A HISTORY OF
THE AMERICAN PEOPLE

A HISTORY OF THE AMERICAN PEOPLE

Paul Johnson

'Be not afraid of greatness'

Shakespeare, Twelfth Night, II, v

HarperCollins*Publishers*

This book was originally published in Great Britain in 1997
by Weidenfeld & Nicolson.

HarperCollins books may be purchased for educational, business, or
sales promotional use. For information please write: Special Markets
Department, HarperCollins Publishers, Inc., 10 East 53rd Street,
New York, NY 10022.

FIRST U.S. EDITION

Library of Congress Cataloging-in-Publication Data

Johnson, Paul, 1928–
     A history of the American people / Paul
Johnson.—1st U.S. ed.
        p.   cm.
     "Originally published in Great Britain in
1997 by Weidenfeld and Nicolson"—T.p. Verso.
     Includes bibliographical references and index.
     ISBN 0-06-016836-6
     1. United States—History.   I. Title.
E178.J675   1998
973—dc21   97-43698

98 99 00 01 02 ❖/RRD 10 9 8 7 6 5 4 3 2 1

This book is dedicated to the people of America—strong, outspoken, intense in their convictions, sometimes wrong-headed but always generous and brave, with a passion for justice no nation has ever matched.

CONTENTS

PART THREE

'A General Happy Mediocrity Prevails'
Democratic America, 1815–1850

PART EIGHT
'We Will Pay Any Price, Bear Any Burden'
Problem-Solving, Problem-Creating America, 1960–1997

PREFACE

This work is a labor of love. When I was a little boy, my parents and elder sisters taught me a great deal of Greek, Roman, and English history, but America did not come into it. At Stonyhurst, my school, I was given a magnificent grounding in English constitutional history, but again the name of America scarcely intruded. At Oxford, in the late 1940s, the School of Modern History was at the height of its glory, dominated by such paladins as A. J. P. Taylor and Hugh Trevor-Roper, Sir Maurice Powicke, K. B. McFarlane, and Sir Richard Southern, two of whom I was fortunate to have as tutors and all of whose lectures I attended. But nothing was said of America, except in so far as it lay at the margin of English history. I do not recall any course of lectures on American history as such. A. J. P. Taylor, at the conclusion of a tutorial, in which the name of America had cropped up, said grimly: 'You can study American history when you have graduated, if you can bear it.' His only other observation on the subject was: 'One of the penalties of being President of the United States is that you must subsist for four years without drinking anything except Californian wine.' American history was nothing but a black hole in the Oxford curriculum. Of course things have now changed completely, but I am talking of the Oxford academic world of half a century ago. Oxford was not alone in treating American history as a non-subject. Reading the memoirs of that outstanding American journalist Stewart Alsop, I was intrigued to discover that, when he was a boy at Groton in the 1930s, he was taught only Greek, Roman, and English history.

As a result of this lacuna in my education, I eventually came to American history completely fresh, with no schoolboy or student prejudices or antipathies. Indeed my first contacts with American history were entirely non-academic: I discussed it with officers of the US Sixth Fleet when I was an officer in the Garrison at Gibraltar, during my military service, and later in the 1950s when I was working as a journalist in Paris and had the chance to meet such formidable figures as John Foster Dulles, then Secretary of State, President Dwight D. Eisenhower, and his successor at SHAPE Headquarters, General Matthew Ridgeway. From the late 1950s I began visiting the United States regularly, three or four times a year, traveling all over the country and meeting men and women who were shaping its continuing history. Over forty years I have grown to know and admire the United States and its people, making innumerable friends and acquaintances, reading its splendid literature, visiting many of its universities to give lectures and participate in debates, and attending scores of conferences held by American businesses and other institutions.

In short, I entered the study of American history through the back door. But I also got to know about it directly during the research for a number of books I wrote in these years: *A History of Christianity, A History of the Jews, Modern Times: the World from the Twenties to the Nineties,* and *The Birth of the Modern: World Society, 1815–1830.* Some of the material acquired in preparing these books I have used in the present one, but updated, revised, corrected, expanded, and refined. As I worked on the study of the past, and learned about the present by traveling all over the world—but especially in the United States—my desire to discover more about that extraordinary country, its origins and its evolution, grew and grew, so that I determined in the end to write a history of it, knowing from experience that to produce a book is the only way to study a subject systematically, purposefully, and retentively. My editor in New York, Cass Canfield Jr of HarperCollins, encouraged me warmly. So this project was born, out of enthusiasm and excitement, and now, after many years, it is complete.

Writing a history of the American people, covering over 400 years, from the late 16th century to the end of the 20th, and dealing with the physical background and development of an immense tract of diverse territory, is a herculean task. It can be accomplished only by the ruthless selection and rejection of material, and made readable only by moving in close to certain aspects, and dealing with them in fascinating detail, at the price of merely summarizing others. That has been my method, as in earlier books covering immense subjects, though my aim nonetheless has been to produce a comprehensive account, full of facts and dates and figures, which can be used with confidence by students who wish to acquire a general grasp of American history. The book has new and often trenchant things to say about every aspect and period of America's past, and I do not seek, as some historians do, to conceal my opinions. They are there for all to see, and take account of or discount. But I have endeavored, at all stages, to present the facts fully, squarely, honestly, and objectively, and to select the material as untendentiously as I know how. Such a fact-filled and lengthy volume as this is bound to contain errors. If readers spot any, I would be grateful if they would write to me at my private address: 29 Newton Road, London W25JR; so that they may be corrected; and if they find any expressions of mine or opinions insupportable, they are welcome to give me their comments so that I may weigh them.

The notes at the end of the book serve a variety of purposes: to give the sources of facts, figures, quotations, and assertions; to acknowledge my indebtedness to other scholars; to serve as a guide to further reading; and to indicate where scholarly opinion differs, directing the reader to works which challenge the views I have formed. I have not bowed to current academic nostrums about nomenclature or accepted the fly-blown philacteries of Political Correctness. So I do not acknowledge the existence of hyphenated Americans, or Native Americans or any other qualified kind. They are all Americans to me: black, white, red, brown, yellow, thrown together by fate in that swirling maelstrom of history which has produced the most remarkable people the world has ever seen. I love them and salute them, and this is their story.

PART ONE

'A City on a Hill'

Colonial America, 1580–1750

The creation of the United States of America is the greatest of all human adventures. No other national story holds such tremendous lessons, for the American people themselves and for the rest of mankind. It now spans four centuries and, as we enter the new millennium, we need to retell it, for if we can learn these lessons and build upon them, the whole of humanity will benefit in the new age which is now opening. American history raises three fundamental questions. First, can a nation rise above the injustices of its origins and, by its moral purpose and performance, atone for them? All nations are born in war, conquest, and crime, usually concealed by the obscurity of a distant past. The United States, from its earliest colonial times, won its title-deeds in the full blaze of recorded history, and the stains on them are there for all to see and censure: the dispossession of a indigenous people, and the securing of self-sufficiency through the sweat and pain of an enslaved race. In the judgmental scales of history, such grievous wrongs must be balanced by the erection of a society dedicated to justice and fairness. Has the United States done this? Has it expiated its organic sins? The second question provides the key to the first. In the process of nation-building, can ideals and altruism—the desire to build the perfect community—be mixed successfully with acquisitiveness and ambition, without which no dynamic society can be built at all? Have the Americans got the mixture right? Have they forged a nation where righteousness has the edge over the needful self-interest? Thirdly, the Americans originally aimed to build an other-worldly 'City on a Hill,' but found themselves designing a republic of the people, to be a model for the entire planet. Have they made good their audacious claims? Have they indeed proved exemplars for humanity? And will they continue to be so in the new millennium?

We must never forget that the settlement of what is now the United States was only part of a larger enterprise. And this was the work of the best and the brightest of the entire European continent. They were greedy. As Christopher Columbus said, men crossed the Atlantic primarily in search of gold. But they were also idealists. These adventurous young men thought they could transform the world for the better. Europe was too small for them—for their energies, their ambitions, and their visions. In the 11th, 12th, and 13th centuries, they had gone east, seeking to reChristianize the Holy Land and its surroundings, and also to acquire land there. The mixture of religious zeal, personal ambition—not to say cupidity—and lust for adventure which inspired generations of Crusaders was the prototype for the enterprise of the Americas.

In the east, however, Christian expansion was blocked by the stiffening resistance of the Moslem world, and eventually by the expansive militarism of the Ottoman Turks. Frustrated there, Christian youth spent its ambitious energies at home: in France, in the extermination of heresy, and the acquisition of confiscated property; in the Iberian peninsula, in the reconquest of territory held by Islam since the 8th century, a process finally completed in the 1490s with the destruction of the Moslem kingdom of Granada, and the expulsion, or forcible conversion, of the last Moors in Spain. It is no coincidence that this decade, which marked the homogenization of western Europe as a Christian entity and unity, also saw the first successful efforts to carry Europe, and Christianity, into the western hemisphere. As one task ended, another was undertaken in earnest.

The Portuguese, a predominantly seagoing people, were the first to begin the new enterprise, early in the 15th century. In 1415, the year the English King Henry V destroyed the French army at Agincourt, Portuguese adventurers took Ceuta, on the north African coast, and turned it into a trading depot. Then they pushed southwest into the Atlantic, occupying in turn Madeira, Cape Verde, and the Azores, turning all of them into colonies of the Portuguese crown. The Portuguese adventurers were excited by these discoveries: they felt, already, that they were bringing into existence a new world, though the phrase itself did not pass into common currency until 1494. These early settlers believed they were beginning civilization afresh: the first boy and girl born on Madeira were christened Adam and Eve.[1] But almost immediately came the Fall, which in time was to envelop the entire Atlantic. In Europe itself, the slave-system of antiquity had been virtually extinguished by the rise of Christian society. In the 1440s, exploring the African coast from their newly acquired islands, the Portuguese rediscovered slavery as a working commercial institution. Slavery had always existed in Africa, where it was operated extensively by local rulers, often with the assistance of Arab traders. Slaves were captives, outsiders, people who had lost tribal status; once enslaved, they became exchangeable commodities, indeed an important form of currency.

The Portuguese entered the slave-trade in the mid–15th century, took it over and, in the process, transformed it into something more impersonal, and horrible, than it had been either in antiquity or medieval Africa. The new Portuguese colony of Madeira became the center of a sugar industry, which soon made itself the largest supplier for western Europe. The first sugar-mill, worked by slaves, was erected in Madeira in 1452. This cash-industry was so successful that the Portuguese soon began laying out fields for sugar-cane on the Biafran Islands, off the African coast. An island off Cap Blanco in Mauretania became a slave-depot. From there, when the trade was in its infancy, several hundred slaves a year were shipped to Lisbon. As the sugar industry expanded, slaves began to be numbered in thousands: by 1550, some 50,000 African slaves had been imported into São Tomé alone, which likewise became a slave entrepot. These profitable activities were conducted, under the aegis of the Portuguese crown, by a mixed collection of Christians from all over Europe—Spanish, Normans, and Flemish, as well as Portuguese, and Italians from the Aegean and the Levant. Being energetic, single young males, they mated with whatever women they could find, and sometimes married them. Their mixed progeny, mulattos, proved less susceptible than pure-bred Europeans to yellow fever and malaria, and so flourished. Neither Europeans nor mulattos could live on the African coast itself. But they multiplied in the Cape Verde Islands, 300 miles off the West African coast. The mulatto trading-class in Cape Verde were known as Lancados. Speaking both Creole and the native languages, and practicing Christianity spiced with paganism, they ran the European end of the slave-trade, just as Arabs ran the African end.[2]

This new-style slave-trade was quickly characterized by the scale and intensity with which it was conducted, and by the cash nexus which linked African and Arab suppliers, Portuguese and Lancado traders, and the purchasers. The slave-markets were huge. The slaves were overwhelmingly male, employed in large-scale agriculture and mining. There was little attempt to acculturalize them and they were treated as body-units of varying quality, mere commodities. At São Tomé in particular this modern pattern of slavery took shape. The Portuguese were soon selling African slaves to the Spanish, who, following the example in Madeira, occupied the Canaries and began to grow cane and mill sugar there too. By the time exploration and colonization spread from the islands across the Atlantic, the slave-system was already in place.[3]

In moving out into the Atlantic islands, the Portuguese discovered the basic meteorological fact about the North Atlantic, which forms an ocean weather-basin of its own. There were strong currents running clockwise, especially in the summer. These are assisted by northeast trade winds in the south, westerlies in the north. So seafarers went out in a southwest direction, and returned to Europe in a northeasterly one. Using this weather system, the Spanish landed on the Canaries and occupied them. The indigenous Guanches were either sold as slaves in mainland Spain, or converted and turned into farm-labourers by their mainly Castilian conquerors.[4] Profiting from the experience of the Canaries in using the North Atlantic weather system, Christopher Columbus made landfall in the western hemisphere in 1492. His venture was characteristic of the internationalism of the American enterprise. He operated from the Spanish city of Seville but he came from Genoa and he was by nationality a citizen of the Republic of Venice, which then ran an island empire in the Eastern Mediterranean. The finance for his transatlantic expedition was provided by himself and other Genoa merchants in Seville, and topped up by the Spanish Queen Isabella, who had seized quantities of cash when her troops occupied Granada earlier in the year.[5]

The Spanish did not find American colonization easy. The first island-town Columbus founded, which he called Isabella, failed completely. He then ran out of money and the crown took over. The first successful settlement took place in 1502, when Nicolas de Ovando landed in Santo Domingo with thirty ships and no fewer than 2,500 men. This was a deliberate colonizing enterprise, using the experience Spain had acquired in its *reconquista*, and based on a network of towns copied from the model of New Castile in Spain itself. That in turn had been based on the *bastides* of medieval France, themselves derived from Roman colony-towns, an improved version of Greek models going back to the beginning of the first millennium BC. So the system was very ancient. The first move, once a beachhead or harbour had been secured, was for an official called the *adelantana* to pace out the street-grid.[6] Apart from forts, the first substantial building was the church. Clerics, especially from the orders of friars, the Dominicans and Franciscans, played a major part in the colonizing process, and as early as 1512 the first bishopric in the New World was founded. Nine years before, the crown had established a Casa de la Contracion in Seville, as headquarters of the entire transatlantic effort, and considerable state funds were poured into the venture. By 1520 at least 10,000 Spanish-speaking Europeans were living on the island of Hispaniola in the

Caribbean, food was being grown regularly and a definite pattern of trade with Europeans had been established.[7]

The year before, Hernando Cortes had broken into the American mainland by assaulting the ancient civilization of Mexico. The expansion was astonishingly rapid, the fastest in the history of mankind, comparable in speed with and far more exacting in thoroughness and permanency than the conquests of Alexander the Great. In a sense, the new empire of Spain superimposed itself on the old one of the Aztecs rather as Rome had absorbed the Greek colonies.[8] Within a few years, the Spaniards were 1,000 miles north of Mexico City, the vast new grid-town which Cortes built on the ruins of the old Aztec capital, Tenochtitlán.

This incursion from Europe brought huge changes in the demography, the flora and fauna, and the economics of the Americas. Just as the Europeans were vulnerable to yellow fever, so the indigenous Indians were at the mercy of smallpox, which the Europeans brought with them. Europeans had learned to cope with it over many generations but it remained extraordinarily infectious and to the Indians it almost invariably proved fatal. We do not know with any certainty how many people lived in the Americas before the Europeans came. North of what is now the Mexican border, the Indians were sparse and tribal, still at the hunter-gatherer stage in many cases, and engaged in perpetual inter-tribal warfare, though some tribes grew corn in addition to hunting and lived part of the year in villages—perhaps one million of them, all told. Further south there were far more advanced societies, and two great empires, the Aztecs in Mexico and the Incas in Peru. In central and south America, the total population was about 20 million. Within a few decades, conquest and the disease it brought had reduced the Indians to 2 million, or even less. Hence, very early in the conquest, African slaves were in demand to supply labor. In addition to smallpox, the Europeans imported a host of welcome novelties: wheat and barley, and the ploughs to make it possible to grow them; sugarcanes and vineyards; above all, a variety of livestock. The American Indians had failed to domesticate any fauna except dogs, alpacas and llamas. The Europeans brought in cattle, including oxen for ploughing, horses, mules, donkeys, sheep, pigs and poultry. Almost from the start, horses of high quality, as well as first-class mules and donkeys, were successfully bred in the Americas. The Spanish were the only west Europeans with experience of running large herds of cattle on horseback, and this became an outstanding feature of the New World, where enormous ranches were soon supplying cattle for food and mules for work in great quantities for the mining districts.[9]

The Spaniards, hearts hardened in the long struggle to expel the Moors, were ruthless in handling the Indians. But they were persistent in the way they set about colonizing vast areas. The English, when they followed them into the New World, noted both characteristics. John Hooker, one Elizabethan commentator, regarded the Spanish as morally inferior 'because with all cruel inhumanity . . . they subdued a naked and yielding people, whom they sought for gain and not for any religion or plantation of a commonwealth, did most cruelly tyrannize and against the course of all human nature did scorch and roast them to death, as by their own histories doth appear.' At the same time the English admired 'the industry, the travails of the Spaniard, their exceeding charge in furnishing so many ships . . . their continual supplies to further their attempts and their active and undaunted spirits in executing matters of that quality and difficulty, and lastly their constant resolution of plantation.'[10]

With the Spanish established in the Americas, it was inevitable that the Portuguese would follow them. Portugal, vulnerable to invasion by Spain, was careful to keep its overseas relations with its larger neighbor on a strictly legal basis. As early as 1479 Spain and Portugal signed an agreement regulating their respective spheres of trade outside European waters. The papacy, consulted, drew an imaginary longitudinal line running a hundred leagues west of the Azores: west of it was Spanish, east of it Portuguese. The award was made permanent between the two powers by the Treaty of Tordesillas in 1494, which drew the lines 370 leagues west of Cape Verde. This gave the Portuguese a gigantic segment of South America, including most of what is now modern Brazil. They knew of this coast at least from 1500 when a Portuguese squadron, on its way to the Indian Ocean, pushed into the Atlantic to avoid headwinds and, to its surprise, struck land which lay east of the treaty line and clearly was not Africa. But their resources were too committed to exploring the African coast and the routes to Asia and the East Indies, where they were already opening posts, to invest in the Americas. Their first colony in Brazil was not planted till 1532, where it was done on the model of their Atlantic island possessions, the crown appointing 'captains,' who invested in land-grants called *donatorios*. Most of this first wave failed, and it was not until the Portuguese transported the sugar-plantation system, based on slavery, from Cape Verde and the Biafran Islands, to the part of Brazil they called Pernambuco, that profits were made and settlers dug themselves in. The real development of Brazil on a large scale began only in 1549, when the crown made a large investment, sent over 1,000 colonists and appointed Martin Alfonso de Sousa governor-general with wide powers. Thereafter progress was rapid and irreversible, a massive sugar industry grew up across the Atlantic, and during the last quarter of the 16th century Brazil became the largest slave-importing center in the world, and remained so. Over 300 years, Brazil absorbed more African slaves than anywhere else and became, as it were, an Afro-American territory. Throughout the 16th century the Portuguese had a virtual monopoly of the Atlantic slave-trade. By 1600 nearly 300,000 African slaves had been transported by sea to plantations—25,000 to Madeira, 50,000 to Europe, 75,000 to Cape São Tomé, and the rest to America. By this date, indeed, four out of five slaves were heading for the New World.[11]

It is important to appreciate that this system of plantation slavery, organized by the Portuguese and patronized by the Spanish for their mines as well as their sugar-fields, had been in place, expanding steadily, long before other European powers got a footing in the New World. But the prodigious fortunes made by the Spanish from mining American silver, and by both Spanish and Portuguese in the sugar trade, attracted adventurers from all over Europe. While the Spanish and Portuguese were careful to respect each other's spheres of interest, which in any event were consolidated when the two crowns were united under the Habsburgs in 1580, no such inhibitions held back other nations. Any chance that the papal division of the Atlantic spoils between Spain and Portugal would hold was destroyed by the Reformation of the 1520s and 1530s, during which large parts of maritime northwest Europe renounced any allegiance to Rome. Protestantism took special hold in the trading communities and seaports of Atlantic France and the Low Countries, in London, already the largest commercial city in Europe, and among the seafaring men of southwest England. In 1561, Queen Elizabeth I's Secretary of State, Sir William Cecil, carried out an investigation into the international law of the Atlantic, and firmly told the Spanish ambassador that the pope had had no authority for his award. In any case there had long been a tradition, tenaciously held by French Huguenot seamen, who dismissed Catholic claims on principle, that the normal rules of peace and war were suspended beyond a certain imaginary line running down the mid-Atlantic. This line was even more vague than the pope's original award, and no one knew exactly where it was. But the theory, and indeed the practice, of 'No Peace Beyond the Line' was a 16th-century fact of life.[12] It is very significant indeed that, almost from its origins, the New World was widely regarded as a hemisphere where the rule of law did not apply and where violence was to be expected.

From the earliest years of the 16th century, Breton, Norman, Basque, and French fishermen (from La Rochelle) had been working the rich fishing grounds of the Grand Banks off Newfoundland and Labrador. Encouraged by their rich hauls, and reports of riches on land, they went further. In 1534 the French seafarer Jacques Cartier, from St Malo, went up the St Lawrence River, spent the winter at what he called Stadacona (Quebec) and penetrated as far as Hochelaga (Montreal). He was back again in 1541, looking for the 'Kingdom of Saguenay,' reported to be rich in gold and diamonds. But the gold turned out to be iron pyrites and the diamonds mere quartz crystals, and his expedition failed. As the wars of religion began to tear Europe apart, the great French Protestant leader Gaspard de Coligny, Admiral of France, sent an expedition to colonize an island in what is now the immense harbor of Rio de Janeiro. This was in 1555, and the next year 300 reinforcements were dispatched to join them, many picked personally by Jean Calvin himself. But it did not prosper, and in 1560 the Portuguese, seeing that the colony was weak, attacked and hanged all its inhabitants. The French also set up Huguenot colonies at Fort Caroline in northern Florida, and at Charles Fort, near the Savannah River, in 1562 and 1564. But the Spaniards, whose great explorer Hernando de Soto had reconnoitered the entire area in the years 1539–42, were on the watch for intruders, especially Protestants. In 1565 they attacked Fort Caroline in force and massacred the entire colony. They did the same at Charles Fort the next year, and erected their own strongholds at St Augustine and St Catherine's Island. Six years later, in 1572, French Catholic militants staged the Massacre of St Bartholomew, in which Admiral Coligny was murdered, thus bringing to an end the first phase of French transatlantic expansion.[13]

Into the vacuum left by the discomfiture of French Protestantism stepped the English, and it from their appearance on the scene that we date the ultimate origins of the American people. The Englishman John Cabot had been off the coast of Labrador as long ago as 1497, and off Nova Scotia the following year. Nothing came of these early ventures, but the English were soon fishing off the Banks in strength, occasionally wintering in Newfoundland. Henry VIII took many Huguenot sea-men and adventurers into his service and under his daughter Elizabeth maritime entrepreneurs like Sir John Hawkins worked closely with French Protestants in planning raids on Spanish commerce 'beyond the line.' The West Country gentleman–seafarer Humphrey Gilbert helped the Huguenots to fortify their harbour-bastion of La Rochelle in 1562, was made privy to their Atlantic schemes, and conceived some of his own. He came of a ramifying family clan which included the young Walter Ralegh, his half-brother, and their cousin Richard Grenville. In 1578 Gilbert obtained Letters Patent in which Queen Elizabeth signi-fied her willingness to permit him to 'discover and occupy' such lands as were 'not possessed by any Christian prince,' and to exercise juris-diction over them, 'agreeable to the form of the laws and policies of England.'[14] He was in touch with various scholars and publicists who did everything in their power to promote English enterprise on the high seas. One was Dr John Dee, the Queen's unofficial scientific adviser; another was the young mathematician Thomas Hariot, friend and fol-lower of Ralegh. The most important by far, however, was Richard Hakluyt.

Hakluyt was the son of a Middle Temple lawyer who had made a collection of maps and manuscripts on ocean travel. What his father followed as a hobby, young Hakluyt made his lifework. His countless publications, ranging from pamphlets to books, reinforced by powerful letters to the great and the good of Elizabethan England, were the biggest single impulse in persuading England to look west for its future, as well as our greatest single repository of information about the Atlantic in the 16th century.[15] Young Hakluyt has some claims to be considered the first geopolitical strategist, certainly the first English-speaking one. What Dr Dee was already calling the future 'British Empire,' and exhorting Queen Elizabeth to create, was to Hakluyt not a distant vision but something to be brought about in the next few years by getting seamen and entrepreneurs and 'planters' of 'colonies'—two new words which had first appeared in the language in the 1550s—to set about launching a specific settlement on the American coast.[16] In 1582, Hakluyt published an account of some of the voyages to the northwest Atlantic, with a preface addressed to the popular young hero Sir Philip Sidney, who had already arranged with Gilbert to take land in any colony he should found. Hakluyt com-plained in it that the English were missing opportunities and should seize the moment:

I marvel not a little that since the first discovery of America (which is now full forescore and ten years) after so great conquest and planting by the Spaniards and the Portingales there, that we of England could never have the grace to set fast footing in such fertile and temperate places as are left as yet unpossessed by them. But again when I consider that there is a time for all men, and see the Portingales' time to be out of date and that the nakedness of the Spaniards and their long-hidden secrets are at length espied . . . I conceive great hope that the time approacheth and now is that we of England may share and part stakes (if we will ourselves) both with the Spaniard and Portingale in part of America and other regions as yet undiscovered.[17]

Gilbert immediately took up Hakluyt's challenge and set out with five ships, one of them owned by Ralegh, and 260 men. These included 'masons, carpenters, smiths and such like requisites,' but also 'mineral men and refiners,' indicating that Gilbert's mind, like those of most of the early adventurers, was still focussed on gold. But he did not survive the voyage: his tiny ship, the *Squirrel*, which was only 10 tons, foundered—Gilbert was last glimpsed reading a book on deck, a typical Elizabethan touch.[18] So Ralegh took his place and immediately secured a new charter from the Queen to found a colony. Ralegh is the first great man in the story of the American people to come into close focus from the documents, and it is worth looking at him in detail.

Ralegh was, in a sense, a proto-American. He had certain strongly marked characteristics which were to be associated with the American archetype. He was energetic, brash, hugely ambitious, money-conscious, none too scrupulous, far-sighted and ahead of his time, with a passion for the new and, not least, a streak of idealism which clashed violently with his overweening desire to get on and make a fortune. He was of ancient family, but penniless, born in Devon about 1554 and 'spake broad Devonshire until his dying day.' He was, wrote John Aubrey, who devoted one of his *Brief Lives* to him, 'a tall, handsome and bold man,' with a lot of swagger, 'damnably proud.' His good looks caught the Queen's eye when he came to court, for she liked necessitous youngsters from good families, who looked the part and whom she could 'make.' But what made her single him out from the crowd of smart-looking gallants who jostled for attention was his sheer brain-power and his grasp of new, especially scientific, knowledge. The court was amazed at his rapid rise in favor. As Sir Robert Naunton, an eyewitness, put it, 'true it is, he had gotten the Queen's ear at a trice, and she began to be taken with his elocution, and loved to hear his reasons to her demands. And the truth is, she took him for a kind of oracle, which nettled them all.'[19] Ralegh was one of the first young courtiers to make use of the new luxury, tobacco, which the Spaniards had brought back from America, and typical of the way he intrigued the Queen was his demonstration, with the help of a small pair of scales, of how you measured the weight of tobacco-smoke, by first measuring the pristine weed, then the ashes. His mathematical friend, Hariot, fed him new ideas and experiments with which to keep up the Queen's interest.[20]

Ralegh was not just an intellectual but a man of action since youth, having fought with the Huguenots, aged fifteen, and taken part in a desperate naval action under his half-brother Gilbert. He had also been twice in jail for 'affrays.' But his main experience of action, which was directly relevant to the American adventure, was in Ireland. The English had been trying to subdue Ireland, and 'reduce it to civility' as they put it, since the mid–12th century. Their success had been very limited. From the very beginning English settlers who planted themselves in Ireland and took up lands to turn into English-style estates had shown a disturbing tendency to go native and join the 'wild Irish.' To combat this, the English government had passed a series of laws, in the 14th century, known as the Statutes of Kilkenny, which constituted an early form of apartheid. Fully Anglicized territory, radiating from Dublin, the capital, was known as the Pale, and the Irish were allowed inside it only under close supervision. The English might not sell the Irish weapons or horses and under no circumstances were to put on Irish dress, or speak the local Gaelic language, or employ 'harpers and rhymers.' Conversely the Irish were banned from a whole range of activities and from acquiring land in the Pale, and staying there overnight. But these laws were constantly broken, and had to be renewed periodically, and even so English settlers continued to 'degenerate' and intermarry with the Irish and become Irish themselves, and indeed foment and lead revolts against the English authorities. One such uprising had occurred in 1580, in Munster, and Ralegh had raised a band of 100 footmen from the City of London and taken a ruthless part in suppressing it. He had killed hundreds of 'Irish savages,' as he termed them, and hanged scores more for treason, and had been handsomely rewarded with confiscated Irish lands which he was engaged in 'planting.' In the American enterprise, Ireland played the same part for the English as the war against the Moors had done for the Spaniards—

it was a training-ground both in suppressing and uprooting an alien race and culture, and in settling conquered lands and building towns. And, just as the money from the *reconquista* went into financing the Spanish conquest of the Americas, so Ralegh put the profits from his Irish estates towards financing his transatlantic expedition.[21]

Ralegh's colonizing venture is worth examining in a little detail because it held important lessons for the future. His first expedition of two ships, a reconnaissance, set out on April 27, 1584, watered at the Canaries and Puerto Rico, headed north up the Florida Channel, and reached the Carolina Banks at midsummer. On July 13, they found a passage through the banks leading to what they called Roanoke Island, 'And after thanks given to God for our safe arrival hither, we manned our boats and went to view the land next adjoining, and to take possession of the same, in the right of the Queen's most excellent Majesty.'[22] The men spent six weeks on the Banks and noted deer, rabbits, birds of all kind, and in the woods pines, cypress, sassafras, sweat gum and 'the highest and reddest cedars in the world.' What struck them most was the total absence of any pollution: 'sweet and aromatic smells lay in the air.' On the third day they spotted a small boat paddling towards the island with three men in it. One of them got out at a point opposite the English ships and waited, 'never making any show of fear or doubt' as a party rowed out to him. Then:

> After he had spoken of many things not understood by us we brought him with his own good liking aboard the ships, and gave him a shirt, a hat and some other things, and made him taste of our wine and our meat, which he liked very well; and after having viewed both barks, he departed and went to his own boat again, which he had left in a little cove or creek adjoining: as soon as he was two bowshots into the water, he fell to fishing, and in less than half an hour he had laden his boat as deep as it could swim, with which he came again to the point of land, and there he divided his fish into two parts, pointing one part to the ships and the other to the pinnace: which after he had (as much as he might) requited the former benefits received, he departed out of our sight.[23]

There followed further friendly contact with the Indians, and exchanges of deerskins and buffalo hides, maize, fruit, and vegetables, on the one hand, and pots, axes, and tun dishes, from the ship's stores, on the other. When the ships left Roanoke at the end of August, two Indians, Manteo and Wanchese, went with them. All were back in the west of England by mid-September, bringing with them valuable skins and pearls. Ralegh was persuaded by the detailed account of one of the masters, Captain Arthur Barlow, that the landfall of Roanoke was suitable for a plantation and at once began a publicity campaign, using Hakluyt and other scribes, to attract investors. He had just become member of parliament for Devonshire, and in December he raised the matter in the Commons, elaborating his plans for a colony. On January 6, 1585 a delighted Queen knighted him at Greenwich and gave him permission to call the proposed territory Virginia, after her. In April an expedition of seven ships, carrying 600 men, half of them soldiers, assembled at Plymouth. The fleet was put under the command of Ralegh's cousin Sir Richard Grenville, with an experienced Irish campaigner, Ralph Lane, in charge of the troops. It carried aboard Hariot, as scientific expert. He had been learning the local language from the two Indians, and was given special instructions to make scientific measurements and observe flora and fauna, climate and geology. Also recruited was John White, England's first watercolor-painter of distinction, who was appointed surveyor and painter, and a number of other specialists—an apothecary, a surgeon, and skilled craftsmen.

After various misadventures, some losses, prize-taking from the Spaniards, and quarreling between Grenville and Lane, the bulk of the fleet reached the Roanoke area in July. There they discovered, and Hariot noted, one of the main difficulties which faced the early colonists in America. 'The sea coasts of Virginia,' Hariot wrote, 'are full of islands whereby the entrance into the main land is hard to find. For although they be separated with divers and sundry large divisions, which seemed to yield convenient entrance, yet to our great peril we proved that they were shallow and full of dangerous flats.'[24] There are literally thousands of islands off the American coasts, especially in the region of the great rivers which formed highways inland, and early voyagers could spend weeks or even months finding their way among them to the mainland, or to the principal river-system. And when they occupied a particular island, relief and reinforcements expeditions often found immense difficulty in identifying it. Moreover, the topography of the coast was constantly changing. Ralegh's Virginia lies between Cape Fear and Cape Henry, from latitude 33.50 to 36.56, mainly in what is now North Carolina, though a portion is in modern Virginia. The Carolina Banks, screening the Roanoke colony, are now greatly changed by wind and sea-action, though it is just possible to identify the 16th-century outlines.

No satisfactory harbor was found, though a fort was built on the north of Roanoke Island. Lane was left with 107 men to hold it, while Grenville returned to England in August to report progress. On the return voyage, Grenville took a 300-ton Spanish vessel, the *Santa Maria*, which had strayed from the annual treasure convoy, and brought it into Plymouth harbor on October 18. The prize and contents were valued at £15,000, which yielded a handsome dividend for all who had invested in the 1585 expedition. But the fact that Grenville had allowed himself to be diverted into commerce-raiding betrayed the confusion of aims of the Ralegh enterprise. Was its object to found a permanent, viable colony, with an eye to the long term, or was it to make quick profits by preying on Spain's existing empire? Ralegh himself could not have answered this question; or, rather, he would have replied 'Both,' without realizing that they were incompatible.

Meanwhile Lane had failed to find what he regarded as essential to a settlement, a proper harbor, had shifted the location of the colony, fallen foul of the local Indians and fought a pitched battle; and he had been relieved by a large expedition under Sir Francis Drake, which was cruising up the east coast of America after plundering the Spanish Caribbean. Lane was a good soldier and resourceful leader, but he knew nothing about planting, especially crop-raising. The colonists he had with him were not, for the most part, colonists at all but soldiers and adventurers. Hariot noted: 'Some also were of a nice bringing up, only in cities or towns, and such as never (as I may say) had seen the world before.' He said they missed their 'accustomed dainty food' and 'soft beds of down and feathers' and so were 'miserable.' They thought they would find treasure and 'after gold and silver was not to be found, as it was by them looked for, had little or no care for any other thing but to pamper their bellies.' Lane himself concluded that the venture was hopeless as the area had fatal drawbacks: 'For that the discovery of a good mine, by the goodness of God, or a passage to the south sea, or some way to it, and nothing else can bring this country in request to be inhabited by our nation.' Lane decided to bring his men back to England, while he still had the means to do so. The only tangible results of the venture were the detailed findings of Hariot, published in 1588 as *A Briefe and true report of Virginia*, and a number of high-quality watercolor drawings by White, now in the British Museum, which show the Indians, their villages, their dances, their agriculture, and their way of life. White also made a detailed map, and elaborate colored sketches of flora and fauna, including a Hoopoe, a Blue Striped Grunt Fish, a Loggerhead Turtle, and a plantain.[25]

A further expedition of three ships set out for Roanoke on May 8, 1587, with 150 colonists abroad, this time including some women and children, and John White in charge as governor. His journal is a record of the expedition. Again there were divided aims, for Captain Simon Fernandez, master of the fleet, was anxious to engage in piracy and so quarreled with White. Roanoke was reached, and on August 18 John White's daughter, Elenora, who was married to his assistant Ananias Dare, gave birth to a girl, who was named Virginia, 'because this child was the first Christian born in Virginia.' But there was more trouble with the Indians, and Fernandez was anxious to get his ships away to prey on the Spaniards while their treasure fleet was still on the high seas. So 114 colonists, including Elenora and little Virginia, sixteen other women, and ten children, were left behind while White sailed back with Fernandez to persuade Ralegh to send a back-up fleet quickly. White reached Southampton on November 8 and immediately set about organizing relief. But he found the country in the midst of what was to be its first global conflict, preparing feverishly to resist the Spanish invasion-armada, which was expected in the spring. All shipping was stayed by government order in English ports, to be available for defensive flotillas, and when Ralegh and Grenville got together eight vessels in Devon in March 1588, with the object of equipping them for Roanoke, the Privy Council commanded Grenville 'on his allegiance to forbear to go his intended voyage' and to place them under the flag of Sir Francis Drake, to join his anti-Armada fleet. White's attempt to set out himself, with two small pinnaces, proved hopeless.[26]

As a result of the Armada campaign and its aftermath, White found it impossible to get his relief expedition to Virginia until August 17, 1590. He anchored at Roanoke Island at nightfall, lit by the lurid flickers of a forest fire. He recorded: 'We let fall our grapnel near the shore, and sounded with a trumpet and call, and afterwards many familiar English tunes and songs, and called to them friendly. But we had no answer.'[27] When they landed the next day, White found no sign of his daughter or granddaughter, or anyone else. Five chests were found, broken open, obviously by Indians. Three belonged to White himself, containing books, framed maps, and pictures with which he had intended to furnish the governor's mansion, to be built in the new town he had planned and called Ralegh. They were all, he said, 'rotten and spoyled with raine.' They found three letters, 'CRO,' carved on a tree, and nearby the full word 'Croatoan,' on a post, 'in fayre Capital letters.' White had agreed with the colonists that, if forced to quit Roanoke, they would leave behind a carved signpost of their destination; and in the event of trouble they were to put a Maltese cross beside it. There was no cross. But all the other evidence—the defensive palisade and the cabins overgrown with weed—indicated a hasty departure. And where the colonists went to was never discovered, though White searched long and anxiously. But he failed to get to Croatoan Island, and whether the frightened colonists reached it can never be known. To this day, no further trace of the lost colony has ever been found. Ralegh himself tried to sail past Virginia in 1595, on his way home from a voyage to Guyana, and he sent another search-party in 1602. But nothing came of either attempt. The most likely explanation is that the colony was overwhelmed by Indians on their way from Roanoke to Croatoan, the males killed, the women and children absorbed into the tribe, as was the Indian custom. So the bloodline of the first Virginians merged with that of the Indians they intended to subdue.

In 1625 Sir Francis Bacon, no friend of Ralegh—who in the meantime had been executed by King James I—wrote an essay, 'On Plantations,' in which he tried to sum up the lessons of the tragic lost colony. He pointed out that any counting on quick profits was fatal, that there was a need for expert personnel of all kinds, strongly motivated in their commitment to a long-term venture, and, not least, that it was hopeless to try to win over the Indians with trifles 'instead of treating them justly and graciously.' Above all, back-up expeditions were essential: 'It is the sinfullest thing in the world to forsake or destitute a plantation once in forwardness; for besides the dishonour, it is the guiltiness of blood of many commiserable persons.'[28]

There are two points which need to be added. First, as the historian A. L. Rowse has pointed out, the failure of the Roanoke colony may have been a blessing in disguise. Had it taken root, the Spanish would certainly have become aware of this English intrusion in a continent all of which they claimed. They would have identified its exact location and strength and have sent out a powerful punitive expedition, as they did against the French in Florida in the 1560s. At that stage in the game they were still in a military and naval position to annihilate any English venture on the coast. Moreover, they would almost certainly have built forts in the vicinity to deter further English ventures and have laid specific claim to the entire coast of what is now the eastern seaboard of the United States, and so made it much less likely that the English would have returned, after the turn of the 17th century and in the new reign of James I. James was anxious to be on peaceful terms with Spain and would, in those circumstances, have forbidden any more attempts to colonize Virginia. So English America might never have come into existence.[29]

Secondly, in listing the reasons why Roanoke failed, Francis Bacon omitted one important missing element. It was an entirely secular effort. It had no religious dimension. This was in accordance with Ralegh's own sentiments. Though he was for form's sake an oath-taking, church-attending Protestant, like anyone else who wanted to rise to the top in Elizabethan England, religion meant nothing to him. It is not even clear he was a Christian. It was darkly rumored indeed, by his enemies at the court, that he and his friend Hariot, and others of their circle, were 'atheists'—though the term did not then necessarily imply a denial of God's existence, merely a rejection of the Christian doctrine of the Trinity: in our terminology he was a deist of sorts. At all events, Ralegh was not the man to launch a colonizing venture with a religious purpose. The clergy do not seem to have figured at all in his plans. There was no attempt on his part to recruit God-fearing, prayerful men.

In these respects Ralegh was unusual for an Elizabethan sea-venturer. Most of the Elizabethan seadogs were strict Protestants, usually Calvinists, who had strong religious motives for resisting Spanish dominance on the high seas and in the western hemisphere. Drake was typical of them: his family were victims of the papist persecution under Queen Mary, and Drake had been brought up in a Thames-side hulk in consequence, educated to thump his Bible and believe in double-predestination and to proselytize among the heathen and the benighted believers in Romish superstition. He held regular services on board his ships, preached sermons to his men, and tried to convert his Spanish prisoners. Next to the Bible itself, his favorite book was Foxe's *Book of Martyrs*, that compendium of the sufferings of English Protestants who resisted the Catholic restoration under 'Bloody Mary' and died for their faith. Foxe's vast book, published early in Elizabeth's reign, proved immensely popular and, despite its size and expense, had sold over 10,000 copies before the end of it, an unprecedented sale for those times. It was not just a history of persecution: it also embodied the English national-religious myth, which had been growing in power in the later Middle Ages and came to maturity during the Reformation decades—the myth that the English had replaced the Jews as the Elect Nation, and were divinely appointed to do God's will on earth.[30]

This belief in divine appointment was to become an important factor in American as well as English history, because it was transmitted to the western side of the Atlantic when the English eventually established themselves there. At the origin of the myth was the widely held belief that the Christian faith had been brought to Britain directly by Joseph of Arimathea, on the express instructions of the Apostles. Some thought the agent was St Paul; others that Christ himself had paid a secret visit. It was through Britain that the Roman Empire had embraced the faith: for the Emperor Constantine had been British—his mother Helena was the daughter of the British King Coilus. So, wrote Foxe, 'by the help of the British Army,' Constantine 'obtained . . . peace and tranquillity to the whole universal Church of Christ.'

In the reign of Elizabeth the myth became a historical validation of England's role in resisting the Counter-Reformation and the Continental supremacy of the Catholic Habsburgs. The Elect Nation had imperative duties to perform which were both spiritual and geopolitical. In the second year of the Queen's reign, John Aylmer wrote in his *An Harborow for faithful and true subjects* that England was the virgin mother to a second birth of Christ:

> God is English. For you fight not only in the quarrel of your country, but also and chiefly in defence of His true religion and of His dear son Christ. [England says to her children:] 'God hath brought forth in me the greatest and excellentest treasure that He hath for your comfort and all the world's. He would that out of my womb should come that servant of Christ John Wyclif, who begat Huss, who begat Luther, who begat the truth.'[31]

The most strident in proclaiming the doctrine of the English as the Chosen Race were the explorers and navigators, the seamen and merchant adventurers, and the colonizers and planters. It is they who gave the myth its most direct geopolitical thrust by urging England's divinely appointed right to break open Spain's doomed empire of the Scarlet Woman, the popish Whore of Babylon, and replace it with an English Protestant paramountcy. One of them, John Davys, put the new English ideology thus:

There is no doubt that we of England are this saved people, by the eternal and infallible presence of the Lord predestined to be sent into these Gentiles in the sea, to those Isles and famous Kingdoms, there to preach the peace of the Lord; for are not we only set on Mount Zion to give light to all the rest of the world? It is only we, therefore, that must be these shining messengers of the Lord, and none but we![32]

It is curious that this powerful religious motivation, so strongly marked in seafaring men and others involved in overseas ventures, was made so little use of by the Englishmen controlling or plotting the attempted settlement of North America in the closing decades of the 16th century and even in the first decade of the 17th century. But so it was. It is part of the larger mystery of why the English, and the French too for that matter, were still so reluctant to settle across the Atlantic a whole century after Columbus' first discoveries, during which the Spanish and Portuguese created vast empires there and possessed themselves of enormous fortunes.

France was totally absorbed in a long and bitter religious–civil war until the 1590s, when the Protestant leader Henri IV reluctantly accepted conversion to Catholicism to end the struggle, and gave the Protestants tolerance by the Edict of Nantes of 1598. Once at peace, eager French minds quickly conceived geopolicies of European and global expansion. The English avoided civil war but in the 1590s and the first decade of the 17th century they were embroiled in desperate struggles to subdue the 'wild Irish,' which they finally achieved—for the moment—in the last year of old Elizabeth's reign. Thereafter their colonial energies were absorbed with 'planting' the conquered country, especially Ulster, until then the wildest part of it. Early in the 1600s Ulster was made the theater of the largest transfer of population ever carried out under the crown—thousands of Scots Presbyterians being allocated parcels of confiscated Catholic land along a defensible military line running along the Ulster border: a line which is still demographically significant to this day and explains why the Ulster problem remains so intractable. This major Ulster planting took root because it was based on agriculture and centered round hard-working, experienced Scots lowland farmers who were also ready to take up arms to defend their new possessions.

In transatlantic expeditions, however, the English maritime intellectuals like Ralegh, Hakluyt, and Hariot were still obsessed with the possibility of quick riches and refused to accept the paramount importance of food-growing capacity to any successful settlement. The Indians could and did grow food, especially maize, but not for cash. Once their own needs were satisfied there was little left over. Colonists had either to grow their own or be dependent on continuing supplies from England—that was the great lesson of Roanoke. And the only way to insure that settlers grew food systematically and successfully was to send them out as entire families. This emerged as the leading principle of English colonization. Hakluyt, in his practical book on planting, wrote in terms of commerce and trading posts. He even recognized that religion could be important and he accepted the need to grow food. But he did not discuss the need to send out independent families and he thought agricultural labor could be supplied by criminals, civil debtors, and the like, sent out to regain their freedom by work.[33]

The notion of using overseas colonies for getting rid of 'human offal,' as it was termed, was coming to be accepted. A generation before, Gilbert had thought of using persecuted and discontented papists as settlers, but did nothing about it. In the 1590s, increasingly, life in England was made hard for the Presbyterians and other Nonconformists, but to begin with they migrated to Calvinist Holland. Certainly, by the turn of the century, there were many ill-fitting groups in society for whom the new business of exporting humans seemed the obvious solution. Population was rising fast, the number of 'sturdy beggars,' as parliament referred to them, was growing. In 1598 the House of Commons laid down banishment beyond the realm as one punishment for begging. The same year the French founded their first overseas penal colony. It was only a matter of time before the English state recognized that North America had the answer to many social problems.

Then too, international trade was increasing steadily. In the later Middle Ages trade outside Europe had been falling off as Europe's own meager gold and silver mines became exhausted and the Continent was gradually stripped of its specie to pay for imports. The discovery by the Spanish of precious metals in the Americas had a profound effect on world trade. Once and for all, Europe became a money-economy. Merchants began to operate on an ever-increasing scale. The huge quantities of silver brought to Europe pushed up commodity prices, and since wages and rents lagged behind, those involved in commerce made handsome profits, built themselves grand houses, and upgraded their importance in society. As trade spread throughout the world, and its quantity rose, the importance of colonizing ventures to expand the system became obvious. And finally there was North Atlantic fishing, increasing all the time. By the turn of the century both English and French had semi-permanent fishing settlements off what is now Labrador, Newfoundland, and Canada. Sable Island in the Atlantic was the first French permanent post. They set up another at Tadoussac at the mouth of the Sanguenay River. Their great explorer–entrepreneur Samuel de Champlain came there in 1603 and his party moved into Acadia, Cape Breton Island, and Canada itself. In 1608 Champlain established Quebec. Much of this early French enterprise was conducted by Huguenots, though when the French crown took over in the 1620s, Catholic paramountcy was established. It was now the French, rather than the Spanish, who caused forward-looking Englishmen uneasiness and spurred them to move out across the Atlantic themselves, before it was too late.[34]

All these threads began to come together in the early years of the 17th century. James I was keen on colonization, provided it could be carried out without conflict with either Spain or France. As in Elizabethan times, the method was for the crown to issue charters to 'companies of adventurers,' who risked their own money. The Ulster plantation, which began in earnest in 1606, absorbed most of the available resources, but the same year the Virginia Company was refounded with a new charter. It had a Plymouth-based northern sector, and a southern sector based on London. The Plymouth men settled Sagadahoc on the Kennebec River, but abandoned it in 1608. A related, Bristol-based company founded settlements in southwest Newfoundland two years later. Meanwhile the Londoners followed up the old Roanoke settlement by entering the Chesapeake Bay in 1607 and marking out a city they called Jamestown, after their sovereign, 40 miles up the Powhatan River, renamed the James too.

The Jamestown settlement is of historic importance because it began the continuous English presence in North America. But as a colony it left much to be desired. This time, the men who ran the Virginia Company from London did not leave out the religious element, though they saw their divine purpose largely in terms of converting Indians. The company asserted that its object was 'to preach and baptise into the *Christian Religion* and by propagation of the Gospell, to recover out of the arms of the Divell, a number of poure and miserable soules, wrapt up into death, in almost invincible ignorance.'[35] The true benefits of colonization, wrote Sir George Peckham in a pamphlet, would accrue to the 'natives,' brought by the settlers 'from falsehood to truth, from darkness to light, from the highway of death to the path of life, from superstitious idolatry to sincere Christianity, from the Devil to Christ, from Hell to Heaven.' He added: 'And if in respect of all the commodities [colonies] can yield us (were they many more) that they should but receive this only benefit of Christianity, they were more fully recompensed.'[36]

There was also the 'human offal' argument. The *New Britannia*, published at the time of the Jamestown foundation, justified it by urging that 'our land abounding with swarms of idle persons, which having no means of labor to relieve their misery, do likewise swarm in lewd and naughtie practises, so that if we seek not some ways for their foreign employment, we must supply shortly more prisons and corrections for their bad conditions. It is no new thing but most profitable for our state, to rid our multitudes of such who lie at home [inflicting on] the land pestilence and penury, and infecting one another with vice and villainy worse than the plague itself.'[37]

Converting Indians, getting rid of criminals and the idle poor—that was not a formula for a successful colony. The financing, however, was right: this was a speculative company investment, in which individuals put their cash into a joint stock to furnish and equip the expedition, and reinforce it. The crown had nothing to do with the money side to begin with. Over the years, this method of financing plantations turned out to be the best one and is one reason why the English colonies in America proved eventually so successful and created such a numerous and solidly based community: capitalism, financed by private individuals and the competitive money-market, was there from the start. At Jamestown, in return for their investment, each stockholder received 100 acres in fee simple (in effect perpetual freehold) for each share owned, and another 100 acres when the grant was 'seated,' that is, actually taken up. Each shareholder also received a 'head right' of 50 acres for each man he transported and paid for. That was the theory. But in practice the settlers, who were adventurers rather than farmers— most were actually company employees—did not know how to make the most of their acres.

It was on May 6, 1607 that three ships of the Virginia Company, the *Godspeed*, the *Discovery*, and the *Sarah Constant*, sighted the entrance to Chesapeake Bay. The settlers numbered 105, and they built a fort, a church, and huts with roofs of thatch. None of the original settlement survives but an elaborate reconstruction shows us what it looked like, and it was extremely primitive. It was in fact more like a Dark Age settlement in western Europe during the 6th or 7th centuries than a neat township of log cabins—as though the English in establishing a foothold on the new continent had had to go back a thousand years into their past. As it was, lacking a family unit basis, the colony was fortunate to survive at all. Half died by the end of 1608, leaving a mere fifty-three emaciated survivors.

The rest might have perished too had it not been for the leadership of Captain John Smith (c.1579–1631). Smith was a Lincolnshire man, who had had an adventurous career as a mercenary fighting the Turks. He joined the Jamestown expedition not as an investor but as a hired soldier. His terms of engagement entitled him to a seat on the Jamestown council, set up immediately the colony was formed, but he was denied it for brawling on board ship. He accordingly spent the winter of 1607 mapping the Chesapeake Bay district. In the course of it he was taken by the Indians, part of a tribal grouping Thomas Jefferson was later to call the Powhatan Confederacy. He put this to good advantage by establishing friendly relations with the local inhabitants. When he returned to the colony he found it in distress. Since he was the only man who had a clear idea of what to do, he was elected president of the council in September 1608, the earliest example of popular democracy at work in America. He imposed military discipline on the remaining men, negotiated with the Indians for sufficient food to get the colony through the winter, and in fact kept the mortality rate down to 5 per cent—a notable achievement by the standards of early colonization across the Atlantic. He got no thanks for his efforts. A relief convoy which arrived in July 1609 brought the news that changes in the company's charter left him without any legal status. So he returned to England two months later. Smith continued to interest himself in America, however. In 1614 he conducted a voyage of discovery around the Cape Cod area, and published in 1616 *A Description of New England*, which was to be of importance in the next decade—among other things, it was the first tract to push the term 'New England' into common use.[38]

Meanwhile Jamestown again came close to collapse. Under its new charter, the Virginia Company tried to recruit new settlers from all levels of society by promising them land free in return for seven years of labor. It attracted about 500 men and put them aboard the relief convoy, under a temporary governor, Sir Thomas Gates. Gates' ship (one of nine) was wrecked off Bermuda, where he spent the winter of 1609–10, thus providing England's first contact with a group of islands

which are still a British crown colony at the end of the 20th century—
and Shakespeare with his setting for *The Tempest*. The rest of the fleet
deposited 400 new settlers at Jamestown. But, in the absence of both
Smith and Gates, the winter was a disaster. When Gates and his co-sur-
vivors, having built two small ships in Bermuda—no mean feat in
itself—finally arrived at Jamestown in May 1610, scarcely sixty settlers
were still alive. All the food was eaten, there was a suspicion of canni-
balism, and the buildings were in ruins. The Indians, moreover, seeing
the weakness of the colony, were turning hostile, and it may be that a
repetition of the Roanoke tragedy was pending. An immediate decision
was taken to give up the colony, but as the settlers were marching
downriver to reembark, a further relief convoy of three ships arrived,
this time under the leadership of the titular governor of the Virginia
Company, the grandee Lord De La Ware (or Delaware as the settlers
wrote it). Under his rule, and under his successor Gates, a system of
law was established in 1611.

We have here the first American legal code, what Gates called his
'Lawes Divine, Moral and Martiall.' They are known as 'Dale's Code,'
after his marshal, Thomas Dale, who had the job of enforcing them.
Unlike Smith's ordinances, they were civil not martial law, but they had
a distinctly Puritan tone. Sabbath observance was strictly enforced,
immodest dress was forbidden and idleness punished severely. The
colony was not yet self-supporting even in food, however, and had
nothing to export to England. But, the year after the code was promul-
gated, a settler called John Rolfe, fearing prosecution for idleness,
began experiments with tobacco. After trying various seeds, he produced a satisfactory crop, the first sweet-tasting Virginian tobacco, and
by 1616 it was already exportable. In the meantime, in 1614, he had
married an Indian princess, Pocahontas, who had been in and out of
the colony since its inception, when she was twelve. The marriage produced offspring and many in Virginia to this day are proud of their
descent from the princess. At the time the union produced a precarious
peace with the local tribes.

The year 1619 was significant for three reasons. In order to make
the Virginia colony more attractive to settlers, the company sent out a
ship carrying ninety young, unmarried women. Any of the bachelor
colonists could purchase one as a wife simply by paying her cost of
transportation, set at 125 pounds of tobacco. Second, the company
announced that it would give the colonists their 'rights of Englishmen.'
A fresh governor, Sir George Yeardley, was sent out to introduce the

new dispensation. On July 30, 1619 the first General Assembly of Virginia met in the Jamestown Church for a week. Presided over by Yeardley, flanked by his six fellow-councillors, constituting the government, it also included twenty-two elected burgesses. They sat in a separate 'House,' like the Westminster Commons, and their first task was to go over Dale's Code, and improve it in the light of experience and the popular will, which they did, 'sweating and stewing, and battling flies and mosquitoes.' The result of their deliberations was approved by Yeardley and his colleagues, constituting an Upper House, and both houses together, with the governor representing the King, made up a miniature parliament, as in England itself. Thus, within a decade of its foundation, the colony had acquired a representative institution on the Westminster model. There was nothing like it in any of the American colonies, be they Spanish, Portuguese or French, though some of them had now been in existence over a century. The speed with which this piece of legislative machinery had developed, at a time when its progenitor was still battling with King James and his theory of the divine right of kings, in London, was a significant portent for the future.

Three weeks later, on August 20, John Rolfe recorded in his diary the third notable event of the year: 'There came in a Dutch man-of-warre that sold us 20 negars.' He did not state the price, but added that fifteen of the blacks were bought by Yardley himself, for work on his 1,000-acre tobacco plantation in Flowerdew Hundred. These men were unfree though not, strictly speaking, slaves. They were 'indentured servants.' Theoretically they became free when their indentures expired at the end of five years. After that, they could buy land and enjoy all the rights of free citizens of the colony. White laborers arrived from England under the same terms, signing their indentures, or making their mark on them, in return for the passage to America. But in practice many indentured men acquired other financial obligations by borrowing money during their initial period of service, and thus had it extended. It is doubtful if any of this first batch of blacks from Africa ended up free farmers in the colony. Most white servants, when they struggled free of their indentures, found themselves tenant farmers on the Jamestown River. But it was not impossible for a black to become a free man in early Virginia: some are recorded as having done so. What was more ominous, however, was the success with which Yeardley and other landowners used blacks to work their tobacco plantations. Soon they were buying more men, and not indentured laborers either, but chattel slaves. Thus in 1619 the first English colony in America

embarked on two roads which bifurcated and led in two totally different directions: representative institutions, leading to democratic free-
doms, and the use of slave-labor, the 'peculiar institution' of the South,
as it was to be called. It is important to bear in mind that large numbers
of black chattel slaves did not arrive in North America until the 18th
century. All the same, the bifurcation was real, and it eventually produced a society divided into two castes of human beings, the free and
the unfree. These two roads were to be relentlessly and incongruously
pursued, for a quarter of a millennium, until their fundamental incom-
patibility was resolved in a gigantic civil war.[39]

The very next year occurred the single most important formative event
in early American history, which would ultimately have an important
bearing on the crisis of the American Republic. This was the landing, at
New Plymouth in what was to become Massachusetts, on December
11, 1620, of the first settlers from the *Mayflower*. The original Virginia
settlers had been gentlemen–adventurers, landless men, indentured servants, united by a common desire to better themselves socially and
financially in the New World. The best of them were men cast in the
sturdy English empirical tradition of fair-mindedness and freedom,
who sought to apply the common law justly, govern sensibly in the
common interest, and legislate according to the general needs of the
Commonwealth. They and their progeny were to constitute one principal element in American tradition, both public and private—a useful,
moderate, and creative element, good for all seasons. The *Mayflower*
men—and women—were quite different. They came to America not
primarily for gain or even livelihood, though they accepted both from
God with gratitude, but to create His kingdom on earth. They were the
zealots, the idealists, the utopians, the saints, and the best of them, or
perhaps one should say the most extreme of them, were fanatical,
uncompromising, and overweening in their self-righteousness. They
were also immensely energetic, persistent, and courageous. They and
their progeny were to constitute the other principal element in the
American tradition, creative too but ideological and cerebral, prickly
and unbending, fiercely unyielding on occasions to the point of self-
destruction. These two traditions, as we shall see, were to establish
themselves firmly and then to battle it out, sometimes constructively,
occasionally with immense creative power, but sometimes also to the
peril of society and the state.

The *Mayflower* was an old wineship, used to transport barrels of claret from Bordeaux to London. She had been hired by a group of Calvinists, all English and most of them from London, but including some who had been living in exile in Holland. Thirty-five of the settlers, who were led by William Bradford and William Brewster, were Puritan Nonconformists, dissenters whose Calvinist beliefs made them no longer prepared to submit to the episcopal governance and Romish teachings (as they saw it) of the established Church of England. They were going to America to pursue religious freedom, as a Christian body. In this sense they were not individuals but a community. They were also traveling as families, the first colony to sail out on this basis. They obtained from the Virginia Company an 80,000-acre grant of land, together with important fishing rights, permission to trade with the Indians, and authority to erect a system of self-government with wide powers. They brought with them sixty-six non-Puritans, and the settlers as a whole were grouped into forty-one families. Many carried books with them, in addition to a Bible for each family. The captain, Miles Standish, had *Caesar's Gallic War* and a *History of Turkie*. There were enough beds, tables, and chairs carried on board to furnish a score of family huts, plus dogs, goats, sheep, poultry, and quantities of spices, oatmeal, dried meat and fish, and turnips. One passenger, William Mullins, brought with him 126 pairs of shoes and 13 pairs of boots. Others, carpenters, joiners, smiths, and the like, brought their tools of trade.[40]

An important event occurred on the voyage, when the *Mayflower* was two months out from England, and the discomforts of a crowded voyage were leading to dissension. On November 21, the colony's leaders assembled in the main cabin and drew up a social compact, designed to secure unity and provide for future government. In effect it created a civil body politic to provide 'just and equal laws,' founded upon church teaching, the religious and secular governance of the colony to be in effect indistinguishable. This contract was based upon the original Biblical covenant between God and the Israelites. But it reflected also early–17th-century social-contract theory, which was later to receive such notable expression in Thomas Hobbes' *Leviathan* (1655) and John Locke's *Treatise of Civil Government* (1690). It is an amazing document for these earnest men (and women) to have agreed and drawn up, signed by all forty-one 'heads of households' aboard the tiny vessel in the midst of the troubled Atlantic, and it testified to the profound earnestness and high purpose with which they viewed their venture.[41]

What was remarkable about this particular contract was that it was not between a servant and a master, or a people and a king, but between a group of like-minded individuals and each other, with God as a witness and symbolic co-signatory. It was as though this small community, in going to America together, pledged themselves to create a different kind of collective personality, living a new life across the Atlantic. One of their leaders, William Bradford, later wrote a history, *Of Plymouth Plantation*, in which he first referred to them as Pilgrims. But they were not ordinary pilgrims, traveling to a sacred shrine, and then returning home to resume everyday life. They were, rather, perpetual pilgrims, setting up a new, sanctified country which was to be a permanent pilgrimage, traveling ceaselessly towards a millenarian goal. They saw themselves as exceptions to the European betrayal of Christian principles, and they were conducting an exercise in exceptionalism.[42]

Behind the Pilgrims were powerful figures in England, led by Sir Robert Rich, Earl of Warwick, who in 1612 at the age of twenty-five had become a member of the Virginia Company, and was later to be Lord High Admiral of the parliamentary forces during the English Civil War. Warwick was an adventurer, the Ralegh of his age, but a graduate of that Cambridge Puritan college, Emmanuel, and a profoundly religious man. Together with other like-minded Puritan gentry, he wanted to reform England. But if that proved impossible he wanted the alternative option of a reformed colony in the Americas. Throughout the 1620s he was busy organizing groups of religious settlers, mainly from the West Country, East Anglia and Essex, and London—where strict Protestantism was strongest—to undertake the American adventure. In 1623 he encouraged a group of Dorset men and women to voyage to New England, landing at Cape Ann and eventually, in 1626, colonizing Naumkeag.[43] John White, a Dorset clergyman who helped to organize the expedition, insisted that religion was the biggest single motive in getting people to hazard all on the adventure: 'The most eminent and desirable end of planting colonies is the propagation of Religion,' he wrote. 'This Nation is in a sort singled out unto this work, being of all the States enjoying the liberty of the Religion Reformed, and are able to spare people for such an employment, the most orthodox in our profession.' He admitted: 'Necessity may oppress some: novelty draw on others: hopes of gain in time to come may prevail with a third sort: but that the most sincere and Godly part have the advancement of the Gospel for their main scope I am confident.'[44]

The success of this venture led to a third Puritan expedition in 1628 which produced the settlement of Salem. A key date was March 4, 1629 when the organizers of these voyages formed the Massachusetts Bay Company, under royal charter, which had authority to transfer itself wholly to the American side of the ocean. It promptly dispatched six ships with 350 people and large supplies of provisions, tools, and arms. But that was as nothing to a great fleet which set out in 1630, with 700 settlers aboard. This was the first of a great series of convoys, numbering 200 ships in all, which throughout the 1630s transported 20,000 Englishmen and women to New England. Thus in 1634 William Whiteway noted in his Dorchester diary: 'This summer there went over to [New England] at least 20 sail of ships and in them 2,000 planters, from the ports of Weymouth and Plymouth alone.' It was the greatest outward movement, so far, in English history.

The most important of these early convoys, as setting a new pattern, was the one in 1630 under the leadership of John Winthrop. He was the outstanding figure of the Puritan voyages, the first great American. Son of a Suffolk squire, and neighbor and friend of Warwick, he was tall and powerful, with a long, lugubrious, stern, impressive face, penetrating eyes, prominent nose, and high brow. He was another Cambridge man, trained as a lawyer in Gray's Inn, sat as a justice of the peace, and took up a job in the Court of Wards but lost it because of his uncompromising Puritan views. He was a sad but exalted man, who had buried two beloved wives and reasoned to himself that 'The life which is most exercised with tryalls and temptations is the sweetest and may prove the safest.'[45] He came to the conclusion that over-crowded, irreligious, ill-governed England was a lost cause, and New England the solution, setting his views down fiercely in his *General Observations for the Plantation of New England*:

> All other Churches of Europe are brought to desolation and it cannot be but that the like Judgement is coming upon us . . . This land grows weary of its Inhabitants, so as man, who is the most precious of all Creatures, is heere more vile and base than the earth they tread upon . . . We are grown to the height of intemperance in all excess of Ryot, as no mans estate almost will suffice to keep sayle with his equals . . . The fountains of Learning and Religion are corrupted . . . Most children, even the best wittes and fayrest hopes, are perverted, corrupted and utterly overthrowne by the multitude of evil examples and the licensious government of those seminaries.

Previous colonies had failed, Winthrop argued, because they were 'carnall not religious'. Only an enterprise governed in the name of the reformed religion stood a chance.[46]

Winthrop joined the new company at the end of July 1629, when it was decided that the proposed new colony should be self-governing and not answerable to the backers in England. Under their charter they had power to meet four times annually in General Courts, to pass laws, elect new freemen or members, elect officers, including a governor, deputy governor, and eighteen 'assistants,' make ordinances, settle 'formes and Ceremonies of Government and Magistracy,' and 'correct, punish, pardon and rule' all inhabitants of the plantation, so long as nothing was done 'contrary to English lawe.'[47] The decision to make the colony self-governing persuaded Winthrop to sell up his estate at Groton, realizing £5,760, and put all his assets into the venture. He impressed everyone connected with the venture by his determination and efficiency, and in October he was elected governor, probably because other major shareholders said they would not go except under his leadership.[48]

Winthrop proved extremely successful in getting people and ships together over the winter, thus forming the largest and best-equipped English expedition yet. As the fleet set off, on Easter Monday 1630, Winthrop was in a mood of exaltation, seeing himself and his companions taking part in what seemed a Biblical episode—a new flight from Egypt into the Promised Land. To record it he began to keep a diary, just as he imagined Moses had made notes of the Exodus. These early diaries and letters, which are plentiful, and the fact that most important documents about the early American colonies have been preserved, mean that the United States is the first nation in human history whose most distant origins are fully recorded. For America, we have no ancient national myth or prescriptive legends but solid facts, set down in the matter-of-fact writings of the time. We know in considerable detail what happened and why it happened. And through letters and diaries we are taken right inside the minds of the men and women who made it happen. There can be no doubt then why they went to America. Among the leading spirits, those venturing out not in the hope of a quick profit but to create something new, valuable, and durable, the overwhelming thrust was religious. But their notions of religious truth and duty did not always agree, and this had its consequences in how they set about emigrating.

The original Pilgrim Fathers of Plymouth Rock were separatists.

They thought the church back in England was doomed, irrecoverably corrupt, and they wanted to escape from it. They came to America in the spirit of hermits, leaving a wicked world to seek their own salvation in the wilderness. John Winthrop saw things quite differently. He did not wish to separate himself from the Anglican church. He thought it redeemable. But, because of its weakness, the redemptive act could take place only in New England. Therefore the New England colony was to be a pilot church and state, which would create an ideal spiritual and secular community, whose example should in turn convert and save the Old World too. He set out these ideas to his fellow-travelers in a shipboard sermon, in which he emphasized the global importance of their mission in a striking phrase: 'We must consider that wee shall be as a Citty upon a Hill, the eyes of all people are upon us.'[49] Winthrop observed to his fellows, and set down in his diary, numerous Old Testament-style indications of godly favor which attended their voyage. Near the New England coast, 'there came a smell off the shore like the smell of a garden.' 'There came a wild pigeon into our ship, and another small land-bird.' He rejoiced at providential news that the Indians, within a range of 300 miles, 'are swept away by the small-pox . . . so God hath hereby cleared our title to this place.' He warned the colonists of the coming harsh winter, telling them: 'It hath been always observed here, that such as fell into discontent, and lingered after their former condition in England, fell into the scurvy and died.'[50]

The Winthrop-led reinforcement was the turning-point in the history of New England. He took over 1,000 colonists in his fleet, and settled them in half a dozen little towns ringing Boston harbor. In Boston itself, which became the capital, he built his town house, took a farm of 600 acres, Ten Hills, on the Mystic River, and other lands, and he built a ship, *Blessing of the Bay*, for coastal trading. Throughout the 1630s more ships arrived, to make good losses, swell the community, and form new towns and settlements.

The land God gave them, as they believed, was indeed a promised one. Of all the lands of the Americas, what is now the United States was the largest single tract suitable for dense and successful settlement by humans. The evidence shows that human beings function most effectively outdoors at temperatures with a mean average of 60–65 degrees Fahrenheit, with noon temperature 70 average, or a little more. Mental activity is highest when the outside average is 38 degrees, with mild frosts at night. It is important that temperature changes from one day to the next: constant temperatures, and also great swings, are unfavorable—the ideal conditions are moderate changes, especially a cooling of the air at frequent intervals.[51] The territory now settled and expanded met these requirements admirably, with 40–70 degrees average annual temperatures, a warm season long enough to grow plenty of food, and a cold season severe enough to make men work and store up food for the winter. The rainfall averages were also satisfactory. Until the development of 'dry' farming, wheat was grown successfully only if the rainfall was over 10 inches annually and less than 45: the average United States rainfall is 26.6, and east of the Appalachians, in the area of early settlement, it is 30 to 50 inches a year, almost ideal. Variations of rainfall and temperature were greater than those in Europe, but essentially it was the same general climate. Odd, then, that the English, who came to the Americas comparatively late, got their hands on those parts where Europeans were most likely to flourish.[52]

Successive generations of settlers discovered that almost anything can be grown in America, generally with huge success. Central North America has the best soil in the world for growing regular food-crops. Only 40 per cent may be arable but it has the best combination of arable soil, natural transport, and exploitable minerals. The soil makes a remarkable variety of crops possible, and this is one reason why there has never been a famine in this area since Europeans arrived. The effect of the Ice Age glaciers on North America, which once covered New England, was to scrape some areas to the bare rock but to leave ample valleys with rich deposits. Thus the Connecticut Valley, soon penetrated by the English, proved the most fertile strip in New England, and became in time prodigiously rich not just in settlements but in colleges, publishing houses, and the first high-quality newspapers in the Americas.[53] The colonists brought with them, in addition to livestock of all kinds, most of the valuable plants they grew. In New England, the Pilgrim settlers never made the mistake of the Jamestown people, of looking for gold when they should be growing crops to feed themselves. But they found maize, or 'Indian Corn,' a godsend. It yielded twice as much food per acre as traditional English crops. It was less dependent on the seasons, could be cultivated without plowing using the crudest tools, and even the stalks could be used as fodder. It was the ideal cheap and easy food for an infant colony and it is no wonder the corn-cob became a symbol of American abundance—as did the turkey, a native of North America which the Puritans found much to their taste. The settlers also discovered chestnuts, walnuts, butternuts, beech, hazel, and hickory nuts in abundance, and also wild plum, cherry, mulberry, and persimmon, though most fruit trees were imported. In addition to maize the colonists had pumpkins, squash, beans, rice, melons, tomatoes, huckleberries, blackberries, strawberries, black raspberries, cranberries, gooseberries, and grapes, all growing wild or easily cultivated.[54]

To European arrivals, the wildlife, once they learned to appreciate and hunt it, was staggering in its fecundity. The big game were the deer and the buffalo. But of great importance to settlers were the smaller creatures whose fur and skin could be exported: weasel, sable, badger, skunk, wolverine, mink, otter, sea-otter, beaver, squirrel, and hare. Then there were the fish and seafood. The waters of Northeast America abounded in them, and once the New England colonists built their own ships—as they began to do, with success, almost immediately—there was a never-ending source of supply. John Josselyn, in his *New England Rareties*, published in 1672, lists over 200 kinds of fish caught in the area.

The mineral resources were without parallel, as the settlers gradually discovered. If we can look ahead for a minute, exactly 300 years after John Winthrop's fleet anchored, the United States was producing, with only 6 percent of the world's population and land area, 70 percent of its oil, nearly 50 percent of its copper, 38 percent of its lead, 42 percent each of its zinc and coal, and 46 percent of its iron—in addition to 54 percent of its cotton and 62 percent of its corn.[55] What struck the first New Englanders at the time, however, was the abundance and quality of the timber, to be had for the simple effort of cutting it down. In western Europe in the early 17th century wood for any purpose, including fuel, was increasingly scarce and costly. The ordinary family, which could not afford 'sea coal,' could never get enough of it. So the colonists fell on the wood with delight. Francis Higginson, minister to the settlers at Cape Ann, wrote in 1629: 'Here we have plenty of fire to warm us . . . All Europe is not able to make so great fires as New England. A poor servant here, that is to possess but 50 acres of land, may afford to give more wood for timber and fire as good as the world yields, than many noblemen in England can afford to do. Here is good living for those that love good fires.'[56] William Wood, the first American naturalist, who explored the forests in the years 1629–32, and published his findings two years later in *New England's Prospects*, delightedly listed all the varieties of trees available, virtually all of which could be used for furniture, though also for charcoal, dyes, and potash for soap.[57] He too was amazed by the sheer quantity, as well he might be. It has been calculated that the original forests of what is now the United States covered 822 million acres in the early 17th century. This constituted a stand of marketable saw-timber of approximately 5,200 billion square feet. Early America was a timber civilization, growing out of its woods just as Anglo-Saxon England grew out of its primeval forests. In the first 300 years of their existence, the American people consumed 353 million acres of this huge area of forests, being over 4,075 billion square feet of saw-timber. Washington and Lincoln with their axes drew attention to the archetypal American male activity.[58]

The New Englanders fell upon this astonishing natural inheritance with joy. They were unable to decide whether the Indians were part of this inheritance or competitors for it. They developed, almost from the first, a patriarchal attitude towards the Indians, and the habit, to us distasteful, of referring to them as their children. It is true that the North American Indians, compared to the Indians of Central and parts of South America, were comparatively primitive. They were particularly backward in domesticating animals, one reason why their social organizations were slow to develop. That in turn helps to explain why their numbers (so far as we can guess them) were small compared to the Indians of the south.[59] Being so few, and occupying so large and fertile an area, the Indians did not replenish their cultivated land—they had of course no animal manure—and moved on to fresh fields when it became exhausted. But their agricultural skills, at least in some cases, were not contemptible. The early French and Spanish explorers—Cartier and Champlain on the St Lawrence, De Soto on the Mississippi, Coronado in the southwest—all reported seeing extensive cornfields. Henry Hudson said the Indians built houses of bark and stored them with corn and beans for winter. When the first settlers reached the Ohio Valley they came across cornfields stretching for miles. In 1794 General Wayne said he had 'never before beheld such immense fields of corn in any part of America, from Canada to Florida.'

There were wide differences between the various Indian peoples. Most of them farmed a little, and hunting and fighting tended to be suspended during planting and harvest. The Indians of the southwest, presumably because they had closer contact with the advanced Indians of South and Central America, irrigated their crops from reservoirs and had actual towns. The Pueblo Indians had permanent villages near their fields. The Iroquois villages were semi-permanent. The Indians the New Englanders came across were usually farmers. The settlers noted the way they cleared land of trees and grew corn and beans, pumpkins and squash; in some cased they imitated Indian methods, for instance in the use of fish-fertilizer. The Indians seem to have been low-grade farmers but produced at least a million bushels of crops a year, drying and storing. They also produced poor-quality tobacco.

In most cases the New Englanders began by following Indian practice in sowing, growing, and storing, but then improved on their methods. They also got from the Indians the white potato, which had arrived from Peru in South America, though this was surprisingly little eaten until the Irish arrived in New Hampshire. Where the early New Englanders benefited most from the Indians was in taking over cleared fields, left unclaimed when the tribes were wiped out by smallpox. The early New England farm, cleared of trees, with rows of corn twined with beans arranged as vines, and with squash and pumpkins growing in between, was not very different from the fields of the Indians. William Bradford, for one, testified to the help the Pilgrims received from Indian example, especially in growing corn, 'ye manner, how to set it, and after, how to dress & tend it.'[60]

The importance of livestock was critical. All thrived, pigs especially. One of the earliest exports was barrels of pork. Flocks of sheep were soon common in Massachusetts and Rhode Island. The colonists raised hardy horses and exported them to the West Indies. They brought in seed for turnips, carrots, buckwheat, peas, parsnips, wheat, barley, and oats—all raised with success. New England apples were soon doing particularly well. One commentator, writing in 1642, said they now 'had apples, pears and quince tarts, instead of their former Pumpkin Pies.' Apples were 'reckoned as profitable as any other part of the Plantation.'[61]

New England farming standards were much higher than Indian ones but, by the best European standards, wasteful. All knowledgeable observers noted how the plentiful supply of land, and the shortage of labor, led to 'butchering.' One account said tillage was 'weakly and insufficiently given; worse, ploughing is nowhere to be seen, yet the farmers get tolerable crops; this is owing, particularly in the new settlements, to the looseness and fertility of the old woodlands, which with very bad tillage will yield excellent crops.'[62] Another claimed that New England farmers were 'the most negligent ignorant set of men in the world. Nor do I know of any country in which animals are worse treated . . . they plough cart and ride [horses] to death . . . all the nourishment they are like to have is to be turned loose in a wood.'[63] Visitors were making the same complaints in the mid–18th century, though by then the supply of new land to take in was running out, at least on the eastern side of the Appalachians.

The New Englanders were, on the whole, much less wasteful than the Virginians. Indeed, without tobacco it is doubtful whether the Virginia colony could have survived at all. Initially all the authorities, at home and abroad, were against tobacco farming, largely because King James I hated the 'weed,' thinking it 'tending to general and new Corruption both of Men's Bodies and Manners.' Governor Dale actually legislated against it in 1616, ordering that only one acre could be laid down to tobacco for every two of corn. It proved impossible to enforce. By the next year tobacco was being laid down even in Jamestown itself, in the streets and market-place. Men reckoned that, for the same amount of labor, tobacco yielded six times as much as any other crop. It had a high cash value. Everything conspired to help it. It was grown close to the banks of many little rivers, such as the James, the York, and the Rappahannock. Every small plantation had its own riverside wharf and boat to get the crop to a transatlantic packet. Roads were not necessary. Land would yield tobacco only for three years: then a fresh set of fields had to be planted. But the real problem was labor—hence slavery. The increasing supply of cheap, high-quality slave-labor from Africa came (as the planters would say and believe) as a Godsend to America's infant tobacco industry. So it flourished mightily. James I himself signaled his capitulation as early as 1619 when he laid a tax of a shilling in the pound (5 percent) on tobacco imports to England, though he limited the total (from Bermuda as well as Virginia) to 55,000 lb a year. But soon all such quantitative restrictions were lifted and tobacco became the first great economic fact of life in the new English-speaking civilization growing up across the Atlantic. It continued to be counted a blessing over four centuries until, in the fullness of time, President Bill Clinton brought the wheel back full circle to the days of James I, and in August 1996 declared tobacco an addictive drug.

New England had no such crutch as tobacco to lean on. It had to work harder, and it did. Under John Winthrop, whose first spell as governor lasted 1630–4, it got the kind of firm, even harsh, government a new colony needs. In effect it was a theocracy. This meant that government was conducted by men chosen by all the full members of the congregation. These were the freemen, and they were recruited in batches on account of their 'Godly behaviour.' Thus in May 1631 Winthrop added 118 men to the freeman ranks. More were added from time to time as he and the congregational elders saw fit. In effect, he ran a dictatorship. He summoned his General Court only once a year, not four times as the company's charter stipulated. Everyone, not just the freemen, had to swear an oath of loyalty to his government. He was quite ruthless in dealing with any kind of dissent or (as he saw it) antisocial behavior. In August 1630, in the first weeks of the settlement, he burned down the house of Thomas Morton of Boston for erecting a maypole and 'revelling.' Morton was kept in the stocks until he could be shipped home in the returning fleet. The following June, Philip Radcliffe was whipped and had both his ears cut off for, in the words of Winthrop's *Journal*, 'most foul, scandalous invectives against our church and government.' Sir Christopher Gardiner was banished for bigamy and papism. Again, 'Thomas Knower was sett in the bilbowes for threatening the Court that, if hee should be punished, hee would have it tryed in England whether hee were lawfully punished or not.'[64]

However, Winthrop was not the only man in New England who had a lust for authority and a divine mandate to exercise it. The new American colonies were full of such people. James I pettishly but understandably remarked that they were 'a seminary for a fractious parliament.' Men with strong religious beliefs tend to form into two broad categories, and constitute churches accordingly. One category, among whom the archtetypal church is the Roman Catholic, desire the certitude and tranquility of hierarchical order. They are prepared to entrust religious truth to a professional clergy, organized in a broad-based triangle of parish priests, with an episcopal superstructure and a pontifical apex. The price paid for this kind of orthodox order is clericalism—and the anticlericalism it provokes. There was never any chance of this kind of religious system establishing itself in America. If there was one characteristic which distinguished it from the start—which made it quite unlike any part of Europe and constituted its uniqueness in fact—it was the absence of any kind of clericalism. Clergymen there were, and often very good ones, who enjoyed the esteem and respect of their congregations by virtue of their piety and preachfulness. But whatever nuance of Protestantism they served, and including Catholic priests when they in due course arrived, none of them enjoyed a special status, in law or anything else, by virtue of their clerical rank. Clergy spoke with authority from their altars and pulpits, but their power ended at the churchyard gate; and even within it congregations exercised close supervision of what their minister did, or did not, do. They appointed; they removed. In a sense, the clergy were the first elected officials of the new American society, a society which to that extent had a democratic element from the start—albeit that such electoral colleges were limited to the outwardly godly.

Hence Americans never belonged to the religious category who seek certainty of doctrine through clerical hierarchy: during the whole of the colonial period, for instance, not a single Anglican bishop was ever appointed to rule flocks there. What most Americans did belong to was the second category: those who believe that knowledge of God comes direct to them through the study of Holy Writ. They read the Bible for themselves, assiduously, daily. Virtually every humble cabin in Massachusetts colony had its own Bible. Adults read it alone, silently. It was also read aloud among families, as well as in church, during Sunday morning service, which lasted from eight till twelve (there was more Bible-reading in the afternoon). Many families had a regular course of Bible-reading which meant that they covered the entire text of the Old Testament in the course of each year. Every striking episode was familiar to them, and its meaning and significance earnestly discussed; many they knew by heart. The language and lilt of the Bible in its various translations, but particularly in the magnificent new King James version, passed into the common tongue and script. On Sunday the minister took his congregation through key passages, in carefully attended sermons which rarely lasted less than an hour. But authority lay in the Bible, not the minister, and in the last resort every man and woman decided 'in the light which Almighty God gave them' what the Bible meant.

This direct apprehension of the word of God was a formula for religious excitement and exaltation, for all felt themselves in a close, daily, and fruitful relationship with the deity. It explains why New England religion was so powerful a force in people's lives and of such direct and continuing assistance in building a new society from nothing. They were colonists for God, planting in His name. But it was also a formula for dissent. In its origins, Protestantism itself was protest, against received opinion and the exercise of authority. When the religious monopoly of the Roman Catholic Church began to disintegrate, in the 1520s and 1530s, what replaced it, from the start, was not a single, purified, and reformed faith but a Babel of conflicting voices. In the course of time and often by the use of secular force, several major Protestant bodies emerged: Calvinism in Geneva and Holland;

Anglicanism in England; Lutheranism in northern Germany. But many rapidly emerging sects were left outside these state churches, and more emerged in time; and the state churches themselves splintered at the edges. And within each church and sect there were voices of protest, antinomians as they were called—those who refused to accept whatever law was laid down by the duly constituted authorities in the church they belonged to, or who were even against the idea of authority in any form.

We come here to the dilemma at the heart of the perfect Protestant society, such as the Pilgrims and those who followed them wished to create. To them, liberty and religion were inseparable, and they came to America to pursue both. To them, the Roman church, or the kind of Anglicanism Charles I and his Archbishop of Canterbury, William Laud, were creating in England, were the antithesis of liberty, the essence of thraldom. They associated liberty with godliness because without liberty of conscience godliness was unattainable. But how to define liberty? When did the exercise of liberty become lawlessness? At what point did freedom of conscience degenerate into religious anarchy? All the leaders of opinion in New England tackled this point. Most of them made it clear that liberty had, in practice, to be narrowly defined. Nathaniel Ward, who came to Massachusetts Bay in 1634 and became pastor of Ipswich, wrote a tract he entitled *The Simple Cobbler of Aggawam in America*, and boldly asserted: 'I dare take it upon me to be the herald of New England so far as to proclaim to the world, in the name of our colony, that all Familists, Antinomians, Anabaptists and other enthusiasts shall have free liberty to keep away from us; and such as will come, to be gone as fast as they can, the sooner the better . . . I dare aver that God does nowhere in his world tolerate Christian states to give toleration to such aversaries of his truth, if they have power in their hands to suppress them.'[65] John Winthrop himself gave what he termed a 'little speech,' on July 3, 1645, on the whole vexed question of the authority of magistrates and liberty of the people—a statement of view which many found powerful, so that the words were copied and recopied and eventually anthologized. Man, he laid down, had:

> a liberty to that only which is good, just and honest . . . This liberty is maintained and exercised in a way of subjection to authority, it is of the same kind of liberty whereof Christ hath made us free . . . If you stand for your natural, corrupt liberties, and will do what is good in your own eyes, you will not endure the least weight of authority . . . but if you will

be satisfied to enjoy such civil and lawful liberties, such as Christ allows you, then you will quietly and cheerfully submit under that authority which is set over you . . . for your good.[66]

That was all very well in theory. But the difficulty of applying it in practice was illustrated by the vicissitudes of Winthrop's own career, as a political and religious leader of the colony. It is early American history in microcosm. Winthrop had natural authority, a kind of charisma: that is why he had been picked as governor in the first place. But his stern and at times brutal exercise of it led strong-minded spirits—and there were plenty of them in the Bay Colony—to feel he had exceeded its legitimate bounds.

Moreover, in what Winthrop no doubt felt was a justified stratagem, such as Joshua and David and Solomon had indulged in from time to time, he had cheated. He obliged all colonists, including non-freemen, to swear an oath of loyalty to his government, in accordance with the charter. But, according to the charter, the General Court should meet four times a year, and Winthrop called it only once. After four years of what some called his 'tyranny,' many colonists, freemen and non-freemen alike, demanded he show them the charter, to see what it said. He reluctantly did so. It was generally agreed he had acted *ultra vires*. The colonists brought with them from England a strong sense of the need to live under the rule of law, not of powerful individuals. That was what the developing struggle in the English parliament was all about. The colonists had been promised, by the founding company of Massachusetts, all 'the rights of Englishmen.' Winthrop, no doubt from high motives, had taken some of those rights from them, by flouting the charter. So he was publicly deposed at a general meeting. The freemen of the colony set up what was in effect a representative system of government, with each little town sending deputies 'who should assist in making laws, disposing of lands etc.' This body confirmed Winthrop's dismissal, and replaced him with his deputy, Thomas Dudley. Thus the first political coup in the history of North America was carried out, in 1634, when the colony was still in its infancy. And it was carried out not by force of swords and firearms but by arguments and speeches, and in accordance with the rule of law.[67]

However, the colonists soon discovered that to change a government by popular mandate does not necessarily mean to improve it. During the next three years, 1634–7, the colony was shaken by a series of arguments over rebellious and antinomian figures, such as Roger Williams and Anne Hutchinson. We will come to these two in a minute, because they are important in their own right. From the point of view of good government, they needed to be handled with a mixture of firmness, common sense, and fairness. The feeling grew in the colony that Winthrop's successors lacked all three. Some felt that the authorities were becoming antinomian themselves. It was a fact that the church in Boston itself tended to be antinomian, all the rest orthodox. The antinomians held that the only thing which mattered in religion was the inner light of faith, which was a direct gift of God's grace. The more orthodox held that good works and exemplary behavior were also necessary, and were visible, outward evidence of true faith and godliness. This argument was raging in England and Holland and other countries where Calvinism was strong. But it was fiercer in Massachusetts than anywhere. One contemporary wrote: 'It began to be as common here to distinguish between men, by being under a Covenant of Grace or a Covenant of Works, as in other countries between Protestants and Papists.'[68]

The argument came to a head in the first contested election on American soil, May 17, 1637, an important date in the development of American democracy. The issue was religious; but behind it was the question of good, orderly government. If the antinomians had their way, it was argued, religion and government would cease to be based on reasoned argument, and learning, and the laws of evidence, and would come to rest entirely on heightened emotion—a form of continuous revivalism with everyone claiming to be inspired by the Holy Spirit. The issue was settled at a crowded outdoor meeting in Cambridge. 'There was a great danger of a tumult that day. [The antinomians] grew into fierce speeches; and some laid hands on others; but seeing themselves too weak [in numbers], they grew quiet.'[69] Winthrop was triumphantly reelected governor, the antinomians being 'quite left out' in the voting. So from 1637 Winthrop was free to resume his clear and insistent policy of imposing orthodoxy on the colony by punishment, exclusion, and banishment.[70]

But doctrinal orthodoxy was the not the only measure of a man's fitness to govern, as Winthrop learned to his cost. His natural authority was based to some extent on his descent from the old squirearchy of England, and his manifest possession of means to keep up his station in the New World. In 1639, however, Winthrop discovered that his English agent had cheated him, and that his affairs were in a muddle. The agent was convicted of fraud and sentenced to have his ears cut off. But that did not get the governor out of his financial difficulties. He found himself £2,600 in debt—a formidable sum—and was forced to sell land on both sides of the Atlantic. His financial plight became obvious. Friends and political supporters rallied round. They collected £500 to tide him over. They donated 3,000 acres to his wife. But Winthrop's opponents pointed fingers. The Puritans did not exactly insist that poverty was a sign of wickedness. But there was a general assumption that the godly flourished and that if a man persistently failed to prosper—or if financial catastrophe suddenly struck him—it was because he did not, for some reason, enjoy God's favor. This idea was very potent and passed into the mainstream of American social consciousness. Winthrop was its first victim. In 1640 he was demoted to deputy governor. Some purists even proposed to ban him, and another unsuccessful man, from office for life, 'because they werte growne poor.' But this measure did not pass.[71]

Indeed, Winthrop struggled back into the governorship two years later, when his earthly fortunes revived a little. Embittered and soured by his political vicissittude and 'the unregeneracy of man,' he dealt ever more severely with dissenters. He fell foul of an unorthodox preacher called Samuel Gorton, and commanded him to shut up or leave the colony. Gorton's congregation sent a 'weird and impudent letter' to the Massachusetts government, comparing the Blessed Samuel to Christ and Winthrop to Pontius Pilate—it referred to him as 'the Great and Honoured Idol General, now set up in Massachusetts' and to his supporters as 'a Generation of Vipers.' Almost beside himself with rage at these 'horrible and detestable blasphemies, against God and all Magistry,' Winthrop sent three commissions of forty soldiers to arrest them. He had them tried and put in irons, but they continued preaching, until he took off their shackles and simply dispatched them into the wilderness.[72] As a result of this high-handedness, Winthrop was again demoted to deputy governor in 1644. He wrote in his journal that he feared rule by the rabble, an actual democracy, 'the meanest and worst of all forms of government.' The amount of arguing and political maneuvering was intense. Early Massachusetts was a remarkably argumentative and politically conscious society—reflecting of course the Civil War then raging in England, which was a battle of words as well as arms. Winthrop published a treatise defending his actions, saying that a wise magistrate had no alternative but to stamp out a firebrand like Gorton before he set fire to the whole house. He said that wise men had to be given discretionary power to follow God's law as they saw it.

One of the deputies attacked this view as outrageous: he said the tract should be 'burnt under the gallows' and added that 'if some other of the magistrates had written it, it would have cost him his ears, if not his head.'[73] But Winthrop survived this controversy too, won back public favor, regained the governorship in 1646, and held it to his death three years later.

Winthrop's career and views raised fundamental issues at the time, which have continued to reverberate through American history and political discourse. Where does freedom end and authority begin? What was the role of the magistrate? And how should he combine the need for order and the commands of justice with the Christian virtue of mercy? There is no doubt that Winthrop himself thought deeply about these issues, communing and agonizing with himself in his journal. Generations of American historians have been sharply divided on his civic merits. In the 1830s, George Bancroft depicted him as a pioneer in laying down representative government in America. Later in the century Brooks Adams and Charles Francis Adams emphasized his authoritarian character and his propensity to persecute intellectual opponents—blaming him for the bigotry in the colony which later produced the witchhunting catastrophe in Salem. In the 1930s Perry Miller and Samuel Eliot Morison stressed that Winthrop was a man of religion first, that his political philosophy was a projection of his Christian beliefs, thus indicating to what extent New England was a kind of theocracy, a deliberate attempt to erect a system of government in conformity with Christ's teaching. Yet another historian, Edmund Morgan, went further and argued that Winthrop's magistracy was a continuing struggle, in fashioning this Christian–utopian society, to prevent the separatist impulse, so strong among New England settlers from undermining corporate responsibility, and instead to harness the colonists' sense of righteousness to the cause of social justice.[74]

Winthrop emerges from the chronicle of events as a severe and often intolerant man, and that is how his critics saw him. He regarded himself as a man chosen by God and the people to create a new civil society from nothing by the light of his religious beliefs, and he prayed earnestly to discharge this mandate virtuously. He admitted his shortcomings, at any rate to himself. His political theory was clear. Man had liberty not to do what he liked—that was for the beasts—but to distinguish between good and evil by studying God's commands, and then to do 'that only which is good.' If, by God's grace, you were given this liberty, you had a corresponding duty to obey divinely sanctioned authority. In the blessed colony of Massachusetts, freemen chose their rulers. But, once chosen, the magistrate's word must be obeyed—it was divine law as well as man's. If his commands were not just and honest, his authority was not genuine, 'but a distemper thereof.' Man was sinful, and struggling with his sinful nature. So sometimes magistrates had to exercise mercy and forgiveness. But, equally, final impenitence and stiff-necked obstinacy in sin had to be deal with ruthlessly. Conversely, the people should forgive magistrates their occasional errors of judgment. And if these errors were persisted in, then the people had the right of removal. Winthrop could claim that he was freely elected governor of the colony, not just once but four times, and that therefore he embodied representative government. Moreover it has to be said, on his behalf, that he implanted this system of government firmly in American soil, so that at the end of twenty years the colony had been built up from nothing to a body politic which was already showing signs of maturity, in that it was reconciling the needs of authority with the needs of liberty.[75]

The success of the Bay Colony in this respect would not have been possible without the sheer space America afforded. America had the liberty of vast size. That was a luxury denied to the English; the constraints of their small island made dissent a danger and conformity a virtue. That indeed was why English settlers came to America. A man could stand on Cape Cod with his face to the sea and feel all the immensity of the Atlantic Ocean in front of him, separating him, like a benevolent moat, from the restrictions and conformities of narrow Europe. And, equally, he could feel behind him—and, if he turned round, see it—the immensity of the land, undiscovered, unexplored, scarcely populated at all, a huge, experimental theater of liberty. In a way, the most important political fact in American history is its grandeur and its mystery. For three centuries, almost until 1900, there were crucial things about the interior of America which were unknown to its inhabitants. But what they were sure of, right from the start, was that there was a lot of it, and that it was open. Here was the dominant geopolitical fact which bore down upon the settlers from their first days on the new continent: if they did not like the system they found on the coast, and if they had the courage, they could go on. Nothing would stop them, except their own fear.

This was the point made by Roger Williams (1603–83), the second great American to emerge. Williams was a Londoner, of Welsh descent, who was ordained a minister in 1628 and came to the Bay Colony three years later. His original intention was to be an Indian missionary. Instead he became pastor in Salem. He was clever, energetic, and public-spirited, and promptly made himself a well-known figure in local society. Whereas Winthrop stood for the authority-principle, Williams represented the liberty-principle, though curiously enough the two men liked and respected each other. Williams loved the vastness of the New World and took the opportunity to explore the hinterland of the Bay. He liked the Indians, made contact with them, established friendships. He tried to learn their tongue; or, as he quickly discovered, tongues. In the early 17th century the 900,000 or so Indians of what is now the United States and Canada were divided by speech into eight distinctive linguistic groups, all unrelated, which in turn branched out into fifty-three separate stocks and between 200 and 300 individual languages. Of these, the most widely spoken, used by about 20 percent of all Indians, was Algonquilian. Williams learned this, and noted other Indian languages, and eventually published his findings in *A Key into the Languages of America* (1643), the first and for long the only book written on the subject.[76] His friendships with individual Indians led him to conclude that there was something fundamentally wrong with settler–Indian relationships. The Europeans had come to bring Christianity to the Indians, and that was right, thought Williams. Of all the things they had to impart to the heathen, that was the most precious blessing—far more important than horses and firearms, which some settlers were keen to sell, and all Indians anxious to buy. But in practice, Williams found, few New Englanders took trouble to instruct Indians in Christianity. What they all wanted to do was to dispossess them of their lands and traditional hunting preserves, if possible by sheer robbery. Williams thought this profoundly unChristian. He argued that all title to Indian land should be validated by specific negotiations and at an agreed, fair price. Anything less was sinful.[77]

None of this made Williams popular among right-thinking freemen of Boston. But his religious views, and their political consequences, were far more explosive. He did not believe, as Winthrop's Anglicans held, and as even the Pilgrim Fathers had accepted, that God covenanted with a congregation or an entire society. God, he held, covenanted with each individual. The logic of this was not merely that each person was entitled to his own interpretation of the truth about religion, but that in order for civil society to exist at all there had to be an absolute separation between church and state. In religion, Williams was saying, every man had the right to his individual conscience, guided by the inner light of his faith. In secular matters, however, he must submit to the will of the majority, determined through institutions shorn of any religious content. So to the Massachusetts elders Williams was not merely an antinomian, he was a secularist, almost an atheist, since he wanted to banish God from government.[78] When Williams began to expose these views in his sermons, the authorities grew alarmed. In October 1635 they decided to arrest him and deport him to England.

Winthrop was out of office at the time, or he might perhaps have taken a similar view. But, nursing his own wrongs, he concluded that the treatment of Williams was unjust and ungodly. In tiny England, there was no alternative but to suppress him. In vast America, he should be given the choice of planting himself elsewhere. So Winthrop, who knew what had been decided in council, secretly warned Williams of the plan to send him back to England and advised him to slip off from Salem, where Williams was established, into the Narrangansett Wilderness. As Williams himself related it: '[Winthrop] privately wrote to me to steer my course to Narrangansett Bay and the Indians, for many high and heavenly and public ends, encouraging me, from the freeness of the place from any English claims and patents.' So Williams fled, with his wife Ann, their children, and their household servants. It was the beginning of the harsh New England winter, and Williams and his family had to spend it traveling through the forest, in makeshift shelters, until in the spring of 1636 he reached an Indian village at the head of Narrangansett Bay. To his dying day—and he lived to be eighty—Williams believed that his family's survival was entirely due to divine providence, a fact which confirmed him in the rightness of his views. And he retained correspondingly bitter memories of his 'persecutors.' He negotiated a land purchase with two Indian tribes, and set up a new colony on a site he named Providence. He let it be known that his new settlement on Rhode Island welcomed dissidents of all kinds, fleeing from the religious tyranny of the Bay Colony. As he put it, 'I desired it might be a shelter for persons distressed for conscience.' By 1643, Portsmouth, Newport, and Warwick had been founded as further towns.

Williams may have been an extremist. But he was also a man of business. He knew the law of England and the ways of government. He had no title to his colony and the Bay authorities would not give him one: they called Providence 'the Sewer of New England.' But he knew that parliament and the Puritans now ruled in London, so he went there. On March 24, 1644 parliament, at his request, transformed the four towns into a lawful colony by charter, endorsing an Instrument of Government which Williams had drawn up. He took the opportunity, London being for the time being an ultra-libertarian city where the most extreme Protestant views could be circulated, to write and publish a defense of religious freedom, *The Bloudy Tenet of Persecution for the Cause of Conscience discussed*. And his new Instrument declared that 'The form of government established in Providence Plantations is DEMOCRATICAL, that is to say, a government held by the free and voluntary consent of all, or the greater part, of the free inhabitants.' Williams listed various laws and penalties for specific transgressions but added: 'And otherwise than this, what is herein forbidden, all men may walk as their consciences persuade them, every one in the name of his God. And let the Saints of the Most High walk in this colony without molestation, in the name of Jehovah their God, for ever and ever.'[79]

Williams' rugged angularity and unbiddable nature emerge strongly from his voluminous writings and correspondence. His new colony was by no means popular. Public opinion in Massachusetts was against it. It was believed to be a resort of rogues. Nor was it easy to get title to land there, as Williams insisted that any acquired from the Indians must be paid for at market rates. He opposed force, and was virtually a pacifist: 'I must be humbly bold to say that it is impossible for any man or men to maintain their Christ by the sword, and to worship a true Christ.'[80] The Rhode Island towns were stockaded and fortified. But Williams managed to avoid any conflict at all with the Indians until the disaster of King Philip's War in 1675–6. His Rhode Island colony thus got the reputation of being a place where the Indians were honored and protected. Then again, Williams was accused with some justice of being a man of wild and volatile opinions, an eccentric, an anti-establishment man. A few colonists liked that sort of thing. Most did not. Williams was actually governor of his colony only in the years 1654–7, but he was always the power behind the scenes. He set the tone. So he attracted the antinomians, but not much else. Even as late as 1700, the colony numbered only 7,000 inhabitants, 300 of them slaves.

When Charles II was invited back to govern the British Isles in 1660, and the Puritan ascendancy ended, there was some doubt about the lawfulness of such colonies as Rhode Island. So Williams hastened to England and on July 18, 1663 he obtained from the King a charter confirming the privileges granted in 1644. This made the principle of religious freedom explicit and constitutes an important document in American history. 'No person,' it read, 'within the said colony, at any time hereafter, shall be in any wise molested, punished, disquieted or called in question, for any difference in opinion in matters of religion, and who do not actually disturb the civil peace of our said colony; but that all . . . may from time to time, and at all times hereafter, freely and fully have and enjoy his and their own judgments and consciences in matters of religious concernments.' Rhode Island was thus the first colony to make complete freedom of religion, as opposed to a mere degree of toleration, the principle of its existence, and to give this as a reason for separating church and state. Its existence of course opened the doors to the more angular sects, such as the Quakers and the Baptists, and indeed to missionaries from the Congregationalists of the Bay Colony and the Anglicans of Virginia. Williams himself was periodically enraged by what he saw as the doctrinal errors of the Quakers, and their stiff-necked obstinacy in refusing to acknowledge them. He was almost tempted to break all his own principles and have them expelled. But his sense of tolerance prevailed; the colony remained a refuge for all. The creation of Rhode Island was thus a critical turning-point in the evolution of America. It not only introduced the principles of complete religious freedom and the separation of church and state, it also inaugurated the practice of religious competition. It thus accepted the challenge the great English poet John Milton had just laid down in his pamphlet *Areopagitica* appealing for liberty of speech and conscience: 'Though all the winds of doctrine were let loose to play upon the earth, so Truth be in the field, we do injuriously by licensing and prohibiting to misdoubt her strength. Let her and Falsehood grapple: who ever knew Truth put to the worse, in a free and open encounter?' Who indeed? Rhode Island was now in existence to provide a competitive field in which the religions—or at any rate the varieties of Christianity—could grapple at will, the first manifestation of that competitive spirit which was to blow mightily over every aspect of American existence.

Yet it must be said that, in the 17th century at least, the way of the rebellious individual in New England was a hard one, especially if she was a woman. The case of Anne Hutchinson (1591–1643) is instructive. She was the first woman to achieve any importance in North America, the first to step forward from the almost anonymous ranks of neatly dressed, hard-working Puritan wives and mothers, and speak out with her own strong voice. Yet we know very little about her as a person. Whereas Winthrop and Williams left books and papers, often highly personal, which between them fill a dozen thick volumes, Mrs Hutchinson left behind not a single letter. She published no books or pamphlets—for a woman to do so was almost, if not quite, impossible in the first half of the 17th century. If she kept a journal, it has not survived. The only real documentation concerning her is the record of the two trials to which she was subjected, which naturally leave a hostile impression.[81]

She came from Lincolnshire, one of thirteen children of a dissenting minister, Francis Marbury, who encouraged her early interest in theology and taught her everything he knew. She married a merchant, William Hutchinson, and had twelve children by him. But she kept up her religious enthusiasm and attended the charismatic sermons given by John Cotton at St Botolph's in Lincoln. Under the 'Godless tyranny' of Archbishop Laud, Cotton lost his license to preach in 1633, and promptly emigrated to the Bay Colony. Mrs Hutchinson, her husband, and children followed the next year, and she gave birth to another child shortly after they arrived in Boston. She was capable of delivering a child herself, and acted as a midwife on occasions. She also dispensed home-made cordials and 'simples' and gave medical advice to women. A natural leader, she made her house a resort for women in trouble. There is no need to read into her story the overtones of women's rights with which feminist historians have recently embellished it.[82] But it is clear she was formidable, and that she thought it proper and natural that women should participate in religious controversy.

It was Anne Hutchinson's practice, along with her brother-in-law John Wheelwright, to hold post-sermon discussion groups at her house on Sunday afternoons and on an evening in midweek. There, the words of John Cotton and other preachers were analyzed minutely and at length, and everyone present—there were often as many as sixty, half of them women—joined in if they wished. Cotton himself, and Hutchinson and Wheelwright still more passionately, believed in a Covenant of Grace. Whereas most official preachers held that a moral life was sufficient grounds for salvation, Mrs Hutchinson argued that redemption was God's gift to his elect and could not simply be earned by human effort—albeit the constant practice of good works was usually an external sign of inward election. The logic of this doctrine was subversive. The one power the clergy in New England still possessed was the right to determine who should be a full member of the church—and monitoring of his or her good works was the obvious way in which to do it. But the Hutchinson doctrine stripped the minister of this power by insisting that election, or indeed self-election, with which he had nothing to do, was the criterion of church membership. Such a system, moreover, by which divine grace worked its miracles in the individual without any need for clerical intermediary, abolished distinctions of gender. A woman might just as well receive the spirit, and utter God's teachings, as an ordained pastor. Some people liked this idea. The majority found it alarming.

By 1636 the controversy was dividing the colony so sharply that the elders decided on extreme measures. Cotton was hauled before a synod of ministers and with some difficulty cleared himself of a charge of heresy. Then Winthrop was reelected governor in May 1637 and immediately set about dealing with Mrs Hutchinson, whom he regarded as the root of the problem. He put through an ordinance stipulating that anyone arriving in the colony could not stay more than three weeks without the approval of the magistrates. In November he had Hutchinson, Wheelwright, and their immediate followers up before the General Court, and banished. Some seventy-five of their adherents were disenfranchised and disarmed. He followed this up in March 1638 by having Hutchinson and Wheelwright charged with heresy before the church of Boston, and excommunicated. It is clear that Winthrop believed Anne Hutchinson was in some way being manipulated by the Devil—was a witch in fact. He discovered that she had had a miscarriage, which he interpreted as a sign of God's wrath, and that her friend Mary Dyer had given birth to a stillborn, malformed infant—a monster. He even went so far as to have the pitiful body of the 'monster' dug up and examined. All this he recorded in his diaries.[83] He also communicated the results to England so 'all our Godly friends might not be discouraged from coming to us.' Mrs Hutchinson and her supporters, to save their lives, had no alternative but to leave the Bay Colony and seek refuge in Williams' Rhode Island, where most of them settled and flourished. Her husband, a long-suffering man many might argue, died, and in due course the widow and his six youngest children moved further west, to Pelham Bay in what is now New York State. There, all but one daughter were killed by Indians in 1643.[84] The violent death of Mrs Hutchinson and her brood was promptly interpreted as providential, and New England orthodoxy produced pious literature about the 'American Jezebel,' initiated by a violent pamphlet from the pen of Winthrop himself.[85] Anne Hutchinson's vindication, which has been voluminous and imaginative, had to wait for the women's movement of the 1960s.

The Hutchinson story showed that even the most radical dissent was possible, albeit dangerous. The practice in Massachusetts was to warn people identified as religious troublemakers to move on. If they insisted on staying, or came back, they were prosecuted. In July 1641, for instance, Dr John Clarke and Obediah Holmes, both from Rhode Island, were arrested in Lynn by the sheriff for holding an unauthorized religious meeting in a private house, at which the practice of infant baptism was condemned. Clarke was imprisoned; Holmes was whipped through the streets. Again, on October 27, 1659, three Quakers, William Robinson, Marmaduke Stevenson, and Mary Dyer, having been repeatedly expelled from the colony, the last time under penalty of death, were arrested again as 'pestilential and disruptive' and sentenced to be hanged on Boston Common. Sentence on the men was carried out. The woman, blindfolded and with the noose around her neck, was reprieved on the intervention of her son, who guaranteed she would leave the colony forthwith. She did in fact return, and was finally hanged on June 1, 1660. Other women were hanged for witch-craft—the first being Margaret Jones, sentenced at Plymouth on May 13, 1648 for 'administering physics' with the 'malignant touch.' Severe sentences were carried out on moral offenders of all kinds. Until 1632 adultery was punished by death. In 1639, again at Plymouth, an adul-terous woman was whipped, then dragged through the streets wearing the letters AD pinned to her sleeve: she was told that if she removed the badge the letters would be branded on her face. Two years later, a man and a woman, convicted of adultery, were also whipped, this time 'at the post,' the letters AD 'plainly to be sewn on their clothes.'

To buttress orthodoxy in Boston, a college for training ministers of religion was founded on the Charles River at Newtown in 1636, according to the will of the Rev. John Harvard. He came to the colonies in 1635 and left £780 and 400 books for this purpose. Three years later, the college was named after him and the place rechristened Cambridge, after the university where he was nurtured. The event was an index of the way in which the colony was achieving its primary objects. As one of the Harvard founders put it, 'After God had carried us safe to New England, and wee had builded our houses, provided necessaries for our livlihood, rear'd convenient places for God's wor-ship and settled the Civill Government; One of the next things we longed for, and looked after was to advance Learning and Perpetuate it to Posterity.'[86] But the college never had a monopoly of religious educa-tion, when dissenters could move off and found other establishments for teaching, without any need for a crown charter. In April 1638, for instance, the Rev. John Davenport led a congregation of pious Puritans from Boston, a town they claimed had become 'corrupt,' to settle in Quinnipiac, which they renamed New Haven. Davenport brought with him some successful merchants, including Theophilus Eaton and David Yale, the latter a learned gentleman whose descendant, Elihu Yale, was to found another historic college. Two months later, on May 31, 1638, another dissenting minister, Thomas Hooker, arrived at Hartford on the Connecticut, with 100 followers, marking the occasion by preaching them a sermon stating that all authority, in state or religion, must rest in the people's consent. Thus, within New England, there was a continuing diaspora, often motivated by religious dissent and the urgent desire for greater freedom of thought and action.

As far back as 1623, David Thompson had founded a settlement at Rye on the Piscataqua, the nucleus of what was to become New Hampshire. In 1639 Hooker's Hartford joined with two other Puritan–dissenter townships, Windsor and Wethersfield, to form what they called the Fundamental Orders of Connecticut. Neither they nor Davenport's New Haven had royal charters, but they constituted in effect a separate colony. From the 1620s there were further settlements up the coast in what was to become Maine, set up by dissenting fishermen. Annual new fleets from England reinforced all the areas of settlement. Between 1630 and 1660 about 20,000 Puritans came out, with Massachusetts and Connecticut forming the core area of settlement. This was characterized by what have been called 'Christian, utopian, closed and corporate communities.'[87]

Some settlements were formal. New Haven had nine squares, the central square being a market-place, eventually occupied by the meeting-house, court-house, schoolhouse, and jail. An early map survives of Wethersfield, one of the first of the Connecticut towns, showing house-lots, adjacent field-lots, then outlying strips. Some of these early colony-towns were abandoned. But the vast majority survived with continuous occupation to this day. The fact is, the Puritans were successful settlers. They were homogeneous in belief, literate—they could read the many and often excellent printed pamphlets advising colonists—and skilled. Most of them were artisans or tradesmen, some were experienced farmers, and there was a definite sprinkling of merchants with capital. They came as families under leaders, often as entire congregations under their minister. Their unit-plantation was of several square miles, with an English-style village in the middle (in New

England called a town), where all had houses, then lands outside it.

There was, from the start, no egalitarianism. The free-enterprise investment system ensured that leaders and largest investors got bigger units. There was no symmetrical uniformity or pattern since the countryside was rugged and varied, and there was a universal pragmatism in adapting to the physical features of the place. When more space was needed, the congregation met and decreed a formal move to found a new township. This was the New England equivalent of an Old England village—but with no manor-house and tenant cottages. Virtually everyone came from England and Wales. The religious exclusivity of the original settlements rarely lasted more than a decade or so, with dissenters being expelled. Gradually, Anglicans, Baptists, and even Quakers were allowed to settle. Wealth-gaps widened in the second and third generation, rows and splits weakened church authority, the social atmosphere became more secular and mercantile, and the Puritan merged into the Yankee, 'a race whose typical member is eternally torn between a passion for righteousness and a desire to get on in the world.'[88]

Even Catholics were soon living in America in organized communities. This was the work of the Calvert family. George Calvert, born in 1597, was an energetic Yorkshireman who became James I's Secretary of State and did some vigorous 'planting' in Ireland as well as investing largely in the East India Company and the Virginia Company. When he became a Catholic in 1625 he retired into private life, but James made him a peer, Lord Baltimore, and encouraged him to found a colony for his fellow-papists. He looked at Newfoundland twice, but decided it was too cold. Then he visited Jamestown, but felt unable to sign the Protestant Oath of Supremacy there. In the end Charles I gave him a charter to settle the northern Chesapeake area. It was left to his son Cecil, 2nd Baron Baltimore, to organize the actual settlement in 1633. He recruited seventeen younger sons of Catholic gentry to lead and finance the expedition, taking with them about 100 ordinary settlers, mostly Protestants, some of them married, a few farmers. They had two ships, the *Ark*, 350 tons, and the *Dove*, a mere pinnace, but both armed to the teeth. Various attempts were made by religious enemies to sabotage the venture and in the end Baltimore had to stay behind to protect its London end. Two Jesuits were taken aboard secretly at Cowes on the Isle of Wight. Baltimore, who had studied Captain John Smith's account of Virginia, gave handwritten instructions to his brother Leonard, who was to act as governor, and Jerome Halwey and Thomas Cornwallis, Catholic gentlemen, his co-commissioners. They were to get on good terms with the Virginians, using their Protestant passengers as intermediaries. They were not to disperse but to build a town and concentrate all their efforts on feeding themselves and becoming self-sufficient as quickly as possible. They were to train a militia and build a fort but to try and remain at peace with the Indians. The three leaders were to be 'very careful to preserve unity and peace' between the Protestants and the Roman Catholics: the Protestants were to receive 'mildness and favour', the Catholics to practice their faith as quietly as possible.[89]

One of the priests aboard, Father Andrew White, kept a record of the colony's foundation, and found it full of splendid auguries for success. Of the Chesapeake he wrote: 'This baye is the most delightful water I ever saw . . . two sweet landes, firm and fertile: plenty of fish, woods of walnuts, oakes, and cedars.' He noted 'salad herbes and such like, strawberries, raspberries, fallen mulbery vines, rich soil, delicate springs of water, partridge, deer, turkeys, geese, ducks, and also squirrels, eagles and herons'—'the place abounds not only with profit but with pleasure.' Maryland, he concluded, being halfway between the extremes of Virginia and New England, had 'a middle temperature between the two, and enjoys the advantages, and escapes the evils, of each.'[90]

The land was called Maryland after Charles I's French wife, Henrietta Maria, and the township was St Mary's. Father White had a cross set up and said a dedicatory mass on the shore, 'to take solemn possession of the country.' They traded axes, hoes, hatchets, and cloth for 30 miles of land below the Wicomico River.[91] Strictly speaking, Baltimore's colony was, in English law, a feudal fiefdom, with palatine powers like those of the Bishop of Durham. He was answerable only to the King, owned all the territory granted, received rents, taxes, and fees, appointed all officials, exercised judicial and political authority, could build forts and wage defensive war, confer honors and titles, incorporate boroughs and towns, license trade, was head of the church, could erect and consecrate chapels and churches, and these his 'ample rights, liberties, immunities and temporal franchises' were to be enjoyed by him and his heirs for ever.[92] But this was theory, English lawyers' big talk. In practice none of these grandiose baronies based on feudal models worked as they were intended, quickly losing their gilding under the erosion of America's democratic rock. To begin with,

Baltimore did not have perfect title to his land. A Kentish man called William Claiborne, who had been in Virginia since 1621 and had erected a fur-trading stockade on Kent Island, disputed it. In law he was in the right, for Baltimore's title excluded land already settled. Claiborne threatened to cause trouble, and did, once King Charles' power collapsed in the 1640s. The American coast was already dotted with difficult loners like Claiborne, ferociously opposed to authority of any kind, litigious, well armed, and ready to fight if necessary. Then again, Baltimore's charter specifically stated that the colonists were to enjoy 'the full rights of Englishmen.' That proviso was incompatible with the feudal trappings and was far more likely to be turned into a reality. The first assembly of the colony met in January 1635, consisting of all freemen, that is males not bound in service. Within two years, and after some acrimonious arguments paralleling the debates in Westminster, it had won the legislative initiative. Thereafter there was no chance of Baltimore exacting his feudal privileges in full—quite apart from the fact that, from 1640, the Long Parliament in England systematically demolished what was left of the feudal system.

The principal investors in the colony, called 'Adventurers,' who had to provide their own transport out plus five 'able men' between twenty and fifty, got 2,000 acres each. Anyone bringing less than five men got 100 acres, plus a further 100 for each man beside himself. Married settlers got 200 acres plus 100 for each 'servant.' Each child under sixteen got 50. Widows with children got the same grants as the men, and unmarried women with servants got 50 acres for each. The land was freehold but owners had to pay Baltimore a 'quitrent'—20 shilling for a manor, 12 pence for a 50-acre tract, payable 'in the commodity of the country' and due annually for fifty years. If a man wanted to come and could not afford the voyage, he could travel free in return for a four- to five-year indenture of service. He made it with the captain, who sold it on landing to any bidder. The indenture bound the master to furnish transport, 'meat, Drinke, Apparel and Lodging' during the term and, on completion, to supply clothes, a year's provision of corn, and 50 acres.[93] Skilled men earned their freedom earlier.

The actual apportioning of land proceeded swiftly—something Americans learned to do well very early in their history, and which was for 300 years one of their greatest strengths. A settler went to the secretary of the province, recorded his entitlement, and requested a grant of land. The secretary then presented a Warrant of Survey to the surveyor-general, who found and surveyed an appropriate tract. When he reported, the secretary issued a patent, which described the reasons for the grant, the boundaries and the conditions of tenure. The owner then occupied the land and began farming. Compared to the difficulties of acquiring land in England, even for ready money, it was amazingly simple.[94]

The farming went well from the start. The land produced a surplus the very first year and a load of grain was dispatched to Massachusetts for cash. But most farmers quickly went into tobacco, and stayed there. By the mid–1630s, tobacco prices, after a sellers' market in the 1620s, then a glut, had stabilized at about 4 to 6 shillings a pound. The Maryland settlers planted on creeks and rivers on the western shore, with wharves to receive annual tobacco-export ships. They killed trees by 'girdling'—cutting a ring round the base—then planted. By 1639 the Maryland planters were producing 100,000 pounds of the 'sotte weed.' Tobacco planting was never easy. It required 'a great deal of trouble in the right management of it.' It was expert, labor-intensive, and tricky at all times. A plant had to be 'topped,' using the thumbnail. You could always tell a 17th-century tobacco farmer by his hard, green-stained thumb. Everyone worked hard, at any rate in those early days. The laborers and indentured servants did a twelve- to fourteen-hour day, with Saturday afternoon free and Sunday. They could be transferred by sale and corporally punished, and if they ran away they were punished by longer terms of service. They could not marry until their contract expired. In any case, men outnumbered women by two or three to one. There were lots of bastards and heavily pregnant brides—twice as many as in England.

Housing was poor: 'The dwellings are so wretchedly constructed that even if you are close to the fire as almost to burn yourself, you cannot keep warm and the wind blows through them everywhere.'[95] That was in the winter. The problem in summer was malaria. The more settlers who arrived, the more the mosquitoes bred. Those who got it were peculiarly susceptible to smallpox, diphtheria, and yellow fever. Amebic dysentery, known as Gripes of the Gutts, was endemic. Maryland was noticeably less healthy than New England, where a male who survived to twenty lived generally to around sixty-five. In Maryland it was more like forty-three. About 70 percent died before fifty; only 6 percent of fathers lived to see their offspring mature. And half the children died before twenty.[96] Wives worked very hard, in the tobacco fields, as well as by milking the cow, making cheese and butter, raising chickens, tending to the vegetable garden—mainly peas, beans, squash, and pumpkins. The men butchered but the wives cured, usually pork, which was the commonest meat. Corn was ground with pestle and mortar until the family could afford a grist-mill. These vigorous women were more partners than inferiors to their husbands. They lived longer and inherited more property than was usual in England.[97]

Despite the hardship, there was a feeling of nature's bounty, thanks to tobacco. It was everything to the Marylanders. It was, in practice, the local currency. One settler, who wrote an account of the place, the Rev. Hugh Jones, called it 'our meat, drinke, cloathing and moneys.' The highest-priced variety, which was sweet-scented, the 'true Virginia,' would flourish only in a few counties in Virginia itself. Maryland grew mainly Orinoco, from South America. By the end of the 1630s a Maryland planter could produce 1,000 pounds a season, which rose to 1,500 or even 1,700 later in the century. It is true that the soil soon became exhausted, the yield dropped, and planters had to move on. But they did so—land was cheap and plentiful—and thus the colony scattered and spread. Only four of the original gentlemen–adventurers stuck it out. But they became major landowners, with manor-houses, which in the next generation were rebuilt in fine brick. One of them, Thomas Gerard, soon farmed 6,000 acres. Four-fifths of the land worked fell within such manors; only one in five freemen claimed land, preferring to work as tenants or wage-earning landowners. Thus society rapidly became far more stratified than in New England.

Maryland had a difficult time during the English Civil War. It was invaded by a shipmaster–pirate and parliamentary fanatic called Richard Ungle who, in conjunction with the still-smouldering and discontented Claiborne, pillaged the settlement, claiming the authority of parliament to do so. At one point, Claiborne and a fellow-parliament man went to London, ingratiated themselves with the authorities, and were made governor and deputy governor. They came back and sought to wage an anti-papist war of terror. Not only did they ban Catholic worship but they passed an Act outlawing sin, vice, and the most minute infractions of the sabbath. Throughout the 1640s and 1650s, religion as well as the proprietorial form of government were the issues: more particularly, the degree of toleration to be allowed to different faiths, and which exactly were to be excluded from it. But by the late 1650s toleration had won the battle. Maryland's Toleration Act, based upon an Act Concerning Religion first pushed through the Assembly in 1649, not only laid down the principle of the free practice of religion but made it an offence to use hostile language about the religion of others, 'such as Heretick, Schismatic, Idolator, Puritan, Independent, Presbyterian, Popish Priest, Jesuite, Jesuited Papist, Roundhead, Separatist and the like.' But you could also be penalized for denying Christ was the Savior, the doctrine of the Trinity, or the Existence of God. A free-thinking Jew, Dr Jacob Lumbrozo, was later bound over for saying that Christ's miracles were 'magicianship and body-snatching.' Thus toleration did not extend to outspoken Jews and atheists. But, for its time, it was an astonishing measure. Henceforward, no Christian whatever could 'bee any wais troubled, molested or discountenanced for or in respect of his or her religion nor in the free exercise thereof.'[98]

The Toleration Act proved invaluable to the colony. The upheavals in England, followed by the reassertion of royal authority and the imposition of the so-called Clarendon Code against dissenters, brought a rush of refugees of all religious persuasions to America, and large numbers chose to go to Maryland, where they lived perfectly happily together. The population had risen slowly to pass the 2,500-mark about 1660, but in the next twenty years it increased by 20,000. Maryland even took in Quakers. During its brief period of Puritan rule they were fined, whipped, jailed, and banished: they claimed the Indians treated them better than 'the mad, rash rulers of Mariland.' But once the Puritans were pushed out and the Toleration Act came back into force, the Calverts got the Quakers back, arguing that they were good citizens and farmers. One leading Quaker preacher, Wenlocke Christison, who had been whipped in Massachusetts, called his Maryland land-patent 'The Ending of Controversie.' Another, Elizabeth Harris, fleeing Boston where the authorities had stripped her and other Quaker women to find marks of witchcraft, gave sermons throughout the province, and George Fox himself came out there in 1672. By the late 1670s, Quakers held regular meetings in fifteen different places in Maryland.

The colony attracted Dutch and German dissenters too. One group, the Labadists, of Dutch origin, had a remarkable German polymath leader, who drew the first good map of Maryland, had himself naturalized, and built up by 1674 an estate of 20,000 acres, making him the largest private landowner in America. Within ten years 100 Labadists were settled on this beautiful domaine, farmed with fine German–Dutch neatness and efficiency, sloping down to the Bohemia River and Chesapeake Bay. They followed the communal teaching of a

Jesuit-turned-Calvinist called Jean de Labardie, sleeping in single-sex dormitories, eschewing private property, observing silence at meals, and denying themselves fires in winter. It was too strict and eventually dispersed, but it set a pattern of individualist utopian colonies in America which persists to this day and, in its own way, is one of the glories of the New World.[99]

Studying the history of these early settlements one is astonished—and delighted—by the variety of it all, and by the way in which accidents, events, and the stubborn individuality of ordinary men and women take over from the deep-laid schemes of the founders. The Calverts of Maryland attempted to create a perfect baronial society in America, based on status rather than wealth. But such an idea, it was already clear, simply did not work in America. The basic economic fact about the New World was that land was plentiful: it was labor and skills that were in short supply. To get immigrants you had to offer them land, and once they arrived they were determined to become individual entrepreneurs, subject to no one but the law. So the manorial courts rapidly gave place to elective local government. St Mary's, like Jamestown, remained no more than a village. People just spread out into the interior, out of control of everything except the law, which they respected and generally observed. But they had to make the law themselves.[100]

It is important to remember that the area of settlement of North America covered thousands of miles of coastline and islands, from Providence Island off the coast of Central America, settled by Puritans in 1629, right up to Newfoundland, first exploited by two groups of fishermen, one Anglican, one Irish Catholic, who lived in separate areas there. Various towns claim to have 'the oldest street in North America.' The best claimants are Water Street, St John's, Newfoundland, and Front Street in Hamilton, Bermuda—neither of them in what is now the United States. In the 17th century there were in fact many scores of colonies, only a few of which would acquire full historical status. And not all of them were English. Leaving aside the French to the north, in Canada, and the Spanish to the south, there were the Dutch on the Hudson. As early as 1614 they settled upriver at Fort Nassau, opposite modern Albany. New York, or New Amsterdam as they called it, was founded by them on May 4, 1626. During the Anglo-Dutch war of King Charles II's day, it was conquered by Colonel Richard Nicolls on September 7, 1664, on behalf of Charles' brother James, Duke of York, who got a charter to found a proprietary colony there. Despite a brief Dutch reoccupation in 1673–4, the English were able to consolidate their power in the Hudson Valley. One reason was that they left the Dutch settlers alone, with their lands and privileges, or rather encouraged them to enter their system of local government. In North America, the settlement and actual ownership of land came first. What flag you lived under was secondary—it was successful farming and ownership of land which brought you personal independence, the only kind which really mattered. Again, on the Delaware River, there was a mixed Swedish–Dutch–Finnish settlement at Fort Christian, dating from 1638. It called itself New Sweden, and when the English finally got control in 1674 it was the sixth change of flag over the colony in half a century. The settlers, overwhelmingly farmers—and good ones—did not mind so long as they were left in peace.

The English, French, and Dutch, as well as the Spanish, scattered all over the Caribbean and the islands of Central America. Some islands changed hands again and again. The English put in the biggest effort, both of men and money. In the years 1612–46 alone, 40,000 English Puritans emigrated to various West Indian island-colonies. The most important by far was Barbados, not least because it became a springboard for colonization in the Carolinas on the American mainland. Barbados, unlike most of the other islands, was not a volcanic mountain sticking out of the sea but a limestone block with terraced slopes. It was uninhabited when the English arrived in 1627. First they tried planting tobacco, then cotton, both unsuccessfully. Then, during the English Civil War, there was an influx of royalist refugees, bringing capital and grand ideas, and Dutch expelled by the Portuguese from northeast Brazil. The latter knew about sugar-planting, and with the help of English capital they set up a sugar industry. From the start it was a huge commercial success, the first plantation boom-economy in English-speaking America. By the middle years of Charles II's reign, there were 400 households in the capital, Bridgetown, 175 big planters, 190 middling, and 1,000 small ones, 1,300 additional freemen, 2,300 indentured servants, and 40,000 slaves. It was easily the richest colony in North America—its sugar exports were more valuable than those of all the other English colonies combined.[101] But with over 55,000 people on 166 square miles it was also the most congested.

A solution was found in 1663 when Charles II gave the Carolinas, unsuccessfully settled under his father in 1629, to a group of eight proprietors, who invited experienced colonizers, from the islands as well as from Virginia and New England, to take up land on easy terms. The Barbadians responded with enthusiasm. A first group came in 1664 to Cape Fear, but had to abandon it three years later. In 1670 a much larger group tried again, laying out Charlestown. This time it worked. Of course there was the usual nonsense from the proprietors of 12,000-acre 'baronies' and private courts—feudal ideas died hard among the more romantic gentlemen–adventurers. The actual Barbadian planters simply ignored the propaganda and went for the most likely sugar-bearing lands on inlets and creeks. They ignored proprietorial guidance in other ways. The proprietors wanted religious toleration, in order to attract the maximum numbers of settlers. The planters were Anglicans, insofar as they were anything: they agreed with Charles II that it was 'the only religion for gentlemen.' So they made it their business to enforce second-class status on people from other faiths. The proprietors opposed slavery. The planters needed slaves, and got them. In one sense the wishes of the proprietors were carried out—the Carolinas got a stratified society, with three classes: a small ruling class of plantation owners or gentry, a large class of laborers, and an enormous number of slaves.

The settlement of Carolina was by no means exclusively Barbadian. There were also Scotch Presbyterians at Port Royal, Huguenots on the Santee, English dissenters west of the Edisto. And there were new waves of settlers from Ireland and France, as well as England. Nor did sugar do particularly well in Carolina. It can be argued that Carolina was saved by rice just as Virginia was saved by tobacco. The lands backing onto Charleston and the other river systems made perfect rice-fields: it was easy to set up a water-control system, and rice-fields, unlike tobacco plantations, did not have to be moved every few years. But the essence of Carolina was a Barbadian slave-owning colony transported to the American mainland. This gave the place a distinctive social, political, and cultural character quite unlike the rest of the emerging colonies, even Virginia. Indeed, as we shall see, without this Barbadian implant, which became in due course the aggressively slave-owning state of South Carolina, the emotional leader of the South, it is quite possible that the American Civil War would not have taken place.

Indeed, it is fascinating for the historian to observe how quickly different regions of the North American coast developed distinctive and deep-rooted characteristics. In Europe, where national forms go back to the pre-scriptive Dark Ages and beyond, these differences remain mysterious. In America there is no real mystery. The books are open from the start. The earliest origins of each colony are well documented. We know who, and why, and when, and how many. We can see foreshadowed the historical shape of things to come. With remarkable speed, in the first few decades, the fundamental dichotomy of America began to take shape, epitomized in these two key colonies— Massachusetts and Carolina. Here, already, is a North–South divide. The New England North has an all-class, mobile, and fluctuating society, with an irresistible upward movement pushed by an ethic of hard work. It is religious, idealistic, and frugal to the core. In the South there is, by contrast, a gentry-leisure class, with hereditary longings, sitting on the backs of indentured white laborers and a multitude of black slaves, with religion as a function of gentility and class, rather than an overpowering inward compulsion to live the godly life.

Not that the emerging America of colonial times should be seen as a simple structure of two parts. It was, on the contrary, a complex structure of many parts, changing and growing more complex all the time. It was overwhelmingly English, as yet. But it was also already indestructibly multiethnic, preparing the melting-pot to come. It was also, compared with limited England, which was obliged to think small in many ways, already a place which saw huge visions and thought in big numbers. Bigness was the characteristic of Pennsylvania from the start. In 1682 William Penn (1644–1718) arrived at New Castle, Delaware bearing a massive proprietary grant from Charles II. He was the son of a rich and politically influential admiral, to whom Charles II was much indebted, both financially and otherwise. Penn had already dabbled in colonizing in the Jersey region, but his new charter, in full and final settlement of a £16,000 debt to his father, was on a princely scale and actually termed 'Pennsylvania' as a proprietary colony. Penn had become a Quaker in 1666, and suffered imprisonment for his beliefs, and he was determined to create a 'tolerance settlement' for Quakers and other persecuted sects from all over Europe. He called it his 'Holy Experiment.' There were Europeans in the area already—Swedes from 1643 at Tinicum Island, 9 miles south of modern Philadelphia, Dutch and English too. But they were few: Penn brought the many. His first fleet was of twenty-three ships, many of them large ones; and plenty more soon followed.

Everything in Pennsylvania was big from the start. In Philadelphia, his city of brotherly love and capital, he had plans for what was later called a 'Garden City,' which he termed a 'Greene Country Town,' spread out on an enormous scale so that every houseowner would have 'room enough for House, Garden and small Orchard.' In fact this did not happen: Philadelphia grew up tightly on the Delaware waterfront, and was manifestly from the start a city built for high-class commerce. But it was quite unlike Boston, whose narrow, winding streets recalled medieval London. Philadelphia was a proud and self-conscious example of contemporary town-planning, made of brick and stone from the start, and much influenced by the new baroque London of squares and straight streets. It was laid out on a large scale to fill, eventually, its entire neck of the river, with twenty-five straight streets bisected by eight. All these streets had proper paving and curbing, sidewalks and spaced-out trees.[102]

Into this colony radiating from Philadelphia, Penn poured multitudes of Quakers, from Bristol and London, many of considerable property, who bought the best lots in the city, but also from Barbados, Jamaica, New York, and New Jersey, from Wales—forming a separate, Welsh-speaking area which kept its culture for generations—and from the Rhineland, founding a city they called Germanopolis. Penn wanted the settlement dense for cultural as well as economic reasons: 'I had in my view Society, Assistance, Busy Commerce, Instruction of Youth, Government of Peoples' Manners, Conveniency of Religious Assembling, Encouragement of Mechanics, distinct and beaten roads.' He wrote home: 'We do settle in the way of townships or villages, each of which contains 5,000 acres or at least ten families ... Our townships lie square; generally the village in the Center, for nearneighbourhood.'[103] But it rarely proved possible to carry out these schemes. In practice, land was simply sold off in lots of a hundred acres or more. Again and again in early America, planning—good, bad, or indifferent—was defeated by obstinate individualism. The European notion of the docile, contented peasant, living in agricultural villages under a squire, was an anachronism, or becoming one. A new pattern of owner-occupiers, producing food for the market, was already to be found in England, where they were known as yeomen. America was a natural paradise for such a class, where they were called simply farmers. And Pennsylvania, with its rich soil, was particularly well adapted to promote their numbers and interests.[104] These farmers pushed inland from the river valleys into the low hills of piedmont of the interior, and then across the first range or corrugation of the Appalachian mountains in what was known as the Great Valley. Here was 'the best poor man's country,' the ideal agricultural setting for a farmer with little capital to carve out not only a subsistence for his family but, through hard and skillful work, a marketable surplus for cash. So Pennsylvania soon became known as the 'Bread Colony,' exporting a big surplus not only of grain but of livestock and fruit. Huge numbers of immigrants arrived and most did well, the Quakers setting the pattern. They were well dressed, they ate magnificently, and they had money jingling in their pockets.[105]

Amid this prosperous rural setting, it was natural for Philadelphia to become, in a very short time, the cultural capital of America. It can be argued, indeed, that Quaker Pennsylvania was the key state in American history. It was the last great flowering of Puritan political innovation, around its great city of brotherly love. With its harbor at Philadelphia leading up the Delaware to Pittsburgh and so to the gateway of the Ohio Valley and the west, and astride the valleys into the southern back country, it was the national crossroads. It became in time many things, which coexisted in harmony: the world centre of Quaker influence but a Presbyterian stronghold too, the national headquarters of American Baptists but a place where Catholics also felt at home and flourished, a center of Anglicanism but also a key location both for German Lutherans and for the German Reformed Church, plus many other German groups, such as Moravians and Mennonites. In due course indeed it also housed the African Methodist Episcopal Church, the first independent black denomination.[106] With all this, it was not surprising that Philadelphia was an early home of the printing-press, adumbrating its role as the seat of the American Philosophical Society and birthplace of the American Declaration of Independence.

But much, indeed most, of this was for the future. The question has to be asked: was early America a hard-working but essentially a prosaic, uncultured place? Were the attitudes of early Americans, when they were not narrowly religious, equally narrowly mercantile? It is a curious fact that, whereas in England the 17th century was an age of great literature—and the actual language used by those New Englanders, such as Winthrop and Roger Williams, who did set down their thoughts on paper, was often expressive and powerful—the New World was strikingly slow to develop its own literature. There is more than one argument here, however. Cultural Bostonian historians of the late 19th century were inclined to dismiss their forebears as horribly uncivilized. Charles Francis Adams wrote: 'As a period it was singularly barren and almost inconceivably somber.'[107] On the contrary, argued the great Samuel Eliot Morison: the Puritan clergy and many leading layfolk were notable in their anxiety to educate and distinguished by their interest in science. They did everything possible to promote intellectual activity by founding schools and colleges.[108] It is true they disliked individualism, a necessary ingredient in cultural creativity. Perry Miller, the historian of the Puritan mind, argued that they were communalists, who believed that government should interfere and direct and lead as much as it could, in all aspects of life. And when necessary it should discipline and coerce too. Puritans saw the individualist as a dangerous loner, meat for the Devil to feed on. As one of them, John Cotton, put it, 'Society in all sorts of humane affairs is better than solitariness.'[109] The Puritans believed they had the right to impose their will on this communally organized society.

John Davenport of Connecticut summed up the entire Puritan theory of government thus: 'Power of Civil rule, by men orderly chosen, is God's ordinance. It is from the Light and Law of Nature, because the Law of Nature is God's Law.' They did not accept that an individual had the right to assert himself, in religious or indeed in any matters. When in 1681 a congregation of Anabaptists published an attack on the government of Massachusetts Bay and appealed to what they called the 'tolerant spirit' of the first settlers there, Samuel Willard, minister to the Third Church in Boston, wrote a pamphlet in reply, with a preface by Increase Mather, saying: 'I perceive they are mistaken in the design of our first Planters, whose business was not toleration but were professed enemies of it, and could leave the world professing they *died no Libertines*. Their business was to settle and (as much as in them lay) to secure Religion to Posterity, according to that way which they believed was of God.'[110]

In this kind of would-be theocracy, it was difficult for cultural individualism to flourish. But the elites proposed—and the people disposed themselves otherwise. Sermons, tracts, and laws say one thing; town and church records often show that quite different things actually happened. The New England rank and file contained many individualists who would not be curbed by Puritan leaders. If there were enough of them, there followed a heated town-meeting, an unbridgeable difference, a split—and a move by one faction. A study shows that this is exactly what happened in Sudbury, Massachusetts, in the 1650s, leading to the founding of Marlboro, Massachusetts.[111] The head of the conservative faction, Edmund Goodnow, put his case thus: 'Be it right or wrong, we will have [our way] . . . If we can have it no other way we will have it by club law.' To which the leader of the younger generation, John How, replied: 'If you oppress the poor, they will cry out.

And if you persecute us in one city, wee must fly to another.' And so they did.[112] The early Americans were lucky people—they had the space for it.

Individualism did assert itself, therefore, even in Puritan New England. Indeed in a sense it had to, for America was a do-it-yourself society. Potential settlers were warned they would have to depend on their own skills. A London broadsheet of 1622 has survived entitled 'The Inconvenients that have happened to Some Persons which Have transported Themselves from *England* to *Virginia*, without Provisions Necessary to Sustain Themselves.' It advised that settlers should take arms, household implements, and a list of eighteen tools, recommended to be carried in duplicate, from axes to saws and shovels, not excluding a grindstone.[113] Early settlers erected their own huts, and made their own furniture when necessary.

But it was not always necessary, even in the earliest decades. It is a notable fact that America, from the start, had a powerful attraction for skilled men. The reason was clear. One of the original *Mayflower* backers, Robert Cushman, wrote that England was a poor place for an honest man to raise a family. The towns, he said, 'abound with young tradesmen and the hospitals are full of the ancient. The country is replenished with new farmes, and the alehouses are filled with old labourers. Many are those who get their living with leaving burdens; but more are fain to burden the land with their whole bodies. Multitudes get their means of living by prating and so do numbers more by begging.' He complained that 'even the most wise, sober and discrete men go often to the wall, when they have done their best.'[114] He and others pointed out that a skilled young man in England had a poor economic future, and no status at all, since status was entirely dependent on land, which he had virtually no hope of acquiring. In America, he could get higher wages, and his raw materials were cheap. And there was a strong likelihood he could get land too.

So there was no shortage of craftsmen in the colonies. Carpenters and joiners were particularly fortunate. Not only was wood plentiful and in great variety: it was also cheap in sawn lengths. One of America's earliest innovations was the rapid spread of water-driven sawmills. England had no real tradition of mechanical sawing. America by contrast had masses of timber located near fast-flowing streams. So mills were built everywhere but particularly in New England. They saved labor—the biggest item in furniture-making. One small water-powered mill could produce seven times the output of two skilled sawyers. There was waste of material—so what? Timber was plentiful; it was human labor which was scarce. Settlers were moving from an economy of scarcity to an economy of plenty, where men were valuable to a degree unknown in Europe.[115] It was this fact which shaped the early culture.

Excellent furniture was made in 17th-century America, and a surprising amount of it has survived. There were skilled glassmakers from the start, for glass was difficult to transport with safety and had if possible to be manufactured on the spot. We know there were professional glassmakers on the first Virginia voyage of 1608. They flourished because raw materials, particularly wood, were cheap and easily available. The shortage of skilled labor attracted foreigners as well as Englishmen: the first glass-factories in America, both in Jamestown, were run by Venetians and Poles.[116] The same principle applied to pottery. Suppliers were unwilling to ship pots across the Atlantic, saying there was no money in it. So potters went instead. The skilled English ceramicist Philip Drinker was in Charlestown, Massachusetts, by 1635, working away. The Dutch 'pottmaker' Dirck Clausen was turning out ware on Manhattan Island by 1655. Redware was the quintessential American pioneer ceramic, dictated by the clays available (porcelain was not successfully manufactured in America until the 19th century). It came with simple abstract geometric patterns, rather like proto-classical Greek pottery of the 8th century BC, then with written mottoes: 'Mary's Dish,' 'Clams and Oysters,' and 'Pony Up the Cash.' But very little survives from the 17th century.[117]

Records exist of dozens of other categories of craftsmen at work in America by the mid–17th century. Thus, by the 1630s, two expert shoemakers, Henry Elwell and Philip Kirkland, were already settled in Lynn, Massachusetts, later to become a major shoe-manufacturing center. They specialized in women's shoes.[118] Church silver was in demand from early on—even the Puritans liked it. In fact by the end of the 17th century superb silver, almost on a par with the European best, was being made in Boston.[119] One of the interesting points about New England craftsmen is that they came from all ranks—something impossible in 17th-century England. The best Boston silversmith–goldsmith was Jeremiah Dummer, born in 1645, the son of a leading landowner. He was a member of the Boston elite and invested in shipping, but he was not above slaving at his bench making candlesticks for prominent families and Boston churches.[120] That was the kind of social mobility which augured well for America's future.

It is, however, a futile quest to look for much in the way of fine art produced in 17th-century America. Only about thirty paintings have survived from this period, all of them amateurish. We know the names of men described as painters but it is impossible to match them convincingly to the surviving paintings. Any who practiced as painters did so intermittently, it seems.[121] One or two Dutchmen in New York— Gerrit Duyckinck (1660–1712) for instance—combined portrait painting with other crafts, such as glazing. There were virtually no architects in the first decades. Men, even men with many acres, designed their own houses: the tradition of the rich amateur architect began early in America. Thus there is Adam Thoroughgood's house in Norfolk, Virginia, 1636–40, of brick, in a mixture of Elizabethan and Jacobean styles, with a huge, medieval-type chimney. Another early Virginia house, built by Arthur Allen and known as Bacon's Castle, had towers, front and rear, massive chimney-stacks, and Flemish gables.[122] All the writers were amateurs too, be they authors of works of travel, like John Smith or William Bradford, or Puritan poets like the metaphysician Edward Taylor of Westfield, Massachusetts (c.1644–1729) or Michael Wigglesworth, whose theological poem, *Day of Doom*, published in Cambridge in 1662, popularized Puritan dogma in ballad meter.

Yet there was one sense in which early America was abreast of the European world, even ahead of it. It had a deep-rooted, and increasingly experimental, political culture. Here the English tradition was of incomparable value. It was rich and very ancient. By comparison, the French and Spanish settlers knew little of the art of politics. Both France and Spain, as geographical entities with national institutions, were still recent developments in the 17th century, and neither had much experience of representative government, or even at that date of unified legal systems. By contrast, England had been a national unity since the 9th century, with forms of representation going back to that date and even beyond. Its common law began to mature as early as the 12th century; its first statute of the realm, Magna Carta, was enrolled as early as 1215; its parliaments, with their knights of the shire and their burgesses of the towns, had had a continuous history since the 14th century, as an institution which passed laws for all the people and raised revenues from all of them too. Behind the Englishmen who came to settle in Virginia and Massachusetts, in Carolina and Maryland and Pennsylvania, were 1,000 years of political history.

Moreover, it must be said that the period at which this tradition was implanted in America was also of great significance. English America 'took off' as a viable social and economic entity in the three decades 1630–60. That was when its population reached a critical mass large enough to produce self-sustaining growth. And it was during these three decades that there took place in England a veritable explosion of political argument and experiment, in which, perhaps for the first time in history, the fundamentals of participatory and democratic politics were discussed.[123] It could be said, indeed, that modern politics was invented in the England of the 1640s, and the English settlers in America were, in a sense, participants in this process—the coming and going between England and America during this decade was of great political significance. If the English had first settled America in the first half of the 16th century, during the Tudor autocracy, or in the first half of the 18th century, during the long calm of the Whig supremacy, it might have been a very different story. But they settled it during the first half of the 17th century, when the smoldering dispute between king and parliament reached its climax, burst into flames, and was resolved by a parliamentary victory, albeit a qualified one. The English settling in America brought with them this political tradition, just when it was at its most active and fruitful.

The early settlers, then, came from an intensely religious and political background, and most of them were independent-minded, with ingrained habits of thinking things out for themselves. And it was the earliest settlers who counted most. It is almost a law of colonization that the first group, however small, to set up an effective settlement has more effect on the political and social character of the colony than later arrivals, however numerous. Until Charles II's reign, indeed until the 1680s, by which time the English had effectively wrested maritime control of the North Atlantic from the Dutch, the English crown made little attempt to supervise closely what went on across the Atlantic. It awarded charters, then let the colonists get on with it. This was an old tradition in England, applying particularly to local government through justices of the peace (magistrates), sometimes referred to as 'Self-government by the King's Command.' So governors, however appointed or elected, operated independently of England. And every colony, almost from its inception, and in most cases within a year of its foundation, had some kind of representative assembly. Electing people was one of the first things a settler in America learned to do. Moreover, many offices in America which, in England, would have been filled by appointment, by lords lieutenant or even by the crown—key offices in the administration and enforcement of the law—became elective from the start.

The American tradition of electing large numbers of public officers directly took deep root quickly. Those men so chosen might be
very humble people. Forty years after the foundation of Maryland,
governors were complaining that many men chosen as justices or
sheriffs could not even sign their names.[124] Virtually everyone voted
for somebody or other. In orthodox Calvinist New England, voting
rights, to begin with at least, were confined to church members.
Elsewhere all freemen had them, as a rule. In Maryland, for instance,
at least from the 1650s, all freemen voted for four delegates per
county to serve in the Lower House. It was linked to service in the
militia, compulsory for every male over sixteen: if you fought for the
colony, you voted in it. In Carolina, you voted automatically if you
took up a 50-acre plot, though to be a delegate you were supposed
to have 500 acres. These 50-acre men were lowly folk who would
never have been allowed to take part in politics in Europe: Thomas
New, who came out in 1682, described them as 'tradesmen, poor
and wholly ignorant of husbandry . . . their whole business was to
clear a little ground to get bread for their families.'[125] But they voted
all the same. Some of the Carolina elite, who originally tried to call
themselves 'landgraves' and 'cassiques,' grumbled at this. As one of
them put it, 'It is as bad as a state of Warre for men who are in want to
have the making of Laws over Men that have Estates.'[126] But it was a
fact of colonial life, even in Carolina, explained partly by the proprietors themselves being mostly absentee landlords. As an independent-
minded Caroliner wrote in the 1670s: 'By our frame [of government],
noe bodys power, noe not of any of the Proprietors themselves were
they there, is soe great as toe be able to hurt the meanest man in the
Country.'[127]

You may ask: how did the early settlers reconcile their acceptance
that even the meanest had rights—including rights to vote—with the
institution of slavery? The point was to be made with great force by Dr
Samuel Johnson at the time of the American Revolution, and it echoes
through American history: 'How is it that the loudest YELPS for LIBERTY
come from the drivers of Negroes?' The answer is that America was
only gradually corrupted into the acceptance of large-scale slavery. The
corruption entered through Carolina, whence it came from Barbados.
In the West Indian islands, those occupied by the Spanish, Portuguese,
and French under Catholic teaching, slaves were treated as actual or
potential Christians, with souls and rights—not just property. In the
islands occupied by the English and Dutch Protestants, who got their

doctrine about slavery from the Old Testament, slaves were seen as legal chattels, with no more rights than cows or sheep. The Barbadian planters in Carolina, who set the tone, never troubled themselves to Christianize their slaves and even prevented others from doing so. In any case, it made no difference. Early laws laid down that baptism did not change a person's free or unfree status. Such laws spread north. Thus in 1692 a Maryland statute insisted that baptism did nothing to change a black's servile status.

Carolina was the first slave state, properly speaking. From the start it imported black slaves, even before it acquired rice as an agricultural staple. A Carolina promotion pamphlet of 1682 stated flatly: 'without [negro slaves] a planter can never do any great matter.' The same year, a settler told a friend: 'Negroes are more desirable than white servants.'[128] This was because a white indentured servant cost £2–£4 a year in capital investment. A slave cost £18–£30 outright, plus the likelihood of breeding. Young, healthy female slaves were particularly valuable for this reason. In Maryland, slavery grew only slowly. Until the late 1680s, an estate was more likely to be run by indentured labor. Early probates from the years 1658–70 show that only 15 out of 150 estates had slaves. But the treatment and legal status of slaves, especially black ones, declined as the 17th century wore on. A statute of 1663 recognized black service as perpetual, writing of 'Negroes and other Slaves who are incapable of making Satisfaction by Addition of Tyme.' A black had to prove he was under limited contract by producing documents, otherwise the law assumed he (and his children) was a chattel slave.[129]

Much legislation of these years strengthened the hands of the planters against slaves. In Carolina, slavery was an early source of corruption in politics. Slavers were heard to boast that they could 'with a bowl of punch get who they would Chosen of the parliament and afterwards who they would Chosen of the Grand Council.'[130] The Barbadians in Carolina also enslaved numbers of Indians. This was strictly against the law. The policy of Charles II's government was 'to get and continue the friendship and assistance of the Indians and make them useful without force or injury.' It laid down (1672) that enslavement of Indians was forbidden 'upon any occasion or pretence whatsoever.'[131] But the Barbadian planters induced Indian tribes—who did not need much persuasion—to make war on others to produce Indian slaves. An early anti-slaver in Charleston, John Stewart, wrote angry letters home to England, protesting about the behavior of the

Barbadians, or 'Goose Creek Men' as he called them (this was their densest area of settlement). He said that one of their leaders, Maurice Matthews, an important man because he was the official surveyor as well as a planter and slaver, was 'Hell itself for Malice, a Jesuit for Design politick.' Stewart eventually got Matthews sacked for slaving. His predecessor, Florence O'sullivan, was as bad: 'a very siddencious, troublesome Man, an ill-natured buggerer of children,' as other settlers complained.[132]

All the same American slavery was on a small scale, Carolina being the only exception. Large-scale slavery was an 18th-century phenomenon. Even by 1714 there were fewer than 60,000 slaves in the whole of the English colonies on mainland America. Thereafter the numbers grew steadily—78,000 by 1727, 263,000 by 1754, and 697,000 at the first census in 1790. So in Dr Johnson's day the existence of huge, black, servile multitudes in America was a recent development, growing daily—one reason he was so outraged by it. In early settler times, by contrast, over most of the colonies, slavery was very marginal, blacks were almost invisible, and servile work was seen in terms of indentured whites, who served their terms, became freemen and soon owned land and exercised their votes. So the leading settlers, creating their assemblies, were not struck by the paradox of free whites and blacks who had no rights at all. That came later—when it was too late and slavery was deeply entrenched.

The early lack of interest by the English government in the American mainland colonies led, therefore, both to a rapid growth of legislative assemblies, with wide franchises and, rather later, to an unregulated growth of slavery. When the home country first began to take a closer interest, during Charles II's reign, its main concern was with regulated trade. By an Act of 1660, 'enumerated' commodities from the English mainland colonies in America had to be sent direct to England. These included tobacco, cotton, wool, indigo, to which were later added tar, pitch, turpentine, hemp, masts, yards, rice, copper, iron, timber, furs, and pearls. These included all the staples of the South, chiefly tobacco, rice, and indigo. But, further north, leading exports like fish—for a long time the chief staple of New England—grain, and other foods were kept out of England by high tariffs. So the North, especially New England, sent to the West Indies and southern Europe dried fish, pickles and pickled beef and pork, horses and livestock, plus building materials. New York and Philadelphia sent flour and wheat. As, by the closing decades of the 17th century, the West Indies was concentrating largely on sugar and tobacco, it imported food and wood cheaply from the mainland colonies. In return, they got molasses, to turn into rum for the fishing fleet and to buy slaves. If they were lucky, they got gold and silver too.[133]

The cash was welcome, because under this mercantalist system the balance of trade was in England's favor and there was a chronic shortage of coin in America. Any specie they got from the West Indies 'seldom continues six months in the province before it is remitted to Europe.' Business was accounted in pounds, shillings, and pence, but English coin was seldom seen. Large-scale internal trade was done in drafts and bills of exchange, local trade in barter. Termly bills of students at Harvard College were for decades met by produce, livestock, and pickled meat. In 1649 one student is recorded as settling his bill with 'an old cow.' The accounts for the college's first building includes one item: 'Received a goat 30s plantation of Watertown rate, which died.'[134] All kinds of dodgy procedures were used to get round the shortage of specie. Thus in Virginia and Maryland, receipts for tobacco deposited in warehouses circulated as cash. Then colonial governments began to create paper credit—a slippery slope. In 1690 Massachusetts created bills of credit for payment to militia soldiers. This example was eagerly followed, and such paper money acceptable, at a discount, for silver and marked with a date of payment soon spread. But this was followed by larger, weaker issues, which discredited it. So American paper money tended to be rotten from the start, and distrust of it—followed by distrust of the banks which circulated it—became deep-rooted in Americans from an early date, and was to have very long-term consequences. Primitive 'loan banks,' issuing credit on the security of real estate, made the financial system even more suspect. Parliament in England, instead of solving the problem by ensuring that America got enough coin, stamped on the consequences of the shortage as an abuse. In 1751 it forbade issue of further bills of credit as legal tender in New England, and in 1764 it extended the ban to all the colonies. This both infuriated the Americans and proved ineffective, since by then an estimated $22 million of unlawful paper was already in circulation. It was an early example of the way in which government from both sides of the Atlantic would not work.[135]

Irritation with England, whenever the home government exercised any authority at all, was an early American characteristic. It is a curious fact that the first printed work ever published in America, put out by Stephen Daye in Cambridge, Massachusetts, in January 1639 (he had come out only the year before) was a broadside, *The Oath of a Free-Man*, attacking the oath of allegiance all settlers had to swear to the English crown. Taxes in goods were exacted on settlers, without much success to be sure, and some of this got back to the English crown. But in return it was hard for the colonies to see what they got, other than notional protection. The home government certainly did nothing to defend outlying farms or plantations from occasional Indian raids—for that, the settlers were left entirely to their own devices. On the whole, relations between English settlers and Indians were good and there was surprisingly little conflict. When it occurred, the settlers were usually to blame. But not always. The Indians were capable of unpredictable changes of mood, and downright bellicosity under a certain type of leader. Settler–Indian relations were complicated by disputes among Indian tribes, which were often in a state of perpetual warfare with each other.

This was the origin of the Pequot War in the 1630s. It began with a dispute between the Pequots and the Mohicans in the Connecticut River area, over the valuable shoreline, whose shells and beads were collected for wampum, the Indian form of exchange. Neither the English nor the nearby Dutch would come to the aid of the Mohicans and they were beaten. The Pequots, 'grown arrogant,' attacked an English sea-captain, John Stone, and his seven companions, who were trading upriver. They were murdered. Two years later, there was another murder, on Block Island, of a New England trader, John Oldham. In response, the Massachusetts governor, John Endecott, sent three armed vessels, which destroyed the two Indian villages believed to have been guilty of these crimes. In May 1637, the Pequots retaliated by attacking Wethersfield, Connecticut, killing nine people and abducting two. This in turn provoked a combined operation by all the militia forces of Massachusetts and Connecticut, accompanied by several hundred Narragansett and Niantic Indians, who together surrounded the main Pequot fort on June 5, 1637, and slaughtered 500 Indians, men, women, and children, within it. The village was set on fire and most of those who tried to escape were shot or clubbed to death.[136] This bloody war against the Pequots, which seems to have ended Indian raiding in New England for a generation, was conducted without any assistance from England.

Further south, in the Hudson Valley and Virginia, wars among the Indians, and with settlers over fur and trading, continued sporadically. In June 1644 as many as 350 settlers south of the James River were massacred by the warriors of a chieftain called Opechancanought. This led to large-scale counterattacks by the governor of Virginia, William Berkeley, and the acting governor Richard Kemp. Again, only the local militia was employed. There was a major flare-up in Dutch territory near the Hudson the same year. Near New Amsterdam, 120 Algonquins, fleeing from their Mohawk enemies, were massacred by the Dutch in retaliation for early murders. Various Algonquin tribes then united for a vengeful raid against Dutch settlements, but were defeated when 150 heavily armed Dutch killed 700 Indian warriors near Stamford, Connecticut, in February 1644.

In Virginia it was a constant complaint among settlers pushing into the interior that the authorities never provided them with any protection from hostile Indians. The trouble with Virginia, as indeed with other colonies, is that although its latitude, that is its extent along the coast, was fairly accurately determined by original charters, its extension inland was indefinite. There was an early conflict of interest between the large plantation-owners of the Tidewater, who dominated the assembly and ran the government, and the smaller farmers who penetrated into the foothills, or piedmont, of the Appalachian ridges, and beyond them. In fact almost from the start two very different societies began to emerge. On the coast, there was a characteristically 'Southern' civilization, slave-owning, tobacco-growing, cultured, elitist, leisured, and there was a much more rugged farming society in the interior—a bifurcation which was eventually to find constitutional expression when West Virginia hived off from the rest during the Civil War and formed a separate state.

Early in 1676, the small farmers up the James River became convinced that the plans of Sir William Berkeley, the royal governor, were inadequate, and that this sprang from the fact that they were under-represented in the House of Burgesses, dominated by the Tidewater aristocracy. They got a wealthy planter, Nathaniel Bacon, to lead them, both against the Indians and to remonstrate with the governor. Berkeley accused Bacon of treason and had him arrested when he arrived in Jamestown with 500 men on June 6. Then having—as he thought—asserted his authority, he let Bacon go, and the result was an angry confrontation in which Bacon demanded a commission of inquiry into the government's failure to police the Indians, and authorization to raise an army. The governor then fled, to the eastern shore, and Bacon rampaged for three months in the capital, raising volunteers and plundering the estates of Tidewater grandees. He denounced

Berkeley and his 'clique' as 'sponges' who 'sucked up the Publik Treasure.' But on October 26, 1676 he abruptly succumbed to what was called 'a severe attack' of 'the bloody flux.' Without his leadership, the rebellion collapsed. When a party of English redcoats, summoned by the governor three months before, finally arrived in November from England, only eighty slaves and twenty 'servants' still defied the authorities, turning a serious white man's revolt into a servile one, which was soon suppressed.[137]

Bacon's Rebellion showed how fragile authority in America was in these early times. In the same year there was another demonstration of its fragility in New England. The Puritans had not been particularly assiduous in converting Indians to Christianity. But one of them, John Eliot (1604–90), had done his best from 1646 onwards, preaching widely to the tribes and translating the Bible into Algonquin. His converts were known as 'praying Indians,' and since they often became detribalized he settled them in what were known as 'praying towns.' One of these converts, Sassamon, actually attended Harvard, though he seems to have lapsed afterwards and became a follower of 'King Philip,' also known as Metacom, who was a chief and sacham, or holy man. Sassamon was murdered early in 1676, and since he had once again become a Christian before this occurred, three men of the Wampanoag tribe, who were heathen, were held to be guilty and executed by the Plymouth authorities. That was the ostensible cause of King Philip's War, a conflict between Christianity and Indian religious culture, but it is likely that increasing pressure on Indian land by the rapidly expanding Massachusetts colony was the real reason.

Throughout the summer and autumn of 1679, Philip and his men destroyed white farms and townships over a large area, and at one point came within 20 miles of Boston itself. If Philip had been able to organize a grand coalition of the Indians, it is quite possible he could have extinguished the entire colony. However, the inability to unite against their white enemies was always the fatal weakness of the Indians. The Massachusetts governor, another Winthrop, raised the militia, which was dispatched in parties of armored dragoons, from 10 to 150 and more, to meet the danger whenever large parties of Indians were reported to be gathering. This warfare continued throughout the winter of 1676–7 and the spring, until in August Philip himself was cornered and killed. Thereafter it was a question of isolating small groups of Indians, or hunting them down in the backwoods, though some fighting went on in New Hampshire and Maine until 1678. The casualties on both sides were very heavy. Every white family in New England was involved in one way or another. It is probably true to say that no war in American history produced so many killed and wounded in proportion to the total population. It goes without saying that no assistance was forthcoming from England. Without the local militia, which proved itself in the end a formidable fighting machine, far superior to its English counterparts, the Indians could not have been held at bay.[138] The war was fought with great bitterness. When Philip was finally killed, his head was hacked off and sent for public display in Boston, his hands to Plymouth. It left deep scars among the survivors, and it had a profound effect on the Puritan ministry, who felt that the near-disaster indicated divine displeasure with New England.[139] It had been, as they put it, 'So Dreadful a Judgment.'

The ravages of King Philip's War, the break-up of families it brought about, and the widespread feeling that the godly people of New England had somehow become corrupted and were being punished in consequence, was the long-term background to the Salem witchcraft hysteria of 1692. The immediate background, however, was a prolonged disruption in the normal government of the colony. From 1660 onwards, the authorities in England had been taking an increasing interest in America, and were endeavoring to recover some of the power that had been carelessly bestowed on the colonists during the early decades. This tendency increased sharply in the 1680s. In 1684 the crown revoked the original charter of Massachusetts, which gave it self-government, and in 1686 appointed Sir Edmund Andros (1637–1714) governor. Andros was a formidable public official who had been sent by James, Duke of York, in 1674 to run his proprietary colony of New York, seized from the Dutch. He made the place the strategic focus of England's North American empire, enlarging the anchorage, building warehouses, establishing an exchange, laying down regulations to foster commerce, and building forts. It has been said, 'He found New York a village, he left it a city.'[140]

That was all very well, but Massachusetts wanted to run its own affairs, and the arrival of Andros as 'Governor of Our Dominion of New England,' coinciding as it did with the accession of the Duke of York, now James II and an open Catholic, to the throne, was not welcome in Boston. It was clear that King James wanted to unite all the northern colonies into one large New England super-colony, and that Andros was his instrument. When a group of Whig nobles invited

William of Orange to England, to become its Protestant king, and James fled, the New England elite took the opportunity to stage their own 'Glorious Revolution,' put Andros behind bars, and resumed their separate existences. The president of Harvard, Increase Mather (1639–1723), was sent to London to negotiate a new settlement and charter. It was while he was away doing this that the witch hysteria broke out. It is important to grasp that what, in retrospect, was a breakdown in the rule of law occurred when the entire political frame of New England was in a state of suspension and uncertainty.

There was nothing new about witchcraft, or the suspicion of it, in New England. Religious dissidents, such as Quakers, were regularly stripped and examined for its marks. The fear of the witch was linked to fear of the Devil, his or her master, and the Devil was omnipresent in the moral theology of the 17th century. Conviction and hanging of witches was not common in Massachusetts, but it occurred from time to time. In Connecticut we know of ten cases of witches being hanged for 'familiarity with the Devil.' Rhode Island alone had an unblemished record in this respect. Nor were Calvinists the only people who believed strongly in the reality of sorcery. Witches were prosecuted in Anglican Virginia. There was a case in Catholic Maryland too, where a 'little old woman,' suspected of being a witch, was cast into the sea to appease an inexplicably violent storm.[141] What made the Salem case in 1692 unique was the scale and suddenness of the accusations, the sinister farce of the trials, and the severity of the punishments.

There may be an explanation for this too. The huge religious controversies and wars which had convulsed Europe from over a century since the outbreak of the Reformation in the 1520s, came to a climax in the first half of the 17th century, with the appalling Thirty Years War in central Europe and such marginal catastrophes as the Civil War in Britain. But with the Peace of Westphalia in 1648, the world slowly turned to secularity. It was as though the volcanic spirit of religious intolerance had exhausted itself and men were turning to other sources of dispute.[142] But there were nonetheless periodic convulsions of the dying beast of fanaticism. In the 1680s Louis XIV, at the urging of Catholic extremists, revoked the toleration for Protestants accorded by the Edict of Nantes. The same decade, in Protestant London, there was a violent mob-led hunt for Catholic subversives led by the renegade Titus Oates. The Salem hysteria was part of this irrational, recidivist pattern.

The ostensible facts of the Salem case are not in dispute.[143] Early in 1692, two children in the household of the vicar of Salem, Samuel Parris—his daughter Betty, aged nine, and his niece Abigail, eleven— began to be taken with hysterical fits, screaming and rolling on the floor. Their behavior affected some of their friends. Neither girl could write and they may not have been able to read. They were fond of listening to the tales of Tituba, a black female slave who formed part of the household. When the girls' behavior attracted attention, they were medically examined and closely questioned by their credulous father and local busybodies. The girls finally named Tituba as the source of their trouble and she, under pressure to confess witchcraft, admitted she was a servant of Satan, and spoke of cats, rats, and a book of witchcraft, 'signed by nine in Salem.' Two names of local women were screamed out by the girls, and this set off the hunt.

It soon attracted a great deal of interest, not just in Salem but in the entire neighborhood, including Boston. One of those who took a hand in it was Increase Mather's learned minister-son, Cotton Mather (1663–1728), a young but already prominent member of the Boston elite. The authorities, such as they were, also took a hand. In mid-May, the temporary governor, William Philips, arrived in Salem and, impressed by what he heard—perhaps horrified is a better word—set up a special court, under William Stoughton, to get to the bottom of the matter. This of course was a serious error. The ordinary law might or might not function fairly in sorcery cases, but a special court was bound to find culprits to justify its existence. And so it did. Its proceedings were outrageous. Accused persons, men and women, who confessed to using witchcraft, were released—as it were rewarded by the court for 'proving' the reality of the Devil's work in Salem. The more sturdy-minded among the accused, who obstinately refused to confess to crimes they had not committed, were judged guilty. The hysteria raged throughout the long New England summer. By the early autumn, fourteen women and five men, most of them respectable people with unblemished records, had been hanged. One man, who refused to plead at all, was pressed to death with heavy stones, the old English *peine forte et dure*, for contempt of court—the only time it was ever used in America's history. Over 150 people were awaiting trial in overcrowded jails, and some of them died there. The reaction set in during October, when prominent people, including the governor's wife, were 'named.' The authorities then came to their senses. The special court was dissolved. Those under arrest were released.

The Salem trials can be seen as a throwback to an early age of credulity. In a sense they were. But they were more complicated than that. Belief in witches and the modern, sceptical mind were not opposite polarities. Cotton Mather who, at the climax of the hysteria in October, published a tract, *Wonders of the Invisible World*, 'proving' the existence of witchcraft and its connection with the Devil, was not an obscurantist opponent of science. Quite the contrary. He was descended from the Cottons and Mathers, two of Boston's leading intellectual families since the inception of the colony. In the late 17th century, the new empirical science and older systems of belief overlapped. Isaac Newton, greatest scientist of the age, was an example. He was fascinated by all kinds of paranormal phenomena and his library contained large numbers of books on astronomy. Cotton Mather was a learned man and a keen scientist. He was not only awarded an honorary Doctorate of Divinity by Glasgow University but was also elected a Fellow of the Royal Society, then the leading scientific body in the world. He popularized the Copernican system of astrophysics in the colonies.[144] He regarded the empirical study of nature as a form of worship, a notion pursued by the New England Transcendentalists in the 19th century. For him, his numerous scientific interests were in no way opposed to his religious beliefs but were an extension of them. He argued that the existence of witches was a collateral proof of the life to come: 'Since there are witches and devils,' he wrote, 'We may also conclude that there are immortal souls.'

It was in fact precisely Cotton Mather's scientific interests which made him such an enthusiastic witchhunter. He believed that the trials, if pursued vigorously enough, would gradually expose the whole machinery of witchcraft and the operations of the Devil, thereby benefiting mankind enormously. But here he disagreed with his own father, another learned, scientific gentleman. Increase Mather held that the very operation of hunting for witches might be the work of the Devil, and characteristic of the way the Great Deceiver led foolish men into wickedness. His return from England in the autumn of 1692 was one factor in the ending of the witchhunt, and the following spring he published a book, *Cases of Conscience Concerning Evil Spirits*, which drew attention to the risks of public delusions and suggested that the real work of the Devil was the hanging of an innocent old woman. Increase Mather was instrumental in persuading the General Court of Massachusetts to pass a motion deploring the action of the judges. Members of the jury signed a statement of regret, indemnities were granted to the families of the victims who had been hanged. Some of those who had made false statements later confessed to them, though in one case not till many years had passed. These events, and Increase Mather's book, virtually put an end to trials for witchcraft in America.

The Salem trials, then, can be seen as an example of the propensity of the American people to be convulsed by spasms of self-righteous rage against enemies, real or imaginary, of their society and way of living. Hence the parallels later drawn between Salem in 1692 and the 'Red Scare' of 1919–20, Senator McCarthy's hunt for Communists in the early 1950s, the Watergate hysteria of 1973–74, and the Irangate hunt of the 1980s. What strikes the historian, however, is not just the intensity of the self-delusion in the summer of 1692, by no means unusual for the age, but the speed of the recovery from it in the autumn, and the anxiety of the local government and society to confess wrongdoing, to make reparation and search for the truth. That indeed is uncommon in any age. In the late 17th century it was perhaps more remarkable than the hysteria itself and a good augury for America's future as a humane and truth-seeking commonwealth. The rule of law did indeed break down, but it was restored with promptness and penitence.

The real lesson of the affair, a contemporary historian may conclude, is not the strength of irrationality but the misuse of science. Cotton Mather had been trained as a doctor before he decided to follow his father into the ministry, and was a pioneering advocate of smallpox vaccination, especially of the young. He was a keen student of hysterical fits as a medical as well as a religious phenomenon. Both he and his father took an interest in the behavior of children under extreme psychic and physical stress, though they reached different conclusions. He, and the interrogators of the children, were manifestly anxious to hear tales of possession and sorcery and devilish activity to confirm their preconceptions, and the children intuited their need and supplied it. We have here a phenomenon by no means confined to the 17th century. Perhaps the best insight into the emotional mechanism which got the Salem trials going can be provided by examining some of the many cases of child-abuse hysteria, and cases in which children were alleged to have been abused by Satanist rings, occurring in both the United States and Britain in the 1980s and 1990s. The way in which children can be encouraged, by prosecuting authorities, to 'remember' imaginary events is common to both types of case. The Salem of the 1690s is not so far from us as we like to think.[145]

Cotton Mather himself is a significant and tragic figure in American history. When he was born in 1663, New England still seemed outwardly the religious commonwealth the Pilgrim Fathers had wished to create, through Congregationalism was already losing its physical grip on the machinery of government and signs of a growing secularization were manifest. He was born to the Puritan purple: both his Christian and his surname proclaimed it. He entered Harvard, its school for clerical princes, at the age of twelve—the youngest student ever enrolled there. He seemed destined to inherit not merely his father's mantle as president of the college, but leadership of the entire New England intellectual and religious community. But neither happened. Indeed he was publicly defeated for the presidency of Harvard, and when he was finally asked, in 1721, to become head of its rival, Yale, which had been founded in 1701 and moved to New Haven in 1716, it was too late. He was too old and he refused.

Mather spent his entire life industriously acquiring knowledge and regurgitating it. He learned seven languages, well. He was a living rebuke to the proposition that New England was uncultured and lacked authors. He wrote 450 books. Many more remained unpublished but those that saw print were enough to fill several shelves—and all this in addition to seven thick volumes of diaries.[146] He stood for the proposition that, in America, religion was the friend of enlightenment. He promoted schemes of public charity for the poor and infirm. His books, *The Family Well Ordered* and *The Ornaments for the Daughters of Zion*, put forward sensible and in some ways surprisingly modern views about the role of parents in education, and especially of girls. He wrote about the rights of the slaves and the Indians. He tried to bring order and sense to the medical and legal professions—something high-minded American intellectuals have been trying to do ever since. He was not exactly a man for all seasons—he was too opinionated and cantankerous for that—but he was a man for all disciplines, an all-purpose American do-gooder and right-thinker who adumbrated Benjamin Franklin. But he lacked Franklin's chance to operate on a world stage. He was damned at the time and for posterity (until recent scholarship came to his rescue) as the man behind the Salem witch-trials. He appeared to move effortlessly from Young Fogey to *Laudator Temporis Actae.*

Long before his death Cotton Mather recognized that the times had moved against him and that the kind of religion the Puritans had brought to America was changing beyond recognition. In 1702 he published his masterpiece, the *Magnalia Christi Americana*, which despite its prolixity has a strong claim to be considered the first great work of literature produced in America. It is a primary source-book because it gives lives of the governors of New England and leading divines, a history of Harvard and various churches, and valuable details about early Indian wars. But in essence it is an epic history of the New England religious experiment—the attempt to create the Kingdom of God in the New World—and an inquiry into what went wrong. He proclaims: 'I write the wonders of the Christian religion, flying from the depravations of Europe to the American strand,' and his tone is often wondering; but it is also querulous and elegiac. He put his bony finger on the inherent contradiction in the Puritan mission. Their Protestant ethic, their intensity of religious endeavor which was the source of their law-abiding industry, contained the seeds of its own dissolution. As he put it, '*Religion* brought forth *Prosperity*, and the *daughter* destroyed the *mother*.' He could see what had happened to Boston in his own lifetime: the mercantile spirit flourishing in its busy streets, and its conformist preachers in the well-filled churches goading on their complacent congregations to amass yet more wealth as an outward symbol of inward grace. Thus America's success was undermining its divine mission: 'There is danger lest the *enchantments* of this world make them forget their *errand into the wilderness*.'

Here is rich food for thought about the whole American experiment, secular as well as religious. It is worth noting that when Cotton Mather died, full of warnings and fulmination, in 1728, Benjamin Franklin, so like him in his universality, so unlike him in his objectives, was already a young man of twenty-two, making his way purposefully in Philadelphia. Whereas Mather was obsessed by the need to save one's soul for the next world, Franklin was preoccupied—like the vast majority of his fellow-countrymen—with getting on in this one. To move from one man to the other is to cross a great watershed in American history.

We are now in the 18th century and the final pieces of the jigsaw of early America are beginning to fit into place. From its growth-points on the coasts of New England and Virginia, now joined by the middle colonies of Maryland, Pennsylvania, and New York, settler America was moving north, and south, but above all west. The frontier was already a physical reality and a powerful metaphysical concept by the year 1700. The overwhelming dynamic was the lust to own land. Now, for the first time in human history, cheap, good land was available to the multitude. This happy prospect was now open, and it remained so for the best part of the next two centuries; then it closed, for ever. In the early 18th century, the movement to acquire land outside the original settlements and charters, and dot it with towns, was just getting into its stride, which was not to relax until the frontier ceased to exist in the 1890s. The advance from the shoreline and the tidewater into the piedmont was what might be called America's first frontier. It took place everywhere in English America. Thus there was a push up the Housatonic Valley into the Berkshires, leading to the foundation of Litchfield in 1719, Sheffield in 1725, Great Barrington in 1730, Williamstown in 1750. In 1735, four closely linked townships were founded to bridge the gap between these Housatonic settlements and the Connecticut River itself. Governor Benning Wentworth (1696–1770), who was instrumental in separating New Hampshire from Massachusetts, then granted lands west of the Connecticut in what was to become Vermont.[147]

This northern push consisted mainly of Ulster Protestants, provoked into seeking a new, transatlantic life by an Act prohibiting the export of Irish wool to England, by the enforced payment of tithes to Anglican churches, and by the expiry of the original Ulster plantation leases in 1714–18. So here were hardy frontier farmers, after three generations of fighting and planting to defend the Protestant enclave against the Catholic-Irish south of Ireland, moving to expand the new frontier in North America. They came in organized groups, and for the first time the authorities had the resources to take them direct to the frontier, where they founded Blandford, Pelham, and Warren, or settled in Grafton County in New Hampshire, and Orange, Windsor, and Caledonia counties in Vermont. These were first-class colonists: law-abiding, church-going, hard-working, democratic, anxious to acquire education and to take advantage of self-government. We heard little of them: always a good sign.

This was only the beginning of the Ulster-Scotch migration. From 1720, for the next half-century, about 500,000 men, women, and children from northern Ireland and lowland Scotland went into Pennsylvania. A similar wave of Germans and Swiss, also Protestants, from the Palatinate, Württemberg, Baden, and the north Swiss cantons, began to wash into America from 1682 and went on to the middle of the 18th century, most of them being deposited in New York, though 100,000 went to Pennsylvania. For a time indeed, the population of Pennsylvania was one-third Ulster, one-third German. Land in Pennsylvania cost only £10 a hundred acres, raised to £15 in 1732 (plus annual quitrents of about a halfpenny an acre). But there was plenty of land, and the rush of settlers, and their anxiety to start farming, led many to sidestep the surveying formalities and simply squat. The overwhelmed chief agent of the Penn family, James Logan, complained that the Ulstermen took over 'in an audacious and disorderly manner,' telling him and other officials that 'it was against the laws of God and nature that so much land should be idle while so many Christians wanted it to labor on and raise their bread.'[148] How could he answer such a heartfelt point, except by speeding up the process of lawful conveyance?

The further south you went, the cheaper the land got. Indeed it was often to be had for nothing. From the 1720s onwards, Germans, Swiss, Irish, Scotch, and others, moved down from the northeast along the rich inland valleys of the mountain area—the Cumberland, Shenandoah, and Hagerstown valleys, then through the passes east into what is now North Carolina, Kentucky, and Tennessee. Shortly after the mid-century they were getting into Georgia this way. As F. J. Turner was later to note, in *The Frontier in American History*, this moving mass of people contained children with names like Daniel Boone, John Sevier, and James Robertson, and the forebears of Andrew Jackson, Sam Houston, Davy Crockett, John C. Calhoun, James K. Polk, Jefferson Davis, Abraham Lincoln, and Stonewall Jackson. This was when Andrew Jackson's father set up in Carolina piedmont and Thomas Jefferson's built his home on the frontier at Blue Ridge.

South of the Chesapeake, the framework of government became weaker. In the Carolinas there was constant bickering between north and south, as well as between Tidewater grandees and inland settlers in the piedmont. In 1691 the Carolina proprietors recognized the *fait accompli* of a northern region by dividing the colony into two provinces, with a deputy governor living in the town of Albemarle, capital of what was already being called North Carolina. On May 12, 1712 the separation was completed and North Carolina became a colony on its own. It had already run its own legislature, be it noted, for forty-seven years—five years longer than South Carolina's in Charleston. This did not solve the problem in either half, for the proprietors were absentee landlords—the absentee grandee was the curse of the early South, as it always was in Ireland—and that meant there was a lack of control and purpose in the governor's mansion, leading to tardy and inadequate response to Indian raids, a poorly led and equipped militia, and other evils. The settlers petitioned London for help—it is significant that, even in the 1720s, colonists still had the 'look homeward' reflex and saw the crown as their father and savior. The crown responded: South Carolina became a royal colony on May 29, 1721 and North Carolina followed eight years later on July 25, 1729. But that did not mean the arrival of royal soldiers or assured protection from London.

Nor were the Indians the only threat. In 1720, for instance, South Carolina had only 7,800 whites, as opposed to 11,800 black slaves—the largest ratio of blacks to whites, about 60 percent, in any colony. And it was bringing in more slaves fast; another 2,000 in the years 1721–5 alone. Many slaves escaped, and these maroons, as they were called, tended to organize themselves into gangs to break out of British territory into Spanish Florida, which issued a decree in 1733 stating that slaves who defied the British and managed to reach land under the Spanish flag would be considered free. The result, in 1739, was a series of slave revolts. A band of Charleston slaves set out for Spanish St Augustine and freedom, killing all whites they met on the way, a total of twenty-one; forty-four of these maroons were rounded up and executed on the spot. On the Stono River, a black firebrand called Cato led an even bloodier uprising—thirty whites and about fifty blacks were killed before order was restored. There was a third revolt in St John's Parish, in Berkeley County.

Violence between blacks and whites was by no means confined to Carolina, of course, as the number of blacks imported from Africa and the West Indies steadily increased. In 1741 a series of mysterious fires in New York City, where blacks were a fifth of the population, led to rumors that a negro conspiracy was to blame and that the slaves were planning to take over the city. Many blacks were arrested, eighteen were hanged, and eleven burned at the stake, though a public prosecutor, Daniel Horsemanden, later admitted that there was no evidence such a conspiracy ever existed. But in the Carolinas, especially towards the south and in the back-country, security was much more fragile. Stability was not established until a first-class royal governor, James Glen, took over in 1740.[149] He was even able to get some action from the crown: early in 1743 General James Oglethorpe, with a fierce body of Scottish Highlanders, as well as local militia, thrashed a Spanish force four times its size at the Battle of Bloody Marsh.

James Oglethorpe (1696–1785) was a fascinating example of the bewildering cross-currents and antagonisms which make early American history so confusing at first glance. He was a rich English philanthropist and member of parliament, who came to America as a result of his passionate interest in prison reform. He was particularly interested in helpless men imprisoned for debt and believed they ought to be freed and allowed to work their way to solvency on American land. In 1732, George II gave him a charter to found such a colony between the Savannah and the Altama rivers, to be named Georgia after himself. Oglethorpe himself went out with the first band of settlers in January 1733. This was another utopian venture, though with humanitarian rather than strict religious objectives—an 18th-century rationalist as opposed to a 17th-century doctrinaire experiment. Oglethorpe and his supporters wanted to avoid extremes of wealth, as in South Carolina, to attract victims of religious persecution and the penal system, and to create a colony of small landowners, with total landholdings limited by law, and slavery prohibited. He was also a military man and he intended, with British government backing, to set up Georgia as a defended *cordon sanitaire* against Spanish troublemaking in the South. He built forts, recruited a militia, and attracted fighting Highlanders for a defensive colony on the Altamaha frontier, which they called New Inverness. His victory at Bloody Marsh not only put an end to the Spanish threat but was a warning to the Indians too—though he made it clear his approach to them was essentially friendly by setting up Augusta as an advance post for the Indian trade. In every respect, Georgia was intended to be a model colony of the Age of Enlightenment. Oglethorpe planned to introduce silk-production, and in Savannah, his new capital, he even set up what was called the Trustees Garden, an experimental center for plants.[150]

The colony itself prospered; but the experiment in reason, justice, and science failed. Just as in North Carolina, attempts to ban slavery came up against the ugly facts of economic interest and personal greed. Georgia was too near to the rambunctious but undoubtedly flourishing planter economy of South Carolina to remain uncorrupted. Oglethorpe's regulations were defied. Slaves were smuggled in. So was rum—another banned item. Then the Savannah assembly legitimized widespread disobedience by changing the law. Rum was officially admitted from 1742. Five years later the laws against slavery were suspended and in 1750 formally repealed.[151] These changes brought a flood of newcomers from north of the Savannah, including experienced planters and their slaves, taking up Georgia's cheap land. The utopian colony was Carolined. Oglethorpe was already in trouble with the English authorities for muddling the military finances. So the man who, in the words of Alexander Pope, went to America 'driven by strong benevolence of soul,' returned to England disillusioned and disgusted, surrendering his charter in 1752.

By mid-century all the original Thirteen Colonies were in actual, though not always legal, existence, and all were being rapidly transformed by unequal, sometimes patchy, but on the whole overwhelming prosperity. It was already a region accustomed to dealing in millions—'the land of the endless noughts.' In 1746 a New Hampshire gentleman, John Mason, sold a tract of land totalling 2 million acres, which had been in his family for generations, to a group of Portsmouth businessmen for a planned settlement of new towns. This was merely the largest single item in a continuing process of buying and selling farms, estates, and virgin soil, which had already made British America the biggest theater in land-speculation in human history. Everyone engaged in it if they could—a foreshadowing of the eagerness with which Americans would take to stock-market speculation in the next century.

Four years later in 1750, the population of the mainland colonies passed the million mark too. The British authorities of course saw North America as a whole, and missed the significance of this figure. But whereas at mid-century Barbados had a population of 75,000 and Bermuda–Bahamas 12,000 and Canada, Hudson Bay, Acadia, and Nova Scotia, plus Newfoundland, had a further 73,000, Massachusetts and Maine together were approaching a quarter-million, Connecticut had 100,000, Rhode Island and New Hampshire had 35,000 each, there were 34,000 in East Jersey, 36,000 in West Jersey, 75,000 in New York, 165,000 in Pennsylvania and the Lower Counties, 130,000 in Maryland, 135,000 in the Carolinas—plus 4,000 in infant Georgia—and a massive 260,000 in Virginia. Greater New England had 400,000, Greater Virginia 390,000, Greater Pennsylvania 230,000, and Greater Carolina nearly 100,000.[152] These four major self-sustaining growth-centres were the main engines of demographic increase, attracting thousands of immigrants every year but also ensuring high domestic birth-rates with a large proportion of children born reaching adulthood, in a healthy, well-fed, well-housed family system.

Noting all these facts, Benjamin Franklin, writing his *Observations Concerning the Increase of Making, Peopling of Countries etc* (1755), felt that the country had doubled in population since his childhood and calculated it would double again in the next twenty years, which it did—and more.[153] In attracting yet more people, to keep up the impetus of growth, local authorities did not worry too much about boundaries, an early indication of how the whole territory was beginning to meld together. Thus in 1732 Maryland invited Pennsylvanian Germans to take up cheap 200-acre plots in the difficult country between the Susquehanna and Patapsco, which became an inland district for the new and soon flourishing town of Baltimore. Equally, in the 1750s there was a large movement from Pennsylvania at the invitation of the Virginia government into the western region of the colony, where large blocks in the Shenandoah Valley were offered at low prices. This created, from an old Indian tract, the famous Great Philadelphia Waggon Trail, which became a major commercial route too. Thus Greater Pennsylvania merged into Greater Virginia, creating yet more movement and dynamism. As settlement expanded inland from the tidewaters, colonies lost their original distinctive characteristics and became simply American.[154]

The historian gets the impression, surveying developments in the first half of the century, that so many things were happening in America, at such speed, that the authorities simply lost touch. Their information, such as it was, quickly got out of date and they could not keep up. Strictly speaking, in an economic sense, the colonies were supposed to exist entirely for the home country's benefit. A report to the Board of Trade sent by Lord Cornbury, governor of New York 1702–8, reveals that all governors were instructed 'To discourage all Manufactures, and to give accurate accounts of any Indications of the Same,' with a view to their suppression.[155] One member of the Board of Trade stated flatly in 1726, that certain developments in a colony were *eo ipso* unlawful whether or not there was a specific statute forbidding them:

> Every act of a dependent provincial government ought to terminate to the advantage of the Mother State unto whom it owes its being and protection in all valuable privileges. Hence it follows that all advantageous projects or commercial gains in any colony which are truly prejudicial to and inconsistent with the interests of the Mother State must be understood to be illegal and the practice of them unwarrantable, because they contradict the end to which the colony has a being and are incompatible with the terms on which the people claim both privileges and protection ... for such is the end of colonies, and if this use cannot be made of them it is much better for the state to do without them.[156]

This was hard doctrine, manifestly unjust and equally clearly unenforceable. There were of course many legislative efforts to turn it into reality. An Act of 1699 forbade the colonies to ship wool, woolen yarn, or cloth. Another in 1732 vetoed hats. An Act of 1750 admitted entry of bar-iron into England but banned slitting or rolling mills, plat-force, or steel furnaces. But iron casting was not specifically forbidden and so the colonies produced such things as kettles, salt-pans, and kitchen utensils, as well as cannon. According to Board of Trade economic doctrine these must be inherently unlawful. But they continued to be made. And what about shipbuilding? The sea was Britain's lifeblood and ships were made, competitively, in yards all over England and Scotland. But with wood so cheap and accessible, America had a huge competitive advantage in shipbuilding before the age of iron and steam. By mid-century New England yards were turning out ships at an average cost of $34 a ton, 20 to 50 percent cheaper than in Europe. They had vigorously promoted shipbuilding from the 1640s and as early as 1676 were turning out thirty a year for the English market alone; this rose to 300 to 400 a year by 1760. By this time fully a third of the British merchant fleet of 398,000 tons was American-built, and the colonies were turning out a further 25,000 tons a year. The reason for permitting this obvious anomaly was the British need for cheap timber. A British merchant could sail his ship to Boston, sell his cargo, then with the proceeds build an additional ship, and load both with timber. The British authorities unwittingly encouraged this procedure, paying substantial bounties on timber-related products such as pitch, tar, rosin, turps and water-rotted hemp, to reduce its dependence on supplies from the Continent.[157]

This cheapness of wood, and so of ships, also encouraged the development of an enormous fishing fleet which again, strictly speaking, was a challenge to British interests. As early as 1641 figures show that New England was exporting 300,000 cod a year plus halibut, mackerel, and herring. By 1675, 4,000 men and 600 ships were involved in the industry. By 1770 its exports were worth $225,000 a year. The largest and most difficult-to-cure fish were eaten locally; small, damaged, or tainted fish were sent to the West Indies to be eaten by slaves; the best smaller fish were cured and sent to Britain.[158] All this stimulated a large cooperage industry, again encouraged by cheap wood—New England farmers often increased their incomes by turning out barrels on the side. As New England made bigger and better ships, it went into worldwide deep-sea whaling, already important by 1700 and growing rapidly. For its own mysterious reasons the home government again favored this activity, paying a pound bounty per ton (1732) on whalers of 200 tons of more, and raising it (1747) to 2 pounds a ton. By mid-century America had the most skillful whalers in the world, 4,000 of them from New Bedford and Provincetown, Nantucket and Marblehead, operating over 300 ships.

The fact is, though America's was largely an agricultural economy, far more so than Britain's, it was stealthily catching up in manufactures of all sorts. When the Board of Trade wrote to colonial governors, asking for figures of goods produced locally, the governors, with their eye on local opinion, deliberately underestimated output. A lot of phony statistics passed across the Atlantic in the 18th century—not for the last time, either.[159] Comptroller Weare wrote anxiously to the Board of Trade c.1750: 'The Planters throughout all New England, New York, the Jersies, Pennsylvania and Maryland (for south of that province no knowledge is here pretended) almost entirely clothe themselves in their own woollens, and generally the people are sliding into the manufactures proper to the Mother Country, and this not through any spirit of industry or economy, but plainly for want of some returns to make to the shops.'[160] Another report at the same time suggested that American producers were competing successfully with English ones, even in exports, in cotton yarn and cotton goods, hats, soap and candles, woodwork, coaches, chariots, chairs, harness and other leather, shoes, linens, cordage, foundry ware, axes, and iron tools.[161]

American spokesmen, like Benjamin Franklin, were anxious to play down how well the colonies were doing in this respect, for fear of arousing the wrath of the jealous Mother Country. As Agent of Pennsylvania, he informed a House of Commons committee in 1766 that his colony imported half a million pounds' worth of goods from Great Britain but exported only £40,000 in return. Asked how the difference was made up he replied: 'The balance is paid in our produce to the West Indies, or sold in our own island, or to the French, Spaniards, Danes or Dutch; by the same carried to other colonies in North America, or to New England, Nova Scotia, Newfoundland, Carolina and Georgia; by the same carried to different parts of Europe, as Spain, Portugal and Italy: in all which places we receive either money, bills of exchange or commodities that suit our remittance to Britain; which together with all the profits on the industry of our merchants and mariners, arising in those circuitous voyages, and the freights made by their ships, center finally in Britain, to discharge the balance and pay for British manufactures . . . '[162] Separating 'visibles' from 'invisibles,' distinguishing between all the different elements in triangular or quadrilateral trading patterns—it was all too difficult for an amateur group of parliamentary gentlemen, and all too easy for Franklin to bamboozle them, though it is very likely that his own figures were inaccurate and many of his assumptions misleading. The truth is, by the mid–18th century, mercantilism was on its last legs, overwhelmed by the complexity of global trade and the inability to distinguish what was in the true long-term interests of a country with burgeoning self-sustaining dominions. Entrepreneurial capitalism, spanning the Atlantic, was already too subtle and resourceful for the state to manage efficiently.

In any case, the British economic strategists—if that is not too fancy a name for classically educated Whig country gentlemen advised by a handful of officials who had never been to America (or, in most cases, to the Continent even)—were slow to grasp the speed with which the American mainland colonies were maturing. The conventional wisdom in London was to treat them as poor and marginal. They had played little part in the great wars of King William and Queen Anne's day. Tobacco was the only thing they produced of consequence. In the early 18th century they accounted for only 6 percent of Britain's commerce, less than one-sixth of the trade with northern Europe, two-thirds or less of that with the West Indies, even less than the East Indies produced. Almost imperceptibly at first this situation changed. By 1750 the mainland American colonies had become the fastest-growing element in the empire, with a 500 percent expansion in half a century. Britain, with the most modern economy in Europe, advanced by 25 percent in the same period. In 1700 the American mainland's output was only 5 percent of Britain's; by 1775 it was two-fifths. This was one of the highest growth-rates the world has ever witnessed.[163]

It seems as though everything was working in America's favor. The rate of expansion was about 40 percent or even more each decade. The availability of land meant large family units, rarely less than 60 acres, often well over 100, huge by European standards. Couples could marry earlier; a wife who survived to forty gave birth on average to six or seven children, four or five of whom reached maturity. Living standards were high, especially in food consumption. Males ate over 200 pounds of meat a year, and this high-protein diet meant they grew to be over two inches taller than their British counterparts. They ate good dairy food too. By 1750 a typical Connecticut farm owned ten head of cattle, sixteen sheep, six pigs, two horses, a team of oxen. In addition the farm grew maize, wheat, and rye, and two-fifths of the produce went on earning a cash income, spent on British imports or, increasingly, locally produced goods. It is true that widows might fall into poverty. But only 3 to 5 percent of middle-aged white males were poor. One-third of adult white males held no appreciable property, but these were under thirty. It was easy to acquire land. Over the course of a life-cycle, any male who survived to be forty could expect to live in a household of median income and capital wealth. In short by the third quarter of the 18th century America already had a society which was predominantly middle class. The shortage of labor meant artisans did not need to form guilds to protect jobs. It was rare to find restriction on entry to any trade. Few skilled men remained hired employees beyond the age of twenty-five. If they did not acquire their own farm they ran their own business. In practice there were no real class barriers. A middle-aged artisan usually had the vote and many were elected to office at town and county level. These successful middle-aged men were drawn not just from the descendants of earlier settlers or from the ranks of the free immigrants but from the 500,000 white Europeans who, during the colonial period, came to America on non-free service contracts running from four to seven years.[164] White servitude, unlike black slavery, was an almost unqualified success in America.

The policy, begun in 1717, of transporting convicts to the American mainland, for seven years as a rule, worked less well—far less well than it later did in Australia. This was subsidized by the British state, which wanted to get rid of the rogues, but was also a private business tied to the shipping trade. The convicts left Britain in the spring, were landed in Philadelphia or the Chesapeake in the summer, and the ships which transported them returned in the autumn loaded with tobacco, corn, and wheat. In half a century, 1717–67, 10,000 serious criminals were dumped on Maryland alone. They arrived chained in groups of ninety or more, looking and smelling like nothing on earth. Marginal planters regarded them as a good buy, especially if they had skills. They went into heavy labor—farming, digging, shipbuilding, the main Baltimore iron-works, for instance. In 1755 in Baltimore, one adult male worker in ten was a convict from Britain.[165] They were much more troublesome than non-criminal indentured labor, always complaining of abuses and demanding 'rights.' People hated and feared them. Many were alcoholics or suicidal. Others had missing ears and fingers or gruesome scars. Some did well—one ex-thief qualified as a doctor and practiced successfully in

Baltimore, attracting what he called 'bisness a nuf for 2.' But there were much talked-about horror-stories—one convict went mad in 1751 and attacked his master's children with an axe; another cut off his hand rather than work. From Virginia, William Byrd II wrote loftily to an English friend: 'I wish you would be so kind as to hang all your felons at home.'[166] There were public demands that a head-tax be imposed on each convict landed or that purchasers be forced to post bonds for their good behavior. But the British authorities would never have allowed this. As a result of the convict influx we hear for the first time in America widespread complaints that crime was increasing and that standards of behavior had deteriorated. All this was blamed on Britain.

Indeed the historian notes with a certain wry amusement, as the century progressed, an American tendency to attribute everything good in their lives to their country and their own efforts, and to attribute anything which went wrong to Britain. Certainly, America showered blessings on its people, as English newcomers noticed. One visitor said that 'Hoggs in America feed better than Hyde Park duchesses in England.' Another called the country 'a place of Full Tables and Open Doors.'[167] Miss Eliza Lucas, much traveled daughter of an English army officer, discoursed eloquently in a letter home of 'peaches, nectarines and mellons of all sorts extremely fine and in profusion, and their oranges exceed any I ever tasted from the West Indies or from Spain or Portugal.' There were many more, and better, vegetables than were available in England. German immigrants were particularly good at producing in quantity and for market, at low prices, apples, pears, quinces, chestnuts, and a wide range of strawberries, raspberries, huckleberries, and cherries for preserves. Ordinary people filled their stomachs with beef, pork, and mutton, as well as 'jonny cake' and 'hoe cake.' A contributor to the *London Magazine* in 1746 thought the American country people 'enjoy a Life much to be envied by Courts and Cities.' And there were always new evidences of nature's bounty to those who looked hard enough for it. Clever Miss Lucas, left in charge of a South Carolina plantation, took advantage of a parliamentary bounty on indigo, raised to sixpence a pound in 1748, to experiment successfully with a crop. Thanks to her, the Carolinas were exporting 1,150,662 pounds of it in 1775, and it became the leading staple until displaced by cotton after the Revolution.

While the pioneers pushed inland, opening up new sources of wealth, and gradually creating the demographic base from which America could take off into an advanced industrial economy, the cities of the coast were coining money and spending it. The queen of the cities was Philadelphia, which by mid-century had become the largest in the entire British Empire, after London. Its Philosophical Society (1743) was already famous and its Academy (1751) burgeoned into the great University of Pennsylvania. New York City was also growing fast and was already the melting-pot in embryo. By 1700 the English and the Huguenots outnumbered the original Dutch inhabitants: half a century later, many of the Dutch had become Anglicans and all were bilingual or English-speaking. They had been joined by multitudes of Walloons and Flemings, Swedes, Rhineland Protestants, Norwegians, and North Germans, as well as Scotch and English Calvinists and Quakers, freed slaves, Irish, and more Dutch. By mid-century the Lower Hudson, including East and West Jersey, joined as the royal colony of New Jersey in 1702, was a collection of communities—Dutch in Harlem and Flatbush, lowland Scots in Perth Amboy, Baptist settlers from New Hampshire in Piscataway, New England Quakers in Shrewsbury, Huguenots in New Rochelle, Flemings in Bergen, New Haven Puritans in Newark and Elizabeth, and pockets of Scotch, Irish, and Germans upriver, as well as many Dutch—Albany was a Dutch town then, though English-speaking. It was already competing with French Montreal for the Indian and wilderness trade in furs, with an offshoot at Fort Oswego on Lake Ontario.

The economic and political freedom enjoyed in English America, with its largely unrestricted enterprise, self-government, and buccaneering ways, was already reflected in growth-rates which made Canada, in which the French state had invested a huge effort but also a narrow system of controls, seem almost static. By 1750 there were well over 100,000 in the Hudson Valley alone, compared to only 60,000 in the vast St Lawrence basin, and New York City was four times the size of Quebec. And unlike inward-looking and deadly quiet Quebec, New York and its politics were already noisy, acrimonious, horribly faction-ridden, and undoubtedly democratic.[168]

The venom of New York politics led to America's first trial for seditious libel in 1735, when John Peter Zenger, who had founded New York's *Weekly Journal* two years earlier, was locked up for criticizing the governor, William Cosby, and finally brought to trial after ten months behind bars. Zenger was by no means America's first newspaper publisher. That honor goes to the postmaster of Boston, William Campbell, who set up the *News-Letter* in 1704 to keep friends scattered around the Bay Colony informed of what was going on in the great world. By mid-century more than a score of newspapers had been started, including the *Philadelphia American Weekly Mercury* (1719), the Boston *New England Courant* (1721), started by Benjamin Franklin's elder brother James, and Franklin's own *Pennsylvania Gazette*, which he acquired in 1729. There was also an Annapolis paper, the *Maryland Gazette* (1727), and the *Charlestown South Carolina Gazette* (1732).[169] It is significant that Zenger, or rather his lawyer, Andrew Hamilton of Philadelphia, put forward truth as his defense. That would not have been admitted in an English court where anything was criminally libelous, whether it was true or not, which fostered 'an ill opinion of the government.' Indeed, it was an axiom of English law, in seditious libel, that 'the greater the truth, the greater the libel.' In Zenger's case the judge tried to overrule his defense, but the jury acquitted him all the same—and that was the last of such prosecutions. This in itself was an indication of what critics of society could get away with in the heady air of colonial America—prosecutions for criminal libel continued in England until the 1820s and even beyond.[170]

Not all these cities were booming or bustling. Charleston, the only city in the South for more than a century, had little over 8,000 people in 1750, but it was spacious, tree-shaded, elegant, and free-spending, with a recognizable gentry living in town mansions and parading in their carriages. Annapolis was another gentry town, though even by 1750 it had only 150 households. It was brick-built with paved streets, as good as any in Boston, and had fine shops selling silverware, gold, well-made furniture, and paintings. Not only did it have its own newspaper, it also sported a bookstore–publisher from 1758. By the 1740s it was holding regular concerts and claimed its own gifted composer, the Rev. Thomas Bacon (1700–68), who also compiled *The Laws of Maryland*.[171] In June–July 1752 it had a theatrical season in which visiting professional players performed Gay's *Beggar's Opera*, the great London hit, and a piece by Garrick. A permanent theater was opened in 1771, the first in all the colonies to be brick-built. Its opening night was attended by a tall young colonel called George Washington. Its Tuesday Club, attended chiefly by clerics and professional men, was the center of scientific inquiry. Williamsburg, which became the capital of Virginia Colony in 1699, developed into a similar place, small, elegant, select, with a conscious air of cultural superiority, generated from its William and Mary College, the second oldest in the colonies (1693).

Its main building was designed by Sir Christopher Wren, architect of St Paul's Cathedral in London.

These miniature red-brick cities were adorned by the rich of the Chesapeake with fine town houses. Many of them were modeled on one built in Annapolis by the secretary of Maryland Colony, Edmund Jennings, a magnificent building set in 4 acres of gardens at the foot of East Street. Another with splendid gardens—and no fewer than thirty-seven rooms—was built by William Pace. The chimneys of James Brice's mansion towered 70 feet above street-level. Many of the finest houses were the work of the local architect–craftsman William Buckland, credited with turning the place into the 'Athens of America.' Annapolis had an English-style Jockey Club from 1743, which supervised the regular race-meetings and was the meeting-place of local breeders. By the third quarter of the century over 100 English-bred horses of Arab strains had reached the Chesapeake and the gentry could attend races held near both these elegant cities—they were within commuting range. City artisans had cockfighting. But, as in England, the artisans went to the races if they could afford it, and the gentry certainly attended cockfights.

For boom you went to Baltimore, then the fastest-growing city in America, probably in the world. In 1752 it was nothing much—twenty-five houses, 200 people. Less than twenty years later it was the fourth-largest city in America. Its jewel was its magnificent harbor, which made it the center for Virginia–Maryland tobacco exports to Glasgow (European end of the trade), all-purpose trade with the West Indies, and ships loaded with imports from all over Europe. On top of the hill overlooking the harbor, enormous flags, each from a different shipping company, announced major arrivals from the ocean. Fells Point was one of the most crowded shipping wharves on earth, backed by 3,000 houses, most of brick, two or three stories high. Later, the haughty French aristocrat François Alexis de Chateaubriand conceded that entering Baltimore harbor was like 'sailing into a park.' There was a downside to all this bustle, needless to say. Land values shot up astronomically and people complained the cost of living was higher than London's and much higher than in Paris. There was a terrific stench from the harbor at low tide and the streets near it were crowded with Indian, black, and white whores—also said to be high-priced and insolent. On the other hand there were not one but two theaters, and the Indian Queen Hotel on the corner of Market and Hanover streets was one of the best in the western hemisphere by the 1790s: excellent food, boots and shoes polished by assiduous blacks if left outside the rooms, and slippers provided free for guests.[172]

There also grew up in colonial America, beginning in the last decades of the 17th century and progressing in stately fashion and growing confidence in the 18th century, a country-house culture, modeled on England's but with marked characteristics of its own. To begin with, these baroque–Georgian–Palladian houses were almost invariably at this date built on navigable rivers and creeks, to serve the plantation export economy. The wharf was as important as the drawing-room—indeed, without the wharf the elaborate furniture, imported from London or Paris or made in New England, could not be afforded. These grand houses arose naturally from the economic activities which sustained them and were not plonked artificially in the midst of a cap-doffing countryside like Blenheim or Chatsworth or Althorp in England. Nor, until the arrival of the plutocracy after the Civil War, were American country houses on anything like the same scale as the English aristocracy's. Except when the Dutch *patroons* built them, they were rarely of stone. But in the deployment of brick the American house-builders, amateur and professional, have rarely been excelled.

The greatest early 18th-century house in America was Rosewell, erected by Mann Page (1691–1730) in 1726 on the York River. Page married a Carter, of the family of 'King' Carter (1663–1732), the famous and rapacious agent of Lord Fairfax, proprietor of the Northern Neck of Virginia. Carter amassed 300,000 acres of prime land, and he gave his fancy son-in-law, Page, 70,000 of them. Page had this superb house built using the designs in Colin Campbell's *Vitruvius Britannicus*—published in London 1715–25 and quickly shipped across the Atlantic—as models. Page overspent, his grand house was unfinished when he died in 1730, and his debts exceeded the value of all his property, slaves included. Moreover, Rosewell, having triumphantly survived the horrors of the Civil War, was burned down in 1916. But its ruins compel one to believe that, in its day, 'there [was] nothing like it in England.'[173]

Almost as grand, and still in excellent condition (and open to the public) is Hampton, near Baltimore. This was built (from 1783) by Charles Ridgeley (1733–90) and testifies to the failure of the British authorities to carry out their intention of preventing the American colonies from becoming a major iron-producer. Ridgeley was not only a planter with 24,000 slave-worked acres, but the owner of a large ironworks. That was where most of his money came from. Maryland not only had rich iron-ore deposits but plentiful hardwoods for making charcoal and fast streams for power. As early as 1734–7 it shipped 1,977 tons of pig-iron to England; by the 1740s it had a huge forge and made bar-iron as well as pigs; by the 1750s it had multiple furnaces and forges: in 1756 there were six ironworks in Baltimore County alone. Then it began to push inland, with rich members of the local elite, like Daniel Dulany, Benjamin Tasker, and Ridgeley, buying up gigantic blocks of iron-bearing land by patent, then moving in Swiss-German and Scots-Irish workmen, as well as slaves for the heavy work. This glorious iron-master's house was held by six generations of Ridgeleys until, in 1948, it was bought by the National Park Service and made available to visitors.

There were equally fine, and many more, country houses built in the 18th century in Virginia, by members of the 100 leading families—Byrds, Carters, Lees, Randolphs, Fitzhughes, and so on—of which many, such as Westover, Stratford, and Shirley, survive. Drayton Hall, built 1738–42, on the Ashley River, a good example of the way local American architects used classical models, is based on Palladio's Villa Pisani, happily survived the Revolutionary and Civil wars and is now part of the American Trust for Historical Preservation. Another, rather later masterpiece, now part of Johns Hopkins University, is Hoewood, a Baltimore classical villa erected by the famous Charles Carroll of Carrolltown (1737–1832), grandest of the Revolutionary politicians. These houses and mansions sometimes contained fine libraries of ancient and modern works. A visitor described William Byrd II's library at Westover as 'consisting of nearly 4,000 volumes, in all Languages and Faculties, contained in 23 double-presses in black walnut . . . the Whole in excellent Order.' He added, admiringly: 'Great Part of the books in elegant Bindings and of the best Editions and a considerable Number of them very scarce.'[174] This opulent pile also still exists, though the interiors have been remodeled.

The men who owned these country houses, and others like them along the James, the Connecticut, and the Hudson, and the neat and in some cases spacious city houses in Boston and New Haven, Albany and New York, Philadelphia, Charleston, Williamsburg, Annapolis, and Baltimore, would in England have sat in the House of Commons, 'to keep up the consequence of their families,' as Dr Samuel Johnson put it. In some cases they would have sat in the House of Lords, with a writ of individual summons to parliament. In the American colonies they played a similar role, the main difference being that they were usually forced to consort with a host of lesser folk, many of whom could barely write their names, in helping to run the country. American colonies had their elites everywhere. There were immense differences in wealth and social customs, especially in the South and notably between the Tidewater grandees and the farmers of the piedmont and the inland valleys. Sometimes these grandees behaved as if they owned the place. Thus, in early South Carolina, the Tidewater elite did not even have a House of Assembly but met in one another's houses, just as if they were Whig dukes holding Cabinet dinners in London. But that kind of thing did not last. Rich Americans who got too uppity or tried to pull a rank they did not in law possess were soon reminded that America was a society where all freemen were equal, or liked to think they were any-way. One of the effects of slavery was to make even poor whites assertive about their rights. They felt they were of consequence because they were complacently aware of a huge servile class below them.

To 18th-century Frenchmen or Spaniards, who were familiar with the uniform manner in which their own colonies were directed, with an omnipresent state, a professional bureaucracy, and only the most nom-inal element of local representation, the British colonies in America must have seemed bewildering, chaotic, and inconsistent in the way they were run. The system was empirical and practical rather than coherent. It evolved almost organically, in the way English institutions had always evolved. No two colonies were quite the same. The system is worth examining in a little detail both because of its bearing on the events leading up to the American Revolution, and because of its influ-ence on the way the American Republic developed thereafter.

Originally all the colonies had been divided into two categories: trading or commercial companies, run like primitive joint-stock corpo-rations, or proprietary companies run by one or more great landed estate-owners. All had charters issued directly by the crown. Without these two forms of ownership, which involved a high degree of self-government, the colonies would never have got going at all, because the English crown, unlike the crowns of France and Spain, was simply not prepared to pour out the prodigious amounts of cash needed. So the English state got its colonies largely for nothing and this successful stinginess continued to condition the thinking of British governments throughout the 18th century. They did not expect to have to pay for the empire or, if they did, they expected those who lived in the empire to refund the money through taxes. However, having set up these quasi-independent and self-supporting colonies in the early and mid–17th century, the crown began to wrest back some degree of control before the end of it. From Charles II to William III, they revoked or refused to renew charters—there was always a perfectly good excuse—and turned both commercial and proprietary colonies into crown ones. By 1776 only two commercial colonies (Connecticut and Rhode Island) and two proprietary colonies (Maryland and Pennsylvania) were left. It is true that Massachusetts was also still operating under a royal charter, but it was governed as a royal province.

This ought to have given the crown a great deal of power, at any rate in the nine colonies it controlled directly. In practice, English meanness in colonial matters again frustrated London's ability to control what happened. In each colony, the governor constituted the apex of the pyramid of power—and it is characteristic of the profound constitutional conservatism America has inherited from England that the fifty members of the United States are still run by governors. But the actual power of the colonial governors was less than it looked in theory, just as today the state governors of the federal republic are severely limited in what they can do. In the crown colonies the governors were appointed by the King on the advice of his ministers. In the proprietary colonies they were chosen by the proprietors, though the King had to approve. In the charter colonies they were elected, though again royal approval was needed. All were treated in some ways with deference, as viceroys. But whereas in the Spanish and French colonies they had not only enormous legal powers but the means to enforce them, in America they were not even paid by the crown. In every case except Virginia, their salaries were determined and paid by the colonial assemblies, who, true to the tradition of British meanness, kept these stipends small. They were often grudgingly and tardily paid too. Nor did they have many valuable perks and privileges. Most of them seem to have been able and—amazingly for the 18th century—honest. But they were not, on the whole, great, forceful, self-confident, or masterful men. That in itself made a difference to the degree of authority exercised over the people of the colonies.

The governors were caught between two quite different and often opposed forces. On top of them, but exercising power from distant London, was the crown. Colonies were supervised by the Privy Council, which operated through a Commission, variously called for trade or plantations and, from the days of William III, the Board of Trade and Plantations (1696–1782), which continued to be in charge of American policy until the end of the Revolutionary War. It was handicapped by the fact that it did not actually pay the governors (or in many cases appoint them) and it was very rare indeed for any of its members, or officials, to have set foot in America. The instructions it issued to governors were not always clear, or sensible, or consistent, and were often beyond their power to carry out. On the other hand, the crown tended to see governors as weak, ineffectual, demanding, and 'expensive servants,' always quarreling with the planters, provoking rebellions, or getting themselves involved in Indian wars through needless brutality and insensitive actions. The Crown usually sided with the Indians in cases of dispute and sometimes even with white rebels. When Governor Berkeley, who had run away from Nathaniel Bacon and his followers, turned on them savagely after Bacon died, Charles II exclaimed in exasperation: 'That old fool has taken away more lives in that naked country than I did here for the murder of my father.'

The governors, of course, did not rule alone. Each had some kind of council, which formed the executive or administrative body of the colony and constituted the upper chamber (like the House of Lords) of its assembly. They were appointed by the crown (in royal colonies) or by the proprietors, and their numbers varied—ten in Rhode Island, twenty-eight in Massachusetts. They also had judicial functions and (with the governor) served as courts of appeal, though certain important cases could be appealed again to the Privy Council in London. A good, firm-minded governor could usually get his council solidly behind him.

It was a different matter with the Houses of Burgesses (or whatever they were called), the lower chambers of the assemblies. The first one dated from as far back as 1619. All the colonies had them. Most of them were older than any working parliaments in Europe, apart from Britain's. They aped the House of Commons and studied its history assiduously, especially in its more aggressive phases. Most of these assemblies kept copies of one or more volumes, for instance, of John Rushworth's Historical Collections, which documents the struggles of the Commons against James I and Charles I and was regarded by royalists as a subversive book. Whenever the Commons set a precedent in power-grabbing or audacity, one or other assembly was sure to cite it.

However, there was an important different between the English parliament and the colonial assemblies. England had never had a written constitution. All its written constitutional documents, like Magna Carta or the Bill of Rights, were specific ad hoc remedies for crises as they arose. They were never intended, nor were they used, as guides for the present and future. All the English had were precedents: their constitutional law operated exactly like their common law, organically. The Americans inherited this common law. But they also had constitutions. The Fundamental Orders of Connecticut (1639) was the first written constitution not only in America but in the world. Written constitutions were subsequently adopted by all the colonies. It is vital to grasp this point. It was the constitutions as much as the assemblies themselves which made the colonies unique. In this respect they could be seen as more 'modern' than England, certainly more innovatory. Its constitution was what made Connecticut, for instance, separate from and independent of Massachusetts, its original 'Mother.' Having a constitution made a colony feel self-contained, mature, almost sovereign. Having a constitution inevitably led you to think in terms of rights, natural law, and absolutes, things the English were conditioned, by their empiricism and their organic approach to change, not to trouble their heads about. That was 'abstract stuff.' But it was not abstract for Americans. And any body which has a constitution inevitably begins to consider amending and enlarging it—a written constitution is a signpost pointing to independence.

The early establishment of assemblies and written constitutions—self-rule in fact—arose from the crown's physical inability, in the first half of the 17th century, to exercise direct control. The crown was never able to recover this surrender of power. Nor could the English deny the Americans the fruits of their own past. Their parliament had waged a successful struggle against the crown in the 1640s and acquired powers which could never subsequently be taken away. The colonial assemblies benefited from this. In 1688 the Glorious Revolution turned a divine-right monarchy into a limited, parliamentary one. The colonies participated in this victory, especially in New York, Massachusetts, and Maryland, which overthrew the royal government of James ii and replaced it by popular rule. When William III, the beneficiary of the Glorious Revolution, sought to reorganize the English colonies on Continental lines, he found it impossible and was forced to concede their rights to assemblies. These were all further milestones along a road which led only in one direction, to ultimate independence.[175]

In constitutional terms, the story of the first half of the 18th century in the American mainland colonies is the story of how the lower, elected houses of each assembly took control. The governor had the power of veto over legislation and he was expected, using his council members sitting in the upper chamber, to take the lead, with the elected assemblies deferentially subordinate. The reverse happened. In 1701 the Pennsylvania elite extracted from William Penn a charter of privileges which made it the most advanced representative body in America. When South Carolina ceased being a proprietary colony and became a crown one in the 1720s, which might in theory have led to a diminution of popular power, the House of Burgesses exploited the handover to increase its influence. In the first three decades of the 18th century, lower houses not only in Pennsylvania and South Carolina but in New York and Massachusetts also waged constitutional battles with governors, councils, and the crown, blocked orders, and, on the whole, determined the political agenda. In every colony, the lower houses increased their power during the first fifty years of the 18th century, sometimes very substantially. They ordered their own business, held elections, directed their London agents, and controlled the release of news to the press. They claimed and got the sole right to frame and amend money Bills, and so to raise or lower taxes. They controlled expenditure by specific allocations—something the British parliament could not do because of the huge power of the Treasury—and this meant they appointed and paid money commissioners and tax-collectors, regulated the fees of the administration, and subjected all officials, including the governor, to annual salary regulations. In fact, unlike the House of Commons, they gradually acquired all kinds of executive responsibilities and began to think of themselves as government.

It was not a one-way struggle by any means. Governors, on behalf of the crown, tried to cling on to their prerogative powers—to appoint judges and regulate the courts, to summon, dissolve, or extend assemblies. They made efforts to build up 'court parties' or buttress conservative factions among the burgesses, especially in New Hampshire, Maryland, and Massachusetts. In Virginia and New York, governing councils managed to retain power over land policy, an important source of patronage. As elsewhere in the British Empire, they tried divide and rule. Squabbles between coastal elites and up-country men were perennial. Franchises were heavily weighted in favor of property owners. So were constituency boundaries. For instance, in Pennsylvania, the three 'old' counties of Chester, Bucks, and Philadelphia elected twenty-six deputies to the legislature, the five frontier ones only ten. The young Thomas Jefferson, himself frontier-born, complained that 19,000 men 'below the falls' legislated for more than 30,000 elsewhere. But in most cases the majority of adult males had votes. The further from the coast, the more recent the settlement, the more the franchise became democratic.[176] In practice, it was impossible to enforce any regulation which most people did not like. In the towns mobs could form easily. There were no police to control them. There was the militia, to be sure. And most members of the mob belonged to it!

But there was no real need for mobs. People were too busy, making money, pushing themselves upwards. A growing number got experience of local government, being elected to one office or another, sometimes several. If Americans, in an economic sense, were already predominantly middle class by 1760, the colonies were also in many respects a middle- class democracy too.[177] But this applied more to New England, especially Massachusetts, than to, say, Virginia, where a good deal of deference remained. It is a fact that most people elected to the assemblies, especially in the South, and certainly most of the men who set the tone in them and took a leading legislative and executive part, were vaguely gentry. They were fluent orators, by virtue of their education, and spoke the language of political discourse—very significant in the 18th century, on both sides of the Atlantic. Lesser men, even those proud to call themselves 'free-born Americans,' looked up to them. This was important, because it gave such members of the political elite self-confidence, made them feel they 'spoke for the people' without in any way being demagogs.[178]

Bearing all these factors in mind, it was inevitable that the lower houses would eventually get the upper hand in all the colonial assemblies. And so they did, but at different speeds. The chronological scorecard reads as follows. The Rhode Island and Connecticut Houses of Representatives were all-powerful from well before the beginning of the 18th century. Next came the Pennsylvania House, building on its 1701 Charter of Privileges, and so securing complete dominance in the 1730s, despite the opposition of governors. The Massachusetts House of Representatives actually shared in the selection of the council under its new charter of 1691—that was unique—and in the 1720s it became paramount in finance. By the 1740s it was dominant in all things. The South Carolina Commons and the New York House of Assembly came along more slowly, and trailing behind them were the lower houses in North Carolina, New Jersey, and Virginia—in fact Virginia's burgesses did not get on top until about the mid–1750s. In Maryland and New Hampshire the victory of the lower house had still not been achieved by 1763. But every one had got there by 1770 except remote and under-populated Nova Scotia. The movement was all in one direction—towards representative democracy and rule by the many.[179]

This triumph of the popular system had one very significant consequence for everyday life. It meant that the American mainland colonies were the least taxed territories on earth. Indeed, it is probably true to say that colonial America was the least taxed country in recorded history. Government was extremely small, limited in its powers, and cheap. Often it could be paid for by court fines, revenue from loan offices, or sale of lands. New Jersey and Pennsylvania governments collected no statutory taxes at all for several decades. One reason why American living standards were so high was that people could dispose of virtually all their income. Money was raised by fees, in some cases by primitive forms of poll-tax, by export duties, paid by merchants, or import duties, reflected in the comparatively high price of some imported goods. But these were fleabites. Even so, there was resentment. The men of the frontier claimed they should pay no tax at all, since they bore the burden of defense on behalf of everyone. But this argument was a self-righteous justification of the fact that it was hard if not impossible to get them to pay any tax at all. Until the 1760s at any rate, most mainland colonists were rarely, if ever, conscious of a tax-burden. It is the closest the world has ever come to a no-tax society. That was a tremendous benefit which America carried with it into Independence and helps to explain why the United States remained a low-tax society until the second half of the twentieth century.

By the mid–18th century, America appeared to be progressing rapidly. It was unquestionably a success story. It was to a large extent self-governing. It was doubling its population every generation. It was already a rich country and growing richer. Most men and women who lived there enjoyed, by European standards, middle-class incomes once the frugality and struggles of their youth were over. The opportunities for the skilled, the enterprising, the energetic, and the commercially imaginative were limitless. Was, then, America ceasing to be 'the City on the Hill' and becoming merely a materialistic, earthly paradise? Had Cotton Mather's Daughter Prosperity destroyed her Mother Religion? A visitor might have thought so. In Boston itself, with its '42 streets, 36 lanes and 22 alleys' (in 1722), its 'houses near 3,000, 1,000 brick the rest timber,' its massive, busy 'Long Wharf' which ran out to sea for half a mile and where the world's biggest ships could safely berth in any tide, the accumulation of wealth was everywhere visible.[180] True, the skyline was dotted with eleven church spires. But not all those slender fingers pointing to God betokened the old Puritan spirit. In 1699 the Brattle Street Church had been founded by rich merchants, who observed a form of religion which was increasingly non-doctrinaire, was comfortably moral rather than pious, and struck the old-guard Puritans as disgustingly secular. A place like Philadelphia was even more attached to the things of this world. It had been founded and shaped by Quakers. But the Quakers themselves had become rich. A tax-list of 1769 shows that they were only one in seven of the town's inhabitants but they made up half of those who paid over £100 in taxes. Of the town's seventeen richest men, twelve were Quakers. The truth is, wherever the hard-working, intelligent Quakers went, they bred material prosperity which raised up others as well as themselves. The German immigrants, hard-working themselves but from a poor country still only slowly recovering from the devastation of the Thirty Years War, were amazed at the opportunities the Quaker colony presented to them. One German observer, Gottlieb Mittelberger, summed it up neatly in 1754: 'Pennsylvania is heaven for farmers, paradise for artisans, and hell for officials and preachers.'[181] Philadelphia may have already acquired twelve churches by 1752. But it had fourteen rum distilleries.[182]

However, though Puritanism was in decline in 18th-century America, and the power of the old Calvinist dogmas—and the controversies they bred—were declining, religion as a whole was not a spent force in the America of the Enlightenment. Quite the contrary. In fact American religious characteristics were just beginning to mature and define themselves. It could be argued that it was in the 18th century that the specifically American form of Christianity—undogmatic, moralistic rather than credal, tolerant but strong, and all-pervasive of society—was born, and that the Great Awakening was its midwife. What was the Great Awakening? It was, is, hard to define, being one of those popular movements which have no obvious beginning or end, no pitched battles or legal victories with specific dates, no constitutions or formal leaders, no easily quantifiable statistics and no formal set of beliefs. While it was taking place it had no name. Oddly enough, in the first major history of America, produced in the middle decades of the 19th century, George Bancroft's *History of the United States* (1834–74), the term Great Awakening is never used at all. One or two modern historians argued that the phrase, and to some extent the concept behind it, was actually invented as late as 1842, by Joseph

Tracey's bestselling book, *The Great Awakening: a History of the Revival of Religion in the Times of Edwards and Whitefield*.[183]

Whatever we call it, however, there was a spiritual event in the first half of the 18th century in America, and it proved to be of vast significance, both in religion and in politics. It was indeed one of the key events in American history. It seems to have begun among the German immigrants, reflecting a spirit of thankfulness for their delivery from European poverty and their happy coming into the Promised Land. In 1719, the German pastor of the Dutch Reformed Church, Theodore Frelinghuysen, led a series of revival meetings in the Raritan Valley. 'Pietism,' the emphasis on leading a holy life without troubling too much about the doctrinal disputes which racked the 17th century, was a German concept, and this is the first time we find non-English-speaking immigrants bringing with them ideas which influenced American intellectual life. It is also important to note that this Protestant revival, unlike any of the previous incarnations of the Reformed Religion, began not in city centers, but in the countryside. Boston and Philadelphia had nothing to do with it. Indeed to some extent it was a protest against the religious leadership of the well-fed, self-righteous congregations of the long-established towns. It was started by preachers moving among the rural fastnesses, close to the frontier, among humble people, some of whom rarely had the chance to enjoy a sermon, many of whom had little contact with structured religion at all. It was simple but it was not simplistic. These preachers were anxious not just to deliver a message but to get their hearers to learn it themselves by studying the Bible; and to do that they needed to read. So an important element in the early Great Awakening was the provision of some kind of basic education in the frontier districts and among rural communities which as yet had no regular schools.

A key figure was William Tennent (1673–1745), a Scotch-Irish Presbyterian who settled at Neshaminy, Pennsylvania, in the 1720s, where he built what he called his Log College, a primitive rural academy teaching basic education as well as godliness. This was 'Frontier Religion' in its pristine form, conducted with rhetorical fireworks and riproaring hymn-singing by Tennent and his equally gifted son Gilbert, but also in a spirit of high seriousness, which linked knowledge of God with the spirit of knowledge itself and insisted that education was the high road to heaven.[184] Many of Tennent's pupils, or disciples, became prominent preachers themselves, all over the colonies, and his Log College became the prototype for the famous

College of New Jersey, founded in 1746, which eventually settled at Princeton.

As with most seminal religious movements in history, news of these doings spread by word of mouth and by ministers—some of them unqualified and without a benefice—traveling from one small congregation to another, rather than through the official religious channels. The minister at the Congregationalist church in Northampton, Massachusetts, Jonathan Edwards (1703–58), was intrigued by what he heard. Edwards was a man of outstanding intellect and sensibility, the first major thinker in American history. He was the son and grandson of Puritan ministers, and had gone to Yale almost as young (not quite thirteen) as Cotton Mather went to Harvard. Yet he graduated first in his class and made a name for himself there as a polymath, writing speculative papers on the Mind, Spiders, the Theory of Atoms, and the Nature of Being. His ability was such that, at the age of twenty-one, he was already head tutor of the college—virtually running it, in fact. But, when his grandfather died, Edwards took over his church in Northampton and labored mightily in what was a rather unrewarding vineyard until he learned to base his message not so much on fear, as the old puritan preachers did, as on joy.[185]

It was not that Edwards neglected the element of 'salutary terror,' as he called it. He could preach a hellfire sermon with the best. He told sinners: 'The God that holds you over the pit of hell, much as one holds a spider, or some loathsome insect, over the fire, abhors you, and is dreadfully provoked.' This particular sermon, published (in 1741) under the title *Sinners in the Hands of an Angry God*, was avidly read all over the colonies and committed to heart by many lesser evangelists who wanted to melt hardened hearts. But it was Edwards' nature, as an American, to stress not just God's anger but also his bounty to mankind, and to rejoice in the plenty and, not least, the beauty, of God's creation. Edwards put an entirely new gloss on the harsh old Calvinist doctrine of Redemption by stressing that God did not just choose some, and not others, but, as it were, radiated his own goodness and beauty into the souls of men and women so that they became part of him. He called it 'a kind of participation in God' in which 'God puts his own beauty, ie his beautiful likeness, upon their souls.' In a riveting discourse, *God Glorified in the Work of Redemption*, first published in 1731, he insisted that the happiness human beings find in the 'the Glorious Excellences and Beauty of God' is the greatest of earthly pleasures as well as a spiritual transformation. Through God we love beauty and our joy in beauty is worship. Moreover, this joy and knowledge of beauty, and through beauty God, is 'attainable by persons of mean capacities and advantages as well as those that are of the greatest parts and learning.'[186] It was part of Edwards' message that knowledge of God was education as well as revelation, that it was an aesthetic as well as a spiritual experience, and that it heightened all the senses. Edwards was not a simple evangelist but a major philosopher, whose works fill many thick volumes.[187] But the core of his message, and certainly the secret of his appeal, then and now, and to the masses as well as to intellectuals, is that love is the essence of the religious experience.

In *A Treatise Concerning the Religious Affections* (1746) Edwards lists in detail the twelve signs by which true religious love and its false counterfeit can be distinguished, the most important of which is the ability to detect 'divine things' by 'the beauty of their moral excellency.' It is from 'the sense of spiritual beauty' that there arises 'all true experimental knowledge of religion' and, indeed, 'a whole new world of knowledge.'[188] Through this doctrine of love, Edwards proceeds to liberate the human will by demolishing the old Calvinist doctrine of determinacy and double-predestination. In his *The Freedom of the Will* (1754) Edwards insists that human beings are free because they act according to their perception and conviction of their own good. That will can be corrupted, of course, leading men and women to find the greatest apparent good in self and other lesser goods rather than in God. But earnest teaching can restore the purity of the will. At all events, all can choose: they are responsible for their choices and God will hold them accountable for it. But nothing is determined in advance—all is to be played for. What Edwards in fact was offering—though he did not live long enough to write his great *Summa Theologica*, which was to have been called *A History of the Work of Redemption*—was a framework for life in which free will, good works, purity of conduct, the appreciation of God's world and the enjoyment of its beauties, and the eventual attainment of salvation, all fitted, blended and fused together by the informing and vivifying energy of love. Here was indeed a frontier religion, for persons of all creeds and backgrounds and ethnic origins, native-born Americans and the new arrivals from Europe, united by the desire to do good, lead useful and godly lives, and help others to do the same in the new and splendid country divine providence had given them.

Edwards' earliest published sermons were widely read and discussed. What particularly interested fellow-evangelists, in England as well as America, was his remarkable account, *A Faithful Narrative* (1737), of the conversions his methods brought about in his own parish. One of the Englishmen he thus stirred was John Wesley, over in Georgia in the years 1735–8, to help General Oglethorpe evangelize the colonists and Indians. Another was George Whitefield (1714–70), also a member of the general's mission. Wesley was the greatest preacher of the 18th century, or certainly the most assiduous, but his preoccupation was mainly with the English poor. Whitefield, however, was a rhetorical and histrionic star of spectacular gifts, who did not trouble himself, as Wesley did, with organization. He simply carried a torch and used it to set alight multitudes. He found America greatly to his taste. In 1740 he made the first continental tour of the colonies, from Savannah in Georgia to Boston in the north, igniting violent sheets of religious flame everywhere. It was Whitefield, the Grand Itinerant as he was known, who caused the Great Awakening to take off. He preached, as he put it himself, 'with much Flame, Clearness and Power' and watched hungrily as 'Dagon Falls Daily Before the Ark.' He seems to have appealed equally to conventional Anglicans, fierce Calvinists, German pietists, Scotch-Irish, Dutch, even a few Catholics. A German woman who heard him said she had never been so edified in her life, though she spoke not a word of English. He enjoyed his greatest success in the Calvinist fortress of Boston, where the established churches did not want him at all. There he joined forces with Gilbert Tennent, and an angry critic described how 'people wallowed in snow, night and day, for the benefit of their beastly brayings.'[189]

When Whitefield left, others arose to 'blow up the Divine Fire lately kindled.' John Davenport (1716–57), a Yale man from Long Island, was perhaps the first of the new-style American personal evangelists. At public, open-air meetings in Connecticut he called for rings, cloaks, wigs, and other vain personal adornments to be thrown on the bonfire, together with religious books he denounced as wicked. He thus fell foul of the colony's laws against itinerant preaching, was arrested, tried by the General Assembly, judged to be mentally disturbed, and deported to Long Island. That did not stop him, or anyone else. Denied churches, the new evangelists preached in the open, often round camp-fires. Indeed they soon began to organize the camp-meetings which for two centuries were to be a salient part of American frontier religion. But many clergy welcomed these wild and earnest men. Even Anglican Virginia—its piedmont anyway—joined in the revival.[190] People went to revival meetings, then started attending regularly in their own parish church, if there was one. If not, they clubbed together to set one up. Whitefield attracted enormous crowds—10,000 was not uncommon for him. It may be, as critics claimed, that only one in a hundred of his 'converts' stayed zealous. But he returned again and again to the attack—seven continental tours in the thirty years from 1740—and all churches benefited from his efforts, though the greatest gainers were the Baptists and the stranger sects on the Protestant fringes.

The curious thing about the Great Awakening is that it moved, simultaneously, in two different directions which were in appearance contradictory. In some ways it was an expression of the Enlightenment. One of the most important of the Anglican Awakeners, Samuel Johnson (1693–1772), who had been with Edwards at Yale—was his tutor in fact—was a typical Enlightenment clergyman. He said that reading Francis Bacon's *Advancement of Learning* left him 'like one at once emerging out of the twilight into the full sunshine of open day.' The experience, he said, freed him from what he called the 'curious cobweb of distributions and definition'—17th-century Calvinist theology—and from Bacon he went on to the idealism of the great Anglo-Irish philosopher Bishop Berkeley, who taught him that morality was 'the same thing as the religion of Nature,' not indeed discoverable without Relevation but 'founded on the first principles of reason and nature.' Johnson became the first president of King's College. The Awakening indeed had a dramatic impact on education at all levels. The Congregationalist minister Eleazar Wheelock (1711–79), one of the New England Awakeners, went on to operate a highly successful school for Indians, and this in turn developed into Dartmouth College (1769), which specialized in the classics. Charles Chauncy (1705–87), pastor of the First Church in Boston, originally opposed Edwards and his missions, setting out his views in *Thoughts on the State of Religion in New England* (1743) and other pamphlets. But the Awakening had its effect on him nonetheless, turning him away from the traditional structures of Christianity to what became Unitarianism. He lived just long enough to see the Anglicans of King's Chapel, Boston, adopt a non-trinitarian theology in 1785 and so become America's first Unitarian church. Ebenezer Gay (1696–1787), of Dedham, followed a similar trajectory. And in America, as in England, Unitarianism was, for countless intellectuals, a halfway house on the long road to agnosticism. Paradoxically, as a result of the Awakening, splits arose in many churches between those who endorsed it enthusiastically and those who repudiated its emotionalism, and the second group captured many pulpits and laid the foundations of American religious liberalism.[191]

But if the Awakening, in itself and in the cross-currents it stirred up, was a movement towards a rational view of life, it was also a highly emotional experience for most of those who participated in it—perhaps three out of four of the colonists. It was not just the fainting, weeping, and shrieking which went on at the mass meetings and round the campfires. It was the much less visible but still fundamental stirring of the emotions which Edwards aimed to produce. He urged a rebirth of faith, to create a New Man or a New Woman, rather as Rousseau was to do in France a generation later. He was fond of quoting the Cambridge Platonist John Smith: 'A true celestial warmth is of an immortal nature; and being once seated vitally in the souls of men, it will regulate and order all the motions in a due manner; as the natural head, radicated in the hearts of living creatures, hath the dominion and the economy of the whole body under it . . . It is a new nature, informing the souls of men.'

This, and similar ideas, as presented by Edwards, had undoubted political undertones. Just as in France, rather later in the century, the combination of Voltairean rationalism and Rousseauesque emotionalism was to create a revolutionary explosion, so in America, but, in a characteristically religious context, the thinking elements and the fervid, personal elements were to combine to make Americans see the world with new eyes. There was a strong eschatological element in Edwards and many other preachers. Those who listened to him were left with the impression that great events were impending and that man—including American man—had a dramatic destiny. In his last work, going through the presses at the time of his death in 1758, he wrote: 'And I am persuaded, no solid reason can be given, why God, who constitutes all other created union or oneness, according to his pleasures . . . may not establish a constitution whereby the natural posterity of Adam, proceeding from him, much as the buds or branches from the stock or root of a tree, should be treated as *one* with him.' Man was thus born in the image of God and could do all—his capacities were boundless. In human history, Edwards wrote, 'all the changes are brought to pass . . . to prepare the way for the glorious issue of

things that shall be when truth and righteousness shall finally prevail.'
At that hour, God 'shall take the kingdom' and Edwards said he
'looked towards the dawn of that glorious day.'

The Great Awakening was thus the proto-revolutionary event, the
formative moment in American history, preceding the political drive
for independence and making it possible.[192] It crossed all religious and
sectarian boundaries, made light of them indeed, and turned what had
been a series of European-style churches into American ones. It began
the process which created an ecumenical and American type of religious devotion which affected all groups, and gave a distinctive
American flavor to a wide range of denominations. This might be
summed up under the following five heads: evangelical vigor, a tendency to downgrade the clergy, little stress on liturgical correctness,
even less on parish boundaries, and above all an emphasis on individual experience. Its key text was Revelations 21:5: 'Behold, I make all
things new'—which was also the text for the American experience as a
whole.

If, then, there was an underlying political dimension to the Great
Awakening, there was also a geographical one. It made not only parish
boundaries seem unimportant but all boundaries. Hitherto, each
colony had seen its outward links as running chiefly to London. Each
tended to be a little self-contained world of its own. That was to
remain the pattern in the Spanish colonies for another century, independence making no difference in that respect. The Great Awakening
altered this separateness. It taught different colonies, tidewaters and
piedmonts, coast and up-country, to grasp and appreciate what they
had in common, which was a very great deal. As a symbol of this,
Whitefield was the first 'American' public figure, equally well known
from Georgia to New Hampshire. When he died in 1770 there was
comment from the entire American press.

But even more important than the new geographical sense of unity
was the change in men's attitudes. As John Adams was to put it, long
afterwards: 'The Revolution was effected before the War commenced.
The Revolution was in the mind and hearts of the people: and change in
their religious sentiments of their duties and obligations.' It was the marriage between the rationalism of the American elites touched by the
Enlightenment with the spirit of the Great Awakening among the masses
which enabled the popular enthusiasm thus aroused to be channeled
into the political aims of the Revolution—itself soon identified as the
coming eschatological event. Neither force could have succeeded with-

out the other. The Revolution could not have taken place without this religious background. The essential difference between the American Revolution and the French Revolution is that the American Revolution, in its origins, was a religious event, whereas the French Revolution was an anti-religious event. That fact was to shape the American Revolution from start to finish and determine the nature of the independent state it brought into being.

PART TWO

'That the Free Constitution Be Sacredly Maintained'

Revolutionary America, 1750–1815

If the great awakening prepared the American people emotionally for Revolution and Independence the process was actually detonated by the first world war in human history. And, curiously enough, it was an American who struck the spark igniting this global conflict. George Washington (1732–99) was born on the family estate, Wakefield, in Westmoreland County, Virginia. Much of America's history was written in his antecedents. His founding ancestor was a clergyman expelled from his Essex living for drunkenness, who landed in Virginia in 1657 and married the prosperous Anne Pope. He was Washington's great-grandfather, remembered by the Indians as 'town-taker,' Caunotaucarius. Washington's father, Augustine or Gus, was a blond giant, living evidence of the fact that men grew taller in America than in England—though Gus sent his eldest son Lawrence, Washington's adored half-brother, to school in Appleby, England, to give him a bit of class. Gus had a large family and was only a moderately successful planter. He died when Washington was eleven, leaving 10,000 acres in seven parcels, with a total of forty-nine slaves. The core of it was Ferry Farm with 4,360 acres and ten slaves, in which his mother was left a half-interest and which she decided to keep and run. The Washingtons were so characteristic of the modest gentry families who carried through the Revolution that it is worth detailing the inventory of Gus's possessions on the eve of it. He had little plate—one soup spoon, eighteen small spoons, seven teaspoons, a watch, and a sword, total value £125 10s. The glassware was worth only £5 12s. The china-ware, which included two teasets, was valued at a mere £3 6s. There was a fine looking-glass in the hall, a 'screwtoire' (escritoire), two tables, one armchair, eleven leather-bottomed chairs, three beds in the parlour, an old table, three old chairs, an old desk, window curtains, and in the hall two four-poster beds with two more in the back room. In the chamber above the parlor were three old beds—making a total of thirteen beds in all (Gus had ten children by two wives). There were six good pairs of sheets, ten inferior ones, and seventeen pillow-cases, thirteen table-cloths and thirty-one napkins. Thirteen slaves were attached to the house, but only seven of them were able-bodied.[1] These were the material circumstances in which Washington was nurtured.

Like his father, George Washington was big, six feet two. He had enormous hands and feet, red or auburn hair, a huge nose, high forehead, wide hips, narrow shoulders, and he used his height and bulk to develop an impressive presence which, with his ability to stay calm in moments of crisis, was the key to his ability to rule men, both soldiers and politicians.[2] He always took trouble with his appearance. He never wore a wig, which he thought unbecoming, but he dressed and powdered his hair carefully and tied it with a neat velvet ribbon called a solitary. He broke his teeth cracking nuts and replaced them by false ones of hippopotamus ivory and was self-consciously aware that they fitted badly. He would not venture on an expedition into the woods, as a young man, without nine shirts, six linen waistcoats, seven caps, six collars, and four neckcloths.[3] His instincts were aristocratic and in time became regal. He rejected the new American habit, growing throughout the 18th century, of shaking hands with all and sundry, and instead bowed. He did not hesitate to use his physical strength to exert his will: he 'laid his cane over many of [his] officers who showed their men the example of running.' He could throw stones an immense distance and liked to demonstrate this gift to impress. His mother was a strong woman and he esteemed her. His father meant nothing to him. About 17,000 of Washington's letters have survived, and the father is mentioned in only two of them. He was, from an early age, his own father-figure.[4]

Unlike his half-brother, Washington had only the most elementary education. His envious and critical Vice President John Adams was to write: 'That Washington was not a scholar was certain. That he was too illiterate, unread, unlearned for his station is equally past dispute.'[5] Washington himself said he suffered from 'consciousness of a defective education.' That was why he never attempted to write his memoirs. He said that young men of the gentry class brought up in Virginia, and 'given a horse and a servant to attend them as soon as they could ride' were 'in danger of becoming indolent and helpless.' But Washington was in no such temptation. He wanted to get on. There was a powerful drive in this big young man to better himself. He developed a good, neat, legible hand. To improve his manners, he copied out 110 maxims, originally compiled by a French Jesuit as instructions for young aristocrats. Thus: 'Sing not to yourself with a humming noise nor drum with your fingers and feet.' 'Kill no vermin, as fleas, lice, ticks etc in the sight of others.' 'When accompanying a man of great quality, walk not with him cheek by jowl but somewhat behind him, but yet in such a manner he may easily speak with you.'[6]

Alas! It was Washington's misfortune and grievance that he knew no one 'of great quality.' He 'lacked interest,' as they said in the 18th century. 'Interest' was one of the key words in his vocabulary. Men were driven by it, in his opinion. He wrote of *interest, the only bonding cement.'* It applied equally to men and nations. 'Men may speculate as they will,' he wrote, 'they may talk of patriotism ... but whoever builds upon it as a sufficient basis for conducting a long and bloody war will find themselves deceived in the end ... For a time it may of itself push men to action, to bear much, to encounter difficulties, but it will not endure unassisted by interest.' He thought it was 'the universal experience of mankind' that 'no nation can be trusted further than it is bound by interest.'[7] It is important to grasp that Washington saw both the Revolution and the constitution-making that followed as the work of men driven mainly by self-interest. It was always his dynamic, and he felt no shame in it, following it until his own interest was subsumed in the national interest. The nearest he came to possessing interest himself was the marriage of his half-brother Lawrence to a daughter of Colonel William Fairfax, head of a branch of one of the grandest families in Virginia. Washington made every use he could of this connection. His brother-in-law George Fairfax, a young man with great expectations and a touch of Indian blood (like many Americans) was a role-model.

Washington discovered, aged sixteen, that for a young man of his background and modest education the next best thing to owning a lot of land was to become a land surveyor. A neat hand and the ability to draw maps, take measurements, and make calculations were all that was required. The fascination all Americans had in land, the constant speculation in it, the vast amount there was still to be occupied further west, ensured there would be no lack of occupation. His first job was to survey part of the Fairfax estate west of the Blue Ridge. This took him into the frontier district for the first time and he found he liked the life, the opportunities, even the danger. He joined the militia and found he liked that too. He was a natural soldier. In 1753, when he was twenty-one, the governor of Virginia, Robert Dinwiddie, sent Washington, with the rank of major, into the Ohio Valley, on behalf of the Ohio Company, a private-enterprise venture set up with government backing to develop the frontier districts. Washington's orders were to contact any French he found there and warn them they were straying on to British territory.

The following year was the critical one. Washington, with the rank of lieutenant-colonel and a force of Virginia volunteers and Indians, was sent back to the Ohio and instructed to build a fort at the Ohio Forks, near what is now Pittsburgh. He kept a detailed journal of this expedition. At the Forks he found the French had been before him and constructed Fort Duquesne. He built his own, which he called Fort Necessity—he was having an administrative battle with Governor Dinwiddie over pay and supplies—at Great Meadows. Then he fell in with a French detachment, under Lieutenant de Jumonville, and when the French ran for their muskets, 'I ordered my company to fire,' Washington reported. His Iroquois Indians attacked with their toma-hawks. Before Washington could stop the killing and accept the surren-der of the French, ten of them were dead, including their commander.[8] This incident, *l'affaire Jumonville*, led to massive French retaliation and the outbreak of what was soon a world war. It raged in North America for six years, 1754–60, in Central and South America, in the Caribbean and the Atlantic, in India and the East, and not least in Europe, where it was known as the Seven Years War (1756–63). Detonating such a conflict made Washington famous, even notorious. Artlessly, he wrote to his brother Jack that he had not been daunted by his first experience of action: 'I heard the bullets whistle and, believe me, there is some-thing charming in the sound.' This, together with material from Washington's reports and diaries, was published in the *London Magazine*, where King George II read it. The King was rather proud of his battlefield experience and snorted: 'By God, he would not think bullets charming if he had been used to hear many.' Voltaire summed it up: 'A cannon shot fired in America would give the signal that set Europe in a blaze.' In fact there was no cannon shot. Horace Walpole, in his *History of the Reign of George II*, was more accurate: 'The volley fired by a young Virginian in the backwoods of America set the world on fire.'[9]

This global conflict finally brought to a head the competition between France and Britain to be the dominant power in North America. It was a conflict Britain was bound to win in the end because its American colonies, with their intensive immigration over many decades, their high birth-rate and natural population increase, their booming economy and high living-standards, had already passed the take-off point and were rapidly becoming, considered together, one of the fastest-growing and richest nations in the world. By comparison, the French presence was thinly spread and sustained only by continued military and economic effort from the French state. But it did not quite look like that at the time. The British colonists thought of themselves as encircled by French military power. It stretched from the mouth of the St Lawrence into Canada, down through the region of the Great Lakes, and then along the whole course of the Mississippi to New Orleans, which the French, with much effort, were developing as a major port. There were conflicting claims everywhere. Under the Treaty of Utrecht of 1713, concluding an earlier contest that Britain had won, the French renounced their claim to Hudson's Bay in Canada, but in fact had continued to trade in the region and build forts. Again, Utrecht had given the British 'Nova Scotia or Arcadie with its ancient boundaries,' and to them it meant all territory east of St Croix and north up to the St Lawrence. But the French contested this and in 1750 built more forts to back up their point. Most important of all, from the viewpoint of the British colonies, the French claim to the whole of the Mississippi Basin was in flat contradiction to the claims of the colonies to extend their boundaries along the latitudes indefinitely in a western direction. In the south there were endless conflicting claims too, and a genuine fear that the French would gang up with the Spanish in Florida to attack Georgia.

Fear of France was the great factor which bound the American colonies to Britain in the mid–18th century. They regarded falling under the French flag as the worst possibility that could befall them. On the Atlantic coast, people from numerous nations had found themselves coming under British suzerainty—Spanish, French, Swedes, Dutch, Germans, Swiss—partly by conquest, partly by immigration, and none had found any difficulty in adjusting. By Continental standards Britain was a liberal state with a minimalist government and a tradition of freedom of speech, assembly, the press, and (to some extent) worship. These advantages applied *a fortiori* in the colonies, where settlers often had little or no contact with government from one year's end to another. But for a British subject to shift from the Union Jack to the fleur-de-lys was a different matter. France still had a divine-right absolutist monarchy. Its state was formidable, penetrative, and demanding, even across the Atlantic. It conscripted its subjects and taxed them heavily. Moreover, it was a Catholic state which did not practice toleration, as thousands of Huguenot immigrants in the British colonies could testify.

The American colonies had played little part in the war against France during the reign of William III and Queen Anne. But since then the French military presence in North America had grown far more formidable. When war with Spain, quickly followed by war with France, broke out in 1740 (the War of the Austrian Succession, as Europe called it), the colonies were in the forefront of the action in North America. Not only did Oglethorpe's Georgians invade Spanish Florida but colonial militias, with Massachusetts and New York supplying most of the manpower, took the offensive against France and succeeded in capturing Louisburg. New England and New York were disgusted when the British agreed, at the Treaty of Aix-la-Chapelle in 1748, to hand the fortress back. Nor was colonial opinion impressed by British strategy and grip during the first phases of the world war which Colonel Washington inadvertently started. The British effort in North America was ill provided and ineffective, marked by many reverses. With William Pitt in power from 1758 things changed totally. He had close ties with London mercantile interests and he switched the war from a Continental one in Europe to an imperial one all over the world. He amassed big fleets and raised effective armies, he picked able commanders like General James Wolfe, and he enthused public opinion on both sides of the Atlantic. His armies not only pushed north up the Hudson and down the St Lawrence, but along the Ohio and the Allegheny too. Suddenly, with the fall of Quebec in 1759, French power in North America began to collapse like a house of cards. The Peace of Paris, 1763, confirmed it.[10]

The treaty was one of the greatest territorial carve-ups in history. It says a lot for the continuing ignorance of European powers like Britain and France, and their inability to grasp the coming importance of continental North America, that they spent most of the peace process haggling over the Caribbean sugar-islands, which made quick returns in ready cash. Thanks to its command of the sea, Britain used the war to seize St Vincent, the Grenadines, Tobago, Dominica, St Lucia, Guadeloupe, and Martinique. The British sugar lobby, fearing overproduction, objected to keeping them all, so Britain graciously handed back Guadeloupe, Martinique, and St Lucia. In return, the French made no difficulty about surrendering the whole of Canada, Nova Scotia, and their claims to the Ohio Valley—'Snow for Sugar,' as the deal was called. Moreover, Britain, which now had no fear of a Spain evidently in irreversible military decline, was quite happy to hand Spain back its other conquests, Cuba and Manila. As part of a separate deal France gave Spain all of Louisiana to compensate it for losses in Florida to Britain. Thus more American territory changed hands in this settlement than in any other international treaty, before or since.[11] The net result was to knock France out of the American hemisphere, in which it retained only three small Caribbean islands, two in the fisheries, and a negligible chunk of Guyana. This was a momentous geopolitical shift, a huge relief to British global strategists, because it made Britain the master of North America, no longer challenged there by the most formidable military power in Europe. The hold Spain had on the lower Mississippi was rightly regarded as feeble, to be loosened whenever Britain saw fit. Suddenly, in the mid–1760s, Britain had emerged as proprietor of the largest empire the world had seen since Roman times—larger, indeed, in terms of territorial extent and global compass.

Did this rapid expansion bring a rush of blood to the heads of the British elite? One can put it that way. Certainly, over the next two decades, the characteristic British virtues of caution, pragmatism, practical common sense and moderation seemed to desert the island race, or at any rate the men in power there. There was arrogance, and arrogance bred mistakes, and obstinacy meant they were persisted in to the point of idiocy. The root of the trouble was George III, a young, self-confident, ignorant, opinionated, inflexible, and pertinacious man, determined to be an active king, not just in name, like his grandfather George II, but in reality. George II, however, was a sensible man, well aware of his considerable intellectual and constitutional limitations. He had employed great statesmen, when he could find them, like Sir Robert Walpole and William Pitt the Elder, who had helped to make Britain the richest and most successful nation in the world. George III employed second-raters and creatures of his own making, mere court-favorites or men whose sole merit was an ability to manage a corrupt House of Commons. From 1763 to 1782, by which time the American colonies had been lost, it would be hard to think of a more dismal succession of nonentities than the men who, as First Lords of the Treasury (Prime Minister), had charge of Britain's affairs—the Earl of Bute, George Grenville, the Marquis of Rockingham, the Duke of Grafton, and Lord North. And behind them, in key jobs, were other boobies like Charles Townshend and Lord George Germaine.

This might not have mattered quite so much if the men they faced across the Atlantic had been of ordinary stature, of average competence and character. Unfortunately for Britain—and fortunately for America—the generation that emerged to lead the colonies into independence was one of the most remarkable group of men in history—sensible, broad-minded, courageous, usually well educated, gifted in a variety of ways, mature, and long-sighted, sometimes lit by flashes of genius. It is rare indeed for a nation to have at its summit a group so variously gifted as Washington and Benjamin Franklin, Thomas Jefferson, Alexander Hamilton, James Madison, and John Adams. And what was particularly providential was the way in which their strengths and weaknesses compensated each other, so that the group as a whole was infinitely more formidable than the sum of its parts. They were the Enlightenment made flesh, but an Enlightenment shorn of its vitiating French intellectual weaknesses of dogmatism, anticlericalism, moral chaos, and an excessive trust in logic, and buttressed by the English virtues of pragmatism, fair-mindedness, and honorable loyalty to each other. Moreover, behind this front rank was a second, and indeed a third, of solid, sensible, able men capable of rising to a great occasion. In personal qualities, there was a difference as deep as the Atlantic between the men who led America and Britain during these years, and it told from first to last. Great events in history are determined by all kinds of factors, but the most important single one is always the quality of the people in charge; and never was this principle more convincingly demonstrated than in the struggle for American independence.

Poor quality of British leadership was made evident in the immediate aftermath of the collapse of French power in Canada by an exercise of power thoroughly alien to the English spirit—social engineering. Worried by the concentration of French settlers in Nova Scotia, British ministers tried to round up 10,000 of them and disperse them by force to other British colonies. This was the kind of thing which normally took place in Tsarist Russia, not on British territory. The Protestant colonies did not want the papist diaspora. Virginia insisted on sending its allotment, 1,100, to England. Some 3,000 escaped and went to Quebec, where in due course the British deported them—plus several thousand others—to a reluctant France. The spectacle of these wretched people being marched about and put into ships by redcoats, then replaced by phalanxes of Ulster Protestants, Yorkshire Methodists, and bewildered Scotch Highlanders—themselves marched from ship to inland allotments as though they were conscript members of a military colony—was repugnant to the established colonists. Might not the British authorities soon start to shove them around too, as though they were loads of timber or sacks of potatoes?[12]

By contrast, in American eyes, the British showed a consideration and delicacy towards the Indians which, the colonists felt, was outrageous. In their eyes, the management of the Indians was one field in which social engineering (as they called it 'polity' or 'policing') was not only desirable but essential. In dealing with the heart of America, now to dispose of as they saw fit, the British were faced with a genuine dilemma: how to reconcile three conflicting interests—the fur traders, the colonies with their expanding-westward land hunger, and Britain's Indian allies, such as the Creeks, the Cherokees, and the Iroquois. There were enormous areas involved, none of them properly mapped. Lord Bute, in London, knew nothing about the subject—could not distinguish between a Cherokee and an Eskimo, though he knew all about highland clans—and was entirely dependent on on-the-spot experts like Sir William Johnson and John Stuart, who had Indian interests at heart. The government of Pennsylvania had recently used the term 'West of the Allegheny Mountains,' to denote land reserved for the Indians, presumably in perpetuity. The pro-Indian interest seized on this and persuaded the British government to apply it to the whole of North America. A royal proclamation of October 7, 1763 laid down the new boundary to separate the colonies from land reserved to the Indians. It forbade Americans to settle in 'any lands beyond the heads or sources of any of the rivers which fall into the Atlantic Ocean from the West or Northwest.'[13] In effect this would have created an Atlantic fringe America, inwardly blocked by an Indian interior. That was anathema to the colonies—it destroyed their future, at a stroke. In any case, it was out of date; countless settlers were already over the watershed, well dug in, and were being joined by more every day. The Proclamation noted this point and, to please the Indians, laid down that any 'who have either wilfully or inadvertently seated themselves upon any lands [beyond the line] must forthwith . . . remove themselves.' This was more attempted social engineering, and with heavily armed settlers already scattered and farming over vast distances, there was no question of herding them east except at the point of the bayonet. To make matters worse, British Indian allies were permitted to remain in strength and in large areas well to the east of the line.

The Great Proclamation in short was not a practical document. It enraged and frightened the colonists without being enforceable; indeed it had to be altered, in 1771, to adjust to realities, by conceding settlement along the Ohio from the Great Forks (Fort Pitt or Pittsburgh) to the Kentucky River, thus blowing a great hole in the entire concept, and in effect removing any possible dam to mass westward expansion. The Proclamation was one of Britain's cardinal errors. Just at the moment when the expulsion of the French had entirely removed

American dependence on British military power, and any conceivable obstacle to the expansion of the colonies into the boundless lands of the interior, the men in London were proposing to replace the French by the Indians and deny the colonies access. It made no sense, and it looked like a deliberate insult to American sensibilities.

One American who was particularly upset by the Proclamation was George Washington. He saw himself as a frontiersman as well as a tidewater landowner. Access to land on the frontier was his particular future, as well as America's. The idea of consigning America's interior to the Indians for ever struck him as ridiculous, flying in the face of all the evidence and ordinary common sense. He disliked the Indians and regarded them as volatile, untrustworthy, cruel, improvident, feckless, and in every way undependable. He shared with every one of the Founding Fathers—this is an important point to note—a conviction that the interests of the Indians must not be allowed to stand in the way of America's development. We should not think of Washington as a natural rebel or an instinctive republican. Like most Americans of his class, he was neither. Like most of them, he was ambivalent about England, its crown, its institutions, and its ways. He was fond of using the word 'Empire.' He was proud of England's. If anything his instincts were imperialist. He certainly considered the idea of a career in the British Empire. He had 'had a good war.' In 1756 he had been given the command of the Virginia Militia in frontier defense. In 1758 he led one of the three brigades which took Fort Duquesne. His success as a rising military commander under the British flag helped him to court and win the hand of a wealthy and much sought-after young widow, Martha Dandridge Custis (1732–1802), who had 17,000 acres and £20,000 in money.

It had long been Washington's ambition, which he made repeated efforts to gratify, to get a regular commission in the British Army. This might have changed his entire life because it would have opened up to him the prospect of global service, promotion, riches, possibly a knighthood, even a peerage. He knew by now that he was a first-class officer with the talent and temperament to go right to the top. His fighting experience was considerable and his record exemplary. But the system was against him. In the eyes of the Horse Guards, the headquarters of the British Army in London, colonial army officers were nobodies. American militias were dismissed with special contempt, both social and military. It was a cherished myth in London that they had contributed virtually nothing to winning the war in America and could not be depended on to fight, except possibly against ill-armed Indians. Washington's service actually counted against him, just as, a generation or so later, the young Arthur Wellesley (later Duke of Wellington) found himself dismissed by the Horse Guards as a mere 'sepoy general' because of his service in India. So Washington discovered that his colonial army commission was of no value and that he had no chance of getting a royal one.[14] It was an injustice and an insult and it proved to be the determining factor in his life and allegiance.

Washington's financial experiences also illustrated the way in which the American gentry class were inevitably turning against Britain. The marriage to Martha and the death of his half-brother and his widow made Washington master of Mount Vernon and transformed him from a minor planter into a major landowner. He lived in some style, with thirteen house-servants plus carpenters and handymen about the building. In the seven years from 1768 alone, the Washingtons entertained over 2,000 guests. He did all the things an English gentleman, and the Virginians who aped them, might be expected to do. He bred horses. He kept hounds—Old Harry, Pompey, Pilot, Tartar, Mopsey, Duchess, Lady, Sweetlips, Drunkard, Vulcan, Rover, Truman, Jupiter, June, and Truelove. He set up a library and ordered 500 bookplates from London, with his arms on. He and Martha had no children but he was kind to the step-children she brought with her, ordering fine toys from London: 'A Tunbridge teaset,' reads one list, 'three Neat Tunbridge Toys, a neat book, fashionable tea chest, a bird on bellows, a cuckoo, a turnabout parrot, a grocer's shop, a neat dressed wax baby, an aviary, a Prussian Dragoon, a Man Smoking, and six small books for children.'[15]

But he was not an English gentleman, of course; he was a colonial subject, and he found the system worked against him as a landowner too. He had to employ London Agents, Robert Cary & Co—every substantial planter did—and his relations with them made him anti-British. English currency regulations gave them an advantage over Virginia planters and they tended to keep them in debt with interest mounting up. Any dealings with London were expensive because of the complexity, historical anomalies, and obscurantism of the ancient administration there, which had evolved like a weird organism over centuries. Americans were not used to government. What they had—for instance, the lands offices—was simple, efficient, and did its business with dispatch. London was another universe. The Commission of Customs, the Secretary-at-War, the Admiralty, the Admiralty Courts, the Surveyor-

General of the King's Woods and Forests, the Postmaster-General, the Bishop of London—all were involved in the colonies. The Admiralty alone had fifteen branches scattered all over a city which was already 5 miles wide. There were bureaucratic delays and a five-week voyage added to each end. As Edmund Burke was to put it, 'Seas roll and months pass between order and execution.'[16]

On top of this there was taxation. Like all Americans, Washington paid few taxes before the mid–1760s and resented those he did pay. Now the British government proposed to put colonial taxation on an entirely new basis. The Seven Years War was the most expensive Britain had ever waged. Before it, the national debt had stood at £60 million. It was now—1764—£133 million, more than double. The interest payments were enormous. The British Treasury calculated that the public debt carried by each Englishman was £18, whereas a colonial carried only 18 shillings. An Englishman paid on average 25 shillings a year in taxes, a colonial only sixpence, one-fiftieth. Why, argued the British elite, should this outrageous anomaly be allowed to continue, especially as it was the American colonies which had benefited more from the war?

George Grenville, now in charge of British policy, was a pernickety and self-righteous gentleman determined to correct this anomaly by introducing what he called 'Rules of Right Conduct' between Britain and America. As Burke said, he 'had a rage for regulation and restriction.' His attack was two-pronged. First, he determined to get Americans to pay existing taxes, which were indirect, customs duties and the like. In the English-speaking world, normally law-abiding, evading customs was a universal passion, practiced by high and low, rich and poor. The smugglers who pandered to this passion formed huge armies of rascally seamen, who fought pitched battles on the foreshore and sometimes well inland with His Majesty's Customs Service, who became equally brutal and ruthless in consequence (and still are). But if the English evaded customs duties, the Americans largely got off scot-free because the Colonial Customs Service was inefficient and corrupt. It cost more than it collected. Its officials were almost invariably absentees whose work was done, or not done, by deputies. It was popularly supposed that the duties thus lost amounted to £700,000 annually, though the true figure was nearer £500,000. Grenville's so-called Sugar or Revenue Act of 1764 halved the duty on molasses but provided for strict enforcement. Officials were ordered to their posts. A new Vice-Admiralty Court was set up in Halifax, Nova Scotia, to impose harsh penalties. Suddenly, there were a lot of officious revenue men everywhere. One critic, Benjamin Franklin, reported to the Boston elders that low-born and needy people were being given these jobs as anyone better would not take them:

Their necessities make them rapacious, their offices make them proud and insolent, their insolence and rapacity make them odious, and being conscious they are hated they become malicious; their malice urges them to a continual abuse of the inhabitants in their letters of administration, presenting them as disaffected and rebellious, and (to encourage the use of severity) as weak, divided, timid and cowardly. Government believes all; thinks it necessary to support and countenance its officers; their quarrelling with the people is deemed a mark and consequence of their fidelity . . . I think one may clearly see, in the system of customs now being exacted in America by Act of Parliament, the seeds sown of a total disunion of the two countries.[17]

Franklin's neat summary says it all. But soon there was more. Grenville thought it monstrous that India should pay for itself by having its own taxes and paying its bills, netting large profits for the English gentlemen lucky enough to have posts there, whereas America was run at a thumping loss. So he devised (1765) a special duty for America called the Stamp Act. This was an innovation, which made it horribly objectionable to Americans, who paradoxically were very conservative about such things. It caused exactly the same outrage among them as Charles I's Ship Duty had caused among the English gentry in the years leading up to the Civil War—and this historical parallel did not escape the notice of the colonists. To make matters worse, for some reason which made obscure sense to Grenville's dim calculations, the duty fell particularly hard on two categories of men skilled in circulating grievances—publicans (who had to pay a registration fee of £1 a year) and newspapers (who had to print on stamped paper). Grenville had a gift for doing the wrong thing. His Sugar Act cost £8,000 in administrative costs for every £2,000 raised in revenue.[18] His Stamp Act cost a lot in administration too but raised nothing. It proved unenforceable. Colonial assemblies pronounced it unconstitutional and unlawful. The irresistible popular catchphrase 'No taxation without representation' was heard. The stamps were publicly burned by rioters. One stamp master, Zachariah Hood, had to ride so hard from Massachusetts for protection in the British garrison in New York that he killed his horse under him. Unless the redcoats did it, there was no force prepared to curb the riots. Moreover there were plenty of people in London, led by Pitt, ready to agree with the colonists that parliament had no right to tax them in this way. So the Stamp Act was repealed. That was rightly seen in America as weakness. Parliament then compounded its error by insisting on passing a Declaratory Act asserting its sovereignty over America. That made the dispute not just financial but constitutional.

It is now time to see the origins and progress of the breakdown between Britain and America through the eyes of a man who was involved in all its stages and did his considerable best to prevent it— Benjamin Franklin. One of the delights of studying American history in the 18th century is that this remarkable polymath, visionary, down-to-earth jack-of-all-trades pops up everywhere. There were few contemporary pies into which he did not insert a self-seeking finger. We know a lot about him because he wrote one of the best of all autobiographies.[19] He was born in Boston in 1706, youngest son of a family of seventeen sired by a tallow-chandler immigrant from Oxfordshire. His parents lived to be eighty-four and eighty-seven, and all this was typical of the way America's population was exploding with natural growth—in Philadelphia Franklin met Hannah Miller, who died at 100 in 1769, leaving fourteen children, eighty-two grandchildren and 110 great-grandchildren. Franklin had only two years' schooling, then went to work for his elder brother James' printing business. He became a life-long autodidact, teaching himself French, Latin, Italian, Spanish, maths, science, and many other things. At the age of fifteen he started writing for James' newspaper, the *New England Courant*. His mentor was another self-taught multiple genius, Daniel Defoe, but he learned self-discipline from yet another polymath, Cotton Mather. James was twice in trouble with the authorities for his critical articles, and jailed; Benjamin was a rebel too—'Adam was never called Master Adam,' one of his articles went. 'We never read of Noah Esquire, Lot Knight and Baronet, nor of the Rt Hon. Abraham, Viscount Mesopotamia, Baron of Canaan.' James' paper banned, it reappeared with Benjamin as editor–proprietor, but he soon rebelled against James too and left for Philadelphia.

This was now effectively the capital of the colonies and bigger than Boston. Franklin thrived there. In 1724 the governor of Pennsylvania, Sir William Keith, sent him to England for eighteen months and he returned full of ideas and new technology. By the age of twenty-four he was the most successful printer in America's boom-city, owner of the

Pennsylvania Gazette, and currency-printer to the Assembly, 'a very profitable jobb and a great help to me.' He persuaded other young, self-educating artisans to form a 'Junto or Club of the Leather Aprons,' which set up a circulating library—the first in America and widely imitated—which was notable for its paucity of religious books and its plethora of do-it-yourself volumes of science, literature, technology, and history.[20] Franklin worked hard at improving his adopted city. He helped set up its first police or watch. He became president of its first fire-insurance company and its chief actuary, working out the premiums. He took a leading part in paving, cleaning, and especially lighting the streets, designing a four-sided Ventilated Lamp and putting up whale-oil street-lights. With others, he founded the American Philosophical Society, equivalent of England's famous Royal Society, the city's first hospital, and, not least, the Academy for the Education of Youth, which became the great University of Pennsylvania. It had a remarkable liberal curriculum for its day—penmanship, drawing, arithmetic, geometry, astronomy, 'and even a little Gardening, Planting, Grafting and Inoculating.' It was also used to cultivate English style, especially 'the *clear* and the *concise*.'[21]

Franklin fathered two illegitimate children, took on a common-law wife, kept a bookshop importing the latest pamphlets from London ('Let me have everything, good or bad, that makes a Noise and has a Run'), became postmaster, and, from 1733, made himself a national figure with his *Poor Richard's Almanac*, a calendar–diary with mottoes, aphorisms, and poems. He pinched the idea from Swift but made it his own. It was original in two distinct senses, both highly American. First, it introduced the wisecrack—the joke which imparts knowledge or street-wisdom, as well as makes you laugh. Second, it popularized the notion, already rooted in America, of the Self-Made Man, the rags-to-riches epic, by handing out practical advice. Franklin's 'Advice to a Young Tradesman written by an Old One' (1748) sums up the Poor Richard theme: 'Remember that *time* is money . . . remember that *credit* is money . . . The way to Wealth is as plain as the Way to Market. It depends chiefly on two words, *Industry* and *Frugality*.' The *Almanac* sold 10,000 copies a year, one for every 100 inhabitants, and a quarter-million over its lifespan, becoming the most popular book in the colonies after the Bible. Extracts from it, first printed in 1757 under the title *The Way to Wealth*, have gone through over 1,200 editions since and youngsters still read it.[22] By 1748 Franklin was able to put in a partner to run his business, retiring on an income of £476 a year to devote the rest of his life to helping his fellow-men and indulging his scientific curiosity.

His activities now multiplied. He crossed the Atlantic eight times, discovered the Gulf Stream, met leading scientists and engineers, invented the damper and various smokeless chimneys—a vexed topic which continued to occupy him till the end of his life—designed two new stoves, but refused to patent them from humanitarian principles, invented a new hearth called a Pennsylvania Fireplace, manufactured a new whale-oil candle, studied geology, farming, archeology, eclipses, sunspots, whirlwinds, earthquakes, ants, alphabets, and lightning conductors. He made himself one of the earliest experts on electricity, publishing in 1751 an eighty-six-page treatise, *Experiments and Observations on electricity made in Philadelphia*, which over twenty years went into four editions in English, three in French and one each in German and Italian, giving him a European reputation. As one of his biographers put it, 'He found electricity a curiosity and left it a science.' But he also had fun, proposing an Electricity Party: 'A turkey is to be killed for our dinner by an electric shock, and roasted by an electrical jack, before a fire kindled by the electrified bottle; when the healths of all the famous electricians in England, Holland, France and Germany are to be drunk in electrified bumpers, under the discharge of guns from an electrified battery.'[23]

Honors accrued: a Fellowship of the Royal Society, degrees conferred not only by Yale, Harvard, and William and Mary but by Oxford and St Andrews. He corresponded with sages all over the civilized world and in time belonged to twenty-eight academies and learned societies. As Sir Humphry Davy acknowledged: 'By very small means he established very grand truths.'[24] He came to politics comparatively late. He was elected to the Pennsylvania Assembly in 1751 and two years later was appointed deputy postmaster-general for all the colonies. This made him, for the first time, think of the American continent as a unity. But it was the British Act forbidding new iron forges in the colonies which drew him into the great argument. His *Observations Concerning the Increase in Mankind, People of Countries etc* (published 1754) noted the much higher population increase in America and he predicted that 'within a century' America would have more people, 'a glorious market wholly in the power of Britain.' So it was wrong to restrain colonial manufactures: 'A wise and good mother will not do it.' He added, setting out for the first time the theory of the dynamic frontier: 'So vast is the territory of north America that it will require many ages to settle it fully; and till it is settled, labor will never be cheap here, where no man continues long a laborer for others but gets a plantation of his own, no man continues long a journeyman to a trade, but goes among those new settlers and sets up for himself.'[25]

It was this line of thought, and further experience as a commissioner negotiating with the Indians on the Ohio, and during the war, which led Franklin to propose a general government of the mainland colonies, except for Georgia and Nova Scotia. He thought such a federated government should deal with defense, frontier expansion, and Indian affairs. A Grand Council, elected by delegates from all the colonial assemblies in proportion to tax paid, would have the power to legislate, make peace and war, and pay a president-general. London was not hostile to the idea, but not one of the assemblies showed an interest, so the British government proceeded no further. Franklin later noted, sadly, in his *Autobiography*: 'I am still of the opinion it would have been happy for both sides of the water if it had been adopted. The colonies, so united, would have been sufficiently strong to have defended themselves [against the French]; there would have been no need of troops from England; of course, the subsequent pretence for taxing America, and the bloody contest it occasioned, would have been avoided.' Alas, 'the assemblies did not adopt it, as they all thought there was too much *prerogative* in it, and in England it was judged to have too much of the *democratic*.'[26]

Franklin never abandoned this idea. He was still at this stage (like young Washington) an imperialist, advocating a huge, self-contained Anglo-American empire, pushing to the Pacific by land and sea—a Manifest Destiny man, though under the crown. It was only in the late 1750s, after much wartime experience, when he went to London as representative of the Pennsylvania Assembly (which was at loggerheads with the Penn family, still proprietors) that he began to realize the enormous intellectual and constitutional gap, as wide as the Atlantic itself, which separated Americans from the English ruling class. He had a talk with Earl Granville, Lord President of the Council, who told him, to his astonishment, that 'The King in Council is legislator for the Colonies, and when His Majesty's instructions come there, they are the law of the land.' Franklin continued: 'I told him this was new doctrine to me. I had always understood from our charters that our laws were to be made by our assemblies, to be presented indeed to the King for his royal assent, but that being once given the King could not repeal or alter them. And as the assemblies could not make permanent laws without his assent, [so] neither could he make a law for them without theirs. He assured me I was totally mistaken.'[27] This is a very revealing exchange. There is no doubt that all Americans took exactly the same view of the position as Franklin, and that this view reflected the practice of over a century. There is equally no doubt that ministerial and parliamentary opinion in England, judges, bureaucrats—the lot—took Granville's view. What was to be done?

The constitutional impasse was aggravated by a gradual breakdown in order in some of the colonies, caused by a variety of factors some of which had nothing to do with disagreements between America and London but which nonetheless made them more serious. In 1763 a powerful Indian chief called Pontiac, a former ally of the French who had been exasperated beyond endurance by the consequences of the British conquest, formed a grand confederacy of various discontented tribes, and ravaged over a thousand miles of the frontier, destroying every fort except Detroit and Pittsburgh. The violence ranged from Niagara to Virginia and was by far the most destructive Indian uprising of the century. Over 200 traders were slaughtered.[28] It took three years to put down the uprising, which was achieved only thanks to regular British units, deployed at considerable expense. Only four colonies, New York, New Jersey, Connecticut, and Virginia, made any attempt to assist. On top of this came the violent refusal to pay the Stamp Tax, which many rightly saw as a triumph of mob rule.

There were other outbreaks, some trivial, some serious, but all constituting a threat to a system of government which was clearly outmoded and in need of fundamental reconstruction. For instance, at the end of 1763 a gang of Scotch-Irish frontiersmen, from Paxton and Donegal townships, carried out an atrocious massacre of harmless Indians, some of them Christians, and many of whom had taken refuge in the workhouse at Lancaster. They slaughtered another group of 140 Indians, converted by the Moravians, who had been taken for safety to Province Island on the Schuylkill River. They threatened to march on Philadelphia and slaughter the Quakers too, for they saw them as 'Indian-lovers' who would prevent the development of the frontier and the freeing of land for settlement. Franklin was asked to organize the defence of the city against the 'Paxton Boys,' mustered the militia—six companies of foot, two of horse, and a troop of artillery—and eventually persuaded the rioters to disperse. But there was not the will to punish even the ringleaders, and Franklin, no friend of the Indians but disgusted by what had happened, had to content himself with writing a bitter pamphlet denouncing 'the *Christian White Savages*.'[29] There was yet more violence when Charles Townshend, on behalf of the British government, returned to the financial attack (he was Chancellor of the Exchequer) with a new series of duties, on glass, lead, paint, and tea, in 1767. The colonies responded with what they called Nonimport Agreements, in effect a boycott of British goods. But a considerable amount of tax was collected this time—£30,000 a year, at a cost of £13,000—and this encouraged the British authorities to press on. The port and town of Boston became the center of resistance, which was increasingly violent, with individual attacks on customs officials, and mob raids on customs warehouses and vice-admiralty courts.

The effect of these outrages on British opinion was disastrous. There was a call for 'firmness.' Even those generally sympathetic to the colonists' case called for a strong government line, not ruling out force. Pitt, now Earl of Chatham, laid down: 'The Americans must be subordinate . . . this is the Mother Country. They are children. They must obey, and we prescribe.' The Earl of Shelburne, cleverest and wiliest of the London politicians, wanted the civilian governor of New York, Sir Henry Moore, replaced by 'a Man of a Military Character, who would act with Force or Gentleness as circumstances might make necessary.'[30] The view of the British military men, especially of the foreign mercenary commanders, like Colonel Henri Boughet, who put down the Pontiac Rising, was that the American militias were useless and that, however gifted the colonists might be at playing noisy politics, they were no good at fighting. By the late 1760s Britain had about 10,000 troops in the theater, regulars and German mercenaries, based in Jamaica, Halifax, and the mainland colonies, and costing about £300,000 a year. Why not use them?

Just as the British despised the colonial militias, so they refused to recognize the constitutional or moral legitimacy of the colonial assemblies. Lord North, Prime Minister from 1770, a man dismissed by Dr Johnson in the words 'He fills a chair' with 'a mind as narrow as the neck of a vinegar-cruet,' criticized the Massachusetts constitution as a whole because it depended on 'the democratik part.' His minister in charge of colonial matters, Lord George Germaine, took an even more contemptuous view: 'I would not have men in mercantile cast every day collecting themselves together and debating on political matters.' This view was shared by the generals. General Guy Carleton, governor of Quebec, warned where it was all leading: 'A Popular Assembly, which preserves its full Vigor, and in a Country where all Men appear nearly

on a Level, must give a strong bias to Republican Principles.' General
Gage summed up the conclusion: 'The colonists are taking great strides
towards independence. It concerns great Britain by a speedy and spirited Conduct to show them that these provinces are British Colonies
dependent on her and that they are not independent states.'[31]

The upshot was that the British garrison in Boston, the most 'difficult' of the colonial cities, was suddenly increased by two whole regiments. That, as Franklin put it, was 'Setting up a Smith's Forge in a
Magazine of Gunpowder.' On March 3, 1770, a sixty-strong mob of
Boston youths started to snowball a party of redcoats. There was a
scrimmage. Some soldiers fired, without orders, killing three youths
outright and wounding others, two of whom later died. Britain and its
colonies were under the rule of law and for soldiers to open fire on
civilians without a previous reading of the Riot Act was to invite
charges of murder or manslaughter. Ten years later, indeed, the whole
of central London was given over to the mob because of the timidity of
the military authorities for this reason. In this case the commander of
the redcoats, Captain Preston, was put on trial; so were some of his
men. But there was no conclusive evidence that an order was given, or
who fired the shots, so all were acquitted, though to appease the
Bostonians two of the men were branded. This was to hand the
colonists the first of a whole series of propaganda victories—the story
of the 'Boston Massacre,' as it was called, and the failure of Britain to
punish those responsible. Sam Adams and Joseph Warren skillfully verbalized the affair into a momentous act of deliberate brutality, and Paul
Revere engraved an impressive but entirely imaginary image of the
event for circulation through the eastern seaboard.

The American Revolution was the first event of its kind in which the
media played a salient role—almost a determining one—from first to
last. Americans were already a media-conscious people. They had a lot
of newspapers and publications, and were getting more every month.
There were plenty of cheap printing presses. They now found that they
had scores—indeed hundreds—of inflammatory writers, matching the
fiery orators in the assemblies with every polysyllabic word of condemnation they uttered. There was no longer any possibility of putting
down the media barrage in the courts by successful prosecutions for
seditious libel. That pass had been sold long ago. So the media war,
which preceded and then accompanied the fighting war, was one the
colonists were bound to win and the British crown equally certain to
lose.[32]

Boston was now the center of outright opposition to British colonial rule. We can look at it through the eyes of its most distinguished and certainly most acrimonious son, John Adams (1735–1826), who was then in his thirties and a prominent lawyer of the city. Adams came from Quincy, the son of a fourth-generation Bay Colony farmer, and was as impregnated with the self-righteous, opinionated, independent-minded, and contumacious spirit of Massachusetts as anyone who had ever crossed the Common. He was a Harvard graduate and had the high-minded sense of intellectual superiority of that famous academy, and his sense of importance had been much increased by his marriage in 1764 to Abigail Smith of Weymouth, an able, perceptive, charming, and socially prominent lady. The proto-Republicans of Boston called themselves Whigs, in sympathy with the London parliamentary critics of the British government, such as Edmund Burke and Charles James Fox, and Adams became a prominent Whig at the time of the Stamp Act agitation. He published, anonymously, four notable articles attacking the British authorities in the *Boston Gazette*, and he later brought out under his own name *A Dissertation on the Canon and Feudal Law* (1768), which argued that the tax was unconstitutional and unlawful and so invalid. It says a lot for the fair-mindedness of Britain in these years before the conflict broke out openly that Adams published this philippic in London.[33] But it is important to note that Adams, then and later, was not a man who believed in force if arguments were still listened to. Unlike his cousin Sam Adams, and other men of the mobs, he deplored street violence in Boston and, as a lawyer, was prepared to defend the soldiers accused of the 'Massacre.' The breaking-point for him came in 1773–4 when North, by an extraordinary act of folly, made British power in Boston look not only weak, vindictive, and oppressive, but ridiculous.[34]

The origins of the Boston Tea Party had nothing to do with America. The East India Company had got itself into a financial mess. To help it to extricate itself, North passed an Act which, among other things, would allow the company to send its tea direct to America, at a reduced price, thus encouraging the 'rebels' to consume it. Delighted, the harassed company quickly dispatched three ships, loaded with 298 chests of tea, worth £10,994, to Boston. At the same time, the authorities stepped up measures against smuggling. The American smuggling interest, which in one way or another included about 90 percent of import–export merchants, was outraged. John Hancock (1737–93), a prominent Boston merchant and political agitator, was a respectable large-scale smuggler and considered this maneuver a threat to his liveli-hood as well as a constitutional affront. He was one of many substan-tial citizens who encouraged the Boston mob to take exemplary action.

When the ships docked on December 16, 1773, a crowd gathered to debate what to do at the Old South Meeting House. It is reported 7,000 people were jammed inside. Negotiations were held with the ship-masters. One rode to Governor Hutchinson at his mansion on Milton Hill to beg him to remit the duties. He refused. When this news was conveyed to the mob, a voice said: 'Who knows how tea will min-gle with salt water?' Sam Adams, asked to sum up, said 'in a low voice,' 'This meeting can do nothing more to save the country.' The doors were then burst open and a thousand men marched to the docks. There had been preparations. An eyewitness, John Andrews, said that 'the patriots' were 'cloath'd in blankets with the heads muffled, and copper-color'd countenances, being each arm'd with a hatchet or axe, and a pair of pistols.' The 'Red Indians' ran down Milk Street and onto Griffith's Wharf, climbed aboard the *Dartmouth*, chopped open its tea-chests, and then hurled the tea into the harbor, 'where it piled up in the low tide like haystacks.' They then attacked the *Eleanor* and the *Beaver*. By nine in the evening all three ships had been stripped of their cargo. Josiah Quincy (1744–75), one of the leading Boston pamphle-teers and spokesmen, said: 'No one in Boston will ever forget this night,' which will lead 'to the most trying and terrific struggle this country ever saw.' John Adams, shrewdly noting that no one had been injured, let alone killed, saw the act, though one of force, as precisely the kind of dramatization of a constitutional point that was needed. As he put it: 'The people should never rise without doing something to be remembered, something notable and striking. This destruction to the tea is so bold, so daring, so firm, intrepid and inflexible, and it must have so important consequences, and so lasting, that I can't but con-sider it an epoch of history.'[35]

Adams was quite right. The episode had the effect of forcing every-one on both sides of the Atlantic to consider where they stood in the controversy. It polarized opinion. The Americans, or most of them, were exhilarated and proud. The English, or most of them, were out-raged. Dr Johnson saw the Tea Party as theft and hooliganism and pro-duced his maxim: 'Patriotism is the last refuge of a scoundrel.' In March 1774, on the invitation of the government, parliament closed the port of Boston to all traffic and two months later passed the Coercive Acts. These punitive measures, paradoxically, were accompanied by the Quebec Act, a highly liberal measure which gave relief to the Canadian Catholics and set Upper and Lower Canada firmly on the road to self-government and dominion status. It was designed to keep the Canadians, especially the French-speaking ones, loyal to the crown, and succeeded; but it infuriated the American Protestants and made them suspicious that some long-plotted conspiracy was afoot to reimpose what John Adams called 'the hated despotism of the Stuarts.' In the current emotional atmosphere, anything could be believed. At all events, these legislative measures, which included the compulsory quartering of troops on American citizens in Boston and elsewhere, were lumped together by the American media under the term the 'Intolerable Acts.' They mark the true beginning of the American War of Independence.

We must now shift the eyewitness focus yet again and see how things appeared to Thomas Jefferson (1743–1826), then in his early thirties and already a prominent politician in Virginia. He came from the same background as George Washington, and was related to many families in the Virginia gentry, such as the Randolphs and the Marshalls. His father, Peter Jefferson, was a surveyor who mapped the Northern Wilderness part of Lord Fairfax's great domain. Jefferson was one of ten children and owed a great deal to his devoted elder sister Jane, who taught him to read books and, equally important, to love music. He learned to play the violin well and carried a small instrument with him on all his travels. He delighted to sing French and Italian songs. When he went to William and Mary College, aged sixteen, he was already fluent in Latin and Greek, and could ride, hunt, and dance well. He had a gift for friendship and became a devoted pupil of his Scots teacher, William Small, as well as a disciple of the gifted Virginia jurist George Wythe, seventeen years his senior. Small secured for the college the finest collection of scientific instruments in America and the two together, said Jefferson, 'fixed the destinies of my life.' Wythe was another of the enterprising polymaths whom America produced in such numbers at this time and had many clever guests at his house. Jefferson was in some ways the archetypal figure of the entire Enlightenment, and he first learned to blossom in Wythe's circle.[36] In terms of all-round learning, gifts, sensibilities, and accomplishments, there has never been an American like him, and generations of educated Americans have rated him higher even than Washington and Lincoln. A 1985 poll of members of the Senate showed that conservative and liberal senators alike regarded him as their 'favorite hero.'[37]

We know a great deal about this remarkable man, or think we do. His *Writings*, on a bewildering variety of subjects, have been published in twenty volumes. In addition, twenty-five volumes of his papers have appeared so far, plus various collections of his correspondence, including three thick volumes of his letters to his follower and successor James Madison alone.[38] In some ways he was a mass of contradictions. He thought slavery an evil institution, which corrupted the master even more than it oppressed the chattel. But he owned, bought, sold, and bred slaves all his adult life. He was a deist, possibly even a sceptic; yet he was also a 'closet theologian,' who read daily from a multilingual edition of the New Testament. He was an elitist in education—'By this means twenty of the best geniuses will be raked from the rubbish annually'—but he also complained bitterly of elites, 'those who, rising above the swinish multitude, always contrive to nestle themselves into places of power and profit.' He was a democrat, who said he would 'always have a jealous care of the right of election by the people.' Yet he opposed direct election by the Senate on the ground that 'a choice by the people themselves is not generally distinguished for its wisdom.' He could be an extremist, glorying in the violence of revolution: 'What country before ever existed a century and a half without rebellion? . . . The tree of liberty must be refreshed from time to time with the blood of patriots and tyrants. It is its natural manure.' Yet he said of Washington: 'The moderation and virtue of a single character has probably prevented this revolution from being closed, as most others have been, by a subversion of that liberty it was intended to establish.'

No one did more than he did to create the United States of America. Yet he referred to Virginia as 'my country' and to the Congress as 'a foreign legislature.' His favorite books were *Don Quixote* and *Tristram Shandy*. Yet he lacked a sense of humor. After the early death of his wife, he kept—it was alleged—a black mistress. Yet he was priggish, censorious of bawdy jokes and bad language, and cultivated a we-are-not-amused expression. He could use the most inflammatory language. Yet he always spoke with a quiet, low voice and despised oratory as such. His lifelong passion was books. He collected them in enormous quantity, beyond his means, and then had to sell them all to Congress to raise money. He kept as detailed daily accounts as it is possible to conceive but failed to realize that he was running deeply and irreversibly into debt. He was a man of hyperbole. But he loved exactitude—he noted all figures, weights, distances, and quantities in minute detail; his carriage had a device to record the revolutions of its wheels; his house was crowded with barometers, rain-gauges, thermometers and anemometers. The motto of his seal-ring, chosen by himself, was 'Rebellion to tyrants is obedience to God.' Yet he shrank from violence and did not believe God existed.[39]

Jefferson inherited 5,000 acres at fourteen from his father. He married a wealthy widow, Martha Wayles Skelton, and when her father died he acquired a further 11,000 acres. It was natural for this young patrician to enter Virginia's House of Burgesses, which he did in 1769, meeting Washington there. He had an extraordinarily godlike impact on the assembly from the start, by virtue of his presence, not his speeches. Abigail Adams later remarked that his appearance was 'not unworthy of a God.' A British officer said that 'if he was put besides any king in Europe, that king would appear to be his laquey.' His first hero was his fellow-Virginian Patrick Henry (1736–99), who seemed to be everything Jefferson was not: a firebrand, a man of extremes, a rabble-rouser, and an unreflective man of action. He had been a miserable failure as a planter and storekeeper, then found his metier in the law-courts and politics. Jefferson met him when he was seventeen and he was present in 1765 when Henry acquired instant fame for his flamboyant denunciation of the Stamp Act. Jefferson admired him no doubt for possessing the one gift he himself lacked—the power to rouse men's emotions by the spoken word.

Jefferson had a more important quality, however: the power to analyze a historic situation in depth, to propose a course of conduct, and present it in such a way as to shape the minds of a deliberative assembly. In the decade between the Stamp Act agitation and the Boston Tea Party, many able pens had set out constitutional solutions for America's dilemma. But it was Jefferson, in 1774, who encapsulated the entire debate in one brilliant treatise—*Summary View of the Rights of British America*. Like the works of his predecessors in the march to independence—James Otis' *Rights of the British Colonists Asserted* (1764), Richard Bland's *An Inquiry into the Rights of the British Colonists* (1766), and Samuel Adams' *A Statement of the rights of the Colonies* (1772)—Jefferson relied heavily on Chapter Five of John Locke's *Second Treatise on Government*, which set out the virtues of a meritocracy, in which men rise by virtue, talent, and industry. Locke argued that the acquisition of wealth, even on a large scale, was neither unjust nor morally wrong, provided it was fairly acquired. So, he said, society is necessarily stratified, but by merit, not by birth. This doctrine of industry as opposed to idleness as the determining factor in a just society militated strongly against kings, against governments of nobles and their placemen, and in favor of representative republicanism.[40]

Jefferson's achievement, in his tract, was to graft onto Locke's meritocratic structure two themes which became the dominant leitmotifs of the Revolutionary struggle. The first was the primacy of individual rights: 'The God who gave us life, gave us liberty at the same time: the hand of force may destroy, but cannot disjoin them.' Equally important was the placing of these rights within the context of Jefferson's deep and in a sense more fundamental commitment to popular sovereignty: 'From the nature of things, every society must at all times possess within itself the sovereign powers of legislation.'[41] It was Jefferson's linking of popular sovereignty with liberty, both rooted in a divine plan, and further legitimized by ancient practice and the English tradition, which gave the American colonists such a strong, clear, and plausible conceptual basis for their action. Neither the British government nor the American loyalists produced arguments which had a fraction of this power. They could appeal to the law as it stood, and duty as they saw it, but that was all. Just as the rebels won the media battle (in America) from the start, so they rapidly won the ideological battle too.

But they had also to win the emotional battle—the war for men's hearts—before they could begin the battle of bayonets. In the events leading up to the fighting, ordinary men and women in America were roused by a number of factors. There was the desire for a republic—the commitment to place each selfish and separate interest in the search for the *res publica*, 'the public thing,' the common good. Let us not underestimate this. It was strongly intuited by a great many people who could barely write their names. It was vaguely associated in their minds with the ancient virtue and honor of the Romans. When James Otis gave the address at the public funeral of 'the fallen' of the Boston Massacre, he wore a toga. And republicanism was a broad concept—every man could put into it the political emotions he felt most keenly.[42] But there was also fear. The early 1770s were marked by recession throughout the English-speaking world. There were poor crops in England in 1765–73, with a primitive cyclical downturn, 1770–6. A fall in English purchasing-power hit American exports in most colonies, and this came on top of economic disruption caused by boycotts. Exports from New England hit the 1765 high only twice in the decade 1765–75, after many years of uninterrupted increases. Exports from Virginia and Maryland fell below the 1765 high every year until 1775.[43] There was distress in England, which stiffened the resolve of parliament to 'make the Americans pay.' But there was profound unease among Americans that the exactions of the British government were bringing the good times—most colonists had never known anything else—to a close.

There was another fear, and a more deep-rooted one. Next to religion, the concept of the rule of law was the biggest single force in creating the political civilization of the colonies. This was something they shared with all Englishmen. The law was not just necessary—essential to any civil society—it was noble. What happened in courts and assemblies on weekdays was the secular equivalent of what happened in church on Sundays. The rule of law in England, as Americans were taught in their schools, went back even beyond Magna Carta, to Anglo-Saxon times, to the laws of King Alfred and the Witanmagots, the ancient precursor of Massachusetts' Assembly and Virginia's House of Burgesses. William the Conqueror had attempted to impose what Lord Chief Justice Coke, the great early 17th-century authority on the law, had called 'The Norman Yoke.' But he had been frustrated. So, in time, had Charles I been frustrated, when he tried to reimpose it, by the Long Parliament. Now, in its arrogance and complacency, the English parliament, forgetting the lessons of the past, was trying to impose the Norman Yoke on free-born Americans, to take away their cherished rule of law and undermine the rights they enjoyed under it with as much justice as any Englishman! Lord North would have been astonished to learn he was doing any such thing, but no matter: that is what many, most, Americans believed.[44] So Americans now had to do what parliamentarians had to do in 1640. 'What we did,' said Jefferson later, 'was with the help of Rushworth, whom we rummaged over for revolutionary precedents of those days.' So, in a sense, the United States was the posthumous child of the Long Parliament.

But Americans' fears that their liberties were being taken away, and the rule of law subverted, had to be dramatized—just as those old parliamentarians had dramatized their struggle by the Grand Remonstrance against Charles I and the famous 'Flight of the Five Members.' Who would play John Hampden, who said he would rather die than pay Ship Money to King Charles? Up sprang Jefferson's friend and idol, Patrick Henry. As a preliminary move towards setting up a united resistance of the mainland colonies to British parliamentary pretensions, a congress of colonial leaders met in Philadelphia, at

Carpenters Hall, between September 5 and October 26, 1774. Only Georgia, dissuaded from participating by its popular governor, did not send delegates. Some fifty representatives from twelve colonies passed a series of resolutions, calling for defiance of the Coercive Acts, the arming of a militia, tax-resistance. The key vote came on October 14 when delegates passed the Declarations and Resolves, which roundly condemned British interference in America's internal affairs and asserted the rights of colonial assemblies to enact legislation and impose taxes as they pleased. A common American political consciousness was taking shape, and delegates began to speak with a distinctive national voice. At the end of it, Patrick Henry marked this change in his customary dramatic manner: 'The distinction between Virginians and New Englanders are no more. I am not a Virginian but an American.' Not everyone agreed with him, as yet, and the Continental Congress, as it called itself, voted by colonies rather than as individual Americans. But this body, essentially based on Franklin's earlier proposals, perpetuated its existence by agreeing to meet again in May 1775. Before that could happen, on February 5, 1775, parliament in London declared Massachusetts, identified as the most unruly and contumacious of the colonies, to be in a state of rebellion, thus authorizing the lawful authorities to use what force they thought fit. The fighting had begun. Hence when the Virginia burgesses met in convention to instruct their delegates to the Second Continental Congress, Henry saw his chance to bring home to all the revolutionary drama of the moment.

Henry was a born ham actor, in a great age of acting—the Age of Garrick. The British parliament was full of actors, notably Pitt himself ('He acted even when he was dying') and the young Burke, who was not above drawing a dagger, and hurling it on the ground to make a point. But Henry excelled them all. He proposed to the burgesses that Virginia should raise a militia and be ready to do battle. What was Virginia waiting for? Massachusetts was fighting. 'Our brethren are already in the field. Why stand we here idle? What is it that gentlemen wish? What would they have?' Then Henry got to his knees, in the posture of a manacled slave, intoning in a low but rising voice: 'Is life so dear, our peace so sweet, as to be purchased at the price of chains and slavery? Forbid it, Almighty God!' He then bent to the earth with his hands still crossed, for a few seconds, and suddenly sprang to his feet, shouting, 'Give me liberty!' and flung wide his arms, paused, lowered his arms, clenched his right hand as if holding a dagger at his breast, and said in sepulchral tones: 'Or give me death!' He then beat his breast, with his hand holding the imaginary dagger. There was silence, broken by a man listening at the open window, who shouted: 'Let me be buried on this spot!' Henry had made his point.[45]

By the time the Second Continental Congress met, the point of no return had been reached. Benjamin Franklin, who saw himself—rightly—as the great intermediary between Britain and America, better informed than any other man of attitudes and conditions on both sides of the Atlantic, had been in London in 1774 trying to make peace and in particular presenting a petition to the Privy Council to have the unpopular Governor Hutchinson of Massachusetts removed. He still believed in a negotiated compromise. But he got no thanks for his pains. His petition coincided with the Boston Tea Party and the inflaming of English opinion. He was fiercely attacked by Alexander Wedderburn, North's Attorney-General, a man typical of the British hardliners who made a deal impossible. Wedderburn, to Franklin's amazement, attacked him as 'the leader of disaffection,' a rebel 'possessed with the idea of a great American republic.' The petition was dismissed as 'groundless, vexatious and scandalous,' and to add insult to injury Franklin was peremptorily fired from his job as deputy postmaster-general. He saw Burke, and agreed with him that the British Empire was 'an aggregate of many states under a common head;' but he agreed with Burke also that the notion was now out of date—'the fine and noble China Vase, the British Empire,' had been shattered.[46] He saw Chatham, but found the old man degenerated into a windbag, who talked but did not listen, and counted for little now. Sadly, Franklin set sail for Philadelphia on March 20, 1775, convinced there was nothing more he could do in London to make the peace.

When Franklin got to Philadelphia on May 5—five days before the Second Continental Congress was due to meet—the first shots had been fired. On April 19, sixteen companies of redcoats were dispatched on what one of their officers called 'an ill-planned and ill-executed' expedition to seize patriot arms-dumps in Lexington and Concord. They failed to get the arms, and in a series of confused engagements got the worse of it, losing seventy-three dead and over 200 wounded or missing (American casualties were forty-nine dead, thirty-nine wounded, and five missing). John Adams was profoundly disturbed at the losses. It was 'the most shocking event New England ever beheld.' He saw it as the microcosm of all the tragedy of civil war—'the fight was between those whose parents but a few generations ago were brothers. I shudder at the thought, and there is no knowing where these calamities will end.' But his cousin Sam, hearing the first gunfire, called out: 'What a glorious morning this is—I mean, for America.' The patriotic media machine seized on the skirmishes with delight and presented them as a major victory, and proof that colonial militias could stand up to veterans.

Adams, Franklin, Jefferson, and Washington met on May 11 in Philadelphia, when the Second Continental Congress assembled. Franklin had known Washington twenty years before, during the Seven Years War. But most of the rest were strangers, many of them young men. He noted that 'the Unanimity is amazing.'[47] But that was unanimity for resistance. Only a minority yet thought in terms of outright independence. The rich John Dickinson of Maryland (1732–1808) wanted a direct appeal to King George to give Britain a last chance, and drafted an Olive Branch Petition. But even former moderates thought this pointless. John Adams, with characteristic *ad hominem* bitterness, dismissed it as 'the product of a certain great fortune and piddling genius' giving 'a silly cast to our doings.' He thought 'Power and artillery are the most efficacious, sure and infallible conciliatory measures we can adopt.'[48] Franklin sadly agreed with him. Knowing what he did of British political opinion, he was moving to the view that independence was the only solution, and he busied himself preparing for a long war, seeing to the printing of currency, the manufacturing of gunpowder, and the designing of an independent postal system. He drew up Articles of Confederation and Perpetual Union, which carried his defense union scheme a great deal further and was an early blueprint for the United States Constitution itself. This was to include besides the Thirteen Colonies (Georgia had now joined the Congress), Canada, the West Indies, and even Ireland if it wished. Though sad about the break with Britain, he was confident that America's huge economic and demographic strength—he was one of the few people on either side who appreciated its magnitude—would make it a certain victor, though he thought it should look for allies immediately. He wrote confidently to the English radical Joseph Priestley: 'Britain, at the expense of 3 millions, has killed 150 Yankees this campaign, which is £20,000 a head. During the same time, 60,000 children have been born in America.'[49]

In the meantime, though, everyone agreed that an army was needed to bring Britain to the negotiating table. Dr Joseph Warren of Massachusetts, president *pro tempore* of the Congress, who was soon to pay for his patriotism with his life at Bunker's Hill, put it succinctly:

'A Powerful Army on the side of America is the only means left to stem the rapid Progress of a Tyrannical Ministry.'[50] But who was to command it? Since the clashes at Lexington, the large, imposing delegate from Virginia, General Washington, had taken to appearing in the uniform of an officer in the Fairfax Militia. He was the only member of the Congress in martial attire. He had been a leading critic of British rule since the Great Proclamation. He called the Stamp Act 'legal thievery.' He blamed Britain for falling tobacco prices, which was his 'interest.' He refused to buy British-made articles for his estate. His wife and step-children no longer got presents from London. He set his people to manufacture substitutes. As long ago as 1769 he had advocated forming an American army, though only as 'a last resort.' He strongly disapproved of the Boston Tea Party, which seemed to him a disorderly affair, a needless provocation which gave Britain an excuse to 'rule with a high hand.' But the 'Intolerable Acts' resolved his doubts. The last straw was a British ruling that generous land-grants to officers who served in the Seven Years War applied only to regulars—this invalidated his large claims to Western lands. If ever a man now had an 'interest' in going to war, he did. He told John Adams: 'I will raise one thousand men, subsist them at my own expense, and march myself at their head, for the relief of Boston.'[51] He made it plain he was enthusiastic for fighting. He told fellow-delegates that he regarded the Indians as a sufficient menace—'a cruel and bloodthirsty enemy on our backs.' But this told in his favor. The delegates were experienced, serious men. They did not want to be led by a hothead. They liked the look of Washington. He was described as 'Six foot two inches in his stockings and weighing 175 pounds . . . His frame is padded with well-developed muscles, indicating great strength.' And again: 'In conversation he looks you full in the face, is deliberative, deferential and engaging. His demeanor at all times composed and dignified. His movements and gestures are graceful, his walk majestic.'[52] Moreover, he was 'generally beloved.'

Adams gives us a blow-by-blow account of how a commander-in-chief was chosen. He himself was by now in a fever of martial emotions: 'Oh, that I were a soldier,' he recorded in his diary. '[But] I will be! I am reading Military Books!' Washington, he said, 'by his great experience and abilities in military matters, is of great service to us.' Adams tried to maintain, twenty-seven years later, that his foresight was responsible for Washington's election. Actually there was not much choice. His only rivals were Israel Putnam, now serving as a major-general, who was too old at fifty-seven; and Artemus Ward, in temporary command of the provisional army at Cambridge, described as 'a fat old gentleman.' According to the Congressional minutes, Washington was chosen unanimously.[53] Washington, who whatever his faults was never arrogant or pushy, was so overwhelmed by his selection that he was unable to write his letter of acceptance, but dictated it to Isaac Pemberton, in whose hand it is, apart from the signature. He refused a salary and asked only for expenses. This was received with great approval, and it is clear from the minutes that the delegates intended him to be treated as more than a mere general. He was to be leader. 'This Congress,' they read, 'doth now declare that they will maintain and assist him and adhere to him, the said George Washington Esquire, with their lives and fortunes in the same cause.'[54]

On June 14 Congress agreed to raise six companies on the Pennsylvania, Maryland, and Virginia frontier, to be paid for by itself (as opposed to any individual state), and to be termed the 'American Continental Army.' Washington was instructed to draw up regulations for the new force. By July 3, the general was at Cambridge, taking charge. One of the reasons the New Englanders had been so keen to choose him was that they had, so far, borne the brunt of the fighting. They were anxious that Virginia, the most populous state, should be fully committed too. By his prompt move to the Boston theater of war, Washington showed he accepted the logic of this and that he intended to fight a continental struggle for an entire people and nation.

But was it a nation yet? Three days after Washington took over the army, Congress issued a formal Declaration of the Causes and Necessity for Taking Up Arms. This rejected independence. As late as January 1, 1776, when the first Grand Union flag was raised over Prospect Hill in Boston, it consisted of thirteen alternating red and white stripes with, in the left-hand corner, a red, white, and blue Union Jack. But the measures taken by Congress, far from compelling Britain to negotiate, as they hoped, had the opposite effect. General Gage, the last royal governor of Massachusetts, wrote home: 'Government can never recover itself but by using determined measures. I have no hopes at present of any accommodation, the Congress appears to have too much power and too little inclination [and] it appears very plainly that taxation is not the point but a total independence.'[55] Acting on his advice, George iii proclaimed all the colonies in a state of rebellion.

At this point an inspired and rebellious Englishman stuck in his oar. Thomas Paine (1737–1809) was another of the self-educated polymaths the 18th century produced in such large numbers. He was, of all things, a customs officer and exciseman. But he was also a man with a grudge against society, a spectacular grumbler, what was termed in England a 'barrack-room lawyer.' In a later age he would have become a trade union leader. Indeed, he was a trade union leader, who employed his fluent and forceful pen on behalf of Britain's 3,000 excisemen to demand an increase in their pay, and was sacked for his courage. He came to America in 1774, edited the Pennsylvania Magazine, and soon found himself on the extremist fringe of the Philadelphia patriots. Paine could and did design bridges, he invented a 'smokeless candle'—like Franklin he was fascinated by smoke and light—and at one time he drew up a detailed topographical scene for the invasion of England. But his real talent was for polemical journalism. In that, he has never been bettered. Indeed it was more than journalism; it was political philosophy, but written for a popular audience, with a devastating sense of topicality, and at great speed. He could pen a slashing article, a forceful, sustained pamphlet, and, without pausing for breath, a whole book, highly readable from cover to cover.

Paine's pamphlet *Common Sense* was on the streets of Philadelphia on January 10, 1776, and was soon selling fast all over the colonies. In a few weeks it sold over 100,000 copies and virtually everyone had read it or heard about it. Two things gave it particular impact. First, it was a piece of atrocity propaganda. The first year of hostilities had furnished many actual instances, and many more myths, of brutal conduct by British or mercenary soldiers. Entire towns, like Falmouth (now Portland, Maine) and Norfolk, had been burned by the British. Women, even children, had been killed in the inevitable bloody chaos of conflict. Paine preyed on these incidents: his argument was that any true-blooded American who was not revolted by them, and prepared to fight in consequence, had 'the heart of a coward and the spirit of a sycophant.' Crude though this approach was, it went home. Even General Washington, who had read the work by January 31, approved of it. Second, Paine cut right through the half-and-half arguments in favor of negotiations and a settlement under British sovereignty. He wanted complete independence as the only possible outcome. Nor did he try to make a distinction, as Congress still did, between a wicked parliament and a benign sovereign. He called George III 'the royal brute.' Indeed, it was Paine who transformed this obstinate, ignorant, and, in his own way, well-meaning man into a personal monster and a political tyrant, a bogey-figure for successive generations of American schoolchildren. Such is war, and such is propaganda. Paine's *Common Sense* was by no means entirely common sense. Many thought it inflammatory nonsense. But it was the most successful and influential pamphlet ever published.[56]

It was against this explosive background that Thomas Jefferson began his finest hour. By March, Adams noted that Congress had moved from 'fighting half a war to three quarters' but that 'Independence is a hobgoblin of so frightful a mien that it would throw a delicate person into fits to look it in the face.' By this he was referring to opponents of outright independence such as John Dickinson and Carter Braxton, who feared that conflicts of interest between the colonies would lead to the dissolution of the union, leaving America without any sovereign.[57] But the logic of war did its work. The British introduced not just German but—heavens above!—Russian mercenaries, allegedly supplied by the Tsar, the archetypal tyrant, who had equipped them with knouts to belabor decent American backs. More seriously, they were inciting slaves to rebel, and that stiffened the resolve of the South. On June 7 the Virginia Assembly instructed Richard Henry Lee to table a resolution 'That these United Colonies are, and of right ought to be, free and independent States,' which was seconded by Adams on behalf of Massachusetts. At this stage Pennsylvania, New York, South Carolina, and New Jersey were opposed to independence. Nonetheless, on June 11 Congress appointed a committee of Franklin, Adams, Roger Sherman, Robert Livingston, and Jefferson to draft a Declaration of Independence 'in case the Congress agreed thereto.'

Congress well knew what it was doing when it picked these able men to perform a special task. It was aware that the struggle against a great world power would be long and that it would need friends abroad. It had already set up a Committee of Correspondence, in effect a 'Foreign Office,' led by Franklin, to get in touch with France, Spain, the Netherlands, and other possible allies. It wanted to put its case before 'the court of world opinion,' and needed a dignified and well-argued but ringing and memorable statement of what it was doing and why it was doing it. It also wanted to give the future citizens of America a classic statement of what their country was about, so that their children and their children's children could study it and learn it by heart. Adams (if he is telling the truth) was quite convinced that Jefferson was the man to perform this miracle and proposed he be chairman of the Committee, though in fact he was the youngest member of it (apart from Livingston, the rich son of a New York judge). He recorded the following conversation between them. Jefferson: 'Why?' Adams: 'Reasons enough.' 'What can be your reasons?' 'Reason first: you are a Virginian, and a Virginian ought to appear at the head of the business. Reason second: I am obnoxious, suspect and unpopular. You are very much otherwise. Reason third: you can write ten times better than I can.'[58] All this was true enough.

Jefferson produced a superb draft, for which his 1774 pamphlet was a useful preparation. All kinds of philosophical and political influences went into it. They were all well-read men and Jefferson, despite his comparative youth, was the best read of all, and he made full use of the countless hours he had spent poring over books of history, political theory, and government. The Declaration is a powerful and wonderfully concise summary of the best Whig thought over several generations. Most of all, it has an electrifying beginning. It is hard to think of any way in which the first two paragraphs can be improved: the first, with its elegiac note of sadness at dissolving the union with Britain and its wish to show 'a decent respect to the opinions of mankind' by giving its reasons; the second, with its riveting first sentence, the kernel of the whole: 'We hold these truths to be self-evident, that all men are created equal, that they are endowed by their Creator with certain inalienable rights, that among these are life, liberty and the pursuit of happiness.' After that sentence, the reader, any reader—even George III—is compelled to read on. The Committee found it necessary to make few changes in Jefferson's draft. Franklin, the practical man, toned down Jefferson's grandiloquence—thus truths, from being 'sacred and undeniable' became 'self-evident,' a masterly improvement.[59] But in general the four others were delighted with Jefferson's work, as well they might be.

Congress was a different matter because at the heart of America's claim to liberty there was a black hole. What of the slaves? How could Congress say that 'all men are created equal' when there were 600,000 blacks scattered through the colonies, and concentrated in some of them in huge numbers, who were by law treated as chattels and enjoyed no rights at all? Jefferson and the other members of the Committee tried to up-end this argument—rather dishonestly, one is bound to say—by blaming American slavery on the British and King George. The original draft charged that the King had 'waged a cruel war against human nature' by attacking a 'distant people' and 'captivating and carrying them into slavery in another hemisphere.' But when the draft went before the full Congress, on June 28, the Southern delegates were not having this. Those from South Carolina, in particular, were not prepared to accept any admission that slavery was wrong and especially the acknowledgment that it violated the 'most sacred rights of life and liberty.' If the Declaration said that, then the logical consequence was to free all the slaves forthwith. So the slavery passage was removed, the first of the many compromises over the issue during the next eighty years, until it was finally resolved in an ocean of tears and blood. However, the word 'equality' remained in the text, and the fact that it did so was, as it were, a constitutional guarantee that, eventually, the glaring anomaly behind the Declaration would be rectified.

The Congress debated the draft for three days. Paradoxically, delegates spent little time going over the fundamental principles it enshrined, because the bulk of the Declaration presented the specific and detailed case against Britain, and more particularly against the King. The Revolutionaries were determined to scrap the pretense that they distinguished between evil ministers and a king who 'could do no wrong,' and renounce their allegiance to the crown once and for all. So they fussed over the indictment of the King, to them the core of the document, and left its constitutional and ideological framework, apart from the slavery point, largely intact. This was just as well. If Congress had chosen to argue over Jefferson's sweeping assumptions and propositions, and resolve their differences with verbal compromises, the magic wrought by his pen would surely have been exorcized, and the world would have been poorer in consequence. As it was the text was approved on July 2, New York still abstaining, and on July 4 all the colonies formally adopted what was called, to give it its correct title, 'The Unanimous Declaration of the Thirteen United States of America.' At the time, and often since, Tom Paine was credited with its authorship, which did not help to endear it to the British, where he was (and still is) regarded with abhorrence. In fact he had nothing to do with it directly, but the term 'United States' is certainly his. On July 8 it was read publicly in the State House Yard and the Liberty Bell rung. The royal coat of arms was torn down and burned. On August 2 it was engrossed on parchment and signed by all the delegates. Whereupon (according to John Hancock) Franklin remarked: 'Well, Gentlemen, we must now hang together, or we shall most assuredly hang separately.'[60] Interestingly enough, Cromwell had made the same remark to the Earl of Manchester at the beginning of the English Civil War 136 years earlier.

It is a thousand pities that Edmund Burke, the greatest statesman in Britain at that time, and the only one fit to rank with Franklin, Jefferson, Washington, Adams, and Madison, has not left us his reflections on the Declaration. Oddly enough, on July 4, the day it was signed, he noted that the news from America was so disturbing 'that I courted sleep in vain.' But Burke was at one with Jefferson, in mind and still more in spirit. His public life was devoted to essentially a single theme—the exposure and castigation of the abuse of power. He saw the conduct of the English Ascendancy in Ireland as an abuse of power; of the rapacious English nabobs in India as an abuse of power; and finally, at the end of his life, of the revolutionary ideologues who created the Terror in France as an abuse of power. Now, in 1776, he told parliament that the crown was abusing its power in America by 'a succession of Acts of Tyranny.' It was 'governing by an Army,' shutting the ports, ending the fisheries, abolishing the charters, burning the towns: so, 'you drove them into the declaration of independency' because the abuse of power 'was more than what ought to be endured.' Now, he scoffed, the King had ordered church services and a public fast in support of the war. In a sentence which stunned the Commons, Burke concluded: 'Till our churches are purified from this abominable service, I shall consider them, not as the temples of the Almighty, but the synagogues of Satan.'[61] In Burke's view, because power had been so grievously abused, America was justified in seeking independence by the sword. And that, in essence, is exactly what the Declaration of Independence sets forth.

With Independence declared, and the crown dethroned, it was necessary for all the states to make themselves sovereign. So state constitutions replaced the old charters and 'frameworks of government.' These were important not only for their own sake but because they helped to shape the United States Constitution later. In many respects the colonies—henceforth to be called the states—had been self-governing since the 17th century and had many documents and laws to prove it. Connecticut and Rhode Island already had constitutions of a sort, and few changes were needed to make them sovereign. Then again, many states had reacted to the imposition of parliamentary taxation from 1763 by seizing aspects of sovereignty in reply, so that the total gestation period of the United States Constitution should be seen as occupying nearly thirty years, 1763–91.[62] The first state to act, in 1775, was Massachusetts, which made its charter of 1691 the basis. Others followed its lead: New Hampshire and South Carolina in 1775, then

Virginia, New Jersey, New York, Pennsylvania, Delaware, Maryland, and North Carolina in 1776, and Georgia early in 1777. New York was the first to adopt a reasonably strong executive. Massachusetts decided it liked the idea, and drafted a revised constitution. The new draft was submitted to a popular referendum in March 1777, the first in history, but rejected 9,972–2,083. Then elections were held for a constitutional convention, which produced the final version of 1780, adopted by a two-thirds vote.

The Massachusetts constitution (as amended) was the pattern for others. All but two, Pennsylvania and Georgia, were bicameral, and these two changed their minds, 1789–90. In all the lower house was elected directly, and the upper house was elected directly too in all but one, Maryland, which had an electoral college. All but one, South Carolina, had annual elections for the lower house, and many had popularly elected executives and governors. Twelve required electors to own property, usually 50 acres, which was nothing in America. In three you had to prove you paid taxes. All but one required property qualifications from office-seekers. The percentage of white adult males enfranchised varied from state to state but the average electorate was four times larger than in Britain. All in all, they amounted to popular sovereignty and were very radical indeed for the 1770s.[63] They had an immediate and continuing impact all over Europe and Latin America. One constitution, Pennsylvania's, initially went further in a radical direction. Franklin claimed parentage (though it was probably written by Paine's follower, James Cannon) and took it proudly with him when he went to France, where the liberal bigwigs gasped in admiration: as Adams put it, 'Mr Turgot, the Duc de la Rochefoucauld, Mr Condorcet and many others became enamored with the constitution of Mr Franklin.'[64] But it proved 'inconvenient' and was deradicalized in 1790. But by that time it had done its insidious work in French Revolutionary heads.

While the states were making themselves sovereign, the Continental Congress had also to empower itself to fight a war. So in 1776–7, it produced the Articles of Confederation, in effect the first American Constitution.[65] In drafting it, delegates were not much concerned with theory but were anxious to produce practical results. So, oddly enough, although Americans had been discussing the location of sovereignty with the British for over ten years, this document made no effort to locate it in America and nothing was said of states' rights. It was unanimously agreed that the Congress should control the war and foreign policy, and the states the rest—what it called 'internal police.' Thomas Burke of North Carolina proposed an article stating that each state 'retained its sovereignty, freedom and independence, and every power, jurisdiction and right, which is not by this Confederation expressly delegated to Congress.' This was approved by eleven of thirteen delegations and became Article II. But Burke himself later stated: 'The United States ought to be as one Sovereign with respect to foreign Powers, in all things that relate to War or where the States have a common interest.' So the question was left begged.[66] The whole thing was done in a hurry and finished on November 15, 1777. But ratification was slower; in fact Maryland did not ratify till March 1, 1781, by which time experience had demonstrated plainly that a stronger executive was needed, and that in turn made the case for a new, and more considered, constitution.[67]

In the meantime, the urgent work of liberating and building the new country had necessarily passed from the men of the pen to the men of the sword. The War of Independence was a long war, lasting in effect eight and a half years. It was a war of attrition and exhaustion. The issue was: could the Americans hold out long enough, and maintain an army in the field of sufficient caliber and firepower, to wear out and destroy Britain's willingness to continue a struggle, and pay for it, which was actually begun in order to save the British taxpayer money? Here was the basic paradox of the war, which in the end proved decisive. The British had no fundamental national interest in fighting the war. If they won, it merely brought more political problems. If they lost, it hurt little but their pride. Few, outside London, were interested in the outcome; it made remarkably little impact on the literature, letters, newspapers, and diaries of the time. Certainly, no one volunteered to fight it. A few Whigs were passionate in opposing the war. But they had no popular support. Nor had the King and his ministers in waging it. There were no mass meetings or protests. No loyal demonstrations either. It was a colonial war, an imperialist war, which in a sense had more in common with the future Vietnam War or the Soviet war in Afghanistan, than with the recent Seven Years War. It was the first war of liberation.

In view of this, the American patriots were fortunate in their commander-in-chief. Washington was, by temperament and skills, the ideal commander for this kind of conflict. He was no great field commander. He fought in all nine general actions, and lost all but three of them. But he was a strategist. He realized that his supreme task was to train an army, keep it in the field, supply it, and pay it. By doing so, he enabled all thirteen state governments, plus the Congress, to remain functioning, and so to constitute a nation, which matured rapidly during the eight years of conflict. Somehow or other, legislatures functioned, courts sat, taxes were raised, the new independent government carried on. So the British were never at any point fighting a mere collection of rebels or guerrillas. They were up against an embodied nation, and in the end the point sank home. It was Washington who enabled all this to happen. And, in addition, he gave the war, on the American side, a dignity which even his opponents recognized. He nothing common did, or mean, or cruel, or vengeful. He behaved, from first to last, like a gentleman.

His resources were not great. At no point did his total forces number more than 60,000 men, subject to an annual desertion rate of 20 percent. He was always short of everything—arms, munitions, cannon, transport, clothes, money, food. But he managed to obtain enough to keep going, writing literally hundreds of begging letters to Congress and state governments to ensure there was just enough. He was good at this. In some ways running an army was like running a big Virginia estate, with many things in short supply, and make-do the rule. He remained always calm, cool, patient, and reassuring with all. As Jefferson testified, he had a hot temper—what red-haired man does not?—but he kept it mostly well under control. He had to take on many of the administrative responsibilities which Congress should have handled but, being weak executively, did not. He got through a vast amount of paperwork. He had some good people to help him. Friedrich Wilhelm von Steuben did the drill side of the army, and in effect served as Washington's chief-of-staff. From early 1777 he had as his secretary and principal ADC a brilliant young New Yorker from the West Indies, Alexander Hamilton (1755–1804). Colonel Hamilton had already served with distinction as an artillery officer, and he proved to be the most effective aide any American commander-in-chief has ever had. But essentially Washington had to do it all himself.

He was much criticized, then and later. Adams asked: 'Would Washington ever have been Commander of the Revolutionary Army or President of the United States if he had not married the rich widow of Mr Curtis?'[68] General Charles Lee was amazed anyone called him great: 'He is extremely prodigal of other men's blood and a great economist of his own.'[69] One close observer, Jonathan Boucher, summed him up:

'He is shy, silent, stern, slow and cautious, but has no quickness of parts, extraordinary penetration, nor an elevated style of thinking . . . He seems to have nothing generous or affectionate in his nature.'[70] A French observer, Ferdinand Bayard, said he lacked human animation: 'He moved, spoke and acted with the regularity of a clock.' But another Frenchman, General Marqui de Barbé-Marbois, testified: 'I have never seen anyone who was more naturally and spontaneously polite.'[71] He could be compassionate, and a great actor too. Elias Boudinot, in charge of prisoners-of-war, who went to Washington in 1778 to plead for clothing for them, reported: 'In much distress, with Tears in his Eyes, he assured me that if he was deserted by the Gentlemen of the Country, he would despair. He could not do everything—he was General, Quartermaster and Commissary. Everything fell on him and he was unequal to the task. He gave me the most positive Engagement, that if I could contrive any mode for their support and Comfort, he would confirm it as far as it was in his Power.'[72] But he could also relax, an eyewitness reporting from his HQ: 'He sometimes throws and catches a ball for whole hours with his aides-de-camp.'

With Washington deliberately fighting a war of endurance, the British strategy made no sense. Indeed it is arguable that Britain had no discernible, and certainly no consistent, strategy from beginning to end. It is a mystery that the British, with their political genius, and their very uncertain touch with military affairs, should have rejected a political solution and put all their trust in a military one. Lord George Germaine, placed in charge of the war by North, had no military gifts. But then he had no political gifts either. He believed that the American militias could never be any good, and that the Tory loyalists greatly outnumbered the revolutionary patriots. How could he possibly know? He had never set foot in America. And it never occurred to him to go there and see for himself what needed to be done, or whether an honorable compromise could be negotiated. No member of the government ever thought of crossing the Atlantic on a fact-finding mission. At various times, generals were given powers to treat, but only after the rebels had agreed to lay down their arms. What good was that? In fact the generals were frequently changed, a sure sign of mismanagement. First Gage came and went, then Admiral Richard Howe and his brother General William Howe shared a joint command—an absurd arrangement—then General Burgoyne and Marquis Cornwallis were given separate and unrelated armies—another absurdity—both of which they lost. Far from getting the chance to negotiate after a rebel surrender, the British generals in fact were instructed to make concessions only after they were involved in disasters—exactly the other way round. Much of the fault for these egregious errors lay with George III, a man who had never seen a shot fired in anger, who had never been abroad—and who never even saw the sea until he was an old man.

The British commanders were not starved of manpower. Some 30,000 mercenaries were sent out. But this was probably counter-productive, since their conduct outraged even the Tory loyalists. When the two Howes were operating in New York in 1776, they had no fewer than seventy-three warships, manned by 13,000 seamen, and transports loaded with 32,000 troops. That was a big expedition by British standards. But none of the large resources Britain put into the war produced long-term results, or indeed any at all. It might have been different if George III or North had picked one really first-class general, and given him unlimited military and political authority, on the spot. But such a person would almost certainly have concluded that the war was folly, and negotiated an end to it. As it was, all the generals (not the admirals) were second-rate, and it showed.

The course of the war is soon told.[73] The first winter 1775–6, when the conflict was concentrated around Boston, was inconclusive and enabled Washington to organize his army. The Howes' strategy in New York in 1776 was to take the city, cut off New England from the south, then destroy the rebellion at its heart, in Massachusetts. To frustrate this, Washington ferried his army from Manhattan to Brooklyn and dug in on the Heights. Howe outflanked him and Washington lost 1,500 men to Howe's 400. Washington was lucky to get 9,000 men back to Manhattan. But Howe failed to surround the American army and destroy it, and Washington escaped to New Jersey and across the Delaware. These were the times that 'tried men's souls,' as Tom Paine wrote in his topical tract, *The Crisis*. In fact, Washington fought a successful winter campaign, killing or capturing 1,000 German mercenaries at Trenton, defeating the garrison at Princeton, then retiring in good order to Morristown in late January 1777. Howe now moved south, descending on Philadelphia, and beating Washington at Brandywine on September 11, 1777.

Meanwhile General John Burgoyne, commanding in Canada, had defeated a second American army, under Richard Montgomery, which had moved north in the hope of raising allies along the St Lawrence. But the Canadians, whether British-descended Protestants or French-descended Catholics, were not interested. They had got a good deal from Britain in 1774 and they remained loyal, now and later. So Burgoyne was able to move onto the offensive. But he was a rash man. In June 1777 he shipped 7,000 men, British, loyalists, Indians and Brunswickers, across Lake Champlain and then down the Hudson. The aim should have been to catch Washington in a pincer with Burgoyne forming one arm and Howe the other. But no such plan was concocted. Instead, Burgoyne soon got into difficulties. He lost two minor actions on September 19 and October 7, then found himself surrounded and surrendered at Saratoga on October 17, 1777. That led to the first genuine British offer of terms—turned down, naturally. Washington's army managed to survive another winter. It shrank during the cold weather, which he spent mainly in winter quarters, but expanded again in the spring, and each year it was better. He and it learned from their mistakes and he gradually secured longer terms of service for his men, better pay for them, stricter articles of war, which allowed him to hang men in extreme cases, more artillery, better transport, and reliable supplies.[74]

By February 1778, Franklin's mission to Europe to secure allies was bearing fruit. In France he was perhaps the most successful of all American envoys. When he had been in England, the English ruling class, perhaps put off by his rustic clothes, plebeian manners, and artisanal background (and accent) would not admit him to their homes, with one or two exceptions. The French aristocracy, whether from Anglophobia, intellectual snobbery—they were much more familiar with his learned work—or sheer curiosity, treated him as a lion. He seemed to them another Rousseau, and a more piquant one, being an American exotic rather than a mere Swiss. He was sponsored by Jacques-Donatian de Chaumont, a rich businessman with extensive American interests who spent 2 million livres of his own fortune in aid to the patriots. The Comte de Ségur found positive virtue and nobility in his mean appearance: 'His clothing was rustic, his bearing simple but dignified, his language direct, his hair unpowdered. It was as though the simplicity of the classical world, the figure of a thinker of the time of Plato, or a republican of the age of Cato or Fabius, had suddenly been brought by magic into our effeminate and slavish age, the 18th century.' By an extraordinary conjunction, the notion of the Americans as the new Romans hit a culturally fashionable note—just at this precise moment, the rococo was suddenly yielding to the classical revival and Franklin seemed a man of the new wave. As a matter of fact, his style of living was not all that modest, either. The Duc de Croy might

enthuse over the humble dinners he served to his high-born guests—
'Everything breathed simplicity and economy as befitted a philosopher'—but Franklin had 1,041 bottles of wine in his cellar in 1778, rising to 1,203 before he left. He had nine indoor servants and spent
freely, justifying his luxuries with a typical American moral: 'Is not the
hope of one day being able to purchase and enjoy luxuries a spur to
labor and industry?' Adams, parsimonious and puritanical, snorted
that 'The life of Dr Franklin [in Paris] was a Scene of continual dissipa-
tion,' and he suspected, probably with reason, that Franklin was enjoying women as well as good food and drink.[75]

So what? The mission was a success, even more so with popular than
with official opinion and *le gratin*. Jacques Necker, the great banker
who was put in charge of the finances in 1776, was against involvement. He predicted it would prove financially disastrous, as it did. So
was Louis XVI, on the grounds that 'it is my profession to be a royalist.'
But they were overruled by the Duc de Choiseuil, the chief minister, the
Comte de Vergennes, the Foreign Minister, and leaders of public opinion like de Beaumarchais, author of fashionable comedies like *The
Barber of Seville* and *The Marriage of Figaro*, who organized public
subscriptions to buy 'arms for America,' as well as pushing the government to provide more.

So from the spring of 1778 America was no longer alone. Nantes
became the American supply base in Europe. Nearby, the Department
of Marine built a special foundry to cast cannon for America. In July
1778 alone one wealthy merchant sent ten ships to Boston, loaded with
munitions. In 1782 he sent thirty. The news of Saratoga spurred the
signing of a Treaty of Amity and Commerce. Louis XVI graciously
received Franklin, who was wigless, swordless, and wearing the rusty
old brown coat in which he was savaged by Wedderburn—sweet
revenge![76] And France's decision persuaded the Spanish and Dutch to
join in, though Spain merely backed France in the hope of recovering
Gibraltar and was never a formal ally of men it considered rebels who
might corrupt its own colonies.

French intervention, by land and sea, raised the stakes for Britain
but brought no early end to the war. Admiral Comte d'Estaigne
appeared with a fleet off the American coast in the summer of 1778
but failed to beat Admiral Howe. The next year he made an attempt,
in conjunction with an American force, to take Savannah in October,
but failed again. After another indecisive winter, Sir Henry Clinton,
who had replaced Howe, took Charleston and 5,500 American pris-

oners under Benjamin Lincoln, the biggest single loss the patriots suffered in the entire war. This was in May 1780. Three months later, on August 16, Lord Cornwallis beat another American force, under General Horatio Gates, at Camden. Clinton now returned to New York, his main base, leaving Cornwallis to command in the south. Cornwallis invaded North Carolina, but his loyalist force was destroyed at King's Mountain on October 7, 1780. The following January 1781, Banastre Tarleton's Tory Legion was beaten at Cowpens, losing 900 men, by General Daniel Morgan. Cornwallis also suffered heavy casualties at Guildford Courthouse two months later, though he held the field. None of these battles was decisive or even particularly important, but they had a cumulative effect in eroding Britain's will to continue the war.

Then Cornwallis made a strategic mistake. He decided to concentrate his forces at Yorktown on the coast. Clinton strongly disagreed with this move, which made Cornwallis' army vulnerable if ever the French were able to concentrate their naval forces and so deprive Britain, for the first time, of command of the sea. That is exactly what happened. The French by now had a substantial force of 5,500 under the Comte de Rochambeau, based on Rhode Island. More important, the Comte de Barras had a naval force operating from Newport. In the summer of 1781 Admiral de Grasse hurried up from the West Indies with twenty ships of the line and a further 3,000 soldiers. He arrived in time to transport Washington's army, plus a French force under the Marquis de Lafayette, from the Chesapeake to the James River, thus concentrating an enormous conjunction of land- and sea-power around Cornwallis' armed camp. To make matters worse for the British, it was joined by De Barras' naval squadron from Newport. The waters around Yorktown were now controlled by the French fleet, and an attempt by Admiral Thomas Graves, sent from New York to break the blockade, failed. He was obliged to return to New York. Britain no longer could reinforce its armies by sea, at any rate in the western North Atlantic, and this was catastrophic for its whole method of fighting the war. Cornwallis, with 8,000 men, faced a Franco-American army of 17,000, well provided with artillery. He was short of supplies, but it was his exposure to the guns which persuaded him to surrender on October 19, 1781.

So the British, who had begun the war with an enormous superiority in trained men and guns and with complete control of the sea, ended it outnumbered, outgunned, and with the French ruling the waves. They

still controlled New York, Savannah, and Charleston, but the catastrophe at Yorktown knocked the stuffing out of the British war-party. On
March 19, 1782, North resigned, making way for a peace coalition
which contained Shelburne, Fox, and Burke. Happily for all concerned,
a series of brilliant British victories against France and Spain—the lift-
ing of the Spanish siege of Gibraltar, success in India, and, above all,
Lord Howe's destruction of De Grasse's fleet at the Battle of the Saints
on April 12, thus saving the British West Indies and restoring Britain's
absolute command of the seas, made it easier for Britain to swallow its
pride and accept an independent America.

Franklin was sent back to Paris to open negotiations with
Vergennes, on behalf of France, and Thomas Grenville, the clever, eru-
dite Foxite son of the 'Stamp Act' Grenville, on behalf of Britain.
Franklin was both the architect and the hero of the Peace of Paris. The
'four points' he set out in July 1782 became the basis of the agreement:
first, outright independence of the United States, and withdrawal of all
British forces; second, Canada to remain British and a definitive boundary to be drawn; third, agreement on the boundaries of all Thirteen
States; and, finally, freedom for fishing off Newfoundland—the first
international fisheries agreement.

What is so fascinating about these talks, and the background to
them, is the ambivalence of British attitudes to America and vice versa.
A short time before, Britain's back had been against the wall, with not
only France, Spain, and the Netherlands actively making war against it,
but with the League of Armed Neutrality—Russia, Denmark, and
Sweden—also hostile and poised to attack. Franklin was planning with
the French Ministry of Marine a series of attacks on the English coasts,
with John Paul Jones in charge of the naval forces and Lafayette of an
invasion army. British resources were stretched thin all over the world.
The French had two divisions, totalling 40,000 men, ready for the invasion, and sixty-four Franco-Spanish ships, mounting 4,774 guns, to
escort them. Against this, Sir Charles Hardy's Channel Fleet had only
thirty-eight ships with 2,968 guns. Lord Barrington, the British War
Secretary, said there was no one in England—all the best generals were
overseas—fit to command an anti-invasion army. It was wind and sickness, hitting the Franco-Spanish fleet hard, which probably averted a
Channel crossing.

Then, in no time at all, there were persistent rumors of a dramatic
renversement d'alliances, in which Britain would concede the United
States sovereignty and both powers would attack France and Spain,

driving them out of North America completely. America had already found them treacherous and unreliable allies, and after all Britain was its main trading partner, and Britain's control of the oceans a precondition of American prosperity. Once America was recognized as independent by the British, the two nations had far more to agree about than to dispute, and in these rumors—nothing came of them of course—we can trace the distant foreshadowing of the Monroe Doctrine forty years later. At the peace talks, the French were surprised at the readiness of the British to make concessions to America. Vergennes declared: 'The British buy peace rather than make it. Their concessions exceed all that I could have thought possible.' That was Franklin's doing: he persuaded the British to be generous to America and in return he abandoned France and signed a separate peace on November 30, 1782. During the celebrations at Passy an exchange between a French guest and one of the British delegates, Caleb Whitefoord, made the point. The flattering Frenchman stressed the growing greatness of America, predicting that 'the thirteen United States would form the greatest empire in the world.' Whitefoord: 'Yes, Monsieur, and they will all speak English, every one of 'em.' From the fires of the war, phoenix-like, sprang that mysterious and long-lived creature, still with us, the Anglo-American Special Relationship.[77]

The consequences of this second world war were profound and reverberated for years. It is worth looking at them for a moment from a global perspective, because of their effects on subsequent American history. All things considered, Britain emerged from the long conflict comparatively unscathed. The people were not emotionally involved, there were no soul-scars. Many interests, not just the Whigs, had been against it all along and the merchants in particular were anxious to end it and get on with the Atlantic trade. If anything, the war boosted the British economy, which entered the decade of the 1780s—the take-off point of the first Industrial Revolution—in roaring form. The war ended mercantilism once and for all. The ideas of Adam Smith—who had been strongly opposed to a coercive policy throughout—triumphed, and with the formation of the great peacetime ministry by William Pitt the Younger at the end of 1783, Smith was now a welcome visitor in 10 Downing Street and his free-trade ideas began to take over British policy. Britain was now in the process of becoming the world's first great industrial power and the victory for Smith's notions of free-enterprise capitalism and a world market was good news both for America's farmers and for its infant manufactures.

The war was a disaster for the old-style European monarchies. Spain emerged from the Peace of Paris (1783) with nothing, with its crown poorer and weaker and its great viceroyalties in Central and South America looking increasingly to the North for example and inspiration. The big loser was France. It, too, got nothing from the peace. The war cost it a billion livres and ruined its credit with the bankers of Europe. As Necker predicted, it did irretrievable damage to France's public finances and compelled the bankrupt monarchy to take the road which led to the calling of the Estates General, the Fall of the Bastille, the Terror, the Republic, military dictatorship, and two decades of disastrous wars. All the wealthy aristocrats and leading merchants who had helped America with their personal fortunes lost everything too, and one or two had to be put on a pauper's payroll by a grudging Congress. The French ruling class learned the hard way not to meddle with republicanism. The Comte de Ségur, who served in America, summed it up: 'We walked gaily over a carpet of flowers which concealed from us the abyss.' But he was comparatively lucky. Admiral d'Estaigne, who brought the first French fleet to the American coast, died by the guillotine.[78]

The war brought to the Thirteen States, now united after a fashion, immense miseries, losses, benefits, and unexpected blessings. There were winners and losers. Chief among the losers, especially in the long term, were the Indians. At the time of the Declaration of Independence, about 200,000 Indians lived east of the Mississippi, grouped in eighty-five nations. Their instinct was to stay neutral. One Iroquois chief told the governor of Connecticut in March 1775: 'We are unwilling to join on either side . . . for we love you both—Old England and New.'[79] Once the war started, however, both sides sought Indian help and it was usually the British who got it. They had defended Indian interests in the past and the Indians' intuition told them that an independent America would be unrestrained in permitting western white expansion. So about 13,000 fought for Britain, and if Sir William Johnson, greatest of Britain's Indian agents, who was 'Honorary Chief of the Six Nations,' had not died in 1774, the Indian alliance would have been much more effective. His son John and his nephew Guy did their best, however, and the Indians felt they had fought a hard and successful war on the whole. Their dismay at the Peace of Paris was all the more bitter, therefore. Britain abandoned them. At Niagara, the British envoys were told by the Indian chiefs: 'If it were really true that the British had basely betrayed them by pretending to give up [our] country to the Americans, without our consent or consulting them, it was an Act of Cruelty and Injustice that only the Christians were capable of doing.'[80] The Americans interpreted the treaty as giving them the right of conquest, and set to with a will. Federal agents told the Delawares and Wyandots in 1785: 'We claim the country by conquest, and are to give, not to receive.'[81] The Indians of the great plains lost too. They had originally been protected from western expansion by French claims to the Mississippi. That barrier had gone in 1763. Then the British came to their rescue by the Great Proclamation. Now that, too, was null and void. They were on their own.

For the slaves, the consequences of the Revolution were mixed. By forcing the Thirteen States to pool their resources and miminize their differences, the war necessarily obliged the New Englanders, who were growing increasingly restive about the 'organic sin' of slavery, a phrase coined during the Great Awakening, to overlook it for the time being. Hence even Adams, already passionately opposed to slavery, agreed without argument to omit the slavery passage from the Declaration of Independence. That was clearly a defeat for the slaves, and worse was to come in the process of constitution-making. Moreover, the number of slaves actually increased and their distribution spread during and immediately after the conflict. It is a melancholy fact that the number of slaves in Virginia actually doubled between 1755 and the end of the war in 1782. That was mainly through natural increase and longevity—though this in itself testified to the healthy and in some ways comfortable conditions they enjoyed in the South: slaves lived there twice as long as in Africa and 50 percent longer than in South America. Despite this, most of the South emerged from the war impoverished as a result of military occupation, naval attacks, civil war between patriots and Tories, war with the Indians, and the flight of thousands of slaves to the British Army and freedom. Many Southerners felt the only way they could restore their fortunes was by the strict restoration of slavery and its rapid expansion. That is indeed what happened. Slave-owners, pushing westwards, as they could now do with increasing freedom, took their slaves with them into Kentucky and Tennessee, South Carolina and Georgia. So, even before the great cotton revolution, the slave South was expanding. With the demand for slaves rising, more were imported direct from Africa—100,000 in the years 1783–1807.[82]

On the other hand, the new climate of liberty and even equality undoubtedly caused many people, especially in the North, to look afresh at the extraordinary anomaly of holding men and women in perpetual slavery in a land which had just won its freedom. The movement to end slavery in some states began even before the crisis. In 1766 Boston instructed its representatives in the Assembly to 'move for a law to prohibit the importation and purchasing of slaves for the future,' and other towns in New England did the same. In 1771 a prohibitory law did pass, but Governor Hutchinson would not sign it. However, in December that year Lord Mansfield, Chief Justice, delivered in London his famous ruling that slavery was 'so odious' an institution that nothing 'could be suffered to support it but positive law.' That made slavery unlawful in England under the common law, and since most American colonies adhered to the common law too, the legal drift was plainly against it. In 1773 Pennsylvania (under the influence of Quakers) and in 1774 Rhode Island and Connecticut passed laws prohibiting the slave trade. The General Articles of Association adopted by the First Continental Congress in 1774 had an anti-slave-trade clause which pledged the members 'neither [to] import nor purchase any slave imported after the first day of December next,' after which time it was agreed 'wholly [to] discontinue the slave trade' and 'neither [to] be concerned in it ourselves' nor to 'hire our vessels, nor sell our commodities or manufactures to those who are concerned in it.' Laws permitting manumission or removing existing restraints on it were passed by five states between 1786 and 1801, and these included slave states like Kentucky and Tennessee. Virginia had allowed manumission even before, in 1782, and 10,000 were freed almost immediately. Maryland followed in 1783 and a generation later over 20 percent of its blacks were free.

During the Revolutionary War, and as a direct result of the climate it produced, all the Northern states except New York and New Jersey, following Britain's lead, took steps to outlaw slavery itself. In 1780 Pennsylvania enacted the first (gradual) emancipation law in American history and others followed, New Jersey being the last to do so before the Civil War. In addition to positive laws, the common law worked in favor of the slaves during these years, as it had in England. In 1781, in *Brom and Bett* v. *John Ashley*, Elizabeth Freeman, called Bett, argued that a phrase in the new 1780 Massachusetts constitution saying that all individuals were 'born free and equal' applied just as much to blacks as to whites. She won the case, and this and other decisions brought slavery in Massachusetts to an end. On top of all this, the constitutional struggle and the war gave birth to mass agitation in England, which soon spread to America and elsewhere, in the organized antislavery movement, led by Samuel Wilberforce and the 'Clapham Sect.' They ultimately drove the law through parliament, which outlawed the international slave trade in 1807.[83]

The consequence for white Americans were mixed too. Like most 'wars of liberation' the American War of Independence was a bitter civil war too. One contemporary guess divided the people into three: the patriots, one-third, the Tory loyalists, one-third, and the remainder prepared to go along with either party. It is likely, however, that those who declined to take an active part were fully half the nation, the militants being almost equally divided, though the Tories, by their very nature, lacked leaders and the extremism which drove the liberators. They looked to leadership from England and were poorly served. They were the biggest losers of all. Indeed, they lost everything—jobs, houses, estates, savings, often their lives. Some families were severed for ever, a typically tragic example being Franklin's. His son William, governor of New Jersey, stayed loyal, and Franklin cut him out of his will, which stated: 'The part he acted against me in the late war, which is of public notoriety, will account for me leaving him no more of an estate he endeavored to deprive me of.' William died in exile, destitute, in 1813. Another, typical, loser was Philip Richard Fendall, one of the fourth generations of Fendalls who farmed on the lower Western Shore of Charles County, Maryland. When he boarded ship for England, he surrendered his career as a merchant, his income as county clerk, his profits from a 700-acre tobacco plantation on the Potomac, his 'large, elegant brick dwelling house' which was in 'a beautiful healthy situation, and commands an extensive view up and down the river'—everything in fact.[84] What became of him we do not know.

Most loyalists in the Thirteen States had no alternative but to stay where they were and swallow their feelings. Of course this worked both ways. In Jamaica, Barbados, and Grenada, the local assemblies declared their sympathy with the patriots, but British naval supremacy prevented them from doing anything more. Bermuda and the Bahamas remained formally loyal but would have shifted if the patriots had been able to offer military help. Florida was loyal because it needed British protection from Spain. Recent research shows that the loyalists were strongest, in proportion to population, in Georgia, New York, and South Carolina (in that order); in absolute terms, New York contained

the largest number of loyalists, having three or four times more supporters of the crown than any other colony/state. Royalists remained
relatively strong in New Jersey and Massachusetts, weaker in Rhode
Island, North Carolina, Connecticut, Pennsylvania, and New
Hampshire, and impotent in Virginia, Maryland, and Delaware.[85]

Loyalty or patriotism was determined to some extent by ethnic origins and religion. People of English origins were divided by temperament, as they had been in the English Civil War 130 years before. The
Scots Highlanders who were clannish and had recently arrived with
generous land-grants, were fanatically loyal. Scots Lowlanders were
also loyal, though less so. Irish and Scotch-Irish (Ulstermen) were
fanatically anti-British, if they were Catholic and still more (at this
stage) if they were Presbyterian. The Dutch were divided. The Germans
were neutral, tending to go along with the prevailing mood in their
locality. The Huguenots were patriots. Religion was a big factor, as it
was and is in everything connected with America. The Quakers of
Pennsylvania were inclined to side with the King but were prevented by
their pacifism from fighting for him. In Philadelphia Benjamin Franklin
found great difficulty in persuading Quakers to serve in the civil
defense forces, man the fire-brigade, or tend the wounded. The Roman
Catholics were patriots. The first Catholic bishop in America, Bishop
John Carroll of Baltimore, actually went on a mission to Canada to try
to persuade Canadian Catholics to help or at least remain neutral.

The Anglicans, the religious group least affected by the Great
Awakening, were predominantly loyalist, except in Virginia. They were
particularly loyal in New York, the biggest single center of support for
the crown. Of course this was due to some extent to the fact that New
York was a major British base and many had a direct economic motive
for supporting the crown—Washington's 'interest' at work again. The
same was true to a limited extent of Boston and Newport, Rhode
Island. But New York was also, at that time, an Anglican stronghold.
Charles Inglis, a leading New York Anglican, called the war 'the most
causeless, unprovoked and unnatural [rebellion] which ever disgraced
any country.'[86] But the Anglicans were weakened and frustrated by the
failure to introduce bishops and a hierarchy. And it is important to
remember that America, as a whole, was a religious breakaway from
Anglicanism. Research reveals that in 1780, the Anglicans had 406
churches. The Presbyterians had 495, and the Congregationalists were
by far the largest with 749. In this sense, Anglican arrogance in the
early 17th century came home to roost in the 1770s, and James I had

not been so far wrong when he called the settlement of America 'a sem-inary for a seditious parliament.' If you equate the Congregationalists with the Presbyterians (both being Calvinist), George III was not far wrong either when he called the Revolution 'a Presbyterian Rebellion.'[87] The English church and state lost the political and military battle because they had already lost the religious battle.[88]

On the whole the loyalists were not successful in organizing resis-tance to the rebellion. In North Carolina, the loyalist David Fanning led an effective guerrilla war for a time against the patriotic leader Governor Thomas Burke, both engaging in terror and counter-terror. In South Carolina, Thomas Brown, who had been tortured by the extremist republican Sons of Liberty, also gave the patriots a hard time. Another successful loyalist leader was Joseph Galloway of Philadelphia. The loyalists fought hard in Georgia and in the North Carolina back-country, where 1,400 Highland Scots seized upper Cape Fear but were badly beaten by the patriot militia, which had cannon. They were let down by the second-rate British commanders, as were most of the other loyalist bands. Other loyalists were discouraged by the bad behavior of mercenaries and British troops—in New Jersey, for instance, a center of loyalism, 2,700 signed a loyalty oath to the King but were put off from doing more by military looting. Other loyalists were silenced by patriot terror-leaders like Colonel Charles Lynch of Virginia, who invented lynching, which in his day was thirty-nine lashes rather than hanging. All the loyalists felt betrayed by the British at the peace. The fate of the loyalist blacks was pitiful. Some 800 Virginia slaves fled north following Governor Lord Dunmore's promise of freedom. They went to New York, where they joined thousands of others who worked for the British garrison. When the troops departed in 1783 they were left to their own resources and most of them fled fur-ther, to Nova Scotia. About 1,000 black loyalists were shipped to Sierra Leone, the first of many attempts to repatriate ex-slaves to West Africa. Thousands of loyalists went to England and tried to file claims for compensation. A total of 3,225 claims were eventually dealt with in London and 2,291 awards of compensation made—a miserable total compared with the vast numbers who lost all.[89]

The most important consequence of the loyalist diaspora was felt in Canada. The total number of loyalists who left the United States may have been as high as 80,000. Some went to England; others to crown colonies in the West Indies. But the vast majority emigrated north to Canada, where they caused a radical shift in its demography. Until then British Upper Canada had been thinly held and the total English-speaking population was outbalanced by French-speaking Lower Canada. Both remained loyal to the crown during the struggle but the influx of fierce loyalists was crucial in binding Canada to the crown and also in making it a predominantly English-speaking country. So if Britain lost America it gained Canada, a point reinforced in the war of 1812.[90]

The overwhelming majority of loyalists remained in the United States, but not necessarily in their old localities. Large numbers moved from Virginia, Maryland, the Carolinas, and Georgia to more northerly states, especially Pennsylvania and New York. Others moved west, into and across the Appalachians, into Tennessee, Kentucky, and the Ohio Valley. The war, then, diluted the pure English stock of the American population somewhat but, more importantly, it mixed everyone up more, dissolving old patterns and forming new ones, and so adding heat to the melting-pot process which was already at work transforming people from innumerable ethnic and religious backgrounds into full-blooded American citizens.

The war, indeed, was a transforming drama, which left deep psychological and physical scars on a much tried people, as well as enobling ones. The women were the big sufferers in this long, divisive, bitter conflict, bearing the brunt of the poverty to which hundreds of thousands were reduced, at least for some years. We hear of Betsy Ross stitching her flag, Abigail Adams writing to her husband in Philadelphia in 1776, 'remember the ladies,' and the clever black girl Phyllis Wheatley writing her poetry. Modern feminist historians pick on certain highlights of the women's struggle to aid the patriots, such as the pro-Revolutionary statement signed by all the women of Edenton, North Carolina, which declared: 'We the ladys of Edenton do hereby Solemnly Engage not to Conform to that Pernicious Custom of Drinking tea, or that we the aforesaid Ladys will not promote the wear of any Manufacture from England until such times as all Acts which tend to Enslave this our Native Country shall be Repealed.'[91]

But that was window-dressing. Most women had a hard, tragic war, losing brothers and sons and fathers and homes and sometimes seeing their families bitterly divided for ever. A more typical story, one suspects, than that of the non-tea-drinking ladies of Edenton was Mrs Elizabeth Jackson's of Waxhaw Settlement, South Carolina. She was from Carrickfergus and her husband Andrew from Castlereagh, both part of a big Ulster immigration of 1765. She had three sons, the last, Andrew, being posthumous, for her husband died soon after they settled. She raised them in grim poverty and stern, English-hating rectitude. Andrew, aged six, remembered crying. 'Stop that, Andrew,' Mrs Jackson admonished him. 'Do not let me see you cry again. Girls were made to cry, not boys.' 'What were boys made to do, Mother?' 'To fight.' All three sons were encouraged by their mother to serve in the Revolutionary War, Andrew being only twelve when he enlisted in 1779. The eldest son, Hugh, died on active service, aged sixteen. Andrew and his second brother, Robert, were both flung into a prisoner-of-war camp and slashed with the sword of a British officer: 'The sword reached my head, and has left a mark there as durable as the skull, as well as on the fingers'—all this for refusing to clean the officer's boots. Andrew Jackson carried these scars to his dying day as a reminder of English brutality. His brother was more seriously wounded and died soon after being released. Finally, in 1782, Mrs Elizabeth Jackson, who had been nursing the American wounded in an improvised hospital, contracted an infection and died too, leaving young Andrew an orphan of the war. He remembered every word of the dying advice of this grim woman: 'Avoid quarrels as long as you can without yielding to imposition. But sustain your manhood always. Never bring a suit in law for assault or battery or defamation. The law affords no remedies for such outrages that can satisfy the feelings of a true man [but] if you ever have to vindicate your honor, do it calmly.'[92]

This kind of suffering, and the bitterness it engendered, makes us thankful for the French intervention, which helped to bring the war to an end. Without it, the civil war–guerrilla war phase might have dragged on for many years, further envenomed by British-inspired servile revolts and Indian raids. That is what happened in South and Central America a generation later, when the wars between the rebels and Spain and between pro- and anti-royalist elements went on for decades, leading to Caesarism, military rule, army mutinies and revolts, and every variety of cruelty. The nature of the revolutionary struggle in Latin America helps to explain the weaknesses and instability of the independent civil societies which arose from it and the political role played by the military almost to this day. The United States has been spared this. But it was touch and go. There were some ugly incidents during and immediately after the war. Congress, with its weak, indeed virtually non-existent executive, was a thoroughly bad war-manager. There was no proper currency and, in effect, rapid inflation. Washington in practice managed the war as well as commanded, and without him there would have been a social as well as a military breakdown. Yet at one point, at the end of 1777, there were rumors he would be replaced by Gates, who had beaten Burgoyne. Washington himself thought that there was a plot, led by Thomas Conway (the 'Conway Cabal') and that Gates was privy to it. But nothing came of it.

After Yorkstown, feelings among some officers about the undersupplying of the army owing to Congressional weakness and negligence led to pressures on Washington to take power—exactly the kind of movement which was to ruin independence in Latin America. Colonel Lewis Nicola, an Irish-born Huguenot, wrote to Washington urging him to 'take the crown.' The general wrote that the letter 'left him with painful sensations.' He admitted the army was short of supplies, but he said he would work to remedy things 'in a constitutional way.' With the war effectively over but many men still under arms and pay in arrears, there was a near-mutiny on March 10, 1783 at Washington's camp. It was led by twenty-four-year-old Major John Armstrong, who wrote the Newbrugh Addresses, protesting at Congress's treatment of its army and urging the officers not 'to be tame and unprovoked when injuries press down hard upon you.' But only the younger officers were involved in this business, which has been described as 'the only known instance of an attempted coup in American history.' But Washington called all the disaffected officers together and persuaded them to put their complaints through regular channels. He was a great persuader. Three months later, in June, the last in a string of mutinies occurred when several hundred angry soldiers actually surrounded the State House in Philadelphia where both Congress and the Executive Council of Pennsylvania State were meeting. But this military mob dispersed on the approach of a regular army unit under General Robert Howe.[93]

The least suspicion of Caesarism was finally scotched by the prompt and decisive manner in which Washington himself terminated his military duties. On December 23, 1783 he presented himself at the Philadelphia State House, where Congress sat, drew a note he had written from his pocket and held it with a hand that visibly shook. He read out: 'Mr President [he was addressing the presiding officer, Thomas Mifflin], the great events on which my resignation depended having at length taken place, I have now the honor of offering my sincere congratulations to Congress and of presenting myself before them in order to surrender into their hands the trust committed to me, and to claim the indulgence of retiring from the service of my country . . . and bidding an Affectionate Farewell to this August body under whose orders I have so long acted, I here offer my commission, and take my leave of all the employments of public life.' At this point, he took his commission from his uniform coat, folded the copy of his speech and handed both papers to Mifflin. He was self-consciously imitating Cincinnatus handing back his sword. He then shook hands with every member of Congress, mounted his horse, and rode through the night to Mount Vernon, reaching it the next morning.[94]

With the British gone and Washington back in Mount Vernon, how was America to govern itself? It did not miss monarchy. The British crown was only a parliamentary monarchy anyway; 18th-century Britain was a semi-republic in many ways. When Benjamin Rush, the radical doctor who ran Washington's army medical services, was on a prewar visit to England, the attendant at the House of Lords (parliament was in recess) allowed him to disport himself on the throne 'for a considerable time' and Rush found himself 'seized with a kind of horror.'[95] The Americans, wrote Jefferson, 'shed monarchy with as much ease as would have attended their throwing off an old and putting on a new suit of clothes.' Monarchy did not make much practical sense in a country without an aristocracy. America had a sort of ruling class—in Virginia, about one man in twenty-five was a 'gentleman,' further north one in ten and the distinction mattered less. In Virginia about 8 percent of the population controlled a third of the land, so there was a class divide based on wealth.[96] Distinctions in status were reflected in careless speech. Even Washington spoke of ordinary farmers as the 'Grazing Multitude.' His aide, Hamilton, referred to the 'unthinking populace.' John Adams termed them the 'common herd of mankind,' and Gouverneur Morris felt ordinary people 'had no morals but their interests.' But this was just club talk. Virtually all American landowners engaged in trade. For once they rejected a saying of Locke's, 'Trade is wholly inconsistent with a gentleman's calling.' In fact New Yorkers stood the adage on its head, merchants listing themselves in directories as 'gentlemen,' if they were prosperous enough. And, since it cost only £400 to set yourself up as a merchant in New York, as opposed to £5,000 in England—that was why so many emigrated—there were plenty of 'gentlemen' in Manhattan. In country districts, money was short, credit hard to get, monetary instruments crude, so rich landowners, if they had it, lent money out—Charles Carroll of Anapolis lent £24,000 to neighbors. This worked as a kind of bastard feudalism, supplemented by family links, so that a really rich man, especially in Maryland and the South, had a following. But the kind of clientage taken for granted by English dukes in their districts, supplemented by pocket boroughs, simply did not exist. There was no top tier in white society—no bottom tier either.[97]

Then again, anyone deciding how America was to be governed had to take account of what was perhaps the most pervasive single characteristic of the country—restlessness. Few people stayed still for long. Mostly they were moving upwards. And vast numbers were moving geographically too. A British observer noted, wonderingly, that Americans moved 'as their avidity and restlessness incite them. They acquire no attachment to Place; but wandering about Seems engrafted in their Nature; and it is weakness incident to it that they Should forever imagine the Lands further off, are Still better than those upon which they are already Settled.'[98] This mobility acted as an economic dynamic—restlessness was one reason the American economy expanded so fast, as new, and often better, land was brought into production and new economic growth-centers created almost overnight in frontier districts. But constant moving broke up settled society, worked against hierarchy and 'respect,' and promoted assumptions of equality.

There is a lot of evidence that farmers' money incomes rose during the war as food was sold to armies for cash. Spending habits grew more luxurious—farmers' wives demanded not only tea but tea-sets. Merchants 'set up their carriage.' The same thing was happening in England—read Jane Austen's novels—but in America it started lower down the socioeconomic scale. And in America there were fewer of the moralists who, in England, deplored the spread of luxuries among the common people. On the contrary: America was already developing the notion that all were entitled to the best if they worked hard enough, that aiming high was not only morally acceptable but admirable. Silk handkerchiefs, feather mattresses, shop-made dresses, imported bonnets—why shouldn't people have them? 'The more we have the better,' enthused James Otis, 'if we can export enough to pay for them.' Ebenezer Baldwin, a little more sharply, agreed: 'We have no such thing as a common People among us. Between Vanity and Fashion, the Species is utterly destroyed.'[99]

It was a short step from admitting ordinary folk had a right to the best to giving them a full share in government—and giving it to them not grudgingly but eagerly. Words like 'husbandman,' 'yeoman,' 'esquire' quickly dropped out of use, being replaced by 'citizen'—a decade before the French Revolutionaries took it up. Collectively, the citizens were the 'Publick,' a new word coming into fashion. 'Cato' wrote: 'Ordinary people [are] the best Judges, whether things go well or ill for the Publick.' Cato thought: 'Every ploughman knows a good government from a bad one.' Jefferson agreed: 'State a problem to a ploughman and a professor. The former will decide it often better than the latter, because he had not been led astray by artificial rules.' John Adams invented a hick-farmer archetype, Humphrey Ploughjogger, and extolled his sense and shrewdness in newspaper articles. He was 'made of as good a Clay as the so-called Great Ones of the world.' 'The mob, the herd and the rabble, as the Great always delight to call them,' were, wrote Adams, 'by the unalterable laws of god and Nature, as well entitled to the benefit of the air to breathe, light to see, food to eat, clothes to wear, as the nobles or the king.'[100] All that was necessary was to educate them, to add knowledge to their native wit.

It was the great merit of the new egalitarian spirit in America that it consciously placed education right at the front of national priorities. Adams wrote that the settlement of America was part of a providential plan 'for the illumination of the ignorant and emancipation of the slavish part of mankind,' first in America, then all over the world. Stanhope Smith, president of Princeton, believed that a combination of 'republican laws' and education would effect a general moral improvement in the population and create a 'society of habitual virtue.' Virtue, said Ezra Styles, could be taught, like any other art.[101] And it was education, said Adams, which made the gentleman, not birth or privilege. He and most of the key men in the Revolution were first-generation gentlemen, made such by their ability to read and make use of books and by their mastery of the pen—Adams' cousin Sam, Jefferson, Rush, John Marshall, James Madison, David Ramsay, John Jay, James Wilson, Benjamin Franklin. Adams' father had been 'an ungenteel farmer'—he himself had become a gentleman by going to Harvard. Jefferson, though from a much higher starting-place in society, had also been the first in his family to go to college. Ultimately, all would do so: then indeed America would be a republican commonwealth of taste, art, manners, and above all virtue. It was education which would make the republican structure and the democratic content of the new union of states engines of peaceful progress. In the 1830s Macaulay was to say that, in England, education was engaged in a race to civilize democracy before it took over. But it is worth remembering that the American elite grasped this point—and did something about it—half a century before.

In the meantime, the republican structure had to be created as a matter of urgency. The wartime system was a series of improvisations and obviously not good enough. The original idea of the United States was a coming together of the states, as sovereign bodies, to create an umbrella-state over them, to do certain things as the states should delegate to it. The people did not come into this process, except insofar as they elected state legislatures. It is important to grasp the point: the original revolution, a military and political one, which produced this improvised form of Congressional government, was followed by a second revolution, this time a constitutional one, which produced the United States Constitution as we know it. The second revolution began during the war and it was (in the old English tradition) an organic development, in response to need. In October 1777 Congress decided it had to create Boards of War, Treasury, and the Admiralty, with professional staffs, simply to get things done. This was the beginning of executive government. Courts had to be created to hold Admiralty appeals from state courts. This was the beginning of the federal judiciary. In September 1779, the doctrine of US citizenship began to emerge. There was the first suggestion, as war supplies ran out, that Congress had the power to coerce mean or uncooperative states—the doctrine which ultimately was to enable President Lincoln lawfully to coerce the Confederacy.

The biggest formative force was financial need. The improvised currency broke down under the pressures of war. Inflation started to accelerate. These were the evils which, in Latin America during the next generation, were to poison the youth and malform the maturity of the Spanish-speaking republics. The men of New York, already emerging as a center of 'sound,' that is expert, finance, were determined not to allow this to happen. Gouverneur Morris, Philip Schuyler, Alexander Hamilton, and James Duane got together to propose what would later be termed a 'federalist solution,' that is a strong government pledged to an honest currency. They believed in government deliberately creating the framework in which the economy could develop and expand rapidly by sponsoring an advanced banking system, managing credit, and promoting fiscal efficiency. They got their ideas from Britain and Adam Smith, and the man who advanced them most confidently, Alexander Hamilton, first gave them expression in his 'Continentalist Letters,' published in 1781–2.[102]

That was the beginning of the debate on the Constitution. So who was this Hamilton, who began it? He was born in 1755 in the small West Indian island of Nevis, and it is vital to remember that he was not an American, except by adoption, and could never have become president, though he was in some respects better fitted for the job than any other of the Founding Fathers. In a sense he was the archetypal self-made man of American mythology—born out of wedlock, deserted by a no-good father, left an orphan at thirteen by the death of his mother, he was helped by friends and relatives to find his way to New York where, at seventeen, he entered King's College (later Columbia University). There he thrived and absorbed a mass of political, historical, constitutional, and forensic knowledge which made him one of the sharpest lawyers of his generation. He was soon in the thick of the Revolutionary agitation as a speaker and churner-out of pamphlets, having a gift for rapid writing for print unequaled by any of the time, even Paine and Franklin. He joined the army, found himself in the artillery, where he quickly mastered the art of gunnery, became a lieutenant, saw action repeatedly, attracted the attention of the Commander-in-Chief, and so served on Washington's staff, as his best and closest ADC for five years. Washington was his hero, his 'aegis,' as he put it. Washington, in turn, found him the best executive officer in the army, a man who could be trusted to carry out the most difficult staff-duty with efficiency and speed, who was full of ideas, brave to a fault, and absolutely loyal.

Hamilton left Washington's staff to command a battery at Yorktown (it was the guns which made Cornwallis' surrender unavoidable), and he undoubtedly saw more military action than any other of the great Constitution-makers. But he never quite strikes one as a typical American, or even an extraordinary one. He might perhaps have been more at home in the House of Commons and in Pitt's Cabinet of the 1780s. He had no fear of kingship as such: if it worked, use it. He was an empiric, a pragmatist on the English model, and his instinct was always to look at how England did things and see if America would be well advised to follow suit, *ceteris paribus*. He was a disciple not so much of Locke as of Hobbes, a man who believed that society was inherently chaotic and in need of a strong Leviathan-figure (whether a man or an institution) to 'keep them all in awe.'[103] In 1780 he married Elizabeth Schuyler, daughter of a major-general and large-scale Hudson Valley landlord. Once demobilized, he started a highly successful law practice, served in Congress 1782–3, and put himself in the forefront of those demanding a stronger national government.

Within the government it was Hamilton's ally, Robert Morris, Superintendent of Finance, who pushed for reforms. In 1781–2 he produced a tax and finance program to provide funds and a stable currency, and he went outside government to organize support, in Congress, in business, and even in the army. Morris and Hamilton realized that, now all British impediments to westward expansion had been removed, selling land to eager farmers was one way the federal government, or the general government as they still called it, could finance itself—but only if individual states relinquished to the federal center control over Western lands. Up to the Revolutionary War, the states had admitted no western limit to their claims. In 1780, however, they had agreed in principle that all western territory would be 'settled and formed into distinct republican states, which shall become members of the federal union, and have the same rights of sovereignty, freedom and independence as the other states.'

The 1783 Peace of Paris doubled the size of the United States, adding the western territories to the Atlantic states. But the size, number, and boundaries of the new states had to be determined, together with the constitutional procedure to bring them into the Union. A committee of Congress under the chairmanship of Jefferson was appointed to settle this, and in 1784 it reported that the Western territories should be divided into fourteen new states—including Assenisipia, Cherroonesus, Metropotamia, Miochigania, and Washington. Congress did not like these weird names and dropped them. But in its ordinance of 1784 it laid down that the territories should have temporary governments (managed by Congress) until each had a free population equal to that of the least populated of the existing states. At that point it could apply for admission. This came into effect only after each state had formally ceded all its Western claims. The Land Ordinance of 1785 defined how this new federal land was to be surveyed and sold. Finally, the Northwest Ordinance of 1787 dealt with the northern sector of the West and made more specific the process of state-creation. First, a governor, secretary, and judges would be appointed to the territory by Congress. The second stage began when a district acquired 5,000 adult free males; it could then elect an assembly and nominate a list of candidates from which Congress chose a governing council, though it retained the right of veto over legislation and still appointed the governor. The third stage began when population passed the 60,000-free-inhabitants mark, when it could petition to become a state.

This ordinance or law was the last passed under the old Articles of Confederation, and many objected to it on the ground that it lodged political power in the hands of Eastern legislators or company promoters rather than Western squatters—it was centralist rather than democratic. So it was. But then the whole question of Western lands inevitably tended to strengthen the power of the federal government, as Hamilton spotted, because it gave it direct authority over a huge spread of territory as big as the existing states—much bigger as it turned out—which it could rule like an imperial power, and support by selling off bits to settlers. That was a geographical fact, which made it inevitable that the federal center would strengthen itself as time went by. It was the states themselves which sold the pass on state sovereign rights when they renounced their sovereignty over Western lands and handed it over to Congress. However, for the time being individual states carried out all kinds of sovereign acts which logically belonged to a central authority—they broke foreign treaties and federal law, made war on Indians, built their own navies, and sometimes did not trouble themselves to send representatives to Congress. They taxed each other's trade while failing to pay what they had promised to the Congressional coffers. That, of course, was at the root of the collapse of credit and the runaway inflation. All agreed: things could not go on this way.[104]

At this point, yet another Founding Father emerged from the shadows into the bright lights of national prominence. James Madison (1751-1836) was born in 1751 in Virginia, the son of a fairly prosperous planter, who had him educated by private tutors before dispatching him to Princeton in 1771. There he was a classmate of Aaron Burr (1756-1836) and with two budding authors, Hugh Henry Brackenridge (1748-1816) and Philip Freneau (1752-1832), produced a remarkable 'Poem of the Rising Glory of America,' reflecting the view of their generation of educated elitists that leadership in culture was inevitably passing westward from Europe to America, which would be the theater of 'the final stage . . . of high invention and wond'rous art, which not the ravages of time shall waste.' Freneau indeed has often been called, with justice, 'the poet of the American Revolution.'[105] Madison, however, can be called, with equal justice, the constitutionalist of the Revolution, for he did more than Jefferson or even Hamilton to ensure that the United States got a workable system of government. He had read Francis Bacon's famous essay, 'Of Honor and Reputation,' which discussed the hierarchy of 'categories of fame and honor,' placing at the top of it 'founders of states and commonwealths, such as Romulus, Cyrus and Caesar.' It was his good fortune, as John Quincy Adams was to write a few years later, to join this select company, being 'sent into life at a time when the greatest lawgivers of antiquity would have wished to live. How few of the human race have

ever enjoyed an opportunity of making an election of government—
more than of air, soil or climate—for themselves and their children?
When, before the present epoch, had three millions of people full
power and a fair opportunity to form and establish the wisest and happiest government that human wisdom can contrive?'[106] It was, Madison
congratulated himself, 'a period glorious for our country and, more
than any preceding one, likely to improve the condition of man'—
hence to be privileged to write the constitution was 'as fair a chance of
immortality as Lycurgus gave to that of Sparta.'[107]

Madison was a frail man, whose physique prevented him from serving in the army. In 1776 he was elected to the Virginia state convention
where, in the drafting of the new state constitution, he made his first
gift to the vernacular of American constitutional law by suggesting that
the phrase 'toleration of religion' be given a positive twist by being
changed to 'the free exercise of religion'—an important improvement,
with many consequences. That year, as a member of the state executive
council, he first met Jefferson, formed a friendship with him which
lasted for the rest of Jefferson's life, and produced an exchange of letters of which over 1,250 survive, one of the great correspondences of
history and by a long way the most important series of political letters,
between two leading statesmen, ever to have been written. There is no
more agreeable way of learning about how history was made during
this half-century than by browsing in the three grand volumes in which
these letters are printed.[108] It is important to remember, in judging the
contributions made by each of these two great men, the extent to which
one influenced the other, at all times.

The stages by which the United States Constitution was created were as
follows. The efforts by Morris and Hamilton to reform the existing
Confederation, especially in finance, had produced no fundamental
response. In 1783 Madison turned his hand to the problem, producing
a three-point plan of reform, less radical in some ways than the
Morrison–Alexander scheme, but introducing the concept of popular
elections for the first time (with the slave element in the population of
states counting as three-fifths of whites, per capita—the formula eventually adopted). Nothing much came of this either, at the time. Then
accident intervened, as it often does in great historical events. Virginia
and Maryland were rowing over the navigation of the Potomac, which
both claimed the legal right to direct. In this confusion, importers were
taking the opportunity to evade customs-dues. Matters came to a head

at the end of 1783 and Madison, who looked after national affairs for the Virginia government, proposed that negotiating commissioners be appointed by both states. Washington, a born conciliator, was delighted to give them hospitality at Mount Vernon, from March 25, 1785. There they ranged well beyond their mandate, settling not only navigational and naval differences between the two states but customs, currencies, regulation of credit, and many other topics.

The conference was so successful that Pennsylvania was roped in on the Potomac issue, and Madison skillfully brought the agreements to the attention of Congress, which ratified them. He then put through a motion for Virginia to invite commissions from all states to meet and discuss 'such commercial regulations as may be necessary to their common interest and their permanent harmony.' The meeting took place at Annapolis on three days in September 1786 and only five states actually sent commissioners. But it did some important preparatory work and lobbying, and it enabled Madison to get to know Hamilton and both to put their heads together to see how to proceed further. Madison was a cautious, deliberative man, Hamilton a plunger, an audacious adventurer. He built onto Madison's tentative scheme of constitutional revision, which dealt only with economic issues, a broader plan, inviting state delegates to Philadelphia in May 1787, 'to devise such further provisions as shall appear to them necessary to render the constitution of the Federal Government adequate to the exigencies of the Union.' It set no limit on the things the Convention might discuss.

However, if Hamilton gave the momentum for constitutional reform a decisive push, it was Madison who provided the Convention's agenda, by presenting the Virginia Plan. The new element in this, of fundamental importance, was that the national government ought to operate directly on the people (rather than through the mediating agency of the states) and that it ought to receive its authority directly from the people (rather than from the states). In other words the sovereign people—it was Madison who coined the majestic phrase 'We, the People'—delegated authority both to the national government and to the states, thereby giving it the power to act independently in its own sphere, as well as imposing restrictions on the actions of the states. This could be described as the most important constitutional innovation since the Declaration of Independence itself. Madison proposed that the limiting power should be exercised by a federal power of veto on state laws. This was rejected, as smacking too much of the old royal veto. But the principle was accepted, and limitations on the power of the states imposed by the federal Constitution have been accepted as a fundamental mechanism of the federal system. In Madison's scheme, such power was legitimized by the federal government drawing authority directly from the votes of the people. The positive point was of comparable importance to the negative point of limiting state authority because it knocked the bottom out of the subsequent states' rights argument (of John C. Calhoun and others) that only the states themselves conferred power on the federal government, and could remove it just as comprehensively. But the people conferred power too, and it was on that basis that President Lincoln was later able to construct the moral and legal case for fighting a war to hold the Union together. All this was Madison's doing.[109]

The Convention met in Philadelphia again and sat for four months, breaking up on September 17, 1787, its work triumphantly done. Its success owed a lot to the fact that all the states had been writing or improving their own constitutions over the past decades and so many of the men attending were experts at the game. Of those attending, forty-two had sat in the Continental Congress or the congresses held under the Articles of Association. Most were planters, landowners or merchants; a number had served in the army; there were twenty-six college graduates—nine from Princeton alone—but probably the most important single element were the lawyers. It was Hamilton who pointed out the significance of this, both at the time and later in the *Federalist* (number 35), in one of his newspaper essays. All the Constitution-makers distinguished between private interests and an autonomous public interest, representing republican ideals—the *res publica* itself. Washington, who presided over the Convention, but who wisely confined his activity to insuring order and decorum, stuck to his view that most men were guided by their own private interests: to expect ordinary people, he said, to be 'influenced by any other principles but those of interest, is to look for what never did and I fear never will happen ... The few, therefore, who act upon Principles of Disinterestedness are, comparatively speaking, no more than a drop in the ocean.' That was true, agreed Hamilton; nonetheless, there was a class of people in society who, as a 'learned profession,' were disinterested—the lawyers. Unlike farmers, planters, and merchants, they had no vested economic interest to advance and therefore formed a natural ruling elite and ought to form the bedrock of public life. Madison complemented this argument by asserting (a point he also repeated in the

Federalist (number 10) that, whereas the states represented local interests, the federal government and Congress represented the national or
public interest, and would mediate between them. Hence, concluded
Hamilton, it would be natural and right for state legislatures to be
dominated by planters, merchants, and other interest-groups but for
the Congress to be dominated by the lawyers. Though America's ruling
elite, insofar as it still existed in the 1780s, intended for the new
Constitution to provide rule by gentlemen, what it did in fact produce
was rule by lawyers—a nomiocracy.[110]

There was a fair spectrum of opinion in the Convention. There were
extreme federalists, who wanted a centralized power almost on the
lines of a European state like Britain—Gouverneur Morris of New
York, James Wilson of Pennsylvania, Rufus King of Massachusetts,
and Charles Pinckney of South Carolina. On the other hand, there were
some states' rights extremists like Luther Martin of Maryland. The
existence of these two opposing groups had the effect of making
Hamilton (the pro-federalist) and Madison (who was closer to
Jefferson) seem moderates, and therefore to strengthen their influence.
But the atmosphere of the Convention was positive, constructive, and
reasonable at all times. Even those who formed, as it were, the opposition—such as Elbridge Gerry, who refused to sign the Constitution,
and Edmund Raldolph, who likewise declined to sign though, unlike
Gerry, he supported ratification—were helpful rather than obstructive.
These were serious, sensible, undoctrinaire men, gathered together in a
pragmatic spirit to do something practical, and looking back on a thousand years of political traditions, inherited from England, which had
always stressed compromise and give-and-take.[111]

The Convention moved swiftly because these practical men were
aware of the need to get the federal power right as quickly as possible.
The previous autumn, a dangerous revolt of debt-ridden farmers, many
of whom had fought in the Continental Army and were well provided
with crude weapons, had developed in rural Massachusetts. Under the
leadership of Daniel Shays (1747–1825), a bankrupt farmer and former
army captain, they had gathered at Springfield in September and forced
the state Supreme Court to adjourn in terror. In January Shays led
1,200 men towards the Springfield arsenal to exchange their pitchforks
for muskets and seize cannon. They were scattered, though many of
them were still being hunted in February 1787, shortly before the
Convention met. The net effect of Shays' Rebellion was to force the
Massachusetts legislature to drop direct taxation, lower court fees, and

make other fiscal concessions to the mob. But it also reminded every-one attending in Philadelphia that the Confederation, as it stood, was powerless to protect itself, or any of the states, from large-scale domes-tic violence, and that this absence of a central power was itself a limita-tion on state sovereignty, as the humiliating climb-down of the Massachusetts legislature demonstrated. The pressure, then, was on to get a federal constitution written—and adopted.[112]

Hence the Convention set to with a will. An analysis of the voting shows that the mechanics of compromise operated throughout—in 560 roll-calls, no state was always on the losing side, and each at times was part of the winning coalition. Broadly speaking, the Virginia Plan was adopted, and in this sense Madison can be called the author of the United States Constitution. A rather weaker version, from New Jersey, was rejected. On the other hand, the federalists, led by Hamilton, could make no real progress with their proposal for a strong central govern-ment on European lines. Amid many compromises, there were three of particular importance. In early July, the so-called Connecticut Compromise was adopted on the legislature. This gave the House of Representatives, directly elected by popular votes in the localities, the control of money Bills, and a senate, particularly charged with foreign policy and other matters, to represent the states, with two senators for each state, chosen by the individual legislatures.

In August the Convention turned its attention to the knotty problem of slavery, which produced the second major compromise. The debating was complex, not to say convoluted, since the biggest slave-holder attending, George Mason, attacked the institution and especially the slave-trade. Article 1, section 9, grants Congress the power to regulate or ban the slave-trade as of January 1, 1808. On slavery itself the Northerners were prepared to compromise because they knew they had no alternative. Indeed, as one historian of slavery has put it, 'It would have been impossible to establish a national government in the 18th cen-tury [in America] without recognising slavery in some way.'[113] The con-vention did this in three respects. First, it omitted any condemnation of slavery. Second, it adopted Madison's three-fifths rule, which gave the slave states the added power of counting the slaves as voters, on the basis that each slave counted as three-fifths of a freeman, while of course refusing them the vote as such—a masterly piece of humbug in itself. Third, the words 'slave' and 'slavery' were deliberately avoided in the text. As Madison himself said (on August 25), it would be wrong 'to admit in the Constitution the idea that there could be property in men.'[114]

The third compromise, in early September, was perhaps the most important of all in the long run, dealing as it did with the election of the president. Although federalists like Hamilton lost the general battle about the nature of the state, which remained decentralized rather than concentrated, they won a significant victory over the presidency. Hamilton won this by tactical skill, compromising on the election procedure—if no candidate got a majority of the popular vote, the House elected one from among the top three, voting by states, not as individuals. Each state was further given the right to decide how to choose its electoral college. This appeared to be a gesture to the states, balancing the fact that the president was directly elected by the people. But it left open the possibility of popular participation. Thus in practice the president was elected independently of the legislature. Moreover he was given a veto (offset by a two-thirds overriding rule) over Congressional legislation, and very wide executive powers (offset to a limited degree by the requirement that the Senate should 'advise and consent').

Almost by accident, then, America got a very strong presidency—or, rather, an office which any particular president could make strong if he chose. He was much stronger than most kings of the day, rivaled or exceeded only by the 'Great Autocrat,' the Tsar of Russia (and in practice stronger than most tsars). He was, and is, the only official elected by the nation as a whole and this fact gave him the moral legitimacy to exercise the huge powers buried in the constitutional thickets. These powers were not explored until Andrew Jackson's time, half a century on, when they astonished and frightened many people; and it is perhaps fortunate that the self-restraint and common sense of George Washington prevented any display of them in the 1790s, when they would certainly have led to protest and constitutional amendment. As it was, the new republic got a combined head of state and head of government entrusted with formidable potential authority.

Although the Convention worked with some speed, which was necessary, and desirable for its own sake—too long debates on constitutions lead to niggling and confusion of issues—it worked deliberatively. The making of the United States Constitution ought to be a model to all states seeking to set up a federal system, or changing their form of government, or beginning nationhood from nothing. Alas, in the 200 and more years since the US Constitution was drawn up, the text itself has been studied (often superficially) but the all-important manner in which the thing was done has been neglected. The French Revolutionaries in the next decade paid little attention to how the

Americans set about constitution-making—what had this semi-bar-barous people to teach Old Europe? was the attitude—and thirty years later the Latin Americans were in too much of a hurry to set up their new states to learn from the history of their own hemisphere. So it has gone on. The federal constitutions of the Soviet Union (1921) and of Yugoslavia (1919) were enacted virtually without reference to the American experience, and both eventually provided disastrous and bloody failures. It was the same with the Central African Federation, the Federation of Malaysia, and the West Indies Federation, all of which had to be abandoned. The federal structure of the European Union is likewise being set up with no attempt to scrutinize and digest the highly successful American precedent, and attempts to persuade the European constitution-makers to look at the events of the 1780s are contemptuously dismissed.[115]

Just as important as the process for drawing up the Constitution was the process of ratifying it. In some ways it was more important because it went further to introduce and habituate the country to the democratic principle. Article VII of the Constitution provided for the way it was to be adopted, and resolutions passed by the Convention on September 17, 1787 set out a four-stage process of ratification. The first was the submission of the document to the Congress of the old Confederation. This was done on September 25, and, after three days of passionate debate, federalists (who supported ratification) and anti-federalists (who wanted it rejected) agreed to send the Constitution to the individual states, the second stage, without endorsing or condemning it. The third stage was the election of delegates in each state to consider the Constitution, and the fourth was ratification by these conventions of at least nine of the Thirteen States. When the ninth state signified its acceptance, the Constitution then became the basic law of the Union, irrespective of what other states did.

This introduction of the rule of majority, as opposed to unanimity, itself signified the determination of the federalists to create a forceful and robust government. Majority rule made fast action possible. It reflected the desire that the ratification process proceed briskly, and the hope that quick ratification by key states early in the day would stampede the rest into acquiescence. It was a high-risk strategy, obviously. If any of the four biggest states, Virginia, Massachusetts, New York, and Pennsylvania, let alone all of them, rejected the Constitution, ratification by all the rest would be meaningless. But the federalists thought they could be pretty sure of the Big Four. Again, the Constitution took an even bigger risk in insisting ratification had to be by popular, specially elected conventions rather than by state legislatures. This was to introduce the people—democracy indeed—with a vengeance. But it was felt that approval by state legislators was not enough. Here was a fundamental law, affecting everyone in the nation and their children and grandchildren and generations to come. The people ought to participate, as a nation, in deciding whether to endorse it, and the ratification process itself would encourage them to look beyond the borders of their own states and consider the national interest as well as their own. This was a wise decision, again with momentous consequences, because once the people had thus been invited onto the political stage, and asked their opinion, they could never be pushed into the wings again.

Ratification by convention also had the effect of inviting a grand public debate on the issue, and in a way this was the most significant aspect of the whole process. If Jefferson, Madison, and Adams were right in believing that education, virtue, and good government went together, then there was a positive merit in getting not just state legislatures but the people themselves to debate the Constitution. The wider the discussions, the more participants, the better—for public political debate was a form of education in itself, and a vital one. If, in the 1760s and early 1770s, the Americans, or their representatives, had been allowed to debate with the British, or their representatives, on the proper relationship between the two peoples, the Revolution might have been avoided. Words are an alternative to weapons, and a better one. But a debate was refused, and the issue was put to the arbitrament of force. The Americans had learned this lesson (as indeed had the British by now) and were determined to give words their full play. In the next decade the French were to ignore the lesson, at the cost of countless lives and ideological bitterness which reverberates to this day.

So that ratification process was a war of words. And what words! It was the grandest public debate in history up to that point. It took place in the public square, at town meetings, in the streets of little towns and big cities, in the remote countryside of the Appalachian hills and the backwoods and backwaters. Above all it took place in print. America got its first daily newspaper in 1783 with the appearance of the Philadelphia *Evening Post*, and dailies (often ephemeral) and weeklies were now proliferating. Printing and paper, being completely untaxed, were cheap. It cost little to produce a pamphlet and the stages carried packets of it up and down the coast. Americans were already developing the device (eventually to be called the syndicated column) of getting articles by able and prominent writers, usually employing pseudonyms like 'Cato,' 'Cicero,' 'Brutus,' 'Publius,' 'A Farmer,' 'A Citizen of New York,' and 'Landholder,' circulated to all newspaper editors, to use as they pleased. So literally thousands of printed comments on the issues were circulated, and read individually or out loud to groups of electors, and then discussed and replied to. It was the biggest exercise in political education ever conducted. An important issue was felt to be at stake, which went beyond the bounds of the Constitution as such. As Hamilton, writing as 'Publius,' put it, the process was to determine 'whether societies of men are really capable or not, of establishing good government by reflection and choice, or whether they are forever destined to depend, for their political constitutions, on accident and force.'[116]

The federalists were led by Alexander Hamilton, the most active of all, James Madison, who came second, John Jay, John Marshall, James Wilson, John Dickinson, and Roger Sherman. They had the initial advantage that George Washington was known to favor ratification, and his name carried weight everywhere. Franklin was also a declared supporter, and he counted for a lot in Philadelphia, the biggest city. Hamilton, Madison, and Jay produced jointly the *Federalist*, a series of eighty-five newspaper essays, much reproduced and printed in book form in 1788. Hamilton was the principal author and collectively they represent the first major work of political theory ever produced in America, discussing with great clarity and force such fundamental questions of government as the distribution of authority between the center and the periphery, between government and people, and the degree to which the constituent elements of government, executive, legislature, and judiciary, ought to be separate. It is the one product of the great debate which is still widely read.[117] How widely it was read, and understood, at the time is debatable. It certainly served as a handbook for speakers on the federalist side before and during the ratification conventions. In that sense it was very important.

The most popular publication on the federalist side was John Jay's *Address to the People of the State of New York*, which was reprinted many times, and another bestseller, as a pamphlet, was the major speech made by James Wilson on November 24, 1787 to the Pennsylvania convention. It was Wilson who put the stress on election and representation as the core of the constitution. That, he argued, was what distinguished this new form from the ancient orders of Athens and Rome and the curious mixture of voting and inherited right which made up the British Constitution. 'The world,' he wrote, 'has left to America the glory and happiness of forming a government where *representation* shall at once supply the basis and the cement of the superstructure. For representation, Sir, is the true chain between the people and those to whom they entrust the administration of the government.' After Madison, Wilson's was the most important hand in shaping the Constitution, and after Hamilton's his was the most important voice in getting it accepted.[118]

The anti-federalists, such as Patrick Henry, Richard Henry Lee, George Mason, John Hancock, James Monroe, Elbridge Gerry, George Clinton, Willie Jones, Melancton Smith, and Sam Adams, were formidable individually but lacked the cohesive force of the federalists. Their objections varied and they appeared unable to agree on an alternative to what they rejected. The *Letters of Brutus*, probably written by Robert Yates, Otis Warren's *Observations on the New Constitution*, the anonymous *Letters from the Federal Farmer to the Republican* and Luther Martin's *General Observation* contradict each other and leave a negative impression. One pamphleteer, signing himself 'A Republican Federalist,' equated the proposed Congress with the British: 'The revolution which separated the United States from Great Britain was not more important to the liberties of America, than that which will result from the adoption of a new system. The *former* freed us from a *foreign subjugation*, and there is too much reason to apprehend that the *latter* will reduce us to a *federal domination*.' This fear of Big Government was allied to a widespread conviction, which the anti-federalists articulated, that the new federal congress and government would quickly fall into the hands of special interests and groups who would oppress the people. Hamilton's notion of lawyers as a disinterested class formed by nature to run the center did not impress. As Amos Singeltary of Massachusetts put it, 'These lawyers, and men of learning, and monied men, that talk so finely, and gloss over matters so smoothly, to make us poor illiterate people swallow down the pills, expect to get into Congress themselves: they expect to be the managers of this Constitution, and get all the power and the money into their own hands, and then they will swallow up all us little folks, like the great Leviathan.'[119]

But the alternative some anti-federalists proposed, of Small Government on the lines of the Swiss cantons, did not go down well. After all, America had experienced small government already, during the war and since, and most people knew it had not worked well—would not have worked at all without Washington. The problem, during the war and since, had not been too much government but too little. That was a very general view, in all states; and fear of Big Government was further mitigated by a general assumption that, once the new Constitution was in force, Washington would again be summoned to duty and would prevent its power from being abused just as once he had made good its lack of powers. Where the anti-federalists struck home was in stressing that the new Constitution said little or nothing about rights, especially of the individual. But the federalists admitted this defect, and they agreed that, once the Constitution was ratified, the first thing was to draw up and pass a Bill of Rights which (as a constitutional amendment) would require the consent of three-quarters of the states and would thus be sure to satisfy the vast majority.[120]

With this qualification in mind, the ratification procedure began. The first five ratifications took place December 1787–January 1788: Delaware (unanimous), Pennsylvania (46–23), New Jersey and Georgia (unanimous), and Connecticut (128–40). In Massachusetts, the two leading anti-federalists, Sam Adams and John Hancock, negotiated a rider to ratification under which the state agreed to accept the Constitution on condition it was amended with a Bill of Rights. This went through in February 1788 (187–168). All the other states adopted this device, and insured the acceptance of the Constitution, though making it imperative that the rights provisions be adopted quickly. Maryland ratified in April (63–11), South Carolina in May (149–73), New Hampshire (57–47) and Virginia (89–79) in June, and New York in July (30–27). That made eleven states and insured the Constitution's adoption. North Carolina's ratification convention adjourned in August 1788 without voting, and Rhode Island refused to call a convention at all. But the virtual certainty that amendments would be introduced guaranteeing rights persuaded both states to change their minds: North Carolina ratified November 1789 (195–77) and Rhode Island May 1790 (34–32). Thus, in the end, the ratification by states was unanimous, and the Constitution was law. Benjamin Franklin, who had attended every session of the Constitutional Convention and who had actually fathered the idea that the House should represent the people and the Senate the states, hailed the adoption of the Constitution with a memorable remark: 'Our Constitution is an actual operation,' he wrote to a friend in Europe, 'and everything appears to promise that it will last: but in this world nothing can be said to be certain but death and taxes.'[121]

Congress now had to enact rights. Some states had already done so, so there were precedents. The federalists who wrote the Constitution were chary on the subject. Individual rights were presumed to exist in nature—that was the basis on which the Declaration of Independence had been drawn up—and a formal, legal statement of them might imply the extension of government into spheres in which it did not and should not operate. 'The truth is,' Hamilton wrote in the *Federalist*, 'the Constitution is itself, in every rational sense, and to every useful purpose, a bill of rights.' That was a shrewd point and it may be that enacting individual rights formally has proved, especially in the 20th century, a greater source of discord than of reassurance. But Hamilton and the others went along with the general feeling, very strong in some states and especially in the backwoods and country districts, that rights must be enumerated and spelt out.

Hence Madison, who had originally opposed what he called 'parchment barriers' against the tyranny of interests or of the majority, relying instead upon structural arrangements such as the separation of powers and checks and balances, now set about the difficult task of examining all the amendments insisting on rights put forward at the ratifying conventions, and various bills of rights enshrined in state constitutions, and coming up with a synthesis. He also had a complete model in the shape of the Virginia Declaration of Rights (1776), written by the anti-federalist George Mason. Early in the first session of the new Congress in 1789, Madison produced drafts of ten amendments. The first amendment, the most important, prohibits legislative action in certain areas, giving citizens freedom of religion, assembly, speech, and press, and the right to petition. The next seven secure the rights of property, and guarantee the rights of defendants accused of crimes. The ninth protects rights not specifically enumerated. The tenth, reinforcing this, insists that 'the powers not delegated to the United States by the Constitution, nor prohibited by it to the states, are reserved to the states respectively, or to the people.' The ratification proceeded smoothly and on December 15, 1791, when Virginia ratified, the Bill of Rights became part of the Constitution.[122]

Two more matters remained to be determined. Should representatives be paid? They never had been in England, except sometimes by localities. The states varied. Franklin, who was rich, argued before the convention of 1787 that no salaries be paid—in his self-made-man way he thought the right to represent should be earned and paid for by the ambitious individual. But he was turned down. Even the Pennsylvania Assembly paid 'compensation' for loss of earnings. There was no issue on which the Founding Fathers were more divided. Many 'gentlemen,' such as lawyers, found they could not hold office and make a living, so they demanded salaries, and then complained they were too low. Hamilton, though rich, spoke for them. John Adams had a high view of the dignity of public officials. When he was first sent to England as minister he refused to take a hand with the ship's pumps, like everyone else, 'arguing it was not befitting a person who had public status.' This claim, so un-American (one might think), makes one suppose that Adams would be against salaries. But he was not. He thought it was perks and privileges which produced evil in public men. Without salaries, he said, public office would become the monopoly of the rich. He thought it disgraceful that Washington had been allowed to serve as commander-in-chief without being paid. Jefferson shared Washington's view, adhering to what he called the 'Roman principle.' 'In a virtuous government,' he said, 'public offices are what they should be, burthens to those appointed to them, which it would be wrong to decline, though foreseen to bring with them intense labor, and great private loss.'[123] In general, the Southerners were against salaries, the Northerners in favor. The North won, and it was decided even senators should be paid. The amount was left to Congress, which fixed on $6 a day. It seemed high to critics, but then the first Congress met in New York City, where the cost of living was 'outrageous.' In any event congressmen were soon grumbling it was too little, as were senators, who thought they should be paid more than mere members of the House.

What nobody seems to have bothered much about was the cost of electioneering. This could be enormous in 18th-century England, up to £100,000 for a single contest, sometimes even more. Nor was it just an English problem. When George Washington was first elected a Virginia burgess in 1758, it cost him £40 for 47 gallons of beer, 35 gallons of wine, 2 gallons of cider, half a pint of brandy, and 3 barrels of rum-punch.[124] These electioneering costs were going up in both countries all the time and in England parliament was slowly coming to grips with the problem and disqualifying MPs for bribing electors with drink and money. It is curious, and disappointing, that the gentleman–politicians who created the United States did not tackle the problem of election-costs right at the start, and thus save their successors a great deal of trouble—and cash.

By agreeing to let each state send two senators to Congress, the Founding Fathers built states' rights into the representational process. The House, on the other hand, was to represent the people, and it was agreed that each state was to have at least one Congressman and not to exceed one for every 30,000 persons (excluding Indians not paying taxes and allowing for the three-fifths rule for slaves). A census was to take place every ten years to determine the numbers and thus the total and distribution of congressmen. In 1787, for the first Congress, there were sixty-five congressmen, Rhode Island and Delaware getting one each, Georgia and New Hampshire three each, New Jersey four, Connecticut and North and South Carolina five each, New York and Maryland six, Massachusetts and Pennsylvania eight each, and Virginia ten. But America was changing and expanding so fast that this allocation was out of date within a year or two. For one thing, more territories were clamoring to get statehood. Vermont had been declared independent in 1777 by delegates from areas originally called New Connecticut and it pinched bits of New Hampshire and New York, neither of which was ready to yield them. Settlers who wanted to get a valid title for their lands did not know which state to apply to. Vermont was virtually neutral during the Revolutionary War, though Britain withdrew any claim to its territory, and it considered signing a separate treaty with Britain and claiming a Swiss-style neutral status. It remained aloof until New Hampshire (1782) and New York (1790) withdrew their land claims. Then it applied to and joined the Union in 1791. So when the Congressional structure was reordered in 1793, as a result of the 1790 census, Vermont was given two seats.

There was a long and acrimonious row over the Virginia back-country—'that dark and bloody land' as it was (perhaps unfairly) called—eventually resolved when Virginia withdrew its claims and the new state of Kentucky was admitted in 1792 and given two seats. The Pennsylvania back-country, organized as the independent state of Franklin, and regarded by North Carolina as a rebellious, land-grabbing illegality, collapsed in 1788, and had to be reorganized by Congress as the Southwest Territory in 1790. Settlers poured in and it soon passed the 60,000 mark and was admitted as the state of Tennessee, though not till 1796. Hence, in the 1793 reconstruction, fifteen states were represented in Congress and the number of House seats was raised to 105, Virginia now getting nineteen, Massachusetts fourteen, Pennsylvania thirteen, and New York ten. The 1790 census revealed that the population of the United States was increasing even faster than optimists like Franklin guessed—it was now 3,929,827. Ten years later, at the end of the century, the census shows a jump to 5,308,483, which was a 35 percent growth in a decade, and double the 1775 estimate.

This rapid growth gratified many but alarmed some, including the elite. Franklin, who worried himself about the dangers of over-population a generation before Malthus systematized them, did not object to settlers of English descent breeding fast but was disturbed by the prospect of the Englishness of America being watered down by new, non-English, and non-white arrivals. It was one reason he objected to the slave-trade and slavery itself: 'Why increase the sons of Africa by planting them in America,' he asked, 'where we have so fair an opportunity, by excluding all blacks and tawnys, of increasing the lovely white and red?' His mind reaching forward as always, he feared a future world in which the white races, and especially the English, would be swamped:

> The number of purely white people in the world is proportionately very small. All Africa is black or tawny; Asia chiefly tawny; America (exclusive of the newcomers) wholly so. And in Europe the Spaniards, Italians, French, Russians and Swedes [sic] are generally of what we call a swarthy complexion; as are the Germans also, the Saxons only excepted, who with the English make the principal body of white people on the face of the earth. I would wish their numbers were increased . . . But perhaps I am partial to the complexion of my country, for such kind of partiality is natural to mankind.

He was not at all happy about the number of Germans coming to America, especially to Pennsylvania, where they tended to vote en bloc, the first instance of ethnicity in politics. 'Why should the Palatine boor be suffered to swarm into our settlements and, by herding together, establish their language and manners to the exclusion of ours? Why should Pennsylvania, founded by the English, become a colony of aliens, who will shortly be so numerous as to Germanise us, instead of us Anglicising them?' He wanted language qualifications 'for any Post of trust, profit or honor.' He also considered monetary rewards to encourage Englishmen to marry the German women, but dismissed the idea for 'German women are generally so disagreeable to an English eye that it wou'd require great portions to induce Englishmen to marry them.'[125] These views were by no means unusual among the founders. Neither Washington nor Jefferson wanted unlimited or even large-scale immigration.

Defining what constituted an American citizen was not easy. As early as 1776, New Hampshire and South Carolina, writing their new constitutions, laid down that all state officers must swear an oath 'to support, maintain and defend' the provisional constitution.[126] Six months later, Congress, in adopting independence, replaced loyalty to the crown by loyalty to the nation: 'All persons residing within any of the United Colonies, and deriving their protection from the laws of the same, owe allegiance to the said laws, and are members of such colony . . . [and] all persons, members of or owing allegiance to any of the United Colonies . . . who shall level war against any of the said colonies . . . or be adherents to the King of Great Britain . . . are guilty of treason against any such colony.'[127]

This did not settle what citizenship was, however. Indeed the term was then new and little understood. The assumption was that everyone belonged to his or her particular state and thence derived their citizenship of the United States, a view later categorized by Justice Joseph Story (1779–1845) of the Supreme Court, who laid down that 'Every citizen of a State is *ipso facto* a citizen of the United States.' Most states had citizenship rules of one kind or another. But what of immigrants coming to the country from outside? The federal Constitution of 1787 laid down a national standard of neutralization by Act of Congress. Several Acts were passed, in 1795, 1798, and again in 1802, before Congress felt it had got the formula right, the main difference being the length of residence required before the applicant got nationality—the first criterion, two years, was considered too short, the next, fourteen, too long, and finally five years was judged right. The federal Constitution, and the states, reserved citizenship to whites, implicitly excluding blacks (even if free) and still-tribalized Indians, regarded as belonging to foreign nations. White women were citizens except for voting purposes, a rule which was not changed till 1920. Blacks did not get automatic citizenship till 1868, Indians not till 1924.[128] But the most important point was that the new country, like the old colonies, continued to admit immigrants virtually without restriction, and they continued to come, in ever growing numbers.[129]

After five years, most immigrants got the vote, for, as a result of the Revolution, America was rapidly becoming democratic. The Founding

Fathers might insist on checks and balances and take precautions against 'the tyranny of the majority,' but though constitutions are made by educated elites, what actually happens on the ground is usually determined by ordinary people. Their demands, as citizens and taxpayers, turned on its head the Revolution slogan 'No taxation without representation.' If the King of England was not allowed to tax Americans without giving them representation, why should states tax any American citizen without giving him a vote about how his taxes were raised and spent? Most states readily agreed. In New York State the federalists, who generally opposed what one of their leaders, Chancellor James Kent, called 'the evil genius of democracy,' fought a determined rearguard action to retain a freehold property qualification, at any rate for the electors of the state Senate. Kent argued that, while everyone else was worshiping 'the idol of universal suffrage,' New York should set an example and maintain property as qualification because it was 'a sort of moral and independent test of character in the electorate, which we could get at in no other practicable mode,' and only voters of sound character could defend society against 'the onrushing rabble.' But he was answered that making distinctions between one set of Americans and another, especially one based on ownership of land, was 'an odious remnant of aristocracy,' a system of 'privilege,' running directly contrary to the principle that in a true republic 'there is but one estate—the people.' Kent was thus driven to fall back on the argument that property qualifications were needed to protect 'the farmers.' But that made farmers into a mere interest, and why should farming, as an interest, get more protection than any other?[130] Manning the barriers against democracy was a losing cause as early as the 1780s and by 1800 was a lost one. By 1790 five states permitted all males (in some of them only white males) the vote for some or all offices, provided they paid tax. These states, and others, increasingly recognized residency, rather than land-ownership, as the qualification for 'attachment' to the state, and most set the period as two years (some, one).

It struck Europeans as amazing that, after arriving, penniless, from a country where they could never have a vote at all, even if their ancestors had lived there a thousand years, and however rich they grew, they could get off a ship in New York, cross the Hudson to New Jersey, and exercise a vote the following year—in five they would be voting for the president. New Jersey was particularly free and easy. From 1776 it had given the vote to all 'worth' 50 pounds after a year's residence and election officials even permitted women to vote if they thus qualified (until 1809). The wartime inflation made the old property qualification pretty meaningless anyway, and states like North Carolina and New Hampshire, with poll-taxes and taxpayer qualifications, adopted near-universal male suffrage as a matter of course. By 1783 the eligible electorate in the states ran from 60 to 90 percent, with most states edging towards the 100 percent mark. New states, like Kentucky, automatically embraced universal white male adult suffrage when they were admitted, if not before. But while states rapidly enfranchised white males, they usually disenfranchised free blacks at the same time.[131] Rhode Island, true to its tradition of being odd man out, alone resisted the democratic flood. Its qualification of a $134 freehold—the dollar had been fixed by law in 1792—was enforced increasingly fiercely and half the male citizens were disenfranchised.

A remarkable letter has survived which gives an indication of how the arrival of democracy was seen by one highly intelligent American. It was written in 1806 to the Italian nationalist Philip Mazzei by Benjamin Latrobe, an Englishman who had settled in Philadelphia ten years before and had become America's first professional architect. He wrote:

> After the adoption of the federal constitution, the extension of the right of Suffrage in all the states to the majority of all the adult male citizens ... has spread actual and practical democracy and political equality over the whole union ... The want of learning and science in the majority is one of those things which strike foreigners who visit us very forcibly. Our representatives to all our Legislative bodies, National as well as of the States, are elected by the majority *unlearned*. For instance from Philadelphia and its environs we sent to Congress not one man of letters. One of them indeed is a lawyer but of no eminence, another a good Mathematician but, when elected, he was a Clerk in a bank. The others are plain farmers. From the next county is sent a Blacksmith, and from just over the river a Butcher. Our state legislature does not contain one individual of superior talents. The fact is, that superior talents actually excite distrust.

But Latrobe was not discouraged. America was about 'getting on,' and he was getting on very well. He admitted that 'to a cultivated mind, to a man of letters, to a lover of the arts [America might] present a very unpleasant picture.' But 'the solid and general advantages are undeniable.' 'There is no doubt whatsoever,' he concluded, 'that [democracy] produces the greatest sum of human happiness that perhaps any nation ever enjoyed.'[132]

Since the arrival of democracy made the 'tyranny of the majority,' feared by Jefferson, Madison, and others, a real threat, who was to protect minorities—or indeed the ordinary citizen confronted by the federal Leviathan? The Bill of Rights went some way. But that depended for its efficacy on enforcement by the courts. Considering the importance the Founding Fathers attached to the separation of powers, and their insistence that the judiciary, along with the executive and legislature, was one of the tripods on which government must rest, the Convention paid little attention to it. Indeed, perhaps the most important provision in the Constitution dealing with the judiciary came about by accident, and is a classic example of Karl Popper's Law of Unintended Effect. Luther Martin, the great states' rights champion, proposed that instead of a federal veto on state laws, federal laws and treaties should be 'the supreme law of the individual states,' whose courts 'bound thereby in their decisions, anything in the respective laws of the individual states to the contrary notwithstanding.' This obscure formulation was accepted unanimously and would have made state courts the authority, in each state, on questions of federal law. This would have been a decisive victory for the states, and altered the whole course of American history. But in subsequent wrangling over the judiciary, especially the provisions for inferior federal courts, the proposal was amended to make state constitutions, as well as law, subordinate to the federal Constitution and the laws and treaties enacted by Congress. This made all the difference in the world, though its importance does not seem to have been grasped at the time.

Indeed the Constitution really left the detailed provision for a judiciary to the first Congress, which in 1789 enacted the Judiciary Act. This law, written mainly by Oliver Ellsworth (1745–1807), the agile Connecticut lawyer who had earlier put together the 'Connecticut Compromise,' is a remarkable piece of work because it has remained virtually unchanged for over two centuries. It created a bottom tier of federal district courts, usually matching state lines, and a middle level of three circuit courts, composed of two Supreme Court justices plus a district judge, who traveled to hear cases twice a year. They heard appeals from district courts and gave a first hearing to cases involving different states—a system which endured until 1891. The Act also formally set up the Supreme Court, as envisaged in the Constitution, with one chief and five associate justices, nominated by the president and confirmed by the Senate. (It had changed size repeatedly, being reduced from six to five in 1801, increased to seven in 1807, to nine in 1833, to ten in 1863, reduced to eight in 1866 and increased to nine in 1869, but otherwise functioning in the same manner.) But Ellsworth's Act, probably inadvertently, gave the Supreme Court an additional right of great importance, the executive power of ordering federal officials to carry out their legal responsibilities.[133]

These aspects of the judiciary's role, however, were little pondered at the time. It is a serious criticism of the Founding Fathers that they devoted insufficient attention to the role judges might play in interpreting a written constitution, and took no steps either to encourage or to inhibit judicial review. The truth is, they were brought up in the English tradition of the common law, which the judges were constantly modifying as a matter of course, to solve new problems as they arose. They did not appreciate that, with a written constitution, which had never existed in England, judge-made law assumed far greater significance, with almost limitless possibilities of expansion, and should have been dealt with in the Constitution. As it was, and is, the American federal judiciary have always been, in a sense, a law unto themselves, evolving organically as, in their wisdom, they saw fit. The process began shortly after the Constitution came into effect. In England, law and politics had always been closely enmeshed, and America had followed that pattern. Until the second half of the 16th century, English governments had always been presided over by the Lord Chancellor, the head of the law, and only gradually had the judiciary and the government bifurcated, and even then incompletely, with the Lord Chancellor continuing to sit in the Cabinet, as he still does. The Founding Fathers decided on a complete and formal separation of powers but they did not follow the logic of this course and insist on separating, at a personal level, judicial sheep from political goats. Thus the early chief justices tended to be professional lawyer–politicians, who saw running the court as merely a step on a public ladder which might lead to higher things rather than as the culmination of a legal career placing the occupant high above all political temptations.

The first Chief Justice, John Jay, was primarily a politician, who resigned in 1795 to run for the governorship of New York. The second, John Rutledge (1739–1800), resigned before he could even be confirmed by the Senate, in order to take up what was then regarded as a higher post on the South Carolina Supreme Court. The third, Oliver

Ellsworth himself, served 1796–1800 but then resigned to take up a diplomatic post in Paris. One of the Supreme Court justices, Samuel Chase, engaged openly in politics while sitting on the bench. This applied lower down too. Of the twenty-eight judges on the federal district courts during the 1790s, only eight had held high judicial office, but all had been prominent politicians. There was, however, a strong desire, first articulated by Alexander Hamilton, that federal judges should stand above the political battle, should be primarily experts, dedicated to interpreting the law as the ultimate protection of the citizen's rights, rather than politicians engaged in the hurly-burly of making it. There was a complementary feeling among the judges themselves that they should be the new priests of the Constitution, treating it as the secular Ark of the Covenant and performing quasi-sacramental functions in its service. That meant a withdrawal from politics, into a kind of public stratosphere. This hieratic notion was gradually gaining ground in the 1790s, displacing the more robust view of the Revolutionary democrats that, in a republic, any citizen was fit to discharge any public duty, if voted into it. The federal judges, it began to be mooted, were 'special,' remote, godlike defenders of the public interest and the private rights of all, who sat in the empyreum.[134] But for this to become generally accepted doctrine, confirmed by events, we have to await the arrival of Chief Justice John Marshall in 1801. We will deal with that shortly.

In the meantime, what of the real priesthood, the real religion of the people? We have said nothing, so far, about the part played by the churches, or by Christianity, as such, in the constitution-making. As we have seen, America had been founded primarily for religious purposes, and the Great Awakening had been the original dynamic of the continental movement for independence. The Americans were overwhelmingly church-going, much more so than the English, whose rule they rejected. The Pilgrim Fathers had come to America precisely because England had become immoral and irreligious. They had built the 'City on the Hill.' Again, their descendants had opted for independence and liberty because they felt their subjugation was itself immoral and irreligious and opposed to the Providential Plan. There is no question that the Declaration of Independence was, to those who signed it, a religious as well as a secular act, and that the Revolutionary War had the approbation of divine providence. They had won it with God's blessing and, afterwards, they drew up their framework of government with

God's blessing, just as in the 17th century the colonists had drawn up their Compacts and Charters and Orders and Instruments, with God peering over their shoulders. How came it, then, that the Constitution of the United States, unlike these early documents in American history, lacks a religious framework, as well as a religious content? The only reference to religion in the document is in Article VI, Section 3, which bans any 'religious Test' as a 'Qualification to any Office,' and the only mention of God is in the date at the end—'in the Year of our Lord one thousand seven hundred and eighty seven.' Even the wretched irreligious English had an established church and a head of state crowned in a sacramental ceremony and a parliament which began its proceedings, each day, with a prayer. The American Constitution's first susbtantial reference to religion comes only in the First Amendment, which specifically rejects a national church and forbids Congress to make 'any law respecting an establishment of religion or prohibiting the free exercise thereof.' How do we explain this seeming anomaly?

There is no doubt that if the United States Constitution had been drawn up in 1687 it would have had a religious framework and almost certainly provided for a broad-based Protestantism to be the national religion. And if it had been drawn up in 1887 it would have contained provisions acknowledging the strong spirit of religious belief and practice in America and the need for the state to nurture and underpin it. As it happens, by a historical accident, it was actually drawn up at the high tide of 18th-century secularism, which was as yet unpolluted by the fanatical atheism and the bloody excesses of its culminating storm, the French Revolution. Within a very few years, this tide began to ebb, and the religious spirit to flood back. In France this was marked by Chateaubriand's epoch-making book *Le Génie du Christianisme* (1802), in Britain by the formation of the Clapham Sect in the early 1790s, and, the same decade in the United States, by the start of the Second Great Awakening. But in 1787, the new religious impulses, which were to make the 19th century into one of the great ages of religious activity and commitment, were not yet felt. Thus the actual language of the Constitution reflects the spirit of its time, which was secular.

It also reflects the feelings of some of the most prominent of the Founding Fathers. Washington himself, who presided at the convention, was probably a deist, though he would have strenuously denied accusations of not being a Christian, if anyone had been foolish enough to make them. He rarely used the word 'God,' preferring 'Providence' or 'the Great Ruler of Events.' He was not interested in doctrine. Sometimes he did not trouble himself to go to church on Sunday, rare in those days. He wrote of immigrants, whom he did not much like in general: 'If they are good workmen, they may be of Asia, Africa or Europe. They may be Mohammedans, Jews or Christians of any sect, or they may be atheists.'[135] He regarded religion as a civilizing force, but not essential. Later hagiographers, such as Parson Weems and Bishop William Meade, tried to make out Washington as more religious than he was—Weems relates that he was found praying in a wood near Valley Forge, by Quaker Poots, and Meade has him strongly opposed to swearing, drinking, dancing, theater-going, and hunting—all untrue. In fact Washington's adopted son, Parke Curtis, in his book about his father, has chapters on hunting and on balls and theater-visits.[136] The most notable aspect of Washington's approach to religion was his tolerance—again, unusual for the time.

Franklin was another deist, though much more interested in religion than Washington was. His approach to it reflected America's rising impatience with dogma and its stress on moral behavior. He wrote to his father in 1738: 'I think vital religion has always suffered when orthodoxy is more regarded than virtue; and the scriptures assure me that at the last day we shall be examined not on what we *thought* but on what we *did*; and our recommendation will be that we did good to our fellow creatures.'[137] In his characteristically American desire to hustle things along, he felt that religious practices simply took up too much time. He particularly disliked long graces before meals—one should be enough for the whole winter, he felt. He took the trouble to abridge the Book of Common Prayer, producing much shorter services—the time saved on Sunday, he argued, could be then spent studying improving books. His *Articles of Belief and Acts of Religion* (1728) contains a form of religious service he invented whose climax is the singing of Milton's 'Hymn to the Creator,' followed by readings from a book 'discursing on and exciting to *Moral Virtue*.' He summed up his faith six weeks before he died in a letter to Ezra Stiles, saying he followed the precepts of Christ while doubting his divinity, that he believed in a Supreme Being and 'doing Good to his other Children.'[138]

Of the Founding Fathers, the man least affected by religion was Jefferson. Some people indeed classified him not just as a deist but as an atheist. In 1800 the *New England Paladin* wrote that 'Should the infidel Jefferson be elected to the Presidency, the *seal of death* is that moment set on our holy religion, our churches will be prostrated and some infamous prostitute, under the title of the Goddess of Reason, will preside in the Sanctuaries now devoted to the worship of the Most High.'[139] But this was electoral propaganda. Jefferson was no more an atheist than the much maligned Walter Ralegh, whom he resembled in so many other ways too. And, strongly as he sympathized with the French Revolution, at any rate for a time, he deplored its anti-religious excesses. He believed in divine providence and confided to John Adams, in spring 1816: 'I think it is a good world on the whole, and framed on Principles of Benevolence, and more pleasure than pain dealt out to us.'[140] Jefferson and his follower Madison certainly opposed Patrick Henry's attempt to get the Virginia legislature to subsidize the churches, but in the whole of their long and voluminous correspondence, amounting to 2,000 printed pages, it is impossible to point to any passage, by either of them, showing hostility to religion. What they both hated was intolerance and any restriction on religious practice by those who would not admit the legitimacy of diverse beliefs.

Madison, unlike Jefferson, saw an important role for religious feeling in shaping a republican society. He was a pupil of John Witherspoon (1723–94), president of the New Jersey College at Princeton, and author of a subtle and interesting doctrine which equated the religious polarity of vice/virtue with the secular polarity of ethics/politics—politics understood in their Machiavellian sense.[141] Witherspoon seems to have given Madison a lifelong interest in theology. Letters to friends (not Jefferson) are dotted with theological points—he advised one to 'season' his studies 'with a little divinity now and then'—and his papers include notes on the Bible he made in the years 1772–5, when he undertook an extensive study of Scripture. He carried around with him a booklet, *The Necessary Duty for Family Prayer, with Prayers for Their Use*, and he himself conducted household prayers at his home, Montpelier. Deist he may have been, but secularist—no.[142]

The same can be said for the great majority of those who signed the Declaration of Independence, who attended the Constitutional Convention, and who framed the First Amendment. An investigation by the historian W. W. Sweet revealed that, of the last group, eight were Episcopalians, eight Congregationalists, two Roman Catholics, one Methodist, two Quakers, one a member of the Dutch Reformed Church, and only one a deist. Daniel Boorstin discovered that of the Virginians who composed the State Constitutional Convention, over a hundred, only three were not vestrymen. Among the Founding Fathers and First Amendment men were many staunch practicing Christians: Roger Sherman and Oliver Ellsworth of Connecticut, Caleb Strong and Elbridge Gerry of Massachusetts, William Livingston of New Jersey, Abraham Baldwin of Georgia, Richard Bassett of Delaware, Hugh Williamson of North Carolina, Charles Pinckney of South Carolina, John Dickinson and Thomas Mifflin of Pennsylvania, Rufus King of Massachusetts, David Brearley of New Jersey, and William Few of Georgia.[143]

Even the doubting and the unenthusiastic were quite clear that religion was needed in society, especially in a vast, rapidly growing, and boisterous country like America. Washington served for many years as a vestryman in his local Anglican church, believing this to be a pointed gesture of solidarity with an institution he regarded as underpinning a civilized society.[144] Franklin wrote to Tom Paine, rebuking him for dismissing religion as needless: 'He who spits in the wind spits in his own face . . . If men are wicked with religion, what would they be without it?'[145] Both men constantly brought providence into their utterances, especially when talking of America. They may not have thought of Americans as the chosen people, like the Pilgrim Fathers, but they certainly believed that America was under some kind of divine protection. John Adams shared this view. The day the Declaration of Independence was signed, Adams wrote to his Abigail: 'The second day of July 1776 will be the most memorable epoch in the history of America . . . it will be celebrated by succeeding generations as a great anniversary festival. It ought to be commemorated as the day of deliverance, by solemn acts of devotion to God Almighty.'[146] Adams had been deflected from a career in the church by a spasm of rationalism in 1755, but he never changed his opininon that belief in God and the regular practice of religion were needful to the good society: 'One great advantage of the Christian religion,' he wrote, 'is that it brings the great principle of the law of nature and nations, love your neighbour as yourself, and do to others as you would that others should do to you—to the knowledge, belief and veneration of the whole people. Children, servants, women as well as men are all professors in the science of public as well as private morality . . . The duties and rights of the citizen are thus taught from early infancy to every creature.' Madison held exactly the same view, and even Jefferson would have endorsed it. All these men believed strongly in education as essential to the creation of a workable republic and who else was to supply the moral education but the churches?

The Founding Fathers saw education and religion going hand in hand. That is why they wrote, in the Northwest Ordinance of 1787: 'Religion, morality and knowledge, being necessary to good government and the happiness of mankind, Schools and the means of education shall forever be encouraged.'[147]

It is against this background that we should place the opening sentence of the First Amendment, 'Congress shall make no law respecting an establishment of religion or prohibiting the free exercise thereof.' This guarantee has been widely, almost willfully, misunderstood in recent years, and interpreted as meaning that the federal government is forbidden by the Constitution to countenance or subsidize even indirectly the practice of religion. That would have astonished and angered the Founding Fathers. What the guarantee means is that Congress may not set up a state religion on the lines of the Church of England, 'as by law established.' It was an anti-establishment clause. The second half of the guarantee means that Congress may not interfere with the practice of any religion, and it could be argued that recent interpretations of the First Amendment run directly contrary to the plain and obvious meaning of this guarantee, and that for a court to forbid people to hold prayers in public schools is a flagrant breach of the Constitution. In effect, the First Amendment forbade Congress to favor one church, or religious sect, over another. It certainly did not inhibit Congress from identifying itself with the religious impulse as such or from authorizing religious practices where all could agree on their desirability. The House of Representatives passed the First Amendment on September 24, 1789. The next day it passed, by a two-to-one majority, a resolution calling for a day of national prayer and thanksgiving.

It is worth pausing a second to look at the details of this gesture, which may be regarded as the House's opinion of how the First Amendment should be understood. The resolution reads: 'We acknowledge with grateful hearts the many signal favors of Almighty God, especially by affording them an opportunity peacefully to establish a constitutional government for their safety and happiness.'[148] President Washington was then asked to designate the day of prayer and thanksgiving, thus inaugurating a public holiday, Thanksgiving, which Americans still universally enjoy. He replied: 'It is the duty of all nations to acknowledge the providence of Almighty God, to obey His will, to be grateful for His mercy, to implore His protection and favor ... That great and glorious Being who is the beneficent author of all the good that was, that is, or that ever will be, that we may then unite in rendering unto Him our sincere and humble thanks for His kind care and protection of the people.'[149]

There were, to be sure, powerful non- or even anti-religious forces at work among Americans at this time, as a result of the teachings of Hume, Voltaire, Rousseau, and, above all, Tom Paine. Paine did not see himself as anti-religious, needless to say. He professed his faith in 'One god—and no more.' This was 'the religion of humanity.' The doctrine he formulated in *The Age of Reason* (1794–5) was 'My country is the world and my religion is to do good.'[150] This work was widely read at the time, in many of the colleges, alongside Jefferson's translation of Volney's skeptical *Ruines ou Méditations sur les revolutions des empires* (1791), and similar works by Elihu Palmer, John Fitch, John Fellows, and Ethan Allen. *The Age of Reason* was even read by some farmers, artisans, and shopkeepers, as well as students. As one Massachusetts lawyer observed, it was 'highly thought of by many who knew neither what the age they lived in, nor reason, was.'[151] With characteristic hyperbole and venom, John Adams wrote of Paine: 'I do not know whether any man in the world has had more influence on its inhabitants or affairs for the last thirty years than Tom Paine. There can be no severer satire on the age. For such a mongrel between pig and puppy, begotten by a wild boar on a bitch wolf, never before in any age of the world was suffered by the poltroonery of mankind, to run through such a career of mischief. Call it then The Age of Paine.'[152]

As it happened, by the time Adams wrote this (1805), Paine's day was done. His 'age' had been the 1780s and the early 1790s. Then the reaction set in. When Paine returned to America in 1802 after his disastrous experiences in Revolutionary France, he noticed the difference. The religious tide was returning fast. People found him an irritating, repetitive figure from the past, a bore. Even Jefferson, once his friend, now president, gave him the brush-off. And Jefferson, as president, gave his final gloss on the First Amendment to a Presbyterian clergyman, who asked him why, unlike Washington and Adams (and later Madison), he did not issue a Thanksgiving proclamation. Religion, said Jefferson, was a matter for the states: 'I consider the government of the United States as interdicted from intermeddling with religious institutions, their doctrines, disciplines, or exercises. This results from the provision that no law shall be made respecting the establishment of religion, or the free exercise thereof, but also from that which reserves to the states the powers not delegated to the United States. Certainly no power over religious discipline has been delegated to the general government. It must thus rest with the states as far as it can be in any human authority.'[153] The wall of separation between church and state, then, if it existed at all, was not between government and the public, but between the federal government and the states. And the states, after the First Amendment, continued to make religious provision when they thought fit, as they always had done.

With the enactment of the Bill of Rights, the process of constitution-making was completed and it now remained to operate it. That had begun on the first Wednesday in January 1789, when presidential electors were chosen in the different states. They met on the first Wednesday in February to elect, and the first Wednesday in March was chosen 'for commencing proceedings under the said Constitution.' New York was the chosen place and that is where the first permanent government of the new nation began. Electors were chosen on the assumption that they would cast their votes for Washington, and that he was prepared to accept the duty. Where contests were staged they were for Congressional seats. The anti-federalists did not oppose Washington for president, who was elected unanimously. They did consider putting up George Clinton for vice-president, but in the event John Adams was easily elected. Washington was notified of his election in April and immediately set off for New York, though not before confiding to a friend: 'from the moment when the necessity [of accepting the presidency] had become more apparent, and as it were inevitable, I anticipated in a heart filled with distress, the ten thousand embarrassments, perplexities and troubles to which I must again be exposed in the evening of a life already nearly consumed in public cares ... none greater [than those produced] by applications for appointments ... my apprehension has already been too well justified.'[154]

Actually the patriarch protested too much. He was quite prepared to be president and made an excellent one. His disloyal and acerbic vice-president, Adams, might call him Old Muttonhead, but Washington knew very well what he was doing. And the first thing he had to do was to get the national finances in order. That meant appointing Hamilton the first Secretary of the Treasury, and giving him a free hand to get on with the job. The financial mess into which the new nation had got itself as a result of the Revolutionary War and the subsequent failure to create a strong federal executive can be briefly summarized. In 1775 Congress authorized an issue of $2 million of bills of credit called Continentals to finance the war. By 1779 (December) $241.6 million of

Continentals had been authorized. This was only part of the borrowing, which also included US Loan Certificates, foreign loans, bills of credit issued by the states, and other paper debts. Together they produced the worst inflation in United States history. By 1780 the Continentals were virtually valueless. When the war died down in 1782, Congress sent commissioners round the country to investigate claims against Congress and the army, and revalue them in terms of hard money. This produced a figure of $27 million. Under the Articles of Confederation Congress had no power to raise revenue. The states did, but were reluctant to come to Congress's aid. So throughout the 1780s interest payments on the debt were met only by issuing more paper. The new Constitution of 1787 of course gave Congress the power to tax, but by the beginning of 1790 the federal government's debt had risen to $40.7 million domestic and $13.2 million foreign. The market price of government paper (that is, proof of debt) had fallen to from 15 to 30 cents in the dollar, depending on the relative worthlessness of the paper. This consequence of inflation and improvidence was precisely the kind of disaster which was to hit all the Latin American republics when they came into being in the next generation, and from which some of them have never recovered to this day. Somehow, the United States, which sprang from the stock of England, whose credit rating was the model for all the world, had to pull itself out of the pit of bankruptcy.[155]

That was Hamilton's contribution to the founding of the nation. It was of such importance that it ranks him alongside Washington himself, Franklin, Jefferson, Madison, and Adams as a member of the tiny elite who created the country. All these men derived from John Locke the notion that security of one's property was intimately linked to one's freedom. Inflation, by making federal and state paper money valueless, was a direct assault on property and therefore a threat to liberty. John Adams wrote: 'Property must be secured, or liberty cannot exist.' Hamilton made the same point: 'Adieu to the security of property, adieu to the security of liberty.'[156] Believing this, Hamilton acted quickly. In January 1790 he submitted his 'Report on the Public Credit' to Congress. This was accepted after a lot of debate and one curious by-product of the negotiations was that the government accepted the proposal of Jefferson and his followers that the new national capital should be on the banks of the Potomac, in return for their support of Hamilton's proposals. Hamilton solved the problem of the Continentals, now valueless, by giving one dollar for every hundred, the embittered people who held them counting themselves lucky to get anything at all. The rest of the domestic debt, and the whole of the foreign debt, was fully funded, being rescheduled as long-term securities payable in gold.

Hamilton also had the federal government, as part of his scheme, shoulder the burden of the debts of the states, on the same terms. This was denounced as unfair, because some states had already paid their debts, and the less provident ones seemed to profit from their tardiness. But that could not he helped; the all-important object was to get rid of the burden of debt once and for all and start afresh with sound credit. That was also Hamilton's reply to those who said the scheme was expensive. So it was—but not in the long run. The United States was already a rich country. It was probably already, in per-capita terms, the richest country in the world, even though Britain was emerging as the world's first great industrial power. Being rich, it could afford to pay to restore its creditworthiness, which meant that in future America could borrow cheaply and easily on world markets to finance its expansion. Congress took Hamilton's word for it, the scheme was adopted, and events proved him right. In 1791, when the plan came into effect, American dept per capita (adjusted to 1980s dollars) was $197, a figure it was not to reach again until during the Civil War. By 1804 it had fallen to $120 and in 1811 to $49. As a result, when America wanted to borrow $11.25 million in 1803, to finance the Louisiana Purchase, and thus double the size of the country, it had no trouble at all in raising the money, at highly favorable rates. By then, of course, poor Hamilton was a back number (he was killed in 1804). But he had made the United States solvent and financially respected, and set it on the greatest arc of growth in history.[157]

The debt-funding was the first of Hamilton's policies to be put forward because it was the most urgent. But he followed it with three other reports to Congress, on the excise, on a national bank, and on manufactures. To raise money to fund the debt and pay the expenses of the federal government, he had already imposed, in 1789, an import tariff on thirty commodities, averaging 8 percent *ad valorem* and a 5 percent rate on all other goods. Added to this, he proposed in 1791, and Congress agreed, to an excise tax, chiefly on whiskey. This was a dangerous move. The frontiersmen, all of whom made whiskey, and who treated it as a kind of currency—almost their only cash—saw this as an attack on their very existence. They did not see why they should pay taxes anyway, since they regarded their intermittent warfare with the Indians, fought on behalf of all, as discharging their duties to the nation in full. They were armed, aggressive, and self-righteous—most of them were poor too—and they hated the Excise Act just as much as their fathers had hated the Stamp Act. Violence and refusal to pay began in 1791 and became habitual. In July 1794 law officers tried to summon sixty notorious tax-evaders to trial at the federal court in Philadelphia. The result was a riot: the mob burned the chief tax-collector's home and killed a United States soldier. There were open threats to leave the Union. Governor Mifflin of Pennsylvania refused to send in the militia, as Hamilton had requested. The Treasury Secretary, backed by the President, decided to treat the violence as treason-rebellion, and Mifflin's behavior as a defiance of federal law and a challenge to the new constitutional order. Hamilton demanded, and the President agreed, that 15,000 militiamen from not only Pennsylvania but Maryland, Virginia, and New Jersey, be called up and deployed. Under the command of General Henry Lee, and with Hamilton breathing fire and slaughter in attendance, 12,900 men—a larger force than Washington had ever commanded—marched across the Alleghenies in the autumn. The rebels, faced with such an enormous army, naturally melted away, and Hamilton had great difficulty in rounding up a score of insurgents for punishment. According to Jefferson, who poured scorn on the entire proceedings, this was a case of 'the Rebellion that could never be Found.' Two 'ringleaders' were convicted of treason, but Washington spared them from hanging. Hamilton thought he had made his point and that the government had gained 'reputation and strength.'[158]

Following his reports on the debt and the excise, Hamilton introduced two more in 1791, on a national bank and on manufacturing industry. The bank was not a new idea. England had created a national bank in the 1690s, which had successfully acted as a lender of last resort and an underwriter of the national money supply. In 1781 Congress had chartered the Bank of North America as the first private commercial bank in the country and the first to get government incorporation. This was a scheme of Robert Morris, who, as superintendent of finance, had been Hamilton's predecessor. The Bank opened in Philadelphia in 1782 with Franklin, Jefferson, Hamilton, James Monroe, and Jay among its original stockholders and depositors. It paid Washington's army and buttressed the faltering finances of the government. Hamilton's plan was more ambitious. His Bank of the United States was more like the Bank of England, a true central bank chartered for twenty-one years, with a board of twenty-five, a main office, and eight branches, serving as the government's fiscal agent. Most of its stock was held by the government, which was also its principal customer. Jefferson protested that the Constitution made no provision for a central bank, and that in creating such a federal institution the government was acting *ultra vires*. He also protested, even more vehemently, against Hamilton's fourth report on manufactures. In effect, Hamilton, building on Adam Smith's *Wealth of Nations*, but going well beyond it, proposed that the federal government should deliberately and systematically promote the industrialization of the United States. Smith had opposed such state interference in the free-enterprise, capitalist economy as a throwback to mercantalism. Hamilton did not disagree in general, but thought that 'priming the pump' was necessary for a small, new nation, overshadowed by the manufacturing power of its former imperial ruler, Great Britain. He intended such help to be temporary, until American industry could stand on its own feet.

Jefferson and his friends protested against the scheme not on grounds of economic theory but for much more fundamental reasons. He believed that the new republic would flourish only if the balance of power within it was held by its farmers and planters, men who owned and got their living from the soil. His reasoning was entirely emotional and sentimental, and had to do with the Roman republic, where Cicero had made the same point. Farmers, he believed, were somehow more virtuous than other people, more staunch in their defense of liberty, more suited to run a *res publica*. Deliberately to create a huge manufacturing 'interest,' with thousands of money-grubbing manufacturers and merchants, clamoring for special privileges and tariffs, seemed to him the road to moral ruin. Hamilton scoffed at such (to him) puerile reasoning. But many important politicians, especially in the South, agreed with Jefferson. Patrick Henry, for instance, who was opposed to the centralization inherent in the Constitution anyway, linked the proposals for the creation of the central bank to what he called 'a monied interest.' 'In an agricultural country like this,' he remonstrated, 'to erect, and concentrate, and perpetuate a large monied interest [must be] fatal to American liberty.' It was 'the first symptom of a spirit which must either be killed, or will kill the Constitution of the United States.'

The farmers and planters of the South hated Philadelphia and its rich Quakers, they hated New York and its rich lawyers, and, most of all, they hated Boston and its rich merchants and shipowners, many of whom were already joining with the Northern churches in calling for an end to slavery throughout the United States. They noted that the Boston rich—the Cabots, the Lowells, the Jacksons, the Higginsons—were right behind Hamilton. These were the clever gentry who had bought the public paper at 15 or 20 percent and, thanks to Hamilton, had it redeemed at par. Farmers, large or small, had a long history of hatred for banks in the United States, which went back to the times when specie or currency of any kind had been hard to get hold of, and the British government had frustrated local attempts to create credit. Now, almost everything that Hamilton did further inflamed them. They were not impressed by Hamilton's triumphant claim that government issues were not floating over par—whom did that benefit, except the money-men? Nor did they think much of his promise that federal effort would be put into industrializing the South as well as the North—Eli Whitney's 1792 invention of the cotton gin, which immediately revolutionized cotton-planting, made such changes, in their view, unnecessary and undesirable. So two parties began to form in the new state—North versus South, farmers versus manufacturers, Virginia versus Massachusetts, states' rights men versus federalist centralizers, old versus new. Jefferson protested that he had no wish to found a party: 'If I could not go to heaven but with a party, I would not go there at all.' But that is what, in the 1790s, he did.[159]

It may be asked: was Jefferson the Leader of the Opposition then? No: he was the Secretary of State in Washington's administration. Strictly speaking, he was Hamilton's superior in the government pecking-order. In fact, Hamilton had more power. At this stage in the evolution of government, the Treasury ran everything not specifically covered by other departments. It ran the Post Office, for instance. It employed 325 people, more than half the federal civil service. Hamilton was always thinking of additional reasons for bureaucratic empire-building. Jefferson was jealous of him. Just as, in England, Pitt was a high-powered financial statesman, cold, hard, unemotional, and interested chiefly in efficiency, beloved of the City of London and the Stock Exchange, and Charles James Fox was a libertarian romantic, who did not care a damn for the price of Consols or the credit of the pound sterling, but who watered the Tree of Liberty with his copious tears, so Hamilton, America's Pitt, and Jefferson, America's Fox, were at opposite poles of the political temperament. It was characteristic of Hamilton that he deplored the revolutionary events in Paris, and entirely typical of Jefferson that he applauded them.

The difference between Hamilton and Jefferson was as much temperamental as intellectual. Jefferson came from a secure background of landowning privilege, going back generations. Hamilton's background was so insecure and, to him, mysterious that we know more about it than he did. He thought he was born in 1757; in fact it was 1755. What happened was this. His mother, Rachel Faucette or Faucitt or Fawcette or Fawcet or Foztet—it was spelt in a score of different ways, just as Ralegh, the founder of Virginia, was spelt in ninety-six different ways—married at sixteen an old fellow, John Leweine, Levine, Lavien, Lawein etc., said to be 'a smallish Jew.' At the age of twenty-one she left her husband and set up house with an itinerant Scotsman called James Hamilton, a drifter and failure, who promptly drifted away. In 1759 Levine sued for divorce, alleging 'several illegitimate children.' Divorce was granted but, under Danish law, which was then the common law of Nevis and the Leeward Islands, Rachel did not have the right to remarry. So Hamilton was never legitimized. As we have seen, his career as a self-made man was spectacular, but the illegitimacy ate into his soul. He hated poverty, which he equated with the forces of darkness, and therefore he avoided or tried to ignore or despised the poor, who reminded him of it.

Small, red-haired, blue-eyed, Hamilton had an intensity about him which made him both admired and genuinely feared. He gave his opinions with a frankness which, in America, was already becoming a political liability. '*Every man ought to be supposed a knave,*' he wrote, 'and to have no other end, in all his actions, but private interest. *By this interest we must govern him* and, by means of it, make him cooperate to public good, notwithstanding his insatiable avarice and ambition.' This was the gutter-philosophy of the West Indies, where the racial mix was a minestrone of buccaneering and sly skulduggery, and where it was war of every man against every man—and woman. It was distinctly unAmerican, where the inherent goodness and perfectibility of human nature was taken for granted. Hamilton despised this as 'hogwash.' He was infuriated by rich, well-born, secure men like Jefferson paying court to the poor, saying everyone was equal and acting upon it—or, more likely, pretending to act upon it. To Hamilton this was dangerous moonshine. He wanted an elite, an aristocracy, to keep 'the turbulent and uncontrollable masses' in subjection. But the elite had to be tough-minded, motivated by its own self-interest. The state had to conciliate it, as in England, by 'a dispensation of regular honors and emoluments,' to give it 'a distinct, permanent share of the government,' to keep 'the imprudence of democracy on a leash.'

Believing this, Hamilton wanted a permanent senate, elected indirectly and serving for life—rather like a House of Lords composed of life peers. He admired many other aspects of the British Constitution, the only one, he once said, which 'united public strength with personal security.' He was thus labeled 'reactionary,' and in a sense he was. But he was also a man of the future. He thought the state system a ridiculous relic of the past which might prevent America becoming 'a great empire.' Tiny states like Rhode Island and Delaware made no sense to him. He knew, from his wartime experience as Washington's right hand, how selfishly and stupidly the states could behave even in moments of great crisis.

Hamilton, like Jefferson, was a mixture of contradictions—a hater of democracy who fought for the republic; a humbly born colonial who loved aristocracy, a faithful servant of Washington who insulted the 'great booby' behind his back, a totally honest man who winked at the peculation of his friends, a monarchist who helped to create a republic, a devoted family man who conducted (and admitted) an amorous adventure. He told General Henry Knox, his Cabinet colleague and Secretary of War, 'My heart has always been the master of my judgment.' This was true in a sense: Hamilton was impulsive—why else would a man who hated dueling finally get himself killed in a duel? But his heart and Jefferson's were different. Hamilton's heart beat warmly in opposition to his deeply cynical view of mankind; Jefferson's was wholly in tune with his rosy, almost dewy-eyed idealization of human nature. Hamilton had been called 'a Rousseau of the right.' Jefferson admitted that Hamilton was 'a host in himself,' that he was 'of acute understanding, disinterested, honest, and honorable in all private transactions, amiable in society and duly prizing virtue in private life.' But he was, said Jefferson, 'so bewitched and perverted by the British example as to be under the conviction that corruption was essential to the government of a nation.' The truth is, Hamilton was a genius—the only one of the Founding Fathers fully entitled to that accolade—and he had the elusive, indefinable characteristics of genius. He did not fit any category. Woodrow Wilson was later to define him, with some justice, as 'A very great man, but not a great American.' But, if unAmerican, he went a long way towards creating, perhaps one should say adumbrating, one of the central fixtures of American public life—the broad conjunction of opinion which was to become the Republican Party.[160]

Equally, Jefferson's growing opposition to the whole trend of Hamilton's financial and economic policy and his constitutional centralism, gave birth to what was to become in time the Democratic Party, although in its first incarnation it was known, confusingly to us, as the Republican Party. The early 1790s were, in a sense, the end of American innocence, the undermining of the confident if unrealistic belief that the government of a vast, prosperous country could be conducted without corruption. Hamilton never had any illusions on that score—to him, man was always a fallen creature; he was a true conservative in that sense. But to the Jeffersonians it came as a shock. It should be said that Jefferson, true to his divided nature, was a man of pacts and compromises and deals. It was he who brokered the deal on funding the debt whereby the Southerners, in exchange for their votes, got the federal capital located on the Potomac. But, he would reply, there was no *personal* gain in this.

The first shocked awareness of personal corruption is reflected in the diaries of Senator William Maclay of Pennsylvania, who recorded the earliest instances of deliberate leaking of sensitive government information to favored individuals on the day the 'Report on the Debt' was published, January 14, 1790: 'This day the "Budget" as it was called was opened in the House of Representatives. An extraordinary rise of certificates has been remarked for some time past. This could not be accounted for, neither in Philadelphia nor elsewhere. But the report from the Treasury [proposing that certificates be repaid at par] reveals all.' The next week he noted: 'Hawkins of North Carolina said as he came up he passed two expresses with very large sums of money on their way to North Carolina for purposes of speculation in certificates. Wadsworth has sent off two small vessels for the Southern states on the errand of buying up certificates. I really fear that members of Congress are deeper in this business than any others.'[161] To members of the American political class, especially Southerners, this was the first real proof of the existence and unscrupulousness of the 'money power,' the huge, occult, octopus-like inhuman creature associated with banks—especially the central bank—New York, Boston, the North, England and the City of London, and unrepublican, unAmerican attitudes of every kind. This nightmare conspiracy would haunt generations of Democratic politicians in years to come, and it was in the 1790s that it made its first appearance.

Thus Washington's first administration, the earliest true government in America's history, was an incompatible coalition. Washington saw nothing wrong in this, at first. He was head of state as well as head of government and felt that his administration should reflect all the great interests in the nation, North and South, agriculture, commerce, and manufactures—should be in fact a geographical amalgam of the new nation. Of course there would be conflicts: how could it be otherwise with such a vast country? Washington agreed with South Carolina's William Loughton on the new state: 'We took each other with our mutual bad habits and respective evils, for better for worse. The northern state adopted us with our slaves, and we adopted them with their Quakers.' The United States was like a marriage. It was better, in Washington's view, to have interests reconciled and disputes mediated in Cabinet, than to have open warfare between parties, and government and opposition, as in England. Besides, the American system was different. Because of the separation of powers, members of the administration were not also members of Congress, answerable to it in person, as in the House of Commons. Washington found, in practice, that the more separate the powers were, the better. One aspect of government he handled personally was the making of treaties. When he was in the process of negotiating his Indian Treaty he agreed to appear before the Senate. This was a goodwill gesture because he did not need to under the Constitution. He was mortally offended when his explanations of what he was doing, instead of being accepted, were greeted by a decision to refer it all to a select committee, before which he was expected to appear again. He 'started up in a violent fret,' exclaiming 'This defeats every purpose of my coming here.' And he refused to do so, ever again. In future he referred treaties to Congress only when they were completed—as the Constitution provided.

With the powers separated, then, Washington judged it better to contain all the main factions within his administration. In practice, with Adams, as vice-president, speaking for New England, this meant he balanced Hamilton (New York) and the War Secretary Henry Knox (1750–1806), a vast, happy, fat man who had started out as a Boston bookseller but had become Washington's most reliable and trustworthy general—both of them ardent federalists—against Jefferson, Secretary of State, and Edmund Randolph (1753–1813), also from Virginia, who were both states' rights men. Those were the six men who met to decide government policy. These gatherings were called Cabinet meetings, as in England, though, as in England, they had no legal or constitutional standing. They took place at Washington's house, 39 Broadway, just round the corner from Wall Street.[162] It would be hard to overemphasize the informality and small scale of this first administration. Washington had to create it from scratch. That did not worry him, because he had had to do exactly the same thing with the army in 1776. The scale of the job was nothing: until the second half of the 1790s he employed more people on his Mount Vernon estate than in the whole of the central executive of his government.

We think of Washington as old when he became President but in fact he was only fifty-seven. He was a bit of an actor, however, and liked to play the Old Man card when convenient. Thus, with an awkward Cabinet meeting he would pretend to fumble for his glasses and say: 'I have already grown grey in the service of my country—now I am growing blind.'[163] He would also pretend to lose his temper. He was 'tremendous in his wrath,' wrote Jefferson, who was taken in. When his integrity was impugned at a Cabinet meeting he would 'by God them,' saying 'he had rather be on his farm than to be made Emperor of the World, by God! etc.' Jefferson said: 'His heart was warm in its affections, but he exactly calculated every man's value and gave him a solid esteem proportional to it.' He wrote 'better than he spoke' being 'unready.' Jefferson thought Washington pessimistic—he would give the Constitution a fair trial but was so distrustful of men and the use they would make of their liberty that he believed America would end up with something like the British Constitution.[164]

Jefferson argued that Washington's distrust of the people led him to erect ceremonial barriers between himself and the public—'his adoption of the ceremonies of levees, birthdays, pompous meetings with Congress [was] calculated to prepare us gradually to a change he believed possible.' That strikes the historian as nonsense, especially if he compares it with the fantastically elaborate preparations Bonaparte was to make a decade later for precisely that end. Washington did not have an elaborate household—only fourteen in all. His secretariat was tiny. He had to borrow money to set the whole thing up, as it was. It is true he bowed instead of shaking hands. But that was his nature—he had always done it. Jefferson later accused Washington, at a public ball, of sitting on a sofa placed on a dais, almost like a throne. But he had this only on hearsay and it was probably untrue.[165] It is true also that, as President, he gave grand, dull dinners, of many courses. The sharp-tongued Senator Maclay recorded: 'No cheering ray of convivial sunshine broke through the cloudy gloom of settled seriousness. At every interval of eating or drinking, he played on the table with a knife and fork, like drumsticks.'[166] But then Maclay had a nasty word about everyone—Adams was 'a monkey just put into breeches,' Gouverneur Morris was 'half-envoy, half-pimp,' Madison (only five feet four inches) was 'His Littleness.'

And, finally, it is true that when traveling as President—he made two extensive progresses, to the North and to the South—Washington cut an unusual figure by American standards. His white coach was second-hand but had been recently rebuilt by Clarke Brothers of Philadelphia for $950, his coachman was a tall, well-built Hessian called John Fagan, who sat on a leopard skin-covered box, and he traveled with Major Jackson, his ADC, his valet, two footmen, a mounted postillion riding behind, plus a light baggage waggon and five saddle-horses, including his favorite charger, Prescott, a magnificent white mount of sixteen hands who had been with the President on many a bloody and dangerous occasion. This equipage arrived in localities and towns at a cracking pace with many a trumpet blast, to the delight of the locals, for whom it was their only glimpse of a president in the whole of their lives. Jefferson seems, in retrospect, more of a New England puritan killjoy than a Virginia gentleman for protesting at this modest display. Nor did it save the President from occasional great discomfort and even peril to his life on several occasions during these official journeys, including a near-drowning on crossing the Severn a mile from Baltimore—'I was in imminent danger from the unskilfulness of the hands and the dullness of her sailing,' he recorded crossly—and a plunge, white coach and all, into the Ocquoquam Creek.[167] Traveling around rough-hewn America in the 1790s it was impossible for anyone, however grand, to keep his dignity for long. The wonderful thing about Washington was that, even in the midst of travel, or while listening to an endless series of fifteen toasts (plus speeches) at a rustic dinner in Maryland, he retained the respect of all. One of his staff, Tobias Lear, said he 'was almost the only man of an exalted character who does not lose some part of his respectability on an intimate acquaintance.'[168]

Despite the differences within the Cabinet, and the stealthy emergence of two great parties in the state—both of them represented in it—there was general agreement that Washington's presidency had been a success. Both Adams and Jefferson, on behalf of North and South, and both factions, strongly urged Washington to stand again. That might not have been decisive, for in 1792 Washington was almost painfully anxious to return to Mount Vernon. But he was persuaded by the ladies. Washington responded strongly to intelligent, perceptive women. He preferred them even to clever, able young men like Hamilton. His favorites were Henrietta Liston, the sweet and intuitive wife of the Scotsman Robert Liston, the British envoy, and Eliza Powell, wife of the former mayor of Philadelphia, Sam Powell. In 1790 the national capital had been removed from New York to Philadelphia (where it remained until Washington itself began to emerge in 1800) and Mrs Powell wanted her grand friend to preside there in state. So she persuaded the President to lean to the side of duty rather than inclination, and her wiles tipped the balance.[169] Mrs Liston may have helped to sway him too—she took the view, as did most of the British elite, that Washington was a 'sensible' man, unlike some of the Revolutionaries, a man whose 'good feelings' and 'bottom' added 'respectability' to America as a negotiating partner and possible future friend.

During his second term, Washington leaned more heavily on the federalists and took less trouble to conciliate the others. A break with Jefferson was probably inevitable, as Washington's monumental patience wore thin. Towards the end of Washington's first term, Madison, identified as Jefferson's closest political associate, had emerged virtually as leader of the opposition in Congress. In 1791, even before the election, the two men had gone on a so-called 'botanizing expedition' up the Hudson, where they had conferred with a motley group of malcontents—Aaron Burr (1756–1836), a sharp-faced New York lawyer, enemy and opponent of Hamilton there, who was using an organization called the Sons of St Tammany to build up a factional city machine; George Clinton (1739–1812), son of an Irish immigrant, the fiercely oppositional governor of the state; and various members of the Livingston family, a grand New York dynasty who, for reasons mysterious to Hamilton and the President, aligned themselves with the 'rabble.' This was the first party-political convention in American history, for the opposition New Yorkers formed a coalition with the states' rights Virginians and one result of the new alliance was a decision to bring Madison's old classmate, Philip Freneau, to Philadelphia to run the opposition newspaper, the *National Gazette*. His editorials infuriated the President.

What brought matters to a head was the outrageous behavior of the increasingly radical and bloodthirsty government in France, and in particular of their irresponsible ambassador. On November 19, 1792, before Washington's second term had even begun, the sansculottes in

Paris issued a Revolutionary decree declaring, 'War with all kings and peace with all peoples.' Edmond Charles Genet, an excitable *enragé*, as the Paris extremists were labeled, arrived to implement it so far as America was concerned. When Britain and France went to war soon afterwards Washington hastily declared America's neutrality. But that was not Genet's idea or, at first, Jefferson's. Genet arrived in Philadelphia with a clap of broadsides from the Revolutionary frigate *L'Ambuscade*, a dwarfish, dumpy man with dark red hair, coarse features, and a huge mouth from which issued forth a constant stream of passionate oratory in seven languages. Without even waiting to present his credentials he summoned the Americans to 'erect the *Temple of Liberty* on the ruins of *palaces and thrones*.' The mistake was characteristically French, to assume they are always the first to think of anything new. Genet forgot that America had already erected its own temple of liberty and had no palaces and thrones left to ruin.

Of course there were extremists in America—transatlantic Jacobins. Oliver Wolcott, Hamilton's Assistant Secretary at the Treasury and a federalist pillar, sneered at 'our Jacobins' who 'suppose the liberties of America depend upon the right of cutting throats in France.' Such people made up the patriotic French Society, one of over thirty such organizations which sprang up. Freneau's newspaper office at 209 Market Street, Philadelphia, almost under Washington's indignant nose, was a kind of headquarters to all of them, and to Genet's posturings. The French envoy set about recruiting men to join the French armed forces and to man privateers to prey on British commerce. He boasted to his Paris superiors: 'I excite the Canadians to break the British yoke. I arm the *Kentukois* and propose a navel expedition which will facilitate their descent upon New Orleans.' Annoyed by Washington's indifference to his cause, soon turning into active hostility, he threatened to 'appeal from the President to the People.'[170]

Jefferson, who had at first welcomed the 'French monkey,' now turned from him in embarrassment, found himself with a migraine—a recurrent complaint of Jefferson's in moments of crisis and perplexity—and took to his bed. Washington, outraged by Genet's threats, found his Secretary of State, instead of administering an instant rebuke to the envoy of France and demanding his recall, unavailable and engaged in what looked like malingering. He wrote to Jefferson in fury: 'Is the minister of the French Republic to set the Acts of this Government at defiance, *with impunity*, and then threaten the executive with an appeal to the people? What must the world think of such conduct, and of the government of the United States for submitting to it?' Jefferson found himself obliged to offer his resignation, just before a Cabinet meeting which decided that Genet must be recalled and during which the President exploded in fury at a satire in Freneau's newspaper entitled 'The funeral of George Washington' and depicting 'a tyrannical executive laid low on the guillotine.' Washington 'By-Godded' them all, said he would 'rather be in his grave' than President, and accused the opposition—eying Jefferson—of 'an impudent desire to insult him.'[171]

As it happened, Genet never left. The purging of the Girondins and the triumph of the 'Mountain' in Paris suddenly put him in danger of the guillotine himself and he begged to be allowed to stay. Washington gave him grudging permission and he promptly married the daughter of George Clinton, became a model citizen in upstate New York, and lived to read the first volume of George Bancroft's monumental *History of the United States* in 1834. Jefferson was replaced by Randolph, originally supposed to be a supporter of the deposed Secretary of State, now increasingly (according to him) a mere creature of the President: 'the poorest chameleon I ever saw, having no color of his own and reflecting that nearest to him. When he is with me, he is a Whig. When with Hamilton, he is a Tory. When with the President he is what he thinks will please him.' But Randolph did not last long. An intercepted French diplomatic dispatch, deliberately fed to Washington by his Secretary of State's enemies, appeared to reveal him soliciting French bribes in return for bending American policy in the direction of Paris. Washington fell into the trap, treated Randolph with great deviousness and duplicity—he could be very two-faced when he chose—and suddenly pounced and accused him of treason: 'By the eternal God . . . the damndest liar on the face of the earth!' Randolph had no alternative but to go, instantly, though it shortly became clear—and historians have since confirmed—that he was guiltless of anything except a little boasting to the French that he was the man in the administration who called the tunes. Washington realized too late he had made a mistake and inflicted an injustice on an old colleague, and the whole episode sickened him of politics.[172] As his second term drew to an end, there was no doubting the finality of his determination to retire for good.

Although Washington's administration demonstrated, especially towards its close, that the rise of party was irresistible, that bipartisan politics, however desirable, simply did not work and that, in the utopian republic, it would 'never be glad confident morning again,' it was on the whole a remarkable success. Not only did it restore the nation's credit, repay its debts, construct a workable financial system, and install a central bank, it also steered the country through a number of tricky problems. In 1789 the nation for the first time was alerted to possible responsibilities in the Pacific northwest when an Anglo-Spanish dispute over fur-trading rights on Vancouver Island ended in the Nootka Sound Convention (1790). Washington, while keeping the country neutral, laid down the policies which were to become America's norm in this part of the world and eventually to lead to a peaceful partition of the northwest, between the United States and British Canada, which eliminated Spain (and Russia) completely.

Washington also pursued a cautiously neutral policy during the first phase of the great war which pitted the crowns of Europe against Revolutionary France from 1793. He took the opportunity to send Chief Justice Jay to London to tie up the loose ends left by the Treaty of Paris a decade before. Jay's Treaty (1794) was treated by Washington's critics—including Jefferson—as an absurd victory for British diplomacy. It was nothing of the sort. It provided for British evacuation of the northwest posts, which had allowed Canadian traders to control the fur-routes and prevented full settlement of the Ohio Valley; it opened a limited West Indies trade for American vessels; it gave America a 'most favored nation' status in British trade; and, in general, it gave a boost both to America's own exports and commercial trading and to British exports into the United States, thus swelling Hamilton's revenues from import duties. It was one of those commercial treaties which enormously benefited both signatories while hurting neither, and the opposition outcry in Congress—mainly inspired by pro-French sentiment—makes little sense to the historian today. On the basis of Jay's Treaty, Washington sent his minister in England, Thomas Pinckney (1750–1828), to Madrid to negotiate a comparable arrangement with Spain. Pinckney's treaty secured major concessions—Spain's acknowledgment of America's boundary claims east of the Mississippi and in East and West Florida and, equally important, America's right of access to and transit through New Orleans, the strategic port at the mouth of the Mississippi. By these two treaties, in fact, all the last remaining obstacles to full-scale American westward expansion into the Ohio and Mississippi valleys were removed.

At the same time, the last years of the 1780s and the Washington administration saw an enormous increase in the maritime commerce of the United States. American ships penetrated the West Indies on a large scale, first trading with the Dutch and French Islands, then after Jay–Pinckney with the Spanish and above all the British colonies. In 1785 the *Empress of China*, the first American trader to penetrate the Far East, returned from Canton to New York, followed two years later by the Salem-based *Grand Turk*. This coincided with the opening up of the New England–northwest (Oregon) route by Captain Robert Gray (1755–1806), the great American trader and circumnavigator in 1787–90, whose pioneering activities in Oregon were the foundation for all American's subsequent claims to the area. It started a valuable triangular commerce: New England manufacturers to the northwest Indians, their furs to China, and then China tea to Boston. When Washington took office in 1789, an observer noted that of forty-six ships in Canton, eighteen were American; when Washington stood for a second term, the China trade had doubled and, when he finally left office, it had trebled.

Internal economic activity boomed correspondingly in the Washington years. Hamilton's policy of encouraging manufactures was not built on nothing. When Franklin got back to Philadelphia from Paris in 1785, he was astonished at the changes—new stagecoach routes, coal, iron, and woolen industries flourishing, frantic speculation everywhere. The states issued major charters to thirty-three companies—and huge enterprises were set up to build key bridges, turnpikes, and canals. In 1787 the first American cotton factory was built at Beverley, Massachusetts. The next year the first woolen factory followed at Hartford, with a £1,280 capital raised on the open market in £10 shares. Steam was coming and in 1789, already, John Fitch was experimenting in Philadelphia with a working steamship. Washington did not want America to become a manufacturing country like Britain any more than Jefferson did, and for the same reasons, but he was a realist and knew it was coming. He was also a military man who knew how important it was for the United States to have modern military equipment, including the latest warships and cannon, and how closely this was linked to military capacity. So, with all due misgivings, he backed Hamilton's industrial policy, and it was during his presidency that America achieved takeoff into self-sustaining industrial growth.[173]

Washington gave a public valediction to the American people by means of a farewell address, the text of which filled an entire page of the *American Daily Advertiser* on September 19, 1796. There is a bit of a mystery about this document. Washington wrote a rough draft of his declaration, intended as his political testament and considered advice to the nation, in May, and sent it to Hamilton for his approval. Hamilton rewrote it, and both men worked on the text. So it is a joint venture, from two men who had been intimately associated for twenty years and knew each other's thoughts. Some of the phrases are clearly Hamilton's. But the philosophy as a whole is his master's. The result is an encapsulation of what the first President thought America was, or ought to be, about.

He has three main points. He pleads at length, and passionately, against 'the baneful effects of the Spirit of Party.' America, he says, is a country which is united by tradition and nature: 'With slight shades of difference, you have the same Religion, Manners, Habits and Political Principles. The economies of North and South, the eastern seaboard and the western interior, far from dividing the nation, are complementary.' Differences, arguments and debates there must be. But a common devotion to the Union, as the source of 'your collective and individual happiness,' is the very foundation of the state. Central to this is respect for the Constitution: 'The Constitution which at any time exists, till changed by an explicit and authentic act of the whole People, is sacredly obligatory on all.' The fact that the people have 'the power and right to establish Government' presupposed 'the duty of every individual to obey it.' Hence, 'all obstructions to the execution of the Laws, all combinations and associations, under whatever plausible character, with the real design to direct, control, counteract or awe the regular deliberation and action of the constituted authorities, are destructive of this fundamental principle, and of fatal tendency.'

This is a very strong statement of the moral obligations of all citizens to comply with the decisions of duly constituted government, enforcing the laws constitutionally enacted by Congress. It was a solemn reminder by Washington, as the result of eight years' experience as chief executive, that America was a country under the rule of law. With the law, it was everything; without the law, it was nothing. And it was well that Washington made it in such forceful terms. Future presidents were able to take courage from it when dealing with powerful acts of defiance—Andrew Jackson when confronted with South Carolina's claim to the right to nullify federal law and Abraham Lincoln when faced with the unconstitutional act of secession by the South. The statement was typical of Washington's understanding of American government—its range was severely limited but, within those limits, its claims (under God) were absolute.

Second, Washington stressed the wisdom of keeping clear of foreign entanglements. He was proud of the fact that he had kept the United States out of the great war engulfing Europe, though under pressure from both sides to join in. America must seek 'harmony' and 'liberal intercourse' with all nations. It must trade with all on terms of equality. It must maintain 'a respectable defensive posture,' underwritten 'by suitable establishments' (of force). It might form 'temporary alliances for extraordinary emergencies.' But in general the United States must pursue its global course with friendship—if reciprocated—to all, enmity and alliance with none. Isolation? Not at all. Independence—yes.

Finally, Washington—in the light of the dreadful events which had occurred in Revolutionary France—wished to dispel for good any notion that America was a secular state. It was a government of laws but it was also a government of morals. 'Of all the dispositions and habits which led to political prosperity,' he insisted, 'Religion and Morality are indispensable supports.' Anyone who tried to undermine 'these firmest props of the duties of Men and Citizens' was the very opposite of a patriot. There can be no 'security for property, for reputation, for life if the sense of religious obligation *desert* the oaths which are the instruments of investigation in the Courts of Justice.' Nor can morality be maintained without religion. Whatever 'refined education' alone can do for 'minds of peculiar structure'—he was thinking of Jefferson no doubt—all experience showed that 'national morality' cannot prevail 'in exclusion of religious principle.' In effect, Washington was saying that America, being a free republic, dependent for its order on the good behavior of its citizens, cannot survive without religion. And that was in the nature of things. For Washington felt, like most Americans, that his country was in a sense chosen and favored and blessed. Hence he would 'carry to the grave' his 'unceasing vows' that 'Heaven may continue to you the choicest tokens of its beneficence—that your Union and brotherly affection may be perpetual—and that the free Constitution, which is the work of your hands, may be sacredly maintained.'[174]

The whole stress of Washington's presidency, underlined by his farewell, was on the absolute necessity to obey the Constitution. As he said on many occasions, he did not seek or want any more power than the Constitution gave him; but, when needful, he did not want any less either. It should be obeyed in letter and spirit. America was the first major country to adopt a written constitution. That Constitution has survived, where so many imitations all over the world have failed, not only because it was democratically constructed and freely adopted by the people, but precisely because it has been obeyed—by both government and people. All kinds of paper constitutions have been drawn up, perfect in design and detail—the Constitution of the Soviet Union is the classic example—but have become nugatory because the government has not obeyed them and the people have therefore lost faith in their reality. Washington insisted that the executive must follow the Constitution in all things, and he expected Congress and people to do likewise. It was in this respect, above all, that the first President led America to an auspicious start.[175]

When Washington retired there were still fundamental aspects of the Constitution waiting to be brought to life, in particular the role of the judiciary. That began under the second President, John Adams. Cantankerous, unloved, and quarrelsome, Adams was not the best choice to succeed the eirenic and universally respected general. But he was very senior. He had been through it all. He had served as vice-president. He was also from New England, awaiting its 'turn.' In Philadelphia a kind of caucus of federalist politicians, mostly Congressmen, decided it had to be Adams. They added Pinckney's brother to the slate, partly because he was from South Carolina, and therefore balanced the slate, partly because his treaty was popular. Hamilton, neither eligible nor willing to run himself, did not like Adams and believed he would be difficult to manage. He preferred Pinckney and engaged in a furtive plot to have Southern votes switched and get him in ahead of Adams. But it misfired, and as a result the New Englanders dropped Pinckney. Adams won, by seventy-one electoral votes; but Jefferson, who 'stood' for the Republicans—he refused to allow the word 'ran' as undignified, preferring the English term—got almost as many, sixty-eight, and therefore became vice-president. Adams, quite liking Jefferson despite their differences, but not wishing to have him aboard, labeled Hamilton, whom he held responsible, 'a Creole bastard'—Adams' wife Abigail, more decorously, called him 'Cassius—trying to assassinate Caesar.'[176] Adams, despite his low opinion of Old Muttonhead, tried hard to maintain the continuity of his government, keeping on Washington's old crony Timothy Pickering as secretary of state (though eventually obliged to sack him) and promoting Hamilton's able deputy, Oliver Wolcott, to the Treasury. Adams even went so far as to keep up Washingtonian pomp, dressing for his inauguration in an absurd pearl-coloured suit adorned with a sword and a huge hat with cockade. But he was a fat little man, who 'looked half Washington's height.' For the first but by no means the last time in presidential history, his best physical and social asset was his splendid spouse.

Adams' presidency was dominated by one issue—peace or war? Could America stay out of the global conflict? On this point he was at one with Washington: at almost any cost, America should stay neutral. Adams underlined this section in the Farewell Address, and caused the whole to be read out every February in Congress, a tradition maintained until the mid–1970s when, in the sudden collapse of presidential authority after Watergate, it lapsed. It was Adams' great merit as president that he kept America out of the war, despite many difficulties and with (as he saw it) a disloyal Cabinet and vice-president. Jefferson worked hard to have the government come to the aid of France and republicanism. Hamilton, outside the government but with his creatures inside it, hoped to exploit the war by destroying what remained of the Spanish and French empires in North America. He called for an enormous standing army of 10,000 and got the aged Washington to lend a certain amount of support to the idea. Adams accused Hamilton of intriguing to be made head of it and proclaim a dictatorship of what he termed 'a regal government.' This was exaggeration. But it was true that he had visions of personally marching a large professional force through the Louisiana Territory and into Mexico, turning all these 'liberated lands' over to American settlement.[177] Adams thought this was all nonsense. He believed that all North America would fall into the United States' hands, like ripe plums, in the fullness of time, but it would be outrageous, and unrepublican, and anyway expensive, to conquer the continent now. Like England, he believed in 'wooden walls,' a strong navy (to protect New England trade), freedom of the seas, and 'holding the balance.' So he tried to keep the army small and build ships—in New England yards of course.[178]

Adams and his friends believed he was superbly, perhaps uniquely, qualified intellectually to be president. His crony Benjamin Rush wrote in his autobiography that Adams possessed 'more learning, probably, both ancient and modern, than any man who subscribed to the Declaration of Independence.' American children who grew up in the early 19th century were told that, except for Franklin, Adams was without an intellectual superior among the Founding Fathers. This may well have been true, and Adams' writings and letters are a wonderful brantub of sharp *aperçus*, profound observations, and fascinating conjectures.[179] His experience was unique. He had been a commissioner to France, 1777–9, then negotiator in Holland, 1780–2, had negotiated, with Jefferson and Jay, the Treaty of Paris, 1782–3, had been America's first envoy to Britain, 1785–8, and as vice-president had assisted Washington, as far as his short temper would allow, in all things. If ever a man had been trained for the First Magistracy it was Adams. But he was ill suited to the office. Though he earnestly strove to maintain himself above party and faction, he was a man of passionate opinions and even more emotional likes and dislikes, mainly personal. He thought Hamilton 'the incarnation of evil.' He did not believe Jefferson was evil but he considered him a slave to 'ideology.' This was Adams' hate-word. It seems to have been coined by a French *philosophe*, Destutt de Tracy, whom Jefferson admired greatly. At his vice-president's promptings, Adams read the man and had a good laugh. What was this delightful new piece of French rubbish? What did 'ideology' stand for? 'Does it mean Idiotism? The Science of Non Compos Mentisism? The Art of Lunacy? The Theory of Deliri-ism?' He put his finger instantly on the way that—thanks to Jefferson and his ilk—ideology was creeping into American life by attributing all sorts of mythical powers and perceptions to a nonexistent entity, 'The People.' When politicians started talking about 'The People,' he said, he suspected their honesty. He had a contempt for abstract ideas which he derived from the English political tradition but to which he added a sarcastic skepticism which was entirely American, or rather Bostonian.

Adams believed that democracy—another hate-word—was positively dangerous, and equality a fantasy which could never be realized. He had no time for actual aristocracies—hated them indeed—but he thought the aristocratic principle, the rise of the best on merit, was indestructible and necessary. As he put it, 'Aristocracy, like waterfowl, dives for ages and rises again with brighter plumage.'[180] He noted that in certain families, young men were encouraged to take an interest in public service, generation after generation, and that such people naturally formed part of an elite. Unlike European aristocracies, they sought not land, titles, and wealth, but the pursuit of republican duty, service to God and man. He was thinking of such old New England families as the Winthrops and the Cottons—and his own. And of course the Adamses became the first of the great American political families, leaders in a long procession which would include the Lodges, Tafts, and Roosevelts. He brought up his son, John Quincy Adams, to serve the state just as old Pitt had brought up his son William, the Younger Pitt, to sit eventually on the Treasury Bench in the Commons.[181] All this was very touching, and the historian warms to this vain, chippy, wild-eyed, paranoid, and fiercely patriotic seer. But, whatever they think, presidents of the United States should not publicly proclaim their detestation of democracy and equality. That leaves only fraternity, and Adams was not a brotherly man either. He was much too good a hater for that.

The truth is Adams, like his enemy Hamilton, was not made to lead America, though for quite different reasons. Adams was very perceptive about the future. He had no doubts at all that America would become a great nation, possibly the greatest in the world, with a population 'of more than two hundred million.' But he did not want to see it. He hated progress, change, the consequences of science and technology, inventions, innovation, bustle. It was not that he despised science. Quite the contrary. Like most of the Founding Fathers, he admired and studied it. He believed in what he called the 'science of government' and he ingeniously worked into his constitutionalism a variety of scientific metaphors, particularly the principle on which the balance rested. Believing wholeheartedly in educating the young republic, he thought students should be taught science, both theoretical and applied: 'It is not indeed the fine arts our country requires,' he noted, 'but the useful, the mechanical arts.'[182] But he loathed the physical, visual evidence of life in a progressive country. 'From the year 1761,' he wrote to Rush, 'now more than fifty years, I have constantly lived in an Enemy's Country. And that without having one personal enemy in the world, that I know of.'[183]

This tremendously unAmerican dislike of progress was compounded by the purgatory Adams suffered from being dislocated. He was devoted to New England, especially 'the neighborhood of Boston' and his own town, Quincy. Being in Europe, as envoy, was an adventure and in some ways a delight for a man who has a taste for the Old World, but being forced to live outside New England in restless, self-transforming America was punishment. One feels for these early presidents, with their strong local roots, being sentenced to long exile in temporary accommodation before the White House was built and made cozy. Washington hated New York. Philadelphia was marginally better but was then the biggest city in the entire New World, dirty, noisy, and anathema to a country gentleman. Before his presidency was over, Adams was compelled to leave Philadelphia to set up his government shop in the new, barely begun capital of Washington, where the vast, endless streets, which mostly contained no buildings of any kind, were unpaved, muddy cesspools in winter, waiting for summer to transform them into mosquito-infested swamps. Washington in fact is built on a swamp and, then and now, specialized in gigantic cockroaches, which terrified Abigail. She was often ill, and demanded to be sent back to Quincy, and Adams used the excuse of tending her to hurry there himself and try to conduct government from his own house.[184] He found the business of creating a new capital commensurate with America's future profoundly depressing, laying it down that the country would not be 'ready for greatness' in 'less than a century.' One has a vivid glimpse of Adams, towards the end of his presidency, sitting in the unfinished 'executive mansion,' still largely unfurnished and requiring 'thirteen fires' constantly replenished just to keep out the cold and damp, surrounded by packing-cases and festooned with clothes-lines that Abigail used for drying the wash.[185]

However, before leaving the presidency, which, as we shall see, he did most reluctantly despite all its discomforts, Adams made a selection of vital significance, perhaps the most important single appointment in the whole history of the presidency. John Marshall (1755–1835) was a Virginian frontiersman, born in a log cabin on the frontier. Like many early Americans he combined a modest background with honorable lineage, being of old stock, related to the Lees, the Randolphs, and the Jeffersons. His father was prominent in state politics. Marshall fought in the Revolution, but as a result of the crisis he had little formal education apart from a brief spell at William and Mary College. But he set up as a lawyer in Richmond—the Americans were never inhibited by the trade union restrictions of the English Inns of Court system from nailing their name-plates to the wall—and soon showed, by his brilliant advocacy in court, that he was made for forensic life. He and Adams got on well together. They were both confirmed and cerebral federalists, believing in strong government, hierarchy based on merit and no nonsense about states' rights. They did not like nonsense in social life either, beyond the formality needed to keep the executive and the judiciary respected. Marshall, like Adams, was an elitist—but he did not look the part. He was tall, loose-limbed, and raw-boned, badly dressed, none too clean, a great gossip and gregarian. Wit he had too, and charm—in some ways he was a prototype for Lincoln.

Adams, in his desperate struggles to keep America out of the war, and especially to avoid sliding into a war with France by sheer accident and bad luck—the French remained provocative and difficult—sent John Marshall, together with Pinckney and Elbridge Gerry, to Paris on an embassy. They got short shrift from Charles-Maurice Talleyrand, the atheist ex-bishop and aristocrat who was now the hired gun of the Revolutionaries in foreign affairs. He objected strongly to Jay's treaty as pro-British and forced the commissioners to deal with plebeian underlings, whom they referred to contemptuously as X, Y, and Z. The French understrappers demanded a 'loan' of $12 million francs as a condition of opening serious talks, accompanied by a further, personal 'gift' of $250,000 to Talleyrand himself. Pinckney is said to have replied: 'No, not a sixpence—millions for defense but not one cent for tribute.' (The last bit was *esprit d'escalier* and actually coined by Robert Harper, a brilliant dinner-orator and neologist who also named Liberia and its capital Monrovia.) As a result, undeclared war broke out and Adams' new navy—he had thirty-three warships by the end of the century—came in handy in engagements with French commerce-raiders in the West Indies and Mediterranean. Adams had been unhappy about his Secretary of State's handling of the XYZ Affair—he thought Pickering was being manipulated by Hamilton—and in 1800 he sacked him and replaced him by Marshall. Finally, on the eve of his own departure, he decided the best way he could perpetuate his spirit was by making Marshall chief justice. This worked very well, Marshall holding the office for thirty-four years, surviving four of Adams' successors and living to cross swords with the redoubtable Andrew Jackson, a man for whom Adams had a peculiar hatred.[186]

We must now look forward a little to assess the full significance of this remarkable man and his impact on American history. If one man can be said to have wedded the United States indissolubly to capitalism, and particularly to industrial capitalism, it was Marshall. Except for Hamilton, all the Founding Fathers, Adams included, were suspicious of capitalism, or suspicious of banks anyway; some hated banks. And the Southerners hated industry. Even Washington disliked Hamilton's report on manufactures. But Marshall approved of capitalism, he approved of banks, he approved of industry—the lot. He thought they were essential to the future wellbeing of the United States people and that therefore their existence must be guaranteed under the Constitution. It was, as he saw it, his job as chief justice to insure this. Marshall, like the Founding Fathers, put his trust in property as the ultimate guarantor of liberty. But, unlike the Fathers, he did not distinguish morally and constitutionally between types of property.

The Founders, particularly the Virginians, Washington, Jefferson,

Madison, Monroe, *et al.*, equated property, as a moral force, with land. Their views were articulated by John Taylor (1753–1824), like them a Virginia landowner who served in the Senate and published in 1814 a monumental work of 700 pages, *An Inquiry into the Principles and Policy of the United States*. Taylor distinguished between 'natural' property, such as land, and 'artificial property' created by legal privilege, of which banking wealth was the outstanding example. He saw the right to issue paper money as indirect taxation on the people: 'Taxation, direct or indirect, produced by a paper system in any form, will rob a nation of property without giving it liberty; and by creating and enriching a separate interest, will rob it of liberty without giving it property.' Paper-money banking benefited an artificially created and parasitical financial aristocracy at the expense of the hard-working farmer, and this 'property-transferring policy invariably impoverishes all laboring and productive classes.' He compared this new financial power with the old feudal and ecclesiastical power, with the bankers using 'force, faith and credit' as the two others did religion and feudality. What particularly infuriated Taylor was the horrible slyness with which financiers had invested 'fictitious' property, such as bank-paper and stock, with all the prestige and virtues of 'honest' property.[187]

Taylor's theory was an early version of what was to become known as the 'physical fallacy,' a belief that only those who worked with their hands and brains to raise food or make goods were creating 'real' wealth and that all other forms of economic activity were essentially parasitical. It was commonly held in the early 19th century, and Marx and all his followers fell victim to it. Indeed plenty of people hold it in one form or another today, and whenever its adherents acquire power, or seize it, and put their beliefs into practice, by oppressing the 'parasitical middleman,' poverty invariably follows. Taylor's formulation of the theory fell on a particularly rich soil because American farmers in general, and the Southerners and backwoodsmen in particular, already had a paranoid suspicion of the 'money power' dating from colonial times, as we have seen. So Taylor's arguments, suitably vulgarized, became the common coin of the Jeffersonians, later of the Jacksonians and finally of silver-standard Democrats and populists of the late 19th century, who claimed that the American farmer was being 'crucified on a cross of gold.' The persistence of this fallacy in American politics refutes the common assumption that America is resistant to ideology, for if ever there were an ideology it is this farrago.

Fortunately Marshall set his face against it, and he had the power—

or rather he acquired the power—to make his views law. His view of how the American Republic should function was clear and consistent. He had read Edmund Burke's *Reflections on the Revolution in France* as soon as it was published in America and it inspired in him a healthy revulsion against the mob which lasted till his dying day. The people might not always constitute a mob. But they were always to be distrusted as an unfettered political force. The role of the Constitution therefore was to fence the people in. In Marshall's analysis, the popular power in America was essentially vested in the states, for they had been the first, in his own lifetime, to enfranchise the masses. Hence he was not only a federalist but a centralist, who thought the primary role of the general government was to balance the power of the mob which was latent in the states. The Constitution may not have said this explicitly. But the thought was implicit in its provisions, and it was the role and duty of the federal judiciary to reveal the hidden mysteries of the Constitution by its decisions. Thus he asserted, for the first time, the right of the Supreme Court to play its full part in the constitutional process by its powers of interpretation. As he put it in one of his judgments, 'We must never forget it is a Constitution we are expounding . . . something organic, capable of growth, susceptible to change.' Marshall was a graceful persuader with a subtle and resourceful mind, fertile in sinewy arguments, expressed with a silver tongue and a pen of gold. He lived very close to his brethren during the six or eight weeks the court sat in Washington, all of them residing together in the same modest boarding house so that, as his biographer said, Marshall was 'head of a family as much as he was chief of a court.'[188] He was absolutely dominant among his colleagues, though less learned than some of them. During his thirty-four years as head of the court it laid down 1,100 rulings, 519 of which he wrote himself, and he was in a dissenting minority only eight times.[189]

Next to Burke, Marshall revered Adam Smith's *Wealth of Nations*. He was closer to its spirit than Hamilton, believing the state should be chary of interfering in the natural process of the economy. Left to themselves, and with the law holding the ring so that all were free to exert the utmost of their powers, industrious men and women were capable unaided of fructifying America's vast resources and making it the richest country on earth. It was capitalism, not the state, which would conquer, tame, and plant the Mississippi Valley and still further west. All it required was a just, sensible, and consistent legal framework so that entrepreneurs could invest their capital and skills with confidence. Marshall had none of Taylor's reluctance to acknowledge 'artificial' property. It was the market, not sentiment, which defined wealth, provided it were honestly acquired. It was the duty of the court so to interpret the Constitution that the rights of property of all kinds were properly acknowledged, and capitalism thus enabled to do its job of developing the vast territories which Almighty God, in his wisdom, had given the American people just as he had once given the Promised Land to the Israelites.[190]

In this work, Marshall saw it as his primary function to provide property with the security which (in his view) was increasingly threatened by the legislatures of the states with their one-man, one-vote democracy and their consequent exposure to the demagoguery of irresponsible and propertyless men. That meant making the muscle of the Supreme Court felt in every state capital—and indeed in Congress itself. He set the parameters for his work as early as 1803, when in *Marbury* v. *Madison* he asserted the constitutional power of the Court to engage in judicial review of both state and federal legislation and, if needs be, to rule it unconstitutional. Viewing, as he did, the Constitution as an instrument of national unity and safety, he claimed that it not only set forth specific powers but created its own sanctions by implied powers. These sanctions were particularly necessary when, with the spread of the suffrage, politicians made populist assaults on lawful property to curry favor with the mob. To Marshall it made little difference whether an actual rabble stormed the Bastille by force or a legislative rabble tried to take it by unconstitutional statute. His first great blow for property came in 1810 in *Fletcher* v. *Peck*, when he overturned the popular verdict by ruling that a contact was valid whatever ordinary men might think of its ethics.[191] Fourteen years later, in the key case of *Gibbons* v. *Ogden*, he struck a lasting blow for entrepreneurial freedom by ruling that a state legislature had no constitutional right to create a steamboat monopoly. This interpretation of the Commerce Clause (Article I, Section 8) of the Constitution insisted that the US Congress was supreme in all aspects of interstate commerce and could not be limited by state law in that area. He wrote: 'The subject is as completely taken from the state legislatures as if they had been expressly forbidden to act on it.'[192]

In 1819 alone there were three cardinal Marshall rulings in favour of property. Early in February his Court ruled, in *Sturges* v. *Crowninshield*, that a populist New York State bankruptcy law in favor of debtors violated the Constitution on contracts. The same month, in *Dartmouth*

College v. *Woodward*, the Court laid down that a corporation charter was a private contract which was protected from interference by a state legislature. The most important case came in March, in a battle between the state of Maryland and the federal bank, or rather its Maryland branch. In *McCulloch* v. *Maryland* the Court had to rule not only on the right of a state to tax a federal institution but on the right of Congress to set up a federal bank in the first place. The judgment came down with tremendous majesty on the side of the central power, and the lawful status of the federal bank, which thus survived and flourished, until the great populist Andrew Jackson—the rabble incarnate and enthroned, in Marshall's view—destroyed it.[193]

In the light of subsequent history, it is easy for us to applaud Marshall's work as saving the United States from the demagogic legislative and governmental follies which made property insecure in Latin America, and so kept it poor and backward. Marshall's rulings made the accumulation of capital possible on a scale hitherto unimaginable and he can justly be described as one of the architects of the modern world.[194] But it did not seem so at the time to Jefferson and his friends. To Jefferson's delight, John Taylor himself lambasted the Court's ruling in *McCulloch* as an 'outrageous' vindication of 'artificial' property. Taylor's pronouncement, wrote Jefferson, was 'the true political faith, to which every catholic republican should steadfastly hold.' He saw Marshall and his Court as the dedicated enemies of American republicanism: 'The judiciary of the United States is the subtle corps of sappers and miners constantly working underground to undermine the foundations of our confederated fabric. They are construing our Constitution from a general and special government to a general and supreme one alone.'[195]

However, it must not be thought the supporters of a strong central authority had it all their own way. On the contrary. Federalism, as a political movement, was a declining force round the turn of the century, precisely because it was a party of the elite, without popular roots, at a time when democracy was spreading fast among the states and thus beginning to determine the federal executive power too. Adams' valedictory appointment of Marshall as chief justice was a huge blow struck for the federalist principle but Adams was the last of the federalist presidents and he could not get himself re-elected. He was very much in two minds whether to run. Not only did he hate Washington and the horrible, damp presidential mansion, he also thought the job intolerable—the President, he warned his son (also in time an uneasy president), 'has a very hard, laborious and unhappy life.' He laid down: 'No man who ever held the office of president would congratulate a friend on obtaining it.' He ran a second time because he did not want Jefferson to hold the job. There was nothing personal in this: Jefferson was one of the few politicians whom Adams did not actually hold in contempt—liked him, in fact, albeit they were totally different in views and styles of life. But Adams thought Jefferson's view of the Constitution and role of government wholly mistaken—the two men were 'the North and South Poles of the American Revolution'—and he was terrified Jefferson's sentimentality would involve America in a war on France's side which would inevitably lead to conflict with Britain and the destruction of New England's trade.[196]

So Adams ran—and much good it did him. A few weeks before the election, Hamilton, his fellow-federalist and ex-colleague, published an extraordinary pamphlet, *A Letter from Alexander Hamilton Concerning the Public Conduct and Character of John Adams Esq, President of the United States.* It began, 'Not denying to Mr Adams patriotism and integrity and even talents of a certain kind,' and went on to assert that he was 'unfit for the office of Chief Magistrate,' on account of his eccentricity, lack of sound judgment, inability to persevere, 'vanity beyond bounds,' and 'a jealousy capable of discoloring every subject.'[197] The pamphlet was so violent that it has been described as an act of political suicide on Hamilton's part, indicating he was quite unsuited to high office himself. But there is no denying that it harmed Adams too. To be fair to Hamilton, he intended it for private circulation among federalist leaders but (as was foreseeable) it fell into enemy hands, in the shape of Aaron Burr, who promptly insured it had the widest possible circulation.

Adams was in a lot of other trouble in any case. In the age of the French Revolution, which had its unscrupulous agents and credulous sympathizers in every civilized country, America, like Britain, had felt obliged to take steps to protect itself. In 1798 Congress had passed, with Adams' approval, the Alien and Sedition Acts. These four measures limited freedom of the press and speech and restricted the activities of aliens, especially French and Irish. They were part of the paranoia of the decade, which infected both sides of the revolutionary argument and predictably led to ludicrous results. In the first case which came before the courts, Luther Baldwin of New Jersey was convicted and fined $100 for wishing that a wad from the presidential saluting-cannon might 'hit Adams in the ass.'[198] As in England, ordinary people cared little about such measures, which affected only the chattering classes. But Jefferson, albeit a member of the government, and his friend Madison, drafted a series of resolutions, passed by the Virginia legislature and copied in Kentucky, which asserted that the Acts were unconstitutional and that the states 'have the right and are in duty bound to interpose for arresting of the evil.' The proper remedy, they went on, was for individual states to proceed to the 'nullification' of 'such unauthorised acts.' This is the first we hear of the Doctrine of Nullification, which was to haunt the republic for decades to come.[199] At the time it had less public impact than the increases in taxation made inevitable by Adams' construction of a substantial navy, especially a direct tax on houses, slaves, and land, which hit farmers, planters, and city-dwellers alike, and even provoked a feeble insurrection known to historians as Fries' Rebellion.

The 1800 election is often referred to as the first contested presidential election but evidence of the contest is scarce. Jefferson, true to his determination to 'stand' rather than 'run,' remained at his home, Monticello, throughout. Adams, now toothless, was incapable of making a public speech. The issue was decided by Jefferson's standing mate, Burr, whose Tammany organization carried New York, the swing state. So Jefferson beat Adams by seventy-three votes to sixty-five. But Burr also got seventy-three votes and under the Constitution the House had to decide which of them was president. After much skulduggery, the federalists voted for Jefferson, after private assurances that he would allow many federalist office-holders to keep their jobs.[200]

Jefferson, the exalted idealist, thus began his presidency with a bit of a deal. Indeed it was his fate all his public life to be forced—some would say that he chose—to compromise in order to obtain his objectives. He was a means-justifies-the-end casuist. He owed his presidency not just to Burr, who was manifestly a political crook and the first machine-politician in America, but to Elbridge Gerry (1744–1814) of Massachusetts, who was the second, and, as governor of the state, the inventor of gerrymandering. Jefferson raises a lot of difficulties for the historian. He is fascinating because of the range of his activities, the breadth of his imaginative insights, and the fertility of his inventions. But his inconsistencies are insurmountable and the deeper they are probed the more his fundamental weaknesses appear. Jefferson suffered from what were clearly psychosomatic migraines all his life—

and many other ills, real and imaginary, too; he was a monumental hypochondriac—and these tended to increase, as the dislocations in his personality, beliefs, and practices became more pronounced.

Jefferson's fundamental difficulty can be simply explained: he was a passionate idealist, to some extent indeed an intellectual puritan, but at the same time a sybarite, an art-lover, and a fastidious devotee of all life's luxuries. From claret to concubinage, there was no delight he did not sample, or rather indulge in habitually. This set his views and practices in constant conflict. Slavery was a case in point: its dark shadow penetrates every corner of his long life. One should be very careful in judging the Virginia Founding Fathers without making the imaginative leap into their minds on this issue. Slavery, to those involved in it as planters, was not just a commercial, economic, and moral issue: it was an intimate part of their way of life. The emotional vibrations it set up in their lives (and in the lives of their household slaves) are almost impossible for us to understand. But we have to accept that they were subtly compounded of love and fear, self-indulgence and self-disgust, friendship and affection, and (not least) family ties. When Jefferson married the rich widow, Martha Wayles Skelton, and brought her to Monticello, then already a-building, it is likely that he had a black mistress installed there as a household servant. When Martha's father, John Wayles, died, she inherited 11,000 acres and fourteen slaves. Wayles had had a mulatto mistress, Betty Hemings, by whom he had quadroon children who, under the laws of Virginia, were slaves by birth. So Jefferson's wife was in intimate daily contact not only with her own servile half-brothers and sisters, one of whom at least worked in the house, but with her husband's concubine.[201] Some Southern white women put up with this kind of thing, others were deeply grieved, others seemed unconcerned. What Jefferson thought we do not know—in all his voluminous writings he never discusses his own sexual relations with black or colored women. But he was clearly torn in two. We know he came to hate miscegenation, as the source of endless misery for all concerned.[202]

He also hated slavery, feared it, reviled it, privately at least, and sought in vain both to curtail it publicly and to cut it out of his own life. His *Notes on the State of Virginia* (1781) is such an outspoken denunciation of slavery on almost every ground that he told James Monroe that he hesitated to publish it, because 'the terms in which I speak of slavery and of our Constitution [in Virginia] may produce an irritation which will revolt the minds of our countrymen against the reform of these two articles, and thus do more harm than good.' He argued that slavery was not just an economic evil, which destroyed 'industry,' but a moral one which degraded the slave-owners even more than the slave. He wanted outright abolition, and none of the future abolitionists from the North argued more fervently or more comprehensively against the 'peculiar institution.' Friends, including Virginians, urged him to publish and he did so, insuring that a number of copies were put in the library of William and Mary College, so that the young would read it.[203]

Though an emancipationist in theory, however, Jefferson did nothing in practice to end slavery, either as governor of Virginia or as the revisor of its law-code. Nor, as secretary of state, as vice-president, or as a two-term president, did he do anything effective to end the slave-trade. He accepted the Southern contention that emancipated slaves could never be allowed to live as freemen in the Southern states. The liberated blacks would have to form a separate and independent country—preferably in Africa—to which 'we should extend our alliance and protection.'[204] One reason Jefferson shared this Southern view was that he agreed with most Southern whites that blacks were quite different and in some ways inferior. They 'secrete less by the kidneys and more by the glands of the skin, which gives them a very strong and disagreeable odor.' They 'require less sleep.' Their sexual desires are 'more ardent' but lack 'the tender, delicate mixture of sentiment and sensation' displayed by whites. They are 'much inferior' in reason, though equal in memory. Jefferson said he had never heard of a black person who could paint a picture, write a musical composition, or 'discover a truth.' He thought it would not be possible to find any 'capable of tracing and comprehending the investigations of Euclid.' Jefferson, one need hardly say, was not a bigoted racist. One of the grand things about him was that he was always open-minded to new evidence. It is significant that he disagreed with virtually all Americans of his day in rating the Indians as the equals of the whites in ability.[205] And when he was sent specimens of mathematical work by Benjamin Banneker, a free black planter in Maryland, he not only altered his views on this point but gleefully sent the manuscript off to the Marquis de Condorcet, secretary of the Paris Academy of Sciences, saying he was 'happy to inform you that we have now in the United States a negro . . . who is a very respectable mathematician.' Jefferson hoped that more Bannekers would emerge to prove that any apparent inferiority of blacks 'does not proceed from any differences in the structure of the parts on which the intellect depends' but 'is merely the effect of their degraded condition.'[206] He did not, however, change his opinion that freed blacks could not remain in the South.

Nor did Jefferson ever get round to doing anything for his own slaves, such as emancipating them. The reason was pitifully simple: money. Jefferson was never in a financial condition to indulge his conscience. Indeed, in an unsuccessful attempt to increase the income from his estates, he actually bought more slaves. When one of his slaves ran away he offered a reward for his capture. When he was about to return from his embassy to Paris, and his black slave-cook wished to remain there as a freeman, Jefferson persuaded him to come back to Monticello as a slave—he could not afford to lose the cook's 'artistry.' He wrote: 'The whole commerce between master and slave is a perpetual exercise of the more boisterous passions, the most unremitting despotism on the one part, and degrading submission on the other . . . indeed I tremble for my country when I reflect that God is just, that his justice cannot sleep for ever.' But if Jefferson's principles were strong, his appetites were stronger. And his debts were stronger still. Jefferson borrowed money all his life and, however much he hated the English, his indebtedness to two large London banking houses steadily increased. It is a curious and not entirely explicable fact that Southern slave-holding and indebtedness went together. The fact that a ship, from Boston or London itself—or France—could easily call at the plantation wharf and deposit on credit the latest European delicacies and luxuries was a standing temptation few Southern gentlemen could resist. Jefferson's temptations were more complex than most of his peers, for in addition to French wines, brandies, liqueurs and cheeses, hams and pâtés, vintage port from Bristol, coats and shirts from Savile Row, and porcelain from Wedgewood and Doulton, there were endless books, some of them very expensive, accumulating to form the finest library, 15,000 volumes, on the western side of the Atlantic. All these, and the growing interest on the debts, had to be paid for by the sweat of his slaves.[207]

Jefferson's expensive tastes might not have proved so fatal to his principles had he not also been an amateur architect of astonishing persistence and eccentricity. Architecture always tells us a great deal about the political state of a nation. This maxim has never been better illustrated than in America during the last quarter of the 18th century and the first of the 19th. And, in this general illustration, Jefferson and his Monticello provide a vivid particular example. Even more than its growing wealth, the new self-confidence felt in America just before, during, and still more after the Revolution and the securing of Independence, expressed itself in ambitious building-programs by its planter aristocracy (and their city associates) who now saw themselves as a ruling class. As befits their Roman republican principles, their taste was overwhelmingly classical. They went back for models both to antiquity and to Renaissance reinterpretations of classical forms. In particular they looked to Palladio. His *Four Books of Architecture*, published (1738) in English translation, lavishly illustrated by his designs, must have been in more American gentlemen's libraries than any other book of its kind. Palladio popularized a two-story pedimented portico, with ionic columns on the lower level and doric columns above. He also favored the so-called 'colossal portico' where vast columns arise without interruption from the floor of the porch to the pediment and roof.

Classical villas were going up steadily in America in the years just before the break with England—the Longfellow House in Cambridge, Massachusetts (1759) for instance or Mount Pleasant (1763), on the Schuylkill, described by John Adams as 'the most elegant country seat in the Northern colonies.' Also on the Schuylkill was Landsdowne, erected by Governor Penn of Pennsylvania, the first to introduce Palladio's two-storied pedimented portico. It was widely imitated and, when Independence came, this flamboyant architectural device, and the still more impressive colossal portico, became, and remain, symbols of America's triumphant discovery of itself. Some of these swagger-houses were built from scratch. Others, like Washington's own Mount Vernon, had a huge portico added (1777–84). An even bigger swagger-portico was added to Woodlands, the magnificent mansion built by the politician William Hamilton in 1787–90. With the end of the war, the creation of the Constitution, and, still more, the establishment of an efficient central government and the recovery of American credit, the passion for villa-building intensified. The Schuylkill, near America's richest city, Philadelphia, like the Thames to the west of London, was soon dotted with these delectable edifices, every few hundred yards.

The Schuylkill villa-rush became a positive stampede in 1793 when the worst outbreak of yellow fever in America's history killed one in ten of Philadelphia's inhabitants. Between 1793 and 1810 scores of villas emerged, each with its own pleasure-gardens or landscaped park, so that, said a visitor, 'The countryside [near Philadelphia] is very pleasant and agreeable, finely interspersed with genteel country seats, fields and orchards, for several miles around.' That is exactly the impression the new American ruling class wished to convey. None more so than Jefferson, who studied and practiced not only statesmanship but architecture all his life. Unfortunately, his divided nature, the simultaneous existence in his personality of incompatible opposites, his indecisiveness, his open-mindedness and changeability, combined to turn his building activities, especially at Monticello, into a nightmare saga. His plan to create a Palladian villa of his own design first unfolded in 1768 and work continued for virtually the rest of his life, the building being finished, insofar as it ever was, in the winter of 1823-4.

It is just as well that Jefferson had no sense of humor: he constitutes in his own way an egregious comic character, accident-prone and vertiginous, to whom minor catastrophes accrued. Almost from the start, the house was lived in, and guests invited there, though it was, by grandee standards, uninhabitable. When Jefferson became president, work on the house had proceeded for over thirty years, but half the rooms were unplastered and many had no flooring. One guest, Anna Maria Thornton, was surprised to find the upper floor reached by 'a little ladder of a staircase . . . very steep' (it is still there). On the second floor, where she slept, the window came down to the ground so there was no privacy but it was so short she had to crouch to see the view. The entrance hall had a clock perched awkwardly over the doorway, driven by cannon-ball weights in the corners, and with a balcony jutting out at the back.

The house was full of ingenious but amateurish Heath Robinson devices such as this, many of which do not work to this day. The library consisted not of shelves but of individual boxes stacked on top of each other, a weird arrangement. The dining-room looked into the tea-room and was only closed off by glass doors, shut in cold weather. The Dome Room proved an insoluble problem. There was no way to heat it, as a chimney flue would have marred its external appearance—the whole point of its existence—so Jefferson could not install a stove. Hence it was never used. The ice-house, attached to the main building, must have been one of the most awkward structures ever devised. It was filled, unusually, by cisterns but they were riddled by leaks and in Jefferson's day only two out of four ever held water. The chimneys proved too low and blew smoke into the house; the fires smoked too and gave out little heat. Jefferson was too jealous of Count Rumford's fame to install a 'Rumford,' the first really elegant drawing-room fireplace, so much admired by Jane Austen. He insisted on producing his own design, which did not work. The bedrooms were mere alcoves. Jefferson was constantly being delivered the wrong wood, or too much wood, or too little wood, and when he got the right wood one of several fires destroyed the kiln for drying it. As originally built, his bedroom accorded him no privacy at all, a curious oversight considering he had a passion for being alone and unobserved. Thereafter the search for privacy became an obsession in the many changes of design, and in the end he built two large porticoes, which did not fit into the Palladian design at all and were merely screens for his bedroom. Contemporaries assumed they were there so that his alleged mistress, Sally Hemings, could slip in and out of his chamber unobserved. Whether this was so we cannot now judge because they were removed in 1890.[208]

His workmen, Messrs Neilson, Stewart, Chisholm, Oldham, and Dinsmore, required infinite patience as Jefferson changed his mind repeatedly. Often a finished bit had to be redone to accommodate a new gimmick Jefferson had just invented—a concealed miniature lift to haul wine up from the cellars into the dining-room, for example, or a mysterious pulley-system which, in theory, made the tea-room doors open of their own accord. On the other hand, Jefferson conveyed his ever changing instructions to them in copious letters, written on terms of complete equality. And many of the workmen were incompetent anyway. Richard Richardson, his carpenter–columnist, could not get the columns of the swagger-portico straight, despite many tries. Jefferson was very forgiving. He was also good-natured. When he was president, he was expected by Oldham to look after his petty financial affairs in Washington, and Jefferson cheerfully obliged.[209]

The total cost of the house over more than half a century must have been enormous but it is impossible to compute the exact or even an approximate sum. All his life, Jefferson kept accounts, lists, and records in overwhelming quantity, covering all his activities in minute detail. His financial records are particularly copious. Yet, as they do not epitomize or balance, they convey little useful information. With a bit of research, Jefferson could have discovered, down to the last cent, what he had spent on any day of his life. But he never knew what he was worth or how much he was in debt. As he told his secretary, William Short, his true financial position remained a mystery to him. It was in fact deplorable and grew steadily worse from the 1770s onwards. As the editors of his memorandum books put it, 'The daily ritual of recording pecuniary events gave Jefferson an artificial sense of order in his financial world.'[210] In this respect Jefferson's accounts were a microcosm of the present-day federal budget, listing every detail of expenditure in tens of thousands of pages and millions of words, which obscures the fact that the government is adding to the national debt at the rate of $10,000 a second.

The story of Jefferson's financial downfall is a melancholy one. He should have saved money when he was president, living free and earning $25,000 annually for eight years. But he left office more in debt than when he entered it, and over $10,000 more than he thought. He assigned $2,500 a year, the income from his Bedford estates—half his landed profits—to pay the debts off; but they mysteriously rose. In 1815 he negotiated with Congress to sell them his library for $24,000, to form the basis of the Library of Congress. But this cleared less than half his borrowings, which then began to rise again. It was not all his fault. In 1819 William Carey Nicholas, the rascally father-in-law of Jefferson's grandson, Jefferson Randolph, pressured him into endorsing notes for $20,000. The next year Nicholas defaulted and Jefferson became liable for the lot. This coincided with the financial collapse of 1819 which made it impossible for Jefferson to sell lands and slaves, now his only option. In his last years visitors noticed that Monticello was 'old and going to decay,' the gardens 'slovenly.' His attempt to sell it in a lottery failed and when he died his debts were over $100,000. Jefferson's original plan had been to give all his slaves manumission at his death. That had to be scrapped. Jefferson Randolph, as heir, felt he had no alternative but to sell his grandfather's 130 slaves in 1827, splitting up families and separating mothers and children in the process, to achieve the maximum cash total. The next year he tried to sell Monticello itself, but there were no bidders and the house was vandalized. It is a miracle that it survived at all. Happily in 1834 it came into the hands of the Levy family, who maintained it for ninety years until, in the fullness of time (1923), it was bought by the Jefferson Memorial Foundation for $500,000. Now it is restored and glorious and a Historic Home—and a remarkable monument to the divided nature and peculiarities of its illustrious begetter.[211]

In the light of this saga of debt, it is amazing that Jefferson was as good a president as he contrived to be. In fact he managed to reduce the national debt by 30 percent. This was no doubt mainly due to the continuing effects of Hamilton's refunding measures, but Jefferson's minimalist ideas of central government had something to do with it too. Once Jefferson took up office, all the ceremonial grandeur of the

Washington presidency, kept up by Adams, was scrapped. We hear no more of the white coach. Dress swords were discarded. Jefferson traveled on horseback and his clothes were plain, not to say slovenly at times. Not only was he unguarded, his house in Washington was open to all-comers. One visitor reported that he arrived at eight o'clock in the morning, without any letters of introduction, and was immediately shown into the President's study, where he was received with courtesy and left 'highly pleased with the affability, intelligence and good sense of the President of America.'[212]

What is perhaps even more remarkable is that Jefferson let it be known that anyone could write to him with their suggestions, observations, or complaints, and that their letters would receive his individual attention. All they had to pay was the cost of the paper and the ink, as Jefferson agreed to pay the postage on receipt. This was an astonishing concession, for depending on the distance, postage then cost from 8 to 35 cents for each sheet of paper, at a time when laborers worked for a dollar a day.[213] The President's generosity encouraged prolixity, many correspondents writing him letters of a dozen sheets or more. Though Jefferson had a secretary, he insisted on opening, reading, answering, and filing all these letters himself. As he never in his life threw anything away, they are all still in existence and many of them have recently been edited.[214] Jefferson's replies, registered on a smudged copier or traced by a more efficient polygraph of his own devising, have also survived.

The letters the President received were political ('Thomas Jefferson, you infernal villun'), supplication for office ('Could it be possible to Give a Youth of my Age the Appointment of a Midshipman in the Navy?'—this purporting to come from four-year-old Thomas Jefferson Gassaway), pleas from widows ('You will no Doubt think Me posest of a Deal of asureance for adressing you, but Neacesary has no Law'), requests for money ('The hope which is kindled from the very ashes of despair alone emboldens me to address you'), appeals from imprisoned debtors and victims of miscarriages of justice ('I Rote to you for assistance not for Relesement'), death threats which read as though they had been written by the young Tom Sawyer ('The retributive SWORD is suspended over your Head by a slender Thread—BEWARE!') and pure abuse ('Thomas Jefferson you are the damnedest fool that God put life into, God dam you'). Taken together, these letters give an extraordinarily vivid glimpse into American lives in the first decade of the 19th century. All except the merely abusive got a reply in Jefferson's hand, even anonymous writers receiving this courtesy provided they gave some sort of address. Some of the replies were long and detailed, some contained money, others embodied careful inquiry into a particular grievance or request. Jefferson was not the only great man to take trouble with correspondents. His contemporary the Duke of Wellington also replied to thousands of letters, most of them from strangers, in his own hand, often by return of post. But Jefferson's conscientious care is without parallel—he was a man of truly heroic civility.

Just occasionally the attention the President paid to his correspondence proved invaluable. He wrote: 'I consider anonymous letters as sufficient foundation for inquiry into the facts they communicate.'[215] On December 1, 1805 he received one such, signed 'Your Friend,' seeking 'to give you a warning about Burr's intrigues . . . be thoroughly persuaded B. is a new Catilina.' Burr, as Jefferson had long known, was an unscrupulous adventurer and he was most embarrassed to have such a rogue as his vice-president during his first term. He forced Burr to keep his distance and the only occasion when the Vice-President came into prominence was when he presided, *ex officio*, at the impeachment of Supreme Court Justice Samuel Chase (1741–1811). It was Jefferson's greatest grievance against his predecessor that Adams had filled up all the court vacancies with ardent federalists, some of them being appointed only days before he left office. Chase was particularly obnoxious to Jefferson's party and his overbearing manner and abusive remarks while judging cases arising out of the hated Alien and Sedition Acts led to a demand for his impeachment in 1804, which Jefferson foolishly encouraged. It is the only time Congress has ever attempted to remove a member of the Supreme Court in this way and the episode demonstrated painfully that impeachment is not an effective method of trying to curb the Court for political reasons. Burr did not distinguish himself and the process failed. He was, accordingly, dropped from the ticket when Jefferson was reelected, George Clinton being chosen instead.[216]

As it happened, even before the election Burr was secretly engaged in various anti-Union intrigues, notably a plan by Senator Timothy Pickering and Massachusetts hardliners to take New England out of the Union. They wanted New York with them too, obviously, and for this purpose it was necessary to get Burr elected governor of the state. But Hamilton frustrated this scheme on the grounds that Burr was 'a dangerous man and one who ought not to be trusted with reins of government.' These remarks got into print and Burr challenged Hamilton to a duel at Weehawken, New Jersey (July 11, 1804). Hamilton strongly disapproved of dueling but felt he could not in honor decline the challenge. His conscience, however, forbade him to shoot at his opponent. Burr killed him without compunction, thus removing from the chessboard of American power one of its most baroque and unpredictable pieces.

Burr went into hiding in Virginia, reemerged, went west, and there embarked on a series of plots to create a new, independent state from Spanish Mexico. Such schemes seem childish to us. But they were not uncommon as the Spanish-American empire disintegrated during these years and romantic adventurers abounded. (Not for nothing did the young Lord Byron consider joining in the scramble for pieces of Spain's rotting imperial flesh.) Burr went further, however, and sought to detach parts of Trans-Appalachian America to join his proposed kingdom. This was treason against the United States, and Jefferson, forewarned, had him arrested and charged. The trial took place in 1807 under Chief Justice Marshall, who, as we have seen, was no friend of the President. It was a highly partisan affair. To embarrass the President, Marshall allowed him to be subpoenaed to appear, and testify under oath. Jefferson refused, invoking, for the first time, executive privilege. Marshall countered by placing a narrow construction on the constitutional law of treason and Burr was acquitted. That was the end of him as a political figure, however, and the episode demonstrated that even a states' rights president like Jefferson was determined to uphold federal authority as far as it legally stretched.[217]

Indeed as president, Jefferson proved himself more assertive and expansionist than he would have believed possible in the 1790s. It was another instance of his contradictions. In the Western Mediterranean, where Barbary pirates from Algiers, Tunis, Morocco, and Tripoli preyed on Western shipping, Jefferson abandoned Adams' policy of following the British example, and paying tribute, and instead sent the ships Adams had built—and which he had opposed—to blockade Tripoli (1803–5) and teach it a lesson. He also sent a land expedition (1804) of American marines and Greek mercenaries across the desert, under the command of William Eaton, the US consul in Tunis—thus producing one theme of the American Marine Corps' marching-song.[218] The Arab beys were the largest-scale slave-merchants (of whites as well as blacks and browns) in the world and hitting them was one way Jefferson could work off his frustrations at not doing anything about American slavery. It certainly aroused the envy of Admiral Nelson, then British naval Commander-in-Chief in the Mediterranean, who was a passionate anti-slaver and was longing to have a crack at the beys. It was also the first example of America's willingness to take the initiative in upholding civilized standards of international behavior—an excellent portent for the future.

More astonishing still is the fact that Jefferson, who saw America's future as that of a medium-sized agrarian republic with no ambitions to great-power status, succeeded in doubling the size of the nation at a stroke. Spain's decision to transfer Louisiana back to France, which was first rumored in Washington early in Jefferson's presidency, immediately rang the alarm-bells. Spain's control of New Orleans and the outlet of the Mississippi was a constant irritant. But Spain was weak and could be bullied. France was the strongest military power on earth and might be tempted to recreate the North American empire it had ceded in 1763. 'Nothing since the Revolutionary War has produced such uneasy sensations through the body of the nation,' Jefferson wrote (April 1802); it was 'the embryo of a tornado.' He added, 'There is on the globe one single spot, the possessor of which is our natural and habitual enemy. It is New Orleans, through which the produce of three-eighths of our territory must pass to market.' His Secretary of State, Madison, agreed. The Mississippi, he wrote, is 'the Hudson, the Delaware, the Potomac, and all the navigable rivers of the United States, formed into one stream.'[219]

Jefferson instructed Robert Livingston, his envoy in Paris, to open immediate negotiations with the Bonapartist government to see whether there was any possibility of France's allowing the United States to mitigate the peril, or at least insure access to the sea through New Orleans, by some kind of territorial bargain or purchase. He sent James Monroe to Paris to assist in the deal—if there was one. The French still held Talleyrand's view that America was a rich cow, which could be milked, and it was the first time Washington was prepared to wave the Almighty Dollar in the greedy faces of foreigners. But Jefferson was gloomy about the outcome; 'I am not sanguine,' he wrote, 'in obtaining a cession of New Orleans for money.' Then, in April 1803, the French Foreign Minister, on Bonaparte's terse instructions, offered America the whole of Louisiana, the entire Mississippi valley, New Orleans— the lot—for $15 million, cash down. Jefferson could hardly believe his luck and immediately set about applying to the hated banks, the masters of 'artificial' property, for the money. The deal was concluded in time for the President to announce it on July 4, 1803, the twenty-sixth anniversary of the Declaration of Independence. Not only did it double the size of America, making it a country as large as Europe, it also removed the last doubts about western expansion and made it virtually certain that America would double in size again in the next few decades.[220] Never before, or since, in history has such an extraordinary territorial cash-bargain been concluded. The Americans were not sure even how much land they had got, but when Livingston asked the French to indicate the exact boundaries of their cession, Talleyrand sourly replied: 'I can give you no direction. You have made a noble bargain for yourselves and I suppose you will make the most of it.'[221] He was, of course, right. As it turned out, America got another 828,000 square miles and 1,000 million acres of good land. Jefferson's only doubt was the constitutionality of the purchase. His federal opponents indeed, reversing their usual view, claimed that the Constitution did not authorize the purchase of foreign territory. But Jefferson for once abandoned his constitutional timidity and begged Congress to accept.

Jefferson admitted privately he was breaking the Constitution, justifying himself in a letter to John Breckinridge in a characteristic means-justifies-the-end manner: If the French kept Louisiana America would have to 'marry ourselves to the British fleet and nation.' Hence:

> I would not give one inch of the Mississippi to any nation, because I see in a light very important to our peace the exclusive right to its navigation . . . the Constitution has made no provision for our holding foreign territory, still less for incorporating foreign nations into our Union. The Executive, in seizing the fugitive occurrence which so much advances the good of their country, have done an act beyond the Constitution. The Legislature, in casting behind them metaphysical subtleties, and showing themselves like faithful servants, must ratify and pay for it, and throw themselves on their country in doing for them unauthorised what we know they would have done for themselves had they been in a situation to do it.[222]

This is a very important statement in American history, showing that even a strict constitutionalist like Jefferson was prepared to dismiss the Constitution's provisions as 'metaphysical subtleties' if they stood between the United States and what would soon be called its Manifest Destiny to occupy the entire northern half of the hemisphere. After Louisiana, the rest of the United States' enormous acquisitions—or depredations, depending on the viewpoint—would follow almost as a matter of course. At all events Congress approved Jefferson's decision on October 20, 1803 and early the following year a territorial government was set up. Eight years later Louisiana was admitted to the Union, the first of thirteen states to be carved from this immense godsend.

That the Louisiana affair was not merely a fortuitous aberration in Jefferson's thinking is proved by his decision, even before the purchase was arranged, to ask Congress secretly to authorize and finance an expedition to explore overland routes to the American Pacific coast. He had nurtured this idea since boyhood and ten years before, as secretary of state, he had tried to persuade the French naturalist Andrew Michaux to explore 'a river called Oregon' and find 'the shortest and most convenient route of communication between the US & the Pacific Ocean.' He now commanded his secretary, Meriwether Lewis (1774-1809), to lead an exploratory team to sort out and map the concourse of huge rivers flowing westward on the other side of the watershed from the Mississippi–Missouri headwaters. Lewis picked his army colleague William Clark (1770-1838) to join him, and they assembled and trained a party of thirty-four soldiers and ten civilians outside St Louis in the winter of 1803, before setting off on a three-year journey. Thanks to a remarkable Shoshone Indian woman, Sacajawea (1786-1812), who acted as guide and interpreter, they crossed the continental divide safely, found the Columbia River and, on November 8, 1805, gazed on the broad Pacific. Lewis went back by the same route (with detours), Clark through the Yellowstone, and they met again at Fort Union, the junction of the Yellowstone and Missouri. They then went down the Missouri, arriving back at St Louis on September 23, 1806. Both reported back in triumph to the President: 'In obedience to your orders we have penetrated the Continent of north America to the Pacific Ocean, and sufficiently explored the interior of the country to affirm with confidence that we have discovered the most practicable route which does exist across the Continent by means of the navigable branches of the Missouri and Columbia rivers.'[223] It was one of the most successful and comprehensive geographical adventures ever undertaken, which brought back a mass of economic, political, military, scientific, and cartographical information recorded in copious journals and maps. Jefferson was delighted, as well he might be: the story of the West had begun. Five years later, John Jacob Astor (1763-1848), a German-born adventurer who had entered America in 1784, became a fur-trader, and formed the American Fur Company (1808) and the Pacific Fur Company

(1810), founded the first trading post, Astoria, on the Pacific itself (1811) in the Columbia estuary. Within a few months it had been reported in the leading St Louis newspapers that 'it appears that a journey across the Continent of N. America might be performed with a wagon, there being no obstruction in the whole route that any person would dare call a mountain.'[224] Thus the concept and the route of the Oregon Trail came into existence.

Since during his presidency Jefferson had in effect created the Deep South and laid the foundations of the West, it is disappointing to relate that his period in office ended in failure and gloom. But so it was, because neither he nor Madison knew how to steer the United States through the troubled waters of the Napoleonic Wars. The truth is, they were emotionally involved, a fatal propensity in geopolitics. In 1803 the renewal of the great war between republican France and the royalist coalition led by Britain made it possible for the United States to get Louisiana cheap but in other respects it was a disaster for a commercial and maritime power such as America had now become. Britain's victory at Trafalgar in November 1805, in which the Franco-Spanish battlefleet was destroyed, made it supreme at sea. Bonaparte's victories against Austria and Russia at Friedland (1807) put the whole of Continental Europe at his mercy. In order to destroy British exports, gold from which financed the resistance to his tyranny, he imposed what was known as the Continental System, a punitive embargo on British goods. The British responded with their Orders in Council which allowed British blockading fleets to impound even neutral ships, caught violating an elaborate set of rules designed to hit France and its allies commercially. Jefferson in turn passed the Non-Importation Act (April 1806), which banned most British goods and embargoed all non-American shipping.[225]

It is important to realize that all three parties were divided on these measures. The mechanics and economics of international trade were little understood. Policies, shaped in ignorance, often produced the opposite effect of that intended. Bonaparte's Continental System led to trouble with most of his allies and satellites and did his cause more harm than Britain's. The Orders in Council, ill understood and difficult to enforce, harmed Britain's trade most. The Non-Importation Act failed completely, though it certainly angered Britain. The commercial clauses of Jay's treaty expired in 1807 and Monroe, now envoy in London, failed to get sufficient backing from Jefferson and Madison to reach a settlement. The result was a series of incidents between British warships and United States vessels culminating in a naval battle off Norfolk in which the British frigate *Leopard*, searching for deserters serving on US ships, forced the American frigate *Chesapeake* to strike its colors, seized four men aboard, and hanged one of them.[226] The fury caused by this incident, visible from America's shore, was such that if Congress had been sitting war must have followed. Jefferson himself was confused. His intellect told him that Britain and America, being both major maritime and trading powers, had a mutual interest in enforcing the freedom of the seas and the free exchange of traffic and goods at all ports—something the Continental System challenged. The two powers should have worked out a sensible joint policy and renewed Jay's treaty on its basis. But all Jefferson's republican emotions tugged him in the direction of France, and his hatred of monarchy blinded him to the fact that Bonaparte's military dictatorship—adumbrating the totalitarian tyrannies of the 20th century—was an infinitely greater threat to individual liberties than Britain's constitutional and parliamentary crown.

Jefferson managed to keep America out of war for the time being, but in order to respond to the war-fever in some way he got Congress, in December 1807, to pass the Embargo Act, virtually without discussion, which effectively ended all American overseas commerce by forbidding US ships to leave for foreign ports. How Congress failed to throw out this absurdity is a mystery. While American ships remained in harbor, their crews idle and unpaid, smuggling flourished and British ships had a monopoly of legitimate trade. By a cunning piece of legal legerdemain Bonaparte impounded $10 million of American goods on the ground that he was assisting Jefferson's embargo. It was the most serious political mistake of Jefferson's entire career because it led the Northern shipping and manufacturing interests to assert, with some plausibility, that the government was being run in the interests of the 'Virginia Dynasty' and its slave-owning planters by a pack of pro-French ultra-republican ideologues.[227] The government was forced to capitulate, backtracking by getting Congress to pass the Non-Intercourse Acts (1809), which got some commerce going again but left everyone's feelings, at home and abroad, raw and inflamed.

The damage to Jefferson's reputation caused by the miseries of the embargo, and the often cruel and disreputable attempts to enforce it, is reflected in the angry letters which poured into his office and which he read with mounting distress. 'Take off the Embargo, return to Carters Mountain and be ashamed of yourself and never show your head in Publick Company again.' 'I am Sir a Friend to Commerce and No

Friend to your Administration.' 'Mr President if you know what is good for your future welfare you will take off the embargo.' 'Look at the Situation in the Country when you Took the Chair and look at it now. I should think it would make you sink with despair and hide yourself in the Mountains.' 'You are bartering away this Countrys rights honor and Liberty to that infamous tirant of the world (Napolien).' 'I have agreed to pay four of my friends $400 to shoat you if you dont take off the embargo.' 'Here I am in Boston in a starving condition . . . you are one of the greatest tirants in the whole world.' Jefferson endorsed some of these letters 'abusive' or 'written from tavern scenes of drunkenness.' But others were detailed and circumstantial accounts of the distress caused, such as one written on behalf of 4,000 penniless seamen in Philadelphia—'Sir we Humbly beg your Honur to Grant us destras Seamen Sum relaf for God nos what we will do.' Some were from destitute seamen's wives claiming that their children were without bread. There were over 300 petitions signed by many hands, multiple threats—one from '300 yankee youths between 18 & 29'—'If I dont cut my *throat* I will join the English and fight against you. I hope, honored sir, you will forgive the abrupt manner in which this is wrote as I'm damn'd mad.' One of many desperate letters says the writer has been forced to steal food to feed his children and intends to 'take to highway robbing.'[228]

Jefferson, who had been an optimist up to the turn of the century, was now gloomy, shaken, and demoralized. During his final months in office, government policy disintegrated, with desperate legislative expedients passing backwards and forwards between the two Houses of Congress, and between Congress and executive, in confused attempts to get off the hook of the embargo. Finally, under the pretense of standing up to both France and Britain, Congress passed the Non-Intercourse Act (1809), which effectively repealed the embargo. Jefferson wearily signed it into law on March 1, writing to his friend Pont de Nemours: 'Within a few days I retire to my family, my books and my farms [at Monticello] . . . Never did a prisoner, released from his chains, feel such relief as I shall on shaking off the shackles of power.'[229] The truth is, he had virtually ceased to be in charge of affairs during the last few months of his presidency, and he left office a beaten man.[230]

But worse was to come. Madison had been preparing for the presidency all his life. First-born son of the wealthiest planter in Orange

County, he came from the summit of the civilized Virginia gentry. He had been elaborately educated, especially under the great Witherspoon at Princeton. He had studied history, political theory, and economics all his life, as well as the classics. He had known Jefferson since 1776 and the two men's intimate correspondence is a political and literary education in itself. Small, industrious, moderate, soft-spoken, seeing all sides of the question, trying always to conciliate and reach the golden mean, he was 'a model of neo-classical self-command,' seeking a dream, as the poet Robert Frost has said, 'of a new land to fulfill with people in self-control.' He had served in the House of Burgesses and in Congress, helped draft the Virginia constitution and its Statute of Religious Freedom, assisted at the Mount Vernon Conference, the Annapolis Convention, and the Constitutional Convention, where, more than any other man there, he was the author of the US Constitution itself. He had written twenty-six of the *Federalist Papers* and was the principal architect of the Bill of Rights. He had been a notable leader of the Jeffersonians in Congress and had served his master as secretary of state. He even, unlike Jefferson, had a sense of humor. When, as secretary of state, he had to entertain the Tunisian envoy, come to Washington to negotiate on behalf of the Barbary pirates, and granted the Arab's request for concubines for his eleven-strong party, he put down the cost as 'appropriations for foreign intercourse' (Jefferson was not amused). His wife, Dorothea or Dolley (1768–1849), was a beautiful girl from North Carolina who made herself the first great Washington hostess. But Madison proved a classic illustration of Tacitus' maxim *omnium consensu capax imperii nisi imperasset.* He was no good.[231]

It was a measure of Madison's executive awkwardness that, in his inaugural address, he set out his aims in a sentence he seemed unable to end and which eventually consisted of 470 words and proved difficult to read. At the reception afterwards in the F Street residence, Dolley was in ravishing beauty—'drest in a plain cambric-dress with a very long train, plain round the neck without any handkerchief, and beautiful bonnet of purple velvet, and white satin with white plumes—all dignity, grace and affability.'[232] Dolley, who was 'very much in Charge of the little Man' (Madison seemed tiny, though he had a large head), finished decorating the new Executive Mansion, spending $2,205 for knives, forks, 'bottle stands and andirons,' $458 for a pianoforte and $28 for 'a guitar.' She soon launched the White House's first 'drawing-room' receptions which, in her day, were celebrated—the men in 'black or blue coat with vest, black breeches and black stockings,' the ladies 'not remarkable for anything so much as for the exposure of their swelling breasts and bare backs.'[233] But, behind the glitter, there was endless confusion about how America should extricate itself from a maritime clash over rights which few people on either side of the Atlantic now understood. Madison wasted precious months, even years, in foolish expectation that the war would end or, more likely, that the parliamentary struggles in Britain would throw up a new ministry which would see things from America's viewpoint and scrap the measures against neutral ships.[234]

Actually, it was not so much the divisions in British politics which made a compromise so difficult as the pressures and increasing sectionalism in America. It might be true, as Washington had said in his Farewell Address, that East, West, North, and South in America had much more in common than points of difference. But, in the short-tempered atmosphere aroused by the long European war and its Atlantic repercussions, the differences seemed insuperable. New England had virtually everything in common with British maritime interests. But the further south and west you traveled the more opinion-leaders you found who wanted a showdown with Britain as the road to expansion. Was not Canada to be had for the taking? And Florida? And the West Indies? And this was Madison's constituency. These states had elected him in 1808, and when he was reelected in 1812 his dependence on the South and West was even more marked. His opponent, Clinton, carried New York (twenty-nine electoral college votes), Massachusetts (twenty-two votes), Connecticut, New Jersey, New Hampshire, and other smaller states, making a total of eighty-nine votes. Madison had Virginia (twenty-five), Pennsylvania (twenty-five), and a group of Southern and Western states, led by the Carolinas, Georgia, and Kentucky, making 128 votes. But the seven states which voted for Madison had a total of 980,000 slaves. These blacks had no voice in government whatever but each group of 45,000 added an electoral vote to the state where they were held, giving the cause of the South—and war—a total of twenty-one electoral votes. Thus the New England federalists claimed that the freemen of the North were at the irresponsible mercy of the slaves of the South.[235]

Even so, war might have been averted. On June 18, 1812 Congress completed the formalities necessary for a declaration of war on Britain. Two days later in Westminster, Henry Brougham's motion for repealing the Orders in Council had elicited from Lord Castlereagh, on behalf of the government, a statement that they were suspended. Unfortunately, an inexperienced American chargé d'affaires in London failed to get the news to Madison with the speed required. To judge by the letters which flew between Madison and his mentor Jefferson throughout 1812, while Madison drifted to war without much passion or eagerness, Jefferson believed that the time had come for a *réglement des comptes* with Britain and that America would make huge and immediate gains, especially 'the conquest of Canada.' With the advantage of hindsight, perhaps, we see both these two pillars of the republic, these upholders of white civilization, as irresponsible and reckless.[236]

When the war began it consisted of three primary forms of hostility: an American invasion of Canada; the naval war on the Great Lakes and on the high seas; and opportunities presented to the South and the American settler interest to despoil the possessions of Britain's ally, Spain, and Spain's and Britain's Indian dependants. Washington pinned its high hopes on the first. But the invasion was based on two misapprehensions. The first was that Canada was a soft target. It consisted of two halves—Lower Canada in the east, overwhelmingly French-speaking, and Upper Canada, to the west and north, English-speaking but thinly settled. Madison and Jefferson believed that the French-speaking Canadians were an oppressed and occupied people, who identified with Britain's enemy, France, and would welcome the Americans as liberators. Nothing could have been more mistaken. The French Canadians were ultra-conservative Roman Catholics, who regarded the French Republic as atheism incarnate, Bonaparte as a usurper and Anti-Christ, and who wanted a Bourbon restoration, one of the prime aims of British war-policy. The Quebec Act of 1774 had given the French community wide cultural, political, and religious privileges and was seen as a masterpiece of liberal statesmanship. They thought that, if the invasion turned Lower Canada into a member-state of the United States, they would be republicanized and Protestantized. In Upper Canada, it is true, there were only 4,500 British troops and a great many recent American settlers. The British Commander-in-Chief, Sir Isaac Brock, thought many of them were disloyal and that his only course was to 'speak loud and think big.' In fact the majority of the English-speaking Canadians were old Tories, anti-republicans, or their sons and grandsons. Canada had resisted the blandishments of American republicanism even in the 1770s; reinforced since then by 100,000 loyalists and their teeming descendants, and by many recent arrivals from Britain, they had no wish to change their allegiance.[237]

The illusions shared by the Virginian Dynasty are summed up in Jefferson's boast to Madison: 'The acquisition of Canada this year [1812] as far as the neighborhood of Quebec, will be a mere matter of marching, and will give us experience for the attack on Halifax the next, and the final expulsion of England from the American continent. Halifax once taken, every cockboat of hers must return to England for repairs.'[238] The second grand illusion was the quality of the American militia, about which Madison had boasted in his inaugural—'armed and trained, the militia is the firmest bulwark of republics.' In the first place, Massachusetts, Connecticut, and New Hampshire flatly refused to send their militias at all. New England did not exactly sit on its hands: it invested its money in London securities and did a good business selling supplies to the British forces. In return, the British declined to impose a blockade on New England, or even on New York until the end of the war. By that stage two-thirds of the beef consumed by the British Army was supplied from south of the border, chiefly Vermont and New York State.[239]

The forces Madison dispatched on his 'march' turned out to be a rabble. The militiamen had done well, in the Revolutionary War, defending their own homes, but outside their native districts their amateurishness became evident. They had no discipline. Every man selected his own ground to pitch a tent. No pickets were posted, no patrols sent out at night. Both the militiamen and the volunteers, who had somewhat stricter terms of service, believed they had no legal duty to fight outside United States territory and at first refused to cross the border. A rumor spread among the Volunteers that, if they did so, they automatically became liable for five years' service. Many had never met Indian fighters before and were terrified of them, believing they tortured and massacred their prisoners. News of Indians in the vicinity led to wholesale desertions and even mutinies.[240] The senior officers were hopeless. Major-General Stephen van Rensselaer of the New York Militia came from one of the oldest Dutch families, had inherited 150,000 acres, let to 900 tenant farmers each with 150 acres under crops, and was known as 'The Patroon,' being 'Eighth in succession.' Grandee he may have been but his men refused to follow him into danger, and his attack from Niagara ended in ignominy. At Frenchdown, General James Winchester contrived to get himself defeated and surrounded, and surrendered his whole army, such as it was.[241] Casualties from Indian attacks, disease, and exposure due to inadequate clothing and tents were high.

The generals blamed each other. General Peter B. Porter accused General Alexander Smyth, in the pages of the *Buffalo Gazette*, of arrant cowardice. They fought a comic-opera duel on Grand Island: no one was hurt but their buffoonery disgusted their men.[242] Smyth was mobbed and his shortcomings posted on handbills. The militias often fought each other with more enthusiasm than they tackled the British. In the camp at Black Rock, Irish Greens from New York waged a pitched battle with the Southern Volunteers, and both turned on the regular troops sent to separate them. The civilian public jeered. The US Light Dragoons, raised in 1808 with the initials USLD on their caps, were branded 'Uncle Sam's Lady Dogs.' By the end of 1813 the invasion of Canada had been effectively abandoned and the British were occupying a large part of Maine.[243]

Madison's forces did better at sea. On the Great Lakes, Oliver Hazard Perry (1785–1819) of Rhode Island built up an efficient little fleet and fought a battle with the British on Lake Erie on September 10, 1813. The *Lawrence*, his flagship, was so badly damaged that he had himself rowed to the *Niagara*, from which he continued the fight until the British squadron surrendered. Afterwards he sent a famous victory dispatch, celebrated for its brevity: 'We have met the enemy and they are ours.'[244] On the high seas, American warships, both regulars and privateers, benefited enormously from the fact that their officers were appointed and promoted entirely on merit—one genuine advantage of republicanism—rather than on 'interest,' as in the Royal Navy. The US ships had all-volunteer crews, too, as opposed to press-ganged British ships. In 1813 and still more in 1814 American privateers did immense damage to British shipping in the western approaches to the British Isles. In an address to the crown, Glasgow merchants, who handled the bulk of the American tobacco trade, complained: 'In the short space of two years, above 800 vessels have been taken by that power whose maritime strength we have hitherto held in contempt.'[245] It is true that the British could play the same game with American coastal shipping. Captain Marryat, later the famous novelist, on the British frigate *Spartan*, sank or captured scores of American vessels in US inshore waters. But what shook the British Admiralty were the successes of American warships against regular units of the Royal Navy. American frigates were bigger, better designed, carried more guns, and had twice as many officers as their British equivalents. Marryat admitted that, ship for ship, the American Navy—manned, he pointed out, largely by British crews—was superior. George Canning, the British statesman, felt he had to tell the House of Commons: 'It cannot be too deeply felt that the sacred spell of the invincibility of the British navy has been broken by these unfortunate [American] victories.'[246]

The naval war against Britain was the first in which Americans were able to demonstrate what was to become an overwhelming passion for high technology. This was the work of Robert Fulton (1765–1815), a genius of Irish ancestry born in Little Britain (now renamed Fulton Township) in Pennsylvania. His father died when he was tiny and his needy childhood was redeemed by an astonishing skill at drawing combined with inventive mechanical gifts—from the age of thirteen he made his own pencils, brushes, paints, and other materials. He studied under the leading Philadelphia portraitist, Charles Wilson Peale, who painted his new pupil, a tough, brooding young man with rage written all over his face. Precise skills in draftsmanship overlapped with scientific passion in those days—Fulton's younger contemporary, Samuel Morse, who was to transform telegraphy, also began as a portrait-painter. Fulton's interest in propulsion began as soon as his art studies. In his teens he made a powerful skyrocket, designed a paddle-wheel, and invented guns.[247]

Fulton had a lifelong hatred of the Royal Navy, which he saw as an enemy not just of American independence but of the freedom of the seas, to him the high road to human advancement. In 1798 he went to France in an attempt to sell to General Bonaparte a design for a submarine for use against the British. Oddly enough, as far back as 1776 a Yankee inventor, David Bushnell, had been awarded £60 for building a submarine—but it did not work when tried against British ships. Fulton's U-boat, with a crew of three, could submerge to 25 feet and was equipped with mines and primitive torpedoes. Like all his marine designs, it imitated the movements of a fish. The French promised him 400,000 francs if he sank a British frigate. But when the sub was tried in 1801 it too failed and the French lost interest.[248] He then had the audacity to go to London and try to sell submarines to the British Admiralty, promising to blow up the French invasion fleet then gathering at Boulogne (1803–4). The British too were keen at first, and one of Fulton's torpedoes actually succeeded in sinking a French pinnace, drowning its crew of twenty-two. But only the French knew this at the time. When Trafalgar ended the invasion scare, the Admiralty gave Fulton the brush-off.[249]

Thus the War of 1812 came to Fulton as an emotional and professional godsend—he could now work for his own government. He was able to buy some powerful steam-engines made by the leading British maker, Boulton and Watt, and he planned to install them in enormous steam-driven surface warships. The project-ship, christened *Demologus* (1813), then *Fulton the First* (1814), was a twin-hulled catamaran with 16-foot paddles between the hulls. It was 156 feet long, 56 wide, and 20 deep and was protected by a 5-foot solid timber belt. With an engine powered by a cylinder 4 feet in diameter, giving an engine-stroke of 5 feet, this would have been the first large-scale armored steam-warship. The British were also working on a steam-warship at Chatham, but it was only a sloop. Fulton's new battleship was planned to carry thirty 32-pound guns firing red-hot shot, plus 100-pound projectiles below the waterline. With its 120 horsepower it could move at 5 miles an hour independent of the winds and, in theory at least, outclassed any British warship afloat. Stories of this monster, launched on the East River June 29, 1814, reached Britain and grew in the telling. An Edinburgh news-paper doubled the ship's size, adding: 'To annoy an enemy attempting to board, it can discharge 100 gallons of boiling water a minute and, by mechanism, 300 cutlasses with the utmost regularity over her gunwales and work also an equal number of heavy iron pikes of great length, darting them from her sides with prodigious force.'[250]

The British were also developing new weapons. In 1803 Colonel Henry Shrapnel invented the hollow-cased shot or 'Shrapnel Shell,' an anti-personnel weapon still in use. It was hoped to combine it with the new chemical rockets developed by William Congreve, son of the man who ran Britain's main arsenal at Woolwich. Whereas Fulton was known as 'Toots' because of the noises he made, Congreve was 'Squibb.' He created the Congreve Rocket in 1808 and by 1812 had developed an advanced version with a 42-pound warhead and a range of 3,000 yards, nearly 2 miles. He had in mind a 400-pounder with a 10-mile range. By 1813, when stories of the American attack on Canada and the burning of towns and villages reached Britain, there was outrage and calls for revenge. Captain Charles Pasley, the leading British geostrategist, proposed bombarding the American coastal towns. Robert Southey, the Poet Laureate, applauded the scheme, espe-cially if put into effect with the new giant Congreves. He wrote to Sir Walter Scott that, if British peace proposals were not accepted, 'I would run down the [American] coast, and treat the great towns with an exhibition of rockets . . . [until] they choose to put a stop to the illu-minations by submission—or till Philadelphia, New York, Baltimore etc were laid in Ashes.'[251]

Southey's suggestion, coming from a man not normally bloodthirsty, reflected the exasperation of the British people with war. In the spring of 1814 Bonaparte's regime collapsed and fighting in Europe ceased. The American war seemed a hangover from the past, an anomaly. The British were obsessed by beating Bonaparte but took no interest in the transatlantic conflict. When Francis Jeffrey, the famous editor of the *Edinburgh Review*, was in Washington in January 1814 and called on Madison, the President asked him what the British thought about the war. Jeffrey was silent. Pressed to reply, he said: 'Half the people of England do not know there is a war with America, and those who did had forgotten it.'[252] But the British government were keen to tidy up loose ends all over the world and, in particular, get the frontier of Canada agreed once and for all, and a settlement in the West Indies. So they put out determined peace-feelers but, at the same time, rushed across the Atlantic forces released by the end of the war in Europe, with the aim of putting pressure on Madison.

In view of America's failure in Canada, Madison should have greeted the news of his French ally's defeat as a spur to get the best peace settlement he could as fast as possible. But he was dilatory and divided in himself, and his administration reflected this division. Monroe, his Secretary of State, was all for peace and thought pursuit of the war madness. But Madison had appointed General John Armstrong (1758–1843) his Secretary for War, with wide powers to direct the field armies and Armstrong was keen on victory. He had been ADC to Horatio Alger in the War of Independence, had political ambitions, and thought a ruthless policy might promote them. Monroe thought him a potential Bonaparte.[253] Armstrong sent an order to General William Harrison, the future President, with instructions to conciliate the Indians, turn them loose on the Canadians, and convert the British settlements on the Thames River into 'a desert.' He also gave General McClure discretion to burn Newark. Madison commanded the Thames order to be revoked, and he disavowed the burning of Newark. Terror was never officially White House policy, and one colonel was court-martialled for a town-burning. Nevertheless, many settlers were murdered and their houses torched. Bearing in mind that Britain was now free to retaliate, with an enormous navy of ninety-nine battleships and countless smaller vessels, and with a large army of Peninsular War veterans, Madison's conduct makes no sense. Moreover, he was warned. The British naval commander, Sir Alexander Cochrane, wrote to Monroe that, unless America made reparations for the 'outrages' in

Upper Canada, his duty was 'to destroy and lay waste such towns and districts upon the [American] coast as may be found available.'[254]

In view of this, the lack of preparations taken by Madison, Armstrong, or any of their commanders is remarkable. The actual landing by the British on the Chesapeake in August 1814 seems to have taken everyone by surprise. The assault ships under Sir George Cockburn succeeded in landing 5,000 troops under General Robert Ross and withdrawing them, largely unscathed, over a month later. When news of the British landing reached the capital, politicians and generals rushed about not quite knowing what they were doing. Madison himself, Monroe, Armstrong, the Navy Secretary William Jones, and the Attorney-General Richard Rush all made off to a hastily devised defensive camp outside the city, 'a scene of disorder and confusion which beggars description.'[255] An eyewitness saw the President's wife, Dolley, 'in her carriage flying full speed through Georgetown, accompanied by an officer carrying a drawn sword.'[256] She seems to have been the only person to have behaved with courage and good sense. She saved Gilbert Stuart's fine portrait of Washington, on the dining-room wall of the President's house, 'by breaking the frame, which was screwed to the wall, and having the canvas taken away.'[257]

The British entered Washington, which was now undefended, on Wednesday, August 24. There was a good deal of cowardice as well as incompetence. Edward Codrington, a British naval officer involved in the operation, reporting events to his wife Jane, wrote: 'the enemy flew in all directions [and] scampered away as fast as possible.' Madison, he added, 'must be rather annoyed at finding himself obliged to fly with his whole force from the seat of government, before 1200 English, the entire force actually engaged.' He said that the Americans had 8,000 troops defending the Washington area but 'they ran away too fast for our hard-fagged people to make prisoners.'[258] Madison himself was a fugitive. Dolley had to disguise herself: one tavern, crowded with homeless people, refused her admittance as they blamed her husband for everything. When she took refuge at Rokeby, the country house of Richard Love, his black cook refused to make coffee for her saying: 'I done heerd Mr Madison and Mr Armstrong done sold the country to the British.'[259] The Middle States, like the West and North, had been strongly for the war. Yet their resistance to invasion was pitiful. As one American historian put it, 'In Maryland, Virginia and Pennsylvania there were living not far from 1.5 million of whites. Yet this great population remained in its towns and cities and suffered 5,000 Englishmen to spend five weeks in its midst without once attempting to drive the invaders from its soil.'[260]

Hence the British were able to take their time about humiliating Washington. They fired a volley through the windows of the Capitol, went inside, and set it on fire. Next they went to the President's house, contemptuously referred to by federalists as 'the palace'—it was conspicuously unfinished and had no front porch or lawn—gathered all the furniture in the parlor, and fired it with a live coal from a nearby tavern. They also torched the Treasury Building and the Navy Yard, which burned briskly until a thunderstorm at midnight put it out. Cockburn had a special dislike for the *National Intelligencer*, which had published scurrilous material about him, and he set fire to its offices, telling the troops: 'Be sure that all the presses are destroyed so that the rascals cannot any longer abuse my name.'[261] The troops pulled out at 9 P.M. the following day, by which time a cyclone and torrential rain had further confused the scattered American authorities and compounded the miseries of thousands of refugees. Madison finally found his wife at an inn at Great Falls and prepared to return to his smoldering capital, though as he confessed to Dolly, 'I know not where we are in the first instance to hide our heads.'[262] In temporary quarters on Eighteenth Street, Madison relieved his despair by sacking Armstrong and accepting the resignations of the Navy and Treasury secretaries. But where was he, and America, to look for a savior?

The savior soon—one might say instantly—appeared, but in a human shape that Madison, his mentor Jefferson, and the whole of the Virginia ruling establishment found mighty uncongenial, the very opposite of the sort of person who, in their opinion, should rule America. By 1814 Andrew Jackson, the twelve-year-old boy who had been marked for life by a British officer's sword, had made himself a great and powerful man, of a distinctively new American type. It is worth looking at him in some detail, because to do so tells us so much about life in the early republic. At seventeen, a hungry, almost uneducated orphan, he had turned to a life in the law. In frontier Tennessee, 'lawyering' was in practice a blend of land-grabbing, wheeler-dealing, office-seeking, and dueling. The frontier was rapidly expanding, rough, violent, and litigious. Jackson became a pleader in a court, attorney-general for a local district, then judge-advocate in the militia. Ten years later he was already deep in land-speculation, the easiest way for a penniless man to become rich in the United States, but he was almost ruined by an associate's bankruptcy. His breakthrough came in 1796, when he helped to create the new state of Tennessee, first as congressman, then as senator. He took office as a judge in the state's Superior Court and founded the first Masonic lodge in Nashville, where he settled in 1801, soon acquiring the magnificent estate of the Hermitage near by. His key move, however, was to get himself elected as major-general of the militia, the power base from which he drove his way to the top.[263]

Jackson was known as a killer. His first duel, fought when he was twenty-one, arose from mutual court-room abuse—a common cause—and ended with Jackson firing into the air. But thereafter, like Burr, he usually shot to kill. Jackson fought many duels on account of his marriage, in 1790, to Rachel Robards, an older divorced woman, a substitute mother, whom he loved passionately and fiercely defended until her death. Rachel's divorce proved invalid and the Jacksons, jeered at, were forced to go through a second marriage ceremony. In 1803, when Jackson was a senior judge in Knoxville, the governor of the state, John Sevier, sneered at Rachel and accused Jackson of 'taking a trip to Natchez with another man's wife.' 'Great God!' responded Jackson, 'do you dare to mention *her* sacred name?' Pistols were drawn—the men being aged fifty-eight and thirty-six respectively—and shots were fired but only a passer-by was injured. Ten days later, however, there was another, bloodier gunfight with various members of Sevier's family. In 1806 Jackson fought a formal duel with Charles Dickinson, being wounded himself and leaving his opponent to bleed to death.[264] In 1813 Jackson was involved in a series of knock-on duels and fights which led to a violent melée in the streets of Nashville, fought with swordsticks, guns, daggers, and bare fists—Thomas Hart Benton, later a famous senator, was another of those involved—the participants rolling, bleeding and bruised, in the dust. Most of Jackson's duels, in which he faithfully carried out his mother's dying injunction, struck a squalid note.[265]

The duels left Jackson's body a wreck. Tall and thin (six feet one, weighing 145 pounds), with an erect body crowned by an upstanding thatch of bright red hair, Jackson had a drawn, pain-lined face from which blue eyes blazed furiously, and his frame was chipped and scarred by the marks of a violent frontier existence. Dickinson's bullet broke two of Jackson's ribs, buried itself in his chest carrying bits of cloth with it, could never be extracted, and caused a lung abscess which caused him pain for decades. In the Benton duel he was hit in the shoulder, barely saved his arm, and, again, the ball could not be prized out, remaining embedded in the bone and provoking osteomyelitis. In 1825 Jackson, who was accident-prone, stumbled on a staircase, ripped the wound open and caused massive bleeding from which he nearly died and which recurred occasionally all the rest of his life. On top of these hideous scars and bits of metal in his anatomy, Jackson had endemic malaria compounded by dysentery, contracted on campaign. For the first, and for his aching wounds, he took sugar of lead, both externally and internally—a horrifyingly drastic remedy—and for the second, huge doses of calomel which rotted his teeth.[266] Jackson met these misfortunes with stoicism, even heroism. He anticipated the hemorrhages by opening a vein; he would 'lay bare his arm, bandage it, take his penknife from his pocket, call a servant to hold the bowl and bleed himself freely.'[267] His acceptance of pain deepened his resolution but left further scars on his psyche and intensified his rages. His unforgettably fierce but frail figure thus became an embodiment of angry will, working for America's grand but ruthless purposes.

The first to reel before the impact of Jackson's bitterness were the Indians. Most people in the West and South wanted war because it would 'solve the Indian problem.' The new republic was ambivalent about Indians. The Constitution ignored them, saying only that Congress had the power 'to regulate commerce with foreign nations, and among the several states, and with the Indian tribes.' Henry Knox, in charge of Indian affairs (as war secretary) in the provisional government and then in Washington's administration, had got Congress in 1786 to pass an ordinance which cut Indian country in two at the Ohio River. North of the Ohio and west of the Hudson was the Northern District; south of the Ohio and east of the Mississippi was the Southern District. Each was under a superintendent who felt some responsibility for his charges, as the British had done. But whereas the crown treated the Indians as 'subjects,' just like the whites (or blacks), the Americans could not regard them as 'citizens'—they were 'savages.'[268]

However, it was one thing to divide the Indians on maps; quite another to get them to do what the government wanted. In the years after the Revolutionary War, the Indians often attacked advancing settlers with success, and efforts by the republic's young and tiny army were liable to end in abject failure. When Washington took over as president, there were only 700-odd regulars of all ranks, and the Creeks alone had between 3,500 and 6,000 warriors. In October 1790 the Indians repulsed General Josiah Harmar's army when it invaded western Ohio; they almost destroyed General Arthur St Clair's force in 1791 near what is now Fort Wayne, Indiana, killing half the 1,400 regulars and militia and sending the rest fleeing in panic. At the Battle of Fallen Timbers, August 20, 1794, Anthony Wayne and his mounted Kentucky Riflemen did something to redress the balance in a vicious engagement which lasted only forty minutes but forced the Shawnees and other tribes to sign the Treaty of Greenville (1795). But outright conquest was never an option. The Indians had to be subdued by treaties, promises, deception, attrition, disease, and alcohol.[269]

The prevailing American view was that the Indians must assimilate or move west. This was a constitutional rather than a racist viewpoint. The United States was organized into parishes, townships, counties, and states. The Indians were organized not geographically but tribally. So organized, they lived in pursuit of game. But the game was gone, or going. They therefore had to detribalize themselves and fit into the American system. If they chose to do so, they could be provided with land (640 acres a family was a figure bandied about) and US citizenship. This was, in fact, the option countless Indians chose. Many settled, took European-type names, and, as it were, vanished into the growing mass of ordinary Americans. In any case there was no clear dividing-line between 'redskins' and whites. There were scores of thousands of half-breeds, some of whom identified with the whites, and others who remained tribal. The bulk of the pure-bred Indians seem to have preferred tribalism when they had the choice. In that case, said the settlers, you must move west, to where there is still game and tribalism is still possible.[270]

The War of 1812 increased the leverage of the settlers—one reason why they favored it so strongly—because the British played the Indian card and therefore justified the most ferocious anti-Indian measures. The British pursued a systematic policy of organizing and arming minorities against the United States. They liberated black slaves wherever they could. In the region of the Apalachicola River, then the boundary between West and East Florida, the British major Edward Nicholas, with four officers and 108 Royal Marines, armed and to some extent trained over 4,000 Creeks and Seminoles, distributing 3,000 muskets, 1,000 carbines, 1,000 pistols, 500 rifles, and a million rounds of ammunition.[271] The Indians themselves were divided on whether to take advantage of all this and attack American settlements. But the leader of their war party, the Shawnee Chief Tecumseh (1768–1813) was in no doubt. With his remarkable oratory, and the predictions of his brother, 'The Prophet,' he had organized a league of Indian tribes and he told their elite (mainly Creeks) in October 1811: 'Let the white race perish! They seize your land. They corrupt your women. They trample on the bones of your dead! Back whence they came, on a trail of blood, they must be driven! Back—aye, back to the great water whose accursed waves brought them to our shores! Burn their dwellings—destroy their stock—slay their wives and children that their very breed may perish! War now! War always! War on the living! War on the dead!'[272]

When the war broke out, the militant Creeks, known as the Red Sticks (they carried bright red war-clubs), joined in enthusiastically, some of them traveling as far north as Canada to massacre the demoralized American invaders in late 1812. On their way home they murdered American settlers on the Ohio, and this in turn led to civil war among the Indians, for the Chickasaw, fearing reprisals, demanded that the southern Creeks punish the murderers. In the wild frontier territory north of the Spanish colonial capital of Pensacola, the American settlers, plus Indian 'friendlies,' attempted a massacre of the Red Sticks, led by the half-breed Peter McQueen, who had his own prophet in the shape of High-Head Jim. The attempt failed, and the whites retreated into the stockade of another half-breed, Samuel Mims, who was pro-white, 50 miles north of Mobile, on the Gulf. It was an acre of ground surrounded by a log fence, with slits for muskets and two gates. Inside were 150 militiamen, 300 whites, half-breeds and friendlies, and another 300 black slaves. Yet another half-breed, Dixon Bailey, was appointed commander. It must be grasped that, at this time, much of the Far South, especially near the coasts, was a lawless area anyway, where groups of men, whites, Indians, half-breeds, escaped slaves, mulattoes, banded together, running their own townships, changing sides frequently. Fort Mims was a typical pawn in this game. A slave who warned Bailey that the Red Sticks were coming was deemed a liar and flogged, and the stockade gates were actually open when 1,000 Sticks attacked. Bailey was killed trying to shut the gates and all except fifteen whites were slaughtered. 'The children were seized by the legs and killed by battering their heads against the stockading, the women were scalped, and those who were pregnant were opened while they were alive and the embryo infants let out of the womb.'[273] The Creeks murdered 553 men, women, and children and took away 250 scalps on poles.

At this point Major-General Jackson was told to take the Tennessee militia south and avenge the disaster. It was just the kind of job he liked and the opportunity for which he had been waiting. On Indians he had exactly the same views as the leader of the anti-British faction in the West, Henry Clay of Kentucky (1777–1852), Speaker of the House and organizer of what were known as the War Hawks. Clay, and John Caldwell Calhoun of South Carolina (1782–1850), the most articulate spokesman of the Southerners, wanted every unassimilable Indian driven west of the Mississippi. Jackson agreed with that. He further argued that the states and the federal government should build roads as quickly as possible, thus attracting settlers who would secure any territory vacated by the Indians immediately. Jackson's Protestant forebears in Ulster had pursued exactly the same strategy against the 'Wild Irish.' When he got his orders, his arm was still in a sling from his latest duel but he hurried south, building roads as he went. With him were his bosom pal and partner in land-speculation, General John Coffee, who commanded the cavalry, and various adventurers, including David Crockett (1786–1836), also from Tennessee and a noted sharpshooter, and Samuel Houston (1793–1863), a Virginia-born frontiersman, then only nineteen.

These men, who were later to expand the United States into Texas and beyond, were bloodied in the Creek War. And bloody it was. On November 3, two months after the massacre, Jackson surrounded the 'hostile' village of Tallushatchee and sent in Coffee with 1,000 men to destroy it. Jackson later reported to his wife Rachel that Coffee 'executed this order in elegant stile.' Crockett put it more accurately: 'We shot them like dogs.' Every male in the village, 186 in all, was put to death. Women were killed too, though eighty-four women and children were taken prisoner. An eyewitness wrote: 'We found as many as eight or ten bodies in a single cabin.' Some had been torched, and half-consumed bodies were seen among the smoking ruins. 'In other instances, dogs had torn and feasted on the bodies of their masters.'[274] A ten-month-old Indian child was found clutched in his dead mother's arms. Jackson, who always had a fellow-feeling for orphans and who was capable of sudden spasms of humanitarianism in the midst of his most ferocious activities, adopted the boy instantly, named him Lyncoya, and had him conveyed to the Hermitage. He wrote to Rachel: 'The child must be well taken care of, he may have been given to me for some valuable purpose—in fact when I reflect that he, as to his relations, is so much like myself, I feel an unusual sympathy for him.'[275]

A week later, Jackson won a pitched battle at Talladega, attacking a force of 1,000 Red Sticks and killing 300 of them. At that point some of his men felt enough was enough. The militia was obliged to provide only ninety days' service. The volunteers had engaged for a year but their term was running out. Both said they wanted to go home. They would either march home under Jackson, or mutiny and go home without him. This was the spirit which had ruined the Canada campaign and was already affecting other forces in the multipronged campaign against the Creeks. But Jackson was not going to let his angry will be frustrated by a few homesick barrack-room lawyers. He used the volunteers to frighten the militia men and his few regulars to frighten both. On November 17 he and Coffee lined the road and threatened to shoot any militiamen who started to march home. Back in camp he faced an entire brigade, his left arm in a sling, his right clutching a musket which rested on the neck of his horse, and said he would personally shoot any man who crossed the line he drew. He held the mob with his fierce glare until regulars with arms ready formed up behind him.[276] When the volunteers, their time up, decided to move off on December 10, Jackson trained two pieces of artillery, loaded with grapeshot, on them, and when they failed to respond to his orders, he commanded the gunners, picked loyalists, to light their matches. At that the mutineers gave way. They hated Jackson, but they feared him more.

He wrote to Rachel that the volunteers had become 'mere whining, complaining Seditioners and mutineers, to keep whom from open acts of mutiny I have been compelled to point my cannon against, with a lighted match to destroy them. This was a grating moment of my life. I felt the pangs of an affectionate parent, compelled from duty to chastise the child.'[277] It is unlikely that Jackson felt any such emotion; he always rationalized his acts of passion in language from a pre-Victorian melodrama. When John Woods, a militiaman of eighteen, refused an order, and grabbed a gun when arrested, Jackson had no hesitation in having him shot by firing-squad, with the entire army watching. He banned whiskey. He made his men get up at 3.30, his staff half an hour earlier, to forestall Indian morning raids. Senior officers who objected were sent home under arrest. The shooting of Woods was decisive. According to Jackson's ADC, John Reid: 'The opinion, so long indulged, that a militiaman was for no offense to suffer death was from that moment abandoned and a strict obedience afterwards characterised the army.'[278]

Jackson thus welded into existence a formidable army, 5,000 strong, which paradoxically attracted volunteers. With this he attacked the Creeks' main fortress, at Horseshoe Bend, an awesome peninsula of

100 acres, almost surrounded by deep water, the land side defended by a 350-yard breastwork 5 to 8 feet high with a double row of firing holes across its neck. It was, wrote Jackson, 'well formed by Nature for defense, & rendered more secure by Art.' Jackson never underestimated the Indians and was impressed by their military ingenuity—'the skill which they manifested in their best work was astonishing.'[279] The Creeks had 1,000 warriors inside the fort. Jackson began with diversions, such as fire-boats, then stormed the rampart, calculating that scaling-ladders, always awkward to use, were not needed. Ensign Sam Houston was the first man to get safely across the breastwork and into the compound. What followed the breach of the wall was horrifying. The Indians would not surrender and were slain. The Americans kept a body-count by cutting off the tips of the noses of the dead, giving a total of 557 in the fort, plus 300 drowned trying to escape in the river. The dead included three leading prophets in full war-paint. The men cuts strips of skin from them for harness. Jackson lost forty-seven whites and twenty-three friendlies.[280]

After that it was simply a matter of using terror—burning villages, destroying crops—until the Indians had had enough. On April 14, 1814, Red Eagle, virtual paramount chief of the Creeks, surrendered. He told Jackson: 'I am in your power . . . My people are all gone. I can do no more but weep over the misfortunes of my nation.'[281] Jackson spared Red Eagle because he was useful in getting other Indians to capitulate. He was given a large farm in Alabama where, like other Indian planters, he kept a multitude of black slaves. Four months later Jackson imposed a Carthaginian peace on thirty-five frightened Indian chiefs. Jackson was an impressive and at times terrifying orator, who left the Indians in no doubt what would be their fate if they failed to sign the document he thrust at them. It forced the Creeks to part with half their lands—three-fifths of the present state of Alabama and a fifth of Georgia. Jackson wrote gleefully to a business partner: 'I finished the convention with the Creeks . . . [which] cedes to the United States 20 million acres of the cream of the Creek Country, opening a communication from Georgia to Mobile.'[282] He knew it was only a matter of time before the Americans got the rest. The Treaty of Fort Jackson was the tragic turning-point in the destruction of the Indians east of the Mississippi.

Jackson now moved swiftly to safeguard his conquests from the Spanish and the British. He was fighting on behalf of the American settler class and he took no notice of orders (or the lack of them) from

Washington. Before the end of August he had occupied Mobile and Fort Bower on the key to the south of it. When British land–sea forces moved into the area in mid-September, they found the fort strongly guarded and failed to take it. On November 7 Jackson occupied the main Spanish base at Pensacola. America and Spain were not at war and Jackson had no authority for this act of aggression, but Washington was still too shell-shocked to protest when his letter arrived telling them what he had done. Jackson's move frustrated the plan of the British force commander, Cochrane, to take Mobile and move inland, cutting off New Orleans. So instead he decided on a frontal assault. This gave Jackson his opportunity to become America's first real hero since Washington. When he reached New Orleans on December 1 he found it virtually undefended. He worked swiftly. He formed the local pirates, who hated the Royal Navy, on whose ships they were periodically hanged, into a defensive unit. Hundreds of free blacks were turned into a battalion under white officers (but with their own NCOs). He paid them well. When his paymaster protested, Jackson told him: 'Be pleased to keep to yourself your opinions ... without inquiring whether the troops are white, black or tea.'[283] He brought as many troops into the city as possible and, using his experience at the Horseshoe, built a main defense line of great strength and height. By the time the British, who had sixty ships and 14,000 troops, mostly Peninsula veterans, were ready to attack on January 8, 1815, New Orleans was strongly defended.

Even so, it could have been outflanked. And that was the British intention. Jackson's main defense was behind Rodriguez's Canal, a ditch 4 feet deep and 10 feet wide, which he reinforced by a high mud rampart. The British land commander, General Sir Edward Pakenham, a stupendously brave but impatient man—'not the brightest genius,' as his brother-in-law the Duke of Wellington put it—planned a two-pronged assault, up the almost undefended left bank of the Mississippi, to take the rampart from the rear, while his troops in front kept the defenders occupied. But the force landed in the wrong place and fell behind schedule. Pakenham decided he could not wait and, relying on the sheer professionalism of his veterans, decided on a frontal assault alone and fired the two Congreves which were the signal for attack. A frontal assault against a strongly defended position not enfiladed from the rear was a textbook example of folly which would have made Wellington despair. It became even more murderous when the leading battalion failed to bring up the fascines to fill the ditch and the ladders to scale the rampart. The result was a pointless slaughter of brave men. The advancing redcoats met a combination of grapeshot, canister-shell, rifles, and muskets, all skillfully directed by Jackson himself. The attack wavered and, in goading on their men, all three British general officers were killed—Pakenham on the spot, Sir Samuel Gibbs, commanding the attack column, fatally wounded, General Keane taken off the field writhing in agony from a bullet in his groin. By the time the reserve commander arrived to take over, the men were running and it was all over. Jackson lost only thirteen killed, the British 291, with another 484 missing and over a thousand wounded. Codrington, watching from HMS *Tonnant*, could only shake his head in disbelief at the débâcle. 'There never was,' he wrote to his wife, 'a more complete failure.'[284]

Thus ended one of the shortest and most decisive battles of history. Three days later the first rumors arrived that Britain and America had made peace. The British expedition continued to fight until formally notified, and on February 11 took Fort Bower, preparatory to occupying Mobile. But by the time Admiral Cochrane was ready to enter the town a dispatch-boat arrived with orders to cease hostilities, and in March his fleet sailed for home. The peace had actually been signed on Christmas Eve, in the 'neutral' town of Ghent, and it had taken six months to negotiate.[285] It might have come sooner, had the American team been less ill assorted—a typical example of Madison's lack of realism. It consisted of the Treasury Secretary, Albert Gallatin, and the federalist Senator from Delaware, James Bayard—two men of opposing viewpoints on virtually all subjects. Then there was John Quincy Adams, the minister to St Petersburg and son of the second President, and Henry Clay, leader of the War Hawks. Clay was a Westerner from Kentucky, a thruster, not quite a gentleman, a drinker, gambler, and womanizer. Adams was a Harvard man and a Boston Brahmin, who had spent his life in embassies, was fluent in foreign languages and the diplomatic arts. He was also argumentative, puritanical and prissy, thin-skinned, quick to take offense, a superb hater, and constant compiler of enemies lists. His final one, drawn up at the end of his life, consisted of thirteen men, including Jackson, Clay, John C. Calhoun, and Daniel Webster, who had 'conspired together [and] used up their faculties in base and dirty tricks to thwart my progress in life.'[286]

Adams got on badly with all his colleagues, who, he claimed, kept late hours and gambled all night, having been corrupted by Clay, whom Adams also accused of making a pass at a chambermaid. They disagreed on most of the peace-issues in dispute, representing as they did quite different regional interests. Happily, the British, having secured Canada, were not too concerned at driving a hard bargain. After their assault on Washington had 'taught the Americans a lesson,' they were all for a quick settlement. The Washington disaster also spurred Madison and Monroe to hurry up the talks. Even before it, the banks in Philadelphia and Baltimore had gone bankrupt. The actual sack of Washington detonated a long-smoldering financial crisis, and the big banks in New York went under. The Treasury was empty, as Gallatin well knew. But in New England the federalists, their own banks sound, watched the ruin of the pro-war states and the confusion of the ruling Republicans with complacency. They held a convention of the New England states at Hartford, Connecticut, in December 1814. Contrary to rumor, they did not actively discuss secession, but they drew up plans to oppose any further war measures, including conscription and further restraints on trade.'[287]

In October a weary Madison instructed a still wearier Monroe, who was looking after the War Department as well as his own—'for an entire month I never went to bed,' he complained—to try to get a settlement as quickly as possible with 'the *status quo ante bellum* as the basis of negotiations.' The Duke of Wellington too thought there was nothing more to be gained by fighting and Lord Castlereagh, British Foreign Secretary, was equally anxious to be 'released from the millstone of the American war.'[288] In fact the *status quo* formula was the simplest solution to a war both sides now silently admitted should never have been started. So it was accepted. Such matters as Newfoundland fishery rights and navigation on the Mississippi were dropped. The actual issues of the war were ignored. All the Treaty of Ghent did was to provide for the cessation of hostilities immediately it was ratified; the release of prisoners; surrender of virtually all territory occupied by either side; the pacification of the Indians, and the more accurate drawing of boundaries, to be handed over to commissioners.[289]

More by accident than design, the Treaty of Ghent proved one of history's great acts of statesmanship. After the signing, Adams remarked to one of the English delegates: 'I hope this will be the last treaty of peace between Great Britain and the United States.'[290] It was. The very fact that both sides withdrew to their prewar positions, that neither could describe the war as a success or a defeat, and that the terms could not be presented, then or later, as a triumph or a robbery—all worked for permanency and helped to erase from the national memory of both countries a struggle which had been bitter enough at the time. And the absence of crowing or recrimination meant that the treaty could serve as a plinth on which to build a friendly, common-sense relationship between the two great English-speaking peoples.

The fact that Jackson's victory at New Orleans came too late to influence the treaty does not mean it was of no consequence. Quite the reverse. It too was decisive in its way for, though the treaty made no mention of the fact, it involved major strategic, indeed historic, concessions on both sides. Castlereagh was the first British statesman of consequence who accepted the existence of the United States not just in theory but in practice as a legitimate national entity to be treated as a fellow-player in the world game. This acceptance was marked by the element of unspoken trust which lay behind the treaty's provisions. America, for its part, likewise accepted the existence of Canada as a permanent, legitimate entity, not just an unresolved problem left over from the War of Independence, to be absorbed by the United States in due course. Henceforth the road to expansion for both the United States and Canada lay not in depredations at each other's expense but in pushing simultaneously and in friendly rivalry towards the Pacific. In return, Britain gave the Americans the green light to expand as they wished anywhere south of the 49th parallel (a line adopted in 1818), at the expense of the Indians and the Spanish alike.

The significance of Jackson's victory was that it determined the way the Treaty of Ghent was interpreted and applied. Britain, along with most other nations, had not recognized the Louisiana Purchase, and acknowledged no American right to be in New Orleans, Mobile, or anywhere else on the Gulf of Mexico. Britain would have been at liberty to hand any of these territories back to Spain if it had been in possession of them, even under the terms of the Treaty of Ghent. And that, Monroe told Madison, was exactly what it would have done, had not Jackson won the battle. The effect of the victory was to legitimize the whole of the Louisiana Purchase in the eyes of the international community. Equally, Britain might have kept Fort Bower and turned it into another Gibraltar. As it was, Britain in effect renounced any such ambition provided America left Canada alone. There were, to be sure, sound economic reasons why Britain wanted friendly relations with the United States in the whole Caribbean area. The financial significance of the rich West Indian sugar islands was fast declining relative to Britain's rapid industrial expansion, based on finished cotton manufactures, for which the American South increasingly supplied the raw material. For America to expand south, placing more square miles under cotton, was in the interests of both countries. But it was the New Orleans victory which clinched Britain's switch of policy.

Equally, the Battle of New Orleans sealed the fate of the Indians of the South. Under Article IX of the Treaty of Ghent, America agreed to end the war against the Indians 'and forthwith to restore to such tribes . . . all possessions . . . which they have enjoyed . . . in 1811 previous to such hostilities.' This clearly made the Fort Jackson Treaty invalid. This was Britain's view and Madison agreed with it. Jackson was told: 'The president . . . is confident that you will . . . conciliate the Indians upon the principle of our agreement with Great Britain.'[291] But Madison had no grounds for such confidence. Jackson had no intention of giving the Indians back anything. And, now that the British forces had left the area, there was no one to compel him. When he simply ignored the Treaty of Ghent, Washington did nothing. Nor did the British. In fact the American settler interest had now received *carte blanche* to pursue its destiny—right to the Pacific. That, too, was the consequence of New Orleans. So Jackson was now the hero, recognized by the South and West as their champion, and by all Americans, who badly needed a successful martial figure to lift their national spirits, as the true successor to Washington. So the Revolutionary Era finally ended and a new figure strode onto America's stage, who was to take the nation into the era of democracy.

PART THREE

'A General Happy Mediocrity Prevails'

Democratic America, 1815–1850

Right at the end of his life, Benjamin Franklin wrote a pamphlet giving advice to Europeans planning to come to America. He said it was a good place for those who wanted to become rich. But, he said, it was above all a haven for the industrious poor, for 'nowhere else are the laboring poor so well fed, well lodged, well clothed and well paid as in the United States of America.' It was a country, he concluded, where 'a general happy mediocrity prevails.'[1] It is important for those who wish to understand American history to remember this point about 'happy mediocrity.' The historian is bound to bring out the high points and crises of the national story, to record the doings of the great, the battles, elections, epic debates, and laws passed. But the everyday lives of simple citizens must not be ignored simply because they were uneventful. This is particularly true of America, a country specifically created by and for ordinary men and women, where the system of government was deliberately designed to interfere in their lives as little as possible. The fact that, unless we investigate closely, we hear so little about the mass of the population is itself a historical point of great importance, because it testifies by its eloquent silence to the success of the republican experiment.

Early in the 19th century, America was achieving birth-rates never before equaled in history, in terms of children reaching adulthood. The 1800 census revealed a population of 5,308,843, itself a 35 percent increase over ten years. By 1810 it had leaped to 7,239,881, up another 36.4 percent. By 1820 it was 9,638,453, close to doubling in twenty years, and of this nearly 80 percent was natural increase. As one Congressman put it: 'I invite you to go to the west, and visit one of our log cabins, and number its inmates. There you will find a strong, stout youth of eighteen, with his Better Half, just commencing the first struggles of independent life. Thirty years from that time, visit them again; and instead of two, you will find in that same family twenty-two. That is what I call the American Multiplication Table.'[2]

But with the end of the world war in 1815 high American birth-rates were compounded by a great flood of immigrants. It is a historical conjunction of supreme importance that the coming of the independent American republic, and the opening up of the treasure-house of land provided by the Louisiana Purchase and the destruction of Indian power by Andrew Jackson, coincided with the beginnings of the world's demographic revolution, which hit Europe first. Between 1750 and 1900 Europe's population rose faster than anywhere else in the world (except North America), from 150 million to over 400 million.[3] This, in turn, produced a huge net outflow of immigration: to South America, Russia, Australasia, Canada, South Africa, and above all the United States. The rush to America began after the Battle of Waterloo in June 1815 and continued right through the autumn and winter, the immigrant ships braving gales and ice. It accelerated in 1816, which in Europe was 'the year without a summer,' with torrential rain and even sleet and snow continuing into July and August and wrecking harvests, sending poor and even starving people to the coast to huddle in the transports. Ezekiah Niles (1777–1839), who ran *Niles's Weekly Register* from 1811 onwards, in many ways America's best journal of record at the time, calculated that 50,000 immigrants reached America in the year, though this figure was later revised downwards. His more careful calculation for 1817, based on shipping lists (the federal government, though it took censuses, did not yet publish statistics), produced a figure of 30,000 up to the end of the main season in September. Of this half went to New York and Philadelphia, though some went straight over the Appalachians into the Ohio Valley.

No authority on either side of the Atlantic was bothered with who was going where or how, though the British limited ship-carrying capacity to one passenger to every 2 tons of registry in their own ships. The sheer freedom of movement was staggering. An Englishman, without passport, health certificate or documentation of any kind—without luggage for that matter—could hand over £10 at a Liverpool shipping counter and go aboard. The ship provided him with water, nothing else, and of course it might go down with all hands. But if it reached New York he could go ashore without anyone asking him his business, and then vanish into the entrails of the new society. It was not even necessary to have £10, as the British provided free travel to Canada, whence the emigrants could bum rides on coastal boats to Massachusetts or New York. There was no control and no resentment. One of them, James Flint from Scotland, recorded in 1818: 'I have never heard of another feeling than good wishes to them.'[4] In the five years up to 1820, some 100,000 people arrived in America without having to show a single bit of paper.

The first check of this inflow—the end of innocence if you like—

came with the catastrophic bank crash of 1819, the first financial crisis in America's history. Such a disaster was inevitable, granted the rate at which the country was expanding. In the years 1816–21 alone, six new states were created; in size and potential power it was like adding six new European countries. The United States was already creating for itself a reputation for massive borrowing against its limitless future. That meant a need for large numbers of banks, and they duly sprang up, good, bad, and indifferent (mostly the last two). The Jeffersonians hated banks, as we have seen, and in 1811 when the First Bank of the United States' charter expired, they controlled Congress and refused to renew it. That was foolish, because the states stepped into the vacuum thus created and happily chartered banks, whose numbers thus rose from 88 in 1811 to 208 two years later. Each state bank was allowed by the state legislature to issue bills up to three times its capital. But in practice there was no check on these issues. Hence, in good times at least, to get a charter to found a state bank was literally a license to print money. As critics like Jefferson and John Taylor claimed, a new kind of money power was coming into existence in America, which ran directly counter to the Founding Fathers' concept of an idyllic rural society based on landed property.[5] During the War of 1812 America was awash with suspect $2 and $5 bills printed by these mushroom banks. Such gold as there was flowed straight into Boston, whose state banks were the most secure. By 1813 Boston notes were at a 9–10 percent premium in Philadelphia. The New England banks refused to take paper notes from the South and West at all. In 1814, with the burning of Washington and the virtual collapse of federal government, every bank outside New England was forced to suspend payment.

The remedy of Congress proved worse than the disease. It created (April 10, 1816) the Second Bank of the United States, bought 20 percent of its stock, and stipulated that the federal government appoint five of its twenty-five directors, but made little provision for supervising its operations. Moreover, its first president, William Jones (1760–1831), a former congressman and Madison's Navy Secretary, knew little about banking; his speciality was having dubious friends. He fitted beautifully into Taylor's demonology. Indeed, he managed to create a fragile boom which was a miniature foretaste of the Wall Street boom of the 1920s leading to the crash of 1929. Jones' boom was in land. From 1815 the price of American cotton rose rapidly and that in turn fed the land boom. At that time public land was sold primarily to raise revenue rather than to encourage settlers, who needed no encouragement anyway. Each was charged $2 an acre in minimum blocks of 160 acres. But they only had to put 20 percent down, borrowing the rest from the banks on the security of the property. The $2 was a minimum; in the South potential cotton land was sold at $100 an acre in the boom years. The SBUS, fueling the boom by easy credit, allowed purchasers to pay even the second installment on credit, again raised on the security of the land, like a second mortgage.[6] Jones, whose only concern seems to have been to pay high dividends, based on the total lent by his bank, ran this federal central bank like a bucket-shop. He actually allowed the SBUS to deal in 'racers,' short for Race Horse Bills. These were bills of exchange paid for by other bills of exchange, which thus raced around rapidly from one debtor to another, accumulating interest charges and yielding less and less of their face value. It was a typical bit of 19th-century ruin-finance, beloved of novelists like Thackeray and Dickens, who used such devices to get their gullible heroes into trouble. This kind of paper explains why needy people actually got so little of the sums they undertook to repay. But then they probably could not repay anyway, which explains why the pyramid was bound to collapse.[7]

Jones' easy-credit policy was further undermined by the activities of the SBUS's branch offices, some of which were run by crooks. In Baltimore the branch was run by two land speculators, James A. Buchanan and James W. McCulloch, who financed their speculations by taking out unsecured loans from their own bank ($429,049 and $244,212 respectively, with the First Teller borrowing a further $50,000). In effect, this was to put their hand in the till. Here was a typical example of the general credit expansion Jones encouraged, raising the debt on public land from $3 million in 1815 to over five times that amount ($16.8 million) three years later. Some of this went into house purchases—it was the first urban boom in US history too. As many of the Latin American goldmines had been shut by their own war of independence against Spain, which was now raging, the relation of paper to gold was astronomical. Moreover, all the other banks followed Jones' example. Sensible men warned of what would happen. John Jacob Astor, who had now used his fur empire to build up a massive holding in Manhattan real estate, accused the SBUS of provoking runaway inflation. In a letter to Albert Gallatin (March 14, 1818) he said the SBUS had made money so cheap 'that everything else has become Dear, & the result is that our Merchants, instead of shipping Produce, ship Specie, so that I tell you in confidence that it is not without difficulty that Specie payments are maintained. The different States are going on making more Banks & and I shall not be surprised if by & by there be a general Blow Up among them.'[8]

Astor was right about the state banks: Hezekiah Niles recorded that in 1815–19 all you needed to start a bank issuing paper money were plates, presses, and paper. It was enough to drive genuine counterfeiters out of business, though they still managed (according to Niles) to produce a lot of forged notes too. He said that counterfeit notes from at least 100 banks were freely circulated in 1819. Many of the new banks were in converted forges, inns, or even churches, thus adding blasphemy to gimcrack finance. By 1819 there were at least 392 chartered banks, plus many more unchartered ones, and the debt on public lands had jumped another $6 million to stand at $22 million. Suddenly, the cotton bubble burst, as Liverpool cotton importers, alarmed by the high prices, started shipping in Indian raw cotton in huge quantities. In December–January 1819 the price of New Orleans cotton halved, and this in turn hit land prices, which fell from 50 to 75 percent. The banks then found themselves with collateral in land worth only a fraction of their loans, which were now irrecoverable. So the banks started to go bust. Jones compounded his earlier errors of inflation by abruptly switching to savage deflation, ordering the branches of the SBUS to accept only its own notes, to insist on immediate repayments of capital as well as interest, and by calling in loans.[9] This immediately doubled and trebled the number of state-chartered banks going bust, and the SBUS, their main creditor, secured their assets—the land-deeds of hundreds of thousands of farmers.

Many congressmen, seeing the future of their electors thus put into the power of a wicked central bank they had never wanted anyway, turned with fury on Jones. A Congressional committee soon discovered the Baltimore business. Jones and his entire board were forced to resign and an experienced money-man, Langdon Cheves (1776–1857), took over in March 1819 to find the SBUS what he called 'a ship without a rudder or sails or mast . . . on a stormy sea and far from land.'[10] Cheves decided that the worst of all outcomes was for the SBUS to go bust too, so he intensified the deflationary policy and contrived, with some difficulty, to keep the SBUS's doors open, thus earning his title 'the Hercules of the United States Bank.' But everyone else had to pay for it. As one contemporary expert, William Goude, put it, 'The Bank was saved but the people were ruined.'[11]

The result of the bank Blow Up was a crisis in manufacturing industry. The Philadelphia cotton mills employed 2,325 in 1816; by autumn 1819 all but 149 had been sacked. In New England the crisis was mitigated by sound banking but it was still acute and unemployment shot up. John Quincy Adams, always quick to strike a note of gloom, recorded in his diary on April 24, 1819: 'In the midst of peace and partial prosperity we are approaching a crisis which will shake the Union to its center.'[12] The news of trouble reached Europe too late to affect the 1819 sailings, so tens of thousands of immigrants continued to arrive, to find no work and rising hostility. One observer, Emanuel Howitt, wrote that 'the Yankees now [1819] regard the immigrant with the most sovereign contempt ... a wretch, driven out of his own wretched country, and seeking a subsistence in this glorious land.'[13] It would 'never be glad confident morning again.' In March 1819 Congress, in a panic attempt to stop ships arriving at New York and other ports, slapped a two-persons-for-5-tons rule on incoming ships, effective from September—the beginning of control. The State Department, in a prescript published in *Niles's Weekly Register*, announced its policy-lines: 'The American Republic invites nobody to come. We will keep out nobody. Arrivals will suffer no disadvantages as aliens. But they can expect no advantages either. Native-born and foreign-born face equal opportunities. What happens to them depends entirely on their individual ability and exertions, and on good fortune.'[14]

There is something magnificent about this declaration, penned by John Quincy Adams himself. It epitomizes the spirit of laissez-faire libertarianism which pervaded every aspect of American life at this time—though, as we shall see, there were state interventionists at large too. Libertarianism was, of course, based upon an underlying, total self-confidence in the future of the country. There was something magnificent too about the speed and completeness with which America recovered from this crisis, which within a year or two seemed a mere mishap, a tiny blip on a rising curve of success. Mass immigration soon resumed, thanks this time to Ireland. Hitherto, America had taken in plenty of Ulster Protestants, but few from the Catholic south. But in 1821, when the Irish potato crop failed, one in an ominous series of failures culminating in the catastrophe of the mid–1840s, the British government tried to organize a sea-lift to Canada. There was panic in Mayo, Clare, Kerry, and Cork, where rumor had it the ships would transport them to convict bondage in Australia. But, once the truth was known, the idea of going to America, at virtually no cost, caught on in the poorest parts of Ireland. When the first letters reached home in 1822, explaining how easy it was to slip from Canada into America, and how the United States, albeit Protestant, gave equal rights to Catholics, the transatlantic rush was on. In 1825 50,000 Southern Irish applied for a mere 2,000 assisted places on a government scheme. It was a foretaste of the exodus which was to transport one-third of the Irish nation to America.[15] This, in turn, was part of the process whereby the continuing English (and Welsh and Scottish) immigration to the United States was now balanced by new arrivals from outside Britain. The number of Continental Europeans rose from 6,000 to 10,000 a year in the early 1820s to 15,000 in 1826 and 30,000 in 1828. In 1832 it passed the 50,000-a-year mark and thereafter fell below it only twice. An Anglicized United States was gradually becoming Europeanized.[16]

Why did the immigrants come? One reason was increasingly cheap sea-passages. Another was food shortages, sometimes widening into famines. The bad weather of 1816, and the appalling winters of 1825-6, 1826-7, and 1829-30, the last one of the coldest ever recorded, produced real hunger. The demographic-catastrophe theories of Thomas Malthus filtered downwards to the masses, in horrifyingly distorted form, and men wanted to get their families out of Europe before the day of wrath came. Then there was the tax burden. At the end of the Bonapartist Wars, all Europe groaned under oppressive tax-ation. A parliamentary revolt in 1816 abolished income tax in Britain, and in the 1820s duties were gradually reduced too. But in Europe it was the same old story of the state piling the fiscal burdens on the backs of poor peasants and tradespeople. This was compounded, on the Continent, by tens of thousands of internal customs barriers, imposing duties on virtually everything which crossed them.

By comparison, America was a paradise. Its army was one-fiftieth the size of Prussia's. The expense of government per capita was 10 per-cent of that in Britain, itself a country with a small state by Continental standards. There were no tithes because there was no state church. Nor were there poor rates—there were virtually no poor. An American farm with eight horses paid only $12 a year in tax. Europeans could scarcely believe their ears when told of such figures. Not only were American wage-rates high, but you kept your earnings to spend on your family. Then there were other blessings. No conscription. No political police. No censorship. No legalized class distinctions. Most employers ate at the same table as their hands. No one (except slaves) called anyone 'Master.'

Letters home from immigrants who had already established themselves were read aloud before entire villages and acted as recruitment-
propaganda for the transatlantic ships. So, interestingly enough, did the
President's annual messages to Congress, which were reprinted in many
Continental newspapers until the censors suppressed them. As the
Dublin Morning Post put it: 'We read this document as if it related
purely to our concerns.'[17]

But the most powerful inducement was cheap land. Immigrants from
Europe were getting cheap land from all the old hunting grounds of the
world's primitive peoples—in Australia and Argentina especially—but
it was in the United States where the magic was most potent because
there the government went to enormous trouble to devise a system
whereby the poor could acquire it. In the entire history of the United
States, the land-purchase system was the single most benevolent act of
government. The basis of the system was the Act of 1796 pricing land
at $2 an acre. It allowed a year's credit for half the total paid. An Act of
1800 created federal land offices as Cincinnati, Chillicothe, Marietta,
and Steubenville, Ohio, that is, right on the frontier. The minimum
purchase was lowered from 640 acres, or a square mile, to 320 acres,
and the buyer paid only 25 percent down, the rest over four years. So a
man could get a big farm—indeed, by Continental standards, an enormous one—for only about $160 cash. Four years later, Congress
halved the minimum again. This put a viable family farm well within
the reach of millions of prudent, saving European peasants and skilled
workmen. During the first eleven years of the 19th century, nearly
3,400,000 acres were sold to individual farmers in what was then the
Northwest, plus another 250,000 in Ohio. These land transfers
increased after 1815, with half a million acres of Illinois, for instance,
passing into the hands of small- and medium-scale farmers every year.
It was the same in the South. In Alabama, government land sales rose
to 600,000 acres in 1816 and to 2,280,000 in 1819. In western
Georgia the state gave 200-acre plots free to lottery-ticket holders with
lucky numbers. In the years after 1815, more people acquired freehold
land at bargain prices in the United States than at any other time in the
history of the world.[18]

Individual success-stories abounded. Daniel Brush and a small group
of Vermonters settled in Greene County, Illinois, in spring 1820. 'A
prairie of the richest soil,' Brush wrote, 'stretched out about four miles
in length and one mile wide . . . complete with pure springs of cold
water in abundance.' Once a cabin, 16 by 24 feet, had been built, they

began the hard task of breaking up the prairie. This done, Brush wrote, 'No weeds or grass sprung up upon such ground the first year and the corn needed no attention with plough or hoe. If got in early, good crops were yielded, of corn and fodder.' He added: 'Provisions in abundance was the rule . . . no one needed to go supperless to bed.'[19] The Ten Brook family moved to what became Parke County, Indiana, in autumn 1822. There were twenty-seven of them altogether—three interrelated families, three single men, two teamsters, thirteen horses, twenty-one cows, two yoke-oxen, and four dogs. Their first priority was to build a strong cabin. The soil was rich but virgin. Working throughout the winter, they had cleared 15 acres by the spring and fashioned 200 fence-rails. They had 100 bushels of corn for winter-feed and spring planting. They put two more acres under potatoes and turnips. The spring brought seven calves, and that first summer they made forty 12-pound cheeses, sold at market for a dollar each. The harvest was good. They not only ground their own corn but made 350 pounds of sugar and 10 gallons of molasses from the same soil they cleared for corn. Their leader, Andrew Ten Brook, recounted: 'After the first year, I never saw any scarcity of provisions. The only complaint was that there was nobody to whom the supplies could be sold.'[20]

The sheer fertility of the soil made all the backbreaking work of opening it up worth while. In the Lake Plains—parts of Indiana, Illinois, and Michigan—a vast glacier known as the Wisconsin Drift had in prehistoric times smoothed off the rocks and laid down a deep layer of rich soil containing all the elements needed for intensive agriculture. The settlers, steeped in the Old Testament, called it Canaan, God's Country, because it yielded a third more than the rest, known as 'Egypt.' Some of the settlements in the years after 1815 became celebrated for quick prosperity. One was Boon's Lick, a belt 60 miles wide on each side of the Missouri River which became Howard County in 1816. It boasted superb land, pure water, as much timber as required, and idyllic scenery. By 1819 the local paper, the *Missouri Intelligence*, produced at the little town of Franklin, offered a spring toast: 'Boon's Lick—two years since, a wilderness. Now—*rich in cotton and cattle!*' It was widely reputed to be the best land in all the West.

Moreover, the tendency was for the land price to come down—in the 1820s it was often as low as $1.25 an acre. The modern mind is astonished that, even so, it was regarded as too high and there was a clamor for cheaper or even free land. Many settlers were termed 'squatters.'

This simply signified they had got there first, paid over money immediately after the survey but before the land was 'sectionalized' for the market. They risked their title being challenged by non-resident purchasers—speculators. By the end of 1828 two-thirds of the population of Illinois were squatters. Their champion was Thomas Hart Benton (1782–1858), Senator 1821–51. He sensibly argued against a minimum price for Western lands, proposing grading by quality, and he insisted that settlers pay compensation for improvements, passing a law to this effect. In frontier areas, speculators were naturally hated and took a risk if they showed their faces. A Methodist preacher recorded at Elkhorn Creek, Wisconsin: 'If a speculator should bid on a settler's farm, he was knocked down and dragged out of the [land] office, and if the striker was prosecuted and fined, the settlers paid the fine by common consent among themselves. [But] no jury would find a verdict against a settler in such a case because it was considered self-defense. [So] no speculator dare bid on a settler's land, and as no settler would bid on his neighbor, each man had his land at Congress price, $1.25 an acre.'[21]

All the same, speculation and land dealing were the foundation of many historic fortunes at this time. And powerful politicians (and their friends) benefited too. When a popular figure like General Jackson bid for a potentially valuable town lot, no one bid against him. He acquired his estate and became a reasonably wealthy man through land sales, though by the end of the war he had ceased to be interested in money. His aide, General Coffee, formed the Cypress Land Company, bought land at Muscle Shoals, and laid out the town of Florence, Alabama, where speculators and squatters bid up the government minimum to $78 an acre.[22] Others in the Jackson camp made fortunes this way. The New York politician Martin Van Buren (1782–1862), who became Jackson's Secretary of State, also grew rich through land deals: he got large parcels of land in Otsego County for a fraction of their true value—one 600-acre parcel he bought for $60.90—and he knocked down land cheap at Sheriff's Auctions when settlers were sold up for non-payment of taxes.[23] Of course some land speculation was parasitical and downright antisocial. But large-scale speculators were indispensable in many cases. They organized pressure on Congress to put through roads and they invested capital to build towns like Manchester, Portsmouth, Dayton, Columbus, and Williamsburg. A lot of speculation was on credit, and speculators went bust if they could not sell land quickly at the right price. That was how big groups like the one organized by Sir William Pulteney, the English politician, acquired huge tracts. His agent spent over $1m building infrastructures—stores, mills, taverns, even a theater. A group of bankers from Amsterdam formed the Holland Land Company, which acquired 4 million acres in northwest New York and western Pennsylvania, put in roads and other services, and eventually (1817) made a profit by selling off land in 350-acre plots at $5 an acre (on ten years' credit).[24] But most settlers preferred cheaper land to the use of an infrastructure which they could create for themselves. Moses Cleveland, agent of the Connecticut Land Company, managed to sell good land at a dollar an acre, with five years' credit, and to found the village named after him which became in time a mighty city. It was from Cleveland that William Henry Harrison (1773–1841) played a major role in creating the new state of Ohio, then moved on to Indiana, and finally became America's ninth President.[25]

There is an important historical and economic point to be noted here. Men always abuse freedom, and 19th-century land speculators could be wicked and predatory. But Congress, true to its origins, was prepared to take that risk. It laid down the ground rules by statute and then, in effect, allowed an absolutely free market in land to develop. It calculated that this was the best and quickest way to get the country settled. And it was proved right—freedom worked. In South Africa, Australia, New Zealand, and Canada, the British authorities interfered in the land market in countless ways and from the highest of motives, and as a result these countries—some of which had even bigger natural advantages than the United States—developed far more slowly. One British expert, H. G. Ward, who had witnessed both systems, made a devastating comparison before a House of Commons committee in 1839. In Canada, the government, fearing speculators, had devised a complex system of controls which actually played straight into their hands. By contrast, the American free system attracted multitudes who quickly settled and set up local governments which soon acted as a restraining force on antisocial operators. The system worked because it was simple and corresponded to market forces. 'There is one uniform price at $1.25 an acre [minimum]. No credit is given [by the federal government]. There is a perfect liberty of choice and appropriation at this price. Immense surveys are carried on, to an extent strangers have no conception of. Over 140 million acres have been mapped and planned at a cost of $2,164,000. There is a General Land Office in Washington with 40 subordinate district offices, each having a Registrar and Receiver . . . Maps, plans and information of every kind are accessible to the humblest persons . . . A man if he please may invest a million dollars in land. If he miscalculates it is his own fault. The public, under every circumstance, is the gainer.'[26]

He was right and the proof that the American free system worked is the historic fact—the rapid and successful settlement of the Mississippi Valley. This is one of the decisive events in history. By means of it, America became truly dynamic, emerging from the eastern seaboard bounded by the Appalachians and descending into the great network of river valleys beyond. The Mississippi occupation, involving an area of 1,250,000 square miles, the size of western Europe, marked the point at which the United States ceased to be a small, struggling ex-colony and turned itself into a major nation.[27]

The speed with which representative governments were set up was an important part of this dynamism. In addition to Kentucky and Tennessee, the first trans-Appalachian states, Ohio became a state in 1803, Louisiana in 1812, Indiana in 1816, Mississippi in 1817, Illinois in 1818, Alabama in 1819, Missouri in 1821, Arkansas in 1836, Michigan in 1837. Insuring rapid progress from territory to state was the best way Washington could help the settlement, though under the Constitution it could also build national roads. The first national road, a broad, hardened thoroughfare across the Appalachians, was open in 1818 as far as Wheeling, whence settlers could travel along the Ohio River. By the early 1830s the road had reached Columbus, Ohio. Further south, roads were built by state and federal government in collaboration or by thrusting military men like General Jackson, who in 1820, as commander of the Western Army, strung a road between Florence, Alabama, and New Orleans, the best route into the Lower Mississippi area. There were also the Great Valley Road, the Fall Line Road, and the Upper Federal Road. They were rough by the standards of the new McAdam–Telford roads in Britain but far superior to anything in Latin America, Australasia, or trans-Ural Russia, other vast territories being settled at this time. In addition there were the rivers, most of them facing in the direction of settlement. Even before the steamers came, there were hundreds, then thousands, of flatboats and keelboats to float settlers and their goods downriver. By 1830 there were already 3,000 flatboats floating down the Ohio each year. In 1825 the completion of the Erie Canal, which linked the Atlantic via the Hudson River to the Great Lakes, made easy access possible to the Great Plains. It also confirmed New York's primacy as a port, especially for immigrants, as they could then proceed, via the Canal, straight to new towns in the Midwest. From that point on steamboats were ubiquitous in the Mississippi Valley, not only bringing settlers in but taking produce out to feed and clothe the people of America's explosive cities—only 7 percent of the population in 1810, over a third by mid-century.[28]

The pattern of settlement varied enormously but salient features were common. With every township, the first structure to be built was a church, to serve farming families already scattered around. Then came a newspaper, running off copies even when townsfolk were still living in tents. Then came traders, doubling up as bankers when required, then proper bankers and lawyers at about the same time. The lawyers lived by riding with the local judge on a horseback circuit, by which they became well known, and they sat in the legislature the moment it was set up—so the grip of the attorneys was firm from the start. Justice was fierce and physical, especially for thieves, above all for horse-thieves. By 1815, the pillory, ears cut off and nailed up, and branding on the cheek were becoming rare. But whippings were universal. The tone of settlers' justice was epitomized in Madison County, Tennessee in 1821 when a local thief, 'Squire' Dawson, was sentenced 'to be taken from this place to the common whipping-post, there to receive twenty lashes well laid on his bare back, and that he be rendered infamous, and that he then be imprisoned one hour, and that he make his peace with the state by the payment of one cent.'[29] There we have it—imprisonment was costly, fining pointless when the miscreant had no money, but there was no shortage of bare back.

A typical growth-point was Indianapolis. It was laid out in 1821. The next year it had one two-story house. By 1823 it still had only ninety families but it had already acquired a newspaper, an important engine of urban dynamics. By 1827 the population topped the 1,000 mark, and twenty-one months later a visitor wrote: 'The place begins to look like a town—about 1,000 acres cut smooth, ten stores, six taverns, a court-house which cost $15,000, and many fine houses.'[30] Elijah Miles, who moved to the Sangamon River country in 1823, left a record of how he founded Springfield. It was then only a stake in the ground. He marked out an 18-foot-square site for a store, went to St Louis to buy a 25-ton stock of goods, chartered a boat, shipped his stock to the mouth of the Sangamon, and then had his boat and goods towed upriver by five men with a 300-foot tow-rope. Leaving his goods on the riverside—'As no one lived near, I had no fear of thieves'—he walked 50 miles to Springfield, hired waggons and teams, and so got his stuff to the new 'town,' where his store was the first to open. It was the only one in a district later divided into fourteen counties, so 'Many had to come more than 80 miles to trade.' Springfield grew up around him. They built a jail for $85.75, marked out roads and electoral districts or 'precinks' as they called them, and levied a tax on 'horses, neat cattle, wheeled carriages, stock in trade and distillery.' By 1824 the town had its own roads, juries, an orphanage, a constable, and a clerk. The key figure in such developments was often the county clerk, who doubled as a schoolteacher, being paid half in cash, half in kind.[31]

Although churches were the first structures to go up in most townships, religion flourished without them if necessary. The Second Great Awakening, which started in the 1790s, was essentially a frontier affair, carried out by traveling evangelists, who often held giant camp meetings. The first of these was at Cane Ridge, near Lexington, Kentucky, in 1801, which became the prototype for many more. It was organized by Barton Stone (1772–1844), a Maryland Presbyterian, who described in great detail the evangelical enthusiasm created by these open-air gatherings, where preachers whipped up the participants into frenzies of worship. Stone divided their antics into what he called 'exercises.' Thus in the 'Falling Exercise,' 'The Subject would generally, with a piercing scream, fall like a log on the floor, earth or mud, and appear as dead.' In 'The Jerks,' 'when the head alone was affected, it would be jerked backwards and forwards, or from side to side, so quickly that the features of the face could not be distinguished. When the whole system was affected I have seen the person stand in one place, and jerk backwards and forwards in quick succession, their head nearly touching the floor behind and before.' Then there was the 'Barking Exercise'—'A Person affected by the jerks, especially in his head, would often make a grunt or bark, if you please, from the suddenness of the jerk.' There was also the Laughing Exercise ('loud, hearty laughter . . . it excited laughter in no one else'), a Running Exercise ('the subject running from fear'), a Dancing Exercise ('the smile of heaven shone on the countenance of the subject'), and the Singing Exercise, the sounds issuing not from the mouth but the body—'such music silenced every thing.'[32] These antics may make us laugh, but the fact is they have set the pattern for one form of revivalism for 200 years and are repeated almost exactly by congregations receiving the Toronto Blessing in the 1990s. And the frontier men and women of Cane Ridge and other camp gatherings had some excuse for indulging in these religious ecstasies: they had no other form of entertainment whatever. Religion not only gave meaning to their lives and was a consolation in distress, it was the only relief from the daily hardship of work.[33]

Lyman Beecher (1775–1863), a New Haven Presbyterian who went west and became president of the Cincinnati Theological Seminary— among his other accomplishments was fathering thirteen children, one of them being Harriet Beecher Stowe, author of *Uncle Tom's Cabin*— believed that this revivalist spirit was essential to the creation of the rapidly expanding American nation. Based upon a free market in land and everything else, it was necessarily driven by a strong current of materialistic individualism, and only religious belief and practice, hot and strong, could supply the spiritual leavening and community spirit—could, in effect, civilize this thrusting people. Religion, politics, and culture all went together, he argued, 'and it is plain that the religious and political destiny of the nation is to be decided in the west.' Revivalism, what is now called fundamentalism, was the only way the scattered frontiersmen and women could be reached and gathered. But when the itinerant preachers passed on, all the churches benefited. Some of the older churches, especially the Episcopalians, sniffed at camp-meetings, saying 'More souls are begot than saved there,' but that was because they failed to adapt their evangelism to the new trends. It was the uninhibited Methodists who profited most from revivalism, keeping up the passionate intensity and drumming it into regular, settled congregations. By 1844 they were the biggest church in the United States.[34] Next came the Baptists, radiating from Rhode Island and its great theological seminary, later Brown University (1764). Like most Calvinist sects, they split from time to time, generating such factions as Separatist and Hard-Shell Baptists, but they were enormously successful in the South and West. By 1850 they had penetrated every existing state and had a major theological college in almost all of them.[35]

But revivalism did more than recruit for the existing churches. It created new ones. Thus one Baptist, William Miller (1782–1849), was inspired by the Second Great Awakening to conduct a personal study of the scriptures for two years, and in 1818 declared that 'all the affairs of our present state' would be wound up by God in a quarter of a century, that is in 1844. He recruited many thousands of followers, who composed a hymn-book, *The Millennial Harp*, survived 'The Great Disappointment' when nothing happened in the appointed year, and even the death of their founder. In 1855 they settled at Battle Creek, took the title Seventh-Day Adventists six years later, and eventually, with 2 million worldwide members, became the center of a vast vegetarian breakfast–cereal empire created by John H. Kellogg (1852–1943), first president of Battle Creek College and one of the earliest modern nutritionists.[36]

The way in which the Adventists popularized cereals throughout the world was typical of the creative (and indeed commercial) spirit of the sects which sprang out of the Second Great Awakening. This kind of intense religion seemed to give to the lives of ordinary people a focus and motivation which turned them into pioneers, entrepreneurs, and innovators on a heroic scale. Kellogg himself was the protégé of Ellen G. Harmon (1827–1915), a simple teenager who conceived her vision of sanctified breakfast-food while in a religious transport. And what could be more American than cornflakes, a nutritious food with moral overtones made from the Indian crop which saved the lives of the Pilgrim Fathers? Another very ordinary young man was Joseph Smith (1805–44), born on a hard-scrabble farm in Vermont, who caught a whiff of spirituality from the Second Great Awakening in Palmyra in upstate New York, where in 1827 the Angel Moroni showed him the hiding place of a set of golden tablets. From behind a curtain and with the aid of seer-stones called Urim and Thurim he translated the mystic utterances they contained, which others transcribed to his dictation. This 500-page *Book of Mormon*, put on sale in 1830 (at which point Moroni removed the original plates), describes the history of America's pre-Colombian people, who came from the Tower of Babel, crossing the Atlantic in barges, but survived only in the form of Mormon and his son Moroni, who buried the plates in AD 384. The language of the *Book* clearly derives from the King James Bible but the narrative, with its tribulations overcome by courage and persistence, fits into frontier life well and the movement attracted thousands.

Smith was murdered by an Illinois mob in 1844 but his successor Brigham Young (1801–77), another Vermonter and a man of immense determination (and appetites) and considerable skills of organization, led the Biblical 'remnant' in a historic trek over the plains and mountains to Salt Lake City, 1846–7, where he virtually created the territory of Utah, of which Washington made him governor in 1850. When he proclaimed the doctrine of polygamy in 1852, taking himself twenty-seven wives who bore him fifty-six children, President James Buchanan removed him from office. The row over polygamy (eventually renounced in 1890) delayed Utah's admission as a state until 1896 but it could not prevent Young and his followers from expanding their Church of Latter Day Saints into a world religious empire of over 3 million souls and making the people of Utah among the richest, best educated, and most consistently law-abiding in the United States. In no other instance are the creative nation-building possibilities of evangelical religion so well illustrated.[37]

Some of the by-products of the Second Great Awakening verged on the cranky. When fervent Americans were stirred up by a camp-meeting or a passing preacher, and they found Baptism or Methodism too tame, they had a wide choice of spicier beliefs. The esoterical reinterpretation of the scriptures produced in thirty-eight huge volumes by the 18th-century philosopher Emanuel Swedenborg became an immense quarry into which American sect-founders burrowed industriously for decades. Mesmerism and homeopathy came from Europe but were eagerly adopted and adorned with rococo additions in America. Spiritualism was definitely home-grown. In 1847, John D. Fox, a Methodist farmer who had been 'touched' by the Second Awakening, moved into a Charles Adams house in Hydesville, New York, and the two youngest daughters quickly established contact with a Rapper, at the command 'Here, Mr Splitfoot, do as I do.' Less than two centuries before, this kind of girlish joke-hysteria might have led to witchhunting as at Salem in the 1690s. In mid–19th-century America, already keen on sensation and media-infested, it led to the two girls being signed up by the circus-impresario P. T. Barnum (1810–91) and Horace Greeley (1811–72), the great editor of the *New York Tribune*. So Spiritualism was born. It seems to have had a strong attraction, right from the start, for political liberals, like Robert Owen, son of the utopian community-founder. Owen read a paper about it at the White House in 1861 which led to Abraham Lincoln's memorable observation at the end: 'Well, for those who like that sort of thing, I should think it is just about the sort of thing they would like.'[38] This did not prevent Mrs Lincoln taking it up after the President's death—with its ability to communicate with the dear departed it had a natural attraction for widows. By 1870 Spiritualism had 11 million followers, not only in America but throughout Europe, and it attracted outstanding intellects, like Victor Hugo and William James.[39]

Many of these new sects, which sprang out of the fervor of the 1810s and 1820s, tackled not only the problem of death, like

Spiritualism, but the even more everyday problem of pain. America was already developing one of its most pronounced characteristics, the conviction that no problem is without a solution. Faith-healing flourished in the American mid-century, and Mary Baker Eddy (1821–1910), who suffered dreadful pain in her youth, for which the doctors could do nothing, believed she had been relieved by a Mesmerist, P. P. Quimby; and from this she created her own system of spiritual healing based upon the belief that mind is the only reality and all else an illusion. After her third marriage to Mr Eddy, a first-class businessman, her creed began to flourish on sound commercial principles. She opened the First Church of Christ Scientist in Boston in 1879, followed by the Metaphysical College in 1881 and what became one of America's greatest newspapers, the *Christian Science Monitor*. It quickly spread into 3,200 branches in forty-eight countries.[40] Here again was overwhelming evidence of the new American phenomenon—the way in which religious belief, often of a strange and (some would say) implausible character, produced hugely creative movements with a strong cultural and educational content. Even the most bizarre of these sects founded schools, training colleges for teachers and evangelists, and even universities. Some of America's greatest institutions of higher education have their origins in the Second Great Awakening. It was, for instance, the leading theologian of the Awakening, Charles Grandison Finney (1792–1875), who created Oberlin College in Ohio. The Awakening gave an impulse to Unitarianism, which had come to America in the 1770s and opened King's College Chapel in Boston. The American Unitarian Association was formed in 1825 and quickly radiated all over America. With its rationalist and undogmatic approach to theology and its low-key ritual it particularly attracted intellectuals and scientists, and those of its members with a romantic and utopian disposition tended to set up rustic communities devoted to high thinking and simple living. Ralph Waldo Emerson (1803–82), who moved into the sect from Calvinism, wrote to the British sage Thomas Carlyle in 1840: 'We are all a little wild here with numberless projects of social reform. Not a reading man but has a draft of a new community in his pocket.'[41] Emerson had a finger in a pie of one such, Brook Farm, in West Roxbury, founded by a Boston Unitarian minister, George Ripley. It included on its agricultural committee the novelist Nathaniel Hawthorne (1804–64)—of whom more later—and had a printing press, a kiln for artistic pottery, and a workshop for making furniture. Needless to say it ended in bankruptcy and was cuttingly dismissed by

Carlyle, who epitomized Ripley as 'a Socinian minister who left the pulpit to reform the world by growing onions.'[42]

One writes 'needless to say' but in fact many of the religious–utopian communities, especially the German ones, flourished mightily as commercial or farming enterprises and survive today as models of moral probity, communal tidiness, and capitalist decorum. But others commercialized themselves out of religion altogether. A group of German Pietists under George Rapp (1757–1847) settled in a community at Harmony, Pennsylvania, in 1804, right at the beginning of the Second Awakening. They practiced auricular confession, among other things, and proved highly successful farmers and traders. But as they strictly opposed marriage and procreation, they eventually ceased to exist. At the other end of the sexual spectrum was Oneida Community in western New York, founded by John Humphrey Noyes (1811–86). This originally began as a socialist community, practicing free love, or what was known as 'complex marriage'—procreation, as distinct from other 'sexual transactions,' was decided communally—and the children were brought up as in a kibbutz. The community made itself rich by manufacturing steel traps but eventually lost its faith and became a prosperous corporation.[43]

It is a curious fact that some of these religious sects had very ancient origins but it was only in the free air and vast spaces of America that they blossomed. Thus a medieval sect which in the 14th century developed a Shaking Dance as a form of its ritual (probably derived, via the Crusades, from a Moslem revivalist group known vulgarly as the Whirling Dervishes), continued to shake as Protestant Huguenots in 16th-century France, were expelled by Louis XIV, came to England, mated with a Quaker sect, and became the Shaking Quakers, and were finally brought to America by 'Mother' Anne Lee (1736–84), the visionary daughter of a Manchester blacksmith. These Shakers took advantage of the Second Awakening to develop a number of highly successful utopian communities, distinguished by separation of the sexes, who lived in distinct dormitories, and amazing Spiritualist seances, leading to apparitions, levitation, and spectral voices. They had a frenzied group dance, distantly derived from the Huguenot *camisard*. It was characteristic of the Shakers in their American manifestation that they took the principle of minimalist government to its ultimate conclusion—their many communities, of 100 or more, lived in happiness and content without taxes, spending nothing on police, lawyers, judges, poor-houses, or prisons. They even dispensed with hospitals, believing they had 'special powers' to cure sickness—that may explain, of course, why they are now extinct.[44] (As their founder Miss Lee, known as Mother Ann, believed herself to be 'The Female Principle in Christ,' Jesus being 'The Male Principle,' and taught that the Second Coming would be marked by an assumption of power by women, the sect, whose full title is 'The United Society of Believers in Christ's Second Coming (The Millenarian Church),' is due in a feminist age for a revival.)

The existence of these angular sects, and many others, in addition to the half-dozen or so great 'imperial' religions of American Protestantism, inevitably raised the question, early in the 19th century if not before, of how, granted America's doctrine of religious toleration, all could be fitted into the new republican society. Curiously enough Benjamin Franklin, far-sighted as always, had thought about this problem as early as 1749 when he published his *Proposal Relating to the Education of Youth* in Pennsylvania. He thought the solution was to treat religion as one of the main subjects in the school/college curriculum and relate it to character-training. A similar view was advanced by Jonathan Edwards when president of Princeton. It was, in effect, adopted by the greatest of all American educationalists, Horace Mann (1796–1859), when he began to organize the public school system in Massachusetts. Mann graduated from Brown, became a Unitarian, and, from 1837, was appointed the lawyer–secretary to the new Massachusetts Board of Education. At such he opened the first 'normal' school in the United States at Lexington in 1839, and thence reorganized the entire primary and secondary education system of the state, with longer terms, a more scientific and 'modern' pedagogy, higher salaries and better teachers, decent, clean, and properly heated schoolhouses, and all the elements of a first-class public school system. Massachusetts' framework served as a model for all the other states and Mann, by propaganda and legislative changes during his period in Congress, 1848–53, led the movement which established the right of every American child to a proper education at public expense.[45] Thus the state took over financial responsibility for the education of the new and diverse millions by absorbing most primary and secondary schools (though not tertiary colleges; in 1819 Marshall's Supreme Court, bowing to the eloquence of Daniel Webster (1782–1852), rejected the right of the New Hampshire legislature to interfere in the running of Dartmouth College, thus establishing once and for all the freedom of all America's privately funded universities).

That meant that the true American public school, in accordance with the Constitution, was non-sectarian from the very beginning. Non-sectarian, yes: but not non-religious. Horace Mann agreed with Franklin and the other Founding Fathers that generalized religion and education were inseparable. Mann thought religious instruction in the public schools should be taken 'to the extremest verge to which it can be carried without invading those rights of conscience which are established by the laws of God, and guaranteed by the constitution of the state.' What the schools got was not so much non-denominational religion as a kind of lowest-common-denominator Protestantism, based upon the Bible, the Ten Commandments, and such useful tracts as Bunyan's *Pilgrim's Progress*. As Mann put it, in his final report to the state of Massachusetts, 'that our public schools are not theological seminaries is admitted . . . But our system earnestly inculcates all Christian morals. It founds its morals on the basis of religion; it welcomes the religion of the Bible; it allows it to do what it is allowed to do in no other system, *to speak for itself*.'[46] Hence, in the American system, the school supplied Christian 'character-building' and the parents, at home, topped it up with whichever sectarian trimmings they thought fit (or none).

Naturally there were objections from some religious leaders. On behalf of the Episcopalians, the Rev. F. A. Newton argued that 'a book upon politics, morals or religion, containing no party or sectarian views, will be apt to contain no distinctive views of any kind, and will be likely to leave the mind in a state of doubt and skepticism, much more to be deplored than any party of sectarian bias.' That might apply to dogmatic theology but in terms of moral theology the Mann system worked perfectly well, so long as it was conscientiously applied. Most Episcopalians, or any other Protestants, would now happily settle for the Mann system of moral character-building if they could. So Newton's objections, which were not widely shared, were brushed aside. A more serious question was: how were Roman Catholics, or non-Christians like the Jews, to fit in?

There had been Catholics in America since the foundation of Maryland (1632), and in 1790 Father John Carroll (1735–1815) had been consecrated Bishop of Baltimore with authority over the 40,000 Catholics then in the United States. The following year he founded Georgetown College, America's first Catholic university. But it was only with the arrival of the southern Irish, and Continental Catholics in large numbers, that Catholicism began to constitute a challenge to Protestant paramountcy. New dioceses testified to its expansion even in the South—Charleston 1820, Mobile 1829, Natchez 1837, Little Rock 1843, Galveston 1847—and in Boston and New York City Irish-dominated communities became enormous and potent.[47] The new Catholics brought with them certain institutions which infringed the American moral consensus in the spirit, if not exactly in the letter, almost as much as Mormon polygamy. One was the convent, which provoked a species of Protestant horror-literature infused almost with the venom of the Salem witch-trials. A journal called the *Protestant Vindicator* was founded in 1834 with the specific object of exposing Catholic 'abuses,' the convent being a particular target. The next year saw the publication in Boston of *Six Months in a Convent* and, in 1836, of the notorious *Maria Monk's Awful Disclosures of the Hotel Dieu in Montreal.* It was written and published by a group of New York anti-Catholics who followed it up with *Further Disclosures* and *The Escape of Sister Frances Patrick, Another nun from the Hotel Dieu nunnery in Montreal.* Unlike Continental anticlerical literature about monks and nuns, a genre going back to Rabelais in the 16th century, the *Maria Monk* saga was not directly pornographic but it had something of the same scurrilous appeal. Maria Monk herself was no fiction—she was arrested for picking pockets in a brothel and died in prison in 1849—but her book had sold 300,000 copies by 1860 and it is still in print today, not only in the United States but in many other countries.[48] Nor was Protestant hostility confined to paperbacks. In 1834, even before Maria made her appearance, a convent of Ursuline nuns was burned down by a Boston mob and those responsible were acquitted—Protestant juries believed Catholic convents had subterranean dungeons for the murder and burial of illegitimate children.

There were also widespread fears of a Catholic political and military conspiracy, fears which had existed in one form or another since the 1630s, when they were associated with the designs of Charles I, and which had been resurrected in the 1770s and foisted on George III. In the 1830s, Lyman Beecher, so sensible and rational in many ways, included in his *Plea for the West* details of a Catholic plot to take over the entire Mississippi Valley, the chief conspirators being the pope and the Emperor of Austria. Samuel Morse, who was not particularly pro-Protestant but had been outraged when, during a visit to Rome, his hat had been knocked off by a papal guard when he failed to doff it as the pope passed, added plausibility to Beecher's theory by asserting that the reactionary kings and emperors of Europe were deliberately driving their Catholic subjects to America to promote the takeover. This, combined with labor disputes brought about the willingness of poor Catholic immigrants to accept low rates of pay, led to the founding in 1849 of a secret-oath-bound society, the Order of the Star-Spangled Banner, which flourished in New York and other cities. It was geared to politics by opposing the willingness of the Democratic Party machine to cater for Catholic votes and when its members were questioned about its activities they were drilled to answer 'I Know Nothing.' The Know Nothing Party had a brief but phenomenal growth in the early 1850s, especially in 1852, when it triumphed in local and state elections from New Hampshire to Texas. In 1856 it even ran ex-President Millard Fillmore as a national candidate, but it was doomed by its pro-slavery Southern leadership.[49]

The Catholics were thus put on the defensive. And some of them, in any event, had reservations about the Horace Mann approach to education. The most incisive Catholic convert of the time, Orestes Brownson (1803–76), argued that the state had no obligation to educate its citizens morally and that to do so on a lowest-common-denominator basis would promote a bland, platitudinous form of public discourse. America, he argued, needed the provocation and moral judgments which only Biblical religion could provide and the stimulation of religious controversy between competing sects.[50] But most American Catholics, then and later, wanted badly to win the acceptance of fellow-Americans by fitting into the citizenship formula. And, less defensively and more enthusiastically, they accepted the fact that America had a free market in religion as well as everything else. From the 1830s they competed eagerly to build the most churches and schools and colleges, to display the largest congregations, win the most converts, and demonstrate that Catholics were more American and better citizens than members of other sects.[51]

The Jews did not proselytize like the Catholics but they competed in other ways and they were just as anxious to demonstrate their Americanism. In 1654 the French privateer *St Catherine* brought twenty-three Jewish refugees from Recife in Brazil to the Dutch colonial town of New Amsterdam. The governor, Peter Stuyvesant, protested to the Dutch West India Company against the settlement of what he called 'a deceitful race' whose 'abominable religion' worshiped 'the feet of mammon.' They were denied all rights of citizenship and forbidden to build a synagogue. But when New Amsterdam fell to the English in 1664 and became New York, the Jews benefited from a decision taken under the English Commonwealth regime, later confirmed by Charles II, to allow them to acquire all the rights of English citizenship 'so long as they demean themselves peaceably and quietly, with due obedience to His Majesty's laws and without scandal to his government.' Some early statutes and proclamations, stressing religious liberty, included only 'those who profess Christianity' in this freedom of worship. But in fact the Jews were never directly persecuted on American soil and the great governor of New York, Edmund Andros, went out of his way to include Jews when he promised equal treatment to all law-abiding persons 'of what religion soever.' As in England, the issue of Jewishness was not raised. Jews simply came, enjoyed equal rights, and, it seems, voted in the earliest elections; they held offices too.[52]

Jews settled in other areas, beginning with the Delaware Valley. Some difficulties arose when the Jews wished to have their own cemetery in New York. But in 1677 one was opened in Newport, Rhode Island—later the subject of one of Longfellow's finest poems—and New York got its own five years later. In 1730 the Shearith Israel Congregation of New York consecrated its first synagogue and a particularly handsome one was built in Newport in 1763, now a national shrine. Even in colonial times, Jews' existence in America was fundamentally unlike the life they lived in Europe. There, they had their own legal status, ran their own courts, schools, shops, paid their own special, heavier taxes, and usually lived in ghettos. In America, where there was no religiously determined law, there was no reason why Jews should operate a separate legal system, except on matters which could be seen as merely internal religious discipline. Since in America all religious groups had equal rights, there was no point in constituting itself into a separate community. All could participate fully in a communal society. Hence from the start the Jews in America were organized not on communal but on congregational lines, like the other churches. In Europe, the synagogue was merely one organ of the all-embracing Jewish community. In America it was the only governing body in Jewish life. American Jews did not belong to the 'Jewish community,' as in Europe. They belonged to a particular synagogue. It might be Sephardi or Ashkenazi and, if the latter, it might be German, English, Polish, or 'Holland,' all of them differing on small ritual points. Protestant groups were divided on similar lines. Hence a Jew went to 'his' synagogue just as a Protestant went to 'his' church. In other respects, Jews and Protestants were simply part of the general citizenry, in which they merged as secular units. Thus the Jews in America, without in any way renouncing their religion, began to experience integration for the first time. And this inevitably meant accepting the generalized morality of the consensus, in which religious education was 'character-training' and part of the preparation for living an adult republican life.[53]

But if even Roman Catholics and Jews could join in the American republican moral consensus, there was one point on which it broke down completely—slavery. One sees why St Paul was not anxious to tackle the subject directly: once slavery takes hold, religious injunctions tend to fit its needs, not vice versa. On the other hand, the general thrust of the Judeo-Christian tradition tended to be anti-slavery, and that was why it had slowly disappeared in Europe in the early Middle Ages. In America the moral and political dilemma over slavery had been there right from the start, since by a sinister coincidence 1619 marked the beginning of both slavery and representative government. But it had inevitably become more acute, since the identification of American moral Christianity, its undefined national religion, with democracy made slavery come to seem both an offense against God and an offense against the nation. Ultimately the American religious impulse and slavery were incompatible. Hence the Second Great Awakening, with its huge intensification of religious passion, sounded the death-knell of American slavery just as the First Awakening had sounded the death-knell of British colonialism.

Religion would have swept away slavery in America without difficulty early in the 19th century but for one thing: cotton. It was this little, two-syllable word which turned American slave-holding into a mighty political force and so made the Civil War inevitable. And cotton, in terms of humanity and its needs, was an unmitigated good. Thus do the workings of mysterious providence balance good with evil. Until the end of the 18th century, the human race had always been unsuitably clothed in garments which were difficult to wash and therefore filthy. Cotton offered an escape from this misery, worn next to the skin in cold countries, as a complete garment in hot ones. The trouble with cotton was its expense. Until the industrialization of the cotton industry, to produce a pound of cotton thread took twelve to fourteen man-days, as against six for silk, two to five for linen, and one to two for wool. With fine cotton muslin, the most sought after, the value-added multiple from raw material to finished product was as high as 900.[54] This acted as a spur to mechanical invention. The arrival of the

Arkwright spinning-machine and the Hargreaves jenny in the England of the 1770s meant that, whereas in 1765 half a million pounds of cotton had been spun in England, all of it by hand, by 1784 the total was 12 million, all by machine. Next year the big Boulton & Watt steam engines were introduced to power the cotton-spinning machines. This was the Big Bang of the first Industrial Revolution. By 1812 the cost of cotton yarn had fallen by 90 percent. Then came a second wave of mechanical innovation. By the early 1860s the price of cotton cloth, in terms of gold bullion, was less than 1 percent of what it had been in 1784, when the industry was already mechanized. There is no instance in world history of the price of a product in potentially universal demand coming down so fast. As a result, hundreds of millions of people, all over the world, were able to dress comfortably and cleanly at last.[55]

But there was a price to be paid, and the black slaves paid it. The new British cotton industry was ravenous for raw cotton. As the demand grew, the American South first began to grow cotton for export in the 1780s. The first American cotton bale arrived in Liverpool in 1784. Then, abruptly, at the turn of the century, American exports were transformed by the widespread introduction of the cotton gin. This was the invention of Eli Whitney (1765–1825). His was a case, common at this time, of a natural mechanical genius. He came from a poor farm in Massachusetts and discovered his talent by working on primitive agricultural machinery. Then he worked his way through Yale as an engineer. In 1793, while on holiday at Mulberry Grove, Savannah, the plantation of Mrs Nathaniel Green, he became fascinated by the supposedly intractable problem of separating the cotton lint from the seeds—the factor which made raw cotton costly to process. Watching a cat claw a chicken and end up with clawfuls of mere feathers, he produced a solid wooden cylinder with headless nails and a grid to keep out the seeds, while the lint was pulled through by spikes, a revolving brush cleaning them. The supreme virtue of this simple but brilliant idea was that the machine was so cheap to make and easy to operate. A slave on a plantation, using a gin, could produce 50 pounds of cotton a day instead of one. Whitney patented his invention in 1794 but it was instantly pirated and brought him in eventually no more than $100,000—not much for one of history's greatest gadgets. But by 1800–10 his gins had made the United States the chief supplier of cotton to the British manufacturing industry's rapidly rising demand.[56] In 1810 Britain was consuming 79 million pounds of raw cotton, of which 48 percent came from the American South. Twenty years later, imports were 248 million, 70 percent coming from the South. In 1860 the total was over 1,000 million pounds, 92 percent from Southern plantations. During the same period, the cost (in Liverpool landing prices) fell from 45 cents a pound to as low as 28 cents.[57]

It testifies to the extraordinary fertility of American genius that the country could produce two such men as Whitney and Fulton in one generation. It is a matter of fine judgment who was the more creative. Whitney is often associated solely with the cotton gin. That is a grave injustice to his genius. Indeed he is a fascinating example of the complex impact one man can have on history. Whitney was a dour, single-minded Puritan type, a lifelong bachelor interested only in his job, a secular hermit, driven by the Calvinist work-ethic. He lived simply in a farmhouse and his 'factory' was never more than a series of crude workshops, a cottage industry, at Mill Rock, New Haven. But he had many assistants and apprentices, some of whom did not like working as hard as he did, until they dropped asleep on the floor. They ran away and he had to chase them to get them back. He built a firearms factory in 1798 but was always short of capital. Congress denied his petition to get his gin patent renewed and, during the war of 1812, he had to go direct to President Madison for money. America had no proper capital market at this time. Whitney thought not merely in terms of single new ideas but of whole processes. He grasped that the way to produce machinery or products in vast quantities at low prices was to achieve interchangeability of parts, uniformity, standardization, on a scale never before imagined. He called this the 'American System.' His firearms factory was the first realization of it.

Whitney's determination to introduce this system was adamantine and was laughed at by the British and French ordnance officers to whom he explained it.[58] They said it denied the craftsman's individuality. Well, of course it did. But labor costs in America were so high that the craftsman was a luxury. Whitney realized that for America to overtake Britain in manufactures it was necessary to bypass the craftsman with a workforce of easily trained, semi-skilled men recruited from the waves of immigrants. America was a place where an industrial worker could save up enough in three years to buy a farm, and no immigrant would stay in the city in manufacturing industry if he could become an independent, landowning farmer. So the thrust to reduce the industrial headcount was enormous, and Whitney showed the way ahead. His

'American System' caught on in the earliest stages of the American Industrial Revolution. As early as 1835, the British politician and industrialist Richard Cobden, visiting America, said that its labor-saving machinery was superior to anything in Britain. By the 1850s, British experts marveled at what they found in the United States—standardized products mass-produced by machine methods including doors, furniture, and other woodwork, boots, shoes, plows, mowing machines, wood screws, files, nails, locks, clocks, small-arms, nuts, bolts—the list was endless.[59] Virtually all this industry was located north of the slave-line. So if Whitney's cotton gin enabled the slave-system to survive and thrive, his 'American System' also gave the North the industrial muscle to crush the defenders of slavery in due course.[60]

The huge growth in the cotton industry made possible by Whitney's genius—it rose at 7 percent compound annually—soon made cotton not only America's largest export but the biggest single source of its growing wealth. It also created 'the South' as a special phenomenon, a culture, a cast of mind. And this in turn was the consequence of General Jackson's destruction of Indian and Spanish power in the lower Mississippi Valley. The Treaty of Fort Jackson was only the first of five in which the Indians were deprived of virtually all the land they had in the whole of this vast area. The Old South—the Carolinas, Virginia, Georgia—was not suited to growing cotton on a large scale; if anything it was tobacco country. The new states Jackson's ruthlessness brought into being, Alabama, Mississippi, and Louisiana, now constituted the Deep South where cotton was king. The population of the states multiplied threefold in the years 1810–30. It was internal migration, settlers moving from New England, where land was now scarce and the Old South, where it was exhausted. James Graham, a North Carolina tobacco planter, wrote to a friend on November 9, 1817: 'The *Alabama Fever* rages here with great violence and has *carried off* vast numbers of our Citizens.'[61]

This migration moved the plantation system from Virginia, the Carolinas, and coastal Georgia to West Georgia, West Tennessee, and the Deep South. But both Old and New South were still linked by chains of slavery. Before the cotton boom, the price of slaves in America had been falling—in the quarter-century 1775–1800 it dropped by 50 percent. In the half-century 1800–50 it rose in real terms from about $50 per slave to $800–$1,000. For every 100 acres under cotton in the Deep South, you needed at least ten and possibly twenty slaves. The Old South was unsuited to cotton but its plantations could and did breed slaves in growing numbers. The US Constitution had prevented Congress from abolishing the slave-trade (as distinct from slavery itself) until 1808. In fact all the states had ended the legal importation of slaves by 1803, and Congress was able to exercise the power to ban the trade from 1808. This had the effect of further increasing the value of the home-bred variety, and slave-breeding now became the chief source of revenue on many of the old tobacco plantations.[62]

The Founding Fathers from Virginia who owned slaves, like Jackson and Madison, and who hated slavery, had taken consolation from their belief that it was an outmoded and inefficient institution which would die out naturally or be easy to abolish. Madison 'spoke often and anxiously of slave property as the worst possible for profit.' He used to say that Richard Rush's 10-acre free farm near Philadelphia brought in more money than his own 2,000-acre one worked by slaves. And it is true that in Russia serfdom, the form of slavery practiced there, was economically outmoded and slowly dying. But in America Madison's views were out of date by 1810. It is a horrible fact that modern economics and high technology do not always work in favor of justice and freedom.[63] Thanks to slavery, a cotton plantation could be laid out and in full production in two years. It was possible to harvest a crop even in one year, and 'a man who stood in a wilderness fewer than 12 months ago now stood at a dock watching his crop load out for the English factory towns.' The frontiersman thus became part of a commercial economy and 'cotton made it possible for a man to hang a crystal chandelier in his frontier log cabin.'[64] Early in 1823 a man in western Georgia planted cleared land with cotton, sold it in May with the crop established, cleared land in Alabama that autumn, planted and sold the farm, and then repeated the process in Mississippi; he ended up with 1,000 good acres freehold, which had cost him $1,250 and two years' work.[65] But of course this rapidity would have been impossible without slavery, which made it easy to carry your workforce with you and switch it at will. Slaves made fortunes for those who owned and skillfully exploited them. There were thousands of small planters as well as big ones. A few plantations were worked by the white families which owned them. But over 90 percent had slaves. By the early 1820s a new kind of large-scale specialist cotton plantation, worked by hundreds of slaves, began to dominate the trade.

The big plantations were in turn supplied by specialist, highly commercial slave-breeding plantations. With monogamous marriages, only

10–15 percent of female slaves produced a child a year. Plantations which sold slaves for the market insured regular provision of sires for all nubile females, so that up to 40 percent of female slaves produced a child each year.[66] The notion that Southern slavery was an old-fashioned institution, a hangover from the past, was false. It was a product of the Industrial Revolution, high technology, and the commercial spirit catering for mass markets of hundreds of millions worldwide. It was very much part of the new modern world. That is why it proved so difficult to eradicate. The value of the slaves themselves formed up to 35 percent of the entire capital of the South. By mid-century their value was over $2 billion in gold; that was one reason compensation was ruled out—it would have amounted to at least ten times the entire federal budget.

With so much money invested in slavery it was not surprising that the South ceased to apologize for slavery and began to defend it. This was a slow process to begin with. In 1816 James Monroe (1758–1831), Madison's right-hand man, succeeded him as president after an easy election. He was the last of the great Virginia dynasty and he shared with his predecessors the anomaly that he owned slaves all his life but wanted to abolish slavery. He had been born in the famous Westmoreland County, attended William and Mary, served in the Revolutionary War, studied law with Jefferson, served with the Virginia legislature and the Continental Congress, had been senator, envoy in France, governor of Virginia, had negotiated the Louisiana Purchase, and then done eight years as secretary of state. As president he was surprisingly popular, if only because he was a change from the unsuccessful Madison, and he was reelected in 1820 almost unanimously. Monroe was a dull man, very conservative in most things, the last President to wear a powdered wig, knee-breeches, and cocked hat, soft-spoken, well-mannered, prudish, and careful. Jefferson said that, if turned inside out, he would be found spotless.[67]

Monroe is known to history chiefly as the author of the Monroe Doctrine (1823), of which more later. He ought to be better known for his promotion of the scheme to solve the slavery problem by repatriating freed blacks to Africa. This was not just a humbugging tactical move on the part of conscience-stricken slave-owners like himself. It was also backed by the powerful Evangelical anti-slavery lobby in Britain, who set up the first repatriation colony in Sierra Leone. So it was a 'liberal' solution, or seemed so at the time. In 1819 Monroe supported Congressional legislation to set up a similar, American-sponsored colony in West Africa, to be called Liberia. Being a Jeffersonian he could not actually bring himself to allow the United States to purchase land for this purpose. But he assisted in other ways, so that when the colony got going in 1824 its capital was named Monrovia after him. Some freed slaves did go to Liberia, where they immediately set themselves up as a ruling caste over the local Africans, a prime source of the country's poverty and its ferocious civil wars, which continue to this day. American blacks seem to have realized instinctively that it would not work, that they were better off in America even as slaves than in Africa. They were scared of being sent there. Ten years after its foundation, Madison sold sixteen of his able-bodied slaves to a kinsman for $6,000, they giving 'their glad consent' because of 'their horror of Liberia.'[68]

By the time of Monroe's presidency, however, many Southern whites, especially their political leaders, were brazenly defending slavery, not as an unavoidable evil but as a positive blessing for blacks and whites alike. Christian churchmen joined in this campaign as best they could. As early as 1822 the South Carolina Baptist Association produced a Biblical defense of slavery. There was a notable closing of Southern Christian ranks after the black preacher Nat Turner led a Virginia slave-revolt in 1831, in which fifty-seven whites were killed. In 1844 Bishop John England of Charleston provided an elaborate theological justification to ease the consciences of Catholic slave-owners.

To understand the level of sophistication, and the passionate sincerity, with which slavery was defended we must look at the case of John C. Calhoun (1782–1850) of South Carolina. Calhoun was one of the greatest of all American political figures, a distinguished member of both Houses of Congress, a superb orator, a notable member of the Cabinet, and a political theorist of no mean accomplishment. His *Disquisition on Government* and *Discourse on the Constitution and Government of the United States* (published together in 1851) deserve to rank with Jefferson on Virginia and the writings of Woodrow Wilson.[69] Calhoun was of Ulster Scots-Irish origin, son of a semi-literate Indian fighter, born a penniless boy with natural good looks, enormous charm, and wonderful brainpower, very much in the tradition of Edmund Burke and Richard Brinsley Sheridan, taking to politics as if he had been born to the purple. In the year of Jefferson's inauguration, he was an eighteen-year-old farmer with virtually no formal education. Ten years later he had graduated brilliantly from Yale, got himself elected to

Congress, and found a beautiful bride, Floride Bouneau, heiress to a large plantation in Abbeville, South Carolina. Studying Calhoun's life gives one a striking picture of the way Americans of strong character transformed themselves in a mere generation. In his childhood, life in the Carolina backwoods was wild—literally: the last panthers were not killed till 1797 and the state paid bounties for their skins, and wolf-skins too. One maternal uncle had been killed by the Tories in cold blood (his mother, like Jackson's, was a bitter hater of the English), another had been 'butchered by thirty sabre-cuts and a third immured for nine months in the dungeons of St Augustine.'[70] His grandmother had been murdered and one of his aunts kidnapped by Indians. There were many ambushes and scalpings, and his father's old hat, with four musket holes in it, was a family treasure. Despite a lack of education, his father became an expert surveyor (like Washington) and built up a holding of 1,200 acres. But they were poor. A contemporary historian, the Rev. Charles Woodmason, said the 'cabins swarmed with children' but 'in many places have naught but a gourd to drink out of, not a plate, knife or spoon, glass, cup or anything.'[71] The Calhouns were among those who organized the church and school, served as justices, and tried to civilize the place a little. In those days justice was do-it-yourself, carried out by bands known as the Regulators, who hanged murderers and thieves. Calhoun's father was a tax-collector who also supervised elections, and served in the state legislature for thirty years. He owned thirty-one slaves and referred to 'my family, black and white.'[72] He died when Calhoun was only thirteen, so the boy had to run the estate as a teenager and at the same time get himself through Yale.

Calhoun had two years at college, where he studied law and revealed his obstinacy, clashing with the president, Timothy Dwight, over politics—Dwight was anti-Jefferson. The boy's dissertation was 'The Qualifications Necessary to Make a Statesman.' He had them. He was probably the ablest public man America ever produced and it is not surprising that there were later rumors that he was the real father of Abraham Lincoln. (But the same was said of John Marshall.) Calhoun got to Washington via Charleston politics. In those days it was an uproarious place, a man's town, the women taking refuge in church. It was also the slave capital of the hemisphere—40 percent of Africans came through Charleston, 'the Ellis Island of Black Americans.'[73]

Calhoun grew up six feet two with stiff black hair standing straight up from his head. He read himself into literature, classical and modern, and used his prodigious memory to good effect. His manners were those of the best kind of 18th-century gentleman. The journalist Ann Royall was dazzled by what she called his 'personal beauty' and 'frank and courteous manners—a model of perfection.' Harriet Martineau called him the 'Iron Man,' inflexible in his principles and conduct—'the only way to get him to change his mind was to appeal to his honor.' His enemy Daniel Webster called him 'a true man.' George Ticknor said he was 'the most agreeable man in conversation in Washington.' Margaret Bayard Smith referred to his 'splendid eye' and his face 'stamped with nature's aristocracy.'

In his life, oratory, and writings, Calhoun tried to tackle one of the great problems of the modern age: how to reconcile centralized and democratic power with the demands of people in unequal and different communities, small and large, to control their own lives. The problem has not been solved to this day, even theoretically. He argued that the political war against the South, and slavery, was being fought mainly by powerful lobbies rather than by the democratic wish of the people: he detected, very early on, the threat to American democracy represented by the lobby system, already growing. Thanks to the slaves he and his wife owned, he could pursue a public career, first in Charleston, then in Washington, of completely disinterested public service. Exactly the same argument had been used in defense of slavery in 5th-century BC Athens.

The incongruities of this defense were revealed in a striking passage written by an Englishman, G. W. Featherstonehaugh, who was in the Carolinas in 1834.[74] A traveling companion told him: 'In the North every young man has to scramble rapaciously to make his fortune but in the South the handing down of slave plantations from father to son breeds gentlemen who put honor before profit and are always jealous of their own, and are the natural friends of public liberty.' The speaker, an educated man from South Carolina College, cited Calhoun as an example of what he meant. He had 'the dignity which had belonged to Southern gentlemen from Washington down to the present time.' He had 'never been known to do a mean action in his life.' In public life he 'never omitted a chance to vindicate the Constitution from the attempts of sordid people to violate its intentions.' This disquisition was interrupted by the arrival of the coach, which had a captured runaway slave in chains on its top. Inside Featherstonehaugh was joined by a white man in chains and a deputy sheriff. The man was about to be hanged for killing a slave in a card-game. The sentence had been passed on him not for murder but for breaking the law against gambling with slaves. A bottle was handed around and they all got drunk, as equals. So 'shut up as I was in a vehicle with such a horrid combination of beings,' he reflected on the cultural paradoxes of the Old South. The traveler was later entertained by Calhoun in his mansion, Fort Hill; it was 'like spending an evening in a gracious Tuscan villa with a Roman senator.'

In Congress 1811–17, Calhoun instantly made his mark as an eloquent War Hawk and was soon chairman of the Committee on Foreign Affairs. After serving Monroe as secretary of war, 1817–25, he won election as vice-president and ran the Senate 1825–32. He favored the War of 1812 because he wanted America to annex Florida and Texas and turn them into slave states. We come here to the key mechanism in the political battle over slavery—the need of the South to extend it, state by state, in order to preserve its share in the power-balance of Congress. The South felt it could not sit still and fight a defensive battle to preserve slavery, because the population of the North was rising much faster and non-slave states were being added all the time. Once the non-slave states controlled not just the House but the Senate too, they could change the Constitution. So the South had to be aggressive, and it was that which eventually led to the Civil War. As we have seen, the Constitution said little about slavery. Article I, stating the three-fifths rule, merely speaks of 'free persons' and 'other persons' (slaves). But more significant was Article IV, Section 2, Paragraph 3: 'No person held to Service or Labor in one State, under the Laws thereof, escaping into another, shall, in Consequence of any Law or Regulation therein, be discharged from such Service or Labor, but shall be delivered up on Claim of the Party to whom such Service or Labor may be due.' This obliged free states to hand over runaways. The South was terrified of a constitutional amendment abolishing this clause which would lead to a mass escape of slaves across its unpoliced borders. The constitutional duty to hand over escaped slaves caused more hatred, anger, and venom on both sides of the slave line than any other issue and was a prime cause of the eventual conflict. And it was fear of losing this constitutional guarantee which determined the tactics of the South in creating new states.

In February 1819 Congress faced a demand from Missouri to become a state, as its population had passed the 60,000 mark. There were then eleven slave and eleven free states. The line between them was defined by the southern and western boundaries of Pennsylvania. This line had been determined by a survey conducted by the English astronomers Charles Mason and Jeremiah Dixon in 1763–7, to settle disputes between Pennsylvania and Maryland. So it was known, then and ever after, as the Mason–Dixon Line, the boundary between freedom and slavery, North and South. By 1819 slavery, though still existing in some places in the North, was rapidly being extinguished. But no attempt had yet been made to extend the dividing line into the Louisiana Purchase territory, let alone beyond it, though the area was being rapidly settled. Missouri already had 10,000 slaves and was acquiring more. It was obviously going to become another slave state if allowed statehood.

A New York congressman now introduced an anti-slavery measure, which prohibited the introduction of more slaves into the Territory, and automatically freed any slaves born after it became a state on their twenty-fifth birthday. In short, this would have turned Missouri from a slave territory into a free state. The measure passed the House, where the free states already had a majority of 105 to 81, but was rejected by the Senate, where the numbers were equal, 22–22. The Senate went further and agreed to statehood being given to Maine, which had long wanted to be separate from Massachusetts, and which of course was free, provided Missouri were admitted as a slave state, thus keeping the balance in the Senate 26–26. This was agreed on March 2, 1820 by a narrow vote in the House. But a further crisis arose when the pro-slavery majority in Missouri's constitutional convention insisted it contain a clause prohibiting free blacks and mulattos from settling in the new state. This infringed Article IV, Section 2, of the Constitution of the United States: 'The Citizens of each State shall be entitled to all Privileges and Immunities of Citizens in the Several States.' Free blacks were citizens of a number of states, including slave states like North Carolina and Tennessee—indeed they even voted there until taken off the rolls in the 1830s.[75]

Might the row over slavery in Missouri have led to a breakdown in the Union? Some people thought it could. Jefferson wrote to a friend: 'This momentous question, like a firebell in the night, awakened and filled me with terror. I considered it at once as the knell of the Union.' There had been some in New England who wanted secession over the War of 1812. Would the South now secede over the refusal of the North to agree to the extension of slavery? John Quincy Adams, now Secretary of State, thought this the logical and even the moral solution. Adams did not see how North and South could continue to live together. He noted grimly in his diaries in March 1820 a conversation he had had with his Cabinet colleague, Calhoun (then War Secretary). Calhoun told him that in his state, South Carolina, 'domestic labor was confined to the blacks and such was the prejudice that if he, who was the most popular man in the district, were to keep a white servant in his house, his character and reputation would be irretrievably ruined . . . It did not apply to all kinds of labor—not for example to farming. Manufacturing and mechanical labor was not degrading. It was only manual labor, the proper work of slaves. No white person could descend to that. And it was the best guarantee of equality among the whites.' Adams commented savagely on Calhoun's admissions: 'In the abstract, [Southerners] say that slavery is an evil. But when probed to the quick on it they show at the bottom their soul's pride and vainglory in their condition of masterdom.'[76]

Adams' moral condemnation of the South ignored the fact that, throughout the North, discrimination against blacks was universal and often enshrined in statutes. In Pennsylvania, for instance, special measures were taken to guard against black crime, the governor of the state insisting that blacks had a peculiar propensity to commit assaults, robberies, and burglaries. Both Ohio and Indiana had a legal requirement that, on entering the state, a black must post a bond for $500 as a guarantee of good behavior. In 1821 New York State's constitutional convention virtually adopted manhood suffrage: anyone who possessed a freehold, paid taxes, had served in the state militia, or had even worked on the state highways could vote—but only if he was 'white.' It actually increased the property qualification for blacks from $100 to $250. Pennsylvania also adopted manhood suffrage in 1838, but on a 'whites only' basis. Anti-black color bars were usual in trade unions, especially craft ones.[77] Adams was well aware that in Europe the North's color bars already shocked educated people. When he was minister in St Petersburg, nobles who were quite happy to beat one of their serfs to death with the knout looked down on Americans as uncivilized because of their treatment of blacks—a foretaste of 20th-century anti-Americanism. He noted (August 5, 1812): 'After dinner I had a visit from Claud Gabriel the black man in the Emperor [Alexander I]'s service, who went to America last summer with his wife and children, and who is now come back [to St Petersburg] with them. He complained of having been very ill-treated in America, and that he was obliged to lay aside his superb dress and saber, which he had been ordered to wear, but which occasioned people to insult and beat him.'[78] It was already known for 'reactionary' European regimes to pay honors to American blacks as a way of demonstrating the hypocrisy of American egalitarianism.

Adams did not deny the humbug of much Northern opposition to slavery but brushed it aside and concentrated on the main issue: the absolute need to end it as a lawful institution. It was his view that slavery made Southerners, who had a sense of masterdom that Northerners did not feel, look down on their fellow-Americans, thus undermining the Union at its very heart. He noted: 'It is among the evils of slavery that it taints the very sources of moral principle. It establishes false estimates of virtue and vice.' Hence, he concluded, 'If the union must be dissolved, slavery is precisely the question on which it ought to break!' Adams was apocalyptic on slavery. He dismissed the African colonization schemes which Madison and other Southern moderates favored as contemptible attempts to pass the responsibility for their crimes onto the federal government—they were, he snarled, 'ravenous as panthers' to get Congress to fund their guilt-ridden schemes. He noted sardonically that, in another of his heart-to-hearts over slavery with Calhoun, the latter admitted that, if the Union dissolved over the issue, the South would have to form a political, economic, and military alliance with Great Britain. 'I said that would be returning to the colonial state. He said Yes, pretty much, but it would be forced upon them.' To the furiously moral Adams, it was only to be expected that the evil defenders of a wicked institution should, in order to perpetuate it, ally themselves with the grand depository of international immorality, the British throne. To him, if the Union could be preserved only at the price of retaining slavery, it were better it should end, especially since in the break-up slavery itself would perish:

> If slavery be the destined sword in the hands of the destroying angel which is to sever the ties of this union, the same sword will cut asunder the bonds of slavery itself. A dissolution of the Union for the cause of slavery would be followed by a servile war in the slave-holding states, combined with a war between the two severed portions of the Union . . . its result must be the extirpation of slavery from this whole continent and, calamitous and devastating as this course of events must be, so Glorious would be the final issue that, as God should judge me, I dare not say it is not to be desired.[79]

With high-placed statesmen talking in the exalted and irreconcilable terms that Adams and (to a lesser extent) Calhoun employed, it is a wonder that the United States did not indeed break up in the 1820s. And if it had done the South would undoubtedly have survived. It was then beyond the physical resources of the North to coerce it, as it did in the 1860s. Moreover, Calhoun was probably right in supposing that Britain, for a variety of reasons, would have come to the rescue of the South, preferring to deal with America as two weak entities, rather than one strong one. The course of American history would thus have been totally different, with both North and South racing each other to the Pacific, recruiting new territory, just as Canada and the United States did on either side of the 49th parallel. However, it must be noted that Adams came from Massachusetts and Calhoun from South Carolina, the two extremist states. Many Americans believed—General Grant was one—that, when Civil War finally came, these two states bore the chief responsibility for it; that, without them, it could have been avoided. These were, on both sides of the argument, the ideological states, the upholders of the tradition of fanaticism which was one part of the American national character, and a very fruitful and creative part in many ways. But there was the other side to the national character, the moderate, pragmatic, and statesmanlike side, derived from the old English tradition of the common law and parliament, which taught that ideological lines should not be pursued to their bitter and usually bloody end, but that efforts should be made to achieve a compromise always.

This second tradition, upheld for so long by the Virginia patricians, the Washingtons and Madisons, now fell firmly into the capable hands of a man from the new West, Kentucky's Henry Clay. Henry Clay is a key figure in the period 1815–50, who three times averted complete breakdown between North and South by his political skills.[80] He was also a man of extraordinary energy and ability, perhaps the ablest man, next to Calhoun, who never quite managed to become president of the United States—though, God knows, he tried hard enough. Clay was from Virginia, as American as they came. The Clays got to Jamestown in 1613 (from Wales) and Clay was fifth generation. His mother was third generation. Clay was born in 1777, to a Baptist preacher and tobacco farmer with a 464-acre homestead and twenty-one slaves in the low-lying marshlands or Slashes. His father hated the British, especially Colonel Banastre Tarleton, who ravaged the area, and they hated him: it was Clay's story that redcoats desecrated his father's grave, looking for treasure. When he was four his father died, and Clay inherited two slaves from him (and one from his grandfather). So he owned slaves until his death in 1852 but called slavery a 'great evil,' imposed by 'our ancestors,' and that it was against the Constitution, which, in his opinion, extended equality to blacks 'as an abstract principle.' If, he said, America could start all over again, slavery would never be admitted. Clay's desperate attempts to hold the balance within the Union on slavery made him a particular favorite of Abraham Lincoln, who, during his famous debates with Stephen A. Douglas, quoted Clay forty-one times.[81]

It was Clay's fate to be born to and live with extraordinary personal tragedy. Of his eight sisters and brothers, only two survived childhood and his widowed mother was an embittered woman in consequence. So was his wife, Lucretia Hart. Of their eleven children, two, Henrietta and Laura, died in infancy, Eliza at twelve, Lucretia at fourteen, Susan at twenty, Anne at twenty-eight; the eldest son, Theodore, spent most of his life in a lunatic asylum, and the second, Thomas, became an alcoholic; the third, Henry Jr, a brilliant graduate of West Point, in whom Clay put all his hopes, was killed in the Mexican War.[82] Clay regarded his background as poverty-stricken and for political purposes exaggerated it. He said he was a 'self-made man,' a term he invented, 'an orphan who never recognised a father's smile . . . I inherited [nothing but] infancy, ignorance and indigence.' He regretted to the end of his days that he never learned Latin and Greek—'I always relied too much on the resources of my genius' (Clay was not a conspicuously modest man). On the other hand, Clay had beautiful handwriting and, after working as an errand boy and drugstore clerk, he went to work in the Virginia Chancellery under the great George Wythe, who had trained Jefferson, Monroe, and Marshall. Wythe turned Clay into a capable lawyer and a polished gentleman. So Clay got into Richmond society, where he broke hearts and made enemies as well as connections.[83] But Virginia swarmed with underemployed lawyers, so Clay went to Kentucky to make his fortune.

'The dark and bloody ground' was an Indian name for Kentucky, but it fitted the origins of this border state. Settlers there were described as 'blue beards, who are rugged, dirty, brawling, browbeating monsters, six feet high, whose vocation is robbing, drinking, fighting and terrifying every peaeable man in the community.'[84] Clay was six feet too and could fight with the best of them. He was slender, graceful, but ugly: 'Henry's face was a compromise put together by a committee' and was distinguished by an enormously wide mouth, like a slash. He used this mouth often, to eat and drink prodigiously, to shape his superbly soft, melodic, caressing voice, and to do an extraordinary amount of kissing. As he put it, 'Kissing is like the presidency, it is not to be sought and not to be *declined*.' His opponents said his prodigiously wide mouth allowed him an unfair advantage: 'the ample dimensions of his kissing apparatus enabled him completely to *rest* one side of it while the other was on active duty.'[85] If women had had the vote Clay would have experienced no difficulty in becoming president every time he chose to run. As it was, having arrived in Kentucky in 1797 at the age of twenty—an excellent time to invest one's youth in this burgeoning territory—he promptly married into its leading establishment family. Within a few years he was the outstanding member of its state legislature, the highest-paid criminal lawyer in the state, a director of its main bank, professor of law and politics at Transylvania University, and the owner of a handsome property, Ashland, his home and solace for the rest of his life. He even served two brief terms in the United States Senate, but it was not until he was elected to the House in 1810 that his national career began.

Clay was probably the most innovative politician in American history, to be ranked with Franklin, Jefferson, Hamilton, and Madison as a political creator. A year after getting to the House he was elected its Speaker. Hitherto, the House had followed the English tradition, whereby the Speaker presided impartially and represented the collective consensus. Clay transformed this essentially non-political post into one of leadership, drilling and controlling a partisan majority and, in the process, making himself the most powerful politician in the country after the President. This made him a key figure in promoting the War of 1812 and also in negotiating the Treaty of Ghent; and, somehow or other, he escaped any blame for the war's disasters and returned from Ghent in triumph. This led him to think he ought to be secretary of state to the new President, Monroe. When the job went to John Quincy Adams instead, Clay organized and led in the House a systematic 'loyal opposition'—another political innovation.[86]

Clay was both principled and unprincipled. That was why other public men found it so difficult to make up their minds about him. (The ladies had no difficulty; they loved him.) His colleagues in the House and later in the Senate saw him as dictatorial and sometimes resented the way he used his authority to promote his views and ambitions. They saw the advantages of the strong leadership he provided. When he was in charge, the House functioned efficiently and fairly. Whenever he chose to stand, he was always voted into the speakership by large majorities. Later, in the Senate, the bulk of his colleagues always looked to him to take the lead. He was extraordinarily gifted in making what was in many ways a flawed system of government to work. He knew more about its nuts and bolts than any of his predecessors. Moreover, he had charm. Men who knew him only by repute were overwhelmed when they came across him face to face. A friend said to Thomas Glascock of Georgia: 'General, may I introduce you to Henry Clay?' 'No, sir, I am his adversary and choose not to subject myself to his fascination.' Calhoun, who became a mortal opponent in rhetorical duels of great savagery, admitted through clenched teeth: 'I don't like Clay. He is a bad man, an imposter, a creator of wicked schemes. I wouldn't speak to him but, *by God, I love him!*'[87]

Like many politicians, Clay tended to confuse his personal advancement with the national interest. But once in Washington he quickly developed, and thereafter extended throughout his life, a body of public doctrine which made him one of the pillars of the new republic. He believed that the liberty and sovereign independence of the hemisphere should be the prime object of American foreign policy. The United States should secure its economic independence from Europe by enlarging its own manufacturing sector. For this reason he got Congress in 1816 to enact the first American protective tariff and pressed for what he termed the 'American System' (of state intervention), under which state and federal governments would build roads, canals, and harbors to hasten industrialization, speed westward expansion, and bind the Union together.[88] There was, to be sure, a large element of self-interest in this. Clay's estate grew hemp, a Kentucky staple, and needed both protection from European hemp imports and good roads to take it east cheaply. Equally he was one of those in Congress who helped to create the Second Bank of the United States. It just so happened he was paid the huge retainer of $6,000 a year to fight the SBUS's cases in Kentucky, and borrowed money from it when necessary—indeed he was also loaned large sums by J. J. Astor, another prime beneficiary of the American System. But Clay clearly believed with every fiber of his being that America could and must become a leading industrial power, and that such expansion would eventually make it the greatest nation on earth. Just as John Marshall laid down the legal basis for American capitalism, so Clay supplied its political foundations.[89]

Clay was a passionate man. That, one suspects, is one reason people liked him. Despite all his poitical skills he could not always keep his temper in check. Tears jumped easily into his eyes. So did rage. When

Humphrey Marshall, cousin and brother-in-law of the Chief Justice, and an even bigger man than Clay (six feet two) called Clay a liar in the Kentucky legislature, Clay tried to fight him on the floor of the House but was separated by a giant man with a strong German accent— 'Come poys, no fighting here, I vips you both'—and the two antagonists crossed the river into Ohio to fight a duel, Clay getting a flesh wound in the thigh during a fusillade of shots (happily pistols were very inaccurate in those days). Clay pursued women relentlessly all his life, drank and gambled heavily ('I have always paid peculiar homage to the fickle goddess'), and, above all, danced. He was probably the most accomplished dancer among the politicians of his generation, with the possible exception of the South American Liberator, Simon Bolivar. Like Bolivar, when Clay was excited he loved to dance on the table at a banquet, and on one Kentucky occasion an eyewitness described how he gave 'a grand Terpsichorean performance . . . executing a *pas seul* from head to foot of the dining table, sixty feet in length . . . to the crashing accompaniment of shivered glass and china.' Next morning he paid the bill for the breakages, $120, 'with a flourish.'[90] Dancing indeed was a frontier craze in the America of the 1820s and 1830s— it was about the only entertainment they had—and laid the foundation for the extraordinary proficiency which enabled the United States, in the late 19th and 20th centuries, to produce more first-class professional dancers than any other country in the world, including Russia. When Clay was in Washington he adopted a different accent, watched his grammar (not always successfully), took delicate pinches of snuff while speaking, played with his gold-rimmed eyeglasses, and generally did his gentleman act. On the frontier, however, he was rambunctious, a true Kentuckian, and dancing was part of the performance.

It seemed to Clay ridiculous that Congress should allow the slavery issue, which it was unwilling to resolve fundamentally by banning it once and for all (as he wished), to obstruct the admission of Missouri, the first territory to be carved out of the Louisiana Purchase entirely west of the Mississippi. It was part of his American System to develop the Midwest as quickly as possible, so that America could continue its relentless drive to the Pacific before anyone else came along. If Missouri could not make itself viable without slavery, so what? He knew that in Kentucky, if he and his wife gave up their slaves, they would have to abandon their estate as uncompetitive and move—just as Edward Coles had had to sell up and move to Illinois, where he became the state's second governor. It was Clay's view that, in God's good time, slavery would go anyway, and developing the West using the American System would hasten the day. Meanwhile, let Missouri be admitted and prosper.

Hence Clay, by furious and skillful activity behind the scenes and on the House floor, ensured that Maine and Missouri were admitted together, along with a compromise amendment prohibiting slavery in the Louisiana Purchase north of latitude 36.30 (March 1820). And by even greater prodigies of skill he resolved the constitutional question provoked by the extremists in the Missouri convention by what is known as the Second Missouri Compromise, the local legislature solemnly pledging never to enact laws depriving any citizen of his rights under the US Constitution (February 1821).[91] As a result President Monroe was able to sign Missouri's admission to the Union in August. This was the first of three compromises Clay brokered (the others were 1833 and 1850) which defused the periodic explosion between North and South and postponed the Civil War for forty years. Indeed Senator Henry S. Foote, who had watched Clay weave his magic spells to disarm the angry protagonists in Congress, later said: 'Had there been one such man in the Congress of the United States as Henry Clay in 1860–1, there would, I am sure, have been no Civil War.'[92]

Clay followed up the Missouri Compromises by encouraging President Monroe to play a positive part in the liberation struggle against Spain in Latin America by giving the revolutionary governments rapid recognition and any diplomatic help they needed. That, too, was part of the American System, in which the United States not only made itself strong and independent in the north, but excluded the rapacious European powers from the center and the south. What Clay did not know was the extent to which the British Foreign Secretary, George Canning, also an enthusiastic supporter of Latin American independence (for British commercial purposes), was pushing Monroe to take the same line and to declare openly that France and Spain were no longer welcome in the hemisphere. On December 2, 1823, as part of his message to Congress, Monroe announced the new American policy. First, the United States would not interfere in existing European colonies. Second, it would keep out of Europe, its alliances and wars. Third, 'the American continents . . . are henceforth not to be considered as subjects for future colonisation by the European powers.' Fourth, the political systems of Europe being different to that of the

United States, it would 'consider any attempt on their part to extend their system to any portion of this hemisphere as dangerous to our peace and safety.' This declaration, which in time was known as the Monroe Doctrine, became progressively more important as America—thanks to Clay's American System—acquired the industrial and military muscle to enforce it.[93]

In the light of his successful supra-party activities, Clay believed he had earned the right to be president. But then so did many other people. The Monroe presidency has been described, at the time and since, as the 'Era of Good Feelings,' the last time in American history when the government of the country was not envenomed by party politics.[94] But a case can also be made for describing the Monroe presidency, and the rule of John Quincy Adams which formed its appendage, as the first great era of corruption in American politics. Many Americans came seriously to believe, during it, that their government, both administration and Congress, was corrupt, and this at a time when in Britain the traditional corruption of the 18th-century system was being slowly but surely extruded. By corruption, Americans of the 1820s did not simply mean bribes and stealing from the public purse. They also meant the undermining of constitutional integrity by secret deals, the use of public office to acquire power or higher office, and the giving of private interests priority over public welfare.[95] But the public thought that plenty of simple thieving was going on too. Indeed, two members of the government, Calhoun at the War Department, and William Crawford (1772-1834), the Treasury Secretary, more or less openly accused each other of tolerating, if not actually profiting from, skulduggery in their departments.[96]

The atmosphere in Monroe's administation during its last years was poisonous, not least because Calhoun, Crawford, and Adams, its three principal members, were all maneuvering to succeed their boss. As a consequence there was particular bitterness over patronage and appointments. Adams' diary records a ferocious quarrel between Crawford and Monroe on December 14, 1825. The two men had met to discuss appointments to the Customs, always contentious because of the rising volume of cash handled and the opportunities for stealing it. Monroe was obstructive and when Crawford rose to go he said contemptuously: 'Well, if you will not appoint the persons well qualified for the places, tell me who you will appoint that I may get rid of their importunities.' Monroe replied 'with great warmth, saying he considered Crawford's language as extremely improper and unsuitable to the relations between them; when Crawford, turning to him, raised his cane, as if in the attitude to strike, and said: "You damned infernal old scoundrel," Mr Monroe seized the tongs at the fireplace in self-defense, applied a retaliatory epithet to Crawford and told him he would immediately ring for servants himself and turn him out of the house . . . They never met afterwards.'[97]

To assist their presidential prospects, both Crawford and Calhoun used agents in their departments, whose duty it was to dispense money, as political campaigners; they were given gold and silver but allowed to discharge payments in paper as a reward. Since each knew what the other was doing and circulated rumors to that effect, these activities became public knowledge. Then again, Crawford allowed one of his senatorial supporters to inspect government land offices, at public expense, during which tour he made speeches supporting Crawford's candidacy. Various members of the administration, it was claimed, had been given 'loans' by businessmen seeking favors, loans which were never repaid. But Congress was corrupt too. Senator Thomas Hart Benton of Missouri served as 'legal representative' of Astor, on a huge retainer, and managed to push through the abolition of the War Department's 'factory' system, which competed with Astor's own posts. The fact that the War Department's system was corrupt, as Benton easily demonstrated, was not the point: what was a senator doing working for a millionaire? Benton was not the only one. The great Massachusetts orator, Daniel Webster (1782–1852), received a fat fee for 'services' to the hated Second Bank of the United States. Webster has been described by one modern historian as 'a man who regularly took handouts from any source available and paid the expected price.'[98] Astor seems to have had financial dealings with other men high in public life; indeed he even loaned $5,000 to Monroe himself; this was eventually repaid though not for fifteen years. He lent the enormous sum of $20,000 to Clay during the panic year 1819 when credit was impossible to come by. Clay, like Webster, served the Bank for money while Speaker. About this time, America's growing number of newspapers began to campaign vigorously about Washington's declining standards. The *Baltimore Federal Republican* announced that navy paymasters were guilty of 'enormous defalcation,' only 'one of innumerable instances of corruption in Washington.' The *New York Statesman* denounced 'scandalous defalcation in our public pecuniary agents, gross misapplications of public money, and an unprecedented laxity in official responsibilities.'[99]

The evident corruption in Washington, coming on top of the finan-cial crisis of 1819, persuaded the victor of New Orleans, General Jackson, that it was high time, and his public duty, to campaign for the presidency and engage in what he called 'a general cleansing' of the fed-eral capital. He did not think it practicable to return to the Jeffersonian ideal of a pastoral America run by enlightened farmers. He wrote in 1816: 'Experience has taught me that manufactures are now as neces-sary to our independence as to our comfort.' But clearly a return to the pristine purity of the republic was imperative. So Jackson became the first presidential candidate to grasp with both hands what was to become the most popular campaigning theme in American history—'Turn the rascals out.' Unfortunately for Jackson, if his own hands were clean—how could they not be? He had barely set foot in Washington—there were quite different objections to his candidature. His victory at New Orleans had enabled him to become a kind of unof-ficial viceroy or pro-consul in the South. As such, he had destroyed Indian power there and, in effect, confiscated their lands. No one objected to that of course. But on March 15, 1818 his troops began an undeclared war against Spain by invading Florida, which the feeble Spanish garrisons were incapable of defending. He even promised Monroe, 'I will ensure you Cuba in a few days' if Washington lent him a frigate—but Monroe refused to oblige.[100] Under pressure from his Secretary of State Adams, an enthusiastic imperialist, Monroe gave Jackson's war against Florida tacit support, though he later denied col-lusion and said he was sick at the time. In a modern context, of course, Jackson's activities—plainly against the Constitution, which gave the right to make peace or war exclusively to Congress—would have been exposed by liberal-minded journalists. In 1818, the general would have seized such reporters and imprisoned or expelled them, or possibly hanged them for treason. In any case there were no liberal-minded journalists then, at least on Indian or Spanish questions. All were belli-cose and expansionary. On February 8, 1819 Congress was happy to endorse the *fait accompli* by rejecting a motion of censure on Jackson, and the territory was formally conveyed by Spain to the United States on July 17, 1821.[101]

Nevertheless, Jackson was not without prominent critics, notably Henry Clay. As part of his opposition campaign against the Monroe administration, Clay accused it of allowing Jackson to behave like a Bonaparte. And when Jackson made it clear he was campaigning for the presidency in 1824, Clay dismissed him as a 'mere military chieftain.' 'I cannot believe that killing 2,500 Englishmen at New Orleans qualifies for the various, difficult and complicated duties of the First Magistracy,' he wrote. Jackson, he said, was 'ignorant, passionate, hypocritical, corrupt and easily swayed by the basest men who surround him.'[102] Jackson brushed aside these charges at the time—though he did not forget them: Clay jumped right to the top of his long enemies lists and remained there till Jackson died—and concentrated on rousing 'the people.' Jackson may have been a military autocrat but what differentiated him from the *caudillos* of Latin America, and the Bonaparte figures of Europe, is that he was a genuine democrat. He was the first major figure in American politics to believe passionately and wholly in the popular will, and it is no accident that he created the great Democratic Party, which is still with us. As governor of Florida territory (thanks to his high-handed methods, Florida did not become a state till 1845), Jackson ruled that mere residence was enough to give an adult white male the vote. In more general terms, he said in 1822 that every free man in a nation or state should have the vote since all were subject to the laws and punishments, both federal and state, and so they 'of right, ought to be entitled to a vote in making them.' He added that every state legislature had the duty to adopt such voting qualifications as it thought proper for 'the happiness, security and prosperity of the state' (1822).[103]

Jackson argued that the more people who had a presidential vote the better since, if Washington was rotten, that gave them the remedy: 'The great constitutional corrective in the hands of the people against usurpation of power, or corruption by their agents, is the right of suffrage; and this when used with calmness and deliberation will prove strong enough—it will perpetuate their liberties and rights.'[104] Jackson thought the people were instinctively right and moral, and Big Government, of the kind he could see growing up in Washington, fundamentally wrong and immoral. His task, as he saw it, was to liberate and empower this huge moral popular force by appealing to it over the heads of the entrenched oligarchy, the corrupt ruling elite. This was undoubtedly a clear, simple political strategy, and a winning one, if the suffrage was wide enough.

It is not clear how far this great innovation in American politics—the introduction of the *demos*—was Jackson's own doing or the work, as Clay said, of the unscrupulous men who manipulated him. His ignorance was terrifying. His grammar and spelling were shaky. The 'Memorandoms' he addressed to himself are a curious mixture of naivety, shrewdness, insight, and prejudice. His tone of voice, in speech and writings, was sub-Biblical. 'I weep for my country,' he asserted, often. Banks, Washington in general, the War Department in particular, and his massed enemies were 'The Great Whore of Babylon.' Hostile newspapers poured on him what he called 'their viols of wrath.' He himself would 'cleans the orgean stables.' By contrast, his aide, Major John Eaton (1790–1854), who became a US senator in 1818, and acted as Jackson's political chief-of-staff and amanuensis, was a skilled writer. He turned the 'clean up Washington' theme into a national campaign, the first modern election campaign, in fact. In early summer 1823, Eaton wrote a series of eleven political articles signed 'Wyoming,' for the *Columbian Observer* of Philadelphia. They were reprinted as a pamphlet, *Letters of Wyoming*, and reproduced in newspapers all over the country. The theme, worked out in specific detail and couched in impressive rhetoric, was that the country had fallen into the 'Hands of Mammon' and that the voters must now insure that it returned to the pure principles of the Revolution.

Jackson was attacked in turn in this newspaper and pamphlet warfare, the most damaging assault coming from the highly respected former Treasury Secretary, Albert Gallatin, who asserted that whenever Jackson had been entrusted with power he had abused it. With the appalling example of Latin America in mind, Gallatin reminded voters: 'General Jackson has expressed a greater and bolder disregard for the first principles of liberty than I have ever known to be entertained by any American.' This line, too, was widely reproduced.[105] Yet Jackson, despite the warnings, proved an outstanding candidate, then and later. Tall, slender, handsome, fierce, but also frail and often ill-looking, he made people, especially women, feel protective. With his reputation for wildness and severity, his actual courtesy, when people finally met him, was overwhelming. Daniel Webster testified: 'General Jackson's manners are more presidential than those of any of the candidates . . . my wife is for him decidedly.'[106] It was the first case, in fact, of presidential charisma in American history.

The presidential election of 1824 was an important landmark for more than one reason. Originally there were five candidates; Crawford, Calhoun, Clay, Adams, and Jackson. But Calhoun withdrew to become vice-presidential candidate on both tickets, and a stroke rendered Crawford a weak runner: he came in a poor third. In the event it was a race between Adams and Jackson. The electoral college system was still a reality but this was the first election in which popular voting was also important. In Georgia, New York, Vermont, Louisiana, Delaware, and South Carolina, the electors of the president were chosen by state legislatures. Elsewhere there were already statewide tickets, though voting by districts still took place in Maine, Illinois, Tennessee, Kentucky, and Maryland. The number of electors was larger than ever before, though with the country prosperous again there was no wrathful rising of the people—and America was already showing a propensity towards low turnouts, or low registrations of eligible voters. In Massachusetts, where Adams was the Favorite Son, only 37,000 votes were cast, against 60,000 for governor the year before. In Ohio, where 76,000 turned out for the governorship race earlier in the autumn, only 59,000 voted for the presidency. Virginia had a white population of 625,000: only 15,000 voted and in Pennsylvania only 47,000 voted, though the population had already passed the million mark.[107] All the same, with 356,038 votes cast, Jackson, with 153,544, emerged the clear leader. Adams, the runner-up, was 40,000 votes behind, with 108,740.[108] Jackson also won more electoral college votes, having ninety-nine, against eighty-four for Adams, forty-one for Crawford, and thirty-seven for Clay. He carried eleven states, against seven for Adams. By any reckoning, Jackson was the winner. However, under the Twelfth Amendment, if no presidential candidate scored a majority of the electoral votes, the issue had to be taken to the House of Representatives, which picked the winner from the top three, voting by states. That, in practice, made Clay, Speaker of the House, the broker. As fourth-runner, he was now excluded from the race. But he determined he would decide who won it, and profit accordingly.

The House was due to meet February 9, 1825. Jackson reached Washington on December 7, 1824, after a twenty-eight-day journey from Tennessee. In a letter to his old army crony John Coffee, he claimed the place was thick with rumors of a deal but he was taking no part in any political talks: 'Mrs Jackson and myself go to no parties [but remain] at home smoking our pipes.' (This was a formidable operation: his wife had clay pipes but Jackson smoked 'a great Powhatan Bowl Pipe with a long stem,' puffing out until the room was 'so obfuscated that one could hardly breathe.')[109] Clay's people put out feelers, asking what office was likely to go to their principal if Jackson was elected. Later, Jackson was asked to confirm this rumor: 'Is that a *fact*?' Jackson: 'Yes, Sir, such a proposition *was* made. I said to the bearer, "Go tell Mr Clay, tell Mr Adams, that if I go to that chair, I go with clean hands." '[110] However, Adams and Clay were less squeamish, though both disliked each other. They met twice, on January 9 and 29, 1825, and the first meeting was probably decisive, though Adams' normally copious diary, while recording it, pointedly omits to say what took place. Possibly the prudish and high-principled Adams could not bring himself to record the deal, if there was a deal. At all events, when the House met, Clay insured that Adams got thirteen states, the winning minimum. The Kentucky vote was particularly scandalous, Clay himself casting it for Adams, though he did not get a single vote there. On February 14, Clay got his part of the bargain: Adams officially appointed him Secretary of State. The office had more significance then than now, since the holder was automatically the next front-runner for the presidency.[111]

Jackson's wrath exploded. He wrote the same evening: 'So you see the Judas of the West has closed the contract and will receive the thirty pieces of silver. His end will be the same. Was there ever witnessed such bare-face corruption?'[112] The cry, 'Corrupt Bargain,' was uttered and taken up all over the country. It became the theme for Jackson's next presidential campaign, which began immediately. The way in which Jackson, having got most suffrages, most electoral votes, and most states, was robbed of the presidency by a furtive deal seemed to most people to prove up to the hilt what he had been saying about a corrupt Washington, which he had been 'elected' to purify. So it was the electorate, as well as Jackson, which had been swindled. Clay did not help matters. Instead of keeping a dignified silence, he produced various, and contradictory, explanations for his giving the presidency to Adams. Jackson exulted: 'How little common sense this man displays! Oh, that mine enemy would write a book! . . . silence would have been to him wisdom.' We shall probably never know whether there *was* a 'corrupt bargain.' Most likely not. But most Americans thought so. And the phrase made a superb slogan.[113]

In spring 1825 the Tennessee legislature nominated Jackson as their president for the race of 1828, and another new tradition in America began: the endless election campaign. The charge of a corrupt bargain went far to undermine the legitimacy of the Adams presidency. Jackson announced that, having hitherto regarded Adams as a man of probity, 'From that moment I withdrew all intercourse with him.'[114] A huge political fissure opened between the administration and the Jacksonites. From that point opposition in Congress became systematic. The modern American two-party system began to emerge. All over what was already an enormous country, and one which was expanding fast, branches of a Jacksonian popular party began to form from 1825. Scores of newspapers lined up behind the new organization, including important new ones like Duff Green's *United States Telegraph*. As the political system polarized, more and more political figures swung behind Jackson. In New York, the master of the Tammany machine, Martin Van Buren (1782–1862), Benton and Calhoun, Sam Houston of the West, the Virginia grandee John Randolph of Roanoke (1773–1833), George McDuffie (1790–1851) of South Carolina, Edward Livingstone (1764–1836), the boss of Louisiana—all these men, and others, assembled what was to become one of the great and enduring popular instruments of American politics, the Democratic Party.[115]

The *Telegraph*, chief organ of the new party, was head of a network of fifty others, in all the states, which reproduced its most scurrilous articles. Those who believe present-day American politics are becoming a dirty game cannot have read the history of the 1828 election. Americans have always taken a prurient interest in what goes on in the White House, particularly if public money is involved. Even the mild Monroe, incensed by an inquiry about spending on interior decoration by Congressman John Cocke of Tennessee, a Jacksonian chairman of one of its committees, 'desired the person who brought him the message to tell Cocke he was a scoundrel and that was the only answer he would give him.'[116] Adams, blameless in this respect anyway, was subjected to still more minute investigations. A White House inventory revealed that it contained a billiard table and a chess-set, paid for (as it happened) out of Adams' own pocket. Congressman Samuel Carson of North Carolina demanded to know by what right 'the public money should be applied to the purchase of Gambling Tables and Gambling Furniture?' That question, parroted in the Telegraph and its satellites, sounded dreadful in New England and the Bible Belt. The Telegraph, anxious to portray Adams as a raffish fellow, instead of the grim old stick he actually was, dragged up an ancient story from his St Petersburg days which had him presenting to Tsar Alexander I an innocent young American girl—he had been 'the pimp of the coalition,' it claimed.[117]

Oddly enough, the one shocking aspect of Adams' tenure of the White House—or so it might seem to us—his daily swims stark naked in the Potomac, attended by his black servant Antoine in a canoe, went unreported. It was by no means a tame river and on June 13, 1825

Adams was nearly drowned when the canoe capsized, losing his coat and waistcoat and having to scramble back to the White House in his pants, shoeless. But a Philadelphia paper complained that, in the humid summer weather, he wore only a black silk ribbon round his neck instead of a proper cravat, and that he went to church barefoot.[118] Adams did not have a happy time in the White House. He dreaded being buttonholed in the street. He seems to have spent several hours every day receiving members of the public, who arrived without appointment or invitation, many with tales of woe. He recorded: 'The succession of visitors from my breakfasting to my dining hour, with their variety of objects and purposes, is infinitely distressing.' He had 'many such visitors' as a Mrs Weedon, who 'said she had rent to pay and if she could not pay it this day, her landlord threatened to distrain upon her furniture.' Of a visit from Mrs Willis Anderson, whose husband was serving ten years for mail robbery, he noted: 'I had refused [to help] this woman three times and she had nothing new to allege. I desired her not to come to me again.' Two weeks after the importunate Mrs Anderson, Adams had a visit from a Mr Arnold, who said he had been traveling and found himself in Washington without money. He would be 'much obliged' if the President would provide him with the cash to get back to Massachusetts, 'which I declined.'[119] There is no indication that President Adams had much time to use his Gambling Furniture.

However, the administration papers were not slow in lashing back at Jackson. The *National Journal* asserted: 'General Jackson's mother was a *Common Prostitute*, brought to this country by British soldiers! She afterwards married a *Mulatto Man*, by whom she had several children, of which number *General Jackson is one!*' Jackson burst into tears when he read this statement, but he was still more upset by attacks on the validity of his marriage to Rachel. He swore he would challenge to a duel, and kill, anyone he could identify being behind the rumors. He meant Clay of course. (On his deathbed, Jackson said the two things he most regretted in his life were that 'I did not hang Calhoun and shoot Clay.') In fact, on a quite separate issue, Clay and Randolph did fight a duel on the Potomac banks, just where the National Airport now stands: neither was hurt but Clay's bullet went through Randolph's coat (he bought the Senator a new one). When Jackson got information that a private detective, an Englishman called Day, was nosing around Natchez and Nashville looking at marriage registers, he wrote to Sam Houston that, when he got information about Clay's 'secret movements,' he would proceed 'to his political and perhaps his actual destruction.' Clay was certainly warned by friends that gunmen were after him. Jackson went so far as to have ten prominent men in the Nashville area draw up a statement, which filled ten columns in the *Telegraph*, testifying that his marriage to Rachel was valid. That did not stop the administration producing a pamphlet which asked: 'Ought a convicted adulteress and her paramour husband be placed in the highest offices in the land?' The *Telegraph* replied by claiming that Mr and Mrs Adams had lived in sin before their marriage and that the President was an alcoholic and a sabbath-breaker.[120]

The Presidential campaign of 1828 was also famous for the first appearance of the 'leak' and the campaign poster. Adams complained: 'I write few private letters ... I can never be sure of writing a line which will not some day be published by friend or foe.' Anti-slavery New England was regaled by a pamphlet entitled *General Jackson's Negro Speculations, and his Traffic in Human Flesh, Examined and Established by Positive Proof*. Even more spectacular was the notorious 'Coffin Handbill,' printed for circulation and display, under the headline 'Some Account of Some of the Bloody Deeds of General Jackson,' listing eighteen murders, victims of duels or executions he had carried out, with accompanying coffins. Harriet Martineau related that in England, where these accusations circulated and were generally believed, a schoolboy, asked in class who killed Abel, replied, 'General JACKSON, Ma'am.'[121] Campaign badges and fancy party waistcoats had made their first appearance in 1824, but it was in 1828 that the real razzmatazz began. Jackson's unofficial campaign manager was Amos Kendall (1789–1869), editor of the *Argus of Western America*, who in 1827 had switched from Clay to Jackson. Jackson had long been known to his troops as Old Hickory, as that was 'the hardest wood in creation.' Kendall seized on this to set up a nationwide network of 'Hickory Clubs.' Hickory trees were planted in pro-Jackson districts in towns and cities and Hickory poles were erected in villages; Hickory canes and sticks were sold to supporters and flourished at meetings. There were Hickory parades, barbecues, and street-rallies. Kendall had the first campaign song, 'The Hunters of Kentucky,' written and set to music. It told of the great victory of 1815 and of 'Packenham [sic] and his Braggs'—how he and his men would rape the girls of New Orleans, the 'beautiful girls of every hue' from 'snowy white to sooty'—and of how Old Hickory had frustrated his dastardly plans and killed him.[122]

Jackson proved an ideal candidate, who knew exactly when to hold his tongue and when to give vent to a (usually simulated) rage. And he had the ideal second-in-command in Martin Van Buren, head of the Albany Regency, which ran New York State, a small, energetic, dandi-fied figure, with his reddish-blond hair, snuff-colored coat, white trousers, lace-tipped orange cravat, broad-brimmed beaver-fur hat, yel-low gloves, and morocco shoes. If Van Buren dressed like the young Disraeli, he had something Disraeli never possessed—a real, up-to-date political machine. Van Buren grew up in the New York of Aaron Burr and De Witt Clinton. New York politics were already very complex and rococo—outsiders confessed inability to understand them—but they were the very air the little man breathed. Burr had turned the old Jeffersonian patriotic club, the Society of Saint Tammany, where mem-bers came to drink, smoke, and sing in an old shed, into the nucleus of a Big City political organization. Clinton had invented the 'spoils sys-tem,' whereby an incoming governor turned out all office holders and rewarded his supporters with their jobs. New York was already politics on a huge scale—a man would hesitate between running for governor and running for president. Van Buren's genius lay in uniting Tammany with the spoils system, then using both to upstage first Burr, then Clinton, and rule the roost himself.[123]

Van Buren was the first political bureaucrat. He came from the pure Dutch backwater of Kinderhook in Albany County, where the Rip Van Winkle stories originated, but there was nothing sleepy about him. His motto was: 'Get the details right.' His Tammany men were called Bucktails by their enemies, because of their rustic origins, but he taught them to be proud of their name and to wear the symbol in their hats, just as the Democrats later flaunted their donkey. Branching out from Tammany, he constructed an entire statewide system. His party news-papers in Albany, and in New York City, proclaimed the party line and supplied printed handbills, posters, and ballots for statewide distribu-tion. The line was then repeated in the country newspapers, of which Van Buren controlled fifty in 1827. The line was set by the party elite of lawyers and placemen. Even by the 1820s, America, and especially New York, was a lawyers' paradise. Frequent sessions in New York's complex court circuit system kept the lawyers moving. Van Buren used them as a communications artery to towns and villages even in remote parts of the state. Officeholders appointed by the governor's council were the basis for party pressure groups everywhere. Van Buren's own views sprang from the nature of his organization. The party identity must be clear. Loyalty to majority decisions taken in party councils must be absolute. All measures had to be fully discussed and agreed, and personal interests subordinated to party ones. Loyalty was rewarded and disloyalty punished without mercy.

When Van Buren's Bucktails took over the state in 1821, he conducted a massacre of major officeholders at the council's very first meeting and thereafter combed through 6,000 lesser jobs removing Clintonians, federalists, and unreliable Bucktails.[124] Clinton, who had invented the spoils system, let out a howl of rage. This kind of punishment-and-rewards system was the very opposite of what the Founding Fathers had envisaged; but it was the future of American politics. And Van Buren, like many American master-politicians since, was quite capable of combining party ruthlessness with high-mindedness. He was a political schizophrenic, admitting he abused power occasionally and vowing never to do it again (he did of course). He supported Clinton's great project of the Erie Canal, because he thought it was in the interests of New York and America, despite the fact that the Canal, triumphantly completed on November 2, 1825, helped Clinton to regain the governorship—and massacre the Bucktails in turn.[125] American political history has since thrown up repeated exemplars of what might be called the Van Buren Syndrome—men who could combine true zeal for the public interest with fanatical devotion to the party principle.

Most of 1827 the assiduous Van Buren spent building up the new Jacksonian Democratic Party, traveling along rotten roads in jolting carriages to win support from difficult men like Benton of Missouri, a great power in the West, the splendid but bibulous orator Randolph, who was often 'exhilarated with toastwater,' down to Georgia to conciliate old, sick Crawford, up through the Carolinas and Virginia and back to Washington. Thus, for the first time, the Democratic 'Solid South' was brought into existence. In February 1828 Clinton died of a heart-attack, clearing the way for Van Buren to become governor of New York State. He spent seven weeks in July and August, electioneering in the sticky heat of grim new villages upstate, taking basic provisions with him in his carriages, for none were to be had en route, complaining of insects, humidity, and sudden storms which turned the tracks into marshes. He brought with him cartloads of posters, Jackson badges (another innovation), Bucktails to wear in hats, and Hickory sticks. He was the first American politician to assemble a team of writers, not just to compose speeches but to draft articles for scores of local newspapers. Artists and writers who supported the Jackson campaign included James Fenimore Cooper, the sculptor Horatio Greenough,

Nathaniel Hawthorne, the historian George Bancroft, William Cullen Bryant, then the leading American poet, and another well-known poet William Leggett. Apart from Ralph Waldo Emerson, most of America's writers and intellectuals seem to have backed Jackson—the first time they ganged up together to endorse a candidate. As Harriet Martineau put it, Jackson had the support of the underprivileged, the humanitarians, the careerists, and 'the men of genius.'[126] Adams confided bitterly in his diary: 'Van Buren is now the great electioneering manager for General Jackson [and] has improved as much in the art of electioneering upon Burr, as the state of New York has grown in relative strength and importance in the Union.'[127] Adams was safe in New England but he could see—already—that the South plus New York made a formidable combination.

This was the first popular election. In twenty-two of the twenty-four states (Delaware and Rhode Island still had their legislatures choose college electors), the voters themselves picked the president. Except in Virginia they were equivalent to the adult male white population. A total of 1,155,340 voted, and Adams did well to get 508,064 of them, carrying New England, New Jersey, and Delaware, and a majority of the college in Maryland. Even in New York he got sixteen out of thirty-six college votes because Van Buren, despite all his efforts, carried the state by a plurality of only 5,000. That gave Adams eighty-three electoral votes in all. But Jackson got all the rest and a popular vote of 647,276.[128] So Jackson went to Washington with a clear popular mandate, ending the old indirect, oligarchical system for ever.

The manner of the takeover was as significant as the result. In those days voting for president started in September and ended in November, but the new incumbent did not take office till March. Washington was then a slow, idle, Southern city. Designed by Pierre L'Enfant and laid out by the surveyor Andrew Ellicott (1764–1820), it was in a state of constant constructional turmoil but contrived to be sleepy at the same time. Its chief boast was its 91,665 feet of brick pavement, though it also had, at the corner of Pennsylvania Avenue and 13th Street, the Rotondo, with its 'Transparent Panoramic View of West Point and Adjacent Scenery.' Banquets for the legislators, which were frequent, began at 5.30 P.M. and progressed relentlessly through soup, fish, turkey, beef, mutton, ham, pheasant, ice cream, jelly, and fruit, taken with sherry, a great many table wines, madeira, and champagne. There was, besides, much drinking of sherry cobblers and gin cocktails, slings made with various spirits, juleps, snakeroot bitters, timber doodly, and eggnogs. Most politicians lived in boarding houses, most of them decorous, a few louche. But there were already hostesses who set the tone, which was oligarchical, elitist, and essentially Virginian Ascendancy.

To Jackson, then, it was a hostile city and he arrived there, President-elect, on February 11, 1829, a sad and bitter man. Early in December his wife Rachel had gone to Nashville to buy clothes for her new position. There, she picked up a pamphlet defending her from charges of adultery and bigamy. Hitherto the General had concealed from her the true nature of the smear campaign waged against her honor, and the shock of discovery was too much. She took to her bed and died on December 22. To his dying day Jackson believed his political enemies had murdered her and he swore a dreadful revenge. He put up at Gadsby's Boarding House. He was not alone. From every one of the twenty-four states his followers congregated on the capital, a 10,000-strong army of the poor, the outlandish, the needy, above all the hopeful. Washingtonians were appalled as these people assembled, many in dirty leather clothes, the 'inundations of the northern barbarians into Rome.' They drank the city dry of whiskey within days, they crammed the hotels, which tripled their prices to $20 a week, they slept five in a bed, then on the floors, spilling over into Georgetown and Alexandria, finally into the fields. Daniel Webster wrote: 'I never saw such a crowd here before. Persons have come 500 miles to see General Jackson and *they really seem to think the country has been rescued from some general disaster.*' But most wanted jobs. Clay joked sardonically about the moment 'when the lank, lean, famished forms, from fen and forest and the four quarters of the Union, gathered together in the halls of patronage; or stealing by evening's twilight into the apartments of the president's mansion, cried out with ghastly faces and in sepulchral tones, "give us bread, give us Treasury pap, give us our reward!"'[129]

The inaugural itself was a demotic saturnalia, reminiscent of scenes from the early days of the French Revolution but enacted against a constitutional background of the strictest legality. It was sunny and warm, the winter's mud, 2 feet deep in places, beginning to dry. By 10 A.M. a vast crowd, held back by a ship's cable, had assembled under the East Portico of the unfinished Capitol. At eleven, Jackson emerged from Gadsby's and, escorted by soldiers, walked to the Capitol in a shambling procession of New Orleans veterans and politicians, flanked by 'hacks, gigs, sulkies and woodcarts and a Dutch waggon full of females.' At noon, by which time 30,000 people surrounded the

Capitol, the band played 'The President's March,' there was a twenty-four-gun salute, and Jackson, according to one critical observer, Mrs Margaret Bayard Smith, bowed low to the People in all its majesty.' The President, with two pairs of spectacles, one on top of his head and other before his eyes, read from a paper words nobody could hear. Then he bowed to the people again and mounted a white horse to ride to his new mansion, 'Such a cortege as followed him,' gasped Mrs Smith, 'countrymen, farmers, gentlemen mounted and dismounted, boys, women and children, black and white, carriages, waggons and carts all pursuing him.'[130]

Suddenly, to the dismay of the gentry watching from the balconies of their houses, it became obvious that the vast crowd in its entirety was going to enter the White House. It was like the sansculottes taking over the Tuilleries. A Supreme Court justice said those pouring into the building ranged from 'the highest and most polished' to 'the most vulgar and gross in the nation—the reign of *King Mob* seemed triumphant.' Soon the ground floor of the White House was crammed. Society ladies fainted, others grabbed anything within reach. A correspondent wrote to Van Buren in New York: 'It would have done Mr Wilberforce's heart good to see a stout black wench eating a jelly with a gold spoon in the president's house.' Clothes were torn; barrels of orange punch were knocked over; men with muddy boots jumped on 'damask satin-covered chairs' worth $150 each to see better; and china and glassware 'worth several thousand dollars' were smashed. To get the mob out of the house, the White House servants took huge stocks of liquor onto the lawn and the *hoi polloi* followed, 'black, yellow and grey [with dirt] many of them fit subjects for a penitentiary.' Jackson, sick of it all, climbed out by a rear window and went back to Gadsby's to eat a steak, already a prime symbol of American prosperity. He declined, being in mourning, to join 1,200 citizens at the ball in Signor Carusi's Assembly Rooms, a more sedate affair, ticket only. The scenes at the White House were the subject of much pious moralizing at Washington's many places of worship that Sunday, the pastor at the posh Unitarian Church preaching indignantly from Luke 19:41—'Jesus beheld the city and wept over it.'[131]

Then came the rewards. One of Van Buren's sidekicks, Senator William L. Marcy, responding to the weeping and gnashing of teeth as the Old Guard were fired, told the Senate that such 'removals' were part of the political process, adding, 'To the victors belong the spoils of the enemy.' The phrase stuck and Jackson will always be credited with bringing the spoils system into federal government. Mrs Smith wrote bitterly of the expulsions: 'so many families broken up—and those of the first distinction—drawing rooms now dark, empty, dismantled.' Adams protested: 'The [new] appointments are exclusively of violent partisans and every editor of a scurrilous and slanderous newspaper is provided for.' It is true that Jackson was the first president to give journalists senior jobs—Amos Kendall for example got a Treasury auditorship. But Jackson partisans pointed out that, of 10,093 government appointees, only 919 were removed in the first eighteen months and over the whole eight years of the Jackson presidency only 10 percent were replaced. Moreover, many of those sacked deserved to be; eighty-seven had jail records. The Treasury in particular was full of useless people and rogues. One insider reported: 'a considerable number of the officers are old men and drunkards. Harrison, the First Auditor, I have not yet seen sober.' One fled and was caught, convicted, and sentenced. Nine others were found to have embezzled. Within eighteen months Kendall and other nosy appointees discovered $500,000 had been stolen, quite apart from other thefts at the army and navy offices and Indian contracts. The Registrar of the Treasury, who had stolen $10,000 but had been there since the Revolution, begged Jackson to let him stay. Jackson: 'Sir, I would turn out my own father under the same circumstances.' But he relented in one case when a sacked postmaster from Albany accosted him at a White House reception and said he had nothing else to live on. He began to take off his coat to show the President his wounds. Jackson: 'Put your coat on at once, Sir!' But the next day he changed his mind and took the man's name off the sackings list: 'Do you know that he carries a pound of British lead in his body?' Jackson's appointments turned out to be no more and no less corrupt than the men they replaced and historians are divided on the overall significance of bringing the spoils system to Washington.[132]

Two of Jackson's appointments turned out disasters. The first was the selection of Samuel Swartwout as collector of customs in New York, which involved handling more cash than any other on earth, $15 million in 1829. His claim to office was that he had backed Jackson in New York even before Van Buren. But he was a crooked old crony of Burr, who gambled on horses, stocks, and fast women. In due course he fled to Europe, taking with him $1,222,705.09, the biggest official theft in US history, worse than all the peculations of the Adams administration put together.[133]

An even more serious mistake was Jackson's sentimental decision to make his old comrade and crony Major John Eaton the War Secretary. The canny Van Buren, who knew Swartwout of old but had been unable to prevent his appointment, was even more uneasy about Eaton, whom he regarded as indiscreet, negligent, and the last man to keep a Cabinet secret. He was even more suspicious of Eaton's wife, a pretty, pert young woman of twenty-nine called Peggy, a known adulteress who had lived in sin with Eaton before Jackson ordered him to marry her. But the President, adoring spirited ladies who stood up to him in conversation, would not hear a word said against her.[134]

This imprudent appointment set in motion a chain of bizarre events which were to change permanently the way in which America is governed. The well-informed Amos Kendall dismissed rumors that Peggy was a whore; she was, he said, merely egotistical, selfish, pushy, and 'too forward in her manners.' But the other Cabinet wives, older and plainer, hated her from the start and insisted she had slept with 'at least' twenty men, quite apart from Eaton, before her second marriage to him. If old Rachel had lived, she might have kept the Cabinet matrons in line (or, more likely, quashed the appointment in the first place). But her place had been taken by the twenty-year-old Emily Donelson, wife of Jackson's adopted son. Emily had been accustomed to managing a huge Southern plantation and was not in the least daunted by running the White House with its eighteen servants. But she would not stay in the same room with Peggy, who, she said, 'was held in too much abhorrence ever to be noticed.' Mrs Calhoun, wife of the Vice-President and a grand Southern lady, would not even come to Washington in case she was asked to 'meet' Mrs Eaton. Adams, for whom the Peggy Eaton row was the first nice thing to happen since he lost the presidency, recorded gleefully in his diary that Samuel D. Ingham, the Treasury Secretary, John M. Berrien, the Attorney-General, John Branch, the Navy Secretary, and Colonel Nathan Towson, the Paymaster-General, had all 'given large evening parties to which Mrs Eaton is not invited . . . the Administration party is slipped into a blue and green faction upon this point of morals . . . Calhoun heads the moral party, Van Buren that of the frail sisterhood.' The fact is, Van Buren was a bachelor, with no wife to raise objections, and he, and the British ambassador, another bachelor, gave the only dinner-parties to which Peggy was invited.[135]

The battle of the dinner-parties, what Van Buren called the 'Eaton Malaria,' was waged furiously throughout the spring and summer of 1829. It became more important than any other issue, political or otherwise. Jackson's first big reception was a catastrophe, as the Cabinet wives cut Peggy dead in front of a delightedly goggling Washington *gratin*. At one point the President laid down an ultimatum to three Cabinet members: they *must* ask Mrs Eaton to their wives' dinner-parties or risk being sacked. He thought Clay had organized it all but patient work by Van Buren showed that the wives, and Emily, had had no contact with Clay. Jackson then referred darkly to a 'conspirasy' organized by 'villians' and 'females with clergymen at their head.' The clergymen were the Rev. Ezra Stile Ely of Philadelphia and the Rev. J. M. Campbell, pastor of the Presbyterian Church in Washington, which Jackson often attended. Both believed the gossip, and Jackson had both of them to the White House to argue them out of their suspicions. He exchanged some striking letters with Ely on the subject and engaged in amateur detective-work, rummaging up 'facts' to prove Peggy's innocence and having investigators consult hotel registers and interview witnesses. At 7 P.M. on September 10, 1829, he summoned what must have been the oddest Cabinet meeting in American history to consider what he termed Eaton's 'alleged criminal intercourse' with Peggy before their marriage. Both Ely and Campbell were bidden to attend. The meeting began with a furious altercation between Campbell and the President on whether Peggy had had a miscarriage and whether the Eatons had been seen in bed in New York or merely sitting on it. The Cabinet sat in speechless embarrassment as the clergyman droned on, often interrupted by Jackson's exclamations: 'By the God eternal!' and 'She is as chaste as a virgin!' Campbell finally rushed out of the room in a rage, saying he would prove his accusations in a law court, and the Cabinet meeting broke up in confusion.[136]

The episode testifies more to Jackson's irrational loyalty than to his common sense. As might have been foreseen, he never contrived to force Mrs Eaton on Washington society. She was a worthless woman anyway. Her black page, Francis Hillery, later described her as 'the most compleat Peace of deception that ever god made, and as a mistres it would be cruelty to put a dum brute under her Command.' Her ultimate fate was pitiful. When Eaton died in 1856, leaving her a wealthy widow, she married an Italian dancing-master, Antonio Buchignani, who defrauded her of all her property and ran off with her pretty granddaughter. But she changed the way America is governed.[137]

One of the most fascinating aspects of history is the way power shifts from formal to informal institutions. The Cabinet system, which itself began in Britain as an informal replacement of the old Privy Council, was adopted by George Washington in the 1790s and was still functioning under John Quincy Adams. Jackson, however, was the first president to be elected by a decisive popular mandate and, in a sense, this gave him the moral right to exercise the truly awesome powers which the US Constitution confers to its chief executive. From the outset, an informal group of cronies began to confer with him in the entrails of the White House. They included Kendall, his old aide Major Lewis, his adopted son Donelson, Isaac Hill, the former editor of the New Hampshire *Patriot*, and two members of the official Cabinet, Eaton and Van Buren. Jackson's enemies called it the 'Kitchen Cabinet' and declared it unconstitutional.

Jackson began to lean more and more on this group as he became slowly convinced that opposition to Peggy was not just moral but political, orchestrated by Calhoun and his wife Floride, who had finally come to Washington to make trouble. Van Buren encouraged this conspiracy theory. Not for nothing was he known as the 'Little Magician.' Behind his spells was the deep, often hidden, but steadily growing antagonism between North and South. Van Buren stood for the commercial supremacy of the industrial North, Calhoun for the extreme version of states' rights. It was not difficult for the Secretary of State to persuade his President that the notion of sovereignty being peddled by Calhoun was a mortal threat to the Union itself, and that the Vice-President, using Mrs Eaton as a pretext, was behind a much wider 'conspiracy' to subvert Jackson's Cabinet. Jackson slowly came to accept this notion and in April 1831 he acted, following a plan of Van Buren's. To avoid suspicion, Van Buren resigned. Then almost all the other Cabinet ministers were sacked and replaced, leaving Calhoun isolated to serve the rest of his term. Van Buren's reward was to be made heir apparent, getting first the vice-presidency (during Jackson's second term), then the reversion of the presidency itself.[138]

Meanwhile the Kitchen Cabinet governed the country. It had no agenda. Its membership varied. Outsiders thought its most important figure was Kendall. He certainly wrote Jackson's speeches. The General would lie on his bed, smoking his fearsome pipe and 'uttering thoughts.' Kendall would put them into presidential prose. Congressman Henry A. Wise termed Kendall 'the President's *thinking* machine, and his *writing* machine—aye, and his *lying* machine.' Harriet Martineau, reporting Washington gossip, said 'it is all done in the dark . . . work of goblin extent and with goblin speed, which makes men look about with a suspicious wonder, and the invisible Amos Kendall has the credit for it all.' Very likely Kendall had much less power than was ascribed to him. But he symbolized what was happening to government. The old Cabinet had been designed to represent interests from all over the Union and its members were a cross-section of the ruling class, insofar as America had one—they were gentlemen. The Kitchen Cabinet, by contrast, brought into the exercise of power hitherto excluded classes such as journalists. Kendall despised Washington society, which he accused of trying to ape London and Paris. He thought 'late dinner' was 'a ridiculous English custom,' drinking champagne instead of whiskey 'uppety,' low-cut evening dresses 'disgusting.'

The idea of men like Kendall helping to rule America was appalling to men like Adams. But there it was. Jackson had successfully wooed the masses, and they now had their snouts in the trough. Jackson not only set up a new political dynasty which was to last, with one or two exceptions, up to the Civil War. He also changed the power-structure permanently. The Kitchen Cabinet, which proliferated in time into the present enormous White House bureaucracy and its associated agencies, was the product of the new accretion of presidential power made possible by the personal contract drawn up every four years between the president and the mass electorate. That a man like Kendall came to symbolize these new arrangements was appropriate, for if Jackson was the first man to sign the new contract with democracy, the press was instrumental in drawing it up.[139]

Ordinary people did not care much whether they were ruled by a formal Cabinet or a kitchen one, as long as that rule was light. And, under Jackson, it was. He let the economy expand and boom. As a result, the revenue from indirect taxation and land sales shot up, the meager bills of the federal government were paid without difficulty, and the national debt was reduced. In 1835 and 1836, it was totally eliminated, something which has never happened before in a modern state—or since.[140] There is no doubt that electors liked this frugal, minimalist, popular style of government, with no frills and no pretensions to world greatness. In 1832 Jackson was reelected by a landslide, the first in American presidential history. The luckless Clay was his main opponent. It is a curious fact that, although Washington, Jefferson, Madison, and Monroe had all been Masons, Clay was the only one whose Masonry was held against him (perhaps because he never went to church), especially in New York. So in 1832 Clay had to face an anti-Mason candidate, Thurlow Weed, who got a popular vote of 101,051, which would have gone mainly to Clay. Clay campaigned frantically, and oscillated between Kentucky and Washington, much of the time with his wife and grandson, four servants, two carriages, six horses, a jackass, and a big shepherd dog—all to no avail. He got 437,462 to Jackson's 688,242 votes, and in the electoral college the margin was even greater—a mere 49 to Jackson's 219.[141]

This was the beginning of the Jacksonian Democratic dominance. Jackson virtually appointed his successor, Van Buren, and though Van Buren failed to get reelected in 1840 because of a severe economic crisis, that was the only blip in the long series of Democratic victories. The Democrats returned with James K. Polk or 'Little Hickory' as he was known, in 1844, with Zachary Taylor in 1848 (who, dying in office, was succeeded by Millard Fillmore), and then by two solid Jacksonians, Franklin Pierce, 1852, and James Buchanan, 1856. In effect, Jackson, or his ideas, ruled America from 1828 to the Civil War.[142]

And what were these ideas? One was Union. No one was ever stronger for the Union than Jackson, not even Lincoln himself. Jackson might be a slave-owner, a small government man, a states' rights man and, in effect, a Southerner, or a Southwesterner, but first and foremost he was a Union man. He made this clear when portions of the South, especially South Carolina, threatened to leave the federal Union, or nullify its decisions, unless Washington's economic policy was tailored to fit Southern interests. The South, being a huge exporter of cotton and tobacco, was strongly in favor of low tariffs. The North, building up its infant industry, wanted tariffs high. Congress had enacted its first protective high tariff in 1816, over Southern protests. In 1828 it put through an even higher one, the 'Tariff of Abominations,' which made US tariffs among the highest in the world and hit Britain, the South's main trading-partner. South Carolina was particularly bitter. From being one of the richer states, it feared becoming one of the poorest. It lost 70,000 people in the 1820s and 150,000 in the 1830s. It blamed high tariffs for its distress. Jackson did his best to get tariffs down and the 1832 Tariff Act was an improvement on the Abominations. But it did not go far enough to satisfy the South Carolinians and their leader, Calhoun. In November 1832 the state held a constitutional convention which overwhelmingly adopted an Ordinance of Nullification. This new constitutional device, inspired by Calhoun, ruled the Tariff Acts of 1828 and 1832 to be unconstitutional and unlawful and forbade all collection of duties in the state from February 1, 1833. Its legislature also provided that any citizen whose property was seized by the federal authorities could get a court order to recover twice its value.

To fight this battle in Washington, Calhoun quit the administration finally by resigning the vice-presidency and was promptly elected senator. In reply, Jackson (with all the authority of a newly reelected president) issued a Nullification Proclamation on December 10 which stated emphatically that 'The power to annul a law of the United States, assumed by one state, [is] *incompatible with the existence of the Union, contradicted expressly by the letter of the Constitution, unauthorised by its spirit, inconsistent with every principle on which it was founded, and destructive of the great object for which it was formed*' (Jackson's italics). The Constitution, he added, 'forms a *government*, not a league.' It was 'a single nation' and the states did not 'possess any right to secede.' They had already surrendered 'essential parts of [their] sovereignty,' which they could not retract. Their citizens were American citizens primarily, and owed a prime obedience to its Constitution and laws. The people, he said, were sovereign, the Union perpetual. This, coming from a man who was born in South Carolina, and had been an anti-federalist all his life, was an amazing statement of anti-states' rights principle, and was to make it infinitely easier for Lincoln to fight for the Union in 1860.

Jackson went further. As chief executive, he had to enforce the laws passed by Congress, and that included collection of the tariffs: 'I have no discretionary powers on the subject; my duty is emphatically pronounced in the Constitution.' He spoke to the people of South Carolina directly. They were being deceived by 'wicked men'—he meant Calhoun—who assured them they would get away with it. He, as president, wanted to disillusion them before it was too late: 'Disunion by force is *treason*' and would be put down with all the strength of the federal government. It would mean 'civil strife' and the necessary conquest of South Carolina by federal forces. Indeed, he rather implied that any ringleaders would be tried for treason and hanged—and in private that is exactly what he threatened to do to his former Vice-President. He requested Congress to pass a Force Bill. He followed this up with a whole series of military measures—moving three divisions of artillery, calling for volunteers, mobilizing militias. He ordered the head of the army, General Winfield Scott, to Charleston Harbor, where Fort Moultrie and Castle Pinckney were reinforced, and a battleship and seven revenue cutters took up station in the harbor. He also organized, within the state, a pro-Union force which he hoped, if it came to war, would act and disarm the traitors. They responded to his proclamation: '*Enough!* What have we to fear? We are right and God and Old Hickory are with us.'[143]

The existence of an armed Unionist party within the state was one reason why the Nullifiers were forced to hesitate. Another was the failure of any other Southern state to join the South Carolina legislature in its measures to defy the tariff. But a third was Henry Clay, the 'Great Compromiser.' On February 12, 1833, just as South Carolina was planning, in effect, to secede, he brought forward an ingenious measure which progressively reduced the tariff to 20 percent by 1842. This was not as much as South Carolina wanted but it was enough to save its face. Jackson signed both the Force Act and the Compromise Tariff on March 1, 1833, and immediately afterwards South Carolina withdrew its Nullification Law. Needless to say, Clay got no thanks from either Jackson or Calhoun for getting them off their respective hooks. But the pending conflict between North and South was put off for another two decades and the power, strength, and rights of the Union publicly vindicated. The South was never quite the same again after this enforced climb-down by its most extreme state.[144] The fact is, Jackson had asserted, as president, that the Union could not be dissolved by the unilateral action of a state (or group of states), and the challenger had been forced to comply, implicitly at least.

If Jackson's democratic America was implacable with Southern separatism, it was even more relentless in destroying the last remnants of Indian power and property east of the Mississippi. Of course Jackson was not alone. White opinion—and black for that matter: the blacks found the Indians harsher masters than anyone—were virtually united in wanting to integrate the Indians or kick them west, preferably far west. Jackson had destroyed Indian power in the Southeast even before he became president. And, under Monroe, Indian power south of the Great Lakes was likewise annihilated by General Lewis Cass (1782–1866), hero of the 1812 War and governor of Michigan Territory 1813–21. In August 1825 Cass called a conference of 1,000 leaders of all the Northwest tribes at Prairie du Chien and told them to settle their tribal boundaries. Once this was done, he made compulsory deals with each tribe separately. In 1826 he forced the Potawatomi to hand over an enormous tract in Indiana. The Miami handed over their lands in Indiana for $55,000 and an annuity of $25,000. Other separate tribal deals were similar. In the years 1826–30 the Indians were forced to surrender not only their old land but their new reservations, as the settlers poured in to take over. There was a substantial Indian uprising in 1829, but it was put down by overwhelming force, Washington for the first time using steam gunboats on the Great Lakes, just as the British were using them to build up their empire all over the world. As a result of this 'Gunboat Diplomacy,' the Indians were pushed across the Mississippi, or left in small pockets, and 190,879,370 acres of their lands passed into white hands at a cost of a little over $70 million in gifts and annuities.[145]

Cass was a sophisticated man, who later held high posts in diplomacy and politics. He was one of the few Indian-fighters who actually set down his views on the subject—an essay entitled 'The Policy and Practice of the United States and Great Britain in their Treatment of Indians,' published in the North American Review, 1827. He said he could not understand why the Indians, after 200 years of contact with the white man, had not 'improved.' It was a 'moral phenomenon'—it had to be—since 'a principle of progressive improvement seems almost inherent in human nature.' But 'the desire to ameliorate their condition' did not seem to exist in 'the constitution of our savages. Like the bear and deer and buffalo in his own forests, the Indian lives as his father lived, and dies as his father died. He never attempts to imitate the arts of his civilised neighbors. His life passes away in a succession of listless indolence, and of vigorous exertion to provide for his animal wants or to gratify his baleful passions ... he is perhaps destined to disappear with the forests.'[146]

In fact the Indians varied enormously. The Creeks, Cherokees, Choctaws, Chickasaws, and Seminoles, who bore the brunt of white aggression, had long been known as the 'Five Civilized Tribes.' John Quincy Adams, who was always hostile to Indians, had to admit that a delegation of Cherokees who came to see President Monroe in 1824 were 'most civilised.' 'These men,' he recorded, 'were dressed entirely according to our manner. Two of them spoke English with good pronunciation and one with grammatical accuracy.'[147] During a Cabinet discussion of what Monroe called 'the absolute necessity' that 'the Indians should move West of the Mississippi,' Calhoun, Secretary of War, argued that 'the great difficulty' was not savagery but precisely 'the progress of the Cherokees in civilisation.' He said there were 15,000 in Georgia, increasing just as fast as the whites. They were 'all cultivators, with a representative government, judicial courts, Lancaster schools and permanent property.' Their 'principal chiefs,' he added, 'write their own State Papers and reason as logically as most white diplomatists.'[148]

What Calhoun said was true. The Cherokees were advancing and adopting white forms of social and political organization. Their national council went back to 1792, their written legal code to 1808. In 1817 they formed a republic, with a senate of thirteen elected for two-year terms, the rest of the council forming the lower house. In 1820 they divided their territory into eight congressional districts, each mapped and provided with police, courts, and powers to raise taxes, pay salaries, and collect debts. In 1826 a Cherokee spokesman gave a public lecture in Philadelphia, describing the system. The next year a national convention drew up a written constitution, based on America's, giving the vote to 'all free male citizens' over eighteen, except 'those of African descent.' The first elections were held in summer 1828. A Supreme Court had been functioning five years. The first issue of the republic's own paper, the *Cherokee Phoenix*, appeared February 28, 1828. Its capital, New Echota, was quite an elaborate place, with a fine Supreme Court building, a few two-story red-brick homes, including one owned by Joseph ('Rich Joe') Van, which is still to be found near what is now Chatsworth, Georgia, and neat rows of log cabins.[149]

The trouble with this little utopia—as the whites saw it—was that it was built as a homogeneous Indian unit. It mattered not to the whites that this self-contained community virtually eliminated all the evils whites associated with Indians. The *Phoenix* campaigned strongly against alcohol and there was a plan to enforce prohibition. The courts were severe on horse thieves. The authorities urged all Indians to work and provided the means. There were 2,000 spinning-wheels, 700 looms, thirty-one grist-mills, eight cotton gins, eighteen schools—using English and a new written version of Cherokee. The 15,000 Indians of this settled community owned 20,000 cattle and 1,500 slaves, like any other 'civilized' Georgians. But its very existence, and still more its constitution, violated both state and federal law, and in 1827 Georgia petitioned the federal government to 'remove' the Indians forthwith. The discovery of gold brought in a rush of white prospectors and provided a further economic motive. The election of General Jackson at the end of 1828 sealed the community's fate. In his inaugural address he insisted that the integrity of the state of Georgia, and the Constitution of the United States, came before Indian interests, however meritorious. A man who was prepared to wage war against his own people, the

South Carolinans, for the sake of constitutional principles, was not going to let a 'utopia of savages' form an anomaly within a vast and growing nation united in a single system of law and government. And of course, with hindsight, Jackson was absolutely right. A series of independent Indian republics in the midst of the United States would, by the end of the 20th century, have turned America into chaos, with representation at the United Nations, independent foreign policies, endless attempts to overthrow earlier Indian treaties and territorial demands on all their white neighbors.

Some whites supported the Cherokee Republic at the time. When Congress, in response to the Georgia petition, decreed that, after January 1, 1830, all state laws applied to Indians, and five months later passed a Removal Bill authorizing the President to drive any eastern Indians still organized tribally across the Mississippi, if necessary by force, a group of missionaries encouraged the Cherokee Republic to challenge the law in the Supreme Court. But in *Cherokee Nation* v. *Georgia*, the Marshall Court ruled that the tribe did not constitute a nation within the meaning of the US Constitution and so could not bring suit. The missionaries then counselled resistance and on September 15, 1831 eleven of them were convicted of violating state law and sentenced to four years' hard labor. Nine had their convictions overturned by submitting and swearing an oath of allegiance to Georgia. Two appealed to the Supreme Court and had their convictions overturned. But Georgia, encouraged by President Jackson, defied the Court's ruling. The end came over the next few years, brought about by a combination of force, harassment—stopping of annuities, cancellations of debts—and bribery. The Treaty of New Echota, signed in December 1835 by a greedy minority led by Chief Major Ridge, ceded the last lands in return for $5.6 million, the republic broke up, and the final Cherokee stragglers were herded across the Mississippi by US cavalry three years later.[150]

If Georgia hated Indians with self-serving hypocritical genuflexions to the rule of law, the humbug of Arkansas was even more striking. It was the keenest of all the states to assert the superiority of white 'civilized' values over the 'savage' Cherokees, what its legislature denounced as a 'restless, dissatisfied, insolent and malicious tribe, engaged in constant intrigues.'[151] Testimony from anyone with a quarter or more Indian blood was inadmissible in Arkansas courts. The state operated a system of apartheid with laws prohibiting dealings between whites and Indians. Yet ironically Arkansas was the most socially backward part of the United States. Its whites tended to be either solitaries—isolated hunters, trappers, and primitive farmers—or clannish, self-sufficient, and extremely violent. Its 14,000 inhabitants got a territorial government in 1819, but its courts and legislature were ruled by duels as much as by law or debate. In 1819 the brigadier-general commanding the militia was killed in a duel and five years later the same happened to a superior court judge, his assassin being his colleague on the bench and the occasion a squalid game of cards. The Flanagan clan 'respected no law, human or divine, but were slaves to their own selfish lusts and brutal habits.' The Wylie clan were illiterate, 'wonderfully ignorant' and 'as full of superstition as their feeble minds were capable of, believing in Witches, Hobgoblins, Ghosts, Evil Eyes . . . They did not farm, had no fences round their shanty habitations and appeared to have lived a roving, rambling life ever since the Battle of Bunker Hill when they fled to this wilderness.'[152] Yet Arkansas was harder on the Indians than any other territory or state.

The sight of Indian families, expelled from Georgia and Arkansas, heading west with their meager possessions was not uncommon in the 1830s, a harsh symbol of the age of mass settlement. In winter 1831, in Memphis, Tennessee, Comte Alexis de Tocqueville, in America to study the penal system on behalf of the French government, watched a band of Choctaws being marshaled across the Mississippi. He wrote: 'The Indians had their families with them, and they brought in their train the wounded and the sick, with children newly-born and old men on the point of death.' He added: 'Three or four thousand soldiers drive before them the wandering race of aborigines. These are followed by the [white] pioneers who pierce the woods, scare off the beasts of prey, explore the course of the inland streams and make ready the triumphal march of civilisation across the desert.' Under President Jackson, he noted, all was done lawfully and constitutionally. The Indians were deprived of their rights, enjoyed since time immemorial, 'with singular felicity, tranquilly, legally, philanthropically, without shedding blood and without violating a single great principle of morality in the eyes of the world.' It was, he concluded, impossible to exterminate a race with 'more respect for the laws of humanity.'[153]

Jackson finished the Indians east of the Mississippi, and effectively laid down the ground rules which insured that they would not survive as substantial units west of it either. But he did not hate Indians: they were simply an anomaly. He did, however, hate banks, and especially the Second Bank of the United States. That was an anomaly too, and he was determined to remove it. It is often said that Jackson knew nothing about banks, and that is why he hated them. That is not true. He is, rather, an example of what Keynes meant when he said that the views of great men of the world, who believe themselves impervious to theory of any kind, are usually shaped by the opinions of 'some defunct economist' which have imperceptibly got into their heads. Jackson once said he had disapproved of banks, and especially central banks, ever 'since I read a book about the South Sea Bubble.' He had already read Adam Smith, and misunderstood him, and Taylor, whom he understood only too well. In the late 1820s his views—and Taylor's—were reinforced by an anti-banking ideologue called William M. Goude, who wrote widely about banking in the *New York Evening Post* and Jackson's favorite paper, the *Washington Globe*. Goude's book *A Short History of Paper Money and Banking in the United States* (1833), which summed up his theories, became one of the great bestsellers of the time. It was a book written against the 'city-slickers', the 'Big Men,' the 'money power,' which contrasted the hard-working farmer, mechanic, and storekeeper with the chartered, privileged banker: 'The practices of trade in the United States have debased the standards of commercial honesty . . . People see wealth passing continuously out of the hands of those whose labor produced it, or whose economy saved it, into the hands of those who neither work nor save.'[154] It was a plea for economic equality before the law, in effect for an end of chartering, and especially of federal chartering.

Jackson made ending the SBUS a major issue in the 1832 election and he felt that the landslide result gave him a clear mandate. It is important to grasp that Jackson spoke from his moral heart as well as his bank-hating head. The nation, he said, was 'cursed' with a bank whose 'corrupting influences' fastened 'monopoly and aristocracy on the Constitution' and made government 'an engine of oppression to the people instead of an agent of their will.' Only the elimination of the 'Hydra' could 'restore to our institutions their primitive simplicity and purity.'[155] So Jackson's love of conspiracy theory and his taste for a moral crusade went hand in hand.

There was also a personal element, as there always was in Jackson's campaigns. He was certain—he 'knew for a fact'—that Clay was paid large sums by the SBUS. So was Daniel Webster, the sophisticated if long-winded Massachusetts orator who aroused all Jackson's suspicions and whom he was also certain—'knew for a fact'—was crooked.

Not least, Nicholas Biddle (1786–1844), president of the Bank since 1822, was just the kind of person Jackson feared and despised, a cultivated, high-minded (that is, humbugging), aristocratic intellectual. Jackson was always wary of 'college men.' Biddle had been to two (University of Pennsylvania and Princeton). He came from an ancient, posh Quaker family of Delaware, and married into another. He patronized the arts and not only collected but actually commissioned paintings of naked women of the kind Amos Kendall felt was an outrage, paying the gifted American artist John Vandelyn (1775–1852) to do him a lubricious *Ariadne*. He had edited a literary and artistic magazine called *Port Folio*, founded the Athenaeum Library in Philadelphia, and commissioned leading architects—at what Jackson believed to be vast expense—to design all the SBUS's buildings in Greek Revival granite and marble. Biddle's favorite architect, Thomas Ustick Walter (1804–87), who built the best of the banks, was also employed by Biddle to enlarge and classify for him his house, Andalucia, on the Delaware, making it into one of the lushest and most beautiful homes in America and (to Jackson) a symbol flaunting the new money power.[156]

Biddle was a first-class central banker, as good at his job as Marshall was at being chief justice, and the two men had similar ideas about how America should be developed, by a highly efficient, highly competitive capitalist system with easy access to the largest possible sources of credit, that access to be maintained by strict fiscal and financial probity. Jackson did not care a damn about that. Marshall had supported the SBUS in one of his most important decisions and Jackson did not 'care a fig' for the reasons he advanced for his ruling. When Marshall finally died in 1835—not before time in Jackson's view—the President appointed as his successor his Attorney-General and crony Rogert Brooke Taney (1777–1864), who conducted his court for thirty years on principles diametrically opposed to Marshall's.[157] When the Senate and the House both reported favorably on the Bank and proposed to renew its charter even before it ran out, Jackson used his veto. The fact that the three greatest orators in the Senate, Clay, Calhoun, and Webster, all pronounced at length and with ornate circumlocution on its merits only reinforced Jackson's determination to destroy it. Brilliant orators they might be, he noted, but they were 'always on the losing side.'

Jackson was one of those self-confident, strong-willed people (one thinks of Ronald Reagan and Margaret Thatcher in our own time) who are not in the least disturbed if the overwhelming majority of 'expert opinion,' the 'right-thinking,' and the intelligentsia are opposed to their own deep-felt, instinctive convictions. He simply pressed on, justifying his veto by producing a curious constitutional theory of his own: 'Each public officer who takes an oath to support the Constitution swears that he will support it *as he understands it, and not as it is understood by others* . . . The opinion of the judges has no more authority over Congress than the opinion of Congress has over the judges, and on that point the President is independent of both.'[158] The fury of the right-thinking was unbounded. Biddle himself described Jackson, in his stupidity and ignorance giving vent to 'the fury of a chained panther biting the bars of his cage.' His statement was 'a manifesto of anarchy such as Marat or Robespierre might have issued to the mobs.' The Jacksonian press hailed it as a 'Second Declaration of Independence' and his organ, the *Globe*, said, 'It is difficult to describe in adequate language the sublimity of the moral spectacle now presented to the American people in the person of Andrew Jackson.'[159]

With the election confirming and endorsing Jackson's standpoint (as he saw it), he proceeded to the next step—withdrawing all federal funds from the SBUS and ending its connection with central government. Whether this act was strictly constitutional was a matter of opinion, but Van Buren (now vice-president) warned him against it on prudential grounds. The SBUS was primarily a Philadelphia financial institution, which performed a useful national role balancing the growing money power of New York. If Jackson pulled the government out of Philadelphia, wasn't he in danger of falling into the hands of Wall Street? But Jackson brushed that aside too, and set Amos Kendall, of all people, busily to work finding alternative banks with which the administration could do business. Kendall fed him the rumor, which Jackson readily believed, that the SBUS's vaults were, in fact, empty of bullion, and that it was not a safe bank to do business with anyway. The fact that Senators Clay and Calhoun put together a committee to inspect the vaults and reported them full did not convince the President, coming from such a source. (He thereby inaugurated an American tradition which continues to this day: every year, the Daughters of the American Revolution send a committee of ladies to visit the vaults of Fort Knox, to ensure that America's gold is still in them.) Nor was Jackson impressed when two Treasury secretaries in turn flatly refused to carry out his orders to remove the deposits. He dismissed them both. Kendall, after a trawl through the financial community, came up with a list of banks willing to dare Biddle's wrath and take the SBUS's place.

Jackson acted. And when, as a result, there were rumblings of trouble in the American economy—more pronounced after its federal charter ran out in 1836 and it was obliged to 'go private'—Jackson was adamant, rejecting Van Buren's plea for 'caution' with a gloriously characteristic reply: 'Were all the worshipers of the golden Calf to memorialise me and Request a Restoration of the Deposits, I would cut my right hand from my body before I would do such an Act. The golden calf may be worshiped by others but, as for myself, I serve the Lord!'[160]

Biddle declared all along that, by forcing the SBUS out of its role as federal banker, Jackson would encourage a fever of speculation fueled by an expansion in the number of banks issuing paper and the quantity and quality of the paper they printed. That is exactly what happened, and the orgy was encouraged still further by Jackson's decision to hand the federal government's cash surplus, which accumulated when the national debt was paid off in 1835, back to the states. This amounted to $28 million, and though described as a loan was understood to be an outright gift, treated as such and spent.[161] The surplus was the result of government land sales jumping from $1.88 million in 1830 to $20 million in 1836, and as the land boom continued the states assumed that federal handouts would continue and increased their borrowing on the strength of it. Banks of all shapes and sizes, many with outright crooks on their boards, poured oil on the smoldering embers of inflation by keeping their presses roaring. In the meantime, nature intervened, as it usually does when men construct houses of straw, or paper. Bad weather in 1835 created a crop failure in many parts of America, and the consequences began to make themselves felt in 1836 with an unfavorable balance of trade against the United States, a withdrawal of foreign credit, and the need to pay suspicious foreign creditors, who did not like American paper, in gold and silver.[162]

Jackson, who was nearing the end of his term, increased the tension by issuing, on July 11, 1836, a Species Circular, which directed that future payments for public lands must be made in specie. This move was made in a simple-minded desire to get back to 'sound' finance, but it had the predictable effect of making gold and silver even more sought after. Characteristically, it was cooked up in the Kitchen Cabinet and announced to the official Cabinet as a *fait accompli*. Most of its members objected, and so did Congress. The new Whig Party, recently formed in opposition to 'King' Jackson, on the lines of the old English Whigs who had opposed Stuart tyranny, objected noisily to this further exercise of presidential prerogative, which Clay said was exactly what a dictator would do, calling the circular an 'ill-advised, illegal and pernicious measure.' It was 'a bomb thrown without warning.'[163] Its effects at end–1836 coincided, almost exactly, with the failure of big financial houses in London, the world financial capital. This in turn hit cotton prices, America's staple export. By the time Jackson finally retired in March 1837, handing over to his little heir apparent, Van Buren, America was in the early stages of its biggest financial crisis to date. By the end of May 1837 every bank in the country had suspended payment in specie. Far from getting back to 'sound money,' Jackson had merely paralyzed the system completely.[164]

Before the panic became obvious Van Buren had squeezed through the presidency with a narrow victory made possible by the fact that the anti-Jackson Whigs fielded three candidates. Van Buren got 764,198 votes against a combined Whig total of 736,147. More important, he won fifteen states, making up 170 votes, while his nearest rival, William Henry Harrison (1773–1841), the victor of Tippecanoe (1811) and the Thames (1813), won only 73.[165] So Van Buren was in the White House at last. His bitter Whig enemy in New York, Thurlow Weed (1797–1882), warned: 'Depend upon it, his Election is to be "the Beginning of the End." ' So, through no fault of the Little Magician— who had opposed Jackson's financial policies throughout, so far as he had dared—it proved. He had worked long and hard for the presidency, being nice to everyone, concealing his intentions, 'rowing to his object with muffled oars' as John Randolph put it, convinced that the great state of New York, whose champions, Hamilton, Burr, De Witt Clinton, and Co., had all failed to get to the White House, was due for its turn at last.

But as president Van Buren never had a chance. The financial panic, which deepened into a real depression, ruined all. The money in circulation (banknotes mainly) contracted from $150 million in 1837 to barely over a third by the end of the decade. An enormous number of people, big and small, went insolvent, so many that Congress, in order not to clog the jails with them, passed a special bankruptcy law under which 39,000 people were able to cancel debts of $441 million. The government itself lost $9 million which, on Kendall's advice, it had deposited in Jackson's 'pet banks,' which now went bust. Worse, the depression lingered for five years. As land sales slumped, the federal government went into sharp deficit, and the national debt began to accumulate again—something it has done ever since. Most of Van

Buren's energies went on an attempt to set up what he called an Independent Treasury—the nearest he could get to a central bank without actually repudiating Jackson's policy. He finally got it through Congress just as he had to run before the voters again, and the Depression made it certain he would lose.[166]

If there were any justice in politics, Clay should have been the beneficiary, since he had opposed the Jacksonians for two decades and his warnings against the Dictator's absurd financial policies had been fully vindicated by events. But at the Harrisburg convention of the Whig Party—more a coalition of personal and local power-groups than a real party based on shared convictions—he was outmaneuvered and hornswoggled in the 'smoke-filled rooms,' the first time that phenomenon made its appearance in American history. Clay's supporters arrived with a plurality of delegates but on the final vote he was beaten by Harrison 148–90, his manager telling him: 'You have been deceived betrayed & beaten [by] a deliberate conspiracy against you.'[167] The election itself was unique in American history, conducted in a carnival atmosphere in which programs and policies were scarcely discussed at all, and all was slogans, gimmickry and razzmatazz. Considering the country was supposed to be, indeed to some extent was, in deep depression, the frivolity was remarkable. But then the mid–19th century was an astonishing age of optimism and America was a resilient nation. Harrison campaigned as a rugged frontiersman, with his running mate John Tyler (1790–1862), a dyed-in-the-wool Virginian and states' rights man who had been alienated by Jackson's high-handed ways, being presented as an experienced and wily professional politician. So the Whig slogan was 'Tippecanoe and Tyler Too.' The Democrats retaliated by ignoring Tyler and branding General Harrison, who liked his noggin—or rather his joram—as 'The Log Cabin and Hard Cyder' candidate. The Whigs turned this to advantage by holding 'Log Cabin Rallies' at which hard cider was copiously served. They also created an electorally effective image of the dapper Van Buren as an effete New York dandy, drinking wine 'from his coolers of silver.' The actual popular vote was fairly close—1,275,000 to Harrison against 1,128,000 for Van Buren—but the college vote was a landslide, 234 to 60. The Whigs thus demonstrated, as the Democrats had already discovered, that picking a general paid electoral dividends.[168]

Harrison was sixty-eight and said he would serve only one term. Clay turned down his invitation to become secretary of state again, saying he would rather remain in the Senate and expecting to succeed

Harrison as president in 1844. So the golden-tongued, distinctly fishy Webster got the job instead. However, Harrison, having formed his Cabinet and celebrated his entry into the White House, was attacked by pneumonia and expired after only a month in office. That put Tyler in the White House and disrupted all Clay's long-term plans. However, he thought that, with Harrison dead, he could now control the Whig Party and dictate to Tyler what he ought to do—in particular to proceed immediately to the creation of a Third Bank of the United States. But Tyler was no pushover. He too was a tall man, with a high 'retreating forehead,' with 'all the features of the best Grecian model,' and such a pronounced Roman nose that two Americans in Naples, present when a bust of Cicero was unearthed at an excavation, exclaimed with one accord: 'President Tyler!' So when Clay called on the new President and unwisely insisted 'I demand a bank *now*!' Tyler replied crushingly: 'Then, Sir, I wish you to understand this—that you and I were born in the same district; that we have fed upon the same food, and have breathed the same natal air. Go you now, then, Mr Clay, to your end of the avenue, where stands the Capitol, and there perform your duty to the country as you shall think proper. So help me God, I shall do mine at this end of it, as *I* shall think proper.'[169] The two men never spoke again.

Nor was this the end of the damage inflicted by the wretched SBUS dispute. In 1841 Biddle's Bank, mortally wounded by the long Depression, finally tottered to its doom and closed its doors. Among its bad debts was a massive $114,000 owed by Webster, the Secretary of State. Biddle was ruined, had to sell his splendid city mansion in Philadelphia, and was able to retain his 'perfect house' Andalucia thanks only to his rich wife's trustees. Even so, it soon wore a neglected look. The spiteful John Quincy Adams dined there and recorded: 'Biddle broods with smiling face and stifled groans over the wreck of splendid blasted expectations and ruined hopes. A fair mind, a brilliant genius, a generous temper, an honest heart, waylaid and led astray by prosperity, suffering the penalty of scarcely voluntary error.' Three years later the ruined banker was dead, old Jackson surviving him by a triumphant year.

Did having a central bank, or not having one, make much difference to America as a whole? It is hard to say. America had financial crises and recessions—all expanding countries do—but it always quickly recovered and went on remorselessly, pushing west, building up industries, creating farms. In 1800 there had been 450,000 American farms. By 1850 there were 1.5 million, a number which would grow steadily until it reached 6.4 million and a peak of 6.95 million in 1935. The Americans, recruited from all the people of Europe, were magnificent rough-and-ready farmers. The great epic of 19th-century America is the internal migration, the occupation and exploitation of the Middle and West. The biggest single factor was that the country was empty, land was cheap or free, credit was easy, the law left them virtually alone, and all was governed by a virtually unrestrained market and by their own ingenuity and energy. Nearly all these internal immigrants had already farmed in the East and they faced conditions they knew they could handle—this was a key point. Soils, shrubs, trees, and grasses were known to them, they were familiar with the weather, there was nearly always plenty of wood close at hand, and the water was there, in lakes and streams or in the table below the surface. They knew that all the old tricks of the farming trade they or their fathers had worked in the East, would work in the West, often better. That was the grand psychological certainty which made them pioneer. They had over two centuries of collective wisdom behind them, and science and machinery were coming along fast. In 1842 Samuel Forry published the first scientific work on the American climate. It was replaced by Lorin Blodget's book on climatology in 1857, and by then the Smithsonian was systematically collecting data on climate.

The Smithsonian, in Washington, was the first institution for scientific research in America, made possible by an enormous gift of about $500,000—more than any university endowment in the US then—by a British chemist, James Smithson. Despite the efforts of the strict constructionists like Calhoun to stop the federal government thus getting involved in research, John Quincy Adams and his friends managed to get the gift accepted, the institute organized, and America's first 'pure' scientist, Joseph Henry (1797–1878), made its founding director in 1846. Thanks to him, Americans began to get short- and long-term weather forecasts, and projections based on historical averages, carefully compiled by the Smithsonian.[170] Farmers were greatly assisted by the thoroughness with which the surveyor-general (originally called the geographer) and his teams worked. The surveyors were not merely geodetic workmen but good field-geographers, and each was obliged to put into his field-books, 'at their proper distances, all mines, salt-springs, salt-licks and mill-seats that shall come to his knowledge; and all water-courses, mountains and other remarkable and permanent things, over and near which such lines shall pass, and also the quality of the lands.'[171] From a surprisingly early stage, considering the sheer size of the Midwest and West, American farmers—and prospectors—got detailed, high-quality maps.

Jefferson had spotted the central principle of American agriculture, that it had to be labor-intensive: 'In Europe the object is to make the most of their land; here it is to make the most of our labor, land being abundant.' Until about 1800, all that most farmers had were a low-quality, often hand-made, plow, harrow, hoe, shovel, fork, and rake. Charles Newbould of New Jersey patented the first American cast-iron plow in 1797. But this was a solid piece, and it was improved on by Jethro Wood of New York, who patented a metal plow with separate interlocking parts each of which could be replaced if broken. These advanced metal plows came into general use in the second half of the 1820s and their impact on productivity was immediate. They were combined, on virgin land, with steel mold-boards, needed to break up the matted grasses of the prairies, made from 1833 by John Lane of Chicago. All farmers had got metal plows by the 1830s. In Pittsburgh two factories were making 34,000 metal plows a year even in the 1830s and mass manufacturing brought prices down. In 1845 in Massachusetts alone there were seventy-three plow-making firms producing 61,334 plows plus other implements. Competition was intense and by 1855 they had merged into twenty-two firms while production had risen to 152,688 and prices had fallen sharply.

In 1833 Obed Hussey produced the first practical reaper, which could do 15 acres a day. But he was a poor businessman and it was the genius of Cyrus McCormick (1809–84) which led to the marketing of the first mass-produced reaper. He was of Ulster origin, who came from Pennsylvania into the Shenandoah Valley, then to Brockport, New York, on the Erie Canal, to be nearer the markets, and finally to Chicago in 1848. He was not merely a great inventor but a remarkable businessman. He sold his first reaper in 1834 and by 1860 was turning out 4,000 machines a year. At the International Exposition in Paris in 1855, to the astonishment of the Europeans, an American reaper cut an acre of oats in twenty-one minutes, a third of the time taken by Continental makes. By then there were already 10,000 machines in use on American farms. Two years later, the United States Agricultural Society held a national trial at the New York State Fair, entered by forty mowers and reapers. This display showed how far and quickly American manufacturers had gone in eliminating the bad features, such as side-draft, clogging, and the inability to get started in standing grain. The quantity of these giant, reliable machines, made possible by intense competition, explains why American grain was so cheap, outselling all European products whenever it could get under the tariff and quota barrier, and why production of large acreages was kept up during the Civil War when the armies took all the young men.[172]

Progress was continual. As early as 1850 American farmer–inventors, after prodigious efforts, managed to attach a separator to a thresher, so the whole process of threshing and winnowing could be done by the same machine—the combine was well on its way. The horse hay-rake, doing the work of ten men, came in during the 1820s, in the 1830s speed-drills for sowing wheat, from 1840 the corn-planter and various types of cultivators. The census of 1860 reported: 'By the Improved Plough, labor equivalent of one horse in three is saved. By means of drills, two bushells of seeds will go as far as three scattered broadcast, while the yield is increased six to eight bushells an acre. The plants come up in rows and may be tended by horse-roes ... The reaping machine is a saving of more than one third the labor when it cuts and rakes ... The threshing machine is a saving of two-thirds on the old hand-flail mode ... The saving in the labor of handling hay in the field and barn by means of horserakes and horse-hayforks is equal to one half.' And American inventors and farmers were the first to power farm-machinery with steam.[173]

This mechanical impulse to labor-saving and scientific farming was backed by an intellectual thrust by no means confined to the Smithsonian. Washington had set the pattern of experimental farming, founding America's mule-raising industry with the help of high-quality asses sent him by Lafayette and the King of Spain. While he was still president, New York's Columbia College, a go-ahead place which had set up a medical school as early as 1767, created a professorial chair of Agriculture, Natural History, and Chemistry (1792). The first true agricultural college, the Dariner Lyceum, was established in Gardfiner, Maine, in 1822, and in 1857 Michigan opened the earliest State College of Agriculture, the first of many. These in turn were backed up by New York's Society for Promoting Agriculture, Commerce, and the Arts (1781), and by the country fairs movement of Elkanah Watson (1758–1842), started in 1807 and inspiring the Berkshire Agricultural Society, the first of hundreds. New York was underwriting fairs with state money ($20,000) as early as 1819. And finally there were the specialist publications of which the most important were the Baltimore weekly, the *American Farmer* (1819), the Albany *Cultivator* (1834), and, for the West, the *Prairie Farmer* of 1840. By mid-century, American farming was, next to Britain's, the most advanced technically in the world, and already overtaking Britain in mechanization.[174]

It is important to realize that, with this successful introduction of capital-intensive farming in the United States, and with the gigantic annual additions of land under cultivation—unprecedented in world history—America remained primarily an agricultural country almost till the end of the 1850s. A number of factors were against industrialization: poor banking facilities, hampered (as we have seen) by political problems; a federal government which, between 1830 and 1860, was heavily influenced by Southern plantation owners who opposed further protection, a central bank, any idea of transcontinental transit systems—roads or rail—built for the North, and free land. After Marshall's death, the Supreme Court was also dominated by Southern interests who tended to be anti-capitalist. On the other hand, a number of factors pushed industrialization. The quarrel with Britain and the War of 1812 in the first two decades of the century gave native manufactures a start, reinforced by the early tariffs. Those of 1816, 1828, 1832–3, and 1842 were in varying degrees strongly protectionist, and undoubtedly benefited US manufactures greatly. Tariffs were scaled down in 1846 and still more in 1857, which put America in the front rank of the free-trading nations. But by then home manufactures were firmly established and an increasingly sophisticated and resourceful capital market had come into being.[175] However, the main forces which industrialized America were the arrival of skilled labor from Europe and, above all, the rapid expansion of a huge domestic market.

It was Samuel Slater, an immigrant, induced to come to America by state bounties, who erected the first Arkwright-type cottom-mill at Pawtucket, Rhode Island, financed by Moses Brown, a Quaker merchant of Providence. Slater had been an apprentice of Arkwright and his arrival is the perfect example of the personal transfer of technology from Britain to the United States, repeated hundreds of times. But, as we have already seen, Americans were from the start highly inventive themselves. From 1790 to 1811 the US Patent Office reported an annual average of seventy-seven registrations. By the 1830s it had jumped to 544 annually, by the 1840s to 6,480 and in the 1850s over 28,000 every year.[176] Americans were using steam widely in Rhode Island and New Jersey even in the 1790s, and in 1803 were the first in the world to apply steam to a sawmill, an important point since wood was universally used for power-fuel in the US until the big coalmines began to come on stream in the 1850s. The steam engines were originally all imported Boulton & Watts from Britain but after Oliver Evans (1755–1819) of Philadelphia introduced a new high-pressure steam engine in 1802, Evans engines competed with imported ones, being used west of the Alleghenies in 1812. Five years later, American engines were being produced at Pittsburgh, Louisville, and Cincinnati, as well as on the coast. The census of 1830 showed that in Pennsylvania 57 out of 161 plants now used steam and 39 out of 169 in Massachusetts—the rest used cheap water-power, which was still the norm in New York, New Jersey, and elsewhere. You might say that, whereas mechanization in agriculture was accelerated by the need to save labor, industry in America was, to some extent, held back by the sheer abundance of nature—by ubiquitous water-power, easily harnessed, by seemingly inexhaustible quantities of wood near by, and by the teeming fisheries of the northwest Atlantic which continued to make sailing ships highly profitable.

It was not that America lacked instances of men operating at the limits of known technology in the first half of the 19th century, or even beyond them. Francis Cabot Lowell (1775–1817), having studied cotton technology in England, employed a mechanical genius called Paul Moddy who designed machines, set up at Waltham in 1814, which for the first time brought together all the processes of spinning and weaving in what became known as the Waltham System. There were other major American innovations—the first sewing machine, made by Elias Howe (1819–67), and the discovery in 1851 by William Kelly (1811–88) of how to decarbonize molten metal by forcing air through it, the so-called Bessemer process. But in general the American metallurgical industry remained basic for a long time because its main market was the do-it-yourself farming population. What they wanted was simply bar-iron which blacksmiths could work into machine parts for agricultural machinery and mills. Not until the late 1850s did iron-producers switch the bulk of their business to serving industry directly. The Singer Sewing-machine factory in New York, and others in Bridgeport and Boston were making 110,000 machines annually by 1860, but were exceptional. A more typical manufactured product was the wood-burning stove, of which 300,000 were made annually in the 1850s, iron axes, springs, bolts, wire, firearms, and locks. Processing foodstuffs soon became important in American manufacturing industry. Cincinnati was the chief town for meat-packing until Chicago took over, and by 1850 tinning meat was conducted on a massive scale, with glue, fertilizers, bristles, candles, and soaps forming byproducts. By 1850 Cincinnati was catering for the largest whiskey market in the world—2 million gallons a year.

There was no doubt about Northern dominance. In terms of capital invested and still more in numbers employed, by 1860 New England, the Middle States and the West outclassed the South by more than ten to one. But the sophistication of America's move into industry should not be exaggerated. Products, ideas, news, and innovation were still essentially spread by peddlers. One observer wrote in the 1820s: 'I have seen them on the peninsula of Cape Cod, and in the neighborhood of Lake Erie, distant from each other more than 600 miles. They make their way to Detroit, 400 miles further, to Canada, to Kentucky and, if I mistake not, to New Orleans and St Louis.'[177] Until the 1850s, the United States was essentially a country of four occupational groups—farmers, planters, fishermen and peddlers.[178]

Peddlers were important because continental America had to overcome the tyranny of distance—the paradox whereby sheer space narrows the lives of people living in scattered communities, far from each other and from urban centers. Happily the American people proved themselves wonderfully adept at overcoming this tyranny. If success in any one field has made America the world's greatest nation, it is transport and communications. The Constitution empowered the federal government to spend money under the 'general welfare' clause as well as under Article I, Section 7 (post offices and post roads, and regulation of interstate commerce). From 1808, when Albert Gallatin reported to Congress recommending sales of public lands to finance federal spending on canals and roads, Washington was fully involved in the transport business, usually in partnership with the states. A typical arrangement was with Ohio, at the time of its admission, 1803. Federal land sold within Ohio's borders was exempt from taxation for five years, and in return the federal government appropriated 5 percent of such sales for roads, three-fifths within the state and two-fifths over the mountains to the east. Similar deals were later made with Indiana, Missouri, and Illinois.

Some of these federal-financed roads were of the highest quality, like Telford's in Britain, often using his designs for road-furniture and bridges. The National Pike was the grandest. One historian wrote: 'Its numerous and stately stone bridges, with handsome stone arches, its iron mile-posts and its old iron gates, attest the skills of the workmen engaged in its construction, and to this day remain enduring monuments of its grandeur and solidity.'[179] When its 834 miles had been completed it had cost $6,821,200 and required thirty Acts of Congress, 1806–38. Such constructions always had to run the hazard of the Constitution, or of self-interested interpretations of it. Thus in 1831 President Jackson vetoed the Bill for a Maysville–Lexington road running for 60 miles entirely within Kentucky on the ground that it was unconstitutional for federal money to be spent for the advantage of a single state—in reality to spite Henry Clay.[180] Turnpike mania was succeeded by canal mania, then by railroad mania in the increasingly successful efforts to reduce freight costs, an important part of the tyranny: it cost $125 per ton on the Pennsylvania–Pittsburgh all-land route, a particularly expensive one, but on average it was $10 per 100 miles in the 1820s, about the same as it cost to get a ton across the Atlantic.

Before rail came, it was water-transport which made America great, especially when steam supplied the driving force. Fulton went into the business of marrying steam to water-transport in 1807 when he got a twenty-year monopoly of routes in New York, followed by a similar one for New Orleans. These monopolies were happily soon destroyed in the Marshall Court but Fulton continued his steam pioneering. In 1811 he built a shipyard at Pittsburgh and launched the *New Orleans*, the first steamer on the Ohio. In 1815 Henry Sheve, probably the greatest of all navigators of the Mississippi, took a steamboat all the way up the river from New Orleans to Louisville in twenty-five days. To Pittsburgh it was originally 100 days. But this was soon reduced to thirty and the New Orleans–Louisville route to a mere five days (upstream). No single fact of nature played a bigger part in American progress than the Mississippi. It was one of the three great rivers of the world, but whereas the Nile is bordered by desert, except for a narrow strip on its lower half, and the Amazon by tropical rainforest which is still largely impassable, the Mississippi runs directly through the largest continuous area of high-quality agricultural land on earth, and is the main artery of this richly productive basin. It is an amazingly changeable river, depositing mud in colossal quantities, changing its shape continually, creating islands and peninsulas, and then destroying them, dragging inland towns directly onto the waterfront, pushing river ports miles inland, and making the business of navigating it one of the most exacting sciences on earth.[181]

Many people, especially foreigners, were dismayed by their first contact with this vast and often terrifying river. Charles Dickens, writing of Cairo, one of its 'dismal towns,' called it:

> the hateful Mississippi, circling and eddying before it, and turning off upon its southern course, a slimy monster, hideous to behold; a hotbed of disease, an ugly sepulchre, a grave uncheered by any gleam of promise ... An enormous ditch, sometimes two or three miles wide, running liquid mud, six miles an hour; its strong and frothy current choked and obstructed everywhere by huge logs and whole forest trees: now twining themselves together in great rafts, from the interstices of which a sedgy, lazy foam works up, to float upon the water's top; now rolling past like monstrous bodies, their tangled roots showing like matted hair; now glancing singly by like giant leeches; and now writhing round and round in the vortex of some small whirlpool like wounded snakes. The banks low, the trees dwarfish, the marshes swarming with frogs, the wretched cabins few and far apart, their inmates hollow-cheeked and pale, the weather very hot, mosquitoes penetrating into every crack and crevice in the boat, mud and slime on everything: nothing pleasant in its aspects but the harmless lightning which flickers every night upon the dark horizon.[182]

But to Mark Twain (1835–1910), it was 'the great Mississippi, the majestic, the magnificent Mississippi, rolling its mile-wide tide along, shining in the sun.' And Twain knew! It was the river which gave him his writer's name, from the words the rivermen called out when they sounded two fathoms. It was as Samuel Clemens, the teenager, that he learned the Mississippi river-pilot's trade in the 1850s, later (1883) setting it down in the best book ever written about a river, *Life on the Mississippi*.

At their peak, there were 6,000 steamers in the Mississippi fleets. They competed ferociously against each other in size, capacity, grandeur, gambling, girls, and, above all, speed. It was the spirit of burgeoning American capitalism, afloat. 'The boats going north,' wrote Twain,

> always left New Orleans between four and five in the afternoon. From three PM onwards they would be burning rosin and pitch-pine to get ready, and a three-mile line of boats would produce a huge mushroom cloud of smoke over the city in consequence. Then the bells rang and

they all slid into the river. This was an amazing sight, not to be seen once the Civil War started or forever after. Races between the two fastest steamers were advertised weeks in advance and watched all along the river, boats being stripped of non-essential weight and loaded exactly to get maximum speed. Wood boats were hitched alongside and towed so refueling could take place in progress.

The *Eclipse* was the fastest, doing the 1,440 miles New Orleans–Louisville in 1853 in four days, nine hours, and thirty minutes. In 1870, the *Robert E. Lee*, in a famous race with the *Natchez*, did the New Orleans–St Louis run, 1,218 miles, in three days, eighteen hours, and fourteen minutes.[183] The craze for speed on the Mississippi was not without its human cost. In February 1830 the *Helen McGregor* was leaving Memphis, Tennessee, when the head of her starboard boiler cracked, and the explosion killed fifty souls, flayed alive or suffocated by the scalding steam. Burst boilers were the most common accidents which beset these steamers, sinking one-third of all boats up to the end of the 1830s. Officially, burst boilers killed 1,400 people up to the year 1850 but the real total was higher, and Dickens was advised to sleep at the stern of the boat if he wished to avoid being scalded to death.[184] Losses on bars and snags led the federal government to spend $3 million, 1820–60, on improving the four main rivers, the Mississippi, Ohio, Missouri, and Arkansas. It was cash well spent—by 1852 the commerce on the Mississippi alone was worth $653,976,000 annually.

Canals were built to link rivers, and early railroads to provide links where canalization was impracticable. Following the tremendous success of the Erie Canal opening in 1825—probably the outstanding example of a human artifact creating wealth rapidly in the whole of history—the states went into canal-making on a prodigious scale, borrowing vast sums (chiefly from Europe) to do it. Most state constitutions had to be rewritten in the 1820s to make this possible. In those days American credit in Europe was high, interest payments being substantial and handed over regularly. Most of these state canals were built and some succeeded in their purpose. But state debts mounted fast, from only $12,790,728 in 1820 to over $170 million in 1838 and $200 million in 1840. Only seven states did not borrow to fuel the canal mania. The 1837 crash made it impossible for the states to pay interest on their debts and six of them—Mississippi, Louisiana, Maryland, Pennsylvania, Indiana, and Michigan—went so far as to repudiate their debts entirely, rather like bankrupt African states today, thus bringing eloquent cries of anguish from innocent European investors like the novelist W. M. Thackeray. At this point most states decided to get out of the transport business and hand over to private enterprise. So state constitutions were again rewritten to forbid state canal mania or any other spending passion.

Hence, though the first railroads were built to supplement the mainly state-owned canal system, rail was a field where private capital and publicly floated companies were dominant almost from the start. Steam-powered railroads had evolved in England from the coal industry, which America did not then have, but the United States was not far behind Britain in introducing major passenger lines. On July 4, 1828 Charles Carroll of Carrollton, the last survivor of those who signed the Declaration of Independence, turned the first spade on the Baltimore & Ohio, the earliest proper line, 13 miles of which were open by 1830. Three years later, the Charleston–Hamburg line in South Carolina, with 136 miles open, was already the longest in the world. The first engine to be built in America, the *Best Friend of Charleston*, traveled this route at 30 miles an hour without freight or at up to 21 miles an hour with four loaded cars.[185] Construction started on what was to become the New York Central in 1830 and the big Pennsylvania line was completed four years later. All the major routes between the East Coast and the Mississippi Valley were completed by the 1840s and the first consolidation of multiple early routes began in 1853 with the New York Central, the same year the first rail service New York–Chicago opened. Benjamin Wright, the great engineer of the Erie Canal, denounced railroads as anti-individual: 'I consider a long line of railroad . . . as being odious in this country, as a monopoly of the carrying, which it necessarily must be. A canal, on the other hand, is open to any man who builds a boat.'[186] No one took much notice. The states even, to some extent, got involved in railroads—especially Pennsylvania, Michigan, South Carolina, and Georgia, though mainly in the early stages—and more money came from counties and towns, and even from the federal government, in the form of land deals with individual states. It is a curious fact that Georgia, for instance, actually ran the Chattanooga Choo-Choo till the 1870s. But private or publicly raised capital was the main provider of finance—$1.25 billion of it in the thirty years 1830–60.[187]

Beyond the rail-line, in the years 1830–60, were the stages, the fast carriers, and the telegraph. All developed with impressive speed and determination. Even before the California gold rush in the late 1840s, the Santa Fe Trail became the first path in advance of the Western frontier. With the rush, stage traffic from Independence, Missouri, to Santa Fe came into regular operation in 1849, with a monthly stage-mail to Salt Lake City the same year. In 1858, John Butterfield got a federal contract to carry mail overland from Memphis and St Louis to California, thus enabling him to run a twice-weekly stage through Preston, El Paso, and Yuma to the Pacific Coast, taking twenty-five days. The Russell, Major & Waddell Com- pany pioneered other routes and by the early 1860s it annually carried 75,000 oxen in 6,250 waggons in the freight business alone. By 1860 Russell's Pony Express, conducted on relays of horseback riders, carried mail from St Joseph, Missouri, to Sacramento, California, in ten days. This ruined the firm, however, and it ended up in the hands of Wells Fargo, which controlled much of the routing until the transcontinental railroads.[188]

In the meantime, Samuel Morse (1791–1872), originally an artist— which was why he was in Rome when a papal guard knocked his hat off and made him fanatically anti-Catholic—and professor of design in New York College, conceived the idea of an electric telegraph in 1832 and built a practical machine in 1837. He had to pester Congress before it appropriated $30,000 for a Washington–Baltimore line, opened in 1844. It was first used that spring to transmit news from the Whig and Democratic conventions, which met in Baltimore, to the capital. Then a private company was formed and moved in, opening its first line in 1846. Ezra Cornell (1807–74) was the organizing genius who formed the Western Union Telegraph Company (chartered 1856) and took the first line through to California in 1861—thus killing the Pony Express stone dead—and generating huge profits used to found Cornell University, which opened in 1868. A little belatedly, the railroads learned the value to them of the telegraph line and helped to finance its extensions running alongside their tracks.[189] Thereafter the telegraph quickly became an indispensable tool of government, commerce, and many kinds of social communication, with the Associated Press, originally the New York AP (1827), taking full advantage to coordinate news and disseminate it nationwide.[190] So, by the 1850s, the United States had an overall transportation system of great versatility and often of high density.

This transportation capacity, and the extraordinary powers of land-digestion shown by American settlers and farmers, combined to make the United States ever greedy for more territory. The vast tracts of the Louisiana Purchase, the conquest of Florida by Jackson—all these were not enough. By the 1830s the notion that America was destined to absorb the whole of the West of the continent, as well as its core, was taking hold. This was a religious impulse as well as a nationalist and ideological one—a feeling that God, the republic, and democracy alike demanded that Americans press on west, to settle and civilize, republicanize and democratize. In 1838 an extraordinary essay in the *Democratic Review*, entitled 'The Great Nation of Futurity,' set out the program:

> The far-reaching, the boundless future will be the era of American greatness. In its magnificent domain of space and time, the nation of many nations is destined to manifest to mankind the excellence of divine principles: to establish on earth the noblest temple ever dedicated to the worship of the Most High—the Sacred and the True. Its floor shall be a hemisphere—its roof the firmament of the star-studded heavens—and its congregation the Union of many Republics, comprising hundreds of happy millions, calling and owning no man master, but governed by God's natural and moral law of equality, the law of brotherhood—of 'peace and goodwill among men'.[191]

The theme was taken up in Congress, especially in the 1840s, the Roaring Forties as they came to be called, certainly distinguished by the roaring of Americans for more lands to conquer. One congressman put it thus in 1845: 'This continent was intended by Providence as a vast theater on which to work out the grand experiment of Republican Government, under the auspices of the Anglo-Saxon race.'[192]

The actual term 'Manifest Destiny' was first used by John L. O'Sullivan in the *Democratic Review* of 1845, complaining of foreign interference and attempts aimed at 'limiting our greatness and checking the fulfillment of our manifest destiny to overspread the continent allotted by Providence for the free development of our yearly multiplying millions.' Representative Duncan of Ohio said he feared federal government centralism, and the answer to this was expansion: 'To oppose the constant tendency to federal consolidation, I know of no better plan than to multiply States, and the further from the center of federal influence and attraction, the greater is our security.'[193] At the New Jersey State Democratic convention of 1844, Major Daveznac rhapsodized on this theme: 'Land enough! land enough! Make way, I say, for the young American Buffalo—he has not yet got land enough! He wants more land as his cool shelter in summer—he wants more land for his beautiful pasture grounds. I tell you, we will give him Oregon for his summer shade, and the region of Texas as his winter pasture!' (*Applause.*)[194]

O'Sullivan repeated his Manifest Destiny demand by predicting that it had to be fulfilled to accommodate 'that riot of growth in population which is destined, within a hundred years, to swell our numbers to the enormous population of *two hundred and fifty millions* (if not more),' a good guess, as it turned out.[195] An editorial in the *United States Journal*, October 15, 1845, asserted: 'It is a truth, which every man may see, if he will but look—that all the channels of communication—public and private, through the schoolroom, the pulpit and the press—are engrossed and occupied with *this one idea*, which all these forces are designed to disseminate—that we, the American people, are the most independent, intelligent, moral and happy people on the face of the earth.' This fact, and most Americans agreed it was a fact, gave ethical justification to the desire to expand the republic which promoted such happiness.

It should be added that an outspoken minority, especially among the churchgoers, opposed western expansion on social and moral grounds. After inspecting Louisville, the New England Unitarian minister James Freeman Clarke expressed alarm that, in the West, man was 'unbridled, undirected and ungoverned,' mothers encouraging their children to fight, women favoring dueling, judges gambling, while vice 'ate into the heart of social virtue.' Cornelius C. Felton of Harvard complained in 1842 that a population was growing up in the West 'with none of the restraints which fetter the characters of the working class in other countries.' Each man in the West considered himself 'a sovereign by indefeasible right and acknowledged no one as his better.'[196] But that was the New England view. In the South, these feelings were considered virtues, to be encouraged. Besides, the South had an additional reason for pushing West—to extend slavery and to found more slave states and so maintain the balance of power in Congress.

With the 49th parallel limiting American expansion to the North, the obvious way to get more land was to dismember Mexico. It had always been menaced by the United States. President Jefferson had claimed it up to the Rio Grande. Then America backed down to the Colorado River, and then to the Sabine, accepted as the frontier by Secretary of State Adams in 1819. But East Texas had already been occupied by the same sort of national and racial oddities who congregated in Florida, and, while they still ruled these parts, the Spaniards had thought of giving up the East and concentrating on West Texas, where the ranching, which was what they most liked, was of high quality. In 1812, a filibuster group of Mexicans and Americans marched in from Louisiana, took San Antonio, and set up what they called the State of Texas. They were wiped out by a Spanish counterattack. A group of Bonapartist exiles set up another Texan republic in 1818, and a third proto-state was proclaimed by a group of Americans the next year. Galveston Island was a pirate base. Then an independent Mexican government, having thrown out the Spanish by force, took over and gradually asserted its authority.[197]

Some Americans in search of land cooperated with the new independent Mexican authorities. Moses Austin (1761–1821), born in Connecticut, moved to Missouri and then, having become a citizen of Spanish Louisiana, was allowed by the new Mexican government to found the Austin Colony, a huge block in the Colorado and Brazos basins, which he passed on to his son Stephen Fuller Austin (1793–1836). This colony consisted of prairies, woodlands, and river-bottom lands of just the kind American settlers liked. The land was allocated in ranches of square leagues, according to the Spanish custom, that is 4,428-acre blocks—much bigger units than you could get in the United States, and for lower prices. By 1830 Austin had settled more than 5,000 Americans on his lands. Importing slaves to work the land was technically unlawful because Mexico had abolished slavery in 1824, but in practice it was allowed.

Then in 1830 the Mexican government suddenly halted immigration, imposed customs duties, reorganized Texas into three departments and set up military forts and garrisons. By this time three-quarters of the 30,000 people in Texas (including slaves) were Americans, and the western margin of America to the north was beyond the Colorado and 250 miles west of the Sabine. So Texas, as part of Mexico, looked increasingly anomalous. Austin behaved like a loyal Mexican citizen, as long as this was feasible. If Mexico had been stable, things might have been different and its power survived. If the United States had been unstable, it might have been less acquisitive. But the historical fact is that Mexico was unstable and America was stable. Various transient Mexican governments made superficially brutal but intrinsically feeble attempts to impose their will and these merely had the effect of irritating the growing body of American settlers and driving them into rebellion. The big blow-up came in 1835, the American insurgents proclaiming the Republic of Texas on December 20 and naming Sam Houston, previously governor of Tennessee, to run their army.[198]

The Mexican dictator, General Santa Ana, who had made his reputation with a military victory over the Spaniards in 1829, led a force of 5,000 regulars across the frontier of the self-proclaimed independent state on February 26, 1836. The same day Colonel William B. Travis (1790–1836) of South Carolina and James Bowie (1799–1836) of Burke County, Georgia—reputedly the inventor of the Bowie knife—retreated into the Alamo, a heavy-walled Spanish mission converted into a fort. With them were 187 Texan desperados of the kind beloved by any healthy schoolboy, including Davy Crockett (1786–1836), a Tennessee congressman, who had served under Jackson in the Creek War and had arrived in Texas only a few days before. General Houston ordered Travis and Bowie to abandon their hopeless position but they declined. And when Santa Ana asked for their surrender, they answered with a cannon shot. The Mexicans then ran up the red flag, a traditional sign that no quarter would be given, and the attack commenced. It is said that over a thousand casualties were inflicted on the Mexicans, but within an hour the fort was in Santa Ana's hands and all the Americans were dead—Bowie bayoneted to death in his cot, where he was suffering from pneumonia, Travis riddled with musket balls next to a cannon, Crockett mutilated amid a pile of bodies of fellow-Tennesseans. The bodies of the dead combatants were thrown together on a pyre and burned.

Susannah Dickinson, wife of a blacksmith, was among a few civilians who survived to tell the tale. Santa Ana told her to inform the Texans that 'fighting is hopeless.' However, on April 21 General Houston mounted a surprise attack on Santa Ana's force on the banks of the San Jacinto River, near Galveston Bay. The Mexicans were scattered, Santa Ana was captured—in the arms of his mistress, Jenny—and he was forced to sign documents surrendering his entire army and acknowledging the independence of Texas. The fighting was thus over in seven weeks. Independent Texas proceeded quickly to hold presidential elections, in which Houston defeated Stephen Austin, receiving 80 percent of the 6,000 votes cast. Houston had lived with the Indians in his youth and married an Indian girl. He was known as 'The Raven' and was said to drink 'a barrel of whiskey a day.' He was also an old friend and fellow-Tennessean of the President, General Jackson.[199]

Jackson wanted Texas to be part of America. He also hated Mexicans. When, in the middle of a Cabinet meeting on June 28, 1836, he received a report from Commodore Dallas, US naval commander in the Mexican Gulf, about the indignities inflicted on the American consul and US citizens in Tampico by the Mexican authorities, he rapped out a characteristic order—without bothering to ask the rest of the Cabinet their views: 'Write immediately to Commodore Dallas, & order him to *blockade* the harbor of Tampico, & to suffer nothing to enter until they allow him to land and obtain his supplies of water & communicate with the Consul, & if they touch a hair of the head of one of our citizens tell him *to batter down and destroy their town & exterminate* their inhabitants from the face of the earth.'[200] However, when he cooled down and consulted Kendall, he decided not to annex Texas as a new state, at the risk of outright war with Mexico, but to let things be for a time. Kendall told Jackson that he had to think of international opinion, which would note and resent land-grabbing by the American republic but would not object if, in due course, all were to drop into its lap: 'The time will come when Mexico will be overrun by our Anglo-Saxon race, nor do I look upon it as a result to be at all deplored. I believe it would lead to the amelioration and improvement of Mexico herself; but as guardians of the peace and interests of the United States we are not permitted to go to war through philanthropy or a design to conquer other nations for their own good.'[201] Old Hickory pondered this pacific advice for a time and, surprisingly, agreed to follow it.

So Texas remained independent for a decade and flourished mightily, though continuing to press for its inclusion in the United States as a slave state. Jackson, now retired, coined in 1843, *à propos* of Texas, the saying that adding to America was 'extending the area of freedom'—although grabbing Texas from Mexico had already meant the legal reimposition of slavery there. In the meantime, President Tyler had been slowly moving towards the annexation of Texas, being anxious to ingratiate himself with the Southerners in order to secure his reelection in 1844, this time in his own right. Early in the election year, on February 28, 1844, a disaster occurred which had a profound political impact. Congress had been persuaded to provide funds to build a revolutionary new warship, the USS *Princeton*, the first to be driven by a propeller-screw, invented by a young Swedish engineer, John Ericsson. It had two enormous wrought-iron 12-inch guns, called 'Oregon' and 'Peacemaker.' To call a huge new gun such a name was an affront to providence, and during a gala trip down the Potomac which President

Tyler had arranged for his Cabinet, diplomats, senators, and numerous grand ladies, 'Peacemaker' exploded, killing Secretary of State Abel Upshur, the Navy Secretary, and a New York State senator, and wounding a dozen others, including Senator Benton. The force of the explosion literally flung into the President's arms the beautiful Julia Gardiner, daughter of the dead state Senator, and she shortly afterwards became his wife. Equally, perhaps more, important, it enabled Tyler to reconstruct his Cabinet, excluding Northerners completely, and bringing in Calhoun as secretary of state. The object was twofold: to get Tyler the Southern ticket and to annex Texas.

The first maneuver did not succeed. The Democrats chose a Jackson protégé from Tennessee, James Knox Polk (1795–1849), who took a strong stand on the Manifest Destiny platform, and beat Henry Clay by 170 electoral college votes to 105 (the popular vote was closer: 1,337,243 to 1,299,062). Clay, having been an expansionist all his life, refused, for reasons which are still mysterious, to back the annexation of Texas.[202] That was the main reason he lost. It was obvious the bulk of the nation, even the North, wanted Texas in the Union, whether or not it was a slave state. Tyler, still president, decided to outsmart Polk by gathering to himself the kudos for Texas' admission. His Secretary of State, Calhoun, had failed to get an annexation treaty approved by the necessary two-thirds vote of the Senate. (Two-thirds, overwhelmingly Northerners, voted against it.) Now Tyler, using the 'verdict' of the election as his justification, recommended that Texas be admitted by a joint resolution of both Houses, for which a simple majority was enough. This was done, February 28, 1845, and on his last day in the White House Tyler dispatched a courier to President Houston inviting Texas to become the twenty-eighth state.[203]

Frustrated over Texas, Polk determined to add the riches and immensities of California to the Union and get the credit for it. And as a makeweight he wanted Oregon too. Polk came from North Carolina and was an expert mathematician. He had migrated to Tennessee, served in Congress, had four years as speaker, and two terms as governor. He was a lawyer, planter, and slave-owner and, now that Van Buren was dead politically, Jackson's heir, known as Young Hickory. But he had nothing in common with Jackson other than determination. He was a sour, stiff, elderly-looking man, with a sad, unsmiling face, who did nothing but work—eighteen hours a day in the White House, it was said. He was the first president killed by the office, though the choice was his. Like J. Q. Adams, whom he resembled, he kept a diary, though not such a nasty and interesting one.[204] It is curious that he was despised in his lifetime and later underrated by historians. Within his self-set limits, he has a claim to be considered one of the most successful presidents. He did exactly what he said he would do. He said he would serve only one term, and he did. He said that, in that one term, he would do four things: settle the Oregon question, acquire California, reduce the tariff, and reestablish Van Buren's Independent Treasury, which the Whigs had abolished. He did all these things.

He also got America into war with Mexico and won it in record time. But first, knowing war with Mexico was likely, he and his Secretary of State, James Buchanan (1791–1868), determined to settle the Oregon question. It had a complicated history and involved an enormous mass of territory, only partly explored and mapped, beginning in the northern Rockies and ending on the Pacific coast. Most people did not even have a name for it until 'Oregon'—presumably of Indian origin—was popularized in a poem, 'Thanatopsis,' published by William Cullen Bryant (1794–1878) in 1817. Since the Treaty of 1814 the British and the Americans had agreed to leave the precise longitudinal frontier between Canada and the United States unresolved. President Monroe had assumed that the best solution was simply to extend the 49th parallel to the Pacific. All subsequent presidents had taken the same line. The area was largely the territory of the ancient Anglo-Canadian Hudson's Bay Company, and they had been operating south of the parallel for generations. On the other hand, American pioneers had been boring into the region and staking claims.[205] Now, in the 'Roaring Forties,' with Americans whipping themselves into a nationalistic frenzy over Manifest Destiny, with 'Oregon Fever' taking settlers into the region by the thousand, with a government formed, a governor appointed, and a state capital, Oregon City, mapped out, the cry was 'All of Oregon or None,' supplemented by a bit of demotic geography, 'Fifty-Four-Forty or Fight.'

This last latitude would have shoved the US frontier right into what is now western Canada and given America the matchless harbor of Vancouver. Polk did not want, or expect, to get so much. He talked big. He told Congress that 'the American title to the country [he carefully did not say the whole of it] is clear and unquestionable.' He gave the Monroe Doctrine a new twist: 'The people of *this continent* alone have the right to decide their own destiny.' He said: 'The only way to treat John Bull is to look him straight in the eye.' But the last thing on earth he intended was to get into a scrap with Britain at a time when war with Mexico loomed. Moreover, it was unnecessary. The fur trade had declined in relative and absolute importance, and the Hudson Bay trapping areas south of the 49th were no longer of great consequence. Sir Robert Peel, the British Prime Minister, was enmeshed in his crowning struggle to repeal the Corn Laws and had no intention of wasting his energies on a strip of largely uninhabited territory in western Canada. The British public did not give a damn. By June 1846 Peel had split his party over Corn Law Repeal and was on his way out of office. One of his last acts was to settle for the 49th parallel and send a draft treaty to this effect to Washington. On June 15 Buchanan signed it for America and three days later it was ratified by the Senate after perfunctory debate. Ignored, the raucous Fifty-Four-Fortiers subsided. Thus are disputes involving vast territories settled, calmly and swiftly, when the two parties are both civilized states with a common language, fundamental common interests, and common sense.[206]

By this time America was at war with Mexico. Looking back on it, it is easy to reach the conclusion that the Mexicans were foolish and the Americans hypocritical. Polk wanted war because he wanted California. But he did not want to start it. The Mexicans played straight into his hands by allowing their pride to overcome their prudence. Two days after Polk got to the White House the Mexican ambassador broke off relations and went home in protest at the annexation of Texas. That was silly, since Texas was a lost cause and, if the Mexicans wanted to retain California, or some of it, it was vital for them to keep up negotiations. Meanwhile Polk made his preparations. As early as June 1845 he got his Navy Secretary to send secret orders to Commodore Sloat, commanding the Pacific Station, that he was to seize San Francisco immediately he could 'ascertain with certainty' that Mexico was at war. In October, the War Secretary was instructed by Polk to write to Thomas O. Larkin, US consul in Monterey: 'Whilst the President will make no effort and use no influence to induce California to become one of the free and independent states of the Union, yet if the people should desire to unite their destiny with ours, they would be received as brethren, whenever this can be done without affording Mexico just cause for complaint.' At this time the numbers of American settlers and Mexican inhabitants were about equal, and the message was calculated to incite the Americans to take over, as they had in Texas.

However, in Polk's favor it has to be said that Mexico was a tiresome neighbor, always asking for trouble. It borrowed huge sums of money and then repudiated its debts. It had periodic civil wars in which the property of foreigners was pillaged. France had taken a much higher line with Mexico in 1839, sending a naval squadron to bombard San Juan de Ulúa, in revenge for outrages. America had submitted its claims for compensation to an independent commission, which had awarded it $3 million. In 1843 Mexico had agreed to pay this, plus accrued interest, in twenty installments, quarterly. But only three deadlines were met. In November 1845 Polk said he would put the whole series of issues on a 'businesslike basis:' America would assume responsibility for the debt if Mexico recognized the Rio Grande as the new border between the two countries; it would pay $5 million for New Mexico; and 'money would be no object' if Mexico ceded California. On January 12, 1846, after another brief civil war, the new Mexican military government, which was violently anti-American, refused even to see the US minister plenipotentiary. The following day Polk ordered General Zachary Taylor (1784–1850), 'Old Rough-and-Ready,' to take up station with his army on the Rio Grande. By May Polk had concluded that war was inevitable and got his Cabinet to approve a war message to Congress. As if on cue, the same evening, May 9, the Mexican army attacked a US unit on the 'American' side of the Rio Grande, killing eleven, wounding five, and taking the rest prisoner. The next day Polk was able to go to Congress boiling with simulated wrath. Even before the murders, he said, 'The cup of forbearance had been exhausted.' Now Mexico 'has passed the boundary of the United States, has invaded our territory, and shed American blood upon American soil.'[207]

One of the few to protest about the provocation and hypocrisy of Polk was a new character on the American public scene, freshman Congressman Abraham Lincoln (1809–65), who argued that Polk had in effect started the war motivated with a desire for 'military glory . . . that serpent's eye which charms to destroy,' and that, as a result, 'the blood of this war, like the blood of Abel, is crying to Heaven against him.'[208] A good many New England intellectuals, antecedents of those who would protest against the Vietnam War in the 1960s, agreed with Lincoln. On the other hand, it is difficult now to conjure up the contempt felt by most Americans in the 1840s for the way Mexico was governed, or misgoverned, the endless coups and pronunciamentos, the intermittent and exceedingly cruel and often bloody civil conflicts, and the general insecurity of life and property. It made moral as well as economic and political sense for the civilized United States to wrest as

much territory as possible from the hands of Mexico's greedy and irresponsible rulers.

The Mexican War of 1846 was important because of its consequences. But it also had a lot of high, and sometimes low, comedy. Polk
tried to play politics with the war from start to finish. In the first place
he allowed the slippery Santa Ana, who was in exile in Cuba, to return
to Mexico, the general having promised him he would usurp power
and give America the treaty it wanted. In fact Santa Ana, who always
broke his promises, broke this one too and provided such serious resistance as the American army encountered. Polk, as Senator Benton
wrote, wanted 'a small war, just large enough to require a treaty of
peace, and not large enough to make military reputations, dangerous
for the presidency.'[209] Polk also wanted to fight the war on the cheap,
starving Taylor of supplies at first, and putting volunteers on short
engagements. Taylor protested, refused to budge until supplies arrived,
then won a brilliant three-day battle at Monterey, taking the city. That
worried Polk, who feared Taylor would get the Whig nomination in
1848. Polk then tried to appoint Senator Benton, of all people, as a
political general to control the army. Congress would not have that. So
he turned instead to General Winfield Scott (1786–1866), general-in-
chief of the army. Scott was a Whig and politically ambitious too, but
he served to balance Taylor and take some of the glory from him. He
was known as 'Fuss-and-Feathers' because of his insistence on pipeclay
and gleaming brass. Scott immediately got into a row with Polk's
Secretary of War, William L. Marcy—the man who had coined the
term the 'spoils system'—again over paucity of supplies, and, in reply
to a quibbling letter from Marcy, he wrote that he had received it in
camp 'as I sat down to take a hasty plate of soup.' This self-pitying
phrase circulated in Washington, and got Scott dubbed 'Marshal
Tureen.'[210]

Fortunately for Polk, both Scott and Taylor were competent generals, and there was a dazzling supporting cast under them—Captains
Robert E. Lee and George B. McClellan, Lieutenant Ulysses S. Grant
and Colonel Jefferson Davis, all of whom distinguished themselves. In
some ways Mexico was a dress-rehearsal for the professional military
side of the Civil War. Taylor was supposed to strike for Mexico City
across 500 miles of desert, with inadequate means. On March 9, 1847,
Scott's army, also starved of equipment, landed at Vera Cruz without
loss, the first big amphibious operation ever mounted by US forces. This
was the short route to Mexico City. On May 15, having taken the sec-

ond city, Puebla, Scott had to let a third of his army return home as their enlistments had run out. He insisted on waiting for more. Thus reinforced, he won four battles in quick succession (Contreras, Churubusco, Molino del Rey and Chapultepec) in August and September and entered Mexico City on September 13, a marine unit running up the flag over 'the Halls of Montezuma.'[211] Meanwhile, in California, John Charles Frémont (1813–90), with a party of sixty American freebooters, had raised a flag with a grizzly bear and star on a white cloth, and proclaimed the Republic of California (June 14, 1846). Frémont was an officer in the US Topographical Corps who had surveyed the Upper Missouri and Mississippi rivers (1838–41), eloped with Jessie Benton, pretty daughter of the US Senator, and headed three expeditions to the West, which involved exploring and mapping more territory than any other American—Western Wyoming (including Fremont Peak), all of California, the routes from Utah to Oregon, and most of Nevada and Colorado.[212]

A month later Commodore John D. Stoat of the Pacific Fleet raised the American flag and proclaimed California US territory. The conquest of California was by no means bloodless, as in the south the Mexican peasants and the Indians revolted against the new American regime and had to be put down by force at the Battle of the Plains of Mesa outside Los Angeles, in January 1847. Nor was it easy to sign a peace treaty as there was no effective government in Mexico, by this stage, to negotiate one. Polk also had trouble with his negotiator, Nicholas P. Trist (1800–74), the Chief Clerk at the State Department, who disobeyed orders and was denounced by the President as 'an impudent and unqualified scoundrel.' However, he did succeed in finding a Mexican government and got it to sign the Treaty of Guadalupe Hildago on February 2, 1848, so Polk swallowed his wrath and accepted the *fait accompli*. By this agreement Mexico accepted the Rio Grande frontier with Texas and handed over California and New Mexico. America agreed to pay off the indemnities and give Mexico an extra $15 million.

It had not exactly been the cheap war Polk planned because he ended up with well over 100,000 men under arms, with 1,721 dead and another 11,155 wiped out by disease, and with a bill for $97.7 million, plus the treaty payments. On the other side of the ledger, America got over 500,000 square miles of some of the richest territory on earth, making an extra million square miles if Texas is counted in. Five years later, Gadsden, by now Secretary of State under President Pierce, nego-

tiated what is known as the Gadsden Purchase, whereby Mexico surrendered another 29,640 square miles on the southern borders of
Arizona and New Mexico, for $10 million. This rounded off the
Manifest Destiny program, but it was essentially complete during
Polk's presidency and he can fairly claim, when Oregon was counted
in, to have added more territory to the United States than any other
president, Jefferson (with the Louisiana Purchase) alone excepted.[213]

California was an even greater prize than Texas. The name goes back
to an imaginary island in a romance by Ordonez de Montalvo published in 1510. It was known to Cortez; Cabrillo made his way to San
Diego in 1542; Drake touched there in 1579. But the permanent settlement by the Spanish did not begin until 1769, when the first of many
presidios and Franciscan missions were established between San Diego
and San Francisco. Considering the benevolence of its climate, the fertility of its soil, and its vast range of obvious natural resources, it is
astonishing that the Spanish, then the Mexicans, did so little to make
use of them. Other great powers had nosed around. In 1807 the
Russians formed a plan to establish settlements in California (and at
the mouth of the Colombia River and in Hawaii too), though nothing
came of it. A few years later, however, the Russian–American
Company was working near the Golden Gate, hunting seals. The
British were interested too and, in the 1820s, had collaborated with the
Americans in chasing the Russians out of the area. American agents in
the area repeatedly warned Washington of the feebleness of the Spanish
(later the Mexican) hold on the area, and the desirability of securing
San Francisco Bay, 'the most convenient, capacious and safe [harbor] in
the world.' Lieutenant Wilkes of the US Navy, there in 1841 as part of a
strategic survey of the Eastern Pacific, again stressed the marvels of San
Francisco, 'one of the most spacious and at the same time safest ports
in the world,' and underlined the vacuum of authority: 'Although I was
prepared for anarchy and confusion, I was surprised when I found a
total absence of all government in California, and even its forms and
ceremonies thrown aside.'[214]

The first American to penetrate California by the overland route had
been Jedediah Strong Smith, 'the Knight of the Buckskin,' who, working for the Rocky Mountain Fur Company, had reached the San
Gabriel mission on the Pacific coast in 1826. The first American settlers
came two years later. But ordinary Americans began to learn of the
wonder of the Far West only in the 1840s, when two gifted and adven-

turous writers reported on them. Richard Henry Dana Jr (1815–82) was a young Harvard man who shipped as a common sailor on a three-master in 1834 for health reasons, voyaged the Pacific, and spent a year gathering hides on the California coast before returning to real life at the Harvard Law School. His *Two Years Before the Mast* (Boston, 1840) gave an unforgettable picture of San Francisco Bay in its pristine state: 'All around was the stillness of nature. There were no settlements on these bays and rivers, and the few ranches and missions were remote and widely separated ... On the whole coast of California there was not a lighthouse, a beacon or a buoy ... Birds of prey and passage swooped and dived about us, wild beasts ranged through the oak groves, and as we slowly floated out of the harbor with the tide, herds of deer came to the water's edge.'[215] This splendid book was widely read and made countless adventurous young men itch to get to the Far West.

Even more remarkable was the work of another Harvard Bostonian, Francis Parkman (1823–93), who set out in 1846 from St Louis to see for himself the reality of unspoiled life in this region, and especially to study the Indians, before the white man overwhelmed it. His travels began in what one modern historian has called 'the year of decision,' the watershed between the old and the new.[216] Parkman carried three books, the Bible, *Shakespeare's Works*, and *The Collected Works of Byron*. He was himself a Byronic young man with an intense desire to see and experience the dangers of the Far West, pioneer trailing, a war between the Dakotas and the Snakes, and the need to move secretly through territory infested with Indian war parties. No one has ever conveyed better the loneliness, the danger, and the immensities of the Western spaces, and the occasional cataclysmic concentrations of wild life:

> From the river bank on the right, away over the swelling prairie on the left, and in front as far as the eye could reach, was one vast host of buffalo. The outskirts of the herd were within a quarter of a mile. In many parts they were crowded so densely together that in the distance their rounded backs presented a surface of uniform blackness; but elsewhere they were more scattered, and from amid the multitude rose little columns of dust where some of them were rolling on the ground. Here and there a battle was going forward among the bulls. We could distinctly see them rushing against each other, and hear the clattering of their horns and their hoarse bellowing.

Parkman is romantic in that he consciously describes a life which he sees is now fragile—the buffalo will be hunted to extermination, the nomadic Indians will be corralled up in reservations, the sparse and primitive settlements will give way to towns and farms—but he is also unsentimental. He shows the Indians as they were: improvident, unreliable, sometimes treacherous, vacillating, above all lazy, with the elderly females doing all the hard work. Thus, in a nomadic party of Ogillallahs,

> The moving spirit of the establishment was an old hag of eighty. You could count all her ribs through the wrinkles of her leathery skin. Her withered face more resembled an old skull than the countenance of a living being, even to the hollowed, darkened sockets, at the bottom of which glittered her little black eyes. Her arms had dwindled into nothing but whipcord and wire. Her hair, half-black, half-grey, hung in total neglect nearly to the ground, and her sole garment consisted of the remnant of a discarded buffalo-robe tied round her waist with a string of hide. Yet the old squaw's meager anatomy was wonderfully strong. She pitched the lodge, packed the horses, and did the hardest labor in the camp. From morning till night she bustled about the lodge screaming like a screech-owl when anything displeased her.[217]

Parkman's marvelous account of his excitements and privations, The Oregon Trail, published in 1849, was an immediate success both with literary New England and with the great public. But by that time the modern world had already overtaken the arcadia he described. The month before the Treaty with Mexico was signed, at Sutter's Mill in the Sacramento Valley, gold was discovered on January 24, 1848. A workman found tiny nuggets of gold in the mill-race. For some time the news was concealed while the few in the secret worked frenziedly to trace the veins and stake claims. By September, the East Coast papers were publishing reports from 'the California goldfields,' telling of 'nuggets collected at random and without any trouble.' The real rush started after President Polk, in his December 1848 message to Congress, boastingly confirmed 'the accounts of the abundance of gold' in 'the territory recently acquired'—by him.

That spring, scores of thousands went to California, from all over the world. Some went direct from Australia, which had had a gold rush of its own in the 1830s. The people of Cutler, in Maine, built and rigged their own ship and sailed her round the Horn to San Francisco

Bay. Some went via the Panama Isthmus. More went over the Rockies by the Oregon and California trails. The early Forty-Niners got their gold by sifting off the gravel and soil using wire mesh—what they called 'panning' or 'placer' mining. Or they ran a stream through a 'long-tom' or sluice- box. That was the easy bit, and inspired the ditty: 'Oh California / That's the land for me / I'm off for Sacramento / With my washbowl on my knee.' But as the surface was worked out it became necessary to sink shafts and build crushing mills to grind the gold from the imprisoning quartz; that needed capital and organization. Many disappointed Forty-Niners went home in disgust, penniless—30,000 a year. But many more stayed because there were ample other opportunities in California, besides gold. Before the first strike, the non-Indian population of the territory was less than 14,000. By 1852 it was over 250,000. San Francisco had become a boomtown of 25,000 people, crowded with gamblers, financiers, prostitutes and wild women, actors and reporters, budding politicians and businessmen. It was free-for-all America at its best and worst.

The atmosphere of the mining camps is wonderfully conveyed in the stories of Bret Harte (1836–1902), a young man from Albany, New York, who was in California in 1854 where he worked on the Mother Lode and later went into printing and journalism in San Francisco. His 'The Luck of Roaring Camp' is the greatest of all mining stories.[218] The prototype rush having taken place, there were plenty of others: Gold Hill, Colorado (1859), Virginia City, Nevada (1860), Orofino, Idaho (1861), Virginia City, Montana (1863), Deadwood, South Dakota (1876), Tombstone, Arizona (1877), Cripple Creek, Colorado (1892), and the great Alaska–Yukon rush, beginning at Nome in 1899. Nevada mining is described in glorious detail in another Mark Twain masterwork, *Roughing It*, which has an exact description of all the mining processes then in use and the skulduggery, violence, greed, and disappointments which surrounded them.

But nothing could beat the original Forty-nine Rush for glamour and riches. The yield of gold in the first decade, 1848–58, was $550 million. In the years 1851–5, California produced over 45 percent of the world's entire output of gold. It was a man's world, for fathers, sons, and brothers left to make their fortunes, telling their womenfolk to wait to be summoned. In 1852 Nevada County contained 12,500 white males, 900 females of various colors, 3,000 Indian coolies, and 4,000 Chinese cooks, laundrymen, and camp-workers. Lola Montez (1818–61), the Irish actress who had been the mistress of Louis I of

Bavaria and had run his government, made her appearance, was a sensational success, and then retired to Grass Valley (her house is still there). When the editor of the *Grass Valley Telegraph* attacked her in print she literally horsewhipped him and he had to slink out of town. Grass Valley and Nevada City became centers of the richest and most continuous goldmining in California, with the North Star Mine, the Eureka, and the Empire setting the pace. Until the opening up of the Rand deep-level mines in South Africa in the 1930s, they were the most successful gold mines in history.[219] Indeed, the California gold rush as a whole was a world-historical event of some importance. Until its gold came on the market, there had been a chronic shortage of specie, especially gold bullion, from which the United States, in particular, had suffered. Until the 1850s in fact there was no true gold standard simply because there was not enough gold to maintain it. Once California gold began to circulate, the development of American capital markets accelerated and the huge expansion of the second half of the century became financially possible.[220] That too (it can be argued) was the work of the 'Unknown President Polk.'

The great California gold rush of 1849, attracting as it did adventurers from all over the world, was the first intimation to people everywhere that there was growing up, in the form of the United States, a materialistic phenomenon unique in history, a Promised Land which actually existed. Not that there was any shortage of routine, detailed information. Josiah T. Marshall's *Farmers and Immigrants Handbook: Being a Full and Complete Guide for the Farmer and Immigrant* (1845) was nearly 500 crammed pages. Minnesota set up a State Board of Immigration in 1855 and other states copied it. By 1864 Kansas was sending emissaries abroad to whip up enthusiasm among would-be immigrants. From the early 1840s railroads began obtaining both state and federal land for the use of immigrants. The Illinois Central advertised abroad; so did the Union Pacific and Northern Pacific. Railroad land departments organized trips for newspapermen and land-seekers and regularly dispatched agents all over Europe.[221] By the 1850s a great deal of public and private money was being spent on telling the world about America.

There was also—more important, perhaps—word of mouth and traveler's tales. By European standards, wage-rates, even for unskilled men, were enormous. After about 1820, no one got less than a dollar a day in the cities. Farmhands got $7.50 to $15 a month, with full board. Thomas Mooney, an Irish visitor, asserted (1850): 'You can, as soon as you a get a regular employment, save the price of an acre-and-a-half of the finest land in the world *every week*, and in less than a year you will have money to start for the West, and take up an 80-acre farm which will be yours for ever.' He calculated that a careful immigrant could save 7–8 English shillings a week.[222] This was irresistible news. Immigration was going up all the time, allowing for fluctuations which reflected the trade cycle. After the first crisis dip in 1819, it rose to 32,000 in 1832 and 79,000 in 1837, then down following the credit panic, then up again to 100,000 in 1842, and then an immense increase, 1845–50, produced by bad winters in Europe, the Irish potato famine, and the revolutions of 1848–9, which caused scores of thousands to flee. Never before or since was immigration so high per capita of the American population. The California gold rush sent it up to a record 427,833 in 1854, then the late-fifties panic sent it down with a crash to 153,640 in 1860. By that date there were 4 million foreign-born settlers in the United States, out of a total population of 27 million. They came from all over Europe, but mostly from Britain, Ireland, and Germany, the Irish staying east of the Alleghenies, the Germans pushing on into the Midwest to farm.[223]

America was also admired for many other things in addition to high wage-rates and cheap land. First was the 'American Cottage,' a hit in Europe about 1800. Then came 'American Gardens.' From about 1815 what struck Europeans most was the size and luxury of American hotels. It is not so surprising that American hotels should have been big and comfortable: entire families lived in hotels for years, and in Washington DC it was rare for congressmen, senators, and Cabinet members to acquire their own houses before 1850. The first luxury hotel was Barnham's City Hotel in the boom-town of Baltimore, built 1825–6, which had no fewer than 200 bedrooms, twice as big as the largest in Europe. The Astor House in New York, built by J. J. Astor from 1832, had 309 bedrooms, plus—amazingly—no fewer than seventeen bathrooms. The Continental in Philadelphia (1858), which housed 800–900 people in suites, doubles, and singles, struck a new high in size and luxury (Europe's largest was then the Queen's, Cheltenham, 'the Grandest Hotel in Europe,' with 110 rooms). American hotels were often distinguished and aggrandized by a central lobby, under a rotunda (the hotel atrium of the 1980s and 1990s is a rediscovery of this feature). The first such was the Exchange Coffee House Hotel in Boston, 1806–9, and the St Louis, in New Orleans, built in 1839, was a replica of this on a larger scale. The Palace Hotel in San Francisco, 1874–6, with its 850 bedrooms and 437 bathrooms, was so big that carriages could actually drive into the center, the coming-and-going forming an amusement for the other guests. It is significant that the influence of monumental American hotels gave rise to the first recorded complaint of Yankee cultural colonization, which came (needless to say) from a Frenchman; in 1870 Edmond de Goncourt lamented that Paris hotels were being 'Americanized.'[224]

The new 'utopian' factories of New England were also much admired. The English novelist Anthony Trollope called Lowell 'the realisation of a commercial Utopia.' Harriet Martineau, the English economist, writing of Waltham, enthused: 'There is no need to enlarge on the pleasure of an acquaintance with the operative classes of the United States.'[225] In fact there was a strong authoritarian atmosphere in some of these 'model' factories, an adumbration of Henry Ford's system 1910–30. At Lowell in 1846, it was reported that operatives worked thirteen hours a day, from dawn till dusk in winter (but this is from a hostile account). Long hours were certainly common.[226] In Rhode Island entire families, including small children, contracted to work for employers. What all observers recorded was the absence of begging. As one of them put it in 1839: 'During two years spent in traveling through every part of the Union, I have only once been asked for alms.' To Europeans, that seemed incredible, the real proof of a benevolent prosperity.[227]

Americans were already associated with 'modernity,' with new ways of doing things. This applied particularly to social welfare and public works. The first big international success was Auburn Prison, New York State, in 1820. This pinched an idea from the big Paris bazaars and applied it to a penitentiary—top-lit galleries with massed stories of cells ranged on either side. Then in 1825, John Haviland joined this idea to Jeremy Bentham's panopticon prison idea of 1791, with a ground-plan formed by the spokes of a wheel and a central observation hall ranked by galleries. (The spokes plan had already been used in the Maison de Force in Ghent, but that had no galleries.) This was typical of American-style utopianism and was so much admired that Haviland was asked to design prisons all over the United States.[228] He specialized in prisons designed to accommodate huge new populations committing more crimes—and new crimes—and young criminals. Typical of his work was the Eastern Penitentiary in Philadelphia. Nearly all 'serious' visitors, such as Charles Dickens, Anthony Trollope, and W. M. Thackeray, who intended to write books about their travels, visited one or more prisons (as well as workhouses, homes for fallen women, and similar dismal but worthy places).

It was prisons which drew to America the most perceptive and influential of the European observers, Alexis de Tocqueville. Of noble descent, born in a Normandy chateau, he was nonetheless a liberal, in some ways a radical, whose object (he said) was 'to abate the claims of the aristocrats' and 'prepare them for an irresistible future'—which he saw to be emerging in America. In 1831 the new French 'liberal' government of Louis-Philippe, which had been delightedly hailed by President Jackson as the first real sign that his kind of democracy was spreading to Europe, gave de Tocqueville an unpaid commission to investigate American penology and write a report, which he published in 1833. He subsequently published his *Democracy in America*, part one in 1835, two in 1840. It has remained in print ever since.[229] The theme of the work is that 'The gradual development of the principle of equality is a Providential fact,' and he traces the implication in American institutions, both in theory and in practice. Volume one is mainly about America, and is tremendously optimistic; volume two is also about France, and tends to pessimism. But this work, and his copious letters, and his subsequent memoirs provide wonderful glimpses of American society in the 1830s.[230]

The sharp-eyed and reflective Frenchman went from Boston to New Orleans with brief forays west of the Alleghenies, and did many of the usual things. He stayed in Boston's Fremont Hotel, built two years before, marveling at the 'private parlor' attached to each room, the slippers supplied while boots were being polished, and the terrific bellboys—though he also noted universal and disastrous bed-sharing in the interior. In Baltimore he dined with Charles Carroll—evidently a public monument to be visited by all, if sufficiently distinguished—and rejoiced at the way such aristocrats, unlike their European counterparts, accepted the new democracy graciously and even managed to get themselves elected by universal suffrage. He had an appalling time in the savage winter of 1831–2. In a letter to his mother he described how he had shared a Mississippi steamship with a crowd of Choctaw warriors being forcibly moved west:

> There was a general air of ruin and destruction in this sight, something which gave the impression of a final farewell, with no going back; one couldn't witness it without a heavy heart. The Indians were calm but

gloomy and taciturn. One of them knew English. I asked him why the Choctaws were leaving their country. 'To be free,' he answered. I couldn't get anything else out of him. Tomorrow we will set them down in the Arkansas wilderness. I must confess it is an odd coincidence that we should have arrived in Memphis to witness the expulsion, or perhaps the dissolution, of one of the last vestiges of one of the oldest American nations.

Shortly afterwards he came across Sam Houston, riding 'a superb stallion,' a man he described as 'the son in law of an Indian chief and an Indian chief himself.'[231]

What makes de Tocqueville's account memorable is the way in which he grasped the moral content of America. Coming from a country where the abuse of power by the clergy had made anticlericalism endemic, he was amazed to find a country where it was virtually unknown. He saw, for the first time, Christianity presented not as a totalitarian society but as an unlimited society, a competitive society, intimately wedded to the freedom and market system of the secular world. 'In France I had almost always seen the spirit of religion and the spirit of freedom pursuing courses diametrically opposed to each other,' he wrote, 'but in America I found that they were intimately united, and that they reigned in common over the same country.' He added: 'Religion . . . must be regarded as the foremost of the political institutions of the country for if it does not impart a taste for freedom, it facilitates the use of free institutions.' In fact, he concluded, most Americans held religion 'to be indispensable to the maintenance of republican institutions.' And de Tocqueville noted on an unpublished scrap of paper that, while religion underpinned republican government, the fact that the government was minimal was a great source of moral strength:

One of the happiest consequences of the absence of government (when a people is fortunate enough to be able to do without it, which is rare) is the development of individual strength that inevitably follows from it. Each man learns to think, to act for himself, without counting on the support of an outside force which, however vigilant one supposes it to be, can never answer all social needs. Man, thus accustomed to seek his well-being only through his own efforts, raises himself in his own opinion as he does in the opinion of others; his soul becomes larger and stronger at the same time.[232]

In de Tocqueville's view, it was education which made this spirit of independence possible. The Rev. Louis Dwight said to him that the Americans were the best-educated people in the world: '[Here] everyone takes it for granted that education will be moral and religious. There would be a general outcry, a kind of popular uprising, against anyone who tried to introduce a contrary system, and everyone would say it would be better to have no education at all than an education of that sort. It is from the Bible that all our children learn to read.'[233] As a result of a liberal system of education and free access to uncensored books and newspapers, there were fewer dark corners in the American mind than elsewhere. Reflecting on his conversations in Boston, he noted: 'Enlightenment, more than anything else, makes [a republic] possible. The Americans are no more virtuous than other people, but they are infinitely more enlightened (I'm speaking of the great mass) than any other people I know. The mass of people who understand public affairs, who are acquainted with laws and precedents, who have a sense of the interests, well understood, of the nation, and the faculty to understand them, is greater here than any other place in the world.'[234]

De Tocqueville, significantly, felt that the American syndrome—morality/independence/enlightenment/industry/success—tended not to work where slavery existed. He was shocked to find the French-speaking people of New Orleans infinitely more wicked and dissolute than the pious French Canadians, and blamed the infection of slavery, anti-freedom. Similarly, he contrasted 'industrious Ohio' with 'idle Kentucky': 'On both sides [of the Ohio River] the soil is equally fertile, the situation just as favorable.' But Kentucky, because of slavery, is inhabited 'by a people without energy, without ardor, without a spirit of enterprise.' He was led, he said, again and again to the same conclusion: leaving aside the slave states, 'the American people, taking them all in all, are not only the most enlightened in the world, but (something I place well above that advantage), they are the people whose *practical, political education is the most advanced.*'[235]

The Americans certainly made tremendous, continuous, and heart-breakingly genuine efforts to become 'enlightened.' Even more than 19th-century Britain, America was a country of conscious self-betterment. The state was trying to make itself better; the people were trying too, not for want of urging. The great orator Daniel Webster took the occasion of the unveiling of the Bunker Hill monument in Boston (June 17, 1825) to intone: 'Our proper business is improvement. Let our age be the age of improvement. In a day of peace, let us advance the arts of people and the works of peace.'[236] The 'works of peace' were proceeding all the time. Boston had gas street-lighting in 1822, almost as soon as London. It came to New York in 1823, to Philadelphia in 1837. But Philadelphia was ahead with piped water, getting it in 1799. By 1822 the Fairmount Waterworks had brought piped water to the entire city. This was amazing even by the standards in England, regarded then as the world pioneer in municipal utilities. Moreover, this magnificent waterworks, in the best classical architecture, expanded from the banks of the Schuylkill, and its grounds embraced a huge area of the country, and in order to preserve it from pollution Philadelphia ultimately created the largest urban park in the world, in the process preserving for posterity all the splendid riverside villas we have already described. There were, to be sure, early signs of skulduggery in the provision of municipal services. Aaron Burr's Manhattan Water Company (1799), the first to build a reservoir in New York, was in reality a front for an unlawful bank competing with Alexander Hamilton's Bank of New York (now the Chase-Manhattan). But, at this stage anyway, most services, public and private, were honest, competitive, and, by world standards, go-ahead. New York got its first omnibuses only a year after Paris and the same year as London, 1828—the first line was Wall Street–Greenwich Village. Philadelphia had buses three years later. America was also quick to imitate Britain's penny post, knocking down the steep prices the generous President Jefferson paid to 5 cents for half an ounce delivered at up to 300 miles (1846). Open competition was driving down prices relentlessly: thus the first penny newspaper dates from 1840, an amazing price by European (even British) standards at that time.[237]

There was no doubt about the determination with which 'enlightenment' was pursued in the field of education, at all levels of American society. Since the colonial period, America had rejoiced in the highest rate of adult literacy in the world, higher even than Germany's. This was due primarily to the school reformers in the big cities. Horace Mann's work in Boston we have already noted, in the context of teaching religion. In 1806 the Public School Society of New York introduced the Lancaster system from England, in which 'pupil teachers' or monitors were used to give basic instruction to the thousands of new city children. From 1815 the society's 'model system' of public schools got state aid, and when New York State finally took over the system in 1853 it was providing education for 600,000 children. In the newer states, Ohio for instance, the sixteenth section of each planned township was devoted to education. But in the Western states, sheer distance made universal education difficult. In Louisiana the population density (1860) was only eleven per square mile; in Virginia (including what is now West Virginia) it was fourteen, by contrast with Massachusetts, where it was 127. Census data show that by 1840 some 78 percent of the total population was literate (91 percent of the white population), and this was mainly due to a rise in national school enrollment rates: from 35 percent in 1830 (ages five to nineteen), to 50.4 percent in 1850 and 61.1 percent in 1860. All the same, there were still 1 million adult illiterates in America in 1850, of whom 500,000 were in the South. Most of these illiterates were not new immigrants (though that too was a problem, because of language) but blacks, an early indication of trouble to come.[238]

At the end of the 1760s, America, on the eve of Independence, had nine colleges, or universities as they were later called. All were denominational, though William and Mary was partly secularized in 1779, when the professorships of Hebrew and Divinity were turned over to law and modern languages. The Presbyterians founded four new colleges in the 1780s, including Liberty Hall, which became the nucleus of Washington and Lee, and Transylvania Seminary, the first institution of higher education beyond the Appalachians. By that date Yale was taking in a freshman class of seventy, Harvard thirty-one, Princeton ten, Dartmouth twenty. Such early foundations bred scores of satellites—sixteen Congregational- ist colleges sprang from Yale and twenty-five Presbyterian ones from Princeton, all before 1860. A total of 516 colleges and universities were scattered over sixteen states by the coming of the Civil War. (Some of these were short-lived: only 104 of this group were still flourishing at the end of the 1920s.) The state universities began with Jefferson's University of Virginia, and some of them had humble beginnings. Thus Michigan had one as early as 1817—the first in the West—but it was really a glorified high school until 1837, when it was moved to Ann Arbor and endowed with state lands proceeds. Another great state university, Wisconsin, was created at Madison in 1836. Curiously enough, such institutions enrolled more students than the big foundations of the East: even by the 1840s, a Western youngster had a better chance of going to college than a contemporary in the Eastern cities (Boston and Philadelphia excepted). Thus New York in 1846, with half a million population, enrolled only 241 new students at its two colleges.

Up to the 1780s, the overwhelming majority of college graduates went into the ministry, though politics claimed a surprising number (thirty-three out of fifty-five men attending the Constitutional Convention were graduates). During the 1790s, however, the balance swung in favor of the lawyers, and by 1800 only about 9 percent went into orders, with 50 percent going into the law. The influence of Germany, whose universities were the best on earth, was enormous. Between 1830 and 1860, for instance, virtually every young professor at Yale had spent a year in a German university. The rise of the Western university was very much influenced by government land policy. If a proceeds-from-land-sales arrangement was in force, a college would spring up overnight, and there was no difficulty in obtaining staff or attracting students. The big breakthrough came with the Morrill Act of 1862, which enabled state agricultural colleges to be founded using federal land funds, and in many cases these were quickly broadened into general universities.

This enlightened Act also benefited women. There were a few women's colleges before that date—Oberlin in Ohio, for instance, dates from 1833 and Georgia Female from 1838. But the Morrill Act encouraged the admission of women to state universities—Wisconsin admitted them from 1867 and Minnesota from 1869. By then some superb women's universities were competing—Vassar (1861), Minnesota (1869), Wellesley (1870). By 1872 women were admitted to ninety-seven colleges or universities and by 1880 they constituted one-third of all students, though over 70 percent of them were condemned to (or chose) teaching. The real shortage was in black higher education: only twenty-eight blacks had graduated by the time of the Civil War. Thereafter a few black colleges came into existence: Atlanta in 1865, Lincoln and Fisk in 1866, and Howard in 1867. By this time one in a hundred American adults was having a college education.[239]

By any statistical standards, America made enormous progress in the first half of the 19th century in making itself 'enlightened.' But not everyone agreed with De Tocqueville that the country had succeeded. Fanny Trollope, herself a novelist and the mother of the more famous Anthony, was in the United States 1827–31, trying to earn a living for herself in Cincinnati and elsewhere. She had been married to a fanatical clergyman who had been unable to support her, and in consequence she took a cynical view of religion: she thought America had far too much of it. The moral point, so important to De Tocqueville, entirely escaped her. What she noted was the manners. She thought it outrageous that the only form of garbage collection in Cincinnati were the pigs (this was true of New York, too, until 1830). She found it was 'petty treason' to call a servant such: 'help' was the only acceptable term, an early example of Political Correctness. Moreover, such was the American mobility of labor that it was impossible to hire a 'help' except for a short term, and in the process of engagement it was the 'help' not the mistress who dictated terms. Thus the first she engaged, when asked what she expected per annum, replied: 'Oh Gimini! You be a downright Englisher, sure enough. I should like to see a young lady engage by a year in America! I hope I shall get a husband before many months, or I expect I shall be an outright old maid, for I be most seventeen already. Besides, mayhap I may want to go to school. You must just give me a dollar and a half a week, and mother's slave, Phillis, must come over once a week, I expect, from the other side of the water to help me clean.'

Mrs Trollope started to write down such things, for her letters home, otherwise her London friends would not believe her; and from this came *Domestic Manners of the Americans*, published in 1832, which was an immediate bestseller on both sides of the Atlantic, made Mrs Trollope the most hated author in America, and still makes American hackles rise today.[240] Her criticisms were all calculated to wound. The Americans were rude, ill-bred, pushy, and coarse. They had no fun and no sense of humor: 'I never saw a population so divested of gaiety: there is no trace of this feeling from one end of the Union to the other.' Americans were totally self-absorbed, uninterested in the outside world, and with a hugely inflated idea of their own importance and merits. The women were ignorant, the men disgusting. She excepted a few bookish men from this censure, adding that America was a signal proof of 'the immense value of literary habits' not only in 'enlarging the mind' but in 'purifying the manners.' She added: 'I not only never met a literary man who was a chewer of tobacco or a whiskey drinker, but I never met any who were not, who had escaped these degrading habits.'[241]

Here we come to it; if there was one thing English visitors could not stand about America, it was the habit of spitting. The English middle and upper classes had cured themselves of public spitting in the 1760s—it was one of the great turning-points of civilization—so that by the 1780s, they were already censorious of the French, and other Continentals, for continuing it: Dr Johnson was particularly severe on this point. In the United States, however, the spitting habit was compounded by the business of chewing tobacco, which in the first half of the 19th century was carried on by three-quarters of the males and even by some females. Hence spitting became an almost continuous process, and where spittoons of enormous size were not provided in large numbers, the results were catastrophic to sensitive souls. It was the first thing all the English, from Dickens to Thackeray, noticed and commented on, and English lady travelers, like Mrs Trollope, were especially offended. When she, and others, were shown round the Senate, their eyes were glued to the gigantic brass spittoons attached to every member's desk.

That was a pity, because the Senate of those days, and for several decades afterwards, was a remarkable institution, perhaps the greatest school of oratory since Roman times. And its finest hour was 1850, when the last Great Compromise on slavery was debated, attacked, defended, and carried. The background was extremely complicated— the reader will have gathered by now that everything to do with slavery in America was complicated—and the Compromise itself was complex. The old Northwest Ordinance of 1787 had prohibited slavery in the new Northwest, and all the states created there were free. In most of the other acquisitions America had made, the whole of the Louisiana Purchase, Florida, and Texas, forms of slavery had existed under the French or Spaniards, so maintaining it there, or reimposing it as in Texas, did not appear so horrific. But when what was eventually to become California, which had always been slave-free, was acquired in 1848, and some of the freebooters who were seizing power there proposed to make it a slave state, the Northern conscience was powerfully aroused. When President Polk submitted a money Bill to the House, asking for funds to make peace with Mexico (in effect to bribe Santa Ana), a Pennsylvania congressman, David Wilmot, added an amendment stipulating, and using the language of the Northwest Ordinance, that in any territory so acquired 'neither slavery nor involuntary servitude shall ever exist.' Furious, Polk got his friends to table a counteramendment, proposing that the Old Missouri Compromise line, running at latitude 36.30, should be extended and divide freedom and slavery in the new territories, as in the old. But the moderates who would have voted for this were denounced as traitors in the South or as Doughfaces (Northerners with Southern principles) in the North. So both were voted down by extremists. Wisconsin got statehood in 1848 with a free constitution, but Polk left office with the issue unresolved in Utah, New Mexico, and California.[242]

From the so-called Wilmot debates, new principles emerged. The first was that Congress had the right to ban slavery wherever its jurisdiction extended—freedom was national, slavery only sectional. That was an important step forward. Both the Free Soil and the Republican Parties were later formed to enforce this doctrine. On the other hand, the Southerners also put forward a new doctrine: not only did Congress have no right to prohibit slavery in the territories, it had a positive duty to protect it there, once established. Calhoun now produced a new theory, reversing the constitutional practice of the past sixty years: newly acquired territories belonged to 'the states united,' not to the United States. Congress, he argued, was merely 'the Attorney to a Partnership' and every partner had an equal right to protection of his property on his territory. He denied that Lord Mansfield's 1772 ruling on slavery in England applied in America, where slaves were 'common law property.' To be sure Congress had prohibited slavery north of 36.30 in 1820—but that was unconstitutional. Slavery followed the United States flag, automatically, wherever it was planted. This doctrine was embodied in resolutions adopted by the Virginia legislature in 1847, later known as the 'Platform of the South.'

It also became the doctrine underlying the Supreme Court's fateful decision in the *Dred Scott Case* in 1857. Describing this takes us a little ahead of the California issue, but it is important to get its implications clear now. Scott was a Missouri slave who was taken (1834) by his master to places where slavery was prohibited by law. In 1846, Scott sued for his freedom in the Missouri courts, arguing that his four-year stay on free soil had given it to him. He won his case but the verdict was reversed in the state supreme court. He then appealed it to the federal Supreme Court, and Taney and his colleagues again ruled against him, for four reasons. First, since Scott was a negro and therefore not a citizen, he could not sue in a federal court. Second, as he was suing in Missouri, what happened in Illinois, under its law, was immaterial. Third, even so, Scott's temporary sojourn on free territory did not in itself make him free. Fourth, the original Missouri Compromise was unconstitutional since it deprived persons of their property (slaves) without due process of law and was therefore contrary to the Fifth Amendment.[243] The *Dred Scott* ruling became of critical importance in the events leading up directly to the Civil War, which we will examine later. Here, it is enough to say that its reasoning followed, and gave constitutional legitimacy to (or appeared to do so), Calhoun's case. However, at this point it is important to remember one thing. Neither

Congressman Wilmot nor Senator Calhoun regarded himself as extremist. Both thought they were putting forward defensive strategies, preemptive strikes as it were, to ward off aggression by the other side. And it is true that there were many more extreme men (and women) in Massachusetts and South Carolina, determined to end slavery, or to maintain it, at literally any cost.

Disgusted by his failure to get a solution to the California admission problem, and worn out anyway, Polk made good his promise not to run again (dying soon after leaving the White House). The Democrats fielded a strong Manifest Destiny candidate in the shape of Lewis Cass (1782–1866), a Michigan senator who favored cheap land, squatters' rights, and all kinds of popular causes. The Whigs countered this by picking General Zachary Taylor (thus confirming Polk's fears of 'political generals'), whose victory at Buena Vista had made him a semi-legendary figure. Neither party had a proper platform, especially on the slavery issue. But Taylor came from Louisiana and had scores of slaves working on his estates. This infuriated three groups of Whigs: Van Buren's New Yorkers who called themselves 'Barnburners,' fanatical Massachusetts anti-slavery men who called themselves 'Conscience Whigs,' and another abolitionist group who called themselves the Liberty Party. They ganged up together, called themselves the Free Soil Party, and nominated Van Buren. In theory this should have split the Whig, anti-slavery vote, and let Cass and the Democrats in. In practice it had the opposite effect. In the election razzmatazz, Taylor was so identified with the South that he carried eight slave states. Cass, the Democrat, could manage only seven. Moreover, the free soilers split the Democratic as well as the Whig vote in New York and handed it to Taylor. He won by 1,360,099 to Cass's 1,220,544 (Van Buren getting only 291,263), and by 163 to 127 college votes.[244]

This confused and confusing election brought to the White House a man whom Clay, who had now missed his last-ever chance to become president, dismissed as 'exclusively a military man,' with no political experience, 'bred up and always living in the camp with his sword by his side and his Epaulettes on his shoulders.' By contrast, Clay characterized his friend Millard Fillmore (1800–74), the Vice-President, an experienced New York Tweed machine-man, as 'able, enlightened, indefatigable and . . . patriotic.'[245] Both these verdicts were soon put to the test. Clay was wrong about Taylor. He was not a mere general, nor was he a pro-slaver, as the South had hoped. He encouraged the Californians, who were anxious to get on with things and achieve constitutional respectability, to elect a free state administration. This was done all the more easily because the miners were overwhelmingly anti-slavery, fearing their jobs would be taken by slaves. On December 4, 1849, in his message to Congress, Taylor asked it to admit California immediately, and to stop debating 'exciting topics of sectional charac-ter'—he meant slavery—'which produced painful apprehensions in the public mind'—that is, talk and fear of secession.

Millard, by contrast, justified Clay's eulogium by presiding fairly and skillfully over the Senate, an important point since Congress, far from heeding the President's advice to steer off slavery, debated virtu-ally nothing else in 1850. Then, on July 4, the President, having presided over the ceremonies, gobbled down a lot of raw fruit, cab-bages, and cucumbers—food 'made for four-footed animals and not Bipeds' as one observer put it—and gulped quantities of iced water (the heat and humidity were intense). It was probably the iced water that did it, though there was talk of poison. Five days later the President died in agony of acute gastroenteritis, and Fillmore took over. The new President, unlike Taylor, favored compromise over the California issue, and Senator Clay, in effect the administration's spokesman in Congress, was able to deliver it to him. By this time the debates had already lasted six months. Students of rhetorical form rate the speeches in the Senate as among the greatest in the entire history of Anglo-Saxon oratory, worthy to rank with the duels of Pitt and Fox, and Gladstone and Disraeli. In fact the three main protagonists were uttering their swansongs. Calhoun was dying, Clay was at the end of his immense career, and Webster became secretary of state in the Fillmore adminis-tration. Readers can consult the record of the debates and decide for themselves who won.[246] The Senate was crowded and enthralled throughout and the spittoons had never been in such continuous use. But it is one of the sad things about congressional or parliamentary democracy that great speeches rarely make much difference to histori-cal outcomes.

What the debates did make clear, however, was that secession by the South, if it did not get its way in making slavery 'safe for ever,' was a real possibility, and that it would not and could not be bloodless. That helped to smooth the road to compromise, which was piloted by old Clay, much assisted by a young Democratic senator, Stephen A. Douglas of Illinois (1813–61). Clay had originally hoped to get all the issues tied up together in one gigantic compromise, what he called an Omnibus Bill. The Senate would not wear it. Then Douglas divided it up into its five component bits, and got them all through separately. Senator Benton explained this by saying that the components 'were like cats and dogs that had been tied together by their tails for four months, scratching and biting, but being loose again, everyone ran off to his own hole and was quiet.' Possibly: there are many irrational and, in the end, inexplicable aspects to the whole controversy over slavery, and between North and South, which baffle historians, as they baffled most people in the middle at the time. The upshot is that Clay carried his last great Compromise in early September and on the 20th of the month Fillmore signed the five Bills into law.[247]

In the Compromise, the most important sop to the South was a new Fugitive Slave Law. This made the capture and return of escaped slaves a matter for federal law and rendered it exceedingly difficult, if not impossible, for Northern states to evade their responsibilities under the Constitution. Second, to balance matters a bit, the anti-slavery lobby in the North was given the minor sop of the District of Columbia becoming an area where slave-trading was made unlawful. It was still possible to keep slaves in Washington, but not to buy or sell them there or hold them for sale elsewhere. If you marched slaves through the street in chains—a common sight up to now, which grievously shocked sensitive Northerners and all foreigners—you were inviting arrest. Third and fourth, both New Mexico and Utah became territories and the acts making them such left their slave- or free-state future vague, beyond insisting that their legislatures were to possess authority over 'all rightful subjects of legislation,' subject to appeal to federal courts. Finally, California entered the Union as a free state. This ended the Senate slave/free balance and ensured that in future Congress would have an anti-slavery majority in both Houses.[248]

The crisis between North and South, having seethed and bubbled for months, suddenly went off the boil, just as it had done after the confrontation of 1819–20. Men on both sides, and still more women, relaxed as the horrific shadow of civil war suddenly disappeared, and they could get on with other things. And there was so much to do in mid–19th-century America, so many blessings to rejoice in and opportunities to seize! America was becoming not merely a wealthy country but in a growing number of ways a civilized and sophisticated one. The year 1850 is remarkable not merely for the apogee of Congressional oratory but for the long-delayed but sure and true beginnings of a great national literature. Considering how assertive politically America was, even in the mid–18th century, it was remarkably slow to assert itself culturally. Speech is a very democratic force: it is the demotic which penetrates upwards into the hieratic, not the other way round. 'Americanisms' had been appearing since the mid–17th century in the way ordinary people spoke, though the term was not coined until 1802, by a Scots immigrant, on the analogy of Scotticism. But Independence was declared, and the Constitution written, debated, approved, and amended entirely in standard English, if anything with a slight touch of archaism, though spelling was already diverging.

In 1783–5 Noah Webster (1758–1843), a Yale-trained lexicographer and philologist from Connecticut, produced *A Grammatical Institute of the English Language*, the first part of which was extracted to form his *Spelling Book*, which gave standard American variations of English spelling forms for use in schools. In 1790 he produced his *Rudiments of English Grammar*, the first book to challenge the linguistic hegemony of Britain, in which he argued '*Now* is the time, and *this* is the country, in which we may expect success, in attempting changes favorable to language, science and government.' But he discovered the hard way that it was easier to turn America from a monarchy into a republic than to force systematic language and spelling reform on a stubborn people who spoke as they felt. The same year he produced a volume of essays in his reformed spelling: 'essays and Fugitive Peeces ritten at various times ... as will appeer by their dates and subjects.' Readers laughed at him.[249] Another language reformer, William Thornton, urged, in *Cadmus, or a Treatise on the Elements of a Written Language* (Philadelphia 1793), addressed to the American people: 'You have corrected the dangerous doctrines of European powers, correct now the language you have imported ... The AMERICAN LANGUAGE will thus be as distinct as its government, free from all the follies of unphilosophical fashion and resting upon truth as its only regulator.' He then gave the text in his new spelling-system. Practical Americans dismissed it as gibberish and went on talking, and changing, the English language as they had learned it from their parents.[250]

Americans were immensely resourceful in making these changes, adapting, translating, inventing, and knocking about words to suit their needs and tastes. Some of these early neologisms were from the French, both from Canada and Louisiana: *depot, rapids, prairie, shanty, chute, cache, crevasse.* Some were from the Spanish of Florida and the Gulf: *mustang* (1808), *ranch* (1808), *sombrero* (1823), *patio* (1827), *corral* (1829), and *lasso* (1831). Americans resurrected obsolete

English words like *talented* and invented ones like *obligate*. They adopted, for instance, the German words *dumm*, which became *dumb*, stupid. Words from their new political customs appeared: *mass meeting*, *caucus*, settlers' words like *lot* and *squatter*. The Lewis–Clark and other expeditions introduced a new crop: *portage, raccoon, groundhog, grizzly, backtrack, medicine man, huckleberry, war party, running-time, overnight, overall, rattlesnake, bowery,* and *moose*. Variant meanings were given to old English terms: *snag, stone, suit, bar, brand, bluff, fix, hump, knob, creek,* and *settlement*.

Then there was the wonderful fertility of the Americans in coining new phrases and amalgams: *keep a stiff upper lip* (1815), *fly off the handle* (1825), *get religion*—an important one, that—in 1826, *knockdown* (1827), *stay on the fence* (1828), *in cahoots* (1829), *horse-sense* (1832), and *barking up the wrong tree* (1833), plus less datable novelties: *take on, cave in, flunk out, stave off, let on, hold on*. As early as the 1820s Americans were trying to *get the hang of a thing* and insisting *there's no two ways about it*. The American thirst added many terms: *cocktail* (1806), *barroom* (1807), *mint julep* (1809), a *Kentucky Breakfast* (1822), defined as 'three cocktails and a chaw of terbacka,' and a *long drink* (1828).[251] At varying speeds most of these new words and expressions crossed the Atlantic. By the time Webster came to produce his *An American Dictionary of the English Language* in two thick volumes in 1828, he was able to list 5,000 words not hitherto included in English dictionaries, including many Americanisms, and using definitions which Americans, rather than the British, recognized. He revised this standard work in 1840 to include 70,000 words instead of the original 38,000 and, suitably amended from time to time, it has become second only to the Oxford English Dictionary as the prime authority on English words.[252]

In the hieratic, as opposed to the demotic, the Americans were slower to become creative. In a notorious article in the *Edinburgh Review* of 1819, the great English wit and reformer the Rev. Sydney Smith hailed some American political innovations but argued that Americans 'during the thirty or forty years of their existence' had done 'absolutely nothing for the Sciences, for the Arts, for Literature or even for the statesman-like studies of Politics and Political Economy.' This was nonsense as regards the sciences, as we have seen, and Smith had obviously never read the *Federalist* or any of the great debates on the Constitution, which rivaled Burke in their penetrating analysis of basic political issues. He was wrong about literature, too, if one considers the works of Jonathan Edwards and Franklin. But it was odd, as he suggested, that independence had not brought about a corresponding *pléiade* of American literary stars.

Many Americans agreed with him. In 1818 the *Philadelphia Portfolio* published an essay by George Tucker, *On American Literature*, drawing attention to the contrast between the literary output of America, with 6 million people, and the performance of tiny countries like Ireland and Scotland—where were the American equivalents of Burke, Sheridan, Swift, Goldsmith, Berkeley, and Thomas Moore from Ireland, and Thomson, Burns, Hume, Adam Smith, Smollett, and James Boswell from Scotland? He pointed out that the two most distinguished novelists were Scott and Maria Edgeworth, both from little Scotland. He calculated that America produced on average only twenty new books a year, Britain (with admittedly a population of 18 million) between 500 and 1,000. In 1823, Charles Jared Ingersoll in an address to the American Philosophical Society, 'A Discourse Concerning the Influence of America on the Mind,' noted that 200,000 copies of Scott's Waverley novels had been printed and sold in the United States, while the American novel was almost nonexistent. The *Edinburgh* and the *Quarterly* were now printed in America and sold 4,000 copies each issue there, whereas the American equivalent, the *North American Review*, was unknown and unobtainable in London.[253]

Even when the first real American literary personality emerged, in the shape of Washington Irving (1783–1859), he seemed to be guilty of the 'Cultural Cringe,' and based himself on English models, chiefly Scott and Moore, to a stultifying degree. When he traveled to Europe from 1815 onwards, he made himself heavily dependent on German literary sources too. His most famous character, Rip Van Winkle, and the Legend of Sleepy Hollow, published in *The Sketch Book* (1820), were taken straight from Christophe Martin Wieland and Riesbeck's *Travels Through Germany*—he merely expanded the Winkle tale and gave it an American setting.[254] Irving was an enormous success in England, precisely because of his cringing and his deference to British cultural idols such as Scott, and also because of his sensible attempts to stop American publishers pirating English copyrights.[255] Irving sold well on both sides of the Atlantic, and seems to have earned from his writings the immense sum of $200,000. Many towns, hotels, squares, steamboats, and even cigars were named after him. He was the first American to achieve celebrity in literature and when he died New

York, his home city, closed down: there were 150 carriages in his funeral procession and 1,000 mourners crowded outside the packed church. President Jackson, who objected to his being made minister in Madrid, snarled: 'He is only fit to write a book, and scarcely that.' Behind the philistinism, one detects a note of all-American truth.[256]

By contrast, the first great American novelist, James Fenimore Cooper (1789-1851), was undoubtedly indigenous in his work and spirit. He grew up in a 40,000-acre tract of land in upper New York State, his father being a land investor and agent who at one time owned 750,000 acres and controlled much more. Cooper Sr wandered at will in what was then largely unexplored country and wrote a *Guide to the Wilderness*. But this was published posthumously in 1810 because when young Cooper was twenty his father was shot dead at a political meeting—not uncommon in those days. Cooper's third novel, *The Pioneers* (1823), first of what became known as the Leatherstocking Tales, introduced his frontiersman hero, Natty Bumppo. The five books of the series, above all *The Last of the Mohicans* (1826), made Cooper world-famous. Natty is the first substantial character in American fiction, a recurrent American ideal-type, putting his own special sense of honor and character above money and position—not so different from the Ernest Hemingway hero who would emerge almost exactly a century later. Cooper used his father's experiences as well as his own to recreate the American wilderness, fast disappearing even as he recorded it. The novels fascinated readers in the big East Coast cities, to whom all this was new and strange. Equally, perhaps more, important it brought home to literally millions of people in Europe what they assumed to be the realities of American frontier life. Germans in particular loved them: they were read aloud at village clubs. *The Pioneers* was published in Britain and France the same year it appeared in America and within twelve months it had found two rival German publishers—eventually thirty Germany publishing houses put out versions of the Leatherstocking tales. In France, where he was *le Walter Scott des sauvages*, eighteen publishers competed. Many Russian translations followed, and his works appeared in Spanish, Italian, and Portuguese and eventually in Egyptian, Turkish, and Persian. By the end of the 1820s, children all over Europe and even in the Middle East were playing at Indians and learning to walk 'Indian file.'[257]

Yet in many respects Cooper was hostile to what America was becoming. He opposed mass immigration. Indian 'removal' was infinitely painful to him. He was backward-looking, conservative, and, in American terms, a hidebound traditionalist, who could not get over the demise of the old federal party. Today he would have been an extreme environmentalist. It was a point he made again and again in his novels that Natty and his friends killed wildlife only to eat, not for sport, still less because they feared the beasts and yearned for 'civility.' He was an elitist, a seer, an aristocrat of sorts, fiercely defending his property, loathing the vulgarity and populism of Jacksonian democracy and egalitarianism, which he assailed in a savagely hostile book, *The American Democrat* (1838). In many ways he was the first critic of the American way of life. The three novels he wrote in the 1840s, known as the Littlepage Trilogy (*Satanstoe*, 1845, *The Chairbearer*, 1845, *The Redskins*, 1846) presented the business of settling the Mississippi Valley as an affair of greed, destroying the pristine morality of the American ideal.[258]

The first American intellectual and writer to go wholly with the mainstream American grain, in some ways the archetypal American of the 19th century, was Ralph Waldo Emerson (1803–82), who consciously set out to reject cultural cringing, to 'extract the tape-worm of Europe from America's body,' as he put it, to 'cast out the passion for Europe by the passion for America.'[259] He too went to Europe but in a critical and rejecting mood. Emerson was born in Boston, son of a Unitarian minister. He followed in his father's footsteps but threw off the cloth when he discovered he could not conscientiously 'administer the Lord's supper.' His skepticism, however, did not make him a critic of the essential moralism and religiosity of American secular life: quite the contrary. In seeking to Americanize literature and thought, he developed a broad identification with the assumptions of his own society which grew stronger as he aged and which was the very antithesis of the hostility of the European intelligentsia to the way things were run. After discovering Kant in Europe he settled in Concord, Massachusetts, where he developed the first native American philosophical movement, known as Transcendentalism, which he outlined in his book *Nature* (1836). It is a Yankee form of neo-platonism, mystical, a bit irrational, very vague, and cloudy. It appealed to some fellow-intellectuals but to very few ordinary people—even the educated found it hard to get the hang of it. All the same, they approved. They thought it grand that America had got its own proper intellectual at last. It was said his appeal rested 'not on the ground that people understand him, but that they think such men ought to be encouraged.'[260] A year after he published *Nature*, he delivered a Harvard lecture, 'The American Scholar,' which Oliver Wendell Holmes (1809–94) was to call 'our intellectual declaration of independence.' The patriotic press loved it. The most influential newspaper, the *New York Tribune* of Horace Greeley (1811–72), promoted Emerson's Transcendentalism as a new kind of national asset, an all-American phenomenon, like Niagara Falls.

There was something a bit too good to be true about Emerson. The Scots critic Thomas Carlyle, who became a dear friend, described him as 'like an angel, with his beautiful, transparent soul.' Henry James later wrote of him, 'his ripe unconsciousness of evil . . . is one of the most beautiful signs by which we know him'—though he added, cruelly: 'We get the impression of a conscience gasping in the void, panting for sensations, with something of the movements of the gills of a landed fish.' He astonished English intellectuals by insisting that Young America was sexually pure: 'I assured [Carlyle and Dickens] that, for the most part, young men of good standing and good education with us, go virgins to their nuptual bed, as truly as their brides.' His own sexual drive seems to have been weak. His first wife called him 'Grandpa.' His second wife's criticism of his lack of marital attentions were naively recorded in his journals. His poem 'Give All to Love' was thought daring but there is no evidence he gave himself. His own great extramarital friendship with a woman, Margaret Fuller, was platonic, or maybe neo-platonic, and not by her desire. His unconsciously revealing journal records a dream, in 1840–1, in which he attended a debate on marriage. One of the speakers, he recorded, suddenly turned on the audience 'the spout of an Engine which was copiously supplied . . . with water, and whisking it vigorous about,' drenched everyone, including Emerson: 'I woke up relieved to find myself quite dry.'[261]

But it is too easy to poke fun at Emerson. He was a good, decent man and his views, on the whole, made excellent sense. He married both his wives for prudential reasons and their property made him independent. Soundly invested, it also brought him an affinity with America's burgeoning enterprise system. He made what eventually became an unrivaled, and never repeated, reputation as a national sage and prophet, not so much by his books as through the lecture circuit. Almost from the earliest days of the century, public lectures became a key feature of American cultural life. As part of Washington Irving's cultural cringing he proposed that the British poet Thomas Campbell be hired to lecture in America to give 'an impulse to American literature and a proper direction to public taste.' Lectures were a popular form of entertainment in New York, Boston, and Philadelphia from 1815 but it was only from 1826, when Josiah Holbrook (1788–1854) founded the Lyceum Movement, that the habit spread everywhere. Holbrook had dabbled in industrial schools and agricultural colleges before he hit upon the lecture form as the best way to educate the expanding nation. Lyceums were opened in Cincinnati in 1830, in Cleveland in 1832, in Columbus in 1835, and then throughout the expanding Midwest and Mississippi Valley. By the end of the 1830s almost every considerable town had one. They had their own weekly newspaper, the *Family Lyceum* (1832), their Young Men's Mercantile Libraries, and they sponsored debating societies, aiming especially at young, unmarried men—bank clerks, salesmen, bookkeepers, and so forth—who then made up an astonishingly high proportion of the population of the new towns. The Movement aimed to keep them off the streets and out of the saloons, and to promote simultaneously their commercial careers and their moral welfare.[262]

Emerson was the perfect star-attraction for this system. He was anti-elitist. He thought American culture must be egalitarian and democratic. Self-help was vital, in this as in all fields. He said that 'the first American who read Homer in a farm-house' performed 'a great service to the United States.' If he found a man out West, he said, reading a good book on a train, 'I wanted to hug him.' His own economic and political philosophy was identical with the public philosophy pushing Americans across the continent to fulfill their manifest destiny. Emerson laid down the maxims of this expansion: 'The only safe rule is found in the self-adjusting meter of demand and supply. Do not legislate. Meddle, and you snap the sinews with your sumptuary laws. Give no bounties, make equal laws, secure life and property, and you do not need to give alms. Open the doors of opportunity to talent and virtue and they do themselves justice; property will not be in bad hands. In a free and just commonwealth, property rushes from the idle and imbecile to the industrious, brave and persevering.'[263]

It would be difficult to think of any doctrine more diametrically opposed to what was being preached in Europe at the same time, notably by Emerson's younger contemporary, Karl Marx. And Emerson's experience in the field repeatedly contradicted the way in which Marx said capitalists not only did but must behave. American owners and managers, said Marx, were bound to oppose their workers' quest for enlightenment. But when Emerson came to Pittsburgh in

1851, for example, firms closed early so the young clerks could go to hear him. His courses were not obviously designed to reinforce the entrepreneurial spirit: 'The Identity of Thought with Nature,' 'The Natural History of Intellect,' 'Instinct and Inspiration,' and so on. But one of the thrusts of his arguments was that knowledge, plus moral character, tended to promote business success. Many who attended expecting to be bewildered by the eminent philosopher found he preached what they thought was common sense. The *Cincinnati Gazette* described him as 'unpretending . . . as a good old grandfather over his Bible.'

Emerson was a marvelous manufacturer of short sayings and pithy *obiter dicta*, many of which—'Every man is a consumer and ought to be a producer,' 'Life is the search after power,' '[Man] is by constitution expensive and ought to be rich,' 'A foolish consistency is the hobgoblin of little minds,' 'Whoso would be a man must be a nonconformist,' 'Hitch your waggon to a star'—struck his listeners as true, and when simplified and taken out of context by the newspapers passed into the common stock of American popular wisdom. It did not seem odd that Emerson was often associated in the same lecture series with P. T. Barnum, speaking on 'The Art of Money Getting' and 'Success in Life.' To listen to Emerson was a sure sign of cultural aspiration and elevated taste: he became, to millions of Americans, the embodiment of Thinking Man. At his last lecture in Chicago in November 1871, the *Chicago Tribune* summed it up: 'The applause . . . bespoke the culture of the audience.' To a nation which pursued moral and mental improvement with the same enthusiasm as money, and regarded both as essential to the creation of its new civilization, Emerson was by the 1870s a national hero (though it is as well to recall his own saying, 'Every hero is a bore at last').[264]

Washington Irving attained success by culture-cringing and getting a condescending nod of approval from the English literary elite. Emerson played the anti-English card and went all out to reflect the basic American ethos. But the first writer who managed to appeal equally both to simple American hearts and to the sophisticated audience of the entire English-speaking world was Henry Wadsworth Longfellow (1807–82). He was a prodigy, born in Portland, Maine, educated privately, publishing his first poem at thirteen, and then at Bowdoin where, while still a student, he was told he was needed to teach languages and literature provided he went to Europe and acquired cultural polish. So he learned French and German and Italian and in time became the most learned (so far) of American literary men, translating Dante, difficult Provençal poets, and German philosophy. He taught not just at Bowdoin but later at Harvard for eighteen years where, thanks to a rich second wife, daughter of a successful cotton-mill owner, he made Craigie House—the mansion his father-in-law provided on their wedding—a center of Cambridge intellectual society.

Longfellow's poems flowed from his pen in steady and stately succession, and his unique gift for resonant lines allowed him to enter into the minds and hearts, and stay in the memories, of the middle class on both sides of the Atlantic. None of his lyric contemporaries, not even Tennyson and Browning, found himself quoted so often: 'I shot an arrow in the air;' 'Life is real / Life is earnest;' 'Footprints in the sands of time;' 'A banner with a strange device;' 'The midnight ride of Paul Revere;' 'A Lady with a Lamp;' 'Ships that pass in the night;' 'Under a spreading chestnut tree;' 'It was the schooner Hesperus;' 'When she was good, she was very, very good;' 'Fold their tents like the Arabs, and silently steal away;' 'Something attempted, something done'—these golden phrases, and the thought behind them, passed into the language. It was Longfellow who attempted, with some success at the time, less since, to write America's first epic poem, *The Song of Hiawatha* (1855), in which he used the Finnish metrics of Kalavala to produce the American equivalent of Tennyson's *Idylls of the King*. Even more ambitious, in a way, was his successful attempt to sum up America's powerful (almost strident) message to the world in one short poem, 'The Building of the Ship:'

> Sail on, O Union, strong and great!
> Humanity with all its fears,
> With all the hopes of future years,
> Is hanging breathless on thy fate!

Longfellow was no stranger to personal tragedy: he lost his first wife when she was still a young bride; his second, loved still more, was burned to death in 1861, and the poet was stricken and silent for a decade. He had many friends, had a sweet, decorous, and benign disposition, and lived a sheltered life in the comfort and safety of university New England. There were no sexual hang-ups in his life, no mysteries, no hidden, smoldering pits to be explored. So he has been largely ignored by 20th-century literary academics. But he was much loved in his day, by ordinary people in clapboard houses and Western cabins, as well as by the Boston literati. When his great poem in unrhymed English hexameters, *The Courtship of Miles Standish*, was published simultaneously (1858) in Boston and London, 15,000 copies were sold on the first day. The English treated him as a member of their grand poetical canon, awarding him degrees at Oxford and Cambridge, and he took tea with Queen Victoria, a privilege hitherto accorded to Tennyson alone. On his death he became the first American to have his niche in Poets' Corner in Westminster Abbey. More important, perhaps, he played a notable role in making Americans familiar with Europe's poetical heritage—he was a transatlantic bridge in himself.[265]

One of the few people who went for Longfellow in his day, in a notorious article called 'Longfellow and Other Plagiarists,' was Edgar Allan Poe (1809–49), who stood right at the other end of the worthiness and acceptability scale. Poe was a natural misfit who crammed an extraordinary quantity of misfortune into his short forty years of life. He was both a throwback to the Gothick Romanticism of the years 1790–1820 and an adumbrator of the Symbolism to come. He was born near Boston, the offspring of strolling players. He had a difficult, orphan childhood under a rich foster-father who starved him of money; rebelled and ran away; got into West Point and was discharged 'for gross neglect of duty;' became a journalist, then an editor, but was sacked for drunkenness; nearly starved to death in a garret; married his thirteen-year-old cousin probably incestuously (that is without getting a license); led a *vie de bohème*, being hired and fired many times, by many publications; got into trouble with women; tried suicide; mourned his wife who died of TB; tried to give up drink; fell in love again and planned marriage; on the way north to bring his bride to the wedding at Richmond (where he was living), he stopped in Baltimore and, five days later, was discovered in a delirious condition near a saloon used as a voting-place—it is possible he was captured in a drunken condition by a mob who used him, as was then common, for the purpose of multiple voting.[266]

Poe aroused rage, derision, contempt, and indignation among right-thinking fellow-Americans. Emerson dismissed him as 'the Jingle Man.' James Russell Lowell (1819–91), his younger poetic contemporary, found his work 'three-fifths genius, two-fifths sheer fudge.' But, like Longfellow, though for wholly different purposes, he stuck thoughts, and still more images, into the minds not just of Americans but of people all over the world. Whether writing short stories or poems, his vivid and often horrific imagination worked powerfully on conscious and subconscious alike: 'The Pit and the Pendulum', 'The Raven' (for which he was paid $2), 'The Premature Burial,' 'The Gold Bug,' 'The Bells,' 'The Fall of the House of Usher,' 'The Murders in the Rue Morgue,' 'A Descent into the Maelstrom,' 'Annabel Lee,' 'A Dream within a Dream'—there are ineffaceable images here. His influence was enormous: he was the first American writer who had a major and continuing impact on Europe. Baudelaire, Verlaine, Bierce, Hart Crane, Swinburne, Rossetti, Rilke—and many others—felt his transforming fruitfulness.

In some ways Poe seemed very unAmerican. Baudelaire wrote: 'America was Poe's prison.' Lacan and Derrida, while purloining him, deAmericanized him also. But it can equally well be argued that Poe was very American: that he both reflected and inspired some of the horrors and fantasies of life in the continental country which was emerging in his day, its mystery and violence and contrasts and silences: also its crowds and loneliness. Cranky and melancholy, a solitary man in a vast continent of space, nostalgic for a smaller, warmer world, but also looking ahead to the future marvels and horrors, Poe did indeed respond to the Gothick side of American life, which grew fast in the 19th century. His work was also a huge depository of ideas and dreams, later to be mined by generations of American popular writers, especially authors of detective stories and crime-thrillers, but also the scriptwriters of Hollywood horror-movies and cartoons. The world of Walt Disney, without the germinating seeds of Poe, would have been tamer, safer, and less threatening. In short, Poe arrived at a time when American culture was suddenly becoming complex, difficult to define, moving out of easy control, and immeasurably more exciting—and he added fundamentally to this new excitement.[267]

It is notable in Poe's work that the hidden recesses of the mind, what might be called the psychological depths, are for the first time broached in American literature. But it is in the novels of Nathaniel Hawthorne that they begin to be thoroughly explored. In a sense Hawthorne was as American as it is possible to be. He was born in Salem. He came from a prominent Puritan family, who spelt the name Hathorne and who provided one of the judges at the witchcraft trials. His father was a New England sea-captain who died young of yellow fever, leaving Hawthorne's mother to lead a long life of eccentric seclusion, which had a profound effect on the young writer's own tender and bizarre imagination. All his life Hawthorne felt overshadowed by his puritan forebears, and by the guilt and secrecy they created, so his genealogy was a grave burden to him, which he sought to exorcize in his novels, the first to reflect the workings of the unconscious and to penetrate human psychology in its hidden recesses. He had a spell with fellow-oddities and writers amid the utopian onion-growing of Brook Farm, then spent most of his life as a customs official and consul. But always there was the shadow of guilt, and Salem.

Hawthorne transformed his Brook Farm experiences into *The Blithedale Romance*, and he used his judgmental ancestor as the villain in *The House with Seven Gables*. But it was the chance discovery of penitential material from the 17th century, found in the Salem Customs House, of which he was controller, that inspired his greatest, deepest, and most moving tale, *The Scarlet Letter*. It contains a key passage, written half a century before Sigmund Freud published his first book, which has been cited as containing the best and shortest summary of what the whole of psychotherapy is about:

> If the doctor possesses native sagacity, and a nameless something more— let us call it intuition; if he show no intrusive egotism, nor disagreeable prominent characteristics of his own; if he have the power, which must be born with him, to bring his mind into such affinity with his patient's, that this last shall unawares have spoken what he imagine himself only to have thought; if such revelations be received without tumult, and acknowledged not so often by an uttered sympathy, but silence, an inarticulate breath, and here and there a word, to indicate that all is understood; if, to these qualifications of a confidant be joined the advantages afforded by his recognised character as physician—then, at some inevitable moment, will the soul of the sufferer be dissolved, and flow forth in a dark, but transparent stream, bringing all its mysteries into the daylight.

In his own day, Hawthorne charmed readers, not merely by his *Tanglewood Tales* and other children's stories, but by his emphasis on the bliss and tenderness of happy married life—the Hawthorne of his own day was cherished for the moral, delicate, spiritual, sentimental, and 'exquisite' qualities of his writing. But, as D. H. Lawrence pointed out in his deeply, perhaps surprisingly, penetrating study of American literature published in 1923, Hawthorne was a complex protomodernist psychologist of the depths whose 'blood knowledge' throbbed beneath the surface of the 'sunbeams' which the readers of his own day loved.[268]

Hawthorne fitted into the kind of cultural gentility for which Longfellow stood, and even the meritocratic values Emerson trumpeted were not so remote from the comfortable middle-class lifestyle, based on family solidity, which Hawthorne seemed to epitomize, however much the hidden depths below them, which he examined, were full of future threats to mainstream American certitudes. But Walt Whitman (1819–92) was altogether harder for 19th-century America to rationalize or digest. He was born in the same year as Queen Victoria herself, from an old 17th-century founding family, with a touch of Dutch blood, which owned slaves until New York State abolished slavery. His father was a patriotic Long Island builder, who named three of Whitman's brothers George Washington, Andrew Jackson, and Thomas Jefferson. But of his father's eight children, one was defective, three were psychic disasters, and Walt was, or became, a homosexual. Homosexual acts were capital crimes in all the Thirteen Colonies, the Connecticut law code actually using the words of Leviticus (20:13) on which the legal condemnation was based: 'If a man also lieth with mankind, as he lieth with a woman, both of them have committed an abomination: they shall surely be put to death; their blood shall be upon them'—this statute, so worded, remained on the law books until well after Whitman's birth. At least five men were executed for sodomy in colonial times. After the Revolution, Jefferson proposed that the death sentence for such behavior be replaced by castration; but most states declined to follow his advice, and North Carolina retained the death penalty for sodomy until 1869. As late as 1897, a court in Illinois described sodomy as a crime 'not fit to be named among Christians.'[269] Fear of the law was one reason Whitman told so many lies about himself—stories of secret marriages, of children, both legitimate and illegitimate, of a mistress kept in New Orleans, of love affairs with women providing keys to his poetry, which have confused biographers.[270]

Whitman's homosexuality led to some furtive self-distancing from mainstream American life. But in many ways he was very much part of it. He took in, and wrote about, most aspects of 'modernity:' industrialization, life in a giant metropolis, working men, clerks, craftsmen trying to make a living, pushing themselves up the ladder in a big city like New York. Whitman was typical of the city's lower-middle-class intellectuals: a journeyman printer by trade, a journalist who worked on no fewer than ten publications, a bit of a schoolmaster. New York did not yet have apartments: it was a city either of houses or of boarding-houses. With a population of 325,000 in 1841, it had enough of these lodgings to accommodate 175,000. That was Whitman's life; and in addition he worked for a time at Tammany Hall, then itself a boarding house with a grubby dining-room as well as the Democratic Party HQ. Whitman began as a proper city clerk with stiff white collars and a full black suit; his name was 'Mr Walter Whitman.' Later he sank into bohemia, became 'Walt Whitman,' dressed down to proletarianize himself, adopted demotic habits and turns of speech, made his friends among laborers, tram-conductors, farm-boys, ferry-sailors, finally left the boarding-house world, and bought a house in a working-class area, 'a little old shanty of my own,' which he filled with disordered documents, a kind of paper nest of indescribable squalor, in the middle of which he sat, keeping his hat on invariably, like a Quaker. He had no tie; his suits were homespun. He was not the first major writer to create a deliberately eccentric image for purposes of systematic self-promotion—that innovation had been Rousseau's—but he set about it with an American thoroughness which was certainly new. Indeed, he was in some ways an early version, in literary guise, of what was to become an American archetype—the commercial salesman.

Whitman first published his central work, *Leaves of Grass*, in 1855, when it consisted of twelve poems and ninety-five pages. He republished it, with as much fanfare as he could muster, in 1856, with additions, and this process of republication continued until the sixth edition, in 1881, had 293 poems and 382 pages.[271] He reviewed his own poetry often, both anonymously and under pseudonyms, wrote articles about himself and promoted biographies. He planted news-stories. He said: 'The public is a thick-skinned beast and you have to keep whacking away on its hide to let it know you're there.' He was his own iconographer, promoting photos and portraits of himself and editing them. He built up his own biographical archive, a practice followed by Bertholt Brecht in the next century. He even designed his own tomb. He was the first American poet to employ free verse on a large scale, as a device for attracting attention, and the first to make a virtue of obscenity, thereby getting himself written about (and prosecuted). He conned Emerson into writing him a letter and then published it to boost himself. Emerson reacted by terming him 'half song-thrush, half alligator.' He described his own body as 'perfect,' a theme taken up by his votaries, who compared him to Christ; actually he was an ungainly youth who became an ugly old man. He got a letter from Tennyson but let it be known that it was so fulsome in his praise that modesty forbade him to publish it. He wrote a sixty-four-page promotional pamphlet to sell his third edition but did not acknowledge authorship till twenty-three years later. As visitors like Henry Thoreau discovered, he 'was not only eager to talk about himself but reluctant to have the conversation stray from the subject for long.' His crude literary behavior was termed by one Boston paper 'the grossest violation of literary Comity and courtesy that ever passed under our notice.'

All the same, Whitman demonstrated (as 'Papa' Hemingway was to do in the 20th century) that literary salesmanship and self-promotion, if pursued relentlessly and skillfully enough, can be as effective as any other kind. The first edition of *Leaves of Grass* sold only ten copies and Whitman had to give the rest away. But by the end of his life he was already a cult figure on both sides of the Atlantic, and his fame, and the interest in his work and personality, have continued to increase. He was, in short, despite his social and sexual heterodoxy, an all-American American, much more so, perhaps, than Longfellow, though unlike Longfellow's his verse has never been learned by heart and quoted—the one exception being his uncharacteristic 'O Captain! My Captain!' Of course Whitman's ascent to fame has been accelerated by the well-organized support of the homosexual community, who have presented him as the literary talisman of inversion, just as Oscar Wilde is in England. But the essence of Whitman's more general appeal is something quite different: he can plausibly be presented as the first apostle of poetic modernity.[272]

A country works hard and long, silently and obscurely, to achieve cultural maturity. But, when at last it comes, it comes suddenly, in a blinding flash, and thereafter all is changed for ever. Curiously enough, Emerson, who was very much part of this maturing process, summed it up brilliantly, in his volume of essays on genius, from Plato to Goethe, *Representative Men*: 'There is a moment in the history of every nation when, proceeding out of this brute youth, the perceptive powers reach their ripeness and have not yet become microscopic, so that man, at that instant, extends across the entire scale and, with his feet still planted on the immense forces of night, converses with his eyes and brain with solar and stellar creation. That is the moment of adult health, the culmination of power.'[273]

American literature's moment of adult health came, with great and unexpected force, in the first half of the 1850s. The key year was 1850, the Year of Debate, when not only *Representative Men* itself but Hawthorne's *The Scarlet Letter* were published, followed that autumn by *White-Jacket*, a novel by a self-made writer just coming into prominence: Herman Melville (1819–91), a New Yorker from an old but impoverished Anglo-Dutch family, who had been in turns bank clerk, store clerk, farmer, teacher, cabin-boy, whaler, naval seaman, and adventurer in the South Seas. *White-jacket* is the story of his life aboard a man-of-war. The next year, 1851, the summation of all his experiences, imagination, and energy, *Moby Dick*, telling the tale of the New England whalers, made its appearance, the first American fictional epic. The same year Hawthorne published his sinister *The House of the Seven Gables*. In 1852 Hawthorne followed it with his Brook Farm tale, *The Blithedale Romance*, and Melville with *Pierre*, both concerned with the American dilemma of combining idealism and practicality, and telling how, often enough, utopia is crushed by materialism.

It is the mark of a mature literature to produce unexpected works which are *sui generis*. This happened to America in 1854, when Henry Thoreau (1817–62), a Concord man of Puritan, Quaker, and Scotch stock, with a dash of Gallic blood, published his masterpiece, *Walden, or Life in the Woods*. Thoreau, a Harvard man, had been a teacher and assistant disciple to Emerson, describing himself as 'a mystic, a transcendentalist and a natural philosopher to boot.' From July 1845 to September 1847 he had lived in a hut he built near Walden Pond in the Concord woods, observing what transpired in nature and 'in the mind and heart of me.' His return to the simplicity of nature was interrupted (as he describes) by a day's imprisonment for refusing to pay a poll-tax to a government that was waging war against Mexico, a war he denounced as a mere scheme by slave-holders to extend slavery and enhance its political power. *Walden* is another book which could have come only from America, a work celebrating pioneering and closeness to nature in wild spaces, written by a tender and sophisticated scholar of Puritan descent.[274] To complete this American *pléiade*, Whitman's *Leaves of Grass* made its first appearance in 1855, as did Longfellow's *Hiawatha*.

Among these remarkable books, however, one stands out not so much for its literary quality as for its political influence. There has never been another book quite like *Uncle Tom's Cabin*. Originally published serially in the *National Era*, it appeared in book form on March 20, 1852, selling 10,000 copies in its first week and 300,000 by the end of the year. The sales in Britain were even higher; 1,200,000 within twelve months.[275] Its author, Harriet Beecher Stowe (1811–96), came from a sprawling Connecticut family of teachers and clerics, married a clerical professor herself, and produced a sprawling family of her own.

She took to magazine- and book-writing, like hundreds of other American and British matrons of her class—Mrs Trollope was a typical example—to make ends meet and give her children a few treats. She had already built up a substantial popular reputation before she exploded the bomb which was *Uncle Tom*. The force of the blast surprised no one more than herself. Curiously enough, she was not really an abolitionist, at least when she wrote the book, and knew little about the South. Her only direct experience of slavery was a short visit to Kentucky, itself a border state. She seems to have got most of her information about slavery from black women servants, especially her cook, Eliza Buck, and from the abolitionist literature.[276] It was not until the factual basis of her novel was challenged by angry Southerners that Mrs Stowe, helped by her brother, combed through newspaper reports of actual legal cases from the South. The result was a 259-page, densely printed compilation, *A Key to Uncle Tom's Cabin*, which Stowe published in 1853, showing that the cruelties and injustices of which the novel complained were, in reality, far more severe than she had imagined.

By then the book was not merely a bestseller; it was also a phenomenon. The sales in Britain were particularly significant. An immense Sunday School edition, at the equivalent of 25 cents, meant that British schoolchildren had their ideas of America shaped by Eliza, Tom, Eva, Topsy, Dinah, Miss Ophelia, Augustine St Clair, and Simon Legree. When Stowe came to Britain in 1853 she was lionized by all classes. She received delegations and thank-offerings from the poor, the leading novelist Charles Kingsley hailed her as the 'founder of American literature' and her book as 'the greatest novel ever written,' and the Duchess of Sutherland presented her with a solid gold bracelet in the form of a slave's shackle.[277] The truth is the British leaped at this opportunity to treat Americans, from whom they had received much preaching about democracy and equality, from visitors such as Senator Webster and Emerson, as morally suspect. So did the rest of the world, the novel being rapidly translated into more than forty languages. In Britain the success of the novel helped to ensure that, seven years later, the British, whose economic interest lay with the South, remained strictly neutral. But in the world as a whole it was the foundation stone of what, in the 20th century, became the mighty edifice of anti-Americanism.[278]

In the United States itself, the impact of the book was multiplied many times by the new American science of boosting and multimedia sales-pitching (a modern expression for what was, by the 1850s, a well-established process). The book was turned into statues, toys, games, handkerchiefs, wallpapers, cutlery, and plates. Its real popularity began when it appeared on the stage, in the form of songs and dramatized versions. 'Tom Shows' toured all the Northern and Western states. One of the highlights of the book, Eliza's escape, carrying her child across the Ohio into free territory, with the slave-catchers in close pursuit, became a key moment in early American drama. When the episode was staged at the National Theater in New York, an immense hush descended on the packed audience and an observer who looked around was astonished to see everyone, including society gentlemen and rough-shirted men in the galleries, in tears. Uncle Tom was the greatest tear-jerker of the 19th century, beating even the death of Little Nell in *The Old Curiosity Shop* (1841) and *Black Beauty* (1877).[279]

Stowe was lucky, in a way, to live at the time when American litera-ture was only just maturing and most of it was still crude or grossly imitative of English fashions. She was no stylist, she loved melodrama (her mentor was Scott), and some of her effects would, and indeed did, make even Dickens blush. But she stuck out because she wrote in the American language and her theme was the great issue which was already beginning to dominate American politics to the exclusion of almost everything else. There was also the additional *frisson* of a woman writing about atrocities hitherto regarded as unspeakable. Readers, especially men, were not sure whether it was proper for a woman novelist to acknowledge that slaves were stripped naked and beaten, that slave women were the sexual property of their masters, and that slave-owners habitually fathered children of all colors. In the South this was precisely the line of attack critics took. One wrote: 'Granted that every accusation brought by Mrs Stowe is perfectly true . . . the pol-lution of such literature to the heart and mind of women is not less.' The *Southern Quarterly* dismissed her work as 'the loathesome rakings of a foul fancy.' Another review read: 'The Petticoat lifts of itself and we see the hoof of the beast under the table.'[280]

Fortunately for Stowe, Northern readers did not think she had gone too far. They found her descriptions more credible, perhaps, just because she was a woman—more so than the highly colored atrocity stories of the emancipationist press, written almost entirely by men, usually clerics. This conviction turned *Uncle Tom* into the most suc-cessful propaganda tract of all time. It was widely believed that Mrs Stowe was responsible for Lincoln's election, and so for the chain of events which led to the bombardment of Fort Sumter. When the towering President received Stowe, who was under five feet, at the White House in 1862—would that we had a photograph of that encounter—Lincoln said to her: 'So you're the little woman who wrote the book that started this great war.'[281] But of course it was more complicated than that.

PART FOUR

'The Almost Chosen People'

Civil War America, 1850–1870

The civil war, in which are included the causes and consequences, constitutes the central event in American history. It is also America's most characteristic event which brings out all that the United States is, and is not. It made America a nation, which it was not so before. For America, as we have seen, was not prescriptive, its people forged together by a forgotten process in the darkness of prehistory, emerging from it already a nation by the time it could record its own doings. It was, rather, an artificial state or series of states, bound together by negotiated agreements and compacts, charters and covenants. It was made by bits of parchment, bred by lawyers. The early Americans, insofar as they had a nationality, were English (or more properly British) with an English national identity and culture. Their contract to become Americans—the Declaration of Independence—did not in itself make them a nation. On the contrary; the very word 'nation' was cut from it—the Southerners did not like the word. Significantly it was John Marshall, the supreme federalist, the legal ideologist of federalism, who first asserted in 1821 that America was a nation. It is true that Washington had used the word in his Farewell Address, but elliptically, and it was no doubt inserted by Hamilton, the other ideologue of federalism. Washington referred to 'the Community of Interest in one Nation,' which seems to beg the question whether America was a nation or not. And even Marshall's definition is qualified: 'America has chosen to be,' he laid down, 'in many respects and for many purposes, a nation.' This leads one to ask: in what respects, and for what purposes, was America not a nation? The word is not to be found in the Constitution. In the 1820s in the debates over the 'National Road,' Senator William Smith of South Carolina objected to 'this insidious word:' he said it was 'a term unknown to the origins and theory of our government.' As one constitutional historian has put it: 'In the architecture of nationhood, the United States has achieved something quite remarkable ... Americans erected their constitutional roof before they put up their national walls ... and the Constitution became a substitute for a deeper kind of national identity.'[1]

Yes; but whose Constitution: that as seen by the North, or the one which the South treasured—or the one, in the 1850s, interpreted by the Southern-dominated Taney Supreme Court? The North, increasingly driven by emancipationists, thought of the Constitution as a document which, when applied in its spirit, would eventually insure that all people in America, whatever their color, black or white, whatever their status, slave or free, would be equal before the law. The Southerners, by which I mean those who dominated the South politically and controlled its culture and self-expression, had a quite different agenda. They believed the Constitution could be used to extend not so much the fact of slavery—though it could do that too—but its principle. Moreover, they possessed, in the Democratic Party, and in the Taney Court, instruments whereby their view of the Constitution could be made to prevail. They were frustrated in this endeavor by their impetuosity and by their divisions—that is the story of the 1850s.

For the South, the decade began well. True, the California gold rush had been, from their point of view, a stroke of ill-fortune, since the slavery-hating miners who rushed there frustrated the South's plan of making California a slave state. But in some other respects the Compromise of 1850 worked in their favor. For one thing it made it possible for them to keep the Democratic Party united, and since 1828 that party had been the perfect instrument for winning elections. All it had to do, to elect a president of its choosing, was to hold the South together and secure a reasonable slice of the North; then, with their own man in the White House, appointing new Supreme Court judges, they could keep the South's interpretation of the Constitution secure too. For the election of 1852 the Democrats were able to unite round a campaign platform which promised 'to abide by and adhere to a faithful execution of the acts known as the Compromise Measures,' and for their candidate they picked a man peculiarly adapted to follow that line, 'a Northerner with Southern inclinations.'

Franklin Pierce (1804–69) was born in Hillboro, New Hampshire, had been to Bowdoin and practiced as a lawyer in Concord. So by rights he should have been an abolititionist and an Emersonian, a political Transcendentalist, and a thorough New Englander. But in reality he was a Jacksonian Democrat, another 'Young Hickory' and an ardent nationalist, all-out for further expansion into the crumbling Hispanic South, and thus to that extent a firm ally of the slavery-extenders. He had been a New Hampshire congressman and senator and had served assiduously in the Mexican War, of which (unusually in the North) he was an enthusiastic supporter, reaching the rank of brigadier-general. At the 1852 Democratic convention he emerged, after many votes, as the perfect Dark Horse compromise candidate, being nominated on the forty-ninth ballot. He is usually described as 'colorless.' When he was nominated, an old farmer-friend from New Hampshire commented: 'Frank goes well enough for Concord, but he'll go monstrous thin, spread out over the United States.' Nathaniel Hawthorne, who had been a close friend of Pierce at Bowdoin, called on Pierce after he was nominated, sat by him on the sofa, and said: 'Frank, what a pity . . . But, after all, this world was not meant to be happy in—only to succeed in.' This story is apocryphal, but Hawthorne said something similar to Pierce in a letter in which he undertook to write Pierce's campaign biography. Horace Mann, who knew both, said of the proposed biography, 'If he makes out Pierce to be a great man or a brave man, it will be the greatest work of fiction he ever wrote.' Hawthorne agreed: 'Though the story is true, it took a romancer to do it.'²

Hawthorne had to conceal two things: Pierce's drinking—it was said he drank even more than Daniel Webster, and he was certainly often drunk—and the fact that he hated Pierce's wife Jane. So did a lot of other people. The Pierces had two sons. Their four-year-old died in 1844; their surviving son was killed a month after the election in an appalling railroad accident, and Jane felt, and said, that the presidency had been bought at the cost of their son's life. Hawthorne burned documents about Pierce which were highly derogatory, commenting: 'I wish he had a better wife, or none at all. It is too bad that the nation should be compelled to see such a death's head in the pre-eminent place among American women; and I think a presidential candidate ought to be scrutinised as well in regard to his wife's social qualifications, as to his own political ones.' Jane was the daughter of the Bowdoin president and sister-in-law of its most distinguished professor: but women of academic families are not always congenial.³ The fact is, Hawthorne hated most women, particularly if they had intellectual pretensions, which Jane certainly did: he said of women writers, 'I wish they were forbidden to write on pain of having their faces scarified with an oyster-shell!'⁴ At any rate, *The Life of Franklin Pierce* duly appeared, the tale of 'A beautiful boy, with blue eyes, light curling hair, and a serene expression of face,' who grew up to be a distinguished military man and a conciliatory politician, anxious to preserve the Union by reassuring the South and appealing to 'the majority of Northerners' who were 'not actively against slavery' to beware of what Hawthorne called 'the mistiness of a philanthropic system.'[5]

Pierce won handsomely. The Whigs selected the Mexican War commander, General Winfield Scott, who like most generals was lost in the complex politics of ethnic America. He not only bellowed out his anti-slavery views, which the Whigs had allowed for, but turned out to be a strident nativist only happy with Americans of Anglo-Saxon stock, so he alienated the Germans and the Irish. In the end he carried only Tennessee, Kentucky, Vermont, and Massachusetts, giving Pierce a landslide in the electoral college, though his plurality over all the other candidates (there were four vote-splitters) was only 50,000.[6] In theory Pierce's Cabinet bridged North and South, since his Secretary of State, William Learned Marcy (1786–1857), was a member of the old Albany Regency, the New York politico who had egged on Jackson to enjoy 'the spoils of victory' in 1829. But Marcy did not care a damn about slavery and, as Polk's Secretary of War, had been a rabid architect of the war against Mexico. Again, Pierce's Attorney-General, Caleb Cushing (1800–79), though a Harvard–Massachusetts Brahmin, was primarily, like Marcy, a 'Manifest Destiny' man, and thus a Southern ally. On the other side, Pierce made Jefferson Davis (1808–89) secretary of war, and Davis was not merely a genuine Southerner but the future President of the Confederation. In practice, then, the Pierce administration was committed to policies which might have been designed to help the South.

The first expression of this policy was the Gadsden Purchase in 1853. This was Davis' idea, significantly. America was then discussing alternative possibilities for transcontinental railways and Davis was determined, for strategic as well as economic reasons, that the South should control one route. This required passage through a large strip of territory in what was then still northwest Mexico. Davis persuaded Pierce to send the South Carolina railroad promoter, Senator James Gadsden (1788–1858), to Mexico to promote the purchase of the strip. This was a dodgy business, as Gadsden had a financial interest in securing the purchase, which was made with US federal money—$10 million for 45,000 square miles—and the Senate agreed to ratify the deal only by a narrow margin, partly because this extra territory automatically became slave soil. Indeed Davis' original idea, that Gadsden should buy not only the strip but the provinces of Tamaulipas, Nuevo Leon, Coahuila, Chihuahua, Sonora, and the whole of Baja (lower) California, was also on the cards but not proceeded with as the Senate knew these vast territories would have been turned into several new slave states, and would never have ratified the deal, the Senate now having a Northern majority, or rather an anti-slave one.[7]

There were other possibilities for the South, however. They wanted Cuba, to turn it into an ideal slave state. 'The acquisition of Cuba,' wrote Davis, 'is essential to our prosperity and security.' He regretted that, in joining the Union, the Southern states had forfeited their right to make treaties and acquire new territories on their own, otherwise Cuba would already be in the Union, and slave soil. James Buchanan (1791–1866), who as Polk's secretary of state had been a leading mover in acquiring Texas, was now minister in London and intrigued and negotiated furiously in 1854 to have Cuba purchased and annexed. But nothing came of it—this was one of many occasions when Northerners in Congress frustrated the South's dream of an all-American, all-slave Caribbean.[8] There were various filibustering expeditions to seize by force what might be more difficult to acquire by diplomacy. Prominent in them was William Walker (1824–60), a Tennessee doctor and populist fanatic, who wanted to annex chunks of Latin America to the US, not to make them slave states but to give their peoples a taste of democracy. The 'gray-eyed man of destiny' entered Lower California in 1853 and proclaimed a republic, but Pierce was not hard-faced enough to allow that. Then Walker took his private army to Nicaragua and actually had himself recognized by the US in 1856. But that aroused the fury of another predator, Cornelius Vanderbilt (1794–1877), whose local transport system was being disrupted by Walker's doings, and as Vanderbilt had more money, he was able to force Walker to 'surrender' to the US Navy. Finally Walker turned to Honduras, but there the British navy took a hand and turned him over, as a nuisance, to a Honduran firing-squad.[9]

Now that the Gadsden Purchase made a Southern railway route to California geographically possible, others were looking for northern routes, and this too had an important bearing on the land strategy of the South. Senator Stephen Douglas of Illinois, who had helped Clay to draft the 1850 Compromise, was now chairman of the Senate Committee on Territories, and in that capacity he brought forward a Bill to create a new territory called Nebraska in the lands west of the Missouri and Iowa, the object being to get rails across it with an eastern terminus in the rapidly growing beef-and-wheat capital of Chicago. To appease the Southerners, he proposed to include in the Bill a popular sovereignty clause, allowing the Nebraskans themselves to decide if they wanted slavery or not. The South was not satisfied with this and Douglas sought to reassure them still further by not only providing for another territory and future state, Kansas, but repealing the old 1820 Missouri Compromise insofar as it banned slavery north of latitude 36.30. This outraged the North, brought up to regard the 1820 Compromise as a 'sacred pledge,' almost part of the Constitution. It outraged some Southerners too, such as Sam Houston of Texas, who saw that these new territories would mean the expulsion of the Indians, who had been told they could occupy these lands 'as long as grass shall grow and water run.' But Douglas, who wanted to balance himself carefully between North and South and so become president, pushed on; and President Pierce backed him; and so the Kansas–Nebraska Act passed by 113 to 100 in the House and 37 to 14 in the Senate, in May 1854.[10]

Backing this contentious Bill proved, for Pierce, a mistake and ruled out any chance of his being reelected. It also led to what might be called the first bloodshed of the Civil War. Nebraska was so far north that no one seriously believed it could be turned into a series of free states. Kansas was a different matter, and both sides tried to build up militant colonies there, and take advantage of the new law which stated its people were 'perfectly free to form and regulate their domestic institutions in their own way, subject only to the Constitution.' The first foray was conducted by the New England Emigrant Aid Society, which in 1855–6 sent in 1,250 anti-slavery enthusiasts. The Southerners organized just across the border in Missouri. In October 1854 the territory's first governor, Andrew H. Reeder, arrived and quickly organized a census, as prelude to an election in March 1855. But when the election came, the Missourians crossed the border in thousands and swamped the polls. The governor said the polls were a fraud but did nothing to invalidate the results, probably because he was afraid of being lynched. Territorial governors were provided by Washington with virtually no resources or money, as readers of Chapter 25 of Mark Twain's *Roughing It*—which describes the system from bitter experience—will know. At all events the slavers swept the polls, expelled from the legislature the few anti-slavers who were elected, adopted a drastic slave-code, and made it a capital offense to help a slave escape or aid a fugitive. They even made orally questioning the legality of slavery a felony.[11]

The anti-slavers, and genuine settlers who wanted to remain neutral, responded by holding a constitutional convention—elected unlawfully—drafted a constitution in Topeka which banned both slaves and freed blacks from Kansas, applied for admission to the US as a state, and elected another governor and legislature. Then the fighting began, a miniature civil war of Kansas' own. The Bible-thumping clergymen from the North proved expert gun-runners, especially of what were known as 'Beecher's Bibles,' rifles supplied by the bloodthirsty congregation of the Rev. Henry Ward Beecher. The South moved in guns too. In May 1856 a mob of slavers sacked Lawrence, a free-soil town, blew up the Free State Hotel with five cannon, burned the governor's house and tossed the presses of the local newspaper into the river. This in turn provoked a fanatical free-soiler called John Brown, a glaring-eyes fellow later described by one who was with him in Kansas as 'a man impressed with the idea that God has raised him up on purpose to break the jaws of the wicked.' Two days after the 'Sack of Lawrence,' Brown, his four sons, and some others rushed into Pottawatomie Creek, a pro-slavery settlement, and slaughtered five men in cold blood. By the end of the year over 200 people had been murdered in 'Bleeding Kansas.'[12]

The Lawrence outrage in turn provoked a breakdown of law in the Congress. The next day, May 22, Senator Charles Sumner (1811–74) of Massachusetts, a dignified, idealistic, humorless, and golden-tongued man who also had a talent for vicious abuse—the kind which causes wars—delivered a philippic in the Senate. One of the weaknesses of Congressional procedure was that, unlike the British parliament, where a speaker must go on until he finishes, senators were allowed an overnight respite then allowed to start again next morning, provoking their antagonized hearers beyond endurance. In his two-day speech, full of excitable sexual images, Sumner said what was going on in Kansas was 'the rape of a virgin territory [sprung] from a depraved longing for a new slave state, the hideous offspring of such a crime.' He made a particular target of Senator A. P. Butler of South Carolina, whom he accused of having 'chosen a mistress who . . . though polluted in the sight of the world, is chaste in his sight—I mean the harlot, slavery.' One cannot help feeling that, in the run-up to the Civil War, sex played a major, if unspoken, part. All Northerners knew, or believed, that male slave-owners slept with their pretty female slaves, and often bought them with this in mind. Abraham Lincoln, aged twenty-two and on his second visit to New Orleans, saw a young and beautiful teenage black girl, 'guaranteed a virgin,' being sold, the leering auctioneer declaring: 'The gentleman who buys her will get good value for his money.' The girl was virtually naked, and the horrific scene made a deep impression on the young man. Southerners denied they fornicated with their female slaves, but they also (contradicting themselves) accused their Northern tormenters of sexual envy, which may have been true in some cases.

In any event Sumner's metaphors were provocative. Butler's nephew, Congressman Preston S. Brooks, fumed over the insults for two days, then attacked Sumner with his cane while he was writing at his desk in the Senate. Sumner was so badly injured, or traumatized, that he was ill at home for two years, his empty Senate desk symbolizing the stop-at-nothing violence of the Southern slavers. Equally significant was that Brooks, having been censured by the House, resigned and was triumphantly reelected, his admirers presenting him with hundreds of canes to mark his 'brave gesture,' though it was in fact a cowardly assault on an unarmed, older man. Here was a case of unbridled and inflammatory Northern words provoking reckless Southern aggression—a paradigm of the whole conflict.[13]

Brooks' attack, and the support it received from the 'gentlemanly South,' reflected the aggressive politics of the slave states. The *Dred Scott* verdict by the Taney Court had given the South hope that the constitutional history of the country could be rewritten in a way that would make slavery safe for ever. All previous arrangements had left the South insecure—insecurity was at the very root of its violence. What the Southern militants, especially in South Carolina, wanted was a 'black code,' enacted by Congress and imposed on the territories. They were not so foolish as to hope they could reinstate slavery in New York and New England but they wanted abolitionism to be made illegal in some way. And they wanted not merely to open new territory in the South and West and outside the present borders of the US to slavery but also to reopen and relegalize the slave trade.

This forward plan received an important boost with the election of 1856. The Kansas–Nebraska Act destroyed the last remains of the crumbling Whig Party. In its place, phoenix-like, came the new Republicans, deliberately designed to evoke the memory of Jefferson, now presented as an anti-slaver, his attacks on slavery being eminently quotable, his ownership of slaves forgotten. At its nominating convention, the Republican Party passed over its chief anti-slaver, William H. Seward (1801–72), as too extreme, and picked John Charles Frémont (1813–90), a South Carolina adventurer who had eloped with the daughter of old Senator Benton and then had innumerable near-death escapes in California, including a capital conviction for mutiny quashed by President Polk.

The Republican slogan was 'Free Soil, Free Speech and Frémont.' The Democratic Party, rejecting Pierce as a sure loser, and Douglas as too all-things-to-all-men, picked James Buchanan, who concentrated on taking all the slave states and as much of the rest as he could. Old Fillmore, with Jackson's son-in-law Donelson as his running mate, popped up from the past as a splitter. That did for Frémont. So Buchanan, with a fairly united Democratic Party behind him, carried all the South plus New Jersey, Pennsylvania, Illinois, Indiana, and California, making 174 college voters, against Frémont's 114. Buchanan was elected on a minority (45.3 percent) of the vote but his plurality over Frémont was wide, 1,838,169 to 1,341,264.

The new President was at heart a weak man, and a vacillating one, but he was not out of touch with the combination of imperialist and Southern opinion which, well led, would have ruled out any prospect of coercion of the South by the North. Whatever he said in public, Buchanan sympathized with the idea of adding new states to the South, even if slavers. In his message to Congress, January 7, 1858, Buchanan criticized Walker's filibustering in Nicaragua not because it was wrong in itself but because it was impolitic and 'impeded the destiny of our race to spread itself over the continent of North America, and this at no distant day, should events be permitted to take their natural course.' He followed this up by asking Congress to buy Cuba, despite the fact that the Spanish were demanding at least $150 million for it (the Republicans blocked the plan). America had absorbed what was once Spanish-speaking territory of millions of square miles in California and Texas: why not the whole of Mexico and Central America? That was all part of the 'North American Continent,' to which the US was 'providentially entitled' by its Manifest Destiny.[14]

Moreover, the price of slaves was rising all the time, despite the efforts of the Virginia slave-farms to produce more, and this in turn strengthened demands for a resumption of the slave-trade. Slave-smuggling was growing, and it was well known, and trumpeted in the South, that merchants in New York and Baltimore bought slaves cheap on the West African coast, and then landed them on islands off Georgia and other Southern states. So why not repeal the 1807 Act and legalize the traffic? That was the demand of the governor of South Carolina in 1856, and the Vicksburg Commercial Convention of 1859 approved a motion resolving that 'all laws, state or federal, prohibiting the African slave trade, ought to be repealed.' The first step, it was argued, was to have blacks captured from slave-ships stopped and searched by the US Navy—the current practice was to send them, free, to Liberia, which most of them did not like—sent to the South and 'apprenticed' to planters with good records. Representative William L. Yancey of Alabama asked: 'If it is right to buy slaves in Virginia and carry them to New Orleans, why is it not right to buy them in Cuba, Brazil or Africa, and carry them there?' If blacks would rather be slaves in the South than free men in Liberia, might it not be that other African blacks would prefer to come to the South, as slaves, rather than remain in the 'Dark Continent,' where their lives were so short and cheap?

Southerners argued that to take a black from Africa and set him up in comfort on a plantation was the equivalent, allowing for racial differences, of allowing a penniless European peasant free entry and allowing him, in a few years, to buy his own farm. The *Dred Scott Case*, by declaring the Missouri Compromise unconstitutional, and the Kansas–Nebraska Act together opened up enormous new opportunities for setting up slave-plantations and ranches, and therefore increased the demand for slaves. Southerners argued that by resuming the slave-trade the cost of slaves in America would be sharply reduced, thereby boasting the economy of the whole country. The aggressive message of the South was: slavery must be extended because it makes economic sense for America. But beneath this aggressive tone was the deep insecurity of Southerners who had no real moral answer to the North's case and knew in their hearts that the days of slavery were numbered.[15]

That sense of insecurity was justified, because in the late 1850s it became obvious that dreams of a vast expansion of slavery to the west and into the Caribbean and other Hispanic areas were fantasies, and the reality was a built-in and continuing decline of Southern political power. Calhoun, in almost his dying words in 1850, had warned the South that if they did not act soon, and assert his theory of states' rights, if necessary by force, they were doomed to a slow death: they would never be stronger than they were, and could only get weaker. That was demonstrated to be good advice; in May 1858 the free state of Minnesota entered the Union, followed by another free state, Oregon, in February 1859, while Kansas, being a slave territory, was denied admission. So the Congressional balance, as Calhoun had foreseen, was destroyed for ever. The South was now outvoted in the Senate 36 to 30 and in the House the gap was enormous, 147 to 90.

Southerners' sense of insecurity was deepened by the fact that, while they boasted publicly that 'Cotton is King' and 'The Greatest Staple in the World,' they were painfully aware of the weaknesses of their cotton-slave economy. Most plantations were in debt or operated close to the margins of profitability. During the 1850s, world cotton prices tended to fall. More and more countries were producing raw cotton—a trend which would knock large nails in the South's coffin when the war began. In the light of economic hindsight, it can be seen that the plantation system, as practiced, was fundamentally unsound, and some planters grasped this at the time. Plantations absorbed good land and ruined it, then their owners moved on. There was an internal conflict in the South, as the newer estates in the Deep South were more scientific and efficient (and bigger), and thus tended to take black slave labor away from the tidewater and border areas, and push up the price of slaves. This, at a time of falling cotton prices, put further pressure on profit margins.[16] As the price of slaves rose, slavery as an institution became more vital to the South: to the Deep South because they used slaves more and more efficiently, to the Old and border South because breeding high-quality, high-priced slaves was now far more important than raising tobacco or cotton. Professor Thomas R. Dew of William and Mary College, in his book *The Pro-Slavery Argument* of 1852, asserted: 'Virginia is a *negro-raising* state for other states: she produces enough for her own supply and 6,000 [annually] for sale.'

Actually, Virginia was living on its slave-capital: blacks formed 50 percent of the Virginia population in 1782, but only 37 percent in 1860s—it was selling its blacks to the Deep South. Virginia and other Old South or border states concentrated on breeding a specially hardy type of negro, long-living, prolific, disease-free, muscular, and energetic. In the 1850s, about 25,000 of these blacks were being sold, annually, to the Deep South.[17] The 1860 census showed there were 8,099,000 whites in the South and 3,953,580 slaves. But only 384,000 whites owned the slaves: 10,781 owned fifty and more; 1,733, a hundred and more. So over 6 million Southern whites had no direct interest in slavery. But that did not mean they did not wish to retain the institution—on the contrary: poor whites feared blacks even more than the rich ones did. By 1860 there were already 262,000 free blacks in the Southern states, competing with poor whites for scarce jobs, and a further 3,018 were manumitted that year. Poor whites were keener than anyone on penal legislation against slaves: they insured no state recognized slave marriage in law, and five states made it unlawful to teach slaves to read and write. In any event, small white farmers in the South were very much at the mercy of the big plantation owners and had to go along with them.[18] Those who produced cotton, rice, sugar, tobacco, and slaves on a large scale were all-powerful. As one historian has put it, 'There was never in America a more perfect oligarchy of business-men.'[19]

Slavery was not the only issue between North and South. Indeed it is possible that an attempt at secession might have been made even if the slavery issue had been resolved. The North favored high tariffs, the South low ones; the North, in consequence, backed indirect taxation, the South direct taxation. It is significant that once the war began, the North, shorn of the South, immediately introduced high tariffs with the Morrill Act of 1861, and pushed through direct federal income tax too. There were huge differences of interest over railroad strategy. Increasingly, the railroad interests of the Northeast and the Northwest came into align-ment in the 1850s, and this in turn led to an alliance between Eastern manufacturers seeking high tariffs and Western farmers demanding low-cost or free lands—both linked by lines of rail. This was the basis of the power of the new Republican Party, and the South saw it as a plot—indeed, it was what finished them. Many Southerners believed deeply in their hearts that the moral indignation of the North was spurious, mask-ing meaner economic motives. As Jefferson Davis put it, 'You free-soil agitators are not interested in slavery . . . not at all . . . It is so that you may have an opportunity of cheating us that you want to limit slave territory within circumscribed bounds. It is so that you may have a majority in the Congress of the United States and convert the govern-ment into an engine of Northern aggrandisement . . . you desire to weaken the political power of the Southern states. And why? Because you want, by an unjust system of legislation, to promote the industry of the North-East states, at the expense of the people of the South and their industry.'[20]

Davis was reflecting a bitter conviction held by all 'thinking' men in the South: that the North, while accusing the South of exploiting the blacks, exploited the whole of the South systematically and without mercy. Their feeling was exactly the same as the resentment felt by the Third World towards the First World today. There was something inherent in a plantation economy which put it in a dependent position, with the capitalist world its master. There was, of course, no control by the state of national production and prices, of cotton or anything else. If world markets were high, profits rose, but there was then a tendency to reinvest them in increased production. If prices fell, the planters had to borrow. In either case, the South lacked liquid capital. So the planters fell into the hands of bankers, ending up dependent on New York or even the City of London.[21] The South lacked its own financial system, like the Third World today. When cotton made big profits, it spent them, as the Arab rulers today dissipate colossal oil revenues. And it was in a real sense milked, like the primary producers today in Africa and Latin America, at the same time accumulating massive debts it had no hope of repaying. In effect, the South had all the disadvantages of a one-crop economy. It had only 8 percent of US manufactures. It should have put up the money to open factories, and so provide employment for poor whites and diversify its economy at the same time. But there was no spare capital in the South itself, and the North had no intention of building factories there and competing against itself with low-wage, low-price products. So the South saw itself as the slave of a Union dominated by Northern capital. As the *Charleston Mercury* put it: 'As long as we are tributaries, dependent on foreign labor and skill for food, clothing and countless necessities of life, *we are in thralldom.*'[22]

The Civil War was not only the most characteristic event in American history, it was also the most characteristic religious event because both sides were filled with moral righteousness for their own cause and moral detestation of the attitudes of their opponents. And the leaders on both sides were righteous men. Let us look more closely at these two paladins, Abraham Lincoln and Jefferson Davis. Lincoln was a case of American exceptionalism because, in his humble, untaught way, he was a kind of moral genius, such as is seldom seen in life and hardly ever at the summit of politics. By comparison, Davis was a mere mortal. But, according to his lights, he was a just man, unusually so, and we can be confident that, had he and Lincoln been joined in moral discussion, with the topic of slavery alone banned, they would have found much common ground.

Both men were also characteristic human products of mid–19th-century America, though their backgrounds were different in important respects. Lincoln insisted he came from nowhere. He told his campaign biographer, John Locke Scripps of the *Chicago Tribune*, that his early life could be 'condensed into a single sentence from Gray's *Elegy*, "The short and simple annals of the poor." ' He said both his parents were born in Virginia and he believed one of his grandfathers was 'a Southern gentleman.' He also believed his mother was illegitimate, probably rightly. He was born in a log cabin in the Kentucky backwoods and grew up on frontier farms as his family moved westwards. His father was barely literate; his mother taught him to read, but she died when he was nine. Thereafter he was self-taught. His father remarried, then took to hiring out his tall lanky (six feet four and 170 pounds) son, for 25 cents a day. He said of his son: 'He looked as he had been rough-hewn with an axe and needed smoothing down with a jackplane.' Lincoln acquired, in the backwoods of Kentucky, Indiana, and Illinois, and on the Ohio and the Mississippi, an immense range of skills: rafting, boating, carpentry, butchering, forestry, store- keeping, brewing, distilling, plowing. He did not smoke, chew tobacco, or drink. He acquired an English grammar, and taught it to himself. He read Gibbon, *Robinson Crusoe*, Aesop, *The Pilgrim's Progress*, and Parson Weems' lives of Washington and Franklin. He learned the *Statutes of Illinois* by heart. He rafted down to New Orleans and worked his way back on a steamer. He visited the South several times and knew it, unlike most Northerners.[23] He listened often to Southerners defend the 'Peculiar Institution' and knew their arguments backwards; what he had personally witnessed made him reject them, utterly, though he never made the mistake of thinking them insincere or superficial. He loved Jefferson, Clay, and Webster, in that order. He was a born storyteller, a real genius when it came to telling a tale, short or long. He knew when to pause, when to hurry, when to stop. He was the greatest coiner of one-liners in American history, until Ronald Reagan emerged to cap him. He was awkward—he always put his whole foot flat down when walking, and lifted it up the same way—but could suddenly appear as if transfigured, full of elegance. With one hand he could lift a barrel of whiskey from floor to counter. He was hypochondriac, as he admitted. He wrote an essay on suicide. He said: 'I may seem to enjoy life rapturously when I am in company. But when I am alone I am so often so overcome by mental depression that I never dare carry a penknife.'[24]

Lincoln was a self-taught lawyer but his instincts were not for the cause. He said 'persuade your neighbors to compromise whenever you can ... As a peacemaker, the lawyer has a superior opportunity of being a good man. There will still be business enough. A worse man can scarcely be found than one who [creates litigation].' As a circuit lawyer, Lincoln fancied himself a Whig and stood for the state legislature. His first elective post, however, was as captain of volunteers in the Black Hawk War (1832), in which he came across five scalped corpses in the early morning: 'They lay heads towards us on the ground. Every man had a round red spot on the top of his head about as big as a dollar where the redskins had taken his scalp. It was frightful. But it was grotesque. And the red sunlight seemed to paint everything over.' But he held no grudge; indeed he saved an Indian from being butchered. He was the first man to refer to Indians as 'Native Americans,' though in the then current usage the term referred to Americans of old Anglo-Saxon stock. He said to those who protested about German immigrants, and claimed the title for themselves: 'Who are the [real] Native Americans? Do they not wear the breechclout and carry the tomahawk? We pushed them from their homes and now turn on others not fortunate enough to come over so early as we or our forefathers.'[25]

He did not win his first political election. And he had bad luck. He bought a store and set up as postmaster too. His partner, Berry, fled with the cash and Lincoln had to shoulder a $1,100 burden of debt. Like Washington, he went into land-surveying to help pay it off. Then he was elected to the state assembly, serving eight years from the age of twenty-five to thirty-two. It met in Vandalia, its eighty-three members being divided into two chambers. Lincoln was paid $3 for each sitting, plus pen, ink, and paper. His first manifesto read: 'I go for all sharing the privileges of government who assist in sharing its burdens. Consequently I go for admitting all whites to the right of suffrage who pay taxes or bear arms (by no means excluding females).' He belonged to a group of Whig legislators who were all six feet or over, known as the Long Nine. He got the state capital shifted to Springfield and there set up a law practice, making his name by winning a case for an oppressed widow. A colleague said: 'Lincoln was the most uncouth-looking man I ever saw. He seemed to have but little to say, seemed to feel timid, with a tinge of sadness visible in his countenance. But when he did talk all this disappeared for the time, and he demonstrated he was both strong and acute. He surprised us more and more at every visit.'[26]

Lincoln's first love, Ann Rutledge, died of typhoid fever. That Lincoln was devastated is obvious enough; that his love for her persisted and prevented him from loving any other woman is more debatable.[27] At all events, it is clear he never loved the woman he married, Mary Todd. She came from a grand family in Kentucky, famous since Revolutionary days for generals and governors. She was driven from it by a horrible stepmother, but never abandoned her quest for a man she could marry in order to make him president. Oddly enough, she turned down Stephen Douglas, then a youngish fellow-member of the Illinois

Assembly, in favour of Lincoln, whom she picked out as White House timber. She said to friends: 'Mr Lincoln is to be president of the United States some day. If I had not thought so, I would not have married him, for you can see he is not pretty.' Lincoln consented, but missed the wedding owing to an illness which was clearly psychosomatic. This led to a sabre duel with Sheilds, the state auditor, which was called off when Lincoln scared his opponent by cutting a twig high up a tree. And this in turn led to reconciliation with Mary, and marriage, he being thirty-three, she twenty-four. His law partner, William H. Herndon, said: 'He knew he did not love her, but he had promised to marry her.'

It was an uncomfortable marriage of opposites, particularly since she had no sense of humor, his strongest suit. He liked to say: 'Come in, my wife will be down as soon as she gets her trotting-harness on.' He was a messy man, disorderly in appearance, she was a duster and polisher and tidier. She wrangled acrimoniously with her uppity white servants and sighed noisily for her 'delightful niggers.' 'One thing is certain,' she said, 'if Mr Lincoln should happen to die, his spirit will never find me living outside the boundaries of a slave state.' She hated his partner, his family, and his so-called office. Herndon said: 'He had no system, no order; he did not keep a clerk; he had neither library, nor index, nor cash-book. When he made notes, he would throw them into a drawer, put them in his vest-pocket, or into his hat . . . But in the inner man, symmetry and method prevailed. He did not need an orderly office, did not need pen and ink, because his workshop was in his head.'

The Lincolns had four sons. Generations of Lincoln-admirers have played down the role of Mary in his life and career, easily finding spicy material illustrative of her shortcomings. But the likelihood is that he would never have become president without her. It took him four years, aged thirty-three to thirty-seven, to get into Congress, and but for her endless pushing he might have become discouraged. For his part, he did his best to behave to her gallantly. There is a touching photograph of her, taken in 1861, arrayed in her inaugural finery, wearing pearls. They were a set which Lincoln had just bought for her, paying $530, at Tiffany's store on 550 Broadway: a seed-pearl necklace and matching bracelets for each arm. They are now in the Library of Congress.[28]

Lincoln won a seat for Congress in 1847, by a big majority. The Whig Party gave him $200 for his expenses. He handed back $199.25, having bought only one barrel of cider. He rode to Washington on his own horse, and stayed with friends. But he served only one term—his opposition to the Mexican War did for him. He recalled that at the foot of the Capitol, within sight of its windows, was 'a sort of negro stable where gangs of negroes were sold, and sometimes kept in store for a time pending transport to the Southern market, just like horses.' Lincoln was broad-minded, tolerant, and inclined to let things alone if possible, but he found this insult to the eye of freedom, literally within sight of Congress, 'mighty offensive.' The first law he drafted was a Bill to Abolish Slavery in the District of Columbia, to be enacted by local referendum (as we have seen it became part of the 1850 Compromise). At the end of his term, he returned contentedly to the law.[29]

But the slavery issue would not let him rest, or stay out of politics. It was even more persistent than Mary Lincoln's pushing. Some notes have survived of his musings:

If A can prove, however conclusively, that he may, of right, enslave B, why may not B snatch the same argument, even prove equally, that he may enslave A? You say A is white and B is black—is it *color* then, the lighter having the right to enslave the darker? Take care—by this rule, you are to be slave to the first man you meet, with a fairer skin than your own. You do not mean *color* exactly? You mean the whites are *intellectually* the superior of the blacks, and therefore have the right to enslave them? Take care again—by this rule you are to be the slave of the first man you meet, with an intellect superior to your own.[30]

As Herndon said, 'All his great qualities were swayed by the despotism of his logic.' There are many memorable descriptions of him lost in thought, turning things over in his mind.

Lincoln did a lot of this musing at home, a place in which he kept a low profile. Mary Lincoln said: 'He is of no account when he is at home. He never does anything except to warm himself and read. He never went to market in his life. I have to look after all that. He just does nothing. He is the most useless, good-for-nothing man on earth.' He replied, in his own way: 'For God, one "d" is enough, but the Todds need two.' He was often driven from his own house by Mary's anger. There are no fewer than six eyewitness descriptions of her furies, one relating to how she drove him out with a broomstick. He was never allowed to ask people to a meal, even or rather especially his parents. He wrote: 'Quarrel not at all. No man resolved to make the most of himself can spare time for personal contention ... Yield larger things to which you can show no more than equal right; and yield lesser ones, though clearly your own.' Mary felt his righteousness as well as his awkwardness: 'He was mild in his manner,' she said, 'but a terrible firm man when he set his foot down. I could always tell when, in deciding anything, he had reached his ultimatum. At first he was very cheerful, then he lapsed into thoughtfulness, bringing his lips together in a firm compression. When these symptoms developed, I fashioned myself accordingly, and so did all others have to do, sooner or later.'[31]

That Lincoln, as his wife implied, had a huge will when intellectually roused to a moral cause is clear. This sprang from a compulsive sense of duty rather than ambition as such. The evidence suggests that he was obliged to reenter politics not because he was an anti-slavery campaigner but because, in the second half of the 1850s, the slavery issue came to dominate American politics to the exclusion of almost everything else. Each time the issue was raised, and Lincoln was obliged to ponder it, the more convinced he became that the United States was uniquely threatened by the evil, and its political consequences. In those circumstances, an American who felt he had powers—and Lincoln was conscious of great powers—had an inescapable duty to use them in the Union's defense. Lincoln did not see slavery in religious terms, as the 'organic sin' of the Union, as the Protestant campaigners of the North put it. Those close to him agreed he had no religious beliefs in the conventional sense. His wife said: 'Mr Lincoln had no faith and no hope in the usual acceptation of those words. He never joined a church. But still, I believe, he was a religious man by nature ... it was a kind of poetry in his nature.' Herndon said Lincoln insisted no personal God existed and when he used the word God he meant providence: he believed in predestination and inevitability.[32]

Lincoln came closer to belief in God, as we shall see, but in the 1850s he was opposed to slavery primarily on humanitarian grounds, as an affront to man's natural dignity; and this could be caused by religious sectarians as well as by slave-owners. In his boyish and youthful reading, he had conceived great hopes of the United States, which he now feared for. He wrote: 'Our progress in degeneracy appears to me to be pretty rapid. As a nation we began by declaring that "all men are created equal." We now practically read of "all men are created equal except negroes." When the Know-Nothings get control it will be "all men are created equal except negroes and foreigners and Catholics." When it comes to this, I shall prefer emigration to some country where they make no pretence of loving liberty—to Russia, for instance, where despotism can be taken pure, without the base alloy of hypocrisy.'³³ The state of America caused him anguish. He said to Herndon: 'How hard it is to die and leave one's country no better than if one had never lived for it! The world is dead to hope, deaf to its own death-struggle. One made known by a universal cry, what is to be done? Is anything to be done? Who can do anything? And how is it to be done? Do you never think of these things?'³⁴

But from this general sense of downward moral plunging, which had to be arrested, the slavery issue, and still more the South's determination to extend and fortify it, loomed ever larger. In an important letter to Joshua F. Speed, the storekeeper with whom he shared some of his most intimate thoughts, Lincoln dismissed the claim that slavery was the South's affair and Northerners 'had no interest' in the matter. There were, he said, many parts of the North, in Ohio for instance, 'where you cannot avoid seeing such sights as slaves in chains, being carried to miserable destinations, and the heart is wrung. It is not fair for you to assume that I have no interest in a thing which has and continually exercises, the power of making me miserable.' Lincoln was as much concerned for the slave-owner as for the slave—the institution morally destroyed the man supposed to benefit from it. It was thus more important, as Lincoln saw it, to end slave-owning than to end slavery itself. He said a Kentuckian had once told him: 'You might have any amount of land, money in your pocket, or bank stock, and while traveling around nobody would be any the wiser. But if you have a darky trudging at your heels, everybody would see him and know you owned a slave. It is the most glittering property in the world. If a young man goes courting, the only inquiry is how many negroes he, or she, owns. Slave-ownership betokens not only the possession of wealth but indicates the gentleman of leisure, who is above labor and scorns it.'³⁵ This image of the strutting slave-owner, corrupted and destroyed by the wretch at his heels, haunted Lincoln. He wept for the South in its self-inflicted moral degradation.

It was because slavery made him miserable, and because he thought it was destroying the nation, not least the South, that Lincoln reentered politics and helped to create the new Republican Party, primarily to prevent slavery's extension. Looking back with the hindsight of history, we tend to assume that slavery was a lost cause from the start and the destruction of the old South inevitable. But to a man of Lincoln's generation, the South appeared to have won all the political battles, and all the legal ones. So long as the Democratic Party remained united, the

South's negative grip on the United States seemed unbreakable, and its power to make positive moves was huge. The creation of the Republican party, from free-soilers, Whigs, and many local elements, was the answer to the Democratic stranglehold on the nation, which had been the central fact of American political life since 1828. Lincoln failed to get into the Senate in 1855 and (as we have seen) Buchanan won the presidency in 1856. But it was by then apparent that the Republican Party was a potential governing instrument, and Lincoln's part in creating it was obvious and recognized.

At Bloomington on May 29, 1856, when the new Illinois Republican Party was inaugurated, Lincoln was called to make the adjournment speech and he responded with what all agreed was the best speech of his life. It was so mesmerizing that many reporters forgot to take it down. Even Herndon, who always took notes, gave up after fifteen minutes and 'threw pen and paper away and lived only in the inspiration of the hour.'[36] Lincoln argued that the logic of the South's case, which was that slavery was good for the negroes, would be to extend it to white men too. Because of the relentless pressure of the South's arguments, Northerners like Douglas, Lincoln warned, were now yielding their case of 'the individual rights of man'—'such is the progress of our national democracy.' Lincoln said it was therefore urgent that there should be a union of all men, of whatever politics, who opposed the expansion of slavery, and said he was 'ready to fuse with anyone who would unite with him to oppose slave power.' If the united opposition of the North caused the South 'to raise the bugbear of disunion,' the South should be told bluntly, *'the union must be preserved in the purity of its principles as well as in the integrity of its territorial parts.'* And he updated the reply of Daniel Webster to the South Carolina nullifiers, as the slogan of the new Republican Party: 'Liberty and Union, now and forever, one and inseparable.'[37] One eyewitness said: 'At this moment, he looked to me the handsomest man I had ever seen in my life.' Herndon recalled: 'His speech was full of fire and energy and force. It was logic. It was pathos. It was enthusiasm. It was justice, equity, truth and right set alight by the divine fires of a soul maddened by the wrong. It was hard, heavy, knotty, gnarly, backed with wrath.'[38]

It was now only a matter of time before Lincoln became the champion of the new Republicans. The Senatorial election of 1858 in Illinois, when he was pitted against Douglas, the 'Little Giant,' provided the opportunity. On June 16 Lincoln, having been nominated as Republican candidate, laid down the strategy at the state convention in Springfield. Together with the Bloomington speech, it represents the essence of Lincoln's whole approach to the complex of political issues which revolved round slavery. He said that all attempts to end both the South's agitation for the right to extend slavery and the North's to abolish it had failed, and that the country was inevitably moving into crisis:

> A house divided against itself cannot stand. I believe this government cannot endure half *slave* and half *free*. I do not expect the Union to be *dissolved*. I do not expect the House to *fall*. But I *do* expect it will cease to be divided. It will become *all* one thing, or *all* the other. Either the *opponents* of slavery will arrest the further spread of it, and place it where the public mind shall rest in the belief that it is in course of ultimate extinction; or its *advocates* will push it forward, till it shall become alike lawful in *all* the states, *old* as well as *new*, *North* as well as *South*. [Emphasis Lincoln's.]

The burden of the speech was a masterly summary of the legal and constitutional threats represented by the *Dred Scott* decisions and the Kansas–Nebraska Act, and Lincoln challenged Douglas—his main opponent in the state—to say clearly where he stood on both these issues. Lincoln said of his speech: 'If I had to draw a pen across my record, and erase my whole life from sight, and if I had one poor gift or choice left as to what I should save from the wreck, I should choose that speech and leave it to the world unerased.'[39]

Lincoln was right to put his finger on Douglas, for he represented the spirit of compromise where it was no longer possible—where further attempts to evade the dread issue would play into the hands of the South and sell the pass. Lincoln objected strongly to Horace Greeley's plan to get Douglas into the Republican Party. He saw Douglas as an unprincipled man motivated solely by ambition. Eventually both North and South came round to Lincoln's view. But in 1858 Douglas was a much weightier politician than Lincoln, albeit a younger man. Only five feet high, but muscular and stocky, he was the son of a doctor but had done many things—laborer in his teens, a teacher at twenty, a lawyer at twenty-one, a state legislator and Secretary of State of Illinois, a judge of its supreme court, then a congressman, a senator before he was forty, a European traveler who had been received by the Tsar of Russia and the Queen of England, a rich man who had married two Southern heiresses. He traveled in princely fashion, by special train or coach, with a truck and field gun behind, which fired a salute when he arrived in any place he was due to speak. He drove to his engagements in a carriage with six horses and with thirty-two outriders. So Douglas was a grand man who looked down his nose at the uncouth Lincoln. But Lincoln was cunning when he wished to be. Annoyed by the conservative *Springfield Journal*, he persuaded it to publish an apology for Southern slavery and so ruined its reputation among right-thinking Illinois readers—it went out of business. Determined to get maximum publicity for his House Divided strategy, he provoked and teased and inveigled Douglas into giving him a series of public debates, from which Lincoln had everything to gain and very little to lose.

The Lincoln–Douglas debates were a series of seven encounters, August–October 1858, conducted throughout the state, with the Senate seat the prize. They were preceded and followed by bands and processions and attracted crowds of 10,000 or more, entire families traveling up to 30 miles to attend them. Both men were good debaters and they made a striking contrast of style, Douglas, meticulously dressed, exuding vigor, Lincoln shambling and awkward in word and gesture, then suddenly, without warning and for brief seconds, becoming godlike in his majestic passion. Douglas won the seat. But the debates eventually finished him, while they transformed Lincoln into a national figure. They were, also, an important process in educating the North in the real issues at stake, and this was of far greater historical importance than the Clay–Webster–Calhoun encounters of 1850.[40]

The strength of Douglas was his warning that the path Lincoln was treading could lead to sectional discord on a scale the country had never known, and possibly civil war. His weakness was that he was never really prepared to say where he stood on slavery and was thus exposed, in debate, as trying to be all things to all men. He said: 'I do not care whether the vote goes on for or against slavery. That is only a question of dollars and cents. The Almighty himself has drawn across this continent a line on one side of which the earth must be for ever tilled by slave labor, whereas on the other side of that line labor is free.' Northerners might accept this—indeed had always accepted it—as a convenient or inescapable fact—but they did not want it spelled out. To do so sounded amoral or even immoral. And most Americans, then as now, wanted to sound moral. Then again, Douglas said: 'When the struggle is between the white man and the negro, I am for the white man. When it is between the negro and the crocodile, I am for the negro.' That too played into Lincoln's hands: it was a remark which would do for a saloon but not for a public platform. Lincoln rightly saw that the debate, the entire controversy, had to be conducted on the highest moral plane because it was only there that the case for freedom and Union became unassailable. He pointed out again and again that even the South was, in its heart, aware that slavery was wrong. The United States had made it a capital offense half a century ago to import slaves from Africa, and that fact, over the years, had wormed its way into Southern attitudes, however much they might try to defend slavery. Hence, even in the South, the slave-dealer was treated with abhorrence. Slave-owners would not let their children play with his—though they would cheerfully see them playing with slave-children. And the South knew that not only slave-dealing was wrong but slavery itself— why else did they manumit: 'Why have so many slaves been set free, except by the promptings of conscience?' As for the *Dred Scott* decision, it was an aberration, which would shortly be set right, at the next presidential election: 'You can fool all the people some of the time, and some of the people all the time, but you cannot fool all of the people all of the time.'[41]

Lincoln's object was not merely to put his name and his case before the American people, as well as Illinois voters. It was also to expose the essential pantomime-horse approach of a man who tried to straddle North and South. He succeeded in both. He put to Douglas the key question: 'Can the people of a United States territory, in any lawful way, against the wish of a citizen of the United States, exclude slavery from its limits prior to the formation of a state constitution?' If Douglas said yes, to win Illinois voters, he lost the South. If he said no, to win the South, he lost Illinois. Douglas' answer was: 'It matters not what way the Supreme Court may hereafter decide as to the abstract question whether slavery may or may not go into a territory under the Constitution; the people have the lawful means to introduce it or exclude it as they please, for the reason that slavery cannot exist a day or an hour unless it is supported by the local police regulations.' This answer won Douglas Illinois but it lost him the South and hence, two years later, the presidency.[42] Lincoln, normally a generous and forgiving man, had no time for Douglas and did not regret destroying his future career. He thought less of Douglas than he did of the Southern leaders. He said: 'He is a man with tens of thousands of blind followers. It is my business to make some of those blind followers see.'

The debates gave Lincoln precisely the impetus he needed. He quoted Clay many times and in a way he inherited Clay's mantle. The rhyme went: 'Westward the star of empire takes its way—the girls link onto Lincoln, their mothers were for Clay.' He was told: 'You are like Byron, who woke to find himself famous.' By 1859 he knew he ought to be president, wanted to be president, and would be president. The campaign autobiography he wrote December 20, 1859 is brief (800 words), plain, and self-dismissive, yet it exudes a certain confidence in himself and his purpose. He sums up his bid for the presidency in two laconic sentences: 'I was losing interest in politics, when the repeal of the Missouri Compromise aroused me again. What I have done since then is pretty well known.'[43] William Henry Seward (1801–72) and Salmon Portland Chase (1808–73) were both initially considered stronger contenders for the Republican nomination than Lincoln. Seward, first governor then Senator for New York, was the leader of the abolitionists, who said he was 'guided by a higher law than the Constitution.' Chase was senator, then governor of Ohio, a free-soiler and Democrat who drafted the first Republican Party set of beliefs.[44] Both had strong claims but Lincoln had a big success in New York. At the Republican State convention in Decatur, Lincoln's cousin John Hanks did a remarkable if unconscious public relations job by holding a demonstration centered around two fence-rails which, he said, were among the 3,000 Lincoln had split thirty years before. He told stories of Lincoln's youth and his pioneering father—entirely fanciful in the latter's case—and made rail-splitting into a national symbol, from which Lincoln hugely benefited. Lincoln was in Springfield when a telegram arrived saying he had been nominated for president at the Republican National Convention in Chicago. He said: 'I reckon there is a little short woman down in our house that would like to hear the news.' He took his acceptance speech to the local school superinten-dent, who corrected a split infinitive.[45]

The Democratic papers dismissed Lincoln as 'a third-rate lawyer,' 'a nullity,' 'a man in the habit of making coarse and clumsy jokes,' one who 'could not speak good grammar,' a 'gorilla.' And we have to remember that most of Lincoln's sayings and speeches, and even his let-ters, have been cleaned up a good deal before coming down to us. The feeling that he was too rough to be president was not confined to the South, or even to Democrats. But William Cullen Bryant (1794–1878), the anti-slavery poet and philosopher, who had helped to found the Republican Party, called him 'A poor flatboatman—such are *the true leaders of the nation.*' Lincoln had the Douglas Debates made into a little pamphlet, which he gave to people who asked his views. It served his purpose well. In dealing with the South's threat that his election would lead them to secede, he had already taken the bull by the horns in his speech at the Cooper Institution in New York City, February 27, 1860: 'You will not abide by the election of a Republican President! In that supposed event, you say, you will destroy the Union; and then, you say, the great crime of having destroyed the Union will be upon us! That is cool. A highwayman holds a pistol to my ear, and mutters through his teeth, "Stand and deliver!—or I shall kill you, and then you will be a murderer!" '[46]

Using the political arithmetic of the previous thirty years, Lincoln should have been defeated. All the South had to do was to retain its links with the North, concentrate on keeping Jackson's old Democratic coalition together, and pick another Buchanan, or similar. But that was increasingly difficult to do, as the anti-slavers of the North raised the political temperature and the South replied with paranoia. Militant abolitionism dated from the early 1830s, when it became obvious that repatriating blacks to West Africa had failed—only 1,420 blacks had been settled in Liberia by 1831 and the number going there was declining. On January 1, 1831 William Lloyd Garrison (1805–79) began publishing the *Liberator* in Boston. It carried its motto on the front page: 'I am in earnest—I will not equivocate—I will not excuse—I will not retreat a single inch—*and I will be heard.*' Garrison said he relied wholly on moral persuasion and condemned force, but some of his fiercest attacks were launched on moderate abolitionists and he began a new round of militancy on the Fourth of July 1854 when he burned a copy of the Constitution with the words, 'So perish all compromises with tyranny.' Meanwhile the American Anti-Slavery Society (1833) had been organized by two New York merchants, Arthur and Lewis Tappan, in conjunction with the most sophisticated and effective of the abolitionist campaigners, Theodore D. Weld (1803–95), whose anonymous tract, *American Slavery As It Is* (1839) furnished the inspiration for *Uncle Tom's Cabin*. Weld organized Oberlin as the first college to admit both blacks and women, and he married Angela Grimke, one of two South Carolina sisters who freed their slaves and moved north to campaign.

Initially there was a lot of opposition to the anti-slavery movement in the North, where most Northerners hated blacks and frequently subjected them to mass violence. But by the end of the 1830s a younger generation who took the morality of abolition for granted began to take up positions and exercise influence. Emerson noticed 'a certain tenderness in the people, not before remarked.' As he put it, 'The young men were born with knives in their brain.' It was the beginning of liberal humanitarianism in the United States, and it took many forms, but slavery was the issue around which it concentrated. Increasingly, direct action of various kinds began to take over from propaganda alone. An underground developed to get escaped slaves across the borders on to free soil and protect them there. It was run by 'conductors' like Harriet Tubman (1821–1913), a Maryland slave who had escaped in 1849, the Quaker Levi Coffin (1789–1877), and the ferocious John Brown. There were about 1,000 conductors in all, and although their successes were numerically insignificant—not more than 1,000 a year after the passage of the Fugitive Slave Act of 1850, which made such operations increasingly risky—their effect on Southern morale was disproportionately great. Moreover, Southern slave-hunters, moving into Northern states in hot pursuit of fugitives, were highly unpopular especially when, as often happened, they grabbed the wrong black. From 1843 in Boston we get the first examples of an abolitionist mob releasing a recaptured fugitive slave by force. Whittier echoed the feelings of many with his lines:

> No slave-hunt in our borders—no pirate on our strand!
> No fetters in the Bay State—no slave upon our land!

During the 1850s, moreover, Northern legislatures passed laws making it exceedingly difficult, and sometimes impossible, to enforce the provisions of the 1850 federal act. The fact is that Southern aggression was all the time pushing Northern moderates into more extreme positions, particularly when the threat to the North's freedom of action became apparent. As William Jay, son of Chief Justice Jay, put it, 'We commenced the present struggle to obtain the freedom of the slave—we are compelled to continue to preserve our own.' James G. Birney (1792–1857), another former slave-owner who favored a modern position and was the Liberty Party candidate in 1840, put the point thus: 'It has now become absolutely necessary that slavery should cease in order that freedom may be preserved in any portion of our land.'[47]

As we have seen, from 1854 Kansas became the battleground of Southern extremists and anti-slavery activists. Indeed, it could be said that the Civil War started there. And it was inevitable, perhaps, that the kind of violence which became a daily occurrence in 'bleeding Kansas'

THE ELECTION OF 1860

should spread. In particular, John Brown, who had received much applause for his 'Pottawatomie Massacre'—'Brown of Pottawatomie' became a slogan of Northern militants—was given money and other help to set up a stronghold in the mountains of western Virginia to assist slaves traveling on the Underground Railroad. Not content with this, on October 16, 1859, with twenty men, he seized the US arsenal at Harpers Ferry. Two days later, Colonel Robert E. Lee and a regular army unit recaptured the post, killed ten of Brown's men, and made him prisoner. He was condemned to death and hanged on December 2. Some, including Lincoln, condemned Brown; others, including Emerson, hailed him as 'The new saint who will make the gallows glorious like the cross.'[48] Brown's violent act completed the process of transforming the South, or at least its leadership class, into a tremulous and excitable body—a case of collective paranoia—which believed anything was preferable to a continuation of the present tension and fear. Some predicted a general rising of the slaves. Others looked to separation as the only safeguard of their property and way of life.

Against this background, the Democrats met for their presidential convention in April 1860 in Charleston, the South Carolina city which was the capital of Southern extremism. The Southerners, in their fear and fury, accused the Northern Democrats of betraying them by failing to present slavery to the North as a positive good. On behalf of the North, George E. Pugh of Ohio replied: 'Gentlemen of the South, you mistake us—you mistake us—*we will not do it*.' When the South failed to get the platform it wanted, the delegations from the Gulf states, South Carolina and Georgia, walked out, splitting the Democratic Party right down the middle. The convention met again at Baltimore on June 18 and finally nominated Douglas on a moderate platform. The Southerners replied by nominating the Vice-President, John C. Breckinridge of Kentucky (1821–75), on a slavery platform. The Whigs reorganized themselves as the constitutional Union Party and nominated John Bell (1797–1869) of Tennessee as, in effect, the candidate of the border states. That meant four candidates. Essentially, however, it was a contest between Lincoln and Douglas in the North and Breckinridge and Bell in the South, since Lincoln could not hope to win Southern votes and Breckinridge had no support north of the Mason–Dixon Line.

In effect, a Lincoln victory was certain provided no untoward events intervened and provided he made no spectacular blunder. Hence all his friends and advisors warned him to keep out of the campaign and let the Republican Party do the work. So Lincoln worked behind the scenes to keep the Republican Party together, and left it to the Democrats, or rather the South, to commit political suicide. His only public appearance in the campaign was at Springfield in August where, pressed to orate, he simply said: 'It has been my purpose, since I have been placed in my present position, to make no speeches.' This gave him an almost Washingtonian detachment and saved him from misrepresentation. On November 6 Lincoln waited in the telegraph office until his victory in New York, signaled at 2 A.M. on the morning of the 7th, made his election certain. He got 1,866,452 votes against Douglas' 1,376,957; there had been 849,781 for Breckinridge and 588,879 for Bell. The result, in terms of electoral college votes, was somewhat different: Lincoln got 180, for he carried all but one of the free states, dividing New Jersey with Douglas (all the latter got, apart from Missouri). Breckinridge won all the slave states except Virginia, Tennessee, and Kentucky in the Upper South, which went to Bell. In ten of the Southern states Lincoln did not receive a single vote. Moreover, he was elected on a minority vote of 39.9 percent, the lowest since J. Q. Adams won the unlucky, ominous election of 1824. The nation was indeed divided.[49]

If we now turn to Lincoln's principal opponent in the duel for the soul of America, we will see why it was that the South, having held so many cards in its hand, allowed itself to be exasperated into throwing away the game in a fit of temper. Jefferson Davis, Calhoun's political heir insofar as he had one, was president of the Confederacy from its reckless birth to its pitiful death-agony. He was flawed and blinkered both as man and as statesman, with huge weaknesses of judgment and capacity. But he was not small in any sense of the word. Six feet tall, slim, ramrod-straight, 'soldierly bearing, a fine head and intellectual face . . . a look of culture and refinement about him,' he 'could infuse courage into the bosom of a coward, and self-respect and pride into the breasts of the most abandoned.' To his cause he brought a passion 'concentrated into a white heat, that threw out no sparks, no fitful flashes, glowing [instead] with an intense but not an angry glare.' These judgments by contemporaries were endorsed even by critics and enemies. Thomas Cobb of Georgia said, 'He is not great . . . [but] the power of will he has, made him all he is.'[50]

The conventional portrait of Davis, the man driven by willpower, is of an old-fashioned Southern gentleman. That is inexact. His middle name was Finis because he was born when his mother was forty-seven, the last of ten. He had a modern-style upbringing: his father rejected any kind of corporal punishment, and the boy was cosseted by big sisters, and taught riding by his adoring big brothers, three of whom were old enough to have fought in the 1812 War. Jeff Davis was brought up to a simple, absolutist patriotism of a kind we would now find incomprehensible. When his father died, Davis' elder brother Joseph, a successful Mississippi cotton planter, took over the role of mentor and guardian. After an education under the Roman Catholic Dominican friars at Wilkinson County Academy and at the famous Transylvania University in Lexington, Kentucky, Davis went to West Point on the nomination of the War Secretary, Calhoun, thereafter his political model and leader. As a frontier officer, he fought the Indians and personally took the surrender of Black Hawk, made peace among the miners and war against his superiors. Stiff-necked and bellicose, he admitted: 'In my youth I was over-willing to fight.' His career was checkered with rows, courts-martial, and frustration at slow promotion. When he married the daughter of General Zachary Taylor, he left the army and Brother Joseph set him up as a planter. This too was frustrating. Joseph owned 11,000 acres and was a wealthy man, but the 800-acre Hurricane Estate he 'lent' or half-gave to Davis was small by Mississippi standards and he remained his brother's dependant.[51]

It is important to grasp that, when Davis spoke of the benevolence of the slave-system in the South, he believed what he said totally and spoke from experience. Joseph, as a planter, was enlightened. None of his slaves was ever flogged. The slaves judged and punished themselves. Families were kept together. One testified: 'We had good grub and good clothes and nobody worked hard.' Another: 'Dem Davises never let nobody touch one of their niggers.' The community at Davis Bend on the river, said General Taylor, was 'a little paradise.' Davis shared to the full his brother's attitudes and was anti-blood sports to boot. He treated his black body-servant, James Pemberton, with exquisite courtesy and put him in charge of his plantation when he was away. He made a point of returning any salute from a black with an elaborate bow: 'I cannot allow any negro to outdo me in courtesy.' Not for him the swaggering society of New Orleans or Charleston. His only genuflection to Southern male habits was a propensity to challenge critics to duels, though he never actually fought any. To sleep with one of his slaves would have been to him an abomination. When his beloved wife Sarah Taylor died of malaria, he acquired a sadness that never left him, though he eventually married again, a beautiful girl, Varina, half his age. His melancholy was aggravated by poor health, including terrifying facial pains and chronic hepatitis which eventually left him blind in one eye. He suffered from insomnia and his chief pleasure was reading—Virgil, Byron, Burns, and Scott.[52]

The overriding weakness of this seemingly civilized and well-meaning man was lack of imagination, compounded by ignorance. America in the 1840s and 1850s was already an immense country, but travel was still difficult, especially in the South, and expensive. It is hard for us to grasp how little Americans knew of the societies outside their region or indeed locality. Davis paid only one visit to New England and was surprised to find the people friendly. Until he became president of the Confederation he knew little of the South beyond his own part of Mississippi. He assumed that the treatment of slaves at Davis Bend was typical and refused to believe stories of cruelty: that was simply Northern malice and abolitionist invention. He was, like so many other well-read and well-meaning people in the South, the victim of its own policy of concentrating its limited media and publishing resources on indoctrinating its own people, and telling the rest of the world to go to hell. Davis was self-indoctrinated too; he had a passion for certitude.

On this narrowness of vision he built up a political philosophy which did not admit of argument. Blacks, he insisted, were better off as slaves in the South than as tribesmen in Africa: 'I have no fear of insurrection, no more dread of our slaves than I have of our cattle . . . Our slaves are happy and contented.' Not only was it in the interests of blacks to be slaves, it was likewise to their benefit that slavery be extended. Davis never possessed more than seventy-four slaves and knew all of them well: it was his policy. He maintained it was wrong for whites to own more slaves than they could personally care for, as he did. If cruelty occurred, it was because sheer numbers undermined the personal owner–slave relationship. So the more slavery spread out geographically, the more humane it would be. This was his argument for dismantling Mexico, turning its territories into new states, and making slavery lawful there and even north of the Missouri Compromise line. Slave-owners must be able to take their slaves with them into new territories just as immigrants had always taken any other form of property with them, such as waggons or cattle. Joseph had dinned into him the fundamental principle: 'Any interference with the unqualified property of the owner in a slave was an abolition principle.'

Davis believed that the Southern case for slavery and its extension rested on firm moral foundations. Indeed he was morally aggressive, accusing the North of hypocrisy: 'You were the men who imported these negroes into this country. You enjoyed the benefits resulting from their carriage and sale; and you reaped the largest profits accruing from the introduction of the slaves.' Abolition was nothing but 'perfidious interference in the rights of other men.' He did not see the agreements of 1820 and 1850 as 'compromises' but as Southern concessions, the limit to which the South could reasonably be expected to go. Further limitations on slavery were merely Northern attacks on the South motivated not by morality but by envy and hatred: 'The mask is off: the question is before us. It is a struggle for political power.' The Constitution was on the South's side. The federal government had no natural authority: 'It is the creature of the States. As such it can have no inherent power; all it possesses was delegated by the States.' If what Davis called 'the self-sustaining majority' continued its oppressive and unlawful campaign against the South, the 'Confederation' as he called it should be dissolved: 'We should part peaceably and avoid staining the battlefields of the Revolution with the blood of a civil war.'[53]

This philosophy, inherited from Calhoun and instilled by Brother Joseph, reexamined by Davis in his lonely musings, polished and consolidated over the years, he regarded as axiomatic. It is significant that he never saw himself as an extremist especially over breaking up the Union. He wrote: 'I was slower and more reluctant than others. I was behind the general opinion of the people [of Mississippi] as to the propriety of prompt secession.' But when his basic assumptions about slavery were challenged, he responded with paranoia. This sprang not just from his Southern conditioning but from a dominant streak of self-righteousness in his character. A variety of incidents in his early life, in the army, in his domestic and public quarrels show that, once he had made up his mind and adopted a position, he treated any attempt to argue him out of it as inadmissible, an assault on his integrity. As he put it to his second wife, Varina: 'I cannot bear to be suspected or complained of, or misconstrued after explanation.' That sentence sums up the tragedy of his life. Senator Isaac P. Walker of Wisconsin noted: 'He speaks with an air which seems to say "Nothing more can be said, I know it all, it must be as I think." ' Davis himself said he ignored press criticism: 'Proud in the consciousness of my own rectitude, I have looked upon it with the indifference which belongs to the assurance that I am right.'[54]

All this suggests that Davis was better suited to a military than a political life. That was Varina's view: 'He did not know the arts of a politician, and would not practice them if understood.' Davis got into politics in his later thirties but the Mexican War gave him the chance to resume his army career. He was elected colonel of a regiment of Mississippi volunteers, had the foresight to equip them with the new Whitney rifle, was favored by his commanding general and former father-in-law, General Taylor, saw action at Monterrey and Buena Vista, and distinguished himself in both these much publicized battles. The Mexican War, as we have noted, was the great proving ground for future American bigshots, both political and military. Davis was described by General Bliss, Taylor's chief-of-staff, as 'the best volunteer officer in the Army,' and President Polk offered him a general's commission. But he had been badly wounded in the foot at Buena Vista and chose instead to be nominated to the Senate.[55]

In politics Davis found it natural to be called the 'Calhoun of Mississippi,' and, when the old fire-eater died, to assume Elijah's Mantle. It was equally natural, when his friend Franklin Pierce became president, to accept office as war secretary, where he became perhaps the most powerful voice in the Cabinet and a forceful administrator. But his weakness quickly made its appearance. He got into a series of arguments with his general-in-chief, Winfield Scott, mostly over trivialities. Scott was arrogant and self-righteous too, but Davis, as his political superior, might have been expected to behave with more sense and dignity. One of Davis' letters to Scott ran on for twenty-seven foolscap pages and was contemptuously described by its recipient as 'a book.' Everything fell into the hands of the press and made amazing reading. Scott closed his last letter: 'Compassion is always due to an enraged imbecile,' to which Davis replied that he was 'gratified to be relieved of the necessity of further exposing your malignity and depravity.'[56] Reading this correspondence helps to explain why the Civil War occurred and, still more, why it lasted so long. It certainly suggests that Davis was not a man fit to hold supreme office at any time, let alone during a war to decide the fate of a great nation.

It was not that Davis was unperceptive. In some ways his views were advanced. He tended to take the progressive line on everything except slavery. That pillar of Bostonian anti-slavery rectitude, John Quincy Adams, commended him warmly for helping to get the Smithsonian set up. And Davis was well aware of some of the South's weaknesses, especially its lack of industry. Its one big industrial complex was the Tredegar Iron Works on the banks of the James River near Richmond. It had been, as it were, replicated from the South Wales Tredegar works in the 1830s, to serve the Southern railroads. It also made cannon, chains, and iron ships, and by 1859 was the fourth-largest ironworks in the United States, employing 800 people. But it was near bankruptcy because it was uncompetitive. It got its iron ore from Pennsylvania because Virginian sources were exhausted, and virtually all its copper and bronze and many parts and machinery had to be bought in the North or from abroad. It had to pay extra wages because white industrial workers hated employment in a slave state. They particularly objected to working alongside slaves, fearing to be replaced by them. The works was notable for high labor turnover, chronic labor shortages, and neglect of innovation. It survived at all only because it gave liberal, risky credit to Southern railroads. It seemed enormous, and so reassuring, to Southerners, but in the nation as a whole it was marginal. There was in the South no central, up-to-date industrial magnet to attract skilled labor and so compensate for the many deterrents.[57]

By contrast, a hundred miles or so to the north there was the beginning of a vast manufacturing complex stretching from Wilmington to New York. From 1840 to 1860 this megalopolis was the most rapidly growing large industrial area in the world—and it was this complex which made inevitable, in military–economic terms, the South's ruin. Davis, knowing the South's weakness, began urging it, from about 1850 on, to start stockpiling arms and ammunition, to encourage immigration from the North, or to build railroads to transport its agricultural products itself, to create an industrial base to manufacture its own cotton goods, shoes, hats, blankets, and so on, and to provide state support for higher education so that its sons were not forced to go to Northern universities and adopt their ideas. What finally happened to the South in the 1950s, Davis was urging in the 1850s. But slavery repelled capital and white skilled labor alike, and Southerners themselves did not want industrialization for many different reasons, most of all because they felt instinctively that it would mean the end of slavery and plantation culture. So Davis got no response to his pleas. In any case they were half-hearted and confused. His wish to 'educate' the South conflicted with his insistence that Southern textbooks be rewritten to eliminate opinions in conflict with the South's view of slavery, his desire that the South's children should learn from books which were 'Politically Correct' and 'indoctrinate their minds with sound impressions and views' and his determination to kick out

'Yankee schoolteachers.' Not for nothing did the *New York Herald* call him 'the Mephistopheles of the South.'[58]

By seceding from the Democratic Party, the Southern states threw away their greatest single asset, the presidency. Then, by seceding from the Union, they lost everything, slavery first and foremost. Bell was right in proclaiming, throughout the election, that the only way the South could retain slavery was by staying in the Union. But that demanded 'a change of heart, radical and thorough, of Northern opinion in relation to slavery.'[59] Up to the beginning of the campaign, Davis, realizing that Lincoln would win, made a desperate effort to get all the other three candidates to withdraw in favor of a compromise figure—a sympathetic Northerner, perhaps. Breckinridge and Bell agreed to stand down and so did Douglas' running mate, Benjamin Fitzpatrick. But Douglas, ambitious and self-centered—and blind—to the end, flatly refused. Thus Douglas made the Civil War inevitable. Or did he? *Was* it inevitable once Lincoln won?

One of the villains was Buchanan, the outgoing President, who in effect did nothing between the beginning of November 1860 and the handover to Lincoln in March. His message to Congress denied the right of secession but blamed the Republicans for the crisis—two incompatible opinions. He was lazy, frightened, confused, and pusillanimous. Thus four vital months were lost. His military dispositions, insofar as he made any, were inflammatory rather than conciliatory. Only two states wanted a civil war—South Carolina and Massachusetts. In the early 1830s over Nullification, the South California extremists failed to carry anyone else with them, the rest of the South being prepared to trust President Jackson, to see the South got justice. But now they would trust nobody. All the same, an armed struggle might have been averted. Had South Carolina persuaded only four or five other states to go with it, the secession would have fizzled out. If all fifteen of the slave states had seceded, the North would have been forced to give way and sue for a compromise. As it was, just enough joined South California to insure war.[60] The real tragedy for America is that Lincoln, the man the South most hated, was exactly the man to get it to see reason, had he been given the chance. If he had been enabled by the Constitution to move into the White House immediately after his election, and assume full powers, all the weight of his intellect, and all the strength of his character, and all the genius of his imagination could have been brought to bear on the problem of exorcizing the South's fears. Instead, he had to sit, powerless (he used the interval to grow a beard), while the Union disintegrated, and by the time he took up command the process of secession was already taking place, and was irrevocable.[61]

As early as November 10, only three days after the election results were received, the South Carolina legislature unanimously authorized the election of a state convention on December 6, to decide 'future relations between the State and the Union.' Eight days later, Georgia followed suit. Within a month every state of the South had taken the initial steps towards secession. When Congress reassembled on December 3, it listened to a plaintive grumble from Buchanan, who said that he deplored talk of secession, but nothing could be done, by him anyway, to prevent it. Three days later South Carolina elected an overwhelmingly secessionist state convention which on December 20 declared that the state was no longer part of the Union. Davis himself tried to promote a compromise, then despaired of it. On January 7 the secession convention of his own state, Mississippi, met and on the 9th voted 84 to 15 to leave the Union. Two days before, the senators from Georgia, Florida, Alabama, Louisiana, Texas, Arkansas, and Mississippi had met in caucus in Washington and decided to meet again in Montgomery, Alabama on February 15 to form a government. Like other senators, Davis made an emotional speech of farewell in Congress. Going south through Tennessee, he was asked to make a speech at his hotel, Crutchfield House, and did so. Whereupon the brother of the hotel's owner, William Crutchfield, told him he was a 'renegade and a traitor . . . We are not to be hoodwinked and bamboozled and dragged into your Southern, codfish, aristocratic, Tory-blooded South Carolina mobocracy.' The crowd, many of them armed, backed these accusations—there was strong Union sentiment in the back-country and in the mountains.[62]

Davis was promptly chosen general in Mississippi's army. Many, including his wife, wanted him to be commander-in-chief of the confederate forces, rather than president. He agreed with Varina. Meeting on February 4, the six states which had already seceded, South Carolina, Mississippi, Florida, Georgia, Louisiana, and Alabama, drew up a new constitution, which was virtually the same as the old except it explicitly recognized slaves as property. Robert Toombs (1810–85), Senator from Georgia, might have got the presidency, but he got publicly drunk several nights running. In the end Davis was chosen more or less unanimously. His journey from his home near Vicksburg to his inauguration in Montgomery was a sinister foretaste of the problems the South faced. The two cities were less than 300 miles apart, along a direct east–west road, but Davis, trying to get there more quickly by rail, had to travel north into Tennessee, then across northern Alabama to Chattanooga, south to Atlanta, and from there southwest to Montgomery, a distance of 850 miles around three-and-one-half sides of a square on half a dozen different railroads using three different gauges. No railway trunk lines bound the rebellious states together. The South had no infrastructure.[63] Its railroad system was designed solely to get cotton to sea for export. There was virtually no interstate trade in the South, and so no lines to carry it. It took five railroad lines to get from Columbia to Milledgeville, for example; the railroads in Florida, Texas, and most of Louisiana had no connection at all with the other Southern states. The functional geography of the South, both natural and manmade, was against secession.

In his inaugural, Davis said the Confederacy was born of 'a peaceful appeal to the ballot box.' That was not true. No state held a referendum. It was decided by a total of 854 men in various secession conventions, all of them selected by legislatures, not by the voters. Of these 157 voted against secession. So 697 men, mostly wealthy, decided the destiny of 9 million people, mostly poor. Davis said he was anxious to show that secession was 'not a rich man's war and a poor man's fight,' but the fact is it was the really rich, and the merely well-to-do, both of whom had a major interest in the struggle, who decided to commence it, not the rest of the whites, who had no direct economic interest at all. And the quality of Southern leadership, intellectually at least, was poor. The reasons for secession, put into the declarations of each states, made no sense, and merely reflected the region's paranoia. Mississippi's said: 'the people of the Northern states have assumed a revolutionary position towards the Southern states.' They had 'insulted and outraged our citizens when traveling amongst them . . . by taking their servants and liberating the same.' They had 'encouraged a hostile invasion of a Southern state to incite insurrection, murder and rapine.' South Carolina's was equally odd, ending in a denunciation of Lincoln, 'whose opinions and purposes are hostile to slavery.' But most presidents of the United States had been hostile to slavery, not least Jefferson, the man whose opinions on the subject Lincoln most often quoted.

The Southern leaders assumed there were absolute differences between the peoples of North and South. In fact allegiances were divided. Mary Lincoln had three brothers in the Confederate Army, all of whom were killed—and her emotional sympathies were certainly with the South. Varina Davis' male relatives, the Howells, were all in the Union Army. Senator John J. Crittenden of Kentucky (1787–1863), who did his best to promote compromise, had two sons, both major-generals, one serving in the Confederate, the other in the Union army. The best Union agent in Europe, Robert J. Walker, was a former senator from Mississippi, while the best Confederate agent, Caleb House, came from Massachusetts. General Robert E. Lee's nephew, Samuel P. Lee, commanded the Union naval forces on the James River, while another Union admiral, David Glasgow Farragut (1801–70), the outstanding maritime commander in the war, was born in Tennessee and lived in Virginia. The examples are endless. The young Theodore Roosevelt was made to pray for the North, the young Woodrow Wilson prayed for the South. There were, literally, millions of divided families, and the number of extremists on both sides probably did not amount to a hundred thousand all told.

It became a necessity, Jefferson Davis wrote to a Northern friend, January 20, 1861, 'to transfer our domestic institutions from hostile to friendly hands, and we have acted accordingly.' Lincoln could not exactly be called friendly towards the South—he was, rather, exasperated and sad. But he was not hostile. Southern leaders like Davis would not accept that Lincoln was hated by many abolitionists, like Wendell Phillips (1811–84), the rich Boston humanitarian ideologue, who called him 'the Slavehound of Illinois.' The most the Lincoln Republicans could do, and proposed to do, was to contain slavery. To abolish it in the 1860s required a constitutional amendment, and a three-quarters majority; as there were fifteen slave states, this was unobtainable. A blocking majority of this magnitude would still have been sufficient in the second half of the 20th century. It is worth noting that, at the time of secession, Southerners and Democrats possessed a majority in both houses of Congress, valid till 1863 at least. If protecting slavery was the aim, secession made no sense. It made the Fugitive Slave Act a dead letter and handed the territories over to the Northerners. The central paradox of the Civil War was that it provided the only circumstances in which the slaves could be freed and slavery abolished.[64]

War was so obviously against the rational interests of the South that Lincoln did not consider it likely. His concern was to prevent the Republicans from appeasing the South by abandoning their platform and embracing Douglas' popular-sovereignty doctrine. Over and over again he repeated his message to Republican congressmen: 'Have none of it. Let there be no compromise on the question of *extending* sovereignty. Stand firm. The tug has to come and better now, than any time hereafter.'[65] By tug, he meant confrontation and crisis, not war. If he had thought in terms of war when appointing his Cabinet, Lincoln would never have made Simon Cameron (1799–1889) his Secretary of War. Cameron was a millionaire banker and railroad tycoon, who was the overwhelming boss of Pennsylvanian Republicanism and he was appointed for entirely political reasons (his handling of army contracts led Lincoln to sack him and to a vote of censure in the House). Nor, probably, would he have made Seward Secretary of State and Chase Treasury Secretary. Lincoln knew a vertiginous time was ahead and he opted for a strong government rather than a warlike one.[66]

Seward, a clever, persuasive man, believed the administration's best strategy was to leave the rebellious Deep South to stew in its own Confederate juice and concentrate on wooing the other slave states to remain faithful to the Union. But that would have meant letting the seven go, and Lincoln was determined to preserve the Union as it was, at all costs. That was the only thing which, at this stage, he could see clear, and he stuck to it. This strategy, in turn, set off the mechanism of the war. On asserting its independence in December 1860, South Carolina called on the custodians of all federal property within the state to surrender it. Major Robert Anderson, the federal commander of the fortifications in Charleston Harbor, concentrated his forces in Fort Sumter and refused to act without instructions from Washington; President Buchanan, lax in most ways, likewise declined to have the federal forces evacuated. General F. W. Pickens of South Carolina thereupon trained his guns on the Fort. When Lincoln took over, the Cabinet deliberated on what to do. The Commander-in-Chief, General Scott, who might have been expected to be anxious to poke his old 'imbecile' enemy Davis in the eye, in fact advised doing nothing. Five out of seven members of Lincoln's Cabinet agreed with him. But Lincoln decided otherwise. His decision to send a relieving expedition by sea, carrying food but no arms or ammunition, and to inform General Pickens of what he was doing, demonstrated his policy of upholding the Union at any cost. The response of the Confederate forces, to fire on the Fort, and the flag, was a decision to secede at any cost. That began the war on April 12, 1861.[67]

As the South was arming and recruiting, Lincoln had no alternative but to take steps too. 'The star-spangled banner has been shot down by

Southern troops,' he said, and on April 15 asked for 75,000 volunteers (answered by 92,000 within days). This move by Lincoln was, curiously enough, the 'last straw' which pushed Virginia (and so North Carolina) into secession. This, too, was undemocratic since the state convention voted 88 to 55, on April 17, to submit an Ordinance of Secession to a popular plebiscite. However, the governor put the state under Confederate command without waiting for the vote. This event was decisive for many reasons. Virginia was the most important of the original colonies, the central element in the Revolutionary War, and the provider of most of the great early presidents, as well as of the US Constitution itself. For the state which had done more than any other to bring the Union into existence to leave it in such an underhand and unconstitutional manner was shabby beyond belief. It is astonishing that the Virginians put up with it. And of course many of them did not. The people of West Virginia, who had no slaves, broke off and formed a separate state of their own, acknowledged by Congress as the State of West Virginia in 1863.

General Lee, the state's most distinguished soldier, had been asked by Lincoln to become commander-in-chief of the Union forces. This was a wise choice and would have been a splendid appointment, for Lee was decent, honorable, and sensible as well as skillful. But Lee was a Virginian before anything else and he waited to see what Virginia did. When Virginia seceded, he reluctantly resigned his commission in the US Army, which he had served for thirty-two years. It seems to us quixotic but he felt he had no other option. He wrote to his sister in Baltimore and his brother in Washington DC: 'With all my devotion to the Union, and the feeling of loyalty and duty of an American citizen, I have not been able to make up my mind to raise my hand against my relatives, my children, my home.'[68]

Arkansas seceded on May 6. The next day Tennessee formed an 'alliance' with the Confederacy, the only decision to be endorsed by popular vote. North Carolina, sandwiched between Virginia and South Carolina, had not much choice and joined on May 20. Missouri was divided but refused to join the Confederacy. Delaware was solid for the Union but shaky on coercion. Maryland too protested against coercion but declined to summon a state convention and so remained in the Union. Kentucky initially refused to send volunteers at Lincoln's request but by the end of 1861 had joined the Union war-effort. So only eleven out of the fifteen slave states formed the Confederacy.[69]

In demographic terms, the Confederacy was at a huge disadvantage.

The census of 1860 showed that the eleven Confederate states had a population of 5,449,467 whites and 3,521,111 slaves. Nearly 1 million of the white males served, of whom 300,000 were casualties. The nineteen Union states had a population of 18,936,579 and the four border states a further 2,589,533, plus 429,401 then-slaves, over 100,000 of whom served in the Union army, which altogether numbered 1,600,000. Moreover, during the war nearly a million further immigrants arrived in the North, of whom 400,000 served in the Union army. Some of the best Northern troops were German, Irish, and Scandinavian, as were some of the smartest officers—Franz Sigel, Carl Schurz (Germans), Philippe de Trobriand (French), Colonel Hans Christian Heg and Hans Matson (Norwegian), and Generals Corcoran and Meagher (Irish). The Union economic preponderance was even more overwhelming. If the North–South ratio in free males aged eighteen to sixty was 4.4:1, it was 10:1 in factory production, iron 15:1, coal 38:1, firearms production 32:1, wheat 412:1, corn 2:1, textiles 14:1, merchant-ship tonnage 25:1, wealth 3:1, railroad mileage 2.4:1, farm acreage 3:1, draft animals 1.8:1, livestock 1.5:1. The only commodity in which the South was ahead was cotton, 24:1, but this advantage was thrown away by overproduction (in the South) and stockpiling (outside the South) in the endless build-up to the crisis. Just before the war, Senator James Henry Hammond of South Carolina boasted: 'Cotton, rice, tobacco and naval stores command the world; and we have the sense to know it, and are sufficiently Teutonic to carry it out successfully. The North without us would be a motherless calf, bleating about, and die of mange and starvation.'[70] The assumption in the South was that the coming of war would lead to an expansion of its economy, and a contraction of the North's. In fact, as was foreseeable, the reverse occurred. The South's economy shrank, the North's expanded, even faster than in the 1850s.[71]

The South compounded its difficulties by weaknesses in its handling of finance, diplomacy, and internal politics, all of which had severe military consequences. First, it is a curious historical fact that most civil wars are lost by one side running out of money, and the American Civil War was an outstanding case in point. The South had no indigenous gold or silver supplies and no bullion reserves, and was entirely dependent on its own paper money. The North had the enormous advantage of a large, well-trained navy and, almost from the start, was able to impose a blockade, often ineffective at first but progressively tighter as the war proceeded. As a result, import and export taxes, the way of raising money traditionally preferred by the South, raised little. Import duties brought in only about $1 million in specie during the entire war, and the Union navy was so vigilant in running down cotton-export ships that only about $6,000 in specie was collected from cotton exports. With its limited capacity to produce armaments, the South was forced to shop abroad. France, always happy to supply arms to dodgy regimes, duly obliged but insisted on being paid in specie (as did independent gun-runners).

As his Treasury secretary, Davis appointed C. G. Memminger, a local South Carolina politician. This was an extraordinary choice: Memminger had virtually no experience of finance and, more important, lacked the creative ingenuity to surmount the almost insuperable difficulties of raising hard cash.[72] An initial war-loan of 8 percent, organized by a consortium of New Orleans and Charleston banks, raised $15 million in specie, all of which was immediately sent abroad to buy arms. But subsequent loans were relative, then total, failures. A cotton-backed foreign loan, organized in London by Erlangers in January 1863, brought in disappointingly little, as a result of high charges and an imprudent attempt to bull the market. Hence Memminger resorted to the device of the improvident through the ages—printing paper. By the summer of 1861, $1 million of Confederate paper currency was circulating. By December it was over $30 million; by March 1862 $100 million; August 1862 $200 million; December 1862 $450 million. In 1863 it doubled again to $900 million and continued to increase, though later figures are mere guesswork. Gold was quoted at a premium over paper as early as May 1861 and was 20 percent premium by the end of the year. By the end of 1862 a gold dollar bought three paper ones and, by the end of 1862, no fewer than twenty.

In July 1846 Memminger, accused of making private profits on cotton-running, resigned in disgust, and Davis then appointed a real economic wizard called George A. Trenholm, a Charleston cotton-merchant who had proved extraordinarily adept at selling the South's staple. But by then it was too late: the South's finances were beyond repair. Inflation became runaway, the gold dollar being quoted at 40 paper ones in December 1864 and 100 shortly thereafter. Inflation, if nothing else, doomed the South. In the second half of the war Southerners showed an increasing tendency to use the North's money, as it inspired more confidence. Towards the end people cut themselves off from paper money altogether, and bought and sold in kind—even the government raised taxes and loans in produce. The only people with means to move around were those who had kept gold dollars. Davis was like everyone else. In the final weeks of the Confederacy he sent his wife Varina off with his last remaining pieces of gold, keeping one five-dollar coin for himself.[73]

The South's diplomacy was as inept as its finance. Davis did not initially see the need for a major diplomatic effort since he believed the economic arguments would speak for themselves. The key country was Britain, because in the 1850s it had imported 80 percent of its cotton from America, and it had the world's largest navy, which could break the Union blockade if it wished. Davis accepted Senator Hammond's assertion: 'You dare not make war upon our cotton. No power on earth dares make war on it. Cotton is King.'[74] But overproduction and stockpiling in anticipation of war led to a 40 percent oversupply of cotton in the British market by April 1861, before the war had properly begun. Britain got cotton from Egypt and India and, later in the war, from the United States itself, via the North. In the years 1860–5 Britain managed to import over 5 million bales of cotton from America, little of which was bought from the South directly. British manufacturers welcomed the opportunity to work off stocks and free themselves from dependence on Southern producers, whom they found difficult and arrogant. It is true the cotton blockade caused some unemployment in Lancashire and Yorkshire—by the end of 1862 it was calculated that 330,000 men and women were out of work in Britain as a result of the conflict. But they had no sympathy for the South. They identified with the slaves. They sent a petition to Lincoln: 'Our interests are identical with yours. We are truly one people . . . If you have any ill-wishers here, be assured they are chiefly those who opposed liberty at home, and that they will be powerless to stir up quarrels between us.' Lincoln called their words 'An instance of sublime Christian heroism.'

The truth is, by opposing slavery and by insisting on the integrity of the Union, Lincoln identified himself and his cause with the two most powerful impulses of the entire 19th century—liberalism and nationalism. He did not have to work at a powerful diplomatic effort—though he did—as world opinion was already on his side, doubly so after he issued his Emancipation Proclamation. It was the South which needed to put an effort into winning friends. It was not forthcoming. Davis hated Britain anyway. The South had many potential friends there—the Conservative Party, especially its leading families, newspapers like *The Times*, indeed a surprisingly large section of the press. But he did not build on this. The envoys he sent were extremists, who bellowed propaganda rather than insinuated diplomacy. The British Prime Minister, Lord Palmerston, was a Whig–Liberal nationalist who played it cool: on May 13, 1861 he declared 'strict and impartial neutrality.' The North's naval blockade caused much less friction with Britain than the South had hoped, because it conformed strictly to British principles of blockading warfare, which the Royal Navy was anxious to see upheld for future use. The one really serious incident occurred in November 1861, when the famous explorer Captain Charles Wilkes (1798–1877), commanding the USS *San Jacinto*, stopped the British steamer *Trent* and seized two Confederate commissioners, John Slidell and James M. Mason. This caused an uproar in Britain, but Seward, as secretary of state, quickly defused the crisis by ordering the men's release, on the ground that Wilkes should have brought the ship into harbor for arbitration.[75]

Added to improvident economics and incompetent diplomacy, the South saddled itself with a political system which did not work. It was a martyr to its own ideology of states' rights. Although Davis and his fellow-Southerners were always quoting history, they did not know it. Had they studied the early history of the republic objectively, they would have grasped the point that the Founding Fathers, in drawing up the Constitution, had to insure a large federal element simply because the original provisional system did not work well, in war or in peace. The Confederacy thus went on to repeat many of the mistakes of the early republic. Each state raised its own forces, and decided when and where they were to be used and who commanded them. To many of their leaders, the rights of their state were more important than the Confederacy itself. Men from one state would not serve under a general from another. Senior commanders with troops from various states had to negotiate with state governments to get more men. Davis had to contend with many of the identical difficulties, over men and supplies and money, which almost overwhelmed Washington himself in the 1770s— and he had none of Washington's tact, solidity, resourcefulness, and moral authority. Everyone blamed him, increasing his paranoia. As a former military man and war secretary, he thought he knew it all and tried to do everything himself. When he set up his office, he had only one secretary. His first Secretary of War, Leroy P. Walker, was a cipher. Visitors noticed Davis summoned him by ringing a desk bell, and Walker then trotted in 'exhibiting a docility that dared not say "nay" to any statement made by his chief.'[76] Congress refused to take account of any of his difficulties and behaved irresponsibly—it was composed mainly of vainglorious extremists. Davis had more trouble with his congress than any Union president, except possibly Tyler. He vetoed thirty-eight Bills and all but one later passed with Congress overriding his veto. Lincoln had to use the veto only three times, and in each case it stuck.

But many of Davis' difficulties were of his own making. His constant illnesses did not help, as during them he became short-tempered and dictatorial. As his absurd row with Scott showed, he could not distinguish between what mattered and what was insignificant. Virtually all his early appointments, both Cabinet and army, proved bad. Davis resumed personal vendettas going back to the Mexican War and even to his West Point days. In the South, everyone knew each other and most had grudges. In picking senior commanders, Davis favored former West Point classmates, war-service comrades, and personal friends. Things were made even more difficult by each state demanding its quota of generals, and by muddles Davis made over army regulations. A lot of his bitterest rows with colleagues and subordinates had nothing to do with the actual conduct of the war. The Navy Secretary Stephen Mallory (1813–73), a Trinidadian and one of the few Confederate leaders who knew what he was doing, deplored the fact that 'our fate is in the hands of such self-sufficient, vain, army idiots.' Davis was not the man to run difficult generals, and he became almost insensate with rage when he was personally blamed for lack of men and supplies, above all lack of success. Varina admitted: 'He was abnormally sensitive to disapprobation. Even a child's disapproval discomposed him . . . and the sense of mortification and injustice gave him a repellent manner.' Faced with criticism he could not bear, he took refuge in illness.[77]

A lot of Davis' strategic difficulties were his own fault. Despite conscripting 90 percent of its able white manpower, the South was always short of troops. In January 1862 its army rolls numbered 351,418, against a Unionist strength of 575,917. It reached its maximum in January 1864, when 481,180 were counted under the Confederate flag. Therafter the South's army declined in strength whereas the North's rose, so that in January 1865 the respective numbers were 445,203 and 959,460.[78] That being so, Davis should have concentrated his smaller forces in limited areas. Instead, he took seriously and followed to the letter his inauguration oath to defend every inch of Confederate territory. This was an impossible task. It involved, to begin with, defending over 3,500 miles of coastline, without a navy to speak of. Texas alone had 1,200 miles of border. If Kentucky had seceded, it would have provided a simple water-border. For a time it kept out both sides, but eventually the Unionists menaced the South from there too. Missouri was also divided but its settled eastern reaches, centered on St Louis, were firmly Unionist, and that left an almost indefensible 300-mile straight-line border in northern Arkansas. Hence a large percentage of the Confederate army, perhaps a third or even more, was always employed on non-combative defensive duties when its active commanders were clamoring desperately for troops. It is true that the Unionists also used vast numbers of men on the gradually extending lines of communication—but then they had more men to use.[79]

Early in the war the Confederate capital was moved from Montgomery to Richmond, mainly to insure that Virginia stayed committed to the fight. This was a mixed blessing. The polished Virginians regarded the South Carolinians, who formed the core of the government, as loudmouthed, flashy, dangerous extremists. They looked down their noses at the Davises. The ladies noted Varina's dark color and thick lips, comparing her to 'a refined mulatto cook' and called her the 'Empress,' a reference to the much-despised Eugénie, wife of the French dictator, Napoleon III. The Georgians, especially Thomas Cobb, were hostile to Davis: he was, said Cobb, as 'obstinate as a mule,' and they dismissed J. P. Benjamin (1811–84), the Attorney-General and by far the ablest member of the Confederate government, as a 'Jew dog.' Senator Louis T. Wigfall of Texas was a strong Davis supporter until their wives fell out, wherupon Charlotte Wigfall, a South Carolina snob, called Varina 'a course, western woman' with 'objectionable' manners, and Wigfall preached mutiny and sedition in the Congress, often when drunk. Confederate Richmond gradually became a snakepit of bitter social and political feuds, and the Davises ceased to entertain.

Once Northern armies began to penetrate Confederate soil, the interests of the states diverged and it was everyone for himself, reflected in Richmond's savage political feuding. It is a curious paradox that ordinary Southerners, who had not been consulted, fought the war with extraordinary courage and endurance, while their elites, who had plunged them into Armageddon, were riven by rancorous factions and disloyalty, and many left the stricken scene long before the end.[80] Davis was too proud, aloof, and touchy to build up his own faction. He thought it beneath him to seek popularity or to flatter men into doing their duty. Hence 'close friends sometimes left shaking their heads or fists, red with anger and determined never to call on him again.'[81] But at least he went down with the stricken cause, ending up in Unionist fetters.

It may be asked: all this being so, why did the South fight so well? Why did the war last so long? In the first place, it has to be understood that Lincoln was operating under many restraints. He did not seek war, want war, or, to begin with, consider he was in any way gifted to wage it. He made a lot of mistakes, especially with his generals, but unlike Davis he learned from them. The South was fighting for its very existence, and knew it; there was never any lack of motivation there. The North was divided, bemused, reluctant to go to war; or, rather, composed of large numbers of fanatical anti-slavers and much larger numbers of unengaged or indifferent voters who had no wish to become involved in a bloody dispute about a problem, slavery, which did not affect them directly. Then there were the four border states, all of them slave-owning, whose adherence to the Union it was essential to retain. Lincoln, beginning with a professional army of a mere 15,000, was fighting a war waged essentially for a moral cause, and he had to retain the high moral ground. But he had also to keep the rump of the Union together. That meant he had to be a pragmatist without ever descending into opportunism. His great gift—perhaps the greatest of the many he possessed—was precisely his ability to invest his decisions and arguments with moral seemliness even when they were the product of empirical necessity. He was asked to liberate the slaves—what else was the war about? He answered: it was to preserve the Union. He realized, he knew for a fact, that if he did preserve the Union, slavery would go anyway. But he could not exactly say so, since four of his states wanted to retain it.

Some of Lincoln's generals, for military purposes, began to issue local emancipation decrees, hoping to get the Southern slaves to rise and cause trouble behind Confederate lines. Lincoln had to disavow these efforts as *ultra vires*. He hated slavery. But he loved the Constitution more, writing to a friend in Kentucky:

> I am naturally anti-slavery. If slavery is not wrong, nothing is wrong. I cannot remember when I did not so think and feel, and yet I have never understood that the presidency conferred on me an unrestricted right to act officially on this judgment and feeling. It was in the oath I took that I would, to the best of my ability, preserve, protect and defend the

Constitution of the United States. I could not take the office without taking the oath. Nor was it my view that I might take an oath to get power, and break the oath in using the power.

He made public his intentions about slavery in an order disavowing an emancipation decree issued by General David Hunter. Declaring it 'altogether void' and rejecting the right of anyone except himself to liberate the slaves, he nonetheless made it publicly clear that such a right might well be invested in his presidential power: 'I further make it known that whether it be competent for me, as Commander-in-Chief of the Army and Navy, to declare the slaves of any State or States free, and whether at any time and in any case, it shall have become a necessity indispensable to the maintenance of the Government to exercise such supposed power, are questions which, under my responsibility, I reserve to myself, and which I cannot feel justified in leaving to the decision of commanders in the field.'[82]

He followed this up by writing a reply to Horace Greeley, who had published a ferocious editorial in the *New York Tribune*, entitled 'The Prayer of Twenty Millions,' accusing Lincoln of being 'strangely and disastrously remiss' in not emancipating the slaves, adding that it was 'preposterous and futile' to try to put down the rebellion without eradicating slavery.[83] Lincoln replied by return of post, without hesitation or consultation, and for all to read:

My paramount object in this struggle is to save the Union and it is not either to save or to destroy slavery. If I could save the Union without freeing any slaves, I would do it; and if I could save it by freeing all the slaves I would do it; and if I could save it by freeing some slaves and leaving others alone I would do that. What I do about slavery and the colored race I do because I believe it helps to save the Union . . . I shall do less whenever I believe that what I am doing hurts the cause, and I shall do more whenever I believe doing more helps the cause.[84]

In seeking to keep the Union together, and at the same time do what was right by the slaves, the innocent victims as well as the cause of the huge convulsive struggle, Lincoln was fully aware that the Civil War was not merely, as he would argue, an essentially constitutional contest with religious overtones but also a religious struggle with constitutional overtones. The enthusiasts on both sides were empowered by primarily moral and religious motives, rather than economic and political ones.

In the South, there were standard and much quoted texts on negro inferiority, patriarchal and Mosaic acceptance of servitude, and of course
St Paul on obedience to masters. In the events which led up to the war,
both North and South hurled texts at each other. Revivalism and the
evangelical movement generally played into the hands of extremists on
both sides.[85] When the war actually came, the Presbyterians, from
North and South, tried to hold together by suppressing all discussion of
the issue; but they split in the end. The Congregrationalists, because of
their atomized structure, remained theoretically united but in fact were
divided in exactly the same way as the others. Only the Lutherans, the
Episcopalians, and the Catholics successfully avoided public debates
and voting splits; but the evidence shows that they too were fundamentally divided on a basic issue of Christian principle.[86]

Moreover, having split, the Christian churches promptly went to
battle on both sides. Leonidas Polk, Bishop of Louisiana, entered the
Confederate army as a major-general and announced: 'It is for constitutional liberty, which seems to have fled to us for refuge, for our hearth-
stones and our altars that we fight.' Thomas March, Bishop of Rhode
Island, preached to the militia on the other side: 'It is a holy and righteous cause in which you enlist . . . God is with us . . . the Lord of Hosts
is on our side.' The Southern Presbyterian Church resolved in 1864:
'We hesitate not to affirm that it is the peculiar mission of the Southern
Church to conserve the institution of slavery, and to make it a blessing
both to master and slave.' It insisted that it was 'unscriptural and fanatical' to accept the dogma that slavery was inherently sinful: it was 'one
of the most pernicious heresies of modern times.'

To judge by the hundreds of sermons and specially composed
church prayers which have survived on both sides, ministers were
among the most fanatical of the combatants from beginning to end.
The churches played a major role in dividing the nation, and it may be
that the splits in the churches made a final split in the nation possible.
In the North, such a charge was often willingly accepted. Granville
Moddy, a Northern Methodist, boasted in 1861: 'We are charged
with having brought about the present contest. I believe it is true we
did bring it about, and I glory in it, for it is a wreath of glory round
our brow.' Southern clergymen did not make the same boast but of all
the various elements in the South they did the most to make a secessionist state of mind possible. Southern clergymen were particularly responsible for prolonging the increasingly futile struggle. Both sides claimed
vast numbers of 'conversions' among their troops and a tremendous

increase in churchgoing and 'prayerfulness' as a result of the fighting.[87]

The clerical interpretation of the war's progress was equally dogmatic and contradictory. The Southern Presbyterian theologian Robert Lewis Dabney blamed what he called the 'calculated malice' of the Northern Presbyterians and called on God for 'a retributive providence' which would demolish the North. Henry Ward Beecher, one of the most ferocious of the Northern clerical drum-beaters, predicted that the Southern leaders would be 'whirled aloft and plunged downward for ever and ever in an endless retribution.' The New Haven theologian Theodore Thornton Munger declared, during the 'March through Georgia,' that the Confederacy had been 'in league with Hell,' and the South was now 'suffering for its sins' as a matter of 'divine logic.' He also worked out that General McClellan's much criticized vacillations were an example of God's masterful cunning since they made a quick Northern victory impossible and so insured that the South would be much more heavily punished in the end.[88]

As against all these raucous certainties, there were the doubts, the puzzlings, and the agonizing efforts of Abraham Lincoln to rationalize God's purposes. To anyone who reads his letters and speeches, and the records of his private conversations, it is hard not to believe that, whatever his religious state of mind before the war again, he acquired faith of a kind before it ended. His evident and total sincerity shines through all his words as the war took its terrible toll. He certainly felt the spirit of guidance. 'I am satisfied,' he wrote, 'that when the Almighty wants me to do or not to do a particular thing, he finds a way of letting me know it.' He thus waited, as the Cabinet papers show, for providential guidance at certain critical points of the war. He never claimed to be the personal agent of God's will, as everybody else seemed to be doing. But he wrote: 'If it were not for my firm belief in an overriding providence it would be difficult for me, in the midst of such complications of affairs, to keep my reason in its seat. But I am confident that the Almighty has his plans and will work them out; and . . . they will be the wisest and the best for us.' When asked if God was on the side of the North, he replied: 'I am not at all concerned about that, for I know the Lord is always on the side of the right. But it is my constant anxiety and prayer that I and this nation should be on the Lord's side.' As he put it, 'I am not bound to win but I am bound to be true. I am not bound to succeed, but I am bound to live up to the light I have.'[89]

Early in the war, a delegation of Baltimore blacks presented him with a finely bound Bible, in appreciation of his work for the negroes.

He took to reading it more and more as the war proceeded, especially the Prophets and the Psalms. An old friend, Joshua Speed, found him reading it and said: 'I am glad to see you so profitably engaged.' Lincoln: 'Yes. I *am* profitably engaged.' Speed: 'Well, I see you have recovered from your skepticism [about religion and the progress of the war]. I am sorry to say that I have not.' Lincoln: 'You are wrong, Speed. Take all of this book upon reason that you can, and the balance on faith, and you will live and die a happier and a better man.' As he told the Baltimore blacks: 'This Great Book . . . is the best gift God gave to man.'[90] After reading the Bible, Lincoln argued within himself as to what was the best course to pursue, often calling in an old friend like Leonard Swett, to rehearse pros and cons before a sympathetic listener.

Thus arguing within himself, Lincoln incarnated the national, republican, and democratic morality which the American religious experience had brought into existence—probably more completely and accurately than a man committed to a specific church. He caught exactly the same mood as President Washington in his Farewell Message to Congress, and that is one reason why his conduct in the events leading up to the war, and during the war itself, seems, in retrospect—and seemed so to many at the time—so unerringly to accord with the national spirit. Unlike Governor Winthrop and the first colonists, Lincoln did not see the republic as the Elect Nation because that implied it was always right, and the fact that the Civil War had occurred at all indicated that America was fallible. But, if fallible, it was also anxious to do right. The Americas, as he put it, were 'the Almost Chosen People' and the war was part of God's scheme, a great testing of the nation by an ordeal of blood, showing the way to charity and thus to rebirth.

In this spirit Lincoln approached the problem of emancipating the slaves. The moment had to be well chosen not merely to keep the border states in the war, and fighting, but because in a sense it marked a change in the object for which the war was being fought. Lincoln had entered it, as he said repeatedly, to preserve the Union. But by the early summer of 1862 he was convinced that, by divine providence, the Union was safe, and it was his duty to change the object of the war: to wash away the sin of the Constitution and the Founding Fathers, and make all the people of the United States, black as well as white, free. Providence had guided him to this point; now providence would guide him further and suggest the precise time when the announcement should be made, so as to bring victory nearer.

Lincoln had weighed all the practical arguments on either side some time before he became convinced, for reasons which had little to do with political factors, that the slaves should be declared free, and laid his decision before the Cabinet on July 22. He told his colleagues he had resolved upon this step, and had not called them together to ask their advice but 'to lay the subject-matter of a proclamation before them.' Their response was pragmatic. Edwin M. Stanton (1814–69), Secretary of War, and Edward Bates (1793–1869), Attorney-General, urged 'immediate promulgation' for maximum effect. Chase thought it would unsettle the government's financial position. Postmaster-General Montgomery Blair (1813–83) said it would cost them the fall elections. Lincoln was unperturbed. The decision was taken: all that was now required was guidance over the timing. 'We *mustn't issue it* until after a victory,' he said, many times. That victory came, as he knew it would, on September 17, with Antietam. Five days later, on September 22, the Emancipation Proclamation, the most revolutionary document in United States history since the Declaration of Independence, was made public, effective from January 1, 1863. Despite an initially mixed reception, the ultimate impact of this move on the progress of the war was entirely favorable—as Lincoln, listening to the heedings of providence, knew it would be.[91]

Political considerations—holding the Union together, putting his case before world opinion, in which emancipation played a key part, satisfying his own mind that the war was just and being justly pursued—were not the only considerations for Lincoln, or even the chief ones. The overriding necessity, once the fighting began, was to win, and that Lincoln found the most difficult of all. His problem was not providing the men and the supplies, or the money to pay for them. The money was spent on a prodigious scale, and soon exceeded $2 million a day. At the outset of the conflict, the US public debt, which had risen slowly since President Jackson wiped it out, was a little under $70 million. By January 1, 1866, when the end of the insurrection was officially proclaimed, it stood at $2,773 million. But Congress was willing to vote heavy taxes including, for the first time, a tax on personal incomes of from 3 to 5 percent (it was phased out in 1872). All the same, payments in specie had to be suspended at the end of December 1861, and in February 1862 Lincoln signed an Act making Treasury notes legal tender. This was followed by the issue of greenbacks, so called on account of their color, both simple paper and interest-bearing.

The fluctuations in the value of government paper against gold were at times frenzied, depending on the military news, and some serious mistakes were made. In attempts to reduce inflation, Treasury Secretary Chase went in person to the Wall Street markets and sold gold, and he got Congress to pass an Act prohibiting contracts in gold on pain of fines and imprisonment. This crude and brazen attempt to interfere with the market proved disastrous. Chase was forced to resign, and his successor, William P. Fessenden (1806–69), quickly persuaded Congress to withdraw it. But on the whole inflation was kept under control and some of the wartime measures—the transformation of 1,400 state banks of issue into a much smaller number of national banks, 1863–4, for instance—were highly beneficial and became permanent.[92]

The problem was generals who would fight—and win. General Scott, head of the army, was not a man of the highest wisdom, as we have seen; he was also seventy-five and ultra-cautious. The overall strategy he impressed on Lincoln was to use the navy to blockade the Confederacy, the number of vessels being increased from 90 to 650, and to divide the South by pushing along the main river routes, the Mississippi, the Tennessee, and the Cumberland. But there was a desire among lesser generals, especially Confederate ones, to have a quick result by a spectacular victory, or by seizure of the enemy's capital, since both Richmond and Washington were comparatively near the center of the conflict. In July 1861 one of Davis' warriors, General P. G. T. Beauregard (1818–93), a flashy New Orleans aristo of French descent, who had actually fired the first shots at Sumter, pushed towards Washington in a fever of anxiety to win the first victory. He was joined by another Confederate army under General Joseph E. Johnston (1807–91), and together they overwhelmed the Unionist forces of General Irvin McDowell (1818–85) at Bull Run, July 21, 1861, though not without considerable difficulty. The new Unionist troops ended by running in panic, but the Confederates were too exhausted to press on to Washington.

The battle had important consequences nonetheless. McDowell was superseded by General George B. McClellan (1826–85), a small, precise, meticulous, and seemingly energetic man who knew all the military answers to everything. Unfortunately for Lincoln and the North, these answers added up to reasons for doing nothing, or doing little, or stopping doing it halfway. His reasons are always the same; not enough

men, or supplies, or artillery. As the North's overwhelming preponderance in manpower and hardware began to build up, McClellan refused
to take advantage of it, by enticing the South into a major battle and
destroying its main army. The War Secretary said of him and his subordinates: 'We have ten generals there, every one afraid to fight . . . If
McClellan had a million men, he would swear the enemy had two million, and then he would sit down in the mud and yell for three.'[93]
Lincoln agreed: 'The general impression is daily gaining ground that
[McClellan] does not intend to do anything.' At one point Lincoln
seems to have seriously believed McClellan was guilty of treason and
accused him to his face, but backed down at the vehemence of the gen-
eral's response. Later, he concluded that McClennan was merely guilty
of cowardice. When Lincoln visited the troops with his friend O. M.
Hatch, and saw the vast array from a high point, he whispered:
'Hatch—Hatch, what is all this?' Hatch: 'Why, Mr Lincoln, this is the
Army of the Potomac.' Lincoln (loudly): 'No, Hatch, no. This is
General McClellan's bodyguard.'

The best thing to be said for McClellan is that he had close links
with Allan Pinkerton (1819–84), the Scots-born professional detective,
who had opened a highly successful agency in Chicago. During
Lincoln's campaign for the presidency, and his inauguration, Pinkerton
had organized his protection, and undoubtedly frustrated at least one
plot to assassinate him. McClellan employed him to build up a system
of army intelligence, part of which worked behind Confederate lines,
with great success. It eventually became the nucleus of the federal secret
service. But Lincoln seems to have known little of this. He believed,
almost certainly rightly, that at Antietam in September 1862,
McClellan, with his enormous preponderance, could have destroyed
the main Confederate army, had he followed up his initial successes
vigorously, and thus shortened the war. So he finally removed his non-
fighting general, and Pinkerton went with him; and the absence of
Pinkerton's thoroughness was the reason why it proved so easy to murder Lincoln in 1865.[94]

First Bull Run had mixed results for the Confederates. It appeared to
be the doing of Beauregard, and so thrust him forward: but he proved
one of the least effective and most troublesome of the South's generals.
In fact the victory was due more to Johnston, who was a resolute, daring, and ingenious army commander. On April 6–7, 1862, in the first
major battle of the war at Shiloh, at Pittsburg Landing in Tennessee,
Johnston hurled his 40,000 troops against General Ulysses S. Grant

(1822–85), who had only 33,000. The first day's fighting brought over-whelming success to the Confederates but Johnston was wounded towards the end of it. That proved a disaster for the South: not only was their best general to date lost, but Grant turned the tide of battle the next day by leading a charge personally and the Confederates were routed. However, Johnston was not the only man brought to the fore by First Bull Run. During the melée, the officer commanding the South Carolina volunteers rallied his frightened men by pointing to the neigh-boring brigade commanded by General Thomas J. Jackson (1824–63) and saying: 'There stands Jackson like a stone wall.' The name stuck and Jackson's fame was assured. But it was inappropriate. Jackson was not a defensive commander but a most audacious and determined offensive one, with the true killer instinct of a great general. There was only one way the South might win the war. That was by enveloping and destroying in battle the main Unionist Army of the Potomac, tak-ing Washington and persuading the fainthearts on the Unionist side—there were plenty of them—that the cost of waging the war was too high and that a compromise must be sought. Had Lincoln thus been deserted by a majority in Congress, he would have resigned, and the whole of American history would have been different.[95]

Jackson was an orphan, the son of a bankrupt lawyer from Allegheny, Virginia. He was about as unSouthern as it was possible for a Virginia gentleman to be. As Grant put it, 'He impressed me always as a man of the Cromwell stamp, much more of a New Englander than a Virginian.' He was a Puritan. There is a vivid pen-portrait of him by Mrs James Chesnut, a Richmond lady who kept a war diary. He said to her dourly: 'I like strong drink—so I never touch it.' He sucked lemons instead and their sourness pervaded his being. He had no sense of humor, and tried to stamp out swearing and obscene joking among his men. He was 'an ungraceful horseman mounted on a sorry chestnut with a shambling gait, his huge feet with out-turned toes thrust into his stirrups, and such parts of his countenance as the low visor of his stocking cap failed to conceal wearing a wooden look.' Jackson had no slaves and there are grounds for believing he detested slavery. In Lexington he set up a school for black children, something most Southerners hated—in some states it was unlawful—and persisted in it, despite much cursing and opposition. His sister-in-law, who wrote a memoir of him, said he accepted slavery 'as it existed in the Southern States, not as a thing desirable in itself, but as allowed by Providence for ends it was not his business to determine.'

Yet, as Grant said, 'If any man believed in the rebellion, he did.' Jackson fought with a ferocity and single-minded determination which no other officer on either side matched. Mrs Chesnut records a fellow-general's view: 'He certainly preferred a fight on Sunday to a sermon. [But] failing to manage a fight, he loved next best a long, Presbyterian sermon, Calvinist to the core. He had no sympathy for human infirmity. He was the true type of all great soldiers. He did not value human life where he had an object to accomplish.' His men feared him: 'He gave orders rapidly and distinctly and rode away without allowing answer or remonstrance. When you failed, you were apt to be put under arrest.' He enjoyed war and battle, believing it was God's work, and he was ambitious in a way unusual for Southerners, who were happy-go-lucky except in defense of their beliefs and ways. Jackson would have liked to have been a dictator for righteousness. But, having won the terrifying Battle of Chancellorsville in May 1863, he was shot in the back by men of one of his own bri- gades, Malone's, who supposedly mistook him in the moonlight for a Yankee. After Jackson's death the Confederacy lost all its battles except Chickamauga.[96]

Jackson was not the only superb commander on the Confederate side. Colonel John Singelton Mosby (1833–1916), who worked behind the Unionist lines, also had the killer instinct. Like many Southern officers, he was a wonderful cavalryman, but he had solid sense too. General Richard Taylor, son of President Taylor, who wrote the best book about the war from inside the Southern high ranks, summed it up: 'Living on horseback, fearless and dashing, the men of the South afforded the best possible material for cavalry. They had every quality but discipline.'[97] Mosby would have none of that nonsense and was the first cavalryman to throw away his saber as useless and pack two pistols instead. He hated the Richmond set-up—'Although a revolutionary government, none was ever so much under the domination of red tape as Richmond'— and that was one reason he chose the sabotage role, remote from the order-chattering telegraph. The damage he did to the Unionist lines of communication was formidable and he was hated accordingly. On Grant's orders, any of his men who were captured were shot. In the autumn of 1864, for instance, General George Custer executed six of them: he shot three, hanged two, and a seventeen-year-old boy, who had borrowed a horse to join Mosby, was dragged through the streets by two men on horses and shot before the eyes of his mother, who begged Custer to treat the boy as a prisoner-of-war. This treatment stopped immediately Mosby began to hang his prisoners in retaliation.

Mosby was 'slender, gaunt and active in figure . . . his feet are small and cased in cavalry boots with brass spurs, and the revolvers in his belt are worn with an air of "business." '[98] He had piercing eyes, a flashing smile, and laughed often but was always in deadly earnest when fighting. He was the stuff of which Hollywood movies are made and indeed might have figured in one since he lived long enough to see *Birth of a Nation*. He became a myth-figure in the North: he was supposed to have been in the theater when Lincoln was shot, masterminding it, and to have planned all the big railroad robberies, long after the war. But he was the true-life hero of one of the best Civil War stories. During a night-raid he caught General Edwin H. Stoughton naked in bed with a floozie and woke him up roughly. 'Do you know who I am, sir?' roared the general. Mosby: 'Do you know Mosby, General?' Stoughton: 'Yes! Have you got the — rascal?' Mosby: 'No, but *he has got you!*'[99]

Jackson and Mosby were the only two Confederate generals who were consistently successful. Jackson's death made it inevitable that Lee would assume the highest command, though it is only fair to Lee to point out that he was finally appointed commander-in-chief of the Southern forces only in February 1865, just two months before he was forced to surrender them at Appomattox. Lee occupies a special place in American history because he was the South's answer to the North's Lincoln: the leader whose personal probity and virtuous inspiration sanctified their cause.[100] Like Lincoln, though in a less eccentric and angular manner, Lee looked the part. He radiated beauty and grace. Though nearly six feet, he had tiny feet and there was something feminine in his sweetness and benignity. His fellow-cadets at West Point called him the 'Marble Model.' With his fine beard, tinged first with gray, then white, he became a Homeric patriarch in his fifties. He came from the old Virginian aristocracy and married into it. His father was Henry Lee III, Revolutionary War general, Congressman and governor of Virginia. His wife, Anne Carter, was great-granddaughter of 'King' Carter, who owned 300,000 acres and 1,000 slaves. That was the theory, anyway. In fact Lee's father was also 'Light Horse Harry,' a dishonest land-speculator and bankrupt, who defrauded among others George Washington. President Washington dismissed his claim to be head of the United States Army with the brisk, euphemistic, 'Lacks economy.' Henry was jailed twice and when Robert was six fled to the Caribbean, never to return. Robert's mother was left a penurious widow with many children and the family's reputation was not improved by a ruffianly stepson, 'Black Horse Harry,' who specialized in adultery.

So Lee set himself quite deliberately to lead an exemplary life and redeem the family honor. That was a word he used often. It meant everything to him. He led a blameless existence at West Point and actually saved from his meager pay at a time when Southern cadets prided themselves on acquiring debts. His high grades meant he joined the elite Corps of Engineers in an army whose chief occupation was building forts. He worked on taming the wild and mighty river Mark Twain described so well. Lee served with distinction in the Mexican War, ran West Point, then commanded the cavalry against the Plains Indians. It was he who put down John Brown's rebellion and reluctantly handed him over to be hanged. He predicted from the start that the 'War between the States,' as the South called, and calls, it, would be long and bloody. All his instincts were eirenic and, the son of an ardent federalist, he longed for a compromise which would save the Union. But, as he watched the Union Washington had created fall apart, he clung to the one element in it which seemed permanent—Virginia, from which both he and Washington had come and to which he was honor-bound. As he put it, 'I prize the Union very highly and know of no personal sacrifice I would not make to preserve it, *save that of honor.*'[101]

Lee was a profound strategist who believed all along that the South's only chance was to entrap the North in a decisive battle and ruin its army. That is what he aimed to do. With Johnston's death he was put in command of the Army of Northern Virginia and ran it for the next three years with, on the whole, great success. He ended McClellan's threat to Richmond (insofar as it was one) in the Seven Days Battle, routed the Unionists at Second Bull Run (August 1862) but was checked at Antietam the following month. He defeated the Unionists again at Fredericksburg in December 1862 and again at Chancellorsville in May 1863. This opened the way for an invasion of Pennsylvania, heart of the North's productive power, which would force it to a major battle. That is how Gettysburg (July 1863) came about. It was what Lee wanted, an encounter on the grandest possible scale, though the actual meeting-point was accidental, both Lee and General George G. Meade (1815–72), the Unionist commander, blundering into it. Lee had strategic genius, but as field commander he had one great weakness. His orders to subordinate generals were indications and wishes rather than direct commands. As his best biographer has put it, 'Lee was a soldier who preferred to suggest rather than order, a general who attempted to lead from consensus and shrank from confrontation. He insisted on making possible for others the freedom of thought and action he sought for himself.' This method of commanding a large army sometimes worked for Lee but at Gettysburg it proved fatal. On the first day the Confederate success was overwhelming, and on the second (July 2), General James Longstreet (1821–1904) led the main attack on the Union right but delayed it till 4 P.M. and so allowed Meade to concentrate his main force on the strongpoint of Cemetery Ridge. Some positions were secured, however, including Culp's Hill. Meade's counterattack on the morning of July 3 retook Culp's Hill and confronted Lee with the crisis of the battle. He ordered an attack on Cemetery Ridge but did not make it clear to Longstreet that he wanted it taken at any cost. Jackson would have made no bones about it—take the hill or face court-martial. The charge was led by the division commanded by General George E. Pickett (1825–75), with a supporting division and two further brigades, 15,000 in all. Longstreet provided too little artillery support and the assault force was massacred by enfilading Union artillery, losing 6,000 men. Only half a company of Pickett's charge reached the crest; even so, it would have been enough, and the battle won, if Longstreet had thrown in all his men as reinforcements. But he did not do so and the battle, the culmination of the Civil War on the main central front, was lost. Lee sacrificed a third of his men and the Confederate army was never again capable of winning the war. 'It has been a sad day for us,' said Lee at one o'clock the next morning, 'almost too tired to dismount.' 'I never saw troops behave more magnificently than Pickett's division . . . And if they had been supported as they were to have been—but for some reason not yet fully explained to me, were not—we would have held the position and the day would have been ours.' Then he paused, and said 'in a loud voice': 'Too bad! *Too bad!* OH! TOO BAD!'[102]

General Meade was criticized for not following up Lee's retreating forces immediately and with energy, but that was easier said than done—his own men had been terribly mauled. But he was a reliable general and with him in charge of the main front on the Atlantic coast Lincoln could be satisfied. Meanwhile, the war in the West was at last going in the Union's favor. Lincoln's strategy was to neutralize as much of the South as he could, divide it and cut it into pieces, then subdue each separately. The naval war, despite the North's huge preponderance in ships, did not always go its way. The South equipped commercial raiders who altogether took or sank 350 Northern merchant ships, but this was no more than minor attrition. When the Union forces abandoned the naval yard at Portsmouth, Virginia, at the beginning of the war, they scuttled a new frigate *Merrimac*. The Confederates raised it, renamed it *Virginia*, and clad it in iron. It met the Union ironclad *Monitor* in Hampton Roads on March 9, 1862 in an inconclusive five-hour duel, the first battle of iron ships in history. But the Confederates were not able to get the *Virginia* into the Mexican Gulf, where it might have served a strategic purpose. They stationed more troops guarding its base than it was worth. The South could run the blockade but they never came near breaking it, and the brilliant campaign of Commodore David Farragut in the Gulf finally sealed the mouth of the Mississippi.

To the north, and in the Western theater, General Grant achieved the first substantial Union successes on land when he took Forts Henry and Donelson; and after Shiloh he commanded the Mississippi as far south as Vicksburg. The North now controlled the Tennessee River and the Cumberland and it took New Orleans and Memphis. But the South still controlled 200 miles of the Mississippi between Vicksburg and Port Hudson, Louisiana. Vicksburg was strongly fortified and protected by natural defenses. Attempts to take it, in May–June 1862 and again in December–January 1863 failed. In May 1863 Grant made a third attempt, and after a fierce siege in which each side lost 10,000, he forced it to surrender the day after Meade won Gettysburg (July 4). Five days later Port Hudson fell, the entire Mississippi was in Union hands, and the Confederacy was split in two.

In Grant Lincoln at last found a war-winning general, and a man he could trust and esteem. Unlike the others, Grant asked for nothing and did not expect the President to approve his plans in advance and so take the blame if things went wrong.[103] Grant was an unprepossessing general. Lincoln said: 'He is the quietest little man you ever saw. He makes the least fuss of any man I ever knew. I believe on several occasions he has been in [the Oval Office] a minute or so before I knew he was there. The only evidence you have that he's in any particular place is that he makes things move.' Grant was born in 1822 at Point Pleasant, Ohio. His father was a tanner. In his day West Point was, as he put it, a place for clever, hard-working boys 'from families that were trying to gain advancement in position or to prevent slippage from a precarious place.'[104]

Lee, an aristocrat of sorts, was unusual. In Grant's class of '43 were Longstreet, McClellan, and Sherman, among other Civil War generals—all of them meritocrats. The chief instructor in Grant's day, Denis Hart Mahan—father of the outstanding naval strategist—taught them that 'carrying the war into the heart of the assailant's country is the surest way of making him share its burdens and foil his plans.' Lee was never able to do this—Grant and Sherman did. Grant was in the heat of the Mexican War, fighting at Palo Alto, Resaca, Monterrey, and Mexico City, and he learned a lot about logistics, later his greatest strength. But he hated and deplored the war, which he regarded as wholly unjust, fought by a Democratic administration in order to acquire more slave states, especially Texas. He saw the Civil War as a punishment on the entire country by God—'Nations, like individuals, are punished for their transgressions. We got our punishment in the most sanguinary and expensive war of modern times.'

Grant was a man with a strong and simple moral sense. He had a first-class mind. He might have made a brilliant writer—both his letters and his autobiography have the marks of genius. He made an outstanding soldier. But there were fatal flaws in his system of self-discipline. All his adult life he fought a battle with alcohol, often losing it. After the Mexican War, in civilian life, he failed as a farmer, an engineer, a clerk, and a debt-collector. In 1861 he was thirty-nine, with a wife, four children, a rotten job, and not one cent to his name, in serious danger of becoming the town drunk. He welcomed the Civil War because he saw it as a crusade for justice. It changed his life. A neighbor said: 'I saw new energies in him. He dropped his stoop-shouldered way of walking and set his hat forward on his forehead in a jaunty fashion.' He was immediately commissioned a colonel of volunteers and, shortly after, brigadier-general. He was not impressive to look at. He was a small man on a big horse, with an ill-kept, scrappy beard, a cigar clamped between his teeth, a slouch hat, an ordinary soldier's overcoat. But there was nothing slovenly about his work. He thought hard. He planned. He gave clear orders and saw to it they were obeyed, and followed up. His handling of movements and supplies was always meticulous. His Vicksburg campaign, though daring, was a model of careful planning, beautifully executed. But he was also a killer. A nice man, he gave no mercy in war until the battle was won. Lincoln loved him, and his letters to Grant are marvels of sincerity, sense, brevity, fatherly wisdom, and support. In October 1863 Lincoln gave Grant supreme command in the West, and in March 1864 he put him in charge of the main front, with the title of General-in-Chief of the Union army and the rank of lieutenant-general, held by no one since Washington and specially revived in Grant's favor by a delighted Congress.[105]

Nevertheless, the war was not yet won, and it is a tribute to the extraordinary determination of people in the South, and the almost unending courage of its soldiers, that, despite all the South's handicaps, and the North's strength, the war continued into and throughout 1864, more desperate than ever. The two main armies, the Army of the Potomac (North) and the Army of Northern Virginia (South) had faced each other and fought each other for three whole years and, as Grant said, 'fought more desperate battles than it probably ever before fell to the lot of two armies to fight, without materially changing the vantage ground of either'—it was, indeed, a murderous foretaste of the impenetrable Western Front of World War One. What to do, then? Grant, after much argument with Lincoln, who steered him away from more ambitious alternatives, determined on a two-pronged strategy. One army under General William T. Sherman (1820–91), who had taken over from Grant as commander-in-chief in the West, would sweep through Georgia and destroy the main east–west communications of the Confederacy. Grant's main army would clear the almost impassable Wilderness Region west of Fredericksburg, Virginia, in preparation for a final assault on Lee's army. The Battle of the Wilderness began on May 5–6, 1864, while on the 7th Sherman launched his assault on Atlanta and so to the sea.

The Wilderness battle proved indecisive, though horribly costly in men, and three days later Grant was repulsed at Spotsylvania with equally heavy loss. At the end of the month Grant again attacked at Cold Harbor, perhaps the most futile slaughter of the entire war. In six weeks Grant had lost 60,000 men. Lee, too, had lost heavily—20,000 men, which proportionate to his resources was even more serious than the North's casualties. Nonetheless, Lincoln was profoundly disturbed by the carnage and failure. The Speaker of the House, Schuyler Colfax, found him pacing his office, 'his long arms behind his back, his dark features contracted still more with gloom,' explaining: 'Why do we suffer reverses after reverses? Could we have avoided this terrible, bloody war? . . . Is it ever to end?' Francis B. Carpenter, who was painting his *First Reading of the Emancipation Proclamation by President Lincoln*, described him in the hall of the White House, 'clad in a long morning wrapper, pacing back and forth a narrow passage leading to one of the windows, his hands behind him, great black rings under his eyes, his head bent forward upon his breast—altogether . . . a picture of the effects of sorrow, care and anxiety.'[106]

All the same, the noose was tightening round the South. Davis himself felt it. Even before Gettysburg, he had personally been forced to quell a food riot of hungry women in Richmond. Unionist troops over-ran his and his brother's property, taking the whites prisoner and allowing the blacks to go. Some 137 slaves fled to freedom leaving, on Davis' own estate, only six adults and a few children. His property was betrayed by a slave he trusted, the soldiers cut his carpets into bits as souvenirs, they drank his wine, stabbed his portrait with knives, and got all his private papers, spicy extracts from which duly appeared in the Northern newspapers. In Richmond, Davis had to sell his slaves, his horses, and his carriage just to buy food—ersatz coffee, pones or corn-cakes, bread, a bit of bacon. Jeb Stuart, Davis' best cavalry comman-der, fell, mortally wounded. He had one good general, Lee, marking Grant; but Lincoln had two—and Sherman now took Atlanta, moved through Georgia, burning and slaughtering, and on December 21, 1864 was in Savannah, having cut the Confederacy in two yet again. By Christmas much of the South was starving. Davis had made Lincoln's job of holding the North together easier by proclaiming, for four years, that he would not negotiate about anything except on the basis of the North admitting the complete independence of the South. Now he again insisted the South would 'bring the North to its knees before next summer.' On hearing this rodomontade, his own Vice-President, Alexander Stephens (1812–83), told him in disgust he was leaving for his home and would not return—it was the beginning of the disintegra-tion of the Confederate government.[107]

Much of the South was now totally demoralized by military occupa-tion. Sarah Morgan of Baton Rouge, who kept a diary, described the sacking of her house:

one scene of ruin. Libraries emptied, china smashed, sideboards split open with axes, three cedar chests cut open, plundered and set up on end; all parlor ornaments carried off. [Her sister Margaret's] piano, dragged to the center of the parlor had been abandoned as too heavy to carry off; her desk lay open with all letters and notes well thumbed and scattered around, while Will's last letter to her was open on the floor, with the Yankee stamp of dirty fingers. Mother's portrait half cut from the frame stood on the floor. Margaret, who was present at the sacking, told how she had saved father's. It seems that those who wrought destruction in our house were all officers![108]

The destruction in Georgia was worse. Like Grant, Sherman was a decent man but a fierce, killer general, determined to end the war and the slaughter as speedily as possible and, with this his end, anxious to demonstrate to the South in as plain a manner as he could that the North was master and resistance futile. He cut a swathe 60 miles wide through Georgia, destroying everything—railroads, bridges, crops, cattle, cotton-gins, mills, stocks—which might conceivably be useful to the South's war-effort. Despite his orders, and the generally tight discipline of his army in action, the looting was appalling and the atrocities struck fear and dismay into the stoutest Southern hearts.[109]

Sherman's capture of Atlanta and his rout of the Southern army in Georgia came in time—just—to insure Lincoln's reelection. During the terrible midsummer of 1864 there had been talk, by 'Peace Democrats,' of doing a deal with Davis and getting control of both armies, thus ending both the rebellion and Republican rule. Many prominent Republicans thought the war was lost and wanted to impose Grant as a kind of president–dictator. He wrote to a friend saying he wanted 'to stick to the job I have'—and the friend showed it to Lincoln. Lincoln observed: 'My son, you will never know how gratifying that is to me. No man knows, when that presidential grub starts to gnaw at him, just how deep it will get until he has tried it. And I didn't know but what there was one gnawing at Grant.' The general put an end to intrigue by stating: 'I consider it as important to the cause that [Lincoln] should be reelected as that the army should be successful in the field.'[110]

Sherman's successes in September, and his continued progress through Georgia, swung opinion strongly back in Lincoln's favor. The increasing desperation of the South, expressed in terrorism, bank-raids, and murder in Northern cities, inflamed the Northern masses and were strong vote-winners for the Republicans. The resentful McClellan fared disastrously for the Democrats. Lincoln carried all but three of the participating states and 212 electoral votes out of 233, a resounding vote of confidence by the people.[111] He entered his second term of office in a forthright but still somber mood, in which the religious overtones in his voice had grown stronger. They echo through his short Second Inaugural, a meditation on the mysterious way in which both sides in the struggle invoked their God, and God withheld his ultimate decision in favor of either:

> Both read the same Bible, and pray to the same God; and each invokes His aid against the other. It may seem strange that any men should dare to ask a just God's assistance in wringing their bread from the sweat of other men's faces; but let us judge not that we be not judged. The prayers

of both could not be answered; that of neither has been answered fully. The Almighty has His own purposes: 'Woe unto the world because of offenses! for it needs be that offenses come, but woe unto that man by whom the offenses cometh!' . . . Fondly do we hope—fervently do we pray—that this mighty scourge of war may pass away. Yet if God wills that it continue, until all the wealth piled by the bond-man's two-hundred-and-fifty years of unrequited toil shall be sunk, and until every drop of blood drawn by the lash shall be paid with another drawn by the sword, as was said three thousand years ago, so still it must be said 'the judgements of the Lord are true and righteous altogether'.

So Lincoln asked the nation to continue the struggle to the end, 'With malice to none, with charity to all, with firmness in the right, as God gives us to see the right.'[112]

The Second Inaugural began the myth of Lincoln in the hearts of Americans. Those who actually glimpsed him were fascinated by his extraordinary appearance, so unlike the ideal American in its massive lack of beauty, so incarnate of the nation's spirit in some mysterious way. Nathaniel Hawthorne wrote (1862):

The whole physiognomy is as coarse a one as you would meet anywhere in the length and breadth of the state; but withal, it is redeemed, illuminated, softened and brightened by a kindly though serious look out of his eyes, and an expression of homely sagacity, that seemed weighted with rich results of village experience. A great deal of native sense, no bookish cultivation, no refinement; honest at heart, and thoroughly so, and yet in some sort, sly—at least endowed with a sort of tact and wisdom that are akin to craft, and would impel him, I think, to take an antagonist in flank, rather than make a bull-run at him right in front. But on the whole I like this sallow, queer, sagacious visage, with the homely human sympathies that warmed it; and, for my small share in the matter, would as lief have Uncle Abe for a ruler as any man that it would have been practical to have put in his place.[113]

Walt Whitman, looking at the President from a height in Broadway, noted 'his perfect composure and coolness—his unusual and uncouth height, his dress of complete black, stovepipe hat pushed back on the head, dark-brown complexion, seam'd and wrinkled yet canny-looking face, black, bushy head of hair, disproportionately long neck, and his hands held behind him as he stood observing the people.' Whitman thought 'four sorts of genius' would be needed for 'the complete lining of the Man's future portrait'—'the eyes and brains and finger-touch of Plutarch and Aeschylus and Michelangelo, assisted by Rabelais.'[114]

There is a famous photograph of Lincoln, taken at this time, visiting the HQ of the Army of the Potomac, standing with some of his generals outside their tents. These officers were mostly tall for their times but Lincoln towers above them to a striking degree. It was as if he were of a different kind of humanity: not a master-race, but a higher race. There were many great men in Lincoln's day—Tolstoy, Gladstone, Bismarck, Newman, Dickens, for example—and indeed master spirits in his own America—Lee, Sherman, Grant, to name only three of the fighting men—yet Lincoln seems to have been of a different order of moral stature, and of intellectual heroism. He was a strong man, and like most men quietly confident of their strength, without vanity or self-consciousness—and also tender. Towards the end of the war, Lincoln went to see Seward, his Secretary of State, a man with whom he often disagreed and whom he did not particularly like. Seward had somehow contrived to break both his arm and his jaw in a carriage accident. Lincoln found him not only bedridden but quite unable to move his head. Without a moment's hesitation, the President stretched out at full length on the bed and, resting on his elbow, brought his face near Seward's, and they held an urgent, whispered consultation on the next steps the administration should take. Then Lincoln talked quietly to the agonized man until he drifted off to sleep. Lincoln could easily have used the excuse of Seward's incapacity to avoid consulting him at all. But that was not his way. He invariably did the right thing, however easily it might be avoided. Of how many other great men can that be said?[115]

Lincoln was well aware of the sufferings of those in the North who actively participated in the struggle. They haunted him. He read to his entourage that terrible passage from *Macbeth* in which the King tells of his torments of mind:

> we will eat our meal in fear, and sleep
> In the affliction of these terrible dreams,
> That shake us nightly; better be with the dead
> Than on the torture of the mind to lie
> In restless ecstasy.[116]

One man who was also well aware of the suffering was Whitman. Too old to fight, he watched his younger brother George, a cabinet-maker, enlist for a 100-day stint which turned into four years, during which time he participated in twenty-one major engagements, saw most of his comrades killed, and spent five months in a horrific Confederate prison. Some 26,000 Union soldiers died in these dreadful stockades, and so great was the Union anger at conditions in them, especially at Andersonville, that its commandant, Major Henry Wirz, was the only Southerner to be punished by hanging.[117] Instead of enlisting, Whitman engaged himself in hospital service, first at the New York Hospital, then off Broadway in Pearl Street, later in Washington DC: 'I resigned myself / To sit by the wounded and soothe them, or silently watch the dead.'

In some ways the Civil War hospitals were bloodier than the battlefield. Amputation was 'the trade-mark of Civil war surgery.' Three out of four operations were amputations. At Gettysburg, for an entire week, from dawn till twilight, some surgeons did nothing but cut off arms and legs. Many of these dismemberments were quite unnecessary, and the soldiers knew it. Whitman was horrified by what happened to the wounded, often mere boys. He noted that the great majority were between seventeen and twenty. Some had pistols under their pillows to protect their limbs. Whitman himself was able to save a number by remonstrating with the surgeons. He wrote:

> From the stump of the arm, the amputated hand,
> I undo the clotted lint, remove the slough, wash off the matter
>   and blood.
> Back on his pillow the soldier bends with curv'd neck and side-
>   falling head,
> His eyes are closed, his face is pale, he dares not look on the
>   bloody stump,
> And has not yet looked on it.

More arms and legs were chopped off in the Civil War than in any other conflict in which America has ever been engaged—but a few dozen fewer than might have been, but for Whitman. A paragraph in the *New York Tribune* in 1880 quoted a veteran pointing to his leg: 'This is the leg [Whitman] saved for me.'[118]

Whitman calculated that, during the war, he made over 600 hospital visits or tours, some lasting several days, and ministered in one way or another to over 100,000 soldiers. His book of poems *Drum-Taps* records some of his experiences. Not everyone welcomed his visits. One nurse at the Armory Square hospital said: 'Here comes that odious Walt Whitman to talk evil and unbelief to my boys.' The scale of the medical disaster almost overwhelmed him—one temporary hospital housed 70,000 casualties at one time. Whitman considered the volume and intensity of the suffering totally disproportionate to any objective gained by the war. Others agreed with him. Louisa May Alcott (1832–88), later author of the famous bestseller *Little Women* (1868), spent a month nursing in the Washington front-line hospitals before being invalided home with typhoid, and recorded her experiences in *Hospital Sketches* (1863). This is a terrifying record of bad medical practice, of the kind Florence Nightingale had utterly condemned a decade before, including lethal overdosing with the emetic calomel. At many points her verdict and Whitman's concurred.[119]

Yet it is curious how little impact the Civil War made upon millions of people in the North. When Edmund Wilson came to write his book on the conflict, *Patriotic Gore: Studies in the Literature of the American Civil War* (1962), he was astonished by how little there was of it. There were hymn-songs, of course: 'John Brown's Body,' Julia Ward Howe's 'Battle Hymn of the Republic,' to rally Northern spirits, Daniel Decatur Emmett's 'Dixie' to enthuse the South. The young Henry James was not there—he had 'a mysterious wound,' which prevented serving. Mark Twain was out west. William Dean Howells was a consul in Italy. It was quite possible to live in the North and have no contact with the struggle whatsoever. It is a notable fact that Emily Dickinson (1830–86), America's greatest poet, lived quietly throughout the war in Amherst without it ever impinging on her consciousness, insofar as that is reflected in her poetry. Of her more than 1,700 poems, not one refers directly to the war, or even indirectly, though they often exude terror and dismay. She was educated at Amherst Academy and spent a year at Mount Holyoke Female Seminary: otherwise her life was passed at home, eventless, and for the last twenty-five years of her life in almost complete seclusion. Only six of her poems were published in her lifetime and evidently she did not consider it part of the poet's job to obtain publication. Effectively, she did not emerge as a writer at all until the 1890s, after her death. In a sense, her poetry is internal exploration and could have been written in almost any country, at almost any period of history, with one exception: the South in the 1860s. Had she lived in, say, Charleston or Savannah, she would have been forced to confront external reality in her verse. That is the difference between North and South.[120]

But not only in cloistered New England was the war distant. In vast stretches of America, it had virtually no effect on the rapid development of the country. Not that Westerners were indifferent to the war. They favored the Union because they needed it. The South was protesting not only against the North's interference in its 'peculiar institution' but against the growth of government generally. But Westerners, for the time being at least, wanted some of the services that only federal government could provide. As the historian of the trails through Oregon and to California put it, 'Most pre-Civil War overlanders found the United States government, through its armed forces, military installations, Indian agents, explorers, surveyors, road builders, physicians and mail-carriers to be an impressively potent and helpful force.'[121] Up to the outset of the Civil War, 90 percent of the US Army's active units were stationed in the seventy-nine posts of the trans-Mississippi—7,090 officers and men in 1860. Withdrawal of many units once the war began made Westerners realize quite how dependent they were on federal power.

Lack of troops raised problems in the West and may have encouraged the Indians to take advantage. There were raids and massacres, and the settlers responded by raising volunteers and using them. They were less experienced at dealing with Indians than the regular units, and their officers were often prone to take alarm needlessly and overreact—all the good commanders were out east, fighting. What was liable to happen was demonstrated at Sand Creek in the Colorado Territory on November 29, 1864, just after Lincoln's reelection. Following Indian atrocities, a punitive column consisting of the Third Colorado Volunteers, under Colonel John M. Chivington, attacked a camp of 500 Cheyennes. Their leaders, Black Kettle and White Antelope, believed a peace treaty was in effect and said they had turned in their arms. The volunteers slaughtered men, women, and children indiscriminately, killing over 150, and returning to Denver in triumph, displaying scalps and severed genitals like trophies. This Sand Creek massacre was later investigated by a joint Committee of Congress, and Chivington condemned, though he was never punished. The Cheyennes retaliated brutally on several occasions, and on December 21, 1866, after the war was over, in combination with Lakotas and Arapahos, they ambushed and slaughtered eighty men under the command of Colonel William J. Fetterman, one of the worst defeats the US Army suffered at Indian hands.[122]

In some ways the Civil War hastened the development of the West because, by removing the Southern–Democratic majority in both Houses of Congress, it ended a legislative logjam which had held up certain measures for decades and impeded economic and constitutional progress. For instance, the Californian engineer–promoter Theodore D. Judah, representing a group of San Francisco bankers and entrepreneurs, contrived in the spring of 1861, immediately after the Southerners had left Washington, to lobby the Pacific Railroad Act through Congress. This was entirely a venture to benefit the North and the Northwest. It involved the railroads receiving from the federal government a 400-foot right of way, ten alternate sections of land for each mile of track, and first-mortgage loans of $16,000 per mile in flat country, $32,000 in foothills, and $48,000 per mile in the mountains—an enormous federal subsidy, in effect, which would only have passed over Southern dead bodies. In the event, the subsidy did not prove enough for this giant undertaking and was increased by a further Act of Congress in 1864, the Southerners still being absent.

In fact the North and West got their revenge, during the Civil War, for the many defeats they had suffered at the hands of Southern legislators in the thirty-two years 1829–60. By 1850 the Southern plantation interests had come to see the cheap-land policy in the West as a threat to slavery. Their senators killed the Homestead Bill of 1852, and in 1860, after Southerners made unavailing efforts to kill a similar Bill, President Buchanan vetoed it. Thus a Homestead Bill became an important part of Lincoln's platform and in 1862 it marched triumphantly through Congress. This offered an enterprising farmer 160 acres of public land, already surveyed, for a nominal sum. He got complete ownership at the end of six months on paying $1.50 an acre, or for nothing after five years' residence.[123] This eventually proved one of the most important laws in American history, the consequences of which we will examine shortly. The removal of Southern resistance also speeded up the constitutional development of the West. Kansas entered the Union as a free state in 1861, Nevada in 1864, and Nebraska soon after the end of the war in 1867. Meanwhile the administration extended the territorial system over the remaining inchoate regions beyond the Mississippi. The Dakotas, Colorado, and Nevada territories were organized in 1861, Arizona and Idaho in 1863, and by 1870 Wyoming and Montana had also become formal territories on the way to statehood.[124]

Out west, then, they just got on with it, and made money. The mining boom, which had cost the South any chance of California becoming a slave state, continued and intensified, thus pouring specie into Washington's war-coffers. The classic boomtown, Virginia City, emerged 7,000 feet up the mountains of Nevada, and was immortalized by Mark Twain. The gold and silver were embedded in quartz, and elaborate crushing machinery—and huge amounts of capital—were needed for the big-pay mines, the Ophir, Central, Mexican, Gould, and Curry. Experienced men from Cornwall, Wales, and the German mountains poured in. The Comstock Lode became the great mineralogical phenomenon of the age. It went straight through Virginia City, from north to south, and laboring men earned the amazing wages of $6 a day, working in three shifts, round the clock. So, as Twain wrote, even if you did not own a 'piece' of a mine—and few did not—everyone was happy; 'Joy sat on every countenance, and there was a glad, almost fierce intensity in every eye, that told of the money-getting schemes that were seething in every brain and the high hope that held sway in every heart. Money was as plentiful as dust; every individual considered himself wealthy and a melancholy countenance was nowhere to be seen.'[125]

Any shots fired in these parts had nothing to do with the Civil War but reflected the normal human appetites of greed, lust, anger, and envy. And, as Mark Twain put it, 'the thin atmosphere seemed to carry healing to gunshot wounds, and therefore to simply shoot your adversary through both lungs was a thing not likely to afford you any permanent satisfaction, for he would be nearly certain to be around looking for you within the month, and not with an opera glass, either.' The miners, most of whom were heavily armed, chased away any Indians who stood between them and possible bullion, ignoring treaties. Gold was found in 1860 on the Nez Percé Indians' reservation at the junction of the Snake and Clearwater rivers. The superintendent of Indian affairs reported: 'To attempt to restrain these miners would be like attempting to restrain the whirlwind.' With Washington's attention on the war, protection of reservations had a low priority and the miners did what they pleased. They created the towns of Lewiston, Boise on the Salmon River, and in 1864 Helena. Idaho was a mining-created state; so was Montana, formed out of its eastern part, and Wyoming Territory. Nor was gold and silver the only lure—it was at Butte, Montana, that one of the world's great copper strikes was made. The miners were almost entirely young men between sixteen and thirty; the women nearly all whores. But it was creative: seven states, California, Nevada, Arizona, New Mexico, Colorado, Idaho, and Montana owe their origins to mining—and the key formation period, in most cases, was during the Civil War.[126]

It was totally different in the South: there, nothing mattered, nothing could occur, but the war. Concern for the war, anxiety to win the war, was so intense that people forgot what it was really about. Davis himself forgot to the point where he was among the earliest to urge that slaves should be manumitted in return for fighting for the South. Resistance to this idea was, at first, overwhelming, on the ground that blacks would not or could not fight—this despite the fact that 180,000 blacks from the North were enlisted in the Union army and many of them fought very well indeed. Arguing with a senator who was against enlisting blacks at any price, Davis in exasperation declared: 'If the Confederacy falls, there should be written on its tombstone, "Died of a theory." '[127] As the Union army sliced off chunks of the South and liberated its slaves, many flocked to join the army—apart from anything else, it was the only way they could earn a living. Slavery itself was breaking down, even in those parts of the South not yet under Union rule. Slaves were walking off the plantations more or less as they chose; there was no one to prevent them, and no one to hunt them once they were at liberty. There was no work and no food for them either. So they were tempted to cross the lines and enlist in the Union forces. Hence Davis redoubled his efforts to persuade the Confederate Congress to permit their enlistment. As he put it, 'We are reduced to choosing whether the negroes shall fight for us, or against us.'

Eventually on March 13, 1865, Congress accepted his arguments, but even then it left emancipation to follow enlistment only with the consent of the owner. Davis, in promulgating the new law, added a proviso of his own making it compulsory for the owner of a slave taken into war service to provide manumission papers. But by then it was all too late anyway. Granted the fact that slaves formed more than a third of the South's population at the beginning of the Civil War, their prompt conscription would have enormously added to the strength of the Confederate armies. And most of them would have been willing to fight for the South, too—after all, it was their way of life as well as that of the whites which was at stake. It is a curious paradox, but one typical of the ironies of history, that black participation might conceivably have turned the scales in the South's favor. But obstinacy and 'theory' won the day and few blacks actually got the chance to fight for their homeland.[128]

* * *

The end of the Confederacy was pitiful. On April 1, 1865, Davis sent his wife Varina away from Richmond, giving her a small Colt and fifty rounds of ammunition. The next day he had to get out of Richmond himself. He went to Danville, to plan guerrilla warfare. By this point General Lee was already in communication with General Grant about a possible armistice, and had indeed privately used the word 'surrender,' but he continued to fight fiercely with his army, using it with his customary skills. He dismissed pressure from junior officers to negotiate, and as late as April 8 he took severe disciplinary action against three general officers who, in his opinion, were not fighting in earnest or had deserted their posts. But by the next morning Lee's army was virtually surrounded. He dressed in his best uniform, wearing, unusually for him, a red silk sash and sword. Having heard the latest news of the position of his troops, and the Union forces, he said: 'Then there is nothing left me but to go and see General Grant and I would rather die a thousand deaths.'[129]

The two generals met at Appomattox Court House, Grant dressed in 'rough garb,' spattered with mud. Both men were, in fact, carefully dressed for the occasion, as they wished to appear for posterity. The terms were easily agreed, Grant allowing that Southern officers could keep their sidearms and horses. Lee pointed out that, in the South, the enlisted men in the cavalry and artillery also owned their horses. Grant allowed those to be kept too. After Lee's surrender on April 9, Davis hurried to Greenboro to rendezvous with General Johnston's army. But in the meantime Johnston had reached an agreement with General Sherman which, in effect, dissolved the Confederacy. Davis gave the terms to his Cabinet, saying he wanted to reject them, but the Cabinet accepted them. Washington, however, did not, and the South had to be content, in the end, with a simple laying down of arms.[130]

By this time Lincoln was dead. He had summoned Grant to hear his account of the surrender at Appomattox, and he beamed with pleasure when the general told him that the terms had extended not just to the officers but to the men: 'I told them to go back to their homes and their families and that they would not be molested, if they did nothing more.' Lincoln expected Sherman to report a similar surrender and he told Grant he expected good news as he had just had one of his dreams which portended such. Grant said he described how 'he seemed to be in some singular, indescribable vessel and . . . he was moving with great rapidity to an indefinite shore.'[131] Lincoln told his wife (April 14), who said to him, 'Dear husband, you almost startle me by your great cheerfulness,' 'And well may I feel so, Mary. I consider *this day*, the war has come to a close.'

On April 15 they went to a performance of the comedy *Our American Cousin* at Ford's Theater. Lincoln was no longer protected by Pinkerton, but Marshal Ward Hill Lamon, who often served as his bodyguard, begged him not to go to the theater or any similar place, and on no account to mingle with promiscuous crowds. That evening was particularly dangerous since it had been widely advertised that Grant, too, would join the President in his theater excursion. Name, date, time, place—all were published. John Wilkes Booth (1838–65), from an acting family of British origins, also noted for mental instability, and brother of the famous tragedian Edwin Booth, was a self-appointed Southern patriot. He had three days to organize the assassination, with various associates. He also planned to kill Seward and Vice-President Andrew Johnson (1808–75), regarded with peculiar abhorrence in the South because he was a Democrat and a Southerner, from Tennessee, the only Southerner who remained in the Senate in 1861—and accordingly rewarded with the vice-presidency in Lincoln's second term.

Booth had no difficulty in getting into the theater, and he obtained entry to the President's box simply by showing Charles Forbes, the White House footman on duty, his calling-card. He barred the door of the box, moved behind Lincoln, who was leaning forward, then aimed his Derringer at the back of the President's head and pulled the trigger. He then drew a knife, stabbing Lincoln's ADC, jumped from the box, breaking his ankle in the process, shouted 'Sic semper tyrannis,' the motto of the State of Virginia, and escaped through the back of the theater. Two weeks later he was shot and killed in Bowling Green, Virginia. Lincoln himself was taken to a nearby house where he lingered for nine hours, never regaining consciousness.

It is clear that Booth had links going back to Richmond but equally clear that Davis knew nothing about the assassination plot and would never have authorized it. But many at the time believed he was involved. His last days of liberty were clouded by rumors, including one that a price of $100,000 was on his head and another that he was dressed as a woman. He was taken on May 10. Almost his last words to his colleagues were that he was glad 'no member of his Cabinet had made money out of the war and that they were all broke and poor.' He himself gave his last gold coin to a little boy presented to him as his namesake. All he then had in his pockets was a wad of worthless

Confederate scrip. His soldiers–captors jeered at him: 'We'll hang Jeff Davis from a sour apple tree.' Their commander, Major-General James Wilson, said later: 'The thought struck me once or twice that he was a mad man.'

Davis was put in heavy leg-irons and taken to Fort Monroe, opposite Norfolk, Virginia, where he was held for 720 days mostly in solitary confinement, and subjected to many humiliations, with bugs in his mattress and only a horse-bucket to drink from. None of this would have happened had Lincoln lived. Johnson, now president, insisted on this to prove to Northern opinion that he was not favoring a fellow-Southerner. On the other hand, he hated the idea, put forward by Stanton, the Secretary of War, and others, that Davis should be tried, convicted, and hanged. So he allowed Dr John J. Craven, who visited Davis many times in his cell and had long conversations with him, to smuggle out his diaries and have them written up by a popular writer, Charles G. Halpine. They appeared as *The Prison Life of Jefferson Davis*, presenting him as a tragic hero, aroused much sympathy, even in the North, and prepared the way for his release. Davis detested the book. He refused to ask for a pardon, demanding instead a trial which (he was sure) would lead to his acquittal and vindicate him totally. Instead, a writ of habeas corpus (which Lincoln had suspended but was now permitted again) got him out in May 1867. He then went to Canada, and wrote rambling memoirs, lived to bury all his sons, and died, full of years and honor—in the South at least—in 1889. His funeral, attended by a quarter of a million people, was the largest ever held in the South.[132]

Lee, by contrast, was broken and tired and did not last long. When he died in 1870 people were amazed to learn he was only sixty-three. He spent his last years in the thankless job of running a poor university, Washington College, believing that 'what the South needs most is education.' He refused to write his memoirs, blamed no one, avoided publicity, and, when in doubt, kept his mouth shut. Legend has it that his last words were 'Tell Hill he *must* come up!' and 'Strike the tent!'. In fact he said nothing.[133]

The end of the Civil War solved the problem of slavery and started the problem of the blacks, which is with America still. Everyone, from Jefferson and Washington onwards, and including Lincoln himself, had argued that the real problem of slavery was not ending it but what to do with the freed blacks afterwards. All these men, and the overwhelming majority of ordinary American whites, felt that it was almost impossible for whites and blacks to live easily together. Lincoln did not regard blacks as equals. Or rather, they might be morally equal but in other respects they were fundamentally different and unacceptable as fellow-citizens without qualification. He said bluntly that it was impossible just to free the slaves and make them 'politically and socially our equals.' He freely admitted an attitude to blacks which would now be classified as racism: 'My own feelings will not admit [of equality].' The same was true, he added, of the majority of whites, North as well as South. 'Whether this feeling accords with justice and sound judgment is not the sole question. A universal feeling, whether well- or ill-founded, cannot be safely disregarded.'[134] He told a delegation of blacks who came to see him at the White House and asked his opinion about emigration to Africa or elsewhere, that he welcomed the idea: 'There is an unwillingness on the part of our people, harsh as it may be, for you free colored people to remain with us.' He even founded an experimental colony on the shores of San Domingo, but the dishonesty of the agents involved forced the authorities to ship the blacks back to Washington. All schemes to get the blacks back to Africa had been qualified or total failures, for the simple reason that only a tiny proportion of them ever had the smallest desire to return to a continent for which, instinctively, they felt an ancestral aversion. Like everyone else, they wanted to remain in the United States, even if life there had its drawbacks.

That being so, what to do? And what to do with the rebellious South? On November 19, 1863, Lincoln had made a short speech at the dedication of the cemetery at Gettysburg. It consisted of only 261 words, and it did not make much impact at the time—the professional orator Edward Everett, president of Harvard, was the chief speaker on the occasion—but its phrases have reverberated ever since, and the ideas those few short words projected have penetrated deep into the consciousness of humanity.[135] Lincoln reminded Americans that their country was 'dedicated to the proposition that all men were created equal' and that the war was being fought to determine whether a nation so dedicated 'can long endure.' Second, he referred to 'unfinished work' and 'the great task remaining before us.' This was to promote 'a new birth of freedom' in America, by which he meant 'government of the people, by the people, for the people.' Lincoln, then, thought the blacks should be treated as equals, politically and before the law; but at the same time he insisted that America was a democracy—and Southern whites, rebels though they might be, had as much right to participate in that democracy as the loyalists. How to reconcile the two?

Lincoln's intentions are known because, while still living, he had to deal with the problem of governing those parts of the South occupied by Union armies. He was clear about two things. First, political justice had to be done to the blacks. Second, the South must be got back to normal government as quickly as possible once the spirit of rebellion was exorcized. He proposed a general amnesty, to qualify for which 'politically accused persons' would have merely to take an oath to abide by the Constitution. A state government would be valid, and recognized by Washington, if not less than 10 percent of the voters who were on the rolls in 1860, and had taken the loyalty oath, voted for it. He wanted the occupying armies withdrawn as soon as possible, but he wanted the blacks on the voting rolls first: 'We must make voters of them before we take away the troops. The ballot will be their only protection after the bayonet is gone.' All this was set down in his Proclamation of Amnesty and Reconstruction, issued December 8, 1863.[136]

His first practical step was to get Congress to pass the Thirteenth Amendment. Its first section banned slavery and 'involuntary service' (except for crimes, after conviction by due process) anywhere in the United States, 'or any place subject to their jurisdiction.' Section Two empowered Congress 'to enforce this article by appropriate legislation.' Lincoln did not live to see the Amendment adopted by the three-quarters majority of the states that it required, but it was clear he was fully committed to the liberation of slaves and to entrusting them with the vote. It was also clear that he was in favor of the spirit of the Fourteenth Amendment, adopted in 1868, which wound up the unfinished business of the Civil War, by dealing with the eligibility for office of former rebels and the debts incurred by the Confederacy, but, above all, by making all born or naturalized citizens of the United States equal politically and judicially, and by making it unconstitutional for any state to 'deny to any person within its jurisdiction the equal protection of the laws.' This very important constitutional provision carried forward Lincoln's policy of justice to the blacks into the future, and became in time the basis for desegregation in the South.[137]

Balancing this, it was abundantly clear that Lincoln wanted to exercise the utmost clemency. He intended to bind wounds. On April 14, 1865, his friend Gideon Welles described him as cheerful, happy, hoping for peace, 'full of humanity and gentleness.' His last recorded words on the subject of what to do with the South and the leaders of the rebellion were: 'No one must expect me to take any part in hanging or killing these men, even the worst of them. Frighten them out of the country, open the gates, let down the bars, scare them off. Enough lives have been sacrificed; we must extinguish our resentments if we expect harmony and union. There is too much disposition, in certain quarters, to hector and dictate to the people of the South, to refuse to recognise them as fellow-citizens. Such persons have too little respect for Southerners' rights. I do not share feelings of that kind.'[138]

However, Lincoln was dead, and the task of reconstruction fell on his successor, Andrew Johnson. Johnson agreed wholly with Lincoln's view that the South, consistent with the rights of the freed slaves, should be treated with leniency. But he was in a much less strong position to enforce such views. He had not been twice elected on a Northern Republican platform, fought and won a Civil War against the rebels, and held the nation together during five terrifying years. Moreover, he was a Southerner—and, until 1861, a lifelong Democrat. The fact that he had defied the whole might of the Southern establishment in 1861 by being the only Southern senator to remain in Washington when the South seceded was too easily brushed aside. So, too, was his profound belief in democracy. Johnson stood for the underdog. He had nothing in common with the old planter aristocracy who had willed the war and led the South to destruction. In many respects he was a forerunner of the Southern populists who were soon to make their entry on to the American scene.

He was born in Raleigh, North Carolina. His background was modest, not to say poor. He seems to have been entirely self-educated. At thirteen he was apprenticed to a tailor but ran away from his cruel master and came to Greeneville, Tennessee, where he plied his trade and eventually became its mayor. He was a typical Jacksonian Democrat, strongly in favor of cheap land for the poor—his passionate belief in the Homestead Act was a major factor in his breach with the Southern leadership in 1860–1. He was state representative and senator and governor, representative and senator in Congress, and finally (in Lincoln's first term) military governor of Tennessee from 1862. He was a brilliant speaker, but crude in some ways, with a vile temper. And he drank. At Lincoln's Second Inaugural, following his own swearing-in, Johnson, who had been consuming whiskey, insisted on making a long, rambling speech, boasting of his plebeian origins and reminding the assembled dignitaries from the Supreme Court and the diplomatic corps, 'with all your fine feathers and gewgaws,' that they were but 'creatures of the people.' Lincoln was disgusted and told the parade marshal, 'Do not let Johnson speak outside.'[139]

* * *

Johnson began his term with a violent denunciation of all rebels as 'traitors' who 'ought to be hanged.' Then he proceeded to change tack and carry out what he believed were Lincoln's wishes and policies. There were three possible constitutional positions to be taken up about the South. The extreme position, urged on the White House and Congress by Senator Charles Sumner, the firebrand who had been caned in the Senate, and by Thaddeus Stevens (1792–1868), chairman of the House Ways and Means Committee, was that secession had, in effect, destroyed the Southern states, which now had no constitutional existence, and it was entirely in the power of Congress to decide when and how they were to be reconstituted. Both men were, first and foremost, good haters, and they hated the South and wanted to punish it to the maximum of their power. And their power, in both Houses of Congress, was enormous. Second, there was the bulk of the Republican majority who took a somewhat more moderate position: the rebellion had not destroyed the Southern states but it had caused them to forfeit their constitutional rights, and it was up to Congress to determine when those rights should be restored, under the article of the Constitution guaranteeing all states a republican form of government. Finally there was the Lincoln–Johnson clemency position: this held that rebellion had not affected the states at all, beyond incapacitating those taking part in it from performing their constitutional duties, and that this disbarment could be removed by executive pardon—as soon as this was done, normal government of the states, by the states, could follow.[140]

Initially, Johnson was in a strong position to make this third position prevail. Not only was it manifestly Lincoln's wish, but he was called on to act alone, since it was against the practice of the United States political system for a Congress elected in the autumn of 1864 to be summoned before December 1865, unless by special presidential summons. He had, then, a free hand, but whether it was wise to exercise it without the closest possible consultation with Congressional leaders is doubtful. On May 29, 1865 Johnson issued a new proclamation, extending Lincoln's clemency by excluding from the loyalty oath-taking anyone in the South with property worth less than $20,000. This was consistent with his general view that the South had been misled by its plantocracy and that it must be rebuilt by the ordinary people. In the early summer, he appointed provisional governors for each rebel state, with instructions to restore normalcy as soon as practicable, provided each state government abolished slavery by its own law, repudiated the Confederation's debts, and ratified the Thirteenth Amendment. This was quickly done. Every state found enough conservatives, Whigs, or Unionists, to carry through the program. Every state amended its constitution to abolish slavery. Most repudiated the Confederate debt. All but Mississippi and Texas ratified the Thirteenth Amendment. When all, including these two sluggards, had elected state officials, Johnson felt able to declare the rebellion legally over, in a proclamation dated April 6, 1866.[141]

The new state governments behaved, in all the circumstances, with energy and sense. But there was one exception. They made it plain that blacks would not be treated as equal citizens—would, in fact, be graded as peons, as in some Latin American countries. They had freedom under the state constitutions, and provisions were made for them to sue and be sued, and to bear testimony in suits where a black was a party. But intermarriage with whites was banned by law, and a long series of special offenses were made applicable only to blacks. A list of laws governing vagrancy was designed to force blacks into semi-servile work, often with their old masters. Other provisions in effect limited blacks to agricultural labor. These Black Codes varied from state to state and some were more severe than others; but all had the consequence of relegating blacks to second-class citizenship. Plantation owners were anxious to get blacks to work as peons. Local black leaders encouraged them to sell their labor for what it would fetch, and so make freedom work. This feeling was encouraged by a new kind of federal institution, called the Freedmen's Bureau, set up under the aegis of the military, which spent a great deal of bureaucratic time, and immense sums of money, on protecting, helping, and even feeding the blacks. It was America's first taste of the welfare state, even before it was established by its European progenitor, Bismarck's Germany. The Bureau adumbrated the countless US federal agencies which were to engage in social engineering for the population as a whole, from the time of F. D. Roosevelt until this day. It functioned after a fashion, but it did not encourage blacks to fend for themselves, and one of the objects of the Black Codes was to supply the incentives to work which were missing.[142]

All this caused fury among the Northern abolitionist classes and their representatives in Congress. They were genuinely angry that the Southern blacks were not getting a square deal at last, and more synthetically so that the Southern whites were not being sufficiently punished. Most Northerners had no idea how much the South had suffered already; otherwise they might have been more merciful. Congress had already passed a vengeful Reconstruction Bill in 1864, but Lincoln had refused to sign it. When Congress finally reassembled in December 1865, it was apparent that this spirit of revenge was dominant, with Sumner and Stevens whipping it up, assisted by most of the Republican majority. It was clear that the President had the backing only of the small minority of Democrats. The majority promptly excluded all senators and representatives from the South, however elected, appointed a joint committee to 'investigate conditions' in the 'insurrectionary states,' and passed a law extending the mandate of the Freedman's Bureau. Johnson promptly vetoed this last measure, lost his temper, and denounced leading Republican members of Congress, by name, as traitors. When Congress retaliated by passing a Civil Rights Bill, intended to destroy much of the Black Codes, especially their vagrancy laws, Johnson vetoed that too. Congress immediately passed it again by a two-thirds majority, the first time in American history that a presidential veto had been overriden on a measure of importance. Thus the breach between the White House and Congress was complete. As Johnson had never been elected anyway, and had no personal mandate, his moral authority, especially in the North, was weak, and Congress attempted to make itself the real ruler of the country, rather as it was to do again in the 1970s, after the Watergate scandal.[143]

The consequence was an unmitigated disaster for the South, in which the blacks ultimately became even greater victims than the whites. By June 1866, the Joint Committee reported on the South. It said that the Johnson state governments were illegal and that Congress alone had the power to reconstruct what it called the 'rebel communities.' It said that the South was 'in anarchy,' controlled by 'unrepentant and unpardoned rebels, glorying in the crime which they had committed.' It tabled the Fourteenth Amendment, already described, and insisted that no state government be accorded recognition, or its senators and representatives admitted to Congress, until it had ratified it. All this became the issue in the autumn 1866 mid-term elections. Johnson campaigned against it, but the vulgarity and abusive language of his speeches alienated many, and he succeeded in presenting himself as more extreme, in his horrible way, than his opponents. So the radical Republicans won, and secured a two-thirds majority in both Houses, thus giving themselves the power to override any veto on their legislation which Johnson might impose. The radicals were thus in power, in a sense, and could do as they wished by law. In view of this, the governments of the Southern states would have been prudent to ratify the Fourteenth Amendment. But, as usual, they responded to Northern extremism by extremism of their own, and all but one, Tennessee, refused.

To break this impasse, the dominant northern Radicals now attacked, with the only weapon at their disposal, the law. In effect, they began a second Reconstruction. Their object was partly altruistic—to give justice to the blacks of the South by insuring they got the vote—and partly self-serving, by insuring that blacks cast their new votes in favor of Republicans, thus making their party dominant in the South too. As it happened, most Republicans in the North did not want the blacks to get the vote. Propositions to confer it in the North were rejected, 1865–7, in Connecticut, Minnesota, Wisconsin, Ohio, and Kansas, all strong Republican states. But the Republican majority insisted nonetheless on forcing black voters on the South. In March–July 1867 it pushed through Congress, overriding Johnson's veto, a series of Reconstruction Acts, placing what they called the 'Rebel States' under military government, imposing rigid oaths which excluded many whites from electoral rolls while insuring all blacks were registered, and imposing a number of conditions in addition to ratification of the Fourteenth Amendment, before any 'Rebel State' could be readmitted to full membership of the Union. It also made a frontal assault on the powers of the executive branch, in particular removing its power to summon or not to summon Congress, to dismiss officials (the Tenure of Office Act) and to give orders, as commander-in-chief, to the army. Fearing obstruction by the Supreme Court, it passed a further Act abolishing its jurisdiction in cases involving the Reconstruction Acts. Much of this legislation was plainly unconstitutional, but Congress planned to make it efficacious before the Court could invalidate it.[144]

This program, characteristic of the tradition of American fundamentalist idealism at its most extreme and impractical, had some unfortunate consequences. In Washington itself it led to a degree of bitterness and political savagery which was unprecedented in the history of the republic. In the debates of the 1840s and 1850s, Calhoun, Webster, Clay, and their colleagues, however much they might disagree even on fundamentals, had conducted their arguments within a framework of civilized discourse and with respect for the Constitution, albeit they interpreted it in different ways. And, in those days, Congress as a whole had treated the other branches of government with courtesy, until the Rebellion, by refusing to accept the electoral verdict of 1860, ruined all. Now the Republican extremists were following in the footsteps of the secessionists, and making a harmonious and balanced government, as designed by the Founding Fathers, impossible.

The political hatred which poisoned Washington life in 1866–7 exceeded anything felt during the Civil War, and it culminated in a venomous attempt to impeach the President himself. Johnson regarded the Tenure of Office Act as unconstitutional, and decided to ignore it by sacking Stanton, the War Secretary. Stanton had always been an unbalanced figure, politically, whom Lincoln had brought in to run the War Department simply because of his undoubted energy, drive, and competence. But with the peace Stanton became increasingly extreme in using military power to bully the South. He also, like the President, had an ungovernable temper and lost it often. Johnson saw him as the Trojan Horse of the Radical Republicans within his own Cabinet, and kicked him out with relish. The Republican majority retaliated by impeaching him, under Article I, Sections 1, 2, and 5, of the Constitution. Article II, Section 4, defines as impeachment offenses 'Treason, Bribery or other High Crimes and Misdemeanors.' This last phrase is vague. One school of thought argues it cannot include offenses which are not indictable under state or federal law. Others argue that such non-indictable offenses are precisely what an impeachment is for—political crimes against the Constitution which no ordinary statute can easily define.

The procedure for impeachment is that the House presents and passes an impeachment resolution and the Senate convicts, or not, by a two-thirds vote. Since 1789, the House has successfully impeached fifteen officials, and the Senate has removed seven of them, all federal judges.[145] Johnson was the first, and so far only, president to be impeached, and the experience was not edifying. Johnson was subjected during the proceedings to torrents of personal abuse, including an accusation that he was planning to use the War Department as a platform for a personal *coup d'état*, and much other nonsense. An eleven-part impeachment resolution passed the House on February 24, 1868. There was then a three-month trial in the Senate, at the end of which he was acquitted (May 26, 1868) by 35 to 18 votes, the two-thirds majority not having been obtained. No constructive purpose was served by this vendetta, and the only political consequence was the discrediting of those who conducted it.[146]

The consequences for the South were equally destructive. The Acts of March 1867 led to a new Reconstruction along Republican, anti-white lines. Registration was followed by votes calling conventions, and these by the election of conventions, the drafting of constitutions, and their approval by popular vote. But those who took part in this process were blacks, guided by Northern army officers, a few Northerners, and some renegade whites. This new electorate was organized by pressure groups called Union Leagues, which built up a Republican Party of the South. In fact, the state constitutional conventions were almost identical with Republican nominating conventions. The new party and the imposed state were one. It was as though the North, with its military power, had imposed one-party dictatorships on all the Southern states. The vast majority of whites boycotted or bitterly opposed these undemocratic procedures. But for the time being there was nothing they could do. Only in Mississippi did they succeed in rejecting the new constitution.

By the summer of 1868 all Southern states except three (Texas, Mississippi, and Virginia) had gone through this second, Congressional-imposed Reconstruction, and by an Omnibus Act seven of them were restored to Congressional participation (Alabama had already passed the test). As a result of the disenfranchisement of a large percentage of Southern white voters, and the addition of black ones, organized as Republicans, the ruling party carried the elections of 1868. General Grant, who had been nominated unanimously by the Republican Convention as candidate, won the electoral college by 214 votes to 80 for the Democrat, Governor Horatio Seymour of New York (1810–86). Without the second Reconstruction, it is likely Grant would have lost, and some of the Republicans, such as Sumner and Stevens, admitted that Congress had recognized the eight Southern states in 1868 primarily to secure their electoral votes. Thus America, after abolishing the organic sin of slavery, witnessed the birth of an organic corruption in its executive and Congress.[147]

These transactions at least had the merit of enabling Congress to bully the South into ratifying the Fifteenth Amendment, which stated that the right of American citizens to vote should not be denied or abridged 'on account of race, color or previous condition of servitude.' On the other hand, in evading its implications, Southerners could later cite, as moral justification, the fact that they had ratified it only under duress—especially true in Georgia, for instance, which had to be placed yet again under military occupation and Reconstructed for the third time. Moreover, the Republican-imposed governments in the Southern states, as might have been expected, proved hopelessly inefficient and degradingly corrupt from the start. The blacks formed the majority of the voters, and in theory occupied most of the key offices. But the real power was in the hands of Northern 'carpetbaggers' and a few Southern white renegades termed 'scalawags.' Many of the black officeholders were illiterate. Most of the whites were scoundrels, though there were also, oddly enough, a few men of outstanding integrity, who did their best to provide honest government. There were middle-class idealists, often teachers, lawyers or newspapermen who, as recent research now acknowledges, were impelled by high motives. But they were submerged in a sea of corruption. State bonds were issued to aid railroads which were never built. Salaries of officeholders were doubled and trebled. New state jobs were created for relatives and friends. In South Carolina, where the prescriptions had been particularly savage, and carpetbaggers, scalawags, and blacks had unfettered power, both members of the legislature and state officials simply plunged their hands into the public treasury. No legislation could be passed without bribes, and no verdicts in the courts obtained without money being passed to the judges. Republicans accused of blatant corruption were blatantly acquitted by the courts or, in the unlikely event of being convicted, immediately pardoned by the governor.[148]

The South, its whites virtually united in hatred of their governments, hit back by force. The years 1866–71 saw the birth of the Ku Klux Klan, a secret society of vigilantes, who wore white robes to conceal their identities, and who rode by night to do justice. They were dressed to terrify the black community, and did so; and where terror failed they used the whip and the noose. And they murdered carpetbaggers too. They also organized race-riots and racial lynchings. They were particularly active at election-time in the autumn, so that each contest was marked by violence and often by murder. Before the Civil War, Southern whites had despised the blacks and occasionally feared them; now they learned to hate them, and the hate was reciprocated. A different kind of society came into being, based on racial hatred. The Republican governors used state power in defense of blacks, scalawags, and carpetbaggers, and when state power proved inadequate, appealed to Congress and the White House. So Congress conducted inquiries and held hearings, and occasionally the White House sent troops. But the blacks and their white allies proved incapable of defending themselves, either by political cunning or by force. So gradually numbers prevailed. The whites, after all, were in a majority, and America, after all, was a democracy, even in the South. Congressional Reconstruction gradually crumbled. The Democrats slowly climbed back into power. Tennessee fell to them in 1869, West Virginia, Missouri, and North Carolina in 1870, Georgia in 1871, Alabama, Texas, and Arkansas in 1874, Mississippi in 1875. Florida, Louisiana, and South Carolina were held in the Republican camp only by military force. But the moment the troops were withdrawn, in 1877, the Republican governments collapsed and the whites took over again.

In short, within a decade of its establishment, Congressional Reconstruction had been destroyed. New constitutions were enacted, debts repudiated, the administrations purged, cut down, and reformed, and taxation reduced to prewar levels. Then the new white regimes set about legislating the blacks into a lowly place in the scheme of things, while the rest of the country, having had quite enough of the South, and its blacks too, turned its attention to other things. Thus the great Civil War, the central event of American history, having removed the evil of slavery, gave birth to a new South in which whites were first-class citizens and blacks citizens in name only. And a great silence descended for many decades. America as a whole did not care; it was already engaged in the most astonishing economic expansion in human history, which was to last, with one or two brief interruptions—and a world war—until the end of the 1920s.

PART FIVE

Huddled Masses and Crosses of Gold

Industrial America, 1870–1912

By the end of the Civil War, the United States and its people were beginning to take on the characteristics with which we are familiar at the end of the 20th century: huge and teeming, endlessly varied, multicolored and multiracial, immensely materialistic and overwhelmingly idealistic, ceaselessly innovative, thrusting, grabbing, buttonholing, noisy, questioning, anxious to do the right thing, to do good, to get rich, to make everybody happy. All the great strengths and weaknesses of the mature republic were already appearing, and the reactions of those who lived there, and those who visited it, were the modern mixture of admiration, astonishment, and shock. Surveying the frantic rush of farmers and mechanics, clerks and schoolteachers out west, to strike it rich in the new gold and silver mines, Henry Thoreau wrote in disgust: 'It matches the infatuation of the hindoos, who cast themselves under the car of Juggernaut.' To Rudyard Kipling, bursting on the world as the new genius from Anglo-Indian civilization, New York was 'the shiftless outcome of squalid barbarism and reckless extravagance.' Its streets were 'first cousins to a Zanzibar foreshore or kin to the approaches of a Zulu kraal.' But Walt Whitman, another great poet, loved it: 'What can ever be more stately and admirable to me than mast-hemm'd Manhattan?' And he sang: 'Stand up, tall masts of Manhattan! Stand up, beautiful hills of Brooklyn!' Henry James, the fastidious Boston Brahmin, was overwhelmed by the vast city too, hailing 'That note of vehemence . . . the appeal of dauntless power—the power of the most extravagant of cities.' Almost against his will he loved 'the diffuse, wasteful, clamoring, *detonations*, the bigness and bravery and insolence.' The new, tall buildings were 'like pins in a cushion already overcrowded.'[1]

People complained that America was not *old* enough. Writing in 1871, John Ruskin, the leading English esthete, protested: 'And, to this day, though I have kind invitations enough to visit America, I could not, even for a couple of months, live in a country so miserable and to *possess no castles.*' That was James' real complaint too, the lacuna which led him, increasingly, to locate himself in Europe, above all England. He wrote in 1879: 'One might enumerate the items of high civilisation, as it exists in other countries, which are absent from the texture of American life, until it should become a wonder to know what was left . . . No sovereign, no court, no personal loyalty, no aristocracy, no church, no clergy, no army, no diplomatic service, no country gentlemen, no palaces, nor castles, nor manors, nor old country houses, no parsonages, no thatched cottages, no ivied ruins, no cathedrals, no abbeys nor little Norman churches; no great universities nor public schools—no Oxford, or Eton, or Harrow; no literature, no novels, no museums, no pictures, no political society, no sporting class—no Epsom or Ascot!'[2] James was grossly exaggerating, as we shall shortly see. But people saw what he meant. Actually, America was not short of the picturesque. There were, for instance its incomparable placenames, subject of Stephen Vincent Benet's poem, 'American Names:'

> I have fallen in love with American names,
> The sharp names that never get fat,
> The snakeskin titles of mining claims,
> The plumed war bonnet of Medicine Hat,
> Tucson and Deadwood and Lost Mule Flat.

And he ended: 'Bury my heart at Wounded Knee.' Then there was the sheer space. As Gertrude Stein put it, 'In America there is more space where nobody is than where anybody is—that is what makes America what it is.'[3]

America was big, and the further you got into America the bigger it became, both the landscapes and the natural wonders and the artifacts. The mere width of the streets was a key indicator of the march of time and the growth of expectations. In Cambridge, Massachusetts, the standard street-width was 30 feet, which corresponded to the latest town-planning ideas of 18th-century England or France. Further south, in Charleston, the broadest went up to 60 feet. In New York, early in the 19th century, what was big for South Carolina became the minimum. Municipal ordinances laid down that, north of 23rd Street, crosstown streets had to be at least 60 feet and north–south avenues 100 feet or more. In the West, 80 feet was the minimum for all streets, even side ones—for instance in Sacramento, California and Cheyenne, Wyoming. Many cities insisted on more. In Omaha, the standard minimum was 100, rising to 120 feet. For Topeka, Kansas, the eight major avenues were 130 feet. There were disadvantages of course. Not all young municipalities could afford to surface and pave these huge civic arteries. George Augustus Sala, reporting critically for the London

Daily Telegraph, called these wide streets 'a dusty desert in summer, a slough of despond in winter.' But these regulations showed foresight, vindicated by the arrival of the first city mass-transit systems and still more by the automobile. Where Europe had to bypass its cities or allow them to be strangled by traffic, urban America was ready for the 20th century. One of the best planned of the 19th-century cities rejoiced in the principle of bigness and took full advantage of it, Salt Lake City. Its streets were 132 feet wide, supplemented by 20-foot minimum sidewalks on each side. Its huge city blocks were 660 feet wide, maximizing privacy. The set-back (space between house and road) was a 20-foot minimum and no house was allowed to front another across the street. Each family had a garden and an orchard.[4]

American authorities thought big, and ahead, because by the late 1860s they were conscious that the United States was expanding, and its people multiplying, faster than any other country in history. At the beginning of the Civil War, the total population, North and South, free and slave, was 31,443,321. That placed it in the forefront of European countries, all of which (except France) were also expanding fast though entirely by natural increase. At the end of the Civil War decade, the US population was 39,818,449, an increase of more than a quarter. By 1880 it had passed the 50 million mark, by 1890 it was 62,847,714, more than any other European country except Russia and still growing at the rate of 25 percent a year and more. By 1900 it was over 75 million, and the 100 million mark was passed during World War One.[5] US birth-rates were consistently high by world standards, though declining relatively. White birth-rates, measured per 1,000 population per annum, fell from 55 in 1800 to 30.1 in 1900 (black rates, first measured in 1850 at 58.6, fell to 44.4 in 1900). But the infant mortality rate, measured in infant deaths per 1,000 live births, also fell: from 217.4 in 1850 to 120.1 in 1900 and under 100 in 1920 (black rates were about two-fifths higher). And life expectancy rose: from 38.9 years in 1850 to 49.6 in 1900, passing the 60 mark in the 1920s (black rates were about twelve years lower). These combined to produce a very high natural rate of population increase.[6]

This natural increase was accompanied by mass-immigration. From 1815 to the start of the Civil War, over 5 million people moved from Europe to the United States, about 50 percent from England, 40 percent from Ireland, and the rest from Continental Europe. Between the end of the Civil War and 1890, another 10 million arrived, mostly from northwest Europe, especially England, Wales, Ireland, Germany, and

Scandinavia. Then, in the twenty-four years 1890 to 1914, another 15 million came, mostly from eastern and southern Europe—Poles, Russian Jews, Ukrainians, Slovaks, Croatians, Slovenes, Hungarians, Greeks, Rumanians, and Italians. In the last four years before World War One, immigration was well over the 1 million a year mark and concern was at last expressed that America as a whole (as distinct from particular cities like New York) was becoming overcrowded. The Naturalization Acts, going back to 1790, had always denied entry and residence to non-whites. On August 18, 1882 Congress passed the Exclusion Act, denying entry to the insane, criminals, paupers, and (for ten years) Chinese contract-laborers, who were further restricted by the Contract Labor Act of 1885. But these restrictions were marginal. Between 1866 and 1915, to summarize, the United States gave a home to 25 million people, overwhelmingly from Europe. After 1883, which witnessed the first massive influx of Russian Jews fleeing the Tsarist pogroms (starting in earnest in 1881), the percentage of northern and western Europeans fell dramatically in proportion to the whole, and after 1895 immigrants from eastern and southern Europe were in a majority—over 9 million had entered the United States by 1914.[7]

The extra tens of millions created by the natural increase and continuing mass-immigration were all fed, clothed, housed, and employed, without much difficulty, by the bountiful country, exploited and developed with ever growing intensity and skill as the century progressed. The role of agriculture was preeminent. We have already noted the 1862 Homestead Act. This was accompanied, the same year, by the appointment of a commissioner of agriculture. That was part of gigantic expansion of government, and the spread of the areas in which it took a direct interest, which the Civil War brought. Thereafter, it may be said, the United States federal government no longer just allowed farming to 'happen'—it had a policy for it. That may well have been the popular wish. It was certainly the wish of the farmers, especially the new farmers and the small homesteaders. They grew rapidly in number from the end of the Civil War, boosted by financial concessions given to veterans, under an amended Homestead Act. The boom in agriculture was made possible by railroads, which not only took bulk food rapidly to the multiplying and growing cities, but made possible the creation of a vast overseas export market. The boom began even before the Civil War started, accelerated during it, and continued relentlessly afterwards. In 1860 the entire population of the United States was not much over 30 million. By 1910, half a century later, over 50 million people were living on US farms or in agricultural villages. The number of farms had grown from 2 million in 1860 to over 6 million by 1910. This was made possible by bringing under cultivation over 500 million acres, an area as large as western Europe.[8]

The growth of farming profoundly altered the geography and demographics of the United States. Though the country was industrializing at breakneck speed in the second half of the 19th century, agriculture held its own as the main source of wealth and work. At late as 1880, 49 percent of those gainfully employed were working in agriculture (this fell to 32.5 by 1910 and 21.4 by 1930). But the center of gravity of this mass of rural labor, like the population itself, was moving west. In 1790 it was near Baltimore. By 1810 it was on the Potomac. In 1820 it was near Woodstock. By 1840 it was near Clarksburg in West Virginia, and by 1850 it was almost across the Ohio. In 1860 and again in 1870 it was still in Ohio, but in the decades 1880–1920 it slowly crossed Indiana. These shifts reflected the fact that the United States was taking in, for farming purposes, about 15 million new acres each year, most of it in the West.[9]

This in turn reflected the Republican policy of providing cheap or even free land. The 1862 Act was the most important, in practical as well as symbolic terms, but it was followed by others. The 1878 Timber and Stone Act enabled 160 acres (a quarter section of a square mile) of land valuable for timber or stone to be sold at appraisal rates of not less than $2.50 an acre. The Dawes Act of 1887 allowed individual Indians, as opposed to tribes, to acquire public land, which meant they could sell it, thus putting more into the market-pool. A 1909 Act raised the Homestead maximum to 320 acres of dry land. In 1912 Homestead land became free after three, as opposed to five, years' residence, and in 1916 the acreage was raised to 640 acres, or a square mile, for stockraising. Never in human history, before or since, has authority gone to such lengths to help the common people to become landowners. There were those who said at the time, and many more since, that it was folly for the state to dispose of its stock of public land so quickly and generously, thus leading to huge waste, food gluts, and falls in Eastern land prices. On the other hand, the Public Land Commission set up by Theodore Roosevelt argued that the 1862 Act and its successors had achieved many of its stated objects: 'It protects the government, its fills the state with homes, it builds up communities and lessens the chance of social and civil disorder by giving ownership of the soil, in small tracts, to the occupants thereof.'[10]

The tracts were not all that small, either. In theory, and perhaps sometimes in practice, a settler could acquire 160 acres under the 1862 Act, another 160 under the old Preemption act, 160 under the Stone and Timber Act, and 640 acres of desert land—making 1,120 acres altogether, all of it free, provided residence qualifications were fulfilled. Critics grumbled that lumber or mining companies, by colluding with greedy individuals, could built up enormous holdings. This was possible under the commuting system authorized by the Acts, and undoubtedly took place. In the quarter-century 1881–1904, it was calculated, about 23 percent of the public land thus sold was commuted. In North Dakota, half the land sales ultimately went to big companies.[11] But that does not seem a particularly high proportion and, it could be argued, was a small price to pay for the immense benefits of having a free market in land—something which had never before occurred at any time, anywhere in the world. Moreover it could also be argued that a mix of big and small holdings was desirable. Large ranching, timber, and mining companies could attract or bring in services which were beyond the means or economic clout of small farmers, but which they could share when they became available. And large-scale farming undoubtedly speeded up the development of agricultural technology from which, in the end, small farmers benefited most of all.[12]

Technology took a wide variety of forms. In the big open country of the West, where land had to be protected against cattle, or cattle from straying, the invention of barbed wire was a godsend, which had a dramatic impact on agricultural history. In the mid–1870s, two Illinois farmers, Joseph F. Glidden and Jacob Haish, took out patents for barbed wire which was both practical and cheap. In 1874, barbed wire cost $20 per 100 pounds and total production was 10,000 pounds. Six years later, production had risen to 80.5 million pounds following dramatic price falls which brought top-quality wire down to $1.90 by 1897. Glidden and Haish had no idea that their invention would cost the lives of millions in World War One (had it been available in 1861, casualties in the Civil War would probably have doubled and the fighting would have prolonged itself to the end of the decade). Barbed-wiring was far cheaper and quicker than wooden-fencing of range land and its mass production made it possible to fence in the back-lands. Shipped into Texas by the trainload, it enabled the west of the state, hitherto disregarded, to be developed rapidly.[13]

Barbed-wire mass production had mixed effects on ranching. Americans took over this form of agricultural production from the

Hispanics of the West and South but carried it on far more efficiently and intelligently. They made, in time, good use of a tremendous area stretching up from Texas to Manitoba, across the Canadian frontier, which had originally been classified as 'desert.' It had grass but little rainfall and was thought to be largely uninhabitable as late as the 1830s. American farmers dug wells to tap the underground water-tables and, by developing drought-resistant strains of wheat and corn, practiced dry-farming until, in the fullness of time, the Boulder and the Grand Coulee dams were built. Water was pumped up from wells by an ingenious small metal windmill, worked by the prevailing winds, of which were was no shortage.[14]

The great discovery, early in the Civil War, was that cattle could survive the harsh winters of the high plains, in Nebraska and elsewhere, and did remarkably well on wild grass. The ranchers took over from the dry-farmers and, operating on an increasingly large scale, made fortunes, as soon as the railroad was close enough to roll out their produce cheaply. It was the great American combination of research into agricultural processes and efficient land-use, plus high technology. Each spring, there was a mass round-up of cattle which had spent the winter on the open ranges. According to the local custom of marks, they were divided up among the various owners. The yearling steers were separated and branded, or rebranded, then driven off to Kansas, Nebraska, or Wyoming to be fattened, the rest being sent back to the ranges. Thus cattle towns grew up—Abilene, Kansas, Dodge City, and Topeka, for example. All these towns were on the new railways, snaking their way across the plains. Steers were fattened in these towns for quick slaughter or entrained for the vast stockyards growing up in Kansas City, Milwaukee, and, above all, Chicago. The 1870s were characterized by long, gigantic cattle-drives, immense herds being driven 4,000 miles from Texas to the north, for slaughter or to restock the high-plains herds. This long-drive existence, the golden age of the cowboy, lasted almost exactly a quarter-century. By the early 1890s it was over, but by then it had been made immortal by the paintings of Frederick Remington (1861–1909), who had actually ridden herd as a cowboy, by the novels of Owen Wister (1860–1938), and, not least, by Wister's political friend, Theodore Roosevelt (1858–1919), who turned the life of a cowboy into a kind of adventure-game for rich city-dwellers.[15]

The years of vast expanses and high profits—J. W. Iliff, once a failed Texan miner, had a herd of 35,000, Joseph G. McCoy took to market, at one time or another, over 2 million head—were ended by the barbed-wire invasion, which enabled farmers to keep cattle off their crops cheaply. In the end, however, barbed-wire benefited the intelligent rancher too because it enabled him to fence in large areas and build up high-quality herds, protected from the diseases and genes of poor-quality Southern breeds. Originally ranching enclosures were unlawful but once barbed wire became cheap everyone had to have them and by 1888 an estimated 8 million acres were enclosed for intensive breeding. Books could be, and indeed have been, written about the excitements and disasters of large-scale cattle-ranching on the high plains—of the diseases and droughts and flash-floods which hit rich farmers and wiped out poor ones, and of the coming of state quarantine laws drawn up against Southern cattle.[16] It was one of the more romantic ways in which the West—the whole of Oklahoma, for instance—was first developed. And, once the cattle could be got to market efficiently, once the railroads were in place and scientific packing and refrigeration developed, beef from the American plains, of high quality and at low prices, was exported all over the world.

Needless to say, there was a destructive element in this colossal new system of exploitation. For the cattle to thrive on the plains, their previous tenants, the immense herds of buffalo which roamed them, had to be destroyed. And they were. That in turn goaded the Indians into their last stand. The true history of the American Indians is only just beginning to be written, and unfortunately it has until now been largely in the hands of enthusiasts who have allowed their sympathies to cloud the objective truth. The Indians were not murderous savages, who ought either to be detribalized and assimilated completely (as scores of thousands indeed were), or exterminated, or penned up in remote reservations—this being the view of the vast majority of 19th-century American whites. Nor were they sophisticated–primitive innocents, living in utopian and preservationist communities, brutally disturbed by cruel and heedless invaders of European extraction, that being the view of 20th-century romantic historians of Indian history. The more they, and the white settlers who displaced them, are studied in detail, and without prejudice on either side, the smaller the differences between them appear. Both Indians and whites were living in the same, often harsh country, and trying their best to master it, in different ways. There seems to have been no discernible difference in intelligence, as is attested not least by the large numbers of Indians who passed imperceptibly into the ranks of the whites and underwent racial and ethnic oblivion in a very few generations.

But the Indians were handicapped by two social characteristics with deep historical roots. First, they were extremely fragmented and the groups in which they were organized were tiny. They tended to distinguish not between Indians and whites but between their own small group and the rest, classified as enemies. Thus Navajo Indian language recognized two categories of humans: the *dine* (themselves) and the *ana'i* (enemies). Whites were not a distinct group but a sub-category of *ana'i*. This taxonomy was characteristic of Indian tongues. From time to time in the late 18th and early 19th centuries, Indian 'prophets,' no doubt influenced by white missionaries, tried to preach the doctrine that Indian ways were fundamentally different and opposed to white ways, and that the Indians should act together on a racial basis. This form of Indian race-theory, itself a development of white racialism, was advocated by Neolin among the Delaware, Handsome Lake among the Iroquois and Tenskwatawa among the Shawnees, and leaders like Tecumseh tried to turn Indian racism to political ends. But on the whole this strategy did not succeed. The Cherokees, for instance, who were one of the most successful of the Indian groups, insisted that the line of difference came between them, on one side of it, and all other Indians, plus whites, on the other side. Ironically enough, the Indian tribes were pushed into alliances with each other by precisely the white institutions designed to undermine their power. It was only after they were allotted reservations and land, and annuities were donated by the whites to individuals, as distinct from tribes, that tribal rolls were compiled and blood-quantums calculated. And it was only when the Indians were forced to learn English that they began to conflate their languages, develop an Indian lingua franca (on the lines of Hindustani in the Indian subcontinent) as an alternative to English, and create the first pan-Indian religion, preached in Oklahoma in 1918.[17]

Second, though Indians were reasonably good settled farmers when they chose, males tended to think agriculture was a female task and hunting the prime activity of the menfolk. Indian males could be persuaded to run farms, but then they nearly always detribalized themselves and joined the white community. What the authorities found exceedingly difficult was to combine settled agriculture with management through tribes. Thus the shooting of large numbers of buffalo by white ranchers was seen by Indian tribes, whether on or off reservations, as a direct assault on their tribal integrity and existence. This was compounded by what they saw, rightly or wrongly, as the consistent unwillingness of whites to stick to the terms of the treaties they signed. White miners and mining interests were particularly ruthless in evading treaties and then persuading or bribing the authorities into sanctioning breaches. White ranchers could be just as lawless. But on the whole— and this is where both the older histories and the myths give a misleading impression—disputes between white pioneers, travelers, and farmers and Indian tribes were extremely rare. The 1851 treaty negotiated by David D. Mitchell, flanked by 270 US soldiers, with 10,000 plains Indians at Fort Laramie, proved extraordinarily successful. Under it, the Indians allowed white wagon-trains to the West Coast safe passage, and permitted the army to construct roads and forts without hindrance. During the next twenty years, over 250,000 pioneers passed through Indian territory and less than 400, it is calculated, were killed in fights, not all of them with Indians. Of these casualties, nine out of ten occurred west of the South Pass.[18]

Most such incidents, including the big massacres of 1864 and 1866, arose out of misunderstandings on both sides. And occasionally Indian groups opted out of treaty diplomacy and preferred to risk open hostility. Thus not all the Lakota bands participated in the renewed Fort Laramie Treaty of 1868, the Hunkpapas being told by their chief, Sitting Bull: 'You are fools to make yourselves slaves to a piece of some fat bacon, some hard tack and a little sugar and coffee.' In 1875, President Grant, who had a low opinion of Indians, ordered the Lakotas to leave the ranges, where they were in conflict with the ranches, and come into the agencies for reorganization. This, coming on top of the systematic destruction of buffalo, led to the Great Sioux War of 1876–7. That was by no means the fiercest of the many Indian wars, but it included the most startling, or best-known, incident of all, the annihilation of General Custer's command on June 25, 1876 near the valley of the Little Bighorn River. Custer was a brave but insensitive, arrogant, and often stupid man, and whenever his military record is examined in detail he emerges badly.[19] The best thing about him was his taste in painting, especially his admiration for Albert Bierstadt (1830–1902), of whom more shortly. In fact shortly before Custer was killed he lunched in Bierstadt's New York studio.[20] The news of the Little Bighorn reached the East on July 4 and led to a massive reaction which did for the plains Indians once and for all. General Philip H. Sheridan, commanding the Division of the Missouri, assured Sherman, now the Commander-in-Chief, 'I will take the campaign fully in hand

and will push it to a successful termination, sending every man that can be spared.' The Indians were reduced to six small reservations, and their ability to use force ended with the slaughter at Wounded Knee on December 29, 1890, when 146 Indians, including 44 women and 16 children, were killed.[21]

The real problem of the Indians, in confronting the whites, and especially their government, was absence of leaders who knew how to manipulate the Washington system. There was no lack of sympathy on the white side. In 1881 Helen Hunt Jackson (1830–85), a notable writer of poetry and novels, including *Ramona* (1884), produced a carefully documented history of white breaches of Indian treaties called *A Century of Dishonor* which shocked the authorities into action. Helen Jackson herself was appointed a special commissioner to investigate conditions among the mission Indians of California. More important, her book led Senator Henry L. Dawes of Massachusetts to put through Congress an Act designed to remedy some of the abuses described in her book. In particular the Dawes Act (1887) sought to turned the nomadic plains Indians, who seemed doomed to be exploited in one way or another, into settled farmers, by allotting each head of family a quarter section (160 acres), plus a one-eighth section to single adults and orphans and a one-sixteenth to each dependent child. These lands were inalienable and held in federal trust for twenty-five years before title was given.

This Act marked a retreat from the reservation policy, and it was subsequently criticized for opening up Indian reserve land to whites, since the quarter sections were taken over from tribal allocations, and after twenty-five years they passed into the general land market. But this criticism ignores the fact that many Indians who took advantage of it, and thus got full citizenship, did not sell their land after twenty-five years and became full members of the American farming community.[22] In material and moral terms, assimilation was always the best option for indigenous peoples confronted with the fact of white dominance. That is the conclusion reached by the historian who studies the fate not only of the American Indians but of the aborigines in Australia and the Maoris in New Zealand. To be preserved in amber as tribal societies with special 'rights' and 'claims' is merely a formula for continuing friction, extravagant expectations, and new forms of exploitation by white radical intellectuals.

Nevertheless, to bring the story into the 20th century, the weaknesses of the Dawes Act were sufficient to justify further efforts to devise a just solution to the Indians who remained tribal. One of the difficulties was that it proved hard to establish exactly how many of them there were. In 1865 it was reckoned there were about 340,000. It is supposed that this total remained relatively unchanged for about eighty years. Those Indians who took advantage of the Dawes Act got citizenship, and rapidly ceased to be, as it were, Indian. By 1900 all Indians who remained tribal were settled on reservations, and in 1924 they were all granted US citizenship. The reservations then numbered over 200, scattered through forty states, though their extent gradually shrank from 147 million acres in 1887 to 54 million by 1960.

By this date, however, general policy had been reversed again. John Collier (1884–1968), a Georgian settlement worker with experience of immigrant ghettos, took up the cause of the Indians after World War One and in 1922 founded the American Indian Defense Association, which criticized official federal policy. In 1932 he was made federal commissioner for Indian affairs, a post he held till 1945, and he introduced what was called the Indian New Deal. He sought to curtail federal interference in Indian affairs, to reduce if not eliminate sales of Indian-held land to whites, to reinvigorate native customs and lifestyles, and to promote economic self-sufficiency. Some of these aims found legal force in the Wheeler–Howard Act (1934).[23] This was a return to the fly-in-amber policy, with the difference that Indians were now encouraged to be more self-assertive and to make full use of their legal rights. It authorized tribes to purchase additional land, made loans available for tribal businesses, and for the first time in United States history introduced the principle of Positive Discrimination, Indians of the right bloodlines being given preference in certain civil service jobs. Only 99 out of 172 listed tribes exercised the option of coming under the Wheeler–Howard Act instead of the Dawes Act, but in the quarter-century following this statutory boost to tribalization, many Indians found it preferable to rejoin tribes which had successfully litigated claims and land rights. Thus the 1960 census revealed that the total number of Indians, or those who now wished to be considered such, had mysteriously risen to over 540,000 (rather as, when Jews, as opposed to other citizens of the Soviet Union, were permitted to leave the USSR in 1988, their numbers abruptly trebled). Litigation has enabled some groups of Indians, in theory at least, to become rich. On the other hand, they find their children, as they grow older, do not want to lead fly-in-amber existences. Meanwhile, the claims made by some Indian leaders, or rather the white academics who speak on their behalf, have alienated majority white sentiment. In short, the problem of the American Indians, like many such created by powerful historical forces, is insoluble, except by time.[24]

Throughout the 1860s, 1870s, and 1880s, the frontier had been advancing westward. It was not just a name but a specific geographical definition, with legal significance in many Acts of Congress. It means land occupied by two or more but less than six persons, on average, per square mile. By 1890 it was gone—there were no more areas which corresponded to this definition. But it lived on in the minds of men and women because it was already romanticized and mythical, demonized too. No other fleeting period in history, enacted in a particular region, has been verbalized and visualized so often. (The only comparable example is the Italian Renaissance, which took place over a similar area and for approximately the same number of years.) It is no easy matter to separate the American West itself from its historiography, which began in earnest in about 1890, just when the frontier as historical fact ended, and just when the movie industry, which supplied the popular image of the West, came into primitive existence.[25]

Until the 1890s, American history was essentially presented as the development of the Eastern states, seen as an extension overseas of English history or an episode in European overseas expansion. Then in 1893 a Midwestern academic, Frederick Jackson Turner (1861–1932), delivered an address before the American Historical Society on 'The Significance of the Frontier in American History.' He showed that the existence of a moving frontier, as a solvent of all America's social and economic problems, constituted an element unknown in Europe and made American history unique. This important paper had an energizing effect on American historians, forcing them to look more closely, and with new eyes, at many aspects of their history, and to try to discover what exactly it was which made the United States quite so distinctive.[26]

It is arguable, for instance, that it was the West which made the United States constitutionally violent, a curious characteristic in a country so devoted to the rule of law and the solution of problems by due process. Until the English-speaking settlers crossed the Appalachians they were content, on the whole, to operate within the framework of the English common law. It is true that, as pioneers, they claimed the right of citizens to bear arms at all times, a right guaranteed by the successful American Revolution, which has survived to haunt the nation ever since. But they respected, on the whole, the common law tradition that, in any dispute threatening violence, the threatened subject had a legal duty to retreat from the scene if possible, and failure to do so was culpable, if homicide followed. One says 'on the whole,' because the tradition of dueling and the concept of 'honor' were at least as strong in America as in Continental Europe. Both were at odds with the common law. But they were particularly strong in the Old South, and in the border states like Tennessee, Kentucky, and Missouri, and from these states they spread westward. The South also stressed the right, and indeed the duty, of an honorable man to seek vengeance, as a form of justice—though Francis Bacon, in a famous essay upholding the English common law tradition, had been at pains to show that vengeance was the very antithesis of justice.[27]

Under the influence of this counter-tradition, once states were founded across the Appalachians, their courts and legislatures scrapped the duty to retreat. In 1876, the supreme court of Ohio, not a state notable for lawlessness, held that a 'true man' was not 'obligated to fly' from an assailant and could kill him in self-defense. The next year the Indiana supreme court ruled approvingly: 'The tendency of the American mind seems to be very strongly against the enforcement of any rule which requires a person to flee when assailed.' The right to stand ground was written in granite by the US federal Supreme Court in 1921. It quashed a Texan conviction for murder of a man who had refused to retreat when attacked by an assailant with a knife and had shot him to death. No less a luminary than Oliver Wendell Holmes (1841–1935), who was, paradoxically, author of a standard textbook, *The Common Law* (1881), delivered the ruling which was, to put it mildly, robust. He insisted that, in Texas above all, 'a man is not born to run away.'[28]

The gunfighting of the American West, therefore, was not just a social phenomenon, still less a Hollywood invention, but was deeply rooted in America's distinctive legal philosophy. It was not, of course, without its critics, even in its heyday. It was assailed by the New England clergy, for instance: gunfighting was one of the aspects of the West they most deplored. It was attacked by, of all people, Kipling, on a visit to the West in 1892. He may have coined the phrase 'the Law of the jungle' in tones of respect, but he expressed the strongest possible disapproval when in Portland, Oregon he discovered that a jury had refused to convict a cowboy who had killed a comrade with his six-shooter on the ground that the fight had been 'fair.' Kipling thought that 'the code of the West' should not apply in 'a civilised city' like Portland. He spoke for the future in the sense that, from about 1900, most courts refused to countenance gunfights, however 'fair.'[29]

Some historians, using a quasi-Marxist type of analysis, have sought to explain Western violence, and the role of gunfighters, in terms of class conflict. According to this analysis, the salaried gunmen, usually on the payroll of a major cattle or railroad baron, tended to be Northerners of Anglo-Saxon extraction, and to vote Republican. Those who fought for the 'little folk' came from what Kipling would have called the 'lesser breeds'—Hispanics, Greeks, Italians, Slavs—or were Southerners and voted Democrat.[30] But there are too many exceptions to this dichotomy—and too many incidents where it is impossible to trace the political, financial, or indeed racial affiliations of the assailants—for this to be a useful historical tool.

Oddly enough, Karl Marx himself took a hand in this diagnosis. On May 11, 1880 there occurred what is known as the Mussel Slough Shoot-out in California. This arose from a dispute over land between pioneer-farmers and the Southern Pacific Railroad created by Collis P. Huntington, Leland Stanford, and Charles Crocker. In most such disputes the railroads usually had the letter of the law on their side, and the farmers natural justice. Sometimes the farmers, if numerous and active enough, and able to pull a bit of electoral weight, won their point and kept their land. On other occasions, the railroads invoked the law and local marshals did any needful shooting. Occasionally, however, they hired gunmen. This particular shoot-out was between two hired guns, one called Walter J. Crow, and five armed farmers. Crow killed all five of the farmers, though the gunmen also died from their wounds. It should be appreciated that the weapons used in Western violence were very inefficient.[31] Farmers and cowboys preferred shotguns and Winchester rifles, not always handy at short range. Professional gunmen used six-shooter revolvers which were indeed short-range weapons, but their barrels were not long enough for much accuracy and at 20 yards or more the chances of hitting anyone, let alone killing them, were remote. The longer the barrel, the more lethal the revolver. A famous photo which has survived shows Jesse James (1847–82), the railroad-bandit and notorious killer, posing for his admirers. He is mounted and holding out his shooter. Both man and horse seem very small indeed. The only object larger than one would expect is James' gun, which has a remarkably long barrel. Evidently James knew what he was doing. And so, it seems, did Crow.[32] Karl

Marx followed the affair closely and rejoiced. 'Nowhere else in the world,' he exulted, was 'the coming class war' arriving 'with such speed' as in California. Well, that was another thing he was wrong about, wasn't it? Actually, had Marx but known it, there was a substantial economic battle going on in California, then and later, between cattle-ranchers and sheep-rangers. The sheep-rangers tended to be 'little people,' recent immigrants, non-whites, Hispanics, Scotsmen, Basques, Mexican Americans, and Mormons. But there was not much violence. And the sheep-owners won decisively in the end, mainly by sheer weight of numbers.

The violent phase in the West's history lasted from 1850 to about 1920, though incidents occurred even later. One of the worst episodes took place at Ludlow in Southern Colorado in April 1914, when gunmen hired by John D. Rockefeller and other mine-owners fought a battle with an armed mob of miners and later burned their tent-city, thirteen women and children being suffocated in what was known as the 'Black Hole of Ludlow.' Violence between militant trade unionists and armed men employed by managements figure largely in the statistics of killings. One historian has compiled a list of forty-two serious cases of violence, 1850–1919, whose origins were, it can be argued, socioeconomic.[33] But it would be possible to compile a much longer list where the causes were individual greed, revenge, personal disputes—especially between families—and criminal gangs. Most violence occurred in the comparatively short period which elapsed between the foundation of a district and the setting up of regular law enforcement and courts. During these lacunae the West was, indeed, lawless, and locals had to organize vigilante bands to protect life and property. Even where law enforcement existed, jurisdiction was usually limited to particular counties or even townships and parishes, and criminals who moved across these boundaries were safe. There were no state police and of course no FBI, authorized to cross state boundaries.

This was the grievance that Allan Pinkerton, who had a strong sense of justice, tried to remedy with his own organization, while pressing Congress to take legislative action. But in many areas, for periods of years or even decades, the vigilante band was the only answer. Over 200 identifiable organizations of this kind have been studied west of the Mississippi. California had the largest, the San Francisco, reputed to have 6,000–8,000 members at one time (1856). Texas had the most. The archetypal state was Montana, whose state capitol is a memorial to frontier vigilantes. The fictional literature sometimes gives a misleading impression. The four novels written about Mussel Slough present it in black-and-white terms, whereas history shows the shades were gray, as one would expect. Walter Van Tilburg Clark, who wrote a famous novel about the vigilante movement, *The Ox-Bow Incident* (1940), presents it as oppressive and murderous. But the evidence is overwhelming that most groups were simply anti-criminal and protected the unarmed against armed robbers and killers. Many congressmen, senators, and governors were vigilantes, and they could not have got themselves reelected if the bands to which they belonged were unpopular.[34]

The worst and often the commonest crime in the West was horse-stealing. Nothing was harder to prevent, detect, or punish, and nothing aroused more anger. Most vigilante bands were directed against horse-thieves, who usually committed other crimes too, of course. Young Theodore Roosevelt, when he lived in North Dakota, was an eager horse-thief vigilante. In Montana, a major vigilante group was led by Granville Stuart, the cattle king of the state, sometimes called 'Mister Montana.' It was known as 'Stuart's Stranglers' and not only arrested horse-thieves but burned their cabins. They also killed—over 100. Stuart, who feared that the noisy young Roosevelt would boast of their doings, kept him out of it, and 'TR' was furious. But the Stranglers, their job done, were peacefully disbanded in 1884. A similar band in Wyoming in 1892 was known as the Regulators, after the South Carolina movement of the 1760s. This was run by the cattle magnate Frank Wolcott, who got Frank Canton, former sheriff of Johnson County, to command the gunmen. His band eventually made itself unpopular and, near Buffalo, was attacked by a huge mob of civilian vigilantes. It was saved from destruction only by the US Cavalry, sent in on the personal orders of President Benjamin Harrison, who was usually on the side of vigilantes if they represented the democracy.[35]

The most violent of all was New Mexico Territory, between the late 1860s and 1900 (it was finally made a state in 1912). This was the hunting ground of 'Billy the Kid'—real name William H. Bonney (1859–81)—who was not a socioeconomic phenomenon at all but a young scoundrel who turned professional criminal after he fancied he had been denied his wages by the territorial governor, Lew Wallace (1827–1905). Wallace was a former Union general in the Civil War, who had presided over the court which hanged Major Witz, commandant of the notorious Andersonville prisoner-of-war prison. He was a man of justice, who became famous as author of *Ben-hur, a Tale of the*

Christ (1880). 'The Kid,' as he was known in those days, killed ten men before being dispatched himself. By contrast, Wild Bill Hickock, whose real given names were James Butler (1837–76), was undoubtedly a Northerner and a Republican voter, and to that extent fits into the thesis of class/racial oppression. But his career as stage-coach driver, US scout, and marshal in various Kansas towns, including Abilene, defies political analysis. He killed various men, mostly criminals, and was himself murdered in Deadwood, Dakota Territory, in August 1876.[36]

What is so fascinating about the West is that it was being mythologized even while it was taking place. The settled East wanted to hear about the unsettled West without all the bother and expense—and danger—of going there. William Frederick Cody (1846–1917) provided for this vicarious appetite for adventure. He was a buffalo hunter who shot meat for the railroad camps, and the name 'Buffalo Bill' was coined for him by his friend Ned Buntline in 1869 when he produced a series of cheap novels, among the first Westerns. From this he gravitated to the stage, and then in 1883 organized his own Wild West show which toured Eastern America and Europe. Wild Bill Hickock made a number of personal appearances in the Wild West Show before returning to the reality, and death.[37] All the famous characters of the West were real enough, and some of them well documented. Indeed, holograph letters and signatures in their handwriting crop up constantly in the salerooms at the end of the 20th century. Buffalo Bill's signature is common. So are those of military leaders like George Crook and Nelson A. Miles, though General Custer's is rare and expensive. It is not hard to get the signature of Sheriff Pat Garrett, who shot Billy the Kid, and the signature of Emmett Dalton of the notorious Dalton Gang can also be obtained. Bat Masterson's signature has come up for sale and so has Wyatt Earp's (he signed himself Wyatt D. Earp), though both are very rare. Annie Oakley, 'Little Miss Sure Shot,' also left signatures, though they too are very rare. Equally rare are Geronimo's and Sitting Bull's, both of whom could sign their names. Geronimo drew the letters of his horizontally. But Crazy Horse never signed his name or if he did it has not survived or come on the market. Signatures by Wild Bill Hickock and Billy the Kid are equally unknown to collectors.[38] All the Wild West was more or less authentic, as the records prove. But the historian is struck by the comparative insignificance of the most famous fights. The most celebrated of all, the showdown at the O.K. Corral in Tombstone, Arizona in 1881, in which Wyatt Earp was the hero, or villain, cost the lives of only three men.[39]

Most of the conflicts of agricultural interests, even those caused by the novelty of barbed wire, were solved peacefully. The notion that the West was preoccupied by battles between intransigent class/ethnic/economic interests is false. The West was preoccupied, like other primary producers all over the United States, by fluctuations in the market. Its swings were particularly hard on ranchers, one reason why California, from the 1890s, took so eagerly to fruit and vegetable farming, where markets were more stable and the weather was the chief foe—and in California a friendly one. Another crop affected by violent swings was cotton; but that had always been true. The notion that the Civil War, which exposed the myth of King Cotton, marked the end of the United States, and the South in particular, as a major cotton supplier, is also false. Parts of the agricultural South were indeed hit by the consequences of the war and the brutal Reconstruction. Many large arable estates were broken up into smallholdings of 20 to 50 acres, operated by freed blacks as tenants or sharecroppers—a sort of Asian system, calamitous for land quality. There was less 'improved' land in 1900 than there had been in 1860.[40] But cotton, which after all was an international staple in a world where population was rising fast and garments in endless demand, continued to do well in good years. In the year before the Civil War, 1860, the South produced 3,841,000 bales of cotton. In 1929 the figure was 14,828,000 bales—still about 65 percent of total world production.[41] The blacks shared in this cotton prosperity. About 200,000 of them eventually owned their own farms, covering 20 million acres, a high proportion of it laid down in cotton. But mechanization at every stage of the process accounted for most of the increase.

In the last four decades of the 19th century, the mechanization of American agriculture accelerated. The first appearance of the Marsh Harvester in 1858, an invention of C. W. and W. W. Marsh which was, in effect, a reaper, sent a thrill of excitement through the more go-ahead American grain-farmers. Even more important was John F. Appleby's 'wire binder' of 1878, which enabled the harvest to be got in at eight times the speed, a vital consideration in the Midwest where the climate demanded instant gathering once the crop was ripe. By raising the amount a farmer could harvest efficiently, it hugely increased the crop he could grow—hence production rose from 5.6 bushels per farmer in 1860 to 9.2 in 1880.[42] About the same time, on the big wheat farms of the West, the combines began to appear, which cut, threshed, cleaned, sacked, and weighed the grain mechanically—originally drawn by a score or more horses, then by gas-tractors. Again, in the early 1880s, appeared the combined planters, coverers, and fertilizers, straddle-row cultivators, the sulky plow, the spring-tooth sulky harrows, and the seeders, together with the Lister, which plowed and planted at the same time. The source of farm energy, to drive the machines and implements, was improving all the time from 1860: first men, then horses, then stronger horses, then steam tractors, then (c.1910) the gasoline engine. The value of agricultural machinery employed more than doubled in the generation 1860–90, then multiplied five times 1890–1930, from half a billion dollars to $2.5 billion, with resulting productivity increases of over 400 per cent.[43]

The benefits brought to the American farming community, to the American consumer, who got cheaper, more plentiful, and higher-quality food in consequence, and indeed to the overseas consumer too, were due above all to the intelligence and industry of the American farmers themselves. But those who invented and marketed the new machinery deserve a handsome share of the credit too. This brings us to an important moral point. The period between the end of the Civil War and the second decade of the 20th century is often categorized as the era of the Robber Barons, of the ruthless, greedy, and selfish men who exploited the large-scale system of industrial capitalism, and the hapless millions employed in it, to enrich themselves and squander their wealth in 'conspicuous consumption.' But a list of American millionaires compiled in 1902 shows that a very large proportion of the new plutocracy, as its critics called it, were those who serviced the farming community, both ending the backbreaking labor of earlier days and bringing cheap food to everyone.

In Illinois, for instance, were Charles H. Deere and John Deere (1804–86), who mass-manufactured modern plows, Edward Wells, the great hog-packer, and Philip Danforth Armour (1832–1901) and his brother Herman Ossian Armour (1837–1901), whose work in meat-packing, refrigeration, canning, rapid transportation, and marketing—plus efficient use of waste-products—transformed the Chicago beef and pork industry. There was Thomas Lynch, the great distiller. Is producing hard-liquor an honorable trade? Thousands of farmers, whose products Lynch turned into hard cash for them, thought so. Another user of Western grain in vast quantities was Adolphus Busch, who brewed beer in St Louis. A rival St Louis millionaire was D. R. Francis, who ran the key options market. There were the Pillsburys of Minneapolis–St Paul, a miraculously thriving twin city whose rapid growth was rhapsodized at the end of Twain's *Life on the Mississippi*—they ran one of the most important wheat markets—and Frederick Weyerhaeser, who provided milled lumber in gigantic quantities for Western farmers. In San Francisco there was the great grain-dealer Henry Pierce, and the farm-machinery pioneer L. L. Baker—not to speak of McCormick himself, blessed more often by farmers for the inventions that made their lives a little easier than any other entrepreneur. All these men made fortunes in a highly competitive world, and all made impressive contributions to the transformation of farming lives and consumer budgets. Where was the robbing?

If we turn to the non-agricultural economy, a different picture emerges. Up to the decade of the Civil War, the United States, though already the wealthiest country in the world, in terms of the living standards of most of its inhabitants, was in many ways what we would now call a Third World country—that is, it exported primary products, such as cotton and tobacco, and imported most of its manufactures. The Civil War, by giving a huge impulse to American industry, changed this position dramatically, and the United States became largely self-sufficient. Between 1859 and 1914, America increased its output of manufactured goods, in value, no less than eighteen times, and by 1919, boosted by World War One, thirty-three times. The decade of the Civil War was the 'takeoff' period for this process, showing a 79.6 percent increase in the number of companies engaged in manufacturing, with a 56.6 percent increase in the number of workers they employed. In 1840, the United States ranked fifth in output among the world's manufacturing countries. By 1860 it was fourth. By 1894, it had taken the first place. By that time America already produced twice as much as Britain, the previous leader in manufacturing, and half as much as all Europe put together. By the turn of the century, the United States' imports of manufactured goods were insignificant and it was already exporting them to the world.[44]

There were six main reasons for the overwhelming and rapid advance in industrial power. First, America (like Britain before it) had liberal patent laws, which gave the maximum incentive to human ingenuity. The number of registered patents had already passed the million mark by 1911. Second, the scarcity and high cost of labor gave the strongest possible motive not only to invent but to buy and install labor-saving machinery, the essence of high productivity, and so mass production. Third, this promoted in turn standardization of machinery and parts. Fourth, the extraordinary success of American agriculture, already noted, was one of the dynamics of industrialization. Already in 1860 production of flour and meal was the biggest single industry in the United States. This was then displaced by slaughtering and meat-packing, which remained the biggest single industry until 1914. In other words, America's ascent to world leadership in manufacturing was driven by its agriculture.[45] Fifth was the abundance and variety of energy sources—first water-power, then steam-power fed by wood and coal, then electricity. America already used 2,346 million horsepower by the end of the 1860s decade, and by 1929 it had risen to 42,931 million. America produced and distributed, or tapped, energy more cheaply than any other country in the world. These natural resources and advantages were reinforced by a unique combination of protection and laissez-faire in federal and state policy. The freedom of interstate commerce, guaranteed by the US Constitution, made America by the 1860s the largest free-trading area in the world. But this was accompanied by high external tariffs, made possible or reinforced by the political ascendancy of the Republicans, from 1861 onwards. Thus the United States enjoyed simultaneously the advantages both of free trade and of protection. As a result the home market was buying 97 percent of its manufactures from domestic producers by 1900.

The transformation of the United States, within five decades, from a primary producer into the world's first industrial superstate was symbolized by the construction of a colossal continental railroad system. This, indeed, was a function of the role of agricultural products in industrialization and a driving force behind the emergence of a vast coal–steel complex and sophisticated financial markets in New York, Chicago, and San Francisco. The railroad was right at the center of America's industrial revolution. It was more than that: it was the physical means whereby Americans mastered a giant continent and began to exploit it with their customary single-minded thoroughness. US railroads began out east in 1825 and followed the frontier, the track often a mere decade behind the first pioneers, enabling the mass settlement to follow. When the admission of Dakota as a state was debated in Congress in 1884, Senator Benjamin Harrison (1833–1901) rightly observed: 'The emigrant who is now seeking a home in the West does not now use as his vehicle a pack-train, a Conestoga Waggon, or even a Broad Horn. The great bulk of the people who have gone into Dakota have gone upon the steam-car, many of them to within sight of the home which they were to take up under the Homestead laws of the

United States, whereas in Indiana it was 30 years after the admission of that state into the Union that a single line of railroad was built on its territory.'⁴⁶

The centrality of railroads in the expansion and development of the United States was used morally and politically to justify the suspension of the laws of laissez-faire and the direct involvement of both federal and state governments in their construction. Similar arguments had been used to justify the National Road. States were involved in promoting and facilitating rail construction right from the start. The federal government began its participation in 1850 by helping the Illinois Central. The real subsidies began, however (as we have already noted), in 1862, as a result of the Civil War, during which it began to seem lawful and natural for Washington to be involved in everything. Lincoln lent $65 million directly to the first transcontinental railroad and during the Civil War decade, 1861–70, the government handed over more than 100 million acres as a further direct subsidy. Washington, of course, was not the only source of subsidy. All the states of the South went into the subsidy business in different ways: Texas alone gave the railroads 27 million acres. Up to the beginning of the 1880s, New York State, Illinois, and Missouri, plus local governments, contributed $70 million in direct subsidies. Before the Civil War, about 30 percent of the financing of the railroads came from the public; after 1870 the proportion slowly fell but in absolute terms it rose, as the scale and cost of the networking increased. It is calculated that the total direct aid of government to railroads in the years 1861–90 was over $350 million.⁴⁷

The railroads were subsidized and legally privileged in six ways. First, they got charters (rather like the original banks) from state legislatures, often—it was claimed—in return for free passes handed out to prominent state politicians. Second, they were given special banking privileges to raise money. Third, they got the right of 'eminent domain,' in effect the legal ability to make compulsory purchases. Fourth, they were given both state and federal tax exemptions. Fifth, they often secured monopoly protection against competitors. Sixth, further capital was raised by federal, state, county, and municipal subscriptions. In the thirty years after 1861, for instance, national bond loans totaled $64.6 million (all repaid). The rails received tariff remissions too.

But the most valuable form of government subsidy was undoubtedly the gift of federal lands, made on a prodigious scale. No other corporations in human history have been endowed in such a profligate manner by a paternal government—they were indeed treated as Eldest Sons.[48] The rails got one-fourth of the states of Minnesota and Washington, one-fifth of Wisconsin, Iowa, Kansas, North Dakota, and Montana, one-seventh of Nebraska, one-eighth of California, and one-ninth of Louisiana. In all 242,000 square miles, a territory larger than Germany or France, was handed over. The biggest recipient was the Northern Pacific with 44 million acres; then came Southern Pacific with 24 million, the Union Pacific with 20 million, and the Santa Fe with 17 million. In addition, individual states donated a total of 55 million acres, bore the cost of surveying in many cases, and contributed to the stock. In New York State alone, for instance, 294 cities, towns, and villages contributed $30 million, and fifty-one counties gave subsidies as well. There was also a very large foreign contribution to the capital needed. By 1857, the British already held $390 million of American railway stock. By the end of the century, $7 billion of stock was owned abroad, the British coming first with $4 billion, the Germans next with $1 billion. All this was wiped out in World War One, by dollar war purchases in Britain's case, seizure in Germany's, so that by 1918 the foreign element had largely disappeared.[49]

The willingness of the state, at various levels, to assist railroad promotion helps to explain why, as the French economist Michel Chevalier put it in 1850, 'the Americans have a perfect passion for railroads.' The second factor was size and scale. In order to lock their continent in a railroad grid, the Americans had to think and build in gigantic terms. By 1840, America had nearly 3,000 miles of track, whereas the whole of Europe had only 1,800 miles, most of it in Britain. During the 1840s, track-mileage tripled in the South, rose by 150 per cent in the Northeast and by 1,200 percent in the old Northwest. Early tracks carried mainly passengers, but by 1850 freight began to dominate. The new-type US locomotive, a swiveled four-wheel truck ahead of four drivers, with cowcatcher, large headlight, and balloon-stack, could pull over long distances, fired by wood, a dozen 10-ton-capacity freight cars. That was a huge load by European standards and explains why hauling freight played such a critical role in promoting rail construction. It is true that, ton for dollar, rails could never haul freight as cheaply as river steam-power downstream. The Mississippi in particular, with its enormous width—at times of flood, for example in 1882, the river was 70 miles wide[50]—permitted sensationally large tows. Mark Twain describes a giant tow from Cincinnati to New Orleans of 600,000 bushels (6 pounds to the bushel) of coal, exclusive of the ship's own fuel, 'being the largest tow ever taken to New Orleans or anywhere else in the world,' the equivalent of 1,800 freight rail-cars. This tow cost $18,000, as opposed to £180,000 by rail.[51] But the Mississippi was one river, navigable from St Paul to New Orleans. Rail would take freight virtually anywhere.

Hence the network continued to grow. By 1850 it had 9,000 miles, but there were many important gaps. By 1860 there was a national network, excluding the still-undeveloped or unsettled West, of 30,000 miles. By this time Iowa, Arkansas, California, and Texas had built their first rails, but there was a massive gap in the middle of the continent. The speed with which the track snaked across the surface of America, breeding and sub-breeding like a living octopus, was startling. Chicago did not get its first locomotive till 1848. Five years later it had a regular rail-service to the East. Seven years after that, it was served by eleven railroads with 100 trains daily. From 1850 to 1870, $2.2 billion was invested in railroads, at which point the network totaled 53,000 miles. During the 1880s, the decade of maximum construction, another 70,000 miles were added, giving a total network of 164,000 miles by 1890, with an investment of $9 billion. During the next twenty years, the expansion-rate, though still very high, slowed appreciably, providing a network of 254,000 miles by 1916, about one-third of total world mileage. By then, the automobile was providing growing competition, followed in the 1920s by aircraft. Total mileage dropped for the first time, in 1920, to 253,000, with a further fall to 246,000 by 1933. (The network in 1987 had shrunk to 163,000.)[52]

US railroads not only absorbed a prodigious amount of capital, about $21 billion by 1916, they also employed vast numbers, which passed the million mark by 1900 and had risen to 2,076,000 by 1920. The climax of railroad construction began on May 10, 1869 when the Union Pacific and the Central Pacific connected at Ogden, Utah, for the first railroad across the continent. Henry Villard (1835–1900) completed the Northern Pacific in 1883 and the same year the Atchison, Topeka, and Santa Fe linked to the Southern Pacific for a south transcontinental route to California. In 1882 the Texas Pacific and the Southern Pacific met at El Paso, linking up with both New Orleans and St Louis. The Kansas Pacific, completed at Denver in 1870, and the Chicago, Burlington, and Quincy, completed in 1882, also opened up extensive areas to the network. By 1885 there were at least four main routes to the Pacific.[53]

The Americans gloried in their great 19th-century railroad network, got excited by it, used it as often as they could, and romanced about it. They also grumbled about it. From first to last, American farmers believed railroads were in the same category as the banks—they made excessive profits out of the farmer's sweat. The *Farmers' Alliance*, August 23, 1890, complained: 'There are three great crops raised in Nebraska. One is a crop of corn, one a crop of freight rates and one a crop of interest. One is produced by the farmers, who by sweat and toil farm the land. The other two are produced by men who sit in their offices and behind their bank counter and farm the farmers.'[54] In fact, though a few made fortunes out of rails by floating companies and selling stock, and later by mergers, railroads as an industry were never very profitable, except for short periods and in limited areas. Despite the part government played in financing them, there were never any overall plans. Railroads tended to be inefficiently designed (in terms of profits) and carelessly run. Some were never run at a profit. Others ran profitably but could never repay their underlying bonded debt. Many were, in realistic terms, bankrupt, right from the start. These disagreeable facts led to problems which were insufficiently understood, often by some of the railway magnates themselves. The public was led to believe that railroads overcharged and made massive profits, and that, if financial difficulties arose, it was because the speculators and managers had their hands in the till.[55]

These public suspicions were intensified by the way in which rail construction contributed to overheating of the economy and periodic panics. The panic of 1857 was caused by a combination of land speculation following the Mexican War and under-regulation of rogue banks, but overbuilding of rails had a big part in it. The panic of 1873 was almost entirely due to unrestrained speculation in rail construction, and in related companies often of a fantasy nature. This was a very serious financial crisis indeed, marked by a ten-day closure of the New York Stock Exchange and the worst depression the nation had yet experienced. More than 18,000 businesses went bust in 1876–7, most railroads went into bankruptcy, and this in turn led to a bitter and further damaging rail-strike in 1877, accompanied by savage violence. The panic of 1883, though less serious, was again caused by railroad overbuilding and was followed by bitter strikes. Rail played less of a part in the panic of 1893, which arose from a loss of European confidence in US securities, but it was extremely serious and again exposed the financial weakness of the US rail system. As, by then, the system employed one-tenth of all US capital and over 5 percent of the workforce, investors who lost their money, whether in rails or anything else, tended to blame the 'railroad millionaires.'[56]

These general charges were given some foundation by the spectacular financial careers of two operators, Jim Fisk (1834–72) and Jay Gould (1836–92). If two men came close to deserving the title of Robber Barons, it was Fisk and Gould. And if a particular episode illustrated the propensity of railway finance to promote skulduggery, it was the struggle of these two men with Commodore Cornelius Vanderbilt (1794–1877) for control of the Erie Railroad. Both Vanderbilt and the Erie road were essentially pre-Civil War phenomena. The Erie had been founded in 1833, and eighteen years later connected Lake Erie to New York, thus competing with and complementing the famous canal. Its 483-mile stretch was then the longest in the world, and during the Civil War it paid handsome dividends to its stockholders.[57] Vanderbilt was the first of the big-time tycoons but he started as a boatman. When he was seventeen his mother lent him $100 to buy a two-masted barge known as a leriauger and he used it as a ferry. From that he went into water-transport. By the age of twenty-one he had $10,000 in capital and he used it to break the monopoly of the Fulton–Livingstone combine on the Hudson, by operating the unlicensed *Bellona*. American businessmen often broke the law when they saw it as monopolistic. They had none of the respect felt by the English for the monopoly charters awarded by parliamentary statute. Vanderbilt came to see the law as something created in favor of one group of businessmen by manipulating members of state legislatures or paying expensive lawyers to win court decisions. So he learned how to use and bribe judges, to square politicians and even whole state legislatures, and in time how to get aid from the federal government—including entire regiments of marines when he was operating in Latin America. It is important to grasp this point. Vanderbilt saw himself as upholding the spirit of the US Constitution in exploiting the comparative liberalism of New Jersey law to break the monopolistic practices of New York State. His pirate *Bellona*'s flag bore the words: 'New Jersey Must Be Free.' When he was finally brought to justice he got the great orator Daniel Webster to defend him. The Supreme Court eventually declared the Hudson water monopoly, and the law which made it possible, unconstitutional—so, as the New York cynics said, 'it was Vanderbilt who brought us the freedom of the seas.'[58]

Vanderbilt was a big, rough, overwhelming man, with 'the loudest voice on the East Coast.' His seamen christened him 'Commodore' because that was how they saw him. From the Hudson he branched out into the Atlantic, running services to Europe and then a 'fast route' to California, which involved in effect buying up the government of Nicaragua, across which the land part of his route lay. That gave him a taste for rail and he moved into the northeast system, buying up the New York Central. He was not elegant. He always took twelve lumps of sugar in his tea. Though not an alcoholic, he liked a tumbler of gin to drink. Parson Weems, mythologist of Washington, said he met him in the street in tears, saying 'I've been swearing again and I'm sorry.' It was his new young wife, more elegant than he was, who did not like the swearing. For her he built the *North Star*, 270 feet long, the biggest and most expensive steam-yacht to date, its staterooms floored with Naples granite and Carrara marble, lined with rosewood and uphol-stered in satin, with portraits of his favorites for decoration: Washington, Franklin, Clay, Calhoun, and Webster (no Jefferson or Jackson—he hated them). There was gilt everywhere and it was said—by some, anyway—to 'evoke the Age of Louis XV.' He took his new wife, ten of his twelve children by her predecessor, various sons and daughters-in-law, plus a chaplain, the Rev. John Overton Chouldes, on a grand tour of Europe, leaving New York Harbor to an explosion of rockets and putting in at London, St Petersburg, and other ports. The chaplain published a deferential account of the voyage, *The Cruise of the Steam Yacht North Star*.

It was natural for Vanderbilt, having the New York Central, to want the Erie too. But he had enemies. One was Daniel Drew (1797–1879). He was born on a farm at Carmel in New York State and like Vanderbilt he started with $100, which he earned as a militia substitute in the War of 1812. He was reported to be unique as the only major capitalist actually to serve in the armed forces. He began trading as a cattle-drover, selling in New York to Henry Astor, butcher-brother of the great fur-trader. He slept in barns or by the side of the road. After narrowly escaping death by lightning he became a devout Christian, of the drink-and-swear-then-repent sort. If fed up, he retired to bed with a bottle of whiskey and a binge could last four days; but he 'never drank in company.' He drove herds of 1,000 head, salting them, and filling them up with water before selling. Hence the phrase 'watered stock,' which he took with him to Wall Street. By 1829 he had acquired a cat-tle yard near 3rd and 24th Street and the nearby Bull's Head Tavern where drovers congregated. Thence he started his own ferry line, forced the Commodore to buy him off, and set up in high finance. He was tall, lanky, rustic, and dressed in black. Men noted he looked like an undertaker and observed 'the cat-like tread of his gait' as he prowled for victims, wearing an old drover's hat and carrying a rusty brolly. He told people: 'I became a millionaire afore I knowed it, hardly,' but his supposed illiteracy was a defensive myth—he knew the Bible well and often quoted it.[59]

Jim Fisk's Big City slickness was complementary to Drew's bumpkin image. Fisk, unlikely though it may seem, came from Calvinist Vermont. His father was a tinware peddler, and Fisk was a born salesman. He ran off to join the circus and became a barker. He loved uniforms and finery of all kinds. He got started in the money game by buying up cotton in occupied areas of the South, then selling it in the North. Later he sold almost worthless Confederate bonds in England. He was said to be 'able to spot a sucker at a hundred yards' and coined the phrase 'Never give a sucker an even break' long before Hollywood was dreamed of. He was also reputed to have been the first man habitually to pinch girls' bottoms. Mae West's earliest partner remembered him well. Fisk emerged from the Civil War with a large stock of surplus blankets which he sold at a handsome profit and set up in Wall Street, soon joining forces with Drew. They specialized in buying up poor stocks cheap and dumping them on the unwary. Their office was open to all with a box of cigars and a bottle of whiskey on the desk. As opposed to Undertaker Drew, Fisk was the spangly type, often dressed in the admiral's uniform he designed for himself as boss of the Fall River Line, or as a colonel of militia. They called him 'Jubilee Jim.'[60]

On October 8, 1867 Drew and Fisk met Jay Gould. Unlike the other two, Gould had 'class.' From Roxbury, New York, he was descended from Nathan Gould, who came to Connecticut in 1647 and knew Roger Williams well. Gould was tiny, silent, with sad dark eyes. Many people 'mistook him for a poet.' He came up through tanning and leather. Then, in Wall Street, he became an expert in railroad stock. When he teamed up with Drew and Fisk, they determined on a stock raid to take over the Erie Railroad, which Vanderbilt was already trying to acquire. Their method was to send Erie stock up and down like a yoyo by sales and purchases, moving in for the kill at the critical low point. Vanderbilt, who had in his pay a notorious judge called George C. Barnard, a member of the Tweed Ring run by Boss Tweed of Tammany Hall, started the rough stuff by getting Barnard to give him an injunction. Gould got another judge, and then started the 'Erie

Panic' by throwing 100,000 shares into the market. They 'exploded like a mine' and, in Drew's phrase, 'Erie went down like a dead heifer.' Much of the tactics were made possible by the fact that smart speculators and their lawyers could play off New York laws against New Jersey, and vice versa. At the climax of the battle, Drew, Fisk, and Gould, who had taken over the Erie HQ in New York, gathered up $8 million of greenbacks there, tied them in bundles, threw them into the back of a hackney cab, drove to the New Jersey ferry, crossed, collected an army of thugs, and fortified Taylor's Hotel on the Jersey City waterfront, renamed Fort Taylor, with their armed men and three cannon. They also had a shore patrol in four lifeboats, each containing a dozen gunmen. All this was to fend off the naval assault of the Commodore, who, it was said, 'could be heard roaring from the New York shoreline.'[61]

That Fisk, Gould, and Drew milked the Erie is undoubted. This once profitable railroad became bankrupt in 1877, was restored to solvency after a series of reorganizations and did not actually pay a dividend until 1942, its first in sixty-nine years. It is true that watering of stock, in which Drew specialized, was common in railroads, though it was also used as a device to raise capital for modernization. All the same, throughout this period Americans enjoyed all the advantages of the largest and most modern railroad system in the world. The rate at which innovations were introduced was high by world standards. As early as 1859, George Mortimer Pullman (1831–97), born in New York and trained as a cabinet-maker, opened a workshop in Chicago to transform rail carriages into sleeping cars. He founded the Pullman Palace Car Company in 1867, introduced his sleeping cars widely the next year, and built an entire industrial town, Pullman, Illinois. Two years later, automatic couplers and air-brakes were introduced. Heavier rails replaced the original pre-steel ones, the entire network was made standard gauge, bridges were built across the Ohio, Mississippi, and Missouri, coal-burning (and, from 1887, oil-burning) locomotives replaced the old wood-burners, and scores of safety devices made travel less hazardous. As engines developed wider fireboxes and became more powerful, with extra drivers to insure continuous running, average speeds increased rapidly, and average train loads grew from 100 tons in 1870 to 500 or more by 1915. Between 1880 and 1916 labor productivity in freight services more than doubled. In addition to dining and sleeping cars (from 1868), passengers got steam-heat in 1881, solid vestibule trains and electric light in 1887, and ultra-safe all-steel coaches in 1904. It may be true that rail stock was more likely to be watered than any other—it is estimated that, of $7.5 billion indebtedness of the network in 1883, $2 billion represented water—but the investment in modernization, safety, and speed was nonetheless prodigious.

The innovations enabled the roads to cut average freight rates from 20 cents a ton–mile in 1865 to as little as 1.75 cents a ton–mile in 1900. These must have been the lowest large-scale freight rates on earth, and help to explain why freight carried rose from 10 billion ton–miles in 1865 to 366 billion in 1916, one important factor in the nation's huge industrial expansion. To put it another way, this represented an annual per-capita increase of freight carried from 285 ton–miles to 3,588 ton–miles. By 1916 the rails were taking 77 percent of the intercity freight traffic and 98 percent of its passenger traffic. That was an amazing performance in such a large country, containing so many natural obstacles. All the same, it was true, as one rail president, Charles Francis Adams Jr, put it, 'The system was indeed fairly honeycombed with jobbery and corruption.' That was why, by 1897, less than a third of railroads paid any dividends at all, and even in the ultra-prosperous war-year 1918, only 58.09 percent paid.[62]

The railroads represented the balance sheet of the American system: an unimaginable freedom to build and serve the public, balanced by an absence of legal safeguards to exclude the spoilers and protect the public. There were fewer inhibitions on enterprise than ever. The Fourteenth Amendment, Section One, stated, 'No state shall make or enforce any law which shall abridge the privileges and immunities of citizens of the United States' and so on, ending 'nor deny to any person within its jurisdiction the equal protection of the law.' This amendment was aimed at slavery and designed to protect the blacks. But it was also immediately and very effectively used by corporation lawyers to prevent state legislatures from interfering in business. It became extremely difficult for states to curb business practices they did not like without falling foul of this Constitutional safeguard. That was one reason why President Grant, victor of the Civil War, so quickly got his administration into trouble. Grant, ironically, had been elected by the poorest black voters. His majority over Seymour was only 307,000 out of a total of 5.7 million votes cast, and nearly all the 500,000 negro voters favored Grant. Grant knew little of politics, and was aware of it. That made him secretive. He would have done better to have admitted his lack of expertise and consulted more widely. As it was, some of his

Cabinet learned of their appointments only by reading the newspapers. He turned out to be a bad picker and some of the better men he picked quickly left, not liking their colleagues. His one pillar was Hamilton Fish (1808–93), a New York lawyer and former governor in the Alexander Hamilton federalist tradition. Grant made him secretary of state and he served the President faithfully throughout both his administrations, becoming one of America's great international statesmen and at least keeping Grant out of trouble abroad.[63]

But Fish could not prevent scandal invading the domestic side. This was where Gould, Drew, and Fisk came in. Having taken over the Erie, spending $700,000 on bribes in a single year through what was known as the 'India Rubber Account,' the trio were out for fresh game. Gould conceived the idea of making a 'corner' in gold, at a time when the new mines were enabling the world to leave the silver standard and climb onto a gold one. The object of a corner was to accumulate gold, raise the price on the markets, then unload your bars before it fell. The government, having a reserve of gold, could at any time foil the plot by unloading itself. So Gould's plan required inactivity by the US Treasury. In May 1869 Gould met Abel Rathbone Corbin, who was married to Grant's sister. Corbin favored farmers. Gould pointed out to him that the farmers were best off during the Civil War inflation when a gold dollar would buy $2.50 of paper. To raise the gold price and lower the dollar price would cause Western grain crops to move rapidly and be sold in Europe. So the scheme was good for the West and at the same time good for individuals who promoted it. He offered Corbin $1.3 million of gold at 133 on credit.

That summer the trio made other efforts to get into Grant's good books. In June Grant was in New York and attended the Opera House as Fisk's guest. The next day the President went aboard SS *Providence*, flagship of Fisk's Narrangansett Steamship Line, being greeted by Fisk himself dressed in his admiral's uniform. Gould was also there. They all attended Boston's Peace Jubilee, which featured Patrick Gilmore (1829–92), the great bandmaster—and author, as Louis Lambert, of 'When Johnny Comes Marching Home'—who conducted a band of 1,000 musicians and 10,000 choristers on the Common. Throughout June, July, and August, Gould bought gold, on his own and his friend's account but also for General Daniel Butterfield, the newly appointed Assistant Secretary of the US Treasury. Gold was 137 in August and the plan was to unload when it hit 180. Gould gave Corbin a check for $25,000, on account of 'profits to come.' In return Corbin told Gould on September 2 that Grant had signed in his presence a specific order to US Treasury Secretary George Sewell Boutwell (1818–1905) not to sell gold without direct orders from the President.

Using funds from the Tenth National Bank, controlled by Tammany Hall, Gould had accumulated $40 million of gold by September 22, representing twice the usual amount in circulation. With the dollar inflating, Horace Greeley's *New York Tribune* demanded the Treasury sell gold. But nothing happened. The next day President Grant became suspicious of Corbin and broke off all connection with him. Gould then began to unload as fast as he could without alerting the market. The following day, September 24, was Black Friday. Gold opened at 142 and went up to 162, with Gould and his chums happily selling. Other stocks fell and the pressure on the Treasury to unload gold became hysterical. News of the crisis reached Grant, who was playing croquet, and he immediately countermanded his written instruction, ordering the Treasury orally to sell without ceasing until the corner was demolished. There is a story that, when Wall Street's Trinity Church started to strike noon, gold was at 160 and that by the time the bells fell silent it was 138. We do not know to this day whether Gould, or Fisk for that matter, made or lost money on the transactions. Congress furiously investigated and virtually called both men evil incarnate, their conduct 'unmitigated by any discernible decency.' Fisk, who was then thirty-five (Gould was only thirty-three), commented on the findings, 'Nothing is lost save honor,' which rather implied he had made on the business.[64]

At all events Fisk continued his high-profile existence, and he used his money to the public benefit in more ways than one. Colonel Fisk restored the 9th Regiment of the New York National Guard, bought its band new instruments and paid its top cornet player $10,000 a year. He raised the regiment's numbers to 700 and took them on splendid outings. Jim Fisk Esq. came to the rescue of Chicago after its fire in October 1871, organizing relief and setting up a fund to restore the burned properties of survivors. He also invested heavily in the New York theater, especially in opera productions, making constant use of the casting-couch. Thereby he met his quietus. On January 6, 1872 he was shot to death by Edward R. Stokes, the pimp who looked after the interests of Fisk's ex-mistress, the vivacious Josie Mansfield. His body lay in state in the Opera House, also known as Castle Erie, and thousands of New Yorkers filed past—they liked him. They also subscribed to a magnificent monument made for Fisk in Brattleboro by Larkin

Mead, which cost $25,000. The stiffer element did not like that. The Rev. Henry Ward Beecher, scourge of Catholics and other undesirables, preached a violent philippic against him from his Brooklyn pulpit, describing Fisk as 'abominable in his lusts and flagrant in his violation of public decency.'[65]

By this time Grant was thinking about getting himself re-elected. The mud flung up by the gold corner and by the subsequent activities of Fisk, Gould, and Drew did not stick to the President or, if it did, there were counterbalancing factors. The Democrats, who were rapidly regaining ground in the South, made a foolish error of judgment by combining with a group calling themselves Liberal Republicans. They formed up around the New York journalistic campaigner Horace Greeley (1811–72), who supported virtually every progressive cause in sight and was a sure loser. He was also the scorned victim of one of the most told stories in American history—too long to repeat here—which cast him in a poor light and was made much of by Mark Twain in *Roughing It*. Except for his own paper, the *New York Herald*, the press subjected him to vituperative abuse and the great cartoonist Thomas Nast (1840–1902), star of *Harper's Weekly* and conqueror of the Tweed Ring, crucified him.[66] Grant won by the century's biggest Republican majority, 3,597,132 to 2,834,125, and by an electoral college vote of 286 to 66, Greeley carrying only six Southern and border states. The wretched loser spent the last week of the campaign sitting by the bedside of his dying wife and, overcome by it all, he died insane three weeks after the result.[67]

As Grant was reelected, however, the scandals were beginning to take their toll of those around him. More railroad skulduggery emerged from the Credit Mobilier affair, involving a construction company which had milked the Union Pacific Railroad for exorbitant fees—they went to the insiders who controlled both companies. Among the beneficiaries were the then Speaker, Schuyler Colfax, later Grant's Vice-President. Thirteen prominent Republican members of Congress were involved, but only two were censured by a compliant legislature.[68] Grant's War Secretary, W. W. Belknap, was forced to resign to escape impeachment for taking bribes from Indian traders at army posts. Grant's private secretary, Orville Babcock, was involved in a tax-collection racket known as the St Louis 'Whiskey Ring.' His Treasury Secretary, W. W. Richardson, was also implicated in a tax swindle. Even Grant's minister in London, General Robert Schenck, had to plead diplomatic immunity, and leave on the next boat, to avoid prison for selling worthless stock in the 'Emma Mines' to foolish Londoners.[69]

Then came the panic of 1873, which brought to light some of the crooks within or close to the Grant administration. The President himself was not implicated—he was much better at losing money than making it—but it damaged the Republican Party, which lost control of the House in the mid-term elections of 1874. The panic also did for Brother Drew, who had been hornswoggled by his friend Gould and went bankrupt. Gould went on cheerfully to invest in the trans-Mississippi railroad network, becoming the sixth-largest rail-stock owner in the nation.

These financiers were not without redeeming features. Drew was a pious Methodist and made huge donations to Madison Theological Seminary in New Jersey, which was renamed Drew Theological Seminary and is now Drew University. Gould was also a large-scale financial benefactor. He was not without taste or a sense of the public responsibilities of wealth. In 1878 he acquired America's finest country house in the Gothick style, Lyndhurst, on a knoll on the East Bank of the Hudson, just below the Tappan Zee Bridge. It had been built by an old New York family, the Paulings, and embellished by George Merritt, the first man to patent a successful railroad-car steel spring-system. Gould bought the astonishing castle-style house, in light pink and gray granite, plus 550 acres, for $250,000, and installed his beautiful wife, Helen Day Miller Gould, plus his six children. He actually lived there, he liked the house so much, and each day his superb yacht *Atalanta* would take him down the Hudson at dawn to work in his New York City office. From Lyndhurst he conducted financial forays which brought him control of the *New York World*, the Western Union Telegraph Company, and the New York Elevated Railway. And at Lyndhurst he raised orchids and read the classics in the magnificent library he built up.

Tuberculosis swept Gould off when he was only fifty-six, but—unusually for an American of his generation—he divided his immense possessions equally among his children. This meant that, while the sons continued to run the Gould empire, his eldest daughter Anna ran Lyndhurst with superb efficiency for forty years, finally handing over to her youngest sister, who had used her fortune to buy herself into one of the grandest families in France, the Talleyrand-Périgords. The sparky Duchesse de Talleyrand-Périgord ran the house until 1961, when it passed, under her will, to the National Trust for Historic

Preservation. Thus, as so often happens in American history, the public were the ultimate beneficiaries of a hard man's graft, and Lyndhurst, excellently preserved and endowed, is one of the noblest artistic delights of the Lower Hudson.[70]

Abuse of financial power was common in expanding America. There was also a strong propensity, among the victims, to hit back, and many ways open to them of doing so. The farmers complained that the railroads overcharged them in order to benefit the big industrial interests, who had more muscle. So from 1869 farmers created the Granger Movement, called after the local granges or lodges of the Patrons of Husbandry in Illinois, Minnesota, Iowa, and Wisconsin. Beginning with Wisconsin in 1874, they persuaded state legislatures to enact what were called Granger Laws, based on the protoype Potter Law. They included schedules of maximum rates, a ban on higher charges for short hauls over long, another ban on free passes for public officials, and anti-consolidation laws to maintain competition.[71] In 1876, the Supreme Court, in *Munn* v. *Illinois*, found for the Grangers, Chief Justice Morrison Remick Waite (1816–88) laying down: 'Property does become clothed with a public interest when used in a manner to make it of public consequence and affect the community at large. When, therefore, anyone devotes his property to a use in which the public has an interest, he in effect grants to the public an interest in that use, and must submit to being controlled by the public for the common good.' This was a very important and wide-ranging legal doctrine. But the Court was inconsistent. In 1886, it reversed itself in the case of *Wabash, St Louis and Pacific Railroad* v. *Illinois*, its decision being that no state could exercise any control of commerce beyond its limits.[72] Congress, led by Senator Shelby M. Cullom of Illinois, then enacted the 1887 Interstate Commerce Act, which met many Granger criteria and set up a commission to carry it out. But the rails fought bitter rearguard actions, the Supreme Court proved obstinate, and of sixteen cases brought 1887–1905 the Commission won only one on behalf of the customers.[73]

It became easier for the railroads to serve the public at a reasonable price *and* do justice to their stockholders after they had been rationalized. That was the work of many ingenious railmen and financiers but chiefly of Edward Henry Harriman (1848–1909). The Harrimans were a merchant family from England. Edward was the son of an unsuccessful clergyman and educated up to a point but chose to leave school at fourteen to seek his fortune in Wall Street. He started as a 'pad shover,' that is a financial scout who gathered information on the curb. By 1870, aged twenty-two, he had a coveted seat on the New York Stock Exchange—it cost him $3,000 borrowed from an uncle, and he repaid it, with interest, within a year. Harriman was straight in that he never cheated, repaid debts promptly, and stuck rigorously within the law. Where the law did not apply, he considered all fair game. He never gave quarter and hated conciliation. He 'always played for keeps.' That was why he was so much hated, notably by Theodore Roosevelt, who called him 'a public malefactor.' He did not consider 'gentlemanly behavior' was appropriate for business. So, though he had much in common with 'TR'—they both hunted, fished, hiked, and were crack shots and expert boxers—they stood at opposite poles of the social ethic. 'TR' was not the only man to hate Harriman. In 1907 many prominent figures banded together to form the Harriman Extermination League.[74] In historical retrospect, however, Harriman was a public benefactor, once fairly started. In early days he indeed played it rough. In 1874, aged twenty-six, he made $150,000 by selling short a railroad speculator called 'Deacon' White. He lost this money by an unsuccessful bear raid on the Delaware & Hudson Rail Company (the Astors beat him). He then bought into the Hudson River Steamers and married into the Ogdensburg & Lake Champlain Railroad, becoming a director in 1880. He made his first serious fortune by buying out his partners and then selling this small railroad, for a handsome sum, to the big Penn Railroad. His strategy was to create value by brainpower, then realize it on the market. The public does not like this because it thinks there is such a thing as absolute value. In fact a property is only worth what someone is willing to pay for it. The Absolute Value Fallacy, linked to the Physical Fallacy (the notion that only farmers and workmen, who actually produce food and objects, are valuable, and that all others are parasites), and similar illusions beloved of 19th-century dogmatists like Karl Marx, was the bane of the lives of the great American wealth-creators. Harriman was an expert at judging relative value, or rather creating it. The fact that he was hated for, as he saw it, staying strictly within the law appealed to his black sense of humor, and he never complained of criticism. He became a real big wheel in railroads in 1883, when he was made a director of the Illinois Central. He came up against J. P. Morgan, the biggest giant of the age, in 1887, and beat him. He lost to Morgan in 1895. Then in 1897 Harriman wormed his way into the Union Pacific, which made him the top railman in the United States.[75]

Harriman's rationalization of the American railroad network showed that he had a dynamic concept of change. Morgan also liked to rationalize, or rather to impose order on competitive chaos, but he had no feeling for rail and simply wanted calm. Harriman, by contrast, had an instinct for the travel business and wanted to make the system better and cheaper. He was an inspired financial manager of the Illinois Central. As his fellow-financier Otto Kahn said, 'Somehow or other [the Illinois] never had bonds for sale except when bonds were in great demand and it never borrowed money except when money was cheap and abundant.' The Union Pacific, the biggest road, went bust in 1893 and Morgan refused to rescue it: too much politics was involved, he said, and it was just 'a rusted streak of iron.' Harriman knew better. He blocked other attempts to sort it out as he wanted to do it himself—and did. By 1908 he had made the Union Pacific the best railroad in the country but he delayed announcing its massive 10 percent dividend until he had bought up more stock, as he admitted to the Interstate Commerce Commission. He had his own personal car put on the front of an engine and slowly toured the entire track, asking questions all the time. Then he put $175 million into improvements and made the Union Pacific the finest railroad property in the world. In short, he ran his properties for the benefit of passengers and businessmen sending freight, as well as for himself. He was also the hero of the 1906 San Francisco earthquake, which did $325 million damage, organizing the rescue effort using his immense rail and financial resources. During the subsequent flood he saved the Imperial Valley from becoming a lake. He saved the Erie Railroad out of his own resources. But people, including great and powerful men like Roosevelt, could not stand him personally, so he became, and to a great extent has remained, a 'male-factor.' The historian takes note of these judgments and passes on.[76]

Men like Gould and Harriman were more important, in their impact for good or ill on the public, than most of the presidents of the time. Indeed, between Lincoln and Theodore Roosevelt all the men who occupied the White House were second rate, with the possible exception of Grover Cleveland. Grant, the great general, turned out a political and administrative booby. He wished to run a third time but his party was not having it and invoked the two-term tradition as an excuse to stand him down. He then went on a two-year tour of the world, during which he got spectacularly drunk and committed many enormities.[77] After that he settled down and invested all his money in a private bank, which collapsed, leaving him bankrupt (1884). To provide for his family and pay his debts, he then set down to write his *Memoirs*, a superb work completed shortly before his death from cancer of the throat in 1885.[78]

The corruption within and close to the Grant administration, following on the disastrous attempt to impeach Andrew Johnson, did a great deal to discredit the presidency itself, following its high point under Lincoln. But worse was to come. For the 1876 election, the Republicans nominated Rutherford B. Hayes (1822–93). He was a lawyer, a former Union general and a respectable three-term governor of Ohio, now emerging as the heartland of the Republican Party. Unfortunately for him, he was a dull campaigner who could not get enough votes. By contrast, the Democratic nominee, Samuel Jones Tilden (1814–86), was a popular campaigner, a member of the Barnburners, the radical and progressive wing of the New York Democratic Party, who had broken the grip of the Boss Tweed Ring on Tammany Hall and settled in as a reforming governor of New York State in 1875. Tilden won the popular vote easily, by 24,284,020 to 4,036,572 for Hayes. He also led in the electoral college, by 184 to 165. But twenty electoral votes were in dispute. Nearly all were in South Carolina, Louisiana, and Florida, which were 'natural' Democratic states but still ruled, in effect by force, by the Republicans. In view of Tilden's plurality, the obvious democratic course was to declare him the winner. The Republicans insisted the disputed returns go to a fifteen-member commission, composed of ten members of Congress and five Supreme Court justices—in effect, of eight Republicans and seven Democrats. It operated, as might have been expected, on party lines and found for the Republicans in each state. The House was divided but the Senate, where the Republicans were in a majority, concurred, so Hayes was declared elected.[79] This was a legalized fraud, a result even more unrepresentative than the 'corrupt bargain' election of 1824–5. It meant Hayes had little moral authority even at the start, since even many Republican members of Congress felt he had no right to be in the White House, and still less after the Republicans lost Congress in 1878.[80]

The 1880 election at least produced a legal result. The Republicans picked another Ohio stalwart, James Abram Garfield (1831–81), born in a log cabin, who worked his way up to major-general in the Civil War, then went into the House. He won a clear though narrow victory, 4,454,416 to 4,444,952, over his Democrat opponent, another general called Winfield Scott Hancock, and a rather more convincing one in electoral college terms, 214 to 155.[81] But four months after entering the White House he was shot by a disgruntled office-seeker, and on September 19 he died. So his Vice-President, Chester Alan Arthur (1830–86), became president. Not only did he have no mandate: he was a thoroughly disreputable character who had been reluctantly put on the ticket to appease the Republican Party bosses. Arthur had originally surfaced in that sink of iniquity the New York Customs House. As recently as 1878 he had been personally sacked by President Hayes as a disgrace to civilization. Curiously enough, Arthur's administration did not turn out to be as corrupt as most feared, and he even presided over the enactment of the Pendleton Act (1883), the first serious attempt to rid the US civil service of the spoils system, from which he and his friends had benefited so conspicuously.[82] But he was clearly not a president anyone could respect nor one capable of wielding moral authority.

In these circumstances of political mediocrity, America tended to look elsewhere for leaders to admire, heed, and follow, and it was natural for it to find paladins in business. America had been founded by adventurers and preachers, and transformed into a republic by gentleman–politicians, but it was businessmen who made it, and its people, rich. Why should citizens not look up to them? Americans were conscious, in the 1870s, 1880s, and 1890s, that they were proud inhabitants of the world's wealthiest country, enjoying living standards unprecedented in the history of humanity. All around them they could see the thriving machinery and infrastructure of this wealth-creating process, and it was inevitable that the men who presided over this pulsating, throbbing, enriching system should inspire confidence and invite emulation.

The archetypal hero of the age was Andrew Carnegie (1835–1919). He was also, in his own way, its most effective economic and political philosopher—he wrote a striking autobiography and an important article, 'The Gospel of Wealth,' in the *North American Review* for June 1889. Carnegie was not an American by birth. He was a Scotsman, born not into Calvinism but into something equally pervasive, the Scotch agnostic tradition. To some extent he remained a Scot, papering his apartments with Scots tartans and, at the end of his life, buying Skibo Castle in the Highlands. But he understood what America was about—none better. It was about the freedom to get rich, and then the duty to give that wealth away. His life was a perfect parabola overarching both objectives.[83] His canny mother taught him thrift. His unhappy father, sacked as a hand-loom weaver to make way for power-looms, taught him the importance of new technology. Carnegie was a child-emigrant in 1848. He was small, thin, spry, with piercing blue eyes. His hair went white early, but he kept it. As a boy he knew hardship. He lived in an Allegheny slum in Pennsylvania, working a twelve-hour day as a bobbin-boy at $1.50 a week. Then he tended boilers. He got about, sleeping in box-cars, cabooses, anywhere. His break came when he got a job as a telegraph messenger and discovered Western Union was the firm of opportunity. At twenty-three he was its Pittsburgh superintendent. He saved his money and invested it.

Carnegie discovered the delights of capitalism when he got his first dividend check for his holding in telegraph stock: ' "Eureka!" I said—here's the goose that lays the golden eggs!' He went into railcars, then oil. At twenty-eight he had an income of $47,860.67, of which only $2,500 was his salary, the rest profits on his savings. He decided that working for other people was unprofitable and looked around carefully to decide which was the line of enterprise in which to specialize as his own boss. He picked steel. It fitted his maxim: 'Capitalism is about turning luxuries into necessities.' When Carnegie started, steel trim was a luxury. When he retired it was standard. Carnegie had other maxims. One was: 'Find out the truth from the experts.' Another was: 'Pioneering don't pay.' (That was also the emphatic view of Commodore Vanderbilt.) Carnegie thought the smart entrepreneur learned from the pioneer's mistakes, then cashed in with the help of the professional boffins.[84]

Steel was waiting for a killing. In 1844 US government surveyors discovered the first of the enormous iron-ore deposits in the Great Lakes region. The abundance of rich iron ore around Lake Superior, the scale of the Pennsylvania anthracite deposits, and the cheap water-transport and power of the region insured that this sector of the Midwest would be the center of America's heavy industry for the foreseeable future, and that Pittsburgh would be its capital. In 1856 Sir Henry Bessemer discovered a new and efficient way of making steel, and two years later the Siemens–Martin open-hearth method supplemented it. What Carnegie quickly found out was that the fundamental chemistry of these, and other, processes was imperfectly understood, and that the US iron- and steelmasters, loudly though they boasted about their expertise as 'practical men,' did not actually know what happened inside a blast-furnace. It was Carnegie who first put the laboratory technicians

(chiefly German) to work, thus marrying chemistry to steel-produc-
tion.[85] It would not be true to say that he created industrial technology
in America, for that had already been done by men like Fulton. But he
was the first to make it an essential part of big business. As a result, he
wiped the floor with the British steel industry, then the world's leaders.
At the beginning of the Civil War, America produced no steel rails—all
had to be imported. By 1873, the US was producing 115,000 tons and
steel was rapidly replacing iron rails over the vast and expanding network. That was Carnegie's doing. Between 1880 and 1900 US steel production rose from 1.25 million tons to over 10 million tons annually—
that was Carnegie's doing too. His furnaces produced nearly one-third
of America's output and they set the standards of quality and price.

Carnegie not only grasped the centrality of steel chemistry. He was
also the first major manufacturer to stress the importance of unit costs.
He insisted that every stage in the steel-making process be properly
accounted. He wrote: 'One of the chief sources of success in manufacturing is the introduction and strict maintenance of a perfect system of
accounting, so that responsibility for money or materials can be
brought home to every man.' Once you understood the principle of
unit costs, he argued, you could isolate the problem of productivity—
output per man–hour employed or capital used—and thus raise productivity at the same time as production (the two were usually organically connected). By raising productivity you could slash prices.
Carnegie argued that steel was at the heart of a modern industrial economy and that, if you could get steel costs down, you ultimately reduced
the price of virtually everything and so raised living standards. The
4,000 men at Carnegie's Homestead works at Pittsburgh made three
times as much steel in any year as the 15,000 men at the great Krupps
works in Essen, supposedly the most modern in Europe. This higher
productivity enabled Carnegie to get the price of steel rails, which cost
$160 a ton in 1875, down to $17 a ton in 1898. These enormous savings worked their way through into every aspect of the economy, with
consequential benefit to the public. No president, by miracles of administration, no Congress, by enlightened legislation, was capable of bringing comparable material benefits to all Americans in this way. So there
was a practical logic in the admiration men like Carnegie inspired.[86]

Carnegie was single-minded and concentrated in his vision. One of
his maxims was 'Put all your good eggs in one basket, and then watch
that basket.' He believed in business simplicity. Like most of America's
industrial pioneers, he hated high finance and, still more, financiers. All

the same, he would have made a good one. He was the first to introduce positive marketing into the sale of steel. He could have run railroads or gone into banking as he had a superb gift for placing railroad
securities, for instance. But he ran an old-fashioned partnership and
never marketed stocks or bonds. The great merit of a partnership, in
Carnegie's view, was that partners could not water the stock—the only
way they could raise money was to sell out. After the 1873 crash, six of
his partners in the great Edgar Thomson works sold out to him and he
ended up with 59 per cent of the firm. As he put it, 'so many of my
friends needed money that they begged me to repay them.' Hating speculation and believing in internal financing—that is, investment from
the firm's savings—he operated all his businesses as partnerships and
went right against the growth of the giant corporation and the stock-
exchange system which raised money for it. He saw countless friends
and colleagues ruined by calls on stock bought on credit. He never
bought on credit and ended up the richest man in the world.

The business cycle, Carnegie argued, was not an accident or anyone's fault: it was a fact of life and provision should be made for it. In
order for the economy to expand, credit has to be created. Hence 'when
the banking system contracts, not all the credit can be honored.'
Absolute virtue, therefore, resides with the man who has cash during
depressions—then he can buy labor and materials rock-low. In any efficient industry, he added, cost- and price-cutting must be continuous,
but depressions provided the finest opportunities. Anybody could make
money in a boom—making it at the bottom of the trough was the real
test of ability. His frequent testimony to Congress added to his *oeuvre*
as a philosopher. He hated humbug and never pretended to be anything
but a capitalist rampant when trading. 'I was in business to make
money,' he told Congress. 'I was not a philanthropist at all. When rails
were high we got the highest prices we could get. When they were low
we met the lowest prices we had to meet.' By buying in depressions,
because he had the cash, he was able to undercut competitors when
business picked up.

Carnegie was also the first big businessman to openly advocate a
philosophy of management which involved picking the best and paying
them commensurately. He had a wonderful gift for spotting and training managers. 'Mr Morgan buys his partners, I grow my own,' he said.
His firm was pure meritocracy. He picked the winners, set aside stock
for them, to be paid for out of their earnings, then made them full partners. He called it 'slave-driving with willing slaves.' He said: 'I am after

the winner. If he can win the race he is our racehorse. If not he goes to the cart.' He paid his managers the highest wages in US industry. He notes in his *Autobiography* that attracting the best men is like buying the best machinery—the most expensive labor tends to be the only sort worth hiring because in a free market its high productivity is the only reason for its high price. He trusted the market and never argued with it. So he concentrated not on profits but on costs. If costs were right, profits would follow: 'The price in the market is not your affair—you must meet the price whatever it is.' Carnegie's discoveries, like Julian Kennedy, Charles Schwab, and William Jones, the 'greatest maker of steel in American history,' all came from nowhere. He made natural foremen like W. E. Corey and A. C. Dinkey top managers while they were still in their twenties. His finds tended to stay with him. The only one who 'escaped,' to go off on his own, was Henry Clay Frick, a man who operated 1,200 coke ovens when he was only thirty-three and whom Carnegie himself called 'a positive genius.'

Carnegie's ability to make first-class steel cheaply was behind the US railroad boom at its most intense, but just when rail-laying began to decelerate he spotted the opportunities presented by the new high-rise buildings, rapidly developing into skyscrapers. They wolfed steel as hungrily as the railroads. He supplied steel girders for the first genuine skyscraper, the Home Insurance Building in Chicago (1883). He issued the first professional handbook of steel shapes for major buildings and he offered a ten-story skyscraper for the price of a seven-story one. He made the steel for the Brooklyn Bridge, for the New York Elevated, and for the modern US Navy.[87] He believed in steel—'The perfect [steel] mill is the way to wealth.' He bought railroads in order to design his own transport system and achieve vertical efficiency, but his heart was in the simplicity of a single product, the key to all the rest. Its perfection could be reached only by ruthless ability to detect the inefficient and eliminate it. Capitalism, he said, was 'creative destruction.' His motto to his managers was 'Well—what can we throw away this year?' It was desirable to carry out major scrapping and reconstruction at depression costs, but the process had to be perpetual whatever the climate.

By 1900 the profits of the Carnegie Steel Company were $40 million a year, almost equal to its supposed net worth. The next year he sold the entire business to J. P. Morgan's new mammoth, the US Steel Company, for the wholly unprecedented sum of $447 million. The modern mind reels at the value this sum then represented, when most forms of taxing wealth, including capital gains tax and income tax, simply did not exist. Carnegie has been accused of being 'the greediest little gentleman that God ever created.' But he was not much interested in money as such. A memo he wrote to himself as far back as 1868, when his career was just blossoming, and which came to light later, read: 'Man must have an idol—the amassing of wealth is one of the worst species of idolatry—no idol more debasing than the worship of money.'[88] The only decent motive for the production of wealth was the betterment of mankind. This could be brought about best in a republican democracy like the United States, where the means existed for money to be poured into progressive purposes with a reasonable chance of it benefiting the man. In *Triumphant Democracy* (1885), he extolled the US system: 'The old nations of the earth creep at a snail's pace. The Republic thunders past with the rush of an express.' The next year, in his essay on 'Wealth,' he argued that it was morally acceptable to become rich: what was reprehensible was to hang onto it—'The man who dies thus rich dies disgraced.'

Carnegie's problem was how to dispose of his money. In 1892 he founded Carnegie Hall for the performing arts. In 1902 he endowed the Carnegie Institution to further research. He set up the Carnegie Institution for the Advancement of Teaching (1905) and the Carnegie Corporation (1911) to support programs in the sciences and humanities. He had some dotty causes, such as spelling reform. In 1919 the Carnegie Endowment for International Peace published *A Manual of the Public Benefactions of Andrew Carnegie*, which showed that $350,695,653.40 had then been spent on a huge variety of projects. They included the construction of 2,811 free public library buildings and the purchase of 7,689 church organs, a curious gift from a lifelong free-thinker—but then Carnegie had endured the childhood agony of unaccompanied singing in the Kirk. By the time he died, in his sleep, in his eighty-fourth year, he had disposed of virtually everything he possessed, and he was buried at Sleepy Hollow, Tarrytown, New York, next to Washington Irving. That Carnegie got his point across is attested by the fact he was granted the freedom of fifty cities spread across the United States—even Gladstone collected only seventeen.[89]

Carnegie was a more important man than any president from Lincoln to Theodore Roosevelt. But he played no part in politics, except indirectly, or in the general running of the economy. It was a different matter for his great contemporary, rival, and colleague, J. Pierpont Morgan (1837–1913). Unlike nearly all the major business and financial figures of the epoch, Morgan was not entirely self-made.

He was fairly old money. His ancestors were of ancient New England stock. In 1854, George Peabody, whose family bank was one of the pioneers among the London–Atlantic houses, which also included Barings, Alexanders, Brown-Shipley, and Dennistouns, invited Morgan's father, Junius Spencer Morgan, to become a partner. Junius Morgan did well. He was the moving spirit in a syndicate which floated $50 million in French government bonds when France was at a low point, and so was able to make a 10 percent profit. He built on Peabody & Co.'s foundations and changed its name to J. S. Morgan and Co. Morgan himself built on his father's work, so he was running a three-generation superstructure, and part of his mesmeric hold over financial men was that they saw him as an aristocrat. They knew he had a code, and stuck to it, and they envied him his background and strong moral convictions.

It is impossible to understand this period of American history unless you grasp that people like Morgan had absolute standards of conduct which they would die rather than repudiate. When he told the US Senate that, in Wall Street, under his aegis, 'character determines credit' he meant it, and it was true. Much about his early life, indeed about his entire life, is obscure because he was a great destroyer of personal letters (like most late Victorians). We know he learned to smoke at seventeen, because he later deplored it. He went to a Swiss school, then Göttingen University. He was over six feet, well built, energetic, but he suffered from headaches, fainting, and skin trouble. He was a sincere Episcopalian, a regular churchgoer who took round the plate at St George's Episcopalian church in New York City. He sang hymns loudly. The opening words of his will—'I commit my soul into the hands of my Savior, in full confidence that having redeemed and washed it in His most precious blood He will present it faultless before my Heavenly Father'—might have been written in Hartford, Connecticut, in the 1840s.[90]

Morgan was in money all his life, beginning as an apprentice in New York's Denton, Sherman & Co. Aged twenty-two, in New Orleans, he bought, on his firm's credit, an entire shipment of coffee, and sold it at a profit. Most of the stories told about his rise by the later muckraking journalists are inventions. He certainly speculated against the federal currency—so did everyone who could. He paid $300 to a substitute to serve in the federal draft. So, also, did everyone with the cash—the Civil War was indeed 'a rich man's war and a poor man's fight.' In 1869, in the 'Susequehanna War,' he got the better of Jim Fisk and Jay

Gould. It is also (probably) true that he knocked the portly Fisk down an outside staircase to prevent him taking over a stockholders' meeting with his thugs. He did a number of big deals, which he managed skillfully. He was in no sense a Robber Baron. His riches were based on standard, respectable margins and incremental accumulation. In the immortal words of John D. Rockefeller, 'Mr Morgan is not even a wealthy man.' But by the early 1880s he stood right at the center of the New York financial community and was the only member of it everyone else trusted. That gave him power and responsibility.[91]

It is a curious fact that the United States, at this time, though already the world's richest nation with its largest economy, had no proper financial system of a public nature. Indeed, of the three fundamental tasks which any government must perform—the securing of external defense, the maintenance of internal order, and the provision of an honest currency—the federal government did none. The United States was protected not by its derisory army or its still-meager navy but by two gigantic oceans. It had no national police force and in large areas was not policed at all. And it had no central bank and was bitterly divided about what constituted its currency. In all these circumstances, especially the last, it was well that unofficial arbiters of behavior and maintainers of standards, like J. P. Morgan, existed.

Modern federal finance dates from the Civil War but it is a murky story. In 1869, David A. Well, the Special Commissioner of the Revenue, reported to Congress that his estimates of the cost of the war 'show an aggregate destruction of wealth, or diversion of industry, which would have produced wealth in the US since 1861 approximating nine thousand millions of dollars . . . It is three times as much as the slavery property of the country was ever worth.' The income tax, imposed in the war, was abolished in 1872. There was a huge excise, in addition to a high tariff. But most of the war was paid for in paper, after the suspension of specie payments. The war finance was deliberately inflationary. Lincoln's Treasury Secretary, Salmon P. Chase, admitted that his policy was 'finance your war costs on borrowed funds, and increase your taxes only for the purpose of covering service on the newly incurred debt.'[92] Only a quarter of the immense cost of the war was financed by taxes. Hitherto the paper issued by the US Treasury had been interest-bearing. Now its paper, greenbacks, were genuine paper notes—or rather, suspect paper notes, since the $450 million originally issued never reached par and fell as low as 39.

After the war, prices fell, and all those who owed money, such as the farmers—they had borrowed heavily to expand fast during the war years—fought a thirty-year battle to reverse the government's deflationary policy and push back prices to the high levels of the 1860s. The Republican government went back to the gold standard in 1879. Farmers and other debtor groups formed the Greenback Movement for a paper currency. Gold continued to go up until 1896, and prices fell. The farmers were badly hit and crucified by interest payments. The pressure on them was increased by the panic of 1873. They then turned to silver (as opposed to gold) as their panacea. But there was great muddle and confusion arising from the inadequacy of technical knowledge, and misapprehension about the way in which Gresham's Law— 'bad money drives good money out of circulation'—actually worked. A further confusing factor was the rapid discovery and exploitation of more mines, both gold and silver, at this time. With a free market in bullion, the total amount of specie available, and the relative supplies of gold or silver, had unpredictable effects on the financial markets.[93]

Public discontent was intensified by the well-grounded belief that politicians did not offer a choice in financial policy. Before the Civil War the Democrats had stood for low tariffs, the Republicans (or Whigs) for high tariffs. The Democrats had been anti-bank, the Republicans pro-bank. Clear choices were offered. Now the Republicans were high-tariff men and hard-money men (supported by the banking interest). So that was clear enough. But what did the Democrats stand for? In the election of 1884, the opportunity at last arose for a Democratic comeback. The Republicans split into three groups—reformers, or Mugwumps, who opposed party and government graft; Stalwarts, or supporters of old General Grant; and Half-Breeds, moderate reformers and hard-money men who stood in the middle, backing the party. They eventually nominated a Maine-man called James G. Blaine (1830-93), a senator and former Secretary of State to Garfield. The Democrats picked Grover Cleveland (1837-1908), a first-class governor of New York who had fought a sturdy battle against his party's Tammany Hall 'pork barrel' politics. During one of the dirtiest campaigns in American electoral history, it emerged that Cleveland, a bachelor, had fathered a bastard—and the Republican mobs chanted, 'Ma! ma! Where's my pa? Gone to the White House Ha! Ha! Ha!' But Blaine was overshadowed by an ancient railroad scandal of the 1870s, in which he was implicated by the 'Mulligan Letters' (including the damning phrase 'Burn this letter!'). Worse, a prominent clerical supporter of his denounced the

Democrats as 'the party of Rum, Romanism and Rebellion.' That just gave Cleveland the margin in New York, where thirty-nine electoral college votes secured him victory 219 to 182 (the popular vote was even closer, 4,911,017 to 4,848,334).[94]

Cleveland, as president, emerged as a highly conservative gentleman, a hard-money man, a defender of high tariffs, on the whole, and a financial stand-patter.[95] So what, asked millions of Democrats, was the point of electing their party? Where was the choice? What was the difference between these two sets of Republican and Democratic office-holders both of whom were bear-led by the bankers and the big businessmen of US industry? Thus cynicism about politicians deepened, and it was from this cynicism and disgust, especially among Democrats and above all among farmers, that there blossomed the extraordinary career of William Jennings Bryan (1860–1925), the 'free silver Democrat' and populist, and friend of the farmers. Bryan's superb speech to the Democratic Convention of 1896, in which he claimed, on behalf of the 'free silver' men, that America was being crucified on a 'Cross of Gold,' won him the nomination. He won it again in 1900, and a third time in 1908, so at least voters had a choice on these economic and financial issues. But as he lost on each occasion it made no difference in practice.[96]

Against this background of disbelief in politics and politicians, of genuine confusion about what was the right financial and monetary policy for the United States to pursue, and in the absence of such elements of regulation as a central bank, the community was probably fortunate to have a man like J. P. Morgan to act as an unofficial guide and lawmaker. One says 'probably' because, in the first place, he was never elected to any public office, which makes the power he wielded in the world's greatest democracy an anomaly. Second, he had views of his own, which were firmly based on experience and cogently argued but not shared by all, even in the financial community over which he held sway. Morgan believed in two things. First, he believed in order. Second, he believed in hard money as the best means of enforcing that order.[97] He took a religious view of economics. The tendency of economic activity in a free society was to produce primeval chaos, in which men fought savagely for supremacy and countless sins were committed. Freedom was needed for economic society to function efficiently but the resulting chaos generated inefficiency as well as sin. So some order was needed, and that was best produced by forms of economic concentration which imposed a degree of order without inhibiting freedom to the point where efficiency was again endangered. The needful concentration was achieved by the corporation and the trust.

Corporations had been made possible in America by the Marshall Court early in the century. Marshall himself had defined a corporation, in *Dartmouth College* v. *Woodward*, in almost metaphysical terms, which Morgan found highly satisfactory:

> A Corporation is an artificial being, invisible, intangible and existing only in contemplation of law. Being the mere creature of law, it possesses only three qualities which the charter of its creation confers upon it, either expressly or as incidental to its very existence ... [the most] important are immortality and, if the expression may be allowed, individuality; properties by which a perpetual succession of many persons are considered as the same, and may act as a single individual.[98]

In short, a corporation was a legal convenience enabling many people to act together as if they formed an individual person, for public purposes to be stated in their charter. The device was originally used to create banks, turnpikes, canals, railroads, and other businesses judged to be generally used by the public. Many states would not allow it for other forms of business. Thus, until 1846, New York banned incorporation under the general law: those wishing to form a corporation had to get a special law through the legislature. Then, during and immediately after the Civil War, the corporation became the normal form of business, especially for large enterprises. All legal barriers were removed. This was natural in a huge country like the United States. A corporation suited bigness and in America a business, to be efficient and serve its public, might have to operate in different places hundreds, even thousands, of miles apart, and in many different states. The advantages of the corporation were, in fact, so great that it became overwhelmingly dominant over a certain size. By the 1890s, Carnegie's partnership company was an anomaly. By 1919, corporations, while forming only 31.5 percent of the total number of businesses, employed 86 percent of the workforce and produced 87.7 percent of total output by value.[99]

On top of the movement to incorporate business was a further consolidation drift into trusts. This again intensified during the Civil War. The case for the corporation was overwhelming—no one has ever been able to devise an alternative which fulfills so many different purposes.

The trust is a much more debatable device. It was a creature of protection. High tariffs, which protected nascent industry against its foreign competitors, was the one mortal sin America committed against the virtuous creed of laissez-faire, and naturally it bred other sins. Henry O. Havemeyer, president of the United States Sugar Trust when it was formed, said complacently: 'The tariff is the mother of trusts.' Once a high tariff was imposed by Congress in response to pressure from domestic industries, it was liable to be maintained long after any absolute need for it had ceased. It taught industries to organize themselves into pressure groups not only against external competitors but, in time, against internal disorder, produced by thrusting newcomers. Just as private companies formed corporations for strength and self-protection, so corporations learned to form themselves into trusts.

Sugar was a good example of how a high tariff created a trust. It did not always happen, of course. The Standard Oil Trust and the American Tobacco Trust were not created by tariffs. But most such trusts were, and it is obvious that the Civil War, which ended Democratic control of Congress and so allowed tariffs to be raised, started the movement towards concentration of which the trust was the most prominent symptom. Two of the early experts on trusts, Jeremiah W. Jenks and Walter E. Clerk, asserted in 1917: 'A calculation of the flat average of the returns from all the leading industrial lines for which figures are given since 1850 gives an almost startling demonstration of industrial concentration in the United States during the past two generations . . . in thirteen leading lines of industry in the United States, the average manufacturing plant in the 60 years from 1850 to 1910 multiplied its capital by more than 39, its number of wage-earners by more than seven, and the value of its output by more than 19.'[100]

Morgan thought that the process of concentration, to produce the highest efficiency including vast economies of scale and marketing, was natural, inevitable, and desirable, and that self-protection against disorder was also, on the whole, welcome and in the public interest. But, like capitalism itself, and like the free market which underpinned it, the trust was an instrument for evil as well as good. That was why it should be placed, wherever possible, in the hands of honest and public-spirited men. Concentration came in many forms—pools, legal trusts, holding companies, to name only three. As early as the 1830s, the salt-producers of West Virginia had combined together to restrict output and control prices. They argued that otherwise the price would collapse and ultimately the public would lack salt. The movement towards

trusts began in earnest as a result of the 1873 panic. Competing railroads formed pools, to apportion business among themselves and so
avoid competition which would ruin all. As we have noted, the
Interstate Commerce Act of 1887 tried to end pools, among other
things regarded as evil, but the railroads created devices to frustrate its
provisions and the Supreme Court aided them. The trust was one such
device, though it had in fact been used before the 1887 Act, in 1879
and 1882 by the Standard Oil Company.

Under a trust agreement, stockholders deposited with a board of
trustees a controlling portion of their stock and received a trust certificate instead. In addition to the Sugar Trust, there was the 1887
Distillers and Cattle Feeders Trust (the so-called Whiskey Trust), the
Lead Trust, and the Cotton Oil Trust. These were in addition to such
internal devices as the De Beers combination, which controlled the output of diamonds (and still does) or the 1902 tobacco agreement under
which the British Imperial Tobacco Co. and the American Tobacco
Company merged into Anglo-American, giving the first the United
Kingdom market unmolested and the second the United States and
Cuba.[101] Each trust had to be judged on its merits to see whether it had
got the balance between producer and consumer right, and the proper
mix of order and competition. By 1889 opposition to trusts had produced anti-trust legislation in various states and the next year the
Sherman Anti-Trust Act began federal regulation. The smashing of the
North River Sugar Refining Company by the New York Court of
Appeals in 1890, and of the Standard Oil Trust by the Ohio courts in
1892 put a brake on the trust movement, not on the ground that they
constituted a monopoly, however, but because their original charters
were judged violated. But it was the panic of 1893 rather than the
courts which put a stop to concentrations—justifying Morgan's belief
that, on the whole, the market was the surest regulator and the one
most likely to serve the public interest best in the end.

When the market revived, a new form of self-protection emerged,
the holding company. This was the device employed during the period
of maximum concentration, 1897–1904. It had already been used by
the American Bell Telephone Company and by the Pennsylvania
Company, but Standard Oil took the leading part in its general adoption. Laws in some states, including New Jersey, actually favored holding companies which were pure financial affairs. New Jersey, being
near Manhattan and Wall Street, was a perfect HQ for a holding company, which had only to produce a simple annual report for a general

meeting of stockholders. In 1904 one investigator, John Moody, listed 318 industrial trusts consolidating 5,300 distinct plants and capitalized at over $7 billion, of which 236, composed five-sixths of the capital, had been incorporated since January 1, 1898, 170 under New Jersey law.[102]

Morgan believed that consolidation, whatever legal device was used, was in the public interest provided those in charge of it were honestly concerned to impose order in accordance with three criteria: (1) lowest possible prices compatible with efficiency and permanency; (2) adequate return to stockholders for lending their capital; (3) regular investment to ensure modernity. Both he and Harriman were engaged in the same business for twenty years—imposing order on the railroad network. But their methods were different and it is significant that Harriman did not meet Morgan's strict criteria on public service. What Morgan disliked about Harriman arose not from their business conflicts as such—Harriman beat Morgan in 1887 but lost to him in 1895—but from Harriman's proclaimed philosophy that only the letter of the law need be observed. Morgan thought the spirit should count too. So he never invited Harriman to dinner at his house. On the other hand, he visited Harriman on his deathbed as that was a Christian duty.[103]

Morgan liked the word 'trust.' To him it meant what it said: a device to spread order and responsibility, and to inspire trust by the public, in a disorderly, inefficient (and so dishonest) industry. From the early 1880s, he was establishing order in the railroads, and after the 1893 panic he controlled the largest group of rails in the United States. The railroad bosses were summoned to his house at 219 Madison Avenue and told what to do. His imposition of order on the rails was distinguished by four characteristics: simplification of corporate structure, elimination of water in the stock, linking of small lines into main systems, and imposition of Morgan management, notable for its honesty, integrity, and professional efficiency. This worked well in railroads. He put the Erie back on its feet again. His rail group was the only one in the US which could always be certain to attract European capital for modernization and investment. It also worked well for his general financial company, Northern Securities, though this was eventually destroyed by anti-trust legislation. In 1901, by buying out Carnegie, he created US Steel, generally regarded as his masterpiece of order-imposing. Some argued that, by the time it was created, it was already an outmoded institution, steadily losing world market share. On the other hand, it was the biggest single productive unit in the US effort in both world wars and in 1960 it was still the world's biggest steel company.[104]

Morgan's greatest moment came in the 1907 panic. It was precipitated by the failures of the Knickerbocker Trust Company of New York and the Westinghouse Electric Company (October 22–23), but in reality was caused by structural weaknesses in the banking system. In 1907 there were 11,469 state banks, making a total of 17,891 commercial banks, all operating on a fractional reserve basis, many issuing paper which was legal tender and all promising to pay their depositors in authorized paper or gold coin on demand. America had gone fully back on the gold standard in 1900, a move made possible by the discoveries of vast new mining deposits, and that was the law. But of course there was no central bank, and no one in charge of the entire operation, and the US Treasury Secretary was a mere official with no all-powerful machine at his command. So the moment there was a serious loss of confidence and people wanted their money, it was certain that some—perhaps many, perhaps nearly all—banks would fail. Without a central bank empowered to create money on its own credit, that is, the future of the nation, the system was bound to seize up to some extent. In its absence in 1907, Morgan was quickly called in to put a brake on the panic and to prevent a significant financial downturn developing into a catastrophic depression. He was used to these emergencies. In 1877 his intervention had enabled the army's payroll to be met. In 1893 he had been called in by President Cleveland to remedy America's depleted gold stocks by forming a syndicate to market US securities in Europe in return for bullion. In effect, he single-handedly stemmed the gold outflow from the United States.

In October 1907 Morgan was coming up to his seventieth birthday and had just moved into the new New York mansion he had built to house his enormous library. On Sunday, October 20 he met first his partners, then members of the New York financial and business communities, in a series of interviews in this tremendous room, which still houses the core of his collection. Having ascertained the magnitude of the danger, he then put together a team of clever young men to go through accounts overnight and see which big houses were too weak to be saved and must go to the wall, and which could be rescued. That done, he set about raising liquidity by summoning the bankers and gouging money out of them, to be loaned immediately at 10 percent. The US Treasury provided its share too. Once the Stock Exchange knew the money to save members was available, albeit at high rates, the panic began to subside, though the man who made the announcement had his coat and even his waistcoat torn to pieces in the scrimmage.[105] So one or two big casualties, like the Knickerbocker, were permitted to fall, but the rest were saved. Morgan's activities in 1907 were the basis on which the US Federal Reserve Bank was finally created seven years later.[106] In 1907 Morgan filled a hole in the structure of society which government did not then cover and one who watched him walk around the financial district during that week said, 'He was the embodiment of power and purpose.'

Morgan terrified people. He was righteous, and the unrighteous trembled in his presence. So did many other people. He liked to give ladies gold trinkets at his dinner parties but they were scared of him too. He was very conscious of the effects of his skin complaint, rhinophyma, which made his nose large, red, and swollen, to the delight of caricaturists. When Mrs Dwight Morrow, the young wife of one of his partners, had to entertain him to tea, she instructed her four-year-old daughter, Euphemia, on no account to say anything about his nose. The little girl dutifully complied, and after she had sat on the great man's knee, her mother thankfully dismissed her to her nursery. Then she started pouring and said: 'Mr Morgan, do you take nose in your tea?' Morgan controlled people with his eye. The photographer Edward Steichen said: 'Meeting his eye is like facing the headlight of an express train bearing down upon you.' At his office he worked in full view of his staff, and they in his. He took no exercise and smoked endless cigars. He liked his yachts, all of which were called *Corsair*. When he arrived by liner from his biannual trips to Europe, his current *Corsair* picked him up in New York Harbor and took him straight to his house, Cragston, overlooking the Hudson. But his real center of activity was the great mansion housing his library and the collections he built up. Left to the nation, and munificently endowed, they now constitute, after the Library of Congress, the British Library and the Bibliothèque Nationale, the finest collection of rare books and manuscripts in the world.

Nor was this all, or even most. In 1888 Morgan became a trustee of the Metropolitan Museum of Art, and in 1904 its chairman. In 1909, under pressure from Morgan, the United States finally abolished its 20 percent import duty on works of art and it became possible for him, and many other American collectors, to bring their choicest artistic possessions into the country and, eventually, to donate them to the Metropolitan. It was under Morgan's influence that trusteeships of the major art collections were awarded to the most generous benefactors and their wives, so that they became, like the honors system in England, the apex of New York society. Thus America's enormous public artistic heritage was accumulated. Almost single-handed, Morgan turned the Metropolitan from a merely notable collection into one of the three or four finest anywhere. In 1905 he appointed the ravishing Virginian, Belle de Costa Greene, then aged twenty-one, his librarian and sent her on purchasing missions all over the world, with *carte blanche* to stay at the Ritz in Paris and Claridge's in London and to dress at the leading Paris couturiers. She found him many treasures and built the library as it now is, and it says a great deal for Morgan's reputation for probity that no one ever questioned the propriety of their relationship.[107] In his collecting, which ranged widely all over the world in a variety of artistic disciplines, Morgan obviously employed experts—including the cantankerous Roger Fry of Bloomsbury, author of many vicious sayings about his master—but it is astonishing how few mistakes he allowed them to make on his behalf. Even in a very difficult area like Egyptology, he enabled the Metropolitan to push itself into the front rank with great dispatch.

Thus the men of affairs bestrode late–19th-century America like colossi. How did they do it? And how, in a democratic country like America, where all had access to political power and where the courts and the people were imbued with the notion that all men are created equal, did the unions allow them to get away with it? There is no easy answer to this question. Take the case of Carnegie's right-hand man, Henry Clay Frick (1849–1919) and the Homestead Strike of 1892. It is a notable fact that whereas Americans took naturally to capitalism, which places the whole stress on individual effort, they have been markedly inept at creating and sustaining trade unions, which demand that men subordinate their individuality to the collective interest. There has been no lack of trade unions in American history, which is strewn with their half-forgotten corpses. Their weakness has been an inability to look to the long term and their propensity to go for quick successes by a fatal use of violence. In the 19th century, labor unions rarely survived financial panics. That was the story of the National Labor Union (1866–73) and the Knights of St Crispin, Milwaukee, founded in 1867 and by 1870 probably the world's strongest labor organization, especially in the Massachusetts shoe-trade. But both were destroyed by the panic of 1873.[108] Then there was the case of the Molly Maguires, a secret organization of Irish miners which terrorized the east

Pennsylvania coalfields for several years (1865–77) by strong-arm methods, including intimidation, arson, and murder. The region was saved from further strife by Allan Pinkerton's detective force, which patiently gathered evidence leading to the conviction and hanging of the ringleaders.

Pinkerton, it should be noted, was not anti-working class. He came from a radical Lowland Scots background, was actually wanted by the British police when he emigrated, was a strong supporter of negro rights, a friend of John Brown, and a Republican democrat of deepest dye. But he hated bullying of the weak from whatever quarter, he upheld the rule of law against all-comers, and he believed passionately in the right of the working man to sell his labor in the open market, as he pleased. Trade unionism as practiced in 19th-century America, went against all these convictions and Pinkerton willingly used his organization to beat undemocratic strikes.[109]

American trade unionism was not without a strong element of idealism in certain cases. One example was the Philadelphia garment-maker Uriah S. Stephens and his Noble Order of the Knights of Labor (1869); both he and Terence V. Powderly, who succeeded him as grand master, set out ringing statements of their utopian aims. But even this organization was marred by its intense secretiveness—indeed, it was essentially a secret society, like the Freemasons, though with overtones of the esthetics of the Pre-Raphaelites and William Morris.[110] When the Knights abandoned their secrecy, they had 700,000 members at one time and staged a successful strike against Jay Gould's railroad system. But from 1888 they were pushed aside by a newcomer (1886), the American Federation of Labor. This was the first union to aim at respectability, experience having shown that resort to violence invariably did fatal damage in the long run with public opinion, even though it might occasionally secure short-term gains. The AF of L got itself recognized by large numbers of employers and was backed by conventional middle-class bodies such as the National Consumers' League (1898), the National Civic Federation (1901), and the National Child Labor Committee (1904). It even had its own political arm, the American Association for Labor Legislation (1906), modeled on the British Labour Party, which was founded by the Trades Union Congress and had just got its critical Labour Disputes Acts through the British parliament. The AF of L was committed to the peaceful and lawful principle of collective bargaining and by 1914 had about 2 million members. Its leader, Samuel Gompers (1850–1924), English-born of

Dutch–Jewish extraction, came to New York, aged thirteen, organized the cigar-makers and was president of the AF of L from 1882 until his death. He hated ideology and theory, despised socialism, and concentrated entirely on wages and conditions. His colleague Adolph Strasser told the Senate Committee on Relations between Capital and Labor (1883): 'We are all practical men, we have no ultimate ends. We are going on from day to day. We are fighting only for immediate objects—objects that can be realized in a few years.'[111]

That was all very well, but it meant that American labor tended to fall between two stools. On the one hand it failed to go the British way and organize its own party, with the object of taking over government and enacting legislation. Thus it never won for itself the privileged legal status enjoyed by British labor from 1906 until the Thatcher union reforms of the 1980s—that was good for the American economy but bad for the power of organized labor. On the other hand, even respectable labor unions in the United States failed to escape entirely from the stigma of violence created by the many militant unions which nonetheless flourished alongside them. Thus the AF of L was opposed by the National Association of Mankind (1895), which used spies, injunctions, and Yellow Dog Contracts, and by the Industrial Workers of the World, led by the radical Eugene V. Debs (1855–1926), which declared itself 'founded on the class struggle and . . . the irrepressible conflict between the capitalist class and the working class' (1905), adding that it would go on 'until the workers of the world . . . take possession of the earth, and the machinery of production, and abolish the wages system' (1908).[112] Most American workers did not want the wages system abolished as they knew they got, by world standards, good wages, and many were still anxious to save from them and buy farms. Ideology was not popular. Nor was union violence.

It was against this background that the strike at Carnegie's vast Homestead plant took place in 1892. Carnegie himself was astounded, shocked, and hurt by the apparent willingness of his men to come out. He thought he paid good wages and that conditions at Homestead, where most of the plant was right up to date, were as good as they could ever be in such a horrific and dangerous trade as steel-making. He thought there had been intimidation. So did Frick, in charge of the plant. He hired 500 Pinkerton detectives to investigate what was actually happening, and when they were being towed upriver on two barges, to be landed at the plant, some of the strikers opened fire. That began the violence and from there on it escalated. In the end, the governor of Pennsylvania was obliged to call up and send in 8,000 militia-men, who broke what was left of the strike by force. We do not know to this day whether most of the men wanted to come out, as no orga-nized strike ballot was taken ('show of hands' meant nothing in the atmosphere of terror). The destruction of the strike was a long-term as well as a short-term victory for steel management. Frick wrote: 'We had to teach our employees a lesson and we have taught them one they will never forget.' Carnegie's biographer wrote forty years later: 'Not a union man has since entered the Carnegie works.' But Carnegie was permanently damaged too. He blamed Frick (quite unfairly) for bungling the strike. He said that, as a result, he himself had been accused of getting Grover Cleveland reelected in November 1892. He said the strike had 'left a hole in me.'[113]

Frick went on to become an associate not only of Morgan, at US Steel, but of Andrew Mellon (1855–1937) and John D. Rockefeller (1839–1937) and to accumulate an enormous fortune. He must have had a collector's eye of exceptional astuteness because he built up a personal art collection of a quality never surpassed in America, before or since, in which virtually every item, ranging from Holbein and Vermeer to Turner and Constable, is a masterpiece. He left the entire assemblage, plus his superb Fifth Avenue mansion where they hang, and an endowment of $15 million, to the people of New York City. How could a simple steelman, admittedly one of genius, in Carnegie's words, bring together what is undoubtedly America's most distin-guished single art collection? It is yet another mystery of the American spirit.

The rapid expansion of the American economy from the Civil War onwards left its physical traces not just in permanent art collections of outstanding quality but in countless other ways, some of them gigantic, even monstrous. The two most obvious and complex, and in them-selves procreative and socially and esthetically provocative, were the new monster-cities, Chicago and New York. Even by the standards of the 19th century, when cities sprang up in a few decades, the phenome-non of Chicago was exceptional. Chicago grew right from the start by using the latest technology to exploit its unique demographical advan-tages. At the beginning of the 1830s it was just a fort surrounded by a few farmhouses, with a population of less than 200. Its location at the head of Lake Michigan made it the natural entry point for the great plains Midwest, but sea-access was blocked by a half-mile sandbar, making it inaccessible for much of the year. The first step then was to get federal army engineers to bring up new digging equipment to drive a canal through the bar and eliminate the need for portage. Then in 1832 a Chicagoan called George Snowe invented the Balloon Frame for house construction. Under this system, milled lumber could be quickly nailed together instead of using mortice and tenon, comparable to inflating a balloon.[114]

The new Balloon Frame helps to explain the astonishingly rapid expansion of the village into a city in the 1830s and 1840s. By 1848 it was an important port equipped with facilities for handling the biggest inland ships in the world, and with 100 trains a day arriving on eleven different railroads. By 1887 it had 800,000 people. Chicago is part of a flat plain virtually the same level as Lake Michigan. So it had to jack itself out of the mud. In 1856 the town council decided to raise the entire city 4 feet by a new jacking-up process. The buildings were mainly timber still, which helped matters, but nothing like this had ever been done before, anywhere in the world. Briggs's Hotel, made of brick, five stories high and weighing 22,000 tons, was jacked up while continuing to function, an example of American determination and ingenuity, driven by the conviction that nothing material is impossible (it could never have happened in Europe). When the buildings were jacked up and the spaces infilled, new roads and sidewalks were laid down.[115]

It was Chicago's willingness to use the latest technology that enabled it to beat St Louis, which in theory ought to have become the economic and commercial capital of the Midwest. But it suffered in the Civil War and by 1870 Chicago had overtaken it, being particularly quick to take advantage of rail communications. Moreover, the appalling Chicago fire of 1871, in immediate terms a catastrophe, made Chicago an engineer's and architect's paradise. The fire lasted twenty-seven hours, destroyed 17,000 buildings—one-third of the total—and rendered 100,000 homeless. Architects descended on the stricken city from all directions to inspect the fire-damage at first hand and devise fireproofing methods. Among them were Peter B. Wright and Stanford Loring, who by 1874 had perfected a new fireproof building system. Among their innovations were steel framing, brick flooring, and ceramic and terracotta cladding.

Architects were also interested in erecting tall buildings, not for land-space reasons—especially in Chicago where land was limitlessly available and two-story tenements were spreading out in all directions—but because they could be made highly profitable in big, concentrated commercial city centers. Peter Brooks, the famous Boston property expert, wrote: 'Tall buildings will pay well in Chicago hereafter and sooner or later a way will be found to erect them.'[116] Ways were found almost immediately. First, caissons developed during the bridge-building boom of the late 1860s and 1870s were found perfect for laying foundations in Chicago's muddy surface. Second, Carnegie's success in bringing steel prices down made large-scale steel-frame buildings almost as cheap to build as any other kind. The third factor was the development of the elevator, on which Chicago seized with passionate relish. A steam elevator had been installed in New York as early as 1857 in the Haughwout Building. It made its appearance in Chicago in 1864, before the fire, in the Charles B. Farewell Company store. Six years later a hydraulic elevator was developed and, in 1887, an electric-powered safety elevator. By 1895 Chicago had over 3,000 electric elevators in its buildings.[117]

Very high buildings began to make their appearance in Chicago in the late 1870s and early 1880s. It is not necessary here to enter into the question of who built the first skyscraper and whether it was put up in Chicago or New York. Chicago was a practical, utilitarian city, and New York a display one too, and if Chicago did indeed invent the skyscraper it was because it yielded higher profits per dollar invested.[118] In 1879 William Le Baron Jenny, who experimented with both caissons and steel-framing, put up the First Letier Building, which incorporated iron pilasters, large windows, and a skeletal facade but was only five stories. His Home Insurance Building of 1884–5 had reinforced iron frames on two façades and was ten stories. In the meantime, the firm of Adler & Sullivan had built the Borden Block, 1879–80, and Burnham & Root the Montauk Block of 1881–2. Both these had more than one elevator and the Montauk was ten stories but it did not have a steel frame. Reinforced iron frames were found in half-a-dozen big buildings of the early 1880s. But not until the late 1880s, or even the early 1890s, did Adler & Sullivan, Burnham & Root, and Holabird & Roche use the full outfit of steel frames clad in masonry—and the only ones of these monsters to survive, alas, are the Monadnock, Reliance, and Fisher buildings.[119]

Behind this was the genius of Louis Sullivan (1856–1924), one of the greatest of all American artists. He came from Boston, where he studied at the superb new Massachusetts Institute of Technology, founded in 1865, which the following year opened America's first school of architecture. His mentor was another great architect, Henry Hobson Richardson (1838–86), and he witnessed the construction of Richardson's twin ecclesiastical masterpieces, Boston's Brattle Street Church (1870) and its Trinity Church (1872). Sullivan also went to the Paris Beaux Arts and he was probably the most broadly educated American architect to date. But he brought to his work a uniquely American sense of what was needed and how to do it—he was an exemplar of the maturing process whereby, in the decades of the Civil War, America was becoming a rich and majestic civilization in its own right.[120] He was influenced by the leading Philadelphia neo-gothic architect Frank Furness and he was lucky enough to work in the office of William Le Baron Jenny, who was not only the leading designer of iron structures but specialized in floating caisson foundations of the kind Chicago's muddy soil required. Then, aged twenty-four, he joined the firm of Dankmar Adler as chief designer, and three years later became a partner. The young Frank Lloyd Wright (1867–1959) joined the firm just before, in 1890, Sullivan burst into the office with the solution, conceived on a walk, to the new Wainwright Building project in St Louis—the moment when, as many architectural historians felt, the true skyscraper was born.[121]

Tall, black-bearded, handsome, romantic, eloquent, Sullivan was a charismatic figure, who could memorably articulate new architectural theory, as well as practice it. He coined the phrase 'form follows function,' though by this he did not mean the anti-ornamental dogma of the 20th-century Modern International Style—quite the contrary.[122] His writings give organic, not mechanical, forms as examples, and his buildings follow the flowing patterns of Art Nouveau. Thus he argued that a waterworks should be designed not merely to be an efficient machine for pumping water but should convey its abstract qualities—should express the essence of flowing water. This was much closer to the thinking of the Aesthetic Movement than to International Modernism, and Wright was justified in calling him a 'lyric poet.'[123]

The extraordinary speed and efficiency, as well as grandeur, with which Chicago was built and adorned is illustrated by the fact that, in the eight years 1887–95, Sullivan's firm received no fewer than ninety major commissions, from theaters to opera houses (as well as huge office blocks). His masterpiece was the Chicago Auditorium Building (1886–90), where his own office was relocated. Ten stories high, with 63,350 square feet of office space, it was the largest building in Chicago, with a 400-room hotel, 136 offices and shops, a sixteen-story tower, a 400-seat theater, and a 4,000-seat concert hall, then the largest in the world. It was thanks largely to the existence of this enormous, luxurious auditorium, and the profits it generated, that Chicago was able to build up its symphony orchestra into one of the greatest in the world, with its own electrifying 'Chicago Sound.' And it was thanks, too, to the success of Chicago's building program, and the profits generated from its efficiency, that Chicago was able to create so many libraries and art collections and endow one of the world's largest and finest universities.

Chicago, indeed, illustrates neatly the splendors and miseries of wealth creation, now proceeding on a greater scale in the United States than ever before in history. Sullivan stood for the splendor principle. In his 1896 essay, 'The Tall Building Artistically Considered,' he argued that because skyscrapers were vertical forms, not horizontal ones like most buildings, they should be designed following the vertical archetype, the column, with its three main features: a base, a vertically accentuated shaft, and a decorative capital. And all his buildings were in fact rich in ornamentation following this principle. But it is significant that the thirty-two-story Fraternity Temple (1891), designed specifically to illustrate his column principle, was never built, and in general Chicago architects tended to be hampered by price-factors and the need to generate the maximum rental income. H. B. Fuller's novel about Chicago skyscrapers, *The Cliff Dwellers* (1993), makes this point: 'Over the whole thing hovers incessantly the demon of 9 percent.' You could indeed get a 9 percent return on your capital in Chicago but only if you built steel cages with the services running through, cheaply filled out in tile, brick, and terracotta. More elaborate buildings, such as the Women's Christian Temperance Union, had to be planned with other motives—its teeming turrets and formidable machicolations were designed to make a splash on behalf of temperance.[124] Chicago skyscraper policy, indeed, was often dictated by wider economic interests as well as short-term cash requirements—commercial interests wanted the city to expand horizontally rather than vertically, so in 1893 buildings over ten stories were banned (maximum heights were raised to 260 feet in 1902, then cut to 200 feet in 1911).[125]

Horizontal expansion in Chicago meant endless and rapid development of working-class suburbs of low, detached tenement houses, often without proper sanitation and connected by ill-maintained roads. Moreover, the sheer pressure of arriving immigrants, beginning with Americans from the East, followed by Irish and Germans, with Russian

Jews and Italians forming the next waves, then Greeks and Bulgarians, and finally Mexicans and blacks, meant that five or six families lived in a tenement built for one. It was from this ill-housed and overcrowded environment that the protection rackets and eventually the highly organized crime syndicates of the Prohibition Era rose, especially in such working-class suburbs as Cicero, Al Capone's stronghold.

On the other hand, it was also from these crowded tenements that there sprang magnificent industries and even first-class applied arts of the highest quality. A description of Chicago, or rather of its progenitor Fort Dearborn, in 1833, noted three muddy streets and a mass of log shanties holding 350 people—but it also noted the existence of a furniture-maker, James W. Reed, who doubled as town constable. The rise of the Chicago furniture industry was as rapid and spectacular as that of the city itself. By 1850 there were thirteen furniture factories, and twenty-six ten years later.[126] Not only did Chicago make furniture of high quality at cheap prices on a colossal scale, to supply not only the expanding city but the entire Midwest, it also (by 1873) produced the highest-quality art furniture. Local millionaires promoted top-class craftsmen. When Potter Palmer built his amazing brick-and-stone castle on Lake Shore Drive in 1885, he furnished it throughout with special sets and items made by local craftsmen. Various residences of the tractor-making McCormick family were decorated in the same way, as was the magnificent chateau put up in the French style by the paint-manufacturer W. W. Kimball. Frank B. Tobey (1833–1913) was probably the greatest furniture-maker on the western side of the Atlantic in the 19th century, as is testified by his company's Dolphin Desk, now in the Chicago Historical Society, or the cherrywood carved dining-room table now in the Metropolitan Museum, New York.[127] Chicago spawned remarkable firms of interior decorators in the 1880s and 1890s, such as Mitchell & Halback and Healey & Millet, as accomplished though not as celebrated as Tiffany and Harter Brothers of New York. The lists of local craftsmen they employed are impressive.

Chicago also had the sense to lay out urban park systems when the land around its center was still cheap and available. Its parkways were conceived in 1849 by John S. Wright and implemented in 1869 when the Illinois General Assembly created the South, West, and North Park Commissions with powers to lay out parks and boulevards. These were highly popular because property values along the parkways rose fast—an excellent example of the American gift for aligning public service with the greed for money. When the system was laid out much of it went through open country, so boulevards could be 200 to 400 feet in width with commercial traffic banned and a middle lane for carriages. The system indicated the sharp eye of the new plutocracy for urban civilization at its best—on one Sunday afternoon in 1881, it was reported that 4,700 carriages moved south in luxurious procession down the Grand Boulevard, not bad for a place which consisted of 200 mud shacks only forty-seven years before.[128]

New York, by contrast, was circumferenced by water and chose to have its park, on a giant scale, in the middle. New York was still second in size to Philadelphia at the time of the 1810 census, with 91,874 to 96,373 people, and the plan for its development laid down the following year provided only minimum public spaces (its original Parade Ground between 23rd and 34th Streets had been long since greedily built over). But when the fashions for laying out big public parks within cities was brought from London and Paris soon after, New York still had plenty of undeveloped land in central Manhattan, and the city fathers were able to set aside an enormous area. The landscape architect F. L. Olmsted (1822–1903), from Hartford, Connecticut, that nursery of genius, together with the Londoner Calvert Vaux (1824–95), designed Central Park as an extraordinary multi-class complex of carriage drives, walks, lakes for fishing, boating, and skating, and boulder-strewn wilderness woods.

By the time the Park was in working order the City was fast growing up around it. Population was then 813,000. Forty years later, thanks largely to immigration, it was nearly 3,500,000 and still growing at breakneck speed. The rise of high buildings meant that the immense flat space of Central Park was increasingly surrounded by a periphery of stone and masonry achieving spectacular effects of precisely the *rus in urbe* appeal which had been the aim of the earliest town planners, like John Nash of London. No other city in the world can produce these skylines. First came four- or five-story structures, developed out of British precedent for shops, factories, and warehouses, the leading spirits being two brilliant iron-founders of the 1850s, Daniel Badger and James Bogardus. From this emerged cage-constructions, whose interiors were self-supporting metal frameworks reinforced by independent masonry walls. Next was skeleton-type construction, in which even the external walls hung off the metal frame. The Equitable Building of 1868–70 is often regarded as the first New York skyscraper: it had a frontage only five bays wide but it rose to 142 feet in eight stories and was serviced by two elevators. (Its replacement, the

Equitable Building of 1913–15, was an entire block, reached forty stories and 542 feet, and had forty-eight elevators making 50,000 trips a day, giving some idea of the leap from large to gigantic in New York City in these four decades.)[129]

Evidently the New York skyline was beginning to assume its characteristic form, and to promote deep thoughts in visitors, as early as 1876, when T. H. Huxley, the leading promoter of scientific ideas in Europe, made his first visit. His verdict was: 'Ah, that is interesting. In the Old World, the first things you see as you approach a great city are steeples; here you see, first, centers of intelligence.' Huxley was in a sense right: the skyscraper represented the application of science at its frontiers and imaginative intelligence in the art of building in precisely the way a great Renaissance architect like Michelangelo would have instantly appreciated. But the men who devoted huge creative intelligence and engineering and mathematical skills to making New York a 'scientific city' did not share Huxley's atheism. Rather the contrary. A characteristic American religiosity tended to enter even the field of the high-rise and the structurally gigantic. John Roebling (1806–69), the German-trained immigrant who designed the Brooklyn Bridge (it was completed by his son Washington in 1883), then the longest suspension bridge in the world, said it was 'proof positive that our mind is one with the Great Universal Mind.'[130]

New York differed from Chicago in key respects. Though less innovative, it was richer in the sense that it was the source of the capital for Chicago as well as itself, and most major firms with immortal longings, who wished to commemorate themselves with the tallest, largest, most expensive skyscraper, had their headquarters in New York. So ultimately New York skyscrapers were not only taller but more decorative. The ten-story Western Union headquarters was put up in 1873–5, followed quickly by the eleven-story Tribune Building, then the sixteen-story World Building in 1889–90 and the twenty-story Manhattan Life Insurance giant of 1893. New York soon surpassed Chicago in height, with ten stories or more added every decade, and it indulged in fantastic and often beautiful accretions of domes, columns, and spires. Most New York skyscrapers were permanent advertisements for their companies. Thus the Singer Building of 1902 paid for its construction by one year's extra sales in Asia alone. Equally, New York's vast insurance industry dictated the construction, regardless of cost, of headquarters buildings which vaunted strength, size, and durability (rather like banks). In the first decade of the 20th century, the Metropolitan Life had insurance in force totaling over $2.2 billion, so it built and occupied, 1909–10, an immense temple in the sky which was 700 feet high, the world's tallest for a time. Another example was the spectacular Woolworth Building of 1911, which for long represented *the* skyscraper. Frank Winfield Woolworth (1852–1919), who established his first five-and-ten-cent store in 1879 and by 1911 had over 1,000 worldwide, told the contractor who put up his building that though it could never make a proper return on capital it had an enormous hidden profit as a gigantic signboard.[131]

By 1903 office rents were four times higher in Manhattan than in central Chicago and that was one reason buildings were taller. High rents also determined the cluster of skyscrapers within easy reach of the Stock Exchange: by 1910 they could be as high as $24,750 a square foot in Wall Street but only $800 in South Street a few blocks away. Then in 1916 came the New York Set-Back ordinance: so long as your architect worked out the set-backs correctly, you could go to any height you liked. Grandeur and display raised the height well above the economic optimum and by 1930 it was averaging sixty-three stories in the best area around Grand Central, with the Chrysler Building (1929–30) pushing up to seventy-seven stories, the extra being the advertising element. The sensation of the 1920s, indeed, was the development of the Grand Central area as an alternative to Wall Street, and New York skyscrapers are still to this day grouped around these two foci.[132]

But we are getting ahead of our story, and above it too, for beneath the towering New York high-rises were the clustering tenements, themselves also multistory, of the burgeoning metropolis of the 1870s, 1880s, and 1890s. New York had begun as a Dutch city, then had expanded as a mainly English city, then in the 19th century had broadened into a multiethnic city, much favored by Germans and, above all, by the Irish. Then came the turn of the Italians, the Greeks, and the Jews from eastern Europe. The outbreak of savage state pogroms in Russia from 1881 had dramatic consequences for New York. In the following ten years Jews were arriving in the city at the rate of 9,000 a year. In the 1890s it jumped to an average of 37,000 a year and in the twelve years 1903–14 it averaged 76,000 a year. In 1886 the French people commemorated the centenary of American Independence by having their sculptor Frederic Auguste Bartholdi fashion a gigantic copper statute of Liberty, which was placed on a 154-foot pedestal on Bedloe's Island in New York Harbor, the whole rising to 305 feet, making it the highest statue in the world. A local Jewish relief worker, Emma Lazarus (1849–87), whose talent had been spotted by Emerson, grasped, perhaps better than anyone else in America at that time, the true significance of the open-door policy to the persecuted poor of Europe. So she wrote a noble sonnet, 'The New Colossus,' celebrating the erection of the statute, in which the Goddess of Liberty herself speaks to the Old World:

> Give me your tired, your poor,
> Your huddled masses yearning to breathe free,
> The wretched refuse of your teeming shore.
> Send these, the homeless, the tempest-toss'd to me.
> I lift my lamp beside the golden door.[133]

The refugees and the huddled masses crowded not just into Manhattan as a whole but in particular into the Lower East Side, one and a half square miles bounded by the Bowery, Third Avenue, Catherine Street, 14th Street, and the East River. In 1894 the density of Manhattan reached 142.2 people an acre, as opposed to 126.9 in Paris and 100.8 in Berlin. They were much higher than the Chicago tenements, perhaps safer—fire escapes had been made obligatory in 1867— and far more crowded. The most infested were the Dumbell Tenements, which get their name from a shape determined by the 1879 municipal regulation which imposed airshafts. They were five to eight stories high, 25 feet wide, 100 feet deep, and with fourteen rooms, only one of which got natural light, on each floor. Over half a million Jews were crowded into the Lower East Side, and the heart of New York Jewry was the Tenth Ward, where, in 1893, 74,401 people lived in 1,196 tenements spread over six blocks. Five years later the population density in Tenth Ward was 747 persons per acre or 478,080 per square mile. By comparison, the modern density of Calcutta is only 101,010 per square mile (1961–3). The New York buildings had more stories of course; even so, the Tenth Ward was probably the most crowded habitation, in the 1890s, in the whole of human history. By 1900 there were 42,700 tenements in Manhattan, housing 1,585,000 people.[134]

So here were luxury skyscrapers surrounded by slums, an image of rich-and-poor America. And the poor were, in a sense, sweated labor, most of them in the 'needle trades.' By 1888 no less than 234 out of 241 New York clothing firms were Jewish. By 1913 clothing was New York's biggest industry, with 16,552 factories, nearly all Jewish, employing 312,245 people. But the apparent rich–poor dichotomy con-cealed a huge engine of upward mobility. The whole of America was upwardly mobile, but New York, for the penniless immigrant, was the very cathedral of ascent. Many immigrants stayed in the Tenth Ward only a matter of weeks or months. The average Jewish residence in the Lower East Side as a whole was fifteen years. Then they moved on, to Brooklyn, to Harlem (once a wealthy German-Jewish quarter), to the Bronx and Washington Heights, then further, and inland. Their chil-dren went to universities. Vast numbers became doctors and lawyers. Others set themselves up as small businessmen; then became big busi-nessmen. One-time Jewish peddlers became mail-order tycoons, epito-mized by Julius Rosenwald's Sears, Roebuck. The family of Benjamin Bloomingdale from Bavaria, who opened a dry-goods store in 1872, had over 1,000 employees in its East Side shop by 1888. The Altman Brothers had 1,600 in their store. Isidore and Nathan Straus took over R. H. Macy. Other families created Gimbels, Sterns, and, in Brooklyn, Abraham & Straus. The immigrant Jews created an enormous Yiddish press, the largest in existence, but they soon dominated the New York printed word in English too. Arthur Hays Sulzberger and Arthur Ochs ran the *New York Times*, Dorothy Schiff and J. David Stern the *New York Post*, and, in time, great Jewish publishing houses emerged— Liveright & Boni, Viking Press, Simon & Schuster, Random House, Alfred A. Knopf.[135]

The ability of America, led by New York, to transform immigrant mil-lions, most of whom arrived penniless and frightened, into self-confi-dent citizens, wealth-creators and social and cultural assets, was the essential strength of the expanding republic, which had now been doing the same for its own people for the best part of three centuries. As the culture of the New World became more complex, so more excit-ing combinations of talent—indeed genius—became feasible. A charac-teristic example, which would have been near-impossible in Europe, was the collaboration between Thomas Alva Edison (1847–1931) and Louis Comfort Tiffany (1848–1933) in the creation of the Lyceum Theater in New York in 1885, a wonderful synthesis of the latest high technology and esthetic innovation. Edison was in many ways the archetypal American. Perhaps to a greater degree than anyone else, he exploited the total freedom the America of his day gave to a man of tal-ents. Never in his long and fruitful life of invention was he frustrated by rules and restrictions and inhibitions.

Edison was born February 11, 1847 in Milan, Ohio, of distant Dutch stock. His forebears were Tories, driven to Canada; but they then joined in Mackenzie's Rebellion (1837) against the Canadian government and were forced back to the United States, to work in a lumber mill. Those were hard times. Edison had only three months in a school and was mainly taught by his mother. But he conducted experiments at home and read Paine's *Age of Reason*. At twelve, a sort of Huckleberry Finn figure, he sold newspapers on the railroads, traveling far. At thirteen he edited a newspaper himself. He learned telegraphy, the great technological frontier of the 1850s. At sixteen he was an itinerant telegraphic expert, and already an inventor. He lived in cheap boarding houses or simply slept on any floor available—he must have slept on floors more often than any other man of his eventual eminence. He was a bit deaf and his hands were scarred with chemicals. He had a country-bumpkin manner and swore that Newton's mathematics and geometry were quite beyond him. But the power and inventiveness of his mind were nonetheless stunning. The book that inspired him, with its lucidity and amazing foresight, was *Experimental Researches into Electricity* by Michael Faraday, the Englishman who came from a similar humble background. Faraday made the first practical electric battery and, having discovered how exactly to transform various kinds of energy into electricity, forecast a whole range of devices from sound-transmission to the fax-machine. Edison's destiny was to make many of these dreams come true.

It was Edison's good fortune to be plunged, as a mere eager boy, into the intense competitive era of the early telegraphic age. Western Union alone multiplied its original handful of offices to 2,250 in 1866, over 4,000 by 1870, 6,500 in 1875, and nearly 10,000 by 1880. The miles of wire spread from 76,000 in 1866 to 234,000 in 1880 and the number of messages from 6 million to 30 million, income rising by 10 percent a year. This expansion was matched by the impressive density and complexity of Edison's work, which began with minor though ingenious improvements to the telegraphic system and eventually embraced the whole gamut of electrical applications.[136] Inventions were central to the success of a telegraph company and Edison supplied hundreds. He registered his first patent in 1869 and by 1910 had 1,328 to his credit. At the height of his career he was putting in a new application every eleven days. Often dirty and nearly always dressed like a tramp, he hated capitalists and financial men almost as much as he detested mathematicians. He wrote on one of his patents: 'Invented by

& for myself and not for any small-brained capitalist.' He hated lawyers too.

Edison developed great intuitive skills in handling his experimental equipment, playing on his electromagnets almost like a musician with a Stradivarius. One assistant described him working like a magus, 'displaying cunning,' as he put it, 'in the way he neutralizes or intensifies electromagnets, applying strong or weak currents, and commands either negative or positive directional currents to do his bidding.' This mechanical genius akin to artistry, even to wizardry, was combined with a relentless energy. No inventor ever worked so hard physically. But he also studied. Another associate wrote: 'I came in one night and there sat Edison with a pile of chemical books that was five feet high when laid one upon another. He had ordered them from New York, London and Paris. He studied them night and day. He ate at his desk and slept in a chair. In six weeks he had gone through the books, written a volume of extracts, made 2,000 experiments and . . . produced a solution, the only one that could do the thing he wanted.'[137]

At Menlo Park, a New Jersey whistle-stop settlement, Edison established the first true industrial laboratory in the world, building up his team of assistants from an original thirteen to fifty, and often forcing them to work round the clock, getting a mere four to six hours' sleep, under the table—just as he did himself. Trade union power, let alone workplace regulations, would have made Edison's output impossible. As it was, he had no restraints on the ruthless exploitation of his men, other than their own lack of willingness to serve him. A few quit, most stayed, and flourished mightily along with their boss. The quality and importance of Edison's thousands of inventions vary. He found the electric phonograph or gramophone (1877) easy. The incandescent lamp (1879) was a triumph of relentless experimental hard work, and in the process he unwittingly produced a 'rectifier' which converted alternating current into direct current, thus making possible a device for receiving telegraph signals and the precursor of the vacuum tube in radio—Edison's one significant discovery in pure science. But he worked hardest of all on electrical lighting systems, from power-generation to the bulb itself. In 1881–2 he constructed in New York City the first central electrical power-plant ever assembled. This, in turn, made possible the widespread use of electricity in downtown Manhattan, and his collaboration with Tiffany involved the design and decoration of New York's—indeed the world's—first electrically lit theater.[138]

Edison's partner in this project was as original as himself, pushing the frontiers of esthetics to the limits of the technology available. If Edison was an artistic scientist, Tiffany was a scientific artist. His father, Charles Lewis Tiffany, was the son of a prosperous textile manufacturer, tracing his ancestry back to Squire Humphrey Tiffany who settled in Massachusetts Bay in 1660. He went into the jewelry business in 1837, aged twenty-five, on a borrowed $1,000 and in thirty years built up his firm (he was a master-salesman) into the most profitable of its kind in the world, selling jade, Venetian glass, Indian ivories, Chinese figures, and, above all, diamonds. He made a cheap killing in 1848, the 'Year of Revolutions,' when the price of diamonds in Europe dropped by half, stocking up for the inevitable revival. He bought and sold the famous Esterhazy Diamonds, bought and broke up Marie Antoinette's jewels, sold masses of gems to Napoleon III and the Empress Eugénie, then when they fell bought the stuff back and sold Eugénie's monstrous necklace of 222 diamonds in four rows to Mrs Joseph Pulitzer, wife of the press tycoon. He bought gems worth $1.6 million from the deposed Queen Isabella of Spain and sold them to the wife of California's Leland Stanford. In a striking bargain in 1877 he acquired for £18,000 from South Africa a gem weighing 128.51 carats and called it the Tiffany Diamond. He made the highest-quality silver, turned out by 500 expert craftsmen in his workshop—indeed he set new standards for contemporary *objets d'art*, rivaling Paris and St Petersburg. Tiffany & Co.'s huge wooden Atlas held over its door (as it still does) a clock which stopped at 7.22 A.M. on April 15, 1865, the exact moment of Lincoln's death, as was only right, for Tiffany Sr owed a great deal to that President. Not only did he sell him Mrs Lincoln's inaugural pearls, he also got the commission to sell the Union forces swords, epaulettes, badges, and cap ornaments—Ohio alone bought 20,000 high-quality cap decorations—and Tiffany & Co. made exemplary profits even by American standards.[139]

Louis Comfort Tiffany, born in 1848, trained as a painter under the eccentric American master George Inness (1825–94) and, at twenty, went to Europe to pick up ideas. In London he watched another egregious American painter, James McNeill Whistler (1834–1903), working on his stunning Peacock Room, studied with Arthur Lazenby Liberty, who founded (1875) the outstanding textile and dress-shop Liberty's, and plunged into the Arts and Crafts movement. In Paris he mixed with the artists who were already creating the movement known as Art Nouveau (the term was coined in 1881 by Octave Maus and Edmond Picard and applied originally to painters). Back in America, he attended the Philadelphia Centennial Exhibition in 1876 and met there Candace Wheeler the embroiderist and Samuel Coleman the textile designer: he immediately formed Tiffany & Associated Artists, supplying the capital himself and making Wheeler and Coleman partners. He himself essentially specialized in colored glass. Indeed his adaptation of opalescent glass to windows was 'one of the greatest contributions to the stained glass repertoire since the Middle Ages.'[140] The great windows he designed and installed are his most substantial contribution to world art but he used opaline techniques, and many others, in creating a vast range of glass objects, all made to the highest standards of craftsmanship and dauntingly expensive even then: in 1906 he charged $750 for a table lamp at a time when you could buy a Sears, Roebuck six-room house for less than $1,000.[141]

What brought Edison and Tiffany together was a common aim: the pursuit, enlargement, and glorification of light. Tiffany designed some of the most remarkable decorative schemes ever created on the western side of the Atlantic, notably the famous Veterans' Room for the Seventh Regiment Armory, the George Kemp House on Fifth Avenue, and Mark Twain's enchanting home in Hartford, Connecticut. He redecorated Ogden Goelt's home at 59th Street and Fifth for the then record fee of $50,000 and charged even more for the Vanderbilt mansion on 58th and Fifth. When Chester Arthur moved into the White House he dispensed with twenty-four horse-van-loads of furniture and got Tiffany to redesign the state rooms, including glass-mosaic sconces in the Blue Room and a floor-to-ceiling glass screen in the State Dining Room.[142] But Tiffany's overriding concern always was the presentation of new forms of colored or pure light by means of glass. He was lucky in that, when he went into colored glass-work, over 4,000 new churches were being built in the United States, most clamoring for stained-glass windows.

But Tiffany put his best ideas into specially designed houses, including his own, Laurelton Hall on Oyster Bay in Long Island, where he produced staggering color effects using glass, water, tiles, and colored domed skylights. This Art Nouveau palace, with copper roofing, mushroom-shaped like a mosque, had eighty-four rooms and twenty-five bathrooms and was perhaps the most photographed private house in American history. It had 'light' and 'dark' rooms, Chinese, Indian, palm-house and tea rooms, and in the drawing-room were some of the finest colored glass windows he ever created. Tiffany held 'light and color' parties at this place—he invented *son et lumière* in the 1890s—

notably his Spring Flowers Party in 1900, when 150 'intellectual men of genius' were invited to 'inspect the Spring flowers' brought by special train and served by girls wearing ancient Greek costumes, or his Quest for Beauty masque of 1916, another *son et lumière* exercise in which he created thirty-three different red lights, the climax being the emergence of a huge bowl of iridescent blown glass containing Beauty as a pearl (the lady being his latest mistress, a dancer from the Metropolitan Ballet).[143]

Edison and Tiffany, working together, equipped the new Lyceum not only with electric sconces throughout but with the first electric footlights. This 1885 innovation, the envy of European impresarios, marked the point at which Broadway entered as a major player in the world game of drama and showmanship. Throughout the second half of the 19th century, America slowly emerged not merely as a great theatrical but as an even greater musical nation. In New York it was merely glees and chorus stuff until 1842 when the foundation of the Philharmonic Society (the oldest in the world after Vienna and London) opened the era of professionalism. It gave the first American performance of Beethoven's Choral Symphony on May 20, 1846, and it soon included among its honorary members Mendelssohn, Jenny Lind, Liszt, and Wagner. The Metropolitan Opera, in its first incarnation, opened on October 22, 1883 with seating for 3,615 and a performance of Gounod's *Faust*. Like the Metropolitan Museum, it became, from the start, a focus for big money and the entry to high society which large donations alone secured. In 1892 Mrs Thurber, founder of New York's National Conservatory of Music, staged a memorable coup by getting Antonín Dvořák, professor of composition at Prague Conservatory, to become her director for three years. He found the splendors of America, where he was shown round by its flourishing Czech community, inspiring and wrote two of his greatest works there, his Cello Concerto, Opus 104, and his Symphony in E-minor Opus 95, 'From the New World.' This great work had its first performance by the New York Philharmonic Society orchestra on December 16, 1893. The year was a notable one for music, witnessing the first performances of Verdi's *Falstaff*, Tchaikovsky's Pathétique Symphony, Puccini's *Manon Lescaut*, and the Karelia Suite by a new genius from Finland, Sibelius. But the 'New World' topped them all, and the fact that it was written and first performed in America, celebrating its vastness, joy, and promise, marked the coming-of-age of American music.[144]

But more was to come. By the turn of the 19th century the

Philharmonic was attracting conductors of the quality of Henry Wood, Édouard Colonne, Felix Weingartner, and Wilhelm Mengelberg. In 1909–11 it was under the baton of Gustav Mahler himself, and thereafter came Stravinsky, Toscanini, and Furtwängler. The Met was even more fortunate, being managed, from 1908, by La Scala's Giulio Gatti-Casazza, who with help from Toscanini and Mahler raised it to the level of one of the top half-dozen opera houses in the world. With the financier Otto Kahn running its board, the Met made the season of 1910 one of the most memorable in the entire history of opera. The company performed at the Châtelet in Paris, May 19–June 25, opening with Toscanini conducting *Aida* and continuing with the first Paris performance of Puccini's *Manon Lescaut*, with Puccini himself present. Then, back in New York, Puccini was again present on December 10 for the world premiere of his *La Fanciulla del West*. Among the California Forty-Niners was the father of David Belasco (1854–1931), the playwright and librettist, whose work *The Girl of the Golden West* was the basis of the opera. The cast included Caruso, Destinn, Amato, Didur, and Peni-Corsi, and it was typical of the electric theater Edison and Tiffany had created a quarter-century before, and of the new-style Broadway super-production, that the elaborate lighting effects were specified in enormous detail. In the third act there were eight live horses on stage. Demand for tickets was so high that the Met made each buyer sign the ticket at the time of purchase and endorse it before entering the theater. The snowstorm for the scene in Minnie's cabin required thirty-two stagehands to operate it. Puccini said he counted a total of fifty-five curtain-calls and the Met gave him a laurel-wreath of solid silver (made by Tiffany of course), which was placed on his head by Destinn, playing Minnie.[145]

That the United States, lifted up by an extraordinary combination of self-created wealth and native talent, became a great cultural nation in the second half of the 19th century is a fact which the world, and even Americans themselves, have been slow to grasp. The lack of recognition has been particularly notable in the field of art. From the 1820s there emerged a school of landscape-painting the like of which the world has never seen. The innovator here was Thomas Cole (1801–48), who came to America in 1819 from Lancashire, England and founded what became known as the Hudson River School. Cole was the first painter to appreciate the immensity of the opportunities offered by the scale and variety of the American landscape. He produced in 1826 the first

American masterpiece of landscape art, *The Falls of Kaaterskill* (now in Tuscaloosa) and the following year *The Clove, Catskills* (now in New Britain, Connecticut), which gave some indication of what could be done. His *The Oxbow* (1836) became an emblem of American topography in art, a performance to inspire every young painter in the country, and the next year he began his great series, *The Course of Empire*, now in the Corcoran, the first considered attempt to achieve monumentality on the highest European scale.[146]

American painters were particularly sensitive to New England coastal light. John F. Kennsett (1816–72) painted the astonishing stillness and depth of the inshore waters and their sands, and Martin Johnson Heade (1819–1904) with *The Coming Storm* (1859; now in the Metropolitan) and *Thunderstorm over Narrangansett Bay* (1868; Fort Worth) made himself a storm-evoker of genius. Jasper F. Cropsey (1832–1900) painted in 1855 a famous view, *Catskill Mountain House* (now in Minneapolis), of one of the earliest of the Catskill mountain resort hotels, known as the Yankee Palace, and opened in 1823, as the climax of the Grand Tour of the mountains, on a site 2,200 feet above sea-level made famous in Fenimore Cooper's *The Pioneers*. This is rivaled by an autumn scene, *Kauteskill Clove* (1862; now in the Metropolitan) by Sandford R. Gifford (1823–80) as the quintessential American landscape painting. Most of the painters of this generation, such as Fitz Hugh Lane, Heade, and Gifford, emphasized the sheer clarity and strength of natural American light, and they have become known as the Luminists.[147] But George Inness (1825–94), the self-taught, epileptic son of a prosperous Hudson grocer, produced miraculous misty effects, for instance in his *The Delaware Water Gap* (1857; now in private hands), and a Claude-like golden light entirely his own. Anything the French Impressionists could do he did better—and earlier.[148]

However, it was Cole's great pupil, Frederick Edwin Church (1826–1900), who took the Hudson River School to the summit of accomplishment. Cole had met and studied Constable and Turner when he was in London in 1829, and he passed onto Church the apostolic succession. Cole recognized his genius from the start, and his imaginative powers, and called him the 'Poet–Painter.' But Church himself approved the description of him by the critic Henry Tuckerman as 'the painter of scientific eloquence.' He not only grasped the immensity of the American landscape and the variety and ferocity of its climate but studied both with a scientific concentration which was peculiarly American. Cole introduced him to the work of the eminent geologist Benjamin Silliman (1779–1864), and Church developed the habit of making literally scores of thousands of sketches of atmospheric, geological, climatological, and topographic subjects, building up in effect an immense archive of visual knowledge of the American natural scene.[149] Church was unsurpassed at realizing subtle difference of texture and form as a draftsman, though Jasper Cropsey came close to him. He studied intently Turner's *Liber Studiorum*, Ruskin's *Modern Painters*, the *oeuvre* of Alexander Humboldt, and a range of other scientific works, preparing digests in exactly the same way as Edison with his chemical books.[150] And on the basis of this knowledge he built up a coordination of the mind, eye, and hand which has rarely been equaled.

Church was already successful and mature by the age of twenty-two, when he took a studio in New York and his first pupil (1848). He treated the American hemisphere as a whole, and like a growing number of American painters explored Latin America in search of subjects.[151] He visited Ecuador and Colombia in 1853 and the paintings he showed on his return made him famous. He then decided to tackle the great showpiece of American topography, the Niagara Falls. There was nothing new in this subject—it was in fact the commonest topographical view in the hemisphere—and it says a lot for Church's thoroughness and ability that he was able to turn this hackneyed theme into one of the sublime moments in 19th-century art. He chose the Canadian angle of the Falls with tremendous skill, made six prolonged visits, produced hundreds of sheets of drawings and twenty-one major oil sketches. The final painting, 7 1/2 feet wide, took him six weeks and was specially exhibited on May 1, 1857 in Manhattan at Messrs Williams & Stevens. It proved to be the most successful picture ever shown in the United States by far and later went on a tour of Europe. Ruskin hailed it as 'the coming of age of American art.' It made Church's worldwide reputation and that of the entire Hudson River School. (It is now in the Corcoran.)[152]

Even more popular was the fruit of his second visit to Latin America in 1857, when he painted *The Heart of the Andes* (now in the Metropolitan). This was the result not only of his own visual inquiries on the spot but of intense study of Humboldt's *Cosmos: An Outline of a Description of the Physical World*, which appeared in English in five volumes in 1848–58. Church paid two visits to the site, with Chimborazo in the distance, nine mules carrying his equipment. The painting, almost 10 feet wide, took him a year and while it was evolving it was spoken of reverently as 'his great canvas' in art circles. Whitelaw, an English decorator, and Church himself together made a magnificent frame, 13 feet high and 14 feet wide, of dark walnut to give the effect of a view seen from a castle window or terrace. It was shown with surrounds of drapery and palm trees, in a specially rented gallery in Lyric Hall on April 27, 1859. Like other great artists, such as Rubens, Church was a professional showman and he had published for the occasion a forty-three-page *Companion to the Heart of the Andes* by Theodore Winthrop, the text being described by one critic as 'the ravings of Ruskin in delirium tremens.' Mark Twain saw the work and wrote to his brother: 'It leaves your brain gasping and straining with futile efforts to take all the wonder in . . . and understand how such a miracle could have been conceived and executed by human brain and human hands. You will never get tired of looking at the picture.'[153]

Church was not the only master-showman. Albert Bierstadt (1830–1902), who came from Germany as a child—his father made barrels in New Bedford, Massachusetts—was another conscientious craftsman who turned, unlike Church, to the heart of the Rockies for inspiration and was the first artist to bring to the sophisticated East Coast public the sheer savagery and magnitude of the high West. He went there in 1859 and again in 1863 and 1871–3. His first great canvas, the same size as Church's *Heart, Lander's Peak* (1863; now in the Metropolitan), nearly cost him his life twice: he was attacked by Indians, and later his obsession with the subject made him run out of food and he almost starved to death. In the difficult Civil War market he turned down a price of $10,000 and finally sold it in 1865, for $25,000, to the railroad financier James McHenry. Thereafter he usually asked $15,000 for his big pictures, such as *Looking Up the Yosemite Valley* (1865–7; now in Stockton) and his masterpiece, *A Storm in the Rocky Mountains: Mount Rosalie* (1866–7; now in the Brooklyn Museum). The setting for this astonishing work is the Chicago Lakes area of Colorado, 490 miles southwest of Denver. It reveals a terrific emotional response to the theme—he named the mountain himself after his tempestuous wife—and a power of depicting distance and scale which no European artist, except Turner, had ever equaled, and no other American one, except Church himself.[154]

Church and Bierstadt, by showing America's natural resources in spectacular detail, gave a powerful impulse to the movement to preserve them through national ownership, at a time when the federal government was selling land cheap, or giving it away, at the rate of millions of acres a year. One of Bierstadt's most notable canvases was *The Great Trees, Maripose Grove*, California (painted 1871–3, exhibited 1876). This stand of trees, the largest on earth, was first discovered in 1833 and reports of their size were simply disbelieved. Bierstadt first saw them in 1863, and thanks to him and others Lincoln was persuaded to make the Grove a National Site in 1864, in the middle of a terrifying war and eight years before the first National Park came into being. Of these trees, 132 were measured, with girths of over 100 feet, bark a foot thick and 350 feet high. Bierstadt picked out one called the Grizzly Giant, a cedar, for his massive picture, 10 feet high. This is now the oldest living *Sequoiadendron giganteum* and is still to be seen at Maripose Grove. As a result of Bierstadt's efforts, the giant sequoias became the center of a National Park of 385,000 acres created in the Sierra Nevada in 1890, with Mount Whitney (14,495 feet) at its eastern border. This was the third of the new National Parks, the first being the Yellowstone (Wyoming), formed in 1872, and the second the Yosemite (1890), which Bierstadt had also explored and painted.[155]

Painters like Bierstadt and Church illustrated the effects of what was already, by the 1880s and 1890s, becoming an important force, for good and evil, in American cultural and social economics—the devastating vagaries of fashion. In 1865 Bierstadt was reported as having earned $120,000 in the previous three years alone, and he kept the best carriage horses in New York. To make his fame seem permanent, he built himself Malkasten, an enormous studio-palace overlooking the Hudson with giant windows and a tower rising 500 feet over the river. But even in the 1860s he was already being attacked by the critics in search of novelty, and in the 1880s his house burned down, his fame collapsed, and many of his paintings later disappeared or were hung in inaccessible places. Only in the 1980s did his reputation reestablish itself.

Church's fame, too, was dimmed as the 20th century began. He too built himself a house overlooking the Hudson, not far from where he had first studied landscape art with Cole. In 1867 he bought the top of a hill, with one of the most spectacular scenic views in the entire Hudson Valley, and in 1870–2, with the help of a professional architect, he designed and built, and then decorated—using fifteen crates of rugs, armor, spears, and ceramics he had bought on his world tours—an eclectic house called Olana, part Saracen, part Moorish, and with a touch of Persian. His first biographer described it as 'a Persianised amalgam of Italian villa, Gothic revival, Ruskinian polychromy and

French Mansard, with an East Indian shingle-style wing.'[156] The colors are superb: brickwork of yellow, red, and black below a slate roof of gray, red, and pale green, the whole surrounded by 300 acres of emerald lawn, with a specially designed terrace-platform or loggia (known as the Ombra) from which to enjoy a view extending into the Vermont Mountains, the highlands at West Point, the Taghanic Range, Albany, and the Catskills.[157] Church was handsome and godlike, his wife tiny but exquisite, and when this celestial couple appeared at the Opera the audience rose to applaud. Oddly enough, Church moved to the Hudson to escape metropolitan lionizing just at the moment when his fame went into precipitous decline, though he happily did not live to see its nadir.

Louis Tiffany was less fortunate. He lived on till 1933 but his high point had been the Art Nouveau Exhibition of 1900. After that his reputation was wiped out by the Modern Movement from Paris—Fauves, Post-Impressionists, Cubists, and the rest. When Theodore Roosevelt, his Oyster Bay neighbor, who hated him because he was reputed to be 'loose with women,' moved into the White House, he told his architect, Charles F. McKim, to 'break in small pieces that Tiffany screen' and, without giving the designer the chance to buy back his artifacts, threw them in the dustbin. Tiffany's stained glass in New York's Cathedral of St John the Divine was deliberately sealed off by the architect Ralph Adam Cram. His father's house on Madison Avenue, which he had elaborately decorated, was destroyed and, after his own death, the real obliteration began. The Tiffany stock was auctioned off at low prices, his 72nd Street mansion, containing many of his masterworks, was razed to the ground (1938), all the contents he lovingly designed and collected at Laurelton Hall were sold off for virtually nothing, and the hall itself burned down in 1958. At that point, almost on cue, his reputation began to revive, a Tiffany lamp regarded as ultra-expensive at $750 in 1906 being auctioned for $1,500,000 in the late 1990s.[158] The Morse Gallery of Art at Winter Park, Florida, rescued the Byzantine chapel he designed for the 1893 Chicago World's Fair and assembled also 4,000 small items of Tiffany-work. But many have disappeared without trace, an example of the extraordinary prodigality with which America raises and destroys, produces and consumes. Church, too, is again an idol and icon. Olana, beautifully restored, maintained and publicly owned, is again one of the most popular historic houses in the United States, his *Home by the Lake*, when auctioned in 1989, fetched the highest price ever achieved by an American work of art, and in the unlikely event of a major work by him coming onto the market, museums and art collectors would bid $20 million or more.

We have looked in some detail at the United States' cultural achievements in the closing decades of the 19th century because the maturing of American civilization needs to be emphasized, not least because it escaped the attention, or was deliberately ignored or downplayed, by observers such as Henry James at the time. Though popular historiography has stressed the 'Age of the Robber Barons' and deplored the gross materialism of the epoch, this hostile view is not borne out by the facts, which display a panorama of general progress in which all classes shared and in which all intellectual and cultural interests were abundantly displayed—a panorama, indeed, highlighted by the emergence of quintessentially American geniuses.

Even when American millionaires of the Gilded Age were spending their money on conspicuous consumption, they did so from a mixture of motives, combining self-satisfaction, competitive swaggering, public service, and a self-conscious desire to exercise cultural leadership and *noblesse oblige*, exactly like Italian dukes of the Renaissance, French and English noblemen of the 18th century, and, indeed, Washington and Jefferson. When George Washington Vanderbilt, grandson of the Commodore, possessor of $10 million of third-generation-almost-old-money, bought 125,000 acres at Asheville in North Carolina and erected America's largest country house on it, his motive was duty as well as display. Biltmore was designed by Richard Morris Hunt, an École des Beaux Arts graduate, and was ultimately based on François I's Château de Blois, though its immediate model was the Rothschild mansion at Waddesdon in Buckinghamshire. Asheville was 2,000 feet high and to create the terrace on which this enormous house stands the brow of a mountain had to be sliced off like an egg. A quarry was opened, a brickworks built, a railroad laid down, and hundreds of thousands of tons of earth were shifted to meet the requirements of Frederick Law Olmsted, the American landscape artist equivalent to Capability Brown or, more nearly, Humphrey Repton.

The project took six years and at the peak of the construction curve it was costing $45,000 a day. It was thought, at the time, to represent, by combining 'the best of all European styles,' the working out of America's Manifest Destiny in architecture.[159] The owner built a village, a church, endless cottages, started industries, and altogether gave permanent employment to 2,000 people, each of whom received a present from his wife Edith at Christmas, and invitations to join her in classes on weaving, needlework, and wood-carving. There was a lot of forestry too and other rural activities of an improving nature copied from the Duke of Westminster's Eaton Hall Estate in Cheshire, England. All this scored high marks from the social critic W. J. Ghent, whose book *Our Benevolent Feudalism* asserted that the United States was now developing a first-class baronial elite. The *Ashville Daily Citizen* called it 'no small shakes of a house.' But the locals hated it. Nor was it lucky. By the time it was opened at Christmas 1895, with 255 rooms, Hunt was already dead and Olmstead was mad, and soon to be put in a padded cell. In 1897 the *New York Times* gleefully reported that Vanderbilt, on his first visit, had asked for a drink of water and had been told the pump was not working—so he went off to India to shoot tigers and escape from Biltmore's problems.[160]

Between 1880 and 1920 more, and bigger, country houses went up in the United States than at any other time in history. There were hundreds on Long Island—at least thirty millionaires, in the 1920s, commuted daily by yacht between Long Island and Wall Street—scores on the Hudson, dozens in the northern half of New Jersey and hundreds more along Boston's North Shore, Chicago's Lake Forest, Philadelphia's Main Line, Cleveland's Chagrin Valley, and Sewicky Heights near Pittsburgh, to mention only a few ghettos of the wealthy.[161] The rubber tycoon Harvey S. Firestone asked: 'Why is it that a man, just as soon as he gets enough money, builds a house much bigger than he needs? I built a house at Akron many times larger than I have the least use for. I have another house at Miami Beach which is also much larger than I need. I do not know why I do it—the houses are only a burden. But I have done it and all my friends who have acquired wealth have done the same.' Actually the answer to Firestone's rhetorical question is that he felt he had to compete with his rival, Frank A. Seiberling, head of the Goodyear Tire Company. Self-made men were told they had to be 'seated.' The fashion had been started by Andrew Jackson Downing, who in 1850 had published a book of design-patterns, *The Architecture of Country Houses*. It took off after the Civil War and accelerated in the 1880s and 1890s just as Europeans were giving up building country houses in boredom and despair. The American rich paid no income tax or death duties. Their children did not expect to live in their fathers' houses as the fashions changed so quickly, so each built a country house of his own. Thus Astors, Vanderbilts, and Rockefellers acquired dozens of properties. They lived like princes of older times because they thought it was their duty, as well as their privilege. The Vanderbilt house at Hyde Park on the Hudson had interiors like a European monarch's of the *ancien régime*, with plenty of marble. Mrs Vanderbilt's bedroom was a replica of the *chambre-boudoir* of Louis XV's queen.[162]

These ostentatious displays were analyzed in 1899 by the Chicago social scientist Thorstein Bunde Veblen (1857–1929) in his *The Theory of a Leisure Class*. He identified a need for 'conspicuous consumption.' But the overwhelming majority of the men who built these houses were not leisured; they were fanatical hard workers, who could not stop however much they acquired. And, being congenitally linked to the entire American experiment in progress and improvement, the houses they built tended, in both construction and functioning, to advance the cause of high technology. At Biltmore, for instance, there was a state-of-the-art laundry room, in which sheets slid into heated chambers for rapid drying, a device later imitated commercially.[163] Nemours, the huge house of Alfred I. du Pont, the chemical millionaire, in Delaware, was notable for its advanced machinery, which included giant compressors for his large-scale ice-freezing room, an amazing bottling room for soft drinks, an electric waterwheel to supply the gardens with running water, and vast furnaces to power a central heating (and air-conditioning) system which would have made English hostesses green with hopeless envy. Moreover, du Pont, an anxious and suspicious man, had two of each machine, so that the second could be switched on instantly in the unlikely event of the first conking out.[164] He had only one organ, however. A firm, the Aeolian Company, came into existence especially to supply the country houses of the rich with this needful piece of cultural equipment, built to state-of-the-art standards.

Behind this mechanization was the desire—perhaps one should say the compulsion—of the American rich to save labor. In 1890 there were 1,216,000 servants in the United States. Ten years later, though the population had increased by 20.7 percent, and the country was perhaps twice as rich, the number had increased to only 1,283,000. To put it another way, in 1880 the percentage of Americans gainfully employed as domestics was 8.4; by 1920 it had shrunk to 4.5 percent.[165] We can trace the gradual reduction of the number of servants by labor-saving devices in the architectural plans of the houses. In the 1880s the servants' wing was usually at least 30 percent of the total area of the house. At Biltmore it was nearly as long and tall as the main block. According to David Phillips, the compiler of that valuable compendium, *The Reign of Gilt* (1905), a millionaire's household required

forty servants, indoor and outdoor, all of whom had to be accommo-
dated. In the 1880s a large country house usually contained about fifty
servants' bedrooms. By the 1920s, despite much greater wealth, the
number had shrunk to ten.[166]

High wages was one factor in this shrinkage of human servitude.
More important were machinery and modernization. If we take the
bathroom alone, there was, between 1870 and 1914, a complete transformation, in which America was often decades ahead of the rest of the
world. The most important advances were: the development of the
washdown, washout, and syphonic jet toilet and watercloset from
1875 on; the manufacture of elaborate porcelain (as opposed to iron)
tub, toilet, and sink shapes by the Mort works around 1895–1900; the
placing of bathrooms *en suite* to bedrooms, and multiple improvements
in plumbing and sewerage systems.[167] Labor-saving in kitchens, store-
rooms, laundry, and house-cleaning were even more spectacular. By
1914 servants in the houses of the American rich had virtually ceased
to be drudges, as they were to remain in Europe for at least another
generation.

More important, however, was the speed and effectiveness with
which luxuries promoted by the wealthy were turned into necessities
for everyone by the process of mass production and mass marketing,
two operations in which America set new standards. The United States
was the first to introduce voting democracy. Almost equally central to
the ethos of the country was market democracy, in which ordinary people voted with their wallets and, in doing so, insured that they got what
they wanted. Salesmanship, market research, advertising, the rapid
response of production machinery to perceived customer requirements—all these forms of materialism which, in their more raucous
aspects, are identified as American failings or rather excrescences, are
in fact central to its democratic strength. The story of Sears, Roebuck,
for instance, is a tale of how high-quality products, once the preserve of
the rich, were humbled and distributed literally everywhere.

This mail-order firm was one of the greatest single benefactors of
mankind in the 19th century because it was targeted on perhaps its
most overworked component—farmers' wives. It made a huge range of
goods readily accessible to isolated households and communities for
whom good shops were remote, and it specialized in products, such as
cooking-stoves, plows and washing-machines which were efficient,
required little or no servicing, and lasted a long time—yet were
nonetheless cheap. Richard Warren Sears (1863–1914) was a

Minnesota farmer's son who worked in railroads and suddenly made $5,000 in six months by selling watches wholesale. What struck him was the selling power, and therefore wealth-creation power, of sheer cheapness. In 1886 he founded the mail-order R. W. Sears Watch Company in Minneapolis, and sold it three years later for $100,000. By 1891 he had set up what became Sears, Roebuck. It represented the conquest of geography by using the postal system—and enormously helping it in the process—and turning the entire nation into a market.

Sears began with watches, then spread to everything and he remained 'firm in his belief that the strongest argument for the average customer was a sensationally low price.'[168] In 1897, for instance, Sears' sewing machines ranged from $15.55 to $17.55 when branded, nationally advertised machines, sold in the shops, were three to six times more highly priced. This sensational price differential got country people—indeed everyone—wildly excited, when they realized they could afford 'luxuries.' By putting relentless pressure on manufacturers, for whom it provided a highly prized market, Sears was able to cut the cost of its $16.55 sewing machine in 1897 by $3.05, which caused pandemonium in the industry. It called itself not just 'the world's cheapest supply house,' which it was, but 'the great price-maker,' because it could push prices down throughout an entire industry or product range. It gave competitors a hard time and housewives a good time. Sears extended the same principle to bicycles, baby-carriages, buggies, harnesses, wagons, stoves, and cream-separators. This last was important to America's 10 million farmers. The standard shop model was $100. Sears had one at $62.50 which was not doing well. He decided to transform the market and in 1903 brought out three models at $27, $35, and $39.50. That caused an uproar, and marked the point at which the separator became standard in the farming dairy. This ruthless and systematic exploitation of the capacity of the capitalist system to cut prices and save back-breaking labor at the same time was the reason why F. D. Roosevelt, when asked what single book he would put into the hands of a Russian Communist, replied: 'The Sears, Roebuck catalog'.[169]

Another fundamental improvement to ordinary lives in which Sears played a notable part was refrigeration, important in a country where high temperatures and humidity were annual burdens and health-hazards. Mass supply of ice for urban markets goes back to the 1820s in Massachusetts. This was kept cool by the use of packing-straw. A Maryland farmer, Thomas Moore, invented an ice-box to take butter

to market. It was an oval-shaped cedar tub with a metal container surrounded by ice. This was not a refrigerator as such, of course.
Americans patented ice-making machines in the 1830s and they
became numerous in the 1850s, the cooling effect depending either
upon the expansion of compressed air or upon the evaporation of very
volatile liquids such as liquified ammonia. But only after the Civil War
did the ice-market take off. Thus, in 1856, New York used 100,000
tons a year, Boston 85,000, New Orleans 24,000. In the year beginning
October 1, 1897, New York used a million tons, Chicago 575,000,
Brooklyn and Philadelphia over 300,000 each, Cincinnati and St Louis
over 200,000 tons each, with total consumption around 5.25 million
tons. About half went to households. Demand for ice tripled in the
years 1880–1905. But warm winters in the 1890s led to shortages. By
1889 in the South alone there were 165 ice plants. All this created a
huge demand for refrigerating machinery, both for making ice and for
low-temperature industries—by 1910 over 2,000 plants in the United
States used coal-fired steam-engines to manufacture ice.

Sears seized on this expansion of the industry to introduce cheap
refrigerators. His 1897 catalog already devoted two pages to fridges or
ice-boxes. But the ice-box system was unsatisfactory because it
required deliveries by an iceman. In 1914 a Detroit firm, Kelvinator,
began experiments to produce a genuine refrigerator based upon an
automatic control device which maintained a constant temperature—
and marketed the first one in 1918. It was bought by General Motors
and sold as the Frigidaire, eventually built at a specially designed and
enormous plant at Mortaine city, 4 miles south of Dayton, Ohio. The
Frigidaire went from 10,000 units in 1920 sold at $600 to 560,000 in
1928 sold at an average of $334—typical of the way in which mass
production reduced prices. By 1939 Sears was selling nearly 300,000 of
its special models for as little as $131.[170]

We are getting ahead of our chronological story—but no matter.
Many fascinating developments in the history of the American people
have deep roots, and then blossom swiftly in overwhelming abundance,
and it is convenient to tell the tale in one narrative. From about 1870
onwards, American industrialists, inventors, and salesmen began to
bring the fruits of the earth, and its man-made mechanical marvels, to
the bulk of the American population in unheard of abundance, so that
not only the 'old' Americans of the farms and townships, but the new,
teeming arrivals of the crowded cities were soon able to participate in
what was the world's first consumer society. Food, housing, warmth,

refrigeration, light, and power—blessings denied in adequate quantities, or even at all, to most of humanity for countless generations of deprivation—were suddenly made available, in less than a lifetime, so that a child born in want lived to enjoy plenty and to see its progeny do far better.

The function of the political system was not so much to promote this process of market democracy as to enable it to take place, and accelerate, by removing obstacles, natural or man-made. It was a case of allowing the ship of state to float downstream under the impulse of a mighty current of innovation and improvement, the government merely having to stretch out an occasional oar to prevent it drifting into the bank. Panics and downturns had to be endured, but, as Carnegie said, they could be turned into blessings in disguise by the smart and the creative—so long as they did not go on too long and demoralize people. Thus Grover Cleveland, in his first spell as president, 1885–9, continued Republican policies of high tariffs and hard money, contenting himself with improving the quality of the civil service by curtailing the spoils system. It was a moot question whether high tariffs, by denying Americans access to cheap imports, but also enabling American industry to flourish and deliver cheap products to its expanding domestic market, benefited the bulk of the population or not. Cleveland was inclined to think it did, but in the 1888 election, when he ran for reelection, he bowed to party pressure from the Solid South and advocated low tariffs. This probably cost him New York, his power base, and the greatest of the swing states.

Cleveland's Republican opponent Benjamin Harrison (1833–1901), an Ohio corporation lawyer and senator, and grandson of President William Harrison, was chosen for the elementary but understandable reason that he was the 'least offensive' of those available. That enabled him to carry New York, so that although Cleveland had 5,540,050 votes to his 5,444,337, he scored 233 electoral college votes to Cleveland's 168. So Republicanism and high tariffs it was.[171] The McKinley Tariff Act identified the Republicans with protection once and for all, though it was softened for some American exporters by ingenious reciprocity provisions. But the main business of the administration appeared to be granting Republican machine-men and supporters, led by the party boss James G. Blaine (1830–93), who became Harrison's Secretary of State, and House Speaker Thomas Brackett 'Tsar' Reed (1832–1902), access to the delights of rule. The Tsar was responsible for 'Reed's Rules' (February 14, 1890), increasing the Speaker's power to expedite legislation favored by the majority party, and a very valuable man to his colleagues. But neither he nor Blaine was much liked by the voters, and their unpopularity rubbed off on the harmless (and probably honest) Harrison when he ran again in 1892. This time Cleveland trounced him 5,554,414 to 5,190,801, despite an intervention by a Populist candidate who probably cost the Democrats votes, and he won the college by a massive 277 to 145.[172]

Cleveland was the best president between Lincoln and Theodore Roosevelt, a man of character and conviction and probity. But he was beset by ill-fortune, notably the panic of 1893 which in turn led to the Wilson–Groman Tariff of 1894, passed over his veto. And his determination to preserve the gold standard against radical opposition in his own party led to the rise of Bryan's 'Silver Democrats.' The rift was widened when Cleveland resolutely used federal troops to break the 1894 Pullman Strike. By 1896 the Silver Democrats controlled the party,[173] nominating Bryan as official Democratic candidate at the 1896 'Cross of Gold' convention. Bryan was wildly popular in the West, and even forced a Republican breakaway of 'Silver Republicans,' who imitated his demand for silver being coined on a sixteen-to-one ratio. But over-concentration on his farming base meant that he was unable to carry a single Northern industrial state, and he even failed to win some agricultural states too, such as Iowa, North Dakota, and Minnesota. That left William McKinley of Ohio (1843–1901), a high-tariff, sound-money lawyer and governor of his estate—who usually did what he was told by party bosses like Mark Hanna (1837–1904)—an easy winner by 7,102,246 to 6,502,925 and 271 college votes to 176.[174] McKinley and Hanna epitomized the principle of what was known as 'Standpatism,' at any rate on domestic affairs—there was some adventure abroad, as we shall see—and the voters, who were doing well, liked it too. The 1900 election was the rerun of the McKinley–Bryan contest of four years before, and the outcome the same, though McKinley increased both his plurality, 7,219,530 to 6,358,071, and his electoral college victory, 292 to 155.[175]

Not all Americans shared in the general prosperity of these decades, or shared in it to the same degree as the majority. Many farmers felt excluded from it, or said they did. Some were fearful of the great concentrations of wealth which were growing up in the United States, and the power which was being vested in giant corporations. In Europe, especially during the 1880s, socialism, whether Marxist or in milder versions, began to attract the workers, especially those organized in trade unions. In Germany the socialists were strong enough to strike a bargain with Chancellor Otto von Bismarck—he gave them the legislative elements of a welfare state in return for their support for his militarism and nationalism. But in the United States socialism was never able to move from the margin of politics. A Labor Reform Party put up a presidential candidate at the election of 1872. Two years later a National Greenback Party emerged from the discontent of farmers in the prairie and Southern states who had found themselves in debt after the panic of 1873. It deliberately campaigned for inflation and sought repeal of the Resumption Act of 1875, which provided for the redemption of greenback notes in gold. Such efforts to flood America with paper resulted in a vote in 1876 of 81,000 votes for its candidate, Peter Cooper. Increasing labor disputes brought about an amalgamation between the Labor Reform Party and the Greenbacks and together they polled over 1,000,000 votes in the 1878 mid-term elections. But in the 1880 presidential election their candidate, James B. Weaver, could scarcely top 300,000 votes and the movement disintegrated.

In 1888 another quasi-socialist organization, the Union Labor Party, emerged; and the following year splinter groups from the Democratic Party, who numbered four senators and over fifty congressmen, began to talk of a Third Party. It eventually held a national convention at Omaha in 1892 under the banner of the People's Party, and the leadership of Ignatius Donnelly (1831–1901), a splendid orator from Minnesota, a writer of utopian fiction and crank pamphlets proving that Francis Bacon wrote Shakespeare's plays. It put up old Weaver again for the presidency, and this time he won over a million votes and carried four high-plains states. But by the next time round, in 1896, Bryan had captured the Democratic machine and incorporated most of the Populists' platform, so the Populists gave him their support and, in effect, abolished themselves. Some of their policies were, indeed, adopted by one or other of the two big parties, and it seems to be the fate of America's numerous third parties to be merely ephemeral, to be cannibalized or plagiarized. In 1898 Eugene Debs and Victor Berger formed the Social Democratic—later Socialist—Party. Debs ran in 1912 and got 900,00 votes and he ran again in 1920 and got a few more, 918,000. Thereafter socialists tended to allow themselves to be drawn into the Democratic Party.[176]

American labor leaders, the left, and the progressives, Debs included, lacked a strong ideological framework adapted to the American scene.

The one man who could, and to some extent did, supply it, Henry George (1839–97), concentrated on a single-policy solution and so never generated the broad appeal needed to win elections in a giant democracy. At sixteen he ran away from his devout Philadelphia family and went to sea before the mast. The ship called at Calcutta and it was there—not in the United States itself—George was struck by the immense contrasts between riches and misery. The impression, reinforced by years back in America working as a printer and newspaperman, eventually produced *Progress and Poverty* (1879), one of the few books ever written on economics which found a popular readership entirely on its own merits. George wondered, like so many other people, at the mysterious coexistence of overwhelming wealth and dire want, and he speculated how the blatant irregularities of the human condition could be smoothed out. He argued that the problem lay in the notion of freehold property in land, which consigned to those who held it the fruits of other men's labors and absolute rights which should never belong to the individual. The book was originally printed in an edition of only 500 copies but it caught on hugely, survived his death in 1897 (when 50,000 people attended his funeral), and by the end of the century had sold over 2 million copies. But George's solution, a single tax on land, to restore to the public what he called the 'unearned increment' of landownership, never really worked, though it was later to inspire savagely high marginal tax rates on investment incomes, which persisted in most Western countries until the 1980s.[177]

Henry George, however, was right to challenge the prevailing view that progress would continue indefinitely 'onward and upward,' and right to insist that ownership of land, and the human occupation of the planet generally, was a trust and ought to be discharged with responsibility. If owners did not conduct themselves in a spirit of social justice, then the state should nudge, and if necessary force, them in the right direction. As he said, this was the only alternative to demands for outright socialism. Henry George, assisted by such popular writers as Henry Demarest Lloyd (1847–1903), popularized the term 'monopoly' as the epitome of all that was wrong with the American system of laissez-faire capitalism. They argued that the state, far from assisting the drift towards monopoly by, for instance, large grants of public land to railroads, should actually break up monopolies when they formed, by specific legislation. By 1872 both main parties were opposed to further land grants to railroads and by 1888 anti-monopoly plans featured in all electoral platforms. By 1890, twenty-seven states and territories had enacted anti-monopoly legislation and fifteen had constitutional provisions against them. It was therefore natural for Congress in 1890 to reinforce the anti-trust movement by a federal Bill, the Sherman Act, which imposed fines and imprisonment on those who exercised monopolies against the public interest and allowed plaintiffs to recover three times the amount of any damages sustained.[178] But it proved ineffective. In its first dozen years the government set in motion eighteen suits but with no success. On the other hand, Sherman was used with spectacular effect against unions judged to be monopolies, in the Pullman Strike (1894) and in many other cases, notably in the Danbury Hatters' case, where leaders of the union were held liable financially to the full extent of their property for losses to business caused by an interstate boycott.[179]

In the period 1880–1914 newspapers and magazines were more effective than either political parties or legislation in curbing the antisocial excesses of capitalism. They grew even faster than the giant industrial corporations. In 1850 there were 260 daily newspapers in the United States. By 1880 there over 1,000. Ten years later the total had jumped to nearly 1,600 and in 1900 there were 2,200. In 1910 the newspaper industry reached its maximum expansion in terms of titles, with almost 2,600 newspapers being produced daily (by 1965 titles had fallen to 1,700, though combined circulations, at 60,000,000 daily, were higher).[180] The existence of powerful and independent newspapers and magazines, with the resources to conduct elaborate investigations into business practices, began to be felt, from about 1880 on, as a salutary check on corporate wrongdoing or, rather, antisocial exploitation of corporate power.

In 1906 Theodore Roosevelt termed these investigative journalists the 'muckrakers,' borrowing a word from Bunyan's *Pilgrim's Progress*, 'the man with a muck-rake in his hand,' who gathered filth rather than look up to nobler things. The original muckraking story was written by Henry Demarest Lloyd in an *Atlantic Monthly* series in 1881, 'The Story of a Great Monopoly,' and it was imitated in a wide range of publications, especially *McClure's, Everybody's, Collier's*, and other weeklies. These three magazines alone published over 1,000 muckrakers in the years 1902–12, on such subjects as slums, racial prejudice and discrimination, child labor, and overpricing. Political weeklies, like the *Nation* (1865) and the *New Republic* (1914), kept up the pressure. Lincoln Steffens (1866–1936) in a general series published in 1902–3 in *McClure's*, called 'The Shame of the Cities,' argued that urban corruption, of which he cited many horror-striking examples, was the symptom of a moral weakness in America, sign of 'a freed people who have not the will to be free:' this 'civil shamelessness' indicated a 'natural process' whereby 'a democracy is made gradually into a plutocracy.'[181] But most investigations concentrated on particular scandals and called for specific legislative remedies. Thus the fictional treatment of the meat-packing industry, *The Jungle* (1906) by Upton Sinclair (1878–1968), led directly to the Pure Food and Drugs Act; and another series produced the Hepburn Act imposing railroad regulation in 1906.[182]

Perhaps the most important of these investigations, or so it seemed at the time, was a series by Ida Tarbell (1857–1944) on Standard Oil. This appeared in *McClure's* in 1904 and was subsequently expanded into a two-volume book which went into many editions.[183] Tarbell, a graduate of Allegheny College, was not exactly objective. Her father had been an independent oil producer and had been put out of business by Standard Oil. The oil industry raised acute problems of size and control because of the huge cost of drilling for oil at deep levels and in remote regions, and the stupendous economies of scale which could be achieved by large refineries. It effectively began in the United States with the drilling of the Drake Well at Titusville, Pennsylvania in 1859. It was such an obvious godsend that, as a result of the Civil War, production lagged seriously behind demand by 1865, with grave shortages of refining machinery and transport. In many fields, where the crude oil lay close to the surface, drilling and extraction were the least of the problems. But thereafter not just access to capital but new managerial skills were imperative.

In 1867 John D. Rockefeller (1839–1937) put together five big refineries into one firm, originally Rockefeller, Andrew & Flagler. Three years later, to get more capital, it was reorganized as the Standard Oil Company of Ohio. Its Clevandon Plant could then refine what seemed the enormous total of 50 barrels a day. But that was only 4 percent of US output and it was not even the biggest plant. Where Standard pushed ahead was in winning the competition for transport facilities and getting favorable rates—it proved itself unsurpassed in tough negotiations with the Erie, the New York Central, and the Pennsylvania Railroad, the three key ones at this stage. In 1871 the Pennsylvania legislature chartered the South Improvement Company, 900 of whose 2,000 shares were held by Rockefeller and his chums. This was a transportation company with unrestricted statutory powers and it made a deal with the three railroads: Standard got favorable rates while its competitors paid full rates. This charter raised such a storm that it had to be revoked after three months. But it was typical of the kind of favored deal Rockefeller managed to strike, and as Standard grew bigger he could put railroads, legislatures—anyone—under more and more pressure to give him breaks. By 1879 Standard controlled 90 to 95 percent of the oil refined, and this fact, plus its new pipeline system, gave it a stranglehold over the railroads—or so it was thought. From 1879 to 1882 Rockefeller created a trust system whereby nine trustees coordinated all their production, refining, transportation, and distribution activities with the main Standard Oil Company, later transferred to New Jersey (1899), where the business regime was looser.[184]

That Standard Oil was ruthless and predatory was obvious to all, and from 1880 onwards it was constantly being investigated by state legislatures and hauled before the courts. In many areas it was manifestly a monopoly. By the time Tarbell wrote it was close to controlling 85 percent of the production of domestic oil products in the United States and 90 percent of exports. On the other hand, its success stimulated exploration for and discovery of oil in various parts of the US and, later, abroad, especially in Rumania, Tsarist Russia, the Middle East, and Latin America. A year before Tarbell wrote, in 1901, perhaps the largest and most significant oil strike in history occurred near Beaumont, Texas, on a mound called Spindletop, a gigantic gusher such as the world had never seen. A year later over 1,500 oil companies had been chartered in consequence, among them future giants like Texas Oil and Gulf Oil. The Texas strike ended any possibility (never very strong) that Standard could retain its near-monopoly.

Moreover, the story of Standard seems to illustrate the argument, now better understood than it was then, that temporary monopolies may benefit the public interest. The per-barrel cost of refined oil at a plant with a 500-barrel daily throughput was $0.06 a gallon. With a 1,500-barrel throughput it fell to $0.03 a gallon. It was as simple as that. In its first big phase of expansion Rockefeller's company was able to reduce the retail price of kerosene, used by every household in the US, by 70 percent. Comparable figures could be quoted for the whole range of products. That Standard had a menacing effect on state legislatures may be true: Lloyd, in his book *Wealth Against Commonwealth* (1894), alleged that Rockefeller, in his efforts to influence the Pennsylvania legal climate, did 'everything to the legislators except refine them.' But whether the total impact of Rockefeller's Standard Oil ran against the public interest is doubtful. The rapid fall in oil prices, first in domestic oils, then in gasoline for transport, was largely its doing.[185]

Nevertheless, Tarbell and other writers stimulated the regulators and the courts to fresh efforts and on May 15, 1911 the Supreme Court dissolved the Standard Oil Trust, though it substituted what it called the 'rule of reason' for an earlier ruling under the Sherman Act which forbade any combination in restraint of trade, thereby allowing trusts to cooperate among themselves so long as they did not unduly interfere with competition. In 1914 the Federal Trade Commission Act, designed to prevent 'unfair methods of competition' was passed, and the Commission conducted studies which eventually led to the Robinson–Patman Act of June 1936, outlawing discounts or rebates given by manufacturers for high-volume purchases and aimed particularly at the big chain-store retailers. The historian has to ask the question: can bigness, which demonstrably and statistically benefits the public as a whole, be said to be antisocial if it can be shown to disadvantage some or to operate, albeit metaphysically, to the public disquiet?

The question was raised, early in the 20th century, by the Great Atlantic and Pacific Tea Company, known as A&P, which went back to 1859 but was transformed in the 1870s by one of its clerks, George Huntington Hartford. The object of the firm, which originally operated through mail-order and clubs, was to bring down the price of tea and coffee which, not being of American origin, were overpriced. By 1900 it had about 200 stores and was the industry leader. It concentrated on selling at the lowest possible prices at a time, 1900–12, when food prices were rising, causing concern, and provoking government investigations. Price-cutting was so successful that in the years 1914–16 alone A&P opened 7,500 stores. By 1936 it had 15,427 stores, with a backwards-integration list of wholly owned suppliers which included 111 warehouses, forty bakeries, thirteen milk-plants, eight coffee-roasting plants, six canneries, and nine general food-factories—even a printing plant. Housewives voted for A&P with their purses—that was obvious in the number of branches which succeeded.[186] But the chain was soon in trouble with anti-trust militants and the courts, and the Robinson–Patman Act was virtually designed around it and against it.[187] As it happened, this Act ended by doing more damage to the independent stores it was designed to protect, since it was interpreted by the courts as preventing them forming trade associations—a good example of Karl Popper's Law of Unintended Effect, which operates most fiercely on occasions when politicians (and journalists) try to reform business.

Fear of size was something new in America, where bigness and scale had hitherto been seen as unmitigated benefits. It was inherent in much of the regulation from the 1880s onwards, by state and eventually by federal government, and it drove on the muckraking journalism. If the reformers were asked what they hated most about Standard Oil, they replied: 'Its size.' There was no answer to that criticism. And the legal doctrine of the oppression of size was finally articulated by one of the ablest of the Supreme Court judges, Louis D. Brandeis (1856–1941): 'I have considered and do consider that the proposition that mere bigness cannot be an offense against society is false, because I believe that our society, which rests upon democracy, cannot endure under such conditions.' He specifically instanced Standard Oil and the chain-stores because of their vertical integration, and he opposed both on principle, whether or not they delivered goods to the consumer more efficiently and therefore more cheaply. To him, 'quantity discounts' were 'fraught with very great evil,' even if accompanied by cost-savings to the supplier, and therefore ultimately with cost-savings to the general public.[188]

A well-paid Supreme Court judge would not, naturally, take the same view of the importance of marginally lower food prices as a working-class housewife operating within a strict budget. And America was about the multitude, not about the elites. That, at any rate, was the assumption of its Constitution. Distinctions between classes were particularly marked in transport, and always had been. Throughout history, the multitude had always had their movement restrained by lack of personal transport. The problem rose in an acute form in the United States because of its sheer scale and what has been called the 'tyranny of distance.' Gentlemen and farmers owned horses, but farmhorses were seldom available for non-essential work. The carriage was always a luxury; hence the distinction between 'carriage folk' and the rest. Thanks to huge concentrations of capital, by 1900 the railroads had made short-distance travel cheap and universal and brought even long-distance journeys within the means of most. But it was not personal, in the way the carriage-and-four was, or a private railroad coach (very common around 1900) or a steam yacht.

Then came the internal-combustion engine. Granted the size of the oil industry in the United States by 1890, and the successful efforts of

Standard Oil to bring down prices, not least of petroleum and gasoline, it is odd that a gigantic country like the United States should have been so slow to develop an automobile industry. Up to 1895 France and Germany were the pioneers. Even in 1899 the American auto industry was ranked 150 in terms of the value of the product, and unranked in terms of wages, workers, costs, and value-added. Edison, always at the forefront of the future trend, shouted in vain that the auto was 'the coming wonder . . . it is only a matter of time before the cars and trucks of every major city will be run by motors' (1895).[189]

Then, within a decade, the world-picture was dramatically changed, very largely as a result of the efforts of one man, Henry Ford (1863–1947). He was born on a farm near Dearborn in Michigan and knew from experience the isolation which lack of personal transport imposed on millions of American farming families. In the years 1879–86 he acquired wonderful skills as a machinist in Detroit, already a center of advanced engineering, and he then had a spell in the Edison Company under the master inventor–entrepreneur himself. He built his first gasoline engine in 1892 and a decade later organized the Ford Motor Company. He illustrated the power, which all historians learn to recognize, of a good but simple idea pursued singlemindedly by a man of implacable will. Ford thought the internal-combustion engine, combined with cheap gas (already available), was the solution to the problem of rural isolation and indeed of suburban and city transport too. The task he set himself was to design and manufacture an automobile which was easy to drive, safe, totally reliable (unlike virtually all the early models), made of the strongest and best materials, and whose price could be progressively reduced by economies of scale. The result was the Model T of 1908. Farmers (and their wives and daughters) found they could handle it and even learn to 'fix' it on the rare occasions when it needed attention. Standard Oil, now in hot competition with Texas, Gulf, and the foreign-owned Shell, supplied the cheap gas.

Ford marketed the first Model T at $850 in 1908, when he sold 5,986. By 1916, when he sold 577,036, he had got the price down to $360.[190] By 1927, when the series was discontinued, he had sold over 15 million of these autos, and made them and their rivals standard equipment for the American family. Ford gave the entire, vast nation a mobility it had never possessed before—and which in time spread throughout the world. Ford grasped that Big Business, with its economies of scale, could lead to even bigger business provided it shared its profits with its workforce by paying handsome wages. In

1914, when industrial workers were averaging about $11 a week, he abruptly decided to pay his men $5 for an eight-hour day, so that they could all buy his Model Ts. The idea was obvious but new, and valid. By 1920 Ford, having bought out all his partners, had contrived to do three things: first, to deliver to the mass consumer a quality product at the lowest possible price; second, to pay the highest wages in the industry and, at that date, in history; and, third, to become (after John D. Rockefeller) the nation's second billionaire—indeed, at the height of his success, he was the richest man on earth.[191] This last made him highly objectionable to many people but he was not easy to prosecute as a monopolist because in 1908, the year of his Model T's birth, General Motors was formed with the quite opposite strategy of offering cars for every taste and pocket. William Durant put together Buick and two other companies, then added Oldsmobile and Cadillac, and finally Chevrolet. By 1920–1 GM was selling 193,275 vehicles a year, against Ford's 845,000. That gave Ford 55.67 percent of the market but with enormous competition (the 253 car-manufacturers of 1908 had narrowed down to 44 by 1929) there was never any possibility of monopoly. But Ford and GM were big—by the 1930s GM was the world's largest industrial company—and that alone made them the focus of the regulators and the investigators.[192]

The anti-bigness emotion, so characteristic of the decades between the Civil War and World War One—and becoming stronger with each—is worth looking at because it was unAmerican. It is necessary to distinguish between Populists, who aimed deliberately at the farming vote and agricultural prejudices, and Progressives, who tended to be highly educated intellectuals aiming at an urban audience. Historians have seen the Populist Era as embracing the years 1880–1900 and the Progressive Era 1900 up to America's entry into World War One. But they overlapped and intermingled throughout. It has been argued that Progressivism was the hostile reaction of the educated middle class to the overwhelming power of Big Business, whose wealth and scale and lure elbowed them out of the political–economic picture entirely, or so they feared. Since the days of the Founding Fathers, the educated elite had guided, if they had not exactly run, the United States, and they felt their influence was being eroded by the sheer quantity of money now sloshing around in the bowels of America's great ship of state.[193] Some of the anti-business reformers, like 'Gold Rule' Jones, Charles Evans Hughes, and Tom Johnson, were self-made men. Others had gilded names like du Pont, Morgenthau, Pinchot, Perkins,

Dodge, McCormick, Spreckels, and Patterson. These earnest scions of wealthy families led some critics to call Progressivism a 'millionaires' reform movement.' But careful prosophological analysis shows that the great majority were solidly middle class, university graduates of old British stock.[194] The 1880s and 1890s had been conservative decades with both Republican and Democratic leaderships reflecting the prevailing mood. By 1900, however, a number of new political fashions had come together to constitute a broad-based educated left—gas-and-water (or municipal) socialism, which was worldwide, trust-busting, conservationism (anti-urban sprawl, pro-wilderness), health fanaticism, the notion of educated purposeful elites as 'guardians' of the people, which was shared by a wide range of elitists, from Walter Lippmann (1889–1974) to Vladimir Ilyich Lenin and Benito Mussolini, and literary and artistic bohemianism.

This was a heady mixture, capable of emitting fumes of both good and evil. And in the background—sometimes in the foreground—were other forces: conspiracy theory and racism. The notion that various sinister gangs of people were plotting to destroy the 'good America' was an old English tradition the Pilgrim Fathers had brought with them and had first made history during the agitation against the Stamp Act. It had been brilliantly exploited by Andrew Jackson in his war against the Bank of the United States. Now it became the folklore of both Populism and the Progressives. Ignatius Donnelly made good use of it. It was brandished especially by women writers, such as Mrs S. E. V. Emery, whose book *Seven Financial Conspiracies Which Have Enslaved the American People* was published in 1887 and sold hugely in states like Kansas. Another female rabble-rouser was Mary E. Lease, author of *The Problem of Civilization Solved* (1895), who coined the 'farmers' slogan' of 'raise less corn and more hell.' Much of this output was anti-semitic (and, by association, Anglophobic, Wall Street being presented as controlled by a Jewish City of London). William Hope Harvey published the most popular of all these insidious books, *Coin's Financial School*, in 1893, following it with a conspiracy novel, *A Tale of Two Nations* (1894), in which Rothschild and Bryan figure, a little disguised. The villains of such novels were all Jews and *Coin's* was illustrated by a drawing of the world in the grip of an octopus, labeled Rothschild, stretching out from London. Not surprisingly, the farmers' campaign for silver, as opposed to gold, linked 'Wall Street and the Jews of Europe,' and an Associated Press reporter who followed the Populist convention at St

Louis in 1896 complained publicly of the ubiquitous anti-semitism there.[195]

There was a point at which Populism, in the broader sense, embraced nationalism, xenophobia, nativism, white racism, and imperialism. Mary Lease's formula for rescuing civilization, for instance, involved a global separation of the races, with white supremacists ruling all. In the second half of the 19th century most members of white races felt, and said more or less openly, that they had a divine mission, or at least a cultural duty, to rule over what Kipling called the 'lesser breeds.' America was the first of the ex-colonial nations, but most Americans, by now, felt the itch too. It was the spirit of the age. Manifest Destiny had made America an acquisitive power. Under the presidency of Andrew Johnson, Secretary of State Seward bought Alaska from Russia at the knock-down price of $7,200,000. At the time, many Americans protested it was too much and called the new purchase (organized as a civil and judicial district in 1884) 'Seward's Icebox.' But the discovery of massive gold deposits in the Klondike, 1897–8, revealed the value of the acquisition and led rapidly to its elevation to constitutional status as a Territory in 1912.[196]

Alaska was the first possession of the expanding United States which was not contiguous. But it was soon not the only one. Americans had been in the Pacific since the 1780s, as navigators, traders, whalers, fishermen, and, not least, missionaries, and from the 1820s US missions were established in Hawaii and other islands in strength. They had helped to promote literacy and liberal political institutions, as well as Christianity and American commerce, and when the local Hawaiian ruler, Queen Liliuokalani, took a sudden anti-progressive lurch in 1891, a local American, Sanford B. Dole, supported by the US minister, organized a 'committee of public safety,' which called in nearby US marines and established a government with Dole as president (1893). Seven years later, with some misgivings—behind Dole were the sugar interests, as well as missionaries—Washington recognized Hawaii as a US territory, with Dole as governor. In return for overcoming its scruples, America got the naval base of Pearl Harbor.[197]

The role played by American missionaries in the acquisition of Hawaii was typical of the way in which American Christianity was acquiring an imperialist persona. The American Board of Commissioners for Foreign Missions (mainly Congregational) had been formed as long ago as 1810, the American Baptist Missionary Board following in 1814. But it was only after the end of the Civil War, seen by many American Protestants not as a Christian defeat, in which the powerlessness or contradictions of the faith had been exposed, but as an American–Christian victory, in which Christian egalitarian teaching had been triumphantly vindicated against renegades and apostates, that the missionary surge gathered momentum. It fitted neatly into a world vision of the Anglo-Saxon races raising up the benighted and ignorant dark millions, and bringing them, thanks to a 'favoring providence,' into the lighted circle of Christian truth. The universalist mission of Christ, which had raised up America to be a 'City on the Hill,' would be triumphantly completed. This Christian mission was essentially, in American eyes, a Protestant one. In these years, when Darwin's *Origin of Species*, popularized by Herbert Spencer as 'the survival of the fittest,' and applied to races as well as species in a vulgarized form, Social Darwinism, the coming Christian triumph was presented as an Anglo-Saxon Protestant one. The American Christian republic was a triumphant success. It was so, essentially, because it was Protestant— failure was evidence of moral unworthiness, of the kind associated with decadent Catholics in southern Europe and Latin America.

In the 1870s, the ultra-Protestant and strongly anti-Catholic preacher Henry Ward Beecher used to tell his congregation in New York: 'Looking comprehensively through city and town and village and country, the general truth will stand, that no man in this land suffers from poverty unless it be more than his fault—unless it be his *sin* . . . There is enough and to spare thrice over; and if men have not enough, it is owing to the want of provident care, and foresight, and industry and frugality and wise saving. This is the general truth.'[198] And a related general truth was that God's will was directly expressed in the destiny of a country where success-breeding virtue was predominant. This was Protestant triumphalism, and its dynamic was American triumphalism. George Bancroft, in his *History of the United States*, began (1876 edition): 'It is the object of the present work to explain the steps by which a favoring providence, calling our institutions into being, has conducted the country to its present happiness and glory.' Sooner or later the entire world would follow suit. It was urged to do so in 1843 by the American missionary Robert Baird, in his *Religion in America*, in which he projected the principle of Protestant Voluntarism onto a global frame. History and interventionist theology were blended to produce a new kind of patriotic millenarianism. Leonard Woolsey Bacon wrote in his *History of American Christianity* (1897): 'By a prodigy of divine providence, the secret of the ages [that a new world lay beyond the sea] had been kept from premature disclosure . . . If the discovery of America had been achieved . . . even a single century earlier, the Christianity to be transplanted to the western world would have been that of the Church of Europe at its lowest stage of decadence.' Hence he saw 'great providential preparation, as for some "divine event," still hidden behind the curtain that is about to rise on the new century.'[199]

The 'divine event' was the Christianization of the world in accordance with American standards of justice and probity. That was to be a racial and national as well as a religious event. In 1885 Josiah Strong, general secretary of the Evangelical Alliance, published a book, *Our Country: Its Possible Future and its Present Crisis*, in which he argued:

> It seems to me that God, with infinite wisdom and skill, is here training the Anglo-Saxon race for an hour sure to come in the world's future . . . *the final competition of races, for which the Anglo-Saxon is being schooled* . . . this race of unequaled energy, with all the majesty of numbers and might of wealth behind it—the representative, let us hope, of the largest liberty, the purest Christianity, the highest civilization—having developed peculiarly aggressive traits calculated to impress its institutions on mankind, will spread itself over the earth. And can anyone doubt that the result of this competition between the races will be the survival of the fittest?[200]

Such theories were not uncommon at the time. What was significant in the United States was that they radiated from a Christian context and were presented as part of an altruistic scheme to Christianize the world. By this stage America was the leading missionary force, in terms of the resources it deployed, especially in Asia and the Pacific, and it was thought that the white races in general, and the Anglo-Saxons in particular, would succeed in bringing to reality Christ's vision of nearly two millennia before—a universal faith. The 19th century had been a period of such astonishing, and on the whole benign, progress that even this great dream now seemed possible. In the 1880s the young American Methodist John Raleigh Mott had coined the phrase 'The evangelizing of the world in one generation.' That was a task for American leadership.[201]

It was against this background that America drifted, or blundered—or perhaps strode—into its one imperialist adventure, the Spanish–American War, 1898. That Cuba had not been annexed by

America long since was itself surprising. As we have seen, some Southerners had wanted it. Others had not. Cuba was the last of Spain's colonies in the Americas, and it was said to be oppressed. In 1868–78, there had been a rising there, followed by a guerrilla war, and America had been begged to intervene by Cuban 'patriots.' But America had not stirred. During the 1880s and early 1890s, however, American investments in Cuba multiplied many times, and when the Cubans again rebelled in 1895 American sympathies, not unaffected by the imperialist spirit of the age, were aroused. There was also the Christian dimension. Catholic Spain, not only in Cuba but more importantly in the Philippines, was adamant in excluding Protestant missionaries, believing them keen to spread democracy as well as heresy. The Populists and the rank-and-file Democrats got very excited over the prospects of Cuban independence. They identified with the oppressed Cuban rebels and threatened to campaign under the slogan 'Free silver and Free Cuba.'[202] The Republicans feared to be outbid in playing the nationalist card, hitherto one of their strong suits.

The United States was involved in the Cuban insurrection from the start because it was the source of various filibustering expeditions, and a Cuban 'National Junta' in New York sold bonds and made noisy propaganda. The financial world, however, urged caution and neutrality, and so long as President Cleveland was in the White House this was official US policy. In the autumn of 1896, however, McKinley was elected on a platform which included a demand for Cuban independence, adopted by the Republicans to prevent the Democrats offering more. McKinley nonetheless tried to act circumspectly, encouraging the Spanish to introduce reforms which would appease both Cuban and American opinion. But an outspoken private letter from the Spanish minister in Washington, Enrique de Lôme, accusing McKinley of being 'weak, and a bidder for the admiration of the crowd,' was stolen and gleefully published (February 9, 1898) in the jingo *New York Journal*, owned by William Randolph Hearst (1863–1951). A week later, the US battleship *Maine*, which had been sent to Havana to protect American citizens, was blown up, with the loss of 260 of her crew, by an underwater mine. Who planted the mine has never been discovered with certainty, but the Hearst press, and the *New York World*, owned by Joseph Pulitzer (1847–1911), blamed Spain. Congress voted $50 million for national defense without a dissenting vote and on April 11 McKinley sent Congress a message demanding 'forceful intervention' in Cuba to establish peace on the island. On the 20th Congress passed a

War Resolution and McKinley signed it, though an amendment was added disclaiming any desire to exercise sovereignty over Cuba. Diplomatic ties were broken and a blockade imposed, and on April 24–25 both sides declared war.[203]

Whether this 'Splendid Little War,' as it was called, was created by Hearst, or by US business interests, or was an accident, or inevitable, is still argued about by historians. The evidence suggests only a few businessmen directly involved wanted war; the rest were for peace or, if peace was unobtainable, what would be now be called a quick 'surgical strike' to restore order.[204] That is what they got. The war was primarily naval. The United States had an army of only 30,000 but its 'steel navy' was modern. The Spanish Pacific squadron was defeated in the Battle of Manila Bay (May 1). The US Fifth Corps landed near the Spanish naval base of Santiago and, after the heights of San Juan Hill were stormed by Theodore Roosevelt's 'Rough Riders,' and the shore batteries, commanding the harbor, passed under American control, the Spanish admiral had no alternative but to surrender or take his ships to sea. He chose battle, and all of them were destroyed (July 3), American casualties being one killed and one wounded. The swift manner in which US naval technology demolished old Spain, once the world's strongest power, was comparable to the annihilation by Britain, at the Battle of Omdurman two months later, of the vast but barbarous army of the Sudanese Mahdi. At a stroke, the United States emerged as a great power, and a global one, with all kinds of new responsibilities, including administration of the 7,100 islands which constitute the Philippines.[205]

At the Treaty of Paris, signed on December 10, Spain ceded to the United States Puerto Rico, Guam, and the Philippines (getting $20 million in return) and gave Cuba its independence. The Senate ratified the treaty by a close vote, many of its members not relishing the idea of the Great Republic acquiring an overseas empire. It was characteristic of the age that McKinley himself placed the retention of the Philippines as a colony firmly in a Christian evangelical context: 'I am not ashamed to tell you, Gentlemen, that I went down on my knees and prayed Almighty God for light and guidance that one night. And one night later it came to me this way . . . There was nothing left for us to do but to take them all and to educate the Filippinos and uplift and civilize and Christianize them, and by God's grace do the very best we could by them, as our fellow men for whom Christ also died.'[206] Thus the United States acquired 7 million subjects, 85 percent of whom, as it happened, were already Christians, albeit lowly Roman Catholics.

With a victorious war under his belt, and a strong economy, McKinley was easily renominated in 1900 and did not trouble himself to campaign actively. Bryan was again his opponent and a very vigorous one, delivering over 600 speeches in twenty-four states, perhaps the most extensive campaign yet carried out by a presidential candidate. But his anti-imperialism was unfashionable and his silver policy dated. He lost by a wider margin than in 1896, McKinley beating him in the popular vote by 7,219,530 to 6,358,071, and carrying the college 292–155. The important change was in McKinley's running mate. Roosevelt, who had been Assistant Navy Secretary in the Cuban War, and led the Rough Riders, succeeded the old Vice-President, Garret A. Hobart, who died in office. His promotion became of critical importance on September 6, 1901, when McKinley, opening the Pan-American Exposition in Cleveland, was shot by a local anarchist, Leon Czolgosz, and died eight days later. At forty-two, Roosevelt became the youngest president in the history of the republic, and the first since Lincoln to combine a coherent body of political, philosophical, and social doctrine with an outstanding personality. At last the great and now overwhelmingly mighty republic got a president of comparable stature. Emerson had written: 'Why should not we also enjoy an original relation to the universe? . . . America is a poem in our eyes.' 'TR' was not exactly a poem, though he could be poetic; prose was more his style. But he was undoubtedly a phenomenon.[207]

The entrance of the Roosevelts on to the center-stage merits a word. America has no kings but it rejoices in its family dynasties, religious, political, and commercial. In the 17th and the 18th centuries, as we have seen, there were the Winthrops and the Mathers. In the 19th century there were the Adamses and the Astors, the Vanderbilts, Morgans, and Rockefellers, and in the 20th century we have had the Roosevelts and the Kennedys. For variety and richness of character—and sheer talent—these clans are well up to the standard of the great English houses, the Russells, Churchills, Cecils, and Cavendishes. It is strange that Henry James, who loved such human continuities so dearly, should have turned his back on the family treasures of his own country to go whoring after anemic Mayfair—and still stranger that Edith Wharton, who knew more about American high society than James did, especially in Manhattan, should have preferred the *menus plaisirs* of the Paris *gratin* to the richer meat of Long Island and Newport. But so it is:

the greatest writers often feel themselves exiles in their own country and prefer Abroad.

The Roosevelts were descended from the old Dutch *patroons*. They had a common trunk but there were two branches, called after their estates. The Oyster Bay branch produced Theodore Roosevelt, President 1901–9, and the Hyde Park branch Franklin Delano Roosevelt, President 1933–45, so that in the first fifty years of the 20th century Roosevelts occupied the White House for nearly twenty of them. The two branches were highly competitive and jealous of each other. They occasionally intermarried—FDR's wife Eleanor was from the Oyster Bay branch and TR's nephew married FDR's niece—though generally relationships were malicious, even hostile. But the presidential Roosevelts were gifted populists in politics and had much in common, including enormous energy, especially under physical affliction, and a zest for life. TR was a radical conservative whereas FDR was a conservative radical. A preference for one over the other is a touchstone of American character. Most American intellectuals would rate FDR the greatest president since Lincoln. But a substantial minority prefer TR, seeing him as the archetypal American 'good guy,' combining the best of the inherited English tradition of gentlemanly honor with a riproaring taste for adventure which is quintessentially American—by comparison judging FDR sly, feline, secretive, and *faux bonhomme*.[208]

TR was the first great American to be dogged by photographers all his life, the results often producing ribald smiles among modern commentators. Aged six, he is seen peering out of a New York brownstone as Lincoln's coffin processed through Manhattan on the way to its last resting place in Illinois. Then he is seen posing against a studio background before heading for Dakota, wearing funny lace-up soft-leather boots, fringed leather jerkin, broad belt, fancy dagger, and a bearskin hat and carrying a carbine in menacing fashion. He posed in the same absurd hat for a reenactment of his capture of Redhead Finnigan, the outlawed horse-thief—the photo was partly fake as one of the 'outlaws' was in fact a member of TR's team. There are further photos of him dressed for war in large leather gloves, side-turned hat, pince-nez and mustaches, or in shirt-and-suspenders, jackboots, huge pistol-holster, and scout-hat, posing with his Rough Riders in Cuba. One has to look through these comic theatricals to the steady, serious, true heart and powerful brain beneath.

TR was a sickly child, an anxious youth and young man, and when his young wife died leaving him a twenty-five-year-old widower with a child, he went out to the Badlands of Dakota, to heal and find himself (1884). Calling it the Badlands was a typical 19th-century topographical moral judgment (though it is true that the French, who were there first, baptized them *mauvaise terres à traverser*). It was indeed a Deathscape, actually strewn with buffalo skulls and bones. Below the buttes were mazes of quicksands and connecting gullies and abysses, into which men and animals simply disappeared. There were mists of steam, and smoldering lignite and coal seams emitted subterranean fires. TR, characteristically, summed up: 'The Badlands looked like Poe sounded.' He noted that volcanic ash covered this part of South Dakota, east of the Black Hills in the southwest of the Territory (it did not become a state till 1889). The ash went to a depth of 300 feet and had in prehistoric times engulfed herds of mammoths, elephants, camels, and other creatures whose bones are still buried there.[209] Natural forces were still scorching, eroding, sandblasting, and freezing, then rock-splitting the surface all the time and TR called it 'As grim and desolate and forbidding as any place on earth could be.' He seems to have seen himself, not quite consciously, as like John the Baptist in the desert, or Jesus Christ spending a time in the wilderness to prepare himself for his ministry.

The Roosevelts were old money and TR was able to buy himself into a cattle-herding business, setting up his headquarters at the Maltese Cross Ranch near Medora, and building a remote lodge called Elkhorn. It was his object to overcome his physical debility by pushing himself to the limit of his resources: he wrote a letter home boasting, 'I have just come in from spending *thirteen* hours in the saddle.' There were still a few buffalo and Sioux Indians around and the frontier was not yet 'closed' in the Jackson Turner sense. There was, now and always, a touch of Hemingway literary-machismo about TR and a longing to play John Wayne roles. (TR, of course, lived before that precious pair and thought more of the scenarios displayed in Twain's *Roughing It*, which had just been published.) In a saloon in Mingusville (called after a couple, Minnie and Gus, who founded the town), he knocked down a drunken cowboy who was shooting off his guns and terrifying the ladies. TR struck the man in the face with both fists, the guns going off harmlessly. He remarked, 'Well, if I've got to, I've got to.'[210] He wrote home: 'I wear a sombrero, silk neckerchief, fringed buckskin shirt, seal-skin *chaparajos* or riding trousers, alligator hide boots, and with my pearl-hilted revolvers and beautifully finished Winchester rifle, I shall be able to face anything.' His silver-mounted Bowie knife came from

Tiffany's, as did his silver belt-buckle with a bear's head and his initialed silver spurs. He duly shot his grizzly—'The bullet hole in his skull was as exactly between his eyes as if I had measured the distance with a carpenter's rule.' He said to a local bully called Paddock, who threatened to hound him off his range: 'I understand you have threatened to kill me on sight. I have come over to see when you want to begin the killing and to let you know that if you have anything to say against me, now is the time to say it.'

It was all very like a future Western, and ended with TR capturing Redhead Finnigan and two fellow-outlaws, for which he was paid a reward of $50. It was entertaining and sometimes tough and even risky but it was play-acting and at the end of two years TR went back east and promptly married, a second time, a girl called Edith, directly descended from Jonathan Edwards. After his return he summed up his philosophy in a Fourth of July oration:

> Like all Americans, I like big things: big prairies, big forests and mountains, big wheat fields, railroads—and herds of cattle too—big factories and steamboats and everything else. But we must keep steadily in mind that no people were ever yet benefited by riches if their prosperity corrupted their virtue. It is more important that we should show ourselves honest, brave, truthful, and intelligent than that we should own all the railways and grain elevators in the world. We have fallen heirs to the most glorious heritage a people ever received and each of us must do his part if we wish to show that this nation is worthy of its good fortune.[211]

There can be no doubt that TR believed every word of these sentiments and did his best to live up to them. He was exactly the same kind of romantic–intellectual–man-of-action–writer–professional-politician as his younger contemporary, Winston Churchill. The two did not like each other, having so much in common, and being so competitive; and TR criticized Churchill severely because 'he does not stand up when ladies come into the room' (such things mattered in those days). TR wanted to have many children, believing 'good blood' should battle with immigrant strains—what he called 'the warfare of the cradle.' He sought action because 'every man must prove himself' and a politician should not send men into battle 'without knowing what it is like.' In Cuba he served with the First Volunteer Cavalry under the official command of General Leonard Wood, the conqueror of Geronimo, but TR was the real mover of the unit. It was like a World War Two command, a mixture of Texas Rangers and working-class adventurers (some of whom later joined the Buffalo Bill show) and Harvard-educated young bloods, Hamilton Fishes, Tiffanys, Astors, and so on. The Cuba campaign was no outing, men dying by disease as well as falling in action. Some drowned in a storm on landing, and when the wounded fell on the beach they were attacked by big land-crabs which tore out their eyes and lips. TR enjoyed the campaign, saying, 'I do not want to be called a Parlor Jingo,' and led repeated charges at San Juan Hill. All the same, there was vote-catching in the trip, and he returned a public hero: the invitation to be vice-president—and the presidency itself—followed.

However, the fact that TR was a big man is proved by the strength of his administration. It included Henry Stimson, Herbert Smith, William Moody, Robert Bacon, Franklin Lane, James Garfield, Charles Prouty, Gifford Pinchot, and other men of mark. Lord Bryce, the great British ambassador, who published one of the best and most comprehensive surveys of the American system,[212] said he had never come across such a high-minded and efficient group of public servants as the Roosevelt administration. They were indeed 'the Best and the Brightest.'[213] Roosevelt himself regretted only that they were ill-read in English literature and, in particular, had never heard of Lewis Carroll. When he quoted to his Navy Secretary, William Moody, 'Mr Secretary *what I say three times is true*,' the aggrieved man replied: 'Mr President, it would never for a moment have occurred to me to impugn your veracity.' So when Roosevelt entertained Edith Wharton to dinner, he greeted her: 'Well, I *am* glad to welcome to the White House someone to whom I can quote "The Hunting of the Snark" without being misunderstood.'[214]

Roosevelt went out of his way, in fact, to make his White House a place where writers and artists felt at home, despite his vandalizing of Tiffany's screen. He even had to dinner Henry James, who had called him 'a dangerous and ominous jingo' and whom he privately dismissed as 'effete' and a 'miserable little snob.' Mrs Wharton at least reciprocated his admiration. She was present in September 1902 when the President's carriage was run down by a trolley car, during a speaking tour; he was hurled, battered and bleeding, onto the sidewalk. His face was badly bruised and one knee so damaged that surgeons almost amputated his leg. But he carried on his engagement and Mrs Wharton, who heard him speak, wrote: 'I think if you could have seen the President here the other day, all bleeding and swollen from that hideous accident, and could have heard the very few quiet and fitting words he said to the crowd gathered to receive him, you would have agreed that he is not all—or nearly all—bronco-buster.'[215]

As it happens, the theme of TR's discourse, repeated over and over again that year, was the wickedness of Big Business. He denounced J. P. Morgan and US Steel, he fulminated against Rockefeller and what he called 'the Bad Trust'—Standard Oil—and he positively excoriated Harriman, whom he treated as a social and moral outcast. There was something petty in this kind of Populism, coming from a man whose family had inherited, and always lived off, wealth accumulated in distant times by methods which would not bear close examination. It was as though TR, representing old and limited money, could not stomach the arrival of new money in unlimited quantities. The fact is, Roosevelt, though president for over seven years, did not have enough to do, or rather sufficient challenges for his powerful brain and boundless energies. There was no grand international crisis for him to grapple with, and the times were not yet ripe—they soon would be—for fundamental changes in the way America was governed.

TR was a firm believer in executive action, as opposed to Congressional legislation. He argued—like Jackson—that the President might do anything in the national interest not expressly forbidden by the Constitution, which, he said, 'must be interpreted not as a straitjacket, not as laying the hand of death upon our development, but as an instrument designed for the life and healthy growth of the Nation.' His motto was 'speak softly but carry a big stick'—the stick being executive power, a new form of royal prerogative. When the United Mine Workers brought the coal industry to a halt in the summer of 1902, and the mine-owners prepared to starve the miners out, TR intervened. The owners' negotiator, George F. Baer, infuriated him by declaring, 'The rights and interests of the laboring man will be protected and cared for not by the labor agitators, but by the Christian men to whom God in his infinite wisdom has given control of the property interests of the country,' and when the two met in the White House TR felt tempted to throw Baer through the window. He said he had the presidential power, and if need be would exercise it, to take over the mines and run them, using the army. This was the first time any president had contemplated an industrial takeover (though one can imagine Lincoln doing it) and the threat worked: the owners accepted mediation, by preference.[216]

TR ran a broad-based regime which found a place for organized labor, a place for Big Business despite his trust-busting (he solicited and received business campaign contributions), a place for the farmers, and a place for blacks. He invited Booker T. Washington (1856–1915), the negro educationist and author of a superb autobiography *Up from Slavery* (1901), to a meal at the White House, for which—amazing as it seems to us now—he was bitterly criticized at the time. Some of the Republican Party bosses thought TR was too liberal and looked around for an alternative to run. But TR found no difficulty in getting himself renominated in 1904 and he walked over his Democratic opponent, Judge Alton Parker (1852–1926), 7,628,461 to 5,084,223, winning the college 336 to 140, Parker carrying Southern states only.[217]

TR had already had one shot at regulating the railroads in 1903 when he got Congress to pass the Elkins Act, outlawing rebates, and now, overwhelmingly elected in his own right, he used his authority to help push through the Hepburn Act (1906) which gave the Interstate Commerce Commission power to set maximum rates. In general, he worked in tandem with the more responsible of the muckrakers to improve conditions in meat-packing, food-processing, drug-manufacturing, and other good causes of the time. He established the Department of Commerce and Labor, with its oversight Bureau of Corporations. As a portent of times to come, he paid special attention to conservation. Gifford Pinchot, whom he made head of the Forestry Division in the Department of Agriculture, was instrumental in securing 172 million acres of treeland for the nation and in instituting a series of national programs to preserve and build up the United States holdings in natural resources. A Public Lands Commission, an Inland Waterways Commission, and a National Conservation Commission were created, and Pinchot struck a fine balance between preserving the wilderness and making needful use of water, land, and timber for the country's rapidly increasing population. During the Roosevelt presidency men and women felt that the White House was inhabited by an intelligent, many-faceted, far-sighted, civilized—though not unduly sensitive—spirit.

TR's most notable achievement, in some ways, was to push through the Panama Canal. That meant softening up with cash the corrupt and venal government of Colombia, which owned the Panama Isthmus. Under the Hay–Herrán Treaty of 1903, America was to build the canal and pay Colombia $10 million down and an annual rental of $250,000. The Colombian government, to the fury of TR, upped its demand to $25 million. Unwilling to be hornswoggled by what he called 'a bunch of dagoes' and 'the foolish and homicidal corruptionists of Bogota,' TR connived at a local conspiracy to set up a separate state of Panama, recognized by his government in November 1903. A new treaty was put through immediately with the Panamanian government, on the same financial terms, but which extended the width of the Canal Zone from 6 to 10 miles and gave the United States 'in perpetuity, the use, occupation and control' of it. When TR asked his Attorney-General, Philander C. Knox, to get an expert legal opinion to uphold the constitutionality of his actions as president, Knox replied: 'No, Mr President, if I were you I would not have any taint of legality about it.' As TR is said to have remarked (the story may be apocryphal): 'What's the Constitution between friends?'[218]

It was Roosevelt's view that America had a right, not exactly God-given but arising naturally out of the circumstances of the hemisphere and the United States' proponderant power within it, to operate as a hemispheric policeman. Just as, in America itself, the federal government had what he called a 'national policing power,' so in American waters the American navy, now the second largest in the world after Britain's, had a duty to uphold democratic republicanism and good government in the interests of all. As a corollary to the Monroe Doctrine, he argued that, since European interference in the hemisphere was unacceptable, and since foreign powers whose nationals had invested heavily in Latin American states might be tempted to subvert them if corrupt governments reneged on debts or permitted plundering, the United States had a prior right and duty to forestall such crisis by acting itself. Hence he made what became known as the Roosevelt Corollary part of his 1904 annual address to Congress: 'Chronic wrongdoing . . . may in America, as elsewhere, ultimately require intervention by some civilized nation, and in the Western Hemisphere the adherence of the United States to the Monroe Doctrine may force the United States, however reluctantly, in flagrant cases of such wrongdoing or impotence, to the exercise of an international police power.' This authority was exercised repeatedly, and on the whole sensibly and to general satisfaction, over the next few decades, especially in the Caribbean area.[219]

Roosevelt had some of the ambivalence towards public life of George Washington (and indeed of Alexander Hamilton, the American hero he most admired). He wrote in his autobiography that his own generation and class shunned politics as vulgar: 'The men I knew best were the men of the clubs of social pretension and the men of cultivated taste and easy life.' They told him to avoid politics and 'assured me that the men I met would be rough and brutal and unpleasant to deal with. I answered that if it were so it merely meant that the people I knew did not belong to the governing class, and that the other people did—and that I intended to be one of the governing class.'[220] But did he? TR enjoyed power and battle but he had many other interests. He liked travel and big-game hunting and international hob-nobbing with the great. He was the first president to leave the country, visiting Panama, but he found the White House restrictive in many ways. It may have been him who first coined the saying, 'Winning the presidency is to condemn yourself to four years of drinking Californian wine' (then in its rude infancy). He disliked many of the people whom his office forced him to consort with, notably businessmen. He thought them timid, unadventurous, what he called 'flub-dubs' and 'mollycoddles,' or just plain dishonest.[221] In 1908, only fifty and still young enough to enjoy himself, he indicated he would resign if he were allowed by the Republican Party to pick his successor. A deal was struck.

William Howard Taft (1857–1930) was an extraordinary choice for a man like TR to make and was proof of the contention that great men should not be allowed to pick their successors. He won the election on Roosevelt's record without difficulty, beating the Democrat no-hoper, Bryan, who was despairingly chosen a third time, by 7,679,006 to 6,409,106 and, in the college, by 321 to 162.[222] As he was driving to the White House after his inaugural, he said to his wife Helen, who wore the pants in their family, 'Now I'm in the White House nobody is going to push me around any more.' In fact nobody exactly pushed him around: he weighed over 300 pounds and was the largest man ever to be president. But it was Helen who led him around and it was entirely her idea that he should go into politics. He was by nature sedentary and judicial. He came from Cincinnati in the Republican heartland, had been solicitor-general in the 1890s, then a federal circuit judge and a much respected governor-general of the Philippines. TR made him secretary of war and he did as he was told. He reached the White House to please his wife but almost immediately after he got there she was stricken by a debilitating illness and was unable to run his affairs any more.

Almost despite himself, Taft contrived to infuriate his predecessor. He did as he thought TR would have wished, finally breaking up Standard Oil and initiating proceedings against US Steel. In fact he started twice as many anti-trust suits as Roosevelt himself. TR denounced all this as 'archaic' and said Taft should have introduced regulation instead. Taft tried to rationalize (and reduce) high US tariffs, again something he thought TR wished; but this merely led Congress to make a fool of him and split the party. He practiced what became known as 'dollar diplomacy' in Central America to boost US trade and he sent the marines into Nicaragua. But his appointment of a customs inspector there without the Senate's 'advice and consent' got him into trouble with Congress.[223] His worst fault, in TR's eyes, was to sack the Chief Forester, Gifford Pinchot, whom Taft found 'insubordinate' and a 'crank.' Pinchot promptly went out to Africa to contact TR on his safari and make his complaints. TR charged home, processing through New York in June 1910 in a fourteen-carriage parade from the Battery to 59th Street, with luggage containing horns, heads, and skins from 13,000 specimens, ranging from elephants and rhinos to the rare dik-dik, an antelope smaller than a jack-rabbit. At first he declined to criticize his successor. But he would not dine at the White House either.

Taft seemed a throwback to the mediocrities who had served in the long, booming decades between Lincoln and Roosevelt when it was the businessmen who seemed to be running the country. Under Taft the country seemed to run itself, as it plunged into the era of the automobile and began to contemplate the air-age. Ford's Model T was now well and truly launched and selling in hundreds of thousands, as Ford himself coined the first of his one-liners: 'A customer can have a car painted any color as long as it's black' (1909). The same month four women, Alice Huyler Ramsey, Nettie R. Powell, Margaret Atwood, and Hermine Jahns, became the first to drive all the way from New York to San Francisco in a Maxwell-Biscoe runabout costing $500. Meanwhile Cadillac developed the first electric self-starter, incorporated in its 1912 models. The Wright Brothers, pioneer aviators, set up the first public company, with a capital of $1 million, to manufacture aircraft for general sale. America was swiftly acquiring a production-line economy, with high wages, high spending, and high output.

Taft seemed irrelevant, especially after the Republicans lost the House in the mid-term elections. He continued to push through TR's policies, as he understood them, beating his anti-trust suit record by eighty to twenty-five and taking over more land for public use in four years than TR had in eight (including withdrawing oil lands from public sale for the fifth time). But, whatever he actually did, he appeared to do nothing, having an ineffaceably statuesque image in the voters' minds. In February 1912, meeting in Chicago, a group of seven Republican governors called on Roosevelt to become a candidate at the forthcoming presidential elections. He responded with gusto, saying, 'I hope that so far as possible the people may be given the chance, through direct primaries, to express their preference.' This referred to the new primary system which, beginning in South Carolina in 1896, reduced the power of party bosses, replacing state conventions by direct elections as the mechanism for selecting party candidates. This change spread over the South and in 1903 was adopted in Wisconsin, thence taking over a growing number of Northern states too. Roosevelt was thus able to put up in thirteen state primaries in 1912 and in all but two he won, beating Taft even in his own state, Ohio.[224]

Hence when TR entered the Republican convention he was only 100 votes short of the nomination, in theory anyway. But the party bosses ran the procedures in exactly the same way as they had done for Roosevelt in 1904 and came out with the same result—the incumbent President and party leader was renominated. TR called it 'naked theft' and bolted the party. In August he held a Progressive Party convention in Chicago and declared himself 'fit as a bull moose,' 'stripped to the buff,' and 'ready for the fight.' TR called his ideas the 'New Nationalism.' He proposed to regulate business by a federal trade commission with formidable powers, to end the tariff rows once and for all by having an objective tariff commission settle them 'on a scientific basis,' and to narrow the extremes between rich and poor by graduated income tax and death duties. These were popular ideas and TR was a great campaigner, even appearing, on one campaign photograph, to be crossing a river on the back of his favorite bull moose. He beat Taft easily by 4,119,582 to 3,485,082 and got 88 electoral college votes against a mere 8 for Taft. But in effect he had split the huge majority Republican vote into two, thus allowing the minority party to get home with 6,293,120 votes. This was considerably fewer than Bryan got in 1908 but with the Republicans split and losing state after Northern state, the Democrats collected a massive total of 435 electoral college votes. So their candidate, Woodrow Wilson, was now president and a new time had come.[225]

PART SIX

'The First International Nation'

Melting-Pot America, 1912–1929

The administration of Woodrow Wilson (1856–1924) is one of the great watersheds of American history. Until this time, America had concentrated almost exclusively on developing its immense natural resources by means of a self-creating and self-recruiting meritocracy. Americans enjoyed a laissez-faire society which was by no means unrestrained but whose limitations to their economic freedom were imposed by their belief in a God-ordained moral code rather than a government one devised by man. The rise of rural Populism, the development of muckraking, the appearance in the big cities of middle-class Progressivism, and, not least, the romantic reformism and altruistic nationalism of Theodore Roosevelt—all these were premonitory symptoms of change. And under Wilson the changes actually began to take place, hastened and accelerated by America's fortuitous involvement in a catastrophic world war which destroyed Old Europe for ever.[1]

Wilson came of Scots-Irish Calvinist stock on both sides of his family. His forebears struck deep roots in the South and, strictly speaking, Wilson was a Virginian, with all that is implied in the title. The South ran in his blood and creaked in his bones and sometimes—not often—went to his head.[2] But by training and temperament and self-formation he was very much a British-American. He once let slip that he wished to pursue Jeffersonian ends by Hamiltonian means—a revealing remark. The statesman he most admired was the great reforming Liberal, William Ewart Gladstone, himself a Liverpool-Scot of Calvinist ancestry. His intellectual mentor was the worldly-wise English banker–editor Walter Bagehot, who ran the *Economist* for many years and wrote the pellucid but laid-back prose Wilson sought to imitate. Wilson's Calvinism went deep. As a youth he experienced a characteristic 'awakening,' believing himself one of the elect. He retained throughout his life what he termed 'faith, pure and simple,' and an accompanying conviction that he was chosen to lead, to teach, and to inspire. He told a White House visitor in 1915: 'My life would not be worth living were it not for the driving power of religion.'[3] Religious certitudes undoubtedly helped to bolster his political certitudes, not to say the confident self-righteousness with which he advanced his aims. The anticlerical and cynical French elite later dubbed him a 'lay pope.' As Disraeli said of Gladstone, it was typical of him 'not merely to keep aces up his sleeve but to insist God put them there,' and similarly Wilson, who played hardball politics always when he felt he needed to, insinuated that he did so not by personal choice but at the urgent direction of providence. No one, not even Lincoln, used the quasi-religious rhetoric of the grand American tradition more effectively, or succeeded so often in conveying the impression that to oppose his policies was not merely unreasonable but downright immoral.

Yet Wilson's family was also notable for teaching, wide reading, and nonconformity, as well as Calvinism, and there were persistent liberal elements in his makeup. He remained too much of a Southerner to do anything for blacks—quite the contrary—but he was almost totally without religious prejudice. Joseph Patrick Tumulty (1879–1960), who was his secretary as both governor and president, and perhaps his closest advisor, was a devout Roman Catholic, which raised sharp eyebrows at the time. And Wilson forced through, against fierce opposition, the appointment of the Boston lawyer Louis Brandeis as the first Jew on the Supreme Court in 1916. Brandeis, for his part, noted after their first meeting in 1912 that Wilson 'has all the qualities for an ideal president—strong, simple and truthful, able, open-minded, eager to learn and deliberate.'[4]

This tribute is worth setting against the usual image of Wilson as inflexible and arrogant. But the truth is, there were many Wilsons, just as there were many Jeffersons. He is not exactly elusive but he is Janus-faced and protean. No more complex personality ever ruled the White House. To begin with, he was not, as might be supposed from his mature career, an example of the relentless drive which his contemporary Max Weber had just (1905–6) described as the 'Protestant Ethic,' springing from the 'Salvation Panic.' Wilson did not even learn to read until he was nine. He may have been dyslexic. More likely he was lazy and unmotivated. Upbraiding letters from his father suggest he long remained reluctant to work hard and, in particular, to become a successful lawyer as Wilson Sr always wished. What he wanted to do, as he eventually discovered almost by accident, was to teach and write, above all to teach and write about the workings of government.[5]

Once Wilson had discovered this calling, and had overcome his father's opposition, his career took off and he worked with staggering dedication. He came to academic life at exactly the right moment. The

American university was coming of age and had entered a period of unprecedented expansion and improvement. At Harvard Charles William Eliot (1834–1926), a mathematician and chemist, who entered his forty-year tenure as president of the college in 1869, transformed the university, widening and adding to its courses, founding graduate schools, establishing exchange professorships with France and Germany, developing an 'elective system' of undergraduate courses in which students played a part in their own curricula, ending sectarianism in the divinity school and introducing professionalism in law and medicine and, not least, founding Radcliffe (1879) as an offshoot for women. Seth Low (1850–1916), a former mayor of Brooklyn and future mayor of New York, turned Columbia, where he was president 1890–1901, into a massive engine of scholarship. At Johns Hopkins, where Wilson taught, Daniel Coit Gilman (1831–1908), having revolutionized scientific teaching at Yale, introduced as its first president the best traditions of German scholarship and built up one of the world's finest graduate schools. Such women's colleges as Bryn Mawr, founded (1880) near Philadelphia as the first non-sectarian place of higher education for women, were rapidly training women for virtually all the professions. Wilson taught there too and proved himself admirably qualified to bring women not only into the circle of academia but into extensive areas of active working life hitherto closed to them.[6]

Wilson was not content with teaching: he wrote prodigiously. The growth of the universities themselves, and the variety of subjects taught there, produced a voracious demand for textbooks, specialist publishers, and expert compilers of handbooks. It would not be quite true to say that Wilson invented Politics as a subject. But he rendered it fashionable and supplied it with much of its working material. He made a name for himself with an expanded PhD thesis (one of the first to employ this career-launch catapult effectively) called *Congressional Government* (1885), still in print nearly 120 years later. He followed it with a number of highly regarded and much reprinted standard works, such as *The State: Elements of Historical and Practical Politics* (1889), a five-part *History of the American People* (1902), and *Constitutional Government in the United States* (1908). To a great extent he became the American Bagehot.[7]

During these years Wilson came to the conclusion that America's system of government, though the best in the world, could be further improved. One way of doing it, he concluded, was by modernizing and purifying the American academy, and by making it into a national seminar for training idealistic young men who wished to work in the public service. When he joined the faculty at Princeton, of which he was a devoted alumnus, in 1890, he gradually came to believe that fundamental changes were needed, and the conviction led him to seek power there. It was an old-fashioned New Jersey institution noted chiefly for the training of Presbyterian clergymen when Wilson became its first lay president in 1902. His vigorous attempts to transform it into America's—and the world's—greatest university, by no means unsuccessful, were ultimately frustrated by what he saw as the malign exercise of the power of money. His grandiose plans met with opposition; and his opponents learned to entice wealthy alumni into providing huge conditional endowments skillfully designed to make Wilson's philosophy of education ineffective.[8]

Indeed it is one of the paradoxes of American college life that at the time of, or shortly after, its greatest period of consolidation and elevation, it developed customs and integuments which made it also peculiarly resistant to moral and civic improvement—or, rather, resistant to changes not sanctioned by the ruling class of which it was the training ground. In 1890 only about 3 percent of the appropriate age-group went to college, one in five of whom went to what a contemporary called the fourteen 'great American universities,' the inner group of which consisted of the Ivy League, who played American football against each other. The Ivy League ascendancy was underpinned by the leading prep schools, or independent private boarding schools, usually denominational, which acted as the equivalent of the English 'public school' system. Henry James, in denouncing America for its lack of social density and interest, chose to ignore these layers of class and culture, though they already existed during his time.

St Paul's School, in Concord, New Hampshire, was opened in 1856, when James was thirteen, and was among the larger ones, with 400 boys. It was Episcopalian and so Anglophile that for many years it played cricket rather than baseball. Its headmaster, Dr Samuel Drury, forced all the boys to learn the Sermon on the Mount and its atmosphere was strict: Charles 'Chips' Bohlen, later a leading American diplomat, was accused of 'bad attitude' and expelled for inflating a condom and playing football with it. Smaller (200 boys) and later (1884), Groton was also High Church Episcopalian. For sixty years its headmaster, called the rector, was Endicott Peabody, who had been educated in England at Cheltenham College and Trinity, Cambridge, and who wore a white bowtie (as well as cap and gown) like an Eton master. One of its founders was J. P. Morgan and it was particularly well endowed. Its declared purpose was 'to develop a manly Christian character.' Its motto was *Cui Servire Est Regnare*. Boys attended chapel every day (twice on Sunday). They were allowed only 25 cents a week pocket-money, of which 5 cents had to go into the collection plate. On the other hand their shoes were polished overnight and there was no bed-making or table-waiting. George Biddle, its 104th graduate, wrote: '95 per cent of the boys came from what they considered the aristocracy of America. Their fathers belonged to the Somerset, the Knickerbocker, the Philadelphia or the Baltimore Club. Among them was a goodly slice of the wealth of the nation.' The school's first 1,000 boys included one president, two secretaries of state, two governors, three senators, and nine ambassadors, though most alumni went into Wall Street. The President was, of course, Theodore Roosevelt and, when it celebrated its first twenty years, he declared to the boys: 'Much has been given you. Therefore we have a right to expect much from you.' TR denounced E. H. Harriman as 'an Enemy of the Republic' and 'a malefactor of great wealth,' perhaps unaware that he had become a notable benefactor of Groton, his son Averell Harriman becoming, in turn, one of its more celebrated alumni—this being precisely the kind of paradox in which Henry James delighted.[9] Besides sons of the rich, Groton liked to take in the offspring of the respectable clergy. Thus Dean Acheson, a future secretary of state, whose father was rector of the Episcopalian Church at Middletown, Connecticut, was there alongside Averell Harriman. There were many 'egalitarian' customs. When Cass Canfield, the future publisher, arrived at Groton, a boy said: 'So you're the new kid,' and punched him in the face. This was known as 'pumping,' and also included being held upside down and drenched. Acheson and Harriman were both 'pumped.'[10]

Other key prep schools were the Hill School, Philadelphia, which was distinguished academically and where all the boys learned Latin and Greek, and English, as opposed to American, history, and the Peddie School, near Princeton, New Jersey, which specialized in Greek history, language, and philosophy. These schools fed all the Ivy League colleges. St Paul's and Groton boys (the latter known as Grotties) tended to go to either Harvard, Yale, or Princeton. In 1906–32, for instance, 405 Groton boys applied to Harvard; only three were rejected. Until the 1960s, the great majority of Yale students came from prep schools and 20 percent of them were the sons of Yale men.

These Ivy League colleges were self-contained and tended to see their own university as all-important. The Yale football coach used to say: 'Gentlemen, today you play Harvard. Never again in your life will you do anything so important.' Yale had its own term for courage/determination—sand. These colleges had internal class-distinctions based on their messing and lodging arrangements. At Princeton the top club was the Ivy, which took in only eleven men a year. The future diplomat George Kennan, who came to Princeton from a military academy, quickly learned 'Princeton manners.' On his first day he asked another student the time. The student puffed on his cigar, blew the smoke in Kennan's face, and walked away. Kennan got into a second-grade club, the Key and Seal, and apologized to his father for the fee. Then his money ran out and he was forced to eat, as he put it, 'among the non-club pariahs' in Upperclass Commons. At Yale, the best club was the Skull and Bones, which took in only fifteen seniors. Henry Stimson, President Taft, Henry Luce, Justice Potter Stewart, and the Bundy clan were among its members. To be 'tapped for Bones' was the greatest honor Yale had to offer, and Tap Day was a red-letter event in the Yale calendar. The happy man was told: 'Skull and Bones. Go to your room!' It had its own rituals in its inner sanctum at '322,' and members spent two evenings a week exploring each other's characters in what was essentially a social therapy-group. They never spoke of its affairs or even admitted they were members. Averell Harriman had three wives without ever mentioning the word 'Bones' to any of them. Just beneath Bones came the Scroll and Key: Dean Acheson was in it, along with Cole Porter. Below came the Turtles, the Grill Room Grizzlies, the Hogans, the Mohicans, and the DKW.[11]

At Harvard there was the Gold Coast, whose clubs existed mainly for boys from church prep schools. It was not essential to join a club: Walter Lippmann and John Reed belonged to none but still made their names at Harvard, as well as in journalism in later life. You joined a 'preliminary club' like the Hasty Pudding; then a 'waiting club' like the Sphinx-Kalumet; then a 'final club.' The Porcellian, founded in 1791, was the most difficult to get into. Franklin Delano Roosevelt, a notable Grottie, said that rejection by 'the Porc' was the bitterest blow of his life. Unlike Bones, election was not recognition of performance: as Lord Melbourne said of the Order of the Garter, 'Thank God, there is no damned merit about it.' Paul Nitze, another high performer in public life, observed: 'The club *prided itself* on not being based on merit in any way'—just money, blood, and, perhaps most important, charm, like 'Pop' at Eton. These institutions and connections were important insofar as they had consequences in public life later, since they bred loyalties and intimacies which meant that one man was chosen for high office when another was overlooked.[12] They also acted as conservative, breaking forces in university reform, since school, college, and club allegiances had an enormous effect on donations by wealthy alumni, the way in which the money was handed over and for what purpose—and the freedom of action of college presidents as a result.

It was the abuse, as Wilson saw it, of money power in the etiolated atmosphere of the senior common room, the trustee boardroom, and, ultimately, the campus itself which turned him into a liberal reformer not just in academia but in a much wider field. The tremendous internal rows at Princeton, leading to his resignation in 1910, caused Wilson to stop just studying politics and become an active performer in the game—and an astonishingly masterful one too. Again, he chose his moment well. The return of Roosevelt to active politics in 1910 introduced, as we have noted, a period of party confusion from which the Democrats in general, and Wilson in particular, profited. Within three years, this austere-seeming Presbyterian college president, who had compiled *Congressional Government* without once having set foot in the Capitol, was installed—via a spell as governor of New Jersey—in the White House. In state politics, too, Wilson took advantage of a trend. The City bosses had been all-powerful since the Civil War, but, in the year before Wilson emerged, bosses like Richard Croker at Tammany Hall, Abe Ruef in San Francisco, T. C. Platt in New York State, and Matthew Stanley Quay in Philadelphia had been stripped of their feathers. New Jersey politics were particularly squalid, but the Democratic bosses there were now running scared and they thought that a distinguished and seemingly incorruptible college president like Wilson would give them camouflage and a bit of class. So they made him governor. But the man they imagined would be their high-minded puppet soon made himself their absolute master, displaying in the process a skill at intrigue, maneuver, and elevated skulduggery—and a nice sense of balance between idealism and *realpolitik*—which left them bitter but powerless.[13]

Wilson's success as governor of New Jersey, the prestige he secured for himself as the installer of an honest regime in a notoriously corrupt state, made him a surprise top-runner for the Democratic presidential nomination in 1912. William Jennings Bryan was still only fifty-two and retained a huge following among the farmers and Westerners, but he accepted the old baseball adage, 'Three strikes and you're out,' and performed a last great service to the Democratic Party by helping to secure Wilson's nomination. In terms of registered voters, the Republicans were still by far the larger organization and must have beaten Wilson had they remained united, the Democrats themselves losing many votes to the socialist Eugene Debs, who scored nearly 900,000. As it was, Wilson scored only 41.8 percent of the votes cast, the lowest percentage for an elected president since Lincoln's 39.9 in 1860. Thus does providence intervene: for the second time in its history, the United States got itself a great president because the ruling party split. Nevertheless, Wilson fought a notable campaign and enormously impressed those who heard him. Though he had been in politics only three years, and never sat in Congress, he quickly discovered that his lecture-room skills served him well for platform oratory. In the days before amplification, his fine voice and admirable, often spontaneous, choice of words could hold audiences of up to 35,000 spellbound.[14] Moreover, as a Southerner of liberal views with a Northern power-base in an urban state, he was able to construct, for the first time, the classic coalition of Southern conservatives and Northern and Western progressives that was to remain the Democratic mainstay till the end of the 1960s.

Wilson's arrival in the White House in 1913 was a perfect instance of Victor Hugo's saying, 'Nothing is more powerful than an idea whose time has come.' Since the Civil War, the United States had become by far the world's richest country, with an industrial economy which made all others on earth seem small, and it had done so very largely through the uncoordinated efforts of thousands of individual entrepreneurs. The feeling had grown that it was time for the community as a whole, using the resources of the United States Constitution, to impose a little order on this new giant and to dress him in suitable clothes, labeled 'The Public Interest.' Theodore Roosevelt had already laid out some of these clothes, and Wilson was happy to steal them. And under the despised President Taft, Congress had been encouraged by the White House to put through two constitutional amendments, Sixteen and Seventeen, both of which were ratified and became law in 1913. The first authorized a federal income tax, and thus placed in the hands of Washington a fiduciary power which was to be used with ruthless and cumulatively overwhelming impact over the next eighty years. The second democratized the Senate by stipulating that senators must be elected directly by the people instead of indirectly by state legislatures.

This was part of the process whereby institutional changes such as the primary were removing power from the party machines and the bosses and turning it over to the voters.[15]

These changes in the Constitution, and the fact that both the House and the Senate were now in Democratic hands, cleared the decks for one of the most comprehensive legislative programs ever enacted by a single US administration. Wilson had to pay one or two political debts by, for instance, making Bryan his Secretary of State. He also found the politicians adamantly opposed to Cabinet rank for one or two of his personal friends. Thus Walter Hines Page (1855–1918), Wilson's fellow-student and radical educational reformer, had to be content with the ambassadorship in London, and Brandeis, whom Wilson wanted to make attorney-general, was told that it would lead to what the *Boston Journal* called 'a general collapse' in banking and trust circles. All the same, Wilson formed a strong administration, much to his liking. The key figure was the Treasury Secretary, William Gibbs McAdoo (1863–1941), who had managed Wilson's campaign and was soon to marry one of his daughters. Another important appointment was Josephus Daniels (1862–1948), the Southern newspaper editor who had run Wilson's publicity, now made navy secretary. Like McAdoo he proved a first-class administrator and the choice was significant, not least because he immediately proposed as his assistant secretary the young Franklin Delano Roosevelt (1882–1945), who being a Hyde Park Roosevelt was a Democrat. Wilson greeted the proposal with one word: 'Capital,' thus establishing on the ladder of promotion one of the most powerful figures in American history. Wilson accepted, being a good student of history, the need to balance interests. Thus he put two political choices, James C. MacReynolds and Albert S. Burleson, into the posts of attorney-general and postmaster-general. This enabled him to make one progressive, Franklin K. Lane, secretary of the interior, and another, Lindley M. Garrison, secretary of war. A third liberal, William Cox Redfield, took up the post of commerce secretary, and for labor secretary Wilson turned to his namesake, William B. Wilson, a congressman who had helped to organize the United Mineworkers of America.[16]

Wilson also had a kitchen cabinet, presided over by Tumulty but whose most important member, Colonel Edward Mandell House (1858–1938), was only a semi-official advisor. House was a Texan who had pushed through Wilson's nomination and was now a big financial wheel. Oddly enough, in 1911 he had published a political novel, *Philip Dru: Administrator*, in which a benevolent dictator imposed a corporate income tax, abolished the protective tariff, and broke up the 'credit trust'—a remarkable adumbration of Wilson and his first term. House would come from New York to sit up with the President late into the night, plotting and arguing. Another, increasingly important member was a young ship's doctor, Captain Cary T. Grayson (1878-1938), whom Wilson made his medical attendant, and much more.[17] Wilson was the first president to establish regular press conferences, holding them twice a week in the East Room. They were attended by hundreds of reporters, but the rules Wilson imposed about quoting him, and the evasiveness of his actual replies, rather defeated their purpose.

Nevertheless, no administration in American history got off to a better start than Wilson's. The enactment of his program involved a revolution in thinking, not only on the part of the Democratic Party but among the progressive intelligentsia generally. It had been long in coming but under Wilson it actually took place in a few short years. In the early 19th century, and for long afterwards, the radical, democratic forces in American society (and not only in America: the same pattern is discernible in Britain) tried to limit the role of government. To Jefferson, and indeed to Jackson, big, heavy-handed government was associated with reactionary forces, with kings and emperors and the federalists and, later, with Wall Street. Heavy taxation, especially through such devices as personal income tax, was a conspiracy to steal money from the hard-working population and squander it among the officeholding elites. A central bank was a mechanism to confer privileges on a banking plutocracy. For Washington to acquire power was merely to take it away from the people and place it in the greedy hands of undemocratic elites. To some extent this view survived the Civil War, in which huge powers were acquired by a Republican federal government to destroy states' rights. What it did not survive was the rapid growth of Big Business and corporate power in the decades which followed. Gradually, the progressive intelligentsia, and the bulk of the Democratic Party, began to see a strong federal government, with wide powers of intervention, as the defender of the ordinary man and woman against the excesses of corporate power. The notion of the Public Sector (good; needs to be expanded) as opposed to the Private Sector (potentially bad; needs to be invigilated and regulated) began to take possession of the minds of the do-gooders. For this purpose, it was necessary for the state to expand its revenues. Therefore a personal income tax, especially if it possessed progressive characteristics, and therefore was income-redistributive as well as revenue-raising, was a desirable institution. At the end of the 20th century we have come to regard the state as, at best, a necessary evil, the only means whereby certain needful tasks can be accomplished, and at worst as an unrivaled oppressor. We have to cast our minds back to the intellectual atmosphere of 1913, when the state, not only in the United States but in many other countries, was seen as a knight in shining armor, coming to the rescue of the poor and the weak and the victimized, and doing with objective benevolence what otherwise would be done selfishly by greedy aggregations of private wealth.

It was Wilson who first introduced America to big, benevolent government. We must not exaggerate the extent of the revolution he carried through. Nor must we think that all his actions, and Congress's, expanded government power. In his first year in office, for instance, the Underwood Tariff Act reversed the protectionist, Republican trend of the last sixty years, removing duties designed to protect industrial vested interests, and thus reducing prices for the many. But, the same year, the Federal Reserve system was created to bring order into the US money market and to provide a mechanism for controlling credit centrally and managing crises when they occurred—or, better still, insuring they did not occur. For the Democrats to bring in a reserve bank was a reversal of everything Andrew Jackson had stood for. But by 1913 virtually everyone agreed it was necessary. Even J. P. Morgan, perhaps one should say especially J. P. Morgan, thought it was preferable that a federal institution should conduct the kind of crisis-management he had performed in his library, more particularly since no one ever again acquired quite the same moral authority in Wall Street that Morgan possessed in 1907. So the Fed came into existence and has been there ever since. But it must be said it took a long time to bring perfect order into America's complex, many-tiered paper-money system. A Treasury report compiled as late as 1942 showed that nine types of officially approved money were then still in circulation—gold certificates, silver dollars, silver certificates, Treasury Notes of 1890, subsidiary metal coins, United States banknotes or greenbacks, national banknotes, Federal Reserve banknotes, and Federal Reserve notes. Creating the Fed was merely the beginning of a long, distracted story of trying to prevent the world's largest economy from spinning itself out of control.[18]

In the following year, 1914, the Wilson administration, working with the Democratic Congress, established the Federal Trade Commission, a non-partisan body of five members appointed by the President for seven-year terms. The FTC was empowered to demand annual reports from corporations and to investigate business practices. It was given a statutory duty to investigate and control monopolistic practices, to prevent adulteration or mislabeling, to frustrate the emergence of combinations formed to fix retail prices or maintain them, and to expose the false claims of patents. It was designed to be preventative rather than punitive and was endowed with the right to issue 'cease and desist' orders to errant firms. By 1920 the FTC had issued nearly 400 such orders and had proved itself (or so it was said) a success. Its work was complemented, also in 1914, by another statute, the Clayton Anti-Trust Act, which filled holes in the Sherman Act and also exempted unions from some anti-trust provisions, thus leading Samuel Gompers to hail it as 'labor's charter of freedom.' These two Acts are often said to have 'brought the era of the Robber Barons to an end,' but the truth is it was ending anyway or had indeed ended, as the sheer size of the American economy, and the degree of natural competition, made cornering and monopolies and oligopolies increasingly difficult to establish, or at least to maintain for long.[19] But the Clayton Act certainly made competition easier and the FTC, besides hunting down the obvious economic malefactor, reassured the public that the financial world now had a sheriff to watch over it, and that the economy was policed, after a fashion.

The Wilson administration did nor forget the farmers or industrial workers either. The Federal Farm Loan Act (1916), which created cheap agricultural credits, and the Adamson Act, which introduced the eight-hour day, were the payoff for these two vital groups in the Wilson–Democratic constituency. The eight-hour day was designed to avert a railroad strike, and was supposed to apply chiefly to railroad workers. But it became a benchmark for all industry. Be it noted, however, that Wilson did not even try to introduce the welfare state, in the primitive version, chiefly consisting of old-age pensions, which David Lloyd George and Winston Churchill were then constructing in Britain. That had to wait until the 1930s, or even later. But the federal government quickly took advantage of the Sixteenth Amendment to levy income tax, which in 1913 was established at 1 percent on taxable net income above $3,000 ($4,000 for married couples), rising slowly to a top rate of 7 percent on incomes above $500,000. This was denounced by Wall Street as class legislation, and in a sense it was. Moreover, the onset of World War One showed the awesome possibilities of the new tax, with rates rising almost vertically to 77 percent, something completely new in American history.

Indeed, it was the impact of the war, even more than Wilson's prewar legislative and administrative program, which helped to build up the great historic watershed in the way America is governed. And the man common to both phases, apart from Wilson himself, was McAdoo. Indeed it is probably right to see him as a key figure in 20th-century American history, who never quite got his deserts in the political arena at the time, or has received the historic accolade he earned.[20] William McAdoo was born in 1863, the year of the Gettysburg disaster, as it appeared in his hometown, Marietta, Georgia. He went to the University of Tennessee, became a lawyer, and soon branched out into company promotion, especially of street cars. As a Southerner, he found it useless to go to Wall Street in search of money, and when his venture failed for lack of cash, he transferred himself bodily to New York City. It was at this point he lengthened his name to William Gibbs McAdoo. Like Lincoln he was immensely tall (six feet two), lanky, ungainly, explosive, and voluble; but he now wore black suits and practiced silences. He proved himself expert at exploiting bankruptcy opportunities, especially in rail, and he built or rather finished the Hudson tubes, under the river, linking Manhattan and New Jersey. He rode the populist wave at a time (1911) when the housewives were demonstrating in the street against rising meat prices, when outrage had been provoked by the wicked Triangle Shirtwaist Factory fire, and when Cornelius Vanderbilt had incited hatred by saying 'The public be damned' (it was the shareholders *he* cared for). McAdoo was a 'public-be-pleased' man. His railroad listened to public complaints. He took the opportunity of linking various unpopular trends together to blame them all on Wall Street's monopoly of credit and make a Southern point. He hailed Wilson as the first Southerner to be elected since Zachary Taylor in 1849. It was a great day for McAdoo when Wilson addressed 1,200 members of the Southern Society, the New York network of Southern gents who met annually on Washington's birthday, in the ballroom of the Waldorf on December 17, 1912—the first Southern President to speak to them since their foundation in 1885. But it was the South with a difference. The Society made a point of cheering the name of Lincoln, whom McAdoo called 'the greatest man God ever created.' Wilson's slogan was 'Sectionalism is dead!' Both men argued that Sherman had done the South a favor by destroying plantation society and opening the way for capitalist development—the point was valid though it did not really begin to produce results until the 1950s.[21]

McAdoo has been described as 'the greatest American Treasury Secretary since Alexander Hamilton,' and he was a true federalist in the sense he shifted the monetary system from Wall Street to Washington. He studied the investigation of the House Committee on Banking and Currency, led by Arsene Pujo of Louisiana, and in particular its interrogation of J. P. Morgan himself. The committee worked out that a group of New York banks held 341 seats on the boards of 112 corporations worth over $22 billion. Morgan assured the committee that his criterion in judging creditworthiness was not wealth but character—'A man I do not trust could not get money [from the banks] on all the bonds in Christendom.' In effect Morgan had been the central banker. But what McAdoo did, in creating the Fed, was not merely to make a geographical shift from New York to Washington but to decentralize the system, with twelve regional banks headed by twelve executives nominally independent of both the Treasury and the bankers. It was, as he put it, 'A blow in the solar-plexus of the money monopoly.'[22]

The passing of the watershed towards a strong, populist state was, therefore, very much a Southern operation. Even the income tax was the work of a Tennessee politician, Cordell Hull (1871–1955), later F. D. Roosevelt's Secretary of State. Without income tax the United States could not in practice have played an active role in international affairs, or begun to address the inequalities in American society. This Southern comeback might have been crowned by the selection of McAdoo to succeed Wilson in due course. But Wilson would never acknowledge him as his crown prince. This may have been because Wilson thought him unfitted for the highest office. McAdoo had some of the spirit of Lloyd George, with whom he was often compared, a skill in taking advantage of events as they unfolded to push his chances, and build up a personal bureaucratic empire as occasion offered. One colleague, Newton D. Baker, Secretary of War from 1916, said: 'McAdoo had the greatest lust for power I ever saw.' Charles Hamlins, Wilson's society friend, was harsher: 'The most selfish man I ever met.' The *Nation* described him as 'a thorough imperialist, eager for a big army and navy and the mailed fist, a sort of American Curzon and Milner without European experience and international vision but with plenty of American vigor and push added.' The *Nation* concluded: 'His election to the White House would be an unqualified misfortune.'[23] But that was never seriously on the cards. Serving Wilson prevented McAdoo from openly campaigning or maneuvering for the presidency, and by the time Wilson was stricken, politics had moved on and it was too late. It would have been a different matter if Wilson had treated McAdoo as his heir apparent. But he was never prepared to do that.

There was indeed a streak of selfish egotism in Wilson, a self-regarding arrogance and smugness, masquerading as righteousness, which was always there and which grew with the exercise of power. Wilson, the good and great, was corrupted by power, and the more he had of it the deeper the corruption bit, like acid in his soul. But there are, as always with Wilson, complexities. There was a hedonistic streak too, a bit of Southern dash. August Heckscher, Wilson's most recent biographer, has traced the stages whereby the original Thomas Wilson, known universally as Tommy, transmogrified himself first into Thomas W. Wilson, then T. Woodrow Wilson and finally the Jovian deity, Woodrow Wilson. The mature Wilson, stern, aloof, almost awesome, high-principled, incorruptible, and Olympian, was to some extent a construct, emerging from an earlier and more meretricious figure, not afraid of being thought dressy. Heckscher has unearthed a memorandum in which the young Wilson itemized his wardrobe, listing 103 articles, including many pairs of spats, pearl-colored trousers, and a blue vest. This earlier Wilson was boisterous, joked, sang songs, and told stories brilliantly. Until his second term, Wilson retained this last gift: along with Lincoln and Reagan, he was the President who used the apt and funny tale to most effect.[24]

Wilson was also fond of women, highly sexed, even passionate, and capable of penning memorable love-letters. His first wife, Ellen, was a proto-feminist, and their marriage was a grand love-affair. But it did not prevent Wilson striking up, in due course, an acquaintance with a frisky widow, whom he met in his favorite vacation-haunt, Bermuda. This developed into a liaison, which led in time to a bit of genteel blackmail. Ellen's death was nonetheless a bitter blow. But Wilson soon recovered and found a second great love, another merry widow, the forty-two-year-old Edith Bolling Galt, like Ellen an emancipated woman, who owned Washington's most fashionable jewelry store and was famous for being the first woman in the city to drive her own car. She was tall, Junoesque, and 'somewhat plump by modern American standards,' as one of the President's secret-servicemen put it. She became Wilson's mistress. A young attaché at the British embassy told the story that 'When the President proposed to Mrs Galt, she was so surprised she fell out of bed.' This tale reached the ears of the White House and a complaint was made to the British ambassador, Sir Cecil Spring-Rice.[25] Be that as it may, when Wilson *did* propose, and secured the consent of this statuesque lady, he was described by an associate as jigging dance-steps on the sidewalk and singing the current vaudeville hit, 'Oh, you beautiful doll, you great big beautiful doll.'[26]

This light-hearted Wilson, however, disappeared, never to return, when the immense and horrific conflict in Europe eventually engulfed the United States too. The Great War of 1914–18 was the primal tragedy of modern world civilization, the main reason why the 20th century turned into such a disastrous epoch for mankind. The United States became a great power in the decades after the Civil War, and even an imperial power in 1898. But it was not yet a world power in the sense that it regularly conferred with the major European states, known as 'the powers,' and took part in their diplomatic arrangements. America was not an isolated power—was never at any time, it can be argued, isolationist—and had always had global dealings from its Republican inception. But it had prudently kept clear of the internal wrangles of Europe (as had Britain until 1903), and held itself well aloof from both the Entente Cordiale of Britain and France, with its links, through France, with Tsarist Russia, an anti-semitic autocracy held in abhorrence by most Americans, and with the militaristic, Teutonic, and to some extent racist alliance of the Central European powers, the German and Austro-Hungarian empires. Thus the United States was a mere spectator during the frantic events which followed the murder of the Archduke Franz-Ferdinand of Austria on June 28, 1914, the Austrian ultimatum to Serbia, the Russian decision to support the Serbs, the French decision to support Russia, the German decision to support Austria and fight a two-front war against Russia and France, and Germany's consequential decision to send its armies through Belgium to enforce quick defeat on the French, and so the involvement of Britain and its dominion allies in support of Belgium. No one in Europe took any notice of Wilson's offer to mediate.[27]

The United States had no quarrel with Germany: quite the contrary. As early as 1785 it had negotiated a commercial treaty with Prussia, at which time immigrants of German origin already constituted 9 percent of the population. The Prussians backed the Union during the Civil War and the United States looked with approval on the emergence of a united Germany. But from the 1870s on, economic, commercial, colonial, and even naval rivalry endangered German–American relations, especially in the Pacific. There was a sharp dispute over Samoa, which did not end until the territory was partitioned in 1899, and which left a legacy of suspicion. America began to dislike Germany for exactly the same reasons as the British did: the arrogance and naive pushiness with which the Germans, latecomers to global naval power and colonialism, sought their own 'place in the sun.' In 1898, for instance, the Americans interpreted the presence of a strong German naval squadron near the Philippines as evidence of designs upon the islands. In the growing Anglo-German antagonism of 1900–14, the Special Relationship of Britain and America operated powerfully in Britain's favor. It is significant that in 1902, when both Germany and Britain punished Venezuela for reneging on its debts, President Roosevelt's administration criticized Germany but not Britain for violating the Monroe Doctrine.[28]

All the same, there was at first no question of America entering the war. On August 4, immediately after it began, Wilson issued a proclamation of neutrality. Two weeks later he urged Americans to be 'impartial in thought as well as in action.' There was no doubt about the sincerity of his pacifism at this stage. He loved and admired Americans but he was also in awe of them. His studies of American history and institutions had led him to recognize that, in addition to the broad-minded, pragmatic, tolerant, prudent, and worldly-wise strain in the American political personality, there was also a utopian, intolerant, and fundamentalist streak which leaped at any opportunity to crusade and impose its creeds. It was that streak which had brought about the Civil War and then waged it with relentless ferocity. So he warned; 'Once lead this people into war and they'll forget there was ever such a thing as tolerance ... The spirit of ruthless brutality will enter into every fiber of our national life'.[29] This warning was echoed by Randolph Bourne (1886–1918), speaking on behalf of the radicals, Pacifists, Progressives, and left-wing Democrats, who rightly saw that war would enormously accelerate the already perceptible growth of Big Government: 'War is the health of the State.'[30]

However, though not physically part of Europe, the United States was the western seaboard of the oceanic continuum which made it an integral part of the North Atlantic community. It was the Atlantic which dragged the United States into the Bonapartist wars, as we have seen, and it was the Atlantic which eventually made the United States an unwilling belligerent in Europe's suicidal conflict. On declaring war against Germany, Britain immediately imposed a blockade on German commerce, which had inescapable consequences for the United States. By 1916 American commerce with Germany had fallen to less than 1 percent of its 1914 value. In the same period, American trade with Britain (and also with its allies France and, from 1915, Italy) more than tripled. While Britain had learned lessons from the War of 1812 and imposed commercial regulations in such a manner as to do the least possible harm to Anglo-American relations, German U-boat warfare had a catastrophic effect on America's view of Germany, at any rate at a popular level. On May 7, 1915 a German U-boat sank the British North Atlantic passenger liner *Lusitania*, without warning. It was an international crime without precedent or mitigating circumstance. Nearly 1,200 passengers drowned, 128 of them American.

In retrospect, this was a clear and adequate pretext for America entering the war, and thus shortening it—or even bringing it to a negotiated conclusion. But Wilson contented himself with securing German assurances that such atrocities would never be repeated Indeed, he fought the 1916 election on a neutrality platform against a formidable opponent in Charles Evans Hughes (1862–1948), a former New York governor who resigned as an associate justice of the Supreme Court to undertake the contest. Hughes had the backing of Theodore Roosevelt and a reunited Republican Party, and Wilson, while he expressed doubts about his ability to keep America out of the war, may have felt that anything less than a formal pledge to remain neutral would have cost him the contest. He even had to be careful in expressing pro-British sentiments. In April 1916 the Easter Rising in Dublin had aggravated relations between Britain and the Irish nationalists, and the American-Irish vote could have cost the Democrats the election. As it was, the vote was close. Wilson got 9,129,606 to Hughes' 8,538,221, a plurality of less than 600,000 (fewer than the Irish vote in New York alone), and the electoral college margin was only 277–254.[31]

Immediately after the election, however, the increasing desperation of the German government led it, on January 21, 1917, to declare unrestricted submarine warfare on all shipping, neutral or belligerent, destined for Britain, and this led almost immediately to an appalling slaughter of seamen and civilian passengers in the Atlantic sea-lanes. Wilson broke off diplomatic relations with Berlin but declined to ask Congress for a declaration of war unless there were what he called 'actual overt acts' against American citizens and property. These came in February–March, when U-boats sank a number of American ships.

At the same time, the press published a secret telegram from the German Foreign Minister Arthur Zimmerman to the Mexican government proposing a German–Mexican offensive alliance against the United States under which Texas and other territories would be handed back to Mexico. The crassness of this attempt to stir up trouble on America's southern border clinched the matter. On April 2 Wilson asked Congress for a declaration of war, and on the 6th Congress complied.[32]

Wilson never considered the United States as a co-belligerent, as that would have implied he was mistaken in not entering the war earlier. He called America an 'associated power,' and his own war-aims, set out as the Fourteen Points on January 8, 1918 in a desperate last-minute attempt to keep Russia in the war, were non-punitive and quite distinct from those of the Entente powers. Nonetheless, the Wilson administration's prosecution of the war was vigorous, not to say enthusiastic. Wilson had never been a man to eschew force when he believed, as he always did, he had moral justice on his side. He used American forces in the Caribbean and Central America more frequently than any other president, before or since. He invaded Mexico twice, once in 1914 when he sent US marines into Veracruz following a fracas which led to the arrest of American sailors, and again in 1916 when, on his orders, General John J. Pershing (1860–1948) chased the Mexican revolutionary leader 'Pancho' Villa back into his own country after he raided New Mexico. This was a serious business, involving many casualties and deep penetration into Mexico's territory, and US troops were not withdrawn until February 1917. Wilson was an aggressive man who, when crossed, was liable to turn vindictive. When a group of senators filibustered his proposal to arm US merchantmen, he issued a statement aimed at producing sensational headlines: 'A little group of willful men, representative of no opinion but their own, have rendered the great government of the United States helpless and contemptible.' The Irish-American leader, Jeremiah A. O'Leary, who accused him of being pro-British, got even rougher treatment: 'Your telegram received. I would be deeply mortified to have you or anybody like you vote for me. Since you have access to many disloyal Americans I will ask you to convey this message to them.'[33]

Wilson's war policy had four aspects, each in its own way ruthless. The first was propaganda. Of the 32 million Americans who were foreign-born or children of foreign parents, 10 million had ties with the Central Powers. Wilson formed the Committee on Public Information, headed by the journalist George Creel (1876–1953), who recruited 75,000 speakers, 'Four Minute Men,' to give short war-aims talks at theater intermissions and other opportune times, to distribute 100 million pamphlets in various languages, to make movies, such as *The Kaiser: Beast of Berlin*, and to hold exhibitions of 'frightfulness' committed by the 'Barbaric Huns.' 'Hamburger' was replaced by 'Liberty Sandwich,' sauerkraut by 'liberty cabbage.' Playing German music and teaching German were prohibited and many private manifestations against German culture were encouraged. Second, important changes in administration, involving new philosophies of government, were instituted, chiefly by and under McAdoo, who set up a personal bureaucratic empire under the name of the War Finance Corporation. This was a continuation of the prewar Wilson program by other means—war. Indeed, during this period, many federal government activities were set in motion, which went underground (or were discontinued) in the 1920s and then reemerged under Roosevelt's New Deal, to become a permanent part of the American system. Under McAdoo the cost of World War One to the federal government was ten times more than the cost of the Civil War and more than twice that of operating the federal government since its inception in 1789. The final direct cost to the US was about $112 billion, not counting £10 billion in US Treasury loans to Allied governments. As a consequence the US Internal Revenue Service became, for the first time, a serious factor in the lives of ordinary Americans.[34]

By the time the United States joined the war, the European powers were in the process of harnessing their entire economies and populations to the effort of winning it. This often involved the state taking over the management and even the ownership of whole industries. In Germany, which was the most efficient in getting down to this task, the new system was called (by the man who was, in effect, the military dictator of the country, General Ludendorff) 'War Socialism.' Its workings were so much admired by Vladimir Lenin that he made its structure and methods the basis for his own Sovietization of the entire Russian economy, after he seized power near the end of 1917. Britain and France followed the same road. So, to some extent, did the United States. McAdoo's War Finance Corporation, whose primary purpose was to help civilian firms to convert to war production, in fact took on many more tasks, and later became the model for the Reconstruction Finance Corporation, set up to combat the Depression by President Herbert Hoover (1874–1964). At the time, Hoover himself worked on the War Industries Board, a McAdoo body transformed from 1918 under 'the wizard of Wall Street,' the self-made financier Bernard Baruch (1870–1965), into a hugely successful boost to production. This in turn became the model for the National Recovery Administration under F. D. Roosevelt's New Deal. And the National War Labor Board, set up to prevent strikes which hindered war output, became the model for the National Labor Relations Act (the Wagner Act) of 1935.[35]

The war, as Wilson had predicted, brought a dramatic curtailment of national liberties and a large element of compulsion. This was the third important aspect of the Wilson war effort. The Espionage Act of June 15, 1917 (as amended by the Sedition Act of May 16, 1918) made it possible for the authorities to prosecute many pacifists and members of left-wing groups such as the IWW and the Socialist Party. The constitutionality of this draconian law was upheld by the Supreme Court in *Schenck* v. *United States* (1919). At the same time, the Selective Service Act of May 18, 1917, which was ruled constitutional in *Arver* v. *United States* (1918), obliged over 23,900,000 men to register for service. Of these, 2,800,000 were inducted (some 16 percent of these did not report for duty, or immediately deserted, and were prosecuted). The draft provided 53 percent of the army's troops and 45 percent of all military personnel, so that the American forces in World War One were about one-half conscript.[36]

The fourth aspect of Wilson's war policy was massive intervention in the conflict to insure the United States played the predominant role in restructuring the world when the Germans were finally overwhelmed. The US Army in January 1917 was 200,000 strong. By the end of the war, Wilson's efforts had expanded it to over 4 million, of whom 2 million served in the American Expeditionary Force and about 75 percent of them experienced actual fighting. The American First Army was formed in France in August 1917 and took over a section of the Western Front near Verdun, where the French forces were demoralized and mutinous. The provisional armistice signed by Lenin and Trotsky with the Germans at the beginning of 1918 meant that Germany could begin transferring divisions from its Eastern to its Western Front, and a devastating offensive was launched on March 21, 1918, which broke through the French and British armies, and enabled the advancing German forces to cross the Marne again. In the circumstances, the British agreed to serve under a French army supremo, Maréchal Foch, and General Pershing, whose orders instructed him to make the AEF 'a distinct and separate component' in the conflict, allowed army and marine units to join the line under Allied command. This helped to arrest the German advance, and in August 1918 the Allies went onto the offensive. Pershing was now able to deploy his formidable and rapidly growing resources as a separate command, both in the St-Mihiel Campaign, 12–16 September, and in the Meuse–Argonne Offensive launched on September 28, in which he committed no fewer than 1,200,000 troops to battle. American casualties, especially in this last encounter, were heavy, and by the time the Armistice was signed on November 11, 1918, US war-deaths totalled 112,432, the vast majority on the Western Front. But in terms of effectives, America now had one of the largest and most powerful armies in Europe, and could convincingly claim that it had played a determining role in ending Germany's ability to continue the war.[37]

Wilson did not regard himself as an international expert, and he told friends just before his inauguration in 1913, 'It would be an irony of fate if my administration had to deal chiefly with foreign affairs.'[38] When he was forced to enter the war, he prudently set up an organization called the Inquiry, in which 150 academic experts, working in the American Geographical Society Building in New York under the leadership of Colonel House and Dr S. E. Mezes, prepared in detail for the peacemaking. As a result, the American delegation was, throughout the peace-process at Versailles, by far the best informed and documented, and indeed was often the sole source of accurate information and up-to-date maps. The British diplomatic historian Harold Nicolson, who was present, commented: 'Had the Treaty of Peace been drafted solely by the American experts, it would have been one of the wisest as well as the most scientific documents ever devised.'[39]

Unfortunately, as the British Foreign Secretary, A. J. Balfour, noted: '[Wilson's style] is very inaccurate. He is a first rate rhetorician and a very bad draftsman.' Wilson's draft plan for peace, the Fourteen Points, was set down on paper by him in a hurry, in January 1918, to counter Lenin's Soviet propaganda that any peace treaty must be based 'on the self-determination of the peoples.' He did not consult Britain and France in forming them. The first five points concerned general principles of the international order. 'Open covenants, openly arrived at'—a characteristic Wilsonian flourish—should replace secret diplomacy (Lenin had just published the texts of all the secret treaties to which Russia had been a signatory). There should be freedom of the seas in peace and war. Barriers to international trade should be lifted. Armaments should be reduced by common agreement. Colonial claims should be adjusted, balancing the interests of the great powers and the aspirations of the subject peoples. The next eight points concerned territorial adjustments. Russia should get back its lost territory. Belgium should have its independence restored. Alsace-Lorraine should go back to France. Italy's frontier with Austria should be redrawn 'along clearly recognizable lines of nationality.' The peoples of Austria–Hungary should have the 'freest opportunity of autonomous development.' The Balkan frontiers should be rearranged 'along historically established lines of allegiance and nationality.' Turkey should keep its independence but the non-Turkish peoples of the Ottoman Empire should be allowed 'autonomous development.' International passage through the Dardanelles should be guaranteed, Serbia and Poland should be given access to the sea, and Poland should have its independence too. The final and fourteenth point called for the creation of 'a general association of nations' with power to guarantee each nation's sovereignty and independence. On February 11, 1918 Wilson produced his 'Four Principles,' which expanded the fourteenth point, and on September 27 what he called the 'Five Particulars,' the first of which promised justice to friends and enemies alike.[40]

The Germans still had an enormous army of 9 million men, which had destroyed Russian power on the Eastern Front and was conducting an orderly withdrawal on the Western. But Ludendorff was scared that, with the continuing American build-up, his army might face what he called a 'catastrophe,' and the Wilson parade of principles seemed to offer a chance for Germany to extricate itself from a lost war with its territory largely intact, except for Alsace-Lorraine. It was on this assumption that the Germans and Austrians agreed to talk to Wilson on October 4 and 7 respectively, and on November 5 the Germans were offered an armistice by Wilson on the basis of the Fourteen Points (as extended), subject only to compensation for war damage and a reserved British interpretation of the meaning of 'freedom of the Seas.' The Germans accepted, and the Armistice followed on November 11. What the Central Powers did not know was that on October 29 Colonel House had had a long secret meeting with the French and British leaders, Georges Clemenceau and David Lloyd George, at which they voiced their reservations about the Wilsonian code and had them accepted. These qualifications were drawn up by House in the form of a 'Commentary,' cabled to Wilson and approved by him. The

'Commentary' made all the difference. It effectively removed all the advantages to the Central Powers which the Fourteen Points seemed to offer and foreshadowed all the features of the subsequent Versailles Treaty to which they took the strongest objection—the dismemberment of Austro-Hungary, the loss of Germany's colonies, the break-up of Prussia by a 'Polish Corridor' to the sea, and the handing over to a 'big' Poland of the German industrial region of Silesia. Moreover, 'compensation' had become 'reparations' on an enormous scale, and the implication of all the terms was that Germany and Austria were atoning for their 'war-guilt.' German war-guilt, it might be argued, was implicit in the Fourteen Points, but the system of 'rewards' for the victors and 'punishments' for the vanquished which the 'Commentary' provided for had been specifically repudiated by Wilson in his twenty-three-point code.

The truth is, from the moment the United States entered the war, Wilson had been on a dramatic learning-curve in international affairs, and his views were changing all the time. There is nothing like war to shatter utopian illusions, turn idealism into *realpolitik*, and transform goodwill into bitterness. The huge American casualties incurred in September 1918 had left their mark on the Commander-in-Chief in the White House, and Wilson's growing anti-German feelings had been stiffened by his first experiences of negotiating with them from October onwards. He was particularly disgusted that on October 12, more than a week after the Germans asked him for an armistice, they sank the Irish civilian ferry *Leinster*, drowning 450 people, including many women and children. In addition, Wilson's preoccupations were shifting. He had originally accepted the British idea for a permanent international forum, or League of Nations, without enthusiasm. But the more he thought about it, the more it appealed to him. Indeed, he became obsessed with turning it into reality, as the formula for an eventual system of world democratic government, with America at its head. This had two consequences, both unfortunate. First, he persuaded himself that he ought to go to Europe in person, to act as midwife to this new child of reason and prince of peace. This meant that, instead of remaining in Washington as an Olympian deity remote from the wrangling at Versailles, to appear as a *deus ex machina* to impose judgment when the impasse was reached, he joined in the jostling just like any other head of government. And, as John Maynard Keynes, an acute observer at the conference, noted, Wilson was less impressive at close quarters. The Europeans were used to seeing him standing, speaking majestically from a podium. With his big head, he looked a bit odd seated at a conference table.

Second, Wilson was advised by his Secretary of State, Robert Lansing (1864–1928), a legal expert, that if the proposal for a League were inserted into the preliminary negotiating document, as Wilson wished, it would still constitute a treaty, under the US Constitution, and would have to be ratified by Congress. Hence Wilson decided to go for a final treaty straight away. That meant the Germans, who had already been deceived over the underlying principles of the peace, were not allowed to take part in the negotiating process, which was carried out solely among the Allies. They were simply handed a *fait accompli*, and asked to sign it—or resume the war. Since the Allies were now occupying strategic parts of German territory, fighting was out of the question. So the Germans signed. But this 'Carthaginian Peace,' as Keynes called it, which the Germans saw as a swindle as well as an out-rageous injustice and an affront to their national dignity, determined them to seek rectification and revenge when opportunity offered, and a leader arose to seize it. Thus Versailles was the impulse behind Adolf Hitler's rise to power, the pretext for his aggressions, and the ultimate cause of World War Two.

It should be added that the American delegation was divided over the treaty. John Foster Dulles (1888–1959), a future secretary of state, thought it just on balance, bearing in mind the 'enormity of the crimes committed by Germany.' Colonel House was behind Wilson's scrap-ping of his own points. Wilson's chief advisor on eastern Europe, Robert H. Lord, was a leading advocate of a 'big' Poland, to Germany the worst single aspect of the treaty. But Lansing rightly recognized that the failure to allow the Germans to negotiate was a cardinal error and he considered Wilson had betrayed his principles in form and sub-stance. His criticisms were a prime reason for Wilson's brutal dismissal of him early in 1920. The younger Americans were particularly bitter. William Bullitt, a future ambassador to France and Russia, wrote Wilson a fierce letter: 'I am sorry that you did not fight our fight to the finish and that you had so little faith in the millions of men, like myself, in every nation, who had faith in you . . . Our government has con-sented now to deliver the suffering peoples of the world to new oppres-sions, subjections and dismemberments—a new century of war.' Other young men on the team, the historian Samuel Eliot Morison, the future Secretary of State Christian Herter, and Adolph Berle, later Assistant Secretary of State, also parted company with Wilson, and Walter

Lippmann, already a heavyweight pundit, wrote: 'In my view, the Treaty is not only illiberal and in bad faith, it is in the highest degree imprudent.' It is only fair to add that most of Wilson's distinguished team stood by him and defended him and the treaty then and later. From the perspective of three-quarters of a century, it does not seem Carthaginian. But, so far as Germany was concerned, it was certainly provocative and, since Germany retained the strongest economy in Europe, that was bound to cause trouble. German anger might not have been so significant if the League, with the United States as its head, had come into being as planned, to defend the settlement, if needs be by force. But, as a result of Wilson's further errors, American never joined it, and it was from the start 'a covenant without a sword.'[41]

The covenant was also ambiguous, bearing the hallmarks of Wilson's increasingly cloudy rhetoric. It was not the kind of international document the United States had ever signed before. It did not codify the law of nations or arrange for peace to be preserved by a system of mediation or arbitration. Wilson himself said that the 'heart of the covenant' was Article X, which asked League members to 'respect and preserve' the territories of all the nations which belonged to it. What exactly did that mean? Wilson compounded the ambiguity by insisting that the covenant was not a legal document but a moral one—and therefore all the more binding on its signatories. There is a historical myth that the European powers wanted the League at all costs, that Wilson tried to give it to them, and that an isolationist Congress refused to ratify it, thus bearing the long-term responsibility for another world war. There is no truth at all in this version. Clemenceau and Foch wanted a mutual security alliance, with its own planning staff, of the kind which had finally evolved at Allied HQ, after infinite pains and delays, in the final months of the war. In short they wanted something like the body which eventually appeared in 1948–9, in the shape of the North Atlantic Treaty Organization, and which has successfully kept the peace for the last half-century and survives to this day. They recognized that a universal system which involved all powers (including Germany) irrespective of their records, and which guaranteed all frontiers, irrespective of their merits, was nonsense. They knew there was small chance of Congress accepting any such monstrosity. Their aims were limited and they sought to involve America by stages, as earlier they had involved Britain, in mutual security arrangements. What they wanted America to accept, in the first place, was a guarantee of the treaty, rather than membership of any League.[42]

That was approximately the position of Senator Henry Cabot Lodge (1850–1924), chairman of the Senate Foreign Relations Committee. He is usually presented as the villain of the piece. That is unfair. Lodge was not an isolationist. He was a Boston Brahmin, and Boston Brahmins had always been, and still are, internationalists. His great mentors were Henry Adams and Theodore Roosevelt, both of them outspoken internationalists, and all the evidence shows that he aimed for the kind of guarantee or League of which they would have approved.[43] He thought that the covenant, though well meaning, was poorly drafted and too wide, and that members of the League would not in practice go to war to enforce the League's decisions since nations eschewed war except when their vital interests were at stake. How could frontiers be indefinitely guaranteed by anything or anybody? They reflected real and changing forces. Would the US go to war to protect Britain's frontiers in India or Japan's in Shantung? Of course not. And arrangements America made with Britain and France must be based upon the mutual accommodation of vital interests—then it would mean something. The American people, as Lodge appreciated, were not opposed to US participation in the League. Polls taken at the time showed that Americans wanted to join a permanent peacekeeping body by ratios of four or five to one.[44]

The Republican majority in the Senate favored a League of some kind. The true isolationists among the Republicans were not more than a dozen, and even some of those could have been won over. Lodge himself wanted the treaty to be ratified overwhelmingly. He wanted the League ratified too but not in its present form. Speaking with all the authority of a Harvard man who was the first ever to take a PhD there in political science, he said, 'It might get by at Princeton but not at Harvard.' That was an unwise remark, a good example of the principle that making jokes in politics is nearly always costly. Wilson was incensed at this *lese-majesté*. He did not like Lodge anyway. Lodge was a decent, clever, and well-meaning man but he was irritating. He had, as someone put it, 'a voice like the tearing of a sheet.' He put down 'Fourteen Reservations' and said that unless these changes—or something like them—were made he would not ask his Senate friends to vote for it. But Wilson refused any changes, despite the fact that the Europeans would have accepted Lodge's reservations and in some cases welcomed them. Having said the covenant was a moral document, Wilson regarded it as scripture on a par with the Old Testament, not to be altered by human hand, and certainly not by senatorial hand. In his intransigence, Wilson was isolated. Many of those associated with the peacekeeping process, like Colonel House and Herbert Hoover, favored the reservations; so did the Democratic leader, William Jennings Bryan. But Wilson determined to defy them all.

In November 1918, before the covenant was drawn up, Wilson had lost the mid-term elections, giving the Republicans control of Congress, including the Senate. The line-up there was twenty-three senators still controlled by Wilson and forty-nine by Lodge; the two groups together were quite enough to insure passage of the treaty and ratification of the League, given a bit of flexibility on Wilson's part. But he was rigid. He thought he would take the issue to the people. That was suicidal, given his growing state of ill-health. In April 1919 he suffered his first stroke, in Paris. It was concealed. In September, back in America, he began a speaking tour, to boost support for the League and bring pressure on the 'Strong Reservationists,' as Lodge's supporters were called. He traveled 8,000 miles by rail in three weeks, speaking constantly. This culminated in a second stroke in the train on September 25. Again there was a cover-up. On October 10 there was a third and incapacitating attack which left his entire left side paralyzed. His physician, Gary Grayson—recently promoted admiral—admitted later: 'He is permanently ill physically, is gradually weakening mentally, and can't recover.'[45] But Grayson refused to declare the President incompetent.

The Vice-President, Thomas Marshall (1854–1925), a former governor of Indiana but a hopelessly insecure man, best known for his remark 'What this country needs is a good five-cent cigar,' declined to challenge the President's inability to perform his duties, especially since this meant challenging the second Mrs Wilson, who guarded the President's sick quarters like a Valkyrie. In practice, Grayson, Tumulty, and Edith Wilson conspired to make her the President, which she remained for eighteen months. Rumors circulated that Wilson was stricken with tertiary syphilis, a raving prisoner in a barred room. Mrs Wilson, who had spent only two years at school, wrote orders in a large, childish hand, 'The President says . . . ,' sacked and appointed Cabinet ministers, and forged the President's signature on Bills. She, as much as Wilson, was responsible for Lansing's departure ('I hate Lansing,' she said) and his replacement by a completely inexperienced lawyer, Bainbridge Colby (1869–1950), who was running the Shipping Board.[46] Senator Albert Fall, who had publicly complained, 'We have petticoat government! Mrs Wilson is president!' was summoned to the White House to 'see for himself.' He found Wilson with a long, white beard but otherwise apparently alert (he could concentrate for five to ten minutes at a time). When Fall said, 'We, Mr President, have all been praying for you,' Wilson foxily replied, 'Which way, Senator?'—a remark interpreted as evidence of his continuing sharpness. Fall was with Wilson only a short time, and being an egoist did most of the talking. So Wilson 'passed muster' and the farce continued.[47]

Thus the great Wilson presidency ended in deception and failure. Wilson's twenty-three supporters, under strict instructions from the paralyzed titan, voted against the League rather than accept the Lodge reservations. The Reservationists voted against it as it stood. The treaty failed in a vote on November 19, 1919 and again, with two more votes, on March 19, 1920. Amazing as it may seem, Wilson hoped at one point to be nominated again and run for a third term. Frustrated in this, he let it be known that he considered the 1920 election to be 'a referendum on the League of Nations.' That compounded his earlier errors, for it suggested to the outside world that the result was a popular endorsement of the Senate's rejection of the Wilson covenant. It was nothing of the sort—the League was hardly an issue at all. The Democrats nominated James M. Cox, governor of Ohio (1870–1957), with Franklin Roosevelt as his running mate. They did not try very hard, and they certainly failed, to make foreign policy seem important. The Republicans nominated an old-fashioned conservative, Warren Harding (1865–1923), an Ohio senator, who was certainly against the Wilson League but was not an isolationist either. If the American people preferred him, it was because he was so tremendously unlike Wilson. He was the 'old America' before the Wilson watershed. His theme was 'Let us return to normalcy.'[48]

Harding was from Ohio, then the American heartland and certainly the Republican center of gravity, which had produced six out of ten presidents since 1865. He had risen from poverty to create a successful small-town paper, the *Marion Star*, and had then become director of a bank, a lumber company, a phone company, and a building society. He was decent, small-town America in person: handsome, genial, friendly to all but quite dignified enough to carry off the presidential role. He thought America was the most wonderful country in the world, in history indeed, and all he wished was to keep it like that. To get elected, he stuck President McKinley's old flagpole in his garden and ran a 'front-porch campaign.' Many famous people made the pilgrimage to Marion to listen to him talk—Al Jolson, Lillian Gish, Ethel Barrymore, Pearl White among them, but also 600,000 ordinary folk too, many of them black—hence a Democratic rumor that Harding had negro blood. Everybody liked Harding. The worst thing about him was his sharp-faced wife, Flossie, known as the 'Duchess,' of whom Harding said (not in her hearing), 'Mrs Harding wants to be the drum-major in every band that passes.' Harding always answered his own front door in person and was not above continuing the practice at the White House. Like many Americans still, he always took a horse-ride on Sunday after church. In May 1920 he told a cheering crowd in Boston: 'America's present need is not heroics but healing, not nostrums but normalcy, not revolution but restoration ... not surgery but serenity.'[49] The Americans liked this. They gave Harding the biggest plurality ever: 16,152,200 to 9,147,353, with an electoral college majority of 404 to 127. Debs, though in prison under Wilson's ferocious wartime legislation, got his 900,000 votes—but then he had them before. Apart from the South, 'solid' for historical reasons, Harding was everyone's choice, all over America, and 'normalcy' was his policy.

What was normalcy? America had moved on since Wilson began his revolution, faster than ever, and the war had accelerated the pace in countless different ways. For one thing, it brought women directly into the industrial workforce over virtually the whole range of occupations. In 1880 there were only 2.6 million employed women in the United States, overwhelmingly in domestic service, teaching, and nursing. Women had been fully qualified doctors since the 1850s, thanks to the pioneering of Elizabeth Blackwell (1821–1910), an Englishwoman who was the first woman in America to receive a medical degree, conferred on her by Geneva Medical College, New York. She and her sister, Dr Emily Blackwell, founded the New York Infirmary for Women and Children (1854), where they gave clinical experience for women aspiring to be doctors who had already graduated from the new schools which trained women: the Female Medical School of Philadelphia (1850) and the Boston Medical School for Women (1852).[50] Rather more slowly, women entered the law; in 1870 the Chicago Union College of Law conferred the first legal degree on a woman. By 1890 there were 4 million employed women, rising to 5.1 million in 1900 and 7.8 million in 1910, and by this date educational facilities for women were available in all the arts and sciences.

Political progress was not so fast. Whereas virtually all women, at least in theory, supported vocational training for women, surprisingly few 19th-century American women were prepared to agitate for the vote. They were much more enthusiastic members of temperance and abolitionist movements, and these were the two main channels through which women entered politics, albeit on single-issue platforms. As we have seen, women had had the vote in New York, 1776–1807, but had lost it. On July 19, 1848, inspired by the revolutionary ferment in Europe during the 'Year of Revolutions,' a gathering of women met at Seneca Falls, in New York, under the leadership of Lucretia Coffin Mott (1793–1880) and Elizabeth Cady Stanton (1815–1902). Mott was a Quaker preacher and abolitionist, Stanton an abolitionist and a temperance preacher. The women present enthusiastically endorsed equal rights in marriage, education, religion, and employment but passed a resolution calling for the vote only by a small majority. However, Stanton recruited to the movement Susan Brownell Anthony (1820–1906), who became the most effective organizer of the first great wave of 'votes for women' campaigning. To do this, Anthony and Stanton broke with the anti-slavery movement, and men, and founded the radical, New York-based National Women's Suffrage Association, denouncing the Fifteenth Amendment because it enfranchised only black men. They accepted only women as members and insisted the votes-for-women issue be kept separate from all the other progressive causes. The rival, Boston-based American Woman Suffrage Association accepted the Fifteenth Amendment as a step in the right direction and welcomed male members.

The two organizations fought each other bitterly for two decades. Anthony tried another tack in the 1870s when she hired lawyers to argue that the Fourteenth Amendment required states to let women vote, but on March 29, 1875, in *Minor* v. *Happersett* the Supreme Court rejected the argument and ruled that 'the Constitution does not confer the right of suffrage on anyone.' The NWSA campaigned for a federal vote, but got nowhere. The AWSA tried campaigning state by state but lost all the referenda they sponsored. However, in Wyoming Territory, women got the vote in 1869, followed by Utah Territory (1870), Washington Territory (1883), Colorado (1893), and Idaho (1896). In 1890 the two organizations finally healed the breach and amalgamated, Anthony taking over the leadership.[51]

Even so, it took another entire generation before American women got the vote, and in the end they were two years behind their British sisters. This is remarkable in view of the ease with which American white males got the vote in the late 18th or early 19th centuries, and in view of the fact that American women who appeared in Britain from the

1820s onwards were variously noted for being 'independent,' 'uppity,' 'self-sufficient,' and 'strong-minded.' British visitors to the United States brought back exactly the same impression. If American women had exerted their strength, they would have secured votes by the mid–19th century. The historian is driven to the conclusion that, for the great majority of American women, voting came low in their order of priorities. American wives in particular preferred to exert evident and satisfying control over their husbands to the infinitesimal chance of determining the selection of a president. Alice James, youngest of the James children and the only girl in a family of brilliant men, complained of male 'cruelty' and 'oppression.' She hotly denied the equality of the sexes when, as she said, it was clear that women enjoyed 'moral superiority.' In practice, she thought, men could be comfortably enslaved by the exercise of feminine intelligence.[52] Edith Wharton, the ablest woman of her generation, equally steered clear of the suffrage movement, believing women were made, as she said, 'for pleasure and procreation' and, skillful by use of these characteristics, would get what they wanted.[53] Moreover, among the minority who did actively seek the vote, sectarianism was rife, as indeed it was in all the radical movements.

The splits of the 1870s and 1880s were followed by further divisions after the turn of the century, when the militant National Women's Party, run by Alice Paul (1885–1977), imitating the British suffragettes, ran hunger strikes and picketed the White House. In July 1917 a large body of them made a determined attempt to storm the White House and a score were seized by the police and transported to the district workhouse. President Wilson, who had now come round to their viewpoint and was shortly to throw his whole weight behind a constitutional amendment, was furious with the women for their needless violence, but still more furious with the police for their overreaction (as he saw it). He issued immediate pardons—whereupon the women, who were even more furious, refused to be pardoned. The Attorney-General advised that a presidential pardon was inoperative unless freely accepted. Eventually the women changed their minds.[54] But the incident helps to explain why the women's suffrage movement was ineffective at persuading opponents, not least among their own sex.

What tipped the balance was unquestionably the same factor which won the battle in Britain—the energy, resourcefulness, and devotion of wartime women workers in occupations hitherto carried out exclusively by men, especially in factories. By the time the women tried to storm the White House, full suffrage had been conceded in fifteen states and partial rights in another thirteen. Both the Democratic and the Republican Parties had endorsed women's suffrage in 1916 and in January 1918 Wilson made it official administration policy. Congress, despite vicious lobbying by both the liquor and the textile trades, which for different reasons were scared of the impact of women voters, passed the legislation by June 4, 1919, when the Nineteenth Amendment was submitted to the states. Section One read: 'The right of citizens of the United States to vote shall not be denied or abridged by the United States or any State on account of sex.' Section Two added: 'The Congress shall have power to enforce this article by appropriate legislation.' Ratified, and proclaimed on August 26, 1920, it allowed women the vote just in time to give Harding his landslide.[55]

Many intelligent and experienced women believed all along that suffrage was not the key issue and that the women's rights issue was very much broader. They had to secure, for instance, equality of pay and equality of opportunity in job selection and promotion, and over a whole range of other matters. That involved another constitutional amendment, and the agitation to secure an Equal Rights Amendment began in 1923. It took half a century, with regular legislative submissions in each Congress, before the ERA was passed on March 22, 1972. But it failed to be ratified despite a three-year extension of the deadline. By July 30, 1982 only thirty-five of the thirty-eight states needed had ratified, and five of them had rescinded their vote. It is only fair to add that, in the meantime, the Equal Pay Act of June 11, 1963 had forbidden sex discrimination in employment by requiring equal pay and benefits for men and women at the same skill level. This was reinforced by Title VII of the Civil Rights Act of July 2, 1964, which has been the basis for most subsequent litigation to secure equality of treatment for women.[56]

It is convenient to pursue this story of women's constitutional entitlements now rather than later. A Woman's Rights Project was set up by the American Civil Liberties Union which, from 1972 to 1980, was directed by an exceptionally able lawyer, Ruth Bader Ginsberg, later a Supreme Court justice. Though she passed out top of her class at both Harvard and Columbia law schools, she was turned down for a Supreme Court clerkship, a celebrated high-road to legal distinction, by no less than Felix Frankfurter (1882–1965), a founding member of the American Civil Liberties Union, who said that—personally—he was not ready to take a woman onto his staff. No law firm offered her a position either. From this experience Ginsberg formed the view that the best way to solve the 'woman problem' was to treat women exactly the same as men, sex rarely being a genuine consideration in practice. The US Constitution is (on the whole) well suited to sustain this philosophy. Ginsburg used the 'equal protection of the law' provision to challenge sex distinction over military benefits, disability programs, parental support obligations, administration of wills and estates, and other matters where gender-based practices were unfair to women. She also litigated on behalf of women in areas where 'real' sex distinctions were involved, rather than irrational distinctions based on sex, such as benefit plans which covered all common disabilities except pregnancy. In three important cases (*Geduldig* v. *Aiello*, *General Electric* v. *Gilbert*, and the so-called *CalFed* case) she lost, the Supreme Court ruling that the distinctions were based not on gender but on the difference between 'pregnant women and non-pregnant persons.' Congress responded by passing the Pregnancy Discrimination Act (1978), which required that pregnant workers be treated the same as others. *CalFed* stirred up feminist sectarianism again, revealing sharp differences between militants who wanted the sexes treated equally and others who argued that workplaces were designed around the needs of men and that 'special treatment' was necessary in law to produce equality. This breach led to the development of 'difference feminism' and 'sameness feminism,' two antagonistic varieties which exactly recalled the sectarianism of the 1880s, and for the same reason. We need not follow the rival arguments into the trackless recesses of feminist jurisprudence, especially after the issue was complicated by the intrusion of militant lesbians, militant black women, and indeed militant black lesbians. Battle raged over complex litigation involving the distribution and sale of pornography, made unlawful by local ordinances (known as the 'Minneapolis Ordinance' in legal shorthand), the argument being that such laws, and others, presented the 'ordinary' or 'essential' woman as white, heterosexual, and professional middle class.[57] The short point to be grasped was that giving women the vote in 1920 did not in itself make much difference to the lives of women: it merely opened a new phase in the quest for justice and equality before the law.

The same principle applied to blacks, though there were fundamental differences too. The end of World War One, in which blacks had fought in large numbers, often with courage and distinction, and sustained heavy losses, drew attention to the fact that the Fourteenth and Fifteenth Amendments (1868–70), constitutionally designed to give blacks, or at any rate black males, legal equality, had not succeeded in doing so though an entire generation had passed since their adoption. In the South blacks had actually lost ground, as a result both of unlawful activities by the white majority and of legal decisions. In 1876, in *United States* v. *Reece*, the Supreme Court ruled that, while the Fifteenth Amendment forbade states to disenfranchise blacks for reasons of race, it left the states with discretionary powers to exclude certain categories of persons for reasons unconnected with race. This enabled Southern states to reduce the number of black voters by literacy tests and poll taxes, whites being exempted from the effects of tests by the so-called Grandfather Clause. This process of exclusion took place in Mississippi (1890), South Carolina (1895), Louisiana (1898), North Carolina (1900), Alabama (1901), Virginia (1902), Georgia (1908), and Oklahoma (1910), while poll taxes eliminated most blacks from voting rolls in Texas, Florida, Tennessee, and Arkansas. A system of 'White Primaries' excluded blacks from voting in Democratic primaries and conventions, which in practice denied them any influence in the South. Despite much litigation before the Supreme Court, the practice was not finally found unconstitutional till *Smith* v. *Allwright* (1944) and *Terry* v. *Adams* (1953). This system of political exclusion was reinforced by terror. From 1882 to 1903, a total of 1,985 blacks were killed by Southern lynch mobs, most being hanged, some burned alive, and for a variety of offenses, real or imaginary.[58] Then there was the Ku Klux Klan. The original Klan was employed to intimidate the Radical Republicans (both black and white) of the South in the Reconstruction Era, and virtually disbanded when the Southern whites regained their ascendancy and drove blacks off the voting rolls. But a second Ku Klux Klan was founded at Atlanta in 1915 and spread during the 1920s to become virtually a national organization. It was actually strongest outside the South, especially in Indiana, and it made itself felt in states like Oregon and Colorado. The Second Klan had many targets: Catholics, Jews, white ne'er-do-wells, and Protestants of Anglo-Saxon origin who indulged in immoral practices, as well as blacks. In effect, it was the enforcement arm of middle-class morality in the Bible Belt. Blacks went in fear of it even when they were not directly threatened by its custom of whipping, torturing, or even murdering its enemies.[59]

Conditions in the South after the restoration of white supremacy detonated one of the greatest internal migrations in American history. The underlying cause was economic: the South had little to offer liberated blacks who wanted to get on and improve their living standards. The first popular movement of blacks out of the South came in the years 1877–81, the so-called Exoduster Movement, when up to 70,000 blacks were encouraged by promoters like Benjamin ('Pap') Singleton to move into the 'Promised Land' of Kansas. This exodus ended in dis-illusion and heartbreak, but it did not stop the wider movement north and west. In 1880 only 12.9 percent of blacks lived in cities, and a majority of blacks continued to live in rural areas until about 1950. But blacks, once they tried urban life, found on the whole that they liked it. Or perhaps one should say that they preferred urban poverty to rural poverty—there was more to do and better chances. The Great Migration of blacks to the North and the cities began in the 1890s and reached a climax in World War One when well-paid wartime emer-gency jobs drew 500,000 blacks from Southern rural areas to the big cities. New York, Philadelphia, Boston, Chicago, Detroit, Cleveland, and St Louis were particular targets. In Chicago, for instance, the black population rose from 44,000 in 1910 to 110,000 in 1920. This migra-tion lasted over sixty years and was huge in cumulative scale. From 1916 until the end of the 1960s over 6 million blacks made the move. Thereafter the flow reversed itself, as the New South offered more jobs and opportunities than the smokestack industries of the North.[60]

The World War One climax was significant in several respects. The war meant a sharp cut in European migration to the US, which in 1914 had totaled over 1.2 million people. By 1918 it was running at the rate of 100,000 a year. The wartime industrial boom meant scores of thou-sands of well-paid jobs for blacks, and for the first time the North needed them more than they needed the North. But their sudden, multi-tudinous arrival caused agonizing social, housing, and cultural prob-lems, and the result was race-riots on a scale never before seen in America. These were mainly white riots against blacks encroaching on whites-only housing districts. In East St Louis in 1917, white rioters killed thirty-nine blacks. There was a similar riot in East Chicago, fol-lowed by two years of sporadic residential violence in which twenty-seven black dwellings were bombed. This culminated in a devastating five-day race war in 1919, when black mobs retaliated against white mobs, twenty-three blacks and fifteen whites were killed, and federal troops were summoned to restore order. Similar, though smaller-scale, riots occurred in twenty other cities in 1919.[61] The Chicago Commission on Race Relations carried out an exhaustive inquiry into this episode, and its report is a model of the kind. The whites were

mainly armed with bricks and blunt sticks and often fought only with
their fists. Use of handguns and rifles was rare. On the whole blacks
had more guns than whites, and knives as well. These were essentially
race riots over territory. But after 1919 it became rare for whites to riot
against blacks. By 1943, indeed, the rioting initiative had passed to
blacks. That year there were serious riots in both Detroit and Harlem,
New York. But whereas the Detroit riot was radical and territorial,
with blacks now taking the offensive, in Harlem it was a new type of
disturbance, termed 'community rioting,' which started not at the
periphery of black settlement but in its heart, and was essentially an
outburst against property, especially shops, whose object was looting.
This was a new pattern of black violence, enacted on a much greater
scale in Harlem and Brooklyn in 1964, in Watts, Los Angeles in 1965,
and in Newark and Detroit in 1967. These in turn detonated the riots
in many cities during the 'Long Hot Summer' of 1968. The urban
whites had taught the blacks how to riot, and the blacks learned the
lesson with a vengeance.[62]

The main reason why white rioting declined after 1919 was that
blacks in the big cities settled down into ghettos and the boundaries
between white and black areas became well defined. Most of these ghettos came into significant existence between 1940 and 1970, when 4 million blacks left the rural South for the urban North.[63] But some ghettos
were older. The creation of black New York's Harlem, the most famous
or notorious ghetto of all, was a human and artistic tragedy of peculiar
poignancy. Until about 1910, Harlem, originally a Dutch village, was
essentially a salubrious white area occupied by people of British, Irish,
German, and Jewish origin. It had a small-town atmosphere with big
city facilities—good restaurants and shops and theaters. In the booming 1880s and 1890s, it was provided with wonderful brownstone
houses, especially Striver's Row on 139th Street, designed by the leading architect Stanford White in 1891. The white 'loss' of Harlem
began in 1904 during an ill-planned housing boom detonated by the
opening of the Lenox Avenue subway. It was intended to accommodate well-to-do whites but they did not turn up.[64] Instead this excess
housing attracted the first black realty speculator, Philip A. Payton Jr,
founder of the Afro-American Realty Company. He was a disciple of
Booker T. Washington and of his doctrine 'Get some property—get a
home of your own.' He helped middle-class black families to move into
the area, down to 110th Street. It was the highest-quality accommodation blacks had ever lived in. Payton has been called the first exponent

of 'black economic nationalism' who 'knew how to turn prejudice into dollars and cents.'[65] He called his apartment buildings after famous black figures—one, for instance, was named after the 18th-century black poetess, Phyllis W. Wheatley. He was accused of fraud and lost his company in 1908, but by his death in 1917 he had housed more blacks in quality housing than anyone, before or since.

The whites fought back, of course. Under the leadership of John G. Taylor of the Harlem Property Owners' Improvement Corporation, local whites campaigned under the slogan 'Drive them out and back to the slums where they belong.' Taylor urged that blacks should be forced to 'colony' on empty land outside New York, rather like an Indian reservation, and in the meantime he urged the erection of 24-foot-high fences to 'protect' white areas. But white landlords saw an opportunity to make a quick profit and sold houses to blacks at up to 75 percent above white rates. That provoked a sudden mass exodus of whites, 'like a community in the middle ages fleeing before an epidemic of the black plague.'[66] In the 1920s, 118,792 whites left Harlem and 87,417 blacks arrived. Restaurants put up signs: 'Just opened for coloreds.' Blacks had always lived in the neighborhood—there had been 1,100 families there in 1902—but the size of the black occupation, from 150th through 125th to 110th Street was the first big black territorial victory in a big city.

The victory turned sour when Harlem, from being first a white, then a black, middle-class settlement, degenerated into a slum, a process completed by the end of the 1920s. This was a result of sheer pressure of numbers. From 1910 to 1920 the black population of New York City increased from 91,709 to 152,467 (66 percent) and, from 1920 to 1930, to 327,706 (115 percent). The white population of Manhattan actually declined in the 1920s by 18 percent as its black population swelled by 106 percent. By 1930 blacks composed over 12 percent of Manhattan, though they were only 4.7 percent of the city as a whole. In Harlem, the area of densest black settlement, they were joined in the 1920s not only by 45,000 Puerto Ricans but, most significantly, by a huge influx of black immigrants from the West Indies. These mainly English-speaking but also Dutch-, French-, Spanish-, and Danish-speaking Caribbeans became notorious for their frugality, thrift, business enterprise, and success within the black community, and were denounced as 'pushy,' 'crafty,' 'clannish,' and 'the Jews of the race.' There was antagonism from native-born blacks, who sang the rhyme: 'When a monkey-chaser dies / Don't need no undertaker / Just throw him in de Harlem River / He'll float back to Jamaica.'

About 40,000 of these immigrants settled in Manhattan, nearly all in Harlem, which thus became America's largest black melting-pot. In turn, Harlem swelled to bursting: by 1930 164,566 blacks, 72 percent of the total in Manhattan, lived in the Harlem ghetto. The pressure of numbers produced skyrocketing rents in 1920s Harlem, and that in turn led to over-occupancy and rapid deterioration of property. It was a 'slum-boom' and vast profits were made by both black and white landlords. Conditions were soon described as 'deplorable,' 'unspeakable,' and 'incredible.' Density was not as great as in the Jewish Lower East Side at the turn of the century but there was a difference: Jews, if they prospered at all, and most did, could escape, to Brooklyn and elsewhere. Blacks were trapped in Harlem: there was in practice nowhere else they were allowed to go. By 1925, Harlem had a population density of 336 to the acre in black districts, against a Manhattan average of 223 (and 111 in Philadelphia, the second most congested black city, and 67 in Chicago). Two streets in Harlem were perhaps the most crowded in the entire world at that time.[67] The result was death-rates 42 percent higher than the city as a whole. Deaths in childbirth, infancy, and from tuberculosis were particularly high and other killers were venereal disease, pneumonia, heart disease, and cancer, all well in excess of New York City averages. Deaths from inter-black violence also increased, by 60 percent, 1900–25—the beginning of a new nightmare for a race which had too many already.[68]

Harlem abounded in specialists who called themselves 'herb doctors,' 'African medicine men,' 'spiritualists,' 'dispensers of snake oils,' 'Indian doctors,' 'faith-healers,' 'layers-on-of-hands,' 'palmists,' and phrenologists. 'Professor Ajapa' sold 'herb juice' guaranteed 'to cure consumption, rheumatism and other troubles that several doctors have failed in.' 'Black Herman the Magician' and 'Sister P. Herrald' sold 'blessed handkerchiefs,' 'potent powders,' love-charms, amulets, and 'piles of roots' which 'keep your wife at home,' 'make women fertile,' and 'keep husbands appealing.' But there were also signs which read: 'Jesus is the Doctor; Services on Sunday.' In 1926, an investigator found 140 black churches, mainly of the store-front variety, in one 150-block area of Harlem. W. E. B. Du Bois (1868–1963), the radical black civil rights leader, observed: 'Harlem is perhaps overchurched.' Only fifty-four of these churches were 'regular' church buildings but they included some of the most magnificent in New York City. Most of the clergy were self-ordained 'jack-leg preachers' or 'cottonfield preachers' representing such affiliations as 'The Church of the Temple of Love,' 'The Church of Luxor,' 'The Live-ever-Die-never Church,' and 'The Sanctified Sons of the Holy Ghost.' As one black clergyman wrote, 'People not only had their worries removed in such places but their meager wordly goods as well.'[69] Home-grown medicine and home-grown religion were but two of the ways in which New York blacks, as in so many other big cities, soothed the pain of their existence.

Although blacks made themselves, by 1930, one of the major racial components of New York City, alongside the Irish, the Italians, the Jews, and the older founding-groups, they were not eligible, in practice, for the melting-pot system. The notion that the United States was a draconian machine into which millions of different ethnic, racial, religious, political, social, and cultural backgrounds were poured, transmogrified under its irresistible pressures, and emerged as Americans—neither more nor less—was as old as the republic itself; older indeed. Rhode Island in Roger Williams' day was already an example of the melting-pot in action. The term is believed to have been coined in 1782 by the naturalized New Yorker M. G. Jean de Crèvecoeur, who wrote: 'I could point out to you a family whose grandfather was an Englishman, whose wife was Dutch, whose son married a Frenchwoman, and whose present four sons have now four wives of different nations. *He* is an American who, leaving behind him all his ancient prejudices and manners, receives new ones from the new mode of life he has embraced . . . Here individuals of all nations are melted into a new race of men.'[70] Some arrivals in America had their doubts about how well the ethnic metallurgy of the American Experience worked. Charles Dickens records that, in a railroad car in the Midwest, he apologized to a steward for misunderstanding something and said, 'You see, I am a stranger here.' The steward replied, 'Mister, in America we are all strangers.' But the melting-pot image appealed particularly to the millions from east Europe, the true 'huddled masses,' who, in the generation after the 1880s, were able to put behind them not just the poverty but the bitter national, racial, and ethnic antagonisms of the Old World, and make themselves indistinguishable from other free citizens of a great and prosperous nation. Oddly enough, it was a Londoner, Israel Zangwill (1864-1926), son of a Russian-Jewish refugee, who brought the metaphor to vibrant life with a play, *The Melting-Pot*, which had a sensational success on Broadway in 1908. David Qixano, a symbolic character based on his father, exults in his escape to New York City: 'America is God's crucible, the great Melting Pot where all the races of Europe are melting and reforming! Here you stand, good folk, think I, when I see them at Ellis Island, here you stand in your fifty groups with your fifty languages and histories, and your fifty blood hatreds and rivalries, but you won't be long like that, brothers, for these are the fires of God you've come to ... A fig for your feuds and vendettas! German and Frenchman, Irishman and Englishman, Jews and Russians—into the Crucible with you all! God is making the American.' Zangwill added, more contentiously: 'The real American has not yet arrived. He is only in the Crucible. I tell you—he will be the fusion of all the races, the coming superman.'[71]

But to a great extent the Crucible was for 'Europeans only' or 'whites only.' It is true that on the rim of the Gulf of Mexico and the Caribbean, countless Americans, whose forebears dated from the period when neither France nor Spain nor England nor America itself exercised effective sovereignty, and the adage 'No Peace Beyond the Line' was the only law, were a racial mixture which defied analysis, so complex it was—and is. It is also true that many Northerners had some Indian blood, and boasted of it. The James family of Boston and both the Oyster Bay and the Long Island Roosevelts were cases in point. But most Americans were wedded to the idea that their birthright, though multinational and non-ethnic, was white. Some pushed the argument further. The Second Ku Klux Klan of 1915 was an attempt to reassert the integrity of the White Anglo-Saxon Protestant community and its dominance. The next year Madison Grant published his bestseller, *The Passing of the White Race*, a quasi-scientific presentation in an American context of European master-race theory. He argued that America, by unrestricted immigration, had already nearly 'succeeded in destroying the privileges of birth; that is, the intellectual and moral advantages a man of good stock brings into the world with him.' The result of an unqualified melting-pot could be seen in Mexico, where 'the absorption of the blood of the original Spanish conquerors by the native Indian population' had produced a degenerate mixture 'now engaged in demonstrating its incapacity for self-government.' The virtues of the 'higher races' were 'highly unstable' and easily disappeared 'when mixed with generalized or primitive characters.'[72]

The concept of the Wasp implied a ruling caste, or rather a racial pecking-order, summed up by Will Hays, campaign-manager to

Warren Harding, when he described his candidate's lineage as 'the finest pioneer blood, Anglo-Saxon, German, Scotch-Irish and Dutch.' Senator Henry Cabot Lodge used the code-phrase 'the English-speaking people.'[73] The war, as Wilson predicted, gave a huge impulse to patriotic xenophobia. But he nevertheless signed the Espionage Act of 1917 and the Sedition Act of 1918. The latter punished expressions of opinion which, irrespective of their likely consequences, were 'disloyal, profane, scurrilous or abusive' of the American form of government, flag, or uniform; and under it Americans were prosecuted for criticizing the Red Cross, the YMCA, and even the budget.[74] Justice Brandeis and his colleague Oliver Wendell Holmes (1841–1935) sought to resist this wave of intolerance. In *Schenk* v. *United States* (1919), Holmes laid down that restraint on free speech was lawful only when the words were of a nature to create 'a clear and present danger;' he dissented from *Abrams* v. *United States*, when the Supreme Court upheld a sedition conviction, and argued that 'the best test of truth is the power of thought to get itself accepted in the competition of the market,' which echoed Milton's point in *Areopagitica*.[75] But these were lonely voices. Patriotic organizations like the National Security League and the National Civic Federation continued their activities into the peace, and the watchword in 1919 was 'Americanization.'

There were two immediate consequences. From the autumn of 1919, with Wilson stricken, there was really no effective government in Washington, and the man in charge, insofar as anyone was, Mitchell Palmer, the Attorney-General (1872–1936), was a xenophobe. He had made foes as Enemy Property Controller during the war and in spring 1919 he was nearly killed when an anarchist's bomb exploded in front of his house. Thereafter he led a nationwide drive against 'foreign-born subversives and agitators.' On November 4 he presented Congress with a report entitled 'How the Department of Justice discovered upwards of 60,000 of these organized agitators of the Trotsky doctrine in the US . . . confidential information upon which the government is now sweeping the nation clean of such alien filth.' 'The sharp tongues of the Revolution's head,' he wrote, 'were licking the altars of the churches, leaping into the belfry of the school bell, crawling into the sacred corners of American homes' and 'seeking to replace marriage vows with libertine laws.' On New Year's Day 1920 his Justice Department rounded up 6,000 aliens, most of whom were expelled. In the 'Red Scare' that followed, five members of the New York State Assembly were disqualified and a Congressman was twice thrown out of the House.

Two Italians, Nicolo Sacco and Bartolomeo Vanzetti, anarchists who had evaded military service, were sentenced to death on July 14, 1921 for the murder, in the course of a payroll raid, of a clerk and a guard at a South Braintree, Massachusetts shoe factory. The evidence was largely circumstantial and the jury may have been prejudiced. On the other hand, the Massachusetts governor of the time, Alvin Fuller, set up a special committee which examined the trial record and pronounced the verdict just, and all the various appeal courts through which the case passed (the men were not executed till August 1927) likewise decided they were guilty. But the organized left decided to make the case a *cause célèbre*, pulling out all the literary stops both in America and in Europe and involving, among others, H. G. Wells, Anatole France, and Henri Barbusse, as well as John Dewey, Walter Lippmann, H. L. Mencken, John Dos Passos, and Katherine Anne Porter. From 1925 the worldwide agitation was directed by Willi Muenzenberg's official Communist International propaganda machine in Paris, and produced spectacular results. When the men were finally executed, there were riots in Pars, Geneva, Berlin, Bremen, Hamburg, and Stuttgart. Artistic memorials included a 'crucifixion' mosaic by Ben Shahn, a notorious play, *Winterset*, by Maxwell Anderson, and a two-volume novel, *Boston*, by Upton Sinclair, who privately admitted that his researches while preparing the book left him in no doubt that the men had committed the murders. The left profited enormously from the case, as they were later to do over the Rosenbergs and Alger Hiss. It laid the first of what were to be many archeological layers of anti-Americanism in the world, and destroyed the faith of many innocent people in the American Dream.[76]

The Red Scare as a whole was counter-productive, as the new President, Harding, was shrewd enough to recognize. Against the advice of his Cabinet and his wife, he insisted on releasing Eugene Debs. He thought Debs was a greater danger to the American people as an imprisoned symbol than as a militant at liberty. He said, 'I want him to eat his Christmas dinner with his wife,' and let him out. He freed twenty-three other prisoners convicted of political offenses the same day, and he commuted death sentences on the 'Wobblies' (Industrial Workers of the World). Long before his death he had virtually cleared the jails of anti-constitutional offenders. But he could do nothing about the other consequence of wartime xenophobia, restrictions on immigration, because this sprang from deep-rooted feelings among many ordinary Americans that the 'open door' policy was no longer tenable. The result was the Emergency Quota Act of 1921, first passed in 1920, pocket-vetoed by Wilson, reenacted and reluctantly signed by Harding. It capped immigration from Europe at 357,000 a year, though it set no limits on Canada or Latin America. This was the first statutory ceiling on immigration. It also created the first quota system by restricting entrants from any country to 3 percent of persons born there as counted by the 1910 census. In 1924 the National Origins or Johnson–Reed Act cut the total in any one year to 164,000, with a final ceiling of 150,000 in 1927, capped entry from any one country to 2 percent, banned all Asians, under the provisions of the old Naturalization Act of 1790, and favoured northern and western Europeans at the expense of the Slavs and the Mediterranean nations. There was a further twist of the screw in 1929. As a result, European immigration to the US fell from 2,477,853 in the 1920s to 348,289 in the 1930s, and total immigration from 4,107,200 to 528,400. The era of unrestricted mass immigration was over.[77]

If America was now trying to pull up the drawbridge, through which the outside world had hitherto entered freely, what kind of people and culture were now the tenants of the castle? The debate about the soul of America had begun. While Mitchell Palmer was hunting Reds, East Coast highbrows—the term was a useful invention of the critic Van Wyck Brooks in 1915—were reading *The Education of Henry Adams*, the posthumous autobiography of the archetypal Boston mandarin, published in October 1918. From then until spring 1920 it was the most popular non-fiction book in America, rejecting the notion of a brazen, uniform national culture, of 'Americanization,' of the matrix Palmer was trying to impose, and favoring instead what Adams called 'multiversity.' Van Wyck Brooks, in a famous essay 'Towards a National Culture,' which he had published in 1917 in his magazine *Seven Arts*, argued that the melting-pot theory was unsound since it turned immigrants into imitation Anglo-Saxons. He argued that Americans ought to aspire to a superior version of European national-ism but should pursue the 'more adventurous ideal' of cosmopoli-tanism and become 'the first international nation.'[78] But one suspects that what Brooks really meant by this was a culture monitored and invigilated by the East Coast elites from Ivy League colleges. In May 1919, hearing that a friend, Waldo Frank, planned to settle in the Middle West, he wrote to him: 'All our will-to-live as writers comes to us, or rather stays with us, through our intercourse with Europe. Never believe people who talk to you about the West, Waldo; never forget that it is we New Yorkers and New Englanders who have the monopoly of whatever oxygen there is in the American continent.'[79]

This view, seen here in its most arrogant and extreme form, is worth quoting since it is still held, though not always openly expressed, by a large section of the American educated class. But it was challenged at the time by one of the leading gurus of the New England intelligentsia, the educationist John Dewey (1859–1952). He was speaking on behalf of the old silver-crusader and Midwestern spokesman William Jennings Bryan, who after vainly opposing the drift of the US into World War One (he resigned as secretary of state in protest) was fighting a last battle on behalf of American religious fundamentalism. When Tennessee enacted a law forbidding teachers in public schools to instruct children in Darwinian evolution, the American Civil Liberties Union financed a test-case involving John D. Scopes of Dayton. He was defended by an expensive specialist in sensational cases, Clarence Darrow (1857–1938), who among his other triumphs saved the murderers Leopold and Loeb from the death-chamber. Bryan, who was a dying man at the time, helped to prosecute Scopes. Thanks to some raucous and tendentious reporting by East Coast scribes, notably H. L. Mencken (1880–1956)— of whom more later—the impression has been left that the trial was a defeat for Bryan and a disaster for the Bible Belt. In fact Scopes was convicted and was fined $100. Fundamentalism survived and flourished and Evolution v. Creationism is still a live issue in large parts of America at the end of the millennium.[80] At the time, the vogue among the Eastern urban elite was to mock the obscurantism of the Midwest which the Scopes trial epitomized (or so it was said). Dewey pointed out that Bryan was in fact not an obscurantist but 'a typical democratic figure.' He might be mediocre but 'democracy by nature puts a premium on mediocrity.' And Bryan spoke for some of the best, and most essential, elements in American society:

> the church-going classes, those who have come under the influence of evangelical Christianity. These people form the backbone of philanthropic social interest, of social reform through political action, of pacifism, of popular education. They embody and express the spirit of kindly goodwill towards classes which are at an economic disadvantage and towards other nations, especially when the latter show any disposition towards a republican form of government. The Middle West, the prairie country, has been the center of active social philanthropy and political progressivism because it is the chief home of this folk ... believing in

education and better opportunities for its own children . . . it has been
the element responsive to appeals for the square deal and more nearly
equal opportunities for all . . . It followed Lincoln in the abolition of
slavery and it followed Roosevelt in his denunciation of 'bad' corporations and aggregations of wealth . . . It has been the middle in every sense
of the word and of every movement.[81]

This first encapsulation of what has come to be known as Middle
America applied, as Americans increasingly came to realize, not just to
the Midwest but to almost every part of the vast nation. The essential
simple 'goodness' of America, and of Americans, was to be found
everywhere. 'Realist' or 'Naturalist' novelists like Theodore Dreiser
(1871-1945) and Sinclair Lewis (1885-1951) did not quite catch its
essence. Concerned with particular tragic-dramas and scandals and
abuses, as in Dreiser's *Sister Carrie* (1900) and *An American Tragedy*
(1925) or Lewis' *Main Street* (1920) and *Babbitt* (1922), they ignored
the enormous satisfactions which countless millions of ordinary
Americans drew from the nation they were continually making and
remaking.

Some of the painters got much closer, not just to the truthful image
of American everyday life, but to its spirit, which at its best had a noble
beauty of its own. Thomas Eakins (1844-1916), the Philadelphia master whose extraordinary devotion to the exactitudes of his craft make
his portraits, of the famous and the obscure, so penetrating, argued
passionately that American painters of genre and popular life ought to
look at America with the same exclusive concentration as Church and
Bierstadt looked at its physical features. He deprecated the fact that
outstanding artists like Whistler, Sargent, and Mary Cassatt
(1845-1926), the ablest of the Impressionists, spent so much of their
time in Europe that they identified with its impulses. 'If America is to
produce great painters,' he said in 1914, 'and if young art students
wish to assume a place in the history of the art of their country, their
first desire should be to remain in America, to peer deep into the heart
of American life.'[82]

One who did so peer was Winslow Homer (1836-1910). It is shocking to record that, when he was a teenager in the 1850s, Boston, then
supposedly the cultural capital of the United States, had no art
school. He learned his trade—insofar as he was taught anything at
all—as a lithographer. This did not prevent him becoming the most
wide-ranging and one of the most accomplished of all American

painters, beginning with on-the-spot action drawings of the Civil War fronts for *Harper's Weekly*, and continuing through series on social life in gardens, on beaches, and in the playing fields, walks in the backwoods, small-town, village, and country life, schoolchildren and teenagers, farming and fishing, the coast and the prairie and the forests, mountain lakes and rushing rivers, always with ordinary Americans going about their daily tasks and pleasures in front of the multifaceted natural background of their country.[83] Homer's eye is accurate, objective, dispassionate, uncommenting, understanding. The original energetic, unsophisticated roughness of his technique, eventually evolving into miracles of virtuosity (especially in his watercolors), allowed him to develop a true national style in painting just as Walt Whitman did in poetry.[84]

It was on the basis laid by Eakins and Homer that Norman Rockwell (1895–1978) provided a unique encapsulation of Middle American life. He began providing cover illustrations for the *Saturday Evening Post* in 1916, the year Eakins died, and for the next forty-seven years produced no fewer than 322 of these windows into the nation in its normalcy, crises and exultation, comedy and tragedy, laughter and sadness. The procedures whereby Rockwell conceived and executed these paintings, on the basis of which the illustration was photoengraved, using sketches, photographs, live models, pencil, ink, gouache, watercolor, crayons, and oils, was extraordinarily complex and arduous, often involving hundreds of preparatory studies. Rockwell was, and indeed still is, described as an illustrator, but it is now possible to predict his emergence as an Old Master, like the Dutch genre painters, especially Jan Steen, or the English moralist William Hogarth. Rockwell painted chiefly from his home at Stockbridge, Massachusetts, where his studio is now a gallery of his work, which attracts vast numbers of visitors from all over the world. Virtually all the neighborhood sat as models at one time or another, and anyone who visits the locality can see the people who inhabit his artistic world actually going about their business in the streets, shops, and offices—though they are increasingly the children, grandchildren, and great-grandchildren of the originals. For inspired verisimilitude, infused with a singular note of geniality, there has been nothing quite like it in the Western artistic tradition. *A Country Editor, Blacksmith's Boy, Heel and Toe, The Ration Board, A Hospital Reception Room, Thanksgiving, The Cleaning Women at the Ballet, Scouts: a Guiding Hand, Mrs O'Leary and Her*

Cow, Lady Drivers, The Veteran—and many others—give truthful glimpses into precisely the magnanimity of Middle America which John Dewey described.[85]

Middle America contrived its own peculiar drama when, largely through its efforts, America embraced Prohibition. Hard liquor, especially rum and whiskey, were entwined with American history from the beginning. Rum was a vital element in the three-cornered or quadrilateral slave-trade. Whiskey was often the only currency of the backwoods, as the Whiskey Rebellion of the 1790s testified. Enormous quantities of spirits were drunk in the United States in the late 18th and still more in the 19th century, when people had more money. But before we look at this in detail it is important to grasp that the soft-drinks industry was also an American phenomenon. Carbonated water itself was not an American invention. Commercial manufacture of artificial seltzer water began in Germany in 1783; five years later Paul, Schweppe and Gossee founded a business to mass-produce it in Geneva, and a decade later Jacob Schweppe moved to Bristol in England and started his great international firm. By 1807, soda-water was being dispensed in draught and in bottles in many cities of the United States. It was listed in the *US Pharmacopoeia* of 1820 as a 'medicated water' and flavored syrups had been added by 1830. The next year was the key one; a patent was issued for machinery to dispense fizzy drinks across the counter. This was the beginning of the soda-fountain, and it duplicated itself across the United States *pari passu* with commercial sellers of medical drugs—the drugstore soda-fountain was well established in America by the 1840s. By the late 1880s, America had no fewer than 1,377 soft-drinks bottling plants producing 17.4 million cases a year.[86]

The driving force behind this national phenomenon was the hot and humid South, which specialized in producing delicious new soft drinks (as well as ingenious hard ones). Not surprisingly, then, it was Atlanta which gave birth to the industry's masterpiece. John Styth Pemberton was a fifty-five-year-old druggist of the old-fashioned sort, who knew everything about dispensing his own preparations—he produced Pemberton's Extract of Styllinger and Pemberton's Globe Flower Cough Syrup. But, as with many druggists, he became more gifted at producing drinks for the soda-fountain in his shop than remedies for his drugs-counter. Like almost all 19th-century druggists, he loathed alcohol and strove hard to produce the perfect soft drink, what he called 'the ideal nerve tonic and stimulant' which was not an intoxicant. He produced first 'the French Wine of Coca,' an extract from the coca leaf, and then experimented with juice from the cola nut, knowledge of which had been brought to the South by slaves from Africa. It took him a long time to eliminate the bitterness, but he eventually managed a combination of the two drinks which was sweet, but not too sweet, and (as he thought) both soothing and stimulating. Then came the name, and 'thinking that the two Cs would look well in advertising,' he hit on Coca-Cola. It is now the second most widely recognized term in the world (the first is 'OK').[87]

Pemberton made his concoction at exactly the right time, for the 1880s were the takeoff decade for proprietary soft drinks, when the number of cases sold rose by 175 percent. By 1905 Coke was advertised throughout America as 'The Great National Temperance Drink.' It is astonishing, looking back on it, that Pemberton was the first to bring cola and coca together; but most great ideas are simple. Much fuss was and is made about the 'secret formula,' and when in 1985 a bored Coca-Cola company tried to change it for the first time in ninety-nine years, the hostile public reaction was devastating.[88] But the formula was easily deducible by expert chemical analysis and anyway did not matter much, as was testified by Alfred Steele, who worked for Coke for ten years before becoming president of Pepsi-Cola in 1950.[89] The war of temperance versus hard liquor was probably not the main factor in Coke's success. Case shipments rose from 113 million annually to 182 million during the Prohibition decade but then, despite Repeal, rose again to 322 million during the Depression decade. The truth is, the success of Coke lay in marketing, salesmanship, organizing skills, and a thousand other things which had not much to do with its actual composition or the morals and habits of those who drank it.

Pemberton sold out to Asa Griggs Candler in 1887, the year after he invented it, for a mere $283.29—the biggest steal since the Dutch bought Manhattan. Candler was a genius whose family had been devastated by the Civil War, which deprived him of a medical career but left him determined to make his mark on the nation's health. By the time he sold the company to a consortium of Atlanta banks, under Ernest Woodruff, in 1919 it was worth $25 million, the biggest business deal yet carried out in the South. Thereafter it was mainly a matter of salesmanship, driven by the idea that there was no such thing as an unsuitable outlet for Coke. Harrison Jones, director of sales, told the bottler in 1923 that forty types of outlet, ranging from bakers' shops to fire-engine houses, could be persuaded to sell Coke and that the job of the sales-force was 'to make it impossible for the consumer to *escape* Coca-Cola . . . Gentlemen, there is no place within reach, by steps, elevator, ladder or derrick, where Coca-Cola can be sold, but what should be reached by a CC salesman, or that salesman should be fired.' He said: 'Salesmen should keep calling unremittingly on their prospects . . . No matter how many times you have talked to a dealer about Coke, there is always something new to say. Repetition convinces a man. A merchant buys so many different things that a persistent salesman wins an opening where a casual order-taker makes no impression.'

Coke had a quasi-religious approach and was determined to use all the devilish lures which tempted men into its rival, the demon drink. Its divisions into regions and districts were reminiscent of the *classis* system of 16th-century Calvinism—they have since been copied by all the main soft-drinks firms—and there were many aspects of Calvinism in its approach. A recruit to Coke, went the message in the 1920s, better still a convert to Coke, was 'born again.' It welcomed 'miracles,' such as the Crown Cork and Seal Bottle-Cap, which had been knocking around since 1892 but was nothing until married to Coke. Between 1850 and 1900 dozens of devices had been patented in America for efficient sealing of soft-drink and beer bottles, but none had really worked. So the bottle-cap, like Coke itself, proved one of the greatest, and simplest, inventions of history. Coke was probably the best single example of the way the American religious spirit was transmuted into a secular force while still keeping its religious overtones. Here is Jones again, to a convention of bottlers: 'Thank God for a Board of Directors and heads of business that came 100 percent clean and said "You need the ammunition and here she is" and they gave us a million dollars more than we have ever had in this world for sales and advertising. And they could have kept it for profits—but they didn't do it, they gave it to us, and believe me, with your help and God's help we are going to get them in 1923'—a perfect illustration of Dr Johnson's dictum: 'Sir, a man is seldom so innocently employed as in making money.' Rather like a Christian sect, but uniquely for a commercial firm, Coke never changed its product. Coke learned to treat itself, and its customers— that is the entire nation—as a church, and above all a Congregational church, run by the pew-folk. The point was made by Robert Guizueta, Coke's chief executive in the 1980s, when the calamitous descent into heresy over the formula occurred: 'It was then that we learned that, if the shareholders think they own this company, they are kidding themselves. The reality is that the American consumer owns Coca-Cola.'

Pepsi, a rival church, was also invented by a Southern druggist whose medical career had been frustrated by the Civil War. Caleb D. Bradham, born in North Carolina, created 'Brad's Drink' in the 1890s but changed its name to Pepsi-Cola because he thought it could cure dyspepsia and bring relief to those suffering from peptic ulcers. Pepsi was less well run than Coke, went bankrupt twice, and competed mainly on price. But it, too, had a genius, Alfred N. Steele, born in 1901, called 'a big-hitter salesman'—'When he came into bat he swung for the fences.' He told the bottlers in 1950: 'I want to take you out of your Fords and put you into Cadillacs;' and again in 1954: 'There are many among you in 1950 who told me yourselves you were afraid of going broke. Today, I am proud to say there are many among you who are millionaires. You don't only own Cadillacs—you can afford them.' He himself lived to marry Joan Crawford—and afford her. His 'cola wars' with Coke were essentially battles between rival churches, with the added interest that they constituted a spectator sport. Roger E. Enrico, chief executive of Pepsi, said: 'At Pepsi we *like* Cola Wars. We know they're good for business—for *all* soft drinks brands. You see, when the public gets interested in the Pepsi-Cola competition, often Pepsi doesn't win at Coke's expense and Coke doesn't win at Pepsi's. Everybody in the business wins. Consumer interest swells the market. The more fun we provide, the more people buy our products—*all* our products.'[90] The 'wars' were, he said, 'a continuing battle without blood,' fought on the pages of magazines and on the airwaves, where the messages were quasi-religious too. The Bible Belt uplift approach to selling soft drinks was especially marked in World War Two when the head of Coke made 'a solemn pledge' that properly iced Coke would be made available to all members of the US armed forces, wherever they were stationed. General Dwight D. Eisenhower (1890–1969), commanding the invasion of Europe, saw to it that ten separate bottling plants, which followed the front, were stretched across his entire command, to provide the maximum possible defense against alcohol. From first to last Coke and Pepsi were commercial rivals—separated churches, as it were—but the real enemy, the Devil as it were, was hard liquor.

That was the position when Middle America embarked on its war against drink—which had both the evangelical spirit of the Pilgrim Fathers and the witchhunting fanaticism of the Salem elders. It was also infused by the feeling that pleasure itself was enviable and sinful. America has always been a land of righteous persecution, whether under the banner of Calvinism, purity, anti-Communism, anti-racism, feminism, or Political Correctness. It has turned out some notable Savonarolas in its time. In the 19th century, Prohibition went hand in hand not only with anti-slavery but with anti-obscenity. Anthony Comstock (1844–1915), for instance, was an abolitionist who fought fanatically for the Union in the Civil War and a noted campaigner against alcohol. But his chief work in life was to be secretary, for forty-three years, to the New York Society for the Suppression of Vice. His personal and public war against obscene publications is told in his books, *Frauds Exposed* (1880), *Traps for the Young* (1883), and *Morals Versus Art* (1887). He virtually founded the Society in 1873, and the same year the US Post Office made him a special agent which allowed him to go into any US Post Office and search for mail he suspected to be obscene. He persuaded Congress to amend an 1865 law, so that it became a crime knowingly to send obscene material through the mails. He also got Congress to amend legislation so that it became criminal to send information about, or to advertise, obscene publications, contraception, or abortion. This was the Comstock Law. In its first six months, Comstock boasted of the seizure of 194,000 obscene pictures and photos, 134,000 pounds of books, 14,200 stereopticon plates, 60,300 'rubber articles,' 5,500 sets of playing cards, and 31,150 boxes of aphrodisiacs. Later, Comstock claimed that over his entire career he had destroyed 'sixteen tons of vampire literature' and convicted on obscenity charges 'enough persons to fill a passenger train of sixty-one coaches—sixty coaches containing sixty passengers each and the sixty-first not quite full'—he was a stickler for precise statistics. He prosecuted booksellers for having Zola, Flaubert, Balzac, and Tolstoy on sale, though most of his work dealt with repellent trash. The Catholic Church, for its own reasons, did not approve of him, but he was backed by the Wasp establishment, his society being subsidized by, among others, J. Pierpont Morgan, William F. Dodge Jr, the car manufacturer, and Samuel Colgate, the toothpaste tycoon.[91]

Comstock argued that obscene publications, prostitution, and the sale of hard liquor were intimately linked by commerce and corruption. That was a common view. There were many Americans who believed, in the 19th century, that if only the sale of alcohol could be made unlawful, not only alcoholism and drunkenness could be stamped out but the country could be morally improved in countless other distinct ways. To some extent this view was shared by many leading statesmen, though they differed about the means. Thomas Jefferson characteristically believed the cultivation of the vine would provide what he called 'a safe alternative to ardent spirits.' James Madison wanted all young men to 'take the pledge.' Abraham Lincoln thought 'intoxicating liquors' came forth 'like the Egyptian angel of death, commissioned to slay, if not the first-born, then the fairest born of every family.' He said he did not drink 'because I like it so much.' US spirits were normally bottled at 80 percent proof, and during the 1830s the per-capita consumption of absolute alcohol in America was calculated at 7.1 gallons annually, an alarming figure considering that many if not most women, and most slaves, did not touch alcohol at all, or consumed it only in small quantities. Hence the name for America, the 'Alcoholic Republic.'[92]

In the 1840s a businessman called Neal Dow in Portland, Maine, made a study of the effects of alcohol there and discovered that an astonishing range of evils, from family violence, crime, and poverty to incompetence and loss of production in factories, were, as he put it, 'alcohol-related.' He thought competition among 'grog shops' was the prime cause of 'excessive consumption' of alcohol. In 1851 he persuaded the state legislature to pass the 'Maine Law,' which banned the sale of alcohol. Thirteen of the thirty states had similar laws by 1855. The rise of the Republican Party, which wanted to broaden its base in the North by recruiting Irish and German Catholics, and German and Scandinavian Lutherans, who were generally opposed to Prohibition, took anti-alcohol off its platform. After the Civil War, however, militant women took up the cause—it was often, as we have seen, an alternative to suffrage-campaigning—and it was women who were chiefly behind the Anti-Saloon League of America, which in 1895 held its first annual convention. Enlisting Protestant congregations as basic campaigning units, the ASL was astonishingly adept at guiding legislatures through a cumulative series of reforms—in the process doing democracy itself a favor—ending in total abolition. It was notable that 'dry' legislatures usually favored women's suffrage too. By 1916 twenty-one states had banned saloons. That year the national elections returned a Congress where dry members outnumbered wets by more than two to one. In December 1917 Congress submitted to the states the Eighteenth Amendment, which, when ratified in 1919, changed the Constitution to ban 'the manufacture, sale or transportation of intoxicating liquors.'[93] The Volstead Act, making America dry, had already been passed: the amendment finally made it constitutional, and Bryan, now an old man, was presented with a vast silver loving-cup in token of his 'prodigious efforts' to insure ratification.

The imposition of Prohibition, and its failure, illustrates perfectly a number of important principles in American history. First, it shows the widespread belief in America that utopia can be achieved in the here-and-now and the millennium secured in this world, as well as the next. Second, it indicates a related belief that 'Americanization' can be achieved by compulsion and law. Third, it draws attention to a weakness in American public opinion and policy—a tendency to will the end without willing the means ('freeing' the blacks was another instance). The Volstead Act was a compromise: if it had provided ruthless means of enforcement it would never have become law. The Prohibition Bureau was attached to the Treasury—efforts to make it part of the Justice Department were defeated. Successive presidents refused to recommend appropriations needed to secure effective enforcement.[94] Fourth, the utopianism inherent in Prohibition came up against the utopianism inherent in the rooted and active American principle that freedom of enterprise must be totally unrestricted. Being one of the least totalitarian countries on earth, America possessed virtually none of the apparatus to keep market forces in check once an unfulfilled need had appeared. What Prohibition did was to transfer the manufacture, sale, and distribution of liquor from legitimate to criminal forces. The speed at which this happened and the illicit system appeared was characteristic of American dynamism. In no time at all—mere months—the liquor gangsters and their backers commanded more physical and financial resources than the law.[95]

Prohibition also illustrated Karl Popper's Law of Unintended Effect, America being the ideal arena in which this law, one of history's great ironies, operates. Ultra-American Prohibitionists proclaimed that it was directed chiefly at the 'notorious drinking habits' of 'immigrant working men.' In fact, far from driving alien minorities into Anglo-Saxon conformity, it enabled them to consolidate themselves. Prohibition was one of the turning-points in a long process whereby the Anglo-Saxon-descended possessing class was driven from its position of preeminence. In New York, for instance, bootlegging was half-Jewish, one-quarter Italian, and one-quarter Polish and Irish. Those of Anglo-Saxon descent were mere consumers, it is true enthusiastic ones—Edmund Wilson delightedly listed the additions that bootlegging made to the English vocabulary, swelling the number of expressions and words for degrees of intoxication to over 300. In Chicago it was the same story: Italians and Irish shared the loot, the old Anglo-Saxon elites merely drank. The Italians proved themselves particularly adept in distributing illicit liquor in an orderly and inexpensive manner, drawing not merely on the experience of the Sicilian, Neapolitan, and Sardinian criminal societies but on revolutionary syndicalism.

Prohibition offered matchless opportunities for 'aliens' to subvert society, particularly in Chicago under the corrupt mayoralty of 'Big Bill' Thompson. John Torrio, who ran large-scale bootlegging in Chicago 1920–4, retired to Italy in 1925 with a fortune of $30 million. No one in the history of the world had made this kind of money from organizing crime before. Then and later it made the young, ambitious, and criminally inclined think furiously about possible career-prospects. Torrio practiced the new Leninist principle of 'total control:' all officials were bribed according to their rank and all elections were rigged.[96] Torrio could deliver high-quality beer for as little as $50 a barrel and his success was based on the avoidance of violence by diplomacy: he secured agreement among gangsters for the orderly assignment of territory. In fact bootleggers were most successful when they conformed most closely to the methods of legitimate business. Torrio's lieutenant and successor, Al Capone, was less politically minded and therefore less successful; and the Irish operators tended to think in the short term and resort to violent solutions. When this happened, gang-warfare ensued, the public became angry, and the authorities were driven to intervene. As a rule, however, bootleggers operated with public approval, at any rate in the cities. Most urban men (not women) agreed with Mencken that Prohibition was the work of 'ignorant bumpkins of the cow states who resent the fact they had to swill raw corn liquor while city slickers got good wine and whiskey.' It had 'little behind it, philosophically speaking, save the envy of the country lout for the city man, who has a much better time of it in this world.'[97]

Attitudes being what they were, cleaning up a city while Prohibition was still in force was virtually impossible. General Smedley Butler of the US Marine Corps, put in charge of the Philadelphia police in 1924 under a 'new broom' administration, gave up the job after less than two years, saying it was a 'waste of time.' Politicians of both parties gave little help. At the 1920 Democratic convention in San Francisco delegates gleefully drank illegal first-class whiskey provided free by the mayor. The Republicans bitterly resented the fact that, at their Cleveland convention of 1924, 'Prohibition agents clamped down on the city with the utmost ferocity' (Mencken again). Mencken claimed: 'Even in the most remote country districts, there is absolutely no place in which any man who desires to drink alcohol cannot get it.'[98] The journalist Walter Ligget, who made himself the greatest living expert on the subject, testified to the House Judiciary Committee in February 1930 that 'there is considerably more hard liquor being drunk than there was in the days before Prohibition and . . . drunk in more evil surroundings,' as he could prove by 'a truckload of detail and explicit facts.' Washington DC, he said, had 300 bars before Prohibition: now it had 700 speakeasies, supplied by 4,000 bootleggers. Police records showed arrests for drunkenness had trebled over the decade. Massachusetts had jumped from 1,000 licensed saloons to 4,000 speakeasies, plus a further 4,000 in Boston alone, where there were 'at least 15,000 people who do nothing except purvey booze illegally.' Kansas had been the first state to go dry; had been dry for half a century; yet 'there is not a town in Kansas where I cannot go as a total stranger and get a drink of liquor and very good liquor at that, within fifteen minutes of my arrival.'[99]

Socially, the experience was a catastrophe for the United States. It brought about a qualitative and permanent change in the scale and sophistication of American organized crime. Running large-scale beer-convoys required powers of organization which were soon put to use elsewhere. From the early 1920s, gambling syndicates used phone-banks to take bets from all over the country. Meyer Lansky and Benjamin Siegel adapted bootlegging patterns to organize huge nation-wide gambling empires. Prohibition generated enormous funds which were then reinvested not only in gambling but in other forms of large-scale crime such as prostitution and drug-smuggling. It was the 'takeoff point' for big crime in America, and of course it continued after the Twenty-first Amendment, which ended Prohibition, was finally ratified in December 1933. Throughout the 1930s organized crime matured, and it was from 1944 onwards, for instance, that the small desert town of Las Vegas was transformed into the world's gambling capital.[100] Prohibition, far from 'Americanizing' minorities, tended to reinforce minority characteristics through specific patterns of crime: especially among Italians, Jews, Irish, and, not least, blacks, where from the early 1920s the West Indian immigrants introduced the 'numbers game' and other gambling rings, forming powerful black-ghetto crime-citadels in New York, Chicago, Philadelphia, and Detroit. Studies by the Justice Department's Law Enforcement Assistance Administration in the 1970s indicate that the beginning of Prohibition in the 1920s was the starting-point for most identifiable crime-families, which continue to flourish and perpetuate themselves today at the end of the 20th century.[101]

*   *   *

Prohibition was a characteristically 20th-century exercise in social engineering which ended by doing unintended, enormous, and permanent damage to society. Reformers were more successful in dealing with other forms of vice, such as prostitution, because their approach was more intelligent and aimed not to stamp out altogether the sale of women's bodies but to suppress its most antisocial institutions and aspects. Thus, Congressional legislation in 1903 and 1907 made unlawful the importation of prostitutes and permitted the deportation of alien women who engaged in prostitution. The Mann Act of 1910, which made the transportation of women across state borders for 'immoral purposes' a crime, was a highly successful statute in itself but also proved unexpectedly useful in dealing with organized crime. Society also succeeded in dealing with America's brothel culture, though there the struggle was more difficult.[102]

America's premier Sin City, from the late 1840s up to the end of the 1920s, was San Francisco, founded by Juan Nautista de Anza (1735–88) in 1776 and called Yerba Buer until christened by its present name in 1847. Two years later the gold rush transformed it from a small village into a lawless frontier town, then a great international port. By 1880, with a population of 233,000, it had become the financial and cultural metropolis of the Far West. It had its magnificent millionaire mansions on Nob Hill but it also had the Barbary Coast, a red-light area which served the whole of the West—and an area where available women were notoriously outnumbered by young men who made high wages not just in the gold and silver mines but in many other occupations. The Barbary Coast was bounded by Broadway, Kierney, Montgomery, and Pacific Avenue and it was a collection of bars, dance-halls, and gambling joints. A typical set-up was the Bull Run on the corner of Jackson and Kierney Streets, which had a low-class dance-hall and bar in the cellar, a middle-class bar and dance-hall on the ground floor, and a brothel upstairs. The joint at the corner of Kierney and California Streets was the first (1885) topless bar in the United States. Other combination joints were the Louisiana, the Rosebud, the Occidental, Brook's Melodium, the Billy Goat and the Coliseum, all of which had professional comedians and dancers, with names like the Dancing Heifer, the Waddling Duck, and the Little Lost Chicken. All they lacked was a Toulouse-Lautrec to record them for posterity. The Bella Union, on the corner of Washington and Kierney Streets, introduced some outstanding performers, such as Eddie Foy, who fathered a great theatrical family, Flora Walsh, Lottie Crabtree, and Harrigan and Hart, who went on to become America's most popular vaudeville team in the Progressive Era.[103] There was even a male whorehouse on Mason Street, with twelve young men and boys available—San Francisco's record as a homosexual centre goes back well into the 19th century.

The King of the Barbary Coast, from 1901 to 1916, was Jerome Bassity (real name Jerry McGlane), who controlled dance-halls, saloons, a pornographic theater, and over 200 prostitutes with names like Emilie, Lucy, Fifi, Madame Weston, Elenore, Madame St Armand, Artemise, Helen, and Lucienne. Bassity introduced a number of notorious dances, banned in other places, such as the Bunny-Hug, the Turkey Trot, the Texas Tommy, and the Hoochy-Koochy, which had been introduced at the *Streets of Cairo Show* at the 1893 Chicago World's Fair.[104] San Francisco was the first town to make popular cheek-to-cheek and belly-to-belly dancing, regarded as the depth of depravity in 1890, still risqué in 1915, just acceptable among the young in 1920, and which has since become normal all over the world.

Curiously enough, the man who cleaned up San Francisco was the monster portrayed in *Citizen Kane*, William Randolph Hearst (1863–1951). When he was twenty-four, his father, a mining millionaire and US Senator from California, gave him the *San Francisco Examiner*. Hearst had been expelled from Harvard for 'riotous behavior' and his father thought the newspaper toy would calm him down. Quite the reverse occurred. Hearst spent $8 million turning the *Examiner* into a huge commercial success, then challenged Pulitzer in New York, helped to start the Spanish-American War, endorsed political assassination as what he called a 'mental exercise,' and was blamed for the slaying of President McKinley. He sat for New York in the House, got 40 percent of the votes in one ballot at the Democratic presidential convention of 1904, and acquired seven dailies, five magazines, two news-services, and a movie company.[105]

The coming of World War One dramatically expanded prostitution in America, and many other forms of 'riotous behavior.' But it also made it possible for drastic action to be taken under wartime legislation. At this stage in his life Hearst was a radical, praised by Upton Sinclair for his 'socialism,' and the darling of the trade union movement, who asked him to found a paper for them. His *Examiner* led a campaign in conjunction with the Central Methodist Church and the Police Department to shut the Barbary Coast down. They got the state to enact the Red Light Abatement Act, eventually declared constitutional by the California supreme court (1916), and enforcement began on January 1, 1917. At that point the San Francisco prostitutes held a mass meeting and marched to the Central Methodist Church, where their leader, Mrs Gamble, had an angry dialog with the minister, the Rev. Paul Smith. It went: 'What are we to do when you close us down?' 'Take refuge in religion.' 'Can we eat that? There isn't a woman here who would be a prostitute if she could make a decent living in any other way. They've all tried it.' 'A woman can remain virtuous on ten dollars a week [*laughter*] and statistics show that families all over the country receive less.' 'That's why there's prostitution—come on, girls, there's nothing for us here.' Three weeks later, on February 14, 1917, the entire Barbary Coast quarter was surrounded by police, over 1,000 women were driven out of their rooms, forty saloons were closed down, and scores of cafés and cabarets shut.[106]

Hearst's campaign against San Francisco's whores was, as it were, a bridge period between his early radicalism and his later mood of somber reaction. He was of no consequence in himself—much less interesting than the fictional Kane—but his life in the Twenties gives us an illuminating glimpse of America's quintessentially utopian state, California, in its most eccentric and endearing period. Hearst was always inheriting estates and then building fantastic structures on them. He was willed, for instance, 50,000 acres at Wyntoon on the Oregon border, and along the McCloud River there he constructed a Fairy House, a Cinderella House, a Bridge House over the waters, and a magic castle, as well as a cemetery for his numerous pets—a personal Disneyland in fact.[107] His exploit in transporting St Donat's Castle from Britain and rebuilding it in the United States has been often described.

Less well known, unfortunately, is his commissioning Julia Morgan (1872–1957), America's greatest woman architect, to erect for him what is arguably the finest building in North America (north of the Mexican border, anyway). Hearst inherited the San Simeon estate in California when his mother died of the influenza epidemic in 1919. He determined to build there a house which united all that was best and grandest in the Spanish-American and Anglo-American traditions, and he picked Morgan because she had already made a reputation for herself locally in doing precisely that. A rare photograph shows the two together—the enormous, genial, and sinister Hearst, and the tiny (five foot) Morgan, in her neat, severely tailored suit and expensive silk blouse, the epitome of fierce proto-lesbianism. The only thing they had in common was a love of architecture. He signed his letters to her 'William Viollet-le-Duc Hearst, Architect,' and she said of him: 'He loves architecturing.'[108] Their letters reveal a close, and in all the circumstances amazingly sweet-tempered, association, which yielded splendid results.

Morgan was an example of what was then called the New Woman. She came from San Francisco and went to Berkeley as an engineering student because there was no degree-course in architecture. She then went to Paris and was the first woman to receive a degree in architecture from the École des Beaux Arts. Her career was a remarkable success, both artistically and commercially, and she designed and saw constructed over 700 buildings, many of them very large indeed. His vast house or houses at San Simeon started out as what Hearst called 'a Jappo-Swisso bungalow.' From that it evolved into a Latino palace surrounded by villas on a hilltop which surveys the world. The cluster-idea came from the eastern American idea of the 'camp.' Rich Americans, trying to return to the backwoods tradition, built themselves luxury wooden shack-mansions in the woods, first in the Adirondacks, then in many wild places. It was the United States adaptation of the English urge, led by Queen Victoria, to find refuge from modernity in the primitive Scottish Highlands. But whereas the English built castles, like Balmoral, the Americans erected camps. The most famous example of this genre is the presidential lodge, Camp David. The archetype was Sagamore Lodge, built for Alfred Vanderbilt by a local Adirondack entrepreneur, William Durant, in 1898–1900.[109] A camp had a central tent for eating surrounded by smaller tents for sleeping. This evolved, in Hearst's case, into a central building with three guest *cottages ornées*, all on a sumptuous scale. It was begun in 1920 and 'finished' in 1926, but work continued thereafter.

The main building or Casa Grande, with its twin Spanish towers, is of reinforced concrete faced with stone. Neither Hearst nor Morgan had forgotten the calamitous San Francisco earthquake of 1907, so the house was made earthquake-proof, insofar as that is possible. It has 127 rooms. One of Morgan's specialities was swimming pools and in the big house she made a Roman pool for the rare rainy or cool days. But the Neptune pool outside, which is more Grecian, is the *clou* of the structure—perhaps the most beautiful swimming pool ever built. The three guest villas have a total of 187 rooms, including two libraries, fifty-eight bedrooms, and forty-seven bathrooms. The detailed accounts survive and they show that the Casa Grande cost Hearst

$2,987,000, the Neptune pool $430,000, and the three 'cottages' $500,000. Morgan's total charges for what in the end amounted to twenty-five years and 558 trips came to a modest $70,755. The house-party atmosphere at San Simeon was communal. When the Calvin Coolidges stayed and asked for room service, they were told that all meals, including breakfast, were eaten in the dining-room. But the story of the house in the 1930s and 1940s was sad. Hearst was running out of money, and before he died in 1951 he even had to borrow the odd $1 million from his mistress, Marion Davies. In 1958 the Hearst Corporation presented San Simeon to the California State Parks system, which has lovingly restored it and opened it to all, another example of how the American plutocracy ultimately benefits the American democracy.[110]

Morgan was one of many gifted architects who flourished in California between the wars and helped to create its Janus-faced southern megalopolis centered round Los Angeles, the Nowhere City which is also the Everywhere City, the place where all styles, fashions, and fads meet. Los Angeles was mission territory until 1822, when it became part of Mexico and 8 million acres were seized from the church and redistributed in the form of 500 land grants which became vast ranches. Richard Henry Dana was there in 1838 compiling material for his *Two Years Before the Mast*: it was 'remote' and 'almost desert' and 'there is neither law nor gospel.' It was seized by the United States in 1847, annexed in 1848 and incorporated as a city in 1850. It was from the start a highly interracial and tense city: in its first big race riot of 1871, nineteen Chinese were murdered. It had 11,000 people when it was linked to the intercontinental railroad system in 1876. Thereafter, competing railroads to the West Coast led to a mass migration in the 1880s. In 1887, when the price of a rail ticket from Kansas City to Los Angeles dropped to $12 and even to a nominal $1, 120,000 people came west on the Southern Pacific alone.[111]

The lure was originally health, rather than just sin. Lorin Blodget, whose 1859 book *The Climatology of the United States* was widely read, compared it to Italy and recommended living there for sufferers from tuberculosis, rheumatism, and asthma. In 1872 Charles Nordoff published a genuine bestseller, *California for Health, Pleasure and Residence*, a promotional book compiled on a Southern Pacific freebie-junket, which detonated the fitness rush. For the first time in American history, a 'frontier' was created, and pioneers developed it, not for land or money or gold but for the sick and the invalid.[112] Americans who did not go there were astonished by the quality of its fruit—at the 1893 Chicago Exposition, the Los Angeles publicist Frank Wiggins gave away 375,000 oranges on the first day. The California Pavilion was the most popular of all, housing an entire citrus grove and a knight on horseback made of prunes. By 1900 Los Angeles had 103,000 people and it was already crowded with fantastic buildings, such as the Bradbury Residence on Bunker Hill.[113] What made it the city it has become was the California Bungalow, an exotic and osmotic variation on the Anglo-Indian one, with mission-style and Chinese-Oriental elements. Many beautiful houses were built, such as the Gamble House (1908) designed by Charles and Henry Greene, who specialized in luxury bungalows. They had large gardens and that dictated a big spread, which began long before the automobile arrived. Los Angeles was bungalow-driven, not car-driven, and it was already a suburban city in 1905, with no real center.[114]

This sprawling, sunny city sucked in people voraciously from all over America (and abroad). In 1900–20 it grew to 575,000 with another 325,000 in the surrounding countryside. In the Twenties the immigration accelerated to 100,000 a year, giving Los Angeles 2.2 million at the end of the decade. It was the largest single internal migration in American history. What made Los Angeles successful was not just the sun and the land. It was also power—and many other forms of energy, including human ingenuity. The first oil strike was made by Edward Doheny in 1892. When his royalties poured in he built himself the first of the Los Angeles monster-mansions, a Gothick-Romanesque-Oriental-Mughal chateau.[115] Oil was struck at La Brea (1902), Fairfax (1904), and in Beverly Hills (1908). There was another, huge strike in 1920–1. One oil tycoon, Alphonso Bell, had 200 acres in what became the Santa Fe field. Royalties brought him in $100,000 a month and with it he bought 2,000 acres on which he built the Beverley Hills Hotel and Bel Air. With oil aplenty, water was scarce and frantic political battles were fought over water-rights and irrigation schemes. An arid tract like the San Fernando Valley became immensely valuable once it was irrigated. By 1903 Los Angeles was already reaching out 250 miles into the mountains for water. With the approval and help of Theodore Roosevelt, the giant aqueduct to the Valley took five years, 5,000 workers, 142 tunnels, 120 miles of railroads, and 500 miles of roads before it was finished in November 1913. All this was a prolegomenon to the giant schemes revolving round the Colorado and Hoover dams. Moreover, harnessing rivers and delivering water were linked directly to power supply and electrification.[116]

Indeed, it was electricity which made California, a decade before the German scientist Karl Ballod published *Der Zukunftsstaat* (1919), advocating the 'all-electric state.' This was the book Lenin read and which led him to pronounce: 'Communism is Soviet power plus electrification of the whole country.' In California, the slogan went 'Electricity is the road to the health, wealth and happiness of mankind.' The long-distance power supply based on harnessed water dates back in California to 1903–6, and in 1909 an engineering genius called Erza F. Scattergood became the chief electrical engineer of Los Angeles. He was an enormously energetic Rutgers graduate who came to Southern California for his health and soon made himself boss of the biggest municipal power system in the world. It was Scattergood, a consummate diplomatic and political maneuverer as well as a fanatical engineering modernist, who brought together the coalition of seven states which got colossal quantities of water and power to share out among them by harnessing the Columbia River system to a vast dam at Boulder Canyon, Nevada. It was his work that lay at the root of the Tennessee Valley Authority concept in the age of F. D. Roosevelt—which was why the President invited him to see the site of the TVA at Muscle Shoals in 1933.[117]

Scattergood made Californian electricity cheap and available virtually everywhere in the region. By 1912 California was next to New York in quantities of electric power used, an astonishing performance. During World War One, the decision was taken to pool all the power of this 1,200-miles-long state through a single, central administration. By 1924, in the United States as a whole, only 35 percent of homes were wired for electricity, but in California the figure was 83 percent. Nationally, the cost per kilowatt-hour averaged $2.17. In California it was $1.42. By generating 10 percent of all the electric power in the US, the state was able to benefit the farmers as well as city-dwellers. In 1924, when 90 percent of all farms in America were still without electricity at all, it was not unusual to see all-electric farms in California, with electrically milked cows, electric pumps irrigating orchards and fields, and the full range of appliances in the farmhouse.

Equally important, especially in Los Angeles and its surrounds, was the all-electric transport system. Henry Edwards Huntington (1850–1927), nephew of the Southern Pacific railroad tycoon Collis Potter Huntington (1821–1900), who sold control of the Southern to

Harriman in 1903, thereafter had two main interests. One was to collect books and art for his great Foundation, the Huntington Library at San Marino near Pasadena. The other was to equip Southern California with the largest, cheapest, and most efficient inter-urban transport system in the world. From 1902 he bought up the various networks in and around Los Angeles and melted them into one, the Pacific Electric. It ran for 1,164 miles and welded into an entity forty-two incorporated cities and towns within a 35-mile radius of central Los Angeles. It was these giant, thundering electric cars, familiar to students of early movies because they appear in many Keystone Kops, Laurel and Hardy, and Hal Roach shorts, which made Southern California a prime growth area. It no longer exists but, almost without exception, the Los Angeles area freeways follow its routes.

And cheap electricity fueled the continuing California boom, through the Twenties, through the difficult Thirties, and beyond. Scattergood's Boulder Canyon project ran into powerful opposition from vested interests, which meant delays, and digging did not start till 1930. But then it became a world phenomenon, attracting highway construction giants like Henry J. Kaiser (1882–1967), who became chairman of the building consortium in 1933. Federal money paid for most of it, including construction of a workers' base in the Nevada Desert called Boulder City—and FDR himself dedicated the completed dam in 1935. It was the biggest dam in the world, followed by three other linked dams, Parker, Bonneville, and Grand Coulee.[118] Its cheap power made possible the vast manufacturing industries which flourished in California during World War Two, and transformed the entire West Coast, enriching it still further. Further north, in Seattle, a disciple of Scattergood, J. D. Ross, also preached the gospel of cheap electricity, and practiced it. By the 1930s Seattle had more electric ranges than any other American city. A few years later it, too, was sharing in the wartime industrial expansion which, today, makes it the center of the US aviation industry.[119]

Cheap electricity was also one of the factors which turned Southern California into the world center of the new movie industry. The other two were sunshine and, above all—and ironically in view of California's reputation today—freedom from litigation. To put it another way, movies were the product of a marriage between California and Ashkenazi Jews. This followed an earlier union between Jewish productive and creative genius and New York. In 1890 there was not a single amusement arcade in New York. By 1900 there were over 1,000, and fifty of them already called Nickelodeons. Eight years later there were 400 of them in New York alone and they were spreading all over the Northern cities. They cost five cents and appealed to the poorest of the urban poor. The hundreds of movies made for them were silent. That was an advantage. Few of the patrons spoke English. It was an immigrant art-form—and so the ideal setting for Jewish enterprise. At first the Jews merely owned the Nickelodeons, the arcades, and the theaters. Most of the copyright processes, and the shorts, were owned by American-born Protestants. An exception was Sigmund Lublin, operating from the great Jewish center of Philadelphia, which he might have turned into the capital of the industry. But when the theater-owners began to go into production, to make the shorts their immigrant patrons wanted, Lublin joined with the other patent-owners to form the giant Patent Company and extract full dues out of the movie-makers. It was then that the Jews led the industry on a new Exodus, from the 'Egypt' of the Wasp-dominated Northeast to the Promised Land of California. Los Angeles had easy laws and, if needs be, a quick escape into Mexico from the Patent Company lawyers and from another litigational killjoy, the New York Film Trust.

The first California movie, *The Count of Monte Cristo*, made by the Selig Polyscope company in 1907, had nothing to do with Hollywood. That was then a stuck-up religious place, founded in 1887 by two Methodists, Horace and Daeida Wilcox, who hoped to turn it into a Bible-thumping district. When Hollywood was incorporated as a city in 1903, it banned not only oil-prospecting and slaughterhouses but sanitariums, liquor, and movie-houses. But it ran out of water, and to get some was forced into incorporation with Los Angeles in 1910, thus losing its autonomy. So the next year it got its first movie-studio, and in 1913 Cecil B. DeMille's crew arrived to film *The Squaw Man*, having been driven out of Arizona by dust-storms. Not that Southern California in general, and Los Angeles in particular, are nature-free. Landslides are ubiquitous. There have been major earthquakes in 1933 and 1971, and everyone now waits for the Big One. There are also specially irritating air currents, known as Santa Ana Winds, which detonate rages. One is described in *Red Wind*, a story by Raymond Chandler, the prose-poet of Los Angeles: 'A time when every booze-party ends in a fight and meek little wives feel the edge of the carving knife and study their husbands' necks.' But most of the time the climate is benign and movie-makers found it cut costs by almost half. They were not liked by the then-inhabitants (not natives: even in the 1990s, half the population of California was born elsewhere). In 1913, over 10,000 citizens of Los Angeles signed a petition to ban movie-making within the city limits. They signed themselves Conscientious Citizens and claimed the movies would bring immorality. Were they so far wrong? But they were turned down and by 1915 the Hollywood pay-roll was already $20 million annually and growing fast—the new arrival was too big to be ejected.[120]

The same year, the first characteristic Hollywood structure, Universal City, was built and a movie a day was turned out in it. This was the work of Carl Laemmle (1867–1939), first of the Jewish movie tycoons. Nearly all of them conformed to a pattern. They were immigrants or of immediate immigrant stock. They were poor, often desperately poor. Many came from families of twelve or more children. Laemmle was an immigrant from Laupheim, the tenth of thirteen children. He worked in clerical jobs, as a book-keeper and a clothing-store manager, before opening a Nickelodeon, turning it into a chain, creating a movie-distribution business, and then founding Universal, the first big studio, in 1912. Marcus Loew (1872–1927) was born on the Lower East Side, the son of an immigrant waiter. He sold papers at six, left school at twelve to work in printing, then furs, was an independent fur-broker at eighteen, had been twice bankrupted by the age of thirty, founded a theater-chain, and then put together Metro-Goldwyn-Mayer. William Fox (1879–1952) was born in Hungary, one of twelve children, and came through New York's Castle Garden Immigration Station as a child. He left school at eleven for the garment industry, set up his own shrinking business, then progressed through Brooklyn penny-arcades to a movie-chain, Twentieth Century-Fox.

Louis B. Mayer (1885–1957) was born in Russia, the son of a Hebrew scholar, and also came through Castle Garden as a child, went into the junk trade at the age of eight, had his own junk business by nineteen, a theater-chain by twenty-two, and in 1915 made the first big adult movie, *Birth of a Nation*. The Warner Brothers were among the nine children of a poor cobbler from Poland. They worked selling meat and ice-cream, repairing bicycles, as fairground barkers and traveling showmen. In 1904 they bought a film-projector and ran their own show, with their sister Rose playing the piano and twelve-year-old Jack singing treble. Joseph Schenck, co-founder of United Artists, ran an amusement park. Sam Goldwyn worked as a blacksmith's assistant and a glove-salesman. Harry Cohn, another Lower East Sider, was a trolley-conductor, then in vaudeville. Jessy Lasky was a cornet player. Sam Katz was a messenger boy but owned three Nickelodeons in his teens. Dore Schary worked as a waiter in a holiday camp. Adolph Zukor, from a family of rabbis, worked as a fur-salesman. So did Darryl Zanuck, who made his first money with a fur-clasp. Not all these pioneers kept the studios and fortunes they built up. Some went bankrupt. Fox and Schenck even went to jail. But Zukor summed up for them all: 'I arrived from Hungary an orphan boy of sixteen with a few dollars sewn inside my vest. I was thrilled to breathe the fresh, strong air of freedom, and America has been good to me.'[121]

Yes; and Hollywood was even better. The scale of the operation was soon as big as Scattergood's power schemes. D. W. Griffith, director of the epoch-making *Birth of a Nation*, followed it with *Intolerance* (1916), an epic which cost $2 million and employed 15,000 extras. Its gigantic set of Babylon was the first to become a tourist attraction. By 1920, some 100,000 people were involved in producing movies in the Los Angeles area and they had become the biggest industry in the city, grossing a billion dollars a year. Movies made millionaires, or people who lived like them, and they in turn built themselves paradisical habitations which gave the city—the entire area indeed—its characteristic physiognomy of sentimental pandemonium. In 1920, Gloria Swanson, having starred in DeMille's *Male and Female* (1919) and *Why Change Your Wife?* (1920), built herself a twenty-two-room, five-bath sort-of-Renaissance palace in Beverly Hills, floored it in black marble, put in golden bathtubs and hung it in peacock silk, saying, 'I will be every inch and every moment a star.' Douglas Fairbanks put up a 'Hunting Lodge' near by on the top of the hill.

That did it—a colony of glittering eccentrics was founded, each in his or her own mini-Babylonian setting. And beyond Beverly Hills, in the valleys and canyons around, and in the lowlands to the sea, and in Los Angeles itself, architects and their indiscriminately picky clients competed to shock, astonish, and display. Egyptian and Mayan temples, Malay longhouses, Chinese and Siamese pagodas, Spanish baroque cathedrals, Romanesque churches, Renaissance palaces, Moorish and Arab mosques—all these forms became the models for houses, shops, office-buildings, and filling-stations. The Los Angeles Theater had a foyer modeled on Versailles' Hall of Mirrors. The lobby of the Tower Theater was a recreation of the Paris Opéra's. Many buildings were copied from pictures in children's story-books. There were Assyrian ziggurats and Babylonian Hanging Gardens. The 1926

Los Angeles Public Library, by Bertram Goodhue, was a combination of Roman, Islamic, Egyptian, and Byzantine styles. The Brown Derby Restaurant was in the shape of a brown derby hat. There was the Hansel and Gretel Cottage, another diner. The Chili Bowl Restaurant was just that—a painted, ferro-concrete chili bowl. Hoo Hoo I Scream was—an ice-cream parlor. A lot of the construction was made to look like food: tomato-sauce roofs, marshmallow igloos, pagodas of peanut butter, Florentine pink nougat, haciendas of chili con carne. There were houses which looked like dogs, fruit, vegetables, owls, pigs, and windmills.[122]

Into this phantasmagoria inevitably stepped Frank Lloyd Wright (1867–1959). He was from Chicago, where he had been a junior in the great firm of Adler & Sullivan. But the artistic, emotional, and cultural climate of Southern California suited him well. The series of important commissions he carried out in Los Angeles began in 1917 with a house for Aline Barnsadall, an oil heiress described as eccentric—egregious would be a better word. The house he built her was made of precast concrete blocks and looked like a Mayan sacrificial temple on the outside, the front door being of concrete. Inside, a moat ran round the fireplace. Construction was punctuated by some classic architect–client rows, a Wright specialty. He used blocks for a number of what he called Mesoamerican houses in Los Angeles. These 'textile blocks' (his name) often went into the interior as decorative walls, for instance in his dramatic Ennis House.[123]

Wright was a Hollywood character in his own dispensation, his life, and the houses he designed for himself, being a series of inexplicable dramas. In August 1914 his Modern Movement abode, Taliesin, was the setting for a massacre when his black servant, Julian Carlton, axed to death Wright's mistress, Mamah Borthwick, and six other people (some of whom he also burned). The murderer starved himself to death in jail. In 1925 Taliesin II, another ultra-modernistic concept, was struck by lightning and burned to the ground. Wright then built Taliesin III, where he installed an architectural community, rather on the lines of Brook Farm, but run by him with an iron, or perhaps one should say a ferro-concrete, hand. The Taliesin Fellowship, as he called it, was part-Arcadian community, part-William Morris enterprise, which involved group-therapy in the mode of Gurdjieff, his terrifying third wife Olgivanna having been a pupil of the sage. The Fellowship was essentially an attempt to solve Wright's insoluble financial problems by employing pupils and associates in endless Ruskinian hard physical work: 'Frank has reinvented slave labor,' as the joke went. The master was notorious for never paying anyone, 'except under the greatest duress.'

Taliesin III was also part-theater, with Wright as chief actor during his drafting sessions. Unitarian services were held on Sunday, Wright having rewritten 'Jesu Joy of Man's Desiring' as the Fellowship hymn ('Joy in work is man's desiring' etc.). There were lots of games, parties, sleigh trips, boating expeditions, and charades, with formal dinner on Sundays, the Wrights sitting on a dais like a king and queen. Recorded or performed music, chiefly Beethoven, was relayed by concealed loudspeakers in every room of the house. Offenders, who were many, were brought before a species of family court, presided over by Wright himself. At the age of eighty-nine, Wright testified in the witness-box he was 'the greatest architect in the world.' His wife told him modesty would have been more effective. Wright replied: 'You forget, Olgivanna, that I was under oath.' Half Taliesin III was built, the other half was falling down, and the whole was never finished. But some apprentices loved it and lived to flourish. And in California, at least, Wright had succeeded, by the end of the 1920s, in creating a prime testing area for the International Modern Movement.[124]

Ordinary Californians, however, continued to indulge their own peculiarities and prejudices. While the great architects thundered and created, the average Los Angeles householder preferred varieties of Spanish-American, just as English families choose mock-Tudor, if they get the chance. And Hollywood itself, after a period of turmoil, became as American as apple pie, albeit one with a lavish cream topping and heavily decorated with crystalized fruit. In 1920–1 the trial of Roscoe 'Fatty' Arbuckle, Hollywood's most highly paid star, the murder of the director William Desmond Taylor, the divorce of Mary Pickford, who played 'good girl' roles, and the death of the actor Wallace Reid from a drugs overdose combined to scare the studio moguls into cleaning up their scenarios. They hired Will H. Hays (1879–1954) of Indiana, former Republican National Chairman and Harding's Postmaster-General, at the then enormous salary of $100,000 to clean up Hollywood and lay down a code for the future. Hays cleared Arbuckle but he drove hundreds of others out of the industry on grounds of sexual depravity, homosexuality, drug abuse, and prostitution, and he inserted 'morals clauses' into stars' contracts. The Hays Code insisted that directors should avoid the following: kisses lasting more than seven feet of film, clergy in comic or villain roles, the 'explicit,' 'attractive,' or 'justified' treatment of adultery and fornication, nudity under any circumstances, sympathy for 'murder, safecracking, arson, smuggling etc in such detail as to tempt amateurs to try their hands,' and 'all low, disgusting, unpleasant though not necessarily evil subjects.' Positively directors should follow 'the dictates of good taste and regard for the sensibilities of the audience.' Hays went into considerable detail: if an actor or an actress were seated or lying on a bed, albeit fully clothed, one and preferably both should have one foot on the ground.[125]

Hays evoked sniggers at the time, and abuse since his death, but his rules were accepted and adhered to for forty years, during which Hollywood movies became a hugely successful industry and one of America's most lucrative and culturally effective exports. From 1923 studios ceased to delete American expressions and slang (and accents) for foreign, especially English-speaking, audiences because they found that Americanisms were part of the attractions of the product. At home, movies stressed patriotism, loyalty, truth-telling, family life, the importance and sanctity of religion, courage, fidelity, crime-does-not-pay, and the rewards of virtue. They also underpinned democracy, Republicanism, the rule of law, and social justice. Their presentation of American life was in all essentials the same as Norman Rockwell's *Post* covers. And the homogenizing effect, the encouragement to accept all-American norms, was far more successful than the crude social engineering of the Red Scare and Prohibition. The power of the movie increased dramatically after sound was introduced in *The Jazz Singer*, October 6, 1927, and the first full-length 'talking picture' was shown in July 1928. In 1928 only 1,300 of America's 20,000 movie houses were equipped for sound. By December 1930, over 10,000 were and the 57 million weekly admissions had doubled to over 100 million. For the next generation, movies, often under the pressure of religious groups (in 1934, at the request of the Catholic Church's Legion of Decency, a stricter Production Code went into effect, especially on choice of subjects), became the most formative influence on American society, and the chief projector of the American Way of Life abroad.[126]

This influence was reinforced by an artist of rare genius, Walt Disney (1901–66), who was also a showman and entrepreneur of unusual force. He came from Illinois, where his father was a farmer and small-time contractor. Disney had a strict Protestant upbringing, against which he rebelled, only to fall back into its assumptions as soon as he became a successful businessman. He did imitations of Hollywood comedians on the halls, then went into cartoon-animals in

Kansas. Once in Hollywood, he created Mickey Mouse, a 'little man' moral figure, and *The Three Little Pigs*, an anti-Depression stimulant, which presented determined hard work as the only protection from the Big Bad Wolf of despair. Like all great children's artists, Disney moralized the animal kingdom, but he Americanized it too, using the resources of high-technology animation, which he revolutionized and extended. His creative ideas were so commercially successful that they became clichés, like all great artistic innovations. Recent research has demonstrated that Disney's more adventurous concepts were discarded because of consumer resistance—*Fantasia*, the most innovative of the feature-length animated movies he drew and produced from 1938 onwards, was an exception—but his willingness to stick close to popular taste enormously increased the force of his moral impact. By weaving animal characters into a moral tale, which was itself underpinned by the Judeo-Christian message of the Decalogue and the Sermon on the Mount—Disney invented a new form of miracle play, a quasi-religious subculture which translated morally based fantasy into screen reality.

In 1951 Disney decided to give his two-dimensional message a third by designing, on a huge scale, the first Disneyland, which opened at Anaheim, California, in 1955. The inspiration for this was the architectural fantasies of Southern California we have already described, but Disney wove them into an optimistic tale about America, in which children took a real-life journey in a little train whose first stop was Main Street, USA, and last Tomorrowland, USA, all American history being briefly covered in the process. These three-dimensional experiences for children proliferated in America and were exported abroad.[127] Building on these creations, in 1965 Disney conceived yet another extension of the concept to what he called 'our Experimental Prototype Community of Tomorrow,' a playland area for children and adults, covering 47 square miles of what was Florida swampland near Orlando, run by 40,000 employees and incorporating fantasies from all over the world, from alps covered in fake snow to Polynesian beaches with fake rollers—Walt Disney World. It was opened in 1971, five years after his death, and again was duplicated in America and elsewhere.[128]

If the commercial system, and its manifestations such as Hollywood family movies and Disney's animated all-American bestiary, were homogenizing the inhabitants of the United States and prodding them onto the gentle slope of upward mobility, there was another force growing up in American society which was beckoning them, for the first time, in the opposite direction. This development has been so important, and is simultaneously so uncharacteristic of what America had hitherto stood for, and yet so emblematic of what America is now becoming, that it merits a little treatment in detail. When visitors from Europe and Asia began to come to the United States in considerable numbers, not as immigrants but as visitors, just after World War One, they came essentially in search of novelty—and they found it. There was so much home-grown novelty already sprouting that no one noticed European blossoms shouting for attention. This was new. In the 1880s, Oscar Wilde's exhibitionism had paid off handsomely. By 1920, when the Surrealist leader, Salvador Dali, arrived to advertise himself, and walked down Fifth Avenue carrying a four-foot loaf under his arm, no one took the slightest notice. 'In America,' he sadly observed, 'Surrealism is invisible for all is larger than life.' He added, 'Each American image I would sniff with the voluptuousness with which you welcome the inaugural fragrances of a sensational meal.'[129] By the 1920s, indeed, America had much to shock, enthrall and fascinate—mass motoring, screaming advertising, endless movies, records sold by millions, twenty-four-hour radio (the new word which America, followed by the world, was adopting to replace the staid English term 'wireless'). It had the 'funnies.' It had photojournalism of a kind Europe had not yet experienced. But, above all, it had jazz. When Arthur Schnabel toured America and was asked what sheet music he wanted to take back, he said, 'Jazz.' When Darius Milhaud was asked on his return what kind of music America had, he said, 'Jazz.' By the 1920s, in countless different ways, America was pulling the world behind it in the mass arts just as France, in the early modern period, had pulled the world behind it in the fine arts. But jazz was the exciting, creative element in this mass conquest, perhaps the only one that non-American creators respected.[130] So what did it mean?

Jazz entered at a tangent an American musical tradition which was rich in song but barren of much else. The Pilgrim Fathers and other settlers, being not only Protestants but Puritans, had no access to the English polyphonic musical tradition, which was maintained almost entirely by Catholics like Byrd, Gibbons, Dowland, Bull, Wilbe, and Weelkes. The Indian tradition was not attractive, at any rate according to Captain John Smith's description:

For their musicke they use a thick Cane, on which they pipe as on a Recorder. For their warres they have a great deepe platter of wood. They cover the mouth thereof with a skinne, at each corner they tie a walnut, which meeting on the backside near the bottome, with a small rope they twitch them together till it be so tough and stiffe, that they may beat upon as upon a drumme. But their chief instruments are Rattles made of small gourds or Pumeons shels. Of these they have a Base, Tenor, Countertenor, Meane and Treble. These mingled with their voyces, sometimes twentie or thirtie together, make such a noise as would rather affright than delight any man.[131]

By contrast black slaves were encouraged by the planters to express themselves in music. Whereas the visual and dramatic arts were banned as distracting, work-songs actually increased production. Their quills and pipes and possibly the banjo they brought from Africa. They proved extraordinarily skillful at using battered folk-fiddles, which took the lead melodic role later assumed by the clarinet, and making instruments out of pickaxes, washboards, knives, cans, and other percussive instruments. Negro spirituals came from European hymns, not from Africa, though some of the themes of the songs may have been African.[132]

From this unpromising mixture came two distinct but connected American traditions. The first, less pervasive, was white. Stephen Foster (1826–64) came from Pittsburgh (later Cincinnati) and was a mother-dominated, introspective, nostalgic genius, who died of drink at thirty-eight. From the English ballad, the French waltz, and Italian opera he wove a music folk-art of innocence and plaintive memory which at its best—'I dream of Jeanie with the light-brown hair' (1854)—is as good as a Schubertian song. His 'Ethiopian Songs' are based not on Negro spirituals but on the songs of the Christy Minstrel Show, the first minstrel being Daddy Rice, who toured the frontier settlements around 1830 with a parody–imitation of an aged rheumaticky negro from the Mississippi. The vogue climaxed just before the Civil War, dying with it. Foster's songs—'My Old Kentucky Home,' 'The Old Folks at Home,' 'The Camptown Races,' are the best—form its epitaph. His last song, 'Beautiful Dreamer,' adumbrates Tin Pan Alley. Foster's natural successor, also from Pittsburgh, was Ethelbert Nevin (1862–1901), a polyglot mixture himself who, in turn, added various elements that came to his hand to produce brilliant songs like 'Mighty Lak a Rose' and 'The Rosary.' His seventy songs are highly professional: indeed it can be said of the 19th-century American composers that they brought to popular music a new kind of professionalism, which incorporated their highly sophisticated approach to dancing. The beat was transcendent. This was particularly marked in the work of the third outstanding melodist, John Philip So, born in Washington, the son of a Spanish trombonist who played in the US Marine Band. So added USA to his name, making Sousa (1854–1932). He trained himself to become the world's greatest writer of marches, evolving a superb professionalism from his natural braggadocio, exhibitionism, and passion to entertain. 'A march,' he wrote,

> speaks to the fundamental rhythm in the human organization, and is answered. A march stimulates every center of activity, awakens the imagination . . . But a march must be good. It must be as free from padding as a marble statue. Every line must be carved with unerring skill—once padded, it ceases to be a march. There is no form of composition wherein the harmonic structure must be more clear-cut. The whole process is an exacting one. There must be a melody which appeals to the musical and unmusical alike. There must be no confusion in the counterpoint . . . A march must make a man with a wooden leg step out.[133]

Sousa was characteristically American in that he combined artistic professionalism, commercial skills, and a strong propensity to promote the feel-good factor. 'The Washington Post,' 'The Stars and Stripes Forever,' 'The Liberty Bell,' *El Capitan*, 'The Power and the Glory:' these are animal-uplift works. Sousa had his own band from 1892 until his death and recruited the trimmings of ceremonial marching, the twirling masses, the bounding girls in high hats, which became an integral part of it, celebrating the sheer delight in disciplined physical activity which is as American as (in another mood) Foster's nostalgia.[134]

All these forms of American music were eclectic. But jazz and the blues were the most eclectic of all in their historical origins. They were a product of the one aspect of the melting-pot in which blacks participated fully—sound (and behind the sound, sentiment). Blues in one sense derived directly from slave-work, from the 'field hollers,' but indirectly from western European ballads, the Bible-inspired spirituals, and the hot gospel-shouts of the camp-meeting. Jazz and rags were something different: they were mockery, criticism, covert protests against the triumphalism of the white man's world. The origin of the word 'rag' or 'ragging' lies in English schoolboy slang—it is the defiance of authority by misbehavior. Just as American idiom and slang achieved its impact and acquired its character by ragging British standard English, so negro American ragged middle-class Anglo-American English. Americans were fond of turning nouns into verbs, something the British thought uncivilized. Mencken, in his book *The American Language*, called it the greatest single characteristic of the American idiom—slum, hog, itemize, burglarize, bug, thumb, goose: all usages which once made Englishmen bristle with outrage (and still do in some cases). The blacks verbalized to an even greater extent—confidence, uglying away, feature, and so on. The charm, the defiance, lay in the flouting of convention.

The piano rag was a music version of this charm-through-defiance. Before the Civil War, blacks rarely had access to pianos. But they could sing. Their status gave them an instinct for parody. Thus Bellini's *La Somnambula*, an 1837 hit in New York, was burlesqued or ragged as *The Roof Scrambler*. Black amateur musicians jazzed or ragged the classics by altering the beat or the stress, as well as the words. Blacks proved themselves brilliant at playing with rhythms to produce ironic versions of music in the Western tradition. The beat, the rhythm was the key. The black singer John Bubbles, with 'Rhythms for Sale,' made the point: 'All the world wants rhythm bad / Blacks are mighty glad / Cause they got rhythm for sale.' From the 1860s onwards, blacks learned the piano, often on battered, second-hand models. As the British had discovered as far back as the 1790s, the piano was a consumer durable which was extraordinarily well adapted to mass production. As early as the 1830s, America was buying and making pianos in vast quantities. Virtuosi pianists were the most popular visitors—Leopold de Meyer in the 1840s, Sigismond Thalberg ('Old Arpeggio') in the 1850s, and then in quick succession Anton Rubinstein, Hans von Bulow, Josef Hofmann, Ferruccio Busoni, Ignacy Paderewski: these were the idols. In the two decades 1870–90, purchases of pianos increased 1.6 times faster even than the rapidly rising population, and in the 1890s the increase was 5.6 times; even this was beaten in the decade 1900–10, when it rose to 6.2 times. In the peak year, 1909, 364,595 pianos were sold in the US. 'There is probably no other country in the world,' wrote one music scholar in 1904, 'where piano-playing is so widespread as in America.'[135] The playing on the piano of classical works was the epitome of Victorian values, involving hard work, producing moral and cultural uplift, and stressing domesticity. In its proper usage, the piano was overwhelmingly a feminine instrument. In

1922, for instance, women formed 85 percent of music students and 75 percent of concert audiences—the piano being the center of both activities. Thus male abuse of the piano to produce rags was a protest against domestic discipline and feminine dominance, just as black ragging of white tunes was an act of defiance.[136]

There was also the question of speed. African songs were slow. The negro spirituals of the Deep South were slow. When the blacks came north after the Civil War, they needed to speed up in response to American hustle. Jazz and rag was their answer. Ragtime was percussive as well as fast—hence the gangster term for a submachine-gun, a Chicago Piano. When the blacks came north and started to use pianos on a large scale, they made them the equivalent of an African drum. They stressed particularly the 'nigger keys,' the five blacks, the African pentatonic or five-note mode as opposed to the 'white keys' of the Western diatonic scale. Ragging was syncopating, putting the stress off the beat or onto the weak beat.[137] It was genuine melting-pot because the left-hand bass performed a steady 2/4 Western-style march time while the right-hand treble did the Afro-syncopation. (It is interesting that Sousa sometimes ragged his marches in performance, though never on record; so no one could prove he had 'sunk so low.') It is possible that the piano rag was a deliberate attempt by some gifted blacks to make a distinctive contribution to 'artistic' music and was designed to be notated. It may have been related to the black dance-form of the cake-walk, with an element of parody, and signified a conscious black attempt to get their own back on white classicism. Scott Joplin, greatest of the piano-raggers, was a sophisticated musician well read in classical music, who even tried to compose a rag-opera, *Treemonisha*, but it failed. Joplin wrote down all his rags. The first accurate description of ragtime dates from 1885, and at the Chicago Fair of 1893 a group of blacks astonished audiences by playing it. The first sheet music of rag was published two years later, about ten years after it became a form, and in contrast to blues, which were not written down for at least half a century after they were first sung.[138]

Rag developed in Northern cities like Chicago and on the borders in places like St Louis and Louisville. Jazz, a melting-pot music from fife-and-drum bands, spirituals and black string bands, with rag as a dash of spice, came from Mobile, Alabama, and above all from New Orleans, especially from its Storyville red-light district. These Gulf towns were egregiously multiethnic and multiracial—always had been—and it is impossible to identify all the various historic strands which went into this close-woven mesh. Ferdinand 'Jelly Roll' Morton on the piano and Charles 'Buddy' Bolden on the cornet—performed at funerals, house-parties, and clubs. It was usual for a New Orleans funeral band to perform 'straight' on the way to the cemetery, then rag, syncopate, and burlesque on the way back to the party. Joe 'King' Oliver and Louis Armstrong—both trumpeters—also came from New Orleans, taking their music in time to Chicago and fame. Edward 'Duke' Ellington, by contrast, came from Washington DC and moved to New York to work in the Harlem Cotton Club. In their origins at least, neither rag nor jazz was very far from the honky-tonk or even the brothel, and their loud, percussive characteristics evolved from the need to be heard over the smoky hubbub. 'Jazz' was a black term for sexual intercourse; so was 'eagle-rocking' and 'boogie-woogie,' a black variation of 'humpy-pumpy.' All this was not incompatible with a religious background too. 'T-Bone' Walker was quoted as saying, 'The blues come a lot from the church. The first time I heard a boogie-woogie piano was the first time I went to church.'[139]

To upright, middle-class American women it all sounded mighty suspect. The suffragists and feminists, led by Carrie Chapman Catt (1859–1947), president of the National American Women's Suffrage Association, were particularly outraged, on behalf of women performers, at what they termed the 'rape' of the piano by the raggers, who played 'nigger whorehouse music,' aided, increasingly, by 'clever, unscrupulous Jews.' The National Federation of Women's Clubs insisted they would wrest American music from 'the hands of the infidel foreigner' and 'black slum-dwellers.' At their insistence, the American Federation of Musicians (1914) pledged they would not play ragtime. The *Musical Observer* of September 1914 wrote that there was no room for ragtime 'in Christian homes where purity of morals are [sic] stressed.' In August 1921 the *Ladies Home Journal* published an article asking 'Does Jazz Put the Sin in Syncopation?' Militant women linked the spread of 'negro music' to the rise in illegitimacy rates. There was a strong feminist element in the opposition to jazz, linked to anti-semitism, xenophobia, and opposition to unrestricted immigration. In New York there was a spectacular increase in the number of professional actresses from 780 in 1870 to 19,905 in 1921. They were strongly feminist. Thus the great stage star Lillian Russell (1861–1922), daughter of the pioneer suffragist Cynthia Russell, who started the Women's Club Movement in Chicago, was not merely a beauty but a brain too. One New York newspaper remarked, 'The beauty of Lillian Russell is as much an institution as Niagara Falls or the Brooklyn Bridge,' and they instanced her, with her pink-and-white looks and direct descent from the Pilgrim Fathers, as evidence that America was producing 'a race of queens.' Though herself notable for debts, multiple husbands, and still more numerous lovers, she campaigned for New York mayor in 1915 on a morality platform linked to strict immigration laws—'Our melting-pot has become too crowded,' as she put it.[140]

Feminine opposition to the black element in the new music led to considerable humbug and downright rewriting of history by white artists who adopted it. One example was Irving Berlin (1888–1989), the greatest white ragger. Born Israel Baline in Russia, the son of a cantor, he crossed the Atlantic in 1893 steerage with five brothers, with eight suitcases between them, and went to live in Cherry Street, the worst part of the Lower East Side. He attended school for only five years, then at twelve went on the streets and into dives to sing for money. He survived to compose an enormous treasure-house of 355 hits, jealously guarded by his own publishing company (1914). He burlesqued Paderewski, the 'long-haired genius,' with his exaggerated hand-gestures at the piano, and he ragged America's welcome to Verdi in 'Watch Your Step.' His 'Alexander's Ragtime Band,' a work which made him 'King of Ragtime' (1912), was fundamentally a straight march but its lyrics were ragged—Berlin found you could rag (or 'slang') a song by syncopating its words and shifting the emphasis on to the weaker beat. Berlin denied the black origin of his music. He argued that Jews had been part of Broadway and New York pop culture since the 1850s, whereas the blacks were post–Civil War newcomers. 'Our popular songwriters,' he claimed, '. . . are not negroes but of pure white blood, many of Russian ancestry.'[141] He came from a parallel anti-authoritarian tradition of mockery, of despised Jews v. mastergoys. Yiddish was an alien-pop version of High German in the same way as black English was an alien-pop version of standard American-English. His first paid job was as a 'singing waiter' at Nigger Mike's in New York's Chinatown (the owner was a Russian Jew, posing as a black), where he did parody acts as an Italian singer, a Jewish one, a German, an Irish, and a black. All was elliptical, inverted, upside down, burlesqued. He called himself Cooney. When he got to the top he kicked away any black rung in the ladder whereby he had climbed.

A similar case was Al Jolson, another cantor's son, born Asa Yelson near St Petersburg in the early 1880s. Exceptionally nervous for a professional, he was told by a black: 'Yous'd be much funnier, boss, if you blacked your face like mine. People always laugh at the black man.' He followed this advice and became 'the blackface with the Grand Opera voice,' whose 1925 movie, *The Jazz Singer*, made him a great star. On Broadway, went the adage, you started black or ethnic and became steadily whiter and more Waspy as you succeeded. The classic case of this ascent was Fred Astaire (1899–1987), a second-generation immigrant from Omaha, Nebraska, born Frederick Austerlitz, who started as a black-style dancer and gradually turned it into what Mencken called 'high-toned shindig.' Astaire went into the new black-into-white style of crooning when the electric microphone arrived in 1925, and he transformed himself from a black–Jewish song-and-dance artiste into a white-tie-and-tails Fifth-Avenue–Park-Lane sophisticate. That involved ladder-kicking too. The most blatant of the whites-only jazz musicians was the aptly named Paul Whiteman (1890–1967), a hugely greedy man weighing over 300 pounds who originally played the viola in the San Francisco Symphony Orchestra, then came to New York to make himself the 'King of Jazz.' His first (1920) hit record was 'Avalon,' a pop version of an aria from Puccini's *Tosca*, and he followed the familiar path of burlesquing, sending up, playing for laughs. He engaged in an elaborate and highly successful effort to make jazz respectable by staging a big concert at the Aeolian Hall on February 12, 1924, at which the *Rhapsody in Blue* by George Gershwin (1898–1937) received its first performance. But he insisted that jazz was a white invention, his band was all-white, and in his book *Jazz* (1926) he does not mention blacks at all.[142]

Ultimately the attempt to write blacks out of the script was a total failure. Indeed, even in the early 1920s, blacks were succeeding in getting some of the credit and the limelight. Black bands were welcome in Paris, as was Florence Mills in London. *The Plantation Review* (1922), *Dixie to Broadway* (1924), and *The Blackbirds* (1926) were New York hits. And Josephine Baker, after her 1925 success in Paris with *La Revue Nègre*, in which she did a bare-breasted *danse sauvage*, returned to triumphs in America too. As early as 1922, in the Ziegfeld Follies, Gilda Gray sang 'It's Getting Dark on Broadway,' with its punchline, 'You must black up to be the latest rage.'[143] What was discovered in the 1920s, for the first time, was that black music had a lure of its own precisely because it was black, and that downward mobility was an important element in the new art-form. Going to nightclubs and joints where jazz was played was one way in which upper-class debutantes themselves ragged society dances they were expected to attend. The Broadway music publisher Edward Marks made the point: 'The best songs come from the gutter.' Talking ungrammatical English was a form of social protest against established mores. Irving Berlin knew when to use correct English, as in his 'God Bless America;' he used the other kind, as in 'Ain't You Going,' for deliberate ragging. It was one of the ironies of history that intensely upwardly mobile figures like himself, coming from nowhere to become multimillionaires and pillars of the establishment (he eloped with the 'society rose' Ellin Mackay in 1926 and spent the last decades of his life assiduously tracking down any infringement of his royalties), were those who also, when it suited their commercial purposes, encouraged the taste for downward mobility among the young. A similar tendency was observable among black writers, such as Langston Hughes (1902–67), poet, playwright, novelist, and journalist, and star of what came to be called the Harlem Renaissance of the 1920s and 1930s: he infuriated the conventionally upwardly mobile black middle class by deliberately moving his idiom downmarket. Black reviewers called him the 'Sewer Dweller,' for pretending to write poetry which was in fact 'trash reeking of the gutter.'[144]

It is important to grasp that downward mobility was, initially, only a tiny fracture in the smooth monolith of America's upwardly mobile society. Jazz, for instance, was an elite taste in the 1920s. Not until it was simplified, purged, and made respectable by the white bands of Glenn Miller and Tommy Dorsey (becoming 'swing' in the process) did it begin to conquer the masses from 1935 onwards. Then in the 1940s came bop, with black players like Charlie Parker (tenor sax), Dizzy Gillespie (trumpet), Thelonious Monk (piano), and singers like Ella Fitzgerald, doing bebop. There followed 1950s cool, hard bop, soul jazz, rock in the 1960s, and in the 1970s blends of jazz and rock dominated by electronic instruments. And all the time pop music was crowding in to envelop the various styles and traditions in the phantasmagoria of commercial music geared to the taste of countless millions of easily manipulated but increasingly affluent young people. And, from the worlds of jazz and pop, the drug habit spread to the masses as the most accelerated form of downward mobility of all.[145]

All this was a cloud no bigger than a child's hand in the 1920s, where critical attention was focussed more on the rise of the American musical, itself a melting-pot amalgam of the Viennese operetta, the French boulevard music-play, the English Gilbert and Sullivan comic opera and music hall—*The Beggar's Opera* of 1728 was the ultimate ancestor—plus the all-American ingredients of burlesque, minstrel-show, vaudeville (and jazz), becoming in the process a completely new and hugely attractive art-form. The prewar talents of Berlin and Jerome Kern were joined, after the end of the war, by a host of newcomers, some of them close to the genius level: in addition to Gershwin, Richard Rogers, Howard Dietz, Cole Porter, Vincent Youmans, Oscar Hammerstein, Lorenz Hart, and E. Y. Harburg. Together they brought the American musical to full flower: Gershwin's *Lady Be Good!*, the first mature example, opened on December 1, 1924 in the Liberty Theater, starring Fred Astaire and his sister Adele. It was an outstanding event of a Broadway season that included forty musicals, such as Kern's *Sitting Pretty*, Irvin Berlin's *Music Box Review*, Youmans' *Lollypop, The Student Prince*, and Sissie and Blake's *Chocolate Dandies*.[146] Twenties New York did not quite have the cultural *réclame* of Weimar Germany at this time but it was the place where the native creator had the widest range of opportunities and where the expatriate artist was most likely to find the freedom, the means, and the security to express himself.

The keynote of the 1920s musical was joy, springing from an extraordinary exuberance in the delight of being alive and American. The central paradox of the 1920s—probably the most enjoyable decade in American history—was that America, having voluntarily saddled itself with Prohibition, took to partying on a scale never seen before. Ernest Hemingway (1898–1961), who fought his way to the top of the young writers' league during the decade, coined the phrase, soon universal: 'Have a drink.' The Twenties also saw the emergence into open society of a secret drink, first made in California (probably San Francisco) in the 1860s, of gin, sweet vermouth, and angostura. In 1920s New York it was made with minute quantities of dry vermouth and rechristened the Dry Martini. Bernard de Voto called it 'the supreme American gift to world culture.' H. L. Mencken thought it 'the only American invention as perfect as a sonnet.' If not sonnets, then ingenious verses were written to it, including a notable one by Ogden Nash and a quatrain by the 1920s Egeria, Dorothy Parker: 'I like to have a Martini / Two at the very most. / After three I'm under the table. / After four I'm under the host.'

Warren Harding did not drink Martinis, however; he drank whiskey. And he chewed tobacco. As Thomas Edison put it, 'Harding is all right—any man who chews tobacco is all right.' (This was said with general male approval. Yes; the Twenties are a long time ago now.) In his own humdrum way, Harding shared in the general joy of the time. He took the press into his confidence and, as he got to know reporters, called them by their Christian names. When he moved, he liked to surround himself with a vast traveling 'family,' many invited on the spur of the moment, who occupied ten cars in his presidential train. He would asked people up to his bedroom for what he called a 'snort' and twice a week he invited his intimates over for 'food and action' (action meant poker). Commerce Secretary Hoover, a stuffed shirt (quite literally; he never wore anything else after 6 P.M.) was the only one who declined to play: 'It irks me to see it in the White House.'[147]

Harding's administration was a strong and in some ways a success-ful one. In addition to Hoover, the Cabinet included Charles Evans Hughes as secretary of state and Andrew Mellon at the Treasury. The Cabinet list was a cross-section of upwardly mobile America: a car manufacturer, two bankers, a hotel director, the editor of a farm jour-nal, an international lawyer, a rancher, and an engineer. It included only two professional politicians. Harding inherited from the comatose Wilson regime one of the sharpest recessions in American history. By July 1921 it was all over and the economy was booming again. Harding and Mellon had done nothing except cut government expendi-ture by a huge 40 percent from Wilson's peacetime level, the last time a major industrial power treated a recession by classic laissez-faire meth-ods, allowing wages to fall to their natural level. Benjamin Anderson of Chase Manhattan was later to call it 'our last natural recovery to full employment.'[148] The cuts were not ill-considered but part of a careful plan to bring the spending of the monster state which had emerged under Wilson back under control. The Budget and Accounting Act (1921) created a Bureau of the Budget, to subject authorizations to sys-tematic central scrutiny and control. Its first director, Charles Dawes, said in 1922 that, before Harding, 'everyone did as they damn well pleased,' Cabinet members were 'comanchees,' Congress 'a nest of cowards.' Then Harding 'waved the axe and said that anybody who didn't cooperate his head would come off.' The result was 'velvet for the taxpayer.'[149]

Yet Harding usually emerges from polls of American citizens and professional historians alike as the least respected of the presidents. He illustrates the adage, demonstrated under Grant and proved again and again in the second half of the 20th century: 'A president cannot be too careful.' Harding was careless in his choice of friends and colleagues. Or, to put it another way, he was generous and unsuspicious. The only specific charge of dishonesty brought against him personally was that the sale of the *Marion Star* was a fix. This was decisively refuted in court, the two men who brought suit collecting $100,000 in libel damages. But Harding made one colossal error of judgment: appointing Albert Fall, the florid Senator for New Mexico, his Interior Secretary. In believing Fall honest he was in good, or at least in numerous company. The Senator sported a handlebar mustache and wore a flowing black cape and broad-brimmed stetson, the very picture of old Southern–Western 'normalcy.' He was so popular that, when his nomination went before the Senate for approval, it was confirmed by immediate acclamation, the only time in American history a Cabinet member has received such a vote of confidence.[150] Harding's second error was believing his Attorney-General, his old Ohio campaign-manager Harry Daugherty, who promised the President he would screen him from the influence-peddlers who swarmed up from his home state. 'I know who the crooks are, and I want to stand between Harding and them,' was Daugherty's boast—which proved an idle one.[151]

The result was a series of blows which came in quick succession from early 1923. In February Harding discovered that Charles Forbes, director of the Veterans Bureau, had been selling off government medical supplies at very low prices. Harding summoned him to the White House, shook him 'as a dog would a rat,' and shouted, 'You double-crossing bastard!' Forbes fled to Europe and resigned, February 15.[152] On March 4 Albert Fall resigned. It was later established he had received a total of $400,000 in return for granting favorable leases of government oilfields at Elk Hills in California and Salt Creek (Teapot Dome) in Wyoming. Fall was eventually jailed for a year in 1929. His leases later turned out well for America, since they involved building vital pipelines and installations at Pearl Harbor. But that was not apparent at the time and Fall's exposure was a disaster for Harding. It was quickly followed by the guilty suicide of Charles Cramer, counsel for the Veterans Bureau.[153]

Finally on May 29 Harding forced himself to see a crony of Daugherty's, Jess Smith, who together with other Ohioans had been selling government favors from what became known as 'the little green house [no. 1625] on K Street.' The 'Ohio Gang,' as the group was soon called, had nothing to do with Harding and it was never legally established that even Daugherty shared their loot (when tried in 1926–7 he refused to take the stand, but was acquitted). But after Harding confronted Smith with his crimes, the wretched man shot himself the following day and this second suicide had a deplorable effect on the President's morale. According to William Allen White (not always a reliable witness), Harding told him: 'I can take care of my enemies all right. But my damned friends, my God-damn friends, White, they're the ones who keep me walking the floors nights.'

Given time, Harding would certainly have managed to stabilize the situation and refute the rumors of guilt by association, as have several presidents since. For his own hands were completely clean, so far as the latest historical research has been able to establish. But the following month he left for a trip to Alaska and the West Coast. Already a prime candidate for a heart attack (autopsy showed his heart was badly enlarged), he tried to exorcise the scandals in the minds of the people by frantic activity. In a Seattle motorcade he 'pumped his arm up and down for hours as he tipped his hat,' straining his heart still further. He collapsed as his train neared San Francisco but, unwilling to disappoint people, he insisted on donning morning dress and walked unaided up the steps of the Palace Hotel, to the cheers of the crowd. As soon as he reached his room he fell head-first across the bed, where he died three days later of what the doctors called 'apoplexy,' which was in fact a massive coronary.[154] Harding's funeral train moving east was the occasion of extraordinary demonstrations of public affection for the man who, unlike Taft and Wilson, 'looked like a president.' In Cheyenne immense crowds stood in a dust-storm, in Chicago they filled the freight-yards until the train could not move: Harding was the kind of president American people of all classes love—kind, genial, decent, ordinary, human, one of them.

The deconstruction of the real Harding and his reconstruction as a crook, a philanderer, and a sleazy no-good was an exemplary exercise in false historiography. It began in 1924 with a series of articles in the *New Republic* by its imaginative and violently anti-business editor, Bruce Bliven. He created the myth that the Ohio Gang, run by Daugherty, had deliberately recruited Harding in 1912 as a front man as part of a long-term conspiracy to hand over America to Andrew Mellon and Big Business.[155] It now seems there was no evidence whatsoever for this invention, and it is not surprising that Bliven went on, in the 1930s, to become a credulous propagandist for the Communist-run Popular Front. Then in 1926, a novel, *Revelry*, describes a guilty president who poisons himself to escape scandals and exposure. It took in the disapproving Hoover, who always thought he would have made a better president than Harding. He read it in manuscript and told a friend it described 'many things which are not known.'

The novel's success in turn prompted Nan Britton, an Ohio girl, daughter of a Marion doctor, to publish in 1927 *The President's Daughter*, asserting she had had a baby girl by Harding in 1919. She claimed that she had been seduced in Harding's then office in the Senate, and that their affair continued, Harding writing her many letters. Even at the time she failed to produce these incriminating letters. Recent research has established that she was the local 'fast' girl, whose embarrassing crush on Harding had led to trouble for the unsuspecting man—including blackmail—though it is likely he was unable completely to resist his 'stalker.' The child did exist, but the father may have been one of many men, and Britton's descriptions of later hotel assignations with Harding have been disproved by research into hotel registers.[156]

The attacks on Harding continued with the publication in 1928 of *Masks in a Pageant* by the inventive William Allen White, who repeated the conspiracy theory in this book and again, ten years later, in his 'life' of Coolidge, *A Puritan in Babylon*. In 1930 a former FBI agent, Gaston Means, produced the bestselling *The Strange Death of President Harding*, portraying wholly imaginary drunken orgies with chorus girls at the K Street house, with Harding prominent in the 'action.' The book has now been shown to be a catalog of ghost-written lies. Equally damaging was the 1933 memoir by TR's daughter, Alice Roosevelt Longworth, *Crowded Hours*, which presented Harding's White House as a speakeasy: 'The air heavy with tobacco smoke, trays with bottles containing every imaginable brand of whiskey stood about, cards and poker chips ready at hand—a general atmosphere of waistcoat unbuttoned, feet on the desk and the spittoon alongside . . . Harding was not a bad man. He was just a slob.' It now emerges that Mrs Longworth, notorious for her sharp tongue and amusing *esprit d'escalier*, bitterly resented the fact that Harding, rather than her bibulous husband, Speaker of the House Nicholas Longworth, got to the White House.

To cap it all, an apparently careful work of research by a *New York Post* writer, Samuel Hopkins Adams, *The Incredible Era: the Life and Times of Warren Gamaliel Harding* (1939), welded together all the inventions and myths, plus a few fibs of his own, into a solid ortho-doxy. By this time, the notion of Harding as the demon king of the

Golden Calf Era had become the received version of events not only in popular books like Frederick Lewis Allen's *Only Yesterday* but in standard academic history—though even in that category some reputable scholars, like Allan Nevins, have now been shown to have had personal scores to settle against Harding.[157] When in 1964 the Harding Papers (which had not been burned, as alleged) were opened to scholars, no truth at all was found in any of the myths, though it emerged that Harding, a pathetically shy man with women, had had a sad and touching friendship with the wife of a Marion store-owner before his presidency. The Babylonian image was a fantasy, and in all essentials Harding had been an honest and shrewd president, prevented by his early death from overwork from becoming, perhaps, a great one. The experience should encourage the historian to look more closely at other accepted presidential myths.

Harding's successor, his Vice-President Calvin Coolidge (1872–1933), came not from a small town of the old Midwest but from rustic Vermont, a district even more closely associated with the pristine values of the American 'City on a Hill.' Vermont was the only New England state without a coastline, and therefore largely untouched by the immoral infiltration of commerce. It was the first state to join the original thirteen, in 1791, and it was by no means unprogressive. In fact its state Constitution was the first to abolish slavery and establish universal manhood suffrage. But it was, and is, rural conservative.[158] In Coolidge's day it lived chiefly by dairy farming and he was brought up on a farm, near the little town of Plymouth. His father, Colonel Coolidge of the militia, worked the farm himself. This was by no means unusual then: in some ways America was still a farmer's country. Indeed, when Vice-President Coolidge was summoned to the White House in August 1923, he was at his father's farm, spending two weeks of his vacation helping to get in the hay, swinging a scythe, handling a pitchfork, and driving a two-horse 'hitch.' This was not done for a photocall either, for no photos were taken. Coolidge never had a press secretary in his life. He would not have dreamed of calling a reporter by his first name, as Harding did, and no reporter, so far as is known, was ever welcome at the Coolidge farm until the presidency descended on him.[159]

The scene when the news penetrated to Plymouth on the night on August 2 that the local boy was now the thirtieth President was indeed arcadian. There was no phone at the farm, the nearest being 2 miles down the hill. The Coolidge family were awakened by a Post Office messenger pounding on the door. He brought two telegrams: one from Harding's secretary giving official notification of the President's death, the second from the Attorney-General advising Coolidge to qualify immediately for the office by taking the oath. So the oath was copied out and Coolidge's father, being a notary public, administered it, by the light of a kerosene lamp, for there was no electricity at the house. It was just a tiny farmhouse sitting-room, with an airtight wood stove, an old-fashioned walnut desk, a few chairs, and a marble-topped table on which stood the old family Bible, open. As he read the last words of the oath, the younger Coolidge placed his hand on the book and said, with great solemnity, 'So help me God.'

Coolidge was not all that remote from our times. He was born in July 1872, a few weeks after Bertrand Russell, whom the author of this work used to know well. That summer, Verdi's *Aïda* was the hit opera and George Eliot's *Middlemarch* was the most talked-about novel. Coolidge saw himself as go-ahead in his own way. He liked to quote Sydney Smith: 'It is a grand thing for a man to find his own line and keep to it—you go so much faster on your own rail.' He declined to follow his father into farming and chose his own line of law and public service—nor did he seek a partnership in an established firm but put out his own sign in Northampton, Massachusetts, at the age of twenty-five, 'Calvin Coolidge, Attorney and Counselor-at-Law.' Two years later he took his first step on the political ladder, as a Republican city councilman, followed by election as city solicitor, two terms in the state legislature, a spell as mayor of Northampton, followed by service in the state Senate as president and then two terms as governor.[160]

Like his predecessor Harding, only more systematically and of set purpose and belief, Coolidge was a minimalist politician. He thought the essence of the republic was not so much democracy itself as the rule of law, and that the prime function of government was to uphold and enforce it. Of course government had an enabling function too. As a city administrator, he took steps to enable local farmers to provide citizens with an adequate supply of fresh milk at competitive prices. He took enormous trouble in supervising railroad Bills to enable the companies to provide reliable and cheap public transport in Massachusetts. He was accomplished at both city and state finance, paying off debt, accumulating surpluses, and so, as a result, raising the salaries of state teachers and attracting the best. Examination of his record in Massachusetts both as legislator and as governor shows in detail that he was not a 'property-is-always-right man.' Quite the contrary. He loathed the pressure-group and lobby system of powerful property interests. He was a 'the-law-is-always-right' man. As governor he made an important statement on the freedom of the elected individual to ignore bullying by the interests and the media. He said: 'We have too much legislating by clamor, by tumult, by pressure. Representative government ceases when outside influence of any kind is substituted for the judgment of the representative.' Voters have the right to vote, but a representative, having been voted into office, must use his judgment. Edmund Burke had said the same thing 150 years before. Coolidge added: 'This does not mean that the opinion of constituents is to be ignored. It is to be weighed most carefully, for the representative must represent, but his oath provides that it must be "faithfully and agreeably to the rules and regulations of the Constitution and laws." Opinions and instructions do not outmatch the Constitution. Against it they are void.' For a state like Massachusetts to pass a law providing for the manufacture of light beer and wines, the so-called 'Two-and-a-Half-Per-Cent Beer Bill,' in defiance of federal Prohibition, was an insult to law. He called it indeed 'nullification,' the unlawful course of the rebellious South, in defiance of the Constitution, which he had no alternative as governor but to veto. 'The binding obligation of obedience [to the law] against personal desire,' he said, was the essence of civilized, constitutional government, without which 'all liberty, all security is at an end' and 'force alone will prevail.' 'Can those entrusted with the gravest authority,' he continued, 'set any example save that of the sternest obedience to law?'[161]

An absolute adherence to the principle of the rule of law, and a meticulous attention to its details, was what distinguished Coolidge's successful handling of the 1919 Boston Police Strike, an event which brought him to the attention of the entire nation. Coolidge's conduct was marked by a willingness to take on any group in society, however powerful—in this case the American Federation of Labor—in defense of the law, by an insistence that the duly constituted authority, in this case the Boston Police Commissioner, be left to exercise his judgment and powers until such time as he publicly confessed that the situation was beyond his control, and then by an equal willingness to exercise the full constitutional powers of the governorship, including his rights as the commander-in-chief of the State Guard, which was called out in its entirety. The policy was minimalist until both the facts of the case and the state of public opinion demanded maximalist measures, which had been carefully and secretly prepared before and were then put into action immediately and in full. It was also backed by a well-formulated and easily grasped expression of political philosophy: 'There is no right to strike against the public safety by anybody, at any place or at any time.' Coolidge's handling of this dangerous strike, at a time when public order was under threat virtually all over the world, became a model for any chief executive to follow at either state or federal level. It was evidently seen as such by both the political class and the whole nation at the time, and prepared the way for Coolidge's nomination as vice-presidential candidate the following year.[162]

However, though propelled to national attention by the vigorous and unhesitating exercise of gubernatorial authority, Coolidge was anxious to reassure the nation that such state intervention was for extreme emergencies only, and that in normal times minimal government must *be* the norm. In his acceptance speech, indeed, he spoke of 'restoring the Lincoln principles' by insisting on 'a government of the people, for the people and by the people.' He made it absolutely clear what he meant by this: 'The chief task which lies before us is to repossess the people of their government and their property.'[163]

Coolidge's minimalism was not just an expression of a political philosophy, though it was certainly that. As a prosperous nation with a largely self-regulating economy and protected by great natural defenses, America was in a position to follow the advice of Lord Salisbury, who had governed Britain when Coolidge was a young man. 'The country is carried comfortably down the river by the current, and the function of government is merely to put out an oar when there is any danger of its drifting into the bank.' That was the Coolidge philosophy too, but it was more than a philosophy, it was a state of mind, almost a physical compulsion. Coolidge, like the great Queen Elizabeth I of England, was a supreme exponent of masterly inactivity. But he was also, unlike that Queen, who could be talkative at times, a person who devoted much thought and a lifetime of experience to strategies of silence. He got this from his father, but whereas the Colonel was silent by instinct, Coolidge turned it into a political virtue. He rejoiced in his nickname, 'Silent Cal'—it often saved him from taking steps or making statements that might prove counterproductive.

A reputation for silence was itself a form of authority. As president of the state Senate in 1914, Coolidge delivered the shortest inaugural on record. It is worth recalling. Here it is, in its entirety: 'Do the day's work. If it be to protect the rights of the weak, whoever objects, do it.

If it be to help a powerful corporation better to serve the people, what-ever the opposition, do that. Expect to be called a standpatter, but don't be a standpatter. Expect to be called a demagogue but don't be a demagogue. Don't hesitate to be as revolutionary as science. Don't hes-itate to be as reactionary as the multiplication table. Don't expect to build up the weak by pulling down the strong. Don't hurry to legislate. Give administration a chance to catch up with legislation.' Good points, well noted. Reelected without opposition, he made his second inaugural even shorter—a mere four sentences. 'Conserve the firm foundations of our institutions. Do your work with the spirit of a sol-dier in the public service. Be loyal to the Commonwealth, and to your-selves. And be brief—above all things, be brief.'[164]

He practiced this brevity. Often he said nothing whatever. Campaigning in 1924, he noted: 'I don't recall any candidate for presi-dent that ever injured himself very much by not talking.' Or again: 'The things I never say never get me into trouble.' When he finally retired he confessed that his most important rule 'consists in never doing anything that someone else can do for you.' He added: 'Nine-tenths of a presi-dent's callers at the White House want something they ought not to have. If you keep dead still they will run out in three or four minutes.' Coolidge was usually silent, but slight twitches in his facial muscles spoke for him. He was described as 'an eloquent listener.'

Yet when he did speak, what he said was always worth hearing. It was direct, pithy, disillusioned, unromantic, and usually true. No one in the 20th century defined more elegantly the limitations of govern-ment and the need for individual endeavor, which necessarily involves inequalities, to advance human happiness. Thus: 'Government cannot relieve from toil. The normal must take care of themselves. Self-govern-ment means self-support . . . Ultimately, property rights and personal rights are the same thing . . . History reveals no civilized people among whom there was not a highly educated class and large aggregations of wealth. Large profits mean large payrolls. Inspiration has always come from above.' It was essential, he argued, to judge political morality not by its intentions but by its effects. Thus, in his 1925 inaugural, the key sentence was 'Economy is idealism in its most practical form.'

Later that year, in an address to the New York Chamber of Commerce, Coolidge produced a classic and lapidary statement of his laissez-faire philosophy. Government and business, he said, should remain independent and separate, one directed from Washington, the other from New York. Wise and prudent men should always prevent the mutual usurpations which foolish men sought on either side. Business was the pursuit of gain but it also had a moral purpose: 'the mutual organized effort of society to minister to the economic requirement of civilization . . . It rests squarely on the law of service. It has for its main reliance truth and faith and justice. In its larger sense it is one of the greatest contributing forces to the moral and spiritual advancement of the race.' That was why government had a warrant to promote its success by providing the conditions of competition within a framework of security. The job of government and law was to suppress privilege wherever it manifested itself and uphold lawful possession by providing legal remedies for all wrongs: 'The prime element in the value of all property is the knowledge that its peaceful enjoyment will be publicly defended.' Without this legal and public defense 'the value of your tall buildings would shrink to the price of the waterfront of old Carthage or corner-lots in ancient Babylon.' The more business regulated itself, he concluded, the less need there would be for government to act to insure competition. It could therefore concentrate on its twin tasks of economy and of improving the national structure within which business could increase profits and investment, raise wages and provide better goods and services at the lowest possible prices.[165]

It was one of the characteristics of America in the 1920s that its chief executive for much of the decade preached and practiced this public philosophy. Virtually everywhere else, the trend was towards the expansion of government, greater intervention, and more power to the center. Of those who came to power at the same time as Coolidge, all the most notable were dedicated to expanding the role of the state. Mussolini, supreme in Italy from 1922, put it bluntly: 'Everything within the state, nothing outside the state, nothing against the state.' Stalin, in power from 1923, began his great series of five-year plans for the entire country. The new nation-creators of the 1920s, Kemal Ataturk, President of Turkey from 1923, Chiang Kai-shek, ruler of China from 1925, Ibn Saud of Saudi Arabia (1926), and Reza Shah of Persia (1925) all took government into corners of their countries it had never before penetrated. Even Poincaré of France and Baldwin of Britain were, by Coolidge's standards, rampant interventionists. Coolidge took a critical view of his masterful Cabinet colleague, Herbert Hoover, who was by training a mechanical engineer and by political instinct a social engineer. Coolidge felt that Hoover was itching to get his hands on the levers of power at the White House so that he could set the state to work to hasten the millennium. He referred to

Hoover with derision as the 'Wonder Boy' and, after he had left office, said of his successor: 'That man has offered me unsolicited advice for six years, all of it bad.' It is possible, though unlikely, that if Coolidge had known in advance that the interventionist Hoover was sure to take over the leadership of the Republican Party, he would have run for another term.[166]

All that we can now say is that, on the facts, Coolidge's minimalism was justified by events. Coolidge Prosperity was huge, real, widespread though not ubiquitous, and unprecedented. It was not permanent— what prosperity ever is? But it is foolish and unhistorical to judge it insubstantial because of what we now know followed later. At the time it was as solid as houses built, meals eaten, automobiles driven, cash spent, and property acquired. Prosperity was more widely distributed in the America of the 1920s than had been possible in any community of this size before, and it involved the acquisition, by tens of millions of ordinary families, of the elements of economic security which had hitherto been denied them throughout history. The Twenties was characterized by the longest housing boom recorded. As early as 1924, some 11 million families had acquired their own homes, and the process was only just beginning. Automobiles gave farmers and industrial workers a mobility never enjoyed before outside the affluent classes. For the first time, many millions of working people acquired insurance—life and industrial insurance passed the 100 million mark in the 1920s—savings, which quadrupled during the decade, and a stake in the economy. An analysis of those buying fifty or more shares in one of the biggest public utility stock issues of the 1920s shows that the largest groups were, in order: housekeepers, clerks, factory workers, merchants, chauffeurs and drivers, electricians, mechanics, and foremen. Coolidge Prosperity showed that the concept of a property-owning democracy could be realized.[167]

Nor was this new material advance essentially gross and philistine, as the popular historiography of the 1920s has it, 'a drunken fiesta,' to use Edmund Wilson's phrase, or as Scott Fitzgerald put it, 'the greatest, gaudiest spree in history.' Middle-class intellectuals are a little too inclined to resent poorer people acquiring for the first time material possessions, and especially luxuries, of a kind they themselves have always taken for granted. Experience shows that, in a democratic and self-improving society like the United States, when more money becomes available the first priority, both for local governments and for families, is to spend it on more and better education. That is certainly what happened in the 1920s. Between 1910 and 1930, but especially in the second half of the period, total education spending in the US rose fourfold, from $426.25 million to $2.3 billion. Spending on higher education rose fourfold too, to nearly a billion a year. Illiteracy fell from 7.7 percent to 4.3 percent.[168] The 1920s was the age of the Book of the Month Club and the Literary Guild, of booming publishing houses and bookshops, and especially of a popular devotion to the classics. Throughout the 1920s *David Copperfield* was rated 'America's favorite novel' and those voted by Americans 'the ten greatest men in history' included Shakespeare, Longfellow, Dickens, and Tennyson.

It is hard to point to any aspect of culture in which the 1920s did not mark spectacular advances. By the end of it, there were over 30,000 youth orchestras in the United States. In 1924 Coolidge got himself reelected in his own right, beating not only the Democrat, John W. Davis, by 15,725,016 to 8,385,586, but a Republican Progressive, Robert La Follette, who got 4,822,856, Coolidge thus getting more votes than both of them combined, and an electoral college margin of 382 to 136.[169] The year, as we have already noted, was a key one in the history of the American musical theater. But it was also fertile in American literature generally. The novels of the period included Scott Fitzgerald's *This Side of Paradise* (1920), Sinclair Lewis' *Main Street* (1920), John Dos Passos' *Three Soldiers* (1921), Theodore Dreiser's *An American Tragedy* (1925), William Faulkner's *Soldier's Pay* (1926), Upton Sinclair's *Boston* (1928), and, in 1929, Hemingway's *A Farewell to Arms* and Thomas Wolfe's *Look Homeward, Angel*. That is, by any standard, a brilliant decade.

During the 1920s, in fact, America began suddenly to acquire a cultural density, or what Lionel Trilling called 'a thickening of life,' which it had never before possessed and whose absence Henry James had plaintively deplored a generation before. It was also learning, like more mature European societies, to cherish its past. It was during the 1920s that the national conservation movement really got under way and restored colonial Williamsburg, for example, while at the same time contemporary painting was brought together in the new Museum of Modern Art, which opened in 1929. A sharp French observer, André Siegfried, following a hundred years later in the steps of De Tocqueville, produced an *aperçu* of the nation in 1927 whose message was presented by its title, *America Comes of Age*. The American people, he declared, 'as a result of the revolutionary changes brought about by modern methods of production . . . are now creating on a vast scale an entirely original social structure.'[170]

This was the blossoming scene Calvin Coolidge chose to leave as abruptly as he had entered it. It was an aspect of his minimalist approach to life and office that he not only refrained from doing whatever was not strictly necessary but also believed it right to stop doing anything at all as soon as he felt he had performed his dutiful service. He was widely read in history, like Woodrow Wilson, but he was much more conscious than Wilson of Lord Acton's warning about the tendency of power to corrupt. He liked the idea of an America in which a man of ability and righteousness emerged from the backwoods to take his place as first citizen and chief executive of the republic and then, his term of office completed, retired, if not exactly with relief, then with no regrets, to the backwoods again.

In one sense Coolidge was a professional politician, in that he had ascended the ladder of office, step by step, for over thirty years. But he was sufficiently old-fashioned to find the concept of a professional politician, making a career of office-seeking and hanging on to the bitter end, profoundly distasteful and demeaning. He had a strong, if unarticulated, sense of honor, and it was offended by the prospect that some people, even in his own party let alone outside it, might accuse him of 'clinging' to power. He had a genuine respect for the American tradition of the maximum two-term presidency. That would not have been infringed, of course, by his offering himself a second time, but he had served two years of Harding's mandate, making six in all, and he felt that was enough. Coolidge was never exactly popular—he lacked both personal charm and the slightest desire to develop winning ways—but he was hugely respected. The Republican nomination was his for the asking and he would have had no difficulty in carrying the country in 1928, probably with a greater plurality than Hoover did. He was only fifty-six. But as he told Associate Justice Harlan Stone, 'It is a pretty good idea to get out when they still want you.'

Coolidge had the sense to follow his own advice. He was not without humor, albeit of a very peculiar kind, and he liked to surprise. In the Oval Office he would sometimes call in his staff by bell, then hide under his desk, observing their mystification with wry pleasure. On August 2, 1927 he summoned some thirty journalists and, when they arrived, told them: 'The line forms on the left.' He then handed each a 2-by-9-inch sheet of paper on which he himself had typed, 'I do not choose to run for president in 1928.' That was it: no questions were allowed. It may be that, the following year, seeing Hoover's triumph, he regretted this decision, but he never made the slightest move to reverse it. At the time, he gave no explanation for it either. Indeed, his last words to the press at the White House were typically negative, snapping at them: 'Perhaps one of the most important characteristics of my administration has been minding my own business.'[171]

When Coolidge ventured to explain himself, in the final chapter of his *Autobiography*, published in 1929, he contented himself with saying that eight years in the White House was enough, perhaps more than enough: 'An examination of the records of those Presidents who have served eight years will disclose that in almost every instance the latter parts of their term have shown very little in the way of constructive accomplishment. They have often been clouded with grave disappointments.' That is true enough.

There may have been a personal reason too. Coolidge was the reverse of a demonstrative man but there were powerful emotions operating under the surface of his laced-in exterior, and the evidence is strong that he was deeply attached to his immediate family. While he was in office as president, he lost both his son Calvin, in 1924, and his beloved father, the Colonel, in 1926. There is no reason to think that Coolidge was particularly superstitious but he seems to have got it into his head that neither death would have occurred had he not occupied the White House. The death of his father he felt deeply and believed that it had come prematurely because the consequences of his own eminence had, as he put it, 'overtaxed his strength.' The loss of Calvin was shattering. 'When he went,' Coolidge wrote, 'the power and the glory of the presidency went with him.' He mused sadly, 'The ways of Providence are often beyond our understanding . . . I do not know why such a price was exacted for occupying the White House.' That last is a curious remark. But Coolidge was not a New England Puritan for nothing, and it may be that he felt, in retrospect, that his son was taken from him as a punishment for his own sins of pride in the exercise of power. There was a particular incident which later haunted him—his firing of a long-serving Secret Service agent, Jim Haley, in a fit of petulance. Haley was blameless but Coolidge thought he had exposed Mrs Coolidge, his much adored Grace, to needless danger. Such an episode would never have given a second's concern to a Franklin Roosevelt or a Winston Churchill who, amid their grander moments, regularly abused power, and in a far more shameless fashion. But it worried Coolidge and he may have come to the conclusion, by August 2, 1927, the fifth anniversary of his accession to power, that his son's death had been a warning.

There is, of course, another explanation, that Coolidge felt in his bones that the good times were coming to an end, and he did not want to be in charge when the bottom fell out of the bull market. This was certainly a factor. By Coolidge's day, the history of the trade cycle was fairly well understood, and Coolidge—by nature a pessimist rather than an optimist—knew perfectly well that the boom would not last. All that was uncertain was when it would end and how dramatically. His closest advisor, Stone, who studied the markets, warned him of trouble ahead. He himself was certain the market would break, probably sooner rather than later. That was his private sentiment, reflected in his wife Grace's remark, 'Poppa says there's a depression coming.' But Coolidge did not feel it was his duty or in America's interests to talk the boom down publicly, to hasten the downturn or to take steps to limit its severity. Among his other lapidary phrases, he might easily have coined the maxim: 'If it works, don't fix it.' That is certainly what he believed. That a depression would come was certain, but Coolidge probably assumed it would be on the scale of 1920, to be cured by a similar phase of masterly inactivity. If, however, something more was required, he felt he was not the man to do it. Grace Coolidge said he told a member of his Cabinet: 'I know how to *save* money. All my training has been in that direction. The country is in a sound financial condition. Perhaps the time has come when we ought to *spend* money. I do not feel I am qualified to do that.'[172]

Thus Coolidge departed, pulling down the curtain on the last genuine capitalist Arcadia. Another myth that has grown up about these times is that the Twenties Boom was a mere drunken spending-spree, bound to end in disaster, and that beneath a veneer of prosperity was an abyss of poverty. That is not true. The prosperity was very widespread. It was not universal. In the farming community it was patchy, and it largely eluded certain older industrial communities, such as the New England textile trade.[173] But growth was spectacular. On a 1933–8 index of 100, it was 58 in 1921 and passed 110 in 1929. That involved an increase in national income from $59.4 to $87.2 billion in eight years, with real per-capita income rising from $522 to $716: not Babylonian luxury but a modest comfort never hitherto thought possible.[174]

The heart of the consumer boom was in personal transport, which in a vast country, where some of the new cities were already 30 miles across, was not a luxury. At the beginning of 1914, 1,258,062 cars were registered in the US, which produced 569,054 during the year. Production rose to 5,621,715 in 1929, by which time cars registered in the US totaled 26,501,443, five-sixths of the world's production and one car for every five people in the country. This gives some idea of America's global dominance. In 1924 the four leading European car producers turned out only 11 percent of the cars manufactured in America. Even by the end of the decade European registrations were only 20 percent of the US level and production a mere 13 percent.[175] The meaning of these figures is that the American working class had acquired the freedom of movement limited hitherto to a section of its middle class, and denied to European workers for another thirty years or more. Meanwhile the middle class was moving into air travel. Air passengers rose from 49,713 in 1920 to 417,505 in 1930 (by 1940 the figure was 3,185,278 and it was nearly 8 million by 1945).[176] What the American Twenties demonstrated was the speed with which industrial productivity could transform luxuries into necessities and spread them down the class pyramid.

In fact, Twenties prosperity was a growing solvent of class and other barriers. Next to automobiles, it was the new electric industry which fueled and reflected the boom. Expenditure on radios rose from a mere $10,648,000 in 1920 to $411,637,000 in 1929, and total electrical product sales tripled in the decade to $2.4 billion.[177] The mass radio audience (followed by the talkies) brought about the Americanization of immigrant communities and a new classlessness in dress, speech, and attitudes. Sinclair Lewis, revisiting 'Main Street' on behalf of the *Nation* in 1924, found two working-class, small-town girls wearing 'well-cut skirts, silk stockings, such shoes as can be bought nowhere in Europe [at the price], quiet blouses, bobbed hair, charming straw hats, and easily cynical expressions terrifying to the awkward man.' One of them served hash. 'Both their dads are Bohemian; old mossbacks, tough old birds with whiskers that can't sling more English than a musk-rat. And yet, in one generation, here's their kids—real queens.'[178] The Twenties marked the biggest advance for American women of any decade, before or since. By 1930 there were 10,546,000 women 'gainfully employed' outside the home; the largest number, as before, were in domestic/personal service (3,483,000) but there were now nearly 2 million in clerical work, 1,860,000 in manufacturing, and, most encouraging of all, 1,226,000 in the professions.[179] Equally significant were the liberated housewives, the 'Blondies,' to whom their appliances, their cars, and their husband's high wages had brought leisure for the first time.

The coming of family affluence was reflected in the decline of radical politics and their union base. A 1929 survey quoted a union organizer: 'The Ford car has done an awful lot of harm to the unions, here and everywhere else [in America]. As long as men have enough money to buy a second-hand Ford and tires and gasoline, they'll be out on the road and paying no attention to union meetings.'[180] In 1915, 1921, and 1922 the unions lost three key Supreme Court actions, and their 1919 strikes were disasters. The American Federation of Labor membership dropped from a high of 4,078,740 in 1920 to 2,532,261 in 1932. 'Welfare capitalism' provided company sports facilities, holidays with pay, insurance and pension schemes, so that by 1927 4,700,000 workers were covered by group insurance and 1,400,000 were members of company unions. The American worker appeared to be on the threshold of a hitherto unimaginable middle-class existence of personal provision and responsibility which made collective action increasingly superfluous.[181]

Reflecting this growth and prosperity, the United States had achieved a position of paramountcy in total world production never before attained during a period of prosperity by any other state: 34.4 percent of the whole, compared with Britain's 10.4, Germany's 10.3, Russia's 9.9, France's 5.0, Japan's 4.0, 2.5 for Italy, 2.1 for Canada, and 1.7 for Poland. The likelihood that Europe, followed by Asia, would soon begin to lean towards what André Siegfried called 'America's original social structure' increased with every year the world economy remained buoyant. Granted another decade of prosperity on this scale, the history not just of America but of the entire world would have been vastly different and far more fortunate. But that was not to be. After the election of Herbert Hoover, and during the interregnum in Washington, President Coolidge was asked for a decision on long-term economic policy. He snapped, 'We'll leave that to the Wonder Boy.'

PART SEVEN

'Nothing to Fear
But Fear Itself'

Superpower America, 1929–1960

The wall street crash of October 1929, and the Great Depression which followed it, lasting effectively till the beginning of World War Two in 1939, remain mysterious, despite more than half a century of economic and historical analysis since. Bear markets, and the economic downturns they provoke, reflect trade-cycles which seem to be an inevitable feature of the modern industrial world economy. They are corrective instruments, restoring overheated markets and economies to reality, thus forming stable platforms from which growth can be resumed, soon reaching higher levels. What is puzzling about the events of the decade 1929–39 is the continuing severity of the market falls and the length and obstinacy of the Depression. What follows is an attempt to make historical sense of a tragic series of events for which no satisfactory explanation has yet been provided.[1]

America was, in general, a laissez-faire country in the 1920s. On the whole businessmen were free to make their own arrangements and workers free to bargain for wages at the market rate. But there was one important and dangerous qualification to this self-regulating economy. American industry was protected from foreign competition by high tariffs. The Republican presidents, Harding, Coolidge, and Hoover, did not resume and intensify the tentative attempts by Wilson to reduce tariffs and move closer to free trade. The Fordney–McCumber Tariff Act of 1922 and, still more, the Smoot–Hawley Act of 1930, which Hoover declined to veto, were devastating blows struck at world commerce, and so in the end at America's own. The fact is that America's presidents, and the Republican Congressional leadership, failed to stand up to the National Federation of Manufacturers, the American Federation of Labor, pressure groups formed by particular industries, and local pressure from industrial states, and so pursue the philosophy of economic freedom they claimed to hold.[2]

Instead, during the 1920s the United States, in conjunction with the British and other leading industrial and financial powers, tried to keep the world prosperous by deliberately inflating the money supply. This was something which had been made possible by the creation of the Federal Reserve Bank system, something which could be done secretly, without legislative enactment or control, and without the public knowing, or the business community caring. Although the amount of money in circulation remained stable—$3.68 billion in dollar bills in circulation at the beginning of the 1920s and $3.64 billion in 1929—credit was expanded from $45.3 billion on June 30, 1921 to $73 billion in July 1929, a 61.8 percent expansion in eight years.[3] The White House, the Treasury under Andrew Mellon (1855–1937), who was in charge for the whole period 1921–32, the Congress, the federal banks, and the private banks too combined to inflate credit. This would not have mattered if interest rates had been allowed to find their own level, that is if the manufacturers and farmers who borrowed money had paid interest at the rates savers were prepared to lend it. But, again, the same combination joined to keep interest rates artificially low. It was the stated policy of the Federal Reserve not only to 'enlarge credit resources' but to do so 'at rates of interest low enough to stimulate, protect and prosper all kinds of legitimate business.'[4]

This deliberately managed inflation of credit applied not just nationally but internationally. The US government demanded the repayment of its war-loans to the European allies, chiefly Britain and France, but it also actively assisted foreign governments and businesses to raise money in New York both by its own cheap money policy and by its own active interference in the foreign bond market. The government made it quite clear that it favored certain loans, and certain governments, and not others. The foreign loan policy foreshadowed, at the level of private enterprise, the official US policy of Marshall Aid in the years after 1947. The aims were the same: to keep the international economy afloat, to support certain regimes, and to promote America's export industries. The administration backed certain loans on condition part of them was spent in the US. The foreign lending boom began in 1921, following a Cabinet decision on May 20, 1921 and a meeting between Harding, Hoover, and US investment banks five days later. It ended in late 1928, thus coinciding precisely with the expansion of the money supply which underlay the boom. America's rulers, it can be argued, rejected the laissez-faire formality of free trade and hard money and took the soft political option of high tariffs and inflation. The domestic industries protected by the tariff, the export industries subsidized by the uneconomic loans, and of course the investment bankers who floated the bonds all benefited. The losers were the population as a whole, who were denied the competitive prices produced by cheap imports, suffered from the resulting inflation, and were the universal victims of the ultimate *dégringolade*.[5]

The architects of the policy were Benjamin Strong, governor of the New York Federal Reserve Bank, who until his death in 1928 was all-powerful in the formation of US financial policy, and Montague Norman, governor of the Bank of England. Their inspiration was the Bloomsbury economist John Maynard Keynes, whose influential *Tract on Monetary Reform* appeared in 1923. One of the myths of the inter-war years is that laissez-faire capitalism made a mess of things until Keynes, with his great book, *The General Theory of Employment, Interest and Money* (1936), introduced 'Keynesianism'—another word for government interference—and saved the world. In fact Keynes' *Tract*, advocating 'managed currency' and a stabilized price-level, both involving constant government interference, coordinated internation-ally, was part of the problem. For most of the Twenties, Strong and Norman directed the currency management. Domestically and interna-tionally, they constantly pumped more money into the system, and whenever the economy showed signs of flagging they increased the dose. The most notorious occasion was in July 1927, when Strong and Norman held a secret meeting of bankers at the Long Island estates of Ogden Mills, the US Treasury Under-Secretary, and Mrs Ruth Pratt, a Standard Oil heiress. This was a form of Long Island power-broking unknown, alas, to Scott Fitzgerald and the characters in *The Great Gatsby* (1925), otherwise we might have heard more about it at the time.

The memoirs of some of those present subsequently described what happened. When, at the meeting, Strong and Norman decided on another bout of inflation, Hjalmar Schacht, the great German financial wizard, protested. He argued that the financial underpinning of the postwar credit system, the so-called gold standard, was in fact a phony: it was merely a gold bullion standard, in which central banks kept tally by transferring gold bars among themselves. The true gold standard, he urged, existed only when banks paid gold coins on demand to people who presented their paper. This was the only means to insure that expansion was financed by genuine, voluntary savings, instead of by bank credit determined by a tiny oligarchy of financial Jupiters.[6] He was supported by Charles Rist, deputy governor of the Bank of France, who objected when Strong told him, 'I will give a little shot of whiskey to the Stock Market.' But the Germans and French were overruled and the New York Fed reduced its rate by another half percent to 3.5 per-cent—an amazing rate in the circumstances. Adolph Miller, a member of the Federal Reserve Board, subsequently described this decision in

Senate testimony as 'the greatest and boldest operation ever undertaken by the Federal Reserve system [which] resulted in one of the most costly errors committed by it or any other banking system in the last 75 years.'[7]

The policy appeared to succeed in the short term. In the first half of the decade, world trade, thanks largely to US protectionism, had failed to return to its prewar level. Indeed in 1921–5, world trade, compared to 1911–14, was actually minus 1.42 percent. But during the four years 1926–9 it achieved a growth of 6.74, a performance not to be exceeded until the late 1950s.[8] But prices remained stable, fluctuating between 93.4 in June 1921 to a peak of 104.5 in November 1925 and then down to 95.2 in June 1929. This suggested that the policy of deliberately controlled growth within a framework of price stability had been turned into reality. Keynes described 'the successful management of the dollar by the Federal Reserve Board from 1923–8' as a 'triumph.'[9]

Yet the inflation was there, and growing all the time. Without the kind of Olympian management the central bankers supplied, both prices and wages should have fallen, the first much faster than the second. Between 1919 and 1929, there was a phenomenal growth of productivity in the US, output per worker in manufacturing industry rising by 43 percent. This was made possible by an unprecedented increase in capital investment, rising at an annual average rate of 6.4 percent a year, and by huge advances in industrial technology.[10] The productivity increase should have been reflected in much lower prices. The fact that they remained stable reflected economic management. It is true that, without this management, wages would have fallen too, but only marginally. Real wages, or purchasing power, would have risen steadily, *pari passu* with productivity, so the workers would have been able to enjoy more of the goods their improved performance was turning out of the factories.

As it was, the workers found it difficult to keep up with the new prosperity. The Twenties, far from being too materialistic, as the myth has it, were not materialistic enough. The Twenties boom was based essentially on automobiles. America was producing almost as many cars in the late 1920s as in the 1950s (in 1929 5,358,000, against 5,700,000 in 1953). The really big and absolutely genuine growth stock of the 1920s was General Motors: anyone who in 1921 had bought $25,000 of GM common stock was a millionaire by 1929, when GM was earning profits of $200 million a year. GM had been built up in the Twenties at the expense of Ford's market share by a genius called

Alfred P. Sloane. In 1920 Ford had a 55.67 percent share of the industry, making 845,000 cars a year. Every other car sold was a Ford. GM came second, selling 193,275 a year. Ford had a mechanical strategy, making the best-value car at the lowest possible price, changing it little, except for mechanical reasons, and offering little choice. Sloane, like Ford, was an engineer, with a degree in mechanical engineering from MIT, something Ford never had, and a Brooklyn accent. His was a consumer strategy. He made his way upwards by becoming an expert in ball-bearings, and his underlings compared him to one—'self-lubricating, smooth, eliminates friction and carries the load.'

While Ford made the product as well as he could, then looked for people to buy it, Sloane did it the other way round. He produced the widest possible range of cars for the maximum spread of customers.[11] He said there was nothing new in it; everyone who made shoes did the same. He had five basic brands of car—Chevrolet, Pontiac, Oldsmobile, Buick, and Cadillac—to cover the major price brackets, but each produced in numerous versions. And the whole range, from 1923, changed every year. He introduced the supremacy of style and turned cars into a mechanical adjunct of the fashion industry. He wrote: 'Today the appearance of a motor-car is a most important factor in the selling end of the business—perhaps *the* most important factor—because everyone knows the car will run.' His cars looked more and more imposing as they expanded in size, guzzled more gas and piled on the chrome. He appealed strongly to the American cult of bigness and the linked cult of variety, he pushed Ford into a bad second place, and he made GM into the largest manufacturing company in the world.

But there was a double price to pay. In the long term Sloane made GM's cars—and all the other companies, including Ford, followed suit—into what have been termed 'overblown, overpriced monstrosities built by oafs for thieves to sell to mental defectives.'[12] By the 1950s, American cars had become 'technologically out of date, impractical and unsafe cathedrals of chrome, manufactured sloppily and sold using methods than can only be described as shameful.' The Buick and Oldsmobile of 1958 were 'Huge, vulgar, dripping with pot metal, and barely able to stagger down the highway. They were everything car people hated about the American automobile.'[13] As a result, the Japanese car industry was able to do to GM what GM had done to Ford in the 1920s.

There was an immediate price to pay too. GM made cars too expensive, or rather more expensive than they need have been. Ford's strategy of getting the price down each year was abandoned, even by Ford; GM, and eventually all the rest, sold not on price but on appearance and gimmickry. The enormous gains in productivity were frittered away on appearances rather than utility, and on increased profits to shareholders with the object of drumming up the share price and making capital gains. This was fine for the stock market but bad for the customer, especially the working-class customer whom Ford had made into car-owners. Hence working-class families could still afford cars— just. GM made it easier by hire-purchase. It was the first big company to establish a wholly owned subsidiary for this purpose. By 1925 GM asked only for one-third down, the rest in installments, and sold three in four of its cars this way. In December 1927, Hoover as secretary of commerce proudly claimed that average industrial wages had reached $4 a day, that is $1,200 a year. But a government agency estimated it cost $2,000 a year to bring up a family of five in 'health and decency.' By the end of the Twenties many working-class families found it hard to keep up with the installments on their car, or to renew. And one disadvantage of building an industrial economy round the automobile is that, when money is short, a car's life can easily be arbitrarily prolonged by a few years. Towards the end of the Twenties, the failure to pass on the fruits of productivity increases to the consumer in terms of lower costs was beginning to take its toll. There is some evidence that the increasing number of women in employment reflected a decline in real incomes, especially among the middle class.[14] As the boom continued, and prices failed to fall, it became harder for the consumer to keep the boom going. The bankers, in turn, had to work harder to inflate the economy: Strong's 'little shot of whiskey' was the last big push; next year he was dead, leaving no one with either the same degree of monetary adventurism or the same authority.[15]

Strong's last push in fact did little to help the industrial economy. It fed speculation. Very little of the new credit went through to the mass consumer. As it was, the spending side of the US economy was unbalanced. The 5 percent of the population with the top incomes had one-third of all personal income, and they did not buy Fords and Chevrolets. The proportion of income received in interest, dividends, and rents, as opposed to wages, was about twice as high as the levels we have become accustomed to in the last half-century.[16] Strong's shot of whiskey benefited almost solely the non-wage-earners—the last phase of the boom was largely speculative. Until 1928, stock-exchange prices had merely kept pace with actual industrial performance. From the beginning of 1928 the element of unreality, of fantasy indeed, began to grow. As Bagehot put it, 'People are most credulous when they are most happy.' People bought and sold in blissful ignorance. In 1927 the number of shares changing hands, at 567,990,875, broke all records. The figure then rose to 920,550,032.[17]

Two new and sinister elements emerged: a vast increase in margin-trading and a rash of hastily cobbled-together investment trusts. Traditionally, stocks were valued at about ten times earnings. During the boom, as prices of stocks rose, dividend yields fell. With high margin-trading, earnings on shares (or dividend yields), running at only 1 or 2 percent, were far less than the interest of 8–12 percent on loans used to buy them. This meant that any profits were on capital gains alone. Over the past 125 years of American history, dividend yields have averaged 4.5 percent. The figures show that, whenever the dividend yield sinks to as low as 2 percent, a crack in the market and a subsequent slump is on the way. That had been true of the last two bear markets before 1929 came, and investors or market analysts who studied historical performance, the only sure guide to prudence, should have spotted this. There were indeed some glaring warnings. Radio Corporation of America, which had never paid a dividend at all, and whose earnings on shares were thus zero, nonetheless rose from 85 to 420 points in 1928. That was pure speculation, calculated on the assumption that capital gains would continue to be made indefinitely, a manifest absurdity. By 1929 some stocks were selling at fifty times earnings. As one expert put it, 'The market was discounting not merely the future but the hereafter.'[18]

A market boom based on capital gains is merely a form of pyramid-selling. The new investment trusts, which by the end of 1928 were emerging at the rate of one a day, were archetypal inverted pyramids. They had what, to use the new 1920s vogue term, was called 'high leverage,' through their own supposedly shrewd investments, and secured phenomenal paper growth on the basis of a very small plinth of real growth. Thus the United Founders Corporation was built up into a company with nominal resources of $686,165,000 from an original investment of a mere $500. The 1929 market value of another investment trust was over a billion dollars, but its chief asset was an electrical company worth only $6 million in 1921.[19] These firms claimed to exist to enable the 'little man' to get a 'share of the action.' In fact they merely provided an additional superstructure of almost pure speculation, and the 'high leverage' worked in reverse once the market broke.

It is astonishing that, once margin-trading and investment-trusting took over, the federal bankers failed to raise interest rates and persisted in cheap money. But many of the bankers had lost their sense of reality by the beginning of 1929. Indeed they were speculating themselves, often in their own stock. One of the worst offenders was Charles Mitchell (finally indicted for grand larceny in 1938), the chairman of National City Bank, who on January 1, 1929 became a director of the Federal Reserve Bank of New York. Mitchell filled the role of Strong at a cruder level, and kept the boom going through most of 1929. The ferocious witchhunt of Wall Street begun in 1932 by the Senate Committee on Banking and the Currency, which served as a prototype for the anti-Communist witchhunts of the 1940s and 1950s, actually disclosed little law-breaking. Mitchell was the only major culprit and even his case exposed more the social mores of finance capitalism than actual wickedness.[20] 'Every great crisis,' Bagehot remarked, 'reveals the excessive speculations of many houses which no one before suspected.'[21] The 1929 crash revealed in addition the naivety and ignorance of bankers, businessmen, Wall Street experts, and academic economists, high and low; it showed they did not understand the system they had been so confidently manipulating. They had tried to substitute their own well-meaning policies for what Adam Smith called the 'invisible hand' of the market and they had wrought disaster. Far from demonstrating, as Keynes and his school later argued—at the time Keynes failed to predict either the crash or the extent or duration of the recession—the dangers of a self-regulating economy, the *dégringolade* indicated the opposite: the risks of ill-informed meddling.

The credit inflation petered out at the end of 1928. In consequence, six months later the economy went into decline. The market collapse followed after a three-month delay. All this was to be expected. It ought to have been welcomed. It was the pattern of the 19th century and of the 20th up to 1920–1. It was capitalist 'normalcy,' a 'market correction.' To anyone who studied historical precedents, the figures suggested the correction would be a severe one, simply because the speculation had been so uninhibited. At the peak of the craze there were about a million active speculators. Out of an American population of 120 million, about 29–30 million families had an active association with the market.[22] The economy ceased to expand in June 1929. The bull market in stocks really came to an end on September 3. The later 'rises' were merely hiccups in a steady downward trend. On Monday, October 21, for the first time, the ticker-tape could not keep pace with the news of falls and never caught up. In the confusion the panic intensified (the first margin calls had gone out on the Saturday before) and speculators began to realize they might lose their savings and even their homes. On Thursday, October 24 shares dropped vertically with no one buying, speculators were sold out as they failed to respond to margin calls, crowds gathered on Broad Street outside the New York Stock Exchange, and by the end of the day eleven men well known in Wall Street had committed suicide. Next week came Black Tuesday, the 29th, and the first selling of sound stocks in order to provide desperately needed liquidity.[23]

Business downturns serve essential purposes. They have to be sharp. But they need not be long because they are self-adjusting. All they require on the part of governments, the business community, and the public is patience. The 1920 recession had adjusted itself, helped by Harding's government cuts, in less than a year. There was no reason why the 1929 fall should have taken longer, for the American economy was fundamentally sound, as Coolidge had said. On November 13, at the end of the immediate four-week panic, the index was at 224, down from its peak at 452. There was nothing wrong in that. It had been only 245 in December 1928 after a year of steep rises. The panic merely knocked out the speculative element, leaving sound stock at about their right value in relation to earnings. If the recession had been allowed to adjust itself, as it would have done by the end of 1930 on any earlier analogy, confidence would have returned and the world slump need never have occurred. Instead the market went on down, slowly but inexorably, ceasing to reflect economic realities—its true function— and instead becoming an engine of doom, carrying into the pit the entire nation and, with it, the world. By July 8, 1932 the *New York Times* industrials had fallen from 224 at the end of the panic to 58. US Steel, selling at 262 before the market broke in 1929, was now only 22. GM, one of the best-run and most successful manufacturing groups in the world, had fallen from 73 to 8.[24] By this time the entire outlook for America had changed, infinitely for the worse. How did this happen? Why did the normal recovery not take place?

The conventional explanation is that Herbert Hoover, President when Wall Street collapsed and during the period when the crisis turned into the Great Depression, was a laissez-faire ideologue who refused to use public money and government power to refloat the economy. As soon as President Franklin Delano Roosevelt succeeded him, in 1933, and—having no such inhibitions about government intervention—started to apply state planning, the clouds lifted and the nation got back to work. There is no truth in this mythology, though there were indeed profound differences of character between the two men, which had some bearing on the crisis. Hoover was a social engineer. Roosevelt was a social psychologist. But neither understood the nature of the Depression, or how to cure it. It is likely that the efforts of both merely served to prolong the crisis.

The new fashion of social engineering—the notion that action from above could determine the shape of society and that human beings could be manhandled and manipulated like earth and concrete—had come into its own in World War One. It had, as we have seen, got a grip even in America, since it fitted so neatly on to the enlarged state laid down by the Wilson administration in the years before America entered the conflict. Some pundits wished to go further and install the engineer himself as king. That was the argument pursued by Thorsten Veblen, the sociologist who was, perhaps, the most influential progressive writer in America in the first quarter of the century. He had first touched on the theme in his *The Theory of the Leisure Class* (1899) and he developed it in *The Engineers and the Price System* (1921). In these years, the vast dams and hydroelectric schemes which were planned or built—the Boulder Canyon Dam being the outstanding example—impressed everyone with the magic power of the mechanical engineer, who could organize such power over nature, harnessing enormous and savage rivers which, a generation ago, had seem untameable. This was a worldwide phenomenon. The two men most impressed by the hydroelectric-dam image of power were Lenin and Stalin, who used their totalitarian authority to carry through colossal distortions of nature, including the switching of immense rivers from the Arctic to other outlets, the tragic consequences of which are only now being measured. Veblen presented the engineer as the new Superman. He saw him as a disinterested and benevolent figure, who was out to replace the Big Businessman, eliminate both the values of the leisure class and the profit motive, and run the economy in the interests of consumers.[25]

Hoover, born in 1874, was a larger-than-life exemplar of the new Superman and was given the opportunity to test Veblen's theory to destruction. He not only believed in a kind of social engineering, he actually was an engineer, and a highly successful one. Born in West Branch, Iowa, of a poor Quaker farming family, he was orphaned at nine and worked himself through school and Stanford University to an engineering degree. His style of dress reflected his Quaker background till the day death finally claimed him in the age of John F. Kennedy. Hoover swore, smoked, and drank, and went fishing on Sundays—but always in a stiff white collar and tie.[26] Between 1900 and 1915 he directed mining projects all over the world and made a fortune of $4 million.[27] He became the outstanding member of Wilson's war-team, absorbed its philosophy of forceful government direction and planning, and then as head of America's postwar Commission of Relief (a foretaste of the later Marshall Aid and Point Four programs) achieved a worldwide reputation for efficient, interventionist benevolence. The notion that Hoover was a hard, unfeeling man, later circulated by the New Dealers, was entirely false. Maxim Gorky wrote to him at the time: 'You have saved from death 3,500,000 children and 5,500,000 adults.'[28] Keynes described him as 'the only man who emerged from the ordeal of Paris with an enhanced reputation . . . [who] imported to the councils of Paris, when he took part in them, knowledge, magnanimity and disinterestedness which, if they had been found in other quarters also, would have given us the Good Peace.'[29] Franklin Roosevelt, who had worked with Hoover as Navy under-secretary and shared his outlook, wrote to a friend: 'He is certainly a wonder and I wish we could make him President of the United States. There could not be a better one.'[30]

As secretary of commerce for eight years during the 1920s, Hoover showed himself a corporatist, an activist, and an interventionist, running counter to the general thrust, or rather non-thrust, of the Harding–Coolidge administrations. His predecessor, Oscar Strauss, told him he needed to work only two hours a day, 'putting the fish to bed at night and turning on the lights around the coast.' In fact his was the only department which increased its bureaucracy, from 13,005 to 15,850, and its cost, from $24.5 million to $37.6 million—quite an achievement in the cost-conscious, minimalist-government Twenties.[31] It was one reason Coolidge disliked him. Hoover came into office at the tail-end of the 1920–1 depression, and immediately set about forming committees and trade councils, sponsoring research programs, pushing expenditure, persuading employers to keep up wages and 'divided time' to increase jobs, and, above all, forcing 'cooperation between the Federal, state and municipal governments to increase public works.'[32] He sponsored working parties and study groups (new vogue terms), got people to produce expert reports, and generated a buzz of endless activity. There was no aspect of public policy in which Hoover was not intensely active, usually personally: oil, conservation, Indian policy, public education, child health, housing, social waste, and agriculture (when he became president he was his own Secretary of Agriculture and the 1929 Agriculture Marketing Act was entirely his work).[33]

Harding did not like this hyperactivity, but was impressed by Hoover's brains and prestige—'The smartest gink I know.'[34] Coolidge hated it but by the time he took over Hoover was too much part of the administration's furniture to be removed. Besides, Hoover's corporatism, the notion that the state, business, the unions and other public bodies should work together in gentle, but persistent and continuous manipulation to make life better for all, was the received wisdom of the day among enlightened capitalists, left-wing Republicans, and non-socialist intellectuals, as well as a broad swathe of Democrats and public-spirited academics. Yankee-style corporatism was the American response to the new ideologies of Europe, especially Mussolini's Fascism, then regarded as a highly promising experiment. It was as important to 'right-thinking people' in the Twenties as was Stalinism in the Thirties.[35] Hoover was its outstanding impresario and ideologue. One of his admirers, interestingly enough, was Jean Monnet, who later renamed the approach 'indicative planning' (as opposed to 'imposed planning,' the 'command economy') and made it the basis both for France's post–1945 planning system and for the planning structure of what was to become the European Union.

Yet Hoover was not a statist, let alone a socialist; and he said he was opposed to any attempt 'to smuggle fascism into America through a back door.'[36] On many issues he was liberal. He wanted aid to flow to what were later to be called 'underdeveloped countries,' which in the 1920s and 1930s were simply called 'backward.' He deplored the exclusion of the Japanese, for racial reasons, from the 1924 immigration quotas. His wife entertained the ladies of black congressmen. Unlike Woodrow Wilson and his wife, or Franklin Roosevelt, he did not make anti-semitic jokes in private.[37] To a wide spectrum of educated Americans in the 1920s, he was exactly what an American statesman ought to be, long before he got to the White House.

His selection as Republican candidate in 1928 was a foregone conclusion. He was up against a strong Democratic candidate, Governor Alfred E. Smith (1873–1944), who was a popular and successful governor of New York and who openly campaigned against Prohibition. But Smith was a Catholic, and anti-Catholic prejudice was still strong, especially in rural areas. Hoover, who stuck up for Prohibition, seemed to stand, despite his social engineering, for old-fashioned moral values, as stiff and unyielding as his starched collars. He promised the voters 'a chicken for every pot and a car for every garage.' He was a substantial victor in a high voter turnout, winning 21,392,190 to Smith's 15,016,443, and sweeping the electoral college 444 to 87. (It was significant, though, that Smith carried all twelve of America's largest cities: a portent.)[38] It was universally believed Hoover would work wonders. The *Philadelphia Record* called him 'easily the most commanding figure in the modern science of "engineering statesmanship." ' The *Boston Globe* hailed the arrival in the White House of a disciple of 'the dynamics of mastery.'[39] They all called him the 'Great Engineer.' Hoover himself said he was worried by these exaggerated expectations: 'They have a conviction that I am a sort of superman, that no problem is beyond my capacity.'[40] But he was not really worried. He knew exactly what to do. He ran the administration like a dictator. (The term 'dictator' did not at this time carry much opprobrium.) He ignored or bullied Congress. He laid down the law, like a comic character from Dickens, and was fond of telling subordinates: 'When you know me better, you will find that when I say a thing is a fact, it *is* a fact.'[41]

When Hoover took over the presidency in March 1929, the mechanism of the Wall Street débâcle was already whirring. The only useful thing he might have done would have been to allow the artificially low interest rates to rise to their natural level, which would have killed off the bull market gradually and at least given the impression that somebody was in charge. That would have avoided the dreadful dramas of the autumn, which had a profound psychological effect. But Hoover did not do so: government-induced cheap credit was the very bedrock of his policy. When the magnitude of the crisis became apparent, late in the year, Andrew Mellon, the Treasury Secretary, at last spoke out to repudiate Hoover's interventionist policy and return to strict laissez-faire. He told Hoover that administration policy should be to 'liquidate labor, liquidate stocks, liquidate the farmers, liquidate real estate' and so 'purge the rottenness from the economy.'[42] It was the only sensible advice Hoover received throughout his presidency. By allowing the Depression to let rip, unsound businesses would quickly have been bankrupted and the sound would have survived. Wages would have fallen to their natural level. That for Hoover was the rub. He believed that high wages were the most important element in prosperity and that maintaining wages at existing levels was essential to contain and overcome depressions.[43]

From the very start, therefore, Hoover agreed to take on the business cycle and stamp it flat with all the resources of government. He wrote: 'No president before has ever believed there was a government responsibility in such cases . . . there we had to pioneer a new field.'⁴⁴ He resumed credit inflation, the Federal Reserve adding almost $300 million to credit in the last week of October 1929 alone. In November he held a series of conferences with industrial leaders in which he exacted from them solemn promises not to cut wages—even to increase them if possible—promises which on the whole were kept till 1932. The American Federation of Labor's journal praised this policy—never before had US employers been marshaled to act together, and the decision marked an 'epoch in the history of civilization—high wages.'⁴⁵ Keynes in a memo to Britain's Labour Prime Minister, Ramsay MacDonald, praised Hoover's moves in maintaining high wage-levels and called federal credit-expansion 'thoroughly satisfactory.'⁴⁶ Indeed in most respects what Hoover did would later have been called a 'Keynesian solution.' He cut taxes heavily. Those of a family man with an income of $4,000 went down by two-thirds.⁴⁷ He increased government spending. He deliberately ran up a huge deficit. It was $2.2 billion in 1931, by which point the government's share of GNP had increased from 16.4 percent in 1930 to 21.5 percent. This was the largest-ever increase in government spending, most of which ($1 billion) was accounted for by transfer-payments.⁴⁸ True, he ruled out direct relief and tried to channel government money through the banks rather than giving it to businesses and individuals. But that he used government cash to reflate the economy is beyond question. Coolidge's advice to angry farmers' delegations which came to him asking for federal cash had been bleak: 'Take up religion.' Hoover's new Agricultural Marketing Act gave them $500 million in federal money, increased by a further $100 million early in 1930. In 1931 he extended this to the economy as a whole with his Reconstruction Finance Corporation (RFC) as part of a nine-point program of government intervention which he produced in December 1931.

This was the real beginning of the New Deal, insofar as it had an objective existence as opposed to propaganda and mythology. More major public works were started in Hoover's four years than in the previous thirty. They included the San Francisco Bay Bridge, the Los Angeles Aqueduct, and the Hoover Dam. A project to build the St Lawrence Seaway was scrapped by Congressional parsimony—the White House favored it. When intervention failed to deliver the goods,

Hoover doubled and redoubled it. In July 1932, the RFC's capital was increased by nearly 100 percent to $3.8 billion and the new Emergency Relief and Construction Act extended its positive role: in 1932 alone it gave credits of $2.3 billion plus $1.6 billion in cash. The essence of the New Deal was now in place. One of its satraps, Rexwell Tugwell, finally conceded in an interview forty years after the event (1974): 'We didn't admit it at the time, but practically the whole New Deal was extrapolated from programs that Hoover started.'[49] By this point, however, Hoover had lost control of Congress, which was horrified by the deficit and insisted that the budget had to be brought back into balance after two years of deficit. The 1932 Revenue Act saw the greatest taxation increase in US history in peacetime, with the rate on high incomes jumping from a quarter to 63 percent. This made nonsense of Hoover's earlier tax cuts, but by this time Hoover was not in a position to pursue a coherent policy.[50]

All he had left was his interventionist rhetoric, and that continued louder than ever. Hoover liked activist military metaphors to describe his interventionism. 'The battle to set our economic machine in motion in this emergency takes new forms and requires new tactics from time to time. We used such emergency powers to win the war; we can use them to fight the Depression' (May 1932). 'If there shall be no retreat, if the attack shall continue as it is now organized, then this battle is won' (August 1932). 'We might have done nothing. That would have been utter ruin. Instead we met the situation with proposals to private business and to Congress of the most gigantic program of economic defense and counter-attack ever involved in the history of the Republic . . . For the first time in the history of Depression, dividends, profits and the cost of living have been reduced before wages have suffered . . . They are now the highest real wages in the world . . . Some of the reactionary economists urged that we should allow the liquidation to take its course until we had found bottom . . . We determined that we would not follow the advice of the bitter-end liquidationists and see the whole body of debtors in the US brought to bankruptcy and the savings of our people brought to destruction' (October 1922).[51]

The effect of this rhetoric was to persuade the financial community that Hoover was pro-labor and anti-business, and this had a further deflationary effect. It was reinforced by his incessant attacks on the stock exchanges, which he regarded as parasitical, and his demands that they be investigated pushed stocks down still further and discouraged private investors. His policy of public investments prevented necessary liquidations. The businesses he hoped thus to save either went bankrupt in the end, after fearful agonies, or were burdened throughout the 1930s by a crushing load of debt. Hoover undermined property rights by weakening the bankruptcy laws and encouraging states to halt auction-sales for debt, ban foreclosures, or impose debt moratoria. This in itself impeded the ability of the banks to save themselves and maintain confidence. Hoover pushed federal credit into the banks and bullied them into inflating, thus increasing the precariousness of their position.[52]

The final crisis came with the collapse of American exports. The punitive Smoot–Hawley tariff of 1930, which sharply increased import duties, helped to spread the Depression to Europe. On May 11, 1931, the collapse of Austria's leading bank, the Credit Anstalt, pushed over a whole row of European dominoes. On June 21 Hoover's plan for a moratorium for reparations and war debts came much too late. All German banks shut on July 13. The British Labour government collapsed on August 24 and on September 21 Britain abandoned the gold standard. Debt-repudiations ensued. No one could now buy America's goods and its policy of foreign loans as a substitute for free trade became meaningless. Foreigners lost confidence in the dollar and, since America was still on the gold standard, began to pull out their gold, a trend that spread to American customers. In 1931–2 some 5,096 banks, with deposits of well over $3 billion, went bust and the process culminated early in 1933 when the US banking system came to a virtual standstill in the last weeks of the Hoover presidency, adding what appeared to be the coping stone to the President's monument of failure.[53]

By this time Hoover's frenzied interventionism had prolonged the Depression into its fourth year. The damage was enormous, though it was patchy and often contradictory. Industrial production, which had been 114 in August 1929, was 54 by March 1933. Business construction, which had totaled $8.7 billion in 1929, fell to $1.4 billion in 1933. There was a 77 percent decline in the production of durable manufactures during this period. Thanks to Hoover's pro-labor policies, real wages actually rose. The losers, of course, were those who had no wages at all.[54] Unemployment had been only 3.2 percent in 1929. It rose to 24.9 percent in 1933 and 26.7 percent in 1934.[55] At one point it was estimated that (excluding farm families) about 34 million men, women, and children were without any income at all—28 percent of the population.[56] Landlords could not collect rents and so could not pay taxes. City revenues collapsed, bringing down the relief system, such as it was, and city services. Chicago owed its teachers $20 million. In some areas schools closed down most of the time. In New York in 1932 more than 300,000 children could not be taught because there were no funds. Among those still attending, the Health Department reported 20 percent malnutrition.[57] By 1933 the US Office of Education estimated that 1,500 higher-education colleges had gone bankrupt or shut. University enrollments fell, for the first time in American history, by a quarter-million.[58] Few bought books. None of Chicago's public libraries bought a single new book for twelve months. Total book sales fell by 50 percent. Little, Brown of Boston reported 1932-3 as the worst year since they began publishing in 1837.[59] John Steinbeck complained, 'When people are broke the first thing they give up is books.'[60]

Thus impoverished, writers and intellectuals generally veered sharply to the left in these years Indeed 1929-33 was a great watershed in American intellectual history. In the 18th century American men of ideas and letters had been closely in tune with the republicanism of the Founding Fathers. In the 19th century they had on the whole endorsed the individualism which was at the core of the American way of life— the archetypal intellectual of the mid-century, Emerson, had been himself a traveling salesman for the spirit of self-help in the Midwest. From the early Thirties, however, the intellectuals, carrying with them a predominant part of academia and workers in the media, moved into a position of criticism and hostility towards the structural ideas of the American consensus: the free market, capitalism, individualism, enterprise, independence, and personal responsibility.

One of those who recognized the importance of this cultural watershed at the time was Edmund Wilson, whose Depression articles were collected as *The American Jitters* (1932). He thought a good time was coming for the American intelligentsia, who hitherto had had no particular function or purpose or direction but were now, like their Russian counterparts in the early 19th century, moving into a position of irrevocable opposition to the regime, becoming true critics of society. Books might not be being bought but more people were reading serious ones than ever before. The age of influence was now dawning for American writers, especially the younger ones 'who had grown up in the Big Business era and had always resented its barbarism, its crowding out of everything they cared about.' For them 'those years were not depressing but stimulating. One couldn't help being exhilarated at the sudden, unexpected collapse of the stupid gigantic fraud. It gave us a new sense of freedom; and it gave us a new sense of power.'[61]

It is a curious fact that writers, often the least organized people in their own lives, instinctively support planning in the public realm. And at the beginning of the 1930s, planning became the new Spirit of the Age. In 1932 it dominated the booklists. Stuart Chase, the popular economist who had fatuously predicted a 'continuing boom' in 1929, now came out with a timely new title, *A New Deal*. George Soule demanded ultra-Hooveresque 'works programs' in *A Planned Society*. Corporatist planning reached its apotheosis in *Modern Corporations and Private Property* by Aldolf Berle and Gardiner Means, which went through twenty impressions as the slump touched its nadir and predicted that the 'law of corporations' would be the 'potential constitutional law' of the new Economic State. America's most widely read historian, Charles Beard, advocated a 'Five-Year Plan for America.'[62] Gerard Swope, head of General Electric, produced his own national plan. Henry Harriman, head of the New England Power Company, insisted: 'We have left the period of extreme individualism . . . Business prosperity and employment will be best maintained by an intelligent planned business structure.' Capitalists who disagreed, he added, 'would be treated like any maverick . . . roped and branded and made to run with the herd.' Charles Abbott of the American Institute of Steel Construction declared the country could no longer afford 'irresponsible, ill-informed, stubborn, and un-cooperative individualism.' *Business Week* under the sneering title 'Do You Still Believe in Lazy-Fairies?' asked: 'To plan or not to plan is no longer the question. The real question is: who is to do it?'[63]

In logic and justice, it should have been the Great Engineer, the Wonder Boy. Had not his time come? But there is no logic and justice in history—only chronology. Hoover's time had come and gone. He had been in power four years, acting frantically, and the result was all around for men and women to see: utter ruin. New words based on his name had entered the vocabulary. What was a 'Hoover blanket?' It was an old newspaper used to keep warm a man forced to sleep in the open. And a 'Hoover Flag?' An empty pocket, turned inside out as a sign of destitution. 'Hoover Wagons' were motor wagons, with no gas, pulled by horses or mules, a common sight by the summer of 1932. And on the outskirts of cities, or in open spaces within them, 'Hoovervilles' were growing up, shantytowns of homeless, unemployed people. In the autumn of 1932 hitchhikers displayed signs reading: 'Give me a ride—or I'll vote for Hoover.' By this time, Republican bosses were telling him: 'Keep off the front page'—the very fact he favored a line of policy discredited it.[64] Oddly enough, he had warned himself in 1929: 'If some unprecedented calamity should come upon this nation, I would be sacrificed to the unreasoning disappointment of a people who had expected too much.' Theodore Roosevelt had put it more bluntly: 'When the average man loses his money, he is simply like a wounded snake and strikes right and left at anything, innocent or the reverse, that represents itself as conspicuous in his mind.'[65] Hoover was anything but inconspicuous. He now became the Depression Made Flesh. He had always been a dour man. Now, almost imperceptibly, he emerged as the Great Depressive. The ablest of his Cabinet colleagues, Henry Stimson, said he avoided the White House to escape 'the ever-present feeling of gloom that pervades everything connected with this administration.' He added: 'I don't remember there has ever been a joke cracked in a single [Cabinet] meeting in the last year-and-a-half.' The bouncy H. G. Wells, who called on him at this time, found him 'sickly, overworked and overwhelmed.'[66]

Politicians whom the gods wish to destroy run out of luck too. The left, which had been crushed, discouraged, and ignored in the 1920s, sniffed the breeze of ruin and began to revive. In 1932 it organized a campaign on behalf of army veterans demanding a 'War Bonus.' A 'Bonus Expeditionary Force' of 20,000 was recruited, persuaded to 'march on Washington,' and set up a shantytown camp in the middle of the city. It was ugly, pathetic, highly political, and, in a horrible way, photogenic; in short, excellent far-left propaganda. Congress flatly refused to provide more money. Hoover, whose policy on the issue was identical to F. D. Roosevelt's when the issue was revived in 1936, ordered the camp to be dispersed on July 28. The police said they could not handle it. So troops were called in under the cavalry commander Major (later General) George S. Patton. Both General MacArthur, then Army Chief of Staff, and his aide Major Dwight D. Eisenhower, played minor roles in the messy episode which followed. Photographs and newsreels did not bear out the assurances by the War Secretary Patrick Hurley that the army treated the Vets 'with unparalleled humanity and kindness.' A War Department official inflamed tempers still further by calling the Vets 'a mob of tramps and hoodlums with a generous sprinkling of Communist agitators.'[67]

No episode in American history has been the basis for more falsehood, much of it deliberate. The Communists did not play a leading role in setting up the camp but they organized the subsequent propaganda with great skill. There were tales of cavalry charges, of the use of tanks and poison gas, of a little boy being bayoneted while trying to save his pet rabbit, and of tents and shelters being set on fire with people, including women and children, still inside. They were published in such works as W. W. Walters' *BEF: the Whole Story of the Bonus Army* (1933) and Jack Douglas' *Veteran on the March* (1934), both almost entirely fiction. A *Book of Ballads of the BEF* appeared, including such slogans as 'The Hoover Diet Is Gas' and 'I have seen the sabres gleaming as they lopped off veterans' ears.' A tract of 1940, by Bruce Minton and John Stuart, *The Fat Years and the Lean*, claimed that Hoover would not suffer the veterans to disband peacefully but ordered the army to attack 'without warning ... The soldiers charged with fixed bayonets, firing into the crowd of unarmed men, women and children.' While the camp was burning, Hoover and his wife—so the account went—who 'kept the best table in White House history,' dined alone in full evening dress off a seven-course meal. Some of these fictions are still being repeated in respectable works of history half a century after the events.[68]

What mattered more at the time was the administration's inept handling of the subsequent investigation, leading to a violent and public disagreement between the US Attorney-General and the superintendent of the Washington police, Brigadier Pelham D. Glassford, who had cultivated friendly relations with the Vets. Hoover, loyally supporting his Cabinet colleague, was made to look a liar and a monster. One of his staff wrote, 'There was no question that the President was hopelessly defeated.'[69] All this took place in the closing stages of the 1932 presidential election campaign. The episode went some way to losing Hoover the support of the Protestant churches, which had hitherto opposed the 'wet' Democrats, Prohibition being the other big issue—perhaps, for most voters, the biggest issue—of the election.

The election of 1932 was also a watershed, since it ended the long period, beginning in the 1860s, when the Republicans had been the majority US party. Between the Civil War and 1932 the Democrats had won four presidential elections, electing Cleveland twice and Wilson twice, but in each case with minorities of the votes cast. Franklin D. Roosevelt, campaigning with Senator John Nance Garner of Texas, now carried the nation by a landslide, winning 22,809,638 to 15,758,901, and taking the electoral college by 472 votes to 59. The Democrats also took both Houses of Congress.[70] The 1932 election saw the emergence of the 'Democratic coalition of minorities,' based on the industrial Northeast (plus the South), which was to last for half a century and turn Congress almost into a one-party legislature. The pattern had been foreshadowed by the strong showing of A. L. Smith in 1928 and, still more, by Democrats in the 1930s mid-term elections. But it was only in 1932 that the Republicans finally lost the progressive image they had enjoyed since Lincoln's day and saw it triumphantly seized by their Democratic enemies, with all that such a transfer implies in the support of the media, the approval of academia, the patronage of the intelligentsia, and, not least, the fabrication of historical ortho-doxy.[71]

The most welcome thing about Franklin D. Roosevelt at the time was that he was not Hoover. *Common Sense*, one of the new left-wing jour-nals, got it right in one sense when it said the election had been a choice between 'the great glum engineer from Palo Alto' and 'the laughing boy from Hyde Park.' Roosevelt laughed. He was the first American President deliberately to make a point of showing a flashing smile whenever possible. By 1932 he was an experienced administrator with seven years in the Navy Department under Wilson and a moderately successful spell as governor of New York. At the beginning of 1932 Walter Lippmann described him as 'a highly impressionable person without a firm grasp of public affairs and without very strong convic-tions . . . not the dangerous enemy of anyone. He is too eager to please . . . no crusader . . . no tribune of the people . . . no enemy of entrenched privilege. He is a pleasant man, without any important qualifications for the office, and would very much like to be presi-dent.'[72] That was a shrewd and accurate assessment, before the reality was obscured by the patina of PR. *Time* called him 'a vigorous, well-intentioned gentleman of good birth and breeding.'

That too was accurate as far as it went. FDR was born in 1882 when the United States was an entity of thirty-nine states with a population of under 50 million. His mother, Sara Delano Roosevelt, had thirteen bloodlines going back to the *Mayflower*.[73] She was described as 'not a woman but a Social Presence.' FDR was an only child and relished the fact, claiming he received in consequence 'a love and devotion that were perfect.' Being an only child, he never in all his life thought of anyone but himself. But he was frightened of his mother: 'Yes, I was afraid of her too.' The Delanos were snobbish, xenophobic, and anti-semitic, and FDR got all these characteristics from his mother. Yet he was not without an affability amounting, in a political sense, almost to populism. As heir to the splendid Hyde Park estate on the Hudson, which had been in the possession of the Jacobus branch of the Roosevelts for generations, FDR was brought up to feel, and felt himself, a young prince, who could afford to condescend. He was the first member of his family to call workers on the estate and local villagers (175 people out of the total population of Hyde Park township were employed by his family) by their Christian names. He found this worked in politics, later, and FDR (who rarely knew or remembered people's surnames) did more than anyone else to launch the American habit—now becoming global—of using first names on the barest acquaintance, or none at all.[74]

FDR liked to think of himself as a countryman: 'I never have been and I hope I never will be a resident of New York City,' he said. He liked to quote Jefferson: 'Those who labor in the earth are the chosen people of God, if ever he had a chosen people, whose breasts he has made the peculiar deposit for substantial and genuine virtue.' Like Jefferson, he made a point of cultivating his estate and claimed he had planted 200,000 trees there 'for my posterity to enjoy.' But, again like Jefferson, he rarely if ever actually worked in the fields, even before he was crippled by poliomyelitis. When he said, 'I am a farmer from a family of farmers,' and when he called himself a 'hayseed' and when he discussed the relative merits of Silver Queen and Country Gentleman, he was making a political point rather than telling the strict truth.[75] But his attachment to the estate and family was quite genuine.

Indeed there was something inbred about him. His marriage to TR's niece Eleanor, another Roosevelt and FDR's cousin (five times removed) was a dynastic, even a family marriage.[76] She was an ugly duckling, whose own mother called her 'Granny,' and FDR married her in the same way as an English royal scion marries a German princess, for reasons of prudence, and with the reserved intention of finding romance elsewhere. FDR's family piety was strong. He inherited from his father a tweed suit made in 1878 and wore it till 1926, when he gave it to his son James, who was still wearing it in 1939.[77] FDR was educated privately at home until fourteen, under his mother's supervision, and he learned to be devious and dissimulating to please her. So he was glad to get away to Groton under the famous Peabody, where he got a quasi-English education. Some claimed his accent was English. He certainly used phrases like 'that's not cricket,' though he never actually played the game. He hated leaving Groton, just as some English boys hate leaving Eton, and he was never quite the same success at Harvard. But he lived on the Gold Coast and was elected to the Hasty Pudding, whose oath was: 'Resolved: that the Lord's Anointed shall inherit the earth. Resolved: that we are the Lord's Anointed.' He also had a good career with the *Harvard Crimson*.

Harvard was a more elitist institution then that it is now and by the time FDR got there it had already provided four presidents: John Adams, John Quincy Adams, Rutherford B. Hayes, and Theodore Roosevelt. FDR's father had been anti-politics—it was then the fashion among the elite, the subject of a chapter 'Why the Best Men Do Not Go into Politics' in James Bryce's famous book *The American Commonwealth*. But FDR was never in any doubt that politics was what he wanted. His politics were hereditary Democrat, rather as certain English families were hereditary Whig. Hence from Harvard he went into a New York law firm which specialized in trust-busting. FDR was always, by family tradition, training, instinct, and conviction, anti-business, or perhaps one should say suspicious and contemptuous of business. It is not surprising, therefore, that he was from the start regarded as good material by Tammany Hall, whose bosses picked him out from the throng.[78]

FDR's rise from state Senator, Assistant Navy Secretary, and New York governor to the Democratic presidential nomination was marked by four characteristics. The first was a capacity to spend public money. At the Navy Department under Josephus Daniels he was the first of the big spenders, a conduit for the lobby-men and congressmen who wanted big contracts, and for crooks-on-the-make like Joe Kennedy, who as a navy supplier dealt with FDR. In many ways FDR was made for Big Government, which first arrived in his youth and which, in maturity, he nourished mightily. He used its patronage to build up his own machine in upper New York State. He told a mass audience in 1920 that in trying to get the navy ready for war he had 'committed enough illegal acts to put [myself] in jail for 999 years.'[79] The second was a capacity to lie. FDR's lies are innumerable and some of those on the record are important.[80] He lied glibly to extricate himself from responsibility for the 'Newport Scandal,' a murky business involving homosexuality at the Naval Training Station in 1919 which might have destroyed him. Even so, the Senate subcommittee investigating it reported that his conduct was 'immoral' and 'an abuse of the authority of his high office.'[81] The third characteristic was a courageous and obstinate persistence, notable in overcoming misfortunes, particularly the attack of polio (1921) which disabled him for the rest of his life.

Polio taught FDR the capacity for the postponement of gratification, which many psychologists list as the hallmark of maturity. He was certainly a more formidable politician as a result of it. His skill in concealing aspects of himself enabled him to hide from the public the full effects of his condition. The affliction also led him to set up the commercial treatment center at Warm Springs, Georgia, which he owned, his most conspicuous act of personal benevolence (though it was racially segregated).[82] The polio business meant that FDR's promiscuity which had estranged him from Eleanor, became more restrained. As a result of his affair with a woman called Lucy Mercer, Eleanor told him that their own sexual relations must cease (they already had six children), a threat to which she stuck; and FDR stuck to Lucy, in his own fashion: she was with him at Warm Springs when he died. FDR continued, however, to have other and discreet mistresses, and Eleanor took refuge in intense friendships with women.[83]

Fourth, FDR developed public-relations skills of all kinds but he was the first US politician to pay particular attention to radio. At the 1928 Democratic convention, as in 1924, FDR was asked to give the speech nominating Al Smith. Afterwards he told the pundit Walter Lippmann: 'I tried the definite experiment this year of writing and delivering my speech wholly for the benefit of the radio audience and press rather than for any forensic effect it might have on the delegates and audience in the Convention Hall.'[84] Then and later, FDR was ahead of his time in anything to do with the presentation of his policies. He also avoided rows, though not necessarily enmities. As governor of New York, he was confronted by an angry parks commissioner, Robert Moses, who shouted: 'Frank Roosevelt, you're a goddammed liar, and this time I can prove it.' But FDR only smiled.[85] He was at the origin of the saying, later often quoted by the Kennedy brothers in the 1960s: 'Don't get mad, get even.'

FDR was not regarded as a strong candidate in 1932. It is likely that any Democrat could have got himself elected against the dismal and discredited Hoover. That was certainly the view of FDR's two closest political associates, Louis Howe, his 'inside man,' who did all the detailed work for his campaigns, but who was regarded by Roosevelt as too ugly and scruffy to appear in public alongside him, and Jim Farley, the polished pro, chairman of the Democratic National Committee in 1932, who stage-managed the campaign from the front and who was made postmaster-general in FDR's first two administrations. They both thought he was lucky, 'the man for the hour.' Like many other people, they underestimated FDR's skill at manipulating the system once he was in charge of it and the tenacity with which he gripped power. Garner, his running mate, who had reluctantly surrendered the speakership of the House in order to help Roosevelt, said that all he had to do to win was to stay alive till election day.

Considering the deplorable state of the nation, the lack of cooperation between the newly elected FDR and the outgoing Hoover during the long interregnum from the election in early November to the inauguration in March was a scandal, reflecting badly on both men but particularly on Hoover. He regarded FDR as beneath contempt, a lightweight, who had allowed his followers during the campaign to make statements he knew to be lies. Hoover had unwittingly embittered FDR by keeping him painfully standing, despite his crutches, during a meeting of governors. Once elected, FDR conceived the absurd notion that Hoover ought to appoint him secretary of state immediately, so that he and his Vice-President could resign and FDR could constitutionally move into the White House. In fact the four months saw only an exchange of chilly letters and a formal call on the eve of the handover by FDR, terminating in an arctic exchange which would have warmed Henry James' heart. When Roosevelt, who was staying at the Mayflower, then Washington's best hotel, said he realized that Hoover was too busy to return his call, the retiring President snarled: 'Mr Roosevelt, when you have been in Washington as long as I have, you will learn that the President of the United States calls on nobody.' FDR took his revenge for this by refusing Hoover, whose life was under constant threat, a Secret Service bodyguard during his retirement.[86]

The lack of cooperation during the interregnum worked in FDR's favor by appearing to draw a clear distinction between the two administrations. FDR's face was a new face at exactly the right time, and it was a smiling face. In 1932 Hoover had asked Rudy Vallee for an anti-Depression song. The curmudgeonly fellow had produced 'Brother, Can You Spare a Dime,' an instant success but not what Hoover wanted. By contrast, FDR's campaign song, actually written for MGM's *Chasing Rainbows* on the eve of the Wall Street crash, struck just the right note: 'Happy Days Are Here Again.' FDR had a knack of coining, or causing others to coin, useful, catchy phrases. The term the 'New Deal' was resurrected for his acceptance speech at the Democratic convention by Sam Rosenman. It had been used before, more than once, but FDR made it seem his alone. In his March 1933 inaugural he had a splendid line: 'Let me assert my firm belief that the only thing we have to fear is fear itself.' That had a useful impact. And FDR was lucky. A weak recovery, which had been under way during Hoover's last six months, became visible in the spring and was promptly dubbed the 'Roosevelt Market.' Luck is a very important element in political success, and FDR usually had it.

No series of events in modern history is surrounded by more mythology than the New Deal, inaugurated by the 'Hundred Days.' Beyond generating the impression of furious movement, what FDR's Treasury Secretary, William Woodin, called 'swift and staccato action,' there was no actual economic policy behind the program.[87] Raymond Moley, the intellectual who helped FDR pick his Cabinet, said future historians might find some principle behind the activities but he could not.[88] The most important figure in the New Deal, especially during the crucial early stages, was Jesse H. Jones (1874-1956), who ran the Reconstruction Finance Corporation under FDR and was, in effect, the New Deal's banker. He came from a Texan tobacco-farming family and turned Houston from a do-nothing town into a great Southern metropolis and port by deepening the Buffalo Bayou to make it possible for ocean-going ships to come up from the Gulf of Mexico, 50 miles away. This scheme was inspired by the Manchester Ship Canal in England and a nice touch was that Jones got the federal government to provide matching funds for local finance. From first to last he was expert at using taxpayers' dollars for combinations of public and private ventures. The new channel had opened in 1914 and Jones followed it by building Houston's first skyscrapers, ranging from Art Nouveau to Art Deco, in the Twenties. He would put up a ten-story building, then add extra stories when business justified it. The new Houston skyline, unmatched except in New York and Chicago, impressed people and drew investments. By 1930 Jones had moved Houston from third place in Texan cities to first, with a population of nearly 300,000, second only to New Orleans in the South. He was known to be a corner-cutter and a sly operator but the citizens said: 'Well, we'd rather have Houston the way it is today, with all Jesse's sharp goings-on, than no Jesse and no Houston.'[89]

During the war, Jones ran the American Red Cross, and played the country hick when he thought it useful. In London, he gatecrashed a party at Buckingham Palace, took off his wet shoes and warmed them in front of King George V's fire. Six feet two, weighing over 220 pounds, he radiated confidence and wealth: every time, said his wife, he passed one of the many buildings he owned, 'he pats and pets it.' He provided Woodrow Wilson, and then his widow, with a $10,000 pension (there was then no official provision) and was celebrated for raising money for the Democratic Party. Critics talked of his 'stalwart avarice and piratical trading spirit.' Jones did well out of the Depression, corraling a crowd of Houston bankers to club together and bail out the system locally. Then in January 1932 he agreed to do the RFC job to duplicate, on a national scale, what he had done in Texas. He despised the East Coasters, who he thought had made a mess of things. One of his sayings was 'Most of the country lies west of the Hudson and none of it east of the Atlantic ocean.'[90]

Jones recognized that banking was the key to recovery, insofar as anything was. The RFC under Hoover had lent over $2 billion to big banks, but they had used the money to straighten out their own affairs, rather than to reinvest. FDR's first important step, under Jones' advice, was to make a virtue of the banks shutting their doors by declaring them shut by law, what was called a 'Banker's Holiday.' This was done, significantly, under an old Wilsonian wartime measure, the Trading with the Enemy Act of 1917. The closure was followed by the Emergency Banking Relief Act (1933), the first and probably the most important measure of the entire New Deal. It was FDR's great advantage that he had a large and subservient—and frightened—Democratic majority in both Houses of Congress. The majority put through this key Act in less than a day, after a mere forty-minute debate interrupted by cries of 'Vote, vote!' This Act allowed banks to reopen with presidential powers to safeguard shareholders and depositors, authority being given to decide which should continue in business. It liquidated just over a thousand, or 5 percent, and gave the rest a federal certificate of soundness. This restored America's confidence in the nation's banks, the first step in getting cash and savings circulating again. The RFC, in return, got bank stock, making it a major shareholder in banks and so moved 'the center of American banking from Wall Street to Washington.'[91]

As chairman of the RFC, Jones acquired, and exercised, wide powers under the 1933 Act. He despised Eastern bankers and enjoyed swearing at them. At the American Bankers Association Congress at Chicago on September 5, 1933, he told them: 'Take the government into partnership with you in providing the credit the country is sadly in need of.' They sat totally silent, radiating disapproval. That evening, at a private party, he told them: 'Half the banks in this room are insolvent,' and reminded them that the 'road to solvency' now lay through Washington.[92] His greatest joy in life was to scold and scare bankers from Ivy League colleges. He put further screws on the banks by making it necessary for them to take out credit insurance from the Federal Deposit Insurance Corporation. This was an old idea of William Jennings Bryan, which Jones had heard the old 'Silver Eagle' expound. FDR was against it, an example of how he was not always in charge of his own New Deal. It was done behind his back by Jones and the Vice-President, Garner. Garner deprecated his own office, saying that the vice-presidency was 'not worth a saucer of warm spit.'[93] But on this and other occasions he proved much more instrumental than FDR knew, or would have wished.[94] Garner, like Jones, came from Texas and was also, through his wife's inheritance, a banker–millionaire. While Jones was big, he was short, with large bushy eyebrows, a white thatch, beaver teeth. Like Jones, he swore constantly and thunderously. They also shared a taste for liquor—whiskey and branch water. 'Cactus Jack' and 'the Emperor Jones' used to meet in a room set aside for Jones in the Capitol building where he could drink and play poker with the legislators, and broker deals.

Jones called himself, and encouraged others to do so, 'Uncle Jesse.' In turn, it was said of him that he was 'the first financial pirate to realize that the new field of opportunity lay in public service'—hence his unofficial title, the 'Economic Emperor of America.' Jones eventually possessed himself of Hoover's old job as secretary of commerce, and piled up other titles, heading the Federal Loan Administration, the RFC Mortgage Company, the Disaster Loan Corporation, the Federal National Mortgage Corporation, the Export–Import Bank, the Federal Housing Administration, the Census Bureau, the Bureau of Standards, the Civil Aeronautics Board, the Patent Office, the Coast and Geodetic Survey—plus another four important posts he added during World War Two. Never before had one man possessed so much public power in a democratic society.[95]

Jones, therefore, incarnated the state capitalism which Hoover prefigured and FDR actually introduced. The President played his own part, which was presentational. At the end of his first week in office he showed his mastery of the new radio medium by inaugurating his 'fireside chats.' In terms of political show-business he had few equals. His regular press conferences were exciting and he boasted that at them he played things by ear. He compared himself to a quarter-back who 'called a new play when he saw how the last one had turned out.'[96] His attempt to restore confidence and good humor received a huge boost at midnight on April 6, 1933 when, after a mere month in office, he had America drinking legally again. His own economic moves were confused and often contradictory. He increased federal spending in some directions and cut it in others. Thus he halved the pensions of totally disabled war veterans from $40 to $20 a month and put pressure on the states to slash teachers' salaries, which he said were 'too high.' He remained devoted to the idea of a balanced budget—he had never heard of Keynes, who, then and later, meant nothing to him—and urged Congress in his first Message to make major cuts in expenditure. One of the first Bills he sponsored was a balanced-budget measure entitled 'To Maintain the Credit of the United States Government.'[97] Nothing made him more angry than journalists' suggestions that his financial policy was 'unsound.'

Roosevelt's most distinctive personal contribution to ending the Depression was to buy gold, in the mistaken belief that by upping the gold price (that is, devaluing the dollar) he helped farm prices. This was delayed by a constitutional argument about whether the President could buy gold anyway. In October, the Attorney-General ruled that the Treasury Secretary had the power to buy gold on the open market. Thereafter, starting on October 25, 1933, there took place every morning, in the President's bedroom, while he ate his breakfast egg lying in state in his vast mahogany four-poster (no sign of Eleanor; she had a very separate bedroom), a curious ceremony called 'setting the price of gold.' This was done by FDR, Jones, and Henry Morgenthau Jr (1891–1967), who succeeded the sick Woodin as Treasury secretary. FDR's aim seems to have been to keep the American gold price ahead of London and Paris, and he delighted in his deviousness. He told his colleagues: 'Never let your left hand know what your right hand is doing.' Morgenthau: 'Which hand am I, Mr President?' 'My right hand, but I keep my left hand under the table.'[98] The downside of FDR's useful cheerfulness in public was what was often an unseemly levity, not to say frivolity, in private, when serious business was to be done. Fiddling with the gold price, though it amused FDR and gave him a sense of power over events, did not help the farmers, or indeed anyone else.

Apart from Jones' banking policy, which succeeded, Roosevelt's legislation, for the most part, extended or tinkered with Hoover policies. The Loans to Industry Act of June 1934 merely extended Hoover's RFC. The Home Owners Loan Act of 1933 added to a similar Act of 1932. The Sale of Securities Act (1933) and the Banking Act (1935), plus the Securities Exchange Act (1934), continued Hoover's attempts to reform business methods. The National Labor Relations Act (1935), the 'Wagner Act,' made it much easier to organize unions and won the

Democrats organized labor for a generation. But it was, in fact, an extension and broadening of the Norris–La Guardia Act passed under Hoover—and it is hard to see what it did to get people back to work, though it greatly increased the power of full-time union officials. FDR's first Agricultural Adjustment Act (1933) actually undermined the reflationary aspects of government policy, curtailed food production, and paid farmers to take land out of cultivation, a policy of despair. It contradicted other government measures to counter the drought and dust-storms of 1934–5, such as the creation of the Soil Conservation Service, the Soil Conservation Act (1935), the Soil Conservation and Domestic Allotment Act (1936), and other measures.[99] FDR's agricultural policy was designed to win votes by raising farming incomes, but it also raised food prices for the consumer and this delayed the general recovery. The National Industrial Recovery Act (1933), which created an agency under General Hugh Johnson, was in essence a Hoover-type shot at 'indicative planning.' But, drawing on FDR's wartime experience, his chief source of inspiration, it had an element of compulsion about it, Johnson warning businessmen that if they ignored his 'voluntary' codes, they would get 'a sock right on the nose.' That led Hoover to denounce it as 'totalitarian.' Johnson's bullying made the scheme counterproductive and there was not much administration regret when the Supreme Court ruled it unconstitutional.[100]

FDR really departed from Hooverism when he revived a World War One Wilson scheme, and extended it, to provide cheap power for the Tennessee Valley. As we have seen, the Californians had pointed the way, though plans to do something about taming Muscle Shoals had been laid well back in the 19th century. The Shoals mark the point at which the Tennessee River plunges 134 feet into northern Alabama, creating rocky stretches of fierce shallow water for 37 miles. FDR decided to go ahead with the operation, appointing an expert on conservation, water control and drainage called A. E. Morgan as head of the new Tennessee Valley Authority. But the real dynamo of the project (Morgan, a difficult man, was eventually sacked) was a public-power man called David Lilienthal, who was the real engineer of the TVA with a rollicking competitive instinct. He picked as chief electrical engineer Llewellyn Evans, who made TVA cheap rates famous. The Wilson Dam was used to provide vast quantities of cheap power to the fury of private sources, which had traditionally overcharged. The TVA rate was $2 to $2.75 a Kw–hour, against a national average of $5.5. This began the industrial and agricultural transformation of a huge area. It was also a spectacular piece of engineering—the flood-control system is so well designed that the turbulent Tennessee River can be shut off instantly like a tap. The project thus received intense national and international coverage, all of it favorable, which persuaded many that state capitalism worked and that it was all FDR's idea.[101]

It was Roosevelt's greatest constructive triumph, and won him the regard of the liberals and progressives, which continues to this day. But as Walter Lippmann pointed out at the time (1935), in all essentials the New Deal continued the innovatory corporatism of Hoover, using public money to bolster private credit and activity, what Lippmann called the 'permanent New Deal': 'The policy initiated by President Hoover in the autumn of 1929 was something utterly unprecedented in American history. The national government undertook to make the whole economic order operate prosperously . . . the Roosevelt measures are a continuous evolution of the Hoover measures.'[102] They were somewhat larger in scale. FDR spent $10.5 billion on public works, plus $2.7 billion on sponsored projects, employing at one time or another 8.5 million people and constructing 122,000 public buildings, 77,000 new bridges, 285 airports, 664,000 miles of roads, 24,000 miles of storm and water-sewers, plus parks, playgrounds, and reservoirs.[103] But then he was in power much longer than Hoover and had a more compliant Congress to provide the money.

If Hoover–Roosevelt interventionism was thus a continuum, the question arises, Why did it not work better? Pro-Roosevelt historians argue that the additional elements FDR brought to the continuum worked the miracle, enabling the New Deal to initiate recovery. Pro-Hoover historians argue that Roosevelt's acts, if anything, delayed what Hoover's were already bringing about.[104] A third possibility is that both administrations, by their meddlesome activism, impeded a natural recovery brought about by deflation: from the perspective at the end of the century, that seems the most probable explanation. The truth is, the recovery was slow and feeble. The only reasonably good year was 1937, when unemployment at 14.3 percent dipped below 8 million; but by the end of the year the economy was in free fall again— the fastest fall so far recorded—and unemployment was at 19 percent in 1938. In 1937 production briefly passed 1929 levels, but soon slipped below again. The real recovery from the boom atmosphere of the 1920s came only on the Monday after the Labor Day weekend of September 1939, when news of the war in Europe plunged the New York Stock Exchange into a joyful confusion which finally wiped out the traces (though not the memory) of October 1929. Two years later, with America on the brink of war itself, the dollar value of production finally passed the 1929 levels for good.[105] If interventionism worked, it took nine years and a world war to demonstrate the fact.

The recovery in other Western economies was also slow and uncertain. The exception was Germany, where the new Nazi regime of Adolf Hitler came to power at almost exactly the same time as Roosevelt. Germany's Depression unemployment figures had been even worse than America's, and its rapid return to full employment was certainly due in part to state intervention, though mainly to the recovery of business confidence, which trusted Hitler to abolish free trade unionism, as he did. In the US business confidence did not recover till World War Two broke out. Wall Street and the business community had disliked Hoover more and more as his presidency progressed. But they positively hated Roosevelt from the start. There were many reasons for this: his benevolence towards Big Labor, the way in which Jesse Jones went out of his way to revile bankers, especially East Coast ones, the composition of FDR's Cabinet and, still more, his kitchen cabinet or 'brains trust,' his anti-business remarks, both in public and in private, which circulated widely, and the general feeling that he was anxious to 'get' men who made fortunes out of the financial and entrepreneurial system. The man whom FDR appointed to run the Securities and Exchange Commission, intended to clean up the financial markets, was Joe Kennedy of Boston, a big Democratic wheeler-dealer, paymaster, and fund-raiser but a crook and known to be such: his personality horrified the stiffer elements in Wall Street. Indeed, FDR himself called Kennedy 'a very dangerous man' and said he had him 'watched hourly.'[106] There was also the feeling that FDR was prepared to pursue vendettas against individual businessmen, making unlawful use of federal agencies, such as the IRS, the Secret Service, and the FBI, as well as the New Deal creations.

FDR's willingness to use the income-tax authorities to harass those on his 'enemies list' is now well established.[107] It was less well known, though suspected, at the time. One of FDR's income-tax vendettas was quite open. For reasons which are obscure, he hated Andrew Mellon— who had been Treasury Secretary under Harding, Coolidge, and Hoover—both personally and as a symbol. Mellon epitomized the 1920 Republican slogan, 'More business in government, less government in business,' and no man was more identified with, or responsible for, Twenties prosperity, with all its strengths and weaknesses. Mellon had amassed a great art collection, which he bequeathed to the nation, founding and endowing in the process the National Gallery of Art in Washington (1937), which became one of the world's most magnificent public museums. Indeed it is probably true to say that Mellon did more for public culture in the United States than any other individual, though his son Paul Mellon, whose largesse created the great school of art history at Yale, ran him close. The donations, made through the A. W. Mellon Educational and Charitable Trust, enabled Mellon to reduce his personal tax liability, though the pictures remained in his custody during his lifetime. FDR seized on this fact to have the IRS witchhunt Mellon through the courts for evasion of tax for the rest of his life. Mellon's son-in-law, David Bruce, though a lifetime Democrat from an old Democratic family, accompanied Mellon to all the court hearings and grew to hate FDR, the 'dictator' as he called him. When a grand jury, mainly of working men, refused to indict Mellon on criminal charges, FDR insisted that the government proceed with a civil case, which dragged on and on until the old man died.[108] Nothing did FDR's reputation with the business and financial elite more harm than this obvious persecution of an upright man who had rendered the state long and faithful service, according to his lights.

If FDR's hostility to business delayed recovery, it did him no harm at all with the intellectual community. He demonstrated the curious ability of the aristocratic *rentier* liberal (as opposed to self-made plebeians like Harding, Coolidge, and Hoover) to enlist the loyalty and even the affection of the clerisy. Many of the newspaper owners hated Roosevelt, but a large majority of their journalists admired him, forgiving his frequent lies and his malicious private injunctions to them to give some of his administration colleagues a 'hard time.' They deliberately suppressed damaging facts about him. The mere suspicion that Harding played poker had damned him. But the fact that FDR played the game with journalists was never printed.[109] There were many dark corners in the Roosevelt White House, a comfortless, cheerless place. Disease left FDR stricken but as handsome and virile as ever. He seems to have needed mothering, which he certainly did not get from Eleanor, and turned instead to the women of his entourage, some of whom became mistresses of a sort.

It was widely believed by Eleanor's many enemies, even at the time (though the suspicion never got into print), that she became a lesbian. If so, she was bisexual, for she was devoted to her bodyguard, Earl

Miller, described as 'large, handsome, athletic and brazen.' She permitted him a lot of physical familiarities even in public and at one time thought of running off with him.[110] But all her most intimate friends were clever, active women, her favorites being Marion Dickerman and, above all, Nancy Cook. She seems to have fallen for Cook simply by talking to her on the phone. When they first met in person, Eleanor presented Cook with a bouquet of violets, then an international symbol of affection among proto-feminist women.[111] One of her lady-friends, Lorena Hickok, actually lived in the White House, sleeping on a day-bed in Eleanor's sitting-room.[112] In 1924 the FBI, under its new and hyperactive director, J. Edgar Hoover (1895–1972), opened a file on Eleanor and it gradually expanded to become one of the fattest in the entire secret archive.[113] All her life she was a passionate supporter of progressive causes, some of them infantile and dangerous, and her propensity to fall into far-left propaganda traps increased with age. President Eisenhower spoke of 'trying to save the United States from Eleanor Roosevelt.' But so long as FDR was alive, and for his benefit, she enjoyed the protection of the press.

FDR's appeal to intellectuals was based on the news that he employed a 'brains trust.' In fact this too was largely myth. Of his kitchen cabinet, only Rexford Tugwell, Harry Hopkins, a social worker and not an intellectual as such, and Felix Frankfurter were radical as well as influential. In any case, Tugwell and Frankfurter disagreed violently, Tugwell being a Stalinist-type large-scale statist, Frankfurter an anti-business trust-buster. They symbolized in turn the First New Deal (1933–6) and the Second New Deal (1937–8), which were flatly contradictory.[114] There was no intellectual coherence to the Roosevelt administration at any time, but it seemed a place where the clerisy could feel at home. Among the able young who came to Washington to serve FDR in one capacity or another were Dean Acheson, who found FDR's first-name-calling 'condescending,' Abe Fortas—FDR could never remember his surname though he was always 'Abe'—Hubert Humphrey, Adlai Stevenson, William Fulbright, Henry Fowler, and, not least, Alger Hiss, who held meetings with four other New Deal members of a Communist cell in a Connecticut Avenue music studio.[115] Another young man on whom FDR's favor fell was Lyndon B. Johnson from Texas, a former schoolteacher who came to Congress full of enthusiasm, cunning, and sharp practice in 1937. FDR loved him and seems to have regarded him as his natural successor at some future date. In 1944 he used his executive authority, unlawfully and unconstitutionally, to save Johnson from going to jail for criminal tax fraud.[116]

Attacks on FDR seemed only to consolidate his hold over the intelligentsia. A case in point was the great libertarian scribe H. L. Mencken of Baltimore (1880–1956), perhaps the most prolific and (in his day, especially the 1920s) the most influential American writer of all. In addition to writing and rewriting *The American Language*, the first systematic catalog of the American idiom, published in 1919 and thereafter reissued in many expanded editions up to 1948, plus thirty other books, editing The *Smart Set* and the *American Mercury*, Mencken produced over 10 million words of journalism, chiefly in the Baltimore *Morning Herald* and the Baltimore *Sun*, and wrote over 100,000 letters (between 60 and 125 per working day)—all this produced with two fingers on a small Corona typewriter the size of a large cigar box.[117]

Mencken believed, along with many Americans of his age—and many now—that all governments tended to become enemies of the people, in that they insisted on doing what ordinary citizens could do far better for themselves. The formative event of his existence was his discovery, aged nine, of *Huckleberry Finn*, which he regarded as a great, realistic poem to American rugged individualism: 'probably the most stupendous event in my whole life.' He was Mark Twain's successor, but a more political animal altogether. He was the first scourge of American religious fundamentalism, and of what he regarded as its progeny, Prohibition, which he took to be an ignorant rural conspiracy against the pleasures of the more sophisticated townsman. But his most persistent venom was reserved for federal Washington and its presidents. From the early years of the century up to the Great Depression he was, in Walter Lippmann's words, 'the most powerful personal influence on this whole generation of educated people.' Most Americans usually spoke respectfully of the President, even if they had voted for someone else, but Mencken's excoriations of the Chief Magistrate were relished as embodying the right of the citizen to commit *lèse-majesté*. And Mencken carried this right to its limits. Theodore Roosevelt was 'blatant, crude, overly confidential, devious, tyrannical, vainglorious and sometimes quite childish.' Taft's principal characteristics were his 'native laziness and shiftlessness.' Wilson was 'the perfect model of the Christian cad' who wished to impose 'a Cossack despotism' on the nation. Harding was a 'stonehead,' Coolidge 'petty, sordid and dull . . . a cheap and trashy fellow . . . almost devoid of any notion of honor . . . a dreadful little cad.' Hoover had 'a natural instinct for low, disingenuous, fraudulent manipulators.'[118] These fusillades enthralled the intelligentsia and helped to undermine the reputations of the men assailed.

Mencken excelled himself in attacking the triumphant FDR, whose whiff of fraudulent collectivism filled him with genuine disgust. He was the 'Führer,' the 'Quack,' surrounded by 'an astonishing rabble of impudent nobodies,' 'a gang of half-educated pedagogues, non-constitutional lawyers, starry-eyed uplifters and other such sorry wizards.' His New Deal was a 'political racket,' a 'series of stupendous bogus miracles,' with its 'constant appeals to class envy and hatred,' treating government as 'a milch-cow with 125 million teats' and marked by 'frequent repudiations of categorial pledges.' The only consequence was that Mencken himself forfeited his influence with anyone under thirty, and was himself denounced in turn as a polecat, a Prussian, a British toady, a howling hyena, a parasite, a mangy mongrel, an affected ass, an unsavory creature, putrid of soul, a public nuisance, a literary stink-pot, a mountebank, a rantipole, a vain hysteric, an outcast, a literary renegade, and a trained elephant who wrote the gibberish of an imbecile.[119] The failure of Mencken's assaults against FDR—and his demotion to a back number in consequence—reflected the feelings of helplessness of most Americans in the face of the appalling and mysterious Depression. Far from trusting to the traditional American ability to fend for oneself, the children of the slump turned trustingly, almost despairingly, to the state, to Big Government, to save, nourish, and protect them. This was a sea-change, and FDR, the embodiment of smiling state geniality, was its beneficiary.

The younger generation, whose spokesmen were the intellectuals of the Thirties, positively relished the paranoia which FDR evoked among the rich and the conventional and laughed at the extraordinary vehemence and fertility of invention with which the President was assailed. Rumor claimed he was suffering from an Oedipus Complex, a 'Silver Cord Complex,' heart-trouble, leprosy, syphilis, incontinence, impotence, cancer, comas, and that his polio was 'inexorably ascending to his head.' He was called a Svengali, a Little Lord Fauntleroy, a simpleton, a modern Political Juliet 'making love to the people from a White House balcony,' a pledge-breaker, a Communist, a tyrant, an oath-breaker, a Fascist, a socialist, the Demoralizer, the Panderer, the Violator, the Embezzler, petulant, insolent, rash, ruthless, blundering, a sorcerer, a callow upstart, a shallow aristocrat, a man who encouraged swearing in his entourage, who himself used 'low slang,' and who was, withal a 'subjugator of the human spirit.'[120] Crossing the Atlantic on the *Europa* shortly before the 1936 election, the novelist Thomas Wolfe said that, when he admitted he proposed to vote for the Monster, 'boiled shirts began to roll up their backs like window-shades. Maidenly necks, which a moment before were as white and graceful as the swan's, became instantly so distended with the energies of patriotic rage that diamond dog-collars and ropes of pearls snapped and went flying like so many pieces of string. I was told that if I voted for this vile Communist, this sinister fascist, this scheming and contriving socialist and his gang of conspirators I had no longer any right to consider myself an American citizen.'[121]

Against this background of rage among the financial and social elites, FDR transformed the Democrats from the minority into the majority party, which it remained for more than a generation. The 1936 election was the greatest victory in Democratic history, FDR carrying the nation by 27,751,612 to 16,681,913. His opponent, Governor Alfred M. Landon of Kansas (1887–1987), could not even carry his own state and had to be content with Vermont and Maine (so much for the old wiseacre slogan, 'As Maine goes, so goes the nation').[122] The Democrats piled up an enormous majority in the House, 334 to 89, so large indeed that some of the victors had to sit on the Republican benches, and they won the Senate 75 to 17.

The scale of the victory was more apparent than real, at least from FDR's viewpoint. It tended to increase the political power of the big-city bosses more than his own. They pandered to his desire to break the old two-term presidential rule, and to run for a third term in 1940 and a fourth in 1944, using him as a vote-getter but consolidating their own power under his waning shadow. In 1940 FDR was pitted against Wendell L. Willkie (1892–1944), head of a utility company and life-long Democrat who emerged suddenly as a prominent critic of the New Deal and was picked by the Republicans to break the Roosevelt spell. He did considerably better than Landon, carrying ten states and increasing the Republican vote to 22,305,198, but FDR still racked up almost as many votes as in 1936, 27,244,160, thus winning by a 5 million plurality with an electoral college majority of 449 to 82. In 1944, FDR ran yet again, but the party bosses were strong enough to force him to abandon his Vice-President, Henry A. Wallace (1888–1965), and to coopt instead an experienced product of the Missouri Democratic machine, Harry S. Truman (1884–1972). The Republicans ran Governor Thomas E. Dewey of New York (1902–71), who hung on to 22,006,285 votes while FDR's total shrank by nearly 2 million to

25,602,504. But the electoral college was still overwhelmingly Democratic, 432 to 99, and the winning formula, the South plus the big cities, remained intact.[123]

The Democratic hegemony enabled FDR, with some diffidence, and the party bosses and his more radical associates with considerable enthusiasm, to lay the foundations of an American welfare state. From the inception of the New Deal, some federal money had gone direct to individuals, for the first time in American history, through the Federal Emergency Relief Administration. In addition, the Civil Works Administration, the Works Progress Administration, and the National Youth Administration provided relief work to the unemployed on projects fully and directly funded by US federal agencies.[124] However, the passage of the Social Security Act (1935) introduced a specific and permanent system of federal welfare, on a two-track basis. One track insured participating employees against unemployment and old age (dependants and the disabled were later brought under the umbrella of secured income), and this track was seen as a legitimate, contributory arrangement, and its participants were not stigmatized. The second track provided federal funds to the states, under matching arrangements, for specific categories of the needy. It was means-tested and its benefits, awarded at levels kept below the prevailing income standards in any given community, were seen as unearned charity rather than entitlements—at any rate in the moral climate of the 1930s and 1940s.[125]

It is useful at this point to look forward and see how this Rooseveltian plinth was used by Congress as the foundation for a huge superstructure of transfer payments to the needy. The original pension program, for instance, reflected FDR's fiscal conservatism: funding for the pensions was to be raised entirely through taxes on employers and employees, and the size of the pension was to reflect individual contributions. Many categories of persons were excluded, for one reason or another, and no pension was to be paid before 1942, to allow funds to accumulate. However, as early as 1939 Congress amended the scheme to authorize the payment of the first pensions in 1940. Later amendments, 1950–72, broadened the coverage enormously, substantially increased the real value of benefits, indexed them against future inflation, and made the federal government the principal guarantor of the scheme, irrespective of its contributory funding. In addition, other forms of welfare were added by Congress. Unemployment insurance was brought into existence by taxes on employer payrolls. More important still, means-tested public assistance was granted, and federally funded, to categories of needy people such as the elderly poor, the handicapped, the blind, and dependent children in single-parent families. In 1950 Congress amended what had become known as ADC to AFDC, to include direct grants to the lone parent (usually a woman) as well as to dependent children. In 1962 an Act of Congress separated funding for AFDC from the Social Security Administration and made it into an independent program, state-administered but operating mainly from federal funds. By 1993 this had become the second-largest public assistance program, covering over 5 million households with 9.6 million children (one in eight of all children). By this date fewer than 1 percent of AFDC parents were required to work and half of AFDC families remained dependent on welfare for ten years or more. No other use of federal money attracted so much comment and criticism, and accusations that it encouraged a 'dependency culture.'[126]

Further tiers were added progressively. Towards the end of FDR's second term, the first food-stamp plan was adopted and tried out at Rochester, New York, on May 16, 1939. This allowed purchasers of 100 stamps to received 150 stamps' worth of surplus agricultural stocks from the Federal Surplus Commodities Corporation. By late 1940 it was operating in over 100 cities, was suspended during World War Two food-rationing, and was then resurrected by the 1964 Food Stamp Act as part of Lyndon B. Johnson's Great Society program. As a result it became even more widely used (and costly) than the AFDC program. By 1982 stamps worth $10.2 billion were being issued. By 1994, over 26 millions in 11 million households—10 percent of the population—received food stamps costing $24 billion, of which the Secret Service estimated over $2 billions' worth were obtained fraudulently.[127]

To the food-stamp program was added a series of measures to provide medical care for the poor, beginning with 'Medicare,' assistance to persons retired under social security (1965). 'Medicaid' payments to providers of health care 'to persons otherwise unable to afford them' (1965) and not entitled to Medicare were also provided by the same Act of Congress. Medicare reimbursed most medical expenses for about 19 million persons over sixty-five and others receiving long-term social security benefits. The program, which came into operation on July 1, 1966, greatly exceeded original cost projections. By 1980 spending on Medicare had risen to $35 billion annually, and it continued to increase rapidly in the 1980s, reaching $132 billion annually by 1992. Combined outlays on Medicare and Medicaid, which had been 5 percent of total federal spending under the original Great Society projections, rose to 17 percent of spending in 1994.[128]

The fact that these additions to the original modest social security program established under FDR were added under the Great Society program of Lyndon Johnson confirmed the accuracy of FDR's prediction that Johnson was his natural successor and placed the Great Society in its true historical context as the logical successor of the New Deal. Both New Deal and Great Society also underlined the historical tendency of welfare programs to expand under their own power and, without any specific decision by the electorate, or the conscious wish of those governing on its behalf, to overwhelm society by the magnitude of the burden imposed and by huge social sequences—such as the creation of lifetime and indeed hereditary dependency—wholly unforeseen and unwished by those who initiated the process.[129]

The creation of the prototype welfare state under FDR and other aspects of the New Deal inevitably ran into legal and constitutional difficulties. The Supreme Court, under Chief Justice Charles E. Hughes (1862–1948), ruled unconstitutional the National Industrial Recovery Act, the Railroad Retirement Act, and the Agricultural Adjustment Act. Furious at what he saw as the political opposition of elderly justices out of tune with the electorate, and flushed by his immense victory in November 1936, FDR submitted to Congress, on February 5, 1937, as the first Act under his new mandate, a Judiciary Reorganization Bill, immediately dubbed by its opponents a court-packing plan. This proposed that the federal judiciary be expanded by adding one new judge for each sitting justice over the age of seventy, creating a total of fifty new judgeships, including six on the Supreme Court, all of them (of course) to be appointed by the President. The Bill was introduced primarily as a measure to streamline judicial action at the federal level, but it was plainly intended to swamp a conservative judiciary with new and radical appointments.

Behind FDR's plan was a new kind of progressive, government lawyer, a breed which had grown up under the influence of Justice Brandeis and Felix Frankfurter. Brandeis, the first Jew on the Supreme Court (served 1916–39), had made himself a millionaire by the age of fifty as a corporation lawyer. But he always urged young lawyers to 'serve the public purpose,' either locally or in Washington, rather than to set out in pursuit of big fees. He was not a centralizer, and therefore was in some respects out of sympathy with the New Deal. But he was the model for an entire generation of young men of law, and his influence was such that FDR, shortly after being elected president in November 1932, paid a call on him to ask advice. More directly influential was the man Brandeis called 'half-brother, half-son,' Felix Frankfurter (1882–1962). Frankfurter had emigrated to America at the age of six, got his law degree at Harvard, taught there for a quarter-century—was indeed perhaps the most influential law professor in US history—and was constantly in and out of Washington jobs. He was part of FDR's inner circle (until his appointment to the Supreme Court in 1939) and perhaps his most important role was to persuade brilliant young lawyers to come to Washington to work in one or other of FDR's new agencies. He was an inspired networker—his wife said, 'Felix has 200 Best Friends'—publicist, and self-publicist.[130] It was Frankfurter's great achievement to establish, at any rate for one generation of the young, the moral superiority of public service over private enterprise. He became and for a decade remained a clever and indefatigable recruiter of able young men, ranging from David Lilienthal and Alger Hiss to Dean Acheson.[131]

Many of these young lawyers, as was pointed out constantly at the time, were Jews, and it is a fact that FDR, despite his constant anti-semitic remarks, befriended and promoted more Jews than any other president before or since. Hence the assertions that his real name was 'Rosenfeld' and that his program was the 'Jew Deal.' It has been worked out that Jews made up about 15 percent of his top appointments. But FDR's reply to criticism was that he would happily appoint more non-Jews if they were suitable and willing. He told a Christian friend: 'Dig me up fifteen or twenty youthful Abraham Lincolns from Manhattan and the Bronx to choose from. They must be liberal from belief and not from lip service. They must have an inherent contempt both for the John W. Davieses and the Max Steurs. They must know what life in a tenement means. They must have no social ambitions.' In fact, most of the New Deal Jews, such as Lilienthal, David Niles, Jerome Frank, and Benjamin Cohen, did not come from poor backgrounds. The only one who did was Abe Fortas.

Among the leading young alumni of law schools at Harvard, Yale, Princeton, and other Ivy League colleges, Washington now presented itself—to Jews and Wasps alike—as the high road to fame, power, and excitement (and eventually to fortune too). So they flocked there, and were recognized and rewarded and promoted, and given huge responsibilities, both open and hidden. The Roosevelt regime stimulated the rise of a new kind of political lawyer who, in pursuit of the public good, specialized in the writing and interpretation and enforcement of federal laws and regulations—thus in turn promoting the rise of a new kind of corporation lawyer, who specialized in compliance with and circumvention of such laws and regulations. In due course lawyers arose who specialized in operating on both sides of the fence, being in turn gamekeepers and poachers. The New Deal, then, while benefiting various categories of society, had a direct and long-term effect in enlarging the power, numbers, and incomes of lawyers.[132] We shall have more to say about this later.

In the meantime, among the young lawyers in and out of the White House in these years, antagonism towards what was seen as a reactionary Supreme Court was strong, and they fueled FDR's own anger. But FDR's Bill was a mistake. He miscalculated, perhaps for the first and last time on this scale, the tenor of public feeling. He was accused of seeking to subvert the Constitution and destroy the independence of the judiciary. Many of those who actually agreed with the President that changes in the federal judiciary were needed thought that a step of this importance should be taken by constitutional amendment, not by a simple statute. FDR antagonized some of his most fervent supporters by declining to consult with them in framing the measure, and by refusing to alter it once it was tabled. In fact it was unnecessary. On May 24, 1937, the Supreme Court showed that it was not against the New Deal as a whole by upholding the Social Security Act in *Steward Machine Company* v. *Davis* and *Helvering* v. *Davis*. Other New Deal measures, such as the National Labor Relations Act, were also found constitutional between March and May 1937.

The drive behind the Bill faltered with the death of its chief advocate, Senator Joseph Robinson. FDR and his Congressional supporters were forced to compromise and leave the number of federal justices unchanged. Instead, in August 1937, a new and harmless measure, the Judicial Procedure Reform Act, was quickly passed by Congress. As it happened, FDR need not have got himself into this mess, his first real defeat and a lasting blemish on his reputation. In 1937–9, death or retirement eliminated four of the most conservative associate justices of the Court. By 1941 seven out of nine justices were FDR appointments, including such liberals as Hugo Black (1886–1971), William O. Douglas (1898–1980), and Felix Frankfurter himself.[133]

Arguments over the Supreme Court, and indeed over the New Deal and the continuing Depression themselves, were overshadowed by the world crisis radiating from Europe, which opened World War Two in September 1939. The United States was an extremely reluctant participant in the Second World War, more so indeed than in the First. At no point did the United States choose to go to war against the Axis powers and Japan. War was forced upon it by the Japanese assault on Pearl Harbor on December 7, 1941 and by declarations of war by Germany and Italy four days later. It is a myth that FDR was anxious to bring America into the war, and was prevented from doing so by the overwhelming isolationist spirit of the American people. The evidence shows that FDR was primarily concerned with his domestic policies and had no wish to join in a crusade against Nazism or totalitarianism or indeed against international aggression. He took no positive steps to involve the United States in the conflict. The war came as much of a surprise—and an unwelcome surprise—to him as to anyone else. There is a persistent myth that he was forewarned about the Japanese aggression at Pearl Harbor, and did nothing to forestall it, being anxious that American participation in the global conflict should be precipitated by this unprovoked act of aggression. That all kinds of warnings were in the air at the time is clear. But an objective survey of all the evidence indicates that Pearl Harbor came as a real and horrifying shock to all the members of the Roosevelt administration, beginning with the President himself.[134]

It is also a myth, however, that America's unwillingness to engage in World War Two—the polls show that about 80 percent of the adult population wished America to remain neutral until the Pearl Harbor assault—sprang from a deep sense of isolationism, which was America's pristine and natural posture in world affairs. This myth is so persistent that it has led in the 1990s to a demand to 'a return to isolationism,' as though that were America's destiny and natural preference. So it is worth examining in a longer historical context. There is nothing unique, as many Americans seem to suppose, in the desire of a society with a strong cultural identity to minimize its foreign contacts. On the contrary; isolationism in this sense has been the norm wherever geography has made it feasible. A characteristic example was ancient Egypt, which, protected by deserts, sought to pursue an isolationist policy for 3,000 years, usually with success. In their ideographs and hieroglyphs, the Egyptians made an absolute distinction between themselves, as Egyptians and human beings, and others, who were not categorized as persons in the same sense. A more recent example of a hermit state is Japan, which tried to use its surrounding seas to pursue a policy of

total isolation, again reflected in its ideograms. China, too, was isolationist for thousands of years, albeit an empire at the same time. The
British were habitually isolationist even during the centuries when they
were acquiring an empire embracing a quarter of the world's surface.
The British always regarded the English Channel as a *cordon sanitaire*
to protect them from what they saw as the Continental disease of war.
The Spanish too were misled by the Pyrenees, and the Russians by the
Great Plains, into believing that isolationism was feasible as well as
desirable.

The United States, however, has always been an internationalist
country. Given the sheer size of the Atlantic (and the Pacific), with its
temptation to hermitry, the early colonists and rulers of the United
States were remarkably international-minded. The Pilgrim Fathers did
not cut themselves off from Europe, but sought to erect a 'City on a Hill'
precisely to serve as an example to the Old World. The original Thirteen
Colonies had, as a rule, closer links with Europe than with each other,
focusing on London and Paris, rather than on Boston or Philadelphia.
Benjamin Franklin had perhaps a better claim to be called a cosmopolitan than any other 18th-century figure on either side of the Atlantic. He
believed strongly in negotiations and in mutually advantageous treaties
between nations. The same could be said of Thomas Jefferson.
America's ruling elite was always far more open towards, interested in,
and knowledgeable about the world (especially Europe) than the French
Canadians to the north and the Spanish- and Portuguese-Americans to
the south. Despite the oceans on both sides, the United States was from
the start involved with Russia (because of Oregon and Alaska), China
(because of trade), Spain, Britain, and other European powers. Isolation
in a strict sense was never an option, and there is no evidence that the
American masses, let alone the elites, favored it, especially once immigration widened and deepened the ties with Europe.

It is true that the United States, through most of the 19th century,
was concerned with expanding its presence in the Americas rather than
with global policies. But exponents of 'America First,' like John Quincy
Adams, Henry Clay, and the 'Manifest Destiny' chorus, were imperial-
ists rather than isolationists. And the only time imperialism was an
issue in an American presidential election was in 1900, when the
Democrats used it to attack what they saw as President McKinley's
expansionist policies. The voters' approval of American imperialism, if
that is what it was, reflected itself in McKinley's convincing victory.
Contrary to a popular belief, held by many in the United States as well

as in Europe, Americans make excellent diplomats, not least because of their thoroughness. This was shown (as we have seen) at Ghent in 1814, during and after the Civil War, and not least at Versailles in 1919. And as we have also seen, the failure of the United States to make a commitment to the League of Nations and collective security in 1919–20 sprang from the obstinacy and cussedness of a sick president rather than from any widespread wish on the part of the American people or their representatives.

Between the two world wars, America sometimes appeared, in theory as well as in practice, isolationist, and much of the tragedy of World War Two is attributed to this. But, despite rejection of the League, America was certainly not isolationist in the 1920s, though its intervention in international affairs was not always prudent, particularly in the Pacific. American interest in Asia had grown steadily throughout the 19th century, and it was not only, or indeed not primarily, commercial. It was religious and cultural too. There was something in Asian culture, it has been argued, that persuaded Americans that they had a mission to intervene and change it, for the better. By the end of the 19th century, there were over 3,000 American missionaries in Siam, Burma, Japan, Korea, and, above all, China. They were joined by educationists, scientists, explorers, and technicians who taught Asians or served their governments as advisors and experts.[135] The one Asian country which resisted Americanization was Japan, and it symbolized this rejection of American cultural notions (though not its technology) by building an ocean-going navy on a large scale.

After the annexation of the Philippines by the United States, and the creation of an American naval base near Manila, America was, in the same sense as Britain, an Asian power as well as a Pacific naval power. That placed it in a potentially confrontational relationship with Japan. It was the same for Britain. But, whereas Britain resolved the dilemma by forming an alliance with Japan, which served an important role in World War One, when Japanese naval units provided protection for convoys transporting Australian and New Zealand forces to the Middle East and European theaters, the United States did nothing to prevent the development of hostile US–Japanese relations. There were reasons for this. In the early 20th century California introduced race laws to prevent the settlement of Japanese immigrants and from 1906–8 the mass migration from Japan had been halted. So the Japanese turned to China, and sought in 1915 to turn it into a protectorate. But it was the Americans who saw themselves as the prime protectors of China, and they succeeded in halting that Japanese option too.[136] American policy in the 1920s tended not merely to perpetuate Japanese–American hostility but to poison the relationship between Japan and Britain too. At Versailles, Wilson antagonized the Japanese by refusing to write a condemnation of racism (which had bearings on the situation in California) into the covenant of the League. Thereafter, America gave the Pacific priority in its naval policy, and in doing so put a sharp question to the British: whom do you want as your friends in the Pacific, us or the Japanese?

When the Anglo-Japanese Treaty came up for renewal in 1922, the Americans wanted it scrapped. The British government wanted to renew it. So did the Australians and the New Zealanders (and the Dutch and the French, who also had colonies in the area). They all agreed that Japan was a 'restless and aggressive power,' as the British Foreign Secretary, Lord Curzon, put it. But they were adamant that the Anglo-Japanese alliance was a stabilizing or 'taming' fact, and ought to be maintained. They agreed to suppress their doubts, however, when the Americans (supported by the South Africans and Canadians) proposed as an alternative a naval conference in Washington to limit armaments, with particular reference to the Pacific. Seen in retrospect, the 1922 Washington Conference was a disaster for all concerned. With heavy misgivings, the British agreed to an American proposal for a 'naval holiday,' with massive scrapping of existing warships, no new warships to be built over 35,000 tons, a 5:5:3 capital-ship ratio for Britain, the US, and Japan, and—to make the last provision acceptable to Japan—British and American agreement to build no main-fleet bases north of Singapore or west of Hawaii. The Japanese saw the agreement as the Anglo-Saxons ganging up on them, and the net result, so far as Britain was concerned, was to turn Japan from an active friend into a potential enemy. Moreover, the limitation on American naval-base construction made it impossible, in effect, for America's fleet to come to the rapid support of British, French, or Dutch possessions if they were attacked. That, in turn, caused the Japanese to regard such an assault, for the first time, as feasible, especially since the limitations on Britain's naval construction meant that its capital-ship presence in the Pacific was token rather than real. At the same time, American–Japanese relations, especially over China, continued to deteriorate.[137]

Under President Hoover, the American government continued to play a world role, with the object of preserving peace. But its actions were usually counterproductive. In 1930, the American government persuaded the semi-pacifist Labour government in Britain to sign the London Naval Treaty, which reduced the British navy to a state of impotence it had not known since the 17th century. The American navy remained comparatively large but increasingly antiquated, and the American army, with 132,069 officers and men, was only the sixteenth-largest in the world, smaller than the armies of Czechoslovakia, Poland, Turkey, Spain, and Rumania.[138] The Chief of Staff, General MacArthur, had the army's only limousine. At the same time, Hoover refused to veto the Smoot–Hawley tariff, which destroyed Japan's American trade, 15 percent of its exports. That, combined with the London treaty, which it signed reluctantly, completed Japan's alienation from the West, and determined its rulers, or at any rate the military cliques which in effect ran Japanese army and naval policy, to go it on their own. There followed the 1931 Japanese occupation of Manchuria and, in 1933, Japan's departure from the League of Nations. Hoover made no positive moves to oppose Japanese expansion.

When Roosevelt took over, he made matters worse. Hoover had helped to plan a world economic conference, to be held in London June 1933. It might have persuaded the 'have not' powers like Japan and Germany that there were alternatives to fighting for a living. But on July 3 Roosevelt torpedoed it. Thereafter the United States did indeed move into isolation, though it was not the only great and civilized power to do so in the 1930s. The French signaled their unwillingness to get involved in further efforts to uphold collective security by building the Maginot Line, a purely defensive gesture of defeatism in the face of German rearmament, indeed a form of military escapism. The British remained largely disarmed and sought to respond to a remilitarized Germany by appeasement. A wound-nursing flight from the demands of the world was the mood of the times, characterizing what W. H. Auden called 'a low, dishonest decade.' Among the victors of World War One, fear of a second, which would invalidate all their sacrifices, was universal. In the United States, the Depression, coming after nearly seventy years of dramatic economic expansion which had made it the richest and most powerful country on earth, abruptly reduced half the population to penury. There was an atmosphere of hysteria in parts of the United States during the middle years of the decade, not least in Washington, marked by outbreaks of that intellectual disease to which Americans are prone: conspiracy-theory.[139]

In this atmosphere, comparatively minor figures were able to exercise disproportionate influence. One such was Dorothy Detzer, secretary of the Women's International League, and an archetypal isolationist.[140] Detzer succeeded in stampeding the Senate towards statutory isolationism by arranging a political marriage between her favorite senator, Gerald Nye of North Dakota, and Senator Arthur Vandenberg of Michigan. The two set up a special committee under Nye to investigate charges that the international arms trade had fomented war. Among other things, the Nye Committee supposedly proved that links between the Wilson administration, the banks, and the arms trade brought America into World War One, and that much the same forces, having failed to push America into the League, were once again plotting wars of profit. Nye was undoubtedly an isolationist of sorts, but Vandenberg was an internationalist by instinct who played a notable role during and after World War Two in creating the United Nations and securing passage of the Marshall Plan. For him, as for Americans as a whole, isolationism in the 1930s was an aberration. At the time, however, the emotional drive to cut America off from what was seen as an incorrigibly corrupt Europe was strong and gave rise to the 1935–9 Neutrality Acts.[141]

These laws, like earlier US policies (that is, in the years 1914–17), limited the exercise of neutral rights as a way of protecting US neutrality. But their most important characteristic was that they deliberately made no distinction between aggressor and victim, both sides being simply characterized as 'belligerents.' That was a complete departure from previous American policy, which had always permitted the US government to make moral distinctions between participants in foreign wars. It had the inevitable effect of favoring the aggressive dictatorships of Europe and Asia at the expense of the pacific democracies and of the victims of aggression. The first Neutrality Act (August 1935) was passed after Italy's attack on Ethiopia the previous May. It empowered the President, on finding a state of war existed, to declare an embargo on arms shipments to the belligerents and (an important point in view of experience in World War One) to declare that US citizens traveling on ships of the belligerents did so at their own risk. This Act was replaced by the Neutrality Act of February 29, 1936, which added a prohibition on extending loans or credits to belligerents. The Spanish Civil War, which broke out in July 1936, was not covered by this legislation. Hence Congress, by joint resolution, January 6, 1937, forbade supplying arms to either side in the conflict. When the 1936 law expired, the Neutrality Act of May 1, 1937, which covered civil as well as foreign wars, empowered the President to add strategic raw materials to the embargo list and actually made travel by US citizens on the ships of belligerents unlawful.[142]

It is curious, looking back on it, that FDR, granted his enormously powerful position with opinion in 1936 and the following three years, made such little effort to prevent this legislation being enacted. The notion that he was a passionate defender of freedom throughout the world, determined to assist the forces of democracy by all the means in his power, but frustrated by an isolationist Congress, is another myth of these times. Efforts by the British government to put pressure on the White House to take a more active role in the defense of freedom against totalitarian aggression, in either Europe or Asia, were quite unavailing. The isolationist spirit in Congress was advanced as an excuse, rather than a reason, for inactivity. In fact the Neutrality Acts allowed the President a good deal of latitude in enforcing them. But the only occasion on which FDR used the discretion given him was in July 1937, when large-scale fighting broke out between China and Japan. Since invocation of the Acts would penalize China, far more dependent on American supplies than its adversary, FDR chose not to categorize the fight as a state of war.

The President's unwillingness to stretch his credit with Congress by trying to force through modification or repeal of the Acts, as the situation in Europe and Asia worsened, was symbolized by the alienation of Bernard Baruch from the administration. Baruch, the World War One chairman of the War Industries Board, was increasingly concerned by the weakness of the former Allies in Europe and by America's lack of preparation for a conflict he saw as inevitable. There was, in theory, a 'Preparedness Program' launched in 1938 but Baruch, regarded as a Jewish alarmist, was pointedly excluded from its policymaking and administration. To draw attention to his disagreements with FDR and his entourage, Baruch told reporters that he had no quarters in the White House or any administration building—his only office was a park bench in Lafayette Square, across the street.[143]

Baruch had, however, considerable influence in Congress. After the outbreak of war in Europe in September 1939, he devised a 'cash and carry' formula and persuaded Congress to incorporate it in a new Neutrality Act (November 4, 1939). Belligerents were once more permitted to buy American war supplies, but they had to pay cash and transport the goods in their own ships. The ostensible reason for this provision was that it would prevent the US from being drawn into war by holding debt in belligerent countries or by violating blockades while transporting war goods. The real reason was that the Baruch formula favored Britain, as it was intended to do. But only up to a point: the cash provision had the effect of stripping Britain of its hard currency reserves and forcing it to liquidate foreign holdings at knock-down prices.

The difficulties of providing war supplies to potential allies like Britain (France surrendered in June 1940 and left the war) was complicated by further Congressional legislation. World War One left behind it a complicated legacy of European Allied war debts, both to each other and to the US, and reparations imposed on Germany by the Versailles Treaty. By the spring of 1931 the spiraling international financial panic made it impossible for the Europeans to continue repaying their debts (or reparations). On June 20, 1931 President Hoover granted a year's moratorium. When this ran out and payments were still in default (only Finland eventually paid in full), largely as a result of the Smoot–Hawley tariff, Congress moved towards sanctions. By the beginning of 1934 the US was owed (allowing for future interest charges) nearly $22 billion. Granted the circumstances in which the debts were acquired—the delay in America's entry into the war—the statesmanlike solution would have been for Roosevelt to request Congress to cancel all the debts, which were clearly not going to be paid anyway, and were merely a source of bitterness and reproaches on both sides. But FDR does not seem to have seriously considered such a move. Moreover, when the isolationist element in Congress insisted on passing the Johnson Debt Default Act (April 13, 1934), he declined to veto it. This made it unlawful for the US government to loan money to any nation 'delinquent in its war obligations.' It was clearly against US interests thus to restrict the hands of its government in a rapidly deteriorating international climate. But FDR made no effort to get the Act repealed.

Not until the beginning of 1941, when Britain was virtually the only democracy still in the ring against the totalitarian powers, did FDR, again at Baruch's urging, persuade Congress to pass the Lend–Lease Act (March 11, 1941), which circumvented both the Debt Default Act and the Neutrality Act of 1939. Lend–Lease authorized the President to transfer arms or any other defense materials for which Congress appropriated money to 'the government of any country whose defense the President deems vital to the defense of the United States.' This allowed help to be sent immediately to Britain and, in due course, to other Allied belligerents, especially Russia and China. When Lend–Lease was terminated on August 21, 1945, over $50.6 billion of aid had been sent to Britain and Russia.[144]

The point has been dwelt upon because it indicates the extent to which the Roosevelt administration was infected by the spirit of isolationism as much as any other element in American society during the 1930s decade. Roosevelt showed himself as lacking in leadership as Baldwin and Chamberlain in Britain, or Daladier in France. It is permissible to speculate that Theodore Roosevelt, with his clearer ideas of America's responsibilities to the world, and his warmer notions of democratic solidarity, would have been more energetic in alerting the American people to the dangers which threatened them and the need for timely preparation and action, thereby saving countless American and Allied lives, and prodigious quantities of US treasure. As it was, not until November 17, 1941, after repeated confrontations with German submarines in the North Atlantic, and the actual torpedoing of the US destroyer *Reuben James*, did Congress amend the Neutrality Acts to allow US merchant vessels to arm themselves and to carry cargoes to belligerent ports. This was only three weeks before Pearl Harbor ended the tragic farce of American neutrality. Thus the United States was finally drawn into the war for the survival of democracy and international law at a time and place not of its own choosing, but of its enemy's.[145]

Confronted with the most momentous decisions of his presidency, Roosevelt seems to have been indecisive and inclined to let events take their course. Granted the nature of the Nazi regime, war with Germany was probably unavoidable. But the Japanese regime was subject to constant osmosis, as power fluctuated between the military and civilian elements, and it is possible that war might have been avoided. On July 26, 1941, FDR was informed that Japanese forces had pushed into the French colony of Indochina, which was virtually undefended by the Vichy regime in Paris. He reacted by freezing all Japanese assets in America, which effectively barred Japan from receiving US oil supplies. Japanese policy oscillated between peace and war during the summer and early autumn, but the consensus in Tokyo was to seek a negotiated settlement. As the Naval Chief of Staff, Admiral Nagano, put it, 'If I am told to fight regardless of consequences, I shall run wild considerably for six months or a year. But I have utterly no confidence in the second or third years.' The ablest of the naval commanders, Admiral Yamamoto, said that, however spectacular its early victories, Japan could not hope to win an all-out war against America and Britain. Colonel Iwakuro, a logistics expert, reported the following differentials in American and Japanese production: steel twenty to one, oil a hundred to one, coal ten to one, aircraft production five to one, shipping two to one, labor force five to one, overall ten to one.

On November 20 a peace offer was made to Washington. Japan promised to move all troops from southern Indochina into the north if, in return, America sold it a million tons of aviation fuel. FDR urged Secretary of State Cordell Hull (1871–1955) to take the offer seriously and wrote out a draft reply himself in pencil. Copies were passed to Winston Churchill and Chiang Kai-shek, both of whom protested strongly and the reply was never sent. Accordingly, six days later news reached Washington of further Japanese troop landings in Indochina, and FDR 'fairly blew up,' as Hull put it. As a result, a peremptory US note was sent demanding that Japan withdraw from China and Indochina immediately. The Japanese authorities treated it as an insult and began urgent preparations for a preemptive strike, which resulted in Pearl Harbor. It is arguable that, if US–Japanese negotiations had succeeded in postponing war over the winter of 1941–2, it might never have taken place at all. By the spring, the failure of the Nazis to take Moscow made it clear that Russia would stay in the war, and this itself might have deterred Japan from throwing in its lot with the Axis powers.[146]

All American doubts and hesitations, however, vanished in response to the Japanese assault and the Axis war-declaration. The Japanese war-preparations were a characteristic combination of breathtaking efficiency and inexplicable muddle. General George Marshall, FDR's principal military advisor, had repeatedly assured the President that the Oahu fortress complex, which included Pearl Harbor, was the strongest in the world and that a seaborne attack was out of the question. The Japanese needed to knock out the main units of the US Pacific Fleet while their armies were racing to occupy British Malaya and the Dutch East Indies, thus providing the Japanese war effort with supplies of rubber and oil. The plan of attack on Pearl Harbor, which involved getting a gigantic carrier force unobserved over thousands of miles of ocean, was the most audacious and complex scheme of its kind in history. It was part of an even more ambitious scheme for the conquest of Southeast Asia, embracing attacks and landings over several million square miles, involving the entire offensive phase of the war Japan intended to launch. Nothing like it had ever been conceived before, in extent and complexity, and it is no wonder that Marshall discounted its magnitude and FDR brushed aside such warnings as he received. On the other hand, the Japanese had no long-term plan to defend their new conquests.

As it was, the Pearl Harbor assault achieved complete tactical surprise but its strategic results were meager. Japanese planes attacked at 7.55 A.M. on Sunday, December 7, a second wave following an hour later. All but twenty-nine Japanese planes returned safely to their carriers by 9.45 and the entire force got away without loss. The attacks destroyed half of America's military airpower in the entire theater, put out of action eight battleships, three destroyers, and three cruisers, and totally destroyed the battleships *Oklahoma* and *Arizona*: 2,323 US servicemen were killed. These results seemed spectacular at the time and served to enrage and inflame American opinion. But most of the warships were only damaged or were sunk in shallow water. Their trained crews largely escaped. The ships were quickly raised and repaired and most returned to active service to take part in major operations. The American carriers, far more important than the elderly battleships, were all out at sea at the time of the attack, and the Japanese force commander, Admiral Nagumo, calculated he had too little fuel to search for and destroy them. His bombers failed to destroy either the naval oil storage tanks or the submarine pens, so both submarines and carriers, now the key arms in the naval war, were able to refuel and operate immediately.

The limited success achieved by the Pearl Harbor attack was a woefully small military return for the political risk of treacherously attacking an enormous, intensely moralistic nation like the United States before a formal declaration of war. The tone of the American response was set by Cordell Hull, who knew all about the attack by the time the Japanese envoys handed him their message at 2.20 P.M. on the Sunday, and had rehearsed his little verdict of history (he was a former Tennessee judge): 'In all my fifty years of public service I have never seen a document that was more crowded with infamous falsehoods and distortions on a scale so huge that I never imagined until today that any government on this planet was capable of uttering them.'[147] Thus America, hitherto rendered ineffectual by its remoteness, its divisions, and its pusillanimous leadership, found itself instantly united, angry, and committed to wage total war with all its outraged strength. Adolf

Hitler's reckless declaration of war the following week drew a full measure of this enormous fury down upon his own nation. Roosevelt supplied the rhetoric, beginning with his speech to a joint session of
Congress in which he proclaimed December 7 'a date which will live in
infamy.'

Japan's plans to annex Southeast Asia and occupy the Philippines
were successfully carried out, given some luck and poor Allied leadership. But that is as far as they got. The invasion of India, the occupation of Australia, and the assault on the United States via the Aleutians
were never seriously attempted. Meanwhile, Japan's first Pacific reverse
occurred much earlier than even its most pessimistic strategists had
expected. On May 7–8, 1942 a Japanese invasion force heading for
Port Moresby in New Guinea was engaged at long range by American
carriers in the Coral Sea and so badly damaged that it had to return to
base. On June 3 another invasion force heading for Midway Island was
outwitted and defeated, losing four of its carriers and the flower of the
Japanese naval air force. The fact that it was forced to return to
Japanese home waters indicated that Japan had already effectively lost
naval air-control of the Pacific.[148]

Meanwhile, the United States had embarked on a mobilization of
human, physical, and financial resources without precedent in history.
All the inhibitions, frustrations, and restraints of the Depression years
vanished virtually overnight. Within a single year, the number of tanks
built in America had been raised to over 24,000 and planes to over
48,000. By the end of America's first year in the war, it had raised its
arms production to the total of all three enemy powers put together,
and by 1944 had doubled it again, while at the same time creating an
army which passed the 7 million mark in 1943.[149] During the conflict,
the United States in total enrolled 11,260,000 soldiers, 4,183,466
sailors, 669,100 marines and 241,093 coastguards. Despite this vast
diversion of manpower to the forces, US factories built 296,000 planes
and 102,000 tanks, and US shipyards turned out 88,000 ships and
landing-craft.[150]

The astonishing acceleration in the American productive effort was
made possible by the essential dynamism and flexibility of the
American enterprise system, wedded to a national purpose which
served the same galvanizing role as the optimism of the Twenties. The
war acted as an immense bull market, encouraging American entrepreneurial skills to fling the country's seemingly inexhaustible resources of
materials and manpower into a bottomless pool of consumption.

Extraordinary exercises in speed took place. One reason the Americans won the Battle of Midway was by reducing what had been regarded as a three-month repair job on the carrier *Yorktown* to forty-eight hours, using 1,200 technicians working round the clock.[151] The construction program for the new defense coordinating center, the Pentagon, with its 16 miles of corridors and 600,000 square feet of office space, was cut from seven years to fourteen months.[152]

The man who emerged as the master of this creative entrepreneurial improvisation was Henry J. Kaiser. As one of the executants of the early New Deal, especially the TVA, he had been outstanding not merely for thinking big but for producing an endless succession of ingenious small ideas too—putting wooden tires on wheelbarrows and having them drawn by tractors, replacing petrol engines in tractors and earth-shovels with diesels, and so on. In building Grand Coulee, he had devised a special trestle, costing $1.4 million, to pour 36 million tons of concrete. He had erected the Parmemente cement plant, then the world's biggest, in six months. In the 1930s, more than any other man, he was building the economic infrastructure of the modern Western USA, and during the early 1940s he put it to the service of the war effort, making the West the principal supplier of mass-produced weaponry and advanced technology. David Lilienthal later remarked: 'The fall of France made it clear that TVA must be converted to war.' The war economy, with the state the biggest purchaser and consumer, was the natural sequel to the New Deal and rescued it from oblivion.

The need for hustle on a prodigious scale also served to put back on his pedestal the American capitalist folk-hero. Henry Kaiser and his colleagues Henry Morrison and John McCone, fellow-creators of the great dams, who had been systematically harassed by FDR's Interior Secretary, Harold Ickes (1874–1952), for breaches of federal regulations, were back in business on a bigger scale than ever. Having built the largest cement plant in the world, they followed it with the first integrated steel mill. The New Deal earth-movers became the creators of the 'Arsenal of Democracy'—once again, FDR's contribution to the effort was a happy phrase. The original Six Company group brought together by Kaiser entered into a partnership with Todd Shipbuilding, and set up new shipyards in Los Angeles, Houston, and Portland, Oregon. The first 'Liberty Ship' they built took 196 days to deliver. Kaiser cut the time to twenty-seven days and by 1943 he was turning one out every 10.3 hours.[153] Over 1,000 of the ships, or 52 percent, came from Pacific yards which had not existed before the war. Kaiser built the West's first steel plant, at Fontana, California, and when the government demanded 50,000 aircraft, he constructed Kaiser Aluminum, then Kaiser Magnesium, most of whose plants went to California too.[154] (It was a tragedy for the South that it had no capitalist leader comparable to Kaiser: that is why its own thrust into the modern world was delayed by nearly two decades.) Other big companies quickly adopted the hustle-style. General Electric, in 1942 alone, raised its production of marine turbines from $1 million worth to $300 million.[155] After the loss of the decisive battle of Guadalcanal, the Japanese Emperor Hirohito asked his Naval Chief of Staff, Admiral Nagano: 'Why was it that it took the Americans only a few days to build an airbase and the Japanese more than a month?' All Nagano could say was 'I am very sorry indeed.' The answer was that the Americans had a vast array of bulldozers and earth-moving equipment, the Japanese little more than muscle-power.[156] America won the war essentially by harnessing capitalist methods to the unlimited production of firepower and mechanical manpower.

There were other important factors. One was the skillful marriage of creative brainpower and new technologies to break enemy codes. Here the Americans were able to build on British foundations. The British had been leading codebreakers for over fifty years—it was their decoding of the notorious 'Zimmerman Telegram' which had helped to bring America into World War One. The British possessed, thanks to the Poles, a reconstruction of the electrical Enigma coding machine, which had been adopted by the German army in 1926 and the German navy in 1928, and which both firmly believed, to the end, produced unbreakable cryptograms. The Enigma machine became the basis for Operation Ultra, a decoding system run from Bletchley in Buckinghamshire, which not only provided the Allies with advanced information of Axis operations but enabled high-level deception schemes to be conducted. Winston Churchill made FDR privy to Ultra immediately America entered the war, and Anglo-American intelligence, based on codebreaking, became one of the main war-winning weapons.

The Americans themselves had broken Japan's diplomatic code as early as 1940, and were much assisted by the belief of Kazuki Kamejama, head of Japan's Cable Section, that such a feat was 'humanly impossible.' From January 1942, the merging of American and British code and intelligence operations led to the early breakthrough in the Pacific War—Midway in June 1942 was in many respects an intelligence victory. Thereafter, the Allies knew the positions of all Japanese capital ships nearly all the time, an inestimable advantage given the size of the Pacific naval theater. Perhaps even more important, they were able to conduct a spectacularly successful submarine offensive against Japanese supply ships. This turned the island-empire the Japanese had acquired in the first five months of the war (10 percent of the earth's surface at its greatest extent) into an untenable liability, the graveyard of the Japanese navy and merchant marine and of some of the best of their army units. Codebreaking alone raised Japanese shipping losses by one-third.[157]

Conversely, the breaking of the German 'Triton' code at Bletchley Park in March 1943 clinched Anglo-American victory in the Battle of the Atlantic, for German U-boats continued to signal frequently, confident in their communications security, and breaking the code allowed the Allies to destroy the U-boats' supply ships too. As a result, victory in the Atlantic came quite quickly in 1943, and this was important, for the U-boat was perhaps Hitler's most dangerous weapon.[158] The Ultra system was also well adapted to the provision of false intelligence to the Axis and was highly successful, for instance, in persuading the Germans that the Allied D-Day landings in Normandy were no more than a feint.[159]

Codebreaking enabled America to kill Japan's ablest admiral, Yamamoto, during his tour of the Solomon Island defenses on April 13, 1943. His flying schedule had been put on air, the Japanese communications office claiming, 'The code only went into effect on April 1 and cannot be broken.' In fact the Americans had done so by dawn on April 2. The shooting down of Yamamoto's plane was personally approved by FDR as a legitimate act of war. After it was accomplished, a signal was sent to the theater commander, Admiral William ('Bull') Halsey (1882–1959), who commanded the South Pacific Fleet in the Solomons Campaign. It read: 'Pop goes the weasel.' He exclaimed: 'What's so good about it? I'd hoped to lead that scoundrel up Pennsylvania Avenue in chains.'[160]

A combination of Allied brainpower and America's matchless ability to concentrate and accelerate entrepreneurial effort was also responsible for the success of the nuclear weapons program. The notion of a man-made explosion of colossal power was implicit in Einstein's Special Theory of Relativity and the splitting of the atom in 1932 brought it within the range of practicality. Over a hundred important scientific papers on nuclear physics appeared in 1939 alone and the most significant of them, by the Dane Nils Bohr and his American pupil J. A. Wheeler, explaining the fission process, appeared only two days before the war began. Next month, at the request of Albert Einstein, who feared that Hitler would get there first and create what he called an 'anti-semitic bomb,' FDR set up a Uranium Committee. This awarded government grants to leading universities for atomic research, the first time federal money had been used for scientific work.

The pace quickened in the autumn of 1940 when two leaders of the British scientific war effort, Sir Henry Tizard and Sir John Cockcroft, went to Washington taking with them a 'black box' containing, among other things, all the secrets of the British atomic program. At that time Britain was several months ahead of any other nation in the race for an A-bomb, and moving faster. Plans for a separation plant were completed in December 1940 and by March 1941 the atomic bomb had ceased to be a matter of scientific speculation and was moving into the zone of industrial technology and engineering. By July 1941 the British believed that a bomb could be made and in operation by 1943, and would prove cheaper, in terms of yield, than conventional high explosives. In fact the British were over-optimistic. The industrial and engineering problems involved in producing pure $U-235$, or $U-238$ plutonium (the alternative fissionable material), in sufficient quantities proved daunting, as did the design of the bomb itself. The success of the project was made possible only by placing behind the British project all the power of American industrial technology, resources, and entrepreneurial adventurism. In June 1941, FDR's administration created the Office of Scientific Research and Development (OSRD) under Dr Vannevar Bush, former dean of engineering at the MIT. This expanded to include teams working at Columbia, Princeton, the University of California, and Chicago University.

By spring 1942 the mechanics of the uranium chain reaction had been worked out and Dr Ernest Lawrence in California made a breakthrough in the production of plutonium. By June 1942 Bush was able to report to FDR that the bomb was feasible, though the demands in scientific manpower, engineering, money, and other resources would be immense. FDR (like Churchill before him) felt that the risk of the Nazis getting a bomb first was so real that he had no alternative but to accord top priority to its manufacture. Accordingly, the Manhattan District within the Army Corps of Engineers was established to coordinate resources and production, and the scheme was henceforth named the Manhattan Project.[161] The first actual chain-reaction was produced by

Dr Arthur Compton's team in Chicago in December 1942. A new laboratory for the purpose of building the bomb was established early in 1942 at Los Alamos, New Mexico, under the direction of Dr J. Robert Oppenheimer, the director of Princeton's Institute for Advanced Studies. Insofar as any one man 'invented' the A-bomb, it was Oppenheimer, though General Leslie R. Groves was almost equally important in supervising Manhattan, establishing vast new plants in Oak Ridge, Tennessee, and Hanford, Washington State, to produce the new materials required. This enormous project, working on the frontiers of technology in half a dozen directions, employed 125,000 people, cost nearly $2 billion, and was a classic exercise in the way high-technology capitalism, as originally envisaged by Thomas Edison, and modified by the experience of the New Deal, could serve the purpose of the state. The first test explosion took place on July 16, 1945 and the bomb was ready for delivery the next month. Only the American system could have produced it within such a time-scale.

Indeed, it could be said that the A-bomb was the most characteristic single product of American entrepreneurial energy. This is not said ironically: the bomb was a 'democratic' bomb, and was spurred on by genuine idealism of a peculiarly American kind. Many of those involved in Manhattan felt that liberty and decency, the right of self-government, independence, and the international rule of law were at stake, and would be imperiled if Hitler got the bomb first. In this sense, the nuclear weapons program of the US was very much part of the immigrant input into American society. Oppenheimer was of Jewish-immigrant origins and believed the future of the Jewish race was involved in the project: that was why he built the first A-bomb. It was equally true that Dr Edward Teller, of immigrant Hungarian origin, who built the first H-bomb, was convinced that by doing so he was protecting American freedom from the Stalinist totalitarian system which had engulfed the country of his forebears. Fear, altruism, the desire to 'make the world safe for democracy,' as much as capitalist method, drove forward the effort. Nuclear weapons were thus the product of American morality as well as of its productive skill.[162]

World War Two was won not just by industrial muscle power and scientific brainpower. The American system may have been near-pacifist in the interwar years, and isolationist in the 1930s, but it contrived to produce a generation of outstanding commanders, most of them born within a few years of each other, who not only organized victory in the field, in the air, and on the oceans, but also helped to establish American internationalism, and concern for the wellbeing of the entire globe, once and for all—a fact demonstrated as much in their postwar civil activities as in their wartime careers. The most important of them by far was General George C. Marshall (1880–1959), who had been Chief of Operations in General Pershing's First Army in World War One, and was later in charge of instruction at the Infantry School at Fort Benning, Georgia. Named Chief of Staff of the Army in 1938, Marshall held this appointment for over six years and was, as Churchill put it, 'the true organizer of victory.' It was Marshall (as well as Churchill) who persuaded FDR that Nazi Germany was America's most dangerous enemy, and that the air war, and later the land war, in Europe and its approaches in North Africa and the Mediterranean, must have priority over the naval war in the Pacific against Japan. This proved the right decision.[163] It involved the preparation and unleashing of an overwhelmingly powerful invasion of German-occupied Europe, at the earliest possible date consistent with its success.

For this enormous enterprise, 'Operation Overlord,' involving all services and forces from many nations, Marshall picked General Dwight D. Eisenhower (1890–1969). He was raised in Kansas, educated at West Point, a World War One captain and later assistant to General MacArthur in the Philippines, 1936–9. It is significant of the shape history was taking by the 1940s, and of the scale of military operations, that neither Marshall or Eisenhower ever commanded armies in battle. Both were strategists, organizers, trainers, coordinators. It was Marshall's gift, very necessary in a man who spent the key years of his military career working with politicians, to make an absolute and clear distinction between the political process of decision-making, and the military execution of policy, and to stick to it. This won him the respect not just of FDR himself, but of Churchill and many other Allied leaders. He instilled in Eisenhower, who had overall command of the invasions of North Africa and Italy, 1942–3, before directing Overlord in 1944, a similar awareness of his function, so that 'Ike,' as he became known to all, was able to exercise his undoubted diplomatic and political skills entirely in the immense effort of coordination needed to keep a multinational armed force functioning smoothly.[164]

It is the function of the historian not only to describe what happened, but also to draw attention to what did not happen. World War Two was marked not by the bitter and destructive rows between politicians and generals which had been such a feature of World War One, especially on the Allied side, but by a general congruity of views. That was due in great part to the characters of these two self-disciplined and orderly men. Their view that war was a business, to be organized, as much as a series of battles, to be fought, did not prevent them picking individual commanders of outstanding battlefield enterprise and energy. General George S. Patton (1885–1945), who, under Eisenhower, commanded the US 2nd Corps in North Africa, the 7th Army in Italy, and then the 3rd Army's spectacular sweep through Brittany and Northern France and across the Rhine in 1944–5 (before being killed in an auto-accident several months after the end of the war) was perhaps the most successful field commander—and certainly the best tank general—on either side during the six-year conflict.[165]

The counterpart of Marshall in the naval war was Admiral Ernest Joseph King (1878–1956), Commander-in-Chief of the fleet when the war opened, then Chief of Naval Operations, 1942–5. King had served as a midshipman in the successful operations against Spain in 1898 and was a staff officer in the Atlantic battle of World War One. From submarines he went on to command the carrier *Lexington* and he was responsible for the vast expansion of navy aviation, as chief of the Bureau of Aeronautics, 1933–6. It was as a carrier man that he went to the top in 1941–2 and it was King's decision to fight a carrier war in the Pacific. Some historians would rate him the greatest naval commander of the 20th century, though his gifts lay in his grasp of global strategy and logistics rather than as a fleet commander.[166] For his Pacific Commander, King picked the Texan Chester Nimitz (1885–1966), though he insisted on conferring with him directly every fortnight and choosing his senior staff. Nimitz, under King's close supervision, himself ran a trio of able commanders, the aggressive 'Bull' Halsey, the highly conservative Admiral Raymond Spruance, and the Marine Corps General Holland M. Smith, known to all as 'Howlin' Mad.'[167]

Nimitz began the offensive war in the Pacific on August 7, 1942, when he landed US Marines on Guadalcanal, and after five months of ferocious combat the Japanese were forced to evacuate. Then followed a strategy of leapfrogging from island to island, drawing closer and closer to Japan. Though the Pacific War against Japan did not have the overall priority of the European theater, progress was steady. By February 4, 1944, Nimitz's forces had reached the western limits of the Marianas. On June 15 two Marine divisions and the 27th Army Division were landed on the key island of the chain, Saipan. The Japanese fleet was ordered to destroy the protective cover, Admiral

Spruance's Task Force 58, but in a decisive naval air battle the Japanese lost 346 planes to Spruance's loss of 50, and the blow to Japanese naval airpower was fatal. This combat was the direct precursor of the Battle of Leyte Gulf, October 25, 1944, the most extensive sea engagement in history, which destroyed virtually all that remained of Japan's strategic naval power.

The naval victories in the central Pacific were paralleled by the second campaign of combined operations under Douglas MacArthur (1880–1964). His gallant rearguard action in the Philippines in 1941–2 had impressed FDR, who gave him the Medal of Honor, ordered him to Australia, and then appointed him Supreme Commander of the Southwest Pacific theater. MacArthur took the hard land–sea route to Japan via New Guinea and the Philippines, which he had sworn to return to, and his success entitled him to the distinction of accepting the surrender of Japan, September 2, 1945, on USS *Missouri* in Tokyo Bay. Thereafter he was given the simple mandate of governing Japan and creating democracy and the rule of law there, a task which he carried out with august proconsular relish and remarkable success.[168]

In Europe, where the D-Day landings of Operation Overlord, June 6, 1944, began an eleven-month continental campaign which ended with the suicide of Hitler and the Nazi surrender in May 1945, the political aspects were far more complicated. Eisenhower himself refused to allow political considerations—that is, the future composition and ideological complexion of Europe after the fall of Hitler—to enter his strategy at any point. Hence he refused to countenance the proposal of his chief British subordinate, General Bernard Montgomery, to throw all the Allied resources into a single, direct thrust at Berlin, which might have ended the war in 1944, but embodied corresponding risks, and instead favored his own 'broad front' policy, of advancing over a wide axis, which was much safer, but also much slower. As a result the Russians got to Berlin first, in the process occupying most of eastern Europe and half of Germany itself. While Russian strategy in the rollback of German forces, 1943–5, was largely determined by Stalin's aim of controlling the maximum amount of territory and natural resources in the postwar period, Eisenhower stuck closely to his military mandate of destroying the Nazi forces and compelling Germany to surrender unconditionally. He resisted any temptation to jockey for postwar advantage. He halted when he felt his task was accomplished.[169]

To some extent Eisenhower, in ignoring the politics of the postwar, reflected the views of his political master, President Roosevelt. FDR, a superb political tactician and a master of public relations, excellent at galvanizing the American people to great purposes, was unclear in his own mind about what those great purposes should be, beyond the obvious one of winning the war. In this respect he lacked Churchill's clarity and vision. FDR was very cosmopolitan at a superficial level: by the age of fourteen he had made eight transatlantic crossings, and he had had varied international experiences during and immediately after World War One. But, while often subtle and shrewd about American domestic politics, he was extremely naive, and sometimes woefully ignorant, about global political strategy. In particular, he tended, like many intellectuals and pseudo-intellectuals of his time, to take the Soviet Union at its face value—a peace-loving 'People's Democracy,' with an earnest desire to better the conditions of the working peoples of the world. He was very badly advised by the businessman and campaign-contributor he sent to Moscow, Joseph E. Davies, who was even more naive than FDR, and whose reports make curious reading. Davies saw Stalin as a benevolent democrat who 'insisted on the liberalization of the constitution' and was 'projecting actual secret and universal suffrage.' The ambassador found this monster, whose crimes were pretty well known at the time, and whom we now realize was responsible for the deaths of 30 million of his own people, to be 'exceedingly wise and gentle . . . A child would like to sit on his lap and a dog would sidle up to him.'[170] Davies told his government that Stalin's notorious show-trials were absolutely genuine and repeated his views in a mendacious book, *Mission to Moscow*, published in 1941. Sometimes FDR believed Davies, sometimes he didn't, but he trusted the *New York Times*, whose reporting from Moscow, by Harold Denny and Walter Duranty, was spectacularly misleading and, in Duranty's case, grotesquely Stalinist. Duranty's favorite saying was 'I put my money on Stalin.'[171]

FDR's unsuspicious approach to dealing with Stalin and the Soviet Union was reinforced by his rooted belief that anti-Communists were paranoid and dangerous people, reactionaries of the worst sort. In this category he included many of his State Department advisors and Churchill himself. FDR particularly distrusted Davies' successor in Moscow, Laurence Steinhard, who took the hardline State Department view of Russia's good intention and strongly urged his government that to appease the Soviets would be fatal to America's policy aims. 'My experience,' he reported, is that the Soviet leaders 'respond only to force, and if force cannot be applied, to straight oriental bartering.'[172] FDR's reaction was to ignore him. Indeed, he went further. The moment Hitler's declaration of war made Russia America's ally, he devised procedures for bypassing the State Department and the American embassy in Moscow and dealing with Stalin directly. Nothing could have been more foolish, as well as catastrophic for the people America was pledged to protect. FDR's intermediary was Harry Hopkins (1890–1946), a former social worker and political fixer whom FDR trusted more than anyone else, insofar as he trusted anybody.

Hopkins reported back that Stalin (naturally) was delighted with the idea: '[he] has no confidence in our ambassador or in any of our officials.' FDR also wanted to bypass Churchill, whom he thought an incorrigible old imperialist, incapable of understanding ideological idealism. He wrote to Churchill, March 18, 1942, 'I know you will not mind my being brutally frank when I tell you that I think that I can personally handle Stalin better than either your Foreign Office or my State Department. Stalin hates the guts of all your top people. He thinks he likes me better, and I hope he will continue to do so.'[173] 'I think,' FDR said of Stalin, 'that if I give him everything I possibly can and ask nothing from him in return, *noblesse oblige*, he won't try to annex anything and will work with me for a world of democracy and peace.'[174] But what FDR, over Churchill's protests, gave to Stalin was not his to give.

Thus it was that, at Roosevelt's insistence, Soviet Russia emerged from World War Two as its sole beneficiary by precisely one of those secret wartime treaties which Woodrow Wilson and the Treaty of Versailles had so roundly condemned. And not only Wilson— Roosevelt himself. The Atlantic Charter of August 14, 1941, which Roosevelt largely drafted—reiterated in the United Nations Declaration of January 1, 1942—stated that the signatories 'seek no aggrandizement, territorial or other . . . they desire to seek no territorial changes which do not accord with the freely expressed wishes of the people concerned.' Yet at the Yalta Conference of February 1945, at which Roosevelt acted in effect as chairman and intermediary between the Communist dictator Stalin and the 'hopelessly reactionary' Churchill, Stalin made a series of precisely such demands. In return for agreeing to enter the war against Japan 'two or three months after Germany has surrendered,' Stalin demanded recognition of Russia's possession of Outer Mongolia, southern Sakhalin, and adjacent islands, outright annexation of the Kuril Islands plus other territorial rights and privileges in the Far East, at the expense of Japan and China and without any reference to the wishes of local inhabitants. FDR agreed to these acquisitive conditions virtually without argument, and Churchill, who was desperate for FDR's support on issues nearer home, acquiesced, since the Far East was 'largely an American affair . . . To us the problem was remote and secondary.'[175]

Even in Europe, however, Roosevelt tended to give Stalin what he wished, thus making possible the immense satellite empire of Communist totalitarian states in eastern Europe which endured until the end of the 1980s. This became apparent as early as the Tehran Conference in November 1943, attended by Stalin, Roosevelt, and Churchill. The chairman of the British Chiefs of Staff, Sir Alan Brooke, summed it up: 'Stalin has got the President in his pocket.'[176] In the period immediately before the decisive Yalta meeting, Churchill was able to save Greece from Soviet domination, and thus bar Stalin's access to the Mediterranean, but only at the cost of sacrificing the rest of the Balkans. He was disturbed by the anxiety of Eisenhower and other American commanders to curtail US military operations in Europe as quickly as possible, once Germany was defeated, so that troops could be transferred to the Far East. Eisenhower thought it wrong to use American armies even as a deterrent to Soviet ambitions in eastern Europe: 'I would be loath to hazard American lives for purely political purposes,' as he put it.[177] At Yalta itself, FDR deliberately rejected Churchill's suggestion that the American and British governments should coordinate their tactics in advance. 'He did not wish,' as Averell Harriman put it, 'to feed Soviet suspicions that the British and Americans would be operating in concert.' When the fate of Poland came to be decided, FDR refused to back the British demand for an international team to supervise the elections which Stalin promised, being content with the Russian assurance that 'all democratic and anti-Nazi parties shall have the right to take part.' He added a typical piece of Rooseveltian rhetoric called 'The Declaration on Liberated Europe,' which verbally committed all its signatories to respect 'the right of all people to choose the form of government under which they will live.' Stalin was happy to sign it, and was delighted to hear from FDR that all American forces would be out of Europe within two years.[178]

It is difficult to surmise what would have happened in Europe if Roosevelt had survived to the end of his fourth term and had continued to direct American policy in the early postwar period. There is some evidence that in the last weeks of his life, in spring 1945, he was becoming increasingly disillusioned about Russian behavior and was beginning to realize that Stalin had betrayed his trust. His views on other issues too were becoming more volatile. A case in point was what to do about the Jewish survivors of Hitler's Holocaust. It was during World War Two that the American Jewish community first developed its collective self-confidence, and began to exert the political muscle its numbers, wealth, and ability had created. David Ben-Gurion, one of the founders of the state of Israel, visited the United States in 1941 and felt what he called 'the pulse of her great Jewry with its five millions.' In the closing stages of the war, and the immediate postwar, it became the best-organized and most influential lobby in America. It was able to convince political leaders that it held the voting key to swing states like New York, Illinois, and Pennsylvania.

FDR was as conscious as any other professional politician of the voting strength of American Jews. He had not been prepared to ask Congress to open America to an emergency influx of Jewish refugees before or during the war. Nor, as America's commander-in-chief, had he been willing to agree to divert Allied resources to physical attempts to prevent the Holocaust, when news of its magnitude and horror trickled through. On the contrary; he was instrumental in persuading the American Jewish leaders that it was in everyone's interests to concentrate on defeating the Nazi forces and winning the war as quickly as possible. FDR had a general commitment to support Jewish efforts to create a national home in Palestine. However, he sounded a distinctly anti-Zionist note when, on his way back to America from Yalta, he had a brief meeting with the King of Saudi Arabia, and they discussed the foundation of the Jewish state. 'I learned more about the whole problem,' he told Congress on his return, '. . . by talking with Ibn Saud for five minutes than I could have learned in an exchange of two or three dozen letters.'[179] David Niles, the passionately pro-Zionist presidential assistant, testified: 'There are serious doubts in my mind that Israel would have come into being if Roosevelt had lived.'[180] But Roosevelt did not live; he died, quite suddenly, of a massive cerebral hemorrhage on April 12, 1945. The United States immediately came under leadership of a quite different kind.

Harry S. Truman—the 'S' stood not for nothing, as his enemies jeered, but for both Solomon and Shippe—proved to be one of the great American presidents, and in some respects the most typical. He was seen at the time of his sudden precipitation into the White House as a nonentity, a machine-man, a wholly parochial and domestic politician from a backward border state who would be lost in the world of international statesmanship in which he now became the leading player. In fact Truman acquitted himself well, almost from the start, and not only the United States but the whole world had reason to be grateful for his simple, old-fashioned sense of justice, the clear distinctions he drew between right and wrong, and the decisiveness with which he applied them to the immense global problems which confronted him from the very first moments of his presidency. Moreover, any careful study of his record shows that he was well prepared, by character, temperament, and experience, for the immense position he now occupied.[181]

Truman's Missouri, admitted as the twenty-fourth state in 1821 as part of the famous Missouri Compromise, remained slave-holding but was never part of the cotton economy of the South. In economic terms it developed as a prairie state with links to the West, and in 1860–1 its Unionists kept it officially loyal to the federal government. All the same, it was the scene of guerrilla fighting throughout the Civil War and thereafter remained thoroughly Democratic in the Southern sense, violent, lawless in many ways, corrupt, and machine-ridden.[182] In the 1920s and 1930s, when Truman made his mark, power in the state was divided between St Louis, comparatively honest thanks mainly to a then-outstanding newspaper, the *St Louis Post-Dispatch*, and Jackson County, which included Kansas City. In the interwar years, Kansas City usually had the upper hand. Since it was entirely controlled by the Democratic Party machine run by T. J. Pendergast, and since Truman came from Jackson County, the future President, if he wanted to go into politics at all, had no alternative but to work for the machine, and be branded for doing so. When he finally reached the Senate in 1935, he was referred to as 'Tom Pendergast's Office Boy,' not least by the *Post-Dispatch*, which long remained his mortal enemy.[183]

But the reality was different. Truman's political career in Missouri demonstrates how he emerged from a corrupt local system but remained honest and his own man. The point was and is important because it helps to show why Truman, not despite his background but because of it, had the character and skills to save the West when the death of Roosevelt gave him the opportunity. Truman came from a family of Baptist farmers, who owned or leased substantial acreages but never seemed to make much money. Jackson County was only semi-tamed in his early days. The old outlaw Frank James still lived there; the veterans of Quantrill's Raiders, the most notorious of Missouri's Confederate guerrillas, held annual reunions in this district; and one of Truman's uncles, Jim 'Crow' Chiles, had been killed in a gunfight with Marshal Jim Peacock in Kansas City's Courthouse Square. Truman's mother Martha 'handled a shotgun as well as most men' and was 'as tough as a barrel of roofing nails.'[184] Despite all this, Truman in many ways had the kind of childhood presented in Norman Rockwell's covers. His family moral training was strict, reinforced by the public schools of his home town, Independence. In some ways his schooling was primitive but its emphasis on character-building was of a kind which American parents, at the end of the 20th century, find it impossible to obtain for their children, however much they pay for it. Family and school combined to give him a religious and moral upbringing which left him with a lifelong conviction that personal behavior, and the behavior of nation, should alike be guided by clear principles which made absolute distinctions between right and wrong. These principles were based on fundamental Judeo-Christian documents, the Old and the New Testaments, and especially the Ten Commandments and the Sermon on the Mount.

In short, Truman was a product of Victorian values, and he also accepted the Victorian conviction that the story of mankind was one of upward progress through industry and high ideals. Truman was brought up to believe that, ever since its inception, America had occupied the moral high ground in the community of nations. Thus formed, he felt this distinction was slipping away from his country as early as the 1930s, confiding in his diary just before he entered the Senate: 'Some day we'll awake, have a reformation of the heart, teach our kids honor and kill a few sex psychologists, put boys in high schools with *men* teachers (not cissies), close all the girls' finishing schools, shoot all the efficiency experts and become a nation of God's people once more.'[185]

Truman was thus imbued with a strong determination to get on in life and to do so honestly in conformity with the good old American tradition. But it did not prove easy. He turned sixteen in the year 1900 and for the next twenty years he engaged, sometimes in conjunction with members of his family, in a variety of farming, business, and mining ventures, always working hard, never making much money, often losing it, and sometimes barely avoiding ruin. When he became a successful politician he was categorized as a former haberdasher, but the truth is he was jack-of-all-trades and a master of none. Patient research has unraveled the history and finances of all Truman's multifarious attempts to get on, but all they seem to prove is that Truman was not born to be a businessman.

It was World War One that made him. It showed he could, and given the chance would, lead from the front—and in the right direction. He enlisted as soon as possible in a spirit of old-style patriotism. Despite poor eyesight, he was commissioned, served in France as a battery commander in the 129th Field Artillery, led his men with conspicuous courage and skill, and was discharged from active service as a major. He loved his gunners, and they loved him. He remained a reserve officer throughout the interwar years, regularly going into camp and on maneuvers, being promoted to full colonel, and finally resigning only in 1945, when he became US president and commander-in-chief. This connection with the US Army was vital to his future performance as president because it led him to take a continuing interest in global strategy, in America's preparedness for war, and in foreign politics. Truman was the very reverse of an isolationist, seeing America as assigned by God and its own circumstances and good fortune, to play a leading role in the world. There can have been few Americans of his generation who took a keener interest in events on the other side of the Atlantic and Pacific.

Truman's thirty years with the US Army was also part of his instinctive Americanism—his desire to join in, to participate in, every aspect of life in his community. In addition to his army reserve service, which involved countless dinners and outings, he joined and rose steadily in the Masonic order, was active in the American Legion, and was a prominent member of the Veterans of Foreign Wars. He joined the Kansas City Club, the Lakewood Country Club, the Triangle Club of young businessmen, and the National Old Trails Association. Any organization he became a member of had the free use of his services, though he sometimes had to resign because he could not afford the dues.

With his war record and his love of community effort, it was inevitable he would go into local politics, care of the Pendergast machine. Having attended Kansas City Law School, he was elected judge of the Jackson County Court in 1922 and from 1926 was its presiding judge. There he discovered his second great gift, for administration. For the court was primarily not a juridical body but a governing one, distantly derived from the old Quarter Sessions, which ran local government in England until the late 19th century. The main function of the court in Truman's day was maintaining and updating the transport system of Kansas City and its environs during a revolutionary period when the whole of America, not least the Midwest, was becoming motorized, and when the building of modern roads was paramount. To carry through a major road-building program as the nominee of the Pendergast machine, whose paymasters specialized in building contracts, was a challenge to any man's honesty. Truman was under no illusions about his political associates. It was his habit, throughout his life, to keep private notes or diaries in which he let off steam and wrote away his frustrations. Apart from Pendergast himself, whom Truman insisted was, despite his faults, 'a real man' who 'kept his word,' albeit 'he gives it very seldom and usually on a sure thing,' all the other machine men were worthless. Robert Barr was 'a dud, a weakling, no ideals, no nothin';' Thomas B. Bash led Truman to ask: 'I wonder what the B stands for—Bull or Baloney?' Leo Koehler's 'ethics were acquired in the north-end precinct . . . he can't be stopped.' Spencer Salisbury 'used me for his own ends, robbed me;' Fred Wallace was 'my drunken brother-in-law;' Cas Welch was 'a thug and a crook of the worst water;' Joe Shannon 'hasn't got an honest appointee on the payroll;' Mike Ross was 'just a plain thief;' as for Willie Ross, who had recently died, 'I suspect his sales of rotten paving have bankrupted the government of Hell by now.'[186]

The background to Truman's road administration was not just corruption but gangland violence. From 1928, the Pendergast machine was enmeshed with the minions of the Kansas City underworld boss John Lazia. He was a poor immigrant boy who had 'made good,' after a fashion, as a gangster-turned-businessman and community godfather—an archetype of the era. He owed his position to a small army of enforcers, which included the notorious Charles 'Pretty Boy' Floyd. They used Kansas City as a sanctuary when things got tough in the more law-abiding world to the north and east. Al Capone himself recognized Lazia as king in Kansas City. Indeed next to Chicago itself, Kansas City in the 1930s was the nation's crime center, specializing in arson, bombing, kidnapping, murder in the street, and official toleration of gambling, prostitution, the illicit sale of liquor, and extortion.[187]

To get his road program through, Truman had to avoid being engulfed by the outrageous rake-off demands of the Pendergast machine—Pendergast himself was a compulsive gambler, which explains his incessant and growing need for cash—and to defy the strong-arm methods of Lazia. So he had to operate the legitimate aspects of machine politics for all they were worth. As late as 1941, when he was one of the most important men in the country, he was still worrying about the nomination of a dentist for the Jackson

County farm, the selection of a shop-foreman for the County garage, and the appointment of road overseers. He undoubtedly felt the strain of working with corrupt men on a day-to-day basis while trying to keep them way from his roads program. Obviously in some cases he was forced to yield simply to get things done. He wrote in one agonizing memo to himself: 'Am I a fool or an ethical giant? I don't know . . . Am I just a crook to compromise just in order to get the job done? You judge it, I can't.'

The most meticulous research has not uncovered any examples of Truman himself profiting from all the easy opportunities open to him. The closest he came to corruption was in 1938 when, to avoid yet another crisis on the family farm, Truman arranged a nine-month loan of $35,000 from the Jackson County School Fund, on a property that was assessed at only $22,680. Pointing this out, the *Kansas City Star* commented that a loan of this size was strictly unlawful, and it later became overdue. But the best proof of Truman's honesty is that he remained poor. Throughout the 1930s he was struggling desperately to pay off the miserable debts of his haberdashery business—and this at a time when he was responsible for spending scores of millions of dollars.[188]

In the end, Truman completed his new highway system on time, within budget, and to the highest quality standards. It was a moral as well as an administrative achievement, which took farmers out of their isolation and increased the value of farmland by an average of $50 an acre, while making the rural beauty of the region accessible to city-dwellers. It was built with a speed which would be unimaginable today. Truman saw it as a triumph of 'planning'—the 1930s vogue word—and celebrated its completion with a little book, *Results of County Planning*. His administrative career was thus built up despite and in some respects against the Pendergast machine. Hence, though he may have been elected senator for Missouri in 1934 as its nominee, he was also an independent figure who could justify himself without the help of Tom Pendergast or anyone else.

His character and skill were put to the test when the Justice Department made Kansas City in general, and Pendergast in particular, its Number One target for 1938. The following year Pendergast was fined $350,000—his entire remaining net assets—and was lucky to get only fifteen months in Leavenworth. All his chief associates were convicted. Truman was able to write to his wife Bess: 'Looks like everyone got rich in Jackson County but me. I'm glad I can still sleep well even if it is a hardship on you and Margie [his daughter Margaret] for me to be so damn poor. Mr Murray, Mr McElroy, Mr Higgins and even Mr P himself probably would pay all the ill-gotten loot they took for my position and clear conscience.'

At the time most observers believed Truman was sure to be convicted too or, if not, wiped out in the 1940 election. The collapse of the Pendergast machine left him exposed to a multitude of enemies. The *St Louis Post-Dispatch* crowed: 'He is a dead cock in the pit.' But the worst that happened is that the county foreclosed on the family farm, and that in turn underlined his own poverty and suggested to most people that Truman was honest. He was already an experienced and effective election campaigner. He now set about building up his own machine and conducted the first of his great barnstorming campaigns. In the Democratic primary—the key election—he actually carried the 'enemy' city of St Louis by 8,391, and won forty-four counties out of seventy-five, five more than he had six years earlier with the Pendergast machine behind him. It was a personal triumph which once and for all established Truman as his own man.[189]

Thereafter Truman became, increasingly, a national figure. During his first term he had stuck to transport, in effect carrying trough a fundamental reorganization of the way in which the US railroad system was regulated. In his second term, with the advent of war and the largest spending program in American history, he created and presided over the Special Committee to Investigate the National Defense Program. Here his experience in carrying through a large-scale and efficient road-building program in the teeth of a corrupt machine was invaluable in insuring that the United States, while arming with all deliberate speed, avoided waste, fraud, and abuse by both Big Business and Big Labor. What became known as the Truman Committee, according to the boast of its chairman, succeeded in saving the country $15 billion, a colossal sum in those days. Whatever the exact figure, the savings were plainly huge and much appreciated. In 1943 Truman's efforts got him on the cover of *Time* magazine as 'a crusader for an effective war effort.' The magazine even turned Truman's Pendergast connection to his advantage since it said he had remained personally uncorrupt in a machine-ridden state but had refused to kick an old political friend just because he was down. The next year a *Look* survey of fifty-two Washington newsmen named Truman as one of the most valuable officials in the nation—and the only one in Congress. Thus the stage was set for Truman's selection as FDR's running mate in 1944, and his accession to the White House less than three months after Roosevelt began his fourth term.[190]

This survey of Truman's previous career has been necessary because it shows he did not simply emerge from nowhere to became an effective president by a kind of natural miracle. He was in every way a product of the American democratic system, just as Lincoln was. His career, character, and experience prepared him to take over control of the greatest power on earth in the concluding stages of a world war and the beginning of a confused and dangerous peace. As a successful adminis-trator, he had been used to making rapid decisions in quick succession. As a former field officer and active reservist he had followed closely the dangers that beset the democratic powers and America's ability to meet them. By presiding over the most important wartime Congressional committee he had acquired an enormous knowledge of the military effort, its cost, efficiency, and outreach. He probably knew more about the nuts and bolts of defense than Roosevelt himself.

Truman was taking decisions of importance from his earliest moments in the White House, but his first act of historic significance was authorizing the use of the new A-bomb against Japan. The promptness and singlemindedness with which he took this step illus-trates, in one sense, aptness for his supreme executive role. But, in another, it demonstrates the way in which total war corrupts even the right-thinking, and imposes a relativistic morality which distorts the judgment even of those, like Truman, who are committed to a set of absolute values. In the 1930s the democracies had looked upon the bombing of cities with horror, and there was outrage when Hitler's Germany began the war by indiscriminate bombing raids on Warsaw and other Polish cities, followed in due course by similar attacks on Rotterdam, Belgrade, and many British cities. Churchill, who was well aware of the moral decay war brings, initiated the strategy of the mass bombing of German cities on July 2, 1940 not so much by way of reprisal as because he was overwhelmed by the prospect of Nazi occu-pation—to him the ultimate moral catastrophe—and saw bombing as the only offensive weapon then available to the British. When the Americans entered the war, with their capacity to build vast numbers of heavy bombers, they fitted naturally into this strategy, and one of the earliest US wartime exploits was the raids on Tokyo and other Japanese cities carried out by B25 bombers, under the command of General James Doolittle on April 8, 1942. The US Army Air Corps' bombers operated in conjunction with the British Royal Air Force in carrying out continual raids, often of a thousand bombers or more each, on German targets, both 'area raids' on cities designed to destroy German civilian morale and pinpoint attacks on military and economic targets, during the years 1942–5.

These horrifying raids culminated in the destruction of Dresden on the night of February 13–14, 1945, a blow agreed upon at Yalta by FDR and Churchill to please Stalin, which was carried out by two waves of British bombers followed by a third, American one. Among other bombs, over 650,000 incendiaries were dropped, the firestorm engulfing 8 square miles, totally destroying 4,200 acres and killing 25,000 men, women, and children. As it was the night of Shrove Tuesday, many of the dead children were still in carnival costumes.[191] Hitler's propaganda chief, Josef Goebbels, claimed: 'It is the work of lunatics.' It was not: it was the response of outraged democracies corrupted by the war he and his Nazi colleagues had started, and now obsessed with what one British military theorist called the 'Jupiter Complex'—the ability granted by the possession of huge air forces, to rain thunderbolts on the wicked. The Jupiter complex was to be with the United States for the rest of the century.

What American bombers could do to Nazi Germany operating from bases in Britain, they began to do to Japan, as soon as it could be brought within their range. This was the object of America's Central Pacific strategy, which began at Tarawa Atoll in November 1943. It consisted of hopping or leapfrogging the islands on the route to Tokyo, using airpower, amphibious landings, and overwhelming firepower.[192] At Tarawa, the desperate resistance of the Japanese army meant the Americans had to kill all but seventeen of the 5,000-strong garrison, and lost 1,000 men themselves. As a result they increased the firepower and lengthened the leapfrogging. At the next island, Kwajalein, the air–sea bombardment was so cataclysmic that an eyewitness said, 'the entire island looked as if it had been picked up to 20,000 feet and then dropped.' Virtually all the 8,500 defenders were killed, but the use of this colossal firepower kept American dead down to 373. These ratios were maintained. In taking Leyte, the Japanese lost all but 5,000 of their 70,000 men, the Americans only 3,500. At Iwo Jima, the Americans sustained their worst casualty ratio, 4,917 dead to over 18,000 Japanese, and in taking Okinawa they had their highest casualty bill of all, 12,520 dead or missing, against Japanese losses of 185,000 killed. Most Japanese were killed by air or sea bombardment, or were cut off and starved. They never set eyes on an American foot-soldier or got within bayonet range of him. Thus Americans came to see overwhelming firepower, often delivered from great distances, as the key both to defeating the Japanese and to keeping their own casualties as low as possible.[193]

The area bombing of Japan, by land-based heavy bombers maintaining a round-the-clock bombardment on an ever-growing scale, was driven by the same understandable motive—to end the war as swiftly as possible with the minimum American casualties. Those who controlled Japan, insofar as anyone really controlled it, knew the war was lost by the autumn of 1942 at the latest, and their culpability in declining to negotiate is obvious and, to Western minds, totally inexplicable. To the American politicians and military leaders in charge of the war against Japan, therefore, it was Japan which bore the moral responsibility for what followed. It started in November 1944, when the captured Guam base came into full use, and B29 Flying Fortresses, each carrying 8 tons of bombs, could attack in 1,000-strong masses with fighter-escorts. It is worth recalling that, as recently as September 1939, FDR had sent messages to all the belligerents begging them to refrain from the 'inhuman barbarities' of bombing civilians. But that was pre-Pearl Harbor, in the distant moral past. From March to July 1945, against virtually no resistance, the B29s dropped 100,000 tons of incendiaries on sixty-six Japanese towns and cities, wiping out 170,000 square miles of closely populated streets. On the night of March 9–10, 1945, for instance, 300 B29s, helped by a strong north wind, turned the old swamp-plain of Musashi, on which Tokyo is built, into an inferno, destroying 15 square miles of the city, killing 83,000 and injuring 102,000. An eyewitness in the nearby prisoner-of-war camp compared it in horror to the cataclysmic 1923 earthquake, which he had also witnessed.[194] Even before the dropping of the A-bombs, Japanese figures show that bombing raids on sixty-nine areas had destroyed 2,250,000 buildings, made 9 million homeless, killed 260,000, and injured 410,000. These raids increased steadily in number and power, and in July 1945 the Allied fleets closed in, using their heavy guns to bombard coastal cities.

It is important to bear the scale of this 'conventional' assault on Japan's cities and population in mind when considering the decision to use nuclear weapons. FDR and Churchill devoted vast resources to the Manhattan Project not only to be first in getting nuclear weapons but in order to use them to shorten the war. As General Groves put it, 'The Upper Crust want it as soon as possible.'[195] A protocol, signed by

Churchill and Roosevelt at the latter's Hyde Park estate on September 9, 1944, stated that 'when the bomb is finally available it might perhaps after mature consideration be used against the Japanese.' By the summer of 1945, this approach to the problem seemed out of date. Though no one in a position of authority in Japan by then believed victory was possible, or eventual defeat avoidable, the consensus among the rulers was that honor demanded resistance to the bitter end. This was precisely the strategy, if so it can be called, of the Japanese Supreme Council, which on June 6, 1945 approved a document, 'Fundamental Policy to Be Followed Henceforth in the Conduct of the War.' Its final plan for the defense of Japan itself, 'Operation Decision,' provided for the use of 10,000 suicide planes (mostly converted trainers), fifty-three infantry divisions, and twenty-five brigades: 2,350,000 trained troops would fight on the beaches, backed by 4 million army and navy civil employees and a civilian militia of 28 million. Their weapons were to include muzzle-loaders, bamboo spears, and bows and arrows. Special legislation was passed by the Diet to form this army.[196]

American intelligence quickly became aware of this fight-to-the-finish strategy, and American commanders were under no illusions, in the light of their experience in conquering the mid-Pacific islands, what it would mean in terms of casualties to themselves, and indeed to the Japanese. By this stage in the war, the Americans had suffered 280,677 combat deaths in Europe and 41,322 in the Pacific, plus 115,187 service deaths from non-hostile causes, and 971,801 non-fatal casualties. In addition, 10,650 US servicemen had died (it was later learned) while prisoners-of-war of the Japanese (out of a total of 25,600). The Allied commanders calculated that, if an invasion of Japan became necessary, they must expect up to a million further casualties. Japanese losses, assuming comparable ratios to those already experienced, would be in the range of 10 million to 20 million.

The continued display of ever increasing firepower, therefore, in the hope of inducing a Japanese surrender, seemed in all the circumstances the logical, rational, and indeed humanitarian solution to an intolerable dilemma produced by the irrational obstinacy of those in charge of Japan's destiny. That was the decision taken, and into it the availability of the A-bomb slotted naturally. When Oppenheimer's test plutonium bomb exploded on July 16, generating a fireball with a temperature four times that at the center of the Sun, its inventor quoted a phrase from the *Bhagavadgita*, 'the radiance of a thousand suns . . . I am become as death, the destroyer of worlds.' He, at least, recognized that a great technological and moral threshold had now to be crossed, or not. But then Oppenheimer had not witnessed a firestorm created in a German or Japanese city by conventional high explosives or incendiaries dropped during a 1,000-plane Allied raid. His colleague Fermi, more prosaically, calculated that the shockwave created by the test-bomb was equivalent to a blast of 10,000 tons of TNT. The news of the successful experiment was flashed to Truman on his way back to Washington from the preliminary Allied peace talks in Potsdam. Truman promptly signed an order to bring the bomb into the offensive program against Japan and use it as soon as possible. There does not seem to have been any prolonged discussion about the wisdom or morality of using the new weapon, at the top political and military level.

Indeed, the figures show that nuclear weapons were merely a new upward notch in a steadily increasing continuum of destructive power which had been progressing throughout the war. The Allied campaign to break Japan's will before an invasion became inevitable was driven forward with relentless energy. On August 1, 820 B29s unloaded 6,600 tons of explosive on five towns in North Kyushu. Five days later, America's one, untested uranium bomb was dropped on Hiroshima, Japan's eighth-largest city, headquarters of the 2nd General Army and an important embarkation port. Some 720,000 leaflets warning that the city would be 'obliterated' had been dropped two days before. No notice was taken by the inhabitants, partly because it was rumored that Truman's mother had once lived near by, and it was thought that the city, being pretty, would be used by the Americans as an occupation center. The bomb was launched from the B29 *Enola Gay*, commanded by Colonel Paul Tibbets, and it caused an explosion equivalent to about 20,000 tons of TNT, three times the power of the August 1 raid. It killed 66,000 to 78,000 people, injured 80,000, and exposed 300,000 more to the effects of radiation.[197]

The Japanese reaction to the Hiroshima bomb does not suggest that one such demonstration would have been enough to compel surrender. Publicly the Japanese government protested about 'the disregard for international law' (which they themselves had totally ignored for twenty years). Privately, they asked Professor Nishina, head of their own atomic program, whether it was a genuine nuclear weapon and, if so, whether he could duplicate it within six months. In the absence of a decisive Japanese reaction to the Allied demand for an immediate and unconditional surrender, the second, plutonium-type bomb was dropped on August 9, not on its primary target, which the pilot could not find, but on its alternative one which, by a cruel irony, was the Christian city of Nagasaki, the nearest thing to a center of resistance to Japanese militarism. Over 74,800 people were killed by it that day. This may have persuaded the Japanese that the Americans had a large stock of such bombs. In fact only two more were ready, and scheduled for dropping on August 13 and 16. At all events, on August 10 the Japanese cabled agreeing in principle to surrender without conditions. This came a few hours before the Russians, who now had 1,600,000 men on the Manchurian border, declared war on Japan, following the agreement made at Yalta. It thus seems likely that the use of the two nuclear weapons was decisive in securing the Japanese surrender. That was the unanimous Allied conviction at the time. Immediately the Japanese message was received, nuclear warfare was suspended, though conventional raids continued, 1,500 B29s bombing Tokyo from dawn till dusk on August 13. The final decision to surrender was taken the next day, the 14th. Truman never had any qualms, at the time or later, that his decision to use both A-bombs had been right, indeed unavoidable, and he believed to his dying day that dropping the bombs had saved countless lives, Allied and Japanese. Most of those who have studied the evidence agree with him.[198]

Using the A-bomb seemed, at least at the time, a comparatively simple decision, and Truman took it readily. What the United States ought to do in Europe was a much more complicated question. Truman inherited an appreciation of America's role in postwar Europe based on FDR's conviction of Stalin's benevolence: US forces were to defeat Germany, then go to the Pacific, or home, as quickly as possible. The UN (set up in October 1945), with America committed to membership, was to do the rest. Truman, unlike FDR, had no illusion about Communism or the nature of the Soviet regime. From the start, he had seen both Nazi Germany and Soviet Russia as two hideous totalitarian systems, with nothing morally to choose between them. When Hitler invaded Russia in 1941, he told a reporter: 'If we see that Germany is winning, we should help Russia and if Russia is winning we ought to help Germany and that way let them kill as many as possible, although I don't want to see Hitler victorious under any circumstances.' Writing to his wife Bess immediately after America entered the war, he told her that Stalin was 'as untrustworthy as Hitler or Al Capone.' He reiterated his brutally realistic view of Russia as a war ally: 'As long as the Russians keep the 192 divisions of the Germans busy in Europe that certainly is a war effort that cannot be sneezed at . . . I am perfectly willing to help Russia as long as they are willing to fight Germany to a standstill.'[199] Between this point and his assumption of the presidency, nothing changed Truman's view that the Soviet Union was essentially a gangster state, and the moment he took up office all the information flowing in strengthened this conviction.

Hence at 5.30 P.M. on April 30 he summoned Stalin's Foreign Minister, Vlacheslav Molotov, to Blair House (he had not yet moved into the White House) and told him that Russia must carry out what it had agreed at Yalta about Poland: 'I gave it to him straight. I let him have it. It was the straight one–two to the jaw.' Molotov: 'I have never been talked to like that in my life.' Truman: 'Carry out your agreements and you won't get talked to like that.'[200] But Truman could not transform American military policy in the last days of the war. General Omar Bradley calculated that it would take an additional 100,000 US casualties to press on and take Berlin. General Marshall advised that capturing Prague was not possible. General Eisenhower was opposed to any US move which might jeopardize the superficially friendly relations between his forces and the Red Army. All wanted Soviet assistance against Japan.[201] By the time Japan had surrendered, the Communist occupation of eastern Europe and most of the Balkans was a *fait accompli*, and the whole of this vast area, including half Germany and what before the war had been nine independent countries, was lost to freedom and democracy for more than a generation.

It was unclear for some time whether western Europe could be saved either. Even at the political and diplomatic level, it took precious weeks and months to reverse the Roosevelt policy. In the first half of 1945, the State Department was still trying to prevent the publication of any material critical of Soviet Russia, even straight factual journalism, such as William White's *Report on the Russians*.[202] At the preliminary Allied peace conference at Potsdam in July 1945, when Truman first met Stalin, he found he had at his elbow Ambassador Davies, now the proud possessor of the Order of Lenin, who urged the President: 'I think Stalin's feelings are hurt. Be nice to him.' In fact, Truman's first view of Stalin was not unfavorable: he thought him a crook but, given firmness, one he could work with. 'Stalin is as near like Tom Pendergast as any man I know' was his verdict.[203] In Britain, Churchill was succeeded by the Labour government of Clement Attlee, obsessed by home problems and Britain's rapidly deteriorating financial plight, vastly increased when Congress abruptly terminated Lend–Lease on August 21, 1945. There were many who thought the game in Europe was up. Harriman, back from Moscow, told the Navy Secretary, James Forrestal, that 'half and maybe all of Europe might be Communist by the end of next winter.'[204]

And so it might have been, had not Stalin overplayed his hand, displayed insatiable greed, and so reversed the process of American withdrawal. And it was greed not only for land and power but for blood: he arrested sixteen leading non-Communist Polish politicians, accused them of 'terrorism,' and set in motion the machinery for the next show-trial. It is worth examining the beginning of the Cold War in some detail, partly because it set the pattern for American foreign and defense policy for the best part of half a century, partly because some historians have attempted to argue that the West, and particularly America, have to share responsibility with Russia for the onset of the Cold War, or even that it was primarily America's doing. But this view does not square with the evidence of Truman's papers, which show clearly that he was extremely anxious to work fairly and honestly with Stalin, not least because he was under the illusion that Stalin was easier to deal with than possible successors, like Molotov. He abandoned this view only reluctantly.

But the evidence on the ground was overwhelming that Soviet armies and agents were enforcing Soviet power or establishing puppet governments wherever they physically could. All American diplomats on the spot and intelligence sources reported in the same way. Maynard Barnes cabled details about a bloodbath of democrats in Bulgaria. Robert Patterson reported from Belgrade that any Yugoslav seen with an American or an Englishman was arrested immediately. Arthur Schoenfeld described in detail the imposition of the Communist dictatorship in Hungary. From Rome, Ellery Stone advised that a Communist *putsch* was likely in Italy. William Donovan, head of the Office of Strategic Services, then America's nearest approach to a global intelligence agency (it was being reorganized as the Central Intelligence Group, which became the Central Intelligence Agency under the National Security Act of July 26, 1947), urged that, in the light of the cumulatively terrifying reports flowing in from American agents all over Europe, measures should quickly be taken to coordinate Western defenses.[205]

Granted his impetuous nature, and his fundamental views, Truman was surprisingly slow to react to Stalin's provocative behavior. He was being advised by his fellow-Southern Democrat James Byrnes (1879–1972), whom he had made secretary of state. Byrnes, a sly and clever man and consummate politician, who might easily have been in Truman's place if events had moved a little differently, had no time for what he called 'those little bastards at the State Department,' and believed he could conduct his own negotiations with the Soviet leadership, not always telling his President what he was doing.[206] Truman complained: 'I have to read the newspaper to find out about American foreign policy'—the beginnings of the suspicion which led him to replace Byrnes by General Marshall in 1947. Truman had no patience with those he designated 'the striped pants boys at the State Department,' but he could not ignore the overwhelming body of evidence in cables and dispatches, and he warned Byrnes to be firm. At the Moscow Conference of Foreign Ministers in December 1945, Stalin's intransigence, transmitted through Molotov, brought matters to a head. Byrnes reported that Russia was 'trying to do in a slick-dip way what Hitler used to do in domineering over smaller countries by force.' When Byrnes reported back, Truman made his mind up (January 5, 1946): 'I do not think we should play compromise any longer . . . I am tired of babying the Soviets.'[207] The next month a well-timed 8,000-word cable arrived from George Kennan in Moscow, which crystalized what most people were beginning to feel about the Soviet threat—the 'Long Telegram' as it came to be known. 'It reads exactly,' its author wrote later, 'like one of those primers put out by alarmed congressional committees . . . designed to arouse the citizenry to the dangers of the Communist conspiracy.'[208]

A fortnight later, on March 5, Winston Churchill, at Truman's invitation, came to Fulton, Missouri, and delivered, with the President's strong approval, his famous speech insisting that 'From Stettin in the Baltic to Trieste in the Adriatic, an iron curtain has descended across the Continent of Europe' and demanding that America and its allies should work together without delay to provide 'an overwhelming assurance of security.' The polls showed that 81 percent of Americans favored his idea of a permanent military alliance. This was a decisive moment in modern world history. Churchill complained that, on the trip, he had lost $75 playing poker with Truman: 'But it was worth it.'[209]

The aggressive behavior of the Communists completed the political education of President Truman and drew from him stronger and stronger reactions. The same month, Russia missed its deadline for the withdrawal of its troops from Iran, and did so only after Truman ordered an angry confrontation at the UN. In August 1946 the Yugoslavs shot down two American transport planes, and the same month Stalin began to put pressure on Turkey. Truman responded by upgrading Donovan's intelligence organization, celebrating the move with a White House party at which he handed out black hats, cloaks, and wooden daggers, and personally stuck a false mustache on the face of Admiral Bill Leahy.[210] American and British intelligence agencies resumed full war-style contact and their air forces began exchanging and coordinating plans again; the US and Canada formed a joint air and anti-submarine defense system. America was still disarming at this stage, and Truman, as he was well aware, did not dispose of much offensive power. Byrnes' complacent assumption that America's possession of nuclear weapons would frighten the Soviets proved unfounded: their information from agents within the US defense establishment probably gave them an accurate idea of the limits of America's nuclear capability. By mid–1946, the US had only seven A-bombs, and a year later the number had risen to thirteen. Moreover, US B29 bombers, based as they then were in Louisiana, California, and Texas, could not fly direct to Russia and it was calculated that it would take a fortnight to drop A-bombs on Russian targets. Not until June 1950 did the United States possess bombers capable of flying on raids to Russia and back.[211]

On the other hand, America's economic and financial power, both absolutely and in relation to the rest of the world, was awesome. Truman remarked in a radio address to the nation in August 1945: 'We tell ourselves we have emerged from this war the most powerful nation in this world—the most powerful nation, perhaps, in all history. That is true, but not in the sense some of us believe it to be true.'[212] Truman was referring essentially to the industrial capacity of the country. In the second half of the 1940s the United States had a productive preponderance over the rest of the world never before attained by any one power, and most unlikely to be experienced ever again. With only 7 percent of the world's population, it had 42 percent of its income and half its manufacturing capacity. It produced 57.5 percent of the world's steel, 43.5 percent of its electricity, 62 percent of its oil, 80 percent of its automobiles. It owned three-quarters of the world's gold. Its per-capita income was $1,450, the next group (Canada, Great Britain, New Zealand, Switzerland) was only between $700 and $900. Calorie consumption per day was about 3,000, some 50 percent more than in western Europe.[213]

It is true that in the United States the distribution of wealth was uneven. In 1947, one-third of American homes still had no running water and two-fifths no flush lavatories. This was largely a rural problem: in 1945, 17.5 percent of the population, or 24.4 million, still lived on the soil and farmed. They had cars and they ate well but many lacked amenities, let alone entertainment, and this was one reason Americans were fleeing the land: the 6 million farmers of 1945 had been halved by 1970, when the farming population had fallen to 4.8 percent of the whole. But in global terms the dollar was almighty, and the federal government had the disposal of huge sums. In 1939 its income had been $9.4 billion only. By 1945 it had risen to $95.2 billion, partly by raising the national debt, from $56.9 billion in December 1941 to $252.7 billion in December 1945, but partly by massive increases in taxation. Income tax rose steadily and steeply, especially after the tax-withholding system from paychecks (copied from the heavily taxed British) was introduced in 1943. Nor did it return to prewar levels after the end of the war. On the contrary: rates peaked in the 1950s with a range of 20 to 91 percent on individuals and a 52 percent corporate rate. Actual federal income fell after 1945, dropping to a postwar low of $36.5 billion in 1948, then rising again until it reached $43.1 billion just before the start of the Korean War.[214] During the 1940s and continuing into the early 1960s, taxation in relation to Gross National Product was higher in the United States than at any other period in the country's history, and this meant that the government had at its disposal the means to bolster, sustain, and reinvigorate the world, and especially Europe, in the face of Soviet encroachments. Truman was the first American statesman to grasp that the United States was physically and financially able to rescue the world not merely in war, but in peace too, and to keep up the effort for the foreseeable future.[215]

Truman felt that this burden could be shouldered without prejudice to the country's future because of the speed at which the US economy had grown and continued to grow. GNP (in constant 1939 dollars) had risen from $88.6 billion in 1939 to $135 billion in 1945. The war had enormously benefited the US economy, raising its productive capacity by nearly 50 percent and its actual output of goods to well over 50 percent.[216] The economy had been growing at the rate of 15 percent annually, a rate never reached before, or since, and much of this was civil production to meet the demands of a nation now enjoying full employment and high wages. This was made possible by very rapid increases in productivity, as new and improved machine-tools and machinery were made and installed. Truman knew that he was president of a country which was now responsible for more than half the world's manufactured goods and a third of the world's production of all kinds. It was by far the world's largest exporter, transporting its goods in US-owned ships constituting half the world's mercantile fleet. With such a preponderance, action to help humanity to survive, and the democracies to retain their freedom and independence, was dictated by moral obligation as well as by political prudence. It was in these circumstances that American isolationism, insofar as it had ever existed, was finally put to rest and interred for ever.

Truman was stimulated into activity by a despairing plea from Britain on February 21, 1947. The British had spent a quarter of their net wealth on the war and accumulated crippling foreign debts. America had made Britain a loan after the end of Lend–Lease but this had soon been swallowed up as the British had spent over $3 billion on international relief to stop Europe from starving. This included large sums devoted to keeping Greece and Turkey out of Stalin's clutches. The winter of 1946–7 in western Europe was one of the harshest in modern history and virtually brought the British economy to a standstill. The British government now said they could support the Greek-Turkey burden no longer and appealed for US help. On February 23 Truman decided he would have to take it on, but first he held a meeting in the Oval Office to outline the idea to leading Congressmen. It was one of the most decisive ever held there. General Marshall, who had just taken over as secretary of state, was still mastering his brief and much of the talking was done by his imperious deputy, Dean Acheson. Acheson said that 'Soviet pressure' on the Near East had brought it to a point where a breakthrough 'might open three continents to Soviet penetration.' Like 'apples in a barrel infected by one rotten one,' the 'corruption' of Greece would 'infect Iran and all the East.' It would 'carry infection to Africa from Asia Minor and Egypt' and 'to Europe through Italy and France.' Soviet Russia was 'playing one of the biggest gambles in history at minimal cost.' It did not need to win them all: 'even one or two offered immense gains.' Only America was 'in a position to break up the play.' These were the stakes which British withdrawal 'offered to an eager and ruthless opponent.' When he finished there was a long silence. Then Senator Arthur Vandenberg, former isolationist and now chairman of the Senate Committee on Foreign Affairs, spoke for all his colleagues: 'Mr President, if you will say that to the Congress and the country, I will support you and I believe most of its members will do the same.'[217]

The result was the appearance of Truman before a special session of both Houses of Congress on March 12, 1947 and the annunciation of what was immediately called the Truman Doctrine: 'I believe it must be the policy of the United States to support free peoples who are resisting attempted subjugation by armed minorities or by outside pressure.'[218] Truman did not say how long such support might be necessary but he indicated that it must be provided for as long as was needed, which might be many years. In short, the US was now undertaking an open-ended commitment, both military and economic, to preserve democracy in the world. It had the means, and it had the will, because it had the men: Truman himself, leading two whole generations of active internationalists, young and old, who had learned from experience and history that America had to take its full part in the world, for the sake of the human race. These men, military and civil, politicians and diplomats, included Eisenhower, Marshall, MacArthur, Dean Acheson, Averell Harriman, George Kennan, John McCloy, Charles Bohlen, Robert Lovett, and many leading senators and congressmen, of whom the 'born again' Vandenberg was representative. In natural abilities and experience, in clarity of mind and in magnanimity, they were probably the finest group of American leaders since the Founding Fathers.[219] And, in their impact upon America, and its role in the world, they were of comparable significance.

Of these men, the most important, after Truman, was Marshall. Truman wrote of him: 'He is the great one of the age. I am surely lucky to have his friendship and support.' Truman often felt his colleagues and subordinates were more talented than he was, in one way or another, but Marshall he freely acknowledged to be the better man too.[220] The general was not a talkative or even a particularly articulate man, and sometimes he could be disconcertingly silent. But no American of his times inspired so much awe and respect, even in great and powerful men like Eisenhower. He was able to command the bipartisan support of Congress because its members always felt he rose above politics and sought the national interest without regard to party or class or lobby. He had the same effect on most of the foreign leaders with whom he had dealings, even 'Stonebottom' Molotov. In April 1947, at a bibulous dinner in Moscow, an inebriated Molotov turned to Marshall and said nastily: 'Now that soldiers have become statesmen in America, are the troops goose-stepping?' Marshall, 'his eyes icy gray,' turned to Bohlen who was translating and said: 'Please tell Mr Molotov that I'm not sure I understand the purport of his remark, but if it is what I think it is, *tell him I do not like it*.' Thereafter, Molotov, who had signed the death warrants of tens of thousands, treated the general with the cowed deference he normally reserved only for his master, Stalin.[221]

However, Marshall was not just a personality, he was an organizer; indeed, he was an administrative genius, as he had proved during the war. What was now required, from America, as it was committed to a global strategy of military, diplomatic, and economic outreach, were institutional and structural changes. Marshall was just the man to advise Truman on their shape and magnitude, and indeed to carry out some of them directly himself. As a result, on July 26, 1947, Congress passed the National Security Act, which amalgamated all political control of the forces in the Department of Defense, set up the new Central Intelligence Agency from an amalgam of its forebears and created an entirely new body, the National Security Council, to give expert advice directly to the President on all matters affecting the defense and security of the nation. Within a decade, the CIA's annual budget, not subjected to Congressional scrutiny, had swollen to an estimated $1 billion and its personnel to over 30,000. The National Security Council, routinely presided over, at the beginning, by the head of the CIA, included at its formal meetings the President, Vice-President, Secretary of State, and Defense Secretary, with the President's Assistant for National Security Affairs directing its staff, which by 1980 had grown to 1,600.[222] These moves, together with the expansion, retraining, and reequipping of the US diplomatic service, completed what might be called the professionalization of America's defense and foreign policy effort.

On June 5, 1947, at the Harvard Commencement, the new Secretary of State unveiled what became the Marshall Plan. It was originally a loosely worded proposal that all the European nations, including Russia, cooperate in bringing about a recovery of the entire continent, with the US providing the pump-priming finance. Stalin rejected the offer on Russia's behalf, and vetoed the desire of Poland and Czechoslovakia to participate. But the nations of western Europe drafted programs submitted to a conference at Paris, June 27–July 2, and on September 22 it was estimated that their joint needs would require between $16 and $22 billion of US aid. On December 19, 1947 Truman submitted to Congress a $17 billion European Recovery

Program, and Senator Vandenberg helped to steer the appropriations through. There was initially some opposition, but once again Stalin came to the rescue. The brutal Communist *coup d'état* he staged in Czechoslovakia in February 1948 helped to push what was to become the first of a huge series of foreign-aid Bills through Congress. In the end, the Marshall Plan channeled about $13 billion of US assistance into the European economies, and it must be regarded as perhaps the most successful scheme of its kind in history. It was particularly effective in reinvigorating the economies of Germany, France, and Italy.[223] It made practical sense for the US too because by the second quarter of 1947 America's export surplus was running at an annual rate of $12.5 billion, and Marshall Aid helped to enable Europe to continue to take US goods.

European economic recovery was one thing, and in the long run the most important thing, but in the short term what was also needed was security from Soviet Communist aggression and subversion, and this could be provided only by a permanent and active US military presence. It took some time before Truman, and the US military, grasped that this was inevitable, though repatriation of American forces slowed down steadily in 1947. But Stalin, as always, was happy to oblige with his greed. Unable to agree on a peace formula for one Germany, the rival blocs had been creating two Germanies in 1946, with Berlin, in which each of the four powers (the US, Britain, France, and Russia) had its own zone, as an isolated enclave in Russian-occupied East Germany. On June 18, 1948, as a purely administrative measure, the three Western Allies announced a new German currency for their zones. Six days later, Stalin took this as a pretext for an attempt by force to extinguish the Berlin enclave, by blocking road access to the Western zones there, and cutting off its electricity.

Truman grasped that the Berlin Crisis, the first large-scale formal confrontation of the Cold War, was an event of peculiar significance. Nikita Khrushchev later characterized Stalin's Berlin move as 'prodding the capitalist world with the tip of a bayonet,' to see what would be the response. Truman was immediately quite clear what the response would be: 'We would stay, period.' He was confirmed in his first reaction by the views of the US Zone Commander, General Lucius Clay. Clay had hitherto been the most reluctant of the Cold Warriors; now he changed decisively, and recommended clearing the approach roads by armed convoys. This was rejected as needlessly provocative, but Truman also rejected the more conciliatory approach of the nervous

Forrestal, the Defense Secretary, who wanted, as the President put it, 'to hedge' and supplied him with 'alibi memos.' Truman noted in his diary, July 19: 'We'll stay in Berlin—come what may. I don't pass the buck nor do I alibi out of any decision I make.'[224]

Forrestal was nervous partly by temperament, partly also because he was aware of what he called 'the inadequacy of United States preparations for global conflict.' Truman had been furious when he discovered, on April 3, 1947, that while materials for twelve A-bombs existed in US arsenals, none was assembled for delivery. He ordered the rapid creation of a 400-bomb stockpile. But not enough had been delivered by mid–1948 to carry through even what the US Air Force termed 'Operation Pincher,' which called for the complete destruction of the Soviet oil industry.[225] Nonetheless Truman sent the first three squadrons of B29 bombers to bases in Britain and Germany. They were not actually equipped to deliver A-bombs but Truman rightly assumed that Stalin would think they were, and it was the closest he came to playing the atomic card during his presidency. He made it clear to his closest colleagues, as Forrestal recorded in his diary, that he was quite prepared, if absolutely necessary, to use A-bombs against Russia. But, in terms of the actual physical response to the Berlin Blockade, he decided to frustrate it by mounting an airlift, which had the additional merit of providing Stalin, and the whole world, with an awesome demonstration of American airpower. And it worked; the airlift was flying in 4,500 tons a day by December 1948 and 81,000 tons a day by spring 1949, as much as had been carried by road and rail when the cut-off came. On May 12, 1949, the Russians climbed down.[226]

This was victory of a sort, but the episode had served to draw everyone's attention, not least Truman's, to the sheer inadequacy of America's, and the West's, military forces on the ground, faced with a Soviet army which had stabilized at 2.5 million, plus an armed police of 400,000, and which seemed to have almost unlimited resources of armor and artillery. The creation of a new West German state, and of a permanent military association of Western powers, was the prime object of American policy in the second half of 1948.

In the meantime Truman needed the endorsement of the electorate for his forceful foreign and defense policies. Truman was an unelected president, though he had never allowed this to inhibit his capacity to take decisions of the highest importance promptly and in masterful fashion. But many people, including a lot of Democrats, assumed he could never get himself elected in his own right, and they toyed seriously with the idea of drafting General Eisenhower, now a retired general and college president. The Democrats finally nominated Truman, at their convention in Philadelphia in July, but they took some time about it and, in the process, adopted civil rights measures which led the South to bolt the party and form the Dixiecrats (the states' rights Democratic Party), which nominated J. Strom Thurmond of South Carolina as presidential candidate. Since Truman was also facing a liberal challenge in Northern states, in the shape of Henry Wallace, an old FDR crony whom Truman had sacked in 1947 as secretary of agriculture, for being 'soft on Communism' and a prize pest, his chances looked slim. The convention would not even allow him to pick his own running mate, foisting on him 'old man Barkley,' as the President called him (Alban Barkley, 1877–1956, former Senate Majority Leader). Truman had wanted the forceful Northern liberal William O. Douglas (1898–1980) from the Supreme Court to help fend off the Wallace challenge. But Douglas retorted woundingly that he 'didn't want to be a number-two man to a number-two man,' which provoked from the President one of his wrathful mixed metaphors: 'I stuck my neck out all the way for Douglas and he cut the limb from under me.' However, he accepted the hand his party had dealt him philosophically and girded himself for battle, telling his partner, in words which soon became famous, 'I'll mow 'em down, Alban, and I'll give 'em hell!'[227]

The 1948 campaign was the last whistle-stop, pre-television American election campaign. Actually it was the first in which the party conventions were televised. As far back as 1932 there were twelve regularly operating TV stations in the US, reaching 30,000 homes. But transcontinental TV broadcasts were not to begin till 1951 and in 1948 only 200,000 American homes had TV. On the other hand, the US still had the world's most extensive (and luxurious) railroad system, with over 200,000 miles of track, and some of the finest transcontinental trains in railroad history, albeit the network was already shrinking, declining by 17,000 miles in the twenty years 1940–60. Truman took full advantage of this huge network for his campaign. He made two air trips, to Miami and Ralegh, October 18–19 (US civil airlines were already carrying 50 million passengers a year), but most of his speechmaking was done on three big railroad swings, September 17–October 2, October 6–16, and October 23–30. The first and busiest covered seventeen states from Pennsylvania to California and involved a dozen rearplatform appearances every day, plus thirteen major set speeches.

Truman abounded in vigorous vignettes: the Republicans were 'gluttons for privilege' who had 'stuck a pitchfork in the farmer's back.' Undeterred by the opinion surveys, which predicted an easy victory for the Republican challenger, again Governor Dewey of New York, Truman barked out that electors would 'throw the Galluping polls right into the ashcan—you watch 'em.' Dewey was a good administrator and a reliable campaigner. But he was short and a little too neat in his dress, and he was damaged by being called the 'Little Man on the Wedding Cake.' The Republicans were complacent, which meant they did not find it easy to raise cash, and the Democrats actually outspent them by $2.7 million to $2.1 million.

Truman's whistle-stop progress began by attracting small crowds but his aides, such as Clark Clifford, caught the headlines by standing at the back and shouting when Truman began to speak: 'Give 'em hell, Harry!' and this was soon taken up by genuine members of the audience. By the end, Truman was pulling bigger numbers than FDR had in 1944. The journalist Robert J. Donovan described the Truman oratory as 'sharp speeches fairly criticizing Republican policy and defending New Deal liberalism, mixed with sophistries, bunkum piled higher than haystacks, and demagoguery tooting merrily down the track.'[228] When a *Newsweek* survey of fifty leading journalists reported that all predicted Truman would lose, he commented, 'I know every one of those fifty fellows and not one of them has enough sense to pound sand into a rathole.' The *Chicago Tribune* won itself a little historical notoriety by going to press on election night with the front-page headline: 'Dewey Defeats Truman,' which the reelected President was able to hold aloft in triumph the next day. In fact Truman got 24,105,812 votes (49.5 percent) to Dewey's 21,970,065 (45.1 percent), with Thurmond and Wallace each getting about 1.2 million. In college terms, the score was Truman 303, Dewey 189, Thurmond 39, Wallace none. Truman replicated FDR's feat of 1944 in carrying all thirteen of America's biggest cities with populations over 500,000, but the chief identifiable reason for his victory was a late-stage swing in the farm vote, then still important.

Thus reassured of popular support, Truman was able to pursue the policies begun with his doctrine. On April 4, the North Atlantic Treaty was signed in Washington by the United States, Belgium, Canada, Denmark, France, Great Britain, Iceland, Italy, Luxembourg, the Netherlands, Norway, and Portugal (Greece and Turkey joined in 1952, West Germany in 1955). The treaty embodied the basic principle of collective security, stipulating that an attack on any one member would be considered an attack upon all, and it set up an Organization, with an integrated command headquarters, based in Paris (SHAPE), to which were assigned forces from all the signatories. General Eisenhower was recalled to duty and served as the first Supreme Allied Commander, 1950–2. Thus collective security, for the lack of which World War Two had to be fought, at last came into being, not just in principle but in fact and practice, and of all the postwar institutions, NATO has proved one of the most durable and, perhaps, the single most effective, keeping the peace for half a century and more.[229]

Simultaneously, America took steps to insure that it possessed the military resources to sustain the new alliance. In February–March 1949, a group of State Department and Defense officials drafted a document called 'National Security Council 68,' which laid down the main lines of American foreign and defense policy for the next forty years, until the collapse of the Soviet regime and empire in the late 1980s. It was based on the proposition that America, as the greatest independent democratic power, had moral, political, and ideological obligations to preserve free institutions throughout the world, and must equip itself with the military means to shoulder them. It must possess adequate conventional forces as well as nuclear striking power. This point was underlined on September 3, 1949 when a B29, on patrol in the north Pacific at 18,000 feet, picked up positive evidence that Russia had exploded its first nuclear weapon at the end of August. The atomic monopoly was over. On January 31, 1950, after long behind-the-scenes argument, Truman authorized the development of the hydrogen bomb or the 'Super.' He was strongly backed by the Joint Chiefs of Staff, especially General Bradley (now his favorite general), and by Dean Acheson. The matter was finally decided when he put the question to his three-man advisory committee of Acheson, Lilienthal (in charge of nuclear development), and Defense Secretary Louis Johnson, 'Can the Russians do it?' and received the unanimous answer: 'Yes.' Truman: 'Then we have no choice. We will go ahead.'[230] The first hydrogen bomb, detonated in November 1952, was a combination of deuterium and tritium and vaporized its entire test island in the Pacific, digging a crater a mile long and 175 feet deep. The second (March 1954) used lithium and deuterium and released still greater energy.

In the meantime, NSC 68 noted that the Russians used 13.8 percent of their GNP for defense, as opposed to America's 6–7 percent, and it recommended that, to secure adequate security, America should be prepared to devote up to 20 percent of its GNP to this purpose. The document was finally approved in April 1950, completing the historic establishment of America's commitment to the outside world. Gradually it produced specific alliances or agreed obligations to forty-seven nations and led American forces to build up or occupy 675 bases and station a million troops overseas, as well as enormous air forces and fleets.[231]

Some of these forces were committed in accordance with treaty organizations like NATO or the Southeast Asian Treaty Organization, signed at Manila on September 8, 1954 by the US, Philippines (which had become independent on July 4, 1946), Thailand, Pakistan, Australia, New Zealand, Britain, and France. This alliance, though modeled on NATO, never had its cohesive power or joint infrastructures: Pakistan withdrew from it in 1972 and France in 1973, and it disbanded by mutual consent in 1977. But it served its purpose in the 1950s and 1960s.[232]

In the Middle East, on whose oil supplies the United States became increasingly dependent during the second half of the 20th century, efforts to create a similar permanent alliance, known as the Bagdad Pact, were eventually to fail (1955), but in the meantime, thanks largely to Truman's sagacity and foresight, the state of Israel came into being, and proved itself a reliable if strong-minded and highly independent American ally. Truman's commitment to a Zionist state was part-emotional, part-calculating. He felt sorry for Jewish refugees. He saw the Jews in Palestine as the underdogs. On the other hand, at the time Israel came into being, 1947–8, Truman was the underdog himself in the coming election, and he needed the Jewish vote. In May 1947, the Palestine problem came before the UN. A special committee, asked to produce a plan, produced two. A minority recommended a federated binational state. The majority favored two states, one Jewish, one Arab. On November 29, 1947, thanks to Truman's vigorous backing, the majority plan, making the birth of Israel possible, was endorsed by the General Assembly, thirty-three votes to thirteen, with ten abstentions. For reasons which are still mysterious, Stalin, who was always anti-semitic and usually anti-Zionist too, was going through a brief philo-semitic phase, and so also helped to bring the new state into existence.

Israel's declaration of independence on May 14, 1948 was preceded by an acrimonious debate in Washington, in which Truman, who favored immediate *de facto* recognition, found himself opposed by both the State Department and the Pentagon. Marshall, in what Clark

Clifford called 'a righteous God-damned Baptist tone,' said that Truman was subordinating an international problem to domestic politics and diminishing the dignity of the presidency. It would lose Truman his own vote in the coming election, he added.[233] The Defense Secretary, James Forrestal, bitterly denounced the Jewish lobby: 'No group in this country should be permitted to influence our policy to the point where it would endanger our national security.'[234] There was also vehement opposition from the oil interests. Max Trornburg of Cal-Tex said that Truman had 'extinguished the moral prestige of America' and destroyed 'Arab faith in her ideals.'[235] Nonetheless, Truman went ahead and accorded Israel *de facto* recognition, upgraded to *de jure* after the US elections. Events have justified this policy. Not only did Israel survive four separate wars for its existence but it contrived to preserve its constitutional integrity and democratic practices so that, half a century later, it remained the only working democracy in the Middle East, with close military, economic, and cultural links with the United States.[236]

Over this half-century, Israel had been the biggest single beneficiary of US foreign aid and overseas assistance programs. But it was only one of many. Truman took a particular interest in what he called 'making the Palestine desert blossom,' but he was just as interested in helping the region as a whole. He told a dinner of businessmen in October 1949 that he wanted US assistance to help restore 'the Mesopotamian Valley' to its fruitfulness 'as the Garden of Eden' in which '30 million people could live.' And he explained how 'the Zambezi River Valley in Africa' could be converted into 'sections comparable to the Tennessee Valley in our own country if the people of those regions only had access to the "know how" which we possessed.'[237] In 1943 the United States had helped to organize the United Nations Relief and Rehabilitation Administration (UNRRA) and had been by far its principal benefactor. Truman had taken a particular interest in its work and he was pleased that, at the end of the war, polls showed that 90 percent of the American people favored continued food rationing if this were necessary to enable food to go to the hungry peoples of Europe and Asia. Coming from the Midwest, where 'help your neighbor if he needs it' was the absolute rule, it seemed to him natural that the United States, out of its plenty, should come to the rescue of the rest of the world, but in a practical spirit of prudence (he strongly dissented from Henry Wallace's proposal for indiscriminate US aid). In June 1947 UNRRA made a final accounting. It had shipped abroad a total of 23,405,978 tons, 44 percent of it foodstuffs, 22 percent industrial equipment, 15 percent clothing, and 11 percent agricultural supplies, the value amounting to $2,768,373,000. In addition America had provided specific sums to meet particular emergencies in a variety of countries. In all, it had provided $9 billion in aid.[238] On top of this there had been the $13 billion spent on the Marshall Plan.

But that was not enough, Truman felt, to satisfy 'the conscience of the American people.' He repeatedly asserted, 'America cannot remain healthy and happy in the same world where millions of human beings are starving.' With his election victory behind him, Truman insisted on inserting, as 'Point Four' of his inaugural address of January 1949, to the surprise and consternation of the State Department, and as part of his 'program for peace and freedom,' a pledge of 'a bold new program for making the benefits of our scientific advances and industrial progress available for the improvement and growth of underdeveloped areas.' Truman was the first statesman to draw attention, at a global level, to the vast disparities between the 'have' and the 'have-not' areas of the world, and to insist that 'More than half the people of the world are living in conditions approaching misery.' He was likewise the first to do anything about it. He considered Point Four the most important peace policy of his entire administration. He told a press conference it had been in his mind 'ever since the Marshall Plan had been inaugurated.' It had 'originated with the Greece and Turkey proposition. Been studying it ever since.' He said he fantasized about a world aid scheme, financed by the US, while studying the huge globe in his office. Knowing Congress, he robustly insisted that the scheme was not wholly disinterested—it was also self-interested. America and other Western countries needed to keep their industrial plant 'going at full tilt for a century.' And the idea was also 'preventing the expansion of Communism in the free world.'[239]

Truman encountered remarkably little opposition from Congress for his plans. The original appropriation was only $34.5 million. In the fiscal year 1952 its budget was increased to $147.9 million.[240] The federal government's generosity was matched by that of many big companies. Westinghouse operated 'its own private Point Four' by licensing foreign manufacturers to use its techniques and processes, helping them to design their plants, and training their operatives. Sears, Roebuck became 'one of Point Four's most aggressive if unofficial vessels' by teaching industrial techniques and merchandising to the Brazilians and others. But taxpayers' money provided the main effort. By 1953 there were 2,445 US technicians working in thirty-five foreign countries, engaged in assisting with food production, railroad efficiency, modern mining techniques, public health, central banks and government administration, every conceivable kind of industrial process, and services ranging from housing to meteorological forecasts. Point Four was enhanced by bilateral aid agreements, and the sums spent by the US government continued to increase throughout the 1950s and 1960s. By the 1970s, when the quantity of US foreign aid began to decline, over $150 billion had been spent, two-thirds of it outside western Europe.[241] This effort, in absolute or relative terms and from whatever viewpoint it is regarded, was wholly without precedent in human history, and is likely to remain the biggest single act of national generosity on record. That much of the aid was wasted is unfortunately true. That most of it helped is equally sure. That, during this unique period of giving, anti-Americanism increased in the world, is likewise undeniable. But then what good deed in history ever went unpunished? Truman, who initiated it all, was content with the teaching of the Judeo-Christian ethic, that virtue is its own reward.

Behind the construction of America's network of worldwide bases, and behind the overseas aid program, was the strategy of 'containment.' This geopolitical philosophy was first set out in an article published in *Foreign Affairs*, July 1947, entitled 'The Sources of Soviet Conduct,' and signed 'X' (George Kennan). It spoke of the need for the US to secure the 'patient but firm and vigilant containment of Russian expansionist tendencies' by 'the adroit and vigilant application of counterforce at a series of constantly shifting geographical and political points.' This was the alternative to the strategy of 'rollback' favored in some Republican circles, which called for the application of military and diplomatic pressure to force the Soviet Empire to 'disgorge.' But rollback risked a third world war, which became increasingly unthinkable as Russia secured its first atomic weapon and then was assumed to be making hydrogen bombs too. Rollback exponents never quite spelled out what they intended to do to make their policy work, and it certainly did not catch on with most of the American public. So containment became standard and permanent American policy, and its application, through the Marshall Plan and NATO, and the response to the Berlin Blockade, worked very well in Europe until, in the fullness of time, the Soviet Empire disgorged of its own volition and from its own weakness, and Communism was indeed rolled back throughout eastern Europe and to the Urals and beyond.[242]

Where containment did not work, initially, was in the Far East, probably because it was applied too late. FDR had backed the corrupt Kuomintang Party leader, Chiang Kai-shek, who had received military aid and Lend–Lease in considerable quantities. Truman had continued this policy. Chiang got a $500 million 'economic stabilization loan' and a total of $2 billion in the years 1945–9. But efforts to reconcile the Kuomintang with the Chinese Communist leader Mao Tse-tung, who had his own battle-hardened peasant army, failed—General Marshall led a mission to China to no effect—and, once civil war started in earnest, all the American aid vanished in a morass of inflation. The collapse of the currency meant that Chiang's originally enormous army disintegrated—much of it was integrated with Mao's forces—and by April 1949 Mao had crossed the Yangtze, taking the capital, Nanking, by the end of the month. Chiang was driven out of mainland China into the island of Taiwan, which he turned into a fortress, protected by the US Seventh Fleet. Truman was bitterly accused by the Republicans and the China Lobby of having 'lost China,' but the truth is China lost itself. The question then was: how was the line of containment now to be drawn in the Far East?

Once again, Stalin came to the rescue of uncertain American strategists. Doubts about where the line lay were reinforced by a foolish speech made by Dean Acheson, now Secretary of State, at the National Press Club in Washington on January 12, 1950. He was concerned to argue that, though China was now Communist, its leader Mao was certain to quarrel with Stalin, just as Marshal Tito, the independent Communist leader of Yugoslavia, had done. But in making this point—which eventually was justified by events—he appeared to exclude Taiwan, Indochina and Korea from the American defensive perimeter. The speech was clearly read and noted by Stalin, who was anxious not to repeat his mistake with Tito and was, at that moment, unknown to Acheson, making conciliatory gestures to Mao. Acheson's reference to an inevitable Russo-Chinese break reminded Stalin of the danger, and his apparent omission of Korea as an American vital interest pointed to the remedy. Stalin decided that a limited proxy war in Korea would be the means to teach China where its true interests lay. If this was indeed Stalin's reasoning, it proved correct: Korea postponed the Soviet–Chinese break for a decade. But in the meantime it brought war. Stalin seems to have agreed with Kim Il-Sung, the North Korean Communist dictator, in the spring of 1950, that in November he could make a limited push across the 38th parallel, which divided

Communist North from non-Communist South Korea, where 500 US troops were stationed as advisors. But Kim was not a cautious or a biddable man. He took Stalin's hint as permission to stage a full-scale invasion, and he launched it on June 25.[243]

When Acheson gave Truman the news on Sunday, June 26, that the invasion was on a huge scale, the President's reply was 'Dean, we've got to stop the sons-of-bitches no matter what.' That was the Truman style, and that is the way he told it.[244] In fact the decision to intervene was a regular deliberative process, but there is no doubt that Truman's first response, taken in ten seconds, was as he later recollected.[245] He felt the US was in a position to act firmly. Its atomic stockpile was now approaching 500 bombs, and 264 aircraft were capable of delivering them on Soviet targets. General MacArthur, the supremo in Japan, took over, and Truman began by allowing him his head, including the launching, on September 15, 1950, of the Inchon landings, a daring and risky venture which produced a rapid and total victory. Truman gave his approval to NSC 81/1, which allowed for operations north of the 38th parallel and the military occupation of North Korea, provided that neither China nor Russia intervened in the war. George Marshall, now Secretary of Defense, telegraphed MacArthur September 29: 'We want you to feel unhampered tactically and strategically to proceed north of 38th parallel.'

Under cover of the crisis, China first swallowed quasi-independent Tibet (October 21, 1950), then as American troops approached its borders, attacked with a massive 'volunteer army' (November 28). Truman had flown 7,500 miles to see MacArthur on Wake Island on October 13, had awarded him the Distinguished Service Medal, citing his 'vision, judgment, indomitable will, gallantry and tenacity,' and then, in a major speech in San Francisco four days later, had proclaimed his complete accord with the proud general.[246] But the Chinese intervention, which MacArthur had told him at Wake could not and would not happen, changed everything. The Americans were sent reeling back and MacArthur's response was to recommend full-scale military action against China: intensive bombing of industrial areas, a full coastal blockade, and support for Taiwanese attacks on the mainland. The Chinese offensive was in fact reversed by MacArthur's subordinate, General Matthew Ridgway, by skillful deployment of the military resources already available in Korea, and without any necessity for direct attacks on China.

But by this time MacArthur had taken to issuing in public his own ideas of what should be done, even if they conflicted with what he knew to be Washington's position. This led to a crisis on March 24, 1951, when a statement by the general forced Truman to cancel a message of his own, calling for negotiations. The problem became more acute on April 5 when the House Minority Leader, Joe Martin, released an exchange of letters with MacArthur in which the general appeared to endorse the Republicans' policy of 'maximum counterforce,' ending 'There is no substitute for victory.' This was a direct intervention in politics, and unacceptable. Truman noted in his diary: 'This looks like the last straw. Rank insubordination.'[247] Again, though his mind was now pretty well made up that MacArthur would have to go, Truman moved with deliberation, consulting with Acheson, Marshall, Harriman, and Bradley (now chairman of the Combined Chiefs of Staff). All supported dismissal. Congressional leaders, when consulted, had more mixed reactions. Nonetheless, on April 9 Truman finally decided, and ordered Bradley to set the dismissal machinery in motion. He later recalled he had told Bradley: 'The son-of-a-bitch isn't going to resign on me, *I want him fired*.'[248]

The reaction to the MacArthur dismissal was even more violent than Truman had expected, and for an entire year majority public opinion ranked itself ferociously against him. He said, characteristically, of the hostile polls: 'I wonder how far Moses would have gone if he had taken a poll in Egypt? What would Jesus Christ have preached if he had taken a poll in the land of Israel? . . . It isn't [polls] that count. It is right and wrong, and leadership—men with fortitude, honesty and a belief in the right that make epochs in the history of the world.'[249] But gradually the rage died down, and MacArthur's own highly emotional appearance before a joint session of Congress was more a valedictory than a gesture of defiance. The conviction gradually spread that Truman had been right, and many now see the episode as his finest hour, a forceful and perhaps long overdue reassertion of the elective, civil power over an undoubted military hero who had ignored the constitutional chain of command.

The truth is, Truman kept in mind, which MacArthur did not, that the object of US intervention in Korea was not to start a third world war, but to prevent one. That is what it did. The war settled down to a stalemate. Negotiations scaled down and eventually ended (July 27, 1953) the fighting, though the country remained divided and the cease-fire line tense. The war was costly. US casualties included 33,629 battle deaths, 20,617 non-hostile deaths, and 103,284 wounded. There were in addition 8,177 missing and, of the 7,140 servicemen made prisoner, only 3,746 were repatriated. Direct military expenses were over $54 billion.[250] All this was in addition to Allied casualties in the UN force, 415,000 deaths in the South Korean forces, and an estimated 520,000 among the North Koreans (Chinese casualties have never been divulged but are believed to have been over 250,000). Nothing was gained by either side. But America had demonstrated its willingness to defend the policy of containment in battle, and at the same time prudence in restraining its superior firepower. For both, Truman was personally responsible, and the verdict on his policy is that the United States has never had to fight another Korea.[251]

All the same, after the MacArthur sacking it was downhill all the way for Truman, who found himself caught up in a series of domestic scandals: the sleaziness of machine-politics came back to haunt the old warrior. By 1950–1, the Democrats had controlled the administration for twenty years and the rot had set in deeply in many branches. Egregiously venal, it turned out, were the Bureau of Internal Revenue and the Tax Department of the Justice Department. The scandals uncovered there in 1951 were particularly damaging to Truman at a time when, to fight the Korean War, he was requesting one round of tax increases after another. The public was furious at successive reports of tax-fixing by corrupt patronage appointees. By the end of the year no fewer than fifty-seven tax officials had been forced into resignation and many were subsequently convicted. Truman eventually introduced civil service status for tax collectors and took them out of the patronage system altogether. But his response to the initial rumblings of scandal had been slow, inadequate, and marked by poor public relations. Ironically enough it was the piddling nature of the sums stolen by comparatively junior officials which diminished the administration, and with it the President. A top Washington columnist, Joseph Alsop, commenting on the misdemeanors of a ridiculous miscreant called T. Lamar Caudle, wrote: 'One would have more respect for men like Caudle if they were big thieves on their own, but in fact they are mainly petty favor takers.'[252]

Truman continued to hold his head high. He invariably rose at 5 A.M., shaved, dressed, then took a vigorous walk at the old army quick-march tempo of 120 paces to the minute, accompanied by panting Secret Servicemen and reporters, who were allowed to put questions if they contrived to keep up with the President. Then back to the White House for a shot of bourbon, a rubdown, and breakfast. He was at his desk by 7 A.M., and he worked hard all day, except for a brief nap after lunch and a swim in the pool, his head bolt upright, wearing his glasses. He and his cherished wife Bess led an exemplary family life. He never seems to have looked at another woman. He said he fell in love with her when he was six and she was five in 1890, later had a long, anxious courtship of seven years, and married in 1919. Their marriage lasted fifty-three years, till Truman's death, aged eighty-eight, in 1972. Bess lived on; the headline in the *New York Times* in October 1982 summed it up perfectly: 'Bess Truman is Dead at 97; Was President's "Full Partner." '253

Their only child, Margaret, was the apple of her father's eye. He strongly backed her musical career (it was when her grand piano crashed through one of the ceilings of the White House that the alarm was raised and the ancient, rotting structure finally given a full repair-job in 1951–2). He supported her desire to become a coloratura soprano. Her first Washington concert, December 5, 1950, got a critical review by Paul Hume in the *Post*. The President dashed off an angry letter, beginning 'Mr Hume—I've just read your lousy review of Margaret's concert. I've come to the conclusion that you are "an eight-ulcer man on four-ulcer pay." ' It ended: 'Some day I hope to meet you. When that happens you'll need a new nose, a lot of beef steak for black eyes, and perhaps a supporter below! [Westbrook] Pegler, a gutter-snipe, is a gentleman alongside you. I hope you'll accept that statement as a worse insult than a reflection on your ancestry. H.S.T.'254 Margaret was at first deeply embarrassed, then touched, by her father's loyalty. The letter got into print, and caused tut-tutting, but Truman never regretted it and said all parents of much loved daughters would understand. They did, eventually at any rate. It is one of those incidents which brought home not only Truman's essential Middle Americanism, but in a curiously reassuring way the democratic nature of the US presidency. No one who reads Truman's diaries, and studies his voluminous papers with care, can be in any doubt that he was outspoken, at times vituperative, a volcano of wrath—though a quickly subsiding one—a good hater, a typical hot-blooded American of his time, but also decent, generous, thoughtful, prudent, and cautious when it came to the point, a constitutionalist and a thorough democrat, and a natural leader too.

When Truman decided not to run again in 1952, the Democratic convention picked, on the third ballot, the successful and much liked governor of Illinois, Adlai E. Stevenson (1900–65). He was a toff, like Roosevelt, but in addition an intellectual. Indeed his shining bald dome led the press to create a new term for the species: 'egghead.' Stevenson was a decent man and would probably have made an above-average president. But the role history allotted to him was that of the respectable loser, once the Republican Party, not normally a good picker, ditched its ideological standard-bearer, Senator Robert Taft of Ohio (1889–1953), 'Mr Republican' as he was known, in favor of General Eisenhower. 'Ike' was bald too, but he was amiable and a smiler—one of the great smilers of all time—and his name lent itself to the most successful button in the whole of US election history: 'I Like Ike.' He was elected by an enormous plurality, 33,936,234 to Stevenson's 27,314,999, with an electoral college majority of 442 to 89. And when, in 1956, the Democrats, who had nothing better to offer, decided on a throwaway rerun, Eisenhower triumphed by an even wider margin: 35,590,472 to 26,022,752, the electoral college margin being 457 to 73.[255] Despite growing health problems, including a serious heart attack, Ike might well have been elected again, had not the Twenty-second Amendment, written mainly by Republicans in the light of FDR's four terms, and limiting any one president to two, been adopted in 1951.

Eisenhower was popular because he was seen as a retired war-hero and because his coming marked the nation's return to peace, prosperity, and wellbeing. He was uncontroversial, non-party, unpolitical (or so it seemed), classless, unsectarian, all things to all men. He was certainly unpartisan. After Truman's unexpected victory in 1948, Eisenhower wrote to him that at no point did the political history of the US 'record a greater accomplishment than yours, that can be traced so clearly to the stark courage and fighting heart of a single man.'[256] On the other hand, Ike was never as simple as he appeared. 'The most devious man I ever came across in politics' was the half-admiring summing-up by Richard Nixon, his vice-president throughout his two terms. Nixon said: 'he always applied two, three or four lines of reasoning to a single problem, and he usually preferred the indirect approach.'[257]

As Supreme Allied Commander in Operation Overlord, Eisenhower had been accustomed to organizing vast masses of men and material in conditions of great stress. Running the US federal government held no terrors for him. He knew how to delegate generously and he knew also how to maintain systems of control. Thus he picked as his secretary of state the experienced and knowledgeable international lawyer, John

Foster Dulles (1888–1959) and gave the impression that Dulles was very much in charge of foreign policy, an impression Dulles did all in his power to confirm. Dulles was an ideologist of a kind, investing his conduct of international relations with a moral dimension, presenting conflict with the Soviet Union and China in terms of absolute rights and wrongs, and 'containment,' which he had reluctantly substituted for 'rollback,' as a righteous crusade against atheistical Marxism. He said openly to the American people that keeping the peace in a nuclear-armed world was a dangerous business: 'The ability to get to the verge [of war] without getting into the war is the necessary art . . . We walked to the brink and we looked it in the face.'[258] This frankness, which led to Dulles' critics coining the term 'brinkmanship' and accusing him of practicing it as a kind of diplomatic 'Russian roulette,' led the unsophisticated to think Dulles was a dangerous man. But the record shows both that he was inherently cautious on all points that mattered, and that he was closely supervised by the President, who insisted that Dulles report to him fully by phone every day even when he was on foreign assignment.[259]

Indeed, it is a curious fact that, as part of his attempt to reassure the American people that all was well, Ike deliberately gave the impression that he was (to use a later term) 'laid back.' He allowed it to be thought that he was a kind of constitutional monarch, who delegated decisions to his colleagues and indeed to Congress, and who was anxious to spend the maximum amount of time playing golf. Many fell for his stratagem, including his rival Taft, who sneered, 'I really think he should have been a golf pro.'[260] His first biographer claimed that 'the unanimous consensus' of 'journalists and academics, pundits and prophets, the national community of intellectuals and critics' had been that Eisenhower had 'elected to leave his nation fly on automatic pilot.' He was seen as well meaning but intellectually ignorant, inarticulate, often weak, and always lazy.[261] The daily digest of his activities and appointments provided by his gravelly voiced press secretary, Jim Haggerty, suggested a light workload.

However, in the late 1970s, the opening up of the secret files kept by his personal secretary, Ann Whitman, phone logs, diaries, and other intimate documents revealed that Eisenhower worked very much harder than anyone, including close colleagues, had supposed. A typical day started at 7.30, by which time he had breakfasted and seen the *New York Times*, *Herald Tribune*, and *Christian Science Monitor*, and it finished close to midnight, though he often worked afterwards. Many of his appointments, particularly those connected with party politics, defense, and foreign policy, were deliberately excluded from the lists put out by Haggerty. Long and vital meetings with the State and Defense Secretaries, the head of the CIA, and other senior figures took place in secret and were unrecorded, before the formal sessions of the National Security Council. The running of defense and foreign policy, far from being bureaucratic and inflexible, as his critics contended, in fact took place in accordance with highly efficient staff principles, which Eisenhower had perfected during his long military career, and which included the military art of deception. These procedures contrastly strongly with the romantic anarchy of the Kennedy regime which followed.

Eisenhower was in charge throughout. He practiced pseudo-delegation. Thus most people thought he left domestic matters largely to his chief of staff, the former governor of New Hampshire, Sherman Adams. Adams himself seems to have shared this illusion. He said that Ike was the last major figure who actively disliked and avoided using the phone.[262] But the logs show he made multitudes of calls about which Adams knew nothing. On foreign policy, he used sources of advice and information about which, equally, Dulles was unaware. He used the industrious Dulles as a superior servant; and Dulles complained that, though he often worked late into the night with the President at the White House, he had 'never been asked to a family dinner.'[263] The notion that Adams and Dulles were prima donnas was deliberately promoted by Eisenhower, since they could be blamed when mistakes were uncovered, thus protecting the office of the presidency.

George Kennan came closer to the truth when he wrote that on foreign affairs Eisenhower was 'a man of keen political intelligence and penetration . . . When he spoke of such matters seriously and in a protected official circle, insights of a high order flashed out time after time through the curious military gobbledygook in which he was accustomed to expressing and concealing his thoughts.'[264] At public press conferences Eisenhower used his gobbledygook to avoid giving answers which plain English could not conceal, and often pleaded ignorance for the same reason. He was Machiavellian enough to pretend to misunderstand his own translator when dealing with persistent foreigners.[265] Transcripts of his secret conferences show the lucidity and power of his thoughts. His editing of drafts by speechwriters and of statements by Dulles reveal an excellent command of English which he could exercise when he chose. Churchill was one of the few men who appreciated him at his worth, and it could be said that they were the two greatest states-men of the mid–20th century.[266]

Eisenhower concealed his gifts and activities because he thought it essential that the masterful leadership which he recognized both America and the world needed should be exercised by stealth. He had three guiding principles. The first was to avoid war, which he had seen at close quarters and hated. Of course, if Russia was bent on destroying the West it had to be resisted and America must be strong enough to do the job. But the occasions of unnecessary war (as he judged Korea) must be avoided by clarity, firmness, caution, and wisdom. He ended the Korean conflict. He avoided war with China. He stamped out the Suez war in 1956, although it involved offending his best ally, Britain, and ending the political career of an old friend, Sir Anthony Eden. He skillfully averted another Middle Eastern war in 1958, by timely action. In Indochina in 1955–6, when the French, who had made a mess of things, refused any longer to carry the burden of resisting Communism, Eisenhower agreed to shoulder it, but he was quite clear that America should not get into a war there. 'I cannot conceive of a greater tragedy for America,' he remarked presciently of Vietnam, 'than to get heavily involved now in an all-out war in any of these regions.' Again: 'There is going to be no involvement . . . unless it is as a result of the constitutional process that is placed upon Congress to declare it.'[267] Congressional authorization and Allied support—those were the conditions he laid down for American military involvement anywhere, and they were reflected in the CENTO and SEATO systems of alliance he added to NATO. These covered the Middle East and Southeast Asia.

Paradoxically, Eisenhower was strongly opposed to generals partici-pating in politics. The 1953 Chicago convention had been so crowded with generals, mainly supporters of Taft or MacArthur, that Eisenhower, in disgust, ordered his chief aide, Colonel Bob Schultz, and his doctor, General Howard Snyder, to leave town. He thought there were too many military men in and around the CIA, and in 1954 he appointed a wily old diplomat, David Bruce, to head a civilian Board of Consultants on Foreign Intelligence Activities. He employed other means to keep the military establishment under his authority. He was the only President who knew exactly how to handle the CIA. He presided skillfully over its operation in Iran, for the removal of the anti-West Moussadeq, and in Guatemala, for the overthrow of an unpopu-lar leftist regime, without any damage to his authority. The 1958 CIA coup in Indonesia failed because, for once, the work was delegated to

Dulles. It is hard to believe that Eisenhower would have permitted the 1961 Bay of Pigs operation to proceed in the form it took.

One of the reasons Eisenhower hated war was because he did not believe that 'limited war' was a viable concept. In war, as he understood it, the object was to destroy your enemy's power as quickly as possible with all the means at your disposal. That was his second guiding principle and it explains why he wound up Korea, to him a 'nonsense,' as quickly as he could, and why he deplored Eden's absurd Suez expedition in 1956, in which the Prime Minister personally approved the weights of bombs to be dropped on Egyptian targets. If Eisenhower had fought the Vietnam War, he would have fought it to a finish with all the power of the American armed forces. To him, Lyndon Johnson's political war, with the White House in effect deciding the timing and level of operations, would have seemed a certain formula for failure. Eisenhower was always clear, from his long experience, about how the business of war, when unavoidable, should be conducted. The President and Congress must decide. The President, as chief magistrate and commander-in-chief, should then state the objects to be achieved clearly, and issue precise and unambiguous orders to the armed forces. Their commanders should then state, equally clearly, the resources they required to carry out the orders and, having received them, do precisely that without political interference. Clear distinctions between the roles of the constitutional power of the Congress and presidency and the executive power of the military were always present in Eisenhower's mind—and it was their absence in the decade which followed which led to disaster.[268]

It was Eisenhower's secret fear, in the tense atmosphere generated by the Cold War, that the government would fall into the grip of a combination of bellicose senators, over-eager brasshats, and greedy arms-suppliers. In his farewell address, broadcast on January 17, 1961, he coined or popularized a new phrase: 'In the councils of government, we must guard against the acquisition of influence, whether sought or unsought, by the military–industrial complex.'[269] His use of this term has often been misunderstood. Eisenhower was not condemning militarism so much as making an important economic point in a military context. Historians can trace the rise of huge executive power in the United States, accompanied by prodigious spending, through the pre-World War One years of the Wilson administration, the vast expansion of federal and military power 1917–18, the revival of large-scale federal industrial projects during the New Deal, their expansion during

World War Two on a gigantic and unprecedented scale, and the way in which the onset of the Cold War made a large-scale, free-spending federal government, linked to an enormous arms industry, and voracious armed forces, a permanent feature of the American system. Eisenhower's third principle, reflected in his private diaries and papers, was that the security of freedom throughout the world depended ultimately on the health and strength of the US economy. Given time, the strength of that economy would duplicate itself in western Europe and Japan—he could see it happening—thus spreading the burden. But the US economy itself could be destroyed by intemperate spending by a greedy, over-large state, generating profligacy and inflation. He said of the military: 'They don't know much about fighting inflation. This country could choke itself to death piling up military expenditures just as surely as it can defeat itself by not spending enough for protection.' Or again: 'There is no defense for any country which busts its own economy.'[270]

For this reason Eisenhower was equally opposed to reckless spending in the domestic field. He was not opposed to deficit finance as a temporary device to fight recession. In 1958, to overcome such a dip, he ran up a $9.4 billion deficit, the largest so far for the US government in peacetime. But that was an emergency. Normally he ran balanced budgets. What he was most opposed to was a massive, permanent increase in federal commitments. He put holding down inflation before social security because he held that price stability was ultimately the only reliable form of social security. He loathed the idea of a welfare state. He was in fact deeply conservative. He admitted in 1956: 'Taft was really more liberal than me in domestic matters.'[271] His nightmare was a combination of excessive defense spending and a runaway welfare machine—a destructive conjunction that became reality in the late 1960s. While he was still in charge, federal spending as a percentage of GNP, and with it inflation, was held to a manageable figure, despite all the pressures. It was a notable achievement and explains why the Eisenhower decade was the most prosperous of modern times in America, and felt to be such, as if the lost Arcadia of the 1920s was being rediscovered. By the end of the 1960s this prosperity was radiating widely all over the world, as the pump-priming by US economic aid took effect. The world was more secure and stable too. In 1950–2, the risk of a major war was acute. By the end of the decade a certain stability had been reached, lines drawn, rules worked out, alliances and commitments settled across the globe. Perhaps only Eisenhower himself knew which of those commitments were real, but the Soviets and Chinese had learned that it was safest to assume that they all were. Thus the containment policy had been successfully applied. Militant Marxism–Leninism, which had expanded rapidly in the 1940s in both Europe and Asia, found its impetuous march slowed to a crawl or even halted entirely. These were tremendous achievements.[272]

Eisenhower's skill and deviousness allowed the country to survive, without too much damage, one of its periodical outbreaks of hysteria, which in this case goes under the heading of 'McCarthyism.' It had long been known that Communist agents had penetrated government at various levels in the 1930s and 1940s. In theory, the United States was protected against subversion by the McCormack Act (1938), which obliged foreign agents to register, and by the Hatch Act (1939) and the Smith Act (1940), under which members of organizations which advocated the overthrow of the US government by force or violence could be prosecuted. But these did not work against covert Communist sympathizers who got into government during the New Deal or the war. In his memoirs, George Kennan admits that 'the penetration of the American governmental services by members or agents (conscious or otherwise) of the American Communist Party' was 'not a figment of the imagination' but 'really existed and assumed proportions which, while never overwhelming, were also not trivial.' He said that those who served in Moscow or the Russian division of the State Department were 'very much aware' of the danger.[273] The Roosevelt administration, according to Kennan and others, was remiss in failing to heed warnings about the extent of Communist activity, 'which fell too often upon deaf or incredulous ears.' The Truman administration was more vigilant. In November 1946 Truman appointed a Temporary Commission of Employee Loyalty and the following March he acted on its recommendations with Executive Order 9835, which authorized inquiries into political beliefs and associations of all federal employees.[274]

A number of prosecutions took place. Some proved very difficult. Alger Hiss, a senior State Department official, accused in 1948 by a self-confessed Communist agent, Whittaker Chambers, of having sent classified documents to Moscow in the 1930s, was indicted for espionage but discharged after a jury failed to reach a verdict on July 8, 1949. He was later convicted of perjury, January 21, 1950, and sentenced to five years in prison, but his guilt on the main charges was never finally established until the mid–1990s. On the other hand, those responsible for betraying atomic secrets to Russia, which allowed Stalin to make his first A-bombs much more quickly than had been expected, were brought to book. Harry Gold was convicted on December 9, 1950, and he and David and Ruth Greenglass confessed their guilt and implicated Julius and Ethel Rosenberg and Morton Sobell, all of whom had been associated with the Communist Party. On March 29, 1951 the three were found guilty by a jury, and all were sentenced, the Rosenbergs, the only ones of the accused to refuse to cooperate, being sentenced to death, and executed June 19, 1953 at Sing Sing prison.[275]

Truman's precautions against future espionage seem to have been effective, on the whole, in eliminating those who spied for ideological reasons, as opposed to the merely mercenary. But, by the time they had been set up, Congress had alerted itself to the danger of subversion which (it was claimed) had led first to the 'loss' of eastern Europe, then to the 'loss' of China. A fortnight after Hiss's conviction, the junior Senator for Wisconsin, Joseph R. McCarthy, made a Lincoln Day speech in Wheeling, West Virginia, in which he caused a sensation by waving a piece of paper naming 'all the men in the State Department' who were 'active members of the Communist Party and members of a spy ring.' He added: 'I have here in my hand a list of 205—a list of names which were made known to the Secretary of State [Dean Acheson] . . . and who nevertheless are still working and shaping policy in the State Department.'[276]

There had been Congressional attempts to remove Communists before, notably by the House Committee on UnAmerican Activities, but McCarthy's speech began the active phase of the anti-Red witch-hunt. By the time it took place, most of the real Communists in public service had been detected, dismissed, indicted, convicted, or imprisoned. There was no 'list of 205 names.' The figure 205 was the result of faulty arithmetic. It arose because James Byrnes, when secretary of state, had written to Congressman Adolph Sabath that 285 alleged security risks had been identified in the State Department and, after investigation, seventy-nine had been fired. After subtraction, the remainder, 206, mistakenly became the figure of 205 McCarthy had used. McCarthy also produced another figure, fifty-seven. This came from a report by Robert E. Lee, chief of staff of the House Appropriations Committee, complaining that fifty-seven State Department employees out of an original list of 108 suspects he had reported to the Department were still on its payroll in March 1948.

But, of these fifty-seven, thirty-five had already then been cleared by the FBI and clearance was later given to others. McCarthy never possessed any actual names, and the only ones he used were those already published in the extremist counter-subversion literature such as Elizabeth Dilling's *Red Network*, Richard Whitney's *Reds in America*, Blair Coan's *Red Web*, and Nesta Webster's *World Revolution*.[277] McCarthy was never a serious investigator of subversion but a politician trying to draw attention to himself. He was first amazed, then unbalanced, and finally destroyed by his success. There is no evidence he ever identified any subversive not already known to the authorities and the only consequence of his activities was to cause trouble and distress for a lot of innocent people and discredit the activities of those genuinely concerned to make America safe.

The response of President Truman and the Democrats was to meet McCarthy head-on. The Senate Democrats appointed a committee under Senator Millard Tidings of Maryland to investigate McCarthy's charges, and the hearings of the Tidings Committee effectively exposed as worthless all specific charges against individuals McCarthy was persuaded to name. Truman told the press that McCarthy was 'the Kremlin's greatest asset in America,' which was true. Truman also commissioned a study of 'hysteria and witchhunting' in American history, which concluded there was a permanent undercurrent of 'hate and intolerance' in America which periodically produced outbreaks such as McCarthyism. This in turn created an academic sub-branch of sociology, leading to a 1954 Columbia University seminar on McCarthyism, during which the historian Richard Hofstadter, using Theodor Adorno's 1950 tract *The Authoritarian Personality*, explained the phenomenon as a projection onto society of the groundless fears of 'pseudo-conservatives.' This was later expressed in a famous essay by Hofstadter, 'The Paranoid Style in American Politics' (1964), which proved hugely influential and became the official liberal explanation of McCarthyism, thus generating even more confusion than the Senator's original accusations.[278]

In the meantime, McCarthy accelerated a process of 'blacklisting' which had already begun before his intervention. As far back as November 1947 a meeting of Hollywood producers had drawn up a blacklist of names of Communists, including one producer, one director, and eight writers. Between 1951 and 1954, the House UnAmerican Activities Committee named 324 Hollywood personalities, who were also blacklisted. There was likewise a broadcast-industry blacklist, dating from April 1947 and constantly added to. Local governments began investigating schools and universities and imposing loyalty oaths as a condition of employment. In 1948 the University of Washington had fired three professors for refusing to answer questions or for admitting they were Communists, and, after McCarthy began the hue and cry, about a hundred professors were fired for similar reasons between 1952 and 1954. Blacklisting was extended and enforced by a variety of groups including the Wage Earners Committee, which specialized in movies, the American Legion, the Catholic War Veterans, a special body called AWARE, created by Godfrey Schmidt, the lawyer of Cardinal Francis Spellman of New York, and an ABC program *Red Channels*.[279]

After the Republicans captured control of the Senate in the November 1952 elections, McCarthy had himself made chairman of the hitherto obscure Investigation Subcommittee of the Senate Committee on Government Operations, and turned it into a forum for his claims, hiring a clever and unscrupulous twenty-five-year-old lawyer, Roy M. Cohn, as his chief counsel. His campaign had long since lost touch with reality. On June 14, 1951 he had subjected the Senate to a three-hour harangue in which he 'named' General Marshall as 'the grim and solitary man' who was at the heart of 'a conspiracy on a scale so immense as to dwarf any previous such venture in the history of man.' He published these charges in a book, *America's Retreat from Victory; the Story of George Catlett Marshall*, and when he got his own subcommittee and staff he turned on the Republicans, attacking Eisenhower for nominating the Harvard president James B. Conant as high commissioner in Germany, and Charles Bohlen as ambassador in Moscow. Eisenhower's response was characteristically devious. He thought that Truman's direct rebuttals of McCarthy's charges were mistaken and that if he himself entered into an argument he would merely provide the Senator with more anxiously sought publicity. He declared privately: 'I just will not—I refuse—to get into the gutter with that guy.'[280] He found it hard not to come to the rescue of his old mentor, Marshall, but he strongly believed that, if he ignored and belittled McCarthy, the Senator would eventually destroy himself.

When Eisenhower, and the National Security Council, blocked an attempt by McCarthy to investigate the CIA, the Senator turned instead on the US Army. That proved his undoing, as the President suspected it would, because in the process of the controversy McCarthy's accusations created, it emerged that Roy Cohn had put pressure on the army to secure favors for his boyfriend, a conscript called David Schine, in defiance of 'good order and military discipline.' This was the kind of scandal the ordinary public could understand, and the suspicions of homosexuality spread to include McCarthy himself, whose relationship to Cohn, and indeed Schine, appeared dubious. These opportunities were skillfully exploited during public hearings by the army counsel, a foxy old Boston lawyer called Joseph N. Welch. McCarthy the bully and the accuser found himself the bullied and the accused, and in his pain he increased his already considerable intake of alcohol (the ability to 'belt a fifth [of a gallon] of bourbon' was a pledge of virility in his circle). His allies in the media, the House, and the Senate deserted him one by one and on December 2, 1954 the Senate voted 67 to 22 to censure him. He died of alcohol-related illness before the end of his term, on May 2, 1957, a burned-out case and almost forgotten, though McCarthyism lingered on as a term of abuse for conservative inquiries into any form of liberal ill-doing, however well substantiated.

The significant fact about McCarthyism, seen in retrospect, was that it was the last occasion, in the 20th century, when the hysterical pressure on the American people to conform came from the right of the political spectrum, and when the witchhunt was organized by conservative elements. Thereafter the hunters became the hunted. It was the liberals and the progressives who, over the next four decades, were to direct such inquisitions as Watergate and Irangate. The Eisenhower decade was the last of the century in which the traditional elements in American society held the cultural upper hand. Eisenhower's America was still recognizably derived from the republic of the Founding Fathers. There were still thousands of small towns in the United States where the world of Norman Rockwell was intact and unselfconsciously confident in itself and its values. Patriotism was esteemed. The flag was saluted. The melting-pot was still at work, turning out unhyphenated Americans. Indeed the 'American Way of Life' was a term of praise, not abuse. Upward mobility was the aim. Business success was applauded and identified with the nation's interests. When Eisenhower appointed Charles ('Engine Charlie') Wilson, who was head of General Motors, as Defense Secretary, and he testified at his nomination hearings before the Senate Armed Services Committee, January 15, 1953, he told them, without apology: 'For years I thought what was good for our country was good for General Motors, and vice-versa. The difference does not exist. Our company is too big. It goes with the welfare of the country.' No senator dissented. The nomination was unanimously approved.[281]

There was, to be sure, plenty of criticism, especially from the new breed of pop-sociologists who flourished at this time. The most influential was C. Wright Mills (1916–62), who seized on the tendency of American society to alienate as well as to embrace people, to exclude as much as to include. He was obsessed by the notion of the powerful few and the powerless multitude. He had been an unsuccessful car salesman, a bad thing to be in America, where Willy Loman, the tragic anti-hero of Arthur Miller's 1950 hit, *Death of a Salesman*, was a figure of pathos. Mills insisted that America was becoming 'a great salesroom, an enormous file, an incorporated brain, and a new universe of management and manipulation.'[282] He pressed this charge in a remarkable trilogy of books, analyzing the various layers of America's giant anonymous structure: *New Men of Power* (1948) on labor and its leaders, *White Collar* (1951) on the new middle class of middle management, and *The Power Elite* (1956), which described the 'higher circles' of corporate, military, and political authority. He wrote of the 'postmodern' society in which democratic 'publics' had been replaced by manipulated 'masses.'[283] The theme was echoed and enlarged by David Reisman and Nathan Glazer in *The Lonely Crowd: A Study of the Changing American Character* (1950), which claimed that Americans were losing their individualism, their 'inner-direction,' and were becoming 'outer-directed,' forced into conformity by pressure from their 'peer groups.' Spreading suburbia and the flight from the inner cities, a phenomenon of the decade, was blamed for this increasing conformity. William H. Whyte in *The Organization Man* (1956) blamed suburban America for stressing 'getting along' or 'belonging' at the expense of the personal, entrepreneurial drives that had made the country remarkable.[284]

However, the most influential writers of the period were not the sociological critics but the dispensers of uplift. This was an old American tradition, with its good and its bad sides, going back at least to Jonathan Edwards, and dispensed in secular form by Emerson. Uplift writing nurtured resolution and self-reliance; it also bred complacency. The latest crop had first sprouted in the Depression, when it was much needed, and overflowed in the Eisenhower era. Dale Carnegie (1888–1955) had started as a speech-trainer, a good entry into the upward-mobility trade, and his *How to Win Friends and Influence People* (1936) was just what millions of Americans felt they needed. His later *How to Stop Worrying and Start Living* (1948) was an updated model for an age of full employment. Another tireless worker in the uplift vineyard was Norman Vincent Peale (1889–1969), an Ohio Methodist who later embraced the Dutch Reformed Church. He started with *You Can Win* (1938), broadened his appeal to include all American religious groups with *A Guide to Confident Living* (1948), and hit the commercial jackpot in 1952 with *The Power of Positive Thinking*, which had sold 3 million copies in hardback by 1974. Its chapter heads, 'I Don't Believe in Defeat,' 'Expect the Best and Get It,' 'How to Get People to Like You,' gave the characteristically American message that there was no problem, personal as well as public, without a solution. There were scores of other quasi-religious preachers reassuring the public along the same lines.[285]

In addition to mere uplift, there was genuine religion in countless varieties, sprouting and flourishing in these years. In 1944, a historian called W. W. Sweet had mistakenly preached a funeral sermon, *Revivalism in America; its Origin, Growth and Decline.* At the very time he was writing, Billy Graham (b. 1918), a Southern Baptist minister ordained in 1939, was gearing up for his successful effort to 'revive revivalism,' beginning with an eight-week tent meeting of 350,000 people in, of all unlikely places, Los Angeles in 1949. Graham concentrated on conservatives within the larger Protestant denominations, or in churches opposed to the ecumenical movement, and he proved remarkably successful and persistent, being still a major figure in evangelical revivalism in the second half of the 1990s. His stress was not on sinfulness so much as opportunity, renewal, and peace of mind, epitomized in his *Peace with God* (1953). This echoed the themes of the conservative Catholic revivalist Monsignor Fulton Sheen (1895–1975), whose *Peace of Soul* (1949) led to a popular TV show, *Life is Worth Living*, with an audience of 30 million, and brought him in a fan-mail of 8,000 to 10,000 letters a day.[286] Successful evangelism was evidence of a much deeper and wider religious revival, reflected in statistics of church affiliation. In 1910, 43 percent of Americans were attached to particular churches. This figure was the same in the 1920s. The Depression, far from alienating ordinary Americans, led to a rise in church attendance, which by 1940 was 49 percent. By 1950 the figure was 55 percent and in 1960 it had risen to 69 percent, probably the highest in America's entire history (it had dropped to 62.4 by 1970).[287] Vast numbers of new churches were built or old ones enlarged to accommodate the expanded congregations: in 1960 alone over $1 billion was spent on church-building. The government took note and gave its sanction, wherever possible, to religious activity. In 1954 the phrase 'under God,' as used by Lincoln in his Gettysburg Address, was added to the national Pledge of Allegiance, sworn by all those gaining citizenship, and 'In God We Trust' was adopted as the country's official motto.

Eisenhower presided benignly over what was termed 'Piety on the Potomac', a generalized form of the Christian religion very much in the American tradition, with no stress on dogma but insistence on moral propriety and good works. He announced in 1954: 'Our government makes no sense unless it is founded on a deeply felt religious faith—and I don't care what it is.'[288] By this he did not mean, of course, that he was indifferent to the articles of faith, far from it, but that he believed sincere faith was conducive to moral conformity, and that religion was the best, cheapest, and least oppressive form of social control. He had made his own personal sacrifice to the demands of conformity. While on active service abroad, 1942–5, he had fallen in love with his Anglo-Irish driver, Kay Summersby, and in June 1945, after Germany was beaten, had even considered divorce and remarriage. But this proposal (according to Truman) had been vetoed by a blistering letter from his military superior, General Marshall, who threatened to 'bust him out of the army' unless he gave up the idea.[289]

But, if the United States was still a conformist and traditional society in the 1950s, the portents of change were present too. In 1948 an Indiana University entomologist, Dr Alfred Kinsey, brought out an 804-page volume, based on 18,000 interviews over many years, called *Sexual Behavior in the Human Male*. He followed it in 1954 with *Sexual Behavior in the Human Female*, which sold 250,000 copies. They revealed that 68–90 percent of American males, and almost 50 percent of females, engaged in premarital sexual intercourse, that 92 percent of males and 62 percent of females had masturbated, and that 37 percent of males and 13 percent of females had experienced homosexual intercourse. The findings also suggested that 50 percent of men and 26 percent of women had committed adultery before the age of forty.[290] Kinsey's findings caused surprise and in some cases rage but they confirmed much other evidence that, even in the 1950s, the Norman Rockwell images no longer told the full story. Hollywood was still trying to hold the lines laid down by the old Hays Code, but it was cracking.[291] The adultery theme had been tackled, to public approval, in *From Here to Eternity* (1953). In 1956 came *Baby Doll*, described by *Time* as 'just possibly the dirtiest American-made motion-picture that has ever been legally permitted.' Hollywood dealt with interracial sex in *Island in the Sun* (1957), homosexuality in *Compulsion* (1958), and abortion in *Blue Denim* (1959). *Playboy* (December 1953) was the herald of the soft-porn magazines, the lubricious *Peyton Place* was the 6-million-copy bestseller of 1956–8, *Lolita* introduced the sexual 'nymphet' in 1958, and in 1959 *Lady Chatterley's Lover* was at last published without prosecution. Movies of these bestselling fictions quickly followed.

Television, anxious at this stage to establish itself as an all-American institution, was more censorious, and modeled itself on US radio. There were 1,200 US radio stations in 1949 and only 28 TV stations, transmitted to the 172,000 (1948) families which owned sets. This figure jumped to 15.3 million in 1952 and 32 million in 1955. By 1960 some 90 percent of households had at least one TV set (and by 1970 over 38 percent had color TV).[292] As the TV habit spread and took deep root, and as the medium made itself indispensable to all the purveyors of mass-consumption goods and services, those who ran the networks and the stations began to flex their cultural muscles and contemplate a society in which all standard measurements of behavior would be up for redefinition, and moral relativism, based on ratings, would rule. Thus the way to the 1960s was prepared, and what began as a sexual revolution was to bring about revolutions in many other areas too.

PART EIGHT

'We Will Pay Any Price, Bear Any Burden'

Problem-Solving, Problem-Creating
America, 1960–1997

In 1960 President Eisenhower was the oldest man ever to occupy the White House, and the universal cry was for youth. The Sixties were one of those meretricious decades where novelty was considered all-important, and youth peculiarly blessed. Normally circumspect men and women, who had once made a virtue of prudence, and were to resume responsible behavior in due course, did foolish things in those years. Such waves of folly recur periodically in history. The wise historian does not seek too assiduously to explain them. He merely notes that they occur, and have baleful consequences. Thus it was with the 1960s.[1] The two candidates for the presidential election both had youth on their side. Eisenhower's Vice-President, Richard Nixon (1913–94) was forty-seven. He was very experienced for his age, having sat in the House from 1947, and the Senate (for California) from 1950, following a distinguished wartime career in the US Navy. During his eight years as vice-president he had been a prominent figure in the Senate, one of Eisenhower's most often-consulted colleagues, and had traveled all over the world as a US spokesman and representative.[2] He had been tested, and emerged with credit. He had been nearly murdered by anti-American radicals in Caracas in 1958 and stood up successfully to Nikita Khrushchev, the Soviet dictator, in a famous public debate in Moscow in 1959. In 1960 the Republicans nominated him for president on the first ballot, virtually without opposition (the actual figure was 1321 to 10) as the obvious man to lead the nation 'in the spirit of Ike.' His running mate, UN ambassador and former Senator Henry Cabot Lodge, was an Eastern establishment man of similar wide experience. The two men constituted an outstanding ticket.[3]

The Democratic candidate, John Fitzgerald Kennedy (1917–63), was also young (forty-three) and experienced. Like Nixon, he had had a successful wartime career in the navy, had served in the House as Congressman from Massachusetts from 1947, and then as senator from 1953. Both men had been fierce anti-Communists in the postwar years. Nixon had played an important part in exposing and convicting Alger Hiss, and had been publicly praised by Kennedy for doing so. It is illuminating that, just before the two men were nominated, Nixon ran into Kennedy's father, FDR's old campaign contributor Joseph P.

('Poppa Joe') Kennedy, outside the Colony Restaurant in New York. Kennedy Sr said: 'I just want you to know how much I admire you for what you've done in the Hiss case and in all the [anti-]Communist activity of yours. If Jack doesn't get it, I'll be for you.'[4] But, though the two men had much in common, the differences were enormous and important. And, still more important—and significant—was the way those differences were perceived, especially by the East Coast media.

We come now to an important structural change in America. America had always been, from the earliest time, a democratic society, in that men (and indeed women) paid little attention to formal rank, even where it existed. Every man felt he had the right to shake hands with every other man, even the President (Washington was the first and last President to deny that right, by bowing). But this democratic spirit was balanced by the tribute of respect to those who, for one reason or another—experience, learning, position, wealth, office, or personality—had earned the title of 'boss.' The balance struck between egalitarianism and deference was one of the most remarkable characteristics of America, and one of its great strengths. The Sixties brought a change. In the space of a decade, the word 'boss' passed almost out of the language, certainly out of universal usage. Deference itself deferred to a new spirit of hostility to authority. It became the fashion to challenge long-established hierarchies, to revolt against them or ignore them. Nowhere was this spirit more manifest than in the media (as it is now convenient to speak of the press, radio, and TV). Television had a powerful impact on the way opinion was formed, not only in the country, but within itself. The growth of the TV personality meant that many of those who appeared in front of the cameras, though originally of little account in the official hierarchy of the state, became famous to millions, valuable commodities, and soon earned more than their superiors up the hierarchy and eventually (in some cases) as much as station-owners. And in time they, rather than the management, let alone the stockholders, began to set the tone of comment and the thrust of opinion.

An early sign of the change was the way in which Edward R. Murrow, presenter of the *See It Now* public affairs program on CBS TV, emerged as a leading American opinion-former. His March 9, 1954 documentary on McCarthy, which played a notable part in breaking the Senator, was entirely conceived by Murrow himself and his producer. Management, board, and owners of CBS had little or nothing to do with it. The gradual but cumulatively almost complete transfer of opinion-forming power from the owners and commercial managers of TV stations to the program-makers and presenters was one of the great new facts of life, unheard of before the 1950s, axiomatic by the end of the 1960s. And it was gradually paralleled by a similar shift in the newspaper world, especially on the great dailies and magazines of the East Coast, where political power, with few exceptions, passed from proprietors and major stockholders to editors and writers. Owners like Hearst and McCormick (of the *Chicago Tribune*), Pulitzer and Henry Luce (of Time-Life), who had once decided the political line of their publications in considerable detail, moved out of the picture and their places were taken by the working journalists. Since the latter tended to be overwhelmingly liberal in their views, this was not just a political but a cultural change of considerable importance. Indeed it is likely that nothing did more to cut America loose from its traditional moorings.

The change could be seen in 1960, in the way the East Coast media (the *New York Times* and *Washington Post*, *Time* and *Newsweek*), handled the contest between Nixon and Kennedy. By all historical standards, Nixon should have been an American media hero. He was a natural candidate for laurels in the grand old tradition of self-help, of pulling yourself up by your own bootstraps. He came from nowhere. His family background was respectable but obscure. He worked his way through an unfashionable college. He had no money except what he earned by his own efforts. He had, to begin with, no influential friends or connections. His life was dominated by a passionate desire to serve in public office, sometimes masquerading as brutal ambition, and by his patriotism and love of country, which knew no bounds. He was an autodidact and voracious reader, always trying to better himself, intellectually as well as professionally. He combined this earnest cultural endeavor with solid campaigning and administrative skills, which brought him early and continuing success, and with a modest private life which was morally impeccable. He was twice elected vice-president and twice president, on the second occasion (1972) by the largest plurality in American history, 47,169,911 to 29,170,382, and by 520 college votes to 17, his opponent, George McGovern, being able to carry only Massachusetts. These were extraordinary achievements in a man from nowhere, in the best American democratic–egalitarian style—the Benjamin Franklin tradition. Yet, from start to finish, the media, especially the 'quality' press, distrusted him, consistently denigrated him, and sought to destroy him, indeed in a sense did destroy him. At every crisis in his career—except the last—he had to appeal above the heads of the media to the great mass of the ordinary American people, the 'silent majority' as he called them.

The origin of the East Coast media hostility to Nixon lay in his 1950 Senate defeat of Helen Gahagan Douglas (1900–80), a left-wing Congresswoman consistently portrayed by the media as a political virgin and martyr, whom Nixon crucified on the cross of anti-Communism. A former singer, married to the movie actor Melvyn Douglas, she was an outstanding example of raucous Hollywood chutzpah, and all the mud thrown at her came from within her own party. During the campaign, Democratic Congressman John F. Kennedy, no less, handed Nixon an envelope from his father containing $1,000, saying he hoped it would help 'to turn the Senate's loss into Hollywood's gain.'[5] Indeed Douglas was beaten primarily by disgusted Democrats switching to Nixon. The Hiss case did Nixon even more damage with the media, which, against all the evidence, tried to turn this undoubted Soviet agent and perjurer into an American Dreyfus in order to portray Nixon as a McCarthyite witchhunter.[6] Only after Nixon's death was Hiss's guilt finally established beyond argument by material from Soviet sources.

By contrast, the media did everything in its power to build up and sustain the beatific myth of John F. Kennedy, throughout his life and long after his death, until it finally collapsed in ruins under the weight of incontrovertible evidence. The media protected him, suppressed what it knew to be the truth about him, and if necessary lied about him, on a scale which it had never done even for Franklin Roosevelt. And this was all the more surprising because Kennedy had most of the characteristics of an American anti-hero. Kennedy was third-generation Boston-Irish upward mobility personified. Both his grandfathers were unscrupulous ward-heeling Boston politicians who well understood what became the running theme of the Kennedy family—how to turn money into political power—though they made their piles in different ways, one through insurance, the other in the liquor trade.[7] It was Kennedy's father, however, Joseph P. Kennedy (1888–1969), who saw that to get national power, if not for himself, then for his sons, it was necessary to amass an enormous fortune. That meant breaking out of the narrow Boston orbit and moving to New York. It also meant grabbing opportunities, including criminal ones, as they arose. So Joe Kennedy created one of the biggest cash piles of the century through banking (or perhaps one should call it money-lending, which is closer to the truth), shipbuilding, Hollywood, stock-jobbing, and bootleg liquor, among many other activities. He set up trust funds of $10 million apiece for his children and squirreled away vast sums for a variety of other purposes, chiefly for buying people, whether politicians, newspaper proprietors, or, indeed, cardinals. His range of contacts and collaborators was enormous and included the mob leader Frank Costello and Doc Stacher, lieutenant of another leading *mafioso*, Meyer Lansky. Joe was used by (and used) FDR, bought an ambassadorship from him, then set about the task of making his eldest son president.

When Joe Jr was killed in the war, Jack became crown prince and president-designate.[8] He inherited his father's values in many respects, though they jostled in uneasy counterpoint with his mother's brand of traditional Catholicism. Jack never had his father's insatiable relish for money, power, and corruption. Most of his life he simply did old Joe's bidding. As he put it, ruefully: 'I guess Dad has decided he's going to be the ventriloquist, so that leaves me the role of dummy.'[9] In one important respect, Jack rejected the family tradition. Money did not interest him and he never troubled to learn anything about it. It was not that he was extravagant. Quite the contrary, he was mean.[10]

But, if he was not interested in making money, in most other respects he happily accepted the family philosophy, especially its central tenet: that the laws of God and the republic, admirable in themselves, did not apply to Kennedys, at any rate male ones. Like his father, he dealt with gangsters, when he felt so inclined. From the start he was taught by his father that bluff, freely laced with money—or outright mendacity if need be—could remove all difficulties. The lies centered on certain areas. One was Jack's health. Old Joe had learned many tricks in concealing the true state of his retarded daughter, Rosemary, buried alive in a home.[11] He used them to gloss over the seriousness of Jack's back problems, and his functional disorder, eventually diagnosed as Addison's Disease. Strictly speaking, Jack was never fit to hold any important public office, and the list of lies told about his body by the Kennedy camp over many years is formidable. The backpain Jack suffered seems to have increased after he became president, and his White House physician, Dr Janet Travell, had to give him two or three daily injections of novocaine. Jack eventually found this treatment intolerably painful. But he did not fire Travell, fearing that, though she had hitherto been willing to mislead the media about his health, she might now disclose his true medical history. Instead, he kept her on the payroll but put himself into the hands of a rogue named Dr Max Jacobson, who later lost his medical license and was described by his nurse as 'absolutely a quack.'[12] Known to his celebrity clients as 'Doctor Feelgood,' because of his willingness to inject amphetamines laced with steroids, animal cells, and other goodies, Jacobson started to shoot powerful drugs into Jack once, twice, or even three times a week. Although he turned down a request to move into the White House, he had succeeded, by the summer of 1961, in making the President heavily dependent on amphetamines.[13]

The other main area of lying centered on Jack's curriculum vitae. In 1940 his thesis was written for him by a number of people, including Arthur Krock of the *New York Times*, and Joe's personal speechwriter, who described it in its original state as 'a very sloppy job, mostly magazine and newspaper clippings stuck together.' But, processed, it not only allowed Jack to graduate *cum laude* but also appeared in book form as *Why England Slept*. Old Joe and his men turned it into a 'best-seller,' partly by using influence with publishers such as Henry Luce, partly by buying 30,000–40,000 copies, which were secretly stored at the family compound in Hyannisport.[14] It was the same story with *Profiles in Courage*, which began as a 'disorganised, somewhat incoherent mélange from secondary sources' and was turned by Theodore Sorensen and a team of academic historians and professional writers into a readable book. By 1958 it had sold nearly 125,000 copies and, after intense lobbying by Joe, Krock, and other Kennedy satraps, it won Jack the Pulitzer prize for biography. Those who suggested the book was ghostwritten were sued for libel or even, at Joe's request, investigated by the FBI.[15]

Old Joe's neatest trick was to turn Jack into a war hero. In view of Jack's health, all Joe's skill at manipulation was needed to get him into the navy, secure him an immediate commission, and advance him in the service, especially since, while a young officer working in naval intelligence, he was detected by the FBI having an affair with a Danish woman suspected of being a Nazi spy. Jack was in charge of a PT boat which was rammed and sunk by a Japanese destroyer. Father Joe's management got him a medal for rescuing a crew member. The death of Joe Jr made the father anxious for Jack's safety. So he was promptly plucked out of the US Navy by the same means he got into it—influence. Maximum use was made of Jack's war career in all his campaigns.[16]

With the peace, Old Joe set about making Kennedy first a congressman, then a senator, then president. This train of events is worth studying because it shows the extent to which money paved the way to polit-

ical power in mid–20th-century America. The young man acquiesced at
first, but gradually his competitive spirit, strong but hitherto devoted to
games and sex, was aroused, and he cooperated with varying degrees of
enthusiasm. Kennedy's political ascent, however, was based essentially
on money and corruption. It is not always clear he was aware of his
father's malpractices. Naturally he knew of the dishonesty and lies that
made him a war hero and a literary celebrity. But getting him into the
House and Senate took enormous sums of money, much of it spent
openly but the rest doled out furtively by his father. In Kennedy's first
House race, large families were each given $50 in cash 'to help out at
the polls'. As Tip O'Neill (later Speaker of the House) put it, 'They
were simply buying votes, a few at a time, and fifty bucks was a lot of
money.'[17] Outsiders found they could not get a mention, or even buy
space, in papers whose owners owed Joe a favor. To get Kennedy into
the Senate, Joe diverted one rival into the governorship race by making
a large campaign donation, gave another to Adlai Stevenson to buy his
endorsement, and loaned the editor of the *Boston Post* half a million to
secure his support. Kennedy certainly knew about that one, since he
later admitted, 'You know, we had to buy that paper or I'd have been
licked.'[18]

After the Senate victory, Joe set his sights higher and, under
Sorensen's guidance, a new John F. Kennedy personality began to
emerge, calculated to appeal to liberals, intellectuals, and 'civilized people.' James MacGregor Burns was employed to manufacture a hagiog-
raphy called *John F. Kennedy: a Political Profile*. Other writers ghosted
a mass of articles signed by Kennedy and published in everything from
Look, *Life*, and the *Progressive* to the *Georgetown Law Review*. The
willingness, then and later, of intellectuals and academics of high
repute to participate in the promotion of Kennedy is worth noting. Self-
deception can only go so far: some of them must have known they were
involved in one of the biggest frauds in American political history. As
one of his biographers puts it, 'No national figure has ever so consistently and unashamedly used others to manufacture a personal reputation as a great thinker and scholar.'[19]

Did Kennedy, or for that matter his even more competitive-minded
younger brother Bobby, much his superior in intelligence, and who
now played an increasing role in the campaigns, have any political convictions? Kennedy shared his father's reverence for the power that
money brings, and the need to hang onto an unearned fortune made
him a fierce anti-Communist. His instincts put him in the same camp as

Senator McCarthy. He was on friendly terms with Nixon. Later he became more liberal. But essentially he was a half-ambitious, half-reluctant, dutiful son-and-heir following the masterful demands of his father to get to the top by fair means or foul. As Tip O'Neill put it: 'Looking back on his Congressional campaign, and on his later campaigns for the Senate and then for the Presidency, I'd have to say that [Jack] was only nominally a Democrat. He was a Kennedy, which was more than a family affiliation. It quickly developed into an entire political party, with its own people, its own approach and its own strategies.'[20] The man who got it right at the time was the British Prime Minister Harold Macmillan. He grasped the important point that electing a Kennedy was not so much giving office to an individual as handing over power to a family business, a clan, almost a milieu, with a set of attitudes about how office was to be acquired and used which at no point coincided with the American ethic. Having paid his first visit after Kennedy's election as President, Macmillan was asked on his return what it was like in Kennedy's Washington. 'Oh,' said he, 'it's rather like watching the Borgia brothers take over a respectable North Italian city.'[21]

By the time Kennedy was in the race for the Democratic nomination, then the White House, he was—or rather his people were—developing certain themes, especially the stress on youth, glamour (cortisone treatment had given him a chunkier, more handsome appearance), and sophistication, including an apparent intellectual elegance. But he was still pushed forward primarily by money. His father used the full force of his fortune to wreck competition from the popular Minnesota Senator, Hubert Humphrey (1911–78) in the West Virginia primary. (Humphrey, a decent and, on the whole, honest man was the great political fall-guy of these years who, like Henry Clay in the 19th century, never quite got to the top: he always played the game, and he always lost it.) The Kennedy disadvantage was, or was thought to be, his Catholicism, which had damned Al Smith a generation before. Cardinal Richard Cushing of Boston, Old Joe's tame 'red hat,' who was a kind of house-chaplain to the Kennedy family, later admitted that he and Joe had decided which Protestant ministers were to receive 'contributions' of $100 or $500 to play down the religious issue. The money used for outright bribes came partly from the Kennedy coffers and partly from the mafia, following a secret meeting between Jack Kennedy and Sam Giancana, the Chicago godfather. FBI wiretaps and documents show that the mafia money went to pay off key election officials, including local sheriffs, who were handed a total of $50,000 to get out the Kennedy vote by any means. In return, Joe promised the mobsters assistance in federal investigations.[22]

To what extent the actual presidential election of 1960 was honest, or whether the real winner was actually awarded the presidency, remains a mystery.[23] The election was certainly fraudulent in one respect. The main issue, insofar as there was one, was Kennedy's charge that the outgoing Eisenhower administration had allowed a 'missile gap' to develop between the strategic forces of the US and Soviet Russia. In fact America had a comfortable lead in missiles. Both Jack and Bobby Kennedy, and Jack's speechwriter Joseph Kraft, all admitted later that the 'gap' was a fiction. When Kennedy, as president, was privately taxed on the point, he laughed: 'Who ever believed in the missile gap anyway?' The shift in power in the media, already noted, enormously helped Kennedy, and his team reinforced their advantage by assiduous courting of media personalities and by making Kennedy available for TV on all occasions, something which could not be taken for granted in 1960. Theodore H. White, covering the campaign for his book *The Making of the President*, reported overwhelming media bias in favor of Kennedy. One of Nixon's aides said to him, of the press, '*Stuff* the bastards. They're all against Dick anyway. Make them work—we aren't going to hand out prepared remarks.' White concluded: 'To be transferred from the Nixon campaign to the Kennedy campaign . . . was as if one were transformed from leper and outcast to friend and battle companion.'[24]

It was the same with TV. Kennedy's greatest success during the campaign was his 'victory' over Nixon in the official TV debates, especially the first, when Nixon was unmade-up, tired, and suffering knee trouble.[25] It was learned later that Kennedy's staff had asked for the two candidates to stand up throughout the debate to cause discomfort in Nixon's weak knee, that they demanded an unusually warm temperature in the studio because they knew Nixon sweated easily, and that during the live transmission they put pressure on the studio director in the control room to show camera close-ups of Nixon mopping his brow and others emphasizing his five o'clock shadow, to confirm accusations that he looked like a 'used-car salesman' and give point to a campaign question: 'Would you buy a used car from this man?' Hence viewers who watched the debates gave the result to Kennedy, though those who heard them on the radio thought Nixon had won.[26]

Kennedy won the election by one of the smallest pluralities in

American history, 34,227,496 to 34,107,646, though more hand-somely in terms of electoral votes, 303 to 219. Nixon had deliberately decided to keep Catholicism out of the campaign, and succeeded in doing so. The day after the result, he said to his aide, Pete Flanigan: 'Pete, here's one thing we can be satisfied about. This campaign has laid to rest for ever the issue of a candidate's religion in presidential politics. Bad for me, perhaps, but good for America.'[27] In fact, Kennedy's Catholicism won him the election: whereas Eisenhower got 60 percent of the Catholic vote, Nixon received only 22 percent, less than any Republican candidate in the 20th century. In the Northern industrial states, where the margins were very close, the swing of the Catholic vote to a Catholic candidate made all the difference.

This was a crooked election, especially in Texas and Illinois, two states notorious for fraud, and both of which Kennedy won. In Texas, which gave its twenty-four college votes to Kennedy by a margin of 46,000 votes (out of 2.3 million), one expert made the calculation that 'a minimum of 100,000 votes for the Kennedy–Johnson ticket simply were non-existent.' As Kennedy's running mate, Senator Lyndon B. Johnson, was a Texan politico of long experience and total lack of scruple, the charge, which was in part substantiated by local results (in one polling station, where only 4,895 voters were registered, 6,138 votes were counted as cast), was almost certainly true.[28] In Illinois, Nixon carried 93 of the state's 102 counties, yet lost the state by 8,858 votes. This was entirely due to an enormous Democratic turnout in Chicago, under the control of the notorious Democratic city boss and mayor, Richard Daley. Daley gave Kennedy the Windy City by the astonishing margin of 450,000 votes, and the evidence was overwhelming that fraud was committed on a large scale in Kennedy's favor.[29] The mafia played an important part in this fraud. Afterwards, its boss Giancana often boasted to Judith Campbell, the mistress he shared with the President, 'Listen, honey, if it wasn't for me your boyfriend wouldn't even be in the White House.'[30]

If Nixon, instead of Kennedy, had carried Texas and Illinois, the shift in electoral votes would have given him the presidency, and the evidence of electoral fraud makes it clear that Kennedy's overall 112,803 vote plurality was a myth: Nixon probably won by about 250,000 votes. Evidence of fraud in the two states was so blatant that a number of senior figures, including Eisenhower, urged Nixon to make a formal legal challenge to the result. But Nixon declined. There had never been a recount on a presidential election and the machinery for one did not exist. A study of procedures in six states where fraud was likely showed that every state had different rules for recounts, which could take up to eighteen months. A legal challenge, therefore, would have produced a 'constitutional nightmare' and worked heavily against the national interest. Nixon not only accepted the force of this argument but he actually pleaded successfully with the *New York Herald Tribune* to discontinue a series of twelve articles giving evidence of the frauds, when only four had been printed.[31]

So Kennedy now had the presidency. What was he to do with it? Bobby became attorney-general, and other members of the Kennedy extended household, the PR men, academics, assorted intellectuals, speechwriters, and fixers, moved into the White House and Washington. One person who knew exactly what to do in the White House, however, was the new President's wife, Jackie Bouvier Kennedy. She wanted to turn the Kennedy White House into what she called an updated version of King Arthur's mythic Camelot, and she succeeded. If the Sixties were an exciting and meretricious decade, Jackie's White House was the volcanic explosion of superficial cultural bustle which launched the age.

Like every other career move, Jack's marriage to Jackie had been decided by Old Joe.[32] The Bouviers were upper class but poor. Jackie inherited a mere $3,000. Her mother was a friend of the socialite wife of Arthur Krock, Joe's pal and Kennedy's ghostwriter, and he urged Joe to bring them together. Joe more or less ordered his son to marry Jackie, whom he recognized would give Jack the social and cultural graces he conspicuously lacked. Before the engagement was announced, Old Joe had a long conversation with Jackie in which he urged that, if she agreed to marry his son, she would never lack for anything money could buy for the rest of her life. After the marriage, Joe also, at various stages, held meetings with Jackie, who was threatening to go to the divorce courts because of Jack's infidelities, to renegotiate the financial conditions under which Jackie would agree to allow the marriage to continue. This covert relationship between Old Joe and Jackie continued until Joe was disabled by a stroke after his son became president, and it sometimes gave the impression that the marriage was a creation of Joe and Jackie rather than of Jack.[33]

The reason why Joe had not only to bring about but assiduously to maintain Kennedy's marriage was that the one respect in which Jack carried on his father's traditions not merely dutifully but with genuine enthusiasm was in his pursuit, seduction, and exploitation of women.

Old Joe went for women ruthlessly all his life, not hesitating to steal, or rather borrow, his sons' girlfriends if he could: co-ownership of women by different generations of males seems to have been a feature of the Kennedy family's sex life. Joe had employed his Hollywood power to get himself actresses and starlets, and had even possessed the redoubtable Gloria Swanson until she discovered he was buying her presents with her own money.[34]

Jack, in turn, used his political glamour to secure political trophies from the movies, including Gene Tierney and Marilyn Monroe, the latter first shared with, then passed onto, Bobby (the behavior of the Kennedy brothers towards this fragile movie-actress, and their cover-up of the traces after her sudden death, was one of the shabbiest episodes in the entire Kennedy story).[35] Most of Jack's affairs, however, were brief, often lasting a matter of minutes, and conducted with whomever was available and willing: girls without even first names, air stewardesses, secretaries, campaign workers, prostitutes if need be. He claimed he needed some kind of sexual encounter, however perfunctory, every day.

He committed adultery with various women throughout the 1960 campaign, preparing for each of the TV debates by having a prostitute, rustled up and paid by his staff, in the afternoon.[36] He had a woman even on the night of his inauguration as president, after Jackie had gone to bed. His tenure of the White House was punctuated by both regular and casual affairs, sometimes in his own room when Jackie was away, sometimes in 'safe' apartments, reached by tunnels under New York's Carlyle Hotel to escape observation. He did not always trouble to conceal what he was doing. Jackie retaliated by going on colossal shopping expeditions and by constantly redecorating all the houses they lived in, something which Kennedy hated anyway but particularly if he had to pay for it.[37]

What Kennedy and his associates brought the presidency was the mastery of public relations. This is a two-edged weapon. In some ways the most dangerous kind of politician is a man who is good at PR and nothing else—and in some respects J. F. Kennedy fitted that description exactly. But Jackie brought style, and gave the White House a *réclame* during the years 1961–3 which it had rarely possessed before and never since. Old Joe, in picking Jackie as his son's wife, had said he wanted 'looks, birth and breeding.' She had all three, and the breeding told. She was culturally omnivorous and made the President's home a center of the arts, at least in appearance. She encouraged all the best sides of her husband, who had not gone through the motions of a first-class education at Choate, Princeton, and Harvard without acquiring some worthy interests. She brought out his reading habits, his liking for intellectual company at times, his Anglophilia and Francophilia, his willingness to go to plays, operas, and the ballet, to entertain and to be gracious to a wide variety of people. She acted as a catalyst to the many clever men whom he brought into his circle and administration. Insofar as Kennedy was a respectable, useful, and valuable president, it was mainly Jackie's doing. Selfish, vain, and extravagant she undoubtedly was, then and later, but as chatelaine of the White House she gave the United States' first family something which it had nearly always lacked, sophistication and glamour, and as the President's wife she brought out in him such qualities as he possessed. And she gave her own creation, Camelot, a certain fierce, if brief and evanescent, reality, so that it burned with an intense flame for a year or two, until dramatically extinguished by murder.

Kennedy's actual presidency was dominated by competition with the Soviet Union, then run by the excitable, reckless but at times formidable Nikita Khrushchev. On January 7, 1961, shortly before Kennedy assumed office, Khrushchev outlined in a speech what he called the new areas of 'peaceful competition' between the two superpowers, 'national liberation wars' and 'centers of revolutionary struggle against imperialism,' all over Asia, Africa, and Latin America. Kennedy took up this challenge, at any rate in rhetoric, using his inaugural as the occasion. Aiming particularly at youth, he declared the time to be an 'hour of maximum danger' for freedom. His generation, he said, had been given the role of defending it. 'I do not shrink from this responsibility,' he said. 'I welcome it.' Under him, America would 'pay any price, bear any burden, meet any hardship, support any friend, oppose any foe, to ensure the survival and the success of liberty.'[38] That was good deal further than any other president in history had been prepared to commit himself abroad, and certainly much further than Truman and Eisenhower. It was also much further than the American people were prepared to go, as they would make abundantly clear long before the end of the decade. Kennedy sought to put flesh on the bones of his rhetoric by a variety of devices: the Peace Corps of young US volunteers to serve abroad, the Green Berets for more forceful activities termed 'counter-insurgency;' campaigns for winning 'hearts and minds' of locals in what were then known as 'non-aligned countries;' the 'Alliance for progress' in Latin America; and increased economic and military aid almost everywhere.[39]

One of Kennedy's ventures played straight into the hands of what the departing Eisenhower had warned against, the 'military–industrial complex.' The Soviet Union, with the help of spies and prodigies of effort and spending, had exploded first A-bombs, then H-bombs, long before American experts had believed possible. On October 4, 1957, thanks to the work of German scientists and engineers recruited by Stalin from Hitler's long-range rocket program, Russia put Sputnik 1, a 184-pound satellite, into orbit round the earth, following it next month by a larger one weighing 1,120 pounds. The first American satellite did not go into orbit until January 31, 1958, and it weighed only 30 pounds. In fact America was building far more efficient rockets than Russia, and was even more ahead in miniaturization, which explained why it was content with low payloads. Eisenhower was not prepared to invest heavily in space beyond the pragmatic needs of the military program. He detested the word 'prestige' and he took no notice of the post-Sputnik panic. With Kennedy the priorities changed totally. He put his V-P, Lyndon Johnson, in charge of the Space Program, with instructions to 'get ahead of Russia,' and the big-spending Texan, who had many links to the aerospace business world, was delighted to oblige. He picked James Webb, a publicity-conscious business operator, as head of the National Aeronautics and Space Administration, and spending, if not yet satellites, skyrocketed.

All the same, on April 12, 1961, less than three months after Kennedy had taken over, the Russians put their first man into orbit, beating America by four weeks. At a frenzied meeting Kennedy held two days later in the White House, he stormed: 'Is there any place where we can catch them? What can we do? Can we go around the moon before them? Can we put a man on the moon before them? . . . can we leapfrog? . . . If somebody can just tell me how to catch up! Let's find somebody, anybody. I don't care if it's the janitor there, if he knows how.'[40] On April 19 Kennedy had a forty-five-minute session with Johnson, followed by an excited directive the next day, ordering him to find out: 'Do we have a chance of beating the Soviets by putting a laboratory in space, or by a trip round moon, or by a rocket to land on the moon, or by a rocket to go to the moon and back with a man? Is there any other space program that promises dramatic results in which we would win?'[41] There is a sense in which Kennedy, who loved to use words like 'beating,' 'results,' and 'in,' was a professional sportsman, a political huckster, and a propagandist rather than a serious statesman.

In May he committed America to the Apollo Program, with its aim of landing a manned spacecraft on the moon 'before this decade is out.' The program got going fully in 1963 and for the next decade America spent up to $5 billion a year on space, a typical Sixties project, with its contempt for finance and its assumption that resources were limitless. Naturally the aim was achieved. On July 20, 1969 Apollo II landed Neil Armstrong and Edwin Aldrin on the moon. There were five more moon landings by 1972, when the program petered out. By then America and Russia had launched something over 1,200 space-probes and satellites, at a combined cost of something like $100 billion. In the more austere conditions of the mid–1970s, the space effort shifted from histrionics and propaganda to pragmatism and science, to space laboratories and shuttles. In 1981 America launched the first true spaceship, the Shuttle. The showbiz era of space travel was over.[42]

Kennedy's biggest test, his biggest apparent triumph, and his biggest mistake, was his handling of Cuba. Cuba was, and is, too close to the United States ever to be a fully independent country. Under the Platt Amendment of March 2, 1901, incorporated into the original Cuba Constitution, the US recognized Cuba's independence and sovereignty on condition Cuba agreed not to make treaties compromising that independence, leased America a base in Guantanamo Bay, and permitted the US to land troops in case of civil disorders, threats to US investments, and so on. The Platt Amendment ended in 1934, but as late as the 1950s the US ambassador in Havana was 'the second most important man in Cuba; sometimes even more important that [its] president.'[43] During the decade, however, under the rule of the so-called strongman, Fulgencio Batista, Cuba dissolved into a morass of corruption and gangsterism. Under Platt, the US would have intervened, legally, and restored decency. Without Platt, the US government could do little other than lecture and exhort.

But there remained the US media, growing steadily in strength, influence, and righteousness. Where Washington hesitated, the New York Times stepped in, in the person of its reporter Herbert Matthews. Matthews picked, as America's democratic candidate to topple Batista and replace him, Fidel Castro, a self-appointed guerrilla leader who had taken to the Sierra with 150 followers. The drawback to this policy was that Castro was not a democrat, but a believer in Marxism–Leninism, and in Stalinist 'democratic centralism,' and other authoritarian methods, and not least in violence. He was the spoiled son of a rich Spanish immigrant who had made a fortune out of a fruit plantation. Castro had flourished as an armed student revolutionary, and had committed a variety of crimes, including murder, in a number of Latin American countries. He was a Communist gangster, bent on achieving personal power at any cost. Matthews had not done his homework and knew little of Castro's record. He saw him as the T. E. Lawrence of the Caribbean, and was able to put the whole resources of the *New York Times* behind him, with the willing cooperation of certain key officials in the State Department. The pattern of muddle, duplicity, and cross-purposes leading to the fall of Batista and his replacement by Castro was one of the worst episodes of the entire Eisenhower administration and recalled FDR's diplomacy at its feeblest (it also foreshadowed the attempts of some State Department officials and US media elements to undermine the Shah of Iran in 1979). At all events it succeeded. Batista fled in January 1959, and Cuba was then at Castro's mercy. He was in effect put into power by the *New York Times*.[44]

At what point Castro acquired, or revealed, his Communist beliefs is still unclear, though from some of his statements it looks as if he was a Marxist–Leninist, of sorts, since early student days. By the second half of 1959 he had signed various agreements with Russia under which he received Soviet arms, advisors, and KGB experts to train his secret police. He had the army commander-in-chief murdered, clapped opponents of all kinds into jail, and staged show-trials of his principal enemies. By the beginning of 1960, Cuba was for all practical purposes a Communist dictatorship and, in a military perspective, a Soviet satellite.[45] At the same time Castro made the first moves to implement his threat, contained in a 4,000-word manifesto published in 1957, that he would actively oppose 'other Caribbean dictators.'[46] This was the beginning of a military–political policy which was to take members of Castro's armed forces into a number of African, Asian, and Latin American countries, as agents of 'anti-imperialist revolution.'

That such a regime, 90 miles from the United States coast, should align itself with America's principal enemy, and begin to export violence, was unacceptable, and the United States would have been well within its moral and legal rights in seeking to overthrow Castro and impose a democratic government. Cubans were already fleeing to the US in very large numbers, and demanding precisely that. But there was an uncertain response, first from Eisenhower, then from Kennedy. Eisenhower had taken a high moral line, during the Suez crisis of 1956, against the action of Britain and France in pursuing what they conceived to be their vital interest by occupying the Suez Canal Zone. The

Cuban case was made to seem too close for comfort. So nothing was done under Eisenhower, though many plans were considered.

When Kennedy took over early in 1961, he found a proposal, apparently supported by the CIA, which had 2,500 agents in the island, and by the chairman of the Joint Chiefs of Staff, for 12,000 armed Cuban exiles, known as the Cuban Liberation Corps, to be landed in an area called the Bay of Pigs and detonate a popular uprising. It is hard to believe that the wily and experienced Eisenhower would have given final approval to this naive scheme. It had all the disadvantages of involving America morally and politically (the first two men to step ashore were CIA operatives) without the guarantee of success provided by open US air and naval participation. Ike, on his record, would have waited for Castro to make the kind of false move which would have allowed America to intervene openly and legally, with its own forces, in a carefully planned professional air, land, and sea operation. One thing he always hated, in military matters, was amateurism, with politicians and generals getting their lines of command mixed up. Oddly enough, Kennedy's first response was similar. As he said to his brother Bobby, 'I would rather be called an aggressor than a bum.'[47] But in the event he lacked the resolution, and he weakly allowed the Bay of Pigs operation to go ahead on April 17, 1961.

The very name of the target area should have rung alarm-bells among Kennedy's PR-conscious inner circle. Its selection was also the one aspect of the plan the Chiefs of Staff most disliked. They were also worried about the CIA's insistence that the operation was to be preceded by an air strike launched from bases in Nicaragua, by US combat aircraft painted to resemble Cuban aircraft acquired by the exiles. Nor were they happy that US destroyers were to accompany the Cuban exile invasion fleet, with US aircraft providing air cover to within five miles of the landing place. In the event, the operation was a total disaster from the start, primarily because Castro was able to read all about it, in advance, in the US media; and, once things went wrong, Kennedy refused to authorize the US carrier *Essex*, cruising 10 miles offshore, to come to the rescue of the stricken men pinned down on the swampy bay. Castro's troops, well prepared for the incursion, killed 114 of the invaders and took prisoner the rest, 1,189, nearly all of whom were executed or later died in Castro's prisons.[48] Kennedy retrieved something from the mess by following advice to take responsibility for himself and publicly pointing out that, had the venture succeeded, no one would have criticized it, using an excellent one-liner produced by his resourceful writers: 'Success has a thousand fathers but failure is an orphan.' Actually Ike had the last laugh, albeit a sad one. Alluding to the title of Kennedy's book, he wrote: 'The operation could be called "A Profile in Timidity and Indecision." '[49]

American opinion was outraged by the Bay of Pigs failure and would have supported direct intervention. One senior policymaker, Chester Bowles, thought a decision by Kennedy 'to send in troops or drop bombs or whatever . . . would have had the affirmative votes of at least 90 percent of the people.' Richard Nixon, consulted, told the President: 'I would find a proper legal cover and I would go in.'[50] But the administration dithered and, lacking a policy, took refuge in plotting, encouraged by the wilder spirits of the CIA. Kennedy's Defense Secretary, Robert McNamara, admitted later: 'We were hysterical about Castro at the time of the Bay of Pigs and thereafter.' Richard Helms, whom Lyndon Johnson had made head of the CIA, testified: 'It was the policy at the time to get rid of Castro, and if killing him was one of the things that was to be done . . . we felt we were acting well within the guidelines . . . Nobody wants to embarrass a President . . . by discussing the assassination of foreign leaders in his presence.'[51]

In fact none of these schemes came to anything. In the event, it was Nikita Khrushchev, the impetuous Russian dictator, who gave Kennedy the opportunity to pull something out of the Cuban mess. Khrushchev had his own 'missile gap,' real or imaginary. He thought the balance of intercontinental missiles was heavily in America's favor, and that he could alter it overnight by stationing medium-range missiles in Cuba. It was an absurdly dangerous move, which was a principal factor in leading Khrushchev's colleagues to remove him, in due course. Castro later claimed he was opposed to it. He told two French journalists: 'The initial idea originated with the Russians and with them alone . . . it was not in order to insure our own defense but primarily to strengthen socialism on the international plane.' He agreed only 'because it was impossible for us not to share the risks which the Soviet Union was taking to save us . . . It was in the final analysis a question of honor.'[52]

It was on October 16, 1962 that Kennedy received incontestable evidence that Russia was preparing to install nuclear missiles in Cuba, that the sites were being prepared, and that the actual missiles and warheads were on the way. The scheme was as crack-brained as the Bay of Pigs venture but infinitely more dangerous, and accompanied by a combination of boasting and deception. Castro said that Khrushchev swanked to him that his move was something Stalin would never have dared. His colleague Anastias Mikoyan told a secret briefing of Soviet diplomats in Washington that it would bring about 'a definite shift in the power relationship between the socialist and the capitalist worlds.' Khrushchev deliberately lied to Kennedy: when questioned, he admitted Russia was arming Castro but gave assurances that only short-range surface-to-air missiles were being installed. The lie was peculiarly childish, since US aerial surveillance instantly revealed the truth. Khrushchev sent forty-two medium-range 1,100-mile nuclear missiles and twenty-four 2,200-mile nuclear missiles (the latter never arrived), together with twenty-four SAM anti-aircraft missile groups and 42,000 Soviet troops and technicians. There was never any possibility of concealing this large-scale strategic military activity, and all the sites were fully photographed by U2 aircraft on October 15, Kennedy being told the next day that, by December, about fifty strategic missiles would be deployed and ready for use, and that they would be capable of destroying the main US defenses in seventeen minutes. In fact by mid-October nine tactical missiles, equipped with nuclear warheads and with ranges of 30 miles, were already operational and their local Soviet commander had authority to use them at his own discretion.[53]

So on October 16 a thoroughly alarmed Kennedy set up an executive committee of the National Security Council, or Ex-Comm, to deliberate and decide policy. A wide range of opinions was canvassed, inside and outside the administration.[54] Adlai Stevenson advised demilitarization of Cuba, the surrender of the US base at Guantanamo, and the removal of US Jupiter missile bases from Turkey, in return for Soviet withdrawal of their missiles. At the other end of the spectrum, hardliners such as Dean Acheson and Vice-President Johnson advised immediate air-strikes to knock out the sites, followed by invasion of the island if necessary. A few of those consulted argued that America's overall strength in strategic weapons was so great that the provocation should be ignored. But, almost from the start, Kennedy rejected any such approach. He had been infuriated by Khrushchev's lying and was determined that the missiles had to go. But he was also persuaded by Secretary of State Dean Rusk, and his Under-Secretary, George Ball, that an unannounced air-strike on the sites would smack of Pearl Harbor and was not in the US tradition. As Rusk put it, 'The burden of carrying the Mark of Cain on your brow for the rest of your lives is something we all have to bear.' The military backed this view, pointing out that not all the Soviet missiles might be knocked out on a first strike and that their commanders might then fire them off at America, with devastating consequences. Bobby Kennedy also observed that Russia might retaliate by occupying Berlin.[55]

In the end, Kennedy and his colleagues decided to hold air-strikes in reserve, and in the meantime impose a 'quarantine' on Cuba, forbidding further Soviet ships to enter the area, on penalty of being fired upon. This was announced directly to Khrushchev in Moscow on October 22, a week after the alert, and at the same time publicly to the American people on prime-time TV. It was characteristic of Kennedy that he decided, having deliberated in secret, to play the crisis publicly and extract the maximum advantage from a tough but flexible and cautious line. He gave a deadline of October 24 for compliance with his quarantine, put in because it was essential to prevent Russians from working on the sites under cover of diplomatic delays. On the 24th Soviet missile-carrying ships approaching the quarantine line stopped, and slowly turned about. Other ships, not carrying war material, agreed to permit inspection. But it remained to get the Soviet missiles out. On the 25th Kennedy contacted Khrushchev again asking for 'a restoration of the earlier situation' (that is, removal of the missiles). Khrushchev demanded, in return, a pledge not to invade Cuba and the removal of the Jupiter missiles from Turkey. Kennedy ignored the second demand but complied with the first, and it was on this basis that Khrushchev agreed to remove the missiles on October 28.

This was what the world knew at the time, and it hailed a Kennedy victory, as did the President himself. Chortling over Khrushchev's discomfiture, he crowed, 'I cut his balls off.' Castro, not having been consulted by Khrushchev and learning the news on the radio, smashed a looking-glass, shouted abuse of his Soviet friend, and called him 'a man with no *cojones*.' The view of Khrushchev's colleagues was different but equally unfavorable. When the Soviet Praesidium dismissed him two years later, it referred to his 'harebrained scheming, hasty conclusions, rash decisions and actions based on wishful thinking.'[56] There is no doubt the world came close to nuclear war, probably closer than at any other time, before or since. On October 22 all American missile crews were placed on maximum alert. Some 800 B47s, 550 B52s, and 70 B58s were prepared with their bomb-bays closed for immediate takeoff from their dispersal positions. Over the Atlantic were ninety B52s carrying multi-megaton bombs. Nuclear warheads were made active on a hundred Atlas, fifty Titan and twelve Minuteman missiles, and on American carriers, submarines, and overseas bases. All commands were in a state of Defcon–2, the highest state of readiness next to war itself.[57] The world was not precisely aware of this at the time but most people imagined that war was close, and were correspondingly relieved when the stand-down came on October 28, and inclined to give Kennedy the credit for it. Then, and for some years afterwards, it was considered the finest hour of the Kennedy presidency.

From the perspective of over thirty years, the Cuban missile crisis looks different and its consequences mixed. In the first place, it now appears Kennedy privately agreed with Anatoly Dobrynin, the Soviet ambassador in Washington, using Bobby as intermediary, that he would pull out the Jupiters from Turkey (and Italy), and later did so. Second, nearly all the 42,000 Soviet troops and experts in Cuba remained and began intensive training of what was to become one of the largest and most mobile armies in the world, which would be exported as politico-military mercenaries, on anti-Western missions, to large parts of Africa and Asia in the late 1960s and throughout the 1970s. Third, Kennedy appears to have agreed with Krushchev to restrain any efforts by Cuban exiles to invade Cuba. They certainly regarded the agreement, at the time and since, as a sell-out. Their view was shared by some at least of the US force commanders, especially General Curtis LeMay, head of Strategic Air Command, who pounded the table and told Kennedy: 'It's the greatest defeat in our history, Mr President.'[58] LeMay was a hardliner, and excitable, and he exaggerated. It was not the 'greatest defeat.' But it was defeat. The missile crisis took place at a time when the strategic nuclear equation was still strongly in America's favor, and in a theater where America enjoyed overwhelming advantages in conventional power. Kennedy was thus in a position to demand an absolute restoration of the *status quo ante*, without any American concessions or pledges. Indeed he could have gone further; he could have insisted on punishment—on Soviet public acceptance of a neutral, disarmed Cuba. As Dean Acheson rightly observed: 'So long as we had the thumbscrew on Khrushchev, we should have given it another turn every day.'[59] Instead, Kennedy not only gave way over the Jupiters but acquiesced in the continuance of a Communist regime in Cuba in open military alliance with Soviet Russia.

But history moves in mysterious ways, its wonders to perform. From the perspective of the end of the century, it is no longer clear that the continuance of the Castro regime in Cuba was wholly to America's disadvantage. It is true that Castro became, for a quarter of a century, America's most persistent and (in some ways) successful minor enemy:

to export revolution to South America in the 1960s and, more cunningly, to Central America in the late 1970s and early 1980s; to vilify American 'imperialism' systematically at Third World gatherings, while posing as a 'non-aligned' nation; and, in the 1970s, to send no fewer than three expeditionary forces to Africa as executants of Soviet policy. With remarkable audacity, Castro posed as a defender of the oppressed in the United States itself, and was rewarded by the adulation of a segment of American liberal opinion. To Saul Landau, Castro was 'steeped in democracy.' To Leo Huberman and Paul Sweezy he was a 'passionate humanitarian.' Other US visitors testified to his 'encyclopedic knowledge.' He made them think of the 'connection between Socialism and Christianity.' He was, said US liberal visitors, 'soft-spoken, shy and sensitive' and, at the same time, vigorous, handsome, informal, undogmatic, open, humane, superbly accessible, and warm. Norman Mailer thought him 'the first and greatest hero to appear in the world since the Second World War.' When Castro stood erect, wrote Abbie Hoffman, 'he is like a mighty penis coming to life, and when he is tall and straight the crowd immediately is transformed.'[60] In the second half of the 1960s, throughout the 1970s, and for most of the 1980s, Castro and his propaganda provided the greatest single input into the machinery of anti-Americanism, both in the United States itself and throughout the world.[61]

But in the end Castro proved a diminishing asset for America's enemies, and increasingly a horrible warning to opponents of the free-enterprise system. As early as 1981 it was calculated that since Castro took over Cuba it had experienced an annual average per-capita growth-rate of minus 1.2 percent. By 1990 the minus growth-rate had increased to an average of over 2 percent. This was because the Soviets, who had been subsidizing the Cuban economy at the rate of $11 million a day in the 1980s, reduced and eventually cut off supplies. By the mid–1990s Cuba was calculated to have the lowest standard of living in the entire western hemisphere, Haiti being the only possible exception. The plight of the unfortunate Cubans, many of whom were reported to be going hungry in 1995–6, served to reinforce to Latin Americans generally the advantages of the free-enterprise economy, which became such a feature of the 1980s and 1990s. The ordinary Cubans had long since rejected Castro by voting with their feet and their outboard motors. In the 1960s alone over a million fled from Castro. By 1980, in which year 150,000 political refugees were added to the total, it was calculated that 20 percent of the nation were living abroad, most of them in the United States. Waves of Cuban refugees continued to reach America's shores in the 1980s and 1990s. Sometimes Castro attempted to poison this outflux by releasing common criminals from his jails (this was partly to create space for his 100,000 political prisoners) and by sending them to the US as 'political refugees.' But this was a policy of despair. In fact, the Cuban community in the United States grew and flourished. By the second half of the 1990s, it had founded 750,000 new businesses, become the richest and most influential political lobby after the Jewish Lobby, and its 2 million members generated a Gross Domestic Product eleven times larger than that of Cuba itself, with 11 million inhabitants. Moreover, Miami, center of the new Cuban settlement, forming links with the entire Latin American society of the hemisphere, became in many ways its financial, economic, communications, and cultural center, hugely boosting American exports in goods and, still more, in services throughout the western half of the globe. In the long run, then, the grand beneficiary of the Cuban missile crisis was indeed the United States.

In 1962–3, however, the last year of the presidency, Kennedy basked in the glory of having extricated the world gracefully from the most dangerous crisis in its history. He was popular everywhere. The Europeans liked him. He went to Germany and made a clever speech in West Berlin, June 26, 1963, in which he electrified the citizens by saying, 'All free men, wherever they may live, are citizens of Berlin, and therefore, as a free man, I take pride in the words, *Ich bin ein Berliner.*' This was bad German—it meant 'I am a doughnut'—but it showed good intent. He was able to announce on TV, July 26, 1963, important progress on nuclear weapons negotiations: 'Yesterday, a shaft of light cut into the darkness ... For the first time an agreement has been reached on bringing the forces of nuclear destruction under international control.' During the final months of his presidency, the legend of Camelot, under the skillful ministrations of Jackie Kennedy and the many expert wordsmiths and publicity men working in and surrounding the White House, went round and round the world, growing and glowing. But Kennedy had to remember that his majority in 1960, if it had existed at all, had been wafer-thin, and that much remained to be done to make his reelection in 1964 reasonably certain. In the second half of 1963 he planned many political expeditions into swing and marginal states, not least Texas, at the urging of his Vice-President, Lyndon Johnson, who was unhappy at the administrations's polling there, and by divisions in the Democratic Party.

Among other engagements, Johnson arranged for the President to visit Dallas on November 22, 1963 and process in a 10-mile motorcade through the downtown area. Kennedy remarked to his wife that morning, 'We're heading into nut country today,' a characteristic Bostonian summing up of one of the richest and most enterprising cities in America, the capital of the Bible Belt, surrounded by one of the largest collections of religious universities and training colleges in the world.[62] This was the pre-security era, and Kennedy was scarcely better protected than President McKinley when he was shot by an anarchist on September 6, 1901. The open car in which Kennedy rode had a protective bubble but the day was so fine that he asked for it to be removed so people could see him better. At 12.35, near the end of the motorcade, as it passed along Elm Street, a sniper on the sixth floor of the Texas School Book Depository fired three shots from a rifle. One hit the President in the back below his collar-bone, another in the back of the head, killing him, and a third wounded the Texas governor, John B. Connally.[63] The world was stunned.

The time of death was certified as 1.00 P.M. Ninety minutes later, Dallas police arrested Lee Harvey Oswald, a twenty-four-year-old former marine, who became a Soviet citizen in 1959 before returning to the United States in 1962, and working at the Depository. He was charged with killing a Dallas police officer, J. D. Tippitt, who had tried to detain him on suspicion of the assassination. Nine hours later he was formally accused of killing the President, a charge he denied. Two days later, a strip-joint owner called Jack Ruby, who was known to various police officers and therefore had access to the local station premises, approached Oswald as he was being transferred, pulled out a concealed handgun, and killed him. He said his motive was grief for the murdered President.

The fact that Oswald could never be brought to trial led Lyndon Johnson, who was sworn in as president aboard Air Force One ninety-eight minutes after the certification of Kennedy's death, to appoint a commission of inquiry headed by the Chief Justice, Earl Warren (1891–1974). The inquiry, which included a senior Southern senator, Richard Russell of Georgia, a future president, Gerald Ford (b. 1913), and the former head of the CIA Allen Dulles, reported (September 1964) that the facts had been as the police had charged and that Oswald had committed the crime and acted alone. It stated that Oswald was an evident 'loner,' a self-styled Marxist, a Castro supporter, who tried and failed to murder General Edwin Walker, a notorious Dallas right-winger. The authorities in Russia, where he had lived for thirty-two months, had regarded him as an unstable character and kept him under surveillance. No evidence was found that he acted in concert with anybody or any organization, but the rifle which killed Kennedy, found in the Depository, had Oswald's palm-print on its stock. These findings disappointed the conspiracy theorists though without silencing them. In more than thirty years since the crime took place, however, no further evidence of any significance has emerged, and virtually all historians now accept that Oswald alone was responsible.[64]

Kennedy left behind him a number of growing but unsolved problems: a war in Vietnam, a civil-rights agitation in large parts of the South, and a rising demand, throughout the United States, for better social security provision. His successor, Lyndon Johnson (1908–73), was well qualified to tackle them, better qualified in many ways than Kennedy. The thirty-sixth president had been born on a farm near Stonewall, Gillespie County, Texas, and had moved with his parents to Johnson City five years later. He attended public schools in Blanco County, Texas, graduated from Southwest Texas State Teaching College at San Marcos in 1930, teaching high school for three years, 1928–31. He served in Washington as secretary to Congressman Richard M. Kleberg (1931–5), meanwhile attending Georgetown Law School (1934), and was then appointed by F. D. Roosevelt, who liked and admired him, state director of the National Youth Administration of Texas, 1935–7. He was elected to Congress April 10, 1937, by special election, to fill the vacancy caused by the death of James Buchanan and was subsequently reelected to five successive Congresses, serving April 1937 to January 1949. He was the first member of Congress to enlist after the beginning of World War Two and served as lieutenant-commander in the US Navy 1941–2. He was elected to the US Senate in 1948 and reelected 1954 and 1960, serving as Democratic whip 1951–3, minority leader 1953–5 and majority leader 1955–61, when he was elected vice-president. He had been a member or chairman of a number of important Senate committees and subcommittees.

Johnson had thus been a professional politician virtually all his life, had a thorough grounding in the politics of Texas, one of the largest, richest, and most complex of states, was well versed in Washington executive politics, and was a master of Congressional procedure. Few presidents have ever been better grounded for the post, and none had possessed such skill in piloting legislation through Congress.[65] Johnson appealed strongly to senior fellow-professionals. When he was only twenty-six, FDR had picked him out as a possible successor. His fellow-Texan, Sam Rayburn, Speaker of the House from 1940, had treated him like a son, and in the Senate he had from the start enjoyed the patronage of Senator Richard Russell, leader of the Southern Democratic bloc. Johnson ate, drank, and played recklessly, but he also worked with fanatical industry and singlemindedness, both to enrich himself and to push his way up the political ladder. He regularly lost two stone in weight during his campaigns, and he was perfectly capable of getting through a fifteen- or eighteen-hour day for weeks at a time.[66] He picked fellow-professionals to serve him and exacted the highest standards of devotion and loyalty from them all. He had little but contempt for his predecessor: '[Kennedy] never said a word of importance in the Senate and never did a thing . . . It was the goddamnest thing . . . his growing hold on the American people was a mystery to me.'[67]

There was a dark side to Johnson. He was unscrupulous. In Texas politics, where he acted as a fund-raiser for FDR, he was closely linked to his contractor ally, Brown & Root, for which he had negotiated enormous government contracts to build the Corpus Christi Naval Air Station. The company illegally financed Johnson's unsuccessful 1941 Senate campaign, and from July 1942 IRS agents began to investigate both them and LBJ himself. They found overwhelming evidence not only of fraud and breaches of the Hatch Act in the use of campaign money, but of lawbreaking in many other aspects of Brown & Root's business, including tax evasion of over $1 million. Both LBJ and Herman Brown, head of the firm who had begun life as a two-dollar-a-day rod carrier for a surveyor, could have gone to jail for many years. The investigation was derailed as a result of the direct intervention of FDR himself, January 13, 1944, and the matter ended with a simple fine: no indictment, no trial, and no publicity.[68] After this, LBJ was involved in various Texan political intrigues of a more or less unlawful nature and in building up a personal fortune (most of it in the name of his wife, the long-suffering 'Lady Bird' Johnson) in radio–TV stations and land.

There was also the case of Robert G. ('Bobby') Baker, a gangling South Carolinan who had served Johnson as secretary and factotum in the years when he was Senate leader. Baker was known, on account of his power and influence, as 'the hundred and first Senator,' and LBJ said of him, fondly: 'I have two daughters. If I had a son, this would be the boy . . . [He is] my strong right arm, the last man I see at night, the first one I see in the morning.' In the autumn of 1963, a private suit against Baker in a federal court, alleging that he had improperly used his influence in the Senate to obtain defense contracts for his own vending-machine firm, provoked a spate of similar accusations against his probity, and in a number of them LBJ was involved. The accusations were so serious that, just before his assassination, Kennedy was considering dropping LBJ from his 1964 ticket, even though he feared that to do so would imperil his chances of carrying Texas and Georgia.[69] At Republican urging, the Senate agreed to investigate the case. But by that time LBJ was president and the full weight of his office was brought to bear to avoid the need for testimony either from Johnson himself or from his aide Walter Jenkins, who possessed a good deal of guilty knowledge. The Senate committee, on which Democrats outnumbered Republicans six to three, voted solidly on party lines to protect the President.

To make sure the truth did not emerge, LBJ directed two of his closest allies, Clark Clifford (b. 1906), later Defense Secretary, and Abe Fortas (1910–82), later Supreme Court Justice, to organize a cover-up. (It is notable that the careers of both these men ended in scandals.) Among other things, the most pertinacious of LBJ's pursuers, Senator John J. Williams of Delaware, was subjected to a White House dirty-tricks campaign, which included persecution by the IRS. Baker was finally indicted in 1967 on nine charges, convicted and, after appeals, went to jail in 1971, actually serving seventeen months. LBJ avoided investigation, trial, and jail completely, though the facts of his involvement are now well established.[70] Among the Senators who played important roles in protecting LBJ from exposure were Sam Ervin, Herman Talmadge, and Dan Inouye, all of whom, and especially the first, were assiduous in hounding President Nixon during the Watergate case in 1973.[71]

LBJ was also a man of unpleasant, sometimes threatening, personal habits. He was huge and, despite his height, his head seemed too big for his body. It was equipped with enormous ears, which stood out like those of an angry African elephant. Unlike Kennedy, who normally received guests in the Oval Office, with others present, LBJ saw people alone, in an adjoining closet, with four TV sets, each adjusted to a different channel and transmitting, though as a concession to visitors he would turn the sound down. He propelled himself forward, until his large head and penetrative nose were only inches from the visitor's face. As Doris Kearns, the most intimate of his biographers, put it, he 'invaded your personal space.'[72] Members of Congress, from whom LBJ wanted a service or a vote, thus 'closeted' (a form of political pressure first invented by Louis XIV in the 17th century), emerged shaken and compliant. Some of his staff were frankly terrified of him. He said of one, about to join his entourage, 'I don't want loyalty. I want *loyalty*. I want him to kiss my ass in Macy's window at high noon and tell me it smells like roses. I want his pecker in my pocket.'[73] Actually, what LBJ's pockets contained were not peckers but innumerable bits of paper, which he produced to impress or convince interlocutors, containing statistics, quotes, extracts from FBI or CIA files, and his own medical records. He was particularly anxious to persuade people that, despite his heart attack in 1955, he was fit, and he showed no hesitation in flourishing his cardiographs and other charts of his condition. He had no respect for privacy, his own or anyone else's. He was more devoted to the phone than any other president, before or since, and insisted phones be installed in all rooms, including those of his staff's private houses, the bathroom not excluded. He would summon staff to confer with him, or receive orders, while seated on the toilet, another characteristic he shared with Louis XIV.[74]

LBJ, the large, unrestrained, earthy animal, had a voracious sexual appetite, no more discriminating than Kennedy's, but less interesting. He had a twenty-one-year affair (1948–69) with a Dallas woman called Madeleine Brown, which produced a son, Steven, and countless more transitory encounters, including (so he boasted) intercourse with a secretary at his desk in the Oval Office. He was an inveterate bottom-pincher, especially in swimming pools.[75] In view of LBJ's constant philandering, especially on White House premises, it is curious that no scandal erupted during his five years as president (the only exception was, ironically enough, a homosexual episode, involving LBJ's chief of staff, Walter Jenkins, which broke on October 14, 1964). LBJ's immunity was due to his close relationship to J. Edgar Hoover, whom LBJ, like Kennedy before him, retained as head of the FBI for sound personal reasons. He defended his decision in a memorable and characteristic quip: 'Better to have him inside the tent pissing out, than outside, pissing in.'[76]

Another factor in LBJ's immunity was the acquiescence of Lady Bird, who told a TV producer, long after LBJ's death: 'You have to understand, my husband loved people—all people. And half the people in the world are women. You don't think I could have kept my husband away from half the people?' Lady Bird, like Jackie Kennedy, operated as a civilizing influence in the Johnson White House. It was she who repaired the culinary damage after the highly sophisticated Kennedy chef walked out, soon after the Johnson regime began, in protest at LBJ's insistence that beetroot salad be served at every meal. The Johnsons entertained prodigiously: over 200,000 guests to meals during their five White House years.[77] A Texas woman who had spent much of her childhood in Alabama, Lady Bird, like her husband, used the Southern idiom to great effect, but in contrast to his coarseness, her vernacular brought delight. 'See you tomorrow, if the Lord be willin' and the creek don't rise.' 'I'm as busy as a man with one hoe and two rattlesnakes.' 'The kind of people who would charge hell with a bucket of water.' 'Nosier than a mule in a tin barn.' 'Doesn't the fire put out a welcoming hand?' Praising people she met (very characteristic) she would say: 'I find myself in mighty tall cotton.'[78] She also devoted herself, with remarkable success, to unusual good causes, such as the sowing of seeds of wildflowers on the verges of the intercontinental highway system Eisenhower had built, a project she developed with the help of the English scientist Miriam Rothschild.[79]

But if Lady Bird was the emollient, LBJ was the master, and the White House, during his tenancy, became the most active engine for passing legislation, and spending money, in the entire history of peacetime America. No president, not even Woodrow Wilson in the prewar period, provided so adept at getting Congress to do his bidding. And his bidding was imperiously ambitious. LBJ, despite his occasional boasts of childhood poverty, had come from a reasonably comfortable, though modest, home. But as a young schoolteacher he had become aware of the extensive poverty and deprivation which, in the United States, coexisted with general affluence. He was a generous man, who had done well for himself and wanted to share the animal comforts of affluence with as many people as possible. By temperament and conviction he was a Big Spender and he thought America, in the mid–1960s, after twenty years of uninterrupted growth and prosperity, was in a position to resume the New Deal on a much more ambitious scale— FDR being always his hero and exemplar. He thought nothing of Kennedy's anemic program, the 'New Frontier,' most of which had got stuck in the Congressional quagmire anyway. LBJ insisted that it was time to give all Americans the benefits of a fully financed welfare state. In a speech at the University of Michigan, May 22, 1964, he declared: 'In your time we have the opportunity to move not only towards the rich society and the powerful society, but upward to the Great Society.'

By 'great' he meant magnanimous, and magnanimity, greatness of heart, was LBJ's redeeming, indeed salient, characteristic. He reiterated this pledge at a preelection rally on October 31, 1964: it was the central theme of his bid, after inheriting the presidency from Kennedy, to get a full mandate of his own. And he succeeded in abundant measure. He was assisted by the decision of the Republicans, who did not think they could beat LBJ anyway, to nominate a quixotic ideologue and perceived extremist from a small state, Senator Barry Goldwater of Arizona (b. 1909). Johnson's overwhelming victory gave him 43,128,958 votes to 27,176,873 for Goldwater and a college margin of 486–52.[80]

Thus mandated, as he saw it, LBJ began his legislative and spending spree, calling on Congress to enact his Great Society program on January 3, 1965. Congress complied. The Great Society, as LBJ conceived it, was concerned not merely with raising the poor and ending the economic anxieties of all, but with improving the quality of life, including access to power of the blacks and the ability of all to exercise their civil rights. The program actually began to be enacted before LBJ's stunning electoral victory, and it continued until his retirement. The Civil Rights Act (1964) and the Voting Rights Act (1965) were the most important civil rights enactments, and undoubtedly achieved their main objectives, helped by the adoption in 1964 of the Civil Rights Amendment, the Twenty-fourth, which ruled poll taxes and other tax barriers to voting unconstitutional. Medicare and Medicaid (1965) and the Older Americans Act (1965) went some way to reducing anxieties over health care. The Omnibus Housing Act (1968) provided $6 billion to build housing for poor and middle-income families and introduced rent-supplement allowances. It was expanded by the Housing and Urban Development Act (1968). The Demonstration Cities and Metropolitan Development Act (1966) provided funds to abolish ghettos, improve urban transport and landscaping, plant parks and other urban amenities, initially in six cities, eventually in 150. It supplemented the Mass Transit Act (1964), which made the first major commitment of federal funds ($375 million) to subsidize bus, subway, and rail commuter systems.

LBJ had announced, in a speech to Congress on March 16, 1964, 'For the first time in our history, it is possible to conquer poverty,' and he declared an 'unconditional war on poverty.' This produced legislation in the form of the Equal Opportunity Act (1964) and the Appalachian Regional Development Act (1965), which together allocated over $2 billion a year to various programs, and the Head Start Act (1965) and Higher Education Act (1967), which introduced better educational opportunities for the poor and less affluent. The Great Society program also included important environmental legislation, including the Clean Water Restoration Act (1966), which put $4 billion into anti-pollution control projects, and the Wilderness Areas Act (1964), which banned development from over 9 million acres of public domain; these designated wilderness areas by 1994 had risen to 602, totaling 95.8 million acres. Johnson's measures involved a gigantic bureaucratic expansion, including the establishment of the Department of Housing and Urban Development and the Department of Transportation, as well as such federal-funded bodies as the National Endowments for the Humanities and the Arts, and the Corporation for Public Broadcasting. It also spawned bodies and agencies on the lines of Kennedy's Peace Corps, with names like Volunteers in Service to America, the Jobs Corps, Upward Bound and the Model Cities Program.[81]

All this cost a great deal of money. During the five-year Johnson administration, federal spending on education, for instance, rose from $2.3 billion to $10.8 billion, on health from $4.1 billion to $13.9 billion, and on the disadvantaged from $12.5 billion to $24.6 billion. In current dollars, the rise of federal spending under Johnson was enormous—to $183.6 billion by fiscal 1969. Thanks to Johnson, by 1971 for the first time the federal government spent more on welfare than defense. In the thirty years 1949–79, defense costs rose ten times, from $11.5 billion to $114.5 billion, but remained roughly 4 to 5 percent of GNP. Welfare spending, however, increased twenty-five times, from $10.6 to $259 billion, its share of the budget went up to more than half, and the proportion of the GDP it absorbed tripled to nearly 12 percent.[82] Some of Johnson's programs, being mere fashions, or to use the new 1960s word, 'trends,' petered out by the end of the decade. But most expanded, and some went out of control financially.

Critics of these projects coined the phrase 'throwing money at problems,' and certainly quantities of money were thrown in the second half of the decade by the federal government. And as the cost increased, so did the volume of critical voices, which complained that the prodigious expenditure was not producing results.[83] Whether it did or not, the actual expenditure introduced a revolution in the history of public finance in the United States. We will deal with its long-term implications later, but at this point it is necessary to draw attention to two aspects. First, was it within the intentions of the Founding Fathers that the Constitution should permit the federal government, as opposed to the individual states, to shoulder the main burden of welfare, as opposed to such obviously federal outgoings as national defense? If not, then the whole of the Great Society program was unconstitutional. Second, since the reforms of Alexander Hamilton in the 1790s, American federal finance had been conducted, on the whole, with exemplary prudence. The national debt had risen in times of war (the Revolutionary War, the Civil War, World War One, World War Two) and times of extreme emergency (the Great Depression). But it had been reduced in times of peace. Under Andrew Jackson, in 1840, it had been eliminated altogether, as we have noted. During the Civil War it rose to over $2.76 billion, but in the next twenty-eight years it was reduced by two-thirds, though the GDP in the meantime had more than doubled. The debt rose to $25 billion at the end of World War One, and was then reduced by a third in the 1920s. It rose again during the Great Depression to $48 billion in 1939, and during World War Two it rose again, very fast, to reach $271 billion by 1946. It then fell again, both in constant dollars and as a proportion of GDP, until 1975. Since then, without either war or national emergency, it has risen inexorably.[84]

Strictly speaking, then, the change in the nature of US federal financing did not actually occur under LBJ. But permanent deficit financing did. Under him, the annual deficit jumped to $3.7 billion in 1966, $8.6 billion in 1967, and $25.1 billion in 1968.[85] In March 1968, the Treasury Secretary, Henry Fowler, protested strongly to President Johnson that, with the cost of the Vietnam War soaring, and various domestic programs roaring ahead, the dollar would soon come under serious threat. The result was the enactment of tax increases the same year, which actually produced a surplus of $3.2 billion in fiscal 1969, the last federal budget surplus of the 20th century. But meanwhile the Great Society programs continued to absorb more and more money, and by fiscal 1975 federal spending had risen to $332 billion, causing a huge budget deficit of $53.2 billion. It was at this point that serious efforts to control expenditure and bring the budget back into balance were abandoned (as we shall see) by a weak executive and a triumphalist, irresponsible Congress.[86] Johnson can likewise be held responsible for the critical increase in the federal government's share of GDP, which had peaked at 43.7 percent in 1945, fell to 16 percent in 1950, then slowly rose again. It reached 20 percent under LBJ and, under the impact of his programs, rose again to 22 percent in 1975, stabilizing in the years down to 1994 at between 22 and 24.4 percent.

*   *   *

The Treasury Secretary's alarm in 1968 was actually provoked by fresh administration demands for a larger military commitment in Vietnam, and that is the tragedy we must now examine. America's involvement in Vietnam stretched over twenty years (1954–75). American military personnel began to serve there in 1954 and the last fifty were evacuated on April 30, 1975. In all, 8,762,000 Americans performed Vietnam-era military service: 4,386,000 army, 794,000 marines, 1,740,000 air force, and 1,842,000 navy. Of these, about 2 million servicemen actually fought in Vietnam or operated offshore. There were 47,244 US battle-deaths in all services, 153,329 hospitalized wounded, 150,375 'lightly wounded,' and 2,483 missing. The weight of firepower used by the US military was enormous: the navy and air force carried out 527,000 bombing missions, unloading 6,162,000 tons of explosives, which was three times the tonnage dropped by US bombers in World War Two. Vietnamese losses were calamitous. About 300,000 civilians were killed in South Vietnam and 65,000 in North Vietnam. South Vietnamese forces lost 223,748 killed and 570,600 wounded. North Vietnamese casualties were estimated at 660,000 killed, the number of wounded being unknown. The war's direct expense to the United States was $106.8 billion.[87] These figures do not include America's more limited, but still significant, involvement in the fighting in Cambodia and Laos. But this, too, had long-term consequences for the US. About 1.5 million refugees left the Indochinese states after America finally pulled out in 1975, many of them departing in small craft and becoming known as the Boat People. The United States began allowing large numbers of Boat People to immigrate in 1978. Over the next six years it admitted 443,000 Vietnamese, 137,000 Laotians, and 98,000 Cambodians. The influx slowed to about 30,000 yearly in the late 1980s, and the 1990 US census showed that 905,512 Indochinese were now American citizens, thus adding yet another ethnic layer to America's multiracial mix.[88]

In its duration, the number of Americans involved, its costs, and its consequences, Vietnam was not only the longest but one of the most important wars in American history. Yet strictly speaking it was not a war at all, and certainly was never waged as such by the Americans— otherwise the outcome would have been entirely different. It was an 'involvement,' and a product of the Cold War. America's involvement stretched over seven presidencies and was a unique succession of misjudgments, all made with the best intentions. Indochina was a 19th-

century French colony, ruled by Paris with a variable degree of administrative efficiency, corruption, and altruism. It was occupied by the
Japanese in 1941 and F. D. Roosevelt, knowing nothing about it,
offered it to nationalist China. Generalissimo Chaing Kai-shek turned
it down, remarking to FDR that the Indochinese were quite different
people to the Chinese, and that he wished to have nothing to do with
them. Immediately after FDR's death, the fervent anti-colonialists of
the Office of Strategic Services, predecessor-but-one of the CIA,
worked hard to set up a left-wing nationalist regime, which would prevent the French from returning. Three weeks after the Japanese surrender, the Communist leader Ho Chi Minh, sponsored by the OSS, staged
a *putsch*, known as the 'August Revolution,' which ousted the pro-
French Emperor of Vietnam. The man who, in effect, crowned Ho Chi
Minh as the new ruler was an OSS agent, Archimedes Patti.[89]

It is important to emphasize that America never had any territorial
ambitions in Indochina, either as a base or in any other capacity. But its
policy was often founded on ignorance, usually muddled and invariably indecisive. Truman, on taking office, was advised that the first priority was to get France back on its feet as an effective ally against
Russia in Europe, and to bolster its self-confidence it was convenient
for it to be allowed to resume authority in Indochina. In December
1946, with American approval, the French drove Ho Chi Minh back
into the jungle and brought the Emperor Bao Dai back from exile in
Hongkong. It was at this point that the French created three puppet
nations, Vietnam, Cambodia, and Laos, and gave them independent
status within the French Union, February 7, 1950. America acquiesced
in this step. Russia and China recognized Ho's regime, and began to
arm it. America responded by doing the same for the French regime,
and, with the outbreak of the Korean War, US aid accelerated. In 1951
it was giving $21.8 million in economic and $425.7 million in military
assistance. By 1952 America was paying 40 percent of France's military
costs. Dean Acheson was warned by State Department officials that the
US was drifting into a position in which it would eventually supplant
France as the 'responsible power' in Indochina. But he replied that
'having put our hand to the plough, we could not look back.' By
1953–4 America was paying for 80 percent of the French war effort.[90]

On May 8, 1954 the French suffered a catastrophic defeat when
their 'impregnable' fortress at Dien Bien Phu surrendered. The French
asked for the direct assistance of US airpower, and when Eisenhower
turned the request down they formed a new government and opened

negotiations to withdraw. The ceasefire agreement, signed at Geneva in July, provided for a division of the country along the 17th parallel, the Communists keeping the north, the West the rest, unity to be brought about by elections in two years' time under an international control commission. Eisenhower, having rejected US military intervention, refused to be a party to the Geneva accords either, and created SEATO instead.[91] A protocol attached to the new treaty designated South Vietnam, Cambodia, and Laos as areas whose loss would 'endanger' the 'peace and security' of the signatories. This was the expression of the then-fashionable Domino Theory. Eisenhower had made this theory public in April 1954: 'You have a row of dominoes set up. You knock over the first one, and what will happen to the last one is a certainty that it will go over very quickly. So you could have the beginning of a disintegration that would have the most profound consequences.' He also spoke of 'a cork in a bottle' and 'a chain-reaction.'[92] Beware public men when they use metaphors, especially mixed ones! Ike not only ignored the accords but encouraged the new Prime Minister of the South, Ngo Dien Diem, to refuse to submit to the test of free elections, which if held in 1956 would probably have handed over the country to Ho, until the South was economically stronger and more stable. As a result, the Communists created a new guerrilla movement for the South, the Vietcong, which emerged in 1957 and started hostilities. Eisenhower in effect made America a party to the war by declaring, April 4 1959: 'The loss of South Vietnam would set in motion a crumbling process that could, as it progressed, have grave consequences for us and for freedom.'[93]

It was under Kennedy and Johnson, however, that the American tragedy in Vietnam really began to unfold. When Kennedy reached the White House, Vietnam was already one of America's largest and costliest commitments anywhere in the world, and it is hard to understand why he made no attempt to get back to the Geneva accords and hold free elections, which by that stage Diem might have won. In Paris on May 31, 1961, General de Gaulle, who knew all about it, urged him to disengage: 'I predict you will sink step by step into a bottomless military and political quagmire.'[94] Kennedy had a hunch that Southeast Asia would prove a trap. The Bay of Pigs débâcle made him think twice about further involvement, especially in Laos, where the Communists were threatening. He told Sorenson in September 1962, 'Thank God the Bay of Pigs happened when it did. Otherwise we'd be in Laos by now—and that would be a hundred times worse.'[95] He also said to

Arthur Schlesinger that he was worried about sending in troops to Vietnam. 'The troops will march in; the bands will play; the crowds will cheer; and in four days everyone will have forgotten. Then we will be told to send in more troops. It's like taking a drink. The effect wears off, and you have to take another.'[96]

Nevertheless, in November 1961 Kennedy did send in the first 7,000 US troops to Vietnam, the critical step down the slippery incline into the swamp. That was the first really big US error. The second was to get rid of Diem. Diem was by far the ablest of the Vietnam leaders and had the additional merit of being a civilian. Lyndon Johnson, then Vice-President, described him with some exaggeration as 'the Churchill of Southeast Asia,' and told a journalist, 'Shit, man, he's the only boy we got out there.'[97] But Kennedy, exasperated by his failure to pull a resounding success out of Vietnam, blamed the agent rather than the policy. In the autumn of 1963 he secretly authorized American support for an anti-Diem officers' coup. It duly took place on November 1, Diem being murdered. The CIA provided $42,000 in bribes for the officers who set up a military junta. 'The worst mistake we ever made' was Lyndon Johnson's later verdict.[98]

Three weeks later Kennedy was murdered himself, Johnson was president, and began to make mistakes on his own account. Warning signals ought to flash when leaders engage in historical analogies, especially emotive ones. LBJ compared the risk of Vietnam going Communist to the 'loss' of China in 1949: 'I am not going to lose Vietnam,' he declared. 'I am not going to be the President who saw Southeast Asia go the way China went.' He drew, for the members of the National Security Council, a still more dangerous parallel: 'Vietnam is just like the Alamo.'[99] And again, to the so-called 'Tuesday Cabinet:' 'After the Alamo, no one thought Sam Houston would wind it up so quick.'[100] But this was just big talk. LBJ proved as indecisive over Vietnam as Kennedy was, and in addition he made the further mistake (like Eden over Suez) of imagining he could run the war on political principles.

Johnson continued the war in desultory fashion until August 2, 1964, when North Vietnam attacked the American destroyer *Maddox*, which was conducting electronic espionage in the Gulf of Tonkin. Hitherto Johnson had been reluctant to escalate. But he now summoned Congressional leaders and, without disclosing the nature of the *Maddox* mission, accused North Vietnam of 'open aggression on the high seas.' He then submitted to the Senate a resolution authorizing him to take 'all necessary measures to repel any armed attack against the forces of the United States and to prevent further aggression.' Senator William Fulbright of Arkansas, chairman of the Senate Foreign Affairs Committee, who steered the 'Tonkin Resolution' through Congress, said it effectively gave Johnson the right to go to war without further authorization. Only two senators voted against. Later, when more information became available, many in Congress argued that Johnson and his advisors had deliberately misled senators into supporting the expansion of the war. In fact Johnson did nothing for six months. He was fighting an election campaign against Senator Goldwater, who openly advocated the bombing of North Vietnam. Johnson sensed the public would be unhappy about getting into 'another Korea' and played down the war during the campaign. In fact if anything he advocated a 'peace' line during the campaign, just as Wilson had done in 1916 and Roosevelt in 1940. Then, having won his overwhelming victory, he did the opposite—again like Wilson and Roosevelt. In February 1965, following heavy US casualties in a Vietcong attack on a barracks, he ordered the bombing of the North.[101]

This was the third critical American mistake. Having involved itself, America should have followed the logic of its position and responded to aggression by occupying the North. LBJ should have put the case to the American people, and all the evidence suggests that, at this stage, the people would have backed him. They were looking for leadership and decisiveness. The military were quite open to the politicians about the problem. The Joint Chiefs reported on July 14: 'There seems to be no reason we cannot win if such is our will—and *if that will is manifested in strategy and tactical operations.*' The emphasis was in the original.[102] When Johnson asked General Earl Wheeler, of the JCS, 'Bus, what do you think it will take to do the job?,' the answer was 700,000 to a million men and seven years.[103] Johnson was not prepared to pay this bill. He wanted victory but he wanted it on the cheap. That meant bombing; it was the old reflex of the Jupiter Complex, the belief that America, with its superior technology, could rain punishment on the evildoers from the air, without plunging wholly into the mud of battle. LBJ believed the SAC commander, Curtis LeMay, who said that Vietnam could be 'bombed back into the Stone Age.'[104]

But there was no attempt to bomb Vietnam back into the Stone Age. Even in selecting the cheap bombing option, LBJ was indecisive. He was not misled by the generals. The air force told him they could promise results if the offensive was heavy, swift, repeated endlessly without pause, and without restraint. That was the whole lesson of World War Two. They promised nothing if it was slow and restricted. The navy took exactly the same view. Admiral Ulysses S. Grant Sharp, naval commander in the Pacific, said: 'We could have flattened every war-making facility in North Vietnam. But the handwringers had center stage . . . The most powerful country in the world did not have the will-power to meet the situation.' Instead it was a case of 'pecking away at seemingly random targets.'[105] From start to finish, the bombing was limited by restrictions on quantity, targets, and timing which were entirely political and bore little relation to tactics or strategy. Every Tuesday Johnson had a lunch at which he determined targets and bomb-weights. Johnson was not always the ruthless man he liked to impersonate. He could be paralyzed by moral restraints. As his biographer Doris Kearns shrewdly observed, to him 'limited bombing was seduction, not rape, and seduction was controllable, even reversible.'[106] So the bombing was intensified very slowly and the North Vietnamese had time to build shelters and adjust. When Soviet Russia moved in defensive ground-to-air missiles, American bombers were not allowed to attack while the sites were under construction. There were, in addition, sixteen 'bombing pauses,' none of which evoked the slightest response, and seventy-two American 'peace initiatives,' which fell on deaf ears.[107] Unlike the Americans, the North Vietnamese leaders never once wavered in their determination to secure their clear political aim—total domination of the entire country—at any cost. They do not seem to have been influenced in the smallest degree by the casualties their subjects suffered or inflicted.

There was thus a bitter irony in the accusations of genocide which came to be hurled at the Americans. An examination of classified material in the Pentagon archives revealed that all the charges made against American forces at the 1967 Stockholm 'International War Crimes Tribunal' were baseless. For instance, evacuation of civilians from war zones to create 'free fire' fields not only saved civilian lives but was actually required by the 1949 Geneva Convention. The heavy incidence of combat in civilian areas was the direct result of Vietcong tactics in converting villages into fortified strongholds, itself a violation of the Geneva agreement. And it was the restrictions on American bombing to protect civilian lives and property which made it so ineffective. The proportion of civilians killed, about 45 percent of all war deaths, was about average for 20th-century wars. In fact the population increased steadily during the war, not least because of US medical programs. In the South, the standard of living rose quite fast.[108]

All this went for nothing. The experience of the 20th century shows that self-imposed restraints by a civilized power are worse than useless. They are seen by friend and enemy alike as evidence not of humanity but of guilt and of lack of moral conviction. Despite them, indeed because of them, Johnson lost the propaganda battle, at home and abroad. And it was losing the battle at home which mattered. Initially, the war had the support of the media, including the moderate liberal consensus. Two of the strongest advocates of US involvement were the *Washington Post* and the *New York Times*. The *Post* wrote, April 7, 1961, 'American prestige is very much involved in the effort to protect the Vietnamese people from Communist absorption.' The *New York Times* argued, March 12, 1963, that 'The cost [of saving Vietnam] is large, but the cost of Southeast Asia coming under the domination of Russia and Communist China would be still larger.' On May 21, 1964 the *Times* urged: 'If we demonstrate that we will make whatever military and political effort [denying victory to Communism] requires, the Communists sooner or later will also recognize reality.' The *Post* insisted, June 1, 1964, that America continue to show in Vietnam that 'persistence in aggression is fruitless and possibly deadly.' But the *Times* deserted Johnson early in 1966, the *Post* in summer 1967.[109] About the same time the TV networks became neutral, then increasingly hostile.

There were three reasons why first the media, then public opinion, turned against the war. The first was that the bombing campaign, chosen as an easy option, proved to be not so easy after all. To bomb the North, a huge airbase had to be created at Da Nang. Once the bombing started, the base had to be protected. So on March 3, 1965, 3,500 marines were landed at Da Nang. By April, the total US troop level in Vietnam had risen to 82,000. In June a demand came for forty-four more battalions. On July 28 Johnson announced: 'I have today ordered to Vietnam the Airmobile division and certain other forces which will raise our fighting strength ... to 125,000 men almost immediately. Additional forces will be needed later, and they will be sent as requested.'[110]

The more Americans became involved in ground defense and fighting, the more of them were wounded and killed. The term 'body bags' came into general and ominous use. Selective Service, or the draft, reimposed at the time of the Korean War and extended periodically by Congress since, became increasingly unpopular. It was widely believed to be unfair and easily avoided by the rich and well connected. Of the 2 million conscripted during the Vietnam War, providing 23 percent of all military personnel and a massive 45 percent of the army, no fewer than 136,900 refused to report for duty, an unprecedented number. Draft calls had been only 100,000 in 1964. By 1966 they had risen to 400,000. Draftees made up the bulk of the US infantry riflemen in Vietnam, as many as 88 percent by 1969, and accounted for over half the army's battle deaths. Because of student deferments, the draft fell disproportionately on working-class white and black youths. Blacks, 11 percent of the US population, made up 16 percent of the army's casualties in Vietnam in 1967, and 15 percent for the entire war. So opposition mounted, accompanied by clergy-led protests, the burning of draft-cards, sit-ins at induction centers, and burglaries of local draft-board premises, with destruction of records.[111]

The government responded, between 1965 and 1975, by indicting 22,500 persons for violations of the draft-law. Some 6,800 were convicted and 4,000 imprisoned. But the Supreme Court expanded the basis on which the draft could be refused from purely religious to moral and ethical reasons, and 'Conscientious Objections' grew, in relation to those summoned to the forces, from 8 percent in 1967 to 43 percent in 1971. In the last five years of the 1960s, about 170,000 of those registered qualified as conscientious objectors. That made those who actually served even angrier. In addition, about 570,000 evaded the draft illegally. Of these, 360,000 were never caught, 198,000 had their cases dismissed, and about 30,000–50,000 fled into exile, mainly to Canada and Britain. The worst thing about the draft was that the rules were frequently changed or evaded so that young people felt that the way in which they were being recruited and sent to an unpopular war was unAmerican and unfair. There is no doubt that the draft was behind the unprecedented rise in anti-war protest among American youth in the late 1960s.[112]

The second reason why America turned against the war was not so much editorial criticism as tendentious presentation of the news. The US media became strongly biased in some cases; more often it was misled, skillfully and deliberately, or misled itself. A much publicized photograph of a 'prisoner' being thrown from a US helicopter was in fact staged. Accounts of American 'tiger cages' on Con Son island were inaccurate and sensationalized. Another widely used photo of a young girl burned by napalm gave the impression, in fact quite wrong, that many thousands of Vietnamese children had been incinerated by Americans. A photo, used many times, of a youth being shot in cold

blood by one of America's Vietnamese 'allies'—and apparently genuine—gave the misleading impression that captured Vietcong were
habitually executed. Once the TV presentation of the war became daily
and intense, it worked on the whole against American interests. It generated the idea that America was fighting a 'hopeless' war. Not only
did the media underplay or ignore any US successes, it tended to turn
Vietcong and North Vietnamese reverses into victories.

Media misrepresentation came to a decisive head in the handling of
the Vietcong 'Tet Offensive' on January 30, 1968. The military position at this time was that the Americans and their Vietnamese allies,
having strongly established themselves in all the urban centers of the
South, were winning important successes in the countryside too. That
persuaded the Communists to change their tactics, and try their first
major offensive in the open. On the first day of Tet, the lunar New
Year holiday previously observed as a truce, their units attacked five of
Vietnam's six cities, most of its provincial and district capitals, and fifty
hamlets. The Vietnam forces and units of the Army of the Republic of
Vietnam (ARVN), though taken by surprise, responded quickly. Within
a week they had regained all the ground the attackers had won, except
in one town, Hué, which was not retaken until February 24. Media
coverage concentrated on the fact that the Vietcong enjoyed initial successes in attacking the Government Palace in Saigon, the airport, and
the US embassy compound, and the cameras focused on the continued
fighting in Hué rather than US successes elsewhere. In military terms,
the Tet Offensive was the worst reverse the Vietcong suffered throughout the war: they lost over 40,000 of their best troops and a great number of heavy weapons.[113] But the media, especially TV, presented it as a
decisive American defeat, a Vietcong victory on the scale of Dien Bien
Phu. An elaborate study of the coverage, conducted in 1977, showed
exactly how this reversal of the truth, which was not on the whole
deliberate, came about.[114]

The image not the reality of Tet was probably decisive, especially
among influential East Coast liberals. In general American public opinion backed forceful action in the war. According to the pollsters, the
only hostile category was what they called the 'Jewish sub-group.'
Johnson's popularity rating rose whenever he piled on the pressure: it
leaped 14 percent when he started the bombing.[115] He was criticized for
doing too little, not for doing too much: what the polls showed was
that Americans hated the indecisiveness in Washington. Despite the
draft, support for intensifying the war was always greater among the

under-thirty-fives than among older people, and young white males were the group most consistently in favor of escalation. Among the people as a whole, support for withdrawal never rose above 20 percent until after the November 1968 election, by which time the decision to get out had already been taken. The American citizenry were resolute, even if their leaders were not.[116]

The crumbling of American leadership began in the last months of 1967 and accelerated after the media reaction to Tet. The Defense Secretary, Clark Clifford, turned against the war; so did old Dean Acheson. Even Senate hardliners began to oppose further reinforcements. Finally Johnson himself, diffidently campaigning for reelection, lost heart on March 12, 1968, when his vote sagged in the New Hampshire primary. He said he had decided not to run for reelection and would spend the rest of his term trying to make peace. It was not the end of the war by a long way, but it was the end of America's will to win it. The trouble with the Washington establishment was that it believed what it read in the newspapers—always a fatal error for politicians. New Hampshire was presented as a victory for peace. In fact, among the anti-Johnson voters, the hawks outnumbered the doves by three to two.[117] Johnson sagged in the primary, and so lost the war, because he was not tough enough.

The faltering of Johnson's once strong spirit in the Vietnam War was the result of media criticism, especially from its East Coast power-centers, which itself reflected a change in American attitudes towards authority. One of the deepest illusions of the Sixties was that many forms of traditional authority could be diluted—the authority of America in the world, and of the President within America—without fear of any consequences. Lyndon Johnson, as a powerful and in many ways effective president, stood for the authority principle. That was, for many, a sufficient reason for emasculating him. Another was that he did not share East Coast liberal assumptions, in the way that F. D. Roosevelt and Kennedy had done. He had been doubtful about running for president even in 1964 for precisely that reason: 'I did not believe that . . . the nation would unite definitely behind any Southerner. One reason . . . was that the Metropolitan press would never permit it.'[118] The prediction proved accurate, though its fulfillment was delayed. By August 1967, however, the Washington correspondent of the *St Louis Post-Dispatch*, James Deakin, reported: 'the relationship between the President and the Washington press corps has settled into a pattern of chronic disbelief.'[119] If media misrepresentation of the Tet Offensive was the immediate cause of LBJ's decision to quit, both office and Vietnam, more fundamental still was its presentation of any decisive and forceful act of the Johnson White House as in some inescapable sense malevolent. The media was teaching the American people to hate their chief executive precisely because he took executive decisions.

This sinister development in American history became more pronounced when Johnson yielded office to Richard Nixon. Nixon had suffered some reverses since he lost—or at any rate conceded—the 1960 election. But he never gave up. Nor did the East Coast media stop loathing him, or he reciprocating the feeling. In 1962 he ran for governor of California, and largely because of the Cuban missile crisis lost the race to a weak left-wing Democratic candidate, Pat Brown, who turned out to be one of the worst governors in California's history. The media campaign, led by the East Coast reporters, had been particularly unfair, and Nixon snapped out to the press afterwards: 'Just think how much you're going to be missing. You won't have Nixon to kick around any more because, gentlemen, this is my last press conference.'[120] In fact it was not the end, but the beginning, and kicking around of Nixon by the media was to go on for another decade and more. The catastrophic defeat of Goldwater in 1964 persuaded the Republican bosses that they would have to field an experienced, mainstream candidate against Johnson in 1968. So Nixon made his comeback without much difficulty; his only mistake was to accept, as running mate, the crass and dishonest, but right-wing, Spiro Agnew, governor of Maryland.

The stepping down of Johnson plunged the Democrats into confusion. A strong contender for the succession was Bobby Kennedy, for the Camelot myth was then still powerful; but he was assassinated by Sirhan Sirhan, a Marxist Palestinian refugee, during the California primary on June 6. That made LBJ's Vice-President, Hubert Humphrey, the front runner, and he duly got the Democratic nomination. He was an experienced campaigner and might conceivably have won if LBJ had given him the loyal support he had a right to expect and, in particular, made it easier for him to advocate withdrawal from Vietnam. But Johnson was a bitter man by now and not averse to seeing the Republicans take over. In addition, the South ran a breakaway candidate, Governor George Wallace of Alabama, then still a segregationalist. As a result, Nixon won comfortably, by 31,710,470 votes to Humphrey's 30,898,055, with Wallace collecting 9,466,167: the electoral college votes went 302 (Nixon), 191 (Humphrey), 46 (Wallace).[121]

All the same, the Wallace intervention meant that Nixon was a minority victor. He got 43.4 percent of the popular vote, the lowest percentage since Woodrow Wilson won the three-man race of 1912. As the poll was low (61 percent), it meant that Nixon got only 27 percent of all voters. What sort of a mandate was that? asked the hostile media, pointing out that Nixon did not carry a single big city.[122] In parts of the media, there was an inclination to deny the legitimacy of Nixon's presidency and to seek to reverse the verdict by non-constitutional means. 'Remember,' Nixon told his staff, 'the press is the enemy. When news is concerned, nobody in the press is a friend. They are all enemies.'[123] That was increasingly true. As one commentator put it, 'The men and the movement that broke Lyndon Johnson's authority in 1968 are out to break Richard Nixon in 1969 . . . breaking a president is, like most feats, easier to accomplish the second time round.'[124]

It was something new for the American media to wish to diminish the presidency. Hitherto, opposition to a strong chief executive had come, as was natural, from Congress, and especially from the Senate. As FDR put it, 'The only way to do anything in the American government is to bypass the Senate.' That was one point on which his opponent, Wendell Wilkie, agreed with him: he spoke of devoting his life to 'saving America from the Senate.'[125] Under both FDR and Truman, the press and academic commentators had strongly supported firm presidential leadership, especially in foreign policy, contrasting it with Congressional obscurantism. During the McCarthy era, Eisenhower had been fiercely criticized for failing to defend executive rights against Congressional probing, the *New Republic* complaining (1953): 'The current gravitation of power into the hands of Congress at the expense of the Executive is a phenomenon so fatuous as to be incredible if the facts were not so patent.'[126] When Eisenhower finally invoked 'executive privilege' to deny information about government activities to the UnAmerican Activities Committee, he was warmly applauded by the liberal media. The committee, said the *New York Times*, had no right 'to know the details of what went on in these inner Administration councils.' Eisenhower, wrote the *Washington Post*, 'was abundantly right' to protect 'the confidential nature of executive conversations.' Until the mid–1960s, the media continued to support resolute presidential leadership, on civil rights, on social and economic issues, and, above all, on foreign policy, endorsing Kennedy's dictum (1960): 'It is the President alone who must make the major decisions on our foreign policy.'[127]

It was the Tonkin Gulf Resolution which accelerated the change, and the accession of Nixon which confirmed it. Despite these handicaps, Nixon's first administration was, on the whole, successful in clearing up the problems left by the Kennedy and Johnson administrations. His senior White House men, Bob Haldeman and John Erlichman, were able and devoted—too devoted, perhaps. He had a brilliant National Security Advisor in Henry Kissinger. He brought in one or two clever and original Democrats, such as Daniel Patrick Moynihan. His speechwriters, who included William Safire, Pat Buchanan, Ray Price, David Gergen, and Lee Huebner, were probably the best team of its kind ever assembled.[128] Nixon and Kissinger between them developed the first clear geopolitical strategy for America since the retirement of Eisenhower. Nixon did not want to devote too much time to Vietnam: he wanted to 'wind it down' as it was essentially a 'short-term problem.' He wanted to concentrate on the things that really mattered: the Atlantic Alliance, relations with China and Russia. He believed that friendly relations with China could be brought about by intelligent diplomacy, that China would then be separated from Russia, and that, in the end, the regime in Russia, which he regarded as fundamentally inefficient, could be fatally undermined—something which indeed came about in the 1980s. He and Kissinger were at one on all these matters.[129] And they had considerable success. Nixon called Vietnam by Ike's phrase, 'a cork in a bottle,' and he did not want to pull the cork out too quickly. But he began the process of disengagement. He said, 'We seek the opportunity for the South Vietnamese people to determine their own political future without outside interference,' and so long as he was fully in charge of US policy he stuck to this resolve. He regarded the existence of the Vietcong and North Vietnam Army (NVA) sanctuaries in Cambodia as such 'interference' and he decided to bomb them. He regarded such an act as neither an extension of the war nor an unwarranted invasion of another country's neutrality. The neutrality had already been invaded by the Communists, and it was right to bomb them. But, not wishing to incur the charge that he was 'extending' the war, he had the bombing conducted secretly. Indeed, it was not even noted in the military's own records.[130]

At the same time, Nixon scaled down the US presence. In four years, he reduced American forces in Vietnam from 550,000 to 24,000. Spending declined from $25 billion a year under Johnson to less than $3 billion. This was due to a more intelligent use of US resources in the area. They became more flexible, being used in Cambodia in 1970, in Laos in 1971, in more concentrated bombing of North Vietnam in 1972, all of which kept the determined men in Hanoi perplexed and apprehensive about America's intentions.[131] Nixon actively pursued peace negotiations with the North Vietnamese, without much optimism, but he also did something neither Kennedy nor Johnson had dared: he exploited the logic of the Sino-Soviet dispute and reached an understanding with China. It was Nixon's Californian background which inclined him towards Peking. He saw the Pacific as the world arena of the future. He began his new China policy on January 31, 1969, only eleven days after he started work in the White House. The policy was embodied in National Security Memorandum 14 (February 4, 1969). He was much impressed by a conversation he had with the French politician and Sinologist André Malraux, who told him it was a 'tragedy' that 'the richest and most productive people in the world' should be at odds with 'the poorest and most populous people in the world.'[132] Reports of serious clashes on the long Sino-Soviet border in central Asia offered an opportunity for Nixon to offer China a friendly hand. He warned the Soviet leaders secretly (September 5, 1969) that the US would not remain indifferent if they were to attack China: Kissinger rightly described this move as 'perhaps the most daring step of his presidency.'[133] His first moves towards China were also secretive and he went to considerable lengths to get pledges of silence from the Congressional leaders he consulted. He told his staff: 'A fourth of the world's people live in Communist China. Today they're not a significant power, but 25 years from now they could be decisive. For the US not to do what it can at this time, when it can, would lead to a situation of great danger. We could have total *détente* with the Soviet Union but that would mean nothing if the Chinese are outside the international community.'[134]

The new China policy, and the change in US military strategy, made possible peace with Hanoi. On January 22, 1973, in Paris, Nixon's Secretary of State William Rogers and Le Duc Tho of North Vietnam signed 'An Agreement on Ending the War and Restoring Peace in Vietnam.' The merit of this understanding, which made it possible for America to leave Vietnam, was that it reserved Nixon's right to maintain carriers in Vietnamese waters and to use aircraft stationed in Taiwan and Thailand if the accords were broken by Hanoi.[135] So long as Nixon held power, that sanction was a real one. Granted the situation he had inherited, and the mistakes of his predecessors, Nixon had performed a notable feat of extrication.

But America, and more tragically the peoples of Indochina, were denied the fruits of this successful diplomacy because, by this point, the Nixon administration was already engulfed in the crisis known as Watergate. This was the culmination of a series of assaults on authority which had its roots in the Sixties culture. Indeed in some respects the challenge to authority went back to the 1950s and, in its early stages at least, had the approbation, if not the outright approval, of the federal establishment. It started among blacks (and their white liberal allies), mainly in the South, and emerged from the frustration felt by many blacks at the slow pace of their acquisition of civil rights, especially educational and voting ones, through court process and legislative change. The first outbreak of physical activity—we will look at the legal position later—began in Montgomery, Alabama on December 1, 1955. Rosa Parks, a black woman, refused to give up her bus seat to a white rider, thereby defying a Southern custom which required blacks to yield seats at the front to whites. When she was jailed, a black boycott of the company's buses was begun, and lasted a year. 'Freedom riders' and sit-in movements followed in a number of Southern states. The boycott, which ended in a desegregation victory for blacks (December 1956) was led by Martin Luther King (1929–68), from Atlanta, Georgia, who was pastor of a black Baptist church in Montgomery. King was a non-violent militant of a new brand, who followed the example of Mahatma Gandhi in India in organizing demonstrations which used numbers and passive resistance rather than force and played the religious card for all it was worth. In 1957 he became the first president of a new umbrella group, the Southern Christian Leadership Conference. King's house was bombed and he and some colleagues were convicted on various conspiracy charges, but King slowly emerged as a natural leader whom it was counterproductive for Southern sheriffs and courts to touch, and whose outstanding oratory was capable of enthusing enormous black crowds.[136]

The Montgomery boycott was followed by the first anti-segregation sit-in, February 1, 1960, when four black college students asked to be served at Woolworth's whites-only lunch-counter in Greensboro, North Carolina, were turned down, and refused to leave. Sit-ins spread rapidly thereafter. On May 4, 1961 the tactic was reinforced when the first Freedom Riders, seven blacks and six whites, tested equal access to services at bus terminals from Washington DC to New Orleans. These tactics proved, on the whole, highly successful, with private companies usually ending segregation after initial resistance (and some white violence), and the courts ruling that discrimination was unlawful anyway. Such activities almost inevitably involved the use or threat of force, or provoked it, and, as the Sixties wore on, urban violence between blacks and whites grew, and King himself came under competitive threat from other black leaders, such as the black racist Malcolm X and the proponent of 'black power,' Stokely Carmichael. It was important for King to be seen to succeed in getting civil rights legislation through Congress. He backed the campaign of James Meredith (1961–2) to be enrolled at the whites-only University of Mississippi, and Meredith succeeded in getting himself admitted (and graduating). But his appearance on campus led to riots, September 30, 1962, in which two died, and Meredith himself was later shot and wounded on an anti-segregation 'pilgrimage' from Memphis to Jackson, Mississippi (June 6, 1966).

In the spring of 1963 King's organization embarked on a large-scale desegregation campaign in Birmingham, Alabama, which produced scenes of violence, widely shown on TV, and some memorable images of the local white police chief, Eugene 'Bull' Connor, directing his water cannon and police dogs at the black protesters. King was briefly jailed and later an attempt to bomb him led to the first substantial black mob riot of the campaign (May 11, 1963). Mass demonstration in black communities throughout the nation culminated on August 28, 1963 in a march of 250,000 protesters, led by King, to the steps of the Lincoln Memorial in Washington DC, where King delivered a memorable oration on the theme of 'I have a dream that my four little children will one day live in a nation where they will be judged not by the color of their skin but by . . . their character.'[137] This demonstration was part of the process which led to the passing of the Civil Rights Act (1964). The Act restored the federal government's power to bar racial discrimination for the first time since the 19th century. Title II requires open access to gas stations, restaurants, lodging houses and all 'public accommodations' serving interstate commerce, and places of entertainment or exhibition. Title VI forbids discrimination in programs accepting federal funds. Title VII outlaws any employment discrimination and creates the Equal Employment Opportunity Commission.[138]

The Civil Rights Act gave the blacks much more than they had ever had before, but not everything, and the rise of the Black Panther party, with its strategy of 'picking up the gun,' inspired militancy in many inner cities and serious rioting. King was under pressure from both the advocates of violence, to hit harder at Washington, and from the administration, to cool things down. His relations with Lyndon

Johnson, once warm, deteriorated and he went in fear of his life from black extremists (his rival, Malcolm X, had been murdered by a black Muslim on February 21, 1965). King was involved in a bitter sanitation workers' strike in Memphis when he made his last prophetic speech, April 3, 1968: 'We've got some difficult days ahead, but it doesn't matter with me now because I've been on the mountain-top.' The next day he was assassinated.[139]

One reason King and LBJ moved apart was that King was lending his name and presence increasingly to anti-Vietnam demonstrations. Black leaders were against the war because of the high incidence of casualties (and draft service) among blacks. Black demonstrations and anti-Vietnam demonstrations thus tended to merge, especially when linked to the new forms of student protest which emerged in the Sixties. Black militants had been active among students, forming the Student Non-Violent Coordinating Committee (1960), the work of King's colleague in the Christian Leadership Conference, Ella Baker, who was one of the organizers of the early sit-ins. This committee, under James Forman, Bob Moses, and Marion Barry (later mayor of Washington DC), was behind many of the Southern protests in Mississippi and Alabama, drawing students into the protest movement, which theoretically eschewed force but became involved in scenes of violence, sometimes ending in death. The same year, mainly white student activists formed Students for a Democratic Society, the anchor-body for what became known as the New Left. With the student population growing by hundreds of thousands every year during the Sixties, the opportunities for mass demonstrations grew, especially under the stimulus of the Vietnam War. The New Left's first success was the Free Speech Movement in the University of California at Berkeley (1964), in protest at restrictions on student involvement in politics. The collapse of the authorities at Berkeley was followed by further student riots at Columbia University (1968), at Harvard (1969), and on many other campuses, in virtually all of which student leaders won huge concessions. If the 1960s saw the degradation of authority, it was in many instances a self-degradation, the men in authority breaking and running at the first whiff of student grapeshot.[140]

As the Sixties turned into the Seventies, the trend was for campus demonstrations to acquire coordination and specific political purpose, and so to become correspondingly more violent and alarming to those in power. When President Nixon appeared on national TV, April 30, 1970, to announce draft-extensions on account of the trouble in

Cambodia, there was an organized series of demos at campuses all over America, some of which degenerated, or were pushed, into riots. At Kent State University in Ohio, students set fire to the local army cadet building, and this act of arson led the governor of Ohio to send in 900 National Guardsmen, who occupied the campus. President Nixon had made his feelings plain the day after his TV address, when he was accosted in the Pentagon by the wife of a serving soldier, and told her he admired men like her husband: 'I have seen them. They are the greatest. You see these bums, you know, blowing up the campuses. Listen, the boys that are on the college campuses today are the luckiest people in the world, going to the greatest universities, and here they are burning up the books . . . Then out there [in Vietnam] we have kids who are just doing their duty. And I have seen them. They stand tall and they are proud.'[141] The media reporting of these remarks, which accused Nixon of calling all students 'bums,' further inflamed the campus rage. On May 4, some of the young guardsmen on duty at Kent reacted to rocks thrown at them by students by firing a volley into the crowd, killing four students, two of them girls, and injuring eleven more. This in turn detonated student riots in many campuses: over 200 incidents involving the burning, ransacking, and destruction of university property were recorded. The National Guard were called out to restore order at twenty-one campuses in sixteen states. Over 450 colleges had to be closed down for a spell. And Nixon's contrast between privileged students engaged in nihilism and hard-working kids from poor families getting on with life struck home. On May 7 in New York a crowd of construction workers stormed City Hall and beat up students who were occupying it—the first 'hard hat' demo against the New Left. This was Nixon's 'silent majority' beginning to react, adumbrating his historic landslide victory of November 1972.[142]

The hostility of the media towards Nixon and his administration, which became more and more intense in 1970–2, mingled with the attacks by the new youth culture on authority of any kind, gave a misleading impression that Nixon was in trouble. It led the Democrats, in 1972, to permit themselves the indulgence of a candidate who was popular with the students and the liberal media, George McGovern of South Dakota. His platform was an immediate and unconditional withdrawal from Vietnam and an increase in welfare spending. Nixon was delighted. He told his staff: 'Here is a situation where the Eastern Establishment media finally has a candidate who almost totally shares their views.' The 'real ideological bent of the *New York Times*, the *Washington Post*, *Time*, *Newsweek* and the three TV networks' was 'on the side of amnesty [for draft-dodging], pot, abortion, confiscation of wealth (unless it is theirs), massive increases in welfare, unilateral disarmament, reduction in our defenses and surrender in Vietnam.' At last, he concluded, 'the country will find out whether what the media has been standing for during these last five years really represents the majority thinking.'

The election of 1972 was not just a contest between an unelected media and the 'silent majority,' though certainly the East coast media did its best to lose it for Nixon. There were many solid grounds for an administration victory. The reopening of links to China, which was Nixon's personal policy, went down well. Summiting in Moscow, which he had also practiced, was popular too. The return of many troops from Vietnam and the reduction in drafting were felt and welcomed. Inflation came down to 2.7 percent in the summer of 1972; GNP was currently growing at 6.3 percent annually; real incomes were rising by 4 percent annually; and since 1969 federal taxes had been cut by 20 percent for the average family. Stocks were rising and passed the 1,000 mark for the first time just before the election. Hence the resulting landslide. Nixon's own score of 60.7 percent of the vote was not quite as high as LBJ's in 1964, but McGovern's 29.1 percent was the lowest score ever recorded by a major party candidate.[143] Nevertheless, this election underlined an increasing tendency in America for voters to split their tickets. While losing the presidency by a landslide the Democrats actually gained one Senate seat, thus strengthening their hold there, and while losing twelve House seats they comfortably retained control there too. The Democrats had now dominated Congress, with brief periods of Republican rule under Truman and Eisenhower, for an entire generation, and their continued grip on the Senate, in particular, was to prove fatal to Nixon.

But at the time it was the triumphant Nixon who seemed to be in control, and his success not only humiliated the media liberals but actually frightened them. As one powerful editor put it, 'There's got to be a bloodletting. We've got to make sure that nobody ever thinks of doing anything like this again.'[144] The aim was to use the power of the press and TV to reverse the electoral verdict of 1972 which was felt to be, in some metaphorical sense, illegitimate—rather as conservative Germans, in the 1920s, had regarded the entire Weimar regime as illegitimate, or Latin American army generals, in the 1960s and 1970s, regarded elected but radical governments as illegitimate. The media in the 1970s, rather like the Hispanic generals, felt that they were in some deep but intuitive sense the repository of the honor and conscience of the nation and had a quasi-constitutional duty to assert it in times of crisis, whatever the means or the consequences.

This view was given some spurious justification by what was coming to be called the 'Imperial presidency.' That the power of the executive had been growing since Woodrow Wilson's time, with dips in the Twenties and again in the late Forties and Fifties, was undeniable. And, as already noted, the media was beginning to conceive its duty to be the critical scrutiny of an over-active presidency rather than the goading of a comatose legislature. The war in Vietnam, which necessarily increased the activity, spending power, and decision-making of the presidency, made it seem more formidable than it actually was. Kennedy, in taking over, had been shaken by the number of things over which the President had no power. This had led him to accelerate a process which was already under way—the expansion of the White House bureaucracy. Lincoln had paid for a secretary out of his own pocket. Hoover had had to struggle hard to get three. Roosevelt appointed the first 'administrative assistants,' and World War Two brought a big increase in staff. All the same, the Truman White House was not overmanned, and lack of enough people probably accounted for the inadequate supervision which Truman blamed for corruption within the lower levels of the administration uncovered in 1950–2.[145] Eisenhower reorganized the White House staff and expanded it. But it was under the Kennedy regime that the real inflation began. He had, for instance, twenty-three administrative assistants alone, and at the time of his death the White House personnel numbered 1,664. Under LBJ it jumped to forty times its size in Hoover's day. Under Nixon it rose again, to 5,395 in 1971, the cost jumping from $31 million to $71 million.[146] There was nothing necessarily alarming in this. The administration as a whole had grown in size, enormously, and there was more for the White House to supervise. Congress, too, was growing, perhaps faster, as senators and representatives took on more 'staffers' and 'research assistants'—its cost was to top a billion a year by the end of the 1970s. But the increase in security, following the Kennedy killing, made more notable and visible the portentous manner in which the presidency now carried on and moved around. Kissinger, in particular, promoted to secretary of state, liked to travel with a huge entourage, which raised eyebrows. In a curious way, the position of the presidency in the early 1970s recalled the supposed rise of royal power under George III in the 1760s, which provoked the famous House of Commons motion, 'The power of the Crown has increased, is increasing and ought to be diminished.'

Nixon's reciprocal hostility to the media and his unwillingness to trust them, even more pronounced than under LBJ, persuaded some editors that 'something was going on,' which fitted into their other critical assumptions on what they termed the 'Nixon regime.' And of course something was going on. The White House was a power center engaged in all kinds of activities which would not always bear scrutiny. It necessarily engaged, in a wicked, actual world, in the *realpolitik* which was theoretically banned by an idealistic Constitution. This was a problem which had plagued presidents since Washington's day. But some presidents had taken positive pleasure in unlawful skulduggery. Bad habits had set in under FDR. He had created his own 'intelligence unit,' responsible only to himself, with a staff of eleven and financed by State Department 'Special Emergency' money.[147] He used J. Edgar Hoover's FBI, the IRS, and the Justice Department to harass his enemies, especially the press and business, and to tap their phones, the mineworkers' leader John L. Lewis being one victim. FDR's use of the IRS to 'get' names on his 'enemies' list' was particularly scandalous and unlawful. FDR had made persistent efforts to penalize the *Chicago Tribune*, which he hated, in the courts, and to get the *New York Times* indicted for tax fraud.[148] He even used the intelligence service to bug his wife's hotel room.[149] Though Truman and Eisenhower, who hated underhand dealings, kept clear of clandestine activities by their staff and the CIA, as a rule, they were generally aware of them and considered that, in dealing with Soviet Russia and other totalitarian terror regimes, they were unavoidable. Kennedy and his brother Bobby positively revelled in the game, and Kennedy's chief regret was that he had not made his brother Bobby head of the CIA, to bring it under family control. Kennedy had been privy to CIA assassination plots and had been a party to the coup which led to the killing of his ally Diem, though he had opposed the murder itself.[150] At the Justice Department, Bobby Kennedy in 1962 had agents carry out dawn raids on the homes of US Steel executives who had opposed his brother's policies.[151] In their civil rights campaign, the Kennedy brothers exploited the federal contract system and used executive orders in housing finance (rather than legislation) to get their way.[152] They plotted against right-wing radio and TV stations.[153] They used the IRS to harass 'enemies.' Under

Kennedy and Johnson, phone-tapping increased markedly.[154] So did 'bugging' and 'taping.' JFK's closest aides were stunned to learn in February 1982 that he had taped no fewer than 325 White House conversations. The large-scale womanizing of Martin Luther King was taped and played back to newspaper editors. The efforts made by LBJ to protect himself from the Bobby Baker scandal, already mentioned, included the unlawful use of secret government files, the IRS, and other executive devices.

Until the Nixon era, the media was extremely selective in the publicity it gave to presidential wrongdoing. Working journalists protected Roosevelt on a large number of occasions, over his love affairs and many other matters.[155] They did the same—and more—for Kennedy. The fact that Kennedy shared a mistress, while he was president, with a notorious gangster, though known to several Washington journalists, was never published in his lifetime. In Johnson's struggle to extricate himself from the Bobby Baker mess, the *Washington Post* actually helped him to blacken his chief accuser, Senator Williams. Nixon enjoyed no such forbearance. On the contrary. The anti-Nixon campaign, especially in the *Washington Post* and the *New York Times*, was continual, venomous, unscrupulous, inventive, and sometimes unlawful. This was to be expected, and though it lowered the standard of US journalism, it was something Nixon was prepared to put up with. What was more serious, and a matter which could not be ignored, was the theft, purchase, or leaking of secret material to these two papers (and others) and its subsequent appearance in print. Under the First Amendment, legislation designed to protect military security, such as the British Official Secrets Act, was generally thought to be unconstitutional. The absence of such an Act was deplored by senior American diplomatic and military officials. Kissinger was particularly concerned that leakages would imperil his Vietnam negotiations.[156] The appearance of secret material in newspapers shot up in spectacular fashion after Nixon assumed the presidency. In his first five months in office, twenty-one major leaks from classified National Security Council documents appeared in the *New York Times* and *Washington Post*.[157] Later that year, the CIA sent to the White House a list of forty-five newspaper articles which were regarded as serious violations of national security.[158] It is not known how many US lives were lost as a result of these leaks, but the damage to US interests was in some cases considerable.

Then on June 13, 1971 the administration was startled by the publication of what became known as the 'Pentagon papers' in the *New*

York Times.[159] This was a 7,000-word survey of American involvement in Vietnam from the end of World War Two till 1968, which had been commissioned by Robert McNamara, Defense Secretary under JFK and Johnson, and based on an archive of documents from Defense, State, the CIA, the White House, and the Joint Chiefs of Staff.[160] Many of these documents were classified Top Secret. Despite this, the *New York Times* not only published their contents but in some cases the originals. The author of the leak was Daniel Ellsberg, a forty-year-old Rand Corporation employee who had been a researcher on the study McNamara had commissioned.[161] The administration discovered that publication of the source notes of the Pentagon papers, if analyzed by KGB experts, could jeopardize a whole range of CIA codes and operations. One of the most sensitive of these was the CIA's device for recording the car-phone conversations of Politburo members. So serious were the security breaches that at one point it was thought Ellsberg was a Soviet agent.[162]

Nixon himself thought it best to ignore these leaks but Kissinger warned him: 'It shows you're a weakling, Mr President. The fact that some idiot can publish all the diplomatic secrets of this country on his own is damaging to your image, as far as the Soviets are concerned, and it could destroy our ability to conduct foreign policy. If the other powers feel that we can't control internal leaks, they will never agree to secret negotiations.'[163] Since all the Vietnam talks, and most of those with Russia and China, were being conducted in the greatest secrecy, Kissinger had a point. An anti-leak unit was formed by one of Erlichman's assistants, Egil 'Bud' Krogh, and Kissinger's administrative assistant David Young, and they recruited various helpmates, including G. Gordon Liddy, a former FBI agent now working for the Justice Department and a romantic 'cloak and dagger' enthusiast. They were called the Plumbers after Young's grandmother, hearing he was running a unit to stop White House leaks, wrote to say that her husband, a New York plumber, would be proud that 'David is returning to the family trade.'[164]

The Plumbers were engaged in a variety of activities of an entirely justifiable nature. But in view of the seriousness of the Ellsberg case, they got from Erlichman authorization to engage in a 'covert operation' to obtain the files of Ellsberg from his psychiatrist's office. This break-in was the point at which the Nixon administration, albeit quite unknown to the President, overstepped the bounds of legality. But at least it could be claimed that the infraction was dictated by national security. During the election campaign, however, the Plumbers broke into Democratic Party headquarters, in Washington's Watergate building, on two occasions, late May 1972 and again on June 17. The second time the Plumbers were caught. The published details of the break-in made it sound like a low-grade farce or a Mack Sennet movie. The police burst in with drawn guns, to a cry of 'Don't Shoot!' and five men, three of them Cubans, emerged with their hands, wearing surgical gloves, over their heads. Nixon, who knew nothing about it and who read about it in a small item in the *Miami Herald*, while weekending in Key Biscayne, thought it some sort of joke. It was referred to as a 'caper.'[165] Political espionage, even theft, had never hitherto been taken seriously in America. This one was particularly absurd since it was already clear that Nixon was going to win easily in any case. The Plumbers seemed to have been engaged in a fishing expedition, or were breaking in just for the hell of it, and no one has ever produced a plausible political justification for the burglary, though many have tried.[166]

But election-year dirty tricks were common. Johnson had certainly 'bugged' Democratic Party headquarters during the Goldwater campaign—he loved bedtime reading, usually supplied to him by the FBI's Hoover, of transcripts from phone-taps and tapes of his political enemies. Both the *New York Times* and the *Washington Post*, before and after Watergate, purloined material. Sometimes it was of an extremely valuable nature (the Haldeman and Kissinger memoirs, for example) and clearly involved criminal activities on someone's part. However, in the paranoid atmosphere generated by the media's anti-Nixon vendetta, anything served as ammunition to hurl against the 'enemy.' The *Washington Post*'s editor, Ben Bradlee, was particularly angry, not to say hysterical, because he believed (without any warrant) that the authorities, at Nixon's insistence, were maliciously opposing the *Post*'s application for broadcasting licenses. Hence, unlike the rest of the press, the *Post* had Watergate stories on its front page seventy-nine times during the election and from October 10 began publication of a series of 'investigative' articles seeking to make the Watergate burglary a major moral issue.

This campaign might have had no impact but the *Post* was lucky. A publicity-hungry judge, John Sirica, known as 'Maximum John' from the severity of his sentences—and not a judge under any other circumstances likely to enjoy the approval of the liberal media—gave the burglars, when they came before him, provisional life-sentences to force them to provide evidence against members of the administration. That he was serious was made clear by the fact that he sentenced the only man who refused to comply, Liddy, to twenty years in prison, plus a fine of $40,000, for a first offense of breaking and entering, in which nothing was stolen and no resistance offered to the police. Moreover, Sirica directed that Liddy's sentence be served in a prison where he had good reason to believe Liddy's life might be in danger from inmates.[167] This sentence was to be sadly typical of the juridical vendetta by means of which members of the Nixon administration were hounded and convicted of various offenses, chiefly obstructing justice—a notoriously easy charge to press home, granted a prejudiced judge. In some cases the accused had no alternative but to plea-bargain, pleading guilty to lesser offenses, in order to avoid the financial ruin of an expensive defense. Some of the sentences bore no conceivable relation to the gravity, or non-gravity, of the original offenses.[168]

Thus the Watergate scandal 'broke,' and allowed the machinery of Congressional investigation, where of course the Democrats enjoyed majority control, to make a frontal assault on the 'Imperial presidency.' Matters were made easy for the witchhunters by the admission, on Friday, July 13, 1973, by one of the White House staff, that all Nixon's working conversations were automatically taped. This of course had been routine for many years, though few outside the White House were aware of it. Oddly enough, one of Nixon's first acts, when he moved into the White House, was to have the taping system installed by Johnson, an inveterate taper, ripped out. Then in February, worried that liberal historians of the future would misrepresent his Vietnam policy, he ordered a new system installed. His chief of staff, Haldeman, picked one which was indiscriminate and voice-activated, 'the greatest single disservice a presidential aide ever performed for his chief.'[169] These transcribed tapes, which the courts and Congressional investigators insisted Nixon hand over, were used to mount a putative impeachment of the President. The witchhunt in the Senate was led by Sam Ervin, the man who had successfully covered up LBJ's crimes in the Bobby Baker affair, a shrewd and resourceful operator who concealed his acuteness and partisanship under a cloud of Southern wisecracking.[170]

To make matters worse for Nixon, in a quite separate development, his Vice-President, Spiro Agnew, was accused of accepting kickbacks from contractors while governor of Maryland. There were over forty counts in the potential indictment against him, of bribe-taking, criminal conspiracy, and tax fraud, and the evidence looked solid. Agnew resigned on October 9, 1973, the Justice Department offering him favorable terms for a *noli contendere* plea on one count of tax evasion. General Alexander Haig, now the White House chief of staff (Haldeman had been driven into resignation by the witchhunters) was in favor of Agnew going without fuss as he privately feared that the Democratic Congress, baying for Republican blood, would impeach both President and Vice-President, leaving the succession to the constitutional next-in-line, Speaker of the House Carl Albert, known to be a serious alcoholic under psychiatric care.[171]

By this stage, Ervin, and the special Watergate prosecutor, Archibald Cox, had between them over 200 lawyers and special assistants working for Nixon's downfall, and feeding all the damaging material they could muster to an eager anti-Nixon media. It became difficult for the President to conduct the ordinary business of government, let alone to handle an international crisis. On October 6, 1973, a treacherous attack was launched on Israel, without warning, by Egypt and Syria, which had picked Yom Kippur, the holiest day in the Jewish calendar, for this Pearl-Harbor-type strike. Both the CIA and Israel's secret intelligence service, Mossad, were caught napping, and the results were devastating. The Israelis lost a fifth of their air force and a third of their tanks in four days, and it became necessary to resupply them. The American media did not let up in its hunt for Nixon's scalp, but he actually had to deal with the crisis and save Israel from annihilation. Nixon acted with great courage and decisiveness, cutting through red-tape, military and diplomatic obstructiveness and insisting that Israel be resupplied. Within seventy-two hours an airlift was operating, delivering daily over 6,400 miles more than 1,000 tons of military supplies and equipment, and continuing with over 566 missions by the USAF over the next thirty-two days. (By contrast, the Soviet resupply of the Egyptians and Syrians never exceeded 500 tons a day, despite much shorter lines of communication.) Without the resupply, which transformed Israel's sagging morale, it is likely that the Israeli army would have been destroyed and the entire Israeli nation exterminated. Indeed it is probable that this is precisely what would have happened, had Nixon already been driven from his post at this stage. As it was, he was still around to save Israel. In many ways, October 1973, his last major international achievement, was his finest hour.[172]

As a result of the Yom Kippur War, the Organization of Petroleum Exporting Countries (OPEC) raised its charges to $11.65 a barrel, a price 387 percent higher than before the conflict.[173] That had desirable consequences in the long run because it forced the advanced countries, including the US, which were becoming increasingly dependent on cheap Middle Eastern crude, to find alternative supplies of oil, to develop alternative sources of energy, and to set in motion energy-conservation and fuel-efficiency programs. But in the short term it set off a long period of turbulence in the international economy and an inflationary recession in the United States which made the rest of the decade difficult years for most Americans. Wholesale prices increased by 18 percent in 1974 alone, unemployment rose, reaching a postwar high of 8.5 percent in 1975, and the GDP actually fell by 2 percent in 1974 and 3 percent in 1975.[174]

With darker times coming, the pressures on Nixon increased, and he strove desperately to combine two objects: to preserve his presidency and to do what was in the national interest. To replace the disgraced Agnew, his wish—the appointment being in his gift—was to hand the vacant vice-presidency to John Connally, the former governor of Texas, who had been wounded at the time of Kennedy's murder. He believed Connally would make a great leader and had already settled in his mind to try his best to make him his successor in 1976. Now he wanted Connally's strength by his side during the difficult months ahead. But with the Senate in a rabid and destructive mood, it became apparent that there was no chance of Connally's nomination being confirmed. Indeed inquiries soon established that the only prominent Republican nominee likely to get through the nomination procedure was Congressman Gerald L. Ford (b. 1913) of Michigan, the Minority Leader in the House. Ford was unblemished, harmless, moderate, and loyal, and Nixon had no real hesitation in choosing him, even though he was aware that his acceptability to the Democratic majority arose from their belief that he could easily be beaten in 1976, and that his selection would therefore feed the Democrats' appetite for his own head.[175] So Ford resigned from Congress, December 6, 1973, being the first Vice-President to be nominated by the President and confirmed by Congress, in accordance with the Twenty-fifth Amendment.

In the meantime, and while Nixon was struggling to save Israel, he had been fanged by the man he called 'the viper we planted in our bosom,' Archibald Cox, special prosecutor charged with the Watergate inquiry. This novel and clumsy arrangement had been conjured up in order to preserve the separation of powers and insure the executive, under legal attack, would not unduly influence the judiciary. In fact it had the opposite effect. Cox proved much more anxious to please a

Democratic Congress than to give the benefit of the doubt to a Republican chief executive. On October 12 Cox won a legal battle to secure the right of access to the tapes recording all Nixon's White House conversations. Nixon determined to fire Cox, and did so, though not without considerable obstruction from the Department of Justice.

It was at this point that the hysteria usually associated with American witchhunts took over, and all reason, balance, and consideration for the national interest was abandoned. It was an ugly moment in America's story and one which future historians, who will have no personal knowledge of any of the individuals concerned and whose emotions will not be engaged either way, are likely to judge a dark hour in the history of a republic which prides itself in its love of order and its patient submission to the rule of law. As one of Nixon's biographers put it, 'the last nine months of [the] Presidency consisted of an inexorable slide towards resignation.'[176]

Right from the start of the Watergate case, Democratic liberals in both Houses of Congress had been calling for the President's impeachment. After the Cox dismissal, and the leaking of stolen tapes to the *New York Times*, which published them, thus forestalling any possibility of a fair trial according to 'due process,' impeachment became a practical possibility. Nixon had still been ahead in the polls until publication of the tape extracts, with many passages containing marks of 'expletive deleted,' persuaded members of the 'silent majority' that Nixon and his colleagues had habitually used swearwords and obscenities. Rumors, published in the *New York Times* and the *Washington Post*, that the tapes, before being handed over, had been doctored and censored by Nixon's staff—rumors later proved to be unfounded—intensified these suspicions. With the polls and public apparently moving against the President, the machinery of impeachment was brought into play. The process, specified in Article I, Sections 2 and 3, of the Constitution, provides for a president to be impeached for offenses described in Article II, Section 4, as 'Treason, Bribery or other High Crimes and Misdemeanors.' In practice, however, this simply means one lot of politicians sitting in judgment upon another. That is what it meant in English history, from which the system derived, and what it meant in the case of Andrew Johnson's impeachment in 1868.[177]

The machinery of impeachment is as follows. The House Judiciary Committee, acting as a court of first instance, decides if there is a fair bill. The full House then debates, votes, and puts forward a formal accusation known as 'articles of impeachment.' The Senate then tries the accused on these articles. A two-thirds majority is required for conviction, whereupon the convict is removed from office and disqualified from holding any subsequent office under the Constitution. Nixon's view from an early stage in the contest was that the House would vote for impeachment but that it would never get a two-thirds majority in the Senate. The House Judiciary Committee consisted of twenty-one Democrats and seventeen Republicans. Nixon noted that eighteen of the twenty-one Democrats on the Committee were certain to vote for impeachment, whatever the evidence, because they were hard-core partisans from the liberal wing of the party. So the Committee, as he put it, was 'a stacked deck.' As it turned out, all twenty-one Democrats and six of the seventeen Republicans voted to recommend impeachment to the full House, July 27, 1974. There was no doubt that the full House, by a simple majority, would forward the charge to the Senate. There, Nixon was sure, he would at least have an opportunity to defend himself in public, and the chances of the Democrats' assembling the two-thirds majority needed to convict were not strong. Even Andrew Johnson had survived, by a single vote. Of the twelve occasions when the House had insisted on impeaching a public official, the Senate had convicted on only four.

As late as the night of July 30–31, 1974 Nixon's combative instincts told him, as he scribbled on a note-pad, 'End career as a fighter.' But he had to bear in mind that the witchhunt had already lasted eighteen months and had done incalculable damage to the American system and to America's standing in the world. The impeachment process would add many months to the process, during which the executive power of the world's greatest democracy, the leader of the Western alliance, would be in suspended animation, and his own authority would be in doubt. In all these circumstances—and granted a good deal of pusillanimity among his staff and colleagues—he decided it would be in the national interest to resign, rather than stand trial as an impeached official. He was mistaken, as subsequent events showed, but his decision was neither cowardly nor dishonorable. He resigned on August 9, 1974 and Gerald Ford, now president, issued a pardon in September. This spared Nixon the ruinous cost of the his legal defense and any further harassment by his enemies, but it meant he never had an opportunity to put his case. On the other hand, he gradually reestablished his reputation as a political seer among the American political community—he never lost it abroad, where the Watergate hysteria was almost universally regarded as an exercise in American juvenilia—and became in due course one of the most respected American elder statesmen since Jefferson.[178]

Gerald Ford had to take over in the debris of a media *putsch* which had reversed the democratic verdict of a Nixon landslide less than two years earlier, and had left a triumphalist press and a Democratic Congress which had tasted blood in possession of the stage. Instead of the Imperial presidency there was, all of a sudden—like a *deus ex machina*—an Imperial Congress. But it was an empire without an emperor. None of the men who had dragged Nixon down was in a position, or possessed the gifts, to offer responsible leadership, even of the Congressional majorities. And Ford himself was left to defend the ruined fort of the White House without a mandate. He was a man who had never sought office on a wider franchise than a Congressional district. Born in Douglas County, Nebraska, in 1913, he had moved with his family the following year to Grand Rapids, Michigan, and gone to public schools there, graduating from the University of Michigan at Ann Arbor in 1935, then attending Yale Law School. Like his three immediate predecessors, he had rendered distinguished war service in the US Navy, emerging in 1946 as a lieutenant-commander, and was elected to Congress three years later. He then served, without much notice or achievement but with growing approbation from his colleagues, in twelve succeeding Congresses, emerging as minority leader by virtue of seniority and lack of enemies. Indeed, if Ford had enemies, they never made themselves visible. What he had was critics, usually of his intelligence. LBJ dismissed him as 'So dumb he can't fart and chew gum at the same time.'[179] In fact Ford was not dumb, but he was unassertive, sometimes inarticulate, and notably unacademic. He had been an outstanding athlete at Ann Arbor and preferred to discuss politics in terms of football, thus inspiring another cruel LBJ jibe: 'The trouble with Jerry Ford is that he used to play football without a helmet.'[180] He had trouble with his inner ear, as had Mamie Eisenhower. Whereas Mrs Eisenhower's occasional lack of balance led the media to assume she had taken to drink, in Ford's case it produced much-photographed stumbles, which raised derision among the public. But there was nothing wrong with Ford's brain. He had, indeed, a remarkable memory, was able to record the face, name, occupation, and often other data about everyone he ever met. This did not make him a great president, but it made him countless friends, often humble ones.[181]

Ford was much sustained by his wife Betty, a vigorous and outspoken lady, a former model and dancer known as the 'Martha Graham of Grand Rapids.' While Ford made a nervous president, Mrs Ford took to the White House with enthusiasm, a marked contrast to the retiring Pat Nixon, and she reversed the secrecy of the Nixon White House by a display of spontaneous openness, telling the world about her mastectomy, her drinking problems, and her psychiatric treatment. On CBS TV's *60 Minutes* program in August 1975 she gave a spectacularly successful and frank interview about a series of issues, such as abortion, premarital sex, and other controversial matters, which infuriated some moral conservatives but had the effect of reassuring many more that humanity and normalcy had returned to the White House.[182] Among other things, she was the first President's wife since Grace Coolidge to share not only a bedroom but a double bed with the President in the White House.

But there was no normalcy elsewhere in Washington. The fall of Nixon was made the occasion for a radical shift in the balance of power back towards the legislature. Some movement in this direction was, perhaps, overdue. Presidential over-activism is always a constitutional vice, and one from which even Nixon, despite his fundamental conservatism, sometimes suffered. But in the event the swing proceeded much too far in the opposite direction, at heavy cost to America and still more to the world. The attack on the executive's traditional powers by Congress began even before Nixon's departure. On November 7 Congress enacted, over the President's veto, the War Powers Resolution, which required presidents to inform Congress within forty-eight hours if they sent troops overseas or significantly reinforced troops already serving; and if Congress failed to endorse such actions within sixty days the President would be obliged to cease such operations (unless he certified that an additional thirty days were needed to make the withdrawal safely).

Further limitations of presidential foreign policy were imposed by the Jackson–Vanik and Stevenson Amendments of 1973–4. In July–August 1974 Congress paralyzed the President's handling of the Cyprus crisis. In the autumn of 1974 it imposed restrictions on the use of the CIA. In 1975 it effectively hamstrung the President's policy in Angola, thus producing a civil war which led to the deaths of one-fifth of the population and whose effects are still being felt at the end of the 20th century. Later that year it passed the Arms Export Control Act, removing the President's discretion in the supply of arms—a piece of legislation warmly welcomed by the expanding Soviet and French arms-export industries and by the nascent Chinese one. Congress used financial controls to limit severely the system of 'presidential agreements' with foreign powers, over 6,300 of which had been concluded from 1946 to 1974, as opposed to only 411 treaties, which required Congressional approval. It reinforced its aggressive restrictions on presidential authority by enabling no fewer than seventeen Senatorial and sixteen House committees to supervise aspects of foreign policy and by expanding its own expert staff to over 3,000 (the House International Relations Committee staff tripled, 1971–7), to monitor White House activities. By the late 1970s it was calculated that there were now no fewer than seventy limiting amendments on the President's conduct of foreign policy. It was even argued that a test of the War Powers Act would reveal that the President was no longer commander-in-chief, and that the decision whether or not American troops could be kept abroad or withdrawn might have to be left to the Supreme Court.[183]

Ford was obliged to look on helplessly while such freedom as a decade of effort by America had secured for the peoples of Indochina was removed step by step. As US military aid to South Vietnam tailed off from 1973, the balance of armed power shifted decisively to the Communist regime in the North. By the end of the year the North had achieved a two-to-one superiority and, in defiance of all the accords carefully worked out by Nixon and Kissinger, launched a general invasion. Bound hand and foot by Congress, Ford felt powerless to act and simply made verbal protests. In January 1975 the whole of central Vietnam had to be evacuated and a million terrified refugees fled towards Saigon. In an appeal to Congress, Ford warned: 'American unwillingness to provide adequate assistance to allies fighting for their lives could seriously affect our credibility throughout the world as an ally.' Congress did nothing. At his news conference on March 26 Ford again pleaded with Congress, warning of 'a massive shift in the foreign policies of many countries and a fundamental threat . . . to the security of the United States.'[184] Congress paid no attention. Four weeks later, on April 21, the Vietnamese government abdicated. Marine helicopters lifted American officials, and a few Vietnamese allies, from the rooftop of the US embassy in Saigon, an image of flight and humiliation etched on the memories of countless Americans who watched it on TV. It was indeed the most shameful defeat in the whole of American history. The democratic world looked on in dismay at this abrupt collapse of American power, which had looked so formidable only two years before.

But it was the helpless people of the region who had to pay the real price. Nine days after the last US helicopter clattered out to sea, Communist tanks entered the city of Saigon, and the secret trials and shootings of America's abandoned allies began almost immediately. All over Indochina, the Communist elites which had seized power by force began their programs of 'social engineering.' The best-documented is the 'ruralization' conducted by the Communist Khmer Rouge in Cambodia, who entered the capital Phnom Penh in mid-April, the US embassy having been evacuated on the 12th. The atrocities began on the 17th. The object of the plan was to telescope into one terrifying year the social changes carried out in Mao's China over a quarter-century. Details of the plan had been obtained by the State Department expert Kenneth Quinn, who had circulated it in a report dated February 20, 1974.[185] Members of Congress were made well aware of what they were permitting to happen. But they averted their gaze. Between April 1975 and the beginning of 1977, the Marxist–Leninist ideologues ruling Cambodia ended the lives of 1,200,000 people, a fifth of the population.[186] Comparable atrocities took place in Laos, and during Communist efforts to unify Vietnam by force, 1975–7, following which Vietnam invaded Cambodia and occupied Phnom Penh on January 7, 1979. Laos was likewise occupied by Vietnamese troops. By 1980 Vietnam had over 1 million in its armed forces, next to Cuba's the largest, per capita, in the world. Communist colonial rule inevitably provoked a return to guerrilla warfare in the countryside, which continued throughout the 1980s and into the 1990s.

Unable to do anything in Southeast Asia, Ford, who was desperate to achieve some success which would help him get elected in 1976, tried hard to revitalize the Strategic Arms Limitation Talks, begun by Nixon and Kissinger in 1972, in conjunction with the Soviet dictator Leonid Brezhnev. Nixon was careful to call these negotiations 'talks' not a 'treaty.' SALT I, as it was called, succeeded in reaching agreement between Russia and America to reduce the number of their offensive intercontinental ballistic missiles and defensive antiballistic missiles, though not MIRVs (Multiple Independently Targetable Reentry Vehicles), the Senate approving 88–2. Ford's attempt to push on to SALT II met resistance when he tried to link the arms reductions to Soviet treatment of Jews. All he had to report was a minor agreement known as the Helsinki Accords, whereby Russia renounced the right to use force within its satellite empire, though it had the demerit that it appeared to recognize the *status quo* in eastern Europe. As his running mate for the election, Ford—who got the nomination simply in the absence of a major contender—recruited Senator Robert Dole of Kansas (b. 1923), already regarded as a Congressional 'old timer' and reputed to be witty, though this did not emerge during the campaign.

The Democrats, having destroyed a strong and able president, had no masterful alternative to offer in exchange; no one much at all, in fact. In the second half of the 1970s, not surprisingly, Washington insiders were generally held in low regard, and against this background a Democratic 'outsider,' Jimmy Carter, governor of Georgia (b. 1924) was chosen. Carter inherited a peanut farm and did well at it (or so it was said), but his claim to notice was that he represented the new generation of moderate Southern politicians who had accepted—more or less—the civil rights revolution and learned to live with it. 'Jimmy,' as he liked to be called, had a big smile, and his slogan, endlessly repeated, was 'I'll never lie to you.' The truth is he had been recruited and packaged by a clever Atlanta advertising executive, Gerald Rafshoon, who had brushed 'Jimmy' up and left out of the package inconvenient accessories, such as his no-good drunken brother, and the thrusting ambitions of his strong-minded wife, Rosalynn.[187] The polls, which had been heavily anti-Ford, moved against Carter during the campaign and if it had lasted another fortnight it is possible that Ford might have won the presidency in his own right. As it was Carter scraped home with a vote of 40,828,587 to 39,147,613, and an electoral college majority of 297 to 241. He won the presidency by an unconvincing margin against perhaps the weakest incumbent in history and became a still weaker one. The fact that Congress was Democratic too seemed to make little difference, and the succession of American reverses abroad continued.[188]

Carter actually added to American weakness by well-meaning but ill-thought-through ventures. One of them was his 'human rights' policy, based on the Helsinki Accords, under which those who signed them undertook to end violations of human rights everywhere. Carter's object was to force Soviet Russia to liberalize its internal policy, and especially to abolish its imprisonment of political prisoners in psychiatric hospitals. But the consequences were quite different. Within Russia and its satellites, the Helsinki Accords were ignored and groups set up to monitor them were broken up and their members arrested. In the West, America found itself pitted against some of its oldest allies. A human rights lobby grew up within the administration, taking over a whole section of the State Department, which worked actively to enforce the Accords. Thus, it played a major role in the overthrow of the Somoza regime in Nicaragua. An assistant secretary, Voron Vaky, announced on behalf of the US government: 'No negotiation, mediation or compromise can be achieved any longer with a Somoza government. The solution can only begin with a sharp break with the past.' The 'sharp break' took the form, in 1979, of the overthrow of Somoza, a faithful if distasteful ally of America, and his replacement by a Marxist and pro-Soviet regime, whose attitude to human rights was even more contemptuous and which campaigned openly for the overthrow of America's allies in Guatemala, El Salvador, and elsewhere in Central America. In September 1977 Brazil reacted to State Department criticisms of its internal policies by cancelling all its four remaining defense agreements with the US, two of which went back to 1942. Argentina too was alienated.

The next year the State Department's Bureau of Human Rights played a significant part in undermining the position of another old US ally, the Shah of Iran, whose pro-Western regime was overthrown by orchestrated street-mobs in 1979. It was replaced by a Moslem fundamentalist terror regime, which swiftly accumulated an unprecedented record of gross human rights abuses and characterized the US as the 'Great Satan.'[189] On January 20, 1980, in an attempt to retrieve the situation in Iran, Carter publicly proclaimed what he hoped would be known as the 'Carter Doctrine,' an assertion that the oil reserves of the Persian Gulf were of vital interest to the US and that it would be justified in intervening with military force to prevent domination of the region from outside (that is, by Soviet Russia). This declaration, noted by Congress, came in useful in August 1990, under President Bush, when Iraq invaded Kuwait, but at the time the only attempt by the Carter administration to use military power in the area, an ill-planned helicopter operation to rescue American hostages held by the Iran government, which took place in April 1980, ended in a humiliating failure.

During the 1970s the Cold War spread to virtually every part of the globe and was marked by two developments: the contraction of US naval power, and the expansion of Soviet naval power. In 1945, the United States had 5,718 naval vessels in active service, including 98 aircraft carriers, 23 battleships, 72 cruisers, and over 700 destroyers and escorts. As late as June 1968, the US had 976 naval vessels in commission. But in the 1970s the American fleet shrank rapidly to thirteen carriers and their escorts. While America became a major importer not only of oil but of chrome, bauxite, manganese, nickel, tin, and zinc, and therefore became ever more dependent on supply by sea, its ability to keep sea-lanes open declined sharply. Secretary of Defense Donald Rumsfeld, in his budget report for 1977, noted that the 'current [US] fleet can control the North Atlantic sea-lanes to Europe' but only after 'serious losses' to shipping. The 'ability to operate in the eastern Mediterranean would be, at best, uncertain.' The Pacific fleet could 'hold open the sea-lanes to Hawaii and Alaska' but 'would have difficulty in protecting our lines of communication into the Western Pacific.' He warned that in a global war America would be 'hard pressed to protect allies like Japan and Israel or to reinforce NATO.'[190] This was a transformation from 1951, when Admiral Carney, Commander of NATO forces in Southern Europe, had dismissed Soviet naval power in the Mediterranean: 'He said it was possible there were a few "maverick" Soviet submarines in the Mediterranean and they might be able to push in some others in preparation for a war. But they couldn't support them long.'[191]

The big change in Soviet naval policy came in 1962, when the Kremlin, reflecting on the Cuban missile crisis, decided to expand its navy greatly. Over the next fourteen years Soviet Russia built a total of 1,323 ships of all classes (against 302 American ships built), including 120 major surface combat ships. By the same date Russia had accumulated 188 nuclear submarines, 46 of them carrying strategic missiles. By the late 1970s the first genuine Soviet carriers appeared. Even before this, during the 1973 Yom Kippur War, the position of the American fleet in the Mediterranean theater was described by its Commander as 'very uncomfortable,' an expression not used since the destruction of Japanese naval power.[192] By this point Soviet naval power, already predominant in the Northeast Atlantic and Northwest Pacific, was ready to move into the South Atlantic and Indian oceans. This was the background to the Soviet descent on black Africa in the late 1970s, often using Cuban forces as surrogates. In December 1975, under Soviet naval escort, the first Cuban troops landed in Angola. In 1976 they moved into Ethiopia, now in the Soviet camp, and into Central and East Africa. By the end of the 1970s there were ten African states, under Soviet 'protection,' which were proclaiming themselves Marxist–Leninist.

While the Carter administration was adept at damaging friends and allies, it failed to develop any coherent response to this extension of the Cold War. Under Carter, there was a triangular tug of war between his Secretary of State, Cyrus Vance, his security adviser, Zbigniew

Brzezinski, and his Georgian assistant, Hamilton Jordan, much of which was conducted in public, leaving aside the freelance activities of brother Billy Carter, who acted as a paid lobbyist to the anti-American Libyan government. The only point on which Carter's men agreed was on America's declining ability to control events. Cyrus Vance thought that to 'oppose Soviet or Cuban involvement in Africa would be futile.' 'The fact is,' he added, 'that we can no more stop change than Canute could still the waters.' Brzezinski insisted, 'the world is changing under the influence of forces no government can control.' Carter himself said America's power to influence events was 'very limited.' Feeling itself impotent, the Carter administration took refuge in cloudy metaphor, for which Brzezinski had a talent. Vietnam, he said, had been 'the Waterloo of the Wasp Elite' and no such intervention could ever again be undertaken by America. 'There are many different axes of conflict in the world,' he noted, '[and] the more they intersect, the more danger-ous they become.' West Asia was 'the arc of crisis.' But 'the need is not for acrobatics but for architecture.'[193]

It was not that Carter was incapable of doing things when he exerted himself. From September 6 to 17, 1978, he hosted a summit between Egyptian and Israeli leaders at Camp David at which agreements were reached which led to a formal Egyptian–Israeli treaty in March 1979, the first and most crucial step towards bringing peace between Israel and its Arab neighbors. This was a notable achievement, but it remained unique in Carter's record. Most of the time he appeared hyperactive but curiously supine in action. An aide protested on his behalf: 'Look, this guy is at his desk every day by 6 A.M.,' to which the reply came: 'Yes, and by 8 A.M. he has already made several serious errors of judgment.'[194] Things might have been better, looking back on it, had Rosalynn been in charge. She was ubiquitous. The *Washington Star* reported that in her first fourteen months in the White House she had visited eighteen national and twenty-seven American cities, held 259 private and 50 public meetings, made fifteen major speeches, held twenty-two press conferences, given thirty-two interviews, attended eighty-three official receptions, and held twenty-five meetings with spe-cial groups in the White House, and was said to be 'trying to take on all the problems we have.'[195] What it did not report, though it emerged after she and her husband left the White House, was that she occasion-ally attended Cabinet meetings.

Behind the faltering of American power in the 1970s, and giving it a psychological overtone, was the consciousness of the relative economic decline of the United States. During the early stages of the Cold War, the US had been sustained by its consciousness of the 'Baby Boom,' which began in 1945 when the soldiers returned from abroad, and continued through most of the 1950s: the 'Baby Boomers' were to be a new and infinitely numerous generation of an extra 50 million well-fed and well-educated and trained Americans, able to take on the world.[196] Campaigning for election in 1952, Truman had been able to say, with general agreement, 'This is the greatest nation on earth, I think. The greatest nation in history, let's put it that way. We have done things that no other nation in the history of the world has done' (Salem, Oregon, June 11, 1952). The figures could not lie. In metals, the United States produced as much as the combined output of Canada, the Soviet Union, and Chile, in fuel minerals as much as Russia, Germany, Britain, Venezuela, Japan, France, Poland, Iran, the Netherlands, India, Burma, Belgium, and Luxembourg. US production of minerals was about four times larger than that of Russia, the second-largest producer.[197] The symbol of these years was Pan-American Airways, the world-conquering airline which in June 1947 inaugurated the first round-the-world regular flight of 25,003 miles New York–New York, and eleven years later, in October 1958, brought in the first big-jet 707 regular service. America did well throughout the 1950s and into the 1960s. In 1968, the year when LBJ first became painfully aware of the financial pressure on the world's richest power, US industrial production was still more than a third (34 percent) of total world production. The American GDP, which had doubled during World War Two, doubled again by 1957, and yet again by 1969. That was why President Nixon, on December 15, 1970, was able to celebrate the registering, by the Commerce Department's 'GDP Clock,' of a 'trillion-dollar economy,' which moved at the rate of $2,000 a second.

But by the following year, the US proportion of world production had fallen to under 30 percent, and Nixon was warning: 'Twenty-five years ago, we were unchallenged in the world, militarily and economically. As far as competition was concerned, there was no one who could challenge us. But now that has changed.' America's economic leadership was 'jeopardized.'[198] This was reflected in the inability of America to continue managing the world's international monetary system, which it had done since 1945. In 1971 the Nixon administration lost or abandoned control of what was happening. Two years later, in March 1973, Nixon cut the link between gold and the dollar, and thereafter most major currencies floated. The float revealed the weakness of the dollar, which lost 40 percent of its value against the Deutschmark between February and March 1973.[199]

America added to its problems of competing successfully in the world by a continuing spate of regulatory legislation passed by Congress in the 1960s and 1970s. This followed the remarkable success of Rachel Carson's book *The Silent Spring* (1962), which first drew public attention to the long-term dangers of industrial pollution and the poisoning of the environment. In 1964 came the Multiple Use Act and the Land and Water Act; in 1965 the Water Pollution Act and the Clean Air Act; in 1966 the Clean Water Restoration Act. Then came the 'Conservation Congress' of 1968 and a whole series of gigantic Acts which attempted to impose what was called 'Ecotopia' on the US: the Environmental Protection Act, the Toxic Substances Control Act, the Occupational Health and Safety Act, the Clean Air Amendments Act, and a series of Food and Drug Acts. By 1976 it was calculated that compliance with the new legislation was costing US businesses $63 billion a year plus a further $3 billion to the taxpayer for maintaining regulatory services. By 1979 total costs had risen to $100 billion annually.[200]

Much of this legislation was not only well meaning but desirable. But it had a serious effect on the productivity of US business. In the coal industry, for example, where production had been 19.9 tons per worker per day in 1969, it had slipped, by the time the full effects of the 1969 Coal-Mine Health and Safety Act had been felt in 1976, to 13.6 tons, a fall of 32 percent.[201] In 1975, over the whole of American industry, productivity was 1.4 percent lower than it otherwise would have been as a result of meeting federal pollution and job-safety regulations.[202] As a result, in the decade 1967–77, productivity in American manufacturing industry grew by only 27 percent, about the same as in Britain (regarded, by this time, as the 'Sick Man of Europe'), while the corresponding figure for West Germany was 70 percent, for France 72 percent, and for Japan 107 percent. In some years in the 1970s, American productivity actually fell. Detailed analysis of this stagnation or decline in American dynamism suggested the causes were mainly political: failure to control the money supply, excessive taxation, but chiefly government intervention and regulation.[203]

There were many other disturbing indicators. With growing competition from Europe, Japan, and other Far Eastern manufacturers, America's share of total world production of motor vehicles fell in the decade 1972–81 from 32 to 19 percent; in steel it dropped from 20 to

12 percent; in manufacturing as a whole the US share slid to 26 percent by the mid–1970s, 24 percent by the end of the decade, and 20 percent in the early 1980s. The standard of living, measured by total output per person, gave the US an annual figure of $6,000 by 1973; but Switzerland's figure was then $7,000, and the US was also behind Sweden, Denmark, and West Germany. By the end of the 1970s it was behind Kuwait too, and it fell behind Japan in the 1980s.[204] The decline in America's once-strong balance of payments position, a feature of the 1960s and 1970s, reflected the fact that, by the early 1970s, America was importing nearly half its oil and large quantities of tin, bauxite, diamonds, platinum, and cobalt, as well as a growing percentage of its manufactured goods, not only consumer durables but machine-tools. The relative economic decline continued uninterruptedly during the 1970s.[205]

However, there was a pendant to this picture of relative decline which, in time, proved to be more significant than the main image. The declining dynamism of the US economy observable in the 1960s and 1970s was very much a regional phenomenon largely confined to the Northeast, the old manufacturing core, the 'smokestack industries.' Beginning in the 1920s (as we have seen), encouraged by the New Deal's state capitalism, and hugely accelerated by World War Two, was the rise of America's 'Pacific Economy.' In the years after 1960 this was at last followed by the modern industrial development of the Old South and the border states. This rise of new industry was accompanied and made possible by a movement of people, not unlike the wagon-trail march to the West in the 19th century, in search of jobs and (not least) better weather—a move from the 'frost-belt' to the 'sun-belt.'

The shift of America's center of gravity, both demographic and economic, from the Northeast to the Southwest, was one of the most important changes of modern times. In the 1940s, the geographer E. L. Ullman located the core of the US economy in the Northeast. Though only 8 percent of the total land area, it had 42 percent of the population and 68 percent of manufacturing employment.[206] The pattern remained apparently stable through most of the 1950s. The geographer H. S. Perloff, writing in 1960, saw what he called the 'manufacturing belt' as 'still the very heart of the national economy.'[207] But even as he wrote the balance was changing. In 1940–60 the North still gained population (2 million), but this was entirely accounted for by low-income, mostly unskilled blacks from the South. It was already suffering a net loss of whites; this soon became an absolute loss. The change came in the 1960s and was pronounced by the 1970s. In the years 1970–7, the Northeast lost 2.4 million by migration; the Southwest gained 3.4 million, most of them skilled whites. The shift from frost- to sun-belt was reinforced by rising energy prices, as the 1980 census showed. Regional variations in income, once heavily in favor of the old 'core area,' converged, then moved in favor of the Southwest. Investment followed population. The core area's share of manufacturing employment fell from 66 percent in 1950 to 50 percent in 1977. The Southwest's rose from 20 to 30 percent.[208]

The economic-demographic shift brought changes in political power and philosophy. Since the mid–1960s, all America's elected presidents have come from the South and West: Johnson and Bush from Texas, Nixon and Reagan from California, Carter from Georgia, and Clinton from Arkansas. The only Northerner, Ford, was never elected. (It is arguable that Bush was not a genuine Texan, but he claimed to be, in itself significant.) When Kennedy was elected in 1960, the frost-belt had 286 electoral college votes to the sun-belt's 245. By 1980 the sun-belt led by four and by 1984 by twenty-six. The shift marked the end of the old FDR interventionist coalition, dominant for two generations, and the emergence of a South–West coalition, more closely attuned to the free market. Nixon's landslide victory of 1972 was a foretaste of the political consequences of the shift, though that was overshadowed by Watergate and its consequences. But the election of the Georgian Jimmy Carter in 1976 was another indicator. The idea of a weak candidate from a smallish Southern state making it to the White House would have been inconceivable only ten years before. In 1964, however, California ousted New York as America's biggest state. By 1990 it had 29,760,021 people, against 17,990,450 in New York. By 1990, again, Texas had become the country's third-largest state, with 16,986,510 people (it became the second-largest in 1994), and Florida the fourth-largest with 12,937,926. In 1980, for the first time in American history, the election was between a man from the West, Ronald Reagan, and a man from the South, Jimmy Carter.

Ronald Reagan, born in 1911, came from Tampico, Illinois, and was the son of a hard-drinking, wisecracking, intermittently unemployed shoe-salesman. He graduated from Eureka College in 1932, worked briefly as a sports broadcaster, then moved to California, where he prospered in the movies as 'Mister Norm.' He was a true B-movie star, right at the top of the second grade, a 'quick study,' always punctual on set, easygoing, obedient to the director, friendly to fellow-actors, a bankable name.[209] But after the war (during which he worked in government movies) a near-fatal bout of pneumonia, a painful divorce from the actress Jane Wyman, a growing disgust with some aspects of the movie industry, formed during his service in a union representing actors, combined to persuade him to start a new career, a spokesman for General Electric. This in turn led him into politics and changed his mind on the subject of what politics was about. He was a natural Democrat who had voted four times for FDR and in some metaphysical ways he remained a New Dealer. But 'by 1960,' as he wrote, 'I realized the real enemy wasn't Big Business, it was Big Government.'[210] In 1966 he was elected Republican governor of California, the number-one US state with the world's seventh-largest economy, was triumphantly reelected, and established himself as a reliable, cautious, and effective administrator. Despite this, and his proven record as a vote-winner, Reagan was ignored by the Republican establishment in the post-Nixon era, notwithstanding their poverty of talent. He had to fight his way every inch to the Republican nomination in 1980. The East coast media did not hate him, as they had hated Nixon (and Johnson), they simply despised and dismissed him as a 'maverick,' an 'outsider,' a 'California nut-case,' an 'extremist,' 'another Goldwater,' and so on.

As late as August 1980 most Washington pundits agreed that Carter, as incumbent, would have no difficulty in disposing of the challenger. In fact Carter became the first elected President to be defeated since Hoover in 1932. Reagan beat Carter by a huge margin, 43,904,153 to 35,483,883, and this was all the more remarkable as John B. Anderson, an Illinois congressman who had opposed Reagan for the Republican nomination, ran as an independent and received 5,720,000 popular votes. Reagan carried the electoral college by 489 votes to Carter's 49. The fact is, Reagan was one of the great vote-getters of American history, a man who could win over the majority among both sexes, all age groups, virtually all occupation and income groups, and in all parts of the Union. The only categories where he failed to make majority scores were blacks and Jews. In 1984, against Carter's Vice-President, Walter Mondale, who was running with a woman vice-presidential candidate, for the first time—Representative Geraldine Ferraro of New York—Reagan scored an even more remarkable victory. He carried all states, except Mondale's home-state, Minnesota (and the District of Columbia), winning a 525–13 majority in the college, and an overwhelming popular plurality, 54,455,074 votes to Mondale's 37,577,185.

Reagan looked, spoke, and usually behaved as if he had stepped out of a Norman Rockwell *Saturday Evening Post* cover from the 1950s. More important, he actually thought like a Rockwell archetype. He had very strong, rooted, and unshakeable views about a few central issues of political and national life, which he expressed in simple and homely language. He saw America as the Pilgrim Fathers had, as a 'City on the Hill,' as the Founding Fathers had, as the ideal republic, as Lincoln had, as 'the last, best hope for mankind,' and as Theodore Roosevelt had, as a land where adventure was still possible and any determined and brave man could aspire to anything. He also saw America as (in his view) FDR had, where those who had much gave to those who had less, but he saw it too as Andrew Jackson had, as a federation of states where Washington was (as it were) merely the first among equals.

Reagan had been inspired, in his 1980 campaign, by the success of Margaret Thatcher in Britain, who had set about reducing the size and role of the state by her campaign of curbing expenditure and taxation and regulation, and by her privatization of the state sector. Like Thatcher, Reagan saw himself as a radical from the right, a conservative revolutionary who had captured the citadel of the state but, like Thatcher, still treated it as an enemy town. Both these remarkable figures of the 1980s replaced the doubts and indecisions of the 1970s with 'conviction politics,' homely ideologies based on the Judeo-Christian ethics of the Ten Commandments and the Sermon on the Mount, reinforced by such secular tracts as Adam Smith's *The Wealth of Nations*, the writings of Jefferson and John Stuart Mill, and the arguments of 20th-century conservatives like Milton Friedman, Friedrich A. Hayek, and Karl Popper. But whereas Thatcher had read the last three in the original, Reagan had absorbed them in filtered-down versions, via *Reader's Digest* and other popular publications. He was in fact a great and constant reader, though not in the sense academics recognize.

Reagan was sometimes confused in his thinking. But then he knew he was confused and wisecracked about it, saying: 'Sometimes our right hand doesn't know what our far-right hand is doing.' That was where he differed from Thatcher. Like Lincoln, he possessed the secret weapon of the homely politician: humor. Quite incapable of the majesty Lincoln brought to public papers and speeches, Reagan nonetheless had a way with words which was far from negligible in terms of the political dividends it earned. One writer on the presidency has noted: 'Style is the President's habitual way of performing his three political roles, rhetoric, personal relations, and homework.'[211] In this sense, Reagan had a consistent and coherent style, that of light comedy, a comedy which reassured all and threatened none, which was quotidian, ubiquitous, and all-purpose. As one observer put it, he was 'the Johnny Carson of politics, the Joker-in-Chief of the United States.'[212] And whereas, in Lincoln's case, presidential jokes were resented by many as coarse, unbecoming, and vulgar, with Reagan they seemed natural. His childhood had been unhappy and insecure; the family had moved ten times. Their refuge was joking. Reagan got his one-liners from his father, his ability to get them across with masterful timing from his mother, a frustrated actress. He drew a lot and thought in visual terms. He had ambitions, at one time, to be a cartoonist. His omnivorous reading in homely quarters, and his excellent memory, allowed him to build up an extraordinary stock of jokes, metaphors, funny facts, proverbs, and sayings, added to assiduously during his Hollywood and PR years, and polished and categorized for service on all occasions. He liked to have on his desk a jar of jelly-beans which he could delve into and pop into his mouth to calm himself down, and he likewise delved into his formidable joke-arsenal, which never failed him. It is no accident that his two favorite authors were Mark Twain and Damon Runyon, both of whom used humor to make acceptable the agonies of life. Reagan worked not through logic and statistics but through metaphor, analogy, and jokes, and he was good not just at listening but at interpreting body language. Reagan was never afraid of seeking the presidency or enjoying it. He knew he had been a successful governor of a huge state when everyone said he would fail and that he had won the presidency when most people thought it impossible. He was a very secure person, and that (as with Lincoln) enabled him to joke about serious things.[213]

Observers first became aware of the power of Reagan's humor when, during the election, he had a debate with Carter in New York. This was and is heartland territory for a Democratic candidate but Reagan nonetheless won the encounter hands down. He destroyed Carter in the TV debates with a simple aside to the viewers: 'There he goes again.' His 1980 proposed budget plan was the work of the outstanding economist Alan Greenspan, later chairman of the Federal Reserve, but it was Reagan who gave it the formulation which aroused the public: 'A recession is when your neighbor loses his job. A depression is when you lose yours. And recovery is when Jimmy Carter loses his.' Once in the White House, Reagan used jokes to make both himself and everyone else feel at home. Like Lincoln, he used jokes to narrow the daunting psychological gap between the President and the rest. He knew he had been blessed with common sense and he felt his being in politics was in pursuit of a serious purpose—so let them laugh. He laughed with them. When a reporter asked him to autograph an old studio picture showing him with the chimpanzee Bonzo, Reagan did so and wrote, 'I'm the one with the watch.' He regularly joked about not only his age and stupidity but his memory lapses, his second-rate movies, his mad ideologies, and supposed domination by his much cleverer wife—all the areas of vulnerability. He joked about his laziness: 'It's true hard work never killed anyone but I figure, why take the chance?' He was not exactly lazy, in fact, but he was careful of himself and conserved his energy: he spent 345 days of his presidency at his 688-acre property, the Rancho del Cielo, which he had owned in the Santa Ynez Mountains, northwest of California, since 1974. That was one year out of eight in total. He made many of his most important decisions there and felt, probably correctly, that it was in the national interest to use the ranch as much as he did.[214]

Many of his jokes were learned by heart or even written for him. But some of his best one-liners were spontaneous. Of a mob of peace-protestors: 'Their signs say make love not war, but they didn't look as if they could do either.' To a bearded man who shouted, 'We are the future,' Reagan replied, 'I'll sell my bonds.' When he was shot and nearly died in 1981, he let out a string of one-liners. Thus, to his wife Nancy, 'Honey, I forgot to duck,' a recycling of Jack Dempsey's famous 1926 joke about his defeat by Gene Tunney. Just before being wheeled into the operating theater, he said to the doctors, 'Please tell me you're all Republicans,' and when he was in the recovery room he said to the nurses, paraphrasing W. C. Fields during the wagon-train fight against the Sioux: 'All in all, I'd rather be in Philadelphia.' He had a euphemism for his frequent naps: 'Personal staff time' or 'Staff time with Bonzo.'

His jokes often expressed his fundamental ideas and helped to pass off his ignorance in areas solemn people thought vital, such as economics. But most of his anti-economist, anti-lawyer, and anti-clerical jokes were addressed directly to economists, lawyers, and clergy. He liked cheering-up as well as put-'em-at-their-ease jokes. 'How many are dead in Arlington Cemetery?,' a question which evoked bafflement and was answered by 'They all are.' He told the 1984 Gridiron Dinner, 'I'm not worried about the deficit—it's big enough to take care of itself.' One of his great strengths was that he was not ashamed to ask simple questions. Thus: 'What makes the Blue Mountains blue?' (A lot of people want to ask this but fear to reveal their ignorance.) He asked Paul Volcker, chairman of the Fed, 'Why do we need the Federal Reserve at all?' Volcker, six feet seven-and-a-half—half a foot taller than Reagan—sagged in his chair and was 'speechless for a minute.' But that was exactly the kind of question Andrew Jackson asked.[215] Reagan used questions to disarm or sidetrack issues he felt overwhelming. Thus, asked about the 'Butter Mountain,' he replied: '478 million pounds of butter? Does anyone know where we can find 478 million pounds of popcorn?'

Sometimes Reagan forgot names or misremembered them or became confused. He called the Liberian leader, Samuel K. Doe, 'Chairman Mao.' He thought the former British Defense Minister, Denis Healey, was an ambassador. He announced, at the White House, 'It gives me great pleasure to welcome Prime Minister Lee and Mrs Lee to Singapore.' The first time the author met him he said, 'Good to see you again, Paul.' He relied a lot on his cue cards. But he was humble and cooperative with his staff, as he had been as an actor. He regarded the events in his daily diary schedule as stage directions he must obey. He often took more trouble with ordinary citizens than he did with important people. One of his staff, Anne Higgins, selected letters for him to read from his huge mail. Like Jefferson, he replied to as many as he could. He responded to hard-luck stories with advice and sometimes a small personal check. Some of his letters to humble people are wonderfully well phrased, and treasured. He also made frequent phone calls to the relatives of victims, wounded members of the forces, or policemen and the like. He was the 'Great Comforter.' Reagan was accessible to people who never met him, and they felt it. On the other hand, the closer people got to him, the less they knew him. He was an affable monarch, but still a monarch. Seen really close, he was as unresponsive and remote as Louis XIV. Some infighting did take place in his court, but usually well outside his earshot. No one was anxious to disturb his placidity. His best biographer sums up: 'He was a happy president, pleased with his script, his cue cards and his supporting cast.'[216]

The new President's self-confidence, and confidence in America, soon communicated itself. It was not long before the American public began to sense that the dark days of the 1970s were over, and that the country was being led again. There were limits to what Reagan could do to get the country back on the rails. The Democrats still controlled Congress and would not cut welfare spending. During his first term, spending still increased annually by 3.7 percent. But at least it was less than the 5 percent annual increase under Carter. At the same time, Reagan made good his promise to cut taxes. The Economic Recovery Tax Act of 1981, the work of Congressman Jack Kemp and Senator William Roth, with Reagan's full backing, reduced the highest tax rate to 50 percent and included across-the-board tax reductions of 25 percent. There were other reductions in taxes on capital gains, estates, and gifts. It was followed by the Tax Reform Act of 1986, which greatly simplified the entire tax structure. This success, together with a massive program of deregulation, acted as a potent stimulant to business.[217] Reagan took over in January 1981 and by the beginning of 1983 the nation was in full recovery. The growth continued throughout Reagan's second term, then into his successor's and well into the 1990s, the longest continual expansion in American history. Inflation, which had been 12.5 percent under Carter, fell to 4.4 percent in 1988. Unemployment fell to 5.5 percent as 18 million new jobs were created. Interest rates were down too.

Nevertheless, there has been some confusion about the economic recovery of the Reagan years and it is worth while to go into a little detail to sort it out. We have already noted the rapid, indeed alarming growth of social security spending. The sinister aspect of this growth was that it was largely unfunded, and had been from the start. That was FDR's doing. The first social security check was issued to Ida Fuller of Brattleboro, Vermont in 1940. It was for $22.54. At the time she got this check she had paid in only $22.00 in social security taxes. By the time she drew her last check in 1974, she had received $20,944, all except $22.00 unfunded. No wonder Reagan called it 'an inter-generational Ponzi game.' It was a kind of pyramid fraud, played on youth. Congress knowingly promoted the fraud. Between 1950 and 1972 Congress, often over presidential protests, raised social security benefits or extended eligibility eleven times, six of them in election years. The most irresponsible change came in 1972, when Wilbur Mills, the great House Democratic financial panjandrum and big spender, raised it 20 percent and indexed all benefits to the Consumer Price Index. By 1982 the average pensioner was collecting (in real terms) five times in benefits what he/she had paid in taxes. By the time Reagan became president, social security was 21 percent of the total budget and continued to rise at 3.5 percent a year.

Reagan knew this in general terms and hired a budget wizard, David Stockman, to cut social security. But Stockman was incapable of explaining the problem to Reagan in ways he could grasp, through simple graphics and headline numbers. The consequences were disastrous. Reagan turned down Stockman's cutting scheme partly out of old-fashioned New Deal emotionalism but mainly because he could not understand it. It was the one big instance in which his inherent weaknesses really mattered. As a result 'Reaganomics' showed a yawning gap between theory and practice. It was supposed to produce a $28 billion budget surplus by 1986. In fact it produced an accumulated $1,193 billion deficit over the five-year period. Under Reagan the deficit, which essentially dated back to the year 1968 and had got out of hand during the collapse of executive authority in 1975, began to hit the big numbers with a vengeance. He left it running at the rate of $137.3 billion a year. Meanwhile the public debt had risen to $2,684 trillion. In the last year of Carter, when Reagan took over, interest payments on the debt took up the first 10 percent of the budget receipts. By the first year of Reagan's successor, the percentage had risen to 15.

That was the Reagan downside. But we now know it was made to look worse than it actually was. Despite the prolonged expansion of the economy during the 1980s and into the 1990s, figures for US national wealth appeared to show it growing more slowly than its main competitors, indicating that the country was still in relative decline. Other figures suggested that wage rates in the US were failing to keep pace with rates (in real terms) in other leading economies. It was calculated that real average hourly earnings in America were actually falling. A mid–1990s study showed they had declined by 9 percent between 1975 and 1996. Another study indicated that real average family incomes rose by only 2 percent between 1978 and 1995. The fall in real wage rates was attributed to assumptions that the 18 million new jobs created under Reagan (plus 7 million more under President Bush) were mostly poorly paid, unskilled jobs mainly in service industries. The poor figures for median family incomes were quoted as evidence of the continuing, even deepening, problem of poverty in the US.

However, these and other figures did not seem to correspond to what most people regarded as the reality of modern America in the 1980s and 1990s; that it was an increasingly prosperous country, with almost universally and visibly rising living standards, and that even those officially defined as 'poor' were manifestly living better. Then at the end of 1996, a panel of experts led by the Stanford University economist Michael Boskin, and known as the Congressional Advisory Commission on the Consumer Price Index, released figures showing that the Consumer Price Index had been overestimating the inflation rate of the US economy, that it had been doing so steadily but increasingly for twenty years, and that in consequence some of the most fundamental statistics about the US economy were grievously misleading. The effect of exaggerated rises in the CPI over so long a period was to depress the value of real wages (that is, nominal dollar wages adjusted to take account of inflation). At the request of the *New York Times*, the economist Leonard Nakamura conducted a revisionist exercise. He calculated that the CPI had overstated the US inflation rate in the 1970s by 1.25 percentage points (compounded) annually and that the figure had slowly risen to 2.75 in 1996. By feeding a corrected CPI inflation figure into the statistics, Nakamura estimated that real average hourly aggregates, far from falling, had actually risen by 25 percent, 1975–96. Moreover, the GDP had grown by twice the received rate during the period and family incomes had shot up by 19 percent.[218]

This revelation of the frailty of official figures also cast new light on the deficit. Under the Wilbur Mills index-linking Act, the US government was obliged by law to raise social security provisions annually in accordance with the CPI. The Boskin Commission showed that these increases had been consistently added, at rates made more significant by compound interest, appreciably higher than the real increase in the cost of living. The old, benefiting in any event from an unfunded system, were doing even better than critics had supposed. This explains why retailers and advertising agents were paying more and more attention to the elderly as sources of spending power. But it also explained why the deficit had grown so fast, because an exaggerated CPI affected government checks going to all kinds of other categories, in addition to the pensioners. (It is true that taxpayers also benefited from an overestimated inflation rate, but only marginally.) The Boskin Commission estimated that, if the CPI remained uncorrected, an extra $1 trillion would be added to the debt by 2008. On the other hand if the corrections were made, one-third of the budget deficit projected for 2005 would disappear. And, projected backwards, it could be seen that the false CPI calculations were responsible for a large percentage of the deficit under Reagan, and for which he was blamed. Moreover, the revised figures for the growth in national wealth, wage rates, and family incomes made it clear that the 'feel-good' atmosphere which Reagan succeeded in generating was not just a public relations exercise by the

Great Communicator but was solidly based on real improvements. There had been no 'smoke and mirrors' after all!

Leaving aside this exercise in statistical revisionism, the Reagan years brought solid achievements which were manifestly real even at the time. It was Reagan's aim, from the start, to restore the confidence of ordinary Americans in themselves and in their country. Reagan was good at paying attention to things which to more sophisticated people seemed so obvious as to be not worth bothering about. He thought that a good deal of America's pride had rested in the presidency. In his view, the Imperial presidency, from FDR to Nixon, had worked very well, and the diminished presidency of 1974–80 had failed. He rightly blamed Ford and Carter. On August 10, 1974, the day after Nixon left office, Ford ordered the Marine Guard to stand down from the front of the West Wing of the White House in symbolic acknowledgment of the decline in the power of the presidency. Carter, a scruffy person at the best of times, wore sweaters doing business in the Oval Office. Reagan ended all that egalitarianism immediately. He did not need to be told that the stage and democratic statesmanship have a lot in common. As his White House administrative officer he appointed a young student of the presidency, John F. W. Rogers, who was delighted to recreate its solemn grandeur. He brought back 'Hail to the Chief' and the Herald Trumpeters. Presidential symbolism and sumptuary were brushed up. Staff wore suits and ties to the office—by order. Guards of honor were regularly in attendance and properly inspected. The armed forces were delighted. Foreign visitors were impressed. The public was pleased. And Reagan flourished as the chief actor in a show which was again playing to full houses.

But it was more than show. There was substance. Reagan was not content with the relative decline in US military power. He did not think it was necessary. He was sure it was undesirable. He noted that defense spending as a percentage of GDP, which had been 13.2 percent at the peak of the Korean War, had fallen to 8.9 percent at the peak of hostilities in Southeast Asia. By the time he reached office it was only 4.8 percent of GNP. One of Reagan's axioms was that real need in global terms was to be the criterion of defense spending, not financial restraints. He believed, as did Margaret Thatcher, that the flowing tide of Soviet expansion in the 1970s could be reversed by judicious additions to the West's defense efforts, and a renewed will to resist further encroachments. He also believed (as did she) that Russia was a fundamentally flawed power economically, facing growing pressures of all

kinds, and that its will to match the West in global defense would eventually falter and crack provided the West itself reasserted the determination which had been so conspicuously lacking in the 1970s. So he
embarked on a rearmament program to recover the lost ground and
morale. In current dollars, defense spending rose from $119.3 billion in
1979 to $209.9 billion in 1983 and $273.4 billion in 1986. Reagan
said: 'I asked [the Joint Chiefs of Staff] to tell me what new weapons
they needed to achieve military superiority over our potential enemies.'
If it came to a choice between national security and the deficit, 'I'd have
to come down on the side of national defense.'[219]

One of the results of increased spending was the redressing of the
strategic nuclear balance, upset by the fact that Russia had, for nearly
a decade when Reagan took office, been outspending the US by 50
percent annually on missiles. Reagan was particularly concerned by
the large-scale deployment in eastern Europe of intermediate-range,
multiple-warhead SS20 rockets. On June 17, 1980 Margaret Thatcher
had negotiated with President Carter an agreement to counter the SS20s
by deploying American Cruise missiles in Britain. On the basis of this first
move, Reagan and Thatcher were able to persuade other NATO members
to provide sites for the Cruise network and also to deploy Pershing missiles. Deploying Cruise in particular served notice on the Soviet government that the era of indecision in the White House was over.[220]

Reagan was seen at his most decisive and masterful in his adoption
of the Strategic Defense Initiative, a project conceived by his National
Security Advisor, Robert C. McFarlane. This was an attempt to provide
an effective defense against incoming missiles and so to move away
from the horrific concept of Mutually Assured Destruction which had
dominated the Cold War in the 1970s. McFarlane and Reagan agreed
that MAD was a horrible concept anyway and had the additional
demerit that it placed the United States, Russia, and China on the same
moral (and technical) level. The SDI (or 'Star Wars' as it came to be
known), on the other hand, allowed the US to make full use of its
advanced technology, where it held a big (and, as it turned out, growing) lead over Russia. SDI was an example of Reagan's ability to grasp
a big new idea, simplify it, and give it all it was worth, including presenting it to the American people with consummate skill. It was the
most important change in American strategic policy since the adoption
of containment and the foundation of NATO, yet from its inception in
1982 to its adoption (and publication) it took only a year. The old man
could hustle when necessary.[221]

The rearmament program also included the expansion and training of Rapid Deployment Forces, de-mothballing World War Two battleships and equipping them with Cruise missiles, the development of the radar-evasive Stealth bomber, and a vast panoply of high-technology missiles, defensive and offensive, nuclear and conventional. Strategic planning and tactical training of the US forces were redesigned around the use, for both conventional and nuclear purposes, of these new weapons, a change which was to prove of critical importance in the 1991–2 Gulf War. However, Reagan did not hesitate to use armed force, in appropriate measure, during his presidency. On April 2, 1982, without warning or declaration of war, Argentine armed forces invaded and occupied the British Falkland Islands in the South Atlantic. Reagan agreed with Margaret Thatcher that this act of unprovoked aggression must be reversed, and the US provided valuable high-technology intelligence assistance to the British forces deployed to remove the aggressors, an operation which ended in the unconditional surrender of the Argentine forces (and, in consequence, the replacement of the Argentine military dictatorship by a democratic regime). In late October 1983 Reagan took decisive action to reverse the violent installation of a Communist regime in the island of Grenada in the West Indies, a move which had been made with Cuban and Russian support and was designed to lead to Communist takeovers in other West Indian islands. Reagan acted at the request of the leaders of Grenada's neighbors, Jamaica, Barbados, St Vincent, St Lucia, Dominica, and Antigua, and the speed and secrecy with which he moved was an important part of the operation's success. Reagan took the decision to act on October 21, US troops landed on the 25th and restored constitutional authority, and began withdrawing promptly on November 2. Reagan also took effective action against international terrorism, a growing scourge of the 1970s and 1980s. On July 8, 1985, he branded five nations, Iran, North Korea, Cuba, Nicaragua, and Libya as 'members of a confederation of terrorist states,' carrying out 'outright acts of war' against the US. On April 5, 1986, a terrorist bomb exploded in a Berlin disco frequented by US servicemen, killing one and a Turkish woman, and injuring 200. US intercepts established beyond doubt that Libya had a hand in this crime and on April 14 Reagan authorized US F111 bombers to carry out an attack on Gadafy's military headquarters and barracks in Tripoli, Mrs Thatcher giving permission for US aircraft to operate from their bases in Britain. This successful raid got the message across to Gadafy and others.[222]

Reagan's confident simplicity and willingness to take risks and act fast continued to surprise his colleagues. Critics laughed at his supposed naivety, but as 'Bud' McFarlane would say, shaking his head in wonder, 'He knows so little and accomplishes so much.' George Shultz, Reagan's Secretary of State, presented a remarkable portrait of his boss. He wrote: 'Reagan knew more about the big picture and matters of salient importance than most people, perhaps especially some of his immediate staff, gave him credit for or appreciated. He had blind spots and a tendency to avoid tedious detail. But the job of those around him was to protect him from those weaknesses and build on his strengths.' He had, said Shultz, 'a strong and constructive agenda, much of it labeled impossible and unsustainable in the early years of his presidency. He challenged the conventional wisdom on arms-control, on the possibility of movement towards freedom in the Communist-dominated world, on the need to stand up to Iran in the Persian Gulf, on the superiority of market- and enterprise-based economies.' Reagan's great virtue, said Shultz, was that he 'did not accept that extensive political opposition doomed an attractive idea. He would fight resolutely for an idea believing that, if it was valid, he could persuade the American people to support it.'[223]

Reagan's essential achievement was to restore the will and self-confidence of the American people, while at the same time breaking the will and undermining the self-confidence of the small group of men who ran what he insisted on calling the 'Evil Empire' of Communism. In this second part of his task, he was aided by an extraordinary stroke of fortune. From December 1979, the Soviet leaders, from a mixture of fear, greed, and good intentions, plunged into the civil war in Afghanistan which they had helped to promote. They thus became involved in a Vietnam-type guerrilla war which they could never win and which placed a growing strain on their resources. 'The boot,' as Reagan put it, was 'now on the other foot.' In Vietnam, Russia and China, by supplying weapons, had been able to inflict a totally incommensurate degree of damage on US resources and morale. Now America, in turn, by supplying a comparatively small number of weapons to the Afghan guerrillas, was able to put an immense strain on the Soviet economy and, still more, on the resolution of its leadership. The war lasted a decade and Russia never even began to win it, despite deploying 120,000 troops (16,000 of whom were killed and 30,000 wounded) and immense numbers of tanks, aircraft, and helicopters, many of which were destroyed by US-supplied weapons. During the last year of Reagan's presidency, the new Soviet dictator, Mikhail Gorbachev, accepted the inevitable: on February 8, 1988 he announced that Soviet troops would pull out of Afghanistan completely, and the withdrawal was completed by February 15, 1989, just after Reagan handed over to his successor.

The cost of the war to the already strained and declining Soviet economy was unbearable and undoubtedly played a major role in bringing about the changes in Moscow's thinking which began in the mid–1980s and soon accelerated. On top of the debilitating effect of Afghanistan came the Reagan rearmament program, and especially the Strategic Defense Initiative. In a desperate attempt to match the US technological effort in defense, the Soviet leadership did the unthinkable: it attempted to reform the Marxist–Leninist economy in the hope of reinvigorating it. In the process they lost control of the system as a whole, saw the Evil Empire disintegrate, and watched, almost helpless, as the great monolith of the USSR collapsed. So Communism was submerged, the Cold War ended, and the United States became the world's sole superpower. The effect of SDI was to add to the stresses on the Soviet economy and thus eventually destroy the totalitarian states. As Vladimir Lukhim, Soviet foreign policy expert and later ambassador to the United States, put it at the Carnegie Endowment for International Peace in Washington in 1992: 'It is clear that SDI accelerated our catastrophe by at least five years.'[224]

The end of the Cold War began on September 12, 1989 when the first non-Communist government took over in Warsaw, continued with the destruction of the Berlin Wall on the night of November 9–10, and culminated in the first half of 1991, when the Soviet Union was abolished and Boris Yeltsin was elected first democratic President of a non-Communist Russia. All this took place after Reagan had left the White House but it came as no surprise to him. He had never accepted the view, commonly expressed even late in his presidency, that the US was in irreversible decline, especially in relation to Russia. One of the most fashionable books of 1988, much flourished by liberal pundits as a stick to beat the Reagan administration, pointed to the many 'threats' to America's economic wellbeing, especially from Mexico: 'By far the most worrying situation of all . . . lies just to the south of the United States, and makes the Polish "crisis" for the USSR seem small by comparison.'[225] Reagan was content to follow the guidance of his own President's Committee on Integrated Long Term Strategy, which also in 1988 published its long-term forecast, *Discriminate Deterrence*. This argued that there were no economic reasons why the US should not continue to shoulder its superpower burden and continue its limited role of preserving peace up to the year 2010 and beyond. It predicted that, by this date, the USSR would have sunk to the fourth place in the world league (fifth if the European Union was counted as a single power) and that America, with an $8 trillion economy, would still be producing twice as much as its nearest rival, China. The events of the years 1989–91 reinforced the essential truth of this calculation, and no developments which have occurred since the early 1990s have undermined it.[226]

Reagan's Vice-President, George Bush, won the Republican nomination in 1988 by acclamation because he was identified with the Reagan achievement, and he won the election handsomely because he promised that he would carry on in the Reagan spirit, and was generally believed. His opponent, Michael Dukakis, governor of Massachusetts, said he stood on his state record, which he said demonstrated his 'competence,' but he made a mess of organizing his campaign, which suggested otherwise. The result gave Bush (and his running mate Dan Quayle, Senator from Indiana) a comfortable plurality, 48,886,097 to 41,809,074, and a big majority in the electoral college, 426 to 111.[227] It had been forgotten, however, that Bush had originally hotly opposed Reagan in 1980 as a maverick and extremist, and had joined Reagan on the ticket as part of a balanced compromise package. He was not fundamentally in sympathy with Reagan's aims either intellectually or by temperament and, more important, he lacked Reagan's simple clarity and will. Bush was at his best when he had a more resolute foreign ally to stiffen his resolve, and indeed to make up his mind for him.

One of the consequences of the end of the Cold War was that Russia (and to some extent even China, which remained Communist) began to play a more responsible role at the United Nations and so allowed the Security Council to operate, for the first time, as its founders had intended. During the 1980s, the Iraqi dictator Saddam Hussein, who was in bitter dispute with fundamentalist Iran, had received some support and encouragement from the West, not least America, on the basis of the old adage, 'my enemy's enemy is my friend.' Thus encouraged, he had used his country's oil revenues to build up immense armed forces. Suspicion of Saddam's intentions had been growing, however, and on July 27, 1990 the US Senate, without any prompting from the administration, prohibited the sale of any further military technology to Iraq and, for good measure, cut off farm credits. By July 31, over

100,000 Iraqi troops were massed on the border of Kuwait, the small but immensely rich Gulf state which was a major supplier of crude to the United States and many other advanced countries. At the time it was widely reported that the American ambassador in Baghdad had failed to warn Saddam during a meeting at this time that an occupation of Kuwait would be regarded by Washington as a threat to America's vital interests, though this was denied in evidence before the Senate Foreign Affairs Committee in March 1990.[228] At all events, on August 1, 1990 Saddam's army invaded and occupied Kuwait and annexed its oilfields.

Happily, this unprovoked act of aggression coincided with a private international meeting in Aspen, Colorado, attended by both President Bush and Margaret Thatcher, who was then still British Prime Minister. Bush's first inclination was to concentrate on the defense of Saudi Arabia, America's chief ally (and oil supplier) in the area. But Mrs Thatcher, with the Falklands campaign in mind, forcefully persuaded him that the act of aggression must be reversed, not merely for the sake of the unfortunate people of Kuwait, who were being treated with merciless cruelty by the Iraqi troops, but to deter any other potential aggressor. Bush agreed, that was the plan adopted, the Security Council was consulted and signified its consent, and over the next few weeks, under Bush's leadership, an international coalition of over fifty states was assembled. An enormous armed force, chiefly composed of American units but with significant contributions by Britain and France (as well as many Arab states) was assembled in the area. In December 1990 Bush, quoting the 'Carter Doctrine,' secured the consent of the Democratic Congress to use force, and Saddam was told that, if his troops were not withdrawn by January 15, 1991, he would be ejected from Kuwait by force.

By this time, Mrs Thatcher had been ousted by an internal coup in her own party, being succeeded by the featureless John Major. But the momentum of Operation Desert Storm, as it was called, was unstoppable, even in Bush's hands, and when Saddam missed the deadline, intense aerial bombing of Iraqi military targets began. The ground offensive, involving units from the huge coalition army (540,000 US troops, 43,000 British, 40,000 Egyptian, 16,000 French, and 20,000 Syrians) began on February 24 and lasted five days, during which all Iraqi troops had been cleared from Kuwait or surrendered, and Saddam's 600,000-strong army decisively defeated. This brief campaign was a triumph for General Norman K. Schwarzkopf, the

American force commander, and the chairman of the Joint Chiefs of Staff, General Colin L. Powell, who had organized the gigantic build-up and directed the overall strategy. But neither was anxious to press on to Baghdad and impose a democratic alternative to Saddam's military dictatorship, which they saw as a 'political' extension of a UN-authorized campaign which was entirely military. In the absence of pressure from Mrs Thatcher to finish the job, Bush was content with the liberation of Kuwait. So an unrepentant Saddam was left in power, with much of his armed forces intact, and the United States was left with the burden of maintaining large forces in the area and supervising Saddam's adherence to the terms he had agreed for a ceasefire and to UN resolutions.

This unsatisfactory conclusion and aftermath to one of the most brilliant military campaigns in US history, which had demonstrated the power and efficacy of the new generation of sophisticated weapons produced by Reagan's rearmament program, naturally lowered public confidence in President Bush's judgment and resolve. He was increasingly seen as well meaning but ineffectual and indecisive. There was indeed something fundamentally wrong with Bush, to which his curriculum vitae gave a clue. In addition to serving two terms as vice-president, he had been a Texas oilman, a Texas congressman (1967–70), American ambassador to the UN (1971–2) and to China (1974–5), head of the CIA (1976) and chairman of the Republican National Committee (1972–4). But he had done all these jobs for brief spells before moving on, and there was always the suggestion that he was not a stayer. Moreover he had fought two unsuccessful campaigns to get into the Senate, as well as his aborted bid for the presidency in 1980. Though presenting himself as a large-scale Texan operator, he had in fact been born in Massachusetts, had an Ivy League education, lived in Maine, and worshiped in an Episcopalian church: he was, in short, an Eastern Wasp in a stetson. He was curiously inarticulate for a well-educated man, used words clumsily (unlike Reagan, whose verbal touch was always sure). He had an unfortunate habit of abandoning his written text, in speeches, for extempore remarks, and beginning sentences without being sure what his main verb would be, entangling himself in syntactical confusion and thus ending each utterance on an uncertain note.[229]

Bush was always a gentleman and, so long as the Reagan penumbra surrounded him, did moderately well. Then, as the impact of Reagan faded, Bush faded with it. Panama was a case in point. President Carter had concluded a treaty with Panama, ratified by the Senate on March 16, 1978, which in effect handed over the Panama Canal in stages to a sovereign Panamanian government. This retreat of US power had been much criticized, and the feeling that America had got a poor deal was reinforced by the behavior of President Manuel Noriega, who used the Panamanian government to operate a huge drug-smuggling business in the United States. In 1989 Florida grand juries indicted Noriega on a variety of serious drug offenses and demanded his extradition. America imposed economic sanctions when Noriega refused to surrender to the US authorities and, after a US naval officer was killed by Panamanian police on December 16, Bush brought himself to order an invasion of the country by 10,000 American troops. Panamanian dissidents duly overthrew the government; Noriega was arrested on January 3, 1990 and taken to Miami, where he was convicted of drug-smuggling on April 9, 1992 and sentenced to forty years in a top-security prison. But, characteristically, Bush did not follow up this striking, and much approved, success by revising the operations of the Panama Canal, where conditions continued to deteriorate under Panamanian control. Bush's instincts were often right but he did not know the meaning of the word 'thorough.' Thus the fact that he delighted to display a portrait of Theodore Roosevelt in the Cabinet room and tell visitors about his quiet voice and Big Stick, was an ironic comment on his administration.[230]

Nevertheless, it is likely Bush would have been reelected in 1992 had it not been for the intervention of a Texan billionaire, H. Ross Perot, in the race. Perot, born in 1930, was an Annapolis graduate who left the US Navy in 1957 to found Electronic Data Systems, which he built into a firm worth over $2 billion by the mid–1980s. Campaigning on a platform of eliminating the deficit and paying off the national debt, Perot originally attracted more interest, in a poor field, than any third candidate since the Bull Moose Party. However, his indecisiveness—he filed his candidacy on March 18, 1992, pulled out on July 16, and reentered on October 1—ruined his chances of a dramatic upset. Even so, he polled 19 percent of the votes, the largest third-party fraction of the polls since Theodore Roosevelt in 1912 (27.4 percent) and Millard Fillmore in 1856 (21.6 percent). The effect of the high Perot poll was to let in a lackluster Democratic candidate, William ('Bill') Jefferson Clinton, governor of Arkansas. Born in 1946, Clinton came from a modest family background and rose through academic performance, was Rhodes Scholar at Oxford, graduated at Yale Law School in 1973, and taught law at the University of Arkansas. Critics dismissed him as a 'spoiled Sixties rebel' and 'draft dodger' who had consistently taken up anti-American postures from Vietnam to the Gulf War. Others drew attention to his career in the murky waters of Arkansas politics. There, he ran unsuccessfully for Congress at thirty-two, was elected Arkansas attorney-general in 1976, and became the state's youngest governor in 1978. Although defeated in 1980, he won reelection as governor in 1982, 1984, and 1986. In Arkansas he was universally known as 'The Boy,' and when elected president of the United States in 1992, with 43 percent of the vote, he was the first minority President since 1968 and the youngest since Kennedy was chosen in 1960.[231]

Clinton, and his wife Hillary Rodham Clinton, rapidly ran into trouble in the White House. One reason was inexperience. Clinton had great difficulty in completing his administration and getting Senate approval for some of his nominees, especially for the post of attorney-general. When he finally succeeded, his choice, Janet Reno, soon involved the administration in a disastrous episode near Waco, Texas, where federal agents conducted an ill-planned assault on the headquarters of the Branch Davidian religious sect, run by a messianic figure called David Koresh. The raid, by nearly 100 agents from the Bureau of Alcohol, Tobacco, and Firearms, was a failure and led to five deaths, four of them agents. Reno then took over personally and, tiring of negotiations, she ordered a military-style attack with heavy equipment and gas, during which the entire compound went up in flames, in which seventy-eight of Koresh's followers were burned alive. Among the dead were seventeen children.

On the more positive side, Clinton entrusted the presentation of his health-care bill on Capitol Hill to his wife, an outspoken and venturesome lady who impressed the legislators, one of whom, Democrat Dan Rostenkowski, of the House Ways and Means Committee, told her: 'In the very near future, the President will be known as your husband.' This turned into an accurate, if embarrassing, prophecy since the First Lady was soon involved in a complex financial investigation into an Arkansas firm with which she was connected, the Whitewater Development Corporation, and a number of other awkward matters. Indeed she became the first President's wife to be subpoenaed in a criminal investigation.[232] In the meantime the health-care Bill was blocked. Clinton's budget-package attempt to reduce the deficit, as amended by Congress, cut it from 4 percent to 2.4 percent of GDP (from $255 billion in 1993 to $167 billion in 1994), but there was little general coherence to his domestic program and still less to his foreign policy. He ended Bush's disastrous involvement in Somalia; on the other hand, he occupied Haiti for a spell, to little purpose. He was tardy in trying to sort out the series of Balkan conflicts left by the disintegration of the former Yugoslav federation and he used the continuing difficulties with Saddam's Iraq more as an occasional exercise to strike military postures and win votes than for any constructive purpose.[233] Virtually all the rest of Clinton's first term was spent fending off a series of scandals.

The idea of a president being brought down by the exposure of wrongdoing had haunted the White House since Watergate, though it must be said that the media had never since shown quite the same relish for presidential blood. Under Reagan, National Security Council officials Admiral John Poindexter and Colonel Oliver North had been accused of conducting secret arms sales (via Israel) to Iran, hoping for goodwill to free US hostages held by the fundamentalist regime and in order to use profits from the $48 million sales to finance democratic Contra rebels seeking to undermine the Communist regime in Nicaragua. This was said to be in violation of the Boland Amendments, passed by Congress on December 21, 1982, which forbade the CIA or the Defense Department from using any funds 'for the purpose of overthrowing the government of Nicaragua or provoking a military exchange between Nicaragua and Honduras,' and by a further Amendment on October 12, 1984, which extended the prohibition to any agency involved in 'intelligence activities.' There was a general impression in Congress during the 1970s and 1980s that US undercover activities, including the support of democratic insurgents, were the product of the Cold War. Actually, they go back to President Washington, who was voted a 'Contingency Fund' for this purpose and given entire discretion over its use. Madison funded dissidents in Spanish Florida and Jefferson wanted to finance an attempt to burn down St James's Palace in retaliation for the burning of Washington. Andrew Jackson used secret funds to subvert Mexico and James Polk subsidized American insurgents in Texas. Lincoln created an organization in Europe for secret purposes, eventually dignified with the title of the Office of Naval Intelligence, and there were many instances of covert work in almost every subsequent presidency.[234] An attempt was made not only to indict North and Poindexter and crank up a Congressional witchhunt on the scale of Watergate but even to involve President Reagan himself. However, after a series of televised hearings from November 1986, which failed to inflame the public, a Senate–House committee, dominated by Democrats, was obliged to report there was no direct evidence involving Reagan. In 1989 both Poindexter (April 7) and North (May 4) were found guilty of misleading Congress, but an appeals court (July 20, 1990) overturned both convictions and other charges against them were dropped.

The witchhunt served only to reveal a waning popular backing for proceedings against officials who had merely acted in what they thought American interests, even if they had technically broken the law. At the same time, several minor functionaries were forced to plea-bargain and had their savings wiped out by the heavy legal costs involved in defending themselves against a special prosecutor, with unlimited funds, appointed solely to deal with possible unlawful activities by the executive branch. The continual fear of prosecution engendered what has been called a 'culture of mistrust' in American government and led to an unwillingness of public-spirited citizens to run for high political office or to accept presidential invitations to serve. One critic of the intermittent witchhunting of the executive by Congress noted that 'a self-enforcing scandal machine' had come into existence: 'Prosecutors use journalists to publicise criminal cases [involving members of the administration] while journalists, through their news-stories, put pressure on prosecutors for still more action.' 'Scandal politics' were thus part of the process whereby 'popular democracy' was degenerating into a 'media democracy.'[235]

It could be argued that the growing reluctance of gifted men and women to become candidates for national office explained the presidency of Bill Clinton. Certainly he was soon in trouble both for past misdemeanors as governor of Arkansas and for present malfeasance and abuse of power as chief magistrate. The charges against Clinton fell under three heads. First there was his own personal involvement, as his wife's partner, in the Whitewater and related scandals. Second was a series of sexual accusations made against him by a procession of women with names like Gennifer Flowers, Connie Hamzy, Sally Perdue, and Paula Corbin, some of these charges including Clinton's use of Arkansas police as procurers. Third were a miscellaneous series of charges reflecting badly on Clinton's behavior in federal office, the most important of which concerned the mysterious death of his aide, Vince Foster. Since Watergate, the media had developed an irritating habit of adding 'gate' to the name of any episode with a whiff of political scandal. So, in addition to Irangate in the 1980s, the 1990s now saw piling up Travelgate and Troopergate, Whitewatergate, Fostergate, and Haircutgate. This wearied the public, and, granted the unwillingness of newspapers like the *Washington Post* and the *New York Times* to investigate President Clinton's wrongdoing with the same zeal they had once displayed in Nixon's case, or even to report the investigations of others, it is not surprising that the climate of scandal, in the end, did not prevent Clinton getting reelected in 1996. It might have been a different matter if the Republicans had come up with a genuine vote-winner. But General Colin Powell would not run, or perhaps his wife, the strong-minded but fearful-of-assassination Alma, would not let him run. So the Republicans, in despair, at last picked the elderly Senator Robert Dole, who had been demanding to be their candidate as long as most people could remember, and in November 1996 Clinton had no difficulty at all in fending him off and getting himself reelected by a comfortable margin.[236]

The failures and moral inadequacies of Clinton did have one important, indeed historic consequence however. In the mid-term elections of 1994, Clinton's unpopularity enabled a brilliant political tactician, Newt Gingrich, the Republican Minority Leader in the House, to organize a nationwide campaign which ended in the Republican recapture of both Houses of Congress. This victory, the biggest Congressional upset since 1946, was significant for a number of reasons. For all practical purposes, the Democrats had dominated Congress, except for one or two brief intervals, since 1932. There had been no precedent for this long legislative hegemony in American history and, the Democrats tending to be the high-tax, high-spending party, their continued tenure of the Constitution's financial mechanism explained why the federal government had continued to absorb so large a proportion of the GDP, why the deficit had mounted, and why national indebtedness had become such a problem. What was so remarkable about Democratic control of Congress was that it had continued during the period when the nation preferred to elect Republican presidents. Thus in every Congress from 1961 to 1993 (the 87th to the 103rd), the Democrats had a majority in the House, even though the Republicans held the presidency in ten of them. These majorities were sometimes overwhelming and always substantial: even in Nixon's best year, 1973, for instance, the Democrats still had a margin of 47 in the House. Carter, a weak Democratic president, nevertheless had a majority of 149 in 1977. Under Reagan, who twice won Republican landslides, the Democratic majority varied from 51 to 104. Reagan did contrive, however, to exert a pull over senatorial elections and had a small Republican Senate majority in three of his four Congresses.[237] But the situation returned to 'normal' under Bush, who had to work throughout his presidency with Congresses in which his opponents controlled both Houses easily.

This Democratic paramountcy did not come about for political reasons. Expenditure on political campaigning had been rising throughout the 20th century and incumbents found it easier to raise money both because of their higher exposure and because they were able to direct federal funds to their districts. The Federal Election Campaign Act of 1974 attempted to reduce candidates' dependence on money, but the Supreme Court struck down one of its central provisions as unconstitutional (the limitation on contributions by candidates themselves) and as a result it actually worked in favor of the incumbent. By 1990, over 90 percent of the House members were getting themselves reelected, mainly for financial reasons, against 75 percent in the late 1940s.[238] Hence the 1994 defeat, which reversed a strong Democratic majority in the House and secured the Republicans a working majority in the Senate too, was achieved against a long-entrenched structural bias in the system, and came as a profound shock to the Democrats. Its effect on Clinton was devastating. Dick Morris, his chief political strategist, described a scene with the President after the defeat: 'Then I looked straight at him and shook him harshly, violently. I said through clenched teeth: "Get your nerve back. Get your f— nerve back." He looked at me with bloodshot weary eyes and with face downcast, solemnly nodded yes.'[239] Morris persuaded Clinton that the remedy was to move his presidency towards the center and cooperate with Congress in pushing through popular legislation of a more conservative cast. Morris himself fell as a result of a morals scandal in autumn 1996 but before he did so he had largely achieved his object.[240]

In August 1996 Clinton signed into law the Welfare Reform Act, the first major piece of legislation from the new Republican Congress, which dismantled a fundamental part of the welfare-state structure going back to the New Deal in 1935. Under this Aid to Families with Dependent Children ceased to be an open-ended distribution of cash entitlements to individual beneficiaries and became a $16.4 billion lump-sum 'block grant' to the fifty states to allocate as they believed more effectively. This return of financial power from the federal government to the states set up pragmatic competition between them in ways to use money to reduce poverty. It marked the first change of approach to the problem since Lyndon Johnson declared 'unconditional war on poverty' in 1964—a war which had been going on longer than Vietnam, had cost more, and had been even less successful. Clinton's action in signing the Bill into law was the most significant move of his presidency and helped to explain how he contrived to get himself reelected three months later. The Republicans, however, retained control of the House—the first time they reelected a House majority since 1928—and even strengthened their grip on the Senate. This insured that Clinton's move towards the center would be sustained and that Congress would continue to probe into the chinks in his armor.[241]

The Welfare Reform Act reflected a general feeling among Americans in the late 1990s that the country had moved too far in the direction of what might be called federal centralism, and that a redressment was needed. At the end of the 20th century America remained a profoundly democratic country. All its instincts were democratic, and the core of its democracy lay in the fifty states. The states were merely a top tier of a series of devolved representations. In 1987, the last count, there were 83,166 local governments in the United States, with over 526,000 elective posts.[242] No other country in the world had anything approaching this number and diversity of democratic forms. Considerable local power was still exercised, especially by the states, producing a wide variety of laws. Thus, most states levy income tax, but twelve do not. Most states have capital punishment, but twelve do not. Oregon and Hawaii have forms of state planning unknown in the other forty-eight. Utah is dry, whereas Louisiana has virtually no laws about liquor. Up to about 1933, the states defended their rights effectively (except during the Civil War). In the 1930s, however, the federal government began to invade the states' right to handle economic issues. At first the Supreme Court resisted this federal invasion, but since the 1950s the Court, Congress, and the federal government have combined to diminish the powers of the states.[243] Strictly speaking, the Constitution gives the states only four absolute guarantees: equal representation in the Senate; protection from invasion and domestic violence; the right to jurisdictional integrity; and the right to a republican form of government. All the rest is open to interpretation and the interpretations by the Court since the 1940s have gone against the states.

We will come to the role of the Court itself shortly, but in the meantime it is worth noting that the switch from state to federal power tends to work against democracy. There is in America no automatic, nationally organized, and compulsory registration system, and only about 70 percent of those eligible are registered. If one existed, turnout in federal elections would rise by 10–12 percent. As it is, not much more than half those entitled to vote do so. This is a long-term problem. Thus there was a 52.4 percent turnout in 1932, at the height of the Great Depression, and 54 percent in 1992. The highest turnout, 62.6, was in the 1960 presidential election. Next to Switzerland, the United States has the lowest figure for voting as a percentage of the voting-age population than any other advanced state (though it is round about average in the percentage of registered voters who turn out). Local elections are often fought with far more passion and higher participation. The one respect in which federal democracy has improved in the 20th century (apart from the obvious extension of the franchise to women and eighteen-year-olds) is in the participation of blacks, especially in the South. Thus in Mississippi as recently as 1964 only 7 percent of adult blacks were registered; by 1990, black registration was only 8 percent below white, though the turnout was lower.[244]

All aspects of local government, from the states down, are particularly important because the US continued to grow in population throughout the 20th century. It was part of the uniqueness of America that it was the only advanced country with this apparently invincible propensity. The biggest decennial increase was in the 1950s, when population rose by 27,997,377, as big as the entire population of a medium-large European country. During the 1980s the jump upwards was almost as large, 27,164,068, giving the country a population of 249,632,692 in 1990. The census for 2000 is likely to show an even bigger increase, taking the US to a population figure of around 280,000,000. Birth-rates remained high, but a lot of the increase was due to a new surge of immigration. In fact the 1965 Immigration Act produced the greatest legal flood of new arrivals since the years 1900–14: the share of immigration in total US population growth rose from 11 percent in 1960–70 to 35 percent in 1970–80 and to almost 40 percent 1980–90. (These figures leave out illegal immigration, which is large but often transitory.) Demographic projections, especially long-term ones, are notoriously unreliable, but if the trends of the last forty years continued, the total population of the US would rise to 400 million by the end of the 21st century, with whites (as conventionally defined) in a minority as early as 2180.[245] The effect of immigration and demographics on American culture will be examined shortly. For the moment, it is important to register that America has become notably more crowded in the 20th century. It grew geographically at an astonishing rate in the 19th century. In 1790 it had a land area of 864,746 square miles. At the end of this century it is 3,536,342 square miles, more than four times greater. But while area has long since ceased to expand, population growth has continued relentlessly. In 1800 the average population density of the United States was 6.1 persons per square mile. At the beginning of this century it had risen to 25.6 and by the end of it to well over 70 persons per square mile.[246]

Yet America remained, at the end of the 20th century, a country with enormous contrasts between huge crowded cities and almost primeval wilderness. It was still the country of the skyscraper, reaching upward to breathe air into intense city densities, even in places where land is plentiful. Thus Phoenix, Arizona, a typical 20th-century American big city, grew from 17 square miles with a population of 107,000 in 1950, to 248 square miles and a population of 584,000 in 1970, to 450 square miles with a population of 1,052,000 in 1990—but it continued to put up towers of glass and steel.[247] The first phase of skyscraper-building ended with the completion of the Empire State Building in New York in 1931. Up to that point, towers had been growing in size by up to ten stories a year. Then there was a long pause, partly explained by the Great Depression and the war, partly by technical difficulties. The resurgence began in the 1970s, with the construction not only of 'modest' big skyscrapers like the twin-tower World Trade Center in New York but the much higher 441-meter Sears Tower in Chicago. By the 1990s American supremacy and giganticism in skyscraper-building was being challenged all over the world, in Melbourne and Hongkong, for example, and still more in Malaysia, where in 1996 the Kuala Lumpur twin Petronas Towers, at 440 meters, snatched the crown from Chicago. However, even the planned Mori Tower in Shanghai, at 460 meters, will be only one-fifth higher than the Empire State, and techniques now exist to build towers twice as high—800 meters and more.[248] American architects and engineers are working on such projects and the tallest buildings the world has ever seen will rise in America during the first decade of the 21st century. Meanwhile, American architects, from the 1970s, had been introducing various forms of postmodernism, reacting from the plain functional towers to forms of classicism, gothicism, and ornamentalism reminiscent of the earliest skyscrapers. The first such exercise was the 'Chippendale Highboy,' the AT&T headquarters set up on Madison Avenue by Philip Johnson and John Burgee, 1979–84, with its classical broken pediment on a five-story scale set thirty-seven stories above the sidewalk.[249] This was followed by the Johnson and Burgee gothic Republican Bank, Houston, 1981–4, by the baroque Wacker Drive, Chicago, by Kohn Pedersen Fox Associates in 1981–3, and by a growing variety of extravaganzas.

The notion of architecture, and the business enterprise behind it, as large-scale playthings which—as we have noted—was born in California at the beginning of the 20th century continued to flourish in various parts of America at the end of it. This was American vitalism at its most fervid, unacceptable to some elites, evidently appealing to the many. The archetype of the city which grabbed the visitor by the lapels in the second half of the century was Las Vegas, Nevada, gambling center of the world. Las Vegas Style, based on noise, light, gigantic neon signs, and amazing architectural pastiche, first attracted outside attention when the writer Tom Wolfe paid tribute to 'the designer-sculptor geniuses of Las Vegas' in 1964. This was soon followed by a group of professors from the Yale School of Architecture, led by Robert Venturi, who took their students on an analytical tour of the city, producing a key book, *Learning from Las Vegas*.[250] Thus encouraged, the state of Nevada changed the law in 1968 to allow corporations as well as individuals to own casinos, and, with vast amounts of money now available, hotel-gambling hells on an unprecedented scale began to sprout out of the desert, recalling the creation of the world's first large-scale hotels by Americans in the early 19th century—America's earliest major contribution to cultural patterns. 'The Strip' at Las Vegas became, in the words of Alan Hess, 'the outdoor museum of American popular culture.' During the Seventies and Eighties and into the Nineties, Las Vegas acquired the Luxor, a pyramid larger than any in Egypt, made of glass not stone, the Excalibur, a romantic Old-World-style castle, on a scale no European castle-builder ever imagined, and the MGM Grand, a pastiche of Hollywood (itself a pastiche) of a daunting size, with over 5,000 rooms. Opposite was opened on January 3, 1997 the New York, New York, a giant pastiche of the Manhattan skyline, housing thousands of rooms, gambling spaces, and entertainment facilities. Characteristically American, they followed in the tradition of the gargantuan, paddle-wheeled Mississippi gambling steamers, with their Babylonian luxuries and rococo decor, which outraged and fascinated Americans between the Age of Jackson and the Civil War.[251]

Yet the urban extravagances and fantasias of end-of-the-century America coexisted with fiercely independent smaller units which still thought of themselves as mere big villages. In the mid–1990s, for instance, the County of Los Angeles forced itself out of the deep recession which, for the first time in the early 1990s, devastated Southern California, by adopting an urban village or mini-city approach, forming units of under 40,000 inhabitants, with a strongly democratic spirit and powerful sense of local identity of a kind lost in the sprawling wilderness of the city of Los Angeles itself. New or reinvigorated urban villages formed themselves with names like Burbank, Glendale, Beverly Hills, West Hollywood, South Gate, Huntington Park, Culver City, and Torrance. These developments took note of the fact that the United States grew up as an independent village culture. Early New England and Virginia were congregations of places which called themselves towns but which were in fact villages. As one urban historian has put it, 'Each of several hundred villages repeated a basic pattern. No royal statute, no master-plan, no strong legislative controls, no central administrative offices, no sheriffs or justices of the peace, no synods or prelates, none of the apparatus of government then or now.'[252] Modern urban villages can restructure their economies to meet change quickly, can keep the cost of living down more easily, reduce crime, improve education and services and with less difficulty than urban megaliths. Investors find everything is easier, cheaper, quicker, and less bureaucratic.[253]

Bigness is of the essence of America, a big country with a teeming people, but the notion that everything in United States history has tended towards ever-larger units is false. At the end of the 20th century, small towns were flourishing all over America, from the Vermont hills to the Napa Valley in California, and new ones were being born all the time. Nor has America been an inevitable Hegelian progression from the natural to the mechanical and the artificial. The very earliest European settlers in Massachusetts and Virginia made gardens, perhaps inspired by Sir Francis Bacon's observation, in his essay 'On Gardens' (1625), that 'God Almighty first planted a garden; and indeed, it is the purest of human pleasures.' Between 1890 and 1940 some of the finest gardens in the history of the world were created in the United States: Otto Kahn's Oheka in Long Island; Mary Helen Wingate Lloyd's Allgates in Haverford, Pennsylvania; Lila Vanderbilt's Shelburne Farms, Vermont; Mrs Gustavus Kirby's Fountain Garden in Westchester, New York; Millicent Estabrook's Rose Garden in Santa Barbara; Arcade in California; Branford House near Groton in Connecticut; James Deering's Vizcaya in Florida; as well as the vast gardens at Biltmore and the many-generational gardens at Hampton, Maryland, which go back to 1790.[254] In the second half of the 20th century these *plaisances* of millionaires became resorts of the millions as one after the other they were opened to the public, who thus helped to pay for their maintenance. But more gardens were being created. By mid-century American universities had more graduate schools of garden architecture and history than anywhere else in the world, and designers like Fletcher Steele (1885–1971) were performing prodigies to link the traditional with new technologies. Fletcher's great garden at Stockbridge, Massachusetts, called Naumkeag, the original name for Salem, which he created between 1947 and the 1960s, testified to the way in which America had enhanced one of the oldest of the European arts.[255]

While adding to garden history, Americans were also in the 20th century establishing themselves as unmatched preservers and cultivators of the wilderness. As we have seen, the history of the US National Parks goes back to the early 19th century. But it was the Wilderness Act of 1964 and the Federal Land Policy and Management Act of 1976 which made public authority not merely the protector of the wild but an active and advanced administrator of more than 700 million acres of land. By 1993 the Park Service handled almost 60 million visitors annually, who came to see its 49 national parks, 77 national monuments, 112 historic parks or sites, and 104 other designated areas, covering 124,433 square miles. These included the largest stretches of protected wilderness in the world.[256] The Americans had always oscillated between the convivial and the solitary, between the creator of villages, gardens, great cities, and the explorer of the primeval forest and the lonely stranger. If the monolithic skyscraper, with its thousands of inmates, remained the monument to American gregariousness at the end of the second millennium, it is not surprising that America's best painter of the second half of the 20th century, Andrew Wyeth (b. 1917), devoted his life to the depiction of isolation.

Thereby hangs a traditional tale, or rather two tales. Wyeth, seeking to paint privacy in a public land like America, modeled himself on two great American painters who had done exactly that, Winslow Homer and Thomas Eakins. Homer worked chiefly in Maine and Eakins in rural Pennsylvania; Wyeth oscillated between both for his settings. The loneliness of America, distinctive people in a vast land, was the recurrent theme of Homer in, for instance, *The Nooning* (1892) and *Boys in a Pasture* (1974). Eakins, in his attempts to get deeper into American life, which we have noted, again found himself often depicting the solitary.[257] In following in their steps, Wyeth also worked hard to acquire the delicate touch of Thomas Dewing and the watercolor skills of Sargent—two more outstanding American artists who were essentially private men. Wyeth, however, also belongs to the American tradition of illustration, one of the more formidable strengths of American art. The century 1850–1950 was the golden age of American magazine illustration, when literally hundreds of superb draftsmen, operating often at the very limits of their technology, were able to make a decent living from a vast range of pictorial publications, read by millions. Winslow Homer himself was trained, or trained himself, as a magazine illustrator. His skills were passed on to Howard Pyle, another master illustrator, and Pyle taught Wyeth's father, N. C. Wyeth, who brought up his son in this hard school. But Wyeth also looked to Dürer, the first and greatest of the grand European line-illustrators, and to an American prodigy, George de Forest Brush, who traveled widely in the American West in the last quarter of the 19th century and depicted its solitude and the loneliness of the Indian solitary hunter, often with the birds and beasts he had killed: *Out of the Silence* (1886), *The Silence Broken* (1888), *The Headdress Maker* (1890), *The Indian Hunter* (1890).[258]

Twisting together these various strands with growing skill, Wyeth began to produce outstanding paintings soon after World War Two, such as *Christina Olson* (1947), *Roasted Chestnuts* (1956), *Garret Room* (1962), *Adam* (1963), *Up in the Studio* (1965), and *Anna Kuerner* (1971).[259] Many of his works were of nudes, showing a devotion to and inspiration from another powerful American tradition: *Black Water* (1972), *Barracoon* (1976), *Surf* (1978); female nudes, these, though Wyeth also painted the male nude: *Undercover* (1970) and *The Clearing* (1979).[260] Wyeth's passion, which he shared with other major American painters, such as Edward Hopper, Charles Burchfield, Grant Wood, Edward Weston, Charles Sheeler, and Georgia O'Keeffe, was for the single object or subject, carefully isolated and intensely realized. He concentrated on particular models, whom he painted again and again over long periods, exercising astonishing patience, concentration, and singlemindedness. Thus for years he concentrated on Siri Erikson, daughter of a Maine neighbor of Finnish descent.[261] In 1971 he began work on Helga Testorf, a neighbor of Wyeth and his wife Betsy in Chadds Ford, Pennsylvania. Between that year, when Helga was thirty-eight, and 1985, when she was fifty-three and he brought the series to an end, Wyeth created one of the most remarkable series in the history of American art, 240 drawings and paintings in all. Some of the nudes in this series strongly recall the charcoal nudes of Eakins, an inspired academic craftsman of noble skills, who had been trained under the great Gérôme in Paris. Almost the entire collection was bought by the connoisseur Leonard E. B. Andrews (1986) for display in the Andrews Foundation, thus preserving for America a collective artifact of immense dignity and significance from a period when the art of painting has been vitiated (not least in America) by mere commercial fashion, determined by the agents and galleries of New York, the world's biggest art center.[262]

The creation of major works of art amid a welter of kitsch, vulgarity, and empty fashion was merely one of the paradoxes of America as it neared the end of the century. America had always been a land of paradoxes, for Europeans first came there from a mixture of greed and idealism. The two had fought for supremacy through its existence but never more so than on the eve of the third millennium. The watershed of the 1960s, after which America moved away from its founding tradition into uncharted seas, intensified the paradoxes. The Declaration of Independence had held it as 'self-evident' that 'all men are created equal,' yet it ignored the slavery in its midst. In the last half of the 20th century, America strove more earnestly than ever to achieve equality, yet it became more unequal and even the concept of equality before the law came under threat, in theory as well as practice. There was evidence, for instance, that economic inequality, which had declined in the years 1929–69, began to rise thereafter and continued to rise to at least the mid–1990s.

This point should not be overstated. America in the 1990s was essentially a middle-class country, in which over 60 percent owned their own homes, 20 percent owned stocks and bonds (though only half more than $10,000), 77 percent had completed high school, 30 percent had four years in college, and the largest single group, 44 percent, were in professional, technical, and administrative jobs. The true blue-collar working class was only 33 percent and shrinking fast, the remainder, 23 percent, being in service and farming.[263] America had 3,500 universities and colleges of higher education. The Higher Education Acts of 1967 and 1969 and legislation passed in 1972 establishing a federally funded system of need-based student aid combined, with other factors, to raise the number of those enjoying higher education to 6 million in 1965, 7.6 million in 1968, and 9.1 million in 1973. By 1996 it was estimated that 16 million students enrolled in college and that higher education was touching the lives of more than half America's young people. Of these, 9.2 percent were blacks, 5.7 percent Hispanics, and 4.2 percent of Asian origin, with women of all races making up 52 percent of the whole. Of these, 258,000 were at what were termed 'historically black colleges' and 2.2 million at private colleges (1990 figures).[264]

Nevertheless, in the mid–1990s it was claimed that about 12 percent of the population were 'below the poverty level,' that both an 'overclass' and an 'underclass' existed and that the two were drifting further and further apart. Over the twenty-year period the highest 10 percent of incomes rose in real terms by 18 percent and the lowest 10 percent fell in real terms by 11 percent.[265] At the end of the 1980s, the richest 1 percent of the population, numbering 932,000 families, together owned more than was owned by 90 percent of the entire population, and nearly half their income came from 'passive investment' (stocks, bonds, trusts, bank holdings, and property).[266] Incomes from work showed a similar pattern of rising inequality. At the beginning of the 1980s, the average chief executive made 109 times as much as the average blue-collar worker, and in the decade raised his pay by 212 percent against 53 percent for average workers.[267] By 1993 CEOs were making (on average) 157 times as much as shopfloor workers, against 17 times as much in Japan, for instance.[268] Taking the period 1961–96 as a whole, it was possible to show, statistically, that the overclass enjoyed a privileged existence in many ways: far less likely to serve in the draft—in 1962–72, 39,701 men graduated from three typical overclass universities, but only twenty of them died in Vietnam, from a total death figure of 58,000—far more likely to gain admission to an Ivy League college, and far more likely to make a significant contribution to public service.[269] It is notable, for instance, that the federal district judges appointed during Reagan's two terms, 22.3 percent of the whole, had an average net worth of over a million dollars.[270]

Better-off Americans are far more likely to vote and far, far more likely to get into Congress. Thus in the 1992 elections in California, whites, 55 percent of the population, provided 82 percent of the votes. Hispanics, 25 percent of the population, provided only 7 percent of the votes, and Asians, with 10 percent of the population, provided only 3 percent of the votes. As we have noted, incumbency was the key to Congressional membership because of the high cost of electioneering. A congressman spent an average of $52,000 to get elected in 1974, $140,000 in 1980, and $557,403 in 1992. That meant he and his contributors had to spend $5,000 a week for a two-year term. A senator had to spend $12,000 a week for a six-year term. Changes in the law about election funding meant that members of Congress were more likely to succeed if they were wealthy and could fund themselves.[271] The result could be seen, for instance, in the composition of the 103rd Congress, 1993–4. Of its 539 members, 155 were businessmen and bankers and 239 were lawyers, many of the lawyers also being involved in business. There were seventy-seven in education, thirty-three in the media, and ninety-seven in public service.[272]

The fact that the largest group in Congress by far was composed of lawyers (about two-thirds of those classified as businessmen and bankers also had law degrees) pointed to another significant fact about America at the end of the 20th century. America might or might not be a country 'under law,' as the Founding Fathers intended. It was certainly a country where lawyers held disproportionate power and influence compared to the rest of the population, and where the power of the courts was growing dramatically at the expense of the executive and the legislature. This important point needs examination in some detail. In the mid–1990s, both the President and his wife were law graduates, plus 42 percent of the House and 61 percent of the Senate (comparable figures for legislatures in other advanced countries average 18 percent).[273] Lawyers were prominent in the Continental Congress and the whole independence movement, as they had been in the events leading up to the Civil War in England a century before—as was natural. (Indeed, lawyers had a big hand in Magna Carta in 1215.) But the subsequent growth in numbers and power in America, especially in the last quarter-century, had been dramatic. Between 1900 and 1970, the number of lawyers as a percentage of the growing population was fairly constant, at about 1.3 per thousand. Doctors were 1.8 per thousand. After 1970, lawyers outstripped doctors despite the increase in medical services, Medicare and Medicaid, and growing health-consciousness, because the demand for legal services rose still more sharply. By 1987 lawyers were 2.9 per thousand and by 1990 3.0 per thousand.[274] In the quarter-century 1960–85, the population of the US grew by 30 percent and the number of lawyers by 130 percent. Equally significant was the increase in the number of lawyers resident in Washington DC in the quarter-century 1972–87, from 11,000 to 45,000.[275]

The increase in the demand for legal services, and in litigation, was the product mainly of two factors: consciousness of rights, in an age when the courts, especially the Supreme Court, was stressing the importance and availability of rights; and, secondly, the increase in legislation, especially regulatory law demanding compliance and therefore posing complex legal problems for businesses and individuals. This factor can be measure by the number of pages in the Federal Register, which contains the various regulations put forward annually by the federal executive in response to Acts passed by the legislature. There were only 2,411 pages in 1936, at the height of the New Deal. By 1960 the total had crept up to 12,792. During the Kennedy–Johnson era it leaped to 20,036, and during the 1970s, when Congress was on top, it shot up again to 87,012. Reagan reduced the flood to about 53,000 a year, but it rose again in 1991 to 67,716. The number of pages of Federal Reports, that is the transactions of the federal appeal courts, also rose, from 6,138 pages in 1936 to 49,907 pages in 1991. These increases far outstripped both the growth in GDP and, still more, the growth in population.[276] This explained why, despite the huge increase in those entering the legal profession, from 1,000 women and 15,000 men a year in 1970, to 14,000 women and 22,000 men in 1985, average annual legal earnings kept up very well. Inevitably, this increase in lawyers, litigation, and legal work was parasitical on the economy as a whole. One 1989 study indicated that the optimum number of lawyers needed was only 60 percent of the existing total and that each additional lawyer joining the profession above this total reduced America's GDP by $2.5 million.[277]

More important, perhaps, the growth of litigation and of the number and importance of lawyers in society formed the background to a power bid by those right at the top of the legal profession: the justices of the Supreme Court. The growth of the federal judiciary in numbers had proceeded more than *pari passu* with the growth of federal government as a whole. By 1991 there were 600 federal district judges and 150 appeal judges, reflecting the quadrupling to 250,000 district cases, 1970–90, and the (almost) quadrupling of appeals in the same twenty years. Thanks to the rejection of FDR's attempt to pack it, the Supreme Court remained at the same number, nine, and by refusing writs of certiorari the Court kept its cases more or less stable from 4,212 in 1970 to 4,990 in 1989.[278] What was quite new was the importance and nature of some of its decisions, which made law and interpreted the Constitution in a novel and radical fashion.

It is vital to grasp that the judicial tradition of the United States, as understood by the Founding Fathers, was based upon the English common law and statutory tradition, whereby judges interpreted common law and administered statute law. The English judges had interpreted the law in the light of the principle of equity, or fairness, which is a synonym for natural justice. But they had done so heeding the warning by Grotius (1583–1645), the Dutch founder of modern legal science, that equity should be no more than 'the correction of that, wherein the law is deficient.' This warning was amplified in the most solemn fashion by Sir William Blackstone (1723–80), the greatest of all English 18th-century jurists, whose teachings were very much present in the minds of the Founding Fathers when they drew up the Declaration of Independence and wrote the Constitution of the United States. Blackstone said that judges must bear justice in mind as well as the law, but 'The liberty of considering all cases in an equitable light must not be indulged too far, lest thereby we destroy all law, and leave the decision of every question entirely in the breast of the judge. And law, without equity, though hard and disagreeable, is much more desirable for the public good, than equity without law; which would make every judge a legislator, and introduce most infinite confusion.'[279]

The American system, being based (unlike England's) on a written constitution, and needing to reconcile both federal law and laws passed by many states, necessarily involved some degree of judicial review of statutory law. This process was started by the Marshall Court in 1803, in *Marbury* v. *Madison*, which declared Section 13 of the 1789 Judiciary Act unconstitutional and therefore null and void. As we have seen, Marshall and his court played a crucial role in shaping the structure of the American legal system to admit the rapid and full development of capitalist enterprise. However, President Jefferson strongly opposed this role of the Court, President Jackson flatly refused to enforce the Court's mandate in *Worcester* v. *Georgia* (1832), and President Lincoln refused to comply with a variety of Court mandates and writs. Over 200 years, the Court, on the whole, while veering between 'judicial restraint' and 'judicial activism,' that is between a strict construction of what the Constitution means and a creative or equitable one, had been careful to follow Blackstone's warning.[280]

A change, however, came in 1954, and that too had a complex background. Many American liberals had been dismayed by the fact that, despite emancipation, and despite the Fifteenth Amendment, guaranteeing all citizens the right to vote, blacks had failed to participate fully in American political life, especially in those states where they were most numerous, and the black community remained poor, badly educated, downtrodden, and supine. In 1937 the president of the Carnegie Foundation, Frederick Keppel, decided to finance a study of race relations. His colleague Beardsley Rumi persuaded him to get a Swedish politician called Gunnar Myrdal to do it. Myrdal received a grant of $300,000, an enormous sum in those days, and in due course he produced his findings, published in January 1944 as *An American Dilemma*, with 1,000 pages of text, 250 of notes, and ten appendices. Who was Gunnar Myrdal? He was, essentially, a disciple of Nietzsche and his theory of the Superman: Myrdal's belief that 'Democratic politics are stupid' and 'the masses are impervious to rational argument' led him to the social engineering of the Swedish Social Democrat Party, in which the enlightened elite took decisions on behalf of the people for their own good. Myrdal's book had a profound impact on the American intelligentsia. A 1960s survey of leading intellectuals conducted by the *Saturday Review*, asking what book over the past forty years had had the most influence, reported the finding that, apart from Keynes' *General Theory*, *An American Dilemma* topped the list, many of those questioned ranking it with *Uncle Tom's Cabin*, Tom Paine's *Common Sense* and De Tocqueville's *Democracy in America*.[281]

Essentially Myrdal argued, and 'proved' by his statistics, that America was too deeply racist a country for the wrongs of the blacks ever to be put right by Congressional action. Indeed, the book is a sustained attack on the inadequacies of the democratic process. It concludes by exhorting the Supreme Court to step in where democracy had failed and to apply 'the spirit of the Reconstruction Amendments' to end segregation. The book became the bible of Thurgood Marshall (1908–93), head of the National Association for the Advancement of Colored People Legal Defense and Education Fund. Marshall retained as his chief strategist Nathan Margold, a disciple of Justice Frankfurter. Margold belonged to the school of 'legal realism,' arguing that sociological approaches to the law ought to have priority over the 'original intent' of constitutions and statutes, and of precedent, and that all legal decisions are, or ought to be, forms of social policy, to improve society. Margold's approach was to work on the element of legal realism already embedded in the Supreme Court by Frankfurter, and Myrdal's book, studied by the justices, was of critical importance in his campaign, because liberal judges in particular were enthusiastic about the Myrdal approach.

The result was perhaps the most important single Supreme Court decision in American history, *Brown* v. *Board of Education of Topeka* (1954), in which the Court unanimously ruled that segregated schools violated the Fourteenth Amendment guaranteeing equal protection under the law, and thus were unconstitutional. It is notable that, during the arguments, one of the justices, Robert Jackson, laid down: 'I suppose that realistically the reason this case is here is that action couldn't be obtained by Congress.' Thus the judges were consciously stepping in to redeem what they saw as a failure of the legislature. The next year, in a pendant to the case, the Court issued guidelines for desegregating schools and vested federal courts with authority to supervise the process, urging that it take place 'with all deliberate speed.' By setting itself up as an enforcement agency, the Court was thus substituting itself for what it regarded as a failed executive too. In short, in *Brown* the Court not only made law, but enforced it, and the enforcement was no small matter: as late as 1994 over 450 school districts remained under federal court supervision, and entire generations of children grew up in schools run not by the locally elected democratic authority but by judges in distant Washington.[282]

Desegregation in the schools was reinforced by the Civil Rights Act of 1964, promoted by President Johnson in a famous speech to Congress on November 27, 1963: 'We have talked long enough in this country about equal rights. We have talked for a hundred years or more. It is time now to write the next chapter, and to write it in the books of law.' That was what Congress did, exercising its proper constitutional function, and the Act itself was the result of democratic compromise in the legislature. Fearing Southern senators would filibuster the Bill into oblivion, LBJ made an alliance with Everett Dirksen of Illinois, the Republican Minority Leader, to get the measure through, in return for a key amendment. This specifically and plainly forbade racial quotas, thus rejecting the underlying philosophy of the Myrdal school of social engineering, which advocated 'positive discrimination' and 'affirmative action' to remedy the historic legacy of racism.[283] The Act was, on the whole, a notable success. But the manifest intentions of Congress, in enacting it, were distorted by the lawyers of the 'legal realist' school. The Act created the Equal Employment Opportunities Commission as an enforcement agency. As it happened, in its first five years of operation, the Commission had no fewer than four chairmen, mostly absentee, and during this key period its *de facto* head was its compliance officer, a Rutgers law professor called Alfred

W. Blumrosen. The Commission was directed by the Act to respond to complaints. Blumrosen brushed this provision aside and used the agency to take direct action by imposing quotas in defiance of the Act. He openly boasted of what he called his 'free and easy ways with statutory construction' and praised the agency for working 'in defiance of the laws governing its operation.'[284]

This was the real beginning of Affirmative Action and it is worth remembering that it was based on illegality, as indeed was most of the racialism it was designed to correct. However, this defiance of the law was, as it were, legitimized on March 8, 1971 when the Supreme Court, in *Griggs* v. *Duke Power Company*, interpreted the 1964 Act in such a way as to make lawful discrimination in favor of what it termed 'protected minorities.' The ruling gave such minorities an automatic presumption of discrimination, and so gave them standing to sue in court without having to prove they had suffered from any discriminatory acts. Members of the majority, that is whites, and especially white males unprotected by discriminatory preferences in favor of females, were granted no similar standing in court to sue in cases of discrimination. Thus the principle of 'equality before the law,' which in the Anglo-American tradition went back to Magna Carta in 1215, was breached. Victims of reverse discrimination were disadvantaged in law, as was painfully demonstrated by the Randy Pech case in January 1991, when the US Attorney-General argued that Pech, being white, had no standing in court.[285]

Many university courses on constitutional law (for example, at Georgetown, Washington DC), now begin with *Brown*, the line of instruction being that the Supreme Court, and other courts, were compelled, in the light of *Brown*, to take notice of social protest, such as demonstrations and riots—especially when these were ignored by Congress—in shaping decisions. The rise of legal triumphalism was opposed by a growing number of lawyers themselves. One of those who shaped the *Brown* decision, Alexander Bickel, who had been Frankfurter's clerk, repudiated it. Another, the NAACP litigation consultant Herbert Washsler, Columbia law professor, argued that *Brown* was an 'unprincipled decision' that sacrificed neutral legal principles for a desired outcome. Raoul Berger, of Harvard, accused the Court of following the moral fallacy that the end justifies the means, usurping legislative power 'on the grounds that there is no other way to be rid of an acknowledged evil.' Other experts argued that race, and its handling by the courts, was in danger of crowding out the proper political process altogether, and that race, quickly followed by ethnicity, was displacing citizenship as a badge of identity. Yet others pointed out that the *Brown* decision had not even worked as an act of *realpolitik*, since after thirty years the schools deteriorated even in the districts which were the original defendants in the case. In practice, the evidence showed, the judiciary could not produce desirable change: when it tried to do so it merely produced new forms of injustice. 'Contrary to expectation,' as one critic put it, '[*Brown*] turned out to be a prelude to a major step backward in American race relations.'[286]

Once the principle was established that it was lawful and constitutional to discriminate in favor of certain racial groups—and in some cases obligatory—a number of consequences followed. Race quotas were introduced in all government employment at the lower levels and in private industry under threat of government litigation: for example, AT&T increased the number of minority managers from 4.6 to 13.1 percent, 1973–82, and black employees at IBM rose from 750 in 1960 to 16,564 in 1980.[287] An enormous race-quota compliance bureaucracy grew up: the Civil Rights Commission, the Equal Opportunity Employment Commission, the Labor Department's Office of Federal Contract Compliance Programs, the Office of Minority Business Enterprise, and large civil rights offices within federal departments such as Justice, Defense and Health, Education and Welfare. Curiously enough, this burgeoning race relations industry, employing tens of thousands of people, treated itself as exempt from the quotas, based on demographics, which it was imposing on the rest of society, and employed an overwhelming majority of blacks and other minority groups in workforces, though it did employ some whites, especially if they were women.[288] This formidable bureaucratic army became the core of the Civil Rights Lobby in Washington. And the lobby was, in effect, led by the judges themselves, who in some cases were quite frank about what they were doing: 'Certainly a majority of the educated elite,' said Richard Neely, Chief Justice of the West Virginia Supreme Court,

> as reflected by the attitudes of the faculties, trustees and student bodies at major universities, consider affirmative action, though predatory, morally justifiable. It is just as certain, however, that a majority of Americans disapprove. There is no theoretical justification for continued support of affirmative action other than the elitist one that the courts know from their superior education that affirmative action is necessary

in the short run to achieve the generally applauded moral end of equal opportunity in the long run. That is probably not illegitimate, since judges are social science specialists and have available to them more information and have pursued the issue with more thought and diligence than the man in the street.

This blunt statement of judicial triumphalism occurred in a book appropriately titled *How the Courts Govern America.*[289]

Unfortunately, courts cannot govern, because that is not their function, and the result of judicial aggression was to make America not more, but less, governable. Instituting race quotas merely stimulated more forms of direct action, including more rioting. Introducing inequality before the law merely helped to undermine the legal process itself. Mesmerized by the philosophy of quotas and reverse discrimination, majority black juries began to refuse to convict where blacks were accused of murdering whites (especially, in New York, white Jews). Families of whites murdered by blacks were forced to resort to civil processes to secure a remedy in cases where racially determined verdicts deprived them of justice. And where whites (especially white policemen) were acquitted on charges of murdering blacks, huge black riots forced retrials. Both these last two types of case breached another fundamental axion of the law, that no accused may be tried twice for the same crime.

Nor did the disturbing consequences end there. America became in danger of embracing a caste system, like India or, worse still, of obliging itself to set up the juridical infrastructure of a racist state, like Hitler's Germany. With the beginning of large-scale non-European immigration in the late 1960s as a result of the 1965 Immigration Reform Act, entrants from Europe fell from over 50 percent, 1955–64, to less than 10 percent in 1985–90, while Third World entrants rose rapidly. This opened an opportunity for lobbies to create new categories of 'disadvantaged minorities.' Thus the Mexican-American Legal Defense and Education Fund, a powerful interest-group in alliance with the Democratic Party, succeeded in establishing a racial category known as 'Hispanic,' which included latin mestizos, people of predominantly European, black, and American Indian descent, descendants of long-assimilated Californios and Tejanos, and other groups who once spoke Spanish—almost anyone in fact who found it advantageous to belong, so long as they could not be accused of being 'Caucasian' or 'Aryan.' This pseudo-race came into existence as the result of statistical classification by bureaucrats. Nor was it the only one. In 1973 Washington asked the Federal Interagency Committee on Education to produce consistent rules for classifying Americans by ethnicity and race. The FICE produced a five-race classification: American Indian or Alaskan Native, Asian or Pacific Islander, Black, White, and Hispanic. Oddly enough, at the insistence of the Office of Management and Budget, Asian Indians, who had been upgraded to White by a Supreme Court decision in 1913, then downgraded by another Supreme Court decision in 1923, were downgraded still further to the Asian and Pacific Islander category in the late 1970s, and were delighted at the result since it meant that all of them, including high-caste Hindus, could qualify for radical preference.[290] The new racial code was adopted in Statistical Directive 15.[291]

The new American race code began to experience these kinds of difficulties from the 1980s, when membership of certain categories brought immediate financial and other advantages. Thus the City of San Francisco laid down that 80 percent of its legal services had to be conducted by Asians, Latinos, blacks, or women, a process known as set-asides, then discovered that some of these categories were fronting for white firms. But that was a point difficult to prove. Equally, a white who claimed to be five-eighths Indian could not easily be convicted of fraud even if he did so to get a set-aside contract.[292] Moreover, the rise in interracial marriage produced large numbers of mixed-race citizens who, from the early 1990s, were demanding the creation of a statistical mixed-race category, or categories, with corresponding privileges. This was what happened to the originally simple race code of the 18th-century British Caribbean colonies, which ended up with eight groups—negro, sambo (mulatto/negro), mulatto (white male/negress), quadroon (white male/mulatto female), mustee (white male/quadroon or Indian), mustifini (white male/mustee), quintroon (white male/mustifini) and octaroon (white male/quintroon)—before collapsing in confusion.[293] At one time there were 492 different race categories in Brazil before the system was abandoned, as was apartheid, as unworkable.

It was distressing to see the United States, the model democratic and egalitarian republic, go down that retrograde road. But meanwhile the system spread. Political gerrymandering to benefit racial minorities began in the wake of the Voting Rights Act (1965) and was extended by amendments to the Act in 1975.[294] After the 1990 census, black majority Congressional districts rose from seventeen to thirty-two and Hispanic from nine to twenty, and in the 1992 elections membership of the Black Caucus jumped from twenty-six to thirty-nine, as a direct result of gerrymandering on a racial basis.[295] Affirmative Action in universities led to massive lowering of admission standards in such key schools as medicine. Thus, by the early 1990s, less than half of black medical-school graduates passed their National Board Exams for medical certification, compared to 88 percent among whites.[296] From the late 1960s, more and more universities offered courses and/or degrees in Afro-American Studies, African Identity, Afro-Awareness, Afrocentricity, and related topics. Howard University, chartered by Act of Congress in 1867 and long regarded as America's best predominantly black college, formed an enormous Department of African Studies and Research, listing sixty-five courses, including African Political Thought, Africa and the World Economy, Political Organization of the African Village, African Social History, Educational Systems in Africa, plus courses in such African languages as Ibo, Wolof, Lingala, Swahili, Xhosa, Zulu, Amharis, and Algerian Arabic. Some of the courses made academic sense, others did not. There was a good deal of fabricated Afro-centered history, which included reclassifying Hannibal as a 'black warrior,' Cleopatra as a 'black African queen,' categorizing the Ancient Egyptians as black, and assigning to them the ultimate origins of Western civilization.[297]

This racist approach to learning was based upon the assumption, treated as a truth to be 'held as self-evident,' and which underlay both official policies of Affirmative Action and the new university curricula, that blacks were by definition incapable of racism, which was historically a creation of white color prejudice. Hence the Black Caucus in Congress was not merely not racist but actually anti-racist. Equally, a $420 million donation to Yale University in 1995 to finance a course on Western Civilization was rejected, under pressure from interest groups, on the ground that such a course was, by definition, racist. One paladin of the new learning, Professor Leonard Jeffries, head of the Afro-American Studies Department at City College, New York, taught that whites were biologically inferior to blacks, that Jews financed the slave trade, and that the 'ultimate culmination' of the 'white value system' was Nazi Germany. Jeffries' students were informed that white genes were deformed in the Ice Age, producing an inadequate supply of melanin, which in turn made whites capable of appalling crimes, while black genes were enhanced by 'the value system of the sun.'[298] Other self-evident truths were unveiled by Professor Becky Thompson, head of the Women's Studies Department at Brandeis University—the college named after the first Jew to attain the Supreme Court—who described to a meeting of the American Sociological Association her teaching methodology: 'I begin the course with the basic feminist principle that in a racist, classist and sexist society we have all swallowed oppressive ways of being, whether intentionally or not. Specifically, this means that it is not open to debate whether a white student is racist or a male student is sexist. He/she simply is. Rather, the focus is on the social forces that keep these distortions in place.'[299]

The growth of political indoctrination and malign race-theory was part of a phenomenon known as 'Political Correctness,' which swept the American campus in the 1980s and early 1990s, rather as protest swept it in the 1960s. As in the Sixties, some escaped the disease, others were heavily infected, some succumbed. The worst aspect of PC, critics complained, was not its foolishness but its intolerance, and its tendency to stifle free speech. Thus, a classic statement of PC was the 'Policy on Harassment,' drawn up by President John Casteen of the University of Connecticut and printed in its *Students' Handbook, 1989–90*. This urged black, Hispanic, and female students to report all derogatory remarks they heard to the authorities, and it ordered those administering the complaints procedure to 'avoid comments that dissuade victims from pursuing their rights' since 'such behavior is itself discriminatory and a violation of the policy.'[300] Casteen went on to become president of the University of Virginia, founded by Thomas Jefferson. Conduct like that of Casteen might be greeted with derision by the outside world but during this period large numbers of students were expelled from universities for behavior or remarks judged to be politically incorrect, and some university teachers lost their jobs. PC persecutions were in the long American tradition of the Salem witchhunts, the Hollywood blacklist, and other manifestations of religious or ideological fervor overriding the principles of natural justice.

There was also concern that at a time when more American young people were attending university than ever before, the quality of what they were taught was being lowered by ideological factors. One of the characteristics of PC, and its associated theory, Deconstruction—an import from Europe but pursued more relentlessly on American campuses—was the legitimation of popular culture as a subject of study, children's comics being placed on the same level as Shakespeare or Melville. The switch to 'pop' had moral overtones too. Professor Houston Baker of the University of Pennsylvania, founded by Benjamin

Franklin but in the 1990s a center of PC, argued that 'Reading and writing are merely technologies of control' or 'martial law made academic.' Students should be taught to listen to 'the voice of newly emerging peoples' and especially their music, such as Rap. He cited such groups as Public Enemy and Niggaz Wit' Attitude, the last singing lyrics about the need for violence against white people. In the decade 1985–95, Rap became the favorite kind of music for black students on campus (and many white ones also), who liked its tendency to treat the police as enemies to be killed.[301] One 1990s campus hero was Tupac Shakur, whose albums sold millions of copies and whose movies were cult-properties, openly praised crime and urged the killing of police officers. Since 1991 he had been arrested eight times for violence, served eight months in prison for sexual abuse, was accused of shooting two policemen, and was finally shot to death himself on September 13, 1996, by unknown assailants in Las Vegas. In 1991 Rap, especially its most menacing version, Gangsta Rap, accounted for 10 percent of all records sold in America, and Shakur's label, Death Row (its logo portrays a gangster hero strapped into an electric chair) sold 18 million records and grossed more than $100 million.[302]

All this seemed to be part of a process of downward mobility which, as we have seen, first became an American phenomenon in the 1920s and picked up speed from the 1960s onwards. The critical literature of the 1980s and 1990s was crowded with works complaining about the 'dumbing down' of America—a 1933 Hollywood phrase used for preparing screenplays for semi-literates. A 1987 bestseller by Allan Bloom of the University of Chicago was *The Closing of the American Mind*, subtitled 'How higher education has failed democracy and impoverished the souls of today's students.' A complementary work, from 1991, by Roger Kimball, *Tenured Radicals: How Politics Has Corrupted our Higher Education*, drew attention in particular to the omission of facts and entire subjects from university courses for fear that they might 'offend' a particular racial group. A Harvard professor of ichthyology told a meeting organized by AWARE (Actively Working Against Racism and Ethnocentrism), one of the PC enforcement agencies, that a teacher should never 'introduce any sort of thing that might hurt a group' because 'the pain that racial insensitivity can create was more important than a professor's academic freedom.'[303] And emasculation or falsification of courses, from history to biology, for fear of causing offense was merely one example of the way in which education, it was claimed, was being infantilized by politics. The monthly review the *New Criterion*, founded in 1982 precisely to combat both the decay of culture and deliberate assaults on its standards, drew attention repeatedly to the corrosive process in art and architecture, music, and letters.[304]

One target of attack was the American language. The use of English, with its enormous vocabulary, matchless capacity to invent and adapt new words, and subtlety of grammar and syntax, had always been one of the great strengths of American culture and society. The emergence of the United States as the first really big free-trade area in the world, so important to its expansion and industrialization in the 19th century, was due in large part to the coherence provided by a common language, pronounced (and, thanks to Webster, spelled) in the same way over an area the size of Europe. Webster distinguished between the New England dialect, Southern dialect, and General American; and Professor Hans Kurath, working in the 1940s, distinguished between eighteen 'speech areas,' elaborated in the *Linguistic Atlas of the United States and Canada*, publication of which began in the 1950s. But the use of English by Americans was, on the whole, remarkably uniform, and all Americans could usually understand each other well, something which could not always be said for Britain.[305] The effectiveness of American use of English was of growing importance at the end of the 20th century since by the mid–1990s over 2 billion people throughout the world spoke English as their first, second, or commercial language; not only was this number growing but more and more of those learning it were acquiring it in its American idiom.

Acquisition of English was a vital part of the melting-pot process and, equally, non-English speakers arriving in America as immigrants, in the process of learning it, added to its richness, a point powerfully made by Henry James in a talk he gave at Bryn Mawr College on June 8, 1905: 'The thing they may do best is to play, to their heart's content, with the English language [and] dump their mountains of promiscuous material into the foundation of the American ... they have just as much property in our speech as we have and just as much right to do what they choose with it.' However, James pressed strongly for common pronunciation and what he called 'civility of utterance,' warning against 'influences round about us that make for ... the confused, the ugly, the flat, the thin, the mean, the helpless, that reduce articulation to an easy and ignoble minimum, and so keep it as little distinct as possible from the grunting, the squealing, the barking or roaring of animals.' James called the correct pronunciation of English 'the good cause,' and it had to be based on constant reading: a true culture was a literary culture, and a purely aural culture was bound to degenerate into a mindless and unimaginative demotic. One of the characteristics of downward mobility, often reinforced by Political Correctness and Deconstruction, was precisely the stress on the aural as opposed to the written and printed. Attempts to teach children to speak and write 'correct' English were treated as 'elitist' or even 'racist,' since the very notion of one form of English being 'correct' was 'oppressive.'[306]

One form of this aural dumbing down was the proposal by the School Board in Oakland, California, that Ebonics be recognized as a distinct language and used for teaching purposes. Ebonics, a compound from ebony and phonetics, was a fancy name for black street-speech, based partly on contraction ('You know what I am saying?' becomes 'Nomesame?' and 'What's up?' becomes 'Sup?'), and partly on special use of words (woman/lady/girl is 'bitch,' kill is 'wet,' nice is 'butter,' teeth is 'grill,' money is 'cream,' foot/friend/man is dog/dawg/dogg), and partly on inversion (ugly is 'pretty,' intelligent is 'stupid,' beautiful is 'mad'). The proposal, made in January 1997, was attacked by black opinion leaders, as well as white ones, but it sprang from a growing tendency among some educationists to regard the acquisition of 'correct' skills in English as an unjust form of cultural property, to be corrected by Affirmative Action. This took many forms, in the 1980s and 1990s, including the right of ethnic minorities to be taught in their own languages and, more seriously, the right of Spanish, as the language of the Hispanic minority, to be given status as an official alternative to English. Signs in Spanish as well as English at US airports and other places of entry, which came as a shock to Americans returning to their own country, indicated the extent to which this new doctrine was being adopted. Those who advocated a multilingual as well as a multicultural America—a 'Waldorf Salad' as opposed to the melting-pot, to use one popular simile—were seemingly unaware of the fearful cost of internal linguistic disputes in countries unfortunate enough to be plagued by them, from Sri Lanka to Burundi to Belgium.

The prospect of plunging the United States into linguistic civil wars would not have deterred some rights enthusiasts. The bible of the rights lobby, John Rawls' *A Theory of Justice* (1971), argued that justice was categorically paramount, superior to any other consideration, so that one aspect of justice could be sacrificed only to another aspect, not to any other desired goal. Rawls argued that a policy which benefited the entire human race except one should not be adopted (even if that one person was unharmed) because that would be an 'unjust' distribution of the benefits of the policy.[307] This anti-realist approach was attacked most persuasively by America's leading philosopher, Thomas Sowell, himself a black, in his libertarian cult-book, *Knowledge and Decisions* (1980), who argued that rights were more firmly based in practice on easy access to property with its self-interested monitoring. He quoted Adam Smith, who argued that some degree of justice (possibly a high degree) was necessary for any of the other desirable features of society to exist but increments of justice did not necessarily outrank increments of other desirables—such a belief was doctrinaire and counterproductive.[308] That, it seemed, was true of the whole drive for rights: to press everyone's notional rights to their logical conclusion led not to their attainment, since there was not enough justice to go round, but merely to a conflict of rights.

This became more and more apparent as Affirmative Action was applied. Statistics began to accumulate showing clearly that 'racial preferences,' applied by law, benefited well-off blacks and Hispanics at the expense of the poor of all races. And there were many individual cases of injustice. Thomas Wood, executive director of the California Association of Scholars, said he was denied a job solely because he was a white male. He was one of a group of people who drafted the 1996 California Initiative, Proposition 209, to amend the state's Constitution to prohibit use of quotas by California's state institutions. The phrasing of the Initiative was modeled on the 1964 Civil Rights Act and it passed by a substantial majority in November 1996. This democratic decision by a 30-million electorate was promptly blocked by Federal Judge Thelton Handerson, a black with close links to the radical left, who issued an injunction to prevent its enforcement, on the ground there was a 'strong probability' that the law would be proved 'unconstitutional' when it reached the Supreme Court. This in turn provoked a comment from Professor Sowell, a supporter of Proposition 209: 'Is there nothing left other than the choice between quietly surrendering democracy and refusing to obey a court order?'[309]

The likelihood that the usurpation by the judiciary of the powers of the legislature would provoke violence was demonstrated by the consequences of the decision by the Supreme Court in the case of *Roe* v. *Wade*. On January 22, 1973, the Court ruled (seven to two) that the option of choosing an abortion in the first trimester of pregnancy was a fundamental constitutional privilege. The direct implication of the ruling was that all state laws forbidding abortion were unconstitutional and void. The ruling was regarded by many as an extraordinary piece of judicial sophistry since it was based on the assumption that the Fourteenth Amendment, guaranteeing personal liberty as part of due process, encompassed a right to privacy which was violated by anti-abortion laws. Protest against the tenuous logic behind the Court's decision was reinforced by strong moral and religious feelings among people who held abortion to be wrong irrespective of what the Constitution or the courts said, just as in the 19th century millions of Americans held slavery to be wrong although both Court and Constitution held it was within the rights of states to uphold it.

As a result of the Court's decision, abortions became available on request for any reason and 30 million were performed up to the early 1990s, when they were taking place at the rate of 1,600,000 a year. The justification for this wholesale destruction, without any democratic mandate and at the behest of unelected judges, was that the sheer number of terminations indicated that women as a whole, half the population, evidently acquiesced in the process. But an August 1996 study by the Planned Parenthood Association, itself a pro-abortion institution, indicated that many women had at least two abortions and that the 30 million were concentrated in certain groups, especially blacks and unmarried white women cohabitating with a man or men. The great majority of American women had no recourse to abortion at all in their lifetimes.[310]

The strength of anti-abortion feeling, which appeared to be growing as new scanning techniques demonstrated the early point in pregnancy at which the fetus became a recognizable human being, recalled the anti-slavery campaigns of the period 1820–60. As with the efforts to end slavery, the anti-abortion cause became quiescent and underground for a time, then resurfaced with added fury. It was an issue, in the 1970s, 1980s, and 1990s, which would not go away and it tended to be expressed with more violence as time went by, as indeed was the slavery debate. The anti-abortionists, or pro-lifers as they called themselves, often worked in secret and operated the equivalent of the anti-slavery 'underground,' as well as including homicidal extremists like John Brown and his sons. Between 1987 and 1994, over 72,000 protesters were arrested for picketing abortion clinics, and some were jailed for up to two and a half years for non-violent actions chiefly consisting of chanting and prayer. In 1994, under pressure from the Pro-Choice Lobby, Congress passed the Freedom of Access to Clinic Entrances Act, making peaceful picketing a federal crime punishable by ten years in prison. That did not deter the anti-abortionists: far from it. By the end of 1996 there had been 148 fire-bombings of clinics and a number of doctors specializing in late abortions had been shot or stabbed to death with their own instruments; receptionists had been killed too. Protection and rising insurance costs compelled many clinics to shut down, and in some areas (the Dakotas, for instance, and the Deep South) doctors were no longer willing to perform abortions, which tended increasingly to be concentrated in a few large cities. In January 1997 a new level of violence was reached when an abortion clinic in Atlanta was devastated by a large bomb, and a second, booby-trapped bomb was detonated when firefighters, journalists, and federal agents arrived on the scene. The fall in the number of abortions carried out every year, from 1.6 to 1.4 million, proclaimed by the pro-life organizations as the 'saving' of 200,000 lives annually, indicated the effects of the campaign of violence. Like the emancipationists in the 19th century, the pro-lifers discovered that force was more likely to produce results than due process or voting.

That the Supreme Court, albeit inadvertently, should provoke high-principled citizens into resorting to violence was a serious matter in a society already facing a seemingly intractable problem of crime, especially crimes of violence. In the 19th and early 20th centuries, the United States did not keep national crime statistics, but local ones suggest there was a decline in crime in the second half of the 19th century (except for a few years after the Civil War). This decline continued well into the 20th century. There was even a decline in homicides in large cities, where such crimes were most common: Philadelphia's homicide rate fell from 3.3 per 100,000 population in c. 1850 to 2.1 in 1900.[311] US national crime statistics became available in 1960, when the rate for all crimes was revealed as under 1,900 per 100,000 population. That doubled in the 1960s and tripled in the 1970s. A decline in the early 1980s was followed by a rise to 5,800 per 100,000 in 1990. In crimes of murder, rape, robbery, and aggravated assault, the rise was steeper, fivefold from 1960 to 1992.[312] The US Justice Department calculated in 1987 that eight out of every ten Americans would be victims of violent crime at least once in their lives. In 1992 alone one in four US households experienced a non-violent crime.[313] In what was called 'the demoralization of American society,' a number of statistical indicators came together. In the thirty years 1960–90, while the US population rose by 41 percent, there was a 560 percent increase in violent crime,

200 percent in teenage suicide, 200 percent rise in divorce, over 400 percent rise in illegitimate births, 300 percent rise in children living in single-parent homes—producing *in toto* the significant fact that children formed the fastest-growing segment of the criminal population. During this period, welfare spending had gone up in real terms by 630 percent and education spending by 225 percent.[314]

The rapid rise in violent crime inevitably produced a corresponding rise in the prison population. In 1993, for instance, 1.3 million Americans were victims of gun-related crimes. Guns were used in 29 percent of the 4.4 million murders, rapes, robberies, and serious assaults in the US in 1993, an 11 percent increase over the previous year.[315] In the year ending June 30, 1995 the prison population grew by nearly 9 percent, the largest increase in American history. As of June 30, 1994, there were a million people in US jails for the first time, double the number in 1984. By the mid–1990s, one in 260 Americans was imprisoned for terms of over a year, with another 440,000 awaiting trial or serving shorter sentences. Of the 90 percent in state prisons 93 percent were violent or repeat offenders. A quarter of federal and a third of state prisoners came from 'criminal families' (that is, a close family member had already been jailed).[316] The growth in the prison population was itself the result of 'prison as last resort' policies. Courts were not anxious to send anyone to jail. Evidently there were few innocents locked away. In 1992, for instance, of the 10.3 million violent crimes committed in America, only 165,000 led to convictions and of these only 100,000 led to prison sentences. Less than one convicted criminal was sent to prison for every hundred violent crimes. Nearly all those sent to prison were violent offenders, repeat offenders, or repeat-violent offenders.[317]

In the 1970s, 1980s, and 1990s, as crime multiplied, there were many improvements in policing, which sometimes produced results and sometimes were aborted by political factors. Thus the Los Angeles Police Department, so corrupt in the 1930s and 1940s that Raymond Chandler's brilliant fictions about its operations actually underestimated the problem, was transformed into an honest and efficient force under two brilliant chiefs, William Parker and his successor Thomas Redden, but set back decades by the highly politicized 1992 riots and their consequences.[318] In many US big cities, such as Los Angeles, the high proportion of blacks convicted of crimes—the 12 percent of blacks in the total population produced 50–60 percent of homicide arrests, 50 percent of rape arrests, almost 60 percent of robbery arrests, and 40–50 percent of aggravated-assault arrests—meant that the anti-racial lobby tended to treat the police as the 'enemy' and bring political factors to bear against efficient crime-prevention.[319]

In 1990, a Boston police expert, Bill Bratton, was appointed head of New York's subway police and introduced the 'zero tolerance' policy of arresting and prosecuting in even minor cases. This worked so well that in 1993, when the federal prosecutor Rudolph Giuliani was elected mayor of New York, he gave Bratton the opportunity to apply the strategy to the city as a whole, and the results were impressive. In 1993–5, violent and property crimes dropped 26 percent in New York, murders by teenagers dropped 28 percent, car theft 46 percent, robbery 41 percent, and murder 49 percent. New York, with 3 percent of the population, accounted for one-third of the fall in reported crimes in three years. In 1996 the city reported fewer murders than at any time since 1968.[320] But, though more effective policing helped, abetted by underlying demographic factors, such as a rise in the average age of the population, most studies agreed that a radical improvement in the level of crime in America would depend on the return to a more religious or moralistic culture. Historians have always noted that organized religion had proved the best form of social control in Western societies.

In the light of this, it may strike future historians as strange that, in the second half of the 20th century, while public sin, or crime, was rising fast, the authorities of the state, and notably the courts—and especially the Supreme Court—did everything in their power to reduce the role of religion in the affairs of the state, and particularly in the education of the young, by making school prayers unlawful and unconstitutional, and by forbidding even such quasi-religious symbols as Christmas trees and nativity plays on school premises. This raised a historical point of some importance about the status of religious belief in American life. We have seen how important, right from the outset, was the force of religion in shaping American society. Until the second half of the 20th century, religion was held by virtually all Americans, irrespective of their beliefs or non-belief, to be not only desirable but an essential part of the national fabric. It is worth recalling De Tocqueville's observation that Americans held religion 'to be indispensable to the maintenance of free institutions.' Religion was identified first with republicanism, then with democracy, so as to constitute the American way of life, the set of values, and the notions of private and civil behavior which Americans agreed to be self-evidently true and right. In consequence, those who preached such values from the pulpit or who most clearly, even ostentatiously, upheld them from the pews, were acknowledged to be among the most valuable citizens of the country. Whereas in Europe, religious practice and fervor were often, even habitually, seen as a threat to freedom, in American they were seen as its underpinning. In Europe religion was presented, at any rate by the majority of its intellectuals, as an obstacle to 'progress;' in America, as one of its dynamics.

From the 1960s, this huge and important difference between Europe and America was becoming blurred, perhaps in the process of disappearing altogether. It was one way in which America was losing its uniqueness and ceasing to be the City on the Hill. For the first time in American history there was a widespread tendency, especially among intellectuals, to present religious people as enemies of freedom and democratic choice. There was a further tendency among the same people to present religious beliefs of any kind which were held with certitude, and religious practice of any kind which was conducted with zeal, as 'fundamentalist,' a term of universal abuse. There was a kind of adjectival ratchet-effect at work in this process. The usual, normal, habitual, and customary moral beliefs of Christians and Jews were first verbally isolated as 'traditionalist,' then as 'orthodox,' next as 'ultra-orthodox,' and finally as 'fundamentalist' (with 'obscurantist' added for full measure), thought they remained the same beliefs all the time. It was not the beliefs which had altered but the way in which they were regarded by non-believers or anti-believers, not so much by those who did not share them as by those who objected to them. This hostile adjectival inflation marked the changed perspective of many Americans, the new conviction that religious beliefs as such, especially insofar as they underpinned moral certitudes, constituted a threat to freedom. Its appearance was reflected in the extreme secularization of the judiciary, and the academy, and the attempt to drive any form of religious activity, however nominal or merely symbolic, right outside the public sector. Such a change was new and potentially dangerous for it was a divisive force, a challenge to the moral and religious *oikumene* or consensus which had been so central a part of America's democratic unity and strength.

This development produced a hostile atmosphere which specially affected the mainline American Protestant churches, the so-called 'Seven Sisters:' the American Baptist Churches of the USA, the Christian Churches (Disciples of Christ), the Episcopal Church, the Evangelical Lutheran Church of America, the Presbyterian Church (USA), the

United Church of Christ, and the United Methodist Church.[321] In order to exist in the hostile secular atmosphere, the churches liberalized and to some extent secularized themselves. That in turn led them to alienate rank-and-file members. The Seven Sisters remained apparently strong in terms of the American establishment. Every single president came, in theory at least, from the mainliners, except Kennedy (Catholic) and Carter (Southern Baptist). Of the Supreme Court justices from 1789 to 1992, some 55 out of 112 were Episcopalians or Presbyterians. Representation in Congress was also disproportionate: in the 101st, for instance, one-fifth of the senators were Episcopalians, ten times their proportion in the general population. The 10th contained sixty-eight Baptists (mostly mainliners), sixty-four Methodists, sixty-three Episcopalians, fifty-nine Presbyterians, and forty-three Lutherans.

This over-representation at the top, and the fact that some of these churches retained considerable wealth—in American Christianity as a whole, the top income- and property-owners are the Disciples of Christ, the Congregationalists, the Episcopalians, and the Presbyterians—did not disguise the fact that all seven were in rapid decline. Indeed they had been effectively in decline throughout the 20th century, though the rate had increased markedly since the end of the 1960s, when church attendance began to dip below its historical high. One study, for instance, calculated that the Methodists had been losing 1,000 members a week for thirty years.[322] The Seven Sisters as a whole lost between a fifth and a third of their members in the years 1960–90, chiefly because they forfeited their distinguishing features, or indeed any features. After the Episcopal Church's General Convention of 1994, marked by a bitter dispute over the right of practicing homosexuals to become or remain clergy, one official observer commented: 'The Episcopal Church is an institution in free fall. We have nothing to hold onto, no shared belief, no common assumptions, no bottom line, no accepted definition of what an Episcopalian is or believes.'[323]

By 1994, the Seven Sisters were listed as having membership as follows: American Baptist Churches 1.5 million, Disciples of Christ 605,996, Episcopal church 1.6 million, Lutheran Church 3.8 million, Presbyterians 2.7 million, United Church of Christ 1.5 million, and United Methodists 8.5 million. By contrast, the Roman Catholics, who had been America's biggest single denomination since 1890, numbered over 60 million in the mid–1990s, the Southern Baptists 16.6 million, and the Mormons 4.1 million.[324] These churches were still clear about their identities and distinguishing features, taught specific doctrines and maintained, though not without difficulty, especially among the Catholics, their coherence and morale. In addition there were a large number of Christian churches, both new and distantly derived from the various Awakenings, whose followers were difficult to quantify but who collectively numbered more than 50 million. Many of these churches raised their numbers and provided finance by effective use of network, local, and cable TV and radio. Some were represented on the main religious political pressure-group, the Christian Leadership Conference, which had over 1,300 branches by 1996 and was believed to influence and (in some cases) control the Republican Party in all fifty states. This grouping, together with the Catholics, constituted the more assertive face of American Christianity, still powerful in the nation as a whole, the heart of the 'moral majority.' During the late 1980s and 1990s, there was growing cooperation, a form of grassroots ecumenism, between the Protestant churches, the Catholics, the Jews (numbering slightly below 6 million in the mid–1990s), and even the Moslems, Buddhists, and Hindus, in seeking to promote traditional moral values.

Nevertheless, the decline and demoralization of mainstream Protestantism was an important fact of American life at the end of the 20th century. The average age of members of these churches rose rapidly, from fifty in 1983 to about sixty in the mid–1990s, indicating a failure to retain children in the fold (by contrast with the conservative Protestant groups, which pay special attention to instructing families as a whole). One index was the decline of 'giving,' down from 3.5 percent of income in 1968 to 2.97 percent in 1993, lower than it was at the bottom of the Great Depression. In 1992–4, Americans spent less on maintaining Protestant ministries, $2 billion a year, than on firearms and sporting guns, $2.48 billion, illegal drugs, $49 billion, legal gambling, nearly $40 billion, alcohol, $44 billion, leisure travel, $40 billion, and cosmetics, $20 billion. It is true that outright gifts to religious foundations and purposes remained enormous: $58.87 billion in 1994, making it by far the top charitable priority for Americans: gifts to education, at $16.71 billion, were much lower.[325] But routine support, whether by sustaining or attending churches, was waning. This led to large-scale sackings of permanent staff, especially in the Presbyterian, Lutheran, and Episcopal churches. The theologian Stanley Hauerwas of Duke Divinity School summed it up in 1993: 'God is killing mainline Protestantism in America and we goddam deserve it.'[326]

In general, the decline of the mainstream churches was most pronounced in the large cities and the former industrial areas and centers of population. It was an important factor in what, to many analysts of American society, was the decisive single development in America during the second half of the 20th century: the decline of the family and of family life, and the growth in illegitimacy. The enormous number of children born outside the family structure altogether, or raised in one-parent families, appeared to be statistically linked to most of the modern evils of American life: poor educational performance and illiteracy or semi-literacy, children out on the streets from an early age, juvenile delinquency, unemployment, adult crime, and, above all, poverty. The breakdown or even total absence of the marriage/family system, and the growth of illegitimacy, was right at the heart of the emergence of an underclass, especially the black underclass.

One of America's most perceptive social critics, Daniel P. Moynihan (later Senator for New York), drew attention to the growing weakness of the family as a prime cause of black poverty as early as the 1960s, in the so-called 'Moynihan Report.' The theme was taken up, and amplified, in the writings of Charles Murray, whose book *The Bell Curve* was an important—and controversial—contribution to the social analysis of America in the 1990s. This book was sensationalized and attacked on publication because it summarized a mass of statistical material to point out important differences in average IQs between racial categories, and this in turn undermined the permitted parameters of discussion of the race problem laid down by the liberal consensus. But such criticisms missed the main point of the analysis, which was to stress human differences of all kinds, including those arising from marriage and non-marriage and the existence or not of a family structure.[327] Murray drew attention to the rise of what he called a 'cognitive elite,' based on intermarriage between people of high intelligence, producing children far more likely to do well in a period when 'the ability to use and manipulate information has become the single most important element of success.' This structured overclass was accompanied, at the other end of the social spectrum, by an underclass where marriage was rare and illegitimacy the norm.

Up to 1920, the proportion of children born to single women in the United States was less than 3 percent, roughly where it had been throughout the history of the country. The trendline shifted upwards, though not dramatically, in the 1950s. A steep, sustained rise gathered pace in the mid–1960s and continued into the early 1990s, to reach 30 percent in 1991. In 1960 there were just 73,000 never-married mothers between the ages of eighteen and thirty-four. By 1980 there were 1 million. By 1990 there were 2.9 million. Thus, though the illegitimacy ratio rose six times in thirty years, the number of never-married mothers rose forty times. Illegitimacy was far more common among blacks than among whites. The difference in black–white marriage rates was small until the 1960s and then widened, so that by 1991 only 38 percent of black women aged 15–44 were married, compared to 58 percent of white women. A significant difference between the number of black illegitimate children and white ones went back to well before the 1960s, though it increased markedly after the 1960s watershed. In 1960, 24 percent of black children were illegitimate, compared with 2 percent of white children. By 1991, the figures of illegitimate births were 68 percent of all births for blacks, 39 percent for Latinos, and 18 percent for non-Latino whites. At some point between 1960 and 1990, marriage, and having children within marriage, ceased to be the norm among blacks, while remaining the norm among whites (though a deteriorating one).[328] The jump in the illegitimacy rate in 1991 was the largest so far recorded, but it was exceeded by subsequent years. By the end of 1994 it was 33 percent for the nation as a whole, 25 percent for whites, and 70 percent for blacks. In parts of Washington, capital of the richest nation in the world, it was as high as 90 percent.[329]

The alarming acceleration in illegitimacy was probably the biggest single factor in persuading Congress to legislate to end the existing Aid to Families with Dependent Children program, and President Clinton to sign the Bill into law. The AFDC system was widely regarded as an encouragement to women to eschew marriage and have illegitimate children to qualify for federal support, and ending it a precondition of tackling the decline in marriage. However, some critics, notably Senator Moynihan himself, thought it cruel, having created the illegitimacy–dependency culture, to end it suddenly for existing beneficiaries.[330] What had emerged by the second half of the 1990s, however, was that the conventional remedies for racial and class deprivation—improving civil rights and spending money in huge quantities on welfare programs—did not work. The war on poverty had failed. Now the war to restore marriage to its centrality in American life had begun.

In that war, the rise of women to positions of genuine equality in American life might seem, at first glance, an impediment. The more women succeeded in life, the less priority they accorded to marriage, children, and the home. A 1995 study by Claudia Goldin of the National Bureau of Economic Research found that only about 15 percent of women who had received college degrees c. 1972 were still maintaining both career and family by the mid–1990s. Among those with a successful career, indicated by income level, nearly 50 percent were childless.[331] Career and marriage seemed alternatives, not complements. But women had always adopted different strategies in fighting their way to the front, or into happiness. In the 1920s, Dorothy Parker (1893–1967) had created for herself a special upper-class accent of her own—'Finishing-school talk but not the Brearley accent, not the West Side private school accent but a special little drawl that was very attractive'—and used it to become the first American woman who could swear with impunity and make sexual wisecracks: 'She hurt her leg sliding down a barrister;' 'She wouldn't hurt a fly—Not if it was buttoned up;' 'We were playing ducking for apples—There, but for a typographic error, is the story of my life;' 'If all the girls at a Yale prom were laid end-to-end, I wouldn't be at all surprised;' 'She speaks eighteen languages and can't say no in any of them'—and so forth.[332] Her strategy resembled that of Marlene Dietrich: 'See what the Boys in the Back Room will have, and tell them I'm having the same'—a feminine imitation of the male at his own game.

By contrast, Mary McCarthy (1912–89) stressed female superiority. She was a graduate of Vassar, founded by Matthew Vassar, a self-taught Poughkeepsie brewer, who gave as his reason, 'Woman, having received from her Creator the same intellectual constitution as Man, has the same right as Man to intellectual culture and development.' He thought that 'the establishment and endowment of a College for the Education of Young Women would be, under God, a rich blessing to this City and State, to our Country and to the World.' McCarthy described Vassar as 'four years of a renaissance lavishness, in an Academy that was a Forest of Arden and a Fifth Avenue department store combined.' She despised Yale and asserted, 'Yale boys didn't learn anything in the subjects I knew anything about. We looked down on male education.'[333] In a life spent among intellectuals, she never saw reason to revise that opinion.

Most women in mid–20th-century America, however, did not engage in warfare to defeat males by stealth. They subscribed to what was known as the 'Women's Pact,' which recognized that women fell into three groups. The first, unable or unwilling to marry, or afraid of childbirth, devoted their lives to a career (as they had once entered nunneries). The second married and bore children, then left their rearing to others and pursued careers. The third chose marriage as a career and raising children and homemaking as their art.[334] The Women's Pact was broken by the feminists of the 1960s and later, who denounced the marriage-careerists as traitors to their sex. Helen Gurley Brown, founder of *Cosmopolitan* (1965), a Dorothy-Parker-strategy magazine, complete (eventually) with expletives and full-frontal male nudity, denounced the housewife as 'a parasite, a dependent, a scrounger, a sponger [and] a bum.' Betty Friedan (b. 1921) in *The Feminine Mystique* (1974) called femininity 'a comfortable concentration camp,' though she also argued, in *The Second Stage* (1981), that Women's Liberation and 'the sex war against men' were 'irrelevant' and 'self-defeating.'[335] In 1970 *Time* magazine published a notorious essay in which Gloria Steinem castigated 'traditional women' as 'inferiors' and 'dependent creatures who are still children.' This refrain was taken up by innumerable women academics who proliferated in the 1960s, 1970s, and 1980s, and ran 'Women's Studies' departments at many universities.

But, in the meantime, ordinary women were advancing, economically, financially, professionally, and in self-confidence, without much help from the feminist movement, which had particularly opposed women's participation in entrepreneurial capitalism. By 1996, American women owned 7.7 million businesses, employing 15.5 million people and generating $1.4 trillion in sales. In a large majority of cases this was combined with marriage and child-rearing. Indeed women owned 3.5 million businesses based on their homes, employing 5.6 million full-time and 8.4 million part-time. It was established that women-owned businesses were expanding more rapidly than the economy as a whole, and were more likely to continue to flourish than the average US firm.[336] Women also increasingly flourished in American business of all types, including the biggest. Nor, by the second half of the 1990s, were they effectively barred from the higher ranks by the 'Glass Ceiling,' a term invented by the *Wall Street Journal* in 1986 to describe the 'invisible but impenetrable barrier between women and the executive suite.' This notion was confirmed by the report of a Glass Ceiling Commission created by the Civil Rights Act of 1991, which concluded that only 5 percent of senior managers of *Fortune* 2000 industrial and service companies were women, deliberate discrimination being the implied cause.[337] Careful analysis, however, dismissed the 5 percent figures as 'statistically corrupt but rhetorically and politically useful,' and in fact 'highly misleading.' The truth was that few women graduated from management schools in the 1950s and 1960s, and therefore were eligible for top posts in the 1990s; but many more were 'in the pipeline.' During the decade 1985–95, the number of female executive vice-presidents more than doubled and the number of female senior vice-presidents increased by 75 percent. The evidence of actual companies, treated separately, showed that the Glass Ceiling was a myth.[338]

It was also a myth, though a more difficult one to disprove, that women continued to be insignificant in politics. The only woman to be a candidate in a presidential election, Geraldine Ferraro, who ran as vice-president in 1984, got an exceptionally rough handling, largely on account of her husband's business interests; and women continued to be poorly represented in Congress. But, here again, patience was needed. In 1974, for example, 47 women were main-party candidates for Congress, 36 for statewide office, and 1,122 for state legislatures. In 1994 more than twice as many women were running for office: 121 for Congress, 79 for statewide office, and 2,284 for state legislatures. A 1996 study showed negligible differences between men and women in success rates for state or national office. By 1996, 21 percent of state legislators were women, compared with 4 percent in 1968, and with eight senators and forty-seven (of 435) in the House, women were increasing their ratio at the center with gratifying speed, if still a long way from statistical equality.[339] As if to make the point, the reelected President Clinton announced in December 1996 the appointment of America's first woman Secretary of State, Madeleine Albright.

The same month, a special study by the American Enterprise Institute for Public Policy Research announced, effectively, victory for the women's cause in America. The figures now showed that, among younger women, the wage gap had been closed and women were now earning 98 percent of men's wages; that in 1920–80 women's wages grew 20 percent faster than men's and were still growing faster; that earning differentials at all levels were insignificant or non-existent; that the gap was rapidly closing for older women; that more women were now in higher education than men, and that in all the indicators of educational attainment and professional choice and skills—associate degrees, bachelors and masters degrees, doctoral degrees, first professional degrees, qualifications for law, medicine, accountancy, and dentistry, women were approaching the 50 percent mark or had moved ahead. By 1995 women constituted 59 percent of the labor force in work, were on the whole better educated and trained then men, were acquiring professional qualifications more quickly, and were less likely to be unemployed.[340] The authors of the study, Diana Furchtgott-Roth and Christine Stolba, concluded that in education, in the labor force, and in the eyes of the law women had effectively achieved equality. The law was working and being enforced and in this respect the Equal Pay Act of 1963 and the Civil Rights Act of 1964 had succeeded.

It is appropriate to end this history of the American people on a note of success, because the story of America is essentially one of difficulties being overcome by intelligence and skill, by faith and strength of purpose, by courage and persistence. America today, with its 260 million people, its splendid cities, its vast wealth, and its unrivaled power, is a human achievement without parallel. That achievement—the transformation of a mostly uninhabited wilderness into the supreme national artifact of history—did not come about without heroic sacrifice and great sufferings stoically endured, many costly failures, huge disappointments, defeats, and tragedies. There have indeed been many setbacks in 400 years of American history. As we have seen, many unresolved problems, some of daunting size, remain. But the Americans are, above all, a problem-solving people. They do not believe that anything in this world is beyond human capacity to soar to and dominate. They will not give up. Full of essential goodwill to each other and to all, confident in their inherent decency and their democratic skills, they will attack again and again the ills in their society, until they are overcome or at least substantially redressed. So the ship of state sails on, and mankind still continues to watch its progress, with wonder and amazement and sometimes apprehension, as it moves into the unknown waters of the 21st century and the third millennium. The great American republican experiment is still the cynosure of the world's eyes. It is still the first, best hope for the human race. Looking back on its past, and forward to its future, the auguries are that it will not disappoint an expectant humanity.

SOURCE NOTES

PART ONE: Colonial America

1. C. R. Boxer: *The Portuguese Seaborne Empire, 1415–1825* (London 1969), 27.
2. T. B. Duncan: *Atlantic Islands: Madeira, the Azores and Cape Verde in 17th Century Commerce and Navigation* (Chicago 1972), 212.
3. C. A. Palmer: *Slaves of the White God: Blacks in Mexico, 1570–1650* (Cambridge 1976), 10.
4. J. H. Parry: *The Spanish Seaborne Empire* (London 1966), 42.
5. D. W. Meinig: *The Shaping of America: a Geographical Perspective of 500 Years of History,* vol. i: *Atlantic America, 1492–1800* (New Haven 1986), 4–6.
6. J. Lang: *Conquest and Commerce: Spain and England in the Americas* (New York 1975), 28.
7. C. O. Sauer: *The Early Spanish Main* (Berkeley 1966), 150; C. H. Haring: *The Spanish Empire in America* (New York 1947).
8. Meinig, *opus cit.,* 12.
9. C. W. MacLachlan and J. E. Rodriguez: *Forging of the Cosmic Race: a Reinterpretation of Colonial Mexico* (Berkeley 1980), 198; Meinig, *opus cit.,* 14.
10. D. B. Quinn: *The Royanoke Voyages* (Hakluyt Society, London 1990), i 491, ii 717; A. L. Rowse: *The Expansion of Elizabethan England* (London 1955), 219–20.
11. J. A. Rawley: *The Transatlantic Slave Trade* (New York 1981), 25ff.
12. Garrett Mattingly: 'No Peace Beyond What Line?,' *Transactions of the Royal Historical Society* (London 5th series), xiii 145–62.
13. For these encounters see maps in Martin Gilbert: *The Routledge Atlas of American History* (3rd edn London 1993), 6, 10.
14. Quinn, *opus cit.,* i 188–94.
15. G. B. Parke: *Richard Hakluyt and the English Voyages* (London 1900); E. G. R. Taylor: *Writings and Correspondence of the Two Richard Hakluyts* (Hakluyt Society, London), i.
16. H. H. Jones: *O Strange New World: American Culture, the Formative Years* (New York 1967), 164.
17. Taylor, *opus cit.,* i 175.

18. Richard Hakluyt: *Principal Navigations* (Everyman Edition, London), vi 3–35.
19. John Aubrey: *Brief Lives*, ed. Oliver Lawson Dick (Oxford 1960), 'Ralegh;' Robert Naunton: *Fragmenta Regalia* (London 1641), 47.
20. For Ralegh's background see A. L. Rowse; *Ralegh and the Throckmortons* (London 1962), Chapter 8.
21. For Ralegh in Ireland see Robert Lacey: *Sir Walter Ralegh* (London 1973) 5; W. M. Wallace: *Sir Walter Ralegh* (Princeton 1959).
22. Michael Foss: *Undreamed Shores: England's Wasted Empire in America* (London 1974), 135–73.
23. Account of Captain Arthur Barlow in *ibid.*, 138.
24. *Ibid.*, 152.
25. Reproduced in *ibid.*
26. *Ibid.*, 166–8.
27. S. E. Morison: *The European Discovery of America: the Northern Voyages 500–1600* (New York 1971), 675.
28. Sir Francis Bacon: *Essays*, 'On Plantations' (Everyman Edition, London 1960), 104.
29. Rowse, *Expansion*, 221.
30. For the myth see William Haller: *Foxe's Book of Martyrs and the Elect Nation* (London 1963).
31. John Aylmer: *An Harborow for Faithful and True Subjects* (London 1560).
32. John Davys: *The Seaman's Secrets* (London 1594).
33. Richard Hakluyt: 'Discourse of Western Planting,' in *Original Writings of Richard Hakluyt . . .* (London 1935), ii 210–326.
34. W. J. Eccles: *France in America* (Vancouver 1972), 14ff; Morison, *opus cit.*, 600ff.
35. 'A True and Sincere Declaration of the Purposes and Ends of the Plantation Begun in Virginia,' quoted in Alexander Brown: *Genesis of the United States*, 2 vols (New York 1890), i 339.
36. Sir George Peckham: 'A True Report of the Late Discoveries . . . of the New Found Land', reprinted in *Notes and Queries* (London), xvii (1920), 43.
37. *New Britannia*, reprinted in *Colonial Tracts*, i No. 6.
38. For Smith see *Travels and Works*, ed. Edward Arbor, 2 vols (London 1884), and *True Travels, Adventures and Observations, 1630*, ed. J. G. Fletcher (New York 1930); P. L. Barbour: *Three Worlds of Captain John Smith* (New York 1964).
39. L. G. Tyler (ed.): *Narratives of Early Virginia, 1606–26* (Richmond 1907); John Rolfe: *True Relation of the State of Virginia . . . 1616* (New York 1971); T. J. Wertenbaker: *Shaping of Colonial Virginia* (New York 1958).
40. William Bradford: *History of Plymouth Plantation*, ed. S. E. Morison (Cambridge 1952); G. D. Langdon: *Pilgrim Colony* (Boston 1966).
41. For texts of the Compact, see R. E. Moody (ed.): 'Versions of the Mayflower Compact,' *Old South Leaflet*, No. 225 (Richmond 1952).
42. C. H. Lippy in R. Choquette and S. Poole (eds): *Christianity Comes to the Americas* (New York 1992), 460.

43. Frank Thistlethwaite: *The Dorset Pilgrims* (London 1989), 14–21.
44. Quoted in *ibid.*, 63.
45. Quoted in E. S. Morgan: *The Puritan Dilemma: the Story of John Winthrop* (Boston 1958), 8–11.
46. A. B. Forbes (ed.): *The Winthrop Papers, 1598–1649*, 5 vols (Boston 1929–47), ii 114ff.
47. Printed in *Massachusetts Historical Society Proceedings*, xii 262ff.
48. Allen French: *Charles I and the Puritan Upheaval* (London 1955), 331ff.
49. *Winthrop Papers*, ii 293ff.
50. R. C. Winthrop: *Life and Letters of John Winthrop*, 2 vols (Boston 1964–7), *Journal*, i 27ff.
51. Ellsworthy Huntington: *Civilisation and Climate* (New York 1925), 8.
52. H. U. Faulkner: *American Economic History* (6th edn New York 1949), 6–8.
53. C. E. Kellogg: *The Soils That Support Us* (New York 1941).
54. Lyman Carrier: *The Beginnings of Agriculture in America* (New York 1923), 41.
55. *United States Department of Commerce Yearbook* (Washington DC 1930), vol. ii, part i.
56. J. C. Young (ed.): *Chronicles of the First Planters* (New York 1930), 254.
57. See list in B. N. Forman: *American Seating Furniture 1630–1730* (New York 1988), 20.
58. R. S. Kellogg: *Pulpwood and Wood Pulp in America* (New York 1923), 148.
59. See map on Indian tribes of North America in Gilbert, *opus cit.*, 2.
60. Bradford, *opus cit.*, 121.
61. Edward Johnson: 'Wonder Working Providence of Sion's Saviour in New England,' 1628–51, *Original Narratives Series*, 210.
62. *American Husbandry* (London 1775), i 81.
63. *Ibid.*, 80.
64. *Massachusetts Early Records*, i 94; R. S. Dunn, *Sugar and Slaves: the Rise of the English Planter Class in the English West Indies, 1624–1713* (Chapel Hill 1972), 15.
65. Quoted in Choquette and Poole, *opus cit.*, 275.
66. Winthrop, *Journal*, ii 271–82, for full text of speech.
67. *Ibid.*, 198ff, 152ff.
68. Thomas Hutchinson: *History of the Colony and Province of Massachusetts Bay*, ed. L. S. Mayo, 3 vols (Cambridge 1936), i 54.
69. Winthrop, *Journal*, i 254ff, 261ff.
70. For the controversy see Perry Miller: *The New England Mind*, 2 vols (Cambridge reissue 1953); see also Andrew Delbanco: *The Puritan Ideal* (Cambridge 1989).
71. Dunn, *opus cit.*, 20–1.
72. Ebenezer Hazard (ed.), *Historical Collections*, 2 vols (Philadelphia 1972–4), ii 10.
73. K. W. Porter: 'Samuel Gorton: New England Firebrand,' *New England Quarterly*, vii 405ff.

74. For these controversies see Brooks Adams: *The Emancipation of Massachusetts* (Boston 1887); C. F. Adams: *Three Episodes of Massachusetts History* (Boston 1892); George Bancroft: *History of the United States* (New York 1859), i; Perry Miller: *Orthodoxy in Massachusetts* (Cambridge 1933); S. E. Morison: *Builders of the Bay Colony* (Boston 1930).

75. See the entry on Winthrop in R. W. Fox and J. T. Kloppenburg: *Companion to American Thought* (Cambridge 1995), 739–40.

76. Reprinted, Providence, Rhode Island, 1936; see also James Axtell: *The Europeans and the Indians: Essays in the Ethnology of Colonial North America* (New York 1981).

77. For Williams see Cyclone Covey: *The Gentleman Radical: a Biography of Roger Williams* (New York 1966); Harry Chupack: *Roger Williams* (New York 1969); B. F. Swan; 'Light on Roger Williams and the Indians,' *Providence Journal*, magazine section, November 23, 1969.

78. Perry Miller: *Roger Williams: His Contribution to the American Tradition* (Indianapolis 1953).

79. Text in *The Complete Writings of Roger Williams*, 7 vols (New York 1963).

80. G. W. LaFantasie (ed.): *Correspondence of Roger Williams*, 2 vols (Hanover, New Hampshire 1988).

81. The material on Mrs Hutchinson is printed in D. D. Hall (ed.): *The Antinomian Controversy: a Documentary History* (rev. edn New York 1990).

82. See Amy Lang: *Prophetic Woman: Anne Hutchinson and the Problem of Dissent in the Literature of New England* (Berkeley 1987).

83. Winthrop, *Journal*, i 265ff, 292ff, 313ff, 326ff.

84. See F. J. Bremer (ed.): *Anne Hutchinson: Troubler of the Puritan Zion* (New York 1981).

85. John Winthrop and William Welde: *A Short Story of the Rise, Reign and Ruine of the Antinomians, Familists and Libertines* (1643).

86. S. E. Morison: *Three Centuries of Harvard, 1636–1936* (Cambridge 1936), Chapter 1.

87. K. A. Lockridge: *A New England Town: the First Hundred Years: Dedham, Massachusetts, 1636–1736* (New York 1970).

88. S. E. Morison: *The Maritime History of Massachusetts* (Boston 1921), 21.

89. R. J. Brugger: *Maryland: a Middle Temperament, 1634–1980* (Baltimore 1988), 6–7.

90. Andrew White: 'A Brief Relation of the Voyage Unto Maryland' (1634), in C. C. Hall (ed.): *Narratives of Early Maryland, 1634–84* (New York 1910), 7–8, 40, 45, in White's text.

91. M. P. Andrews: *The Founding of Maryland* (New York 1933), 61.

92. Text of charter in Hall, *opus cit.*, 101–10.

93. M. D. Mereness: *Maryland as a Proprietary Province* (New York 1901), 51; White, *opus cit.*, 99–100.

94. Brugger, *opus cit.*, 15–16.

95. Quoted in Gloria Main: *Tobacco Colony: Life in Early Maryland, 1650–1720* (Princeton 1982), 144n.

96. Brugger, *opus cit.*, 20–4.

97. L. G. Carr and L. Walsh: 'The Planter's Wife: the Experience of White Women in 17th Century Maryland', *William and Mary Quarterly*, 34 (1977).

98. W. H. Browne (ed.): *Archives of Maryland* (Baltimore 1883–), i 244ff.

99. Bieter Cunz: *The Maryland Germans: a History* (Princeton 1948), 28.

100. Meinig, *opus cit.*, 151–2.

101. Dunn; *opus cit.*, 165ff.

102. C. Tunnard and H. H. Reed: *American Skyline: the Growth and Forms of Our Cities and Towns* (New York 1956), 33ff.

103. A C. Myers (ed.): *Narratives of Early Pennsylvania . . .* (New York 1912), 263.

104. J. T. Lemon: *The Best Poor Man's Country: a Geographical Study of Early Southeastern Pennsylvania* (Baltimore 1972), 108.

105. *Ibid.*, 216.

106. For the religious role of Philadelphia see S. E. Ahltrom: *A Religious History of the American People* (New Haven 1972).

107. C. F. Adams: *Massachusetts: Its Historians and Its History* (New York 1893), 64.

108. S. E. Morison: *The Intellectual Life of Colonial New England* (New York 1956), vi.

109. Quoted in P. Miller: *Errand Unto the Wilderness* (New York 1964).

110. Quoted in S. Fine and G. S. Brown: *The American Past: Conflicting Interpretations of Great Issues*, 2 vols (New York 1970), i 13.

111. S. C. Powell: *Puritan Village: The Formation of a New England Town* (Middletown, Connecticut 1963).

112. Quoted in Fine and Brown, *opus cit.*, 23.

113. Forman, *opus cit.*, 40.

114. R. B. St George: 'Father, Sin and Identity: Woodworking Artisans in Southeastern New England, 1620–1700,' in I. M. G. Quimby (ed.): *The Craftsmen of Early America* (New York 1984); Alexander Young (ed.): *Chronicles of the Pilgrim Fathers* (Baltimore 1974), 247.

115. Forman, *opus cit.*, 22.

116. Quimby, *opus cit.*, 116.

117. Barbara Perry (ed.): *American Ceramics* (New York 1989), 24.

118. Quimby, *opus cit.*, 235.

119. B. M. Ward: 'Boston Goldsmiths, 169—1720,' in *ibid.*; Katherine Butler: *American Silver, 1655–1825*, 2 vols (Boston 1972).

120. H. F. Clarke and H. W. Foote: *Jeremiah Dummer: Colonial Craftsman and Merchant 1645–1718* (Boston 1970), 3ff.

121. J. Caldwell and O. K. Roque: *American Paintings in the Metropolitan Museum of Art* (Princeton 1994), i.

122. Michael J. Rozbicki: 'Cultural Development of the Colonies,' in J. P. Greene and J. R. Pole (eds): *The Blackwell Encyclopaedia of the American Revolution* (New York 1991), 71ff.

123. See my *The Offshore Islanders: a History of the English People* (rev. edn London 1990), part 4, 171ff.

124. Rozbicki, *opus cit.*, 72. For the 'Law of First Effective Settlement,' see Wilbur Zelinsky: *Cultural Geography of the United States* (Englewood Cliffs 1973), 13–14.

125. J. F. James (ed.): *Narratives of Early American History* (New York 1911), 184.

126. Quoted in M. E. Sirmans: *Colonial South Carolina: a Political History, 1663–1763* (Chapel Hill 1966), 10.

127. Langdon Cheves (ed.): *Shaftesbury Papers ... Relating to Carolina*, South Carolina Historical Society Collections, 5 (1897), 399.

128. Quoted in Sirmans, *opus cit.*, 24–5.

129. Brugger, *opus cit.*, 12ff.

130. Sirmans, *opus cit.*, 38.

131. *Shaftesbury Papers*, 427.

132. 'Letters from John Stewart to William Dunlop,' *South Carolina Historical Magazine* 32 (1931).

133. Faulkner, *opus cit.*, 80–1.

134. D. R. Dewey: *Financial History of the United States* (12th edn New York 1934), 19.

135. See W. J. Schulz and M. R. Caine: *Financial Development of the United States* (New York 1937).

136. For early settler–Indian conflicts, see A. T. Vaughan: *The New England Frontier: Puritans and Indians, 1620–1675* (Boston 1965).

137. W. E. Washburn: *Governor and Rebel: Bacon's Rebellion in Virginia* (New York 1957).

138. For an excellent description of the work of the militia in King Philip's War see Thistlethwaite, *opus cit.*, Chapter 13, 236ff.

139. R. Slotkin and J. K. Folsom (eds): *So Dreadful a Judgment: Puritan Responses to King Philip's War, 1676–1677* (Middletown, Connecticut 1978).

140. S. S. Webb: *1676: the End of American Independence* (New York 1984), 341, 410.

141. G. L. Kittredhe: *Witchcraft in Old and New England* (London 1929).

142. Keith Thomas: *Religion and the Decline of Magic* (Oxford 1971).

143. For a recent account see Frances Hill: *A Delusion of Satan: the Full Story of the Salem Witch Trials* (New York 1995).

144. In his book *The Christian Philosopher* (Boston 1721).

145. Lawrence Wright: *Remembering Satan* (New York 1994).

146. Published by the Massachusetts Historical Society, 1911–12; for Mather, see Kenneth Silverman: *The Life and Times of Cotton Mather* (New York 1985), and K. B. Murdock (ed.): *Selections from Cotton Mather* (Boston 1926).

147. R. N. Hill: *Yankee Kingdom: Vermont and New Hampshire* (New York 1960).

148. Quoted in H. J. Ford: *The Scotch-Irish in the Americas* (New York 1914), 271–2.

149. Sirmans, *opus cit.*, 132–3.

150. Phinizy Spalding: *Oglethorpe in America* (Chicago 1977).

151. H. E. Davis: *The Fledgling Province: Social and Cultural Life in Colonial Georgia, 1733–1776* (Chapel Hill 1976).

152. Table 3 in Meinig, *opus cit.*, 247.

153. L. Labaree (ed.): *The Papers of Benjamin Franklin* (New Haven 1961), iv 227–34.

154. R. D. Mitchell: 'Content and Context: Tidewater Characteristics in the Early Shenandoah Valley,' *Maryland Historian*, 5 (1974).

155. Faulkner, *opus cit.*, 115–16.

156. Quoted in *ibid.*, 78.

157. C. F. Carroll: *The Timber Economy of Puritan New England* (Providence 1973); J. F. Shepherd and G. M. Walton: *Shipping, Maritime Trade and the Economic Development of Colonial North America* (Cambridge 1972).

158. H. A. Innes: *The Cod Fisheries: a History of an International Economy* (rev. edn Toronto 1954).

159. V. S. Clark: *History of Manufactures in the United States, 1670–1860*, 3 vols (rev. edn New York 1929), i 207ff.

160. Massachusetts Historical Society, *Collections*, 1st series, i 74.

161. *Interests of Merchants and Manufacturers of Great Britain in the Present Contests Stated and Considered* (London 1774, reprinted Boston), 12.

162. Quoted in Faulkner, *opus cit.*, 82.

163. E. J. Perkins: 'Socio-economic Development of the Colonies,' in Greene and Pole, *opus cit.*, 53ff.

164. *Ibid.*, 57.

165. A. R. Ekirch: *Bound for America: Transportation of British Convicts for the Colonies, 1718–75* (Oxford 1987).

166. Brugger, *opus cit.*, 87.

167. 'Maryland Hoggs and Hyde Park Duchesses': a brief account of Maryland in 1697, *Maryland Historical Magazine*, 73 (1978).

168. P. U. Bonomi: *A Factious People and Society in Colonial New York* (New York 1971).

169. J. A. Smith: *Printers and Press Freedom: the Ideology of Early American Journalism* (New York 1988). There is a study of one of the early papers: Hennig Cohen: *The South Carolina Gazette* (U. of S. Carolina Press, Charleston 1943).

170. L. W. Levy: *Emergence of a Free Press* (New York 1985).

171. Brugger, *opus cit.*, 81.

172. Raphael Semmes: *Baltimore as Seen by Visitors, 1783–1860* (Baltimore 1953), 4ff; Richard Switzzer (trans. and ed.): *Chateaubriand's Travels in America* (Lexington 1969), 13.

173. See the reconstruction in Roger W. Moss: *The American Country House* (New York 1990), 47.

174. *Ibid.*, 42.

175. See Bernard Bailyn: *Origins of American Politics* (New York 1968); G. B. Nash: *The Urban Crucible: Social Change, Political Consciousness and the Origins of the American Revolution* (Harvard 1979).

176. R. J. Dinkin: *Voting in Provincial America: a Study of Elections in the Thirteen Colonies, 1689–1776* (Westport, Connecticut 1977).

177. R. E. Brown: *Middle-Class Democracy and the Revolution in Massachusetts, 1691–1780* (Ithaca 1955).

178. J. R. Pole: 'Representation and Authority in Virginia from the Revolution to Reform,' *Journal of Southern History*, February 1958.

179. J. P. Green: 'The Role of the Lower Houses of Assembly in 18th Century Politics,' *Journal of Southern History*, November 1961.

180. W. Whitehill: *Boston: a Topographical History* (2nd edn Cambridge 1968), 22ff; the figures come from the map drawn by Captain John Bonner.

181. Gottfried Mittelburger: *Journey to Pennsylvania*, ed. and trans. O. Handlin and J. Clive (Cambridge 1960), 47.

182. Esmond Wright: *Franklin of Philadelphia* (Cambridge 1986), 32.

183. Jon Butler: *Awash in a Sea of Faith: Christianising the American People* (Cambridge 1990); Joseph Conforti: 'The Invention of the Great Awakening, 1795–1842,' *Early American Literature*, 26 (1991).

184. Alan Heimart: *Religion and the American Mind from the Great Awakening to the Revolution* (Cambridge 1966).

185. For Edwards see Perry Miller: *Jonathan Edwards* (Boston 1949).

186. O. E. Winslow (ed.): *Basic Writings of Jonathan Edwards* (New York 1966), 115, 128–9.

187. Jonathan Edwards: *Works* (New Haven 1957–), 10 volumes in print.

188. *Basic Writings*, 256, 275.

189. E. S. Gaustad: *The Great Awakening in New England* (New York 1957); H. S. Stout: *The Divine Dramatist: George Whitefield and the Rise of Modern Evangelicalism* (Grand Rapids 1991).

190. W. M. Gerwehr: *The Great Awakening in Virginia, 1740–90* (Duke University Press 1936).

191. R. J. Wilson: *The Benevolent Deity: Ebenezer Gay and the Rise of Rational Religion in New England, 1696–1787* (Philadelphia 1984); Conrad Wright: *The Beginning of Unitarianism in America* (Boston 1955).

192. The interaction between religious and political fermentation in the decades before the American Revolution is analysed in J. C. D. Clark: *The Language of Liberty: Political Discussion and Social Dynamics in the Anglo-American World, 1660–1832* (Cambridge, England, 1994), esp. 36, 120–1, 148–9, 250, 262–3.

PART TWO: Revolutionary America

1. J. T. Flexner: *George Washington: the Forge of Experience, 1732–75* (New York 1967), 14–15.

2. Thomas A. Lewis: *For King and Country: The Maturing of George Washington, 1748–60* (New York 1993), 6.

3. Richard Norton Smith: *Patriarch: George Washington and the New American Nation* (Boston 1993), 5.

4. *Ibid.*, 16.
5. Quoted in Marcus Cunliffe: *George Washington: Man and Monument* (Boston 1958), 31.
6. *Rules of Civility and Decent Behavior in Company and in Conversation*, ed. Charles Moore (Boston 1926).
7. Smith, *opus cit.*, 18, 23.
8. Lewis, *opus cit.*, 141ff.
9. *London Magazine*, xiii (1954); D. S. Freeman: *George Washington*, 7 vols (New York 1948–57), iii 89.
10. For the wars see H. H. Peckham: *The Colonial Wars, 1689–1762* (Chicago 1964); D. E. Leach: *Roots of Conflict: British Armed Forces and Colonial Americans, 1677–1763* (Chapel Hill 1986).
11. Max Savelle: *Empires to Nations: Expansion in North America, 1713–1824* (New York 1968), 149.
12. G. Gregault: *Canada: the War of the Conquest* (trans. Toronto 1969); G. A. Rawlyk: *Nova Scotia's Massachusetts: a Study of Massachusetts–Nova Scotia Relations, 1630–1784* (Montreal 1973).
13. Louis de Vorsey: *The Indian Boundary in the Southern Colonies, 1763–75* (Chapel Hill 1966).
14. Flexner, *opus cit.*, 142.
15. *Ibid.*, 234, 262.
16. M. Spector: *The American Department of the British Government, 1768–82* (New York 1940).
17. L. W. Labaree *et al.* (eds): *Papers of Benjamin Franklin*, 22 vols (Philadelphia 1959–70), xviii 102–3.
18. W. J. Smith (ed.): *Grenville Papers*, ii 114.
19. L. W. Labaree *et al.* (eds): *Franklin's Autobiography* (New Haven 1964). The best of many biographies of Franklin is Esmond Wright: *Franklin of Philadelphia* (Cambridge 1986).
20. Edwin Wolf: 'Franklin and His Friends Choose Their Books,' *Pennsylvania Magazine of History and Biography*, January 1956.
21. *Franklin Papers*, iii 397–420.
22. J. F. Ross: 'The Character of Poor Richard,' *Proceedings of the Modern Language Association of America*, September 1940; I. G. Willey: *The Self-Made Man in America: the Myth of Rags to Riches* (Princeton 1954).
23. Carl Van Doren: *Benjamin Franklin* (New York 1938), 71.
24. W. C. Bruce: *Benjamin Franklin Self-Revealed*, 2 vols (New York 1917), ii 362.
25. Text in *Franklin Papers*, iv.
26. Wright, *opus cit.*, 90–9; Franklin, *Autobiography*, 210–11.
27. *Franklin Papers*, viii 293.
28. Francis Parkman's classic account, *History of the Conspiracy of Pontiac*, 2 vols (New York 1851), is reprinted in his *Collected Works* (New York 1922).
29. *Narrative of the Late Massacres in Lancaster County*, in *Franklin Papers*, xi.
30. Quoted in Wright, *opus cit.*, 167.
31. Quotations from *ibid.*, 166.

32. B. Bailyn and J. B. Hench (eds): *The Press and the American Revolution* (Worcester, Massachusetts 1980).

33. Printed in G. A. Peek (ed.): *The Political Writings of John Adams* (Indianapolis 1954).

34. For the evolution of Adams' position see J. R. Howe: *The Changing Political Thought of John Adams* (Princeton 1966).

35. For the role of the Tea Party see Pauline Maier: *From Resistance to Revolution: Colonial Radicals and the Development of Opposition to Britain, 1765–1776* (New York 1972).

36. For Jefferson's early background see Dumas Malone: *Jefferson and His Time*, vol. i: *Jefferson the Virginian* (Boston 1948), 27ff.

37. *Washington Dossier*, May 1985, quoted in A. J. Mapp: *Thomas Jefferson: A Strange Case of Mistaken Identity* (New York 1987), 410.

38. A. A. Lipscomb and A. E. Bergh (eds): *Thomas Jefferson's Writings*, 20 vols (New York 1903); J. P. Boyd *et al.* (eds): *The Papers of Thomas Jefferson*, 25 vols (Princeton 1950–); J. M. Smith (ed.): *The Republic of Letters: The Correspondence of Thomas Jefferson and James Madison, 1776–1826*, 3 vols (New York 1995).

39. N. E. Cunningham: *In Pursuit of Reason: the Life of Thomas Jefferson* (Baton Rouge 1987).

40. Isaac Kramknick: 'The Ideological Background,' in J. P. Greene and J. R. Pole (eds): *The Blackwell Encyclopaedia of the American Revolution*, 84ff.

41. Quotations from M. D. Peterson (ed.): *Thomas Jefferson: Selected Writings* (New York 1984), 122, 118.

42. Joyce Appleby: 'Republicanism in the History and Historiography of the United States,' *American Quarterly*, 37 (1985); Jack P. Greene: *Peripheries and Center: Constitutional Development in the Extended Polities of the British Empire and the United States, 1607–1788* (New York 1986).

43. H. V. Faulkner, *American Economic History*, 120–6.

44. For the importance of the rule of law in the American Revolution, see John Philip Reid, *Constitutional History of the American Revolution*, vol. i: *The Authority of Rights* (1986); vol. ii: *The Authority to Tax* (1987); vol. iii: *The Authority to Legislate* (1991); vol. iv: *The Authority of Law* (1993); abridged edn Madison 1995.

45. M. C. Tyler: *Patrick Henry* (Ithaca 1962), 147ff.

46. Edmund Burke, *Works* (London 1893), ii 43.

47. *Franklin Papers*, xxii 94.

48. *John Adams Papers*, iii 89.

49. *Franklin Papers*, xxii 218.

50. *Massachusetts Historical Society Historical Collections*, 72 (1917), 82.

51. Flexner, *opus cit.*, 327.

52. Quotations from Freeman, *opus cit.*, iii 6.

53. *Journal of the Continental Congress*, ii 97.

54. Flexner, *opus cit.*, 340.

55. *Correspondence of Thomas Gage*, 2 vols (New Haven 1931–3), ii 187ff.

56. Jack Fruchtman Jr: *Thomas Paine: Apostle of Freedom* (New York 1994), 59–81.

57. *Adams Papers*, iv 59.

58. *Adams Works*, ii 514n.

59. J. P. Boyd: *The Declaration of Independence: the Evolution of the Text* (Princeton 1945); F. Herbert: *The Declaration of Independence: an Interpretation and Analysis* (New York 1904).

60. This remark is accepted as canonical in Jared Sparks (ed.): *Works of Benjamin Franklin*, 10 vols (Boston 1836–40), i 407.

61. Quoted in Conor Cruise O'Brien: *The Great Melody: a Thematic Biography of Edmund Burke* (London 1992), 161–2.

62. D. S. Lutz: 'State Constitution Making Through 1781,' in Greene and Pole, *opus cit.*, 276ff.

63. For details see W. P. Adams: *The First American Constitutions* (Chapel Hill 1980).

64. *Adams Papers,* iv 65ff.

65. See J. N. Rakove: 'The Articles of Confederation, 1775–83,' in Green and Pole, *opus cit.*, 289ff.

66. Quoted from P. H. Smith: *Letters of Delegates to Congress, 1774–89*, 13 vols (Washington DC 1976–), ix 908.

67. P. S. Onuf: *The Origins of the Federal Republic* (Philadelphia 1983).

68. *Adams Works*, viii 573.

69. *Lee Papers*, iv 9–10.

70. J. Boucher (ed.): *Reminiscences of an American Loyalist* (New York 1925), 49, 146.

71. Quoted in W. A. Bryan: *George Washington in American Literature, 1775–1865* (New York 1952), 46; E. P. Chase (ed.): *Our Revolutionary Forefathers* (New York 1929), ii 3f.

72. C. E. Burnett (ed.): *Letters of Members of the Continental Congress*, 8 vols (Washington DC 1921–6), iii 356.

73. Piers Mackesy: *The War for America, 1775–1783* (Cambridge 1964); Don Higginbotham: *The War of American Independence: Military Attitudes, Policies and Practice, 1763–1789* (New York 1971).

74. Wright, *opus cit.*, 263.

75. *Franklin Writings*, ix 243f; *Adams Diary*, iv 118.

76. E. E. Hale: *Franklin in France*, 2 vols (New York 1887), i 155ff.

77. W. A. Shewins (ed.): *Whiteford Papers* (Oxford 1898), i 87; for the peace, see R. B. Morris: *The Peacemakers: the Great Powers and American Independence* (New York 1965); R. Hoffman and P. J. Albert (eds): *Peace and the Peacemakers: The Treaty of 1783* (Charlottesville 1986).

78. De Segur: *Memoires*, 2 vols (Paris 1824–6), i 35, quoted in Wright, opus cit.; R. Hoffman and P. J. Albert: The Franco-American Alliance of 1778 (Charlottesville 1981).

79. Barbara Greymont: *The Iroquois in the American Revolution* (Syracuse 1972), 58.

80. C. G. Calloway: *Crown and Calumet: British–Indian Relations, 1783–1815* (Norman, Oklahoma 1987), 10f.

81. R. C. Downes: *Council Fires on the Upper Ohio* (Pittsburgh 1940), 294.

82. Allan Kulikov: 'A Prolific People: Black Population Growth in the Chesapeake Colonies, 1700–90,' *Southern Studies*, 16 (1977).

83. Sylvia R. Frey: 'Slavery and Anti-Slavery,' in Greene and Pole, *opus cit.*, 379ff; Ira Berlion and Ronald Hoffman: *Slavery and Freedom in the Age of the American Revolution* (Charlottesville 1983).

84. Jean B. Lee: *The Price of Nationhood: the American Revolution in Charles County* (New York 1994), 124–5.

85. See the findings of Wallace Brown: *The Good Americans: the Loyalists in the American Revolution* (New York 1969), 228–9; see also table in D. W. Meinig: *The Shaping of America*, vol i: Atlantic America, 1492–1800 (New Haven 1896), Table Five, 317.

86. E. S. Gaustad: *Documentary History of Religion in America* (Grand Rapids 1982), 244.

87. See the statistical table on p. 69 in E. S. Gaustad: 'Religion before the Revolution,' in Greene and Pole, *opus cit.*, 64–70.

88. See the section on allegiances in Bonomi, *Under the Cope of Heaven*, 39ff.

89. R. M. Calhoun: *The Loyalists in Revolutionary America, 1760–81* (New York 1983).

90. G. A. Rawlyk: 'The American Revolution and Canada,' in Greene and Pole, *opus cit.*, 497–503.

91. Quoted in Betty Wood: 'The Impact of the Revolution on the Role, Status and Experience of Women,' in Greene and Pole, opus cit., 399–408; R. Hoffman and P. J. Albert: *Women in the Age of the American Revolution* (Charlottesville 1989).

92. Quoted in Aungst C. Buell: *History of Andrew Jackson* (New York 1904), ii 410f. See also R. V. Rimini: *Andrew Jackson and the Course of American Empire, 1767–1821* (New York 1977), 11.

93. See Dan Higginbottom: 'The War for Independence,' in Greene and Pole, opus cit., 315–17.

94. Douglas Southall Freeman: *Washington* (New York 1992), 509–10 (this work is a one-volume abridgment of Freeman's seven-volume biography). See also: Don Higginbotham: *George Washington and the American Military Tradition* (Athens, Georgia 1985).

95. G. W. Corner (ed.): *Benjamin Rush: Autobiography* (Princeton 1948), 198.

96. T. H. Breen: *Tobacco Culture: the Mentality of the Great Tidewater Planters* (Princeton 1985), 161.

97. E. G. Evans: *Thomas Nelson of Yorktown* (Williamsburg 1975), 19ff; Gordon S. Wood: *The Radicalism of the American Revolution* (New York 1992), 69–70, 122.

98. R. A. Billington: *Westward Expansion: a History of the American Frontier* (2nd edn New York 1960), 156; Wood, *opus cit.*, 128.

99. Quoted in Wood, *opus cit.*, 135–6, 137.

100. Quoted in *ibid.*, 237.

101. *Adams Papers*, i 42–3; E. S. Morgan: *The Gentle Puritan: A Life of Ezra Styles, 1727–1795* (New Haven 1962), 167.

102. M. D. Kaplanoff: 'Confederation: Movement for a Stronger Union,' in Greene and Pole, *opus cit.*, 443ff.

103. For Hamilton's background see B. Mitchell: *Alexander Hamilton*, 2 vols (New York 1957–62), i Chapter 1; J. E. Cooke: *Alexander Hamilton* (New York 1982), 10ff.

104. R. B. Morris: *The Forging of the Union, 1781–89* (New York 1987).

105. For Madison's background see R. A. Rutland: *James Madison: The Founding Father* (New York 1987). For Frenau, see P. M. March: *Philip Frenau, Poet and Journalist* (New York 1968); his *Poems* were edited by F. L. Pattee in 3 vols, New York 1902–7.

106. Quoted in Smith, *Jefferson–Madison Correspondence*, i 2.

107. *Federalist Number 38*; see H. T. Colbourn (ed.): *Fame and the Founding Fathers: Essays by Douglas Adair* (New York 1974), 3–26.

108. Smith, *Jefferson–Madison Correspondence*, i Introduction, 1–36.

109. Madison's work before, during, and after the Annapolis meeting can be studied in W. T. Hutchinson *et al.* (eds): *The Papers of James Madison*, 15 vols (Chicago and Charlottesville 1962–), and in Smith, *Jefferson–Madison Correspondence*, i 394–434.

110. Wood, *opus cit.*, 254–6.

111. J. P. Roche: 'The Founding Fathers: a Reform Caucus in Action,' *American Political Science Review*, 55 (1961), 799–816; *The Records of the Federal Convention of 1787* are reprinted in 5 vols, New Haven 1987.

112. David Szatmary: *Shay's Rebellion: The Making of an Agrarian Insurrection* (Amhurst 1980).

113. H. A. Ohline: 'Republicanism and Slavery: Origins of the Three-Fifths Clause in the United States Constitution,' *William & Mary Quarterly* 28 (1971).

114. P. Finkelman: 'Slavery and the Constitutional Convention: Making a Covenant with Death,' in R. Beeman *et al.* (eds): *Beyond Confederation: Origins of the Constitution and American National Identity* (Chapel Hill 1987).

115. 'We Europeans have our own rich history to study, thank you'—Jacques Delors, head of the Brussels bureaucracy of the European Union.

116. See M. Gillespie and M. Liensch (eds): *Ratifying the Constitution* (Lawrence 1989).

117. Jacob Cooke (ed.): *The Federalist* (Cleveland 1961).

118. For Wilson see C. P. Smith: *James Wilson: Founding Father, 1742–98* (Chapel Hill 1956) and R. G. McCloskey (ed.): *Works of James Wilson*, 2 vols (Cambridge 1987).

119. The anti-federalist documents are reprinted in H. J. Storing (ed.): *The Complete Anti-Federalist*, 7 vols (Chicago 1981).

120. For the debates and procedures as a whole, see M. Jenson et al. (eds): *The Documentary History of the Ratification of the Constitution*, 8 vols (Madison 1976–).

121. Letter to Jean Baptiste Le Roy, November 13, 1789, in *Works of Benjamin Franklin* (Philadelphia 1817), Chapter 6.

122. Irving Brandt: *The Bill of Rights: Its Origin and Meaning* (New York 1965).

123. Quoted in Wood, *opus cit.*, 288.

124. Lewis, *opus cit.*, 250–1.

125. *Franklin Papers*, iv 234; v 204–5.

126. J. H. Kettner: *The Development of American Citizenship, 1608–1871* (Chapel Hill 1978), 175.

127. *Journals of the Continental Congress*, v 475–6.

128. Elise Marienstras: 'Nationality and Citizenship,' in Greene and Pole, *opus cit.*, 669–75; Kettner, *opus cit.*

129. Bernard Bailyn: *The Peopling of North America* (New York 1986); Maldwyn Allen Jones: *American Immigration* (New York 1960).

130. Quoted in Wood, *opus cit.*, 268–70.

131. Chilton Williamson: *American Suffrage from Property to Democracy* (New York 1960).

132. M Marchione *et al.* (eds): *Philip Mazzei: Selected Writings and Correspondence* (Prato, Italy 1983), 439, quoted in Wood, *opus cit.*, 295–6; for Latrobe see his *Journal, 1796–1820* (New York 1905).

133. Robert McClockey: *The Supreme Court* (New York 1960); for controversial treatments, see Archibald Cox: *The Court and the Constitution* (New York 1987), and R. H. Bork: *The Tempting of America: the Political Seduction of the Law* (New York 1990).

134. Wood, *opus cit.*, 322–5; J. M. Sosin: *The Aristocracy of the Long Robe: The Origins of Judicial Review in America* (Westport 1989), 280ff.

135. J. C. Fitzpatrick (ed.): *Washington's Writings*, 39 vols (New York 1931–44), xxvii 367.

136. Marcus Cunliffe (ed.): *Mason L. Weems's Life of Washington 1809* (New York 1962); Bishop William Meade: *Old Churches, Ministers and Families of Virginia*, 2 vols (Philadelphia 1857), ii 242–55; George Washington Parke Custis: *Recollections and Private Memoirs of Washington* (New York 1860); W. A. Bryan: *George Washington in American Literature, 1775–1865* (New York 1952).

137. *Franklin Papers*, ii 202–4.

138. Quoted in Wright, *opus cit.*, 49f.

139. Quoted in Mapp, *opus cit.*, 421.

140. L. J. Cappon (ed.): *The Adams–Jefferson Letters*, 2 vols (Chapel Hill 1959), ii 467.

141. R. M. Calhoon: 'The Religious Consequences of the Revolution,' in Greene and Pole, *opus cit.*, 58–9, and his *Evangelicals and Conservatives in the Early South, 1740–1861* (New York 1988).

142. Ralph Ketcham: 'James Madison and Religion: a New Hypothesis,' *Journal of the Presbyterian Historical Society*, July 1960; M. S. Evans: *The Theme is Freedom: Religion, Politics and the American Tradition* (Washington DC 1994), 270–88.

143. W. W. Sweet: *Religion in the Development of American Culture* (London 1963), 50.

144. Daniel Boorstin: *The Americans*, 2 vols (New York 1964), i 131.

145. Rene Williamson: *Independence and Involvement* (Baton Rouge 1964), 213ff; M. E. Bradford: *A Worthy Company* (Plymouth 1982), cited in Evans, *opus cit.*

146. For Washington as vestryman, see Flexner, *opus cit.*; Franklin to Paine quoted in Wright, *opus cit.*

147. *Adams Diary*, August 14, 1796; for Northwest Ordinance see H. S. Commager: *Documents of American History* (8th edn New York 1968), i 131.

148. *Documentary History of the First Federal Congress* (Baltimore 1977), 228, 232, quoted in Evans, *opus cit.*

149. J. D. Richardson (ed.): *Compilation of the Messages and Papers of the Presidents, 1789–1797*, 10 vols (New York 1969), i 64.

150. M. D. Conway (ed.): *The Writings of Tom Paine*, 4 vols (New York 1894–6).

151. Quoted in J. E. A. Smith: *History of Pittsfield, Springfield* (Boston 1876), 145f.

152. Quoted in W. C. Ford: *Statesman and Friend* (Boston 1927), 31; ref. in L. A. Cremin: *American Education: the National Experience, 1783–1876* (New York 1988).

153. *Jefferson Writings*, xi 428.

154. Quoted in Freeman, *opus cit.*, 559.

155. Figures from *Historical Statistics of the United States* (US Department of Commerce, Washington DC 1975); W. G. Anderson: *The Price of Liberty: The Public Debt of the Revolution* (Charlottesville 1983).

156. Quoted in Stuart Bruchey: 'Social and Economic Developments After the Revolution,' in Greene and Pole, *opus cit.*, 559–60.

157. C. P. Nettels: *The Emergence of a National Economy, 1775–1815* (New York 1962).

158. L. D. Baldwin: *Whiskey Rebels: The Story of a Frontier Uprising* (New York 1939).

159. Cecilia M. Kenyon: *Political Science Quarterly*, June 1958.

160. Saul K. Padover: *The Mind of Alexander Hamilton* (New York 1958).

161. E. S. Maclay (ed.): *Journal of William Maclay* (New York 1890), 177–9.

162. Quoted in Smith: *Patriarch*, xvi.

163. *Ibid.*, 19.

164. *Jefferson Writings*, i 254; ix 448ff.

165. R. W. Griswold: *The Republican Court* (New York 1854), 156.

166. *Maclay Journal*, 4.

167. Smith, *Patriarch*, 88–9.

168. Stephen Decatur: *Private Affairs of George Washington, from the Records and Accounts of Tobias Lear* (Boston 1933).

169. For Mrs Powell see Smith, *Patriarch*, 150–1, 184–5.

170. Greville Bathe: *Citizen Genet, Diplomat and Inventor* (Philadelphia 1946); Gilbert Chinard: *George Washington as the French Knew Him* (Princeton 1940), 104; M. D. Peterson: *Thomas Jefferson and the New Nation: a Biography* (New York 1970), 487–8.

171. L. M. Sears: *George Washington and the French Revolution* (Detroit 1960); Smith, *Patriarch*, 173–5.

172. For the Randolph affair see Bonstell Tachau: 'George Washington and the Reputation of Edmund Randolph,' *Journal of American History*, 73 (1986); John Reardon: *Edmund Randolph* (New York 1974), 300ff.

173. For two views of the 1780s and 1790s economically, see John Kiske: *The Critical Period in American History, 1783–9* (New York 1888) and R. A. East: *Business Enterprise in the American Revolutionary Era* (New York 1938), 242; C. P. Nettrels: *The Emergence of a National Economy, 1789–1815* (New York 1962).

174. Victor Huge Paltsis: *Washington's Farewell Address* (New York 1935); Smith, *Patriarch*, 278ff; Barry Schwartz: *George Washington: the Making of an American Symbol* (New York 1987); E. S. Morgan: *The Genius of George Washington* (New York 1980).

175. R. M. Ketchum: *Presidents Above Parties: the First American Presidency, 1789–1829* (Chapel Hill 1984).

176. J. C. Miller: *The Federalist Era* (New York 1952), 198ff.

177. H. C. Syrett *et al.* (eds): *Hamilton Papers*, 15 vols (1961–), xii 388–94, 440–53.

178. For Adams' work for the US Navy see D. W. Knox: *History of the United States Navy* (rev. edn New York 1948); H. and M. Spout: *Rise of American Naval Power, 1776–1918* (New York 1943).

179. See especially J. R. Howe: *Changing Political thought of John Adams* (New York 1966) and Edward Handler: *American and Europe in the Political Thought of John Adams* (New York 1964).

180. Adams to Jefferson, July 9, 1813, in Capon, *opus cit.*, ii 351–2.

181. See L. H. Butterfield (ed.): *John Adams Diary and Autobiography*, 4 vols (New York 1961), and *Supplement* (1966).

182. I. Bernard Cohen: *Science and the Founding Fathers* (New York 1995), 215–36.

183. Quoted in Joseph J. Ellis: *Passionate Sage: the Character and Legacy of John Adams* (New York 1994), 239.

184. For Abigail, see Janet Whitney: *Abigail Adams* (New York 1947); Abigail Adams' letters are printed in Capon, *opus cit.*

185. For early Washington see J. S. Young: *The Washington Community, 1800–28* (New York 1966); Stewart Mitchell (ed.): *New Letters of Abigail Adams, 1788–1801* (Boston 1947), 259f.

186. For Marshall's background and early life see A. J. Beveridge: *Life of John Marshall*, 4 vols (New York 1916–19), i 20ff.

187. John Taylor: *An Inquiry . . .* (Philadelphia 1814), 275; Arthur M. Schlesinger: *The Age of Jackson* (London 1946), 24.

188. Beveridge, *opus cit.*, iv 87.

189. Max Lerner: 'John Marshall and the Campaign of History,' *Columbia Law Review*, 39, no. 3, reprinted in L. W. Levy (ed.): *American Constitutional Law* (New York 1966).

190. D. G. Loth: *Chief Justice John Marshall and the Growth of the Republic* (New York 1949).

191. Beveridge, *opus cit.*, 586ff.

192. George Dangerfield: *The Era of Good Feelings* (London 1953), 165; Felix Frankfurter: *The Commerce Clause Under Marshall, Tainey and Waite* (Cambridge 1937).

193. H. J. Pious and G. Baker: 'McCulloch v Maryland: Right Principle, Wrong Case,' *Stanford Law Review*, 9 (1957).

194. For a range of views on Marshall's significance, see the symposium by Carl B. Swisher *et al.*: *Justice John Marshall: a Reappraisal* (New York 1955) and E. S. Crowin: *John Marshall and the Constitution* (New York 1919).

195. Smith, *Jefferson–Madison Correspondence*, i 33; D. E. Engdahl: 'John Marshall's "Jeffersonian Concept" of Judicial Review,' *Duke Law Journal*, 42 (1992), 279ff.

196. The phrase was Benjamin Rush's; see Ellis, *opus cit.*, 134.

197. Syrett, *Papers of Hamilton*, xxv 186–90.

198. J. M. Smith: *Freedom's Fetters: the Alien and Sedition Laws and America's Civil Liberties* (New York 1956).

199. Patricia Watlington: *The Partisan Spirit* (New York 1972).

200. Daniel Sisson: *The American Revolution of 1800* (New York 1974).

201. Mapp, *opus cit.*, 71ff; C. L. Griswold: 'Rights of Wrongs: Jefferson, Slavery and Philosophical Quandaries,' in M. J. Lacey *et al.* (eds): *A Culture of Rights: the Bill of Rights in Philosophy, Politics and Law* (New York 1991), 144– 51.

202. J. C. Miller: *The Wolf by the Ears: Jefferson and Slavery* (New York 1977), 161ff. For a recent hostile portrait of Jefferson, presenting him not only as a cynical slave-owner but as a racist, see Conor Cruise O'Brien: *The Long Affair: Thomas Jefferson and the French Revolution* (London 1996), 315–25; but see also the review of this book 'Sally and Her Master' by Bernard Bailyn, *Times Literary Supplement* (London), November 15, 1966. On Jefferson's relationship with Sally Hemings, daughter of Betty Hemings, see Douglass Adair: 'The Jefferson Scandals,' in *Fame and the Founding Fathers* (New York 1974).

203. E. S. Morgan: *American Slavery, American Freedom: the Ordeal of Colonial Virginia* (New York 1975).

204. William Peden (ed.): *Jefferson's Notes on the State of Virginia* (New York 1955), Chapters 8 and 14.

205. Peterson, *opus cit.*, 259.

206. Cohen, *opus cit.*, Appendix 8, 'Jefferson's Changing Views Concerning the Abilities of Black People,' 297ff.

207. Roger W. Moss: *The American Country House*, 81–5; R. J. Betts: 'The Woodlands,' *Wintethur Portfolio*, 14 (1979).

208. For these and many other details see Jack McLaughlin: *Jefferson and Monticello: the Biography of a Builder* (New York 1988), 287–327.

209. *Ibid.*, 313. See Fiske Kimball: *Thomas Jefferson Architect* (Boston 1916).

210. J. A. Bear and L. C. Stanton: *Jefferson's Memorandum Books*, 2 vols (Princeton 1990).

211. McLaughlin, *opus cit.*, *passim*.

212. Millicent Sowerby: *Catalogue of the Library of Thomas Jefferson*, 5 vols (Washington DC 1952–9), iv 215. The visitor was John Melish.

213. W. E. Rich: *History of the United States Post Office to the Year 1829* (Cambridge 1924), 137ff.

214. Jack McLaughlin: *To His Excellency Thomas Jefferson: Letters to a President* (New York 1991).

215. Jefferson to David Gelston, November 12, 1802 (Library of Congress), quoted in McLaughlin, *opus cit.*, 4.

216. Milton Lomask: *Aaron Burr: the Years from Princeton to Vice-President, 1756–1805* (New Haven 1979).

217. Milton Lomask: *Aaron Burr: the Conspiracy and the Years of Exile, 1805–1836* (New Haven 1982).

218. Douglas Johnson: 'The Maghreb,' in J. E. Flint (ed.): *The Cambridge History of Africa*, vol. v: *1790–1870* (Cambridge, England, 1976), 99–124; Kenneth Masin: *Gunfire in Barbary* (London 1982), 29ff.

219. P. L. Ford (ed.): *Jefferson's Writings*, 10 vols (1892–99), viii 143–7; Dumas Malone: *Jefferson and His Time*, 4 vols (1948–70), iv 258.

220. Alexander de Conde: *The Affair of Louisana* (New York 1976), 161–75.

221. Edward Channing: *History of the United States*, 6 vols (New York 1905–25), iv 319, takes the view that Bonaparte 'threw the province' at the United States and Livingstone, Monroe, Madison, and Jefferson merely 'caught it.'

222. J. M. Belolavek: 'Politics, Principle and Pragmatism in the Early Republic: Thomas Jefferson and the Quest for American Empire,' *Diplomatic History*, 19 (1991), 599ff; E. S. Brown: *The Constitutional History of the Louisiana Purchase, 1803–12* (Berkely 1920); Jefferson to Breckinridge, August 12, 1803, *Writings*, 1136–9.

223. R. G. Thwaites (ed.): *Original Journals of the Lewis and Clark Expedition*, 8 vols (New York, 1904–5).

224. Quoted in D. W. Meinig: *The Shaping of America: A Geographical Perspective of 500 Years of History*, vol. ii: *Continental America 1800–67* (New Haven 1993), 67.

225. A. L. Burt: *The United States, Great Britain and British North America from the Revolution to the Establishment of Peace after the War of 1812* (London 1940); for the Orders in Council see Chester New: *Life of Henry Brougham to 1830* (Oxford 1961), Chapter 6, 'Repealing the Orders in Council,' 58ff.

226. For the 'Chesapeake Incident' see W. P. Cresson: *James Monroe* (Chapel Hill 1946), 230–5; Burton Spivak: *Jefferson's English Crisis: Commerce, the Embargo and the Republican Revolution* (New York 1979).

227. Smith, *Jefferson–Madison Correspondence*, iii 1503–49; L. W. Levy: *Jefferson and Civil Liberties: the Darker Side* (Cambridge 1963), 93ff.

228. McLaughlin, *opus cit.*, 14–38.

229. Smith, *Jefferson–Madison Correspondence*, ii 1548–54; Richard Mannix: 'Gallatin, Jefferson and the Embargo of 1808,' *Diplomatic History*, 3 (1979), 151–72. This is critical of Jefferson's handling of the crisis.

230. Irving Brandt: *James Madison*, 6 vols (1941–61), iv 306.

231. For two recent views of Madison see J. N. Rakove: *James Madison and the*

Creation of the American Republic (Glenview 1990); D. R. McCoy: *Last of the Fathers: James Madison and the Republican Legacy* (New York 1989).

232. Gaillard Hunt (ed.): *Margaret Bayard Smith: the First Forty Years of Washington Society* (New York 1906).

233. Quoted in William Seale: *The President's House: A History*, 2 vols (Washington DC 1986), i 129.

234. Smith, *Jefferson–Madison Correspondence*, iii 1567.

235. J. B. McMaster: *A History of the People of the United States from the Revolution to the Civil War*, 6 vols (New York 1895), iv 199ff; N. K. Risjord: 'Election of 1812,' in Arthur M. Schlesinger Jr and F. R. Israel (eds): *History of American Presidential Elections, 1789–1968*, 4 vols (New York 1971), i 249ff.

236. Smith, *Jefferson–Madison Correspondence*, iii 1698ff.

237. A. L. Burt: *United States, Great Britain and Canada: from the Revolution to the Establishment of Peace after the War of 1812* (New York 1961).

238. *Jefferson Writings*, ix 366.

239. Edgar McInnis: *Canada: a Social and Political History* (rev. edn New York 1958), 194.

240. William Atherton: *Narrative of the Sufferings and Defeat of the North-Western Army Under General Winchester* (New York n.d.), 25–31, 56–67, etc.; Elias Barnall: *Account of the Hardships etc . . . of those Heroick Kentucky Volunteers and Regulars in the Years 1812–13* (New York n.d.), 36–54.

241. J. C. A. Stagg: *Mr Madison's War: Politics, Diplomacy and Warfare in the Early American Republic, 1783–1830* (Princeton 1983), 225.

242. McMaster, *opus cit.*, iv 7.

243. *Niles Weekly Register*, iii 238, p. 4.

244. C. J. Dutton: *Oliver Hazard Perry* (New York 1935).

245. McMaster, *opus cit.*, iv 116.

246. Quoted in Christopher Loyd: *Captain Marrat and the Old Navy* (London 1939), 148.

247. H. W. Dickinson: *Robert Fulton, Engineer and Artists* (London 1913), 17–21, which reproduces the Peale portrait of Fulton.

248. *Ibid.*, 73–95, 125.

249. *Ibid.*, 182–7, 194–9.

250. *Edinburgh Evening Courant*, August 31, 1815.

251. Southey to Scott, January 13, 1813, quoted in Geoffrey Carnall: *Robert Southey and His Age* (Oxford 1960), 124; for the technology of the rockets, see Kenneth Mason: *Gunfire in Barbary* (London 1982), 185ff.

252. Katherine Cave (ed.): *Diary of Joseph Faringdon*, 16 vols (New Haven 1978–84), xiii 4492, April 18, 1814.

253. S. M. Hamilton (ed.): *The Writings of James Monroe*, 8 vols (New York 1898–1903), v 245ff.

254. Warren M. Hoffnagle: *Road to Fame: William H. Harrison and the North-west* (New York 1959).

255. McMaster, *opus cit.*, iv, 138ff.

256. Anne H. Wharton: *Social Life in the Early Republic* (Philadelphia 1902), 172.

257. L. B. Cutts (ed.): *Memoirs and Letters of Dolly Madison* (New York 1886), 110ff.

258. Lady Bourchier: *Memoire ... of Sir Edward Codrington*, 2 vols (London 1873), vol. i: *The American Campaign*, 315ff.

259. Wharton, *opus cit.*, 172.

260. McMaster, *opus cit.*, iv 155.

261. Bourchier, *opus cit.*, 317.

262. Robert Allen Rutland: *The Presidency of James Madison* (Lawrence 1990), 157–67.

263. For Jackson's early career, see Robert V. Remini: *Andrew Jackson and the Course of American Empire, 1767–1821* (New York 1977), 37–112.

264. *Ibid.*, 120–3.

265. *Ibid.*, 184–5.

266. Robert V. Remini, *Andrew Jackson and the Course of American Freedom, 1822–32* (New York 1981), 1–3.

267. James Parton: *Life of Andrew Jackson*, 3 vols (Boston 1866), iii 63–5.

268. Reginald Horsman: *Expansion and American Indian Policy, 1783–1812* (East Lansing 1967).

269. R. C. Downes: *Council Fires on the Upper Ohio: a Narrative of Indian Affairs on the Upper Ohio Valley until 1795* (Pittsburgh 1940).

270. Dale Van Every: *The Disinherited: the Lost Birthright of the American Indian* (New York 1976).

271. Reginald Horsman: 'British Indian Policy in the North-West 1807–12,' *Mississippi Valley Historical Review*, April 1958.

272. J. F. H. Claiborne: *Mississippi as Province, Territory and State* (Jackson 1880), 3, quoted in Remini, *Jackson*, i.

273. A. J. Pickett: *History of Alabama* (Charleston 1851), ii 275; H. S. Halbert and T. S. Hall: *The Creek War of 1813–14* (Tuscaloosa 1969), 151ff.

274. J. Doherty Jr: *Richard Keith Call, Southern Unionist* (Gainsville 1961), 6, quoted in Remini, *Jackson*, i.

275. Jackson to Rachel Jackson, December 29, 1913, Jackson Papers in the Library of Congress, quoted in Remini, *Jackson*, i 194.

276. J. Reid and J. H. Eaton: *Life of Andrew Jackson* (reprint Tuscaloosa 1974) 63–70; Amos Kendall: *Life of Andrew Jackson* (New York 1844), 216–17.

277. Jackson to Rachel Jackson, December 29, 1813, Jackson Papers, Library of Congress, quoted in Remini, *Jackson*, i 201.

278. Reid and Eaton, *opus cit.*, 142–3, quoted in Remini, *Jackson*, i 212.

279. John Spencer Bassett (ed.): *Correspondence of General Jackson*, 6 vols (Washington DC 1926–33), i 488–9, 490.

280. *Ibid.*, 491–2.

281. Angie Debo: *The Road to Disappearance* (Norman, Oklahoma 1967), 82; Remini, *Jackson*, i 219.

282. Text of treaty in Charles Kappler: *Indian Affairs: Laws and Treaties* (Washington DC 1903), ii 107ff; letter in Remini, *Jackson*, i 240.

283. Remini, *Jackson*, i 154.

284. C. B. Brooks: *The Siege of New Orleans* (New York 1961); H. F. Rankin (ed.): *The Battle of New Orleans, a British View: the Journal of Major C. R. Forrest* (London 1961); Remini, *Jackson*, i 335ff.

285. For Adams' account of the peace talks see Allan Nevins (ed.): *The Diary of John Quincy Adams, 1794–1845* (New York 1951); the 'enemies list' entry is November 23, 1835.

286. Nevins, *opus cit.*, 136–7, 139, 145ff.

287. Stagg, *opus cit.*, 375–80.

288. Smith, *Jefferson–Madison Correspondence*, iii 1753ff.

289. Stagg, *opus cit.*, 500ff; T. A. Bailey: *A Diplomatic History of the American People* (New York 1969), 157ff.

290. Nevins, *opus cit.*, 151.

291. Text of the Ghent Treaty in Fred Israel (ed.): *Major Peace Treaties of Modern History, 1648–1967* (New York 1967), i 704.

PART THREE: Democratic America

1. *Information to Those Who Would Remove to America* (1784), in Benjamin Franklin, *Writings*, viii 603ff.

2. *Congressional Glove*, 29th Congress, 1st Session, January 10, 1846, 211; quoted in D. W. Meinig, *The Shaping of America: Continental America, 1800–67*, 222.

3. Colin Clark: *Population Growth and Land Use* (London 1969), 106f, Table iii 14.

4. James Flint: *Letters from America* (Edinburgh 1822).

5. For the growth of US banks see M. G. Myers: *A Financial History of the United States* (New York 1970).

6. Leon Schur: 'The Second Bank of the United States and Inflation After the War of 1812,' *Journal of Political Economy*, 68 (1960).

7. R. C. H. Catterall: *The Second Bank of the United States* (New York 1903), 28–32, 160n.

8. Quoted in George Dangerfield: *The Era of Good Feelings* (London 1953), 179–80.

9. Murray N. Rothbard: *The Panic of 1819* (New York 1962).

10. Quoted in Catterall, *opus cit.*, 68.

11. William M. Gouge: *Paper Money and Banking*, 2 vols (Philadelphia 1833), ii 109.

12. C. F. Adams (ed.): *Memoirs of John Quincy Adams*, 12 vols (Boston 1874–7), iii 167.

13. Emanuel Howitt: *Selections from Letters . . . written in 1819* (Nottingham 1820), 217.

14. *Niles Weekly Register*, 18 (1820). A complete run of *Niles* was printed in facsimile in 1947.

15. *Select Committee on Emigration from the United Kingdom, Fourth and Fifth Reports, Parliamentary Papers* (London 1826), 1826–7.

16. C. McEvedy and R. Jones: *Penguin Atlas of World Population History* (Harmondsworth 1978), 285–7, 313–14, 327; H. R. Jones: *A Population Geography* (New York 1981), 254; A. W. Crosby: *Ecological Imperialism: the Biological Expansion of Europe* (Cambridge, England 1986), 3–5.

17. Hansen, *opus cit.*, 152, 159–61.

18. R. A. Billington: *Westward Expansion: a History of the American Frontier* (New York 1949), 265ff, 290ff, 310ff.

19. R. J. Rohrbough: *The Transappalachian Frontier: People, Societies and Institutions* (Oxford 1978), 171–2.

20. *Ibid.*, 168–9.

21. Quoted in Daniel J. Boorstin: *The Americans: the National Experience* (New York 1970), 75.

22. Robert V. Remini, *Andrew Jackson and the Course of American Empire, 1767–1821*, 331–2.

23. John Niven: *Martin Van Buren: The Romantic Age in American Politics* (Oxford 1983), 185.

24. P. D. Evans: *The Holland Land Company* (Buffalo 1924).

25. Freeman Cleaves: *Old Tippecanoe: William Henry Harrison and His Times* (New York 1939).

26. *Hansard*, 3rd series, 33, 852.

27. H. H. Bellot: *American History and American Historians* (London 1952), Chapter 4, 'The Settlement of the Mississippi Valley,' 108ff.

28. F. S. Philbrock: *The Rise of the West, 1754–1830* (New York 1965), 314–15; R. G. Albion: *Rise of New York Port, 1815–60* (New York 1939); R. E. Shaw: *Erie Water West: Erie Canal, 1792–1854* (New York 1966).

29. Quoted in Rohrbough, *opus cit.*, 211.

30. *Ibid.*, 361–2; R. C. Buley: *The Old North-West: Pioneer Period, 1815–40*, 2 vols (Indianapolis 1950).

31. Elijah Iles: *Sketches of Early Life and Times in Kentucky, Missouri and Illinois* (Springfield, Illinois 1863); Rohrbough, *opus cit.*, 178ff.

32. Stone's *Autobiography*, from which these quotations come, is reprinted in Rhodes Thompson: *Voices from Cane Ridge* (St Louis 1954).

33. Lyman Beecher: *A Plea for the West* (2nd edn Cincinnati, 1835), 11.

34. W. C. Barclay: *Early American Methodism, 1769–1844* (New York 1949).

35. O. K. Armstrong and M. M. Armstrong: *The Indomitable Baptists: Their Role in Shaping American History* (New York 1967).

36. A. W. Spalding: *Origins and History of Seventh-Day Adventists*, 4 vols (Washington DC 1961–2); D. E. Robinson: *The Story of Our Health Message* (Nashville 1943).

37. T. F. O'Dea: *The Mormons* (Chicago 1957); for the dispute over polygamy see N. F. Furniss: *The Mormon Conflict, 1850–57* (New Haven 1960).

38. Quoted in E. W. Fornell: *The Unhappy Medium: Spiritualism and the Life of Margaret Fox* (Austin 1964).

39. For American Spiritualism see E. Gurney *et al.*: *Phantasms of the Living*, 2 vols (New York 1886); G. W. Butterworth: *Spiritualism and Religion* (New York

1944). Sir Arthur Conan Doyle, a fervent believer, wrote *A History of Spiritualism*, 2 vols (London 1926).

40. R. Peel: *Christian Science: Its Encounter with American Culture* (New York 1958).

41. W. H. Goldman *et al.*: *Ralph Waldo Emerson: Journals and Miscellaneous Notebooks*, 9 vols (Cambridge 1960–72).

42. Charles Crowe: *George Ripley: Transcendentalist and Utopian Socialist* (New Haven 1967).

43. R. A. Parker: *A Yankee Saint: John Humphrey Noyes and the Oneida Community* (New York 1935).

44. E. D. Andrews: *People Called Shakers* (New York 1953).

45. Jonathan Messerli: *Horace Mann: a Biography* (New York 1972).

46. Mann's annual reports are printed in Mary Mann: *Life and Works of Horace Mann*, 5 vols (rev. edn Boston 1891).

47. For the growth of Catholicism see James Hannessy: *American Catholics: a History of the Roman Catholic Community in the United States* (New York 1981).

48. The Maria Monk period is described in Jay Dolan: *The American Catholic Experience: a Social History from Colonial Times to the Present* (New York 1985) and in Philip Gleason: *Keeping Faith: American Catholicism Past and Present* (Notre Dame 1987).

49. Ray Billington: *The Protestant Crusade, 1800–60* (New York 1938); J. P. Dolan: *The Immigrant Church: New York's Irish and German Catholics, 1815–65* (New York 1975).

50. Orestes Brownson: *The American Republic* (New York 1866); his criticism of Mann's Second Annual Report is in the *Boston Quarterly Review*, 2 (October 1839), 394–434.

51. D. J. O'Brien: *Public Catholicism* (New York 1989).

52. J. J. Blau and S. W. Baron: *The Jews in the United States, 1790–1840: a Documentary History*, 3 vols (New York 1963), i Introduction xviiiff.

53. Meyrer Waxman: *American Judaism in the Light of History: Three Hundred Years* (New York 1955).

54. Henry Hobhouse: *Seeds of Change: Five Plants That Transformed the World* (New York 1986), 144ff.

55. *Ibid.*, 142; Samuel Smiles: *Industrial Biography* (London 1863), 322ff; Derry and Williams, *opus cit.*, 287ff.

56. Constance M. Green: *Eli Whitney and the Birth of American Technology* (2nd edn New York 1956).

57. Quoted in Hobhouse, *opus cit.*, 181 n. 20.

58. J. W. Roe: *English and American Tool-Builders* (New York 1916); see his 'Interchangeable Manufacture,' *Newcomen Society Transactions*, 17 (1937), 165ff.

59. H. J. Habbakuk: *American and British Technology in the 19th Century* (New York 1962).

60. J. Mirsky and A. Nevins: *The World of Eli Whitney* (New York 1952).

61. J. G. de R. Hamilton (ed.): *The Papers of Thomas Ruffin*, 4 vols (Releigh, North Carolina 1918–20), i 198.
62. Jan Lewis: *The Pursuit of Happiness, Family and Values in Jefferson's Virginia* (Cambridge, England 1983); Hobhouse, *opus cit.*, 153.
63. Peter Kolchin: *Unfree Labor: American Slavery and Russian Serfdom* (Cambridge 1987),366, Table 11.
64. Billington, *Westward Expansion*, 198–9.
65. Hobhouse, *opus cit.*, 158.
66. *Ibid.*, 183 n. 34.
67. Harry Ammon: *James Monroe: the Quest for National Identity* (repr. New York 1990).
68. Irving Brandt: *The Fourth President: a Life of James Madison* (London 1970), 639–40.
69. Reprinted in the Classics of Liberty Library, New York 1996.
70. According to a 76-page autobiographical pamphlet Calhoun wrote for his abortive presidential campaign in 1843, quoted in Irving H. Bartlett: *John C. Calhoun* (New York 1993).
71. Charles Woodmason: *The Carolina Backwoods on the Eve of the Revolution*, ed. R. J. Hooker (Chapel Hill 1953).
72. Bartlett, *opus cit.*, 27.
73. *Ibid.*, 54; R. M. Weir: *Colonial South Carolina* (New York 1983), 170ff; G. C. Rogers: *Charleston in the Age of the Pinckneys* (Norman, Oklahoma 1969).
74. G. W. Featherstonehaugh: *Excursion Through the Slave States* (New York 1844).
75. Glover Moore: *The Missouri Controversy 1819–21* (Lexington 1953).
76. Alan Nevins (ed.): *Diary of John Quincy Adams, 1794–1845*, 231, March 3, 1820.
77. Lrom Litwack: *North of Slavery: the Negro in the Free States, 1790–1860* (Chicago 1961), 167; R. T. Takaki: *Iron Cages: Race and Culture in 19th Century America* (London 1980), 113–14.
78. Nevins, *opus cit.*, 96.
79. *Ibid.*, 228, 246–7.
80. The best biography of Clay is Robert V. Remini: *Henry Clay: Statesman for the Union* (New York 1991).
81. E. de Witt Jones: *Influence of Henry Clay on Abraham Lincoln* (Lexington 1952), 21.
82. For Clay's ancestry, background, and personal circumstances, see Remini, *Clay*, 1–14.
83. J. F. Hopkins *et al.* (eds): *The Papers of Henry Clay*, 9 vols (Lexington 1959–), vii 511.
84. For Clay and Wythe see Calvin Colton: *Life and Times of Henry Clay*, 2 vols (New York 1846), i 20ff.
85. T. D. Clark: *History of Kentucky* (New York 1937), 60ff.
86. Remini, *Clay*, 155–68.
87. Horace Greely: *Recollections of a Busy Life* (New York 1868), 250; J. M. Rogers: *The True Henry Clay* (Philadelphia 1904), 250.

88. Remini, *Clay*, 225ff.
89. For Clay and the American system see G. G. van Deusen: *Life of Henry Clay* (New York 1937) and Clement Eaton: *Henry Clay and the Art of American Politics* (New York 1957). There is a learned German work on the topic, M. L. Fringst: *Henry Clays American System und die sektionale Kontroverse in den Vereinigten Staaten von Amerika, 1815–29*.
90. L. P. Littel: *Ben Hardin: His Times and Contemporaries* (Louisville 1887), 38ff.
91. Everett Somerville Brown (ed.): *The Missouri Compromises and Presidential Politics, 1820–25* (St Louis 1926), 42; Glover Moore: *The Missouri Controversy, 1819–1821* (Lexington 1953), 156.
92. Henry S. Foote: *A Casket of Reminiscences* (New York 1968), 30.
93. Ernest R. May: *The Making of the Monroe Doctrine* (New York 1975).
94. For the first use of the phrase see *Niles Register*, March 5, 1817; Dangerfield, *opus cit.*, 96.
95. See Morton Borden: *Parties and Politics in the Early Republic, 1789–1815* (New Haven 1967) and Richard Hofstadter: *The Idea of a Party System, 1780–1840* (New York 1969).
96. For Crawford see P. J. Green: *William H. Crawford* (New York 1965).
97. Nevins, *opus cit.*, 353ff.
98. Robert V. Remini, *Andrew Jackson and the Course of American Freedom, 1822–32*, 397 n. 14, lists various villainies of Webster; for Benton see W. N. Chambers: *Old Bullion Benton* (New York 1956).
99. *Baltimore Federal Republican*, September 4, 1822; *New York Statesman*, August 6, 1822; quoted in Remini, *Jackson*, ii 13–14.
100. Letter to Benjamin Austin, January 9, 1816, quoted in Arthur Schlesinger Jr: *The Age of Jackson* (Cambridge 1945), 18.
101. Remini, *Jackson*, i 357–8, 364; P. C. Brooks: *Diplomacy and the Borderlands; the Adams–Onis Treaty of 1819* (Berkeley 1939), 117.
102. Clay to Francis T. Brooke, August 3, 1833, in *Clay Papers*, viii 661–2.
103. H. J. Doherty Jr: 'Andrew Jackson on Manhood Suffrage: 1822,' *Tennessee Historical Quarterly*, 15 (1956), 60.
104. Letter to James Buchanan, June 25, 1825, quoted in Remini, *Jackson*, ii 30–1.
105. Quoted in Remini, *Jackson*, ii 78–9.
106. Letter to Ezekile Webster, February 22, 1824; *Daniel Webster, Private Correspondence* (New York 1902), i 346.
107. Marquess James: *Andrew Jackson: Portrait of a President* (New York 1937), 99ff.
108. E. W. Austin: *Political Facts of the United States Since 1789* (New York 1986), Table 3.1, 92ff. Remini, *Jackson*, ii, gives Jackson 152,901, Adams 114,023.
109. John Spencer Bassett (ed.), *Correspondence of General Jackson*, iii 270; H. A. Wise: *Seven Decades of the Union* (Philadelphia 1881), 110–11.
110. Quoted in James, *opus cit.*, 135.
111. J. F. Hopkins: 'Election of 1824,' in Arthur M. Schlesinger Jr and F. L. Israel (eds), *History of American Presidential Elections, 1789–1968*, i.

112. *Jackson Correspondence*, ii 276.

113. See Remini, *Clay*, Chapter 15, 'The Corrupt Bargian,' 251–72.

114. *Jackson Correspondence*, iii 291.

115. R. V. Remini: *Martin Van Buren and the Making of the Democratic Party* (New York 1959); James, *opus cit.*, 144–5.

116. Nevins, *opus cit.*, entry for April 10, 1824.

117. Remini, *Jackson*, ii 133. The pimping accusation against Adams was made in a campaign biography of Jackson written by Isaac Hill of New Hampshire.

118. Nevins, *opus cit.*, 287, 297, 348–9.

119. *Ibid.*, 368, 378, 382.

120. James, *opus cit.*, 120; John J. Crittenden to Clay, February 15, 1825; *Jackson Correspondence*, iii 325; *United States Telegraph*, June 16, 1827.

121. The Coffin Handbill is reproduced in James, *opus cit.*, 158–9; Harriet Martineau, *Society in America*, 3 vols (London 1837), iii 166.

122. Song quoted in Remini, *Jackson*, ii 134.

123. For the background see Alvin Kass: *New York Politics, 1800–30* (New York 1965).

124. For further details see De Valsa S. Alexander: *A Political History of the State of New York, 1774–1882*, 3 vols (New York 1906–9).

125. Niven, *opus cit.*, 54ff.

126. Martineau, *Society in America*, i 13–14.

127. *Adams Memoirs*, vii 272; Schlesinger, *Age of Jackson*, 369.

128. Austin, *Political Facts of the United States*, Tables 3.1, 3.2, and 3.3, 92ff.

129. Webster, *Private Correspondence* (Boston 1875), i 470; Clay quote is from a Senate speech, 1832.

130. Gaillard Hunt (ed.), *Margaret B. Smith: the First Forty Years of Washington Society*, 484–91.

131. *Life and Letters of Joseph Storey* (Boston 1851), i 563; James Hamilton to Martin Van Buren, March 5, 1829, quoted in S. E. Morison, *A History of the American People*, 3 vols (Oxford 1972), ii 16.

132. Hunt, *opus cit.*, 257; Nevins, *opus cit.*, 396; Remini, *Jackson*, ii 183ff, gives a spirited defense of Jackson's removals and appointments; see also E. M. Eriksson: 'The Federal Civil Service Under President Jackson,' *Mississippi Valley Historical Review*, July 1927; S. H. Aronson: *Status and Kinship in the Higher Civil Service* (Cambridge 1964), who says that the system under Jackson did not change all that much from under Jefferson and Adams; and for a review of the controversy, see F. W. Muggleston: 'Andrew Jackson and the Spoils System: a Historiographical Survey,' *Mid America* 59 (1977).

133. Niven, *opus cit.*, 240–5; Van Buren, *Autobiography*, 268–9; Remini, *Jackson*, ii 198–9.

134. Remini, *Jackson*, ii 160ff; Niven, *opus cit.*, 228.

135. Pauline Wilcox Burke: *Emily Donelson of Tennessee* (Richmond 1941), i 178; Remini, *Jackson*, ii 213.

136. James Parton: *Life of Andrew Jackson*, 3 vols (Boston 1866), iii 186–205; Remini, *Jackson*, ii Chapter 11, 'The Eaton Imbroglio,' 203–16.

137. Remini, *Jackson*, ii 207.
138. Schlesinger, *Age of Jackson*, 66ff; Remini, *Jackson*, ii Chapter 18, 'The Purge,' 300–14; Niven, *opus cit.*, 255ff.
139. Wise, *opus cit.*, 117; Martineau, *opus cit.*, i 257–8; Wise's remarks were in a speech in the House, December 21, 1838; there is an amusing portrait of Kendall in Schlesinger, *Age of Jackson*, 67–72.
140. Robert Heilbroner and Peter Bernstein: *The Debt and the Deficit* (New York 1989); for a useful summary of the history of the US national debt see E. Foner and J. A. Garraty (eds): *The Reader's Companion to American History* (Boston 1991), 771–6, with sources.
141. K. S. Kutolowski: 'Anti-Masonry Reexamined,' *Journal of American History*, lxxxi (1984); Sven Petersen: *A Statistical History of American Presidential Elections* (New York 1963), 20–1.
142. Remini, *Jackson*, iii 8–23.
143. See *State Papers on Nullification* (Boston 1834), 180ff; William W. Freehling: *Prelude to Civil War: the Nullification Controversy in South Carolina, 1816–36* (New York 1965).
144. Remini, *Jackson*, iii 30–4.
145. Billington, *Westward Expansion*, 301.
146. *North American Review*, Spring 1827, 365–442, and January 1830, 64–109; for Cass, see F. B. Woodford: *Lewis Cass, the Last Jeffersonian* (New York 1950).
147. Nevins, *opus cit.*, 313.
148. *Ibid.*, 318–19.
149. For the republic see H. T. Malone: Cherokees of the Old South: a People in Transition (Athens, Georgia 1956), 74–90.
150. Billington, Westward Expansion, 315–16.
151. Quoted in Rohrbough, opus cit., 277.
152. Ibid., 273.
153. A. De Tocqueville: Democracy in America, 2 vols (New York 1945 edn), i 353ff.
154. Gouge, opus cit. See M. G. Madeleine: Monetary and Banking Theories of Jacksonian Democracy (New York 1942).
155. Remini, Jackson, iii 92.
156. For Andalucia, see Roger W. Moss, The American Country House, 105ff; N. B. Wainwright: Andalusia (Philadelphia 1976). For Biddle, see T. P. Govan: Nicholas Biddle, Nationalist and Public Banker, 1786–1844 (Chicago 1959).
157. For Taney see Walker Lewis: Without Fear or Favor: Chief Justice Roger Brooke Taney (New York 1965); R. J. Harris: 'Chief Justice Taney,' Vanderbilt Law Review, 10 (1957).
158. Jackson's veto is described in detail in Remini, Jackson, ii, Chapter 22, 353ff.
159. Biddle to Clay, August 1, 1832, Biddle Papers (Library of Congress), quoted in Remini, *Jackson*, ii, 369; *Globe*, July 12, 1832.
160. Jackson to Van Buren, January 3, 1834, Van Buren Papers (Library of Congress), quoted in Remini, *Jackson*, iii 162.
161. E. G. Bourne: *The Surplus Revenue of 1837* (New York 1885).

162. R. C. McGrane: *The Panic of 1837* (New York 1924).

163. R. H. Timberlake: 'Species Circular and Distribution of Surplus,' *Journal of Political Economy*, 68 (1960), 109ff. Nathan Sargent: *Public Men and Events* (Philadelphia 1875), i 321.

164. G. R. Taylor (ed.): *Jackson v Biddle* (New York 1949). See also documents in the compilation by F. O. Gatell: *Jacksonians and Money Power, 1829–40* (New York 1967).

165. Peterson, *opus cit.*, 22–4.

166. For Van Buren's Treasury plan see D. Kinley: *Independent Treasury* (New York 1910). The Van Buren administration is detailed in J. C. Curtis: *Fox at Bay: Van Buren and the Presidency* (New York 1970).

167. Remini, *Clay*, Chapter 31, 544ff.

168. W. N. Chambers: 'Election of 1840,' in Schlesinger and Israel, *opus cit.*, i 680–2, 690.

169. Remini, *Clay*, 582–3.

170. Quoted in Moss, *opus cit.*, 125.

171. R. V. Bruce: *The Launching of Modern American Science, 1846–1876* (New York 1987); for climate see S. S. Visher: *Climatic Atlas of the United States* (New York 1954), Introduction, and G. H. Kimble: *Our American Weather* (New York 1955), Chapter 1.

172. P. W. Gates: *Farmer's Age: Agriculture 1815–60* (New York 1960).

173. R. M. Wyk: *Steam Power on the American Farm* (New York 1953).

174. C. R. Woodward: *Development of Agriculture in New Jersey, 1640–1880* (New York 1927), 51ff.

175. F. W. Taussig: *The Tariff History of the Unites States* (New York 1923).

176. Faulkner, *opus cit.*, 257.

177. Timothy Dwight: *Travels in New England and New York*, 2 vols (New York 1823), ii 54.

178. H. U. Faulkner, *American Economic History*, 267.

179. T. B. Searight: *The Old Pike: a History of the National Road* (New York 1894), 16.

180. Remini, *Jackson*, ii 252ff.

181. W. J. Petersen: *Steamboating on the Upper Mississippi* (New York 1937); Greville Bathe: *Rise and Decline of the Paddle Wheel* (New York 1962).

182. Charles Dickens: *American Notes* (London 1842), Chapter 12.

183. Mark Twain: *Life on the Mississippi* (New York 1883), Chapters 4, 16.

184. S. A. Howland: *Steamboat Disasters and Railway Accidents in the United States* (New York 1860); see Bourstin, *opus cit.*, 105ff.

185. Edward Hungerford: *Baltimore and Ohio Railroad*, 2 vols (Baltimore 1928); S. M. Derrick: *Centennial History of the South Carolina Railroad* (Charleston 1930).

186. House, executive documents, 18, 1831–2, 22nd Congress, 1st Session, i 174.

187. For the early financing of railroads see L. H. Haney: *Congressional History of Railroads*, 2 vols (Washington DC 1908–10), i, Chapters 2–3.

188. H. P. Walker: *Waggonmasters: High Plains Freighting to 1880* (New York

1966); J. W. Turrentine: 'Wells Fargo, Stagecoaches and Pony Express,' *California Historical Society Quarterly*, 45 (1966), 291ff.

189. Carleton Mabie: *The American Leonardo: the Life of Samuel F. B. Morse* (Boston 1943); Philp Dorf: *The Builder: Ezra Cornell* (Ithaca 1952).

190. Victor Rosewater: *History of Cooperative Newsgathering in the United States* (New York 1930).

191. See A. K. Weidenberg: *Manifest Destiny: a Study of Nationalist Expansion in American History* (Baltimore 1935).

192. *Congressional Globe*, 28th Congress, 2nd Session, App. 161–2, quoted in J. W. Pratt: 'The Origins of Manifest Destiny,' *American Historical Review*, cccii (1927).

193. *Congressional Globe*, 28th Congress, 2nd Session, App. 43.

194. *Young Hickory Banner*, October 15, 1845.

195. *Democratic Review*, xvii (1845).

196. *North American Review*, July 1836, October 1842.

197. D. M. Pletcher: *The Diplomacy of Annexation: Texas, Oregon and the Mexican War* (New York 1973).

198. Stanley Siegel: *A Political History of the Texas Republic, 1836–45* (Austin 1956).

199. M. K. Wisehart: *Sam Houston* (New York 1962).

200. Quoted in Remini, *Jackson*, iii 357ff.

201. Kendall to Jackson, July 30, 1836, in the Jackson Papers, Chicago Historical Society, quoted in Remini, *Jackson*, iii 362.

202. For Clay and the 1844 election see Remini, *Clay*, Chapter 36, 642ff; details of the election are in Schlesinger and Israel, *opus cit.*, i 861ff.

203. J. H. Smith: *Annexation of Texas* (New York 1911).

204. M. M. Quaife (ed.): *Diary of James Knox Polk*, 4 vols (New York 1910); there is a *Selection* edited by Allan Nevins (New York 1929); the best biography is C. G. Sellers: *James K. Polk*, 2 vols (New York 1957–66); for a recent assessment see Joseph Shattan: 'One Term Wonder,' *American Spectator*, October 1996.

205. For the early history of Oregon see C. H. Carey: *A General History of Oregon Prior to 1861*, 2 vols (New York 1935–6).

206. M. C. Jacobs: *Winning Oregon* (New York 1938).

207. For the origins of the war, in addition to Sellers, *opus cit.*, ii, see J. H. Shroeder: *Mr Polk's War* (New York 1973).

208. R. P. Basler: *Collected Works of Abraham Lincoln*, 7 vols (New Brunswick 1953–5), i 439ff.

209. Thomas Hart Benton: *Thirty Years View* 2 vols (New York 1854–6).

210. For Scott see A. D. H. Smith: *Old Fuss-and-Feathers: the Life of Winfield Scott* (New York 1937).

211. For the campaign, see K. J. Bauer: *The Mexican War, 1846–8* (New York 1974).

212. Allan Nevins: *Fremont: The West's Greatest Adventurer*, 2 vols (New York 1928, reissued 1955). His *Memoirs* were published in 1887.

213. For the Treaty and its aftermath, see Pletcher, *opus cit.* See also R. W. Johannsen: *To the Halls of Montezuma: the Mexican War in the American Imagination* (New York 1985).

214. Charles Wilkes: *Narrative of the United States Exploring Expedition,* 5 vols (Philadelphia 1845), v 152, 172; D. W. Meinig, *Continental America,* 142–3.

215. See also R. F. Lucid (ed.): *Journal of Richard H. Dana,* 3 vols (Cambridge 1968).

216. Bernard de Voto: *The Year of Decision* (New York 1943).

217. Francis Parkman: *The Oregon Trail* (Boston 1849). See Howard Doughty: *Francis Parkman* (New York 1962); W. R. Jacobs (ed.): *Parkman Letters,* 2 vols (Boston 1960); Mason Wade (ed): *Parkman Journals,* 2 vols (Cambridge 1947).

218. Richard O'Connor: *Bret Harte: a Biography* (New York 1966).

219. R. W. Paul: *California Gold* (New York 1947), vii; see also his collection of contemporary sources, *California Gold Discovery* (New York 1966); A. L. Rowse: *The Cornish in America* (Truro 1967), 248ff.

220. R. J. Roske: 'World Impact of California Gold Rush, 1849–1857,' *Arizona and West,* 5 (1963), 187ff.

221. G. R. Taylor: *Transportation Revolution, 1815–1860* (New York 1951).

222. Thomas Monney: *Nine Years In America* (London 1850), 37.

223. For wage-rates of immigrants, see N. J. Ware: *The Industrial Worker, 1840–60* (New York 1924).

224. Jefferson Williamson: *The American Hotel* (New York 1930); D. E. King: 'The First Class Hotel and the Age of the Common Man,' *Journal of Social History,* 23 (1957), 172ff.

225. Anthony Trollope: *North America* (London 1864), 245–7; Harriet Martineau: *Society in North America,* 2 vols (London 1937), ii 57ff.

226. R. T. Ely: *The Labor Movement in America* (New York 1925), 49; Ware, *opus cit.*

227. Charles A. Murray: *Travels in North America,* 2 vols (London 1839), ii 297.

228. Orlando F. Lewis: *The Development of American Prisons and Prison Customs, 1776–1845* (New York 1922); D. W. Lewis: *From Newgate to Dannemora; The Rise of the Penitentiary in New York, 1796–1848* (New York 1965).

229. The best recent edition is in the American Classics of Liberty series, New York 1990.

230. For De Tocqueville's letters, see André Jardin: *Tocqueville: a Biography* (trans. London 1988); his memoirs were published posthumously, *Souvenirs* (Paris 1893); for interesting insights into De Tocqueville, see B. J. Smith: *Politics and Remembrance: Republican Themes in Machiavelli, Burke and Tocqueville* (Princeton 1985).

231. Jardin, *opus cit.,* 149, 161, 167, 169.

232. De Tocqueville: *Oeuvres complètes,* 18 vols (Paris 1981–), V, i 89.

233. *Ibid.,* 85.

234. *Ibid.,* v 278–80.

235. Jardin, *opus cit.,* 170–1, 174; *Oeuvres complètes,* V, i 257.

236. All this superb speech is worth reading. See *Great Speeches and Orations of Daniel Webster* (Classic of Liberty Library, New York 1993), 123ff.

237. See L. P. Eisenhart: *Historical/Philadelphia* (Philadelphia 1953) and G. B. Tatum: *Penn's Great Town* (Philadelphia 1961).

238. L. A. Cremin: *American Education: the National Experience, 1783–1876* (New York 1980); C. F. Kaestle: *Pillars of the Republic: Common Schools and American Society, 1780–1860* (New York 1983).

239. L. R. Veysey: *The Emergence of the American University* (New York 1965); B. M. Solomon: *In the Company of Educated Women: a History of Women and Higher Education in America* (New York 1985).

240. For Mrs Trollope see Johanna Johnston: *The Life, Manners and Travels of Fanny Trollope* (London 1979); the best recent edition of *Domestic Manners* is by the Folio Society (London 1974).

241. *Ibid.*, Chapter xxv.

242. The best account of all this is in Samuel Eliot Morison and Henry Steele Commager: *The Growth of the American Republic*, 2 vols (Oxford 1962), vol. i: *1000–1865*, 618ff.

243. Don E. Fehrenbacher: *The Dred Scott Case* (Stanford 1968).

244. Schlesinger and Israel, *opus cit.*, ii 918.

245. Remini, *Clay*, 710.

246. Elbert B. Smith: *The Presidencies of Zachary Taylor and Millard Fillmore* (Lawrence 1988).

247. For the details of the Senate debates see Holman Hamilton: *Prologue to Conflict: the Crisis and Compromise of 1850* (Lexington 1964).

248. Michael F. Holt: *The Political Crisis of the 1850s* (New York 1978).

249. H. D. Babbage (ed.): *Noah Webster: On Being American: Selected Writings, 1783–1828* (Cambridge 1967).

250. *Cadmus*, v–vii, 25. David Simpson: *The Politics of American English, 1776–1850* (Oxford 1986), 22ff.

251. Boorstin, *opus cit.*, 282–4.

252. H. R. Warfel: *Noah Webster: Schoolmaster to America* (New York 1936).

253. Martin Green: 'The God That Neglected to Come: American Literature: 1780–1820,' in Marcus Cunliffe (ed.): *American Literature to 1900* (London 1986), 53ff.

254. W. A. Reichart: *Washington Irving and Germany* (Ann Arbor 1957), 22–3, 42; H. A. Pochmann: 'Irving's German Sources in *The Sketch Book*,' *Studies in Philology*, 1930, 477ff.

255. P. M. Irving (ed.): *Life and Letters of Washington Irving*, 3 vols (London 1962), 28–9; Cunliffe, *opus cit.*, 78.

256. Jackson quoted in Morison, *History of the American People*, ii 162.

257. K. S. House: *James Fenimore Cooper: Cultural Prophet and Literary Pathfinder* (New York 1988); Clarence Golides: 'The Reception of Some 19th-Century American Authors in Europe,' in M. Denny and W. H. Gilman (eds): *The American Writer and the European Tradition* (Minneapolis 1950), 113–14; P. A. Barba: *Cooper in Germany* (Bloomington 1914), 78ff.

258. R. Ruland and M. Bradbury: *From Puritanism to Postmodernism: a History of American Literature* (New York 1991), 100ff.

259. E. Wagenknecht: *Ralph Waldo Emerson: Portrait of a Balanced Soul* (New York 1974), Chapter 6, 'Politics,' 158–201.

260. Thomas Wentworth Higginson: *Every Saturday*, April 18, 1868.

261. *Journals and Miscellaneous Notebooks of Ralph Waldo Emerson*, 14 vols (Cambridge 1960–), viii 88–9, 242, ix 115, vii 544. For Emerson's relationship with Margaret Fuller, see Carlos Baker: *Emerson Among the Eccentrics: a Group Portrait* (New York 1996).

262. Paul Boyer: *Urban Masses and Moral Order in America, 1820–1920* (Cambridge 1978), 109; M. K. Cayton: 'The Making of an American Prophet: Emerson, His Audience and the Rise of the Culture Industry in 19th Century America,' *American Historical Review*, June 1987.

263. Quoted in Wagenknecht, *opus cit.*, 170.

264. Harold Bloom: 'Mr America,' *New York Review of Books*, November 22, 1984, 19ff.

265. The fullest account of Longfellow's life is S. Longfellow: *Life of Henry Wadsworth Longfellow with extracts from his journals and correspondence*, 3 vols (Cambridge 1891); A. R. Hilen (ed.): *Longfellow's Letters*, 2 vols (Cambridge 1967); see also E. Wagenknecht: *Henry Wadsworth Longfellow: Portrait of an American Humorist* (New York 1967).

266. The best life of Poe is K. Silverman: *Edgar A. Poe: Mournful and Never-ending Remembrance* (New York 1991).

267. Michael Allen: *Poe and the British Magazine Tradition* (New York 1969); J. P. Muller and W. J. Richardson: *The Purloined Poe: Lacan, Derrida and Psychoanalytic Reading* (Baltimore 1988).

268. For Hawthorne's life see Edwin Haviland Miller: *Salem Is My Dwelling Place: a Life of Nathaniel Hawthorne* (London 1991); S. Bercovitch: *The Office of the Scarlet Letter* (Baltimore 1991); T. Walter Herbert: *Dearest Beloved: the Hawthornes and the Makings of the Middle-Class Family* (Berkeley 1992).

269. J. Katz: *Gay American History* (New York 1976); J. D'Emilio: *Sexual Politics, Sexual Communities* (Chicago 1983).

270. The best life of Whitman I have read is Justin Kaplan: *Walt Whitman: a Life* (New York 1980), but there are many others; he is now more written about than any other 19th-century writer, including Poe and Hawthorne.

271. Paul Zweig: *Walt Whitman: the Making of the Poet* (New York 1984).

272. M. J. Killingsworth: *Whitman's Poetry of the Body: Sexuality, Politics and the Text* (Chapel Hill 1989); Michael Moon: *Disseminating Whitman: Revision and Corporeality in 'Leaves of Grass'* (Cambridge 1991).

273. R. W. Emerson: *Representative Men* (Boston 1850), Introduction.

274. For Thoreau's life see Walter Harding: *The Days of Henry Thoreau: a Biography* (New York 1982); for Walden, L. Shanley: *The Making of 'Walden'* (Chicago 1957).

275. Joan D. Hedrick: *Harriet Beecher Stowe: a Life* (Oxford 1994), 223ff.

276. Harriet Beecher Stowe: *A Key to Uncle Tom's Cabin* (Boston 1853), 22ff.

277. Hedrick, *opus cit.*, 245.
278. J. S. Van Why and E. French (eds): *Harriet Beecher Stowe in Europe: the Journal of Charles Beecher* (Hartford 1896); F. J. Klinberg: 'Harriet Beecher Stowe and Social Reform in England,' *American Historical Review*, 43 (1937–8); Betty Fladeland: *Abolitionists and Working-Class Problems in the Age of Industrialisation* (Baton Rouge 1984).
279. T. F. Gossett: *Uncle Tom's Cabin and American Culture* (Dallas 1985), 270.
280. Quoted in Hedrick, *opus cit.*, 232.
281. *Ibid.*, 305ff. The remark may be apocryphal, though I doubt it. No record has survived of their conversation, but Mrs Stowe wrote to her twin sister: 'It was a very droll time we had at the White House I assure you. I will tell you all about it when I get home. I will only say now that it was very funny—and we were ready to explode with laughter all the while.'

PART FOUR: Civil War America

1. J. M. Murrin in R. Beeman *et al.* (eds): *Beyond Confederation: Origins of the Constitution and of American National Identity* (Chapel Hill 1987), 346–7.
2. Edwin Haviland Miller, *Salem is My Dwelling Place*, 379–81.
3. R. F. Nichols: *Franklin Pierce* (2nd edn New York 1958), 75.
4. Miller, *opus cit.*, 383–4.
5. There is a facsimile edition of Hawthorne's *Life of Pierce* (Boston 1970), with an introduction by R. C. Robey.
6. Nichols, *opus cit.*, 216.
7. W. C. Davis: *Jefferson Davis: the Man and His Hour* (New York 1991), 251.
8. Robert E. May: *The Southern Dream of a Caribbean Empire* (Baton Rouge 1973), 60ff.
9. Lawrence Greene: *Filibuster: the Career of William Walker* (New York 1937).
10. K. S. Davis: *Kansas: a Bicentennial History* (New York 1976), 47ff; R. W. Johannsen: *Stephen A. Douglas* (New York 1973).
11. P. W. Gates: *Fifty Million Acres: Conflicts Over Kansas Land Policy, 1854–90* (New York 1954).
12. J. A. Rawley: *Race and Politics: 'Bleeding Kansas' and the Coming of the Civil War* (New York 1969); S. B. Oates: *To Purge This Land with Blood: John Brown* (New York 1970).
13. D. H. Donald: *Charles Sumner*, 2 vols (New York 1960–70).
14. Samuel Eliot Morison and Henry Steele Commager, *The Growth of the American Republic*, i 654ff.
15. J. T. Carpenter: *The South as a Conscious Minority* (Baton Rouge 1930); A. O. Craven: *The Growth of Southern Nationalism, 1848–1861* (New York 1953).
16. Louis Hacker: *Triumph of American Capitalism* (New York 1940), 281ff; R. B. Flanders: *Plantation Slavery in Georgia* (Atlanta 1933), 221–3; C. S. Sydnor: *Slavery in Mississippi* (New Orleans 1933), 196ff.

17. Ulrich B. Phillips: *American Negro Slavery* (New York 1918).

18. L. C. Gray: *History of Agriculture in the Southern United States to 1850*, 2 vols (Richmond 1933), i 460ff.

19. W. E. Dodd: *The Cotton Kingdom* (New York 1921), 121.

20. C. and M. Beard: *Rise of American Civilisation*, 4 vols (New York 1927–42), ii 5–6.

21. H. U. Faulkner, *American Economic History*, 320.

22. Quoted in *Niles Weekly Register*, April 19, 1845.

23. David Herbert Donald: *Lincoln* (London 1995), 19–20; for the early Lincoln see E. Hertz (ed.): *The Hidden Lincoln: from the Letters and Papers of William H. Herndon* (New York 1938); L. A. Warren: *Lincoln's Youth: Indiana Years, Seven to 21, 1816–30* (New York 1959); C. B. Strozier: *Lincoln's Quest for Union: Public and Private Meanings* (New York 1982). Beware of unreliable transcripts in Herndon papers and of psychobabble in Strozier.

24. For Lincoln details see M. E. Neely Jr: *The Abraham Lincoln Encyclopaedia* (New York 1982). For his suicide fears, see H. I. Kushner: *Self-Destruction in the Promised Land: a Psychocultural Biology of American Suicide* (New Brunswick 1989), Chapter 5.

25. W. C. Temple and H. E. Pratt: 'Lincoln in the Black Hawk War,' *Bulletin of the Abraham Lincoln Association*, 54 (December 1938), 3ff.

26. For Lincoln as a lawyer see A. A. Woldman: *Lawyer Lincoln* (Boston 1936); J. J. Duff: *A. Lincoln: Prairie Lawyer* (New York 1960); J. P. Frank: *Lincoln as a Lawyer* (Champaign 1961).

27. For Ann Rudledge and her effect on Lincoln, see Donald, *opus cit.*, 608 n. 55.

28. See John Lorring: *Tiffany's 150 Years* (New York 1987), 46–7; for Mrs Lincoln, see R. P. Randall: *Mary Lincoln: Biography of a Marriage* (Boston 1953) and her *The Courtship of Mr Lincoln* (Boston 1957).

29. For Lincoln's spell in Congress, see Donald, *opus cit.*, Chapter 5; D. E. Riddle: *Congressman Abraham Lincoln* (Westport 1979); Paul Findley: *Abraham Lincoln: the Crucible of Congress* (New York 1979).

30. *Sayings and Anecdotes of Lincoln* (New York 1940), 107–8.

31. Jean Baker: *Mary Todd Lincoln* (New York 1960).

32. W. J. Wolf: *The Almost Chosen People: a Study of the Religion of Abraham Lincoln* (New York 1959).

33. For Lincoln's depressions, see Donald, *opus cit.*, 163ff.

34. For the writings and speeches of Lincoln, I have used Don. E. Fehrenbacher (ed.): *Abraham Lincoln: Speeches and Writings*, 2 vols (Classics of Liberty Library, New York 1992).

35. For Lincoln on slavery see Donald, *opus cit.*, 165–7, 180–1, etc.

36. *Ibid.*, 191; *Speeches and Writings*, 365; W. E. Gienapp: *The Origins of the Republican Party, 1852–56* (New York 1987); the supposed full text of the speech published in the September 1896 issue of *McClure's Magazine* had been questioned.

37. Roy P. Basler: *Collected Works of Abraham Lincoln*, 8 vols (New Brunswick 1953), ii 341.

38. *Herndon's Lincoln*, ii 384.

39. *Speeches and Writings*, i 426–34; Don E. Fehrenbacher: *Prelude to Greatness: Lincoln in the 1850s* (Stanford 1962); Donald, *opus cit.*, 206ff.

40. For the debates, see R. A. Heckman: *Lincoln v. Douglas: the Great Debates Campaign* (Washington DC 1967); texts in R. W. Johannsen: *The Lincoln–Douglas Debates of 1858* (New York 1965).

41. Speech at Clinton, September 8, 1858.

42. See David Zarefsky: *Lincoln, Douglas and Slavery: in the Crucible of Public Debate* (Chicago 1990).

43. *Speeches and Writings*, ii 106–8; the long campaign autobiography is ii 160–7.

44. Published in the *New York Times*, January 24, 1854; for Chase see David Donald (ed.): *Inside Lincoln's Cabinet: the Civil War Diaries of Salmon P. Chase* (New York 1954); for Seward, see G. G. Van Deusen: *William Henry Seward* (New York 1967).

45. See the two letters, May 19 and 23, 1860, *Speeches and Writings*, ii 156–7.

46. *Ibid.*, ii 111–30.

47. Among the many recent books on anti-slavery agitation, the best are: Thomas Bender (ed.): *The Antislavery Debate: Capitalism and Abolitionism as a Problem in Historical Interpretation* (Berkeley 1992); Alan M. Kraut (ed.): *Crusaders and Compromisers: Essays on the Relationship of the Antislavery Struggle to the Antebellum Party System* (Westport 1983); L. Perry and M. Fellman (eds): *Antislavery Reconsidered: New Perspectives on the Abolitionists* (Baton Rouge 1979).

48. J. C. Furnas: *The Road to Harper's Ferry* (New York 1959).

49. Elting Morison, 'The Election of 1860,' in Arthur M. Schlesinger Jr and F. R. Israel (eds), *American Presidential Elections*, ii 1097–122. See also W. E. Gienapp: 'Who Voted for Lincoln?,' in J. L. Thomas (ed.): *Abraham Lincoln and the American Political Tradition* (Amherst 1986), 50ff.

50. For Davis see William C. Davis: *Jefferson Davis: the Man and His Hour* (New York 1991), esp. 689ff.

51. For Davis' background see the life by his widow, Varina H. Davis: *Jefferson Davis*, 2 vols (Charleston 1890) and the 'official' collection, *Jefferson Davis, Constitutionalist, His Letters, Papers and Speeches*, ed. Dunbar Rowland, 10 vols (Baton Rouge 1923).

52. For Davis' treatment of slaves etc. see Jefferson Davis: *Rise and Fall of the Confederate Government*, 2 vols (New York 1881), i 518; Varina Davis, *opus cit.*, i 174–9.

53. William C. Davis, *opus cit.*, 125.

54. *Ibid.*, 198–9.

55. *Ibid.*, 127–67.

56. The row is described in Winfield Scott: *Memoirs*, 2 vols (New York 1864), and in William C. Davis, *opus cit.*, 228ff.

57. T. C. Cochran: *Frontiers of Change: Early Industrialism in America* (New York 1981), 73.

58. William C. Davis, *opus cit.*, 258–60.

59. *New Orleans Bee*, December 14, 1860.

60. William C. Davis, *opus cit.*, 283.

61. For Lincoln during this vital period see W. E. Baringer: *A House Dividing: Lincoln as President Elect* (Springfield, Illinois 1945).

62. William C. Davis, *opus cit.*, 296.

63. D. W. Meinig, *Continental America*, 477–8.

64. Morison and Commager, *opus cit.*, 667ff.

65. William C. Davis, *opus cit.*, 270.

66. For the early weeks of the Lincoln presidency see P. S. Paludan: *The Presidency of Abraham Lincoln* (Lawrence 1994), Chapters 2–3.

67. R. N. Current: *Lincoln and the First Shot* (New York 1963).

68. Quoted in Emory M. Thomas: *Robert E. Lee: a Biography* (New York 1995), 188.

69. R. A. Wooster: *Secession Conventions of the South* (New York 1962) for details.

70. Quoted in Drew Gilpin: *James Henry Hammond* (Baton Rouge 1982).

71. See R. L. Andreano (ed.): *Economic Impact of the Civil War* (New York 1962); H. N. Scheiber: 'Economic Change in the Civil War Era: Analysis of Recent Studies,' *Civil War History*, 11 (1965), 396ff.

72. H. D. Capers: *Life and Times of C. G. Memminger* (New York 1893).

73. See the excellent summary of Southern finances by J. C. Schwab: 'The South During the War, 1861–65,' in *Cambridge Modern History* (Cambridge 1934), vii 603ff; Davis, *opus cit.*, 601ff.

74. Elizabeth Merritt: *James Henry Hammond* (Baton Rouge 1923).

75. For a summary of this incident see C. F. Adams: 'The Trent Affair,' *Massachusetts Historical Society Proceedings* 45 (1911), 35ff.

76. Davis, *opus cit.*, 319.

77. *Ibid.*, 366.

78. T. L. Livermore: *Numbers and Losses in the Civil War* (New York 1901).

79. For the influence of geography on the conflict see Meinig, *opus cit.*, 494ff.

80. W. B. Yearns: *The Confederate Congress* (New York 1960); F. L. Owsley: *States Rights in the Confederacy* (New York 1925); Davis, *opus cit.*, 444ff; R. D. Meade: *Judah P. Benjamin* (New York 1943).

81. Davis, *opus cit.*, 447.

82. Donald, *Lincoln*, 362–4; text of Lincoln's proclamation in *Speeches and Writings*, ii 318–19.

83. Quoted in Donald, *Lincoln*, 368.

84. Different texts circulated; see *Speeches and Writings*, ii 357–8; Lincoln's *Collected Works*, v 388–9.

85. See Joseph H. Parks: *General Leonidas Polk, CSA: Fighting Bishop* (New York 1962).

86. J. W. Silver: *Confederate Morale and Church Propaganda* (New Orleans 1957).

87. See the special issue, 'Civil War Religion,' *Civil War History*, 6 (1960).

88. See J. G. Randall and R. N. Current: *Lincoln the President: Last Full Measure* (New York 1955), chapter entitled 'God's Man.'

89. Chester F. Dunham: *Attitude of the Northern Clergy Towards the South, 1860–65* (New York 1942); D. W. Harrison: 'Southern Protestantism and Army Missions in the Confederacy,' *Mississippi Quarterly*, 17 (1965), 179ff.

90. W. J. Wolf: *The Almost Chosen People: a Study of the Religion of Abraham Lincoln* (New York 1959).

91. J. H. Franklin: *The Emancipation Proclamation* (New York 1963); Donald, *Lincoln*, 366ff.

92. *Cambridge Modern History*, vii, Chapter xviii, 'The North During the War: Finance;' B. W. Rein: *Analysis and Critique of Union Financing of the Civil War* (New York 1962); Bray Hammond: *Sovereignty and the Empty Purse: Banks and Politics in the Civil War* (New York 1970); A. M. Davis: *Origins of the National Banking System* (New York 1910).

93. B. P. Thomas and H. M. Hymam: *Stanton* (New York 1962); W. W. Hassler: *General George B. McClellan* (New York 1957).

94. See Allan Pinkerton's autobiography: *Criminal Reminiscences and Detective Sketches* (New York 1879) and *Thirty Years a Detective* (Chicago 1884).

95. For Jackson, see G. F. R. Henderson: *Stonewall Jackson and the American Civil War*, 2 vols (New York 1898); there are many modern books dealing with him.

96. Mrs James Chesnut: *A Diary from Dixie* (New York 1949); General Richard Taylor: *Destruction and Reconstruction: Personal Reminiscences of the Late War* (New York 1879); Edmund Wilson: *Patriotic Gore* (New York 1962), 279ff, 303–4.

97. Wilson, *opus cit.*, 300.

98. John Esten Cooke: *Wearing the Grey* (New York 1867).

99. *Ibid.*

100. Lee is entombed in Douglas Southall Freeman: *R. E. Lee: a Biography*, 4 vols (New York 1934–5); the best life is E. M. Thomas: *Robert E. Lee: a Biography* (New York 1995).

101. Thomas, *opus cit.*, 187ff.

102. See E. B. Coddington: *The Gettysburg Campaign: a Study in Command* (New York 1968). See also J. Luvass and H. W. Nelson (eds): *The US Army War College Guide to the Battle of Gettysburg* (Carlisle 1986); Thomas, *opus cit.*, 287ff.

103. E. S. Miers: *Web of Victory: General Grant at Vicksburg* (New York 1955).

104. Grant wrote one of the great American autobiographies, *Personal Memoirs*, 2 vols (New York 1885–6).

105. For Lincoln's relations with Grant see T. H. Williams: *Lincoln and His Generals* (New York 1952).

106. For these battles and casualties, see R. U. Johnson and C. C. Buel (eds): *Battles and Leaders of the Civil War* (New York 1884–8), iv; for Lincoln's sorrow, Donald, *Lincoln*, 500.

107. Davis, *opus cit.*, 531, 544, 594.

108. Quoted in Wilson, *opus cit.*, 271.

109. J. M. Gibson: *Those 163 Days: Sherman's March* (New York 1961).

110. Bruce Catton: *Grant Takes Command* (New York 1969).

111. H. M. Hyman: 'The Election of 1864,' in Schlesinger and Israel, *opus cit.*, ii; E. C. Kirkland: *Peacemakers of 1864* (New York 1927).

112. Text in *Speeches and Writings*, 686–7.

113. Miller, *opus cit.*, 474.

114. Justin Kaplan: *Walt Whitman: a Life* (New York 1980), 260–1.

115. Donald, *Lincoln*, 580–1. This was April 9, 1865, the date of Lee's surrender.

116. *Macbeth*, Act II, Scene 2. Adolphe de Chambrun: *Impressions of Lincoln and the Civil War: a Foreigner's Account* (New York 1952), 82.

117. Kaplan, *opus cit.*, 262, 297.

118. Stewart Brooks: *Civil War Medicine* (Springfield, Illinois 1966), 97; R. M. Buck: *Walt Whitman* (Philadelphia 1883), 37, quoted in Kaplan, *opus cit.*, 266n.

119. For extracts from Alcott's letters see Edna D. Cheney: *Louisa May Alcott: Her Life, Letters and Journals* (New York 1889).

120. Judith Farr: *The Passion of Emily Dickinson* (Cambridge 1992), for a recent view of all this.

121. John D. Unruh Jr: *The Plains Across: the Overland Emigrants and the Trans-Mississippi West, 1840–1860* (Urbana 1979).

122. For such incidents see Francis Paul Prucha: *The Great Father: the United States Government and the American Indians*, 2 vols (Lincoln, Nebraska 1984) and the trilogy by Robert M. Utley: *Frontier Regulars: the US Army and the Indian, 1866–91* (New York 1973); *The Indian Frontier of the American West 1846–90* (Albuquerque 1984), and *The Last Days of the Sioux Nation* (New Haven 1963).

123. See P. W. Gates and R. W. Swenson: *History of Public Land Law Development* (New York 1968).

124. See E. S. Pomeroy: *The Territories and the United States, 1861–90* (Philadelphia 1947) and J. E. Eblen: *The First and Second United States Empires: Governors and Territorial Governments, 1784–1912* (Pittsburgh 1968).

125. Mark Twain: *Roughing It*; see also C. A. Milner *et al.* (eds): *The Oxford History of the American West* (New York 1994), 201ff.

126. F. L. Paxson: *The Last American Frontier* (New York 1910), 170ff.

127. Davis, *opus cit.*, 598ff.

128. *Rise and Fall of the South*, ii 518ff.

129. Thomas, *opus cit.*, 361ff.

130. For these events see B. H. Liddell Hart (ed.): *Sherman's Memoirs*, 2 vols (New York 1957).

131. For the background to the conspiracy to murder Lincoln, see W. A. Tidwell *et al.*: *The Confederate Secret Service and the Assassination of Lincoln* (Jackson 1988); for the event itself, see W. E. Reck: *Abraham Lincoln: His Last 24 Hours* (Jefferson 1987). The Surratt Society has provided *In Pursuit of . . .*

Continuing Research in the Field of the Lincoln Assassination (New York 1990).

132. For Davis' imprisonment and release, see Davis, *opus cit.*, 640ff.

133. For Lee as college president see Thomas, *opus cit.*, 376ff.

134. For Lincoln's racial views see Don E. Fehrenbacher: 'Only His Stepchildren,' in *Lincoln in Text and Context*, 95–112; G. M. Fredrickson: 'A Man Not a Brother: Abraham Lincoln and Racial Equality,' *Journal of Southern History*, 41 (February 1975), 39ff.

135. For all the circumstances surrounding the address, and the various texts of it see Garry Wills: *Lincoln at Gettysburg: the Words that Remade America* (New York 1992).

136. Text in *Speeches and Writings*, ii 555ff.

137. J. B. James: *Framing of the Fourteenth Amendment* (New York 1956); see also H. J. Graham: 'Antislavery Background of the Fourteenth Amendment,' *Wisconsin Law Review*, 30 (1950), 479ff. See also W. B. Heseltine: *Lincoln's Plan of Reconstruction* (New York 1960).

138. Donald, *opus cit.*, 582–3.

139. G. F. Milton: *The Age of Hate: Andrew Johnson and the Radicals* (New York 1930), 145ff.

140. For the varying positions see A. O. Craven: *Reconstruction: Ending of the Civil War* (New York 1969) and R. W. Patrick: *Reconstruction of the Nation* (New York 1967).

141. For Johnson's executive action, see J. E. Sefton: *Andrew Johnson and the Uses of Constitutional Power* (New York 1980).

142. For conditions in the South immediately after the end of the Civil War, see J. R. Dennett: *The South As It Is, 1865–66*, ed. H. M. Christman (New York 1965) and J. T. Trowbridge: *Desolate South, 1865–66*, ed. G. Carroll (New York 1966). For the working of the Freedman's Bureau, see Eric Foner: *A Short History of Reconstruction 1863–1877* (New York 1990), esp. 31–2, 64–5, 66, and 111–13.

143. For a recent view of Johnson's handling of Congress, see H. L. Trefousse: *Andrew Johnson: a Biography* (New York 1989).

144. H. M. Hyman: *Radical Republicans and Reconstruction, 1861–70* (New York 1967).

145. For the context of impeachment see J. E. Sefton: 'Impeachment of Andrew Johnson: a Century of Writing,' *Civil War History*, 14 (1968), 120ff.

146. David Donald: 'Why They Impeached Andrew Johnson,' *American Heritage*, 6 (1956), 20ff; Michael Benedict: *The Impeachment and Trial of Andrew Johnson* (New York 1972).

147. J. H. Franklin: 'Election of 1868,' in Schlesinger and Israel, *opus cit.*, ii; C. H. Coleman: *Election of 1868* (New York 1933).

148. J. Daniels: *Prince of Carpetbaggers* (New York 1958); R. N. Current: *Three Carpetbag Governors* (New York 1967); O. H. Olsen: 'Scalawags,' *Civil War History*, 12 (1966), 304f. Foner gives a more favorable account of carpetbaggers and Second Reconstruction, *opus cit.*, 129–30, 158, 213, 256.

PART FIVE: Industrial America

1. Thoreau, *Journal*, February 1, 1852; Rudyard Kipling, 'Across a Continent,' in *From Tideway to Tideway* (London 1892); Walt Whitman, 'Crossing Brooklyn Ferry,' from *Leaves of Grass*; Henry James, *The American Scene* (Boston 1907), Chapter 2.

2. John Ruskin: *Fors Clavigera* (London 1871), i, Letter 10; Henry James: *Hawthorne* (Boston 1879), Chapter 2.

3. Stephen Vincent Benet: *Ballads and Poems* (New York 1927); Gertrude Stein: *The Geographical History of America* (New York 1936).

4. C. K. Milner *et al.*, (eds): *Oxford History of the American West*, 538ff.

5. For details see National Academy of Sciences, *Growth of United States Population* (Washington DC 1965).

6. See the table 'Fertility and Mortality in the United States, 1800–1980' in Foner and Garraty (eds), *Reader's Companion to American History*, 104; A. J. Coale and M. Zelick: *New Estimates of Fertility and Population in the United States* (Princeton 1963).

7. Figures from *Reader's Companion to American History*, 533–6 and *Dictionary of American History*, ed. T. L. Purvis (New York 1995), 190–1; Bernard Bailyn: *The Peopling of North America* (New York 1985).

8. G. C. Fite: *Farmer's Frontier, 1865–1900* (New York 1966); F. A. Shannon: *The Farmer's Last Frontier, 1860–1897* (New York 1945).

9. See map in H. U. Faulkner, *American Economic History*, 369.

10. See B. H. Hibbert: *A History of the Public Land Policies* (New York 1924).

11. *Ibid.*, 387.

12. For pros and cons see R. F. Swierenga: *Pioneers and Profits: Land Speculation on the Iowa Frontier* (New York 1968).

13. Walter P. Webb: *The Great Plains* (Houston 1931), 317.

14. *Ibid.*, 322.

15. *Oxford History of the American West*, 252ff.

16. E. W. Hayter: 'Barbed Wire Rending,' *Agricultural History*, 13 (1939), 189ff; R. A. Clemen: *The American Livestock Industry* (New York 1923).

17. R. H. Pearce: *Savagism and Civilisation: the Indian and the American Mind* (New York 1967); Stanley Vestal: *Warpath and Council Fire: Plains Indians, 1851–91* (New York 1948).

18. For some interesting material see the National Park Service compilation, *Soldier and Brave: Indian and Military Affairs in the Trans-Mississippi West* (Washington DC 1963).

19. For Custer see Jay Monaghan: *General George Armstrong Custer* (New York 1959). He wrote an autobiography, *My Life on the Plains* (New York 1874) and his *Letters* were edited by M. A. Merrington (New York 1950).

20. *Oxford History of the West*, 689–90.

21. E. I. Stewart: *Custer's Luck* (New York 1955); R. M. Utler: *Custer Battlefield National Monument, Montana* (Washington DC 1969); Carl C. Rister: *Border Command: Phil Sheridan* (New York 1944).

22. Frederick Hoxie: *A Final Promise: the Plan to Assimilate the Indians, 1880–1920* (Lincoln, Nebraska 1984).

23. See Collier's own account: *From Every Zenith: a Memoir and Some Essays on Life and Thought* (New York 1963).

24. For a general summary, see F. P. Prucha: *The Great Father: the United States Government and the American Indians*, 2 vols (New York 1984).

25. See J. A. Carroll and J. R. Kluger (eds): *Reflections of Western Historians* (New York 1969); W. D. Wyman and C. B. Kroeber (eds): *The Frontier in Perspective* (New York 1957).

26. F. J. Turner: *The Frontier in American History* (Cambridge 1920).

27. E. L. Ayers: *Vengeance and Justice: Crime and Punishment in the 19th-Century South* (New York 1984); B. Wyatt-Brown: *Southern Honor: Ethics and Behavior in the Old South* (New York 1982).

28. Mark DeW. Howe: *Justice Oliver W. Holmes*, 2 vols (Cambridge 1957–63).

29. See *Oxford History of the American West*, Chapter 11, 'Violence,' 393ff.

30. For this and other views see R. M. Brown in 'Historiography of Violence in the American West,' in M. P. Malone (ed.): *Historians and the American West* (Lincoln, Nebraska 1983); Richard White: *'It's Your Misfortune and None of My Own': A New History of the American West* (Norman, Oklahoma 1991).

31. C. P. Russell: *Guns on Early Frontiers* (New York 1957); see also his *Guns, Traps and Tools of Mountain Men* (New York 1967).

32. W. A. Settle: *Jesse James Was His Name* (New York 1966).

33. For the list see *Oxford History of the West*, 412–13.

34. Wayne Garde: *Frontier Justice* (New York 1949); for a contemporary account, see N. P. Langford: *Vigilante Days and Ways* (Chicago 1890).

35. See R. M. Brown: 'The American Vigilante Tradition,' in *A Report to the National Commission on the Causes and Prevention of Violence: The History of Violence in America* (New York 1969).

36. See Wallace's *Autobiography* (New York 1906).

37. See W. F. Cody: *The Adventures of Buffalo Bill* (New York 1904); Don Russell: *Lives and Legends of Buffalo Bill* (New York 1960).

38. Kenneth W. Rendell: *History Comes to Life: Collecting Historical Records and Documents* (Norman, Oklahoma 1995), 121–2.

39. K. L. Steckmesser: *The Western Hero in History and Legend* (New York 1965).

40. D. T. Gilchrist and W. D. Lewis: *Economic Change in the Civil War Era* (New York 1965).

41. H. D. Woodman: *King Cotton and His Retainers, 1800–1925* (New York 1968).

42. T. N. Carver: *Principles of Rural Economics* (New York 1927), 99.

43. For an excellent summary of agricultural progress in America, see Faulkner, *opus cit.*, Chapter 19, 'The Agricultural Revolution,' 375ff.

44. E. C. Kirkland: *Industry Comes of Age 1860–97* (Chicago 1961).

45. See the illuminating table in Faulkner, *opus cit.*, 405.

46. *Congressional Record*, 48th Congress, 2nd session, xvi, Part I, 109, December 9, 1884.

47. Carter Goodrich: *Government Promotion of Canals and Railroads, 1800–90* (New York 1960).

48. For comparative purposes see L. E. Davis and J. Legler: 'Government in the American Economy, 1815–1902,' *Journal of Economic History*, 26 (1966), 514ff.

49. J. C. Bonbright: *Railroad Capitalisation* (New York 1920).

50. See Appendix A from *New Orleans Times-Democrat*, March 29, 1882.

51. Mark Twain: *Life on the Mississippi*, quoting the *Cincinnati Commercial*.

52. John F. Stover: *American Railroads* (Chicago 1978); *Railroad Facts* (Association of American Railroads, New York 1988).

53. J. F. Stover: *The Life and Decline of the American Railroad* (Chicago 1970).

54. Quoted in John D. Hicks: *The Populist Revolt* (New York 1931), 85.

55. For profitability in the climax decades see G. R. Taylor and I. D. Neu: *The American Railroad Network, 1861–90* (New York 1956).

56. E. R. McCartney: *The Crisis of 1873* (New York 1935); Rendigs Fels: *American Business Cycles, 1865–1897* (New York 1959); F. P. Weberg: *The Background of the Panic of 1893* (New York 1929).

57. See E. H. Morr: *Between the Ocean and the Lakes: the Story of the Erie* (New York 1901).

58. A. D. H. Smith: *Commodore Vanderbilt* (New York 1927).

59. For Drew see S. H. Holbrook: *The Age of the Moguls* (New York 1964), 13–35.

60. W. A. Swanberg: *Jim Fisk* (New York 1959).

61. Julius Grodinsky: *Jay Gould: His Business Career, 1867–92* (New York 1957) describes the Erie takeover in detail.

62. See C. F. Adams Jr: *Chapters of Erie* (New York 1866); *The Railroad Problem* (New York 1880), 126.

63. For Fish and Grant's government see Allan Nevins: *Hamilton Fish: the Inner History of the Grant Administration* (New York 1936).

64. J. A. Carpenter: 'Washington, Pennsylvania and the Gold Conspiracy of 1869,' *Western Pennsylvania Historical Magazine*, 48 (1965), 345ff.

65. For Fisk's involvement in the New York theater see Lehman Engel: *American Musical Theater* (New York 1967), and G. C. D. Odell: *Annals of the New York Stage*, 15 vols (New York 1927–49).

66. E. S. Lunde: *Horace Greeley* (New York 1980).

67. For the election of 1872 see W. S. McFeeley: *Grant: a Biography* (New York 1981).

68. F. M. Green: 'Origins of Credit Mobilier,' *Mississippi Valley Historical Review*, 46 (1959), 238ff.

69. See L. E. Guese: 'St Louis and the Great Whiskey Ring,' *Missouri Historical Review* 36 (1942), 160ff; R. C. Prickett: 'The Malfeasance of Belknap,' *North Dakota History* 17 (1950), 5ff; C. C. Spence: 'Schenck and the Emma Mine Affair,' *Ohio Historical Quarterly*, 68 (1959), 141ff.

70. For Lyndhurst, see Roger W. Moss, *The American Country House*, 148–57.

71. S. J. Buck: *The Granger Movement* (Chicago 1913), 205.

72. These cases were 94 US 113 and 118 US 557; see also C. M. Gardner: *The Grange: Friend of the Farmer* (New York 1949).

73. R. W. Harbeson: 'Railroads and Regulation, 1877–1916: Conspiracy or Public Interest?,' *Journal of Economic History*, 27 (1967), 230ff; G. Kolko: *Railroads and Regulation, 1877–1916* (New York 1967).

74. There is a full account of Harriman in George Kennan: *E. H. Harriman*, 2 vols (New York 1922).

75. Michael Conant: *Railroad Mergers and Abandonments* (New York 1965); see also Daggett Stuart: *Railroad Reorganisation* (New York 1908).

76. E. G. Campbell: *Reorganisation of the Railroad System, 1893–1900* (New York 1938).

77. See, for instance, Mary Lutyens: *The Lyttons in India* (London 1979), for Grant's amazing performance at a Viceregal dinner party.

78. W. B. Heseltine: *U.S. Grant, Politician* (New York 1935); the *Personal Memoirs* are two vols (New York 1885–6).

79. S. Pomerantz: 'The Election of 1876,' in Arthur M. Schlesinger and F. R. Israel (eds): *American Presidential Elections*, ii; P. L. Hayworth: *The Hayes–Tilden Disputed Election of 1876* (New York 1906).

80. J. W. Burgess: *The Administration of Hayes* (New York 1916); Harry Barnard: *Rutherford B. Hayes* (New York 1956).

81. H. J. Clancy: *The Presidential Election of 1880* (New York 1958).

82. G. F. Howe: *Chester A. Arthur* (New York 1934).

83. J. W. Wall: *Andrew Carnegie* (New York 1970): H. C. Livesay: *Andrew Carnegie and the Rise of Big Business* (New York 1975).

84. Peter Temin: *Iron and Steel in 19th Century America* (Pittsburgh 1964).

85. T. A. Wetime: *Coming of Age of Steel* (New York 1962).

86. For Carnegie's philosophy see E. C. Kirkland (ed.): *Carnegie's Gospel of Wealth and Other Essays* (New York 1890); J. C. Van Dyke (ed.): *Carnegie's Autobiography* (New York 1920).

87. The most detailed account of Carnegie's business activities is in B. J. Hendrick: *The Life of Andrew Carnegie*, 2 vols (New York 1932).

88. Quoted in Jonathan Hughes: *The Vital Few: American Economic Progress and Its Protagonists* (New York 1965), 252.

89. *The Andrew Carnegie Century* (New York 1935) is a mine of information about this and other matters.

90. F. L. Allen: *The Great Pierpont Morgan* (New York 1949) makes this point.

91. Lewis Corey: *The House of Morgan* (New York 1930).

92. For the Well Report, see Executive Document No. 27, House of Representatives, 41st Congress, 2nd Session, vi. For Chase's statement see W. G. Sumner: *A History of the American Currency* (New York 1876), 197.

93. J. P. Nicols: 'Silver Diplomacy,' *Political Science Quarterly*, 48 (1933), 565ff; J. A. Barnes: 'Gold Standard Democrats and Party Conflict,' *Mississippi Valley Historical Review*, 17 (1930), 422ff.

94. M. D. Hirch: 'The Election of 1884,' in Schlesinger and Israel, *opus cit.*, iii.

95. R. E. Welch Jr: *The Presidencies of Grover Cleveland* (New York 1988).

96. P. E. Coletta: *William Jennings Bryan*, 3 vols (New York 1964–9).

97. For Morgan's philosophy, see Ron Chernow: *The House of Morgan* (New York 1990).

98. *Wheaton's Reports*, iv, 518, p. 636.

99. *Abstract of the Census of Manufactures* (Washington DC 1919), Table 195, 340.

100. J. Jenks and J. Clerk, *The Trust Problem* (7th edn New York 1917), 17. See Faulkren, *opus cit.*, Table 434.

101. John Moody: *The Truth About the Trusts* (New York 1904).

102. C. C. Abbott: *The Rise of the Business Corporation* (New York 1946); W. L. Warner: *The Corporation in Emergent American Society* (New York 1962).

103. For Harriman's dealings with Morgan see Kennan, *opus cit.*

104. G. C. Schroeder: *Growth of the Major Steel Companies, 1900–50* (New York 1953).

105. For a description of the scene and Morgan's moves, see Hughes, *opus cit.*, 445ff.

106. For the origins of the Fed see P. M. Warburg: *The Federal Reserve System*, 2 vols (Washington DC 1930).

107. See Louis Auchincloss: *J. P. Morgan: the Financier as Collector* (New York 1990).

108. J. F. Rhodes: *History of the United States from Hayes to McKinley, 1877–96* (New York 1919), 53ff.

109. J. M. Coleman: *The Molly Maguire Riots* (New York 1936); W. G. Broehl: *The Molly Maguires* (New York 1950).

110. See Powderly's own account, *Thirty Years of Labor* (New York 1889).

111. See Gompers, *Seventy Years of Life and Labor*, 2 vols (New York 1925), ii 105; *Report of Senate Committee on Capital and Labor*, i 460; Philip Taft: *Organized Labor in American History* (New York 1964).

112. Melvyn Dubofsky: *We Shall Be All: Industrial Workers of the World* (New York 1969); for background to the ideological dispute see G. N. Grob: *Workers and Utopia: A Study of the Ideological Conflicts in the American Labor Movement, 1865–1900* (New York 1961).

113. Leon Wolfe: *Lockout: the Homestead Strike of 1892: A Study of Violence, Unionism and the Carnegie Steel Empire* (Pittsburgh 1965), which puts the cases for both sides.

114. A. B. Saarinen: *Proud Possessors: American Art Collectors* (New York 1958).

115. H. M. Meyer and R. C. Wade: *Chicago: Growth of a Metropolis* (Chicago 1969), 94–6.

116. *Ibid.*, 128.

117. C. W. Conduit: *American Building Art: 19th Century* (New York 1960).

118. The point is made in Mark Girouard: *Cities and People* (New Haven 1985), 317ff.

119. Lauline A. Saliga (ed.): *The Sky's the Limit: a Century of Chicago Skyscrapers* (New York 1990).

120. For Sullivan, see Sherman Paul: *Louis Sullivan: Architect in American Thought* (Chicago 1962).

121. Meryle Secrest: *Frank Lloyd Wright: a Biography* (London 1992), 104–5.

122. For Sullivan's writings see Maurice English (ed.): *Testament of Stone: Writings of Louis Sullivan* (New York 1963).

123. Carol Willis: *Form Follows Finance: Skyscrapers and Skylines in New York and Chicago* (New York 1996) analyzes the conflict between aesthetic and economic factors. Wright wrote his own biography of Sullivan: *Genius and the Mobocracy* (Chicago 1949).

124. See the handbook *Chicago and Its Environs* (Chicago 1893).

125. For regulation see V. J. Scully Jr: *American Architecture and Urbanism* (New York 1969).

126. Sharon Darling: *Chicago Furniture: Art, Craft and Industry, 1833–1983* (Chicago 1994), 17.

127. *Ibid.*, 185, 191, for photos.

128. See Andres Simon: *Chicago: the Garden City* (Chicago 1893), 48ff; G. E. Holt: 'Private Plans for Public Spaces: the Origin of Chicago's Park System, 1850–75,' *Chicago History* 8 (1979).

129. Sarah Landau and Carl W. Conduit: *Rise of the New York Skyscraper, 1865–1913* (New Haven 1996), for these and other details.

130. Alan Trachtenberg: *Brooklyn Bridge* (New York 1965).

131. J. K. Winkler: *Five and Ten: the Life of Frank W. Woolworth* (New York 1940).

132. Federal Writers Project: *New York City* (New York 1939).

133. E. Marriam: *Emma Lazarus: Woman with a Torch* (New York 1956).

134. See *New York Tenement House Development Report* (New York 1903), 132ff.

135. For details of the Jewish settlement in New York, see J. S. Blau and S. W. Baron: *The Jews in the United States, 1790–1840: a Documentary History*, 3 vols (New York 1963).

136. Robert Silverburg: *Light for the World: Edison and the Power Industry* (New York 1967).

137. Hughes, *opus cit.*, 162–3.

138. See Malcolm MacLaren: *The Rise of the Electrical Industry During the 19th Century* (New York 1943); J. Bauer and N. Gould: *The Electric Power Industry* (New York 1939).

139. John Loring: *Tiffany's 150 Years* (New York 1987), esp. 43–52.

140. D. B. Burke *et al.*: *In Pursuit of Beauty: Americans and the Aesthetic Movement* (Metropolitan Museum, New York 1987), 185.

141. Only three of these lamps are now known: one was illustrated in *Sotheby's Preview* (London, December 1996), 26–7.

142. Vivienne Couldrey: *The Art of Louis Comfort Tiffany* (London 1989), 61, for photograph.

143. *Ibid.*, 152ff, 160ff.

144. John Erskine: *The Philharmonic-Symphony Society of New York: the First Hundred Years* (New York 1943); Quintance Eaton: *Miracle of the Met: History of the Metropolitan Opera, 1883–1967* (New York 1968).

145. W. Weaver and Simonetta Puccini (eds): *The Puccini Companion* (New York 1994), 214–27.

146. E. J. Hygren: *Views and Visions: American Landscape Before 1830* (Washington DC 1986), deals with pre-Cole landscape and Cole's early work.

147. John Wilmerding: *The Luminist Movement, 1850–75* (Princeton 1989).

148. Nicolai Cikovsky Jr: *George Innes* (Smithsonian, Washington DC 1993).

149. Many are reproduced in Gerald L. Carr: *Frederick Edwin Church: Catalogue Raisonnee of Works of Art at the Olana State Historic Site*, 2 vols (New York 1994).

150. For details of the books that Church read see *ibid.*, note xxxviii.

151. For Church's work in Latin America, and other examples of US artists in Central and South America see K. E. Manthorne: *Tropical Renaissance: North American Artists Exploring Latin America 1839–79* (Smithsonian, Washington DC 1989).

152. There is a full-length study of this painting by Jeremy Elwell Adamanson: 'Frederick Church's *Niagara*: the Sublime as Transcendence,' PhD dissertation, University of Michigan 1981.

153. For Church and his contemporaries, see *American Paradise: the World of the Hudson River School* (Metropolitan, New York 1987).

154. For Bierstadt see N. K. Anderson and L. S. Ferber: *Albert Bierstadt: Art and Enterprise* (Brooklyn Museum 1990).

155. For the parks see Freeman Tilden: *The National Parks* (New York 1968); other forests are dealt with in Orville Freeman *et al.*: *National Forests of America* (New York 1968).

156. Moss, *opus cit.*, 158–71.

157. See Clive Aslet: 'Olana,' in *The American Country House* (New Haven 1990), 35–47.

158. For the Tiffany revival see Robert H. Kock: *Louis C. Tiffany, Rebel in Glass* (3rd edn New York 1982); Kock was the historian most responsible for the revival.

159. Mark Alan Hewitt: *The Architect and the American Country House* (New Haven 1990), 70.

160. Aslet, *opus cit.*, 15.

161. *Ibid.*, 190ff.

162. S. Crowther: *Men and Rubber* (New York 1926), 20ff.

163. Hewitt, *opus cit.*, 133.

164. For Nemours, see Aslet, *opus cit.*, 97ff, where the machines are illustrated.

165. D. I. Sutherland: *Americans and Their Servants* (Baton Rouge 1981), 183.

166. Hewitt, *opus cit.*, 99ff.

167. G. C. Winkler: *The Well-Appointed Bath* (Washington DC 1989), 11ff.

168. L. E. Asher and E. Heal: *Send No Money* (Chicago 1942), 72; see also B. Emmet and J. E. Jeuck: *Catalogues and Counters: a History of Sears, Roebuck & Co* (Chicago 1950), 113.

169. D. M. Potter: *People of Plenty: Economic Abundance and the American Character* (Chicago 1954), 80.

170. O. E. Anderson: *Refrigeration in America: a History of a New Technology and Its Impact* (Princeton 1953), 197ff.

171. R. F. Wesser: 'The Election of 1888,' in Schlesinger and Israel, *opus cit.*, iii.

172. G. H. Knoles: *The Presidential Campaign and Election of 1892* (New York 1942).

173. For Cleveland's loss of party control see Allan Nevins: *Grover Cleveland: a Study in Courage* (New York 1932).

174. R. F. Durden: *The Climax of Populism: the Election of 1896* (New York 1965).

175. W. LeFeber: 'The Election of 1900,' in Schlesinger and Israel, *opus cit.*, iii.

176. Samuel P. Hayes: *The Response to Industrialisation, 1885–1914* (New York 1957).

177. *Progress and Poverty* was reissued by the Robert Schalkenbach Foundation (New York 1981); see E. J. Rose: *Henry George* (New York 1969) and J. L. Thomas: *Alternative America: Henry George, Edward Ballamy, Henry Demarest Lloyd and the Adversary Tradition* (Cambridge 1983).

178. Text of the Act, and related legislation, in J. W. Jenks and W. E. Clark: *The Trust Problem* (New York 1910), Appendix F.

179. For these cases see *In re Debs*, 158 US 564; *Loewe v. Lawlor*, 235 US 522; W. L. Letwin: 'The First Decade of the Sherman Act,' *Yale Law Journal*, 68 (1958), 464ff, 900ff; H. B. Thorelli: *Federal Anti-Trust Policy* (New York 1955).

180. See table, 'Number of Daily Newspapers in the US, 1790–1990,' in *Reader's Companion to American History*, 691.

181. Lincoln Steffens' *Autobiography* was published in New York in 1931; his *Letters* were edited by E. Winter and G. Hicks, 2 vols (New York 1938).

182. A. Weinberg and L. Weinberg (eds): *The Muckrakers* (New York 1961); see also Richard McCormick: 'The Discovery that Business Corrupts Politics: a Reappraisal of the Origins of Progressivism,' *American Historical Review*, 86 (April 1981), 247ff.

183. For instance Ida Tarbell: *The History of the Standard Oil Company*, 2 vols (New York 1925).

184. P. H. Giddens: *The Birth of the Oil Industry* (New York 1939).

185. For pricing and regulation see G. D. Nash: *United States Oil Policy, 1890–1964* (New York 1969); see also D. M. Chalmers: 'Standard Oil and the Business Historian,' *American Journal of Economics and Sociology*, 20 (1960), 47ff.

186. W. I. Walsh: *The Rise and Decline of the Great Atlantic and Pacific Tea Company* (Secaucus 1986).

187. R. A. Pasner: *The Robinson–Patman Act: Federal Regulation of Price Differences* (Washington DC 1976).

188. T. K. McCraw: *Prophets of Regulation* (Cambridge 1984), 102–9.

189. J. J. Flink: *America Adopts the Automobile, 1895–1910* (Cambridge 1971), 21.

190. Other sources give these figures as 10,607 and 730,041 respectively. See Allan Nevins and Frank E. Hill: *Ford*, 3 vols (New York 1954–63) and P. Collier and D. Horrowitz: *The Fords: an American Epic* (New York 1987).

191. W. C. Richards: *The Last Billionaire* (New York 1948), 348ff.

192. James J. Flink: *The Automobile Age* (New York 1988).

193. Richard Hofstadter: *The Age of Reform: Bryan to FDR* (Cambridge 1955).

194. George E. Mowry: *The Era of Theodore Roosevelt, 1900–1902* (New York 1958).

195. But note that W. R. K. Nugent: *The Tolerant Populists: Kansas Populism and Nativism* (Chicago 1963) defends the Populists from charges of xenophobia and anti-semitism.

196. A. W. Sheils: *The Purchase of Alaska* (New York 1967); Ben Adams: *The Last Frontier: Alaska* (New York 1961).

197. W. A. Russ: *The Hawaiian Republic and Annexation* (New York 1961).

198. W. G. McLouchlin: *The Meaning of Henry Ward Beecher: an Essay on the Shifting Values of mid-Victorian America, 1840–1870* (New York 1970).

199. Quoted in S. A. Allstron: *A Religious History of the American People* (New Haven 1972).

200. Josiah Strong: *Our Country: Its Possible Future and Its Present Crisis* (new edn New York 1891).

201. S. C. Neill: *A History of Christian Missions* (London 1964).

202. David Healey: *US Expansionism: the Imperialist Urge of the 1890s* (New York 1970).

203. The fullest account is in F. E. Chadwick: *Relations of the United States and Spain: the Spanish–American War*, 2 vols (New York 1911); for a visual account see Frank Friedel: *A Splendid Little War* (New York 1958), and for the role of the press, J. E. Wisan: *The Cuban Crisis in the New York Press* (New York 1934).

204. J. W. Pratt: 'American Businessmen and the Spanish–American War,' *Hispanic American Historical Review*, May 1934, argues that US business (apart from some few directly involved) preferred trade to colonies and wanted peaceful relations with Spain; Walter LaFeber: *The New Empire: an Interpretation of American Expansion, 1860–98* (Ithaca 1963) argues that many businessmen wanted war, but only to restore order.

205. Ernest R. May: *Imperial Democracy: the Emergence of America as a Great Power* (New York 1961).

206. M. H. Wayne: *William McKinley* (New York 1963).

207. W. H. Harbaugh: *Power and Responsibility: the Life and Times of Theodore Roosevelt* (New York 1961) gives a good general account of the man.

208. Peter Collier: *The Roosevelts: an American Saga* (New York 1994) is a vivacious account of the family by a journalist.

209. For South Dakota in this period see H. R. Lamar: *Dakota Territory, 1861–89* (Chicago 1956).

210. This episode is described in Theodore Roosevelt's big book, *The Winning of the West*, 4 vols (New York 1889–96).

211. Theodore Roosevelt's *Works* are in 24 volumes (New York 1923–6); H. C. Lodge and T. Roosevelt (eds): *Selections from the Correspondence of Theodore Roosevelt*, 2 vols (New York 1925).

212. James Bryce: *The American Commonwealth*, 2 vols (London 1888). The best edition is published by the American Classics of Liberty, New York 1993.
213. See L. L. Gould: *The Presidency of Theodore Roosevelt* (New York 1990).
214. R. E. B. Lewis: *Edith Wharton: a Biography* (New York 1993), 144–5.
215. *Ibid.*, 112–13.
216. For two different views of Roosevelt see E. E. Morris: *The Rise of Theodore Roosevelt* (New York 1979) and G. W. Chessman: *Theodore Roosevelt and the Politics of Power* (New York 1968).
217. W. H. Harbaugh: 'The Election of 1904,' in Schlesinger and Israel, *opus cit.*, iii.
218. R. A. Friedlander: 'A Reassessment of Roosevelt's Role in the Panamanian Revolution of 1903,' *Western Political Quarterly*, 14 (1961), 535ff.
219. For use of the Roosevelt corollary, see D. G. Munro: *Intervention and Dollar Diplomacy in the Caribbean, 1900–21* (New York 1964).
220. T. R. Roosevelt: *An Autobiography* (New York 1913), 56.
221. J. M. Cooper Jr: *The Warrior and the Priest: Woodrow Wilson and Theodore Roosevelt* (Cambridge 1983).
222. Paola E. Coletta: 'The Election of 1908,' in Schlesinger and Israel, *opus cit.*, iii.
223. Paola E. Coletta: *The Presidency of William Howard Taft* (New York 1973).
224. G. E. Mowry: *Theodore Roosevelt and the Progressive Movement* (New York 1946).
225. G. E. Mowry: 'The Election of 1912,' in Schlesinger and Israel, *opus cit.*, iii. For Roosevelt's own account of his program, see his book *The New Nationalism* (New York 1910).

PART SIX: Melting-Pot America

1. For a general survey of the change in the American economy and relation to society see M. J. Sklar: *The Corporate Reconstruction of American Capitalism, 1890–1916: the Market, the Law and Politics* (Cambridge 1988).
2. The standard biography of Wilson is Arthur S. Link: *Wilson*, 5 vols (Princeton 1947–65); Link et al. also edited *The Papers of Woodrow Wilson*, 69 vols (Princeton 1966–93).
3. Wilson's intellectual background is well explored in the best recent biography, August Heckscher: *Woodrow Wilson: a Biography* (New York 1991), 10–46.
4. A. T. Mason: *Brandeis* (Boston 2nd edn 1956).
5. J. M. Mulder: *Woodrow Wilson: Years of Preparation* (Princeton 1978) deals with the religious background and Wilson's relations with his father.
6. R. L. Geiger: *To Advance Knowledge: the Growth of American Research Universities, 1900–40* (New York 1986).
7. For Wilson's constitutional writing and thinking see N. A. Thorsen: *The Political Thought of Woodrow Wilson, 1875–1910* (Princeton 1988).
8. See Hardin Craig: *Woodrow Wilson at Princeton* (Norman, Oklahoma 1960).
9. Frank Ashburn: *Peabody of Groton* (New York 1944) and *Fifty Years On* (New York 1934).

10. Cass Canfield: *Up and Down and Around* (New York 1971).

11. Loomis Havermeyer: *Go to Your Room!* (New Haven 1960); Wilmarth Lewis: *One Man's Education* (New York 1967); Samuel Eliot Morison: *Three Centuries of Harvard* (Cambridge 1936).

12. A good introduction to the effect of education on high office-seekers is Walter Isaacson and Evan Thomas: *The Wise Men: Six Friends and the World They Made* (New York 1986), 39–97.

13. D. W. Hirst: *Woodrow Wilson, Reform Governor: a Documentary Narrative* (Princeton 1965).

14. Woodrow Wilson: *Crossroads of Freedom: 1912 Campaign Speeches*, ed. John W. Davidson (New York 1956).

15. For income tax see Gerald Carson: *The Golden Egg: the Personal Income Tax, Where It Came From, How It Grew* (New York 1977); for the Senate, see G. H. Haynes: *The Senate: History and Practice* (New York 1939).

16. For Wilson's Cabinet-making see Heckscher, opus cit., 262–73.

17. Edward Mandell House: *The Intimate Papers of Colonel House*, 4 vols (Boston 1926–8); Cary T. Grayson: *Woodrow Wilson: an Intimate Memoir* (New York 1960).

18. H. U. Faulkner, *American Economic History*, 561; Elgin Groseclose: *Fifty Years of Managed Money: the Federal Reserve, 1914–63* (Washington DC 1965).

19. L. T. Fournier: 'The Purpose and Results of the Webb–Pomerene Law,' *American Economic Review*, xxii (March 1932), 18ff.

20. William McAdoo's autobiography is called *Crowded Years* (New York 1931); see the important section on McAdoo in Jordan A. Schwattz: *The New Dealers: Power Politics in the Age of Roosevelt* (New York 1993).

21. G. B. Tindall: *The Emergence of the New South, 1913–45* (New York 1967).

22. E. E. Garrison: *Roosevelt, Wilson and the Federal Reserve Law* (New York 1931).

23. *The Nation*, November 30, 1918.

24. Heckscher, opus cit., 56, 46ff, 233.

25. The story is cherished British embassy lore and may be apocryphal. It is not mentioned in Stephen Gwynn (ed.): *The Letters and Friendships of Sir Cecil Spring-Rice*, 2 vols (London 1929). For Wilson's love-letters see Eleanor Wilson McAdoo (ed.): *The Priceless Gift: the Love Letters of Woodrow Wilson and Ellen Axson Wilson* (New York 1962). For Wilson's Bermuda mistress see Mary Allen Hulbert: *The Story of Mrs Peck: an Autobiography* (New York 1933).

26. Heckscher, opus cit., 356.

27. Wilson's involvement in the first two years of the Great War is dealt with in volume 5 of Ray Baker: *Woodrow Wilson: Life and Letters*, 9 vols (New York 1927–39).

28. Manfred Jones: *The United States and Germany: a Diplomatic History* (New York 1984).

29. Quoted in Foster Rhea Dulles: *The United States Since 1865* (Ann Arbor 1959), 263.

30. Randolph Bourne: *Untimely Papers* (New York 1919), 140. Bourne was a victim of the 1918 influenza epidemic and these writings were published posthumously.

31. A. S. Link and W. M. Leary: 'The Election of 1916,' in Arthur M. Schlesinger and F. R. Israel (eds): *American Presidential Elections*, iii; W. M. Leary: 'Woodrow Wilson, Irish-Americans and the Election of 1916,' *Journal of American History*, 54 (1967), 57ff.

32. K. E. Birnbaum: *Peace Moves and U-boat Warfare: Germany's Policy Towards the United States, April 18, 1916–January 9, 1917* (New York 1958); Barbara Tuchman: *The Zimmerman Telegram* (New York 1958).

33. Heckscher, *opus cit.*, 388.

34. For the impact of the war on ordinary Americans, see D. M. Kennedy: *Over Here: The First World War and American Society* (New York 1980).

35. Bernard Baruch later published an account, *American Industry in War*, ed. R. H. Hippelheuser (New York 1941). See also R. R. Himmelberg: 'The War Industries Board,' *Journal of American History*, 52 (1965), 43.

36. H. N. Scheiber: *The Wilson Administration and Civil Liberties, 1917–20* (New York 1960); John Dickinson: *Building of an Army* (New York 1922).

37. See Frederick Palmer: *John J. Pershing, General of the Armies* (New York 1948); Pershing's own account of World War One is in *My Experiences*, 2 vols (New York 1933).

38. P. A. Poole: *America in World Politics: Foreign Policy and Policymakers Since 1898* (New York 1975), 39.

39. Harold Nicolson: *Peacemaking 1919* (London 1945 edn), 21–2; L. E. Gelfand: *The Inquiry: American Preparations for Peace, 1917–19* (New Haven 1963).

40. Nicolson, *opus cit.*, 31–3; for detailed examinations of Wilson's twenty-three assertions, see R. S. Baker: *Woodrow Wilson and the World Settlement*, 3 vols (New York 1922), and H. W. V. Temperley *et al.*: *History of the Peace Conference*, 6 vols (1920–4).

41. See T. A. Bailey: *Woodrow Wilson and the Lost Peace* (New York 1944); N. G. Levin: *Wilson and World Politics* (New York 1968). See also *Foreign Relations of the United States: Paris Peace Conference 1919* (Washington DC 1942–7), xi, 547–9, 570–4; Walter Lippmann, letter to R. B. Fosdick, August 15, 1919, *Letters of the League of Nations* (Princeton 1966).

42. G. Clemenceau: *Grandeur and Misery of a Victory* (trans. London 1930); Andre Tardieu: *The Truth About the Treaty* (trans. London 1921).

43. See W. C. Widenor: *Henry Cabot Lodge and the Search for an American Foreign Policy* (Berkeley 1980) and Lodge's own account, *The Senate and the League of Nations* (Boston 1925).

44. D. F. Fleming: *The United States and the League of Nations* (New York 1932).

45. For the details of Wilson's last phase in office, see Heckscher, *opus cit.*, 611ff; Gene Smith: *When the Cheering Stopped: the Last Years of Woodrow Wilson* (New York 1964).

46. Colby wrote a book about his weird experiences, *The Close of Woodrow Wilson's Administration and the Final Years* (New York 1925).

47. Smith, *Wilson*, 107, 111–13, 126–8; Heckscher, *opus cit.*, 621–2.

48. W. M. Bagby: *Road to Normalcy: the Campaign of 1920* (New York 1962); see also his 'Woodrow Wilson, a Third Term and the Solemn Referendum,' *American Historical Review*, 60 (1955), 567ff.

49. Quoted in Robert K. Murray: *The Harding Era: Warren G. Harding and his Administration* (Minneapolis 1969), 67; see also Randolph Downes: *The Rise of Warren Gamaliel Harding, 1865–1920* (Columbus, Ohio 1970).

50. Thomas Woody: *A History of Women's Education in the United States*, 2 vols (New York 1929); Ishbel Ross: *Child of Destiny: Elizabeth Blackwell, the First Woman Doctor* (New York 1949).

51. Nancy F. Cott: *The Grounding of Modern Feminism* (New York 1987); Ellen Carol DuBois: *Feminism and Suffrage: the Emergence of an Independent Woman's Movement in America* (New York 1978).

52. Jean Straus: *Alice James* (Boston 1980), 215ff.

53. Lewis, *Edith Wharton*, 486.

54. Heckscher, *opus cit.*, 457.

55. Aileen Kraditor: *Ideas of the Woman's Suffrage Movement, 1890–1920* (New York 1965); Eleanor Flexner: *Century of Struggle: the Women's Rights Movement in the United States* (Cambridge 1959).

56. Patricia Smith (ed.): *Feminist Jurisprudence* (New York 1993).

57. K. T. Bartlett and R. Kennedy (eds): *Feminist Legal Theory: Readings in Law and Gender* (Boulder, 1991).

58. J. E. Cutler: *Lynch-Law: an Investigation into the History of Lynching in the United States* (New York 1905), 177; Walter White: *Rope and Faggot: a Biography of Judge Lynch* (New York 1929).

59. D. M. Chalmers: *Hooded Americanism: the First Century of the Ku Klux Klan, 1865–1965* (New York 1965); C. C. Alexander: *The Ku Klux Klan in the Southwest* (Lexington 1965).

60. Nicholas Lemann: *The Promised Land: the Great Black Migration and How It Changed America* (New York 1991); N. I. Painter: *Exodusters: Black Migration to Kansas After Reconstruction* (New York 1988).

61. Chicago Commission on Race Relations: *The Negro in Chicago: a Study of Race Relations and a Race Riot* (Chicago 1922); see also E. M. Rudwick: *The Riot at East St Louis* (Carbondale 1964).

62. Callow, *American Urban History*, 564ff, with important bibliographical references.

63. Studies of ghettos include A. R. Hirsch: *Making the Second Ghetto: Race and Housing in Chicago, 1940–60* (Chicago 1983) and K. L. Kusmer: *A Ghetto Takes Shape: Black Cleveland, 1870–1930* (New York 1976).

64. For the origins of black Harlem see Anne Douglas: *Terrible Honesty: Mongrel Manhattan in the Twenties* (New York 1995), 303ff.

65. Harold Cruse: *The Crisis of the Negro Intellectual* (New York 1967).

66. J. W. Johnson: *Black Manhattan* (New York 1930); see also R. Ottley and W. J. Weatherby (eds): *The Negro in New York: an Informal Social History, 1625–1940* (New York 1969) and Gilbert Osofsky: *Harlem: the Making of a Ghetto* (New York 1971).

67. Callow, *opus cit.*, 389; see sources 397.
68. *Ibid.*, 390–1.
69. J. H. Robinson: *Road without Turning: an Autobiography* (New York 1950), 231.
70. Jean de Crèvecoeur: *Letters from an American Farmer* (New York 1904 edn), 54–5.
71. Israel Zangwill: *The Melting Pot* (New York 1909), 37ff. The Dickens quote is from *American Notes* (1842).
72. Madison Grant: *The Passing of the White Race* (New York 1916), 3–36.
73. Widenor, *opus cit.*; Robert Murray: *The Harding Era* (U. of Minnesota 1969), 64.
74. J. M. Blum: *The Progressive Presidents* (New York 1980), 97.
75. Dulles, *opus cit.*, 295.
76. Francis Russell: *Sacco and Vanzetti: the Case Resolved* (New York 1986); Paul Avrich: *Sacco and Vanzetti: the Anarchist Background* (Princeton 1991).
77. R. A. Divine: *American Immigration Policy, 1924–1952* (New York 1957).
78. J. C. Levenson: *The Mind and Art of Henry Adams* (Boston 1957); J. R. Vitelli: *Van Wyck Brooks* (New York 1968).
79. Quoted in James Hoopes: *Van Wyck Brooks in Search of American Culture* (Amherst 1977), 130.
80. W. B. Gatewood: *Controversy in the Twenties: Fundamentalism, Modernism and Evolution* (New York 1969).
81. *New Republic*, May 10, 1922; for another view, see Alan Ryan: *John Dewey and the High Tide of American Liberalism* (New York 1995), 131–2.
82. *Philadelphia Press*, February 22, 1914, quoted in John Wilmerding (ed.): *Thomas Eakins* (London 1993), 16.
83. The pioneering work on Homer is Gordon Hendricks: *The Life and Work of Winslow Homer* (New York 1979).
84. See the essays in Nicolai Cikovsky and Franklin Kelly (eds): *Winslow Homer* (National Gallery of Art, Washington DC 1996).
85. Arthur L. Gupthill: *Norman Rockwell, Illustrator* (New York, reissue 1975) describes the Rockwell technique, 45ff and 193ff, and reproduces all 322 *Post* covers.
86. J. J. Riley: *A History of the American Soft-Drinks Industry, 1807–1957* (New York 1972), 251ff.
87. Pat Watters: *Coca-Cola: an Illustrated History* (New York 1978), 15ff.
88. T. Oliver: *The Real Coke, the Real Story* (New York 1986).
89. J. C. Louis and H. Z. Yaziian: *The Cola Wars* (New York 1980), 25ff.
90. *Ibid.*, 104.
91. John Tebbel: *A History of Book Publishing in the United States*, 2 vols (New York 1975), ii 611; see also Edward de Grazia: *Girls Lean Back Everywhere: the Law of Obscenity and the Assault on Genius* (London 1992), 3–7, etc.
92. W. J. Rosabaugh: *The Alcoholic Republic* (New York 1979) for more figures.
93. Mark Moore and Dean Gerstein (eds): *Alcohol and Public Policy: Beyond the Shadow of Prohibition* (New York 1981).

94. For details see Albert E. Sawyer: 'The Enforcement of National Prohibition,' *Annals*, September 1932.

95. Herbert Asbury: *The Great Illusion: Prohibition* (New York 1950).

96. *The Illinois Crime Survey* (Chicago 1929), 909–19.

97. Charles Fecher: *H. L. Mencken: a Study in His Thought* (New York 1978), 159.

98. For non-enforcement see Charles Merz: *The Dry Decade* (New York 1931), 88, 107, 123–4, 144, 154.

99. *The Prohibition Amendment: Hearings before the Committee of the Judiciary, 75th Congress, Second Session* (Washington DC 1930), Part 1, 12–31.

100. Rufus King: *Gambling and Organised Crime* (New York 1969).

101. Annalise Graebner Anderson: *The Business of Organised Crime: A Cosa Nostra Family* (Stanford 1979).

102. B. M. Hobson: *Uneasy Virtue: the Politics of Prostitution and the American Reform Tradition* (New York 1987); T. J. Gilfoyle: *City of Eros: New York City, Prostitution and the Commercialisation of Sex, 1790–1920* (New York 1991).

103. Herbert Asbury: *The Barbary Coast* (New York 1933).

104. Bradley Smith: *The American Way of Sex* (New York 1978), 154ff.

105. W. A. Swanberg: *Citizen Hearst* (New York 1961); Pauline Kael: *The Citizen Kane Book* (New York 1971).

106. Doris Muscatine: *Old San Francisco* (New York 1975); Oscar Lewis: *This Was San Francisco* (New York 1962).

107. For descriptions and photos of these buildings see Sarah Holmes Boutelle: *Julia Morgan, Architect* (New York 1988), 216ff.

108. Roger W. Moss, *The American Country House*, 208ff.

109. Alfred L. Donaldson: *History of the Adirondacks*, 2 vols (New York 1921).

110. Full descriptions of San Simeon in Moss, *opus cit.*, 207ff, and in Boutelle, *opus cit.*

111. For early Los Angeles see R. M. Fogelson: *Fragmented Metropolis: Los Angeles, 1850–1930* (Berkeley 1967).

112. John Bauer: *The Health Seekers of Southern California, 1870–1900* (Los Angeles 1960).

113. Sam Hall Kaplan: *Los Angeles Lost and Found: an Architectural History of Los Angeles* (New York 1987), 49.

114. Robert Winter: *The California Bungalow* (Los Angeles 1961).

115. See Kaplan, *opus cit.*, 71, for photo.

116. T. P. Hughes: *Networks of Power: the Electrification of Western Society, 1880–1930* (Baltimore 1983).

117. P. J. Hubbard: *The Origins of the TVA: the Muscle Shoals Controversy, 1920–32* (New York 1961).

118. Norris Hundley Jr: *Water and the West: the Colorado River Compact and the Politics of Water in the American West* (Berkeley 1975).

119. Jordan A. Schwearz: *The New Dealers: Power Politics in the Age of Roosevelt* (New York 1993), 205ff.

120. Lewis Jacobs: *Rise of the American Film: a Critical History* (2nd edn New York 1968); L. A. Griffith: *When Movies Were Young* (New York 1925).

121. Lary May: *Screening Out the Past: the Birth of Mass Culture and the Motion Picture Industry* (Oxford 1980), 253, Table 111a, 'Founders of the Big Eight,' and Table iiib for biographies.

122. Kaplan, *opus cit.*, 91ff.

123. *Ibid.*, 104–5; see also Meryle Secrest, *Frank Lloyd Wright.*

124. Secrest, *opus cit.*, 397ff; Kaplan, *opus cit.*, 106ff; Neil Levine: *The Architecture of Frank Lloyd Wright* (Princeton 1996).

125. Raymond Moley: *The Hays Office* (New York 1945).

126. Hortense Powdermaker: *The Hollywood Dream Factory* (New York 1950).

127. Richard Schickel: *The Disney Version* (New York 1968).

128. Notably in France.

129. Rem Koolhass: *Delirious New York* (New York 1978).

130. Burton Peretti: *The Creation of Jazz* (New York 1992).

131. Quoted in P. L. Barbour: *The Three Worlds of Captain John Smith* (New York 1964).

132. Wilfred Mellers: *Music in a New Found Land: Theme and Developments in the Story of American Music* (rev. edn New York 1987), 264–5.

133. Quoted in *ibid.*, 257–8.

134. David Evan (ed.): *American Popular Songs from the Revolutionary War to the Present* (New York 1966).

135. Louis C. Elton: *A History of American Music* (New York 1904).

136. Craig Roell: *The Piano in America* (New York 1989); Anne Douglas: *Terrible Honesty: Mongrel Manhattan in the 1920s* (New York 1995), 276ff.

137. Douglas, *opus cit.*, 279.

138. *Ibid.*, 390ff.

139. Quoted in *ibid.*, 267.

140. Stanley Green: *The World of Musical Comedy* (New York 1960).

141. David Ewen: *Irving Berlin* (New York 1950).

142. LeRoi Jones: *Blues People: Negro Music in White America* (New York 1963).

143. Lehman Engel: *American Musical Theater* (New York 1967).

144. Arnold Rampersad: *The Life of Langston Hughes*, 2 vols (New York 1986–8); Faith Berry: *Langston Hughes: Before and Beyond Harlem* (New York 1983). For his rows with black critics see *The Nation*, June 23, 1926, and the *Pittsburgh Courier*, April 14, 1927.

145. Mark Gridley: *Jazz Styles* (New York 1985); Martin Williams: *Jazz in Our Own Time* (New York 1980).

146. See the introduction by Edward Jablonski to *Lady, Be Good!* in the Smithsonian Archival Reproduction Series, the Smithsonian Collection R008 (Washington DC 1977).

147. Robert Murray: *The Harding Era* (U. of Minnesota 1969), 117–19.

148. Quoted in Murray N. Rothbard: *America's Great Depression* (Los Angeles 1972), 167.

149. *New York Times* October 14, 1922; see Fritz Marx: 'The Bureau of the

Budget: Its Evolution and Present Role,' *Political Science Review*, August 1945.

150. Murray, *opus cit.*, 112.

151. *Ibid.*, 108.

152. *Investigations of Veterans Bureau: Hearings before the US Senate* (Washington DC 1923).

153. Burt Noggle: 'The Origins of the Teapot Dome Investigation,' *Mississippi Valley Historical Review*, September 1957; M. R. Werner and John Star: *Teapot Dome* (New York 1959), 194–277; Murray, *opus cit.*, 473.

154. Robert H. Ferrell: *The Strange Death of President Harding* (Columbia, Missouri 1996) establishes the facts about Harding's death and demolishes the scurrilous rumors which circulated afterwards.

155. For this journal see David Seideman: *The New Republic: a Voice of Modern Liberalism* (New York 1986).

156. For examination of the Britton story see Ferrell, *opus cit.*

157. Ferrell, *opus cit.*, examines critically all the historiography. Nevins believed Harding responsible for the destruction of his hero Woodrow Wilson's League of Nations policy, and he once worked for Harding's 1920 rival, the Dayton editor James M. Cox.

158. R. N. Hill: *Contrary Country: a Chronicle of Vermont* (Cambridge 1950); see also Vermont Historical Society: *Essays in the Social and Economic History of Vermont* (1943).

159. E. C. Latham: *Meet Calvin Coolidge* (New York 1960).

160. C. M. Fuess: *Calvin Coolidge* (Boston 1943) for Coolidge's early career.

161. Many of Coolidge's sayings are recorded in Donald R. McCoy: *Calvin Coolidge* (New York 1967).

162. For the Boston police strike see E. M. Herlihy *et al.*: *Fifty Years of Boston* (Boston 1932).

163. C. B. Slemp *et al.*: *Mind of a President: President Coolidge's Views on Public Questions* (Washington DC 1926).

164. *Ibid.* See also R. J. Maddox: 'Keeping Cool with Coolidge,' *Journal of American History*, 53 (1967), 772ff.

165. Slemp, *opus cit.*

166. Cyril Clemens and A. P. Daggett: 'Coolidge's "I Do Not Choose to Run," ' *New England Quarterly*, 18 (1945), 147ff.

167. Robert Sklar (ed.): *The Plastic Age, 1917–1930* (New York 1970); George Soule: *Prosperity Decade: from War to Depression, 1917–29* (New York 1947).

168. Kenneth M. Goode and Harford Powell: *What About Advertising?* (New York 1927); Warren Suzman (ed.): *Culture and Commitment, 1929–45* (New York 1973).

169. David Burner: 'The Election of 1924,' in Schlesinger and Israel, *opus cit.*, iii. See also J. L. Bates: 'The Teapot Dome Scandal and the Election of 1924,' *American Historical Review*, 60 (1955), 303ff.

170. George Boas: 'Rediscovery of America,' *American Quarterly*, 7 (1955), 14ff;

M. D. Davison (ed.): *American Heritage History of Colonial Antiques* (Washington DC 1967); H. and M. Katz: *Museums, USA* (New York 1965).

171. Clemens and Daggett, *opus cit.*

172. For Coolidge's views on the economy, see R. H. Ferrell and H. H. Quint (eds): *Talkative President: Coolidge's Press Conferences* (New York 1929).

173. Stuart Chase: *Prosperity: Fact or Myth?* (New York 1930).

174. Soulin, *opus cit.*

175. Faulkner, *opus cit.*, 622; Walt Rostow, *World Economy*, 209 and Table III–38.

176. Faulkner, *opus cit.*, 624.

177. *Ibid.*, 607–8.

178. Sinclair Lewis: 'Main Street's Been Paved,' *Nation*, September 10, 1924.

179. Sophia Breckenridge: 'The Activities of Women Outside the Home,' in *Recent Social Trends in the US* (New York 1930), 709–50.

180. Lewis Lorwin: *The American Federation of Labor: History, Policies and Prospects* (New York 1933), 279.

181. R. W. Dunn: *The Americanisation of Labor* (New York 1927), 153, 193–4.

PART SEVEN: Superpower America

1. Two useful guides to these events are J. K. Galbraith: *The Great Crash of 1929* (3rd edn Boston 1972) and Murray Rothbard: *America's Great Depression* (New York 1963).

2. For the tariffs see F. W. Taussig, *The Tariff History of the United States.*

3. Rothbard, *opus cit.*, 86.

4. Seymour E. Harriss: *Twenty Years of Federal Reserve Policy* (Cambridge 1933), 91.

5. Rothbard, *opus cit.*, 128ff.

6. *Ibid.*, 139.

7. See Lionel Robbins: *The Great Depression* (New York 1934), 53. The term Strong used was given by Monsieur Rist as 'un coup de whiskey.'

8. Walt Rostow, *World Economy*, Table II–7, 68.

9. Rothbard, *opus cit.*, 157–8.

10. Galbraith, *opus cit.*, 180.

11. A. D. Chandler Jr: *Great Enterprise: Ford, General Motors and the Automobile Industry* (Chicago 1964).

12. J. J. Flink, *The Automobile Age*, 281.

13. Richard S. Tetlow: *New and Improved: the Story of Mass Marketing in America* (London 1990), 76; Flink, *opus cit.*, 287.

14. Samuel Schmalhausen and V. F. Calverton (eds): *Women's Coming of Age: a Symposium* (New York 1931), 536–49.

15. Lester V. Chandler: *Benjamin Strong, Central Banker* (Washington DC 1958).

16. Selma Goldsmith *et al.*: 'Size and Distribution of Income Since the Mid-Thirties,' *Review of Economics and Statistics*, February 1954; Galbraith, *opus cit.*, 181.

17. Walter Bagehot: *Lombard Street* (London 1922 edn), 151.
18. For the collected sayings of 'experts,' see Edward Angly: *Oh Yeah?* (New York 1949). For historic earnings on shares percentages, see William Rees-Mogg, 'Wall Street Will Crash,' *The Times* (London), December 12, 1996.
19. Galbraith, *opus cit.*, 57ff.
20. *Securities and Exchange Commission in the Matter of Richard Whitney, Edwin D. Morgan . . .* (Washington DC 1938).
21. Bagehot, *opus cit.*, 150.
22. Galbraith, *opus cit.*, 83.
23. *Ibid.*, 104ff.
24. *Ibid.*, 147.
25. John P. Diggins: *The Bard of Savagery: Thorsten Veblen and Modern Social Theory* (London 1979).
26. George H. Nash: *The Life of Herbert Hoover*, 2 vols (New York 1983–8).
27. Details of Hoover's mining ventures are given in David Burner: *Herbert Hoover: a Public Life* (New York 1979).
28. Quoted in William Manchester: *The Glory and the Dream: a Narrative History of America 1932–72* (New York 1974), 24.
29. J. M. Keynes: *Economic Consequences of the Peace* (London 1919), 257n.
30. The letter was to Hugh Gibson and Hoover preserved it in his files (Hoover Papers).
31. Herbert Hoover: *Memoirs*, 3 vols (Stanford 1951–2), ii 42–4.
32. *Ibid.*, 41–2.
33. Martin Fasault and George Mazuzan (ed.): *The Hoover Presidency: a Reappraisal* (New York 1974); Murray Benedict: *Farm Policies of the United States* (New York 1953).
34. Murray, *The Harding Era*, 195.
35. Ellis Lawler: 'Herbert Hoover and American Corporatism, 1929–33,' in Fasault and Mazuzan, *opus cit.*
36. Eugene Lyons: *Herbert Hoover: a Biography* (New York 1964), 294.
37. Joan Hoff Wilson: *American Business and Foreign Policy, 1920–33* (Lexington 1971), 220; Donald R. McCoy: 'To the White House,' in Fasault and Mazuzan, *opus cit.*, 55; for Wilson's anti-semitism see David Cronon (ed.): *The Cabinet Diaries of Josephus Daniels, 1913–21* (Lincoln, Nebraska 1963), 131, 267, 497; for FDR's, Walter Trohan: *Political Animals* (New York 1975), 99.
38. L. H. Fuchs: 'The Election of 1928,' in Arthur M. Schlesinger Jr and F. R. Israel (eds), *American Presidential Elections*, iii (New York 1971).
39. Quoted in Galbraith, *opus cit.*, 143.
40. Hoover to J. C. Penney, quoted in Fasault and Mazuzan, *opus cit.*, 52–3.
41. Hoover to General Peyton Marsh at the War Food Administration, quoted in Arthur Schlesinger Jr, *The Crisis of the Old Order*, 8.
42. Rothbard, *opus cit.*, 187.
43. Hoover, *opus cit.*, ii 108.
44. *Ibid.*, iii 295.

45. *American Federation*, January–March 1930.
46. Roy Harrod: *Life of John Maynard Keynes* (Cambridge 1950), 47–8.
47. Galbraith, *opus cit.*, 142.
48. Rothbard, *opus cit.*, 233–4.
49. Quoted in *Reader's Companion to American History*, 514. See also Joan Hoff Wilson: *Herbert Hoover: Forgotten Progressive* (New York 1974).
50. J. A. Schwartz: *Interregnum of Despair: Hoover, Congress and the Depression* (New York 1970).
51. Rothbard, *opus cit.*, 187.
52. G. D. Nash: 'Herbert Hoover and the Reconstruction Finance Corporation,' *Mississippi Valley Historical Review*, 46 (1959), 455ff.
53. Rothbard, *opus cit.*, 268.
54. *Ibid.*, 291.
55. Rostow, *opus cit.*, Table III–42, 220.
56. *Fortune*, September 1932.
57. Manchester, *opus cit.*, 40–1.
58. C. J. Enzler: *Some Social Aspects of the Depression* (Washington DC 1939), Chapter 4.
59. Ekirch, *opus cit.*, 28–9.
60. Don Congdon (ed.): *The Thirties: a Time to Remember* (New York 1962), 24.
61. Edmund Wilson: 'The Literary Consequences of the Crash,' in *The Shores of Light* (New York 1952), 498.
62. *Harper's*, December 1931.
63. Charles Abba, *Business Week*, June 24, 1931.
64. Fasault and Mazuzan, *opus cit.*, 10.
65. Quoted in *ibid.*, 80.
66. *Ibid.*, 91, 92; Wells, *An Experiment in Autobiography* (London 1934).
67. Roger Daniels: *The Bonus March: an Episode of the Great Depression* (Westport, Connecticut 1971).
68. For an objective account see D. J. Lision: *The President and Protest: Hoover, Conspiracy and the Bonus Riot* (Columbia, Missouri 1974), esp. 254ff; Louis Liebovich: *Press Reaction to the Bonus March of 1932* (Columbia, South Carolina 1990). See Daniel, *opus cit.*, Chapter Ten, 'The Bonus March as Myth.'
69. Theodore Joslin: *Hoover Off the Record* (New York 1934).
70. Frank Freidel, 'The Election of 1932,' in Schlesinger and Israel, *opus cit.*, iii; R. V. Peel and T. C. Donnelly: *The 1932 Campaign* (New York 1935).
71. Mario Ehandi: *The Roosevelt Revolution* (New York 1959).
72. Quoted in Ekirch, *opus cit.*
73. For the Roosevelts, see Nathan Miller: *The Roosevelt Chronicles* (New York 1979) and Peter Collier: *The Roosevelts* (New York 1995).
74. Ted Morgan: *FDR: a Biography* (New York 1985), 123–4.
75. *Ibid.*, 39ff.
76. For this marriage see Blanche Wiesen Cook: *Eleanor Roosevelt*, vol. I: *1884–1933* (New York 1992), 125ff.

77. Morgan, *opus cit.*, 49.

78. *Ibid.*, 136–7; Rita Halle Kleeman: *Young Franklin Roosevelt* (New York 1946).

79. Quoted in *ibid.*, 238.

80. Morgan, *opus cit.*, lists a number of them, e.g. 175, 176, 179, 252–3, 299.

81. *Ibid.*, Chapter IX, 'The Newport Scandal,' 257ff.

82. Jean Gould: *A Good Fight: The Story of FDR's Conquest of Polio* (New York 1960); Theo Lippmann Jr: *The Squire of Warm Springs: FDR in Georgia 1924–45* (New York 1977).

83. Described in detail in Cook, *opus cit.*, i.

84. Quoted in Ronald Steel: *Walter Lippmann and the American Century* (Boston 1980).

85. Morgan, *opus cit.*, 343.

86. Trohan, *opus cit.*, 83–4.

87. James McGregor Burns: *Roosevelt, the Lion and the Fox* (New York 1956), 148–9.

88. Raymond Moley: *After Seven Years* (New York 1939), 151.

89. B. N. Timmons: *Jesse H. Jones: the Man and the Statesman* (New York 1956).

90. See Jones' own account: *Fifty Billion Dollars: My Thirteen Years with the RFC, 1932–45* (New York 1951).

91. Jordan A. Schwartz: *The New Dealers: Power Politics in the Age of Roosevelt* (New York 1993), 70.

92. *Ibid.*, 73.

93. Sometimes rendered as 'not worth a pitcher of warm piss.' For the FDIC deal see *ibid.*, 74.

94. For Garner see B. N. Timmons: *Garner of Texas* (New York 1948).

95. See Turner Catledge: 'New Deal's Man of Many Jobs,' *New York Times Magazine*, October 6, 1940.

96. Cf. press conferences of March 4 and April 19 and 26, 1933.

97. Cf. letter of Joseph Daniels, March 27, 1933, in Elliot Roosevelt (ed.): *FDR: His Personal Letters*, 4 vols (New York 1947–50), i 339–40; Burns, *opus cit.*, 167, 172.

98. Morgan, *opus cit.*, 439–40, 462.

99. H. U. Faulkner, *American Economic History*, 656ff.

100. Arthur M. Schlesinger: *The Coming of the New Deal* (Boston 1958), 153; Manchester, *opus cit.*, 89; Leverett S. Lyon *et al.*: *The National Recovery Administration* (Washington DC 1935).

101. P. J. Hubbard: *Origins of the TVA* (New York 1961); G. B. Tyndall: *The Emergence of the New South, 1913–45* (Baton Rouge 1967); Judson King: *The Conservation Fight: from Theodore Roosevelt to the TVA* (Washington DC 1959); A. V. Morgan: *The Making of the TVA* (Buffalo 1974).

102. Walter Lippmann: 'The Permanent New Deal,' *Yale Review*, 24 (1935), 649–67.

103. Broadus Mitchell *et al.*: *Depression Decade* (New York 1947).

104. For example William Myers and Walter Newton: *The Hoover Administration:*

a Documented Narrative (New York 1936); see also David Burner: *Herbert Hoover: a Public Life* (New York 1979).

105. Francis Still Wickware in *Fortune*, January 1940; *Economic Indicators: Historical and Descriptive Supplement, Joint Committee on the Economic Report* (Washington DC 1953); Galbraith, *opus cit.*, 173; Rostow, *opus cit.*, Table III–42.

106. Morgan, *opus cit.*, 549.

107. *Ibid.*, 473, 613–5, 621.

108. Nelson D. Lankford: *The Last American Aristocrat: the Biography of Ambassador David K. E. Bruce* (Boston 1996), 92ff. See also Burton Hersh: *The Mellon Family: a Fortune in History* (New York 1978).

109. Trohan, *opus cit.*, 59ff, 67–8, 115.

110. For Miller see Cook, *opus cit.*, i, 429–42.

111. For Cook, see *ibid.*, 318–23, etc.

112. For Hickok, see *ibid.*, 458–92; Morgan, *opus cit.*, 498.

113. For Hoover and Mrs Roosevelt, see Curt Gentry: *J. Edgar Hoover: the Man and the Secrets* (New York 1991), 299–306, etc. This is a work to be used with care.

114. See 'The Hullabaloo Over the Brains Trust,' *Literary Review*, cxv (1933); Bernard Sternsher: *Rexford Tugwell and the New Deal* (New Brunswick 1964), 114–15; Otis Graham: 'Historians and the New Deal,' *Social Studies*, April 1963.

115. Manchester, *opus cit.*, 84.

116. Morgan, *opus cit.*, 615–18.

117. The best edition of Mencken's magnum opus is Raven I. McDavid Jr (ed.): *The American Language* (New York 1963). See also Alistair Cooke (ed.): *The Vintage Mencken* (New York 1990 edn) and Charles A. Fecher: *The Diary of H. L. Mencken* (New York 1989). A collection of his newspaper articles, *A Mencken Chrestomathy* (New York 1948) is still in print.

118. Quoted in Charles Fecher: *Mencken: a Study in His Thought* (New York 1978).

119. Epithets listed in Fecher, *opus cit.*, 179n.; see also Fred C. Hobson Jr: *Mencken: a Life* (New York 1993).

120. Collected in George Wolfskill and John Hudson: *All But the People: President Roosevelt and His Critics* (New York 1969), 5–16.

121. Elizabeth Howell (ed.): *The Letters of Thomas Wolfe* (New York 1956), 551ff.

122. W. E. Leuchtenberg: 'The Election of 1936,' in Schlesinger and Israel, *opus cit.*, iii; H. F. Gosnell: *Champion Campaigner: Franklin D. Roosevelt* (New York 1952).

123. R. W. Burke, 'The Election of 1940,' and Leon Friedman, 'The Election of 1944,' are both in Schlesinger and Israel, *opus cit.*, iv. See also Herbert S. Parmet and Marie B. Hecht: *Never Again: a President Runs for a Third Term* (New York 1968).

124. See Michael B. Katz: *In the Shadow of the Poorhouse: a Social History of Welfare in America* (New York 1989).

125. Blanche D. Coll: *Perspectives in Public Welfare: a History* (New York 1969).

126. Linda Gordon: *Pitied but Not Entitled: Single Mothers and the History of Welfare* (New York 1994); S. McLanahan and G. Sandefur: *Growing Up with a Single Parent* (Cambridge 1990). For criticism see Charles Murray: *Losing Ground: American Social Policy, 1950–80* (New York 1984).

127. For food stamps see James T. Patterson: *America's Struggle Against Poverty, 1900–1980* (New York 1981) and M. B. Katz: *The Undeserving Poor* (New York 1989).

128. For the rising cost of medical care, see Rosemary Stevens: *In Sickness and in Wealth* (New York 1988).

129. For the original ideas of welfare under Roosevelt, see Paul K. Conkin: *FDR and the Origins of the Welfare State* (New York 1967) and T. H. Greer: *What Roosevelt Thought: Social and Political Ideas of Franklin D. Roosevelt* (New York 1958).

130. Liva Baker: *Felix Frankfurter* (New York 1960); Max Freedman (ed.): *Franklin D. Roosevelt and Felix Frankfurter: Correspondence, 1928–45* (New York 1967).

131. Schwartz, *opus cit.*, 126ff.

132. E. O. Smigel: *The Wall Street Lawyer* (New York 1968).

133. Morgan, *opus cit.*, 519–28.

134. *Ibid.*, 688. Roberta Wohlstetter: *Pearl Harbor: Warning and Decision* (New York 1980).

135. Akira Ariye: *Across the Pacific: an Inner History of American–East Asian Relations* (New York 1967).

136. For America's China policy, see Michael H. Hunt: *The Making of a Special Relationship: the United States and China to 1914* (New York 1983).

137. O. J. Clinard: *Japan's Influence on American Naval Power* (New York 1947).

138. Manchester, *opus cit.*, 7.

139. Manfred Jonas: *Isolationism in America, 1935–41* (New York 1966).

140. For Detzer, see David Fromkin: *In the Times of the Americans* (New York 1995).

141. Wayne S. Cole: *Senator Gerald P. Nye and American Foreign Relations* (New York 1962); John E. Wiltz: *In Search of Peace: the Senate Munitions Inquiry, 1934–36* (New York 1963).

142. James M. Seavey: *The Neutrality Legislation* (New York 1939).

143. J. A. Schwartz: *The Speculator: Bernard M. Baruch in Washington, 1917–65* (New York 1981).

144. W. F. Kimball: *The Most Unsordid Act: Lend–Lease, 1939–41* (New York 1969).

145. For views on the origins of World War Two see W. S. Cole: 'American Entry into World War Two,' *Mississippi Valley Historical Review*, 43 (1957), 595ff; Pierre Renouvin: *World War Two and Its Origins: International Relations, 1929–45* (New York 1969); Laurence Lafore: *The End of Glory: an Interpretation of the Origins of World War Two* (New York 1969).

146. For details see John Toland: *Infamy* (New York 1982).

147. J. W. Pratt: *Cordell Hull, 1933–44* (New York 1964).

148. Masatake Okumiya: *Midway: the Battle That Doomed Japan* (Annapolis 1955).

149. See D. Novik and G. A. Steiner: *Wartime Industrial Statistics* (Washington DC 1949).

150. L. P. Adams: *Wartime Manpower Mobilisation* (New York 1951).

151. Toland, *opus cit.*, 327.

152. Susman, *opus cit.*

153. Charles Murphy: 'The Earth-Movers Organise for War,' *Fortune*, August–October 1943.

154. Tom Lilley *et al.*: *Problems of Accelerating Aircraft Production* (New York 1947).

155. See Gilbert Burck, *Fortune*, March 1943.

156. Toland, *opus cit.*, 426.

157. Edward Van Der Rhoer: *Deadly Magic: a Personal Account of Communications Intelligence in World War Two in the Pacific* (New York 1978); W. J. Holmes: *Double-Edged Secrets: US Naval Intelligence Operations in the Pacific During World War Two* (Annapolis 1979).

158. See Ralph Bennett: 'Ultra and Some Command Decisions,' and Vice-Admiral B. B. Schofield, 'The Defeat of the U-Boats During World War Two,' *Journal of Contemporary History*, 16 (1981), 131–51, 119–29.

159. John Masterman: *The Double-Cross System in the War of 1939–45* (New Haven 1972).

160. Burke Davis: *Get Yamamoto* (New York 1969).

161. Ronald Powaski: *March to Armageddon: the United States and the Nuclear Arms Race, 1939 to the Present* (New York 1987).

162. M. J. Sherwin: *A World Destroyed: Hiroshima and the Origins of the Arms Race* (New York 1987); M. Kasku and J. Trainer: *Nuclear Power: Both Sides* (New York 1982).

163. Forest C. Pogue: *George C. Marshall*, 4 vols (New York 1963–87); Marshall's papers, edited by Laddy Bland, with the first two volumes issued in 1981, continue.

164. Stephen E. Ambrose: *Eisenhower, Soldier, General of the Army and President Elect, 1890–1952* (New York 1983).

165. Martin Blumenson (ed.): *The Papers of General George S. Patton, 1885–1940* (New York 1971) covers his early career.

166. See Thomas Buell: *Master of Sea Power* (New York 1979) and King's own account, *Fleet Admiral King, a Naval Record* (New York 1952).

167. E. B. Potter: *Nimitz* (New York 1976); Edwin Hoyt: *How They Won the War in the Pacific: Nimitz and His Admirals* (New York 1970).

168. D. C. James: *The Years of MacArthur*, 3 vols (New York 1970–85).

169. Stephen E. Ambrose: *Eisenhower and Berlin: the Decision to Halt at the Elbe* (New York 1967).

170. Paul Hollander: *Political Pilgrims: Travels of Western Intellectuals to the Soviet Union, China and Cuba, 1928–78* (Oxford 1981).

171. Malcolm Muggeridge: *Chronicles of Wasted Time* (London 1960), 254–5; Duranty's own account is *The Kremlin and the People* (New York 1941).
172. Charles Bohlen: *Witness to History, 1919–69* (New York 1973), 26–9.
173. Robert Sherwood: *Roosevelt and Hopkins*, 2 vols (New York 1950), i 387–423; Winston Churchill: *Wartime Correspondence* (London 1960), 196ff.
174. Cairo Conference, 1943. Quoted in Terry Anderson, *The United States, Great Britain and the Cold War, 1944–47* (New York 1981), 4.
175. Poole, *opus cit.*, 130.
176. Lord Moran: *Churchill: the Struggle for Survival, 1940–44* (London 1968), 154.
177. Anderson, *opus cit.*, 5.
178. Averell Harriman and Elie Abel: *Special Envoy to Churchill and Stalin, 1941–1946* (New York 1975), 390; Winston Churchill: *The Second World War* vi (London 1950) 337.
179. Joseph Schechtman: *The US and the Jewish State Movement* (New York 1966), 110.
180. Quoted in Alfred Steinberg: *The Man from Missouri: the Life and Times of Harry S. Truman* (New York 1952), 301.
181. The most up-to-date biography of Truman is Alonzo L. Hamby: *Man of the People: a Life of Harry S. Truman* (Oxford 1996).
182. Theodore A. Brown: *Politics of Reform: Kansas City's Municipal Government, 1925–1950* (New York 1958); D. D. March: *A History of Missouri* 4 vols (New York 1967).
183. Hamby, *opus cit.*, 198–9. The *New York Times* called him 'a rube from Pendergast Land:' *New York Times*, December 19, 1934.
184. Hamby, *opus cit.*, 6–7.
185. Quoted in *ibid.*
186. Quoted in *ibid*, 158–9.
187. See Lyle W. Dorsett: *The Pendergast Machine* (New York 1968); Brown, *opus cit.*
188. See Richard Lawrence Miller: *Truman: the Rise to Power* (New York 1986).
189. Hamby, *opus cit.*, 232–46.
190. *Time*, March 8, 1943; *Look*, May 16, 1944.
191. H. Rompf: *The Bombing of Germany* (London 1963), 164.
192. Toland, *opus cit.*, 469ff.
193. James, *opus cit.*, 246ff.
194. Lansing Lamont: *Day of Trinity* (New York 1965), 235.
195. For the bomb decision see Martin Sherwin: *A World Destroyed: the Atomic Bomb and the Grand Alliance* (New York 1975), Chapter 8.
196. Toland, *opus cit.*, 756.
197. There is some dispute over the figures of those killed on the day the bomb was dropped and who subsequently died from radiation and wounds. Toland, *opus cit.*, 790n., gives the calculations of Professor Shogo Nakaoka, First Curator of the Peace Memorial in Hiroshima, that about 100,000 died on August 6, and another 100,000 subsequently.

198. Herbert Feis: *The Atomic Bomb and the End of World War Two* (New York 1966); Lester Brooks: *Behind Japan's Surrender* (New York 1968).

199. Hamby, *opus cit.*, 270.

200. Harry S. Truman: *Memoirs*, 2 vols (New York 1995–6), i 81–2.

201. Omar Bradley: *A Soldier's Story* (New York 1951), 535–6; Forrest Pogue: *George C. Marshall: Organiser of Victory* (New York 1973), 573–4.

202. Thomas Campbell and George Herring (eds): *The Diaries of Edward Stettinius Jr, 1943–46* (New York 1975), 177–8.

203. Anderson, *opus cit.*, 69; Hamby, *opus cit.*, 331.

204. Walter Mellis (ed.): *The Forrestal Diaries* (New York 1951), 38–40, 57.

205. Anderson, *opus cit.*, 75–6.

206. David Robertson: *Sly and Able: a Political Biography of James F. Byrnes* (New York 1994), 445; Bohlen, *opus cit.*, 263.

207. Patricia Dawson Ward: *The Threat of Peace: James F. Byrnes and the Council of Foreign Ministers, 1945–6* (Kent, Ohio 1979): Yergin, *opus cit.*, 160–1; George Curry: 'James F. Byrnes,' in R. H. Ferrell and S. F. Bemiss (eds): *The American Secretaries of State and Their Diplomacy* (New York 1965); Hamby, *opus cit.*, 338–9.

208. George Kennan: *Memoirs, 1925–50* (New York 1952), 294.

209. Robert Rhodes James (ed.): *Churchill: Complete Speeches* (London 1974), vii 7283–96; J. K. Ward: 'Winston Churchill and the Iron Curtain Speech,' *History Teacher*, January 1968.

210. *Leahy Diaries.*

211. Ernest May: 'Cold War and Defense,' in K. Nelson and R. Haycock (eds): *Cold War and Defense* (New York 1990).

212. Quoted in Donald W. White: *The American Century: the Rise and Decline of the United States as a World Power* (New Haven 1996), 21.

213. Alan Wolfe: *America's Impasse: the Rise and Fall of the Politics of Growth* (New York 1981), 155.

214. US Department of Commerce: *Historical Statistics of the United States* (Washington DC 1975); *Economic Report of the President* (Washington DC 1989); *Revised Annual Estimates of American GNP, 1789–1889* (New York 1978).

215. For debt in relation to GNP see the chart, 'Ratio of the Gross Federal Debt to GNP 1791–1988,' in Foner and Garraty, *Reader's Companion to American History*, 774.

216. W. Ashworth: *A Short History of the International Economy Since 1850* (London 1975), 268.

217. Yergin, *opus cit.*, 281–2; Dean Acheson: *Present at the Creation* (New York 1969), 219.

218. For the text of the speech see *Public Papers of the Presidents of the United States: Harry S. Truman, 1945–53* (Washington DC 1961–6), 1947, 178–9.

219. Two books which deal with the prosopography of American postwar internationalism are David Fromkin: *In the Time of the Americans: FDR, Truman, Eisenhower, Marshall, MacArthur: The Generation That Changed America's*

Role in the World (New York 1995) and W. Isaacson and E. Thomas: *The Wise Men: Six Friends and the World They Made* (New York 1986).

220. R. H. Ferrell: *Off the Record* (New York 1980), 108–9; Hamby, *opus cit.*, 388.

221. Isaacson and Thomas, *opus cit.*, 402.

222. Melvyn Loffler: *A Preponderance of Power: National Security, the Truman Administration and the Cold War* (Stanford 1992).

223. Michael Hogan: *The Marshall Plan: America, Britain and the Reconstruction of Western Europe, 1947–52* (New York 1987); see also Wilson Miscamble: *George Kennan and the Making of American Foreign Policy, 1947–50* (Princeton 1992).

224. Truman's diary is in his Post-Presidential Papers, Memoirs File; see Hamby, *opus cit.*, 444; F. C. Pogue: *George C. Marshall, Statesman, 1945–59* (New York 1987), 312.

225. D. A. Rosenberg: 'American Atomic Strategy and the Hydrogen Bomb Decision,' *Journal of American History*, June 1979.

226. *Forrestal Diaries*, 460f; W. Phillips Davison: *The Berlin Blockade* (Princeton 1958).

227. *New York Times*, September 18, 1948; Alonso L. Hamby: *Beyond the New Deal: Harry S. Truman and American Liberalism* (New York 1973), 224ff; Irwin Ross: *The Loneliest Campaign: the Truman Victory of 1948* (New York 1968).

228. Robert A. Donovan: *Conflict and Crisis: the Presidency of Harry S. Truman, 1945–48* (New York 1977), 425; see also H. E. Alexander: 'Financing Presidential Campaigns,' in Schlesinger and Israel, *opus cit.*, iv.

229. John J. McCloy: *Atlantic Alliance: Origins and Future* (New York 1969).

230. John Gaddis: *Strategies of Containment: a Critical Appraisal of Postwar National Security Policy* (New York 1982), 82.

231. Anderson, *opus cit.*, 184.

232. A. A. Jordan: *Foreign Aid and the Defense of South-East Asia* (New York 1962).

233. Hamby, *Man of the People*, 415–16.

234. *Forrestal Diaries*, 324, 344, 348.

235. *Petroleum Times*, June 1948.

236. Nadaf Safran: *The United States and Israel* (New York 1963).

237. White, *opus cit.*, 206–7.

238. George Woodbridge (compiler): *UNRRA*, 3 vols (New York 1950).

239. Truman, *opus cit.*, ii 231ff; *Mr President* 225ff, 249ff; *Truman Public Papers*, 1949, 119–20; 1950, 467ff.

240. David M. Potter: *People of Plenty: Economic Abundance and the American Character* (Chicago 1954), 139; *Fortune*, February 1950.

241. US Department of Commerce, *Historical Statistics*, 872ff; *Statistical Abstract*, 1979, 853ff.

242. See John Lewis Gaddis: 'Containment: a Reassessment,' *Foreign Affairs*, July 1977; George Kennan, 'Containment Reconsidered,' *Foreign Affairs*, April 1978.

243. Soon Sung Cho: *Korea, 1940–50: an Evaluation of American Responsibility* (New York 1968).

244. Robert Alan Arthur, 'The Wit and Sass of Harry S. Truman,' *Esquire*, August 1971, 66.   Truman, *opus cit.*, ii 331ff.

245. Burton Kaufman, *The Korean War* (Philadelphia 1986), Chapter One, on the decision to intervene. See also Glenn D. Paige: *Korean Decision, June 24–30, 1950* (New York 1968).

246. L. P. Leffler: *A Preponderance of Power* (Stanford 1992), 306–8, 369–70.

247. *Foreign Relations of the United States*, 1950, 7: 826.

248. Merle Miller: *Plain Speaking: an Oral Biography of Harry S. Truman* (Berkeley 1974), 329.

249. *Truman Diary*, April 10, 1951; Hillman, *Mr President*, 11.

250. John Spanier: *The Truman–MacArthur Controversy and the Korean War* (New York 1965).

251. For a verdict on Truman's handling of Korea, see Michael Lacey (ed.): *The Truman Presidency* (New York 1989).

252. *St Louis Post-Dispatch*, December 17, 1951.

253. P. J. Boller: *Presidential Wives* (New York 1988), 324.

254. Letter given in full in Hamby, *Man of the People*, 478.

255. B. J. Bernstein, 'The Election of 1952,' in Schlesinger and Israel, *opus cit.*, iv; and Malcolm Moos: 'The Election of 1956,' in *idem*, iv. See also S. G. Brown: *Conscience in Politics: Adlai Stevenson* (Chicago 1961) and Heinz Eulau: *Class and Party in the Eisenhower Years* (New York 1962).

256. Quoted in Hamby, *Man of the People*, 464.

257. 'The most devious man'—Nixon to the author, 1984; Richard Nixon: *Six Crises* (New York 1962), 161.

258. In *Life*, January 16, 1956.

259. See the essay by R. D. Challener: 'John Foster Dulles: Theorist/Practitioner,' in L. Carl Brown (ed.): *Centerstage: American Diplomacy Since World War Two* (New York 1990).

260. Trohan, *opus cit.*, 292.

261. Emmet John Hughes: *Ordeal of Power: a Political Memoir of the Eisenhower Years* (New York 1963), 329f.

262. Sherman Adams: *First Hand Report* (New York 1961), 73.

263. Trohan, *opus cit.*, 111.

264. Kennan, *Memoirs, 1950–63*, 196.

265. Vernon A. Walters: *Silent Missions* (New York 1978), 226.

266. For Eisenhower's activism see Fred I. Greenstein: 'Eisenhower as an Activist President: a Look at New Evidence,' *Political Science Quarterly*, Winter 1979–80.   See also Stephen E. Ambrose: *Eisenhower the President* (New York 1984).

267. *Public Papers of Dwight D. Eisenhower 1954* (Washington DC 1960), 253, 206; Robert A. Divine: *Eisenhower and the Cold War* (Oxford 1981).

268. Joseph B. Smith: *Portrait of a Cold Warrior* (New York 1976), 229–40.

269. *New York Times*, January 18, 1961.

270. Sherman Adams, *opus cit.*, Chapter 17, 360ff.

271. Arthur Larsen: *Eisenhower: the President That Nobody Knew* (New York 1968), 34.

272. See Eisenhower's own account, *Waging Peace* (New York 1965), and William Pickett: *Dwight D. Eisenhower and American Power* (Wheeling, Illinois 1995).

273. Kennan, *Memoirs, 1950–63*, 191–2.

274. Alan Harper: *The Politics of Loyalty* (New York 1969).

275. R. Radosh and J. Milton: *The Rosenberg File: a Search for the Truth* (New York 1983).

276. T. C. Reeves: *The Life and Times of Joe McCarthy: a Biography* (New York 1982), 224.

277. Richard Gild Powers: *Not Without Honor: the History of American Anti-Communism* (New York 1995), 239–40.

278. Richard Hofstadter: 'The Pseudo Conservative Revolt,' in Daniel Bell (ed.): *The Radical Right* (New York 1964) and *The Paranoid Style in American Politics and Other Essays* (Chicago 1964).

279. See Ellen Schrecker: *The Age of McCarthyism: a Brief History with Documents* (New York 1994) and *No Ivory Towers: McCarthyism and the Universities* (New York 1986). See also David Caute: *The Great Fear: the Anti-Communist Purge Under Truman and Eisenhower* (New York 1978).

280. Reeves, *opus cit.*, 474.

281. John Steele Gordon: 'The Ordeal of Engine Charlie,' *American Heritage*, February–March 1995.

282. C. Wright Mills: *White Collar* (New York 1951), xv. See Richard Gillam: 'White Collar from Start to Finish: C. Wright Mills in Transition,' *Theory and Society*, 10 (1981), 1–30.

283. Rick Tilman: *C. Wright Mills: An American Radical and His Intellectual Roots* (Philadelphia 1984).

284. Richard Pells: *The Liberal Mind in a Conservative Age: American Intellectuals in the 1940s and 1950s* (New York 1985), 232–48.

285. Donald Meyer: *The Positive Thinkers: Religion as Pop Psychology from Mary Baker Eddy to Oral Roberts* (New York 1980), 258ff.

286. For Sheen's work see George G. Marlin (ed.): *The Quotable Fulton Sheen* (New York 1989); for Graham see William McLoughlin: *Billy Graham: Revivalist in a Secular Age* (New York 1960).

287. Roy E. Eckhardt: *The Surge of Piety in America: an Appraisal by the Yearbook of the American Churches* (New York 1958).

288. Dwight D. Eisenhower in *The Christian Century* 71 (1954).

289. This story first surfaced in Merle Miller: *Plain Speaking: an Oral Biography of Harry S. Truman* (New York 1973), published after Truman's death and based on interviews in 1961–2. The story, if true, had been suppressed in Kay Summersby's memoirs, *Ike Was My Boss* (New York 1948), but was now told, with a good deal of circumstantial detail (including the information that Eisenhower was impotent and that their affair was never consummated) in a

new book by Summersby, *Past Forgetting: My Love Affair with Dwight D. Eisenhower* (New York 1976). There is also a good deal of evidence that both the 'affair' and the divorce plan were inventions. See Lester and Irene David: *Ike and Mamie: the Story of the General and His Lady* (New York 1981); John Eisenhower (ed.): *Dwight D. Eisenhower: Letters to Mamie* (New York 1978). Paul Boller, *Presidential Wives*, 339–41, dismissed the stories.

290. R. M. Morantz: 'The Scientist as Sex Crusader: Alfred C. Kinsey and American Culture,' *American Quarterly*, 29 (Winter 1979).

291. Peter Biskind: *Seeing Is Believing: How Hollywood Taught Us to Stop Worrying and Love the Fifties* (New York 1983), 44ff.

292. Erik Barnouw: *Tube of Plenty: the Evolution of American Television* (New York 1982).

PART EIGHT: Problem-Solving, Problem-Creating America

1. For an overview of Sixties atmosphere, see Morris Dickstein: *Gates of Eden: American Culture in the Sixties* (New York 1977).

2. For Nixon's vice-presidential travels, see Jonathan Aitken: *Nixon: a Life* (London 1993), 225–65.

3. Paul Tillett (ed.): *Inside Politics: National Conventions, 1960* (New York 1962).

4. Quoted in Aitken, *opus cit.*, 269.

5. *Ibid.*, 181–5. Douglas' tone is reflected in her memoirs, *A Full Life* (New York 1982).

6. Aitken, *opus cit.*, 160–74, 175–8, 180.

7. Nigel Hamilton: *JFK: Life and Death of an American President*, vol. i: Reckless Youth (New York 1992), 5–17, 19–24; Thomas C. Reeves: *A Question of Character: A Life of John F. Kennedy* (New York 1991), 17–25; D. K. Goodwin: *The Fitzgeralds and the Kennedys* (New York 1987).

8. David E. Koskoff: *Joseph P. Kennedy: A Life and Times* (Englewood Cliffs, 1974).

9. Reeves, *opus cit.*, 88.

10. *Ibid.*, 115.

11. *Ibid.*, 36–7; Hamilton, *opus cit.*, 83, 507.

12. Reeves, *opus cit.*, 8, 9, 75–6, 92–4, 97, 118, 122–3, 126, 171, 213; Hamilton, *opus cit.*, 793–4.

13. Reeves, *opus cit.*, 295–6; C. David Heymann: *A Woman Named Jackie* (New York 1980), 308, 311, 312–14. (This last book must be handled with care.)

14. Reeves, *opus cit.*, 48–54, 74, 431 nn. 83–5, for sources.

15. The question is from Herbert S. Parmet: *Jack: the Struggles of John F. Kennedy* (New York 1980), 320–33; Reeves, *opus cit.*, 127–8; Drew Pearson: *Diaries, 1949–59* (New York 1974), 407, 420–1.

16. Reeves, *opus cit.*, 55, 63, 66, 73; R. J. Bulkly Jr; *At Close Quarters: PT Boats in the US Navy* (Washington DC 1962), 120–8; Parmet, *opus cit.*, 111–21.

17. Thomas P. O'Neill: *Man of the House: the Life and Political Memories of Speaker Tip O'Neill* (New York 1987), 76; Joan and Clay Blair Jr: *The Search for JFK* (New York 1976), 484–6; Reeves, *opus cit.*, 81–2.

18. Richard J. Whalen: *The Founding Father: the Story of Joseph P. Kennedy* (New York 1966), 419ff; Parmet, *opus cit.*, 242ff; Edwin O. Guthman and Jeffrey Shulman (eds): *Robert Kennedy in His Own Words: the Unpublished Recollections of the Kennedy Years* (New York 1988), 444.

19. Reeves, *opus cit.*

20. O'Neill, *opus cit.*, 45.

21. Macmillan to the author, 1961.

22. Reeves, *opus cit.*, 161–70; Antoinette Giancan and Thomas C. Renner: *Mafia Princess: Growing Up in Sam Giancana's Family* (New York 1985), 278ff, 309ff; Kitty Kelley: 'The Dark Side of Camelot,' *People*, February 29, 1988, 109–11, article based on interview with Judith Campbell Exner, mistress of both Kennedy and Giancana, while she was ill with terminal cancer and anxious to say certain things, 'so that I can die peacefully.' This text should be used with care.

23. Theodore C. Sorensen: 'The Election of 1960,' in Arthur M. Schlesinger Jr and F. R. Israel (eds), *American Presidential Elections*, iv; Theodore H. White: *The Making of the President, 1960* (New York 1961); Lucy S. Dawidowicz and L. J. Goldstein: *Politics in a Pluralist Democracy: Voting in 1960* (New York 1963).

24. Reeves, *opus cit.*, 249; see note on sources, 462 n. 12.

25. White, *opus cit.*, 336–7.

26. *Ibid.*, 288f; Sidney Kraus (ed.): *The Great Debates: Background—Perspective—Effects* (Bloomington, 1962); Richard M. Nixon, *Six Crises*, 340ff; Aitken, *opus cit.*, 277–8, and 597 n. 10 for sources.

27. Interview with Flanigan, quoted in Aitken, *opus cit.*, 280.

28. Earl Mazo and Stephen Hess: *Nixon: a Political Portrait* (New York 1971), 248; Aitken, *opus cit.*, 290.

29. Mazo and Hess, *opus cit.*, 248; Aitken, *opus cit.*, 290; Reeves, *opus cit.*, 213–14 and other sources, 455nn. 110–11; Goodwin, *opus cit.*, 805ff.

30. Judith Campbell Exner: *My Story* (New York 1977), 194–6; Kelley, *People*, February 29, 1988; Davis, *The Kennedys*, 252ff.

31. Aitken, *opus cit.*, 290–1; Mazo and Hess, *Nixon*, 249.

32. There are many books on the Kennedy marriage. See, for example, Christopher Andersen: *Jack and Jackie: Portrait of an American Marriage* (New York 1996) and Edward Kelin: *All Too Human: the Love Story of Jack and Jackie Kennedy* (New York 1966).

33. Noemie Emery: 'JFK: the Great American Novel,' *Weekly Standard* (Washington), October 14, 1996.

34. Reeves, *opus cit.*, 29–30; Gloria Swanson: *Swanson on Swanson* (New York 1980), 306ff, 427, 445f, 457.

35. Reeves, *opus cit.*, 323–7, and 473–4 nn. 48–66 for sources.

36. C. David Heymann: *A Woman Named Jackie* (New York 1989), 242.

37. Reeves, *opus cit.*, 145, and 445 n. 62.

38. Text of the inaugural has often been reprinted, e.g. in William Safire (ed.): *Lend Me Your Ears: Great Speeches in History* (New York 1992), 811–14. The speech was written by Ted Sorensen, after consultations with Adlai Stevenson, John Kenneth Galbraith, Arthur M. Schlesinger Jr, and Walter Lippmann.

39. R. J. Walton: *Cold War and Counter-Revolution: the Foreign Policy of John F. Kennedy* (New York 1972).

40. Eyewitness account by Hugh Sidey: *John F. Kennedy: Portrait of a President* (New York 1964).

41. H. Young *et al.*: *Journey To Tranquillity: the History of Man's Assault on the Moon* (London 1969), 109–10.

42. Walter McDougall: *The Heavens and the Earth: a Political History of the Space Age* (New York 1985).

43. Earl Smith (former US ambassador to Cuba) in testimony to the Senate Judiciary Committee, August 30, 1930. For US–Cuban relations see Hugh Thomas: *Cuba, or the Pursuit of Freedom* (London 1971).

44. Earl Smith: *The Fourth Floor* (New York 1962): Herbert Matthews: *Castro: a Political Biography* (London 1969); Thomas, *Cuba*, 814ff, 946ff, 977ff, 1038–44.

45. The author was present in Cuba in 1960 and observed the final stages of Cuba's transformation into a Communist totalitarian state.

46. Thomas, *Cuba*, 969–70.

47. Arthur Schlesinger Jr: *Robert Kennedy and His Times* (Boston 1978), 452, 445.

48. Trumbull Higgins: *The Perfect Failure: Kennedy, Eisenhower and the CIA at the Bay of Pigs* (New York 1987); John Ranelagh: *The Agency: the Rise and Decline of the CIA* (New York 1986), 381ff.

49. Ambrose, *Eisenhower the President*, 638–9; Reeves, *opus cit.*, 262–75.

50. Schlesinger, *Robert Kennedy* 472; *Reader's Digest*, November 1964.

51. See Schlesinger, *Robert Kennedy*, Chapter 21, and *Alleged Assassination Plots Involving Foreign Leaders* (Washington DC 1975), interim and final reports.

52. Jean Daniel in *L'Express*, December 14, 1963 and *New Republic*, December 21, 1963; Claude Julien: *Le Monde*, March 22, 1963.

53. *Washington Post*, January 14, 1992; Robert McNamara: 'One Minute to Domesday,' *New York Times*, October 14, 1992.

54. For the missile crisis, see Dino Brugioni: *Eyeball to Eyeball: the Inside Story of the Missile Crisis* (New York 1991); Reeves, *opus cit.*, 364–86; Graham Allison: *Essence of Decision: Explaining the Cuban Missile Crisis* (Boston 1971); Theodore Sorensen (ed.): *Robert Kennedy: Thirteen Days, a Memoir of the Missile Crisis* (New York 1969); James Giglio: *The Presidency of John F. Kennedy* (Lawrence 1991); Stephen Ambrose: *Rise to Globalism: American Foreign Policy Since 1938* (New York 1988), 180–200; Michael Beschloss: *The Crisis Years: Kennedy and Khruschev, 1960–63* (New York 1991); Barton Bernstein: 'Reconsidering Khruschev's Gambit: Defending the Soviet Union and Cuba,' *Diplomatic History*, 14 (Spring 1990).

55. See *Boston Globe*, July 28, 1994, quoting from recently released tapes.

56. Beschloss, *Crisis Years*, 542–9; Michael Tatu: *Power in the Kremlin: from Khruschev to Kosygin* (New York 1969), 422.

57. *Newsweek*, October 28, 1963, based on Pentagon sources.

58. Beschloss, *opus cit.*, 543ff.

59. Quoted in Schlesinger, *Robert Kennedy*, 530–1.

60. Quoted in Hollander, *Political Pilgrims*, Chapter 6, esp. 234ff.

61. Paul Hollander: *Anti-Americanism: Critiques at Home and Abroad, 1965–1990* (New York 1992), 39, 127–9, 136–9, 235–43, 292ff, 375ff, etc.

62. Beschloss, *opus cit.*, 670ff; Schlesinger, *Robert Kennedy*, 1020ff.

63. Max Holland: 'After Thirty Years: Making Sense of the Assassination,' *Reviews in American History*, 22 (June 1994).

64. Gerald Posner: *Case Closed: Lee Harvey Oswald and the Assassination of JFK* (New York 1993).

65. Robert Caro: *The Years of Lyndon Johnson: the Path to Power* (New York 1982); *The Years of Lyndon Johnson: Means of Ascent* (New York 1989).

66. Paul Conkin: *Big Daddy from the Pedernales: Lyndon Baines Johnson* (Boston 1986).

67. Johnson to Doris Kearns: *Lyndon Johnson and the American Dream* (New York 1976). See Robert Dallek: 'My Search for Lyndon Johnson,' *American Heritage*, September 1991.

68. Ted Morgan, *FDR*, 615–18.

69. See conflicting testimony on this point in Evelyn Lincoln: *Kennedy and Johnson* (New York 1969), 205, and Benjamin C. Bradlee: *Conversations with Kennedy* (New York 1975), 217–18.

70. See Eric F. Goldman: *The Tragedy of Lyndon Johnson* (New York 1969), 98; C. Vann Woodward (ed.): *Responses of the Presidents to Charges of Misconduct* (New York 1974), 329ff; R. Evans and R. Novak: *Lyndon B. Johnson: the Exercise of Power* (New York 1966), 413–15; *Hearings Before the Select Committee to Study Government Operations with Respect to Intelligence Activities*, US Senate 94th Congress, 1st session, vol. 6 (1975), 728–9. For a good summary of this case, see Victor Lasky: *It Didn't Start with Watergate* (New York 1977), Chapter 11, 127–41.

71. See the speech by Philip M. Crane, *Congressional Record*, July 27, 1973, E–5158–9.

72. Doris Kearns: *Lyndon Johnson and the American Dream* (New York 1976).

73. Quoted in David Halberstam: *The Best and the Brightest* (New York 1972).

74. Joseph Californo: *The Triumph and Tragedy of Lyndon Johnson: the White House Years* (New York 1991), 26ff.

75. A society woman, who had been thus handled, told the author: 'A nip from LBJ was very painful.'

76. Quoted in Halberstam, *opus cit.*

77. Boller, *Presedential Wives*, 387.

78. *Ibid.*, 383. See also Liz Carpenter: *Ruffles and Flourishes* (New York 1970).

79. Information from Miriam Rothschild to the author.

80. J. B. Martin, 'The Election of 1964,' in Schlesinger and Israel, *opus cit.*, iv. For Goldwater's sometimes interesting views, see J. H. Kessel: *The Goldwater Coalition: Republican Strategies in 1964* (New York 1968).

81. James Sundquist: 'The origins of the War on Poverty,' in Sundquist (ed.). *On Fighting Poverty: Perspectives from Experience* (New York 1969); Mark Gelfand: 'The War on Poverty,' in Robert Divine (ed.): *Exploring the Johnson Years* (Austin 1981); Ira Katznelson: 'Was the Great Society a Lost Opportunity?,' in S. Fraser and G. Gerstle (eds): *The Rise and Fall of the New Deal Order, 1930–80* (Princeton 1989), 185ff.

82. *Office of Management and Budget: Federal Government Finances* (Washington DC 1979); for a slightly different calculation, see Walt Rostow, *World Economy*, 272, Table III–65.

83. For example, William O'Neill: *Coming Apart: an Informal History of America in the 1960s* (Chicago 1971); Robert Collins: 'Growth Liberalism in the Sixties: Great Societies at Home and Grand Designs Abroad,' in David Farber (ed.): *The Sixties: from Memory to History* (Chapel Hill 1994); Allan Matusow: *The Unraveling of America: A History of Liberalism in the 1960s* (New York 1984); Daniel Moynihan: *Maximum Feasible Misunderstanding: Community Action in the War Against Poverty* (New York 1967).

84. See the chart in Foner and Garraty (eds): *Reader's Companion to American History*, 774, and the chart in the Special Issue of *Forbes Magazine* on the American Disease, 'Why We Feel So Bad,' September 14, 1992, 180–3; Robert Heilbronner and Peter Bernstein: *The Debt and the Deficit* (New York 1989).

85. Herbert Y. Shandler: *The Unmaking of a President: Lyndon Johnson and Vietnam* (Princeton 1977), 226–9.

86. *Statistical Abstract of the United States, 1994* (Washington DC 1994), 297, 330–3. See James T. Patterson: *Grand Expectations: the United States, 1945–74* (New York 1996), 541 n. 37.

87. George C. Herring: *America's Longest War: the United States and Vietnam, 1950–74* (2nd edn New York 1986); Stanley Karnow: *Vietnam: a History* (New York 1983).

88. For the events which detonated the Boat People phenomenon, see John Barron and Anthony Paul: *Peace with Horror* (London 1977).

89. See Archimedes L. A. Patti: *Why Viet Nam? Prelude to America's Albatross* (Berkeley 1981); but see also Dennis Duncanson, *Times Literary Supplement* (London) August 21, 1981, 965.

90. Acheson, *Present at the Creation*, 675–6.

91. Marylin Young: *The Vietnam Wars, 1945–1990* (New York 1991), 31ff; Townsend Hoopes: *The Devil and John Foster Dulles* (Boston 1973), 220ff.

92. Quoted in Herring, *opus cit.*, 52. Eisenhower, press conferences, April 7, 26, 1954; Leslie H. Gelb and Richard K. Betts: *The Irony of Vietnam: the System Worked* (Washington DC 1979), 59.

93. Eisenhower, *Public Papers*, 1959, 71.

94. De Gaulle, *Memoirs* (Pain 1953), 256.

95. Reeves, *opus cit.*, 283.

96. Schlesinger, *A Thousand Days*, 547.

97. Halberstam, *opus cit.*, 135.

98. Quoted in Henry Graff: *The Tuesday Cabinet: Deliberation and Decision in Peace and War under Lyndon B. Johnson* (New York 1970), 53.

99. Quoted in William Chafe: *The Unfinished Journey: America Since World War Two* (New York 1991), 274ff.

100. Graff, *opus cit.*, 83.

101. Gelb and Betts, *opus cit.*, 117–18; see also Joseph C. Goulden: *Truth Is the First Casualty: the Gulf of Tonkin Affair* (New York 1969), esp. 160.

102. Lyndon Johnson, *Public Papers*, iv 291.

103. Quoted in Halberstam, *opus cit.*, 596.

104. George Herring: 'The War in Vietnam,' in Robert Divine (ed): *Exploring the Johnson Years* (Austin 1981), 27–62; Larry Berman: *Lyndon Johnson's War: the Road to Stalemate in Vietnam* (New York 1989), 12.

105. Gelb and Betts, *opus cit.*, 135ff; Sharp quoted in Patterson, *opus cit.*, 605; see also David Barrett: *Uncertain Warriors: Lyndon Johnson and His Vietnam Advisors* (Lawrence 1993).

106. Kearns, *opus cit.*, 264.

107. Geld and Betts, *opus cit.*, 139–43.

108. See Guenther Lewy: 'Vietnam: New Light on the Question of American Guilt,' *Commentary*, February 1978; but see also Christian Appy: *Working-Class War: America's Combat Soldiers and Vietnam* (Chapel Hill 1993), 16–17, which estimates total Vietnamese deaths 1961–75 at 1.5 to 2 million.

109. Geld and Betts, *opus cit.*, 214–16.

110. *Ibid.*, 120–3.

111. Stephen M. Kohn: *Jailed for Peace: the History of American Draft-Law Violations, 1658–1985* (New York 1986).

112. John Whiteclay Chambers II: *To Raise an Army: The Draft Comes to Modern America* (New York 1987); G. Q. Flynn: *Conscription and American Culture* (New York 1992).

113. Gelb and Betts, *opus cit.*, 171.

114. Peter Braestrup: *Big Story: How the American Press and TV Reported and Interpreted the Crisis of Tet 1968 in Vietnam and Washington*, 2 vols (Boulder 1977).

115. John Mueller: *War, Presidents and Public Opinion* (New York 1973): Gelb and Betts, *opus cit.*, 130.

116. William Lunch and Peter Sperlich: 'American Public Opinion and the War in Vietnam,' *Western Political Quarterly*, Utah, March 1979.

117. Sidney Verba *et al.*: *Vietnam and the Silent Majority* (New York 1970); Stephen Hess: 'Foreign Policy and Presidential Campaigns,' *Foreign Policy*, Autumn 1972.

118. Lyndon Baines Johnson: *The Vantage Point: Perspectives of the Presidency, 1963–69* (New York 1971), 95.

119. Lawrence J. Wittner: *Cold War America: from Hiroshima to Watergate* (New York 1974), 283.

120. *Los Angeles Times*, November 8, 1962.

121. Davis S. Broder: 'The Election of 1968,' in Schlesinger and Israel, *opus cit.*, iv.

122. Wittner, *opus cit.*, 300–1.

123. William Safire: *Before the Fall: an Insider's View of the Pre-Watergate White House* (New York 1975), 70, 75.

124. David Broder, quoted in *ibid.*, 171.

125. Quoted in Arthur Schlesinger: *The Imperial Presidency* (Boston 1973), 123, and Charles Bohlen, *Witness to History*, 210.

126. Thomas Cronin, 'The Textbook Presidency and Political Science,' *Congressional Record*, October 5, 1970; Wilfred Brinkley, *New Republic*, May 18, 1953.

127. *New York Times*, May 18, 1954; *Washington Post*, May 20, 1954; Schlesinger, *Imperial Presidency*, 169.

128. For Nixon's White House, see Aitken, *opus cit.*, 373ff.

129. *Ibid.*, 378ff; Henry Kissinger: *The White House Years* (Boston 1979).

130. Aitken, *opus cit.*, 385–6; Parmet, *opus cit.*, 566ff.

131. Geld and Betts, *opus cit.*, 350.

132. Quoted in Wittner, *opus cit.*, 283.

133. Kissinger, *Diplomacy* (New York 1994), 723.

134. Safire, *Before the Fall*, 375–6.

135. Text of agreement in *State Department Bulletin*, February 12, 1973; Geld and Betts, *opus cit.*, 350.

136. Taylor Branch: *Parting the Waters: America in the King Years, 1954–63* (New York 1988); D. J. Garrow: *Bearing the Cross: Martin Luther King and the Southern Christian Leadership Conference* (New York 1986).

137. For King's speeches etc. see Martin Luther King: *Strength to Love* (New York 1963) and *Where Do We Go from Here?* (New York 1967): the 'I have a dream' speech is in Safire, *Lend Me your Ears*, 497–500.

138. Hugh Davis Graham: *The Civil Rights Era: Origins and Development of National Policy* (New York 1990); S. F. Lawson: *Running for Freedom: Civil Rights and Black Politics in America Since 1941* (New York 1990).

139. George Breitman (ed.): *Malcom X Speaks* (New York 1965); see also Peter Goldman: 'Malcom X,' in *The Dictionary of American Negro Biography* (New York 1982); D. L. Lewis: *King: a Critical Biography* (New York 1970).

140. For the Sixties youth movement see Michael W. Miles: *The Radical Probe: the Logic of Student Rebellion* (New York 1971) and Theodore Roszak: *The Making of a Counterculture* (New York 1969).

141. Nixon, *Memoirs*, 454.

142. Safire, *Before the Fall*, 360.

143. For the election see Theodore White: *The Making of the President, 1972* (New York 1973).

144. Quoted in Safire, *Before the Fall*, 264.

145. Alonzo L. Hamby, *Man of the People*, 584.

146. Charles Roberts: *LBJ's Inner Circle* (New York 1965), 34; Schlesinger, *Imperial Presidency*, 221; see 'The Development of the White House Staff,' *Congressional Record*, June 20, 1972.

147. Richard W. Steele: 'Franklin D. Roosevelt and His Foreign Policy Critics,' *Political Science Quarterly*, Spring 1979, 22 n. 27.

148. Morgan, *opus cit.*, 613, 614, 621, etc. Steele, *opus cit.*, 18; Saul Alindky: *John L. Lewis* (New York, 1970), 238; Safire, *Before the Fall*, 166.

149. Trohan, *opus cit.*, 179; *Daily Telegraph* (London), March 4, 1982.

150. There is argument about the extent to which Kennedy connived in attempts to kill Castro, but the most authoritative account, Thomas Powers: *The Man Who Kept the Secrets: Richard Helms and the CIA* (New York 1980), concludes that JFK knew exactly what the CIA were planning, and encouraged it. For Kennedy and the Diem killing see Reeves, *opus cit.*, 408ff.

151. Schlesinger, *Robert Kennedy*, 403ff; Roger Blough: *The Washington Embrace of Business* (New York 1975).

152. Schlesinger, *Robert Kennedy*, 311–12.

153. Fred Friendly: *The Good Guys, the Bad Guys and the First Amendment* (New York 1976), Chapter 3.

154. Safire, *Before the Fall*, 166.

155. See the *New York Times*, February 5, 1982 and June 24, 1983; Sorensen, *Kennedy*, 295; *Robert Kennedy in his Own Words*, 240; Reeves, *opus cit.*, 260.

156. Schlesinger, *Robert Kennedy*, 362ff; Senate Select Committee on Intelligence Activities (Church Committee), *Final Report* (Washington DC 1976), ii 154, iii 158–60.

157. Trohan, *opus cit.*, 136–7; Morgan, *opus cit.*, 619.

158. Trohan, *opus cit.*, 326; Judith Exner: *My Story* (New York 1977).

159. Alfred Sternberg: *Sam Johnson's Boy* (New York 1968), 671.

160. Aitken, *opus cit.*, 421.

161. For the details and origins of the Watergate affair, see Len Colodny and Robert Getlin: *Silent Coup: the Removal of a President* (London 1991) and Stanley J. Kutler: *The Wars of Watergate: the Last Crisis of Richard Nixon* (New York 1990).

162. Aitken, *opus cit.*, 421–2.

163. H. R. Haldeman, *The Ends of Power* (New York 1980), 112.

164. Aitken, *opus cit.*, 422n. Fred Thompson: *At That Point of Time* (New York 1980). For Watergate etc. see also Stephen E. Ambrose: *Nixon*, vol. iii: *Ruin and Recovery, 1973–1990* (New York 1991); John Blum: *Years of Discord: Politics and Society, 1961–74* (New York 1991); and Joan Hoff: *Nixon Reconsidered* (New York 1994). Virtually all those involved on all sides in Watergate published accounts of it; the best and probably the least self-deceiving is Liddy's *Will: the Autobiography of G. Gordon Liddy* (New York 1981).

165. Nixon, *Memoirs*, 846–7; Aitken, *opus cit.*, 445.

166. Joan Hoff, *Nixon Reconsidered*, thinks the Plumbers may have been looking for evidence of Democratic involvement in a call-girl ring.

167. Liddy, *opus cit.*, 300.

168. Maurice Stans: *The Terrors of Justice: the Untold Side of Watergate* (New

York 1979); James Neutchterlein: 'Watergate: Towards a Revisionist View,' *Commentary*, August 1979; Sirica gave his own account, *To Set the Record Straight* (New York 1979).

169. Anthony Lukas: *Nightmare: the Underside of the Nixon Years* (New York 1976), 375ff; Safire, *Before the Fall*, 292.

170. Ervin's role in the Baker cover-up remains to be fully exposed. It is not mentioned in his own memoirs, *Preserving the Constitution: an Autobiography of Senator Sam Ervin* (Charlottesville, 1984).

171. Haig, as reported in Aitken, *opus cit.*, 504.

172. Nixon, *Memoirs*, 922, 926–7; Henry Kissinger, *Years of Upheaval* (New York 1980), 514–16; Raymond Garthoff: *Detente and Confrontation: American–Soviet Relations from Nixon to Reagan* (Washington DC 1985), 360–85, 404–7.

173. John Blum: *Years of Discord in Politics and Society, 1961–74* (New York 1991), 457.

174. See Frank Levy: *Dollars and Dreams: the Changing American Income Distribution* (New York 1987), 62ff; Patterson, *opus cit.*, 785.

175. See Ford's own account, *A Time to Heal: the Autobiography of Gerald R. Ford* (New York 1979); Ambrose: *Nixon*, iii 238.

176. Aitken, *opus cit.*, 510.

177. For reflections on the impeachment of Andrew Johnson, see J. E. Sefton: 'The Impeachment of Andrew Johnson: a Century of Writing,' *Civil War History*, 14 (1968).

178. Ambrose, *Nixon*, iii.

179. Quoted in Richard Reeves: *A Ford, Not a Lincoln* (New York 1975).

180. Edward L. and Frederick H. Schapsmeier: *Gerald R. Ford's Date with Destiny: a Political Biography* (1989).

181. Personal knowledge; private information.

182. Betty Ford: *The Time of My Life* (New York 1978); Boller, *Presidential Wives*, 417ff.

183. Lee H. Hamilton and Michael H. Van Dusen: 'Making the Separation of Powers Work,' *Foreign Affairs*, Autumn 1978; Georgetown University Conference on Leadership, Williamsburg, Virginia, reported in *Wall Street Journal*, May 10, 1980.

184. Gerald Ford, *Public Papers, 1975* (Washington DC 1977), 119; *State Department Bulletin*, April 14, 1975.

185. *Political Change in Wartime: the Khmer Krahom Revolution in Southern Cambodia, 1970–74*, paper given at American Political Science Convention, San Francisco, September 5, 1975.

186. See the evidence collected from over 300 refugee camps in Thailand, Malaysia, France, and the US and printed in Barron and Paul, *opus cit.*, 10–31, 136–49, 202ff.

187. Paula Smith: 'The Man Who Sold Jimmy Carter,' *Dun's Review* (New York), August 1976.

188. For the Carter presidency, see Erwin C. Hargrove: *Jimmy Carter as President: Leadership and the Politics of the Public Good* (New York 1988).

189. See Jeane Kirkpatrick: 'Dictatorships and Double Standards: a Critique of US

Policy,' *Commentary*, November 1979. This powerful essay was instrumental in getting the policy reversed under President Reagan. See also Michael A. Ledeen and William H. Lewis: 'Carter and the Fall of the Shah: the Inside Story,' *Washington Quarterly*, Summer 1980, 15ff.

190. *Annual Defense Department Report, Financial Year 1977* (Washington DC 1977), section v.

191. James L. George (ed.): *Problems of Sea Power as We Approach the 21st Century* (Washington DC 1978).

192. Admiral Elmo Zumwalt: *On Watch* (New York 1975), 444–5.

193. Quoted in Thomas L. Hughes: 'Carter and the Management of Contradictions,' *Foreign Policy*, 31 (Summer 1978), 34–55; Simon Serfaty: 'Brzezinski: Play It Again, Zbig,' *Foreign Policy*, 32 (Autumn 1978), 3–21; Elizabeth Drew, 'Brzezinski,' *New Yorker*, May 1, 1978; Kirkpatrick, *opus cit.*

194. Exchange between White House aide and the author, spring 1980.

195. Rosalynn Carter: *The First Lady from Plains* (New York 1984), 173–4.

196. Joseph F. Davis: 'Fifty Million More Americans,' *Foreign Affairs*, 28 (April 1950), 109ff.

197. C. K. Leith, J. W. Furness, and Cleona Leith: *World Minerals and World Peace* (Washington DC 1943), 39–41.

198. Nixon, *Public Papers, 1970*, 1134–6; *New York Times*, December 16, 1970 (1.31); Nixon, *Public Papers, 1971*, 669, 895.

199. Charles Coombs: *The Arena of International Finance* (New York 1976), 219.

200. Robert DeFina: *Public and Private Expenditures for Federal Regulation of Business* (St Louis 1977); Murray L. Weidenbaum: *Government Power and Business Performance* (Stanford 1980).

201. Weidenbaum, *opus cit.*

202. Edward F. Denison in *Survey of Current Business* (US Department of Commerce, Washington DC), January 1978.

203. Denison, *ibid.*, August 1979, Part II, and his *Accounting for Slower Economic Growth: the United States in the 1970s* (Washington DC 1980).

204. 'After Many Years as Richest Nation,' *US News and World Report*, September 24, 1973; 'US in the World Economy: the Changing Role,' *US News and World Report*, September 10, 1984; Donald W. White: *The American Century*, 383–4.

205. White, *opus cit.*, 385; P. Bairoch: 'International Industrialisation Levels from 1790 to 1980,' *Journal of European Economic History*, 11 (1982).

206. E. L. Ullman: 'Regional Development and the Geography of Concentration,' *Papers and Proceedings of the Regional Science Associations*, 4 (1958), 197–8.

207. H. S. Perloff *et al.*: *Regions, Resources and Economic Growth* (Lincoln, Nebraska 1960), 50.

208. Robert Estall: 'The Changing Balance of the Northern and Southern Regions of the United States,' *Journal of American Studies* (Cambridge, England), December 1980.

209. The best of many books on Reagan is Lou Cannon: *President Reagan: the Role of a Lifetime* (New York 1991), which replaced his 1982 book, *Reagan*. Other, less sympathetic studies included Laurence I. Barrett: *Gambling with History: Ronald Reagan in the White House* (New York 1984) and J. Mayer

and D. McManus: *Landslide: the Unmaking of the President, 1984–88* (New York 1988).

210. Ronald Reagan: *An American Life: an Autobiography* (New York 1990), 135.

211. James Barber: *The Presidential Character* (2nd edn New Jersey 1977).

212. Cannon, *Role of a Lifetime*, is excellent on Reagan's use of jokes; see esp. 120ff.

213. Owen Ullmann in the *Washington Reporter*.

214. Cannon, *opus cit.*, 528–9.

215. *Ibid.*, 268.

216. *Ibid.*, 147.

217. D. M. Hill *et al.*: *The Reagan Presidency* (Southampton 1990), 39.

218. See Irwin M. Stelzer: 'Lies, Damned Lies and Statistics Revisited,' *Washington Weekly Standard*, December 23, 1996.

219. *An American Life*, 234–5.

220. For the events which led up to the SS20, Cruise, and Pershing deployments see Jonathan Haslam: *The Soviet Union and the Politics of Nuclear Weapons in Europe, 1968–70* (Ithaca 1990).

221. For the origins of SDI see Robert C. McFarlane: *Special Trust* (New York 1994), 227–35.

222. *An American Life*, 517ff.

223. George P. Shultz: *Turmoil and Strife: My Years as Secretary of State* (New York 1993), 1135–6.

224. McFarlane, *opus cit.*, 235.

225. Paul Kennedy: *The Rise and Fall of the Great Powers* (New York 1988), 517.

226. *Discriminate Deterrence: Report of the President's Committee on Integrated Long Term Strategy* (Washington DC 1988).

227. Norman J. Ornstein and Mark Schmitt: 'The 1988 Election,' *Foreign Affairs*, Spring 1989.

228. For the events leading up to the conflict see John Bulloch and Harvey Morris: *Saddam's War: the Origins of the Kuwait Conflict and the International Response* (London 1991).

229. The author first drew Bush's attention to this weakness in 1984, when he promised to remedy it by sticking to prepared texts, but did not do so.

230. For personal details of Bush see Fitzhugh Green: *George Bush: an Intimate Portrait* (New York 1989).

231. For Clinton's background and early career, see Charles F. Allen and Jonathan Portis: *The Comeback Kid: the Life and Career of Bill Clinton* (New York 1992) and John Brummett: *High Wire: from the Back Woods to the Beltway: the Education of Bill Clinton* (New York 1994).

232. For Hillary Clinton's background see Donnie Radcliffe: *Hillary Rodham Clinton: a First Lady for Our Time* (New York 1993).

233. For a review of Clinton's first term, see David Maraniss: *First in His Class: a Biography of Bill Clinton* (New York 1995).

234. For the history of such operations see Stephen F. Knott: *Secret and Sanctioned: Covert Operations and the American Presidency* (Oxford 1997).

235. Suzanne Garment: *Scandal: the Culture of Mistrust in American Politics* (New York 1991), 9ff.

236. For further information about the Clinton White House and the charges against him, see Elizabeth Drew: *On the Edge: the Clinton Presidency* (New York 1994) and Bob Woodward: *The Agenda: Inside the Clinton White House* (New York 1994).

237. See Table 7.2, 'Composition of Congress by Political Party, 1961–1993,' in David McKay: *American Politics and Society* (3rd edn Oxford 1993), 135.

238. Table 7.4, 'House Incumbents Reelected, Defeated or Retired, 1946–90,' *ibid.*, 138.

239. Dick Morris: *Behind the Oval Office* (New York 1996).

240. 'The Man Who Has Clinton's Ear,' *Time*, September 2, 1996.

241. For Clinton's relationship with the 104th Congress see Elizabeth Drew: *Showdown: the Struggle Between the Gingrich Congress and the Clinton White House* (New York 1996).

242. *1987 Census of Governments, Bureau of the Census* (Washington DC 1987), Table A–1.

243. J. P. Zimmerman: *Contemporary American Federalism: the Growth of National Power* (Leicester 1992).

244. See Walter Dean Burnham: 'The Turnout Problem,' in A. J. Reichley (ed.): *Elections, American Style* (Washington DC 1987).

245. Michael Lind: *The Next American Nation* (New York 1995), 132ff.

246. See *Statistical Abstract of the USA* (Washington DC 1991), Tables 1 and 2.

247. 'Make Way for the Urban Confederates,' special issue of *American Enterprise*, November–December 1996.

248. 'Towering Ambitions,' *Daily Telegraph* (London), July 16, 1996.

249. Philip Goldberger: 'A Major Monument of Post-Modernism,' *New York Times*, March 31, 1978.

250. Charles Jenks: *Post Modernism: the New Classicism in Art and Architecture* (London 1987), Chapter VIII, 217ff.

251. Tom Wolfe: 'Las Vegas (What?) Las Vegas (Can't Hear You! Too Noisy) Las Vegas!!!,' *Esquire*, June 1964.

252. Sam Bass Warner: *The Urban Wilderness: A History of the American City* (New York 1972), 8.

253. *American Enterprise*, November–December 1996 for figures, 61ff.

254. See Mac Griswold and Eleanor Weller: *The Golden Age of American Gardens* (New York 1996).

255. Robin Karson: *Fletcher Steele: Landscape Architect; an Account of the Garden-Maker's Life, 1885–1971* (New York 1989), 283ff.

256. Marion Clawson: *The Federal Lands Revisited* (New York 1983).

257. Elizabeth Johns: *Thomas Eakins: the Heroism of Modern Life* (Princeton 1983).

258. See *George de Forest Brush, Master of the American Renaissance*, Exhibition catalogue, Berry-Hill Galleries (New York 1985).

259. Wanda M. Corn: *The Art of Andrew Wyeth*, exhibition catalogue, The Fine Arts Museum (San Francisco 1973).

260. W. H. Gerdts: *The Great American Nude: a History in Art* (New York 1974).

261. *Two Worlds of Andrew Wyeth*, exhibition catalogue, Metropolitan Museum of Art (New York 1976).

262. John Wilmerding: *Andrew Wyeth: the Helga Pictures* (New York 1987).

263. McKay, *opus cit.*, 19ff.

264. See Roger L. Geiger: *Research and Relevant Knowledge: American Research Universities Since World War Two* (New York 1993); L. L. Stevenson in *Companion to American Thought* (Cambridge 1995), 136–7.

265. *The Economist*, November 5, 1994.

266. Quoted in Edward N. Luttwak: *The Endangered American Dream* (New York 1993), 163.

267. M. T. Jacobs: *Short-Term America: Causes and Cures of Our Business Myopia* (Cambridge 1991), 82.

268. Lind, *opus cit.*, 151ff.

269. See James Hallows: 'Low Class Conclusions,' *Atlantic Monthly*, April 1993; Lind, *opus cit.*, 149n.

270. D. M. O'Brien: 'Reagan Judges: His Most Enduring Legacy,' in C. O. Jones (ed.): *The Reagan Legacy: Promise and Performance* (Chatham, New Jersey 1988), Table 3.6, 77.

271. Lind, *opus cit.*, 157.

272. McKay, *opus cit.*, Table 7.1, 134.

273. Richard A. Epstein: *Simple Rules for a Complex World* (Cambridge 1995), Introduction. 'Too Many Lawyers, Too Much Law; America's Parasitic Economy,' *The Economist*, October 10, 1992; Sherwin Rosen: 'The Market For Lawyers,' *Journal of Law and Economics*, 35 (1992); S. P. Magee: 'The Optimum Number of Lawyers,' *Law and Social Inquiry*, 17 (1992).

274. Marc Galanter and Thomas Palay: *Tournament of Laws: the Transformation of the Big Law Firm* (New York 1992), 37.

275. Epstein, *opus cit.*, 3.

276. See *ibid.*, Table, 7.

277. Stephen P. Magee *et al.*: 'The Invisible Foot and the Fate of Nations: Lawyers as Negative Externalities,' in Magee (ed.): *Black Hole Tariffs and Endogenous Policy Theory: Political Economy in General Equilibrium* (New York 1989).

278. *Statistical Abstract of the United States, 1991* (Washington DC 1992), Tables 321–3.

279. W. Blackstone: *Commentaries on the Laws of England* (London 1765–70). This work passed through as many editions in America as in England and inspired the standard work of James Kent (1763–1847), *Commentaries on the American Law* (New York 1826–30).

280. For background see Loren P. Beth: *Politics, the Constitution and the Supreme Court* (New York 1962) and Richard Hodder-Williams: *The Politics of the US Supreme Court* (London 1980).

281. Mark Tushnet: *The NAACP's Legal Strategy Against Segregated Education, 1925–50* (Chapel Hill 1987).

282. Roberts and Stratton, *The New Color Line*; Mark Tushnet: *Making Civil Rights Law* (New York 1994).

283. Sundquist, *Politics and Policy*, 259–71; Califarno, *opus cit.*, 54; Matusow,

opus cit., 92ff; see also Steve Lawson: 'Civil Rights,' in Robert Divine, (ed.): *Exploring the Johnson Years* (Austin 1981), 99–100.

284. For Blumrosen's boasting see his book *Black Employment and the Law* (New Brunswick 1971).

285. Roberts and Stratton, *opus cit.*, 2–4.

286. Irving Kristol: 'How Hiring Quotas Came to the Campuses,' *Fortune*, September 1974; A. M. Bickel: 'The Original Understanding and the Segregational Decision,' *Harvard Law Review*, 1955, and 'The Decade of School Desegration: Progress and Prospects,' *Columbia Law Review*, February 1964; see also his *The Least Dangerous Branch: the Supreme Court at the Bar of Politics* (2nd edn Cambridge 1986); Ivan Hannaford: 'The Idiocy of Race,' *Wilson Quarterly*, Spring 1992; Herbert Wechsler: 'Towards Neutral Principles of Constitutional Law,' *Harvard Law Review*, November 1959; Raoul Berger: *Government by Judiciary: the Transformation of the Fourteenth Amendment* (Cambridge 1977); Raymond Wolters: *The Burden of Brown: Thirty Years of School Desegregation* (Knoxville 1984); Gerald Rosenburg: *The Hollow Hope: Can Courts Bring About Social Change?* (Chicago 1991).

287. Stephen Steinberg: *The Ethnic Myth: Race, Ethnicity and Class in America* (Boston 1989), 294.

288. Hugh Davis Graham: *The Civil Rights Era* (New York 1990), 365, 459–60.

289. Richard Neely: *How the Courts Govern America* (New Haven 1981), 6.

290. Lind, *opus cit.*, 119 and n.

291. For the OMB Statistical Directive 15, see Lawrence Wright: 'One Drop of Blood,' *New Yorker*, July 1994, 46–55.

292. *The Economist*, September 22, 1992.

293. Lind, *opus cit.*, 295; for Caribbean classifications, see E. Braithwaite: *The Development of Creole Society in Jamaica, 1770–1820* (Oxford 1971).

294. Lind, *opus cit.*, 174–5.

295. Bob Zelnick: *Backfire: a Reporter's Look at Affirmative Action* (New York 1996).

296. For many examples see Dinesh D'souza: *Illiberal Education: the Politics of Race and Sex on Campus* (New York 1991). The most significant of the history-rewriting efforts was the book by the white Cornell professor Martin Bernal: *Black Athena: The Afroasiatic Roots of Classical Civilisation* (New Brunswick 1989).

297. On this see Davis Gress: 'The Case Against Martin Bernal,' in H. Kramer and R. Kimball (eds): *Against the Grain* (Chicago 1995), 91ff.

298. See reports in the City University of New York *Campus*, April 15 and 26, 1988, and *Campus Report*, June 1988: 'Racism in Black Studies;' *New York Times*, April 20, 1990; D'souza, *opus cit.*, 7ff.

299. Patricia Collins and Margaret Anderson (eds): *An Inclusive Curriculum: Race, Class and Gender in Sociological Instruction* (American Sociological Association, Washington DC 1987).

300. D'souza, *opus cit.*, 259, n. 21.

301. Michael Collison: 'Fight the Power: Rap Music Pounds Out a New Anthem

for Many Black Students,' *Chronicle of Higher Education*, February 14, 1990.

302. Shakur's first solo album, *Pacalypse Now*, urging violence against police, was the subject of a suit by the widow of a state trooper against its present company, Time-Warner. The trooper's assailant claimed that his crime was a direct result of listening to the album.

303. Roger Kimball: *Tenured Radicals* (New York 1991), 68; see R. R. Detlefsen: 'White Like Me,' *New Republic*, April 10, 198, for other examples of AWARE policies.

304. See the New Criterion Anthology, *Against the Grain*: 'The Academy in the Age of Political Correctness,' 67–193.

305. Albert C. Baugh: *A History of the English Language* (London 1976 edn), Chapter 11, 'The English Language in America,' 406ff. See esp. the list of studies of American regional dialects and pronunciation in n. 2, 436ff.

306. See Cynthia Ozick: 'The Question of Our Speech: the Return to Aural Culture,' in K. Washburn and J. Thornton (eds): *Dumbing Down: Essays on the Strip-Mining of American Culture* (New York 1996), 68–87.

307. John Rawls: *A Theory of Justice* (New York 1971), 3–4.

308. Adam Smith: *The Theory of Moral Sentiments*, part II, ii, Chapter ii, 380–1, in Liberty Classic Library, New York 1976; Thomas Sowell: *Knowledge and Decisions* (New York 1980), 118–22.

309. Ambrose Evans-Pritchard: 'Race Ruling "Threatens Democracy," ' *Sunday Telegraph* (London), January 11, 1997.

310. 'Who Gets Abortions?,' *American Enterprise*, November–December 1996, 18.

311. James Q. Wilson and Richard J. Hernstein: *Crime and Human Nature* (New York 1985), 405; see also H. D. Graham and T. R. Gurr (eds): *The History of Violence in America* (New York 1969).

312. *Crime Index, 1960–92* (US Department of Justice, FBI); *Uniform Crime Reports, 1992* (US Department of Justice, FBI), Table 1, 58.

313. *Highlights from 20 Years of Surveying Crime Victims: The National Crime Victimization Survey, 1973–92* (Washington DC 1993), 5, 7.

314. William J. Bennett: *Index of Leading Cultural Indicators* (New York 1993); see Gertrude Himmelfarb: *The De-Moralisation of Society* (London 1995), 226ff.

315. *Wall Street Journal*, November 9, 1994.

316. *Milwaukee Journal-Sentinel*, December 4, 1995.

317. W. Bennett, J. D. DiLulio Jr, and J. P. Walters: *Body Count: Moral Poverty and How To Win America's War Against Crime and Drugs* (New York 1996).

318. See George Kelling and K. M. Coles: *Fixing Broken Windows: Restoring Order and Reducing Crime in Our Communities* (New York 1996).

319. See figures of black arrests in *Body Count*; see also Jerome J. Miller: *Search and Destroy: African-American Males in the Criminal Justice System* (Cambridge 1996).

320. For figures see *Fixing Broken Windows*.

321. For detailed analysis of America's Protestant churches, see C. Rood and W. McKinney: *American Mainline Religion: Its Changing Shape and Future* (New

Brunswick 1987), 81–90, 110–47. See also P. W. Williams: *America's Religions: Traditions and Culture* (New York 1990), 333ff.

322. B. A. Kosmin and S. P. Lachman: *One Nation Under God: Religion in Contemporary American Society* (New York 1993), 257ff; R. S. Michaelsen and W. C. Roof (eds): *Liberal Protestantism: Realities and Possibilities* (New York 1986), 65ff; Mark Tooley: 'Madness in Their Methodism: the Religious Left Has a Summit,' *Heterodoxy*, May 1995.

323. *Foundations Daily* (a publication of the Synod), September 2, 1994; quoted in Thomas C. Reeves: *The Empty Church: the Crisis of Liberal Christianity* (New York 1996).

324. K. B. Bedell (ed.): *Yearbook of American and Canadian Churches 1996* (Nashville 1996), 250–6. There is dispute about some of the figures. Another source lists Mormon numbers as 4,370,700.

325. *The Empty Church*, 11; *Wall Street Journal*, May 25, 1995.

326. Quoted in *The Empty Church*, 12.

327. Richard J. Hernstein and Charles Murray: *The Bell Curve: Intelligence and Class Structure in American Life* (New York 1994).

328. *Ibid.*, 178ff., 339ff.

329. Charles Murray: 'Bad News About Illegitimacy,' *Washington Weekly Standard*, August 5, 1996.

330. Daniel Patrick Moynihan: *Miles to Go: A Personal History of Social Policy* (Cambridge 1996).

331. Claudia Goldin: 'Career and Family: College Women Look to the Past,' *National Bureau of Economic Research Digest*, December 1995.

332. Marion Meade: *Dorothy Parker: What Fresh Hell Is This?* (London 1988), 11ff.

333. Carol Brightman: *Writing Dangerously: Mary McCarthy and Her World* (New York 1993), 65.

334. F. Carolyn Graglia: 'The Breaking of the Women's Pact,' *Washington Weekly Standard*, November 11, 1996.

335. For feminist extremism, see Alice Echolas: *Daring to Be Bad: Radical Feminism in America, 1967–75* (Minneapolis 1989); Shulamith Firestone: *The Dialectic of Sex: the Case for Feminist Revolution* (New York 1971).

336. See Clair Brown and J. A. Pechman: *Gender in the Workplace* (Washington DC 1987); *Wall Street Journal*, January 29, 1996; US Department of Labor, *1993 Handbook on Women Workers* (Washington DC 1993), 59. See statistics in Diana Furchtgott-Roth: *Women's Figures: the Economic Progress of Women in America* (Washington DC 1996), xii–xiii, and graph on 28, with sources.

337. Federal Glass Ceiling Commission: *Good for Business: Making Full Use of the Nation's Human Capital* (Washington DC 1995).

338. Katherine Post and Michael Lynch: 'Smoke and Mirrors: Women and the Glass Ceiling,' *Pacific Research Institute Fact Sheet*, November 1995; *Women's Figures*, 12–14.

339. *Women's Figures*, 28–34; see publications of the Center for the American Woman in Politics, Eagleton Institute, Rutgers University, *passim*.

340. *Women's Figures*, 13–26.

INDEX